P9-DGZ-873

Principles of Development

Principles of Development

Fifth Edition

Lewis Wolpert | Cheryll Tickle | Alfonso Martinez Arias

Peter Lawrence

Andrew Lumsden

Elizabeth Robertson

Elliot Meyerowitz

Jim Smith

OXFORD
UNIVERSITY PRESS

OXFORD
UNIVERSITY PRESS

Great Clarendon Street, Oxford, OX2 6DP,
United Kingdom

Oxford University Press is a department of the University of Oxford.
It furthers the University's objective of excellence in research, scholarship,
and education by publishing worldwide. Oxford is a registered trade mark of
Oxford University Press in the UK and in certain other countries

© Oxford University Press 2015

The moral rights of the authors have been asserted

Second edition 2001
Third edition 2008
Fourth edition 2011

Impression: 1

All rights reserved. No part of this publication may be reproduced, stored in
a retrieval system, or transmitted, in any form or by any means, without the
prior permission in writing of Oxford University Press, or as expressly permitted
by law, by licence or under terms agreed with the appropriate reprographics
rights organization. Enquiries concerning reproduction outside the scope of the
above should be sent to the Rights Department, Oxford University Press, at the
address above

You must not circulate this work in any other form
and you must impose this same condition on any acquirer

Published in the United States of America by Oxford University Press
198 Madison Avenue, New York, NY 10016, United States of America

British Library Cataloguing in Publication Data
Data available

ISBN 978-0-19-870988-6

Printed in Italy by L.E.G.O. S.p.A.

Links to third party websites are provided by Oxford in good faith and
for information only. Oxford disclaims any responsibility for the materials
contained in any third party website referenced in this work

QR Code images are used throughout this book.
QR Code is a registered trademark of DENSO WAVE INCORPORATED.
If your mobile device does not have a QR Code reader try this website
for advice http://www.mobile-barcodes.com/qr-code-software

Preface

As we pointed out in the preface to the fourth edition of *Principles of Development*, developmental biology is at the core of all of the biology of multicellular organisms. It deals with the process by which the genes in the fertilized egg control cell behavior in the embryo and so determine the character of the animal or plant. Developmental biology is also fundamental to evolution, as organisms that are better adapted to the environment result from changes in development. The four years since the last edition of this book have seen continuing progress in understanding the cellular and molecular basis of embryonic development, and genomics is having an increasing impact. In this fifth edition, we have included many recent advances. Of particular note is the progress in our understanding of cell differentiation (Chapter 8) and in deciphering the developmental changes that underlie evolution (Chapter 14). Throughout this new edition, we have also tried to reflect the increasing emphasis on the medical applications of developmental biology, for example in clinical genetics and in regenerative medicine (Chapter 8).

Principles of Development is designed for undergraduates and aims to provide students with an understanding of the principles that guide development. We have tried to make these principles as clear as possible and to provide numerous summaries, in both words and pictures. We focus on the systems that best illustrate the principles of development, and do not aim to provide a comprehensive text. We have also tried to avoid going into too much detail, as this can be overwhelming and obscure general principles. The details can be found in the many reviews in the literature, which are periodically updated. It is our belief that while the details are likely to change, the principles will remain, and as we understand the general principles better, we should be able to make the book shorter!

We have assumed that students have some familiarity with basic cell and molecular biology and genetics, but all key concepts, such as the control of gene activity, are explained in the text. There is also an extensive glossary, which means that the book is self-contained. The illustrations are a special feature and have been carefully designed and chosen to illuminate both experiments and mechanisms. Many new diagrams and photographs are included throughout the book, together with information about their sources. In providing further reading, our prime concern has been to guide the student to particularly helpful papers and reviews rather than to give credit to all the scientists who have made major contributions: to those whom we have neglected, we apologize. As in previous editions, we have concentrated our attention on vertebrates and *Drosophila*, but include other organisms, such as the nematode and the sea urchin, when they best illustrate a concept. As in the previous edition, we have started the book by considering the process of pattern formation in laying down the body plan in *Drosophila* (Chapter 2). This is because of the central role that *Drosophila* has played, and still plays, in understanding developmental mechanisms.

Chapter 3 describes the embryology and genetics of our vertebrate model organisms, together with some of the main methods used to study them. An outline of human embryonic development is included in this edition, because comparing this, where possible, with embryonic development in other vertebrates will be important for medical applications. The mechanisms involved in pattern formation in the early development of our vertebrate model organisms are then considered in the two subsequent chapters (Chapters 4 and 5). These have been reorganized so that the process of laying down the early body plan is first described in its entirety in *Xenopus* (Chapter 4), the vertebrate in which the general principles were discovered. This is followed by comparisons with the process in zebrafish (Chapter 4) and in chick and mouse (Chapter 5). Chapter 5 also considers how the body plan is completed, which mainly rests on studies in chick and mouse embryos. Chapter 6 now focuses on pattern formation in two invertebrate model organisms, the nematode and the sea urchin. Chapter 7 deals with plant development, which is often neglected in general textbooks of developmental biology, and which is important in its own right. Chapters 8 and 9 focus on the fundamental processes of differentiation and morphogenesis and have been extensively revised, with particular reference to stem cells in Chapter 8. Chapter 10 deals with germ cells and fertilization. Organogenesis (Chapter 11) and the development of the nervous system (Chapter 12) are huge topics, so we have had to be very selective in our coverage, but have included new boxes highlighting examples of medical relevance. In this edition, growth and regeneration are considered together in the same chapter (Chapter 13), which has been reorganized, and the last chapter (Chapter 14) deals with development in relation to evolution.

For this new edition, Alfonso Martinez Arias has joined Cheryll Tickle and Lewis Wolpert as a main co-author, and Andrew Lumsden has also become an author. Each chapter has also been reviewed by a number of experts (see page xxii), to whom we give thanks. The authors made the initial revisions, which were then deciphered, edited, and incorporated by our editor, Eleanor Lawrence. Her involvement has been crucial in the preparation of this edition and her expertise and influence pervades the book. Eleanor's input has also been invaluable in ensuring that the information in the book is readily accessible to students. The new illustrations were brilliantly drawn or adapted by Matthew McClements, who created the illustrations for the first edition.

We are indebted to Alice Roberts and Jonathan Crowe at Oxford University Press for their help and patience throughout the preparation of this new edition.

L. W.

London
September 2014

C. T.

Bath
September 2014

A. M. A.

Cambridge
September 2014

About the Online Resource Centre

www.oxfordtextbooks.co.uk/orc/wolpert5e/

Principles of Development is accompanied by a range of online materials for adopters of the book and their students.

For registered adopters:

Electronic artwork

Figures from the book are available to download, for use in lecture slides.

Journal clubs

Journal clubs consist of discussion questions focused around primary literature articles that relate to topics featured in the book. Use these as an additional learning tool to help your students become more adept at assimilating knowledge from the research literature.

Test bank

A test bank of questions is available for you to use when assessing your students.

For students:

Flashcard glossary

Flashcards, which can be downloaded to mobile devices, can help you test your recall of key terminology.

Multiple-choice questions

Use the extensive bank of multiple-choice questions to check your understanding of concepts introduced in the book, and get instant feedback on your progress.

Answer guidance

The authors have written answer guidance to the long-answer questions found at the end of each chapter, so you can check that you have considered all the appropriate points when responding to each question.

Web links and web activities

Links to websites, with notes to explain how each site relates to concepts featured in the book, are provided to help you explore topics in the book in more detail. Complete the associated activities to get to grips with the material in a hands-on way. *In silico* practicals have also been developed to accompany the book, and include questions to help you think more deeply about the material you have learned.

Movies from real research

Scan the QR code images in the text to access movies showing key developmental processes occurring in real embryos to help you visualize developmental biology in three dimensions.

Signaling pathway animations

Custom-made animations of key signaling pathways, linked to the text via QR code images, break down these complex processes into stages, making them easier to understand and remember.

Online extracts

Further material on the development of ascidians can be found online, in addition to extra topics such as kidney organogenesis and reaction-diffusion mechanisms. QR code images at the relevant points in the text direct you to this extra material.

About the authors

Lewis Wolpert is Emeritus Professor of Biology as Applied to Medicine, in the Department of Anatomy and Developmental Biology, University College London, London, UK. He is the author of *The Triumph of the Embryo, A Passion for Science, The Unnatural Nature of Science,* and *Six Impossible Things Before Breakfast.*

Cheryll Tickle is Emeritus Professor in the Department of Biology and Biochemistry, University of Bath, Bath, UK.

Alfonso Martinez Arias is Professor of Developmental Mechanics at the University of Cambridge, UK.

Peter Lawrence is in the Department of Zoology, University of Cambridge, UK, and Emeritus member of the Medical Research Council Laboratory of Molecular Biology, Cambridge, UK. He is the author of *The Making of a Fly.*

Andrew Lumsden is Professor of Developmental Neurobiology and Emeritus Director of the MRC Centre for Developmental Neurobiology at King's College London, UK. He is the co-author of *The Developing Brain.*

Elizabeth Robertson is a Wellcome Trust Principal Fellow and Professor at the Sir William Dunn School of Pathology at the University of Oxford, Oxford, UK.

Elliot Meyerowitz is the George W. Beadle Professor of Biology and Chair of the Division of Biology at the California Institute of Technology, Pasadena, CA, USA.

Jim Smith is Director of the Medical Research Council National Institute for Medical Research, London, UK.

Eleanor Lawrence is a freelance science writer and editor.

Matthew McClements is an illustrator who specializes in design for scientific, technical, and medical communication.

Summary of contents

Contents

List of boxes

Reviewer acknowledgements

Many thanks to the following who kindly reviewed various parts of the book:

Michael Akam, University of Cambridge

Heather J. Anderson, Winthrop University

Michael Bate, Cambridge University

Jeremy Brockes, University College London

Marianne Bronner, California Institute of Technology

Deborah L. Chapman, University of Pittsburgh

Susan Ernst, Tufts University

Makoto Furutani-Seiki, University of Bath

Peter Holland, University of Oxford

Robert Kelsh, University of Bath

Jane P. Kenney-Hunt, Westminster College

Tetsu Kudoh, University of Exeter

Fang Ju Lin, Coastal Carolina University

Philip Maini, University of Oxford

Bonny Millimaki, Lipscomb University

Tony Perry, University of Bath

Lisa M. Nagy, University of Arizona

Fred Sablitzky, University of Nottingham

James Sharpe, CRG Barcelona (Spain)

Rebecca Spokony, Baruch College, the City University of New York

Ajay Srivastava, Western Kentucky University

Kate Storey, University of Dundee

Vasanta Subramanian, University of Bath

Andrew Ward, University of Bath

Neil Vargesson, University of Aberdeen

Heather Verkade, Monash University

Grant Wheeler, University of East Anglia

History and basic concepts

- The origins of developmental biology
- A conceptual tool kit

The aim of this chapter is to provide a conceptual framework for the study of development. We start with a brief history of the study of embryonic development, which illustrates how some of the key questions in developmental biology were first formulated, and continue with some of the essential principles of development. The big question is how does a single cell—the fertilized egg—give rise to a multicellular organism, in which a multiplicity of different cell types are organized into tissues and organs to make up a three-dimensional body. This question can be studied from many different viewpoints, all of which have to be fitted together to obtain a complete picture of development: which genes are expressed, and when and where; how cells communicate with each other; how a cell's developmental fate is determined; how cells proliferate and differentiate into specialized cell types; and how major changes in body shape are produced. All the information for embryonic development is contained within the fertilized egg. We shall see that an organism's development is ultimately driven by the regulated expression of its genes, determining which proteins are present in which cells and when. In turn, proteins largely determine how a cell behaves. The genes provide a generative program for development, not a blueprint, as their actions are translated into developmental outcomes through cellular behavior such as intercellular signaling, cell proliferation, cell differentiation, and cell movement.

The development of a multicellular organism from a single cell—the fertilized egg—is a brilliant triumph of evolution. The fertilized egg divides to give rise to many millions of cells, which form structures as complex and varied as eyes, arms, heart, and brain. This amazing achievement raises a multitude of questions. How do the cells arising from division of the fertilized egg become different from each other? How do they become organized into structures such as limbs and brains? What controls the behavior of individual cells so that such highly organized patterns emerge? How are the organizing principles of development embedded within the egg, and in particular within the genetic material, DNA? Much of the excitement in developmental biology today comes from our growing understanding of how genes direct these developmental processes, and genetic control is one of the main themes of this book. Thousands of genes are involved in controlling development, but we will focus only on those that have key roles and illustrate general principles.

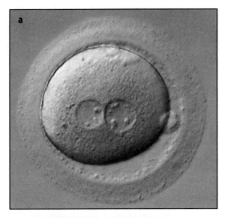

Fig. 1.1 Human fertilized egg and embryo.
(a) Human fertilized egg. The sperm and egg nuclei (pronuclei) have not yet fused. (b) Human embryo at around 51 days' gestation (Carnegie stage 20), which is equivalent to a mouse embryo at 13.5 days post-fertilization. A human embryo at this stage is about 21–23 mm long.

(a) Courtesy of A. Doshi, CRGH, London.
(b) Reproduced courtesy of the MRC/ Wellcome-funded Human Developmental Biology Resource.

Fig. 1.2 The South African claw-toed frog, *Xenopus laevis.* Scale bar = 1 cm.
Photograph courtesy of J. Smith.

Understanding how embryos develop is a huge intellectual challenge, and one of the ultimate aims of the science of **developmental biology** is to understand how we humans develop (Fig. 1.1). We need to understand human development for several reasons. We need to properly understand why it sometimes goes wrong and why a **fetus** may fail to be born or a baby be born with congenital abnormalities. The link here with genetic control of development is very close, as mutations in genes can lead to abnormal development; environmental factors, such as drugs and infections, can affect it too. Another area of medical research related to developmental biology is regenerative medicine—finding out how to use cells to repair damaged tissues and organs. The focus of regenerative medicine is currently on **stem cells**. Stem cells that can proliferate and give rise to all the different tissues of the body are present in embryos. These, and the stem cells with more limited developmental potential that are found in adult tissues, are discussed in Chapter 8. Cancer cells also display some properties of embryonic cells, such as the ability to divide indefinitely, and so the study of embryonic cells and their behavior could lead to new and better treatments for cancer, as many of the same genes are involved.

The development of an embryo from the fertilized egg is known as **embryogenesis**. One of the first tasks is to lay down the overall body plan of the organism, and we shall see that different organisms solve this fundamental problem in several ways. The focus of this book is mainly on animal development, in particular that of vertebrates—frogs, birds, fish, and mammals—whose early development is discussed in Chapters 3 to 5. We also look at selected invertebrates, particularly the fruit fly and the nematode worm, and also the sea urchin. Our understanding of the genetic control of development is most advanced in fruit flies and nematodes and the main features of their early development are considered in Chapters 2 and 6, respectively. The fruit fly is also used throughout the book to illustrate particular aspects of development. In Chapter 7 we look briefly at some aspects of plant development, which differs in many respects from that of animals but involves similar basic principles.

Morphogenesis, or the development of form, is discussed in Chapter 9. In Chapter 10 we look at how sex is determined and how germ cells develop. The differentiation of unspecialized cells into cells that carry out particular functions, such as muscle cells and blood cells, is considered in Chapter 8. Structures such as the vertebrate limb, and organs such as insect and vertebrate eyes, the heart and the nervous system, illustrate the problems of multicellular organization and tissue differentiation in embryogenesis, and we consider some of these systems in detail in Chapters 11 and 12. The study of developmental biology, however, goes well beyond the development of the embryo. Post-embryonic growth and aging, how some animals undergo metamorphosis, and how animals can regenerate lost organs is discussed in Chapter 13. Taking a longer view, we shall consider in Chapter 14 how developmental mechanisms have evolved and how they constrain the very process of evolution itself.

One might ask whether it is necessary to cover so many different organisms in order to understand the basic features of development. The answer is yes. Developmental biologists do indeed believe that there are general principles of development that apply to all animals, but life is too wonderfully diverse to find all the answers in a single organism. As it is, developmental biologists have tended to focus their efforts on a relatively small number of animals, chosen because they were convenient to study and amenable to experimental manipulation or genetic analysis. This is why some creatures, such as the frog *Xenopus laevis* (Fig. 1.2) and the fruit fly *Drosophila melanogaster*, have such a dominant place in developmental biology. Similarly, work with the thale-cress, *Arabidopsis thaliana*, has uncovered many features of plant development.

One of the most exciting and satisfying aspects of developmental biology is that understanding a developmental process in one organism can help to illuminate similar processes elsewhere—for example, giving insights into how humans develop. Nothing illustrates this more dramatically than the influence that our understanding of *Drosophila* development, and especially of its genetic basis, has had throughout developmental biology. The identification of genes controlling early embryogenesis in *Drosophila* has led to the discovery of related genes being used in similar ways in the development of mammals and other vertebrates. Such discoveries encourage us to believe in the existence of general developmental principles.

Amphibians have long been favorite organisms for studying early development because their eggs are large and their embryos are easy to grow in a simple culture medium and relatively easy to experiment on. Embryogenesis in the South African frog *Xenopus* (Box 1A) illustrates some of the basic stages of development in all animals.

In the rest of this chapter we first look briefly at the history of **embryology**—as the study of developmental biology used to be called. The term developmental biology itself is of much more recent origin and reflects the appreciation that development is not restricted to the embryo alone. Traditionally, embryology described experimental results in terms of morphology and cell fate, but we now understand development in terms of molecular genetics and cell biology as well. In the second part of the chapter we will introduce some key concepts that are used over and over again in studying and understanding development.

The origins of developmental biology

Many questions in embryology were first posed hundreds, and in some cases thousands, of years ago. Appreciating the history of these ideas helps us to understand why we approach developmental problems in the way that we do today.

1.1 Aristotle first defined the problem of epigenesis versus preformation

A scientific approach to explaining development started with Hippocrates in Greece in the fifth-century BC. Using the ideas current at the time, he tried to explain development in terms of the principles of heat, wetness, and solidification. About a century later the study of embryology advanced when the Greek philosopher Aristotle formulated a question that was to dominate much thinking about development until the end of the nineteenth century. Aristotle addressed the problem of how the different parts of the embryo were formed. He considered two possibilities: one was that everything in the embryo was preformed from the very beginning and simply got bigger during development; the other was that new structures arose progressively, a process he termed epigenesis (which means 'upon formation') and that he likened metaphorically to the 'knitting of a net'. Aristotle favored epigenesis and his conjecture was correct.

Aristotle's influence on European thought was enormous and his ideas remained dominant well into the seventeenth century. The contrary view to epigenesis, namely that the embryo was preformed from the beginning, was championed anew in the late seventeenth century. Many could not believe that physical or chemical forces could mold a living entity like the embryo. Along with the contemporaneous background of belief in the divine creation of the world and all living things, was the belief that all embryos had existed from the beginning of the world, and that the first embryo of a species must contain all future embryos.

Even the brilliant seventeenth-century Italian embryologist, Marcello Malpighi, could not free himself from preformationist ideas. While he provided a remarkably accurate description of the development of the chick embryo, he remained convinced, against the evidence of his own observations, that the fully formed embryo was present from the beginning (Fig. 1.3). He argued that at very early stages the parts were so small that they could not be seen, even with his best microscope. Other preformationists believed that the sperm contained the embryo, and some even claimed to see a tiny human—a homunculus—in the head of each human sperm (Fig. 1.4).

The preformation/epigenesis issue was vigorously debated throughout the eighteenth century. But the problem could not be resolved until one of the great advances in biology had taken place—the recognition that living things, including embryos, were composed of cells.

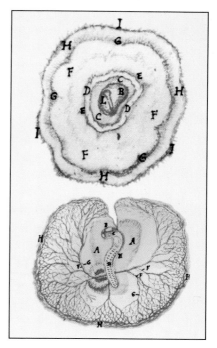

Fig. 1.3 Malpighi's description of the chick embryo. The figure shows Malpighi's drawings, made in 1673, depicting the early embryo (top), and at 2 days' incubation (bottom). His drawings accurately illustrate the shape and blood supply of the embryo. *Copyright The Royal Society.*

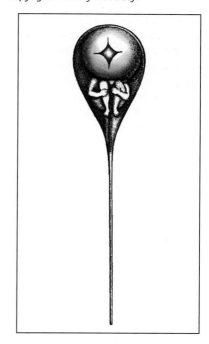

Fig. 1.4 Some preformationists believed that an homunculus was curled up in the head of each sperm.

An imaginative drawing, after N. Harspeler (1694).

BOX1A Basic stages of *Xenopus laevis* development

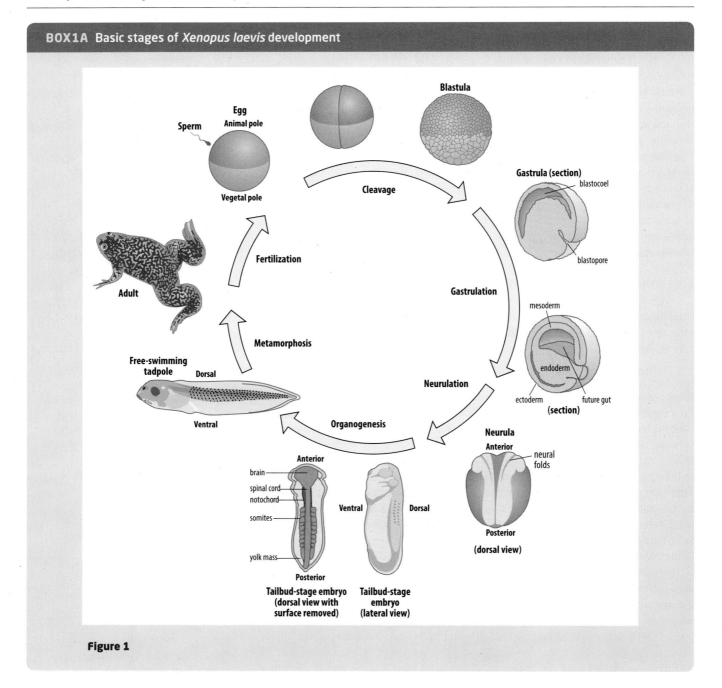

Figure 1

1.2 Cell theory changed how people thought about embryonic development and heredity

The invention of the microscope around 1600 was essential for the discovery of cells, but the 'cell theory' of life was only developed between 1820 and 1880 by, among others, the German botanist, Matthias Schleiden, and the physiologist, Theodor Schwann. It recognized that all living organisms consist of cells, that these are the basic units of life, and that new cells can only be formed by the division of pre-existing cells. The cell theory was one of the most illuminating advances in biology, and had an enormous impact. Multicellular organisms, such as animals and plants, could now be viewed as communities of cells. Development could not therefore be based on preformation, but must be epigenetic, because during development many new cells are generated by division from the egg, and

Although vertebrate development is very varied, there are a number of basic stages that can be illustrated by following the development of the frog *Xenopus laevis* (Figure 1). The unfertilized egg is a large cell. It has a pigmented upper surface (the **animal pole**) and a lower region (the **vegetal pole**) characterized by an accumulation of yolk granules.

After **fertilization** of the egg by a **sperm**, and the fusion of male and female pronuclei, **cleavage** begins. Cleavages are mitotic divisions in which cells do not grow between each division, and so with successive cleavages the cells become smaller. After about 12 division cycles, the embryo, now known as a **blastula**, consists of many small cells surrounding a fluid-filled cavity (the **blastocoel**) above the larger yolky cells. Already, changes have occurred within the cells and they have interacted with each other so that the three **germ layers**: **mesoderm**, **endoderm**, and **ectoderm** are specified (see Box 1C). The animal region gives rise to ectoderm, which forms both the epidermis of the skin and the nervous system. The vegetal region gives rise to the future endoderm and mesoderm, which are destined to form internal organs. At this stage, these cells are still on the surface of the embryo. During the next stage—**gastrulation**—there is a dramatic rearrangement of cells; the endoderm and mesoderm move inside, and the basic body plan of the tadpole is established. Internally, the mesoderm gives rise to a rod-like structure (the notochord), which runs from the head to the tail, and lies centrally beneath the future nervous system. On either side of the notochord are segmented blocks of mesoderm called somites, which will give rise to the muscles and vertebral column, as well as the dermis of the skin (somites can be seen in the cutaway view of the later tailbud-stage embryo).

Shortly after gastrulation, the ectoderm above the notochord folds to form a tube (the **neural tube**), which gives rise to the brain and spinal cord—a process known as **neurulation**. By this time, other organs, such as limbs, eyes, and gills, are specified at their future locations, but only develop a little later, during **organogenesis**. During organogenesis, specialized cells such as muscle, cartilage, and neurons differentiate. By 4 days after fertilization, the embryo has become a free-swimming tadpole with typical vertebrate features.

new types of cells are formed. A crucial step forward in understanding development was the recognition, in the 1840s, that the egg itself is but a single, albeit specialized, cell.

An important advance in embryology was the proposal by the nineteenth-century German biologist, August Weismann, that an offspring does not inherit its characteristics from the body (the soma) of the parent but only from the **germ cells**—egg and sperm. Weismann drew a fundamental distinction between germ cells and the body cells or **somatic cells** (Fig. 1.5). Characteristics acquired by the body during an animal's life cannot be transmitted to the germline. As far as heredity is concerned, the body is merely a carrier of germ cells. As the English novelist and essayist Samuel Butler put it: 'A hen is only an egg's way of making another egg.'

Work on sea-urchin eggs showed that after fertilization the egg contains two nuclei, which eventually fuse; one of these nuclei belongs to the egg, while the other comes from the sperm. Fertilization therefore results in a single cell—the **zygote**—carrying a nucleus with contributions from both parents, and it was concluded that the cell nucleus must contain the physical basis of heredity. The climax of this line of research was the demonstration, towards the end of the nineteenth century, that the

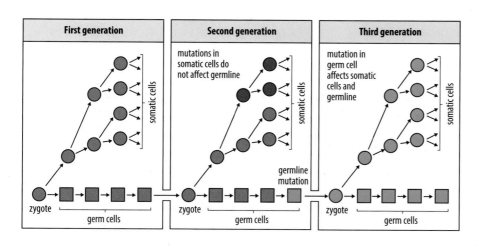

Fig. 1.5 The distinction between germ cells and somatic cells. In each generation a germ cell contributes to the zygote, which gives rise to both somatic cells and germ cells, but inheritance is through the germ cells only (first panel). Changes that occur due to a mutation (red) in a somatic cell can be passed on to its daughter cells but do not affect the germline, as shown in the second panel. In contrast, a mutation in the germline (green) in the second generation will be present in every cell in the body of the new organism to which that cell contributes, and will also be passed on to the third and future generations through the germline, as shown in the third panel.

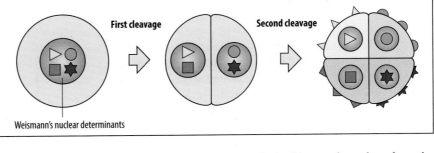

Fig. 1.6 Weismann's theory of nuclear determination. Weismann assumed that there were factors in the nucleus that were distributed asymmetrically to daughter cells during cleavage and directed their future development.

chromosomes within the nucleus of the zygote are derived in equal numbers from the two parental nuclei, and the recognition that this provided a physical basis for the transmission of genetic characters according to the laws developed by the Austrian botanist and monk, Gregor Mendel. The number of chromosomes is kept constant from generation to generation by a specialized type of cell division that produces the germ cells, called **meiosis**, which halves the chromosome number; the full complement of chromosomes is then restored at fertilization. The zygote and the somatic cells that arise from it divide by the process of **mitosis**, which maintains chromosome number (Box 1B). Germ cells contain a single copy of each chromosome and are called **haploid**, whereas germ-cell precursor cells and the other somatic cells of the body contain two copies and are called **diploid**.

1.3 Two main types of development were originally proposed

The next big question was how cells became different from one another during embryonic development. With the increasing emphasis on the role of the nucleus, in the 1880s Weismann put forward a model of development in which the nucleus of the zygote contained a number of special factors, or **determinants** (Fig. 1.6). He proposed that while the fertilized egg underwent the rapid cycles of cell division known as **cleavage** (see Box 1A), these nuclear determinants would be distributed unequally to the daughter cells and so would control the cells' future development. The fate of each cell was therefore predetermined in the egg by the factors it would receive during cleavage. This type of model was termed 'mosaic,' as the egg could be considered to be a mosaic of discrete localized determinants. Central to Weismann's theory was the assumption that early cell divisions must make the daughter cells quite different from each other as a result of unequal distribution of nuclear components.

In the late 1880s, initial support for Weismann's ideas came from experiments carried out independently by the German embryologist, Wilhelm Roux, who experimented with frog embryos. Having allowed the first cleavage of a fertilized frog egg, Roux destroyed one of the two cells with a hot needle and found that the remaining cell developed into a well-formed half-larva (Fig. 1.7). He concluded that the

Fig. 1.7 Roux's experiment to investigate Weismann's theory of mosaic development. After the first cleavage of a frog embryo, one of the two cells is killed by pricking it with a hot needle; the other remains undamaged. At the blastula stage the undamaged cell can be seen to have divided as normal into many cells that fill half of the embryo. The development of the blastocoel, a small fluid-filled space in the center of the blastula, is also restricted to the undamaged half. In the damaged half of the embryo, no cells appear to have formed. At the neurula stage, the undamaged cell has developed into something resembling half a normal embryo.

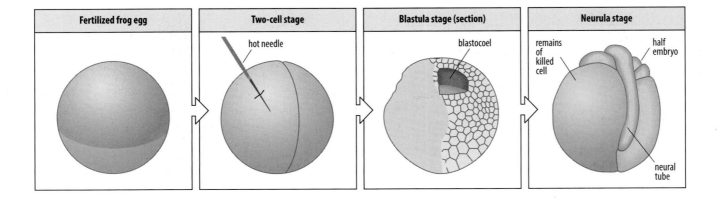

CELL BIOLOGY BOX 1B   The mitotic cell cycle

When a eukaryotic cell duplicates itself it goes through a fixed sequence of events called the **cell cycle**. The cell grows in size, the DNA is replicated, and the replicated chromosomes then undergo mitosis and become segregated into two daughter nuclei. Only then can the cell divide to form two daughter cells, which can go through the whole sequence again.

The standard eukaryotic mitotic cell cycle is divided into well-marked phases (Figure 1). At the M phase, mitosis and cell cleavage give rise to two new cells. The rest of the cell cycle, between one M phase and the next, is called interphase. Replication of DNA occurs during a defined period in interphase, the S phase (the S stands for synthesis of DNA). Preceding S phase is a period known as G_1 (the G stands for gap), and after it another interval known as G_2, after which the cells enter mitosis (see figure). G_1, S phase, and G_2 collectively make up interphase, the part of the cell cycle during which cells synthesize proteins and grow, as well as replicating their DNA. When somatic cells are not proliferating they are usually in a state known as G_0, into which they withdraw after mitosis. The decision to enter G_0 or to proceed through G_1, may be controlled by both intracellular state and extracellular signals such as growth factors. Growth factors enable the cell to proceed out of G_0 and progress through the cell cycle. Cells such as neurons and skeletal muscle cells, which do not divide after differentiation, are permanently in G_0.

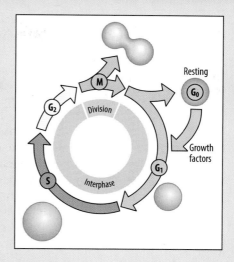

Figure 1

Particular phases of the cell cycle are absent in some cells: during cleavage of the fertilized *Xenopus* egg G_1 and G_2 are virtually absent, and cells get smaller at each division. In *Drosophila* salivary glands there is no M phase, as the DNA replicates repeatedly without mitosis or cell division, leading to the formation of giant **polytene chromosomes**.

'development of the frog is based on a mosaic mechanism, the cells having their character and fate determined at each cleavage.'

But when Roux's fellow countryman, Hans Driesch, repeated the experiment on sea-urchin eggs, he obtained quite a different result (Fig. 1.8). He wrote later: 'But things turned out as they were bound to do and not as I expected; there was, typically, a

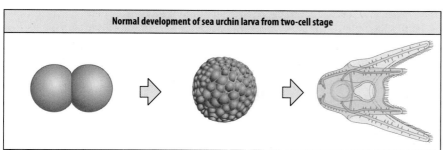

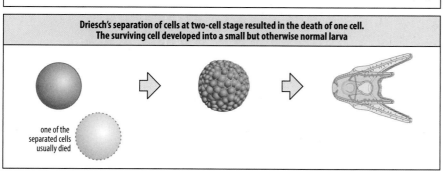

Fig. 1.8 The outcome of Driesch's experiment on sea urchin embryos, which first demonstrated the phenomenon of regulation. After separation of cells at the two-cell stage, the remaining cell developed into a small, but whole, normal larva. This is the opposite of Roux's earlier finding that when one of the cells of a two-cell frog embryo is damaged, the remaining cell develops into a half-embryo only (see Fig. 1.7).

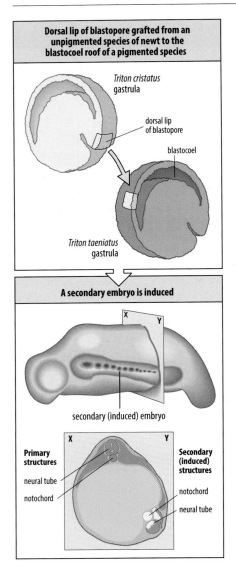

Dorsal lip of blastopore grafted from an unpigmented species of newt to the blastocoel roof of a pigmented species

Triton cristatus gastrula

dorsal lip of blastopore

blastocoel

Triton taeniatus gastrula

A secondary embryo is induced

secondary (induced) embryo

Primary structures

neural tube

notochord

Secondary (induced) structures

notochord

neural tube

Fig. 1.9 The dramatic demonstration by Spemann and Mangold of induction of a new main body axis by the organizer region in the early amphibian gastrula. A piece of tissue (yellow) from the dorsal lip of the blastopore of a newt (*Triton cristatus*) gastrula is grafted to the opposite side of a gastrula of another, pigmented, newt species (*Triton taeniatus*, pink). The grafted tissue induces a new body axis containing neural tube and somites. The unpigmented graft tissue forms a notochord at its new site (see section in lower panel) but most of the neural tube and the other structures of the new axis have been induced from the pigmented host tissue. The organizer region discovered by Spemann and Mangold is known as the Spemann organizer.

whole gastrula on my dish the next morning, differing only by its small size from a normal one; and this small but whole gastrula developed into a whole and typical larva.'

Driesch had completely separated the cells at the two-cell stage and obtained a normal but small larva. That was just the opposite of Roux's result, and was the first clear demonstration of the developmental process known as regulation. The experiment of Roux on frogs was later repeated by the American T. H. Morgan, who separated the two blastomeres instead of killing one of them and leaving it attached, and he obtained the same result as Driesch with sea urchins. This showed the general ability of vertebrate embryos to regulate, that is, to restore normal development, even if some portions are removed or rearranged very early in development. The basis for this phenomenon is explained later in the chapter. The extent to which embryos can regulate differs in different species and we shall see many examples of regulation throughout the book. The existence of regulation does not mean, however, that the unequal distribution of determinants that make two daughter cells different from each other is not important during development. But Weismann was wrong in one crucial respect, in that such determinants are not nuclear but are located in the cell cytoplasm. We shall see many examples of developmentally important proteins and RNAs that act in this way as **cytoplasmic determinants**.

1.4 The discovery of induction showed that one group of cells could determine the development of neighboring cells

The fact that embryos can regulate implies that cells must communicate and interact with each other, but the central importance of **cell–cell interactions** in embryonic development was not really established until the discovery of the phenomenon of **induction**. This is where one cell, or tissue, directs the development of another, neighboring, cell or tissue.

The importance of induction and other cell–cell interactions in development was proved dramatically in 1924 when Hans Spemann and his assistant, Hilde Mangold, carried out a now famous transplantation experiment in amphibian embryos. They showed that a partial second embryo could be induced by grafting one small region of an early newt embryo onto another at the same stage (Fig. 1.9). The grafted tissue was taken from the **dorsal** lip of the **blastopore**—the slit-like invagination that forms where gastrulation begins on the dorsal surface of the amphibian embryo (see Box 1A). This small region they called the **organizer**, as it seemed to be ultimately responsible for controlling the organization of a complete embryonic body; it is now known as the **Spemann–Mangold organizer**, or just the **Spemann organizer**. For their discovery, Spemann received the Nobel Prize for Physiology or Medicine in 1935, the first Nobel Prize ever given for embryological research. Sadly, Hilde Mangold had died earlier, in an accident, and so could not be honored.

1.5 Developmental biology emerged from the coming together of genetics and embryology

When Mendel's laws were rediscovered in 1900 there was a great surge of interest in mechanisms of inheritance, particularly in relation to evolution, but less so in relation to embryology. Genetics was seen as the study of the transmission of hereditary elements from generation to generation, whereas embryology was the study of how an individual organism develops and, in particular, how cells in the early embryo became different from each other. Genetics seemed, in this respect, to be irrelevant to development.

The fledgling science of genetics was put on a firm conceptual and experimental footing in the first quarter of the twentieth century by T. H. Morgan. Morgan chose the fruit fly *Drosophila melanogaster* as his experimental organism. He noticed a fly with white eyes rather than the usual red eyes, and by careful cross-breeding he showed that inheritance of this mutant trait was linked to the sex of the fly. He found three other sex-linked traits and worked out that they were each determined by three distinct 'genetic loci,' which occupied different positions on the same chromosome, the fly's X chromosome. The rather abstract hereditary 'factors' of Mendel had been given reality. But even though Morgan was originally an embryologist, he made little headway in explaining development in terms of genetics. That had to wait until the nature of the gene was better understood.

An important concept in understanding how genes influence physical and physiological traits is the distinction between **genotype** and **phenotype**. This was first put forward by the Danish botanist, Wilhelm Johannsen, in 1909. The genetic endowment of an organism—the genetic information it inherits from its parents—is the genotype. The organism's visible appearance, internal structure, and biochemistry comprise the phenotype. While the genotype certainly controls development, environmental factors interacting with the genotype influence the phenotype. Despite having identical genotypes, identical twins can develop considerable differences in their phenotypes as they grow up (Fig. 1.10), and these tend to become more evident with age.

Following Morgan's discoveries in genetics, the problem of development could now be posed in terms of the relationship between genotype and phenotype: how the genetic endowment becomes 'translated' or 'expressed' during development to give rise to a functioning organism. But the coming together of genetics and embryology was slow and tortuous. The discovery in the 1940s that genes are made of DNA and encode proteins was a major turning point. It was already clear that the properties of a cell are determined by the proteins it contains, and so the fundamental role of genes in development could at last be appreciated. By controlling which proteins were made in a cell, genes could control the changes in cell properties and behavior that occurred during development. A further major advance in the 1960s was the discovery that some genes encode proteins that control the activity of other genes.

Fig. 1.10 The difference between genotype and phenotype. These identical twins have the same genotype because one fertilized egg split into two during development. Their slight difference in appearance is due to nongenetic factors, such as environmental influences.

Photograph courtesy of Josè and Jaime Pascual.

1.6 Development is studied mainly through selected model organisms

Although the embryology of many different species has been studied at one time or another, a relatively small number of organisms provide most of our knowledge about developmental mechanisms. We can thus regard them as 'models' for understanding the processes involved, and they are often called **model organisms**. Sea urchins and amphibians were the main animals used for the first experimental investigations because their developing embryos are easy to obtain and, in the case of amphibians, relatively easy to manipulate experimentally, even at quite late stages. Among vertebrates, the frog *Xenopus laevis*, the mouse (*Mus musculus*), the chicken (*Gallus gallus*), and the zebrafish (*Danio rerio*), are the main model organisms now studied. Among invertebrates, the fruit fly *Drosophila melanogaster* and the nematode *Caenorhabditis elegans* have been the focus of most attention, because a great deal is known about their developmental genetics and they can be easily genetically modified. Two Nobel prizes have been awarded for discoveries about development in *Drosophila* and *Caenorhabditis*, respectively. With the advent of modern methods of genetic analysis, there has also been a resurgence of interest in the sea urchin *Strongylocentrotus purpuratus*. For plant developmental biology, *Arabidopsis thaliana* serves as the main model organism. The life cycles and background details for these model organisms are given in the relevant chapters later in the book. The evolutionary relationships of these organisms are shown in Fig. 1.11.

The reasons for these choices are partly historical—once a certain amount of research has been done on one animal it is more efficient to continue to study it rather than start at the beginning again with another species—and partly a question of ease of study and biological interest. Each species has its advantages and disadvantages as a developmental model. The chick embryo, for example, has long been studied as a model for vertebrate development because fertile eggs are easily available and the embryo withstands experimental microsurgical manipulation very well. A disadvantage, however, was that until very recently little was known about the chick's

Scan here

Scan this QR code image with your mobile device to see the online supplementary material on Ascidians or log on to **http://global.oup.com/uk/orc/biosciences/ devbiol/wolpert5e/qr/qr1a/**

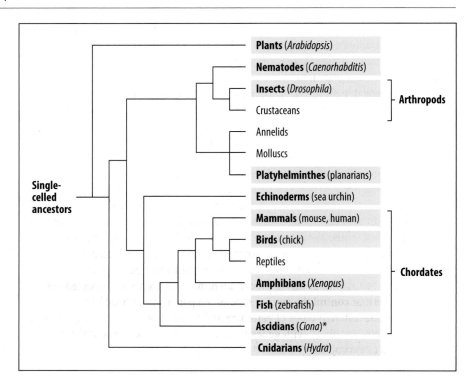

Fig. 1.11 Phylogenetic tree showing the placement of the main developmental model organisms. The organisms discussed in this book are highlighted in blue. *Ascidian development is described online.

developmental genetics. In contrast, we know a great deal about the genetics of the mouse, although the mouse is more difficult to study in some ways, as development normally takes place entirely within the mother. It is, however, possible to fertilize mouse eggs in culture, let them develop for a short time outside the uterus, when they can be observed, and then re-implant the early embryos for them to complete their development. Many developmental mutations have been identified in the mouse, and it is also amenable to genetic modification by **transgenic** techniques, by which genes can be introduced, deleted, or modified in living organisms. It is also the best experimental model we have for studying mammalian development. The zebrafish is a more recent addition to the select list of vertebrate model organisms; it is easy to breed in large numbers, the embryos are transparent and so cell divisions and tissue movements can be followed visually, and it has great potential for genetic investigations.

A major goal of developmental biology is to understand how genes control embryonic development, and to do this one must first identify which genes, out of the many thousands in the organism, are critically and specifically involved in controlling development. This task can be approached in various ways, depending on the organism involved, but the general starting point is to identify gene mutations that alter development in some specific and informative way, as described in the following section. Techniques for identifying such developmental genes and for detecting and manipulating their expression in the organism are described throughout the book, along with techniques for manipulating the genes themselves.

As we have just discussed, some of our model organisms are more amenable to conventional genetic analysis than others. Despite its importance in developmental biology, little conventional genetics has been done on *X. laevis*, which has the disadvantage of being **tetraploid** (it has four sets of chromosomes in its somatic cells, as opposed to the two sets carried by the cells of diploid organisms, such as humans and mice) and a relatively long breeding period of 1–2 years to reach sexual maturity. Using modern genetic and bioinformatics techniques, however, many developmental genes have been identified in *X. laevis* by direct DNA sequence comparison with known genes in *Drosophila* and mice. The closely related frog *Xenopus tropicalis* is a more attractive organism for genetic analysis; it is diploid and can also be genetically manipulated to produce transgenic organisms.

The hereditary information of a particular organism, comprising its complete set of genes and non-coding DNA sequences, is known as its **genome**. In sexually reproducing diploid species, two complete copies of the genome are carried by the chromosomes in each somatic cell—one copy inherited from the father and one from the mother. Genome sequences are available for all our model organisms, although some sequences are more complete than others. In the zebrafish, a genome-duplication event in its evolutionary history means that there are duplicates of at least 2900 out of the estimated 20,000 protein-coding genes in its genome. Thus the zebrafish genome contains a total of over 26,000 protein-coding genes, whereas the mouse genome has only 22,000. Having the complete DNA sequences of the genomes of our model organisms helps enormously in identifying developmental genes and other developmentally important DNA sequences.

In general, when an important developmental gene has been identified in one animal, it has proved very rewarding to see whether a corresponding gene is present and is acting in a developmental capacity in other animals. Such genes are often identified by a sufficient degree of nucleotide sequence similarity to indicate descent from a common ancestral gene. Genes that meet this criterion are known as **homologous genes**. As we shall see in Chapter 5, this approach identified a hitherto unsuspected class of vertebrate genes that control the regular segmented pattern from head to tail, which is represented by the different types of vertebrae at different positions. These genes were identified by their **homology** with genes that determine the identities of the different body segments in *Drosophila* (described in Chapter 2).

1.7 The first developmental genes were identified as spontaneous mutations

Most of the organisms dealt with in this book are sexually reproducing diploid organisms: their somatic cells contain two copies of each gene, with the exception of those on the sex chromosomes. In a diploid species, one copy, or **allele**, of each gene is contributed by the male parent and the other by the female. For many genes there are several different alleles present in the population, which leads to the variation in phenotype one normally sees within any sexually reproducing species. Occasionally, however, a mutation will occur spontaneously in a gene and there will be marked change, usually deleterious, in the phenotype of the organism.

Many of the genes that affect development have been identified by spontaneous mutations that disrupt their function and produce an abnormal phenotype. Mutations are classified broadly according to whether they are dominant or recessive (Fig. 1.12).

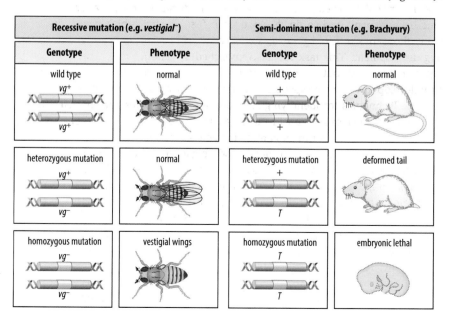

Fig. 1.12 Types of mutations. Left: a mutation is recessive when it only has an effect in the homozygous state; that is, when both copies of the gene carry the mutation. The *vestigial* mutation in *Drosophila* leads to the absence of wings when homozygous. A plus sign (+) denotes the 'normal' allele (wild type), and a minus sign (−) the recessive allele. Right: by contrast, a dominant or semi-dominant mutation produces an effect on the phenotype in the heterozygous state; that is, when just one copy of the gene carries the mutation. The Brachyury mutation in the mouse has an obvious but non-lethal effect in the heterozygous state, but is lethal in the homozygous state. *T* denotes the dominant mutant allele of the *brachyury* gene.

Dominant and **semi-dominant** mutations are those that produce a distinctive phenotype when the mutation is present in just one allele of a pair; that is, it exerts an effect in the **heterozygous** state. By contrast, **recessive** mutations alter the phenotype only when both alleles are of the mutant type; this is known as the **homozygous** state. In recessive mutations, the function of the one wild-type gene is sufficient to confer a normal phenotype in heterozygotes, whereas in dominant or semi-dominant mutations, the effect of the mutant allele completely or partly overrides that of the wild-type gene in the heterozygote.

In general, dominant mutations are more easily recognized, particularly if the phenotype involves alterations in gross anatomy or coloration, provided that they do not cause the early death of the embryo in the heterozygous state. True dominant mutations are rare. However, if one functional allele is not enough to provide the full function of a gene, a heterozygous mutant will have an abnormal phenotype. An example of this is seen in mice that have one copy of the mutant allele of the *brachyury* gene denoted by *T*. These heterozygous mice (*T/+*) have short tails. Because of this phenotype, the gene affected in the mutants was called *brachyury* (from the Greek *brachy*, short, and *oura*, tail). When the mutation is homozygous (*T/T*), it has a much greater effect, and embryos die at an early stage, with very short bodies, indicating that the *brachyury* gene is required for normal embryonic development (Fig. 1.13). Once breeding studies had confirmed that a single gene was involved, the mutation could be mapped to a location on a particular chromosome by classical gene-mapping techniques, and it was found that the *T* allele represents a deletion of the *brachyury* gene.

Identifying genes affected by recessive mutations is more laborious, as the heterozygote has a phenotype identical to a normal wild-type animal, and a carefully worked-out breeding program is needed to obtain homozygotes. In mammals, identifying potentially lethal recessive developmental mutations requires careful observation and analysis, as the homozygotes may die unnoticed inside the mother.

Many mutations in invertebrates have been identified as **conditional mutations**. The effects of these mutations only show up when the animal is kept in a particular condition—most commonly at a higher temperature, in which case the mutation is known as a **temperature-sensitive mutation**. At the usual ambient temperature, the animal appears normal. Temperature sensitivity is usually due to the encoded mutant protein being able to fold into a functional structure at the normal temperature, but being less stable at the higher temperature.

Very rigorous criteria must be applied to identify mutations that are affecting a genuine developmental process and not just affecting some vital but routine 'housekeeping'

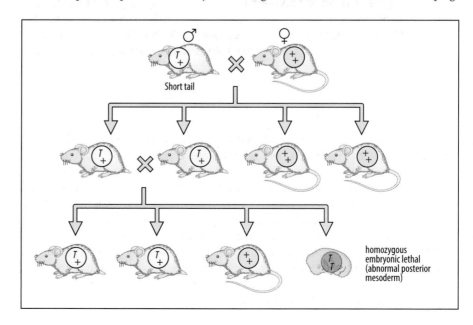

Fig. 1.13 Genetics of the semi-dominant *T* mutation, which deletes the *brachyury* gene in the mouse. A male heterozygote carrying the *T* mutation merely has a short tail. When mated with a female homozygous for the normal, or wild-type, *brachyury* gene (+/+), some of the offspring will also be heterozygotes and have short tails. Mating two heterozygotes together will result in some of the offspring being homozygous (*T/T*) for the mutation, resulting in a severe and lethal developmental abnormality in which the posterior mesoderm does not develop.

Short tail

homozygous embryonic lethal (abnormal posterior mesoderm)

function without which the animal cannot survive. One simple criterion for a developmental mutation is that it leads to the death of the embryo, but mutations in genes involved in vital housekeeping functions can also have this effect. Mutations that produce observable abnormal embryonic development are more promising candidates for true developmental mutations. In later chapters we shall see how large-scale screening for mutations after mutagenesis by chemicals or X-rays has identified many more developmental genes than would ever have been picked up just by looking for rare spontaneous mutations.

In recent years, new techniques for identifying developmental genes have appeared. These depend on knowing the existence and sequence of a gene (for example, from genomic sequences) and working backwards to determine its function in the animal, usually by removing the gene or blocking its function. Such methods are known as **reverse genetics**, as opposed to the traditional methods of **forward genetics** outlined above. Examples of reverse genetics include **gene knock-out**, in which the gene is effectively deleted from the animal's genome by transgenic techniques (discussed in Section 3.10), and **gene knockdown** or **gene silencing**, in which gene expression is prevented by techniques such as RNA interference or antisense RNA (discussed in Box 6B).

SUMMARY

The study of embryonic development started with the Greeks more than 2000 years ago. Aristotle put forward the idea that embryos were not contained completely preformed in miniature within the egg, but that form and structure emerged gradually, as the embryo developed. This idea was challenged in the seventeenth and eighteenth centuries by those who believed in preformation, the idea that all embryos that had been, or ever would be, had existed from the beginning of the world. The emergence of the cell theory in the nineteenth century finally settled the issue in favor of epigenesis, and it was realized that the sperm and egg are single, albeit highly specialized, cells. Some of the earliest experiments showed that very early sea-urchin embryos are able to regulate—that is, to develop normally even if cells are removed or killed. This established the important principle that development must depend, at least in part, on communication between the cells of the embryo. Direct evidence for the importance of cell–cell interactions came from the organizer graft experiment carried out by Spemann and Mangold in 1924, showing that the cells of the amphibian organizer region could induce a new partial embryo from host tissue when transplanted into another embryo. The role of genes in controlling development, by determining which proteins are made, has only been fully appreciated in the past 50 years and the study of the genetic basis of development has been made much easier in recent times by the techniques of molecular biology and the availability of the DNA sequences of whole genomes.

A conceptual tool kit

Development into a multicellular organism is the most complicated fate a single living cell can undergo; in this lies both the fascination and the challenge of developmental biology. Yet only a few basic principles are needed to start to make sense of developmental processes. The rest of this chapter is devoted to introducing these key concepts. These principles are encountered repeatedly throughout the book, as we look at different organisms and developmental systems, and should be regarded as a conceptual tool kit, essential for embarking on a study of development.

Genes control development by determining where and when proteins are synthesized. Gene activity sets up intracellular networks of interactions between proteins and genes, and between proteins and proteins, that give cells their particular properties. One of these properties is the ability to communicate with, and respond to, other cells. It is these cell–cell interactions that determine how the embryo develops; no developmental process can therefore be attributed to the function of a single gene or single protein. The amount of genetic and molecular information on developmental processes is now enormous. In this book, we will be highly selective and describe only those molecular details that give an insight into the mechanisms of development and illustrate general principles.

1.8 Development involves the emergence of pattern, change in form, cell differentiation, and growth

In most embryos, fertilization is immediately followed by cell division. This stage is known as cleavage, and divides the fertilized egg into a number of smaller cells (Fig. 1.14). Unlike the cell divisions that take place during growth of a tissue, there is no increase in cell size between each cleavage division; the cleavage cell cycles consist simply of phases of DNA replication, mitosis, and cell division (see Box 1B). Early embryos are therefore no bigger than the zygote, and there is little increase in size during embryogenesis unless the embryo has access to a substantial source of material it can use to grow—for example, the large yolk available to the chick embryo and the nutrients provided to the mammalian embryo through the placenta.

Development is in essence the emergence of organized structures from an initially very simple group of cells. It is convenient to distinguish four main developmental processes, which occur in roughly sequential order in development, although in reality they overlap with, and influence, each other considerably. They are **pattern formation, morphogenesis, cell differentiation**, and **growth**. Pattern formation is the process by which cellular activity is organized in space and time so that a well-ordered structure develops within the embryo. It is of fundamental importance in the early embryo and also later in the formation of organs. In the developing arm, for example, pattern formation enables cells to 'know' whether to make an upper arm or fingers, and where the muscles should form. There is no universal strategy of patterning; rather, it is achieved by various cellular and molecular mechanisms in different organisms and at different stages of development.

Pattern formation initially involves laying down the overall **body plan**—defining the main body axes of the embryo that run from head (**anterior**) to tail (**posterior**), and from back (**dorsal**) to underside (**ventral**). Most of the animals in this book have a head at one end and a tail at the other, with the left and right sides of the body being outwardly bilaterally symmetrical—that is, a mirror image of each other. In such animals, the main body axis is the **antero-posterior axis**, which runs from the head to the tail. Bilaterally symmetrical animals also have a **dorso-ventral axis**, running from the back to the belly. A striking feature of these axes is that they are almost always at right angles to one another and so can be thought of as making up a system of coordinates on which any position in the body could be specified (Fig. 1.15). Internally, animals have distinct differences between the left and right sides in the disposition of their internal organs—the human heart, for example, is on the left side. In plants, the main body axis runs from the growing tip (the apex) to the roots and is known as the **apical–basal axis**. Plants also have radial symmetry, with a **radial axis** running from the center of the stem outwards.

Even before the body axes become clear, eggs and embryos often show a distinct **polarity**, which in this context means that they have an intrinsic orientation, with one end different from the other. Polarity can be specified by a gradient in a protein

Fig. 1.14 Light micrographs of cleaving *Xenopus* eggs.

Images courtesy of Dr. H. Williams.

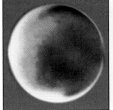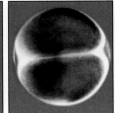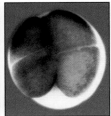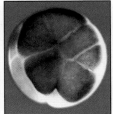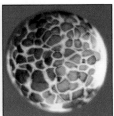

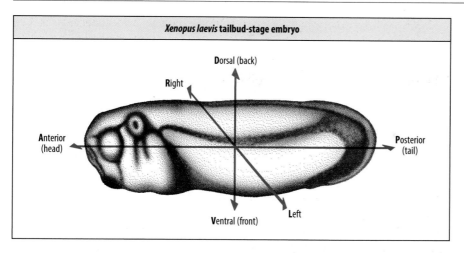

Xenopus laevis **tailbud-stage embryo**

Dorsal (back)
Right
Anterior (head)
Posterior (tail)
Ventral (front)
Left

Fig. 1.15 The main axes of a developing embryo. The antero-posterior axis and the dorso-ventral axis are at right angles to one another, as in a coordinate system.

or other molecule, as the slope of the gradient provides direction. Individual cells within the embryo can also have their own polarity. One type of polarity is called **apico-basal polarity**, and is seen in simple **epithelia**, single-layered sheets of cells in which the two sides of the sheet (the apical and basal sides) are different, and which are building blocks of tissues and organs (discussed in detail in Chapter 9). A second axis of polarity in cells can be discerned along the plane of the tissue and this provides a larger-scale coordinate system for the tissue. This type of individual cell polarity is known as **planar cell polarity** and is discussed further in Chapters 2 and 9. Planar cell polarity is sometimes easily visible in structures—the hairs produced by epidermal cells on the edge of the *Drosophila* wing all point in the same direction.

At the same time that the axes of the embryo are being formed, cells are being allocated to the different **germ layers**—ectoderm, mesoderm, and endoderm (Box 1C).

CELL BIOLOGY BOX 1C Germ layers

The concept of germ layers is useful to distinguish between regions of the early embryo that give rise to quite distinct types of tissues. It applies to both vertebrates and invertebrates. All the animals considered in this book, except for the cnidarian *Hydra*, are **triploblasts**, with three germ layers: the **ectoderm**, which gives rise to the epidermis of the skin and nervous system; the **mesoderm**, which gives rise to the skeletomuscular system, connective tissues, and other internal organs, such as the kidney and heart; and the **endoderm**, which gives rise to the gut and its derivatives, such as the liver and lungs in vertebrates (Figure 1). These are specified early in development. There are, however, some exceptions to the types of tissue that arise from a particular germ layer. The neural crest in vertebrates, for example, is ectodermal in origin, but gives rise not only to neural tissue but also to some skeletal elements that generally arise from mesoderm.

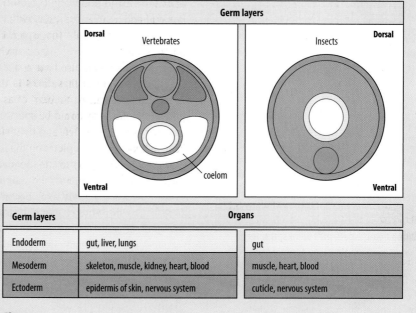

Germ layers

Vertebrates
Dorsal
Ventral
coelom

Insects
Dorsal
Ventral

Germ layers	Organs		
Endoderm	gut, liver, lungs		gut
Mesoderm	skeleton, muscle, kidney, heart, blood		muscle, heart, blood
Ectoderm	epidermis of skin, nervous system		cuticle, nervous system

Figure 1

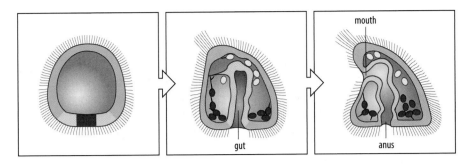

Fig. 1.16 Gastrulation in the sea urchin. Gastrulation transforms the spherical blastula into a structure with a tube—the gut—through the middle. The left-hand side of the embryo has been removed.

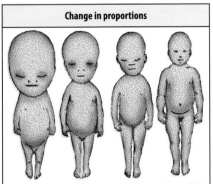

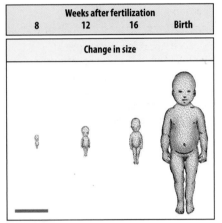

Fig. 1.17 The human fetus changes shape as it grows. From the time the body plan is well established, at 8 weeks, the fetus increases in length some tenfold until birth (upper panel), while the relative proportion of the head to the rest of the body decreases (lower panel). As a result, the proportions and the shape of the fetus change. Scale bar = 10 cm.

After Moore, K.L.: The Developing Human, 1983. Saunders.

During further pattern formation, cells of these germ layers acquire different identities so that organized spatial patterns of differentiated cells finally emerge, such as the arrangement of skin, muscle, and cartilage in developing limbs, and the arrangement of neurons in the nervous system.

The second important developmental process is change in form, or **morphogenesis** (discussed in Chapter 9). Embryos undergo remarkable changes in three-dimensional form—you need only think about the complicated anatomy of structures such as your hands and feet. At certain stages in development, there are characteristic and dramatic changes in form, of which **gastrulation** is the most striking. Almost all animal embryos undergo gastrulation, during which endoderm and mesoderm move inside, the gut is formed, and the main body plan emerges. During gastrulation, cells on the outside of the embryo move inwards and, in animals such as the sea urchin, gastrulation transforms a hollow spherical blastula into a gastrula with a tube through the middle—the gut (Fig. 1.16). Morphogenesis in animal embryos can also involve extensive cell migration. Most of the cells of the human face, for example, are derived from cells that migrated from a tissue called the neural crest, which originates on the dorsal side of the embryo. Morphogenesis can also involve developmentally programmed cell death, called **apoptosis**, which is responsible, for example, for the separation of our fingers and toes from an initially solid plate of tissue.

The third developmental process we must consider here is **cell differentiation**, in which cells become structurally and functionally different from each other, ending up as distinct cell types, such as blood, muscle, or skin cells. Differentiation is a gradual process, cells often going through several divisions between the time at which they start differentiating and the time they are fully differentiated (when many cell types stop dividing altogether), and is discussed in Chapter 8. In humans, the fertilized egg gives rise to hundreds of clearly distinguishable types of cell.

Pattern formation and cell differentiation are very closely interrelated, as we can see by considering the difference between human arms and legs. Both contain exactly the same types of cell—muscle, cartilage, bone, skin, and so on—yet the pattern in which they are arranged is clearly different. It is essentially pattern formation that makes us look different from elephants and chimpanzees.

The fourth process is **growth**—the increase in size. In general there is little growth during early embryonic development and the basic pattern and form of the embryo is laid down on a small scale, always less than a millimeter in extent. Subsequent growth can be brought about in various ways: cell multiplication, increase in cell size, and deposition of extracellular materials, such as, for example, in bone. Growth can also be morphogenetic, in that differences in growth rates between organs, or between parts of the body, can generate changes in the overall shape of the embryo (Fig. 1.17), as we shall see in more detail in Chapter 13.

These four developmental processes are neither independent of each other nor strictly sequential. In very general terms, however, one can think of pattern formation in early development specifying differences between cells that lead to changes in form, cell differentiation, and growth. But in any real developing system there are many twists and turns in this sequence of events.

1.9 Cell behavior provides the link between gene action and developmental processes

Gene expression within cells leads to the synthesis of proteins that are responsible for particular cellular properties and behavior, which in turn determine the course of embryonic development. The past and current patterns of gene activity confer a certain state, or identity, on a cell at any given time, which is reflected in its molecular organization—in particular which proteins are present. As we shall see, embryonic cells and their progeny undergo many changes in state as development progresses. Other categories of cell behavior that will concern us are intercellular communication, also known as **cell–cell signaling**, changes in cell shape and cell movement, cell proliferation, and cell death.

Changing patterns of gene activity during early development are essential for pattern formation. They give cells identities that determine their future behavior and lead eventually to their final differentiation. And, as we saw in the example of induction by the Spemann organizer, the capacity of cells to influence each other's fate by producing and responding to signals is crucial for development. By their response to signals for cell movement or a change in shape, for example, cells generate the physical forces that bring about morphogenesis (Fig. 1.18). The curvature of a sheet of cells into a tube, as happens in *Xenopus* and other vertebrates during formation of the neural tube (see Box 1A), is the result of contractile forces generated by cells changing their shape at certain positions within the cell sheet. An important feature of cell surfaces is the presence of adhesive proteins known as cell-adhesion molecules, which serve various functions: they hold cells together in tissues; they enable cells to sense the nature of the surrounding extracellular matrix; and they serve to guide migratory cells such as the neural crest cells of vertebrates, which leave the neural tube to form structures elsewhere in the body.

We can therefore describe and explain developmental processes in terms of how individual cells and groups of cells behave. Because the final structures generated by development are themselves composed of cells, explanations and descriptions at the cellular level can provide an account of how these adult structures are formed.

Because development can be understood at the cellular level, we can pose the question of how genes control development in a more precise form. We can now ask how genes are controlling cell behavior. The many possible ways in which a cell can behave therefore provide the link between gene activity and the morphology of the adult animal—the final outcome of development. Cell biology provides the means by which the genotype becomes translated into the phenotype.

1.10 Genes control cell behavior by specifying which proteins are made

What a cell can do is determined very largely by the proteins it makes. The hemoglobin in red blood cells enables them to transport oxygen; the cells lining the gut secrete specialized digestive enzymes; and skeletal muscle cells are able to contract because they contain contractile structures composed of the proteins myosin, actin and tropomyosin, as well as other muscle-specific proteins needed to make muscle function correctly. All these are specialized proteins and are not involved in the 'housekeeping' activities that are common to all cells and keep them alive and functioning. Housekeeping activities include the production of energy and the metabolic pathways involved in the breakdown and synthesis of molecules necessary for the life of the cell. Although there are qualitative and quantitative variations in housekeeping proteins in different cells, they are not important players in development. In development we are concerned primarily with those proteins that make cells different from one another and which are often known as **tissue-specific** proteins.

Genes control development mainly by specifying which proteins are made in which cells and when. In this sense the genes are passive participants in development,

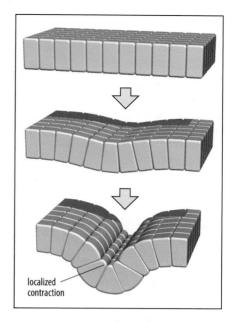

Fig. 1.18 Localized contraction of particular cells in a sheet of cells can cause the whole sheet to fold. Contraction of a line of cells at their apices due to changes in the cells' internal structure causes a furrow to form in a sheet of epithelium.

localized contraction

compared with the proteins they encode. These are the agents that directly determine cell behavior, which includes determining which genes are expressed. To produce a particular protein, its gene must be switched on and transcribed (**transcription**) into **messenger RNA** (**mRNA**). This process is known as **gene expression**. The mRNA must then be translated (**translation**) into protein. The **initiation of transcription**, or switching on of a gene, involves combinations of specialized **gene-regulatory proteins** binding to the **control region** of the gene. The control region is made up of *cis*-**regulatory regions** in the DNA (the *cis* simply refers to the fact that the regulatory region is on the same DNA molecule as the gene it controls). These *cis*-regulatory regions are often modular, and in many developmental genes the time of gene expression and the tissue in which it is expressed are controlled by different *cis*-**regulatory modules**.

Both transcription and translation are subject to several layers of control, and translation does not automatically follow transcription. Figure 1.19 shows the main stages in gene expression at which the production of a protein can be controlled. Even after a gene has been transcribed, for example, the mRNA may be degraded before it can be exported from the nucleus. Even if an mRNA reaches the cytoplasm, its translation may be prevented or delayed. In the eggs of many animals, preformed mRNA is prevented from being translated until after fertilization. mRNAs may also be targeted by specific **microRNAs** (**miRNAs**), short **non-coding RNAs** that can render the mRNA inactive and so block translation (see Box 6C). Some microRNAs are known to be involved in gene regulation in development. Another process that determines which proteins are produced is **RNA processing**. In eukaryotic organisms—all organisms other than bacteria and archaea—the initial RNA transcripts of many genes can be cut and spliced in different ways, called **alternative splicing**, to give rise to two or more different mRNAs; in this way, a number of different proteins with different properties can be produced from a single gene.

Even if a gene has been transcribed and the mRNA translated, the protein may still not be able to function immediately. Many newly synthesized proteins require further **post-translational modification** before they acquire biological activity. One very common permanent modification that occurs to proteins destined for the cell membrane or for secretion is the addition of carbohydrate side chains, or **glycosylation**. Reversible post-translational modifications, such as phosphorylation, can also significantly alter protein function. Together, alternative RNA splicing and post-translational modification mean that the number of functionally different proteins that can be produced is considerably greater than the number of protein-coding genes would indicate. For many proteins, their localization in a particular part of the cell, for example the nucleus, is essential for them to carry out their function.

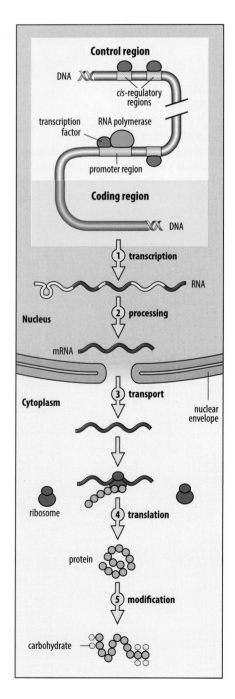

Fig. 1.19 Gene expression and protein synthesis. A protein-coding gene comprises a coding region—a stretch of DNA that contains the instructions for making the protein—and adjacent DNA sequences that act as a control region. The control region includes the promoter region, to which general transcription factors and the enzyme RNA polymerase bind to start transcription, and a *cis*-regulatory region, consisting of one or more modules, at which other transcription factors bind to switch the gene on or off. This latter control region may be thousands of base pairs away from the promoter. When the gene is switched on, the DNA sequence is transcribed to produce RNA (1). The RNA formed by transcription is spliced to remove introns and the non-coding region at the start of the gene (yellow) and processed within the nucleus (2) to produce mRNA. This is exported from the nucleus to the cytoplasm (3) for translation into protein at the ribosomes (4). Control of gene expression and protein synthesis occurs mainly at the level of transcription but can also occur at later stages. For example, mRNA may be degraded before it can be translated, or if it is not translated immediately it may be stored in inactive form in the cytoplasm for translation at a later stage. mRNAs may also be targeted by specific microRNAs so that translation is blocked. Some proteins require a further step of post-translational modification (5) to become biologically active. A very common post-translational modification is the addition of carbohydrate side-chains (glycosylation) as shown here.

An intriguing question is how many genes out of the total genome are **developmental genes**—that is, genes specifically required for embryonic development. This is not easy to estimate. In the early development of *Drosophila*, at least 60 genes are directly involved in pattern formation up to the time that the embryo becomes divided into segments. In *Caenorhabditis*, at least 50 genes are needed to specify a small reproductive structure known as the vulva. These are quite small numbers compared to the thousands of genes that are active at the same time; some of these are essential in that they are necessary for maintaining life, but they provide no or little information that influences the course of development. Many genes change their activity in a systematic way during development, suggesting that they could be developmental genes. The total number of genes in *Caenorhabditis* and *Drosophila* is around 19,000 and 15,000, respectively, and the number of genes involved in development is likely to be in the thousands. A systematic analysis of most of the *Caenorhabditis* genome suggests that approximately 9% of the genes (1722 out of 19,000) are involved in development.

Developmental genes typically code for proteins involved in the regulation of cell behavior: receptors; intercellular signaling proteins such as growth factors; intracellular signaling proteins; and gene-regulatory proteins. Many of these genes, especially those for receptors and signaling molecules, are used throughout an organism's life, but others are active only during embryonic development. Some of the techniques used to find out where and when a gene is active are described in Box 1D and Box 3B. It is also possible to prevent the expression of a given gene and so prevent its protein from being made. This can be done using **antisense RNAs**, which are small RNA molecules of about 25 nucleotides complementary in sequence to part of a gene or mRNA. Such antisense RNAs can block gene function by binding to the gene or its mRNA. Artificial RNA molecules called morpholinos are used, which are more stable than normal RNA. Another technique for blocking gene expression is **RNA interference (RNAi)**, which operates in a similar way, but uses a different type of small RNA called small interfering RNAs (siRNAs) to exploit cells' own metabolic pathways for degrading mRNAs. These techniques are discussed in more detail in Box 6B.

1.11 The expression of developmental genes is under tight control

All the somatic cells in an embryo are derived from the fertilized egg by successive rounds of mitotic cell division. Thus, with rare exceptions, they all contain identical genetic information, the same as that in the zygote. The fact that genetic information is not lost during differentiation was first shown in a landmark series of experiments by John Gurdon working with *Xenopus*. In these experiments, nuclei from differentiated cells were placed into unfertilized eggs in which the nucleus had been destroyed to test the ability of these nuclei to support normal development. In many cases tadpoles were obtained, and in a small number of cases even adult frogs, showing that differentiated cells still contain all the genetic information required to make a complete animal (Fig. 1.20). This is the principle of **genetic equivalence**. The differences between cells must therefore be generated by differences in gene activity that lead to the synthesis of different proteins. This is the principle of **differential gene expression**. John Gurdon was awarded the Nobel Prize in 2012 for this work, together with the Japanese biologist Shinya Yamanaka for his work on mouse stem cells (discussed in Chapter 8).

Turning the correct genes on or off in the right cells at the right time therefore becomes the central issue in development. As we shall see throughout the book, the genes do not provide a blueprint for development but a set of instructions. Key elements in regulating the readout of these instructions are the *cis*-regulatory regions that are associated with developmental genes and genes for specialized proteins (see Fig. 1.19). These are acted on by gene-regulatory proteins, which switch genes on or off by activating or repressing transcription. Some gene-regulatory proteins,

EXPERIMENTAL BOX 1D Visualizing gene expression in embryos

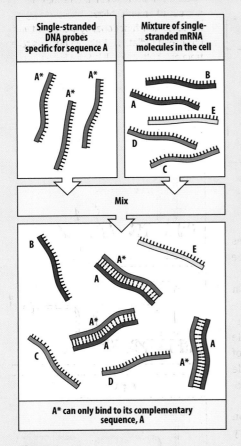

Figure 1

In order to understand how gene expression is guiding development, it is essential to know exactly where and when particular genes are active. Genes are switched on and off during development and patterns of gene expression are continually changing. Several powerful techniques can show where a gene is being expressed in the embryo.

One set of techniques uses *in situ* nucleic acid hybridization to detect the mRNA that is being transcribed from a gene. If a DNA or RNA probe is complementary in sequence to an mRNA being transcribed in the cell, it will base-pair precisely (hybridize), with the mRNA (Figure 1). A DNA probe can therefore be used to locate its complementary mRNA in a tissue slice or a whole embryo. The probe may be labeled in various ways—with a radioactive isotope, a fluorescent tag (fluorescence *in situ* hybridization, FISH), or an enzyme—to enable it to be detected. Radioactively labeled probes are detected by autoradiography, whereas probes labeled with fluorescent dyes are observed by fluorescence microscopy. Enzyme-labeled probes are detected by the enzyme's conversion of a colorless substrate into colored product (see Fig. 5.37a and b). Labeling with fluorescent dyes or enzymes has the advantage over radioactive labeling in that the expression of several different genes can be detected simultaneously, by labeling the different probes with different colored detectors. It also means that researchers do not have to work with radioactive chemicals.

Proteins can also be detected *in situ* by immunostaining with antibodies specific for the protein. The antibodies may be tagged directly with a colored dye or an enzyme, or, more usually, the protein-specific antibody is detected by a second round of staining with an anti-immunoglobin antibody tagged either with a fluorescent dye (as pictured here) or an enzyme (see Fig. 2.34). Figure 2 shows a section through the neural tube of a chick embryo at stage 16 (around 2–2.75 days after laying) which has been immunostained for expression of transcription factors that pattern the neural tube dorso-ventrally. This image was generated by immunostaining adjacent sections separately for each of the five transcription factors and then merging these images. The transcription factor was first labeled *in situ* using an antibody specific for the protein and then visualized by applying a secondary anti-immunoglobulin antibody labeled with a fluorescent dye (purple, Foxa2; red, Nkx2.2; yellow, Olig2; green, Pax6; blue, Pax7). These *in situ* labeling techniques, whether for RNA or protein, can only be applied to fixed embryos and tissue sections.

The pattern and timing of gene expression can also be followed by inserting a 'reporter' gene into an animal using transgenic techniques. The reporter gene codes for some easily detectable protein and is placed under the control of an appropriate promoter that will switch on the expression of the protein at a particular time and place. One commonly used reporter gene is the *lacZ* sequence coding for the bacterial enzyme β-galactosidase. The presence of the enzyme can be detected by treating the specimen after fixation with a modified substrate, which the enzyme reacts with to give a blue-colored product (see Fig. 5.28) or by using a fluorescent-tagged antibody specific for β-galactoside.

Other very commonly used reporter genes are those for small fluorescent proteins, such as green fluorescent protein (GFP), which fluoresce in various colors and were isolated from jellyfish and other marine organisms. These proteins are extremely useful as they are harmless to living cells and are readily made visible by illuminating cultured cells, embryos, or even whole animals with light of a suitable wavelength. Gene expression can therefore be visualized and followed in living embryos and cells (see Figs. 3.32 and 5.37c). The three scientists who discovered and developed GFP and other fluorescent proteins for use in live imaging were awarded the 2008 Nobel Prize for Chemistry.

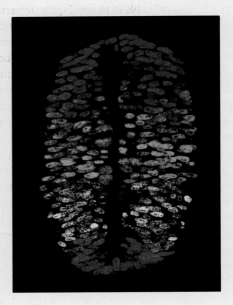

Figure 2

*Photograph courtesy of G. Le Dréau and E. Martí. From Le Dreau, G., Marti, E.: Dev. Neurobiol. 2012, **72**: 1471–1481.*

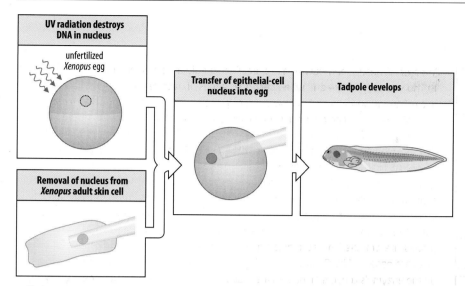

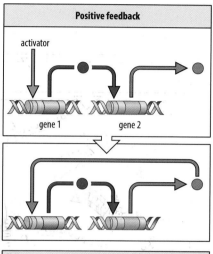

Fig. 1.20 Differentiated cells still contain all the genetic information required to make a complete animal. The haploid nucleus of an unfertilized *Xenopus* egg is rendered non-functional by treatment with ultraviolet radiation. A diploid nucleus taken from an adult skin cell is transferred into the enucleated egg, and can support the development of an embryo until at least the tadpole stage.

called **transcription factors**, act by binding directly to the DNA of the control regions (see Fig. 1.19), whereas others interact with transcription factors already bound to the DNA, and act as **co-activators** or **co-repressors**.

Developmental genes are highly regulated to ensure they are switched on only at the right time and place in development. This is a fundamental feature of development. To achieve this, developmental genes usually have extensive and complex *cis*-regulatory regions composed of one or more modules. Each module contains multiple binding sites for different transcription factors, and it is the precise combination of factors that binds that determines whether the gene is on or off. On average, a module will have binding sites for four to eight different transcription factors, and these factors will in turn often bind other co-activator or co-repressor proteins.

The modular nature of the regulatory region means that each module can function somewhat independently. This means that a gene with more than one regulatory module is usually able to respond quite differently to different combinations of inputs, and so can be expressed at different times and in different places within the embryo in response to developmental signals. Different genes can have the same regulatory module, which usually means that they will be expressed together, or different genes may have modules that contain some, but not all, of the binding sites in common, introducing subtle differences in the timing or location of expression. Thus, an organism's genes are linked in complex interdependent networks of expression through the modules of their *cis*-regulatory regions and the proteins that bind to them. Common examples of such regulation are **positive-feedback** and **negative-feedback** loops, in which a transcription factor respectively promotes or represses the expression of a gene whose product maintains that gene's expression (Fig. 1.21). Indeed, extensive networks of gene interactions for various stages of development have now been worked out for animals such as the sea urchin (discussed in Chapter 6).

As all the key steps in development reflect changes in gene activity, one might be tempted to think of development simply in terms of mechanisms for controlling gene expression. But that would be a big mistake. For gene expression is only the first step in a cascade of cellular processes that lead via protein synthesis to changes in cell behavior and so direct the course of embryonic development. Proteins are the machines that drive development. To think only in terms of genes is to ignore crucial aspects of cell biology, such as change in cell shape, which may be initiated at several steps removed from gene activity. In fact, there are very few cases where the complete sequence of events from gene expression to altered cell behavior has been worked out. The route leading from gene activity to a structure such as the five-fingered hand is tortuous, but progress has been made, as discussed in Chapter 11.

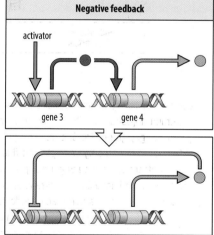

Fig. 1.21 Simple genetic feedback loops. Top: gene 1 is turned on by an activating transcription factor (green); the protein it produces (red) activates gene 2. The protein product of gene 2 (blue) not only acts on further targets, but also activates gene 1, forming a positive-feedback loop that will keep genes 1 and 2 switched on even in the absence of the original activator. Bottom: a negative-feedback loop is formed when the product of the gene at the end of a pathway (gene 4) can inhibit the first gene (gene 3), thus shutting down transcription of both genes. Arrows indicate activation; a barred line indicates inhibition.

1.12 Development is progressive and the fates of cells become determined at different times

As embryonic development proceeds, the organizational complexity of the embryo becomes vastly increased over that of the fertilized egg, as first recognized by Aristotle's 'epigenesis' principle outlined earlier in the chapter. Many different cell types are formed, spatial patterns emerge, and there are major changes in shape. All this occurs more or less gradually, depending on the particular organism. But, in general, the embryo is first divided up into a few broad regions, such as the future germ layers (mesoderm, ectoderm, and endoderm). Subsequently, the cells within these regions have their fates more and more finely determined. **Determination** implies a stable change in the internal state of a cell, and an alteration in the pattern of gene activity is assumed to be the initial step, leading to a change in the proteins produced in the cell. Different mesoderm cells, for example, eventually become determined as muscle, cartilage, bone, the fibroblasts of connective tissue, and the cells of the dermis of the skin.

An important concept in the process of cell determination is that of **lineage**, the direct line of descent of a particular cell. Once a cell is determined, this stable change is inherited by its descendants. Thus, the lineage of a cell has important consequences in terms of what type of cell it can eventually become. For example, once determined as ectoderm, cells will not subsequently give rise to endodermal tissues, or vice versa. In some organisms, such as *Caenorhabditis*, the entire body is derived from the zygote according to very strict lineages (see Chapter 6), but in most organisms, the fate of cells is not determined according to a fixed, invariant lineage.

It is important to understand clearly the distinction between the normal fate of a cell at any particular stage, and its state of determination. The **fate** of a group of cells merely describes what they will normally develop into. By marking cells of the early embryo one can find out, for example, which ectodermal cells will normally give rise to the nervous system, and of those, which, in particular, will give rise to the neural retina of the eye. In these types of experiments one obtains a 'fate map' of the embryo (see Section 3.7). However, the map in no way implies that those cells can only develop into a retina, or are already committed, or determined, to do so in the early embryo.

A group of cells is called **specified** if, when isolated and cultured in the neutral environment of a simple culture medium away from the embryo, they develop more or less according to their normal fate (Fig. 1.22). For example, cells at the animal pole

Fig. 1.22 The distinction between cell fate, determination, and specification. In this idealized system, regions A and B differentiate into two different sorts of cells, depicted as hexagons and squares. The fate map (first panel) shows how they would normally develop. If cells from region B are grafted into region A and now develop as A-type cells, the fate of region B has not yet been determined (second panel). By contrast, if region B cells are already determined when they are grafted to region A, they will develop as B cells (third panel). Even if B cells are not determined, they may be specified, in that they will form B cells when cultured in isolation from the rest of the embryo (fourth panel).

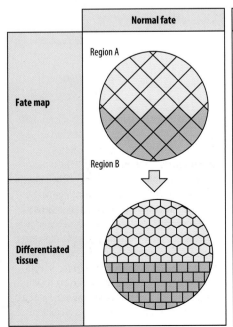
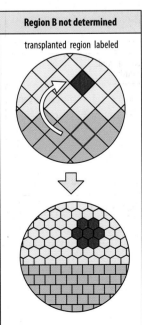
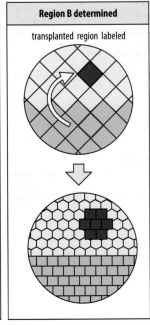
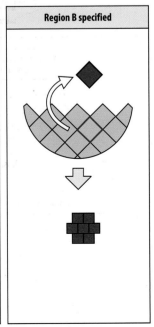

of the amphibian blastula (see Box 1A) are specified to form ectoderm, and will form epidermis when isolated. Cells that are 'specified' in this technical sense need not yet be 'determined', for influences from other cells can change their normal fate. If tissue from the animal pole is put in contact with cells from the vegetal pole, some animal pole tissue will form mesoderm instead of epidermis. At a later stage of development, however, the cells in the animal region have become determined as ectoderm and their fate cannot then be altered. Tests for specification rely on culturing the tissue in a neutral environment lacking any inducing signals, and this is often difficult to achieve, whereas it is much easier to test for determination.

The state of determination of cells at any developmental stage can be demonstrated by transplantation experiments. At the blastula stage of the amphibian embryo, one can graft the ectodermal cells that give rise to the eye into the side of the body and show that the cells develop according to their new position; that is, into mesodermal cells. At this early stage, their potential for development is much greater than their normal fate. However, if the same operation is done at a later stage, then the cells of the future eye region will form structures typical of an eye (Fig. 1.23). At the earlier stage the cells were not yet determined as prospective eye cells, whereas later they had become so, and so the graft development was **autonomous**, irrespective of its new location.

It is a general feature of development that cells in the early embryo are less narrowly determined than those at later stages; with time, cells become more and more restricted in their developmental potential. We assume that determination involves a change in the genes that are expressed by the cell, and that this change fixes or restricts the cell's fate, thus reducing its developmental options.

We have already seen how, even at the two-cell stage, the cells of the sea-urchin embryo do not seem to be determined. Each has the potential to generate a whole new larva (see Section 1.3). Embryos like these, where the developmental potential of cells is much greater than that indicated by their normal fate, are said to be capable

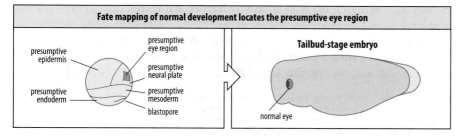

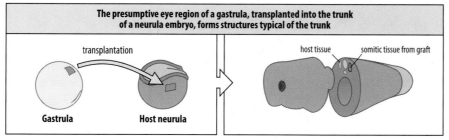

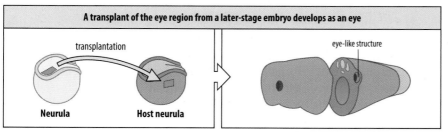

Fig. 1.23 Determination of the eye region with time in amphibian development. If the region of the gastrula that will normally give rise to an eye is grafted into the trunk region of a neurula (middle panel), the graft forms structures typical of its new location, such as notochord and somites. If, however, the eye region from a neurula is grafted into the same site (bottom panel), it develops as an eye-like structure, since at this later stage it has become determined.

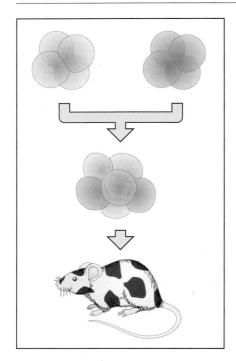

Fig. 1.24 Fusion of mouse embryos gives rise to a chimera. If a four-cell stage embryo of an unpigmented strain of mouse is fused with a similar embryo of a pigmented strain, the resulting embryo will give rise to a chimeric animal, with a mixture of pigmented and unpigmented cells. The distribution of the different cells in the skin gives this chimera a patchy coat.

of **regulation** and are described as **regulative**. Vertebrate embryos are among those capable of considerable regulation. In contrast, those embryos where, from a very early stage, the cells can develop only according to their early fate are termed **mosaic**. The term mosaic has a long history (see Section 1.3). It is now used to describe eggs and embryos that develop as if their pattern of future development is laid down very early, even in the egg, as a spatial mosaic of different molecules in the egg cytoplasm. These cytoplasmic determinants get distributed in a regular fashion into different cells during cleavage, thus determining the fate of that cell lineage at an early stage. The different parts of the embryo then develop quite independently of each other. *Caenorhabditis* is an example of an embryo with significant amounts of mosaic development and is discussed in Chapter 6. In such embryos, cell interactions may be quite limited. But the demarcation between regulative and mosaic strategies of development is not always sharp, and in part reflects the time at which determination occurs; it occurs much earlier in mosaic systems.

The difference between regulative and mosaic embryos also reflects the relative importance of cell–cell interactions in each system. The occurrence of regulation absolutely requires interactions between cells, for how else could normal development take place and how could deficiencies be recognized and restored? A truly mosaic embryo, however, would in principle not require such interactions. No purely mosaic embryos are known to exist.

Underlying the phenomena of regulative and mosaic-type development is the question of whether the effects of a given gene being expressed are restricted purely to the expressing cell or whether they are influencing and are influenced by, other cells not expressing that gene. Effects restricted to the gene-expressing cell are called **cell-autonomous**, while gene expression in one cell that has an effect on other cells is called **non-cell-autonomous**. **Non-autonomous** action is typically due to a gene product such as an intercellular signaling protein that is secreted by one cell and acts on others, whereas cell-autonomous effects are due to proteins such as transcription factors that act only within the cell that makes them.

Cell-autonomous versus non-autonomous genetic effects can be tested in mice produced by fusing together two very early embryos with different genetic constitutions (Fig. 1.24). The resulting mouse will be a **chimera**, a mosaic of cells with different genetic constitutions. If early embryos from a brown-haired and a white-haired mouse are fused, the resulting mouse has distinct patches of brown and white hair, reflecting patches of skin cells derived from the brown and white embryos, respectively. The brown cells still produce brown pigment when put into a white mouse, and they do not have any effect on the white cells; pigmentation is therefore a cell-autonomous character.

1.13 Inductive interactions make cells different from each other

Making cells different from one another is central to development. There are many times in development where a signal from one group of cells influences the development of an adjacent group of cells. This is known as **induction**, and the classic example is the action of the Spemann organizer in amphibians (see Section 1.4). Inducing signals may be propagated over several, or even many, cells, or be highly localized. The inducing signals from the amphibian organizer affect many cells, whereas other inducing signals may be transmitted from a single cell to its immediate neighbor. Two different types of inductions should be distinguished: **permissive** and **instructive**. Permissive inductions occur when a cell makes only one kind of response to a signal, and makes it when a given level of signal is reached. By contrast, in an instructive induction, the cells respond differently to different concentrations of the signal. 'Antagonistic' signal molecules that block induction, for example, by preventing the inducing signal reaching the cell or binding to a cell-surface receptor, are also important in controlling development.

Inducing signals are transmitted between cells in three main ways (Fig. 1.25). First, the signal can be transmitted through the extracellular space in the form of a secreted diffusible molecule, generally a protein. Second, cells may interact directly with each other by means of molecules located on their surfaces, which are also generally proteins. In both these cases, the signal is generally received by receptor proteins in the cell membrane and is subsequently relayed through intracellular signaling systems to produce the cellular response. Third, the signal may pass from cell to cell directly. In most animal cells, this is through gap junctions, which are specialized protein pores in the apposed plasma membranes providing direct channels of communication between the cytoplasms of adjacent cells through which small molecules can pass. Plant cells are connected by strands of cytoplasm called plasmodesmata, through which even quite large molecules such as proteins can pass directly from cell to cell. Some small molecules that act as extracellular signals during development of both animals and plants can pass through the plasma membrane.

In the case of signaling by a diffusible protein, or by direct contact between two
⬚ ceived at the cell membrane
⬚ it is to alter gene expression
⬚ that information has to be
⬚ his process is known gener-
⬚ nd is carried out by relays of
⬚ n the extracellular signaling
⬚ of different types of intracel-
⬚ commonly used intercellular
⬚ . More detailed diagrams of
⬚ We shall encounter these sig-
⬚ in in many different animals

⬚ nother to transmit the signal
⬚ vay proteins by phosphoryla-
⬚ In development, most signal-
⬚ ching transcription on or off.
⬚ lopment is the **cytoskeleton**,
⬚ ment enables cells to change
shape, to move, and to divide. Signaling pathways are also used to alter enzyme activity and metabolic activity within cells temporarily and to initiate nerve impulses in neurons. The various signals a cell may be receiving at any one time are integrated by cross-talk between the different intracellular signaling pathways to produce an appropriate response. An example of what can happen when an important developmental signaling pathway is defective is illustrated in Box 1F. In general, **organizing regions** function by producing extracellular signaling molecules or factors that influence intracellular signaling pathways in neighboring cells to determine their fate. These distinct signal-producing regions are also known as **signaling centers**.

An important feature of induction is whether or not the responding cell is able to respond to the inducing signal. The ability to respond, or **competence**, may depend, for example, on the presence of the appropriate receptor and transducing mechanism, or on the presence of particular transcription factors needed for gene activation. A cell's competence for a particular response can change with time; for example, the Spemann organizer can induce changes in the cells it affects only during a restricted window of time.

In embryos, it seems that small is generally beautiful where signaling and pattern formation are concerned. Most patterns are specified on a scale involving just tens of cells and distances of only 100 to 500 μm. The final organism may be very big, but this is almost entirely due to growth after the basic pattern has been formed.

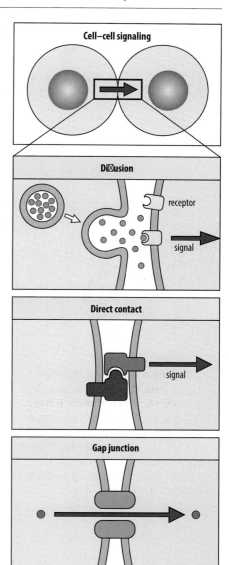

Fig. 1.25 An inducing signal can be transmitted from one cell to another in three main ways. The signal can be a secreted diffusible molecule, generally a protein, which interacts with a receptor on the target cell surface (second panel), or the signal can be produced by direct contact between two complementary proteins at the cell surfaces (third panel). If the signal involves a small molecule it may pass directly from cell to cell through gap junctions in the plasma membrane (fourth panel).

CELL BIOLOGY BOX 1E Signal transduction and intracellular signaling pathways

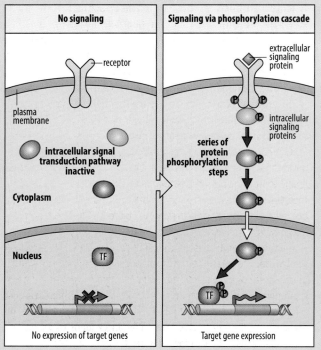
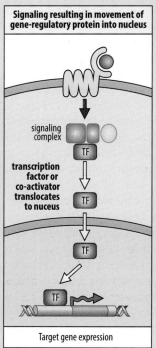
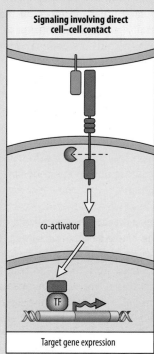

Figure 1

Extracellular signal proteins act on a target cell via receptors located in the cell's plasma membrane. These receptors are transmembrane proteins and are specific for different signal proteins. This means that the receptors a cell expresses determine which signals it can respond to. The signal protein itself does not enter the cell. Instead, when it binds to the extracellular face of its cell-surface receptor this activates a relay of intracellular signaling molecules that results in a cellular response.

For many developmental signals, the end of the pathway is a change in the pattern of gene expression in the responding cell.

1.14 The response to inductive signals depends on the state of the cell

Although an inductive signal can be viewed as an instruction to cells about how to behave, it is important to realize that responses to such a signal are entirely dependent on the current state of the cells, which depends on their developmental history. Not only does the individual cell have to be competent to respond, but the number of possible responses is usually very limited. An inductive signal can only select one response from a small number of possible cellular responses. Thus instructive inducing signals would be more accurately called selective signals. A truly instructive signal would be one that provided the cell with entirely new information and capabilities, by providing it with, for example, new genes, which is not thought to occur during development.

The fact that, at any given time, an inductive signal selects one out of several possible responses has several important implications for biological economy. On the one hand, it means that different signals can activate a particular gene at different stages of development: the same gene is often turned on and off repeatedly during development. On the other, the same signal can be used to elicit different responses in different cells. A particular signaling molecule, for example, can act on several types of cell, evoking a characteristic and different response from each, depending on their

The information that a signal molecule has bound to a receptor on the cell surface has to be transmitted from the receptor through the cytoplasm to the nucleus. This is achieved by an intracellular signaling pathway composed of a series of interacting proteins; different signaling pathways are employed by different receptors. Just as the same families of signal molecules are used by different animal species, the general features of the corresponding signaling pathways in different species are also similar.

There are various general types of intracellular signaling pathway (Figure 1). In the absence of a signal, the receptor is not occupied and many of the intracellular proteins involved in the signaling event are in an inactive state. When the signal binds its receptor, it triggers a series of biochemical events which in the case shown is a cascade of intracellular phosphorylation events. The information that the signaling protein has bound to its receptor is transmitted through the cytoplasm by a protein relay that involves a series of protein phosphorylations (compare the first and second panels of figure 1). Protein phosphorylation, which changes the activity of proteins and their interactions with other proteins, is ubiquitous in intracellular signaling pathways as one means of transmitting the signal onwards, and is carried out by enzymes called protein kinases. There is a large group of receptors that signal in this way. The first step in the pathway involves phosphorylation of the intracellular domain of the receptor by a protein kinase, and the pathway culminates in phosphorylation and activation of a protein kinase that enters the nucleus to phosphorylate and activate a specific transcription factor, thus selectively controlling gene expression. Fibroblast growth factor (FGF, see Box 4E), a secreted signal protein used in many different contexts throughout development, epidermal growth factor (EGF) and many other extracellular signals use pathways of this type.

For some other extracellular signals, activation of the intracellular pathway leads to the movement of a transcription factor or other gene-regulatory protein (such as a co-activator or co-repressor) from the cytoplasm into the nucleus (third panel of figure 1). Extracellular signal proteins that use this type of pathway are the Wnt proteins (see Box 4B), the Hedgehog protein family (see Box 2F), and proteins of the transforming growth factor (TGF)-β family (see Box 4C), which we shall meet throughout the book.

The receptor Notch controls yet another type of signaling pathway. This transmembrane protein is activated by direct contact with its ligands (which are also transmembrane proteins) on another cell (**ligand** is the general term for a molecule that binds specifically to another biomolecule). In this case, binding of the transmembrane ligand to Notch results in cleavage of the intracellular portion of the receptor, which is then free to translocate to the nucleus, where it acts as a co-activator (fourth panel of figure 1). The Notch pathway (see Box 5D) is commonly used to help determine cell fate.

It is important to ensure that intracellular signaling pathways remain inactive when the cell is not receiving a signal. For example, in the absence of a signal, transcription factors and other gene-regulatory regulatory proteins may be bound to other proteins in an inactive complex in the cytoplasm, or be targeted for degradation. Signaling pathways can also contain feedback loops that modulate their own activity. In addition, cells can be receiving several signals at the same time and there can be cross-talk between the intracellular signaling pathways to ensure the appropriate cellular response.

Some small molecules that act as extracellular signals during development (for example, retinoic acid (see Box 5C), steroid hormones, and some plant hormones, such as ethylene) can pass through the cell membrane unaided. They interact with specific receptors present in the cytoplasm or in the nucleus.

current state. As we will see in future chapters, evolution has been lazy with respect to this aspect of development, and a small number of intercellular signaling molecules are used over and over again for different purposes.

1.15 Patterning can involve the interpretation of positional information

One general mode of pattern formation can be illustrated by considering the patterning of a simple non-biological model—the French flag (Fig. 1.26). The French flag has a simple pattern: one-third blue, one-third white, and one-third red, along just one axis. Moreover, the flag comes in many sizes but always the same pattern, and thus can be thought of as mimicking the capacity of an embryo to regulate. Given a line of cells, any one of which can be blue, white, or red, and given also that the line of cells can be of variable length, what sort of mechanism is required for the line to develop the pattern of a French flag?

One solution is for the group of cells to use **positional information**—that is, information on their position along the line with respect to boundaries at either end. Each cell thus acquires its own **positional value**. After they have acquired their positional values, the cells interpret this information by differentiating according to their genetic program. Those in the left-hand third of the line will become blue, those in the middle-third white, and so on. Very good evidence that cells use positional information comes

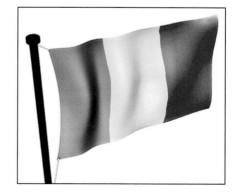

Fig. 1.26 The French flag.

MEDICAL BOX 1F   When development goes awry

The signaling pathway stimulated by the small secreted protein Sonic hedgehog (Shh) (named after the Hedgehog protein in *Drosophila*, which was the first member of this family to be discovered) regulates many developmental processes in humans and other vertebrates from very early in embryogenesis onwards. In adults, Shh controls the growth of stem cells, and misregulation of this signaling pathway leads to tumor formation. Like many other signaling pathways in development (discussed in Box 1E), the end results of Shh signaling are changes in gene expression.

Identifying the components of intracellular signaling pathways can help us understand the reasons for some developmental abnormalities, and Shh is a good example (Figure 1). A relatively common birth defect is a condition known as holoprosencephaly, which is characterized by a spectrum of midline brain and facial abnormalities. The severity of the condition varies greatly: at one end of the spectrum the forebrain (the prosencephalon) fails to divide in the midline to form right and left brain hemispheres, there is complete absence of the midline of the face and the fetuses do not survive; at the other end, the division of the forebrain is only mildly affected and the only observable defect may be a cleft lip or the presence of one incisor tooth rather than two.

The first indications that Shh was involved in brain development came in the early 1990s, when Shh was discovered to be expressed in the ventral midline of the nervous system in mice. In 1996, clinical geneticists identified *Shh* as one of the genes affected in patients with inherited holoprosencephaly. At around the same time, it was found that when the *Shh* gene is completely knocked out experimentally in mice (see Section 3.10 and Box 3D for the technique of gene knock-outs), forebrain division does not occur and the embryos have a condition known as cyclopia, with loss of midline facial structures and development of a single eye in the middle of the face—a condition representing the most severe manifestation of holoprosencephaly.

As clinical geneticists continued to identify genes involved in holoprosencephaly, more genes encoding components in the Shh pathway were implicated. In all cases, the cause of the abnormality could be ascribed to a deficiency in Shh signaling, but different genes were affected in different cases. In some patients, the abnormality was caused by mutations in the *Shh* gene itself, but in others the affected genes encoded either the receptor for Shh (Patched) or one of the transcription factors (Gli2) that moves into the nucleus to control gene expression. In some patients, mutations in the gene for an enzyme required for cholesterol synthesis were found. Cholesterol promotes the activity of a transmembrane protein called Smoothened, which acts together with Patched to control the Shh pathway. Figure 1 shows a simplified version of the Shh signaling pathway and how mutations in various components of the pathway (marked with asterisks) will block the pathway, resulting in a reduction or loss of Shh signaling (the full Shh pathway is illustrated in Box 11D. The region of the face that is affected by reduction in Shh signaling is indicated. Shh signaling in vertebrates takes place on cellular structures called primary cilia, which are immotile cilia present on most vertebrate cell types, and some patients with genetic conditions in which formation of these structures is defective have distinctive facial features. Thus, the discovery of genes that produce similar symptoms in human patients can highlight genes in a common developmental pathway.

Facial defects similar to those produced by mutations in genes for Shh pathway proteins can also be caused by environmental factors. Pregnant ewes grazed in pastures in which the California False Hellebore (*Veratrum californicum*) is growing may give birth to lambs with severe facial abnormalities, such as cyclopia. The chemical in the plant that causes these defects is the alkaloid cyclopamine, and it too disrupts Shh signaling. Chemicals and other environmental factors that cause abnormal development are called **teratogens** (see Box 11A). Cyclopamine interferes with the activity of Smoothened and reduces Shh signaling (Figure 1). This property of cyclopamine is now being explored in a therapeutic context to develop drugs to treat cancers that display abnormal Shh signaling.

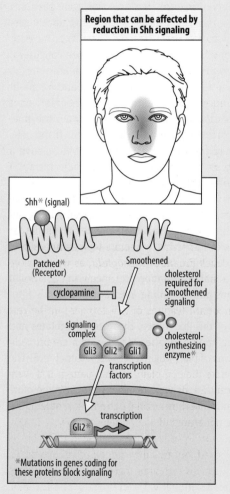

Figure 1

from the regeneration of amphibian and insect limbs, in which the missing regions are precisely replaced (discussed in Chapter 13).

Pattern formation using positional values implies at least two distinct stages: first the positional value has to be related to some boundary, and then it has to be interpreted by the cell. The separation of these two processes has an important implication: it means that there does not need to be a set relation between the positional values and how they are interpreted. In different circumstances, the same positional information and set of positional values could be used to generate the Italian flag or another pattern. How positional values will be interpreted depends on the particular genetic instructions active in the group of cells at the time and thus the cells' developmental history.

Cells could have their positional values specified by various mechanisms. The simplest is based on a gradient of some substance. If the concentration of some chemical decreases from one end of a line of cells to the other, then the concentration in any cell along the line provides positional information that effectively specifies the position of the cell with respect to the boundary (Fig. 1.27). A chemical whose concentration varies and which is involved in pattern formation is called a **morphogen**. In the case of the French flag, we assume a source of morphogen at one end and a sink at the other, and that the concentrations of morphogen at both ends are kept constant but are different from each other. Then, as the morphogen diffuses down the line, its concentration at any point effectively provides positional information. If the cells can respond to **threshold concentrations** of the morphogen—for example, above a particular concentration the cells develop as blue, while below this concentration they become white, and at yet another, lower, concentration they become red—the line of cells will develop as a French flag (see Fig. 1.27). Thresholds can represent the amount of morphogen that must bind to receptors to activate an intracellular signaling system, or concentrations of transcription factors required to activate particular genes. The use of threshold concentrations of transcription factors to specify position is most beautifully illustrated in the early development of *Drosophila*, as we shall see in Chapter 2. Other ways of providing positional information and specifying positional values are by direct intercellular interactions and by timing mechanisms.

The French flag model illustrates two important features of development in the real world. The first is that, even if the length of the line varies, the system regulates and the pattern will still form correctly, given that the boundaries of the system are properly defined by keeping the different concentrations of morphogen constant at either end. The second is that the system could also regenerate the complete original pattern if it were cut in half, provided that the boundary concentrations were re-established. It is therefore truly regulative. We have discussed it here as a one-dimensional patterning problem, but the model can easily be extended to provide patterning in two dimensions (Fig. 1.28).

It is still unclear how positional information is specified: morphogen diffusion is one of the mechanisms that have been proposed, but this may not be reliable enough. We do not know how fine-grained the differences in position are. For example, does every cell in the early embryo have a distinct positional value of its own, and is it able to make use of this difference?

Another way in which a pattern can be generated in a group of cells is by cells that have differentiated in a salt-and-pepper fashion to then 'sort out' so that they occupy a particular position (Fig. 1.29). Cells in the group might initially differentiate in a random spatial arrangement because of differences in sensitivity to a global signal or in response to differences in timing. Sorting out has been demonstrated in various experimental systems in which two or more different tissues are disaggregated into single cells, the cells then mixed together and reaggregated. In many of these mixed reaggregates, like cells tend to associate preferentially. This can result in the formation of separate aggregates, or in cells of one type adopting a particular position within a mixed aggregate. The basis for sorting out is cells having different cell-adhesion molecules on their surfaces or differing amounts of the same cell-adhesion molecules.

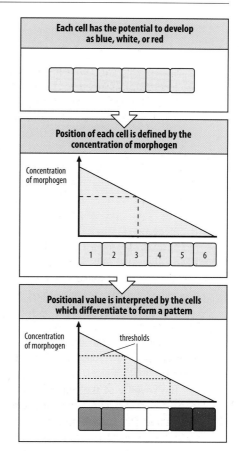

Fig. 1.27 The French flag model of pattern formation. Each cell in a line of cells has the potential to develop as blue, white, or red. The line of cells is exposed to a concentration gradient of some substance and each cell acquires a positional value defined by the concentration at that point. Each cell then interprets the positional value it has acquired and differentiates into blue, white, or red, according to a predetermined genetic program, thus forming the French flag pattern. Substances that can direct the development of cells in this way are known as morphogens. The basic requirements of such a system are that the concentration of substance at either end of the gradient must remain different from each other but constant, thus fixing boundaries to the system. Each cell must also contain the necessary information to interpret its positional value into a developmental outcome. Interpretation of the positional value is based upon different threshold responses to different concentrations of morphogen.

Fig. 1.28 Positional information can be used to generate patterns. A good example, as shown here, is where the people seated in a stadium each have a positional value defined by their row and seat numbers. Each position has an instruction about which colored card to hold up, and this makes the pattern. If the instructions were changed, a different pattern would be formed, and so positional information can be used to produce an enormous variety of patterns.

As we will see, although positional information is the general mode of pattern formation in embryos, there are some instances where this alternative strategy is used, for example in the generation of mesoderm and endoderm in zebrafish embryos and extra-embryonic and embryonic tissue in early mouse embryos. Sorting out has also been invoked as a mechanism for creating sharp boundaries between different cell types.

1.16 Lateral inhibition can generate spacing patterns

Many structures, such as the feathers on a bird's skin, are found to be more or less regularly spaced with respect to one another. One mechanism that gives rise to such spacing is called **lateral inhibition** (Fig. 1.30). Given a group of cells that all have the potential to differentiate in a particular way, for example as feathers, it is possible to regularly space the cells that will form feathers by a mechanism in which the first cells to begin to form feathers (which occurs essentially at random) inhibit adjacent cells from doing so. This is reminiscent of the regular spacing of trees in a forest by competition for sunlight and nutrients. In embryos, lateral inhibition is often the result of the differentiating cell producing an inhibitory molecule that acts locally on the nearest neighbors to prevent them developing in a similar way.

1.17 Localization of cytoplasmic determinants and asymmetric cell division can make daughter cells different from each other

Positional specification is just one way in which cells can be given a particular identity. A separate mechanism is based on **cytoplasmic localization** of factors and **asymmetric cell division** (Fig. 1.31). Asymmetric divisions are so called because they result in daughter cells having properties different from each other, independently of any environmental influence. The properties of such cells therefore depend on their lineage, and not on environmental cues. Although some asymmetric cell divisions are also unequal divisions, in that they produce cells of different sizes, this is not usually the most important feature in animals; it is the unequal distribution of cytoplasmic factors that makes the division asymmetric. An alternative way of making the French flag pattern from the egg would be to have chemical differences (representing blue, white, and red) distributed in the egg in the form of determinants that foreshadowed the French flag. When the egg underwent cleavage, these cytoplasmic determinants would become distributed among the cells in a particular way and a French flag would develop. This would require no interactions between the cells, which would have their fates determined from the beginning.

Although such extreme examples of mosaic development (see Section 1.12) are not known in nature, there are well-known cases where eggs or cells divide so that some cytoplasmic determinant becomes unequally distributed between the two daughter

Fig. 1.29 Two ways in which a pattern can be produced from a group of cells. The upper panel illustrates how the a red stripe can be generated by positional information. The lower panel shows how the same red stripe can be generated by cells in the group differentiating randomly and then associating preferentially with other like cells.

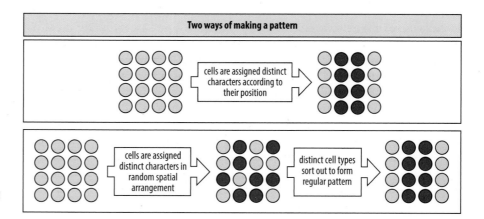

cells and they develop differently. This happens at the first cleavage of the nematode egg, for example, and defines the antero-posterior axis of the embryo. The germ cells of *Drosophila* are also specified by cytoplasmic determinants, in this case contained in the cytoplasm located at the posterior end of the egg. And in *Xenopus*, a key cytoplasmic determinant in the first stages of embryonic development is a protein known as VegT, which is localized in the vegetal region of the fertilized egg. In general, however, as development proceeds, daughter cells become different from each other because of signals from other cells or from their extracellular environment, rather than because of the unequal distribution of cytoplasmic determinants.

A very particular type of asymmetric division is seen in stem cells. When they divide, these self-renewing cells can produce one new stem cell and another daughter cell that will differentiate into one or more cell types (Fig. 1.32). Stem cells that are capable of giving rise to all the cell types in the body are present in very early embryos, and are known as **embryonic stem cells (ES cells)**, whereas in adults, stem cells of apparently more limited potential are responsible for the continual renewal of tissues such as blood, skin epidermis, and gut **epithelium**, and for the replacement of tissues such as muscle when required. The difference in the behavior of the daughter cells produced by division of a stem cell can be due either to asymmetric distribution of cytoplasmic determinants or to the effects of external signals. The generation of neurons in *Drosophila*, for example, depends on the asymmetric distribution of cytoplasmic determinants in the neuronal stem cells, whereas the production of the different types of blood cells seems to be regulated to a large extent by extracellular signals.

Embryonic stem cells, which can give rise to all the types of cells in the body, are called **pluripotent**, while stem cells that give rise to a more limited number of cell types are known as **multipotent**. The hematopoietic stem cells in bone marrow, which give rise to all the different types of blood cells, are an example of multipotent stem cells. Stem cells are of great interest in relation to regenerative medicine, as they could provide a means of repairing or replacing damaged organs and are discussed in Chapter 8. For most animals, the only cell that can give rise to a complete new organism is the fertilized egg, and this is described as **totipotent**.

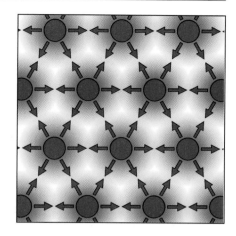

Fig. 1.30 Lateral inhibition can give a spacing pattern. Lateral inhibition occurs when developing structures produce an inhibitor that prevents the formation of any similar structures in the area adjacent to them, and the structures become evenly spaced as a result.

1.18 The embryo contains a generative rather than a descriptive program

All the information for embryonic development is contained within the fertilized egg. So how is this information interpreted to give rise to an embryo? One possibility is that the structure of the organism is somehow encoded as a descriptive program in the genome. Does the DNA contain a full description of the organism to which it will give rise: is it a blueprint for the organism? The answer is no. Instead, the genome contains a program of instructions for making the organism—a **generative program**— that determines where and when different proteins are synthesized and thus controls how cells behave.

A descriptive program, such as a blueprint or a plan, describes an object in some detail, whereas a generative program describes how to make an object. For the same object the programs are very different. Consider origami, the art of paper folding. By folding a piece of paper in various directions it is quite easy to make a paper hat or a bird from a single sheet. To describe in any detail the final form of the paper with the complex relationships between its parts is really very difficult, and not of much help in explaining how to achieve it. Much more useful and easier to formulate are instructions on how to fold the paper. The reason for this is that simple instructions about folding have complex spatial consequences. In development, gene action similarly sets in motion a sequence of events that can bring about profound changes in the embryo. One can thus think of the genetic information in the fertilized egg as equivalent to the folding instructions in origami; both contain a generative program for making a particular structure.

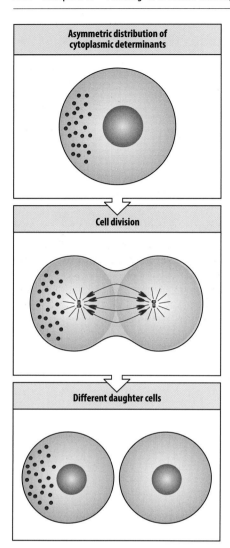

Asymmetric distribution of cytoplasmic determinants

Cell division

Different daughter cells

Fig. 1.31 Cell division with asymmetric distribution of cytoplasmic determinants. If a particular molecule is distributed unevenly in the parent cell, cell division will result in its being shared unequally between the cytoplasm of the two daughter cells. The more localized the cytoplasmic determinant is in the parental cell, the more likely it will be that one daughter cell will receive all of it and the other none, thus producing a distinct difference between them.

1.19 The reliability of development is achieved by various means

Development is remarkably consistent and reliable; you need only look at the similar lengths of your legs, each of which develops quite independently of the other for some 15 years (see Chapter 13). Embryonic development thus needs to be reliable so that the adult organism can function properly. For example, both wings of a bird must be very similar in size and shape if it is to be able to fly satisfactorily. How such reliability is achieved is an important problem in development.

Development needs to be reliable in the face of fluctuations that occur either within the embryo or in the external environment. Internal fluctuations include small changes in the concentrations of molecules, and also changes due, for example, to mutations in genes not directly linked to the development of the organ in question. External factors that could perturb development include temperature and environmental chemicals. However, the mechanisms underlying the developmental processes withstand large changes of these kinds and therefore are said to be **robust**.

One of the central problems of reliability relates to pattern formation. How are cells instructed to behave in a particular way in a particular position in a robust manner? Two ways in which reliability is assured during development are apparent redundancy of mechanism and negative feedback. **Redundancy** is when there are two or more ways of carrying out a particular process; if one fails for any reason another will still function. It is like having two batteries in your car instead of just one. True redundancy—such as having two identical genes in a haploid genome with identical functions—is rare except for cases such as rRNA, where there are often hundreds of copies of similar genes. Apparent redundancy, on the other hand, where a given process can be specified by several different mechanisms, is probably one of the ways that embryos can achieve such precise and robust results. It is like being able to draw a straight line with either a ruler or a piece of taut string. This type of situation is not true redundancy, but may give the impression of being so if one mechanism is removed and the outcome is still apparently normal. Multiple ways of carrying out a process make it more robust and resistant to environmental or genetic perturbation.

Negative feedback also has a role in ensuring consistency; here, the end product of a process inhibits an earlier stage and thus keeps the level of product constant (see, for example, Fig. 1.21). The classic example is in metabolism, where the end product of a biochemical pathway inhibits an enzyme that acts early in the pathway. Yet another reliability mechanism relates to the complexity of the networks of gene activity that operate in development. There is evidence that these networks, which involve multiple pathways, are robust and relatively insensitive to small changes in, for example, the rates of the individual processes involved.

1.20 The complexity of embryonic development is due to the complexity of cells themselves

Cells are, in a way, more complex than the embryo itself. In other words, the network of interactions between proteins and DNA within any individual cell contains many more components and is very much more complex than the interactions between the cells of the developing embryo. However clever you think cells are, they are almost always far cleverer. Each of the basic cell activities involved in development, such as cell division, response to cell signals, and cell movements, are the result of interactions within a population of many different intracellular proteins whose composition varies over time and between different locations in the cell. Cell division, for example, is a complex cell-biological program that takes place over a period of time, in a set order of stages, and requires the construction and precise organization of specialized intracellular structures at mitosis.

Any given cell is expressing thousands of different genes at any given time. Much of this gene expression may reflect an intrinsic program of activity, independent of external signals. It is this complexity that determines how cells respond to the signals they receive; how a cell responds to a particular signal depends on its internal state. This state reflects the cell's developmental history—cells have good memories—and so different cells can respond to the same signal in very different ways. We shall see many examples of the same signals being used over and over again by different cells at different stages of embryonic development, with different biological outcomes.

We have at present only a fragmentary picture of how all the genes and proteins in a cell, let alone a developing embryo, interact with one another. But new technologies are now making it possible to detect the simultaneous activity of hundreds of genes in a given tissue. The discipline of systems biology is also beginning to develop techniques to reconstruct the highly complex signaling networks used by cells. How to interpret this information, and make biological sense of the patterns of gene activity revealed, is a massive task for the future.

1.21 Development is a central element in evolution

The similarity in developmental mechanisms and genes in such very different animals as flies and vertebrates is the result of the evolutionary process. All animals have evolved from a single multicellular ancestor, and it is therefore inevitable that some developmental mechanisms will be held in common by many different species, whereas new ones have arisen in different animal groups as evolution proceeded. The evolution of development is discussed in more detail in Chapter 14.

The basic Darwinian theory of evolution by natural selection is that changes in genes alter how an organism develops, and that such changes in development in turn determine how the adult will interact with their environment. If a developmental change gives rise to adults better adapted to survive and reproduce in the prevailing environment, the underlying genetic change will be selected to be retained in the population. Thus changes in development due to changes in the genes are fundamental to evolution.

Very clear examples of the workings of evolution in development can be seen in the limbs of vertebrates. In Chapter 14 we shall see how analysis of fossil evidence shows that the limbs of land vertebrates originally evolved from fins, and how the genetic and developmental basis of this evolution is being reconstructed (see Section 14.7). We shall also see how the basic pattern of the five-digit limb then evolved to give limbs as different as those of bats, horses, and humans. In bats, the digits of the forelimb grow extremely long and support a leathery wing membrane, whereas in horses, one digit of the 'hand' in the forelimb and the 'foot' in the hindlimb has become a long bone that forms the lower part of the leg and the hoof, while some other digits have been lost (see Fig. 14.25). The fossil record gives many other examples of evolutionary changes. Birds, which have only three digits in the wings, evolved from dinosaur ancestors in which digit reduction had taken place (discussed in Box 14B).

While some major steps in evolution can be identified, the genes involved are not easy to identify. In the few examples so far in which the mutations underlying evolutionary changes have been identified, it has been found that the mutations are not in the protein-coding sequences themselves but in their control regions (see Chapter 14). Thus there is increasing focus on the importance of regulatory sequences as the instruments of evolution. There are also questions as to how big the genetic steps to new forms were, and there are questions about how completely new structures evolve. But there is no doubt that it is all down to changes in embryonic development.

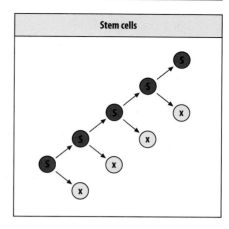

Fig. 1.32 Stem cells. Stem cells (S) are cells that both renew themselves and give rise to differentiated cell types. Thus, a daughter cell arising from stem-cell division may either develop into another stem cell or give rise to a different type of cell (X).

SUMMARY

Development results from the coordinated behavior of cells. The major processes involved in development are pattern formation, morphogenesis or change in form, cell differentiation, growth, cell migration, and cell death. These involve cell activities such as intercellular signaling and turning genes on and off. Genes control cell behavior by controlling where and when proteins are synthesized, and thus cell biology provides the link between gene action and developmental processes. During development, cells undergo changes in the genes they express, in their shape, in the signals they produce and respond to, in their rate of proliferation, and in their migratory behavior. All these aspects of cell behavior are controlled largely by the presence of specific proteins; gene activity controls which proteins are made. Since the somatic cells in the embryo all generally contain the same genetic information, the changes that occur in development are controlled by the differential activity of selected sets of genes in different groups of cells. The genes' control regions are fundamental to this process. Development is progressive and the fate of cells becomes determined at different times. The potential for development of cells in the early embryo is usually much greater than their normal fate, but this potential becomes more restricted as development proceeds. Inductive interactions, involving signals from one tissue or cell to another, are one of the main ways of changing cell fate and directing development. Asymmetric cell divisions, in which cytoplasmic components are unequally distributed to daughter cells, can also make cells different. One widespread means of pattern generation is through positional information; cells first acquire a positional value with respect to boundaries and then interpret their positional values by behaving in different ways. Another strategy for generating a pattern involves cells becoming different and then sorting-out. Developmental signals are more selective than instructive, choosing one or other of the developmental pathways open to the cell at that time. The embryo contains a generative, not a descriptive, program—it is more like the instructions for making a structure by paper folding than a blueprint. Various mechanisms, including apparent redundancy and negative feedback, are involved in making development remarkably reliable and robust. The complexity of development lies within the cells. Evolution is intimately linked to development because heritable changes in the adult organism must inevitably be the result of changes in the genetic program of embryonic development.

Summary to Chapter 1

- All the information for embryonic development is contained within the fertilized egg—the diploid zygote. The genome of the zygote contains a program of instructions for making the organism. The cytoplasmic constituents of the egg, and of the cells it gives rise to, are essential players in working out this developmental program along with the genes.
- Strictly regulated gene activity, by controlling which proteins are synthesized, and where and when, directs a sequence of cellular activities that brings about the profound changes that occur in the embryo during development.
- The major processes involved in development are pattern formation, morphogenesis, cell differentiation, and growth, which are controlled by communication between the cells of the embryo and changes in gene expression.
- Development is progressive, with the fate of cells becoming more precisely specified as it proceeds.
- Cells in the early embryo usually have a much greater potential for development than is evident from their normal fate, and this enables embryos, particularly vertebrate embryos, to develop normally even if cells are removed, added, or transplanted to different positions.
- Cells' developmental potential becomes much more restricted as development proceeds.

- Many genes are involved in controlling the complex interactions that occur during development, and reliability is achieved in various ways.
- Changes in embryonic development are the basis for the evolution of multicellular organisms.
- A full understanding of development is far from being achieved. While basic principles and certain developmental systems are quite well understood, there are still many gaps.

■ End of chapter questions

Long Answer (concept questions)

1. The ultimate goal of the science of developmental biology is to understand human development. In your own words, discuss three areas of medicine that are impacted by the study of developmental biology.

2. The Greek philosopher Aristotle considered two opposing theories for development: preformation and epigenesis. Describe what is meant by these two ideas; refer to the concept of the 'homunculus' in describing preformation, and to the art of origami in describing epigenesis. Which better describes our current view of development?

3. Weismann's concept of 'determinants' was supported by Roux's experiments demonstrating 'mosaic development' in the frog. How would determinants lead to mosaic development? Why does Driesch's concept of 'regulation' during development better explain human twinning?

4. Pattern formation is a central process in the study of development. Describe three examples of pattern formation and explain what 'patterns' are generated.

5. Briefly define the following cell behaviors: intercellular signaling, cell proliferation, cell differentiation, cell movement, changes in cell shape, gene expression, and cell death.

6. Contrast the role of housekeeping genes and proteins during development with the role of tissue-specific genes and proteins. Which class of genes and proteins is more important during development, housekeeping or tissue-specific? Give examples.

7. All cells in an organism contain the same genes; however, it contains different types of cells. How is this paradox of 'genetic equivalence' resolved by the concept of 'differential gene expression'?

8. How do transcription factors work? What is their relationship to control regions in DNA?

9. Contrast cell fate with cell determination. What kinds of experiments allow the distinction between fate and determination to be made?

10. What are three ways in which inducing signals can be transmitted in developing embryos? In what circumstance would an inducing signal be a 'morphogen'?

11. What is the relationship between positional information and pattern formation? Give an example.

Multiple choice (factual recall questions)

NB There is only one right answer to each question.

1. Which of the following is a vertebrate 'model organism' used in the study of development?

a) *Caenorhabditis elegans*
b) *Drosophila melanogaster*
c) *Homo sapiens*
d) *Xenopus laevis*

2. Which of the following is a germ cell?

a) cells of the gonads
b) intestinal cells
c) sperm and eggs
d) zygotes

3. Which statement best describes the relationship between genotype and phenotype?

a) 'Genotype' and 'phenotype' both describe the physical properties of cells and organisms, just using different words or symbols.
b) Phenotype is entirely the consequence of a cell's genotype, which is why we now recognize the principle of 'genetic determinism.'
c) The genes found in a cell are that cell's genotype, and genes directly control a cell's phenotype.
d) The genotype determines what proteins a cell contains, and the proteins in turn determine the cell's properties, or phenotype.

4. Which would qualify as a dominant allele of a gene?

a) an allele that confers a phenotype on the organism when heterozygous; that is, when present in only one copy
b) an allele that confers a phenotype only when both copies of the gene are mutant; that is, when homozygous
c) an allele that is commonly found in wild populations, as opposed to alleles found in laboratory strains or when a pathological condition is present
d) an allele that confers a normal phenotype under some conditions, but confers a different, mutant phenotype when conditions such as elevated temperature are present

5. Which of the following basic developmental processes is most dependent on cellular movements?

a) cell differentiation
b) growth
c) morphogenesis
d) pattern formation

6. Which aspect of cellular regulation operates after a protein has been synthesized?

a) post-translational modification
b) RNA processing
c) transcriptional control
d) translational control

7. Which term could be used to refer to a segment of DNA?

a) *cis*-regulatory module
b) differential gene expression
c) gene regulatory protein
d) transcription factor

8. A morphogen is

a) a cell or group of cells that signal other cells to become determined in a specific way

b) a gene that causes a cell or group of cells to adopt a particular shape

c) a signaling molecule that causes the differentiation of neighboring cells

d) a signaling molecule that confers concentration-dependent positional information

9. What distinguishes stem cells from other cell types?

a) Although all cell types can divide, only stem cells differentiate after division.

b) Stem cells are able to divide to produce populations of cells that differentiate into a related set of cells.

c) Stem cells divide asymmetrically to give rise to one daughter cell that remains a stem cell, and a second daughter cell that differentiates into one or more cell types.

d) Stem cells have a special type of cell division in which they form small stems that then pinch off to give rise to the daughter cell.

10. How does organismal development evolve to give rise to new forms and species?

a) A structure like the fin of a fish evolves into a structure like a human hand through the needs of the organism for a hand rather than a fin.

b) Changes in genes that will be favored by natural selection will be induced, and thereby produce favorable changes in development.

c) Changes in genes will alter an organism's development, and if the new developmental patterns are favored by the environment, they will be retained through natural selection.

d) The characteristics acquired by the organism during its lifetime are passed on to the offspring; if the changes are favorable to the organism, they will eventually come to be encoded in the genes that govern development.

Multiple choice answer key

1: d, 2: c, 3: d, 4: a, 5: c, 6: a, 7: a, 8: d, 9: c, 10: c.

■ Section further reading

The origins of developmental biology

De Robertis, E.M.: **Spemann's organizer and the self-regulation of embryonic fields.** *Mech Dev.* 2009, **126**: 925–941.

Hamburger, V.: *The Heritage of Experimental Embryology: Hans Spemann and the Organizer.* New York: Oxford University Press, 1988.

Harris, H.: *The Birth of the Cell.* New Haven: Yale University Press, 1999.

Milestones in Development [http://www.nature.com/milestones/development/index.html] (date accessed 10 January 2014).

Needham, J.: *A History of Embryology.* Cambridge: Cambridge University Press, 1959.

Sander, K.: **'Mosaic work' and 'assimilating effects' in embryogenesis: Wilhelm Roux's conclusions after disabling frog blastomeres.** *Roux's Arch. Dev. Biol.* 1991, **200**: 237–239.

Sander, K.: **Shaking a concept: Hans Driesch and the varied fates of sea urchin blastomeres.** *Roux's Arch. Dev. Biol.* 1992, **201**: 265–267.

Spemann, H., Mangold, H.: **Induction of embryonic primordia by implantation of organizers from a different species.** *Int. J. Dev. Biol.* 2001, **45**: 13–38. [reprinted translation of their 1924 paper from the original German]

A conceptual tool kit

Alberts, B., *et al.*: *Essential Cell Biology: An Introduction to the Molecular Biology of the Cell.* 4th edn. New York: Garland Science, 2013.

Ashe, H.L., Briscoe, J.: **The interpretation of morphogen gradients.** *Development* 2006, **133**: 385–394.

Barolo, S., Posakony, J.: **Three habits of highly effective signalling pathways: principle of transcriptional control by developmental cell signalling.** *Genes Dev.* 2002, **16**: 1167–1181.

Howard, M.L., Davidson, E.H.: *cis*-**Regulatory control circuits in development.** *Dev. Biol.* 2004, **271**: 109–118.

Istrail, S., De-Leon, S.B., Davidson, E.H.: **The regulatory genome and the computer.** *Dev. Biol.* 2007, **310**: 187–195.

Jordan, J.D., Landau, E.M., Lyengar, R.: **Signaling networks: the origins of cellular multitasking.** *Cell* 2000, **103**: 193–200.

Levine, M., Tjian, R.: **Transcriptional regulation and animal diversity.** *Nature* 2003, **424**: 147–151.

Martindale, M.Q.: **The evolution of metazoan axial properties.** *Nat. Rev. Genet.* 2005, **6**: 917–927.

Martinez Arias A., Nicholls, J., Schröter, C.: **A molecular basis for developmental plasticity in early mammalian embryos.** *Development* 2013, **140**: 3499–3510.

Nelson, W.J.: **Adaptation of core mechanisms to generate polarity.** *Nature* 2003, **422**: 766–774.

Papin, J.A., Hunter, T., Palsson, B.O., Subramanian, S.: **Reconstruction of cellular signalling networks and analysis of their properties.** *Nat. Rev. Mol. Biol.* 2005, **6**: 99–111.

Rudel, D., Sommer, R.J.: **The evolution of developmental mechanisms.** *Dev. Biol.* 2003, **264**: 15–37.

Volff, J.N. (Ed.): **Vertebrate genomes.** *Genome Dyn* 2006, **2**: special issue.

Wolpert, L.: **Do we understand development?** *Science* 1994, **266**: 571–572.

Wolpert, L.: **One hundred years of positional information.** *Trends Genet.* 1996, **12**: 359–364.

Xing, Y., Lee, C.: **Relating alternative splicing to proteome complexity and genome evolution.** *Adv. Exp. Med. Biol.* 2007, **623**: 36–49.

Box 1E Signal transduction and intracellular signaling pathways

Alberts, B., *et al.*: *Molecular Biology of the Cell.* 5th edn. New York: Garland Science, 2008.

Box 1F When development goes awry

Cohen, M.M. Jr.: **Holoprosencephaly: clinical, anatomic, and molecular dimensions.** *Birth Defects Res. A Clin. Mol. Teratol.* 2006, **76**: 658–673.

Geng, X., Oliver, G.: **Pathogenesis of holoprosencephaly.** *J. Clin. Invest.* 2009, **119**: 1403–1413.

Roessler, E., Muenke, M.: **How a Hedgehog might see holoprosencephaly.** *Hum. Mol. Genet.* 2003, **12**: R15–R25.

Zaghloul, N.A., Brugmann, S.A.: **The emerging face of primary cilia.** *Genesis* 2011, **49**: 231–246.

2

Development of the *Drosophila* body plan

- *Drosophila* life cycle and overall development
- Setting up the body axes
- Localization of maternal determinants during oogenesis
- Patterning the early embryo
- Activation of the pair-rule genes and the establishment of parasegments
- Segmentation genes and segment patterning
- Specification of segment identity

The early development of the fruit fly Drosophila melanogaster *is better understood than that of any other animal of similar or greater morphological complexity. Its early development exemplifies clearly many of the principles introduced in Chapter 1. The genetic basis of development is particularly well understood in the fly, and many developmental genes in other organisms, particularly vertebrates, have been identified by their homology with genes in* Drosophila, *in which they were first identified. Like many other animals,* Drosophila's *earliest development is directed by maternal gene products laid down in the egg, and these specify the main axes and regions of the body. We shall see how gradients of maternal morphogens act on the zygote's own genes to specify the antero-posterior and dorso-ventral pattern of the early embryo, to trigger the assignment of cells to particular germ layers, and to start the partitioning of the embryo into segments. Another set of zygotic genes, the Hox genes, acting along the antero-posterior axis then confers on each segment its unique character, as displayed by the appendages the segment eventually bears in the adult, such as wings, legs, or antennae.*

We are much more like flies in our development than you might think. Astonishing discoveries in developmental biology over the past 25 years have revealed that many of the genes that control the development of the fruit fly *Drosophila* are similar to those controlling development in vertebrates, and indeed in many other animals. It seems that once evolution finds a satisfactory way of patterning animal bodies, it tends to use the same strategies, mechanisms and molecules over and over again with, of course, important organism-specific modifications. So, while insect and vertebrate development might seem to be very different, much has been learned from *Drosophila* that has illuminated vertebrate development.

There are around 14,000 protein-coding genes in *Drosophila* as estimated from its genome sequence, which is only twice the number of genes in unicellular yeast and fewer than the 19,000 estimated genes of the morphologically simpler nematode *Caenorhabditis elegans*. However, a recent large-scale analysis of the RNAs transcribed during development and in adult *Drosophila* revealed a high incidence of alternative RNA splicing (see Section 1.10) which would increase the number of different proteins that could be made. In addition, around 1100 genes encoding functional RNAs other than mRNAs, such as tRNAs and microRNAs (see Box 6C), are present in the fly genome. Thus the complete arsenal of genetic information is more than sufficient to specify the large number of different cell types and more complex behavior of the fruit fly compared with the nematode.

The pre-eminent place of *Drosophila* in modern developmental biology was recognized by the award of the 1995 Nobel Prize for Physiology or Medicine for work that led to a fundamental understanding of how genes control development in the fly embryo, only the second time that the prize had been awarded for work in developmental biology. While insect and vertebrate development may seem to be very different, much has been learned that can be applied to vertebrate development; for example, many intercellular signaling pathways are well conserved.

The first part of this chapter describes the life cycle of *Drosophila* and its overall development. The remaining parts look at how the basic body plan of the *Drosophila* larva is established up to the stage at which the embryo becomes segmented, and how the segments are patterned and acquire their unique identities. More about *Drosophila* gastrulation, germ-cell development and sex-determination mechanisms, adult organ development, neural development, growth and metamorphosis, and evolution will be found in Chapters 9, 10, 11, 12, 13 and 14, respectively.

Drosophila life cycle and overall development

The fruit fly *Drosophila melanogaster* is a small dipteran insect, about 3 mm long as an adult, which undergoes embryonic development inside an egg and hatches from the egg as a larva. This then goes through two more larval stages, growing bigger each time, and eventually becomes a pupa, in which metamorphosis into the adult occurs. The life cycle of *Drosophila* is shown in Fig. 2.1.

2.1 The early *Drosophila* embryo is a multinucleate syncytium

The *Drosophila* egg is rather oblong and the future anterior end is easily recognizable by the micropyle, a nipple-shaped structure in the tough external coat surrounding the egg. Sperm enter the anterior end of the egg through the micropyle. After fertilization and fusion of the sperm and egg nuclei, the zygote nucleus undergoes a series of rapid mitotic divisions, one about every 9 minutes, but, unlike in most embryos, there is initially no cleavage of the cytoplasm and no formation of cell membranes to separate the nuclei. The result after 12 nuclear divisions is a **syncytium** in which around 6000 nuclei are present in a common cytoplasm (Fig. 2.2); the embryo essentially remains a single cell during its early development. After nine divisions the nuclei move to the periphery to form the **syncytial blastoderm**. This comprises a superficial layer of nuclei and cytoplasm, and surrounds a central mass of yolky cytoplasm. The syncytial blastoderm is equivalent to the blastula stage in embryos such as *Xenopus* (see, for example, Box 1A). Shortly afterwards, membranes grow in from the surface to enclose the nuclei and form cells, and the blastoderm becomes truly cellular after 14 mitoses. Because of the formation of a syncytium, even large molecules such as proteins can diffuse between nuclei during the first 3 hours of development and, as we shall see, this is of great importance to early *Drosophila* development.

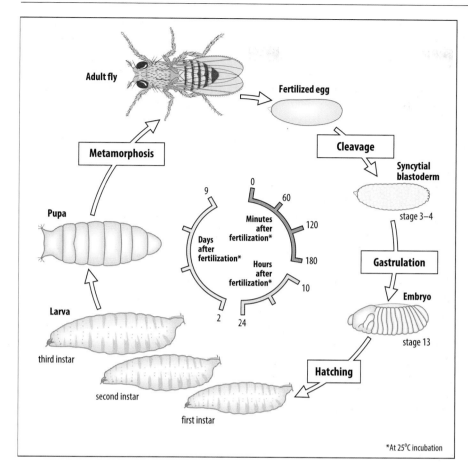

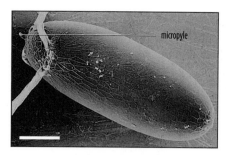

Fig. 2.1 Life cycle of *Drosophila melanogaster*. After cleavage and gastrulation the embryo becomes segmented and hatches out as a feeding larva. The larva grows and goes through two more rounds of growth and molting (instars), eventually forming a pupa that will metamorphose into the adult fly. The photographs show scanning electron micrographs. Top: a *Drosophila* egg before fertilization. The sperm enters through the micropyle. The dorsal filaments (white ribbons) are extraembryonic structures. Center: a first-instar larva. Bottom: a pupa. Anterior is to the left, and larva and pupa are shown in dorsal view. Scale bars = 0.1 mm.

*Top photograph reproduced with permission from Turner, F.R., Mahowald, A.P.: **Scanning electron microscopy of Drosophila embryogenesis. I. The structure of the egg envelopes and the formation of the cellular blastoderm**. Dev. Biol. 1976, **50**: 95–108. Middle photograph reproduced with permission from Turner, F.R., Mahowald, A.P.: **Scanning electron microscopy of Drosophila melanogaster embryogenesis. III. Formation of the head and caudal segments**. Dev. Biol. 1979, **68**: 96–109.*

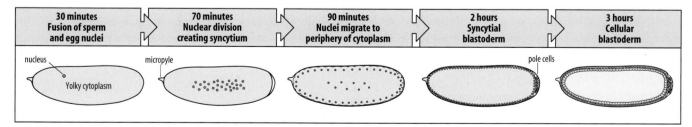

Fig. 2.2 Cleavage of the *Drosophila* embryo. After fusion of the sperm and egg nuclei there is rapid nuclear division but no cell membranes surround the individual nuclei, resulting in a syncytium of many nuclei in a common yolky cytoplasm. After the ninth division the nuclei move to the periphery to form the syncytial blastoderm but some are delayed. About 3 hours after egg laying, cell membranes develop, giving rise to the cellular blastoderm. About 15 pole cells, which will give rise to germ cells, form a separate group at the posterior end of the embryo. Times given are for incubation at 25°C.

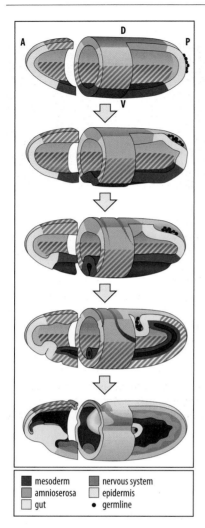

mesoderm · nervous system
amnioserosa · epidermis
gut · germline

Fig. 2.3 Gastrulation in *Drosophila*.
Gastrulation begins when the future
mesoderm (red) invaginates in the ventral
region, first forming a furrow and then an
internalized tube (red). The mesoderm cells
then leave the tube and migrate internally
under the ectoderm (see last panel). The
nervous system comes from cells that leave
the surface of the ventral blastoderm and
form a layer between the remaining ventral
ectoderm and mesoderm. Regions striped
light blue and dark blue represent ectodermal
tissue that will give rise to both nervous
system (neuroectoderm, dark blue) and
epidermis (light blue). The anterior striped
area gives rise to part of the brain and
the epidermis of the head. The gut forms
from two invaginations at the anterior and
posterior end that will fuse in the middle
(yellow). The midgut region is derived from
endoderm, whereas the foregut and hindgut
are of ectodermal origin. The amnioserosa
(green), an extra-embryonic membrane, is
discussed in Section 2.5.

At the syncytial stage, a small number of nuclei move to the posterior end of the embryo and become surrounded by cell membranes to form the **pole cells**, which end up on the outer surface of the blastoderm (see Fig. 2.2). The pole cells eventually give rise to the germ cells from which the gametes—sperm and eggs—develop in the adult insect, whereas the blastoderm gives rise to the somatic cells of the embryo. This setting aside of the germline from the rest of the embryo at a very early stage is a common strategy in animal development.

In this chapter we will be concerned with the development of the embryo to the stage at which it hatches out of the egg as a larva. Embryonic development naturally takes place inside the opaque egg case, and to study it embryos are freed from the egg case by treating with bleach and then placed in a special oil for observation. This can be done at any stage and they will continue to develop and even hatch. The embryo is about 0.5 mm long.

2.2 Cellularization is followed by gastrulation and segmentation

All the future tissues, except for the germline cells, are derived from the single epithelial layer of the cellular blastoderm. This gives rise to the three germ layers—**ectoderm**, **mesoderm**, and **endoderm** (see Box 1C). In *Drosophila*, the prospective mesoderm is located in the most ventral region, whereas the future midgut derives from two regions of prospective endoderm, one at the anterior and the other at the posterior end of the embryo. Endodermal and mesodermal tissues move to their future positions inside the embryo during **gastrulation**, leaving ectoderm as the outer layer (Fig. 2.3). Gastrulation starts at about 3 hours after fertilization, when the future mesoderm in the ventral region invaginates to form a furrow along the ventral midline. The mesoderm cells are initially internalized by formation of a tube of mesoderm, which will be described in more detail in Chapter 9. The mesoderm cells then separate from the surface layer of the tube and migrate under the ectoderm to internal locations, where they later give rise to muscle and other connective tissues. Each of these germ layers becomes a unit of lineage, that is, once a cell has been specified to one of these fates—mesoderm, ectoderm or endoderm—all of its descendants will adopt that fate.

In insects, as in all arthropods, the main nerve cord lies ventrally, rather than dorsally as in vertebrates. Shortly after the mesoderm has invaginated, ectodermal cells of the ventral region, which will give rise to the nervous system, leave the surface individually and form a layer of prospective neural cells (neuroblasts) between the mesoderm and the remaining outer ectoderm (see Fig. 2.3, last panel). At the same time, two tube-like invaginations develop at the sites of the future anterior and posterior midgut. These grow inward towards each other and eventually fuse to form the endoderm of the midgut, while ectoderm is dragged inward behind them at each end to form the foregut and the hindgut. The outer ectoderm layer develops into the epidermis. The ectoderm continues to divide during gastrulation, but the mesoderm does not. Once gastrulation is completed, mesoderm cells start to divide again. The cells of the epidermis only divide two more times before they secrete a thin cuticle composed largely of protein and the polysaccharide chitin.

Also during gastrulation, the central body of the blastoderm or **germ band**, which comprises the main trunk region of the embryo, undergoes a process of extension along the antero-posterior axis (Fig. 2.4). **Germ-band extension** drives the posterior trunk regions round the posterior end and onto what was the dorsal side, as described in Chapter 9. The germ band later retracts as embryonic development is completed. At the time of germ-band extension, the first external signs of **segmentation** can be seen. A series of evenly spaced grooves form more or less simultaneously and these demarcate developmental regions called **parasegments**. As we shall see later, the segments of the larva and adult do not correspond exactly to the embryonic parasegments:

Ventral view, early gastrula	Lateral view, germ-band extension	Lateral view, segmentation
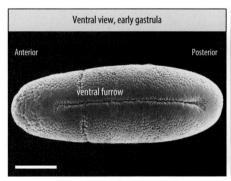	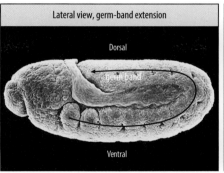	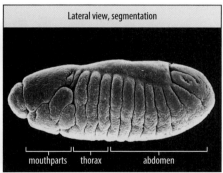

parasegments act as the developmental basis for segments, which result from the fusion of the posterior and anterior halves of adjacent parasegments. There are 14 parasegments: three contribute to mouthparts of the head; three to the thoracic region; and eight to the abdomen.

2.3 After hatching, the *Drosophila* larva develops through several larval stages, pupates, and then undergoes metamorphosis to become an adult

The larva (Fig. 2.5) hatches about 24 hours after fertilization, but the different regions of the larval body are well defined several hours before that. The head is a complex structure, largely hidden from view before the larva hatches. A specialized structure associated with the most anterior region of the head is the acron. Another specialized structure, called the telson, marks the posterior end of the larva. Between the head and telson, three thoracic segments and eight abdominal segments can be distinguished by specializations in the cuticle. On the ventral side of each segment are belts of small tooth-like outgrowths from the epidermis called **denticles,** and other cuticular structures characteristic of each segment. As the larva feeds and grows, it molts, shedding its cuticle. This process occurs twice, each stage being called an **instar.** After the third instar, the larva becomes a **pupa,** inside which **metamorphosis** into the adult fly occurs and there is a major transformation of overall form.

The *Drosophila* larva has neither wings nor legs; these and other organs are formed as the larva grows (discussed in Chapter 13). The primordia for these structures are, however, formed in the embryo and are present in the larva as **imaginal discs,** small sheets of about 20–40 prospective epidermal cells at the time they are formed. These discs grow by cell proliferation throughout larval life and form sacs of epithelia that fold to accommodate their increase in size. There are imaginal discs for each of the six legs, two wings, and the two halteres (balancing organs), and for the genital apparatus, eyes, antennae, and other adult head structures (Fig. 2.6). At metamorphosis, these differentiate and develop into the adult organs. We will discuss the development of the imaginal discs in Chapter 11; they provide continuity between the pattern of the larval body and that of the adult, even though metamorphosis intervenes.

2.4 Many developmental genes were identified in *Drosophila* through induced large-scale genetic screening

Valuable though spontaneous mutations have been in the study of development, suitable developmentally informative mutations are rare. Most of the known developmental genes were identified by inducing random mutations in a large number of individuals through chemical treatments or irradiation with X-rays, and then screening for organisms with developmental defects associated with mutations in individual

Fig. 2.4 Gastrulation, germ-band extension, and segmentation in the *Drosophila* embryo. Gastrulation involves the future mesoderm moving inside through the ventral furrow. During gastrulation, the central body of the blastoderm (the germ band) extends, driving the posterior trunk region onto the dorsal side and segmentation now takes place. Later the germ band shortens. Scale bar = 0.1 mm.

*Reproduced with permission from Turner, F.R., Mahowald, A.P.: **Scanning electron microscopy of Drosophila melanogaster embryogenesis. II. Gastrulation and segmentation.** Dev. Biol. 1977, **57**: 403–416.*

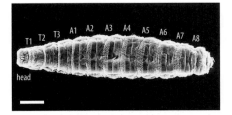

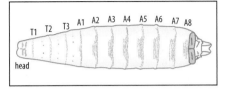

Fig. 2.5 The *Drosophila* first-instar larva. Ventral view. T1 to T3 are the thoracic segments, and A1 to A8 the abdominal segments. The characteristic pattern of denticles can be seen in the anterior region of each abdominal segment. Scale bar = 0.1 mm.

Photograph courtesy of F.R. Turner.

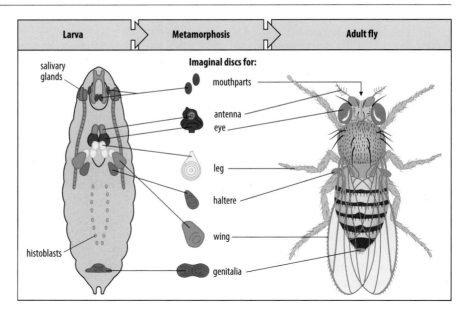

Fig. 2.6 Imaginal discs give rise to adult structures at metamorphosis. The imaginal discs in the *Drosophila* larva are small sheets of epithelial cells; at metamorphosis they give rise to a variety of adult structures. The abdominal cuticle of the adult comes from groups of tissue-forming cells (histoblasts) located in each larval abdominal segment.

genes. This sort of approach is best used in organisms that breed rapidly and can be obtained and treated conveniently in very large numbers, such as *Drosophila*. The overall aim of the experiment is to treat a large enough population so that, in total, a mutation is induced in every gene in the genome.

Many of the developmental mutations that have led to our present understanding of early *Drosophila* development came from a brilliantly successful screening program that searched the *Drosophila* genome systematically for mutations affecting the patterning of the early embryo. Its success was recognized with the award of a Nobel Prize for Physiology or Medicine to Edward Lewis, Christiane Nüsslein-Volhard, and Eric Wieschaus in 1995.

In this screening program thousands of flies were treated with a chemical mutagen, and were bred and screened according to the strategy described in Box 2A. Given the number of progeny involved, it was important to devise a strategy that would reduce the number of flies that had to be examined to find a mutation. So the search was restricted to mutations on just one chromosome at a time and making use of a scheme that would eliminate flies that were of no interest to the screen. As described in Box 2A, the program also incorporated a means of identifying flies homozygous for the male-derived chromosomes carrying the induced mutations and, most important, also included a way of automatically removing from the population flies that could not be carrying a mutated chromosome.

The screen took advantage of the precise and reproducible pattern of the stereotyped pattern of denticles in the cuticle at the end of embryogenesis (see Fig. 2.5) and identified deviations from this. The screens identified deviations in the number of segmental structures as well as in the arrangement of denticles (see Box 2A). In this way, the key genes involved in patterning the early *Drosophila* embryo were first identified. The screens also yielded genes involved in the specification of the germ layers, and even genes affecting mitosis and the cell cycle.

One type of mutation that affects *Drosophila* embryonic development needs a somewhat different type of screen. **Maternal-effect mutations** affect the embryo's development by way of the mother. They are identified as mutations that, when present in the mother, do not affect her appearance or normal physiology but have profound effects on the development of her progeny, independently of the contribution of the

EXPERIMENTAL BOX 2A Mutagenesis and genetic screening strategy for identifying developmental mutants in *Drosophila*

The mutagen ethyl methane sulfonate (EMS) was applied to large numbers of male flies homozygous for a recessive but viable mutation (for example white eyes instead of the normal red eyes, see Fig. 1.12) on the selected chromosome, 'a' in Figure 1.

The treated males, which now produce sperm with various induced mutations on this chromosome (a*), were crossed with untreated females that carried different mutations (*DTS* and *b*) on their two 'a' chromosomes, but were otherwise wild type. These mutations track the untreated female-derived chromosomes and automatically eliminate all embryos carrying two female-derived chromosomes in subsequent generations. *DTS* is a dominant temperature-sensitive mutation that kills flies raised at 29°C while *b* is an embryonic lethal recessive mutation without a visible phenotype that allows the elimination of embryos not bearing the a* chromosome. The female flies also carried a special balancer chromosome (not shown) that prevented the recombination of male and female chromosomes.

To identify the new recessive mutations (a*) caused by the chemical treatment, a large number of the heterozygous males arising from this first cross were outcrossed to *DTS/b* females. Of the offspring of each cross, only a*/*b* flies survive when placed at 29°C; all other combinations die. The surviving siblings of each cross bear the same a* mutation. They were intercrossed and the offspring screened for patterning mutants. There are three possible outcomes: flies homozygous for the induced mutation a*; heterozygous a*/*b* flies; and homozygous *b* flies (which die as normal-looking embryos).

a* patterning mutations of interest are embryonic lethal, and so any culture tubes containing adult white-eyed flies can be discarded immediately, as their a* mutations must have allowed

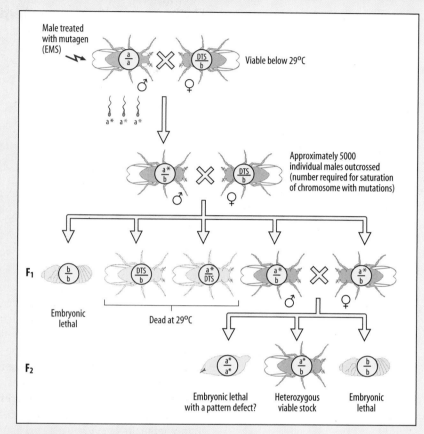

Figure 1

them to develop to adulthood even when homozygous. If no white-eyed flies are present in a tube, then the homozygous a*/a* embryos may have died or been arrested in their development as a result of abnormal development, and the mutation is of potential interest. The embryos or larvae from this cross can then be examined for pattern defects, as in the examples illustrated here (Figure 2). The phenotypically wild-type adults in this tube are heterozygous for a* and are used as breeding stock to study the mutant further. This whole program was repeated for each of the four chromosome pairs of *Drosophila*.

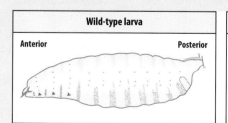

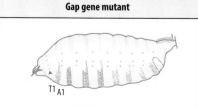

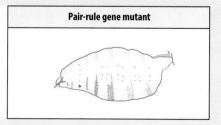

Figure 2

father; that is, the paternal genome cannot rescue the defect from the mother. Screens for these particular types of mutation led to the identification of the genes responsible and the proteins they encode. As we shall see later in the chapter, some of these **maternal-effect genes** or sometimes just **maternal genes** exert their effects by being expressed in the follicle cells of the ovary, which form a bag that contains the germline-derived **oocyte** (the immature egg) and nurse cells. Others are expressed in the nurse cells, which eventually contribute their cytoplasm and nuclei to the oocyte, or in the oocyte itself, producing developmentally important mRNAs and proteins that are deposited in the oocyte in a particular spatial pattern. This initial patterning process is crucial for correct development.

SUMMARY

The embryo of the fruit fly *Drosophila* develops within the egg-case and hatches out as a larva, which goes through several cycles of growth and moulting before pupating and undergoing metamorphosis to produce the adult fly. Maternal proteins and mRNAs laid down in the egg guide the early stages of development, and we shall look next at how these maternal gene products specify and start to pattern the antero-posterior and dorso-ventral axes of the embryo.

Setting up the body axes

The insect body, like that of vertebrates and most of the other organisms described in this book, is bilaterally symmetrical. Like all animals with bilateral symmetry, the *Drosophila* larva has two distinct and largely independent axes: the antero-posterior and dorso-ventral axes, which are at right angles to each other. These axes are already partly set up in the *Drosophila* egg, and we will see here how they become established and start to be patterned in the very early embryo.

2.5 The body axes are set up while the *Drosophila* embryo is still a syncytium

The antero-posterior and dorso-ventral axes become fully established and start to become patterned while the embryo is still in the syncytial blastoderm stage. Along the antero-posterior axis the embryo becomes divided into several broad regions, which will become the **head, thorax,** and **abdomen** of the larva (Fig. 2.7). The thorax and abdomen become divided into segments as the embryo develops, while two regions of endoderm at each end of the embryo invaginate at gastrulation to form the gut (see Fig. 2.3). Each segment, and the head, has its own unique character in the larva, as revealed by both its external cuticular structures and its internal organization.

The dorso-ventral axis of the embryo becomes divided up into four regions early in embryogenesis: from ventral to dorsal these are the **mesoderm**, which will form muscles and other internal connective tissues; the **ventral ectoderm** or **neuroectoderm**, which gives rise to the larval nervous system and epidermis; the **dorsal ectoderm**, which gives rise to larval epidermis; and the **amnioserosa**, which gives rise to an extra-embryonic membrane on the dorsal side of the embryo (see Figs 2.3 and 2.7). Organization along the antero-posterior and dorso-ventral axes of the early embryo develops more or less simultaneously, but is specified by independent mechanisms and by different sets of genes in each axis.

The early development of *Drosophila* is peculiar to certain types of insects, as patterning occurs within a multinucleate syncytial blastoderm (see Fig. 2.2). Fig. 2.8 shows an embryo at the syncytial blastoderm stage, with the nuclei stained for two maternal transcription factors. Only after the beginning of segmentation does the embryo become truly multicellular. At the syncytial stage, many proteins, including those such as transcription factors that are not normally secreted from cells, can diffuse throughout the blastoderm and enter other nuclei. This allows the formation of concentration gradients of transcription factors along the length of the embryo that provide positional information for the nuclei to interpret (see Section 1.15).

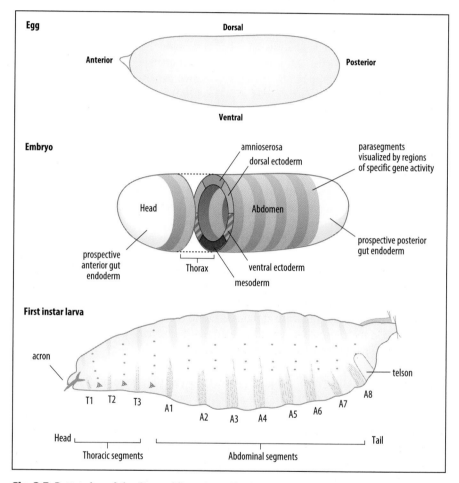

Fig. 2.7 Patterning of the *Drosophila* embryo. The body plan is patterned along two distinct axes. The antero-posterior and dorso-ventral axes are at right angles to each other and are laid down in the egg. In the early embryo, the dorso-ventral axis is divided into four regions: mesoderm (red), ventral ectoderm (neuroectoderm) (dark blue), dorsal ectoderm (light blue), and amnioserosa (an extra-embryonic membrane; green). The ventral ectoderm gives rise to both ventral epidermis and neural tissue, the dorsal ectoderm to epidermis. The antero-posterior axis becomes divided into different regions that later give rise to the head, thorax, and abdomen. After the initial division into broad body regions, segmentation begins. The future segments can be visualized as transverse stripes by staining for specific gene activity; these stripes demarcate 14 parasegments, 10 of which are indicated in this diagram. The embryo develops into a segmented larva. By the time the first-instar larva hatches out, the 14 parasegments have been converted into three thoracic (T1–T3) and eight abdominal (A1–A8) segments, with each segment being made up of the posterior half of one parasegment and the anterior half of the next. The rest of the parasegments contribute to structures in the head. Different segments are distinguished by the patterns of hairs and denticles on the cuticle. Specialized structures, the acron and telson, develop at the head and tail ends of the embryo, respectively.

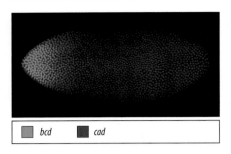

| bcd | cad |

Fig. 2.8 Initial patterning of *Drosophila* occurs at the syncytial blastoderm stage. The photograph shows an embryo about 2 hours after egg laying, in which two translated maternal proteins, the transcription factors Bicoid (green) and Caudal (red), have entered the blastoderm nuclei. These proteins are involved in patterning the antero-posterior axis of the embryo. Bicoid protein prevents the synthesis of Caudal protein at the anterior end of the embryo, thus restricting Caudal to the posterior end. The embryo has been fixed and the proteins stained using fluorescent antibodies.

Reproduced with permission from Surkova, S., et al.: Characterization of the Drosophila segment determination morphome. Dev. Biol. 2008, 313: 844-862.

Scan here

Scan this QR code image with your mobile device to see an online movie showing the Bicoid gradient in *Drosophila* or log on to **http://global.oup.com/uk/orc/biosciences/ devbiol/wolpert5e/qr/qr2a/**

Early development is essentially two-dimensional, because patterning occurs mainly in the blastoderm, which is a single layer of nuclei, and later of cells. But the larva itself is a three-dimensional object, with internal structures organized along the axis. This third dimension develops later, at gastrulation (discussed in Chapter 9), when parts of the surface layer move into the interior to form the gut, the mesodermal structures that will give rise to muscle, and the ectodermally derived nervous system (see Fig. 2.3).

2.6 Maternal factors set up the body axes and direct the early stage of *Drosophila* development

The earliest stage of *Drosophila* embryonic development is guided by **maternal factors**— mRNAs and proteins that are synthesized and laid down in the egg by the mother (see Section 2.4). In all, genetic analysis has identified about 50 maternal genes involved in setting up the two axes and a basic molecular framework of positional information, which is then interpreted by the embryo's own genetic program. In contrast to the maternal-effect genes, those genes expressed in the embryo's own nuclei during development are known as **zygotic genes**. All later patterning, which involves the regulated expression of the zygotic genes, is built on the framework of maternal gene products (Fig. 2.9). After development begins, the maternal RNAs are translated, and the resulting proteins, many of which are transcription factors, act on the embryo's nuclei to activate zygotic genes in a specific spatial pattern along each axis, thus setting the scene for the next round of patterning.

Drosophila illustrates particularly well a general principle of development, namely that an embryo is patterned in a series of steps. Broad regional differences are established first, and within these domains, the expression of specific genes and their interactions produce a larger number of smaller developmental domains, each characterized by a unique profile of gene activity. Developmental genes act in a strict temporal sequence. They form a hierarchy of gene activity, in which the action of one set of genes is essential for another set of genes to be activated, and thus for the next stage of development to occur.

2.7 Three classes of maternal genes specify the antero-posterior axis

We will first look at how maternal gene products specify the antero-posterior axis of the embryo. The expression of maternal genes during egg formation in the mother creates regional differences in the egg along the antero-posterior axis even before it is fertilized. These differences already distinguish the future anterior and posterior ends of the larva. The roles of individual maternal genes can be deduced from the effects of maternal-effect mutations (see Section 2.4) on the embryo. They fall into three classes: those that affect only anterior regions; those that affect only posterior regions; and those that affect both of the terminal regions (Fig. 2.10). Mutations of genes in the anterior class, such as *bicoid*, lead to a reduction or loss of head and thoracic structures in the larva, and in some cases their replacement with posterior structures. Posterior-group mutations, such as *nanos*, cause the loss of abdominal regions, leading to a smaller than normal larva, while those of the terminal class, such as *torso*, affect the acron and telson.

The idiosyncratic naming of genes in *Drosophila* usually reflects the attempts by their discoverers to describe the mutant phenotype—*nanos* is Greek for dwarf, *bicoid* means two ends, and *torso* reflects the fact that both ends of the embryo are missing. In this chapter we meet quite a number of gene names; all these are listed, together with their functions where known, in the table at the end of this chapter.

2.8 Bicoid protein provides an antero-posterior gradient of a morphogen

Maternal *bicoid* mRNA is localized at the anterior end of the unfertilized egg during oogenesis. After fertilization it is translated into Bicoid protein, which diffuses from

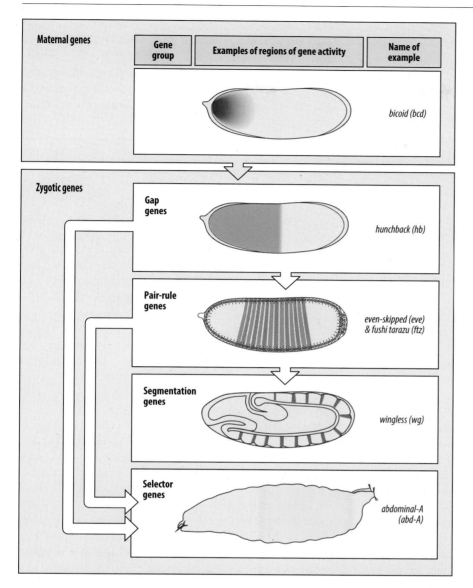

Gene group	Examples of regions of gene activity	Name of example
Maternal genes		*bicoid (bcd)*
Zygotic genes		
Gap genes		*hunchback (hb)*
Pair-rule genes		*even-skipped (eve) & fushi tarazu (ftz)*
Segmentation genes		*wingless (wg)*
Selector genes		*abdominal-A (abd-A)*

Fig. 2.9 The sequential expression of different sets of genes establishes the body plan along the antero-posterior axis. Upper panel: after fertilization, maternal gene products laid down in the egg, such as *bicoid* mRNA (red), are translated. When translated, the Bicoid protein diffuses and its concentration along the main axis provides positional information that activates the zygotic genes (the bicoid protein gradient can be seen in Figs 2.8 and 2.12). Lower panel: the four main classes of zygotic genes acting along the antero-posterior axis are the gap genes, the pair-rule genes, the segmentation genes, and the selector, or homeotic, genes. The gap genes define regional differences, which result in the expression of a periodic pattern of gene activity by the pair-rule genes, which define the parasegments and foreshadow segmentation. The segmentation genes elaborate patterning within the parasegments, and thus the future segments, and the selector genes determine segment identity (whether a segment is thoracic or abdominal, for example, and which specific segment it is). The functions of each of these classes of genes are discussed in this chapter.

the anterior end to form a concentration gradient along the antero-posterior axis. Historically, the Bicoid protein gradient provided the first concrete evidence for the existence of the morphogen gradients that had been postulated to control pattern formation (see Section 1.15).

The role of the *bicoid* gene was first elucidated by a combination of genetic and classical embryological experiments on the *Drosophila* embryo. Female flies that do not express *bicoid* produce embryos that have no proper head and thorax (see Fig. 2.10). In a separate line of investigation into the role of localized cytoplasmic factors in anterior development, normal eggs were pricked at the anterior ends and some cytoplasm allowed to leak out. The embryos that developed bore a striking resemblance to *bicoid* mutant embryos. Together, these observations suggested that normal eggs have some factor(s) in the cytoplasm at their anterior end that is absent in *bicoid* mutant eggs. This was confirmed by showing that anterior cytoplasm from wild-type embryos could partially 'rescue' the *bicoid* mutant embryos—in the sense they developed more normally—if it was injected into their anterior regions (Fig. 2.11). Moreover, if normal anterior cytoplasm is injected into the middle of a fertilized *bicoid* mutant egg, head structures developed at the site of injection and the adjacent segments became thoracic segments, setting up a mirror-image body pattern at the site of injection. The simplest interpretation of these experiments is that the *bicoid* gene is necessary for

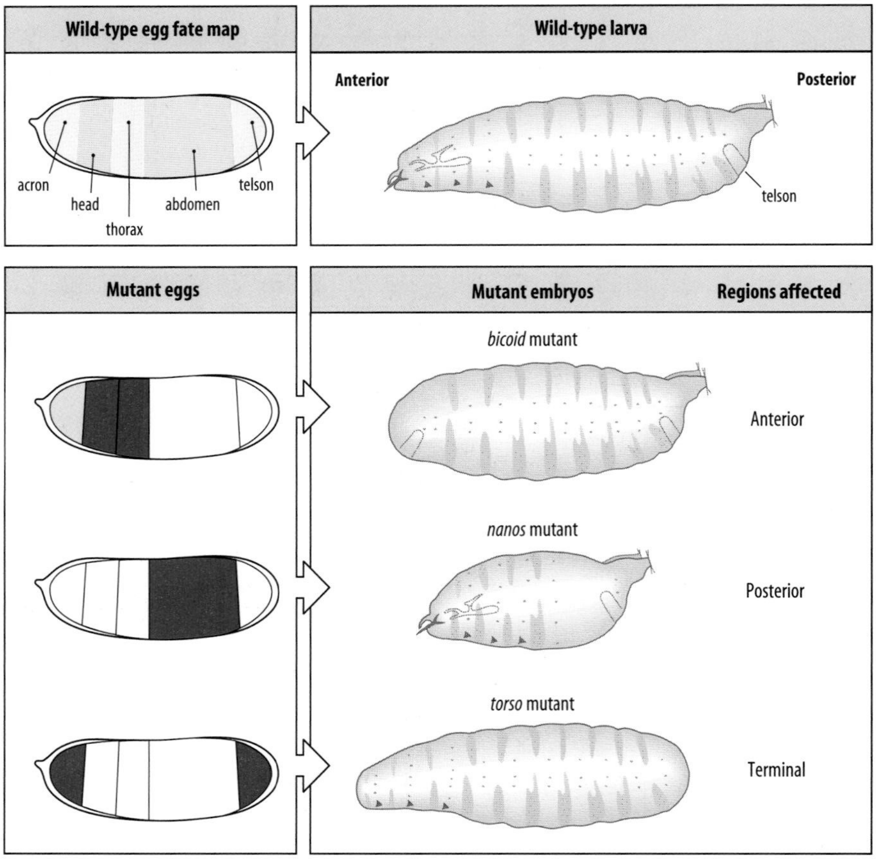

Fig. 2.10 The effects of mutations in the maternal gene system. Mutations in maternal genes lead to deletions and abnormalities in anterior, posterior, or terminal structures. The wild-type fate map shows which regions of the egg give rise to particular regions and structures in the larva. Regions that are affected in mutant eggs, and which lead to lost or altered structures in the larva, are shaded in red. In *bicoid* mutants there is a partial loss of anterior structures and the appearance of a posterior structure—the telson—at the anterior end. *nanos* mutants lack a large part of the posterior region. *torso* mutants lack both acron and telson.

the establishment of the anterior structures because it establishes a gradient of Bicoid protein, whose source and highest level is at the anterior end.

bicoid mRNA was originally shown to be tightly localized to the anterior-most region of the unfertilized egg by the technique of *in situ* hybridization (see Box 1D). Staining with an antibody against Bicoid protein then showed that this is absent from the unfertilized egg, but that after fertilization, *bicoid* mRNA is translated into protein at the anterior pole. The protein then diffuses and forms a gradient with the highest point at the anterior end. *Bicoid* is a transcription factor and so must enter the embryo's nuclei to carry out its function of switching on zygotic genes. By the time the syncytial blastoderm is formed, there is a clear gradient of intranuclear Bicoid protein concentration along the antero-posterior axis, with its high point at the anterior end.

With advances in imaging technology, the dynamics of Bicoid gradient formation have been observed in living embryos by measuring the intranuclear concentrations of Bicoid protein genetically tagged with green fluorescent protein (Bicoid–GFP) (Fig. 2.12). The measurements indicate that the concentration of Bicoid protein inside a nucleus at a given position along the gradient is somehow maintained at a constant level, despite the increase in the number of nuclei at each mitotic cycle and the outflow of Bicoid into the cytoplasm from nuclei when nuclear membranes break down at mitosis.

Bicoid protein acts as a morphogen that patterns the anterior portion of the embryo. As described in more detail later in the chapter, it switches on particular zygotic genes at different **threshold concentrations**, thereby initiating new patterns of gene expression along the axis (the concept of thresholds is discussed in Section 1.15). Thus, *bicoid* is a key maternal gene in early *Drosophila* development. The other maternal genes of the anterior group are mainly involved in the localization of *bicoid*

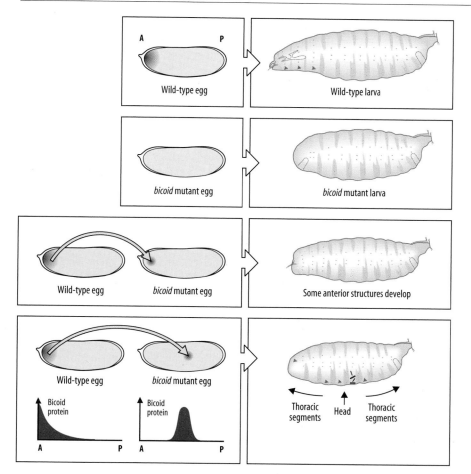

Fig. 2.11 The *bicoid* gene is necessary for the development of anterior structures. Embryos whose mothers lack the *bicoid* gene lack anterior regions (second row). Transfer of anterior cytoplasm from wild-type embryos to *bicoid* mutant embryos causes some anterior structure to develop at the site of injection (third row). If wild-type anterior cytoplasm is transplanted to the middle of a *bicoid* mutant egg or early embryo, head structures develop at the site of injection, flanked on both sides by thoracic-type segments (fourth row). These results can be interpreted in terms of the anterior cytoplasm setting up a gradient of Bicoid protein with the high point at the site of injection (see graphs, bottom left panel). A, anterior; P, posterior.

mRNA to the anterior end of the egg during oogenesis and in the control of its translation after fertilization. Given the crucial importance of *bicoid* in *Drosophila* development, it is worth noting that the *bicoid* gene is only present in a small group of the most recently evolved dipterans—such as fruit flies and blowflies. Later in this chapter we shall briefly discuss different mechanisms of early development that occur in some other groups of insects. It is not surprising that many different developmental mechanisms have evolved in such a large and diverse group of animals as the insects.

2.9 The posterior pattern is controlled by the gradients of Nanos and Caudal proteins

For proper patterning along an axis, both ends need to be specified, and Bicoid defines only the anterior end of the antero-posterior axis. The posterior end is specified by the actions of at least nine maternal genes—the posterior-group genes. Mutations in posterior-group genes, such as *nanos*, result in larvae that are shorter than normal because there is no abdomen (see Fig. 2.10). One function of the products of other posterior-group genes (for example, *oskar*) is to localize *nanos* mRNA at the extreme posterior pole of the unfertilized egg. Another function is to assemble the posterior **germplasm** in the egg, cytoplasm that contains the so-called germline factors; in the embryo this cytoplasm is incorporated into the pole cells (see Section 2.1) and will give rise to eggs or sperm.

nanos mRNA is localized to the posterior end of the egg and, like *bicoid* mRNA, it is translated only after fertilization. It produces a concentration gradient of Nanos protein, in this case with the highest level at the posterior end of the embryo. But unlike

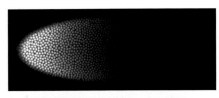

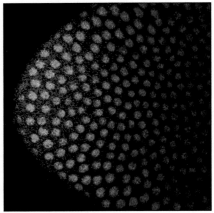

Fig. 2.12 Bicoid protein enters the nuclei and forms an nuclear gradient from anterior to posterior. The top photograph shows the nuclear Bicoid protein gradient in a fixed transgenic *Drosophila* embryo carrying a *bicoid-GFP* fusion gene. In these embryos the concentration of Bicoid protein could be accurately measured either by the intensity of the autofluorescence of the GFP protein itself (not shown) or by the intensity of fluorescence of an anti-GFP antibody tagged with a fluorescent dye, as shown in a surface view of the embryo here. The lower photograph shows the anterior tip of a fixed transgenic *Drosophila* embryo carrying a *bicoid-GFP* fusion gene labeled with fluorescently tagged cDNA probes that bind *bicoid* mRNA molecules. Green color shows autofluorescence of the GFP protein in the nuclei, red color shows packages of multiple *bicoid* mRNA molecules in the anterior cytoplasm.

Upper photograph courtesy of J. O. Dubuis and T. Gregor; lower figure courtesy of S. C. Little and T. Gregor.

Bicoid, Nanos is not a transcription factor and does not act directly as a morphogen to specify the abdominal pattern; it has a quite different action. Its function is to suppress translation of the maternal mRNA of another gene, *hunchback*, in the posterior region of the embryo.

Maternal *hunchback* mRNA is distributed throughout the embryo and starts to be translated after fertilization. However, *hunchback* is also one of the zygotic genes activated by the Bicoid protein as part of its patterning activity in the anterior half of the embryo. Restriction of Hunchback protein to the anterior region is crucial for correct patterning along the antero-posterior axis. To avoid interference by maternal Hunchback protein in the posterior region, and the production of anterior structures there, its translation is prevented in the posterior. This is the sole task of the Nanos protein (Fig. 2.13); it prevents *hunchback* mRNA translation by binding to a complex of the mRNA and Pumilio protein, which is encoded by another of the posterior-group genes. If maternal Hunchback is completely removed from embryos, then Nanos becomes completely dispensable for antero-posterior patterning.

Evolution can only work on what is already there; it cannot 'look' at the whole system and redesign it economically. So, if a gene causes a difficulty by being expressed in the 'wrong' place, the problem may be solved not by reorganizing the gene-expression pattern, but by bringing in a new function, such as *nanos*, to remove the unwanted protein.

Another maternal product crucial to establishing the posterior end of the axis is *caudal* mRNA. Again, this is only translated after fertilization. *caudal* mRNA itself is uniformly distributed throughout the egg, but after fertilization, a posterior-to-anterior gradient of Caudal protein is established by the specific inhibition of Caudal protein synthesis by Bicoid protein, which binds to a site in the 3′ untranslated region of the *caudal* mRNA. This function is quite separate from Bicoid's action as a transcription factor. Because the concentration of Bicoid is low at the posterior end of the embryo, Caudal protein concentration is highest there (see Fig. 2.8). Mutations in the *caudal* gene result in the abnormal development of abdominal segments.

Soon after fertilization, therefore, several gradients of maternal proteins have been established along the antero-posterior axis. Two gradients—Bicoid and Hunchback proteins—run in an anterior to posterior direction, whereas Caudal protein is graded posterior to anterior. We next look at the quite different mechanism that specifies the two termini of the embryo.

2.10 The anterior and posterior extremities of the embryo are specified by activation of a cell-surface receptor

A third group of maternal genes specifies the structures at the extreme ends of the antero-posterior axis—the acron and the head region at the anterior end, and the telson and the most posterior abdominal segments at the posterior end. A key gene in this group is *torso*; mutations in *torso* can result in embryos developing neither acron nor telson (see Fig. 2.10). This indicates that the two terminal regions, despite their topographical separation, are not specified independently but use the same pathway.

The terminal regions are specified by an interesting mechanism that involves the localized activation of a receptor protein that is itself present throughout the plasma membrane of the fertilized egg. The activated receptor sends a signal to the adjacent cytoplasm that specifies it as terminal. The receptor is known as Torso, as mutations in the maternal *torso* gene produce embryos lacking terminal regions. After fertilization, the maternal *torso* mRNA is translated and the Torso protein is uniformly distributed throughout the fertilized egg plasma membrane. Torso is, however, only activated at the ends of the fertilized egg because the protein ligand that stimulates it is only present there.

This ligand for Torso is thought to be a fragment of a secreted protein known as Trunk. *trunk* mRNA is deposited in the egg by the nurse cells and Trunk protein is

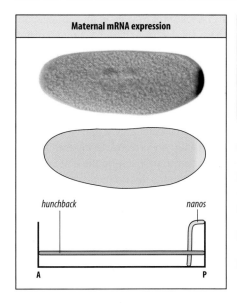
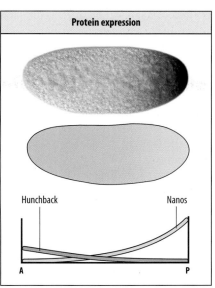

Maternal mRNA expression	Protein expression
hunchback *nanos*	Hunchback Nanos
A P	A P

Fig. 2.13 Establishment of a maternal gradient in Hunchback protein. Left panel: in the unfertilized egg, maternal *hunchback* mRNA (turquoise) is present at a relatively low level throughout the egg, whereas *nanos* mRNA (yellow) is located posteriorly. The photograph is an in situ hybridization showing the location of nanos mRNA (black). Right panel: after fertilization, nanos mRNA is translated and Nanos protein blocks translation of hunchback mRNA in the posterior regions, giving rise to a shallow antero-posterior gradient in maternal Hunchback protein. The photograph shows the graded distribution of Nanos, detected with a labeled antibody.

Photographs courtesy of R. Lehmann, from Griffiths, et al.: An Introduction to Genetic Analysis, *6th edition. New York: W.H. Freeman & Co., 1996.*

secreted into the **perivitelline space** during oocyte development. The perivitelline space is the space between the oocyte plasma membrane and a protective membrane of extracellular matrix called the **vitelline membrane** or **vitelline envelope**, which surrounds the oocyte. The Trunk protein is thought to be present throughout the perivitelline space, but the Trunk fragment that acts as a ligand for Torso is generated only at the poles of the egg because the processing activity that produces it from Trunk is present only in these two regions. Crucial to this processing is the activity of a protein called Torso-like, which is produced exclusively by follicle cells adjacent to the two poles of the egg and is present in the vitelline envelope at the poles of the fertilized egg. By the time development begins after fertilization, small quantities of the Trunk ligand have been produced at the poles and are present in the perivitelline space, where they can bind Torso. Because Trunk ligand is only present in small quantities, most of it becomes bound to Torso at the poles, with little left to diffuse further away. In this way, a localized area of receptor activation is set up at each pole (Fig. 2.14).

Stimulation of Torso by its ligand produces a signal that is transmitted across the plasma membrane to the interior of the developing embryo. This signal directs the activation of zygotic genes in nuclei at both poles, thus defining the two extremities of the embryo. The Torso protein is one of a large superfamily of transmembrane receptors, known as receptor tyrosine kinases, whose cytoplasmic portions have tyrosine protein kinase activity. The kinase is activated when the extracellular part of the receptor binds its ligand, and the cytoplasmic part of the receptor transmits the signal onwards by phosphorylating cytoplasmic proteins.

This ingenious mechanism for setting up a localized area of receptor activation is not confined to determination of the terminal regions of the embryo, but is also used in setting up the dorso-ventral axis, which we consider next.

2.11 The dorso-ventral polarity of the embryo is specified by localization of maternal proteins in the egg vitelline envelope

The dorso-ventral axis is specified by a different set of maternal genes from those that specify the anterior-posterior axis. The basic mechanism is very similar to that described in Section 2.10 for the ends of the embryo. The receptor involved in dorso-ventral axis organization is a maternal protein called Toll, which is present throughout the plasma membrane of the fertilized egg. The ventral end of the axis is determined by

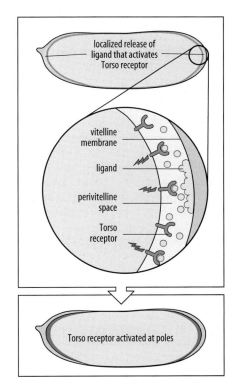

Fig. 2.14 The receptor protein Torso is involved in specifying the terminal regions of the embryo. The receptor protein encoded by the gene *torso* is present throughout the egg plasma membrane. Its ligand is laid down in the vitelline membrane at each end of the egg during oogenesis. After fertilization, the ligand is released and diffuses across the perivitelline space to activate the Torso protein at the ends of the embryo only.

the localized production in the ventral vitelline envelope of the ligand for this receptor. The ligand is a protein fragment produced by processing of a maternal protein called Spätzle. After fertilization, Spätzle itself is uniformly distributed throughout the extra-embryonic perivitelline space. The localized processing of Spätzle is controlled by a small set of maternal genes that are expressed only in the follicle cells that surround the future ventral region—about a third of the total surface of the developing egg. The key gene here is *pipe*, which codes for an enzyme, a heparan sulfate sulfotransferase, that is secreted into the oocyte vitelline envelope by these cells. The subsequent activity of the Pipe enzyme leads, in some way not yet fully understood, to the localization of a protease activity in the vitelline envelope on the ventral side of the embryo. The processed Spätzle fragment is therefore only present in the ventral perivitelline space.

Toll mRNA is laid down in the oocyte and is probably not translated until after fertilization. Although Toll is present throughout the membrane of the fertilized egg, it is only activated in the future ventral region of the embryo, owing to the localized production of its ligand, Spätzle. Toll activation is greatest where the concentration of its ligand is highest, and falls off rapidly on all sides, probably as a result of the limited amount of ligand being rapidly mopped up by the Toll receptors. Activation of Toll sends a signal to the adjacent cytoplasm of the embryo. At this stage, the embryo is still a syncytial blastoderm and this signal causes a maternal gene product in the cytoplasm—the Dorsal protein—to enter nearby nuclei (Fig. 2.15). This protein is a transcription factor with a vital role in organizing the dorso-ventral axis.

2.12 Positional information along the dorso-ventral axis is provided by the Dorsal protein

The initial dorso-ventral organization of the embryo is established at right angles to the antero-posterior axis at about the same time that this axis is being divided into terminal, anterior, and posterior regions. The embryo initially becomes divided into four regions along the dorso-ventral axis (see Section 2.5 and Fig. 2.7) and this patterning is controlled by the distribution of the maternal protein Dorsal.

Unlike Bicoid, Dorsal protein is uniformly distributed in the egg. Initially it is restricted to the cytoplasm, but under the influence of signals from the ventrally

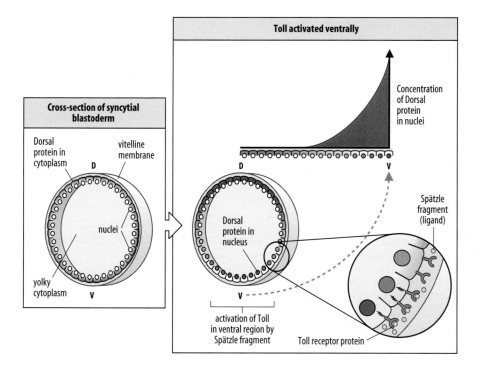

Fig. 2.15 Toll protein activation results in a gradient of intranuclear Dorsal protein along the dorso-ventral axis. Before Toll protein is activated, the Dorsal protein (red) is distributed throughout the peripheral band of cytoplasm. The Toll protein is a receptor that is only activated in the ventral region, by a maternally derived ligand (the Spätzle fragment), which is processed in the perivitelline space after fertilization. The localized activation of Toll results in the entry of Dorsal protein into nearby nuclei. The intranuclear concentration of Dorsal protein is greatest in ventral nuclei, resulting in a ventral to dorsal gradient. D, dorsal; V, ventral.

activated Toll proteins it enters nuclei in a graded fashion, with the highest concentration in ventral nuclei and the concentration progressively decreasing in a dorsal direction, as the Toll signal becomes weaker (see Fig. 2.15). Thus there is little or no Dorsal in nuclei in the dorsal regions of the embryo.

The role of Toll was first established by the observation that mutant embryos lacking it are strongly **dorsalized**—that is, no ventral structures develop. In these embryos, Dorsal protein does not enter nuclei but remains uniformly distributed in the cytoplasm. Transfer of wild-type cytoplasm into *Toll*-mutant embryos results in specification of a new dorso-ventral axis, the ventral region always corresponding to the site of injection. This occurs because in the absence of Toll, the Spätzle fragments produced on the original ventral side diffuse throughout the perivitelline space because there is no Toll protein to bind them. When the wild-type cytoplasm is injected, the Toll proteins it contains enter the membrane at the site of the injection. Spätzle fragments then bind to these receptors, setting in motion the chain of events that defines the ventral region at the site of injection.

In the absence of a signal from Toll protein, Dorsal protein is prevented from entering nuclei by being bound in the cytoplasm to another maternal gene product, the Cactus protein. As a result of Toll activation, Cactus is degraded and no longer binds Dorsal, which is then free to enter the nuclei. The pathway leading from Toll to the activation of Dorsal is shown in Box 2B. In embryos lacking Cactus protein, almost all of the Dorsal protein is found in the nuclei; there is a very poor concentration gradient and the embryos are **ventralized**—that is, no dorsal structures develop.

SUMMARY

Maternal genes act in the ovary of the mother fly to set up differences in the egg in the form of localized deposits of mRNAs and proteins. After fertilization, maternal mRNAs are translated and provide the embryonic nuclei with positional information in the form of protein gradients or localized protein. Along the antero-posterior axis there is an anterior to posterior gradient of maternal Bicoid protein, which controls patterning of the anterior region. For normal development it is essential that maternal Hunchback protein is absent from the posterior region and its suppression is the function of the posterior to anterior gradient of Nanos. The extremities of the embryo are specified by localized activation of the receptor protein Torso at the poles. The dorso-ventral axis is established by intranuclear localization of the Dorsal protein in a graded manner (ventral to dorsal), as a result of ventrally localized activation of the receptor protein Toll by a fragment of the protein Spätzle.

SUMMARY: maternal gene action in the fertilized egg of *Drosophila*

Antero-posterior	**Dorso-ventral**
mRNAs: *bicoid* forms anterior to posterior gradient; *hunchback* and *caudal* uniform; *nanos* posterior	Spätzle protein activates Toll receptor on ventral side
Anterior to posterior gradient of Bicoid protein formed. *hunchback* mRNA translation suppressed in posterior region by Nanos. *caudal* mRNA translation repressed by Bicoid	Dorsal protein enters ventral nuclei, giving ventral to dorsal gradient

Termini: Torso receptor activated by Trunk at ends of egg

CELL BIOLOGY BOX 2B The Toll signaling pathway: a multifunctional pathway

The interaction between the Dorsal and Cactus proteins in the Toll signaling pathway of *Drosophila* is of more than local interest: Dorsal protein is a transcription factor with homology to the Rel/NFκB family of vertebrate transcription factors, which are involved in regulation of gene expression in vertebrate immune responses, and the Toll signaling pathway is also used in the adult fly in defense against infection. The discovery of these innate immune defense pathways in the fly and the recognition of the importance of Toll-like receptors and the NFκB pathway in human immunity was recognized by the award of the Nobel Prize in Physiology and Medicine in 2011 to a researcher in *Drosophila* innate immunity alongside workers in human innate immunity. So what might seem at first sight a rather specialized mechanism for confining transcription factors to the cytoplasm until it is time for them to enter the nucleus during embryonic development is likely to be widely used for controlling gene expression and cell differentiation.

The Toll signaling pathway is thus a good example of a conserved intracellular signaling pathway that is used by multicellular organisms in different contexts, in this case ranging from embryonic development to defense against disease. All members of the Rel/NFκB family are typically held inactive in the cytoplasm until the cell is stimulated through an appropriate receptor. This leads to degradation of the inhibitory protein, which releases the transcription factor. This then enters the nucleus and activates gene transcription (Figure 1). In the *Drosophila* embryonic Toll pathway, Dorsal is held inactive in the syncytium cytoplasm by the protein Cactus. When Toll is activated by binding the Spätzle fragment, its cytoplasmic domain binds an adaptor protein, dMyD88 (or Tube), which in turn interacts with and activates the protein kinase Pelle. Activation of Pelle leads, through several more intermediate steps, which have not yet been fully established, to the

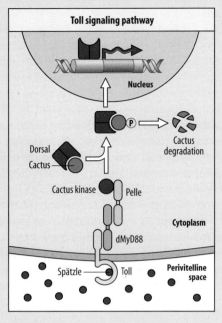

Figure 1

phosphorylation and degradation of Cactus. This releases Dorsal, which is then free to enter a nucleus.

In adult *Drosophila*, the Toll receptor is stimulated by fungal and bacterial infection and the signaling pathway results in the production of antimicrobial peptides. In humans, Toll-like receptors acting by essentially the same pathway are also involved in innate immunity to microbial infection. IRAK, and IκB are the mammalian homologs of Pelle, and Cactus, respectively, and play the same roles in this pathway. In vertebrates, NFκB is also activated in response to signaling through receptors other than Toll.

Localization of maternal determinants during oogenesis

Having seen how localized maternal gene products in the egg set up the basic framework for development, we now look at how they come to be localized so precisely. When the *Drosophila* egg is released from the ovary it already has a well-defined organization. *bicoid* mRNA is located at the anterior end and *nanos* and *oskar* mRNAs at the opposite end. Torso-like protein is present in the vitelline envelope at both poles, and other maternal proteins are localized in the ventral vitelline envelope. Numerous other maternal mRNAs, such as those for *caudal*, *hunchback*, *Toll*, *torso*, *dorsal*, and *cactus*, are distributed uniformly. How do these maternal mRNAs and proteins get laid down in the egg during **oogenesis**—its period of development in the ovary—and how are they localized in the correct places?

The development of an egg in the *Drosophila* ovary is shown in Fig. 2.16. A diploid germline stem cell in the **germarium** divides asymmetrically to produce another stem cell and a cell called the cystoblast, which undergoes a further four mitotic divisions to give 16 cells with cytoplasmic bridges between them; this group of cells

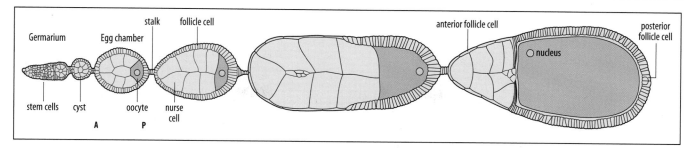

Fig. 2.16 Egg development in *Drosophila*. Oocyte development begins in a germarium, with stem cells at one end. One stem cell will divide four times to give 16 cells with cytoplasmic connections between each other, forming a structure called a cyst. One of the cells that is connected to four others will become the oocyte, the others will become nurse cells. The nurse cells and oocyte become surrounded by follicle cells and the resulting structure buds off from the germarium as an egg chamber. Successively produced egg chambers are still attached to each other at the poles. The oocyte grows as the nurse cells provide material through the cytoplasmic bridges. The resulting chain has an intrinsic polarity, which is reflected in that of the oocyte, with its anterior end being the side in contact with the nurse cells and closer to the germarium. The follicle cells have a key role in patterning the oocyte.

is known as the **germline cyst**. One of these 16 cells will become the **oocyte**; the other15 will develop into **nurse cells**, which produce large quantities of proteins and RNAs that are exported into the oocyte through the cytoplasmic bridges. Somatic ovarian cells make up a sheath of **follicle cells** around the nurse cells and oocyte to form the **egg chamber**. Successively produced egg chambers are joined to each other by 'stalks' derived from the follicle cells. This gives the string of egg chambers (the **ovariole**) a distinct architectural polarity, which is translated into oocyte polarity by signaling between the oocyte and its surrounding follicle cells. In the diagram in Fig. 2.16, the future anterior end of each egg chamber is to the left.

The follicle cells have a key role in patterning the egg's axes. During oogenesis, they become subdivided into functionally different populations at various locations around the oocyte; these subpopulations express different genes and thus have differing effects on the parts of the oocyte adjacent to them. Follicle cells also secrete the materials of the vitelline envelope and egg case that surround the mature egg. Eventually the oocyte fills the whole of the egg chamber, the stalk breaks down, and the mature egg is released from the ovary and laid. *During* most of the stages of development discussed here the oocyte is arrested in the first prophase stage of meiosis (see Fig. 10.8). Meiosis is completed after fertilization.

2.13 The antero-posterior axis of the *Drosophila* egg is specified by signals from the preceding egg chamber and by interactions of the oocyte with follicle cells

The antero-posterior axis is the first axis to be established in the oocyte. The first visible sign of symmetry breaking is the movement of the oocyte from a central position surrounded by nurse cells to the posterior end of the developing egg chamber, where it is in direct contact with follicle cells. This rearrangement occurs while the egg chamber is becoming separated from the germarium, and is caused by preferential adhesion between the future posterior end of the oocyte and the adjacent posterior follicle cells. Oocyte antero-posterior polarity is the result of signaling from the anterior of the older egg chamber to the posterior of the younger chamber (Fig. 2.17).

The older germline cyst first signals to its anterior follicle cells through a widely used signaling pathway in animal development, the Delta–Notch pathway, which will be described in detail later in the book (see Box 5D). The germline cells produce the ligand Delta, which interacts with the receptor Notch on the follicle cells. This signaling event specifies several of the anterior follicle cells to become specialized **polar follicle cells**. The polar cells in turn secrete a signal molecule called Unpaired

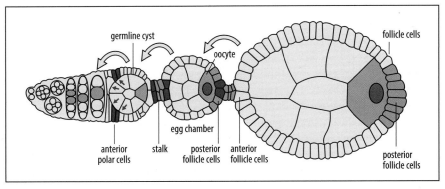

Fig. 2.17 Signals from older to younger egg chambers initially polarize the *Drosophila* oocyte. As a germline cyst buds from the germarium, it signals through the Delta–Notch pathway (small red arrows) to induce the formation of anterior polar cells (red). These in turn signal to the adjacent cells anterior to them and induce them to become stalk (green). Signals from the stalk induce adjacent follicle cells in the younger cyst to become posterior polar cells (red), and induce the younger cyst to round up and the oocyte to become positioned at the posterior of the egg chamber. Signals from anterior and posterior polar cells then specify adjacent follicle cells as posterior and anterior follicle cells, respectively (shown in more detail in Fig. 2.18). The yellow arrows indicate the overall direction of signaling from older to younger egg chambers.

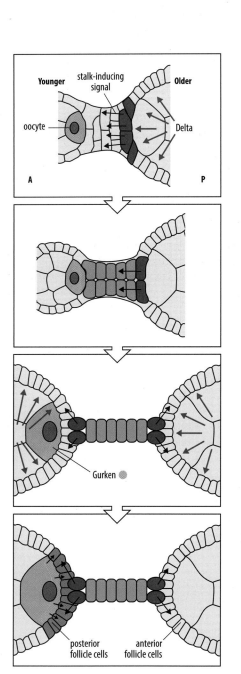

that acts on receptors on adjacent follicle cells. On binding, it stimulates the JAK–STAT signaling pathway (Box 2C), which induces the follicle cells to form a stalk between the two egg chambers. The signals that induce stalk formation and the specification of anterior and posterior follicle cells are shown in more detail in Fig. 2.18. The stalk cells upregulate production of the adhesion molecule cadherin, and adhesive interactions between the oocyte in the younger cyst and the stalk cells immediately adjacent to it position the oocyte firmly at the posterior end of its egg chamber. (The way in which cadherins and other adhesion molecules work is explained in Box 9A). Thus, an antero-posterior polarity is propagated from one egg chamber to the next one to be formed.

The posterior localization of the oocyte is accompanied by a movement of maternal mRNAs from the future anterior end of the oocyte, where they have been deposited by nurse cells, to its future posterior end (see Fig. 2.18). This movement depends on a reorientation of the oocyte microtubule cytoskeleton, on which the maternal mRNAs are transported. At this stage of oocyte development the nucleus is also positioned at the posterior end of the oocyte. Reorientation of the oocyte microtubule cytoskeleton depends on proteins known as PAR proteins, for 'partitioning defective'. Among other

Fig. 2.18 Stalk formation and specification of anterior and posterior follicle cells. First panel: in the older egg chamber, Delta (red arrows) signals from the germline cells to specify anterior polar cells (red). Second panel: these produce the signal molecule Unpaired, which signals adjacent pre-stalk cells to become stalk cells (green). Adhesive interactions anchor the oocyte to the posterior of the younger egg chamber and maternal mRNAs move to the posterior of the oocyte. Third panel: the most posterior follicle cells of the younger cyst are specified as polar cells and secrete Unpaired. mRNA in the oocyte encoding the protein Gurken (see text) is translated and Gurken signals to the adjacent follicle cells. A second round of Delta signaling from the germline initiates differentiation of the follicle cells. Those exposed to both Unpaired and Gurken signals differentiate into posterior follicle cells; those exposed to unpaired in the absence of Gurken differentiate into anterior follicle cells.

*After Figure 4 from Roth, S. and Lynch, J. A.: Symmetry breaking during **Drosophila oogenesis**. Cold Spring Harb. Perspect. Biol. 2009, 1: a001891.*

CELL BIOLOGY BOX 2C The JAK-STAT signaling pathway

This relatively simple signaling pathway was first identified in mammals as an element in the cellular response to cytokines, including the interferons and many other cytokines crucial to the development and function of the immune system. In *Drosophila* the pathway was identified genetically in the context of a variety of functions, including segmentation of the embryo, growth and pattern formation of the imaginal discs, and oogenesis (see Section 2.13) and in antimicrobial responses in adult flies. The central component of the JAK-STAT pathway is a cytoplasmic tyrosine protein kinase of the Janus kinase (JAK) family. This protein family takes its name from the mythical Roman two-headed god Janus, as JAKs have two kinase domains. JAKs are associated with the cytoplasmic tails of the cytokine receptors, and act as a bridge between the receptor and the transcriptional effectors of the pathway, the STAT proteins (signal transducers and activators of transcription) (Figure 1). *Drosophila* has a single receptor for the JAK-STAT pathway called Domeless (Dome), a single JAK called Hopscotch, and a single STAT. The ligands for Dome are the Unpaired proteins. Mammals have multiple receptors, ligands, JAKs and STATs.

Like the mammalian cytokine receptors, Dome consists of two transmembrane protein chains, each associated with a Hopscotch protein. When Unpaired binds to the receptor, the JAKs are activated, and phosphorylate each other and the receptor tails at selected tyrosine residues. This creates binding sites for STAT proteins, which in the absence of ligand binding appear to shuttle between cytosol and nucleus. The phosphorylated STATs dimerize and translocate to the nucleus to effect transcription of

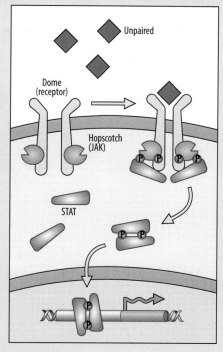

Figure 1

a variety of target genes. In *Drosophila*, the pathway has a negative feedback in the form of the protein SOCS, whose gene is a target of STAT and which interferes with the activation between Hopscotch and STAT.

functions, PAR proteins are involved in determining antero-posterior polarity of cells in many different situations in animal development. They were discovered in *Caenorhabditis elegans*, and are discussed in more detail in Chapter 6 in connection with their role in controlling the first asymmetric division of the fertilized nematode egg.

Once the oocyte has become located posteriorly in its egg chamber, the next steps in its antero-posterior polarization are mediated by a protein called Gurken, which is a member of the transforming growth factor-α (TGF-α) family. Examples of the most commonly used growth factor families in *Drosophila* are given in Fig. 2.19; we shall meet many of these again in vertebrates and other animals. At this stage of oocyte development, maternal *gurken* mRNA is located posteriorly.

Cells at the anterior end of the stalk have now also become specified as polar cells and secrete the ligand Unpaired, which specifies about 200 adjacent follicle cells at the posterior end of the younger egg chamber as 'terminal cells' (see Fig. 2.18). At the same time, *gurken* mRNA is translated at the posterior end of the oocyte close to the nucleus. This produces a local posterior concentration of Gurken protein, which is secreted across the oocyte membrane. Gurken induces the adjacent terminal follicle cells to adopt a posterior fate by locally stimulating the receptor protein Torpedo, which is present on the surface of follicle cells. Torpedo is a receptor tyrosine kinase and is the *Drosophila* equivalent of the epidermal growth factor (EGF) receptor of mammals. This receptor signals by the same type of pathway as the FGF receptor

Common intercellular signals used in *Drosophila*			
Signaling pathway and signal proteins (ligands)	Receptors	Nuclear effector	Examples of roles in *Drosophila* development
Hedgehog (see Box 2F)			
Hedgehog	Patched	Cubitus interruptus (Cu)	Patterning of insect segments (this chapter), positional signaling in imaginal discs (Chapter 11)
Wingless (Wnt) (see Box 4B)			
Wingless and five other Wnt proteins	Frizzled	β-catenin/TCF complex	Patterning of segments (this chapter), patterning of imaginal discs (Chapter 11), development of nervous and muscle systems
Delta/Notch (see Box 5D)			
Delta, Serrate	Notch	Nintra (cleaved intracellular domain of the Notch receptor)/Supressor of hairless (Su(H))	Lateral inhibition in the development of the nervous (Chapter 12) and muscle systems, multiple roles in tissue patterning and cell-type specification
Transforming growth factor (TGF-α)/epidermal growth factor receptor			
Gurken, Spitz, Vein	Torpedo (EGF receptor)	Pointed	Polarization of oocyte (this chapter), eye development, wing vein differentiation
BMP and transforming growth factor-β (TGF-β) (Box 4C)			
Decapentaplegic, screw, glass bottom boat (BMP homologs)	Type I (e.g. Thickveins) and type II (e.g. Punt) subunits form heterodimeric receptor serine-threonine kinases	SMADs (Mad, Medea)	Patterning of the dorso-ventral axis (this chapter), patterning of the antero-posterior axis and growth of imaginal discs (Chapter 11)
Fibroblast growth factor (FGF) (Box 4E)			
Branchless, Pyramus, Thisbe	Breathless, Heartless	Not known if pathway activates transcription in *Drosophila*	Migration of tracheal cells (Chapter 11) and of mesoderm
JAK-STAT (Box 2C)			
Unpaired	Domeless (associates with cytoplasmic serine-threonine kinase, Hopscotch)	STAT	Polarization of oocyte (this chapter), segmentation, patterning of imaginal discs

Fig 2.19 Common intercellular signals in *Drosophila*. These signaling pathways are universally used by all animals and were configured largely through genetic screens in *Drosophila*.

(see Box 4E). In response to Gurken signaling through Torpedo, the posterior follicle cells produce an as-yet-unidentified signal that induces another reorientation of the oocyte's microtubule cytoskeleton, so that about midway through oogenesis, the minus ends of most microtubules are directed towards the anterior, and the plus ends towards the posterior of the oocyte.

2.14 Localization of maternal mRNAs to either end of the egg depends on the reorganization of the oocyte cytoskeleton

Microtubule reorientation is essential for the final localization of maternal mRNAs, such as *bicoid* and *oskar* at either end of the egg. *bicoid* mRNA is originally made by nurse cells located next to the anterior end of the developing oocyte, and is transferred from them to the oocyte. It is subsequently transported to the posterior as a result of the microtubule reorganization that accompanies posterior positioning of the oocyte. After a second microtubule reorganization (described in Section 2.13), *bicoid* mRNA is transported along the reoriented microtubules, probably by the motor protein dynein, to its final location at the anterior-most end of the egg (Fig. 2.20, left panel). *oskar* mRNA is delivered into the oocyte by nurse cells and transported towards the posterior end of the oocyte by another motor protein, kinesin, which transports its cargo along microtubules in the opposite direction to dynein (Fig. 2.20, right panel). Both these localizations require the RNA-binding protein Staufen, which can itself be observed to localize to the posterior and anterior ends of the oocyte (Fig. 2.21).

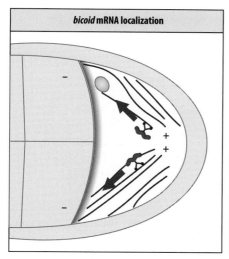

bicoid mRNA localization

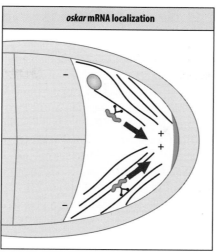

oskar mRNA localization

Fig. 2.20 *bicoid* and *oskar* mRNAs are localized to the anterior and posterior ends of the oocyte respectively. Localization of maternal mRNAs delivered to the oocyte by the nurse cells is by transport along microtubules. The motor protein dynein transports *bicoid* mRNA (red) towards the minus ends of the microtubules. *oskar* mRNA (green) is transported towards the plus ends of the microtubules by the motor protein kinesin.

*Illustration after St Johnston, D.: **Moving messages: the intracellular localization of mRNAs**. Nat. Rev. Mol. Cell Biol. 2005, **6**: 363–375.*

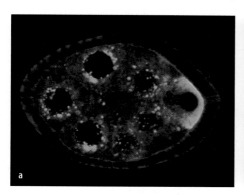

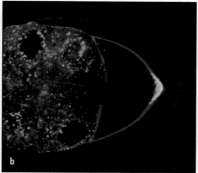

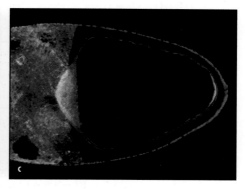

Fig. 2.21 **The RNA-binding protein Staufen localizes to each pole of the *Drosophila* oocyte.** The location of the Staufen protein at different stages of oocyte development is made visible by linking its gene to the coding sequence of green fluorescent protein (GFP) and making flies transgenic for this construct. a, In a stage 6 egg chamber, Staufen (green) accumulates in the oocyte. The fixed egg chambers were counterstained with rhodamine-phalloidin, which labels actin filaments (red). b, In a stage 9 egg chamber, Staufen localizes to the posterior of the oocyte. c, In a stage 10b egg chamber, Staufen is localized to both the anterior and posterior poles of the oocyte.

*From Martin, S.G., et al.: **The identification of novel genes required for** Drosophila **anteroposterior axis formation in a germline clone screen using GFP-Staufen**. Development. 2003, **130**: 4201–4215.*

Observations of *oskar* mRNA particles showed that they moved in all directions, but have a bias towards the posterior end that is sufficient to send the bulk to the posterior end. One role of *oskar* mRNA and protein is to nucleate the assembly of the germplasm at the posterior end of the egg. In the embryo this cytoplasm is incorporated into the pole cells (see Section 2.1) and directs them to form primordial germ cells.

The posterior localization of *oskar* mRNA and the production of Oskar protein there is also required for the subsequent localization of *nanos* mRNA to the posterior end (see Section 2.9), as *nanos* RNA is recruited to wherever Oskar protein is made.

The Torso-like protein, which distinguishes the termini of the *Drosophila* embryo, is synthesized and secreted by follicle cells at both the posterior and anterior poles, but not by the other follicle cells. Torso-like protein is thus deposited exclusively in the vitelline envelope at both ends of the egg during oogenesis. After fertilization it acts together with other proteins to cause the processing of Trunk and production of the ligand for Torso (see Section 2.10).

2.15 The dorso-ventral axis of the egg is specified by movement of the oocyte nucleus followed by signaling between oocyte and follicle cells

The setting-up of the egg's dorso-ventral axis involves a further set of oocyte–follicle cell interactions, which occur after the posterior end of the oocyte has been specified. In response to a signal from the posterior follicle cells, the oocyte nucleus is released from its location at the posterior of the oocyte and moves to a site on the anterior margin (Fig. 2.22). This movement results from a pushing force generated by the growth of astral microtubules—microtubules that anchor the cellular structures called centrosomes to the cortex. Gurken protein is translated at this new site from mRNA that is has been transported on microtubules from its original posterior location.

The locally produced Gurken protein acts as a signal to the adjacent follicle cells, specifying them as dorsal follicle cells; the side away from the nucleus thus becomes the ventral region by default. Figure 2.23 shows the definition of the dorsal side in terms of specific gene expression in the dorsal anterior region only. The ventral follicle cells produce proteins, such as Pipe (see Section 2.11), that are deposited in the ventral vitelline envelope of the oocyte and are instrumental in establishing the ventral side of the axis.

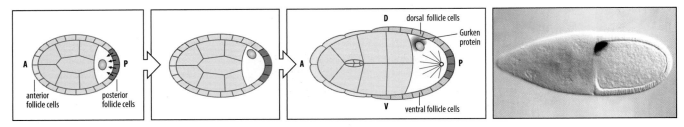

Fig. 2.22 The dorso-ventral axis of the egg chamber and oocyte are specified by further interactions between the oocyte and follicle cells. Signaling from the newly specified posterior follicle cells results in the oocyte nucleus moving to a position on the anterior margin of the oocyte. *gurken* mRNA coming into the oocyte from the nurse cells is now transported on microtubules to the region surrounding the nucleus. *gurken* mRNA is translated and the local release of Gurken protein from the oocyte specifies the adjacent follicle cells as dorsal follicle cells and that side of the oocyte as the future dorsal side. The photograph shows *gurken* mRNA at its new location.

Photograph courtesy of Dr. D. St. Johnston, Wellcome Images.

Illustration after González-Reyes, A., Elliott, H., St Johnston, D.: **Polarization of both major body axes in** Drosophila **by gurken-torpedo signalling.** Nature *1995,* **375**: *654–658.*

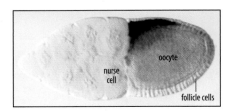

Fig. 2.23 *Drosophila* oocyte development. A developing *Drosophila* oocyte is shown attached to its 15 nurse cells and surrounded by a monolayer of 700 follicle cells. The oocyte and follicle layer are cooperating at this time to define the future dorso-ventral axis of the egg and embryo, as indicated by the expression of a gene only in the follicle cells overlying the dorsal anterior region of the oocyte (blue staining).

Photograph courtesy of A. Spradling.

SUMMARY

Drosophila oocytes develop inside individual egg chambers that are successively produced from a germarium, which contains germline stem cells that give rise to the oocyte and nurse cells, and stem cells that give rise to the somatic follicle cells that surround the oocyte. The nurse cells provide the oocyte with large amounts of mRNAs and proteins, some of which become localized in particular sites. As a result of signals from the adjacent older egg chamber, an oocyte becomes localized posteriorly in its egg chamber as a result of differential cell adhesion to posterior follicle cells. Subsequently the oocyte sends a signal to these follicle cells, which respond with a signal that causes a reorganization of the oocyte cytoskeleton that localizes *bicoid* mRNA to the anterior end and other mRNAs to the posterior end of the oocyte, thus setting up the beginnings of the embryonic antero-posterior axis. The dorso-ventral axis of the oocyte is also initiated by a local signal from the oocyte to follicle cells on the future dorsal side of the egg, thus specifying them as dorsal follicle cells. Thus, directly or indirectly, follicle cells on the opposite side of the oocyte specify the ventral side of the oocyte by deposition of maternal proteins in the ventral vitelline envelope. Follicle cells at either end of the oocyte similarly specify the termini by localized deposition of maternal protein in the vitelline envelope.

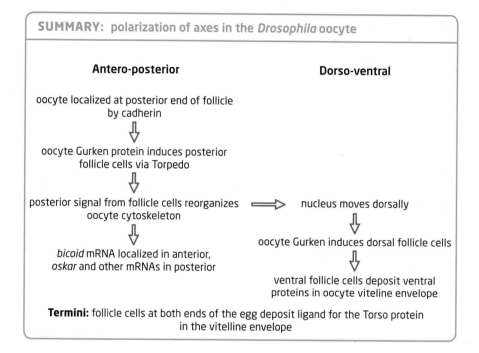

SUMMARY: polarization of axes in the *Drosophila* oocyte

Antero-posterior

oocyte localized at posterior end of follicle by cadherin

⇩

oocyte Gurken protein induces posterior follicle cells via Torpedo

⇩

posterior signal from follicle cells reorganizes oocyte cytoskeleton ⟹

⇩

bicoid mRNA localized in anterior, *oskar* and other mRNAs in posterior

Dorso-ventral

nucleus moves dorsally

⇩

oocyte Gurken induces dorsal follicle cells

⇩

ventral follicle cells deposit ventral proteins in oocyte viteline envelope

Termini: follicle cells at both ends of the egg deposit ligand for the Torso protein in the vitelline envelope

Patterning the early embryo

We have seen how gradients of Bicoid, Hunchback, and Caudal proteins are established along the antero-posterior axis, and how intranuclear Dorsal protein is graded along the ventral-to-dorsal axis. This maternally derived framework of positional information is, in essence, the start of a transcriptional cascade that is interpreted and elaborated on through the regulatory regions of the zygotic genes, many of which do themselves encode transcription factors, to give each region of the embryo an identity. In this part of the chapter we will describe how the embryo is patterned along the antero-posterior and dorso-ventral axes. This patterning occurs simultaneously along both axes but involves different sets of transcription factors and signaling proteins. We will first consider patterning along the dorso-ventral axis, which involves specifying the mesoderm and ectoderm, and the sub-division of the ectoderm to specify the neuroectoderm, which will give rise to the central nervous system.

2.16 The expression of zygotic genes along the dorso-ventral axis is controlled by Dorsal protein

Dorsal protein entering the nuclei of the syncytial blastoderm forms an intranuclear concentration gradient from ventral to dorsal (see Section 2.12). This gradient lasts for about 2 hours, during which the blastoderm becomes cellular. Dorsal is a transcription factor and its gradient divides the dorso-ventral axis into at least four well-defined regions by the expression of key target genes in response to different concentrations of Dorsal (Fig. 2.24). This is also the time at which the germ layers become outlined and

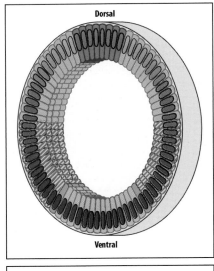

Dorsal

Ventral

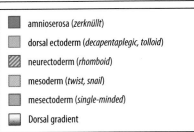

■ amnioserosa (*zerknüllt*)

□ dorsal ectoderm (*decapentaplegic, tolloid*)

▨ neurectoderm (*rhomboid*)

▨ mesoderm (*twist, snail*)

▨ mesectoderm (*single-minded*)

▨ Dorsal gradient

Fig. 2.24 The subdivision of the dorso-ventral axis into different regions by the interactions between Dorsal and its target genes. In the dorsal regions, where nuclear Dorsal protein is absent, the genes *tolloid, zerknullt,* and *decapentaplegic,* for which Dorsal acts as a repressor, can be expressed. In the most ventral region, the high concentration of Dorsal protein activates the genes *twist* and *snail,* which have low-affinity binding sites for Dorsal. The prospective neuroectoderm is characterized by expression of the gene *rhomboid*. The Twist protein helps maintain expression of both its own gene and *snail,* while the Snail protein helps to form the boundary of mesoderm with the neuroectoderm by repressing *rhomboid* in the ventral region. The boundary is characterized by the expression of the gene *single-minded,* whose expression is controlled by a high-affinity site for Dorsal protein in conjunction with sites for other transcription factors.

Dorsal acts to specify the boundaries of these. Going from ventral to dorsal, the main regions are mesoderm, ventral ectoderm, dorsal ectoderm, and prospective amnioserosa. The mesoderm gives rise to internal soft tissues such as muscle and connective tissue; the ventral ectoderm becomes the neuroectoderm, which gives rise to the nervous system as well as ventral epidermis; the dorsal ectoderm gives rise to epidermis. The third germ layer, the endoderm, is located at either end of the embryo and we do not consider it here; it gives rise to the midgut (see Fig. 2.3).

Patterning along the axes poses a problem like that of patterning the French flag (see Section 1.15). It is estimated that 50 genes are direct targets of the Dorsal protein gradient and it is remarkable that the graded expression of Dorsal results in a precise and reproducible pattern of expression of these genes. How this is achieved is now understood, and provides a good example of how a gradient of a transcription factor can directly impart positional information to a field of nuclei that are initially equivalent. Intranuclear Dorsal protein falls off rapidly in the dorsal half of the embryo and as a result, little Dorsal protein is found in nuclei above the equator. Dorsal can act as either an activator or a **repressor** of gene expression and, as a result of its actions, specific genes are expressed in particular regions of the dorso-ventral axis.

The Dorsal protein gradient is one of the best and simplest examples known of how the activity of a single transcription factor can generate different patterns of gene expression, and thus different developmental fates, in the cells in which it acts. Control sites in the target genes for Dorsal along the axis have differences in binding strength, or **affinity**, for the Dorsal protein, and also differ in whether the binding of Dorsal tends to activate or repress the gene, depending on its interactions with adjacent factors. In addition, the control regions of the Dorsal target genes have binding sites for other sets of transcription factors, which act in combination with Dorsal to regulate target-gene expression.

The most ventral region (a strip 12–14 cells wide) is defined by the highest intranuclear concentrations of Dorsal and its interactions with genes that have low-affinity binding sites for Dorsal in their regulatory regions. Such genes will only be expressed where Dorsal is at high concentration. Expression of the genes *snail* and *twist*, both of which encode transcription factors, characterizes this region (see Fig. 2.24). Genes expressed in the ventral region also have binding sites for the Twist protein, a transcriptional regulator that reinforces and maintains their expression.

The neuroectoderm is induced by intermediate levels of Dorsal through the recognition of high-affinity binding sites on target genes in combination with other transcription factors that interact with Dorsal to maintain it on the DNA. A boundary one cell wide between mesoderm and neuroectoderm is characterized by the expression of the gene *single-minded*, whose expression is controlled by a high-affinity site for Dorsal protein in conjunction with sites for other transcription factors. This boundary region is prospective **mesectoderm**, which will give rise to specialized glial cells of the nervous system. The neuroectoderm is defined by expression of the gene *rhomboid*, which encodes a membrane protein involved in regulating cell–cell signaling by the epidermal growth factor (EGFR) pathway. Because *single-minded and rhomboid* are activated by Dorsal, they could, in principle, be expressed in the most ventral domain, but their expression is prevented there by the products of the more ventral genes induced by Dorsal, such as *twist* and *snail*. For example, the regulatory regions of the *rhomboid* gene contain binding sites for both Dorsal and Snail proteins; in this gene, Dorsal activates whereas Snail represses (Fig. 2.25). Thus, *rhomboid* can only be expressed in regions where there is no Snail but where there are sufficient levels of Dorsal, which restricts it to the ventrolateral domain that gives rise to the neuroectoderm.

In the more dorsal regions of the embryo, there is virtually no Dorsal protein in the nuclei. Genes that have control sites at which Dorsal protein acts as a repressor can therefore be expressed in these regions (see Fig. 2.25). Key genes expressed here are

Fig. 2.25 Binding sites in Dorsal target genes. Dorsal can act as an activator or a repressor depending on its concentration and its partners on the DNA. High levels lead to the activation of genes such as *twist* and *snail*, whose products then reinforce their own expression by binding to the same DNA regulatory regions as Dorsal. Twist and Snail also act as repressors of genes such as *rhomboid*, which means that these genes, whose expression also depends on Dorsal, are repressed where Twist and Snail are coexpressed. In some genes, such as *decapentaplegic* (*dpp*) and *zerknüllt* (*zen*), Dorsal acts as a repressor by interacting with co-repressors and recruiting repressors such as Groucho. These genes can only be expressed in the absence of Dorsal, and this is why their expression is restricted to the most dorsal region of the embryo.

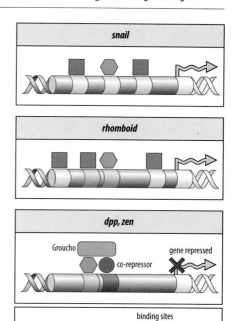

decapentaplegic, tolloid, and *zerknüllt*. *zerknüllt* (*zen*) is expressed most dorsally and specifies the amnioserosa. *decapentaplegic (dpp)* and *tolloid* are expressed throughout a broader dorsal region (see Fig. 2.24). Decapentaplegic protein is a secreted signaling protein with a key role in the specification of pattern in the dorsal part of the *Drosophila* dorso-ventral axis and is considered in detail in Section 2.17.

Mutations in the maternal genes that regulate the formation and readout of the Dorsal gradient can cause dorsalization or ventralization of the embryo (see Section 2.12). In dorsalized embryos, Dorsal protein is excluded uniformly from the nuclei. This has a number of effects, one of which is that the *dpp* gene is no longer repressed and is expressed everywhere. In contrast, *twist* and *snail* are not expressed at all in dorsalized embryos, as they need high intranuclear levels of Dorsal protein to be activated. The opposite result is obtained in mutant embryos in which the Dorsal protein is present at high concentration in all the nuclei, and the embryos are ventralized; *twist* and *snail* are expressed throughout and *dpp* is not expressed at all (Fig. 2.26).

An approximately twofold difference in the level of Dorsal determines whether an unspecified embryonic cell forms mesoderm or neuroectoderm. For simplicity, we have concentrated here on the broad division into mesoderm and neuroectoderm, but within this region, a number of different thresholds in Dorsal concentration pattern

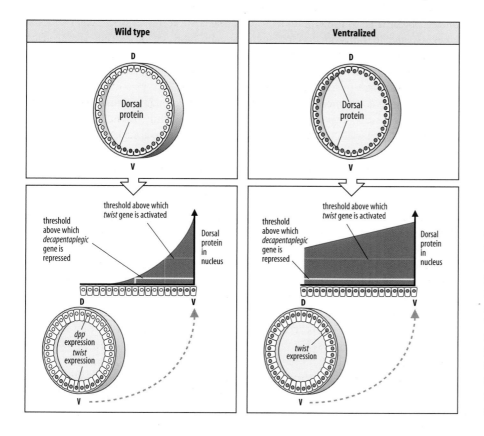

Fig. 2.26 Changes in expression of *twist* and *decapentaplegic* in ventralized embryos. Left panels: in normal embryos, the *twist* gene is activated above a certain threshold concentration (green line) of Dorsal protein, whereas above a lower threshold (yellow line), the *decapentaplegic* (*dpp*) gene is repressed. Right panels: in embryos ventralized as a result of mutations in maternal genes, the Dorsal protein is present in all nuclei; *twist* is now also expressed everywhere, whereas *dpp* is not expressed at all, because Dorsal protein is above the threshold level required to repress it everywhere.

the future ventral midline and the neuroectoderm. The neuroectoderm, for example, is subdivided into three regions along the dorso-ventral axis, which will form three distinct columns of neurons in the future nerve cord (we shall return to this topic in Chapter 12). This subdivision is primarily the result of activation of genes for three different transcription factors at distinct thresholds of the Dorsal gradient.

The gradient of Dorsal protein is therefore effectively acting as a morphogen gradient along the dorso-ventral axis, activating specific genes at different threshold concentrations, and so defining a dorso-ventral pattern. The regulatory sequences in these genes can be thought of as developmental switches, which when thrown by the binding of transcription factors activate new genes and set cells off along new developmental pathways. The Dorsal protein gradient is one solution to the French flag problem. But it is not the whole story—yet another gradient is also involved.

2.17 The Decapentaplegic protein acts as a morphogen to pattern the dorsal region

The gradient of Dorsal protein, with its high point in the ventral-most region, specifies the initial pattern of zygotic gene activity along the dorso-ventral axis, and also further patterns the mesoderm and neuroectoderm by activating other target genes, such as those specifying different parts of the central nervous system, at different thresholds (see Section 2.16). But the further patterning of the dorsal region is not simply specified by a low level of Dorsal, but by a gradient in the activity of another morphogen, the Decapentaplegic protein (Dpp). Thus the combination of low levels of Dorsal and the activity of Dpp specifies the dorsal ectoderm and the most dorsal region, the amnioserosa. As we shall see in Chapter 4, Dpp is a homolog of bone morphogenetic protein-4 (BMP-4), a TGF-β growth factor in vertebrates that is also involved in patterning the dorso-ventral axis (the general signaling pathway for these proteins is shown in Box 4C).

Soon after the gradient of intranuclear Dorsal protein has become established, the embryo becomes cellular, and transcription factors can no longer diffuse between nuclei. Secreted or transmembrane proteins and their corresponding receptors must now be used to transmit developmental signals between cells. Dpp is one such secreted signaling protein (see Fig. 2.19). Dpp is also involved in many other developmental processes throughout *Drosophila* development, including the patterning of the wing imaginal discs (discussed in Chapter 11).

The *dpp* gene is expressed throughout the dorsal region where Dorsal protein is not present in the nuclei. Dpp protein then diffuses ventrally away from this source, forming a gradient of Dpp activity with its high point in the dorsal region. The cells have receptors that allow them to accurately assess how much Dpp is present and to respond by activating transcription of the appropriate genes, thus dividing up the dorsal region into different sub-regions characterized by different patterns of gene expression.

Evidence that a gradient in Dpp specifies dorsal pattern comes from experiments in which *dpp* mRNA is introduced into an early wild-type embryo. As more mRNA is introduced and the concentration of Dpp protein increases above the normal level, the cells along the dorso-ventral axis adopt a more dorsal fate than they would normally. Ventral ectoderm becomes dorsal ectoderm and, at very high concentrations of *dpp* mRNA, all the ectoderm develops as amnioserosa. A most important target of Dpp is the gene *zen*, which is repressed by Dorsal (see Fig. 2.25). Initially, *dpp* and *zen* are expressed in similar domains, but as the gradient of Dpp protein activity emerges, *zen* gene expression becomes restricted to the region with higher levels of Dpp signaling and defines the amnioserosa.

Dpp protein is initially produced uniformly throughout the dorsal region just as cellularization is beginning, but within less than an hour its activity becomes restricted to a dorsal strip of some five to seven cells and is much lower in the adjacent prospective dorsal ectoderm (Fig. 2.27). This sharp peak of Dpp concentration is not a matter of simple diffusion, however; instead, it illustrates how a gradient in activity can be derived from a fairly uniform initial distribution of the protein by interactions with other proteins. In addition, Dpp activity is also thought to be modified by interactions with different forms of its receptors, further refining the activity gradient. It also interacts with collagen in the extracellular matrix, which restricts its movement and thus restricts its signaling range. Dpp is a good example of how the formation of a signal gradient is usually more complicated than simple diffusion.

The sharp transition between the most dorsal ectoderm cells and those just lateral to it involves the proteins Short gastrulation (Sog), Twisted gastrulation (Tsg), and Tolloid. Sog and Tsg are BMP-related proteins that can bind Dpp and prevent it binding to its receptors, and thus inhibit its action. Tolloid is a metalloproteinase that degrades Sog when it is bound to Dpp, thus releasing Dpp.

The gene *sog* is expressed throughout the prospective neuroectoderm. Binding of Dpp by the Sog protein prevents Dpp activity from spreading into this region. Sog protein diffusing into the dorsal region is, in turn, degraded by Tolloid, which is expressed throughout the dorsal region; this sets up a gradient in Sog with a high point in the neuroectoderm and a low point at the dorsal midline and ensures that Sog, along with its bound Dpp, will be shuttled towards the dorsal region. Degradation of Sog or Tsg by Tolloid releases free Dpp, which helps to generate a gradient of Dpp activity with its high point in the dorsal-most region (see Fig. 2.27).

Another factor proposed to sharpen the gradient is the rapid internalization and degradation of Dpp when it binds to its receptors. Dpp activity is also affected by

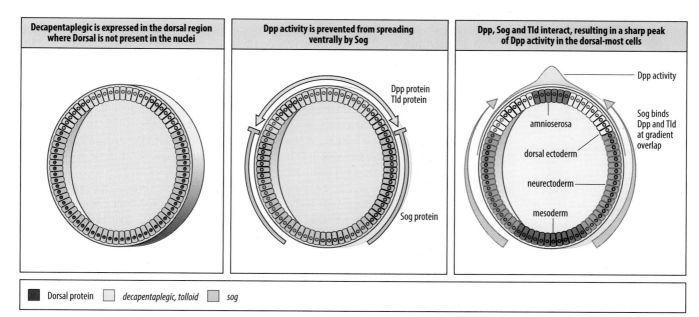

| Decapentaplegic is expressed in the dorsal region where Dorsal is not present in the nuclei | Dpp activity is prevented from spreading ventrally by Sog | Dpp, Sog and Tld interact, resulting in a sharp peak of Dpp activity in the dorsal-most cells |

■ Dorsal protein ☐ *decapentaplegic, tolloid* ☐ *sog*

Fig. 2.27 Decapentaplegic protein (Dpp) activity is restricted to the dorsal-most region of the embryo by the antagonistic activity of the Short gastrulation protein (Sog). Left panel: the gradient of intranuclear Dorsal protein leads to the expression of the gene *short gastrulation* (*sog*) throughout the prospective neuroectoderm. *decapentaplegic* (*dpp*) and *tolloid* are expressed throughout the dorsal ectoderm. Middle panel: *sog* encodes a secreted protein, Sog, that forms a ventral to dorsal gradient, while the secreted Dpp protein is initially present throughout the dorsal region. The protease Tolloid (Tld) is also expressed in the same region as Dpp. Where Sog meets Dpp, it binds it and thus prevents Dpp from interacting with its receptors and Dpp signaling from spreading ventrally into the neuroectoderm. Right panel: Dpp bound to Sog is also carried towards the dorsal region by the developing Sog gradient, thus helping to concentrate its activity in a sharp peak in the dorsal-most region. Tld is involved in creating the final sharp dorsal peak of Dpp activity. It binds to the Sog-Dpp complex and cleaves Sog. This releases Dpp, which can then bind to its receptors. The dorsal region therefore becomes subdivided by Sog activity into a region of high Dpp signaling, which will become the amnioserosa, and a zone of lower Dpp signaling, which is the dorsal ectoderm.

*After Ashe, H.L., Levine, M.: **Local inhibition and long-range enhancement of Dpp signal transduction by Sog**. Nature 1999, **398**: 427-431.*

its synergistic action with another member of the TGF-β family, Screw, which forms heterodimers with Dpp that signal more strongly than homodimers of Dpp or Screw alone; these are likely to be responsible for most of the Dpp activity. Experimental observations and mathematical modeling suggest that heterodimers are formed preferentially in the dorsal-most region, which helps to explain the high Dpp activity here. Homodimers of Dpp and of Screw are responsible for lower levels of signaling elsewhere in the prospective dorsal ectoderm.

The Dpp/Sog pairing has a counterpart in vertebrates, the BMP-4/Chordin partnership (Chordin is the vertebrate homolog of Sog), which is also involved in patterning the dorso-ventral axis. The nerve cord runs dorsally in vertebrates, however, not ventrally as in insects, and thus the dorso-ventral axis is reversed in vertebrates compared with insects; we shall see in Chapter 4 how this reversal in anatomical position is reflected in a reversal of the pattern of activity of BMP-4 and Chordin during early development.

2.18 The antero-posterior axis is divided up into broad regions by gap-gene expression

Patterning along the antero-posterior axis starts, like the patterning of the dorso-ventral axis, when the *Drosophila* embryo is still acellular. The **gap genes** are the first zygotic genes to be expressed along the antero-posterior axis (see Fig. 2.9), and all code for transcription factors. Gap genes were initially recognized by their mutant phenotypes, in which quite large continuous sections of the body pattern along the antero-posterior axis are missing (see figure in Box 2A). Although the mutant phenotype of a gap gene usually shows a gap in the antero-posterior pattern in more-or-less the region in which the gene is normally expressed, there are also more wide-ranging effects. This is because gap-gene expression is also essential for later development along the axis.

Gap-gene expression is initiated by the antero-posterior gradient of Bicoid protein while the embryo is still essentially a single multinucleate cell. Bicoid primarily activates anterior expression of the gap gene *hunchback*, which in turn is instrumental in switching on the expression of the other gap genes, among which are *giant*, *Krüppel*, and *knirps*, which are expressed in this sequence along the antero-posterior axis (Fig. 2.28) (*giant* is in fact expressed in two bands, one anterior and one posterior, but its posterior expression does not concern us here).

The overall mechanism of patterning the antero-posterior axis through transcriptional regulation by *bicoid* and the various gap genes is very similar to that we have just seen for Dorsal: transient gradients of transcription factors are formed; these transcription factors have different affinities for different target genes; and gap-gene proteins act in a combinatorial manner with other transcription factors. As the blastoderm is still acellular at this stage, the gap-gene proteins can diffuse away from their site of synthesis. They are short-lived proteins with half-lives of minutes. Their distribution, therefore, extends only slightly beyond the region in which the gene is expressed, and this typically gives a bell-shaped protein concentration profile. The overlaps at these boundaries are used to sharpen the domains of the different regions of the embryo through cross-regulatory interactions. Zygotic hunchback protein is an exception, as it is expressed fairly uniformly over a broad anterior region with a steep drop in concentration at the posterior boundary of its expression. The control of zygotic *hunchback* expression by Bicoid protein is best understood and will be considered first.

2.19 Bicoid protein provides a positional signal for the anterior expression of zygotic *hunchback*

The Bicoid protein induces expression of the zygotic *hunchback* genes over most of the anterior half of the embryo. This zygotic expression is superimposed on low

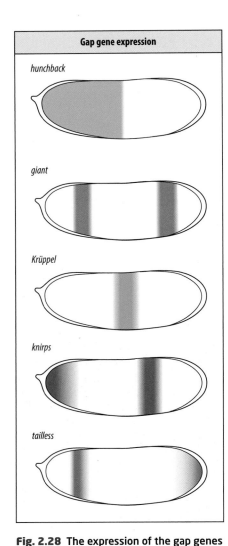

Gap gene expression

hunchback

giant

Krüppel

knirps

tailless

Fig. 2.28 The expression of the gap genes *hunchback, Krüppel, giant, knirps,* **and** *tailless* **in the early** *Drosophila* **embryo.** Gap-gene expression at different points along the antero-posterior axis is controlled by the concentration of Bicoid and Hunchback proteins, together with interactions between the gap genes themselves. The expression pattern of the gap genes provides an aperiodic pattern of transcription factors along the antero-posterior axis, which delimits broad body regions.

Fig. 2.29 Maternal Bicoid protein controls zygotic *hunchback* expression. If the dose of maternal *bicoid* is increased twofold, the extent of the Bicoid gradient also increases. The activity of the *hunchback* gene is determined by the threshold concentration of Bicoid, so at the higher dose its region of expression is extended toward the posterior end of the embryo because the region in which Bicoid concentration exceeds the threshold level also extends more posteriorly (see graph, bottom panel). The position of the head furrow, shown, is also shifted posteriorly at increasing doses of *bicoid*, showing the disruption to the body plan.

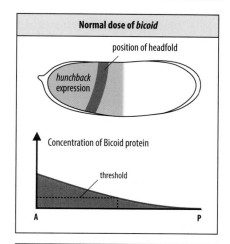

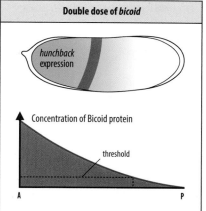

levels of *hunchback* mRNA, whose translation is suppressed posteriorly by Nanos (see Section 2.9).

The localized anterior expression of Hunchback is an interpretation of the positional information provided by the Bicoid protein gradient. The *hunchback* gene is switched on only when Bicoid, a transcription factor, is present above a certain threshold concentration. This level is attained only in the anterior half of the embryo, close to the site of Bicoid synthesis, which restricts *hunchback* expression to this region.

The relationship between Bicoid concentration and *hunchback* gene expression can be illustrated by looking at how *hunchback* expression changes when the Bicoid concentration gradient is changed by increasing the maternal dosage of the *bicoid* gene (Fig. 2.29). Expression of *hunchback* then extends more posteriorly because the region in which the concentration of Bicoid is above the threshold for *hunchback* activation is also extended in this direction.

A smooth gradient in Bicoid protein is translated into a sharp boundary of *hunchback* expression about halfway along the embryo. Neighboring nuclei along the antero-posterior axis experience Bicoid concentrations that are very similar, differing by only about 10% from one cell to the next. This can be seen by measuring the gradient of Bicoid protein and the pattern of *hunchback* expression in many different embryos. It is then possible to see the very tight correlation between a value of Bicoid and the onset of *hunchback* expression (Fig. 2.30). How do nuclei distinguish such small differences and form a sharp boundary of gene expression against a background of biological 'noise' (that is, fluctuations in the concentration of transcription factors at the region of the threshold) such as that resulting from small random variations in the dynamics of Bicoid behavior. This question has been addressed by using expression of a Bicoid–GFP construct (see Section 2.8) to directly observe and measure the distribution of Bicoid protein and its concentrations in nuclei at different points along the antero-posterior axis. Bicoid distribution, intranuclear concentrations at different positions, and the sharp boundary of *hunchback* expression all turn out to be highly reproducible over a large number of embryos, which suggests that background noise is in fact kept low by precise controls in the embryo. These might, for example, include communication between nuclei.

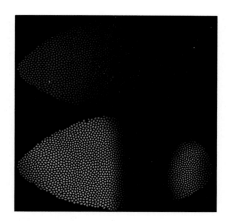

Fig. 2.30 Quantitative relation of Bicoid and Hunchback (Hb) expression. The photographs show the gradient of maternal Bicoid protein (top) in nuclei and the pattern of zygotic Hb protein (bottom) in nuclei in a late syncytial blastoderm embryo. To distinguish the overlapping patterns of expression, the embryo is stained for both proteins using specific antibodies that are visualized by secondary antibodies carrying different fluorescent tags. The fluorescence intensity is then recorded separately for each protein. These images have been false-colored. The graph plots the concentration of Bicoid against that of Hb along the length of the embryo. From the anterior end up to about 45% of embryo length, Bicoid concentration is at or above the threshold level required to activate the *hb* gene. Note the sharp cut-off of Hb expression as the Bicoid concentration in nuclei reduces to below the threshold level. (The *hb* gene has a second domain of expression at the most posterior region of the embryo that is independent of Bicoid.)

Original scans courtesy of J. Reinitz, M. Samsonova, and A. Pisarev. Reproduced with permission from Reinitz, J. **A ten per cent solution**. *Nature. 2007,* ***448****: 420–421.*

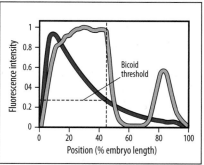

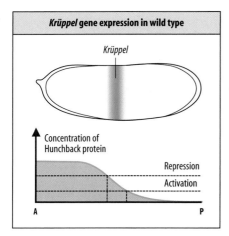

Krüppel gene expression in wild type

Krüppel

Concentration of
Hunchback protein

Repression

Activation

A   P

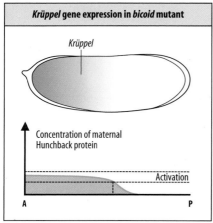

Krüppel gene expression in bicoid mutant

Krüppel

Concentration of maternal
Hunchback protein

Activation

A   P

Fig. 2.31 *Krüppel* gene activity is specified by Hunchback protein. Top panel: above a threshold concentration of Hunchback protein, the *Krüppel* gene is repressed; at a lower concentration, above another threshold value, it is activated. Bottom panel: in mutants lacking the *bicoid* gene, and thus also lacking zygotic *hunchback* gene expression, only maternal Hunchback protein is present, which is located at the anterior end of the embryo at a relatively low level. In these mutants, *Krüppel* is activated at the anterior end of the embryo, giving an abnormal pattern.

Bicoid activates the *hunchback* gene by binding to regulatory sites within the promoter region. Direct evidence of *hunchback* activation by Bicoid was originally obtained by experiments in which a fusion gene constructed from the *hunchback* promoter regions and a bacterial reporter gene, *lacZ*, was introduced into the fly genome using transgenic techniques such as P-element-mediated transformation (Box 2D; some other related techniques for visualizing gene expression and forcing new patterns of gene expression in *Drosophila* are shown in Box 2E). The large promoter region required for completely normal gene expression can be whittled down to an essential sequence of 263 base pairs that will still give almost normal activation of *hunchback*. This sequence has several sites at which Bicoid can bind, and it seems likely that the threshold response involves **cooperativity** between the different binding sites: that is, binding of Bicoid at one site makes binding of Bicoid at a nearby site easier and so facilitates further binding.

The regulatory region of a gene such as *hunchback* is yet another example of a developmental switch that directs nuclei along a new developmental pathway. Transcriptional switches such as this are a universal feature of embryonic development.

2.20 The gradient in Hunchback protein activates and represses other gap genes

The Hunchback protein is a transcription factor and acts as a morphogen to which the other gap genes respond. The other gap genes are expressed in transverse stripes across the antero-posterior axis (see Fig. 2.28). The stripes are delimited by mechanisms that depend on the control regions of these genes being sensitive to different concentrations of Hunchback protein, and also to other proteins, including Bicoid, which now contributes to the sharpening of the pattern. Expression of the *Krüppel* gene, for example, is activated by low levels of Hunchback, but is repressed at high concentrations. Within this concentration window *Krüppel* remains activated (Fig. 2.31, top panel). But below a lower threshold concentration of Hunchback, *Krüppel* is not activated. In this way, the gradient in Hunchback protein locates a band of *Krüppel* gene activity near the center of the embryo (Fig. 2.32). Refinement of this spatial localization is brought about by repression of *Krüppel* by other gap-gene proteins.

Such relationships were worked out by altering the concentration profile of Hunchback protein systematically, while all other known influences were eliminated or held constant. Increasing the dose of Hunchback protein, for example, results in a posterior shift in its concentration profile, and this results in a posterior shift in the posterior boundary of *Krüppel* expression. In another set of experiments on embryos lacking Bicoid protein (so that only the maternal Hunchback protein gradient is present), the level of Hunchback is so low that the *Krüppel* gene is even activated at the anterior end of the embryo (Fig. 2.31, bottom panel).

Hunchback protein is also involved in specifying the anterior borders of the bands of expression of the gap genes *knirps* and *giant*, again by a mechanism involving thresholds for repression and activation of these genes. At high concentrations of Hunchback, *knirps* is repressed, and this specifies its anterior margin of expression. The posterior margin of the *knirps* band is specified by a similar type of interaction with the product of another gap gene, *tailless*. Where the regions of expression of the gap genes overlap,

Fig. 2.32 Localization of *Krüppel* expression by the Hunchback protein gradient. An embryo stained with fluorescent antibodies to simultaneously detect the Hb protein gradient (red) and the expression of *Krüppel* mRNA (green) early in nuclear cycle 14.

*From Yu, D., and Small, S.: **Precise registration of gene expression boundaries by a repressive morphogen in Drosophila.*** Curr. Biol, *2008, **18**: 868–876.*

EXPERIMENTAL BOX 2D P-element-mediated transformation

Transgenic fruit flies have contributed greatly to *Drosophila* developmental genetics. They are made by inserting a known sequence of DNA into the *Drosophila* chromosomal DNA, using as a carrier a **transposon** that occurs naturally in some strains of *Drosophila*. This transposon is known as a **P element**, and the technique as P-element-mediated transformation (Figure 1).

P elements can insert at almost any site on a chromosome, and can also hop from one site to another within the germ cells, an action that requires an enzyme called a transposase. As hopping can cause genomic instability, carrier P elements have had their own transposase gene removed. The transposase required to insert the P element initially is instead provided by a helper P element, which cannot itself insert into the host chromosomes and is thus quickly lost from cells. The carrier and helper elements are injected together into the posterior end of the egg where the germ cells are made.

As well as the gene to be inserted, an additional marker gene, such as the wild-type *white⁺* gene, is added to the P element. When *white⁺* is the marker, the P element is inserted into flies homozygous for the mutant *white⁻* gene (which have white eyes rather than the red eyes of the wild-type *Drosophila*). Red eyes are dominant over white, and so flies in which the P element has become integrated into the chromosome, and is being expressed, can be detected by their red eyes.

In the first generation, all flies have white eyes, as any P element that has integrated is still restricted to the germ cells. But in the second generation, a few flies will have wild-type red eyes, showing that they carry the inserted P element in their somatic cells.

This technique has been used to increase the number of copies of a particular gene, or to introduce a mutated gene that has its control or **coding regions** altered in a known way, or to introduce new genes. It is also possible to introduce genes that carry a marker coding sequence such as *lacZ* (encoding the bacterial enzyme β-galactosidase), whose expression is detectable by histochemical staining (see Box 1D). The P element itself has also been used as a mutagen, as its insertion into a gene usually destroys that gene's function.

This approach has been adapted to large-scale screens that look systematically for genes whose overexpression or misexpression in a particular tissue causes a mutant phenotype (see Box 2E). This is known as misexpression screening. In this case, flies expressing Gal4 in the tissue of interest are crossed with large numbers of different lines of target flies carrying random insertions of the Gal4-binding site, and the progeny screened for a mutant phenotype. This approach is a useful complement to the more conventional genetic screens described in Box 2A, which generally detect loss-of-function mutations. If the target flies also carry a mutation in a known gene, misexpression screening can be used to identify genes whose overexpression enhances or suppresses the mutation. This approach can identify genes whose products interact directly or are part of the same pathway.

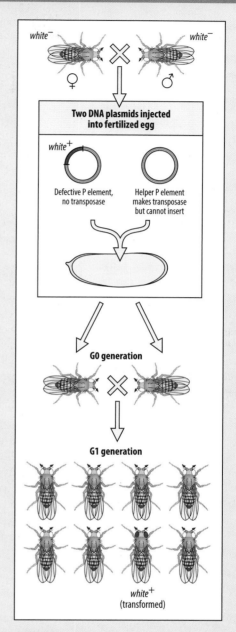

Figure 1

EXPERIMENTAL BOX 2E Targeted gene expression and misexpression screening

The ability to turn on the expression of a gene in a particular place and time during development is very useful for analyzing its role in development. This is called targeted gene expression and can be achieved in several ways. One approach is to give the selected gene a heat-shock promoter using standard genetic techniques such as P-element-mediated transformation (see Box 2D). This enables the gene to be switched on by a sudden rise in the temperature at which the embryos are being kept. By adjusting the temperature, the timing of expression of genes attached to this promoter can be controlled; the effects of expressing a gene at different stages of development can be studied in this way.

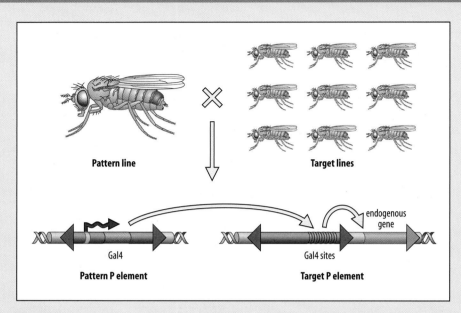

Figure 1

Another approach to targeted gene expression uses the transcription factor Gal4 from yeast. This protein can activate transcription of any gene whose promoter has a Gal4-binding site. In *Drosophila*, genes with Gal4-responsive promoters can be created by inserting the Gal4-binding site. To turn on the target gene, Gal4 itself has to be produced in the embryo. In one approach, Gal4 can be produced in a designated region or at a particular time in development by introducing a P-element transgene in which the yeast Gal4-coding region is attached to a *Drosophila* regulatory region known to be activated in that situation (Figure 1).

A second, more versatile, approach is based on the so-called **enhancer-trap** technique. The Gal4-coding sequence is attached to a vector that integrates randomly into the *Drosophila* genome.

The Gal4 gene will come under the control of the promoter and enhancer region adjacent to its site of integration, and so Gal4 protein will be produced where or when that gene is normally expressed. A large number of *Drosophila* lines with differing patterns of Gal4 expression for different purposes have been produced by this technique. The selected target gene will be silent in the absence of Gal4. To activate it in a particular tissue, for example, flies that express Gal4 in that tissue are crossed with flies in which the target gene has Gal4-binding sites in its regulatory region.

A novel pattern of gene expression can also be obtained using the Gal4 system. Using the Gal4 system, for example, the pair-rule gene *even-skipped* was expressed in even-numbered rather than odd-numbered parasegments, and this led to changes in the pattern of denticles on the cuticle.

there is extensive cross-inhibition between them, their proteins all being transcription factors. These interactions are essential to sharpen and stabilize the pattern of gap gene expression. For example, the anterior border of *Krüppel* expression lies four to five nuclei posterior to nuclei that express *giant*, and is set by low levels of Giant protein.

The antero-posterior axis thus becomes divided into a number of unique regions on the basis of the overlapping and graded distributions of different transcription factors. This particular method of delimiting regions can only work in an embryo such as the acellular syncytial blastoderm of *Drosophila*, where the transcription factors are able to diffuse freely throughout the embryo. But, as we shall see in other organisms, the general principle of first dividing the early embryo into broad regions, which are then further subdivided and patterned, is universal, even though the mechanisms for doing it differ.

In *Drosophila*, the regional distribution of the gap-gene products provides the starting point for the next stage in development along the antero-posterior axis—the activation of the pair-rule genes, cellularization, and the beginning of segmentation.

SUMMARY

Gradients of maternally derived transcription factors along the dorso-ventral and antero-posterior axes provide positional information that activates zygotic genes at specific locations along these axes. Along the antero-posterior axis the maternal gradient in Bicoid protein initiates the activation of zygotic gap genes to specify general body regions. Interactions between the gap genes and their products, all of which code for transcription factors, help to define their borders of expression. The dorso-ventral axis becomes divided into four regions: ventral mesoderm, ventral ectoderm (neuroectoderm), dorsal ectoderm (dorsal epidermis), and amnioserosa. A ventral to dorsal gradient of maternal Dorsal protein both specifies the ventral mesoderm and defines the dorsal region; a second gradient, of the Decapentaplegic protein, patterns the dorsal ectoderm. Patterning along the dorso-ventral and antero-posterior axes divides the embryo into a number of discrete regions, each characterized by a unique pattern of zygotic gene activity.

SUMMARY: early expression of zygotic genes

Antero-posterior	Dorso-ventral
Bicoid protein gradient switches *hunchback* on at high concentration	ventro-dorsal gradient of intranuclear Dorsal protein forms
Hunchback activates and represses gap genes such as *Krüppel, knirps, giant*	ventral activation of *twist, snail,* and repression of *decapentaplegic*
gap-gene products and gap genes interact to sharpen expression boundaries	Decapentaplegic expressed dorsally
axis is divided into unique domains containing different combinations of transcription factors	gradient of Decapentaplegic activity patterns dorsal region
	dorso-ventral axis divided into prospective mesoderm, neuroectoderm, epidermis, and amnioserosa

Activation of the pair-rule genes and the establishment of parasegments

The most obvious feature of a *Drosophila* larva is the regular segmentation of the larval cuticle along the antero-posterior axis, each segment carrying cuticular structures that define it as, for example, thorax or abdomen. This segmental cuticle pattern reflects the patterning of the underlying epidermis, which produces the cuticle and its structures. More generally, it reflects the segmentally repeating organization of the larval body, in which elements such as the tracheal breathing system (discussed in Chapter 11), the larval nervous system, and the muscles of the body wall are organized in a modular fashion.

Each segment first acquires a unique identity in the embryo, and the pattern and identity of segments is carried over into the adult at metamorphosis. Adult appendages such as wings, halteres, and legs are attached to particular segments, but the discernible segments of the newly hatched larva are not in fact the first units of segmentation along this axis. The basic developmental modules, whose definition we will follow in some detail, are the parasegments, which are specified in the embryo and from which the segments derive.

2.21 Parasegments are delimited by expression of pair-rule genes in a periodic pattern

The first visible signs of segmentation are transient grooves that appear on the surface of the embryo after gastrulation. These grooves define the 14 parasegments. Once each parasegment is delimited, it behaves as an independent developmental unit, and in this sense at least the embryo can be thought of as being modular in construction. The parasegments are initially similar to each other in their developmental potential, but each will soon acquire a unique identity. The parasegments are out of register with the definitive segments of the late embryo and hatched larva by about half a segment; this means that each segment is made up of the posterior region of one parasegment and the anterior region of the next. The relation between parasegments, larval segments, and segment identity in the adult fly is illustrated in Fig. 2.33. In the anterior head region the segmental arrangement is lost when some of the anterior parasegments fuse.

The parasegments in the thorax and abdomen are delimited by the action of the **pair-rule genes**, each of which is expressed in a series of seven transverse stripes along the embryo, each stripe corresponding to every second parasegment. When pair-rule gene expression is visualized by staining for the pair-rule proteins, a striking zebra-striped embryo is revealed (Fig. 2.34).

The positions of the stripes of pair-rule gene expression are determined by the pattern of gap-gene expression—a non-repeating pattern of gap-gene activity is converted into repeating stripes of pair-rule gene expression. We now consider how this is achieved.

2.22 Gap-gene activity positions stripes of pair-rule gene expression

Pair-rule genes are expressed in stripes, with a periodicity corresponding to alternate parasegments. Mutations in these genes thus affect alternate segments, and the pair-rule genes were identified originally through mutations that resulted in a discontinuous and periodic loss of tissue in the larva (see Box 2A) Some pair-rule genes (for example, *even-skipped* (*eve*)), define odd-numbered parasegments, whereas others (for example, *fushi tarazu*) define even-numbered parasegments (see Fig. 2.34). The striped pattern of expression of pair-rule genes is present even before cells are formed, while the embryo is still a syncytium, although cellularization occurs soon after expression begins. Each pair-rule gene is expressed in seven stripes, each of which is only a few cells wide. For some pair-rule genes, such as *eve*, the anterior margin of the stripe corresponds to the anterior boundary of a parasegment; the domains of expression of other pair-rule genes, however, cross parasegment boundaries.

The striped expression pattern appears gradually; the stripes of the *eve* gene are initially fuzzy, but eventually acquire a sharp anterior margin. At first sight this type of patterning would seem to require some underlying periodic process, such as the setting up of a wave-like concentration of a morphogen, with each stripe forming at the crest of a wave. It was surprising, therefore, to discover that each stripe is specified independently.

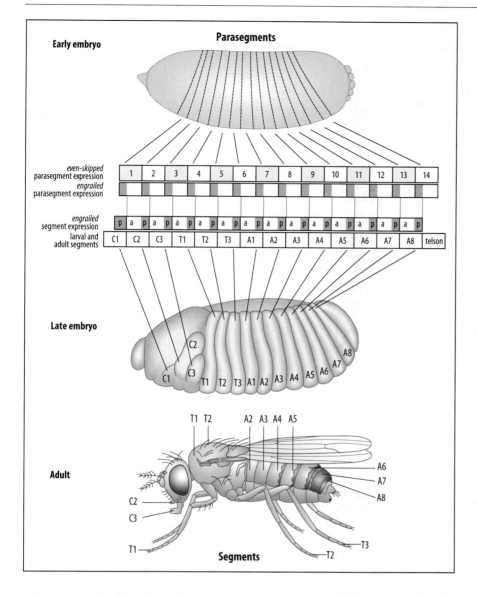

Fig. 2.33 The relationship between parasegments and segments in the early embryo, late embryo, and adult fly. Pair-rule gene expression specifies the parasegments in the future thorax and abdomen, and some head parasegments in the embryo. *even-skipped* (yellow), for example, specifies the odd-numbered parasegments. The segmentation selector gene *engrailed* (blue) is expressed in the anterior region of every parasegment, and delimits the anterior margin of each parasegment. When segmentation occurs, the anterior region of a parasegment becomes the posterior portion of a segment. Each larval segment is therefore composed of the posterior region of one parasegment and the anterior region of the next, and segments are thus offset from the original parasegments by about half a segment. *engrailed* continues to be expressed in the posterior region of each segment throughout all larval stages and in the adult. In this figure, a and p refer to the anterior and posterior regions of the final segments. The segment specification is carried over into the adult and results in particular appendages, such as legs, wings and mouthparts, developing on specific segments only. C1, mandibular segment, C2, maxillar segment, and C3, labial segment. The adult mandibular segment lacks an appendage. T, thoracic segments; A, abdominal segments.

Illustration after Lawrence, P.: The Making of a Fly. Oxford: Blackwell Scientific Publications, 1992.

As an example of how the pair-rule stripes are generated we will look in detail at the expression of the second *eve* stripe (Fig. 2.35). The appearance of this stripe depends on the expression of Bicoid protein and of the three gap genes *hunchback*, *Krüppel*, and *giant* (only the anterior band of *giant* expression is involved in specifying

Fig. 2.34 The striped patterns of activity of pair-rule genes in the *Drosophila* embryo just before cellularization. Parasegments are delimited by pair-rule gene expression, each pair-rule gene being expressed in alternate parasegments. Expression of the pair-rule genes *even-skipped* (blue) and *fushi tarazu* (brown) is visualized by staining with antibody for their protein products. *even-skipped* is expressed in odd-numbered parasegments, *fushi tarazu* in even-numbered parasegments. Scale bar = 0.1 mm.

From Lawrence, P.: The Making of a Fly. Oxford: Blackwell Scientific Publications, 1992.

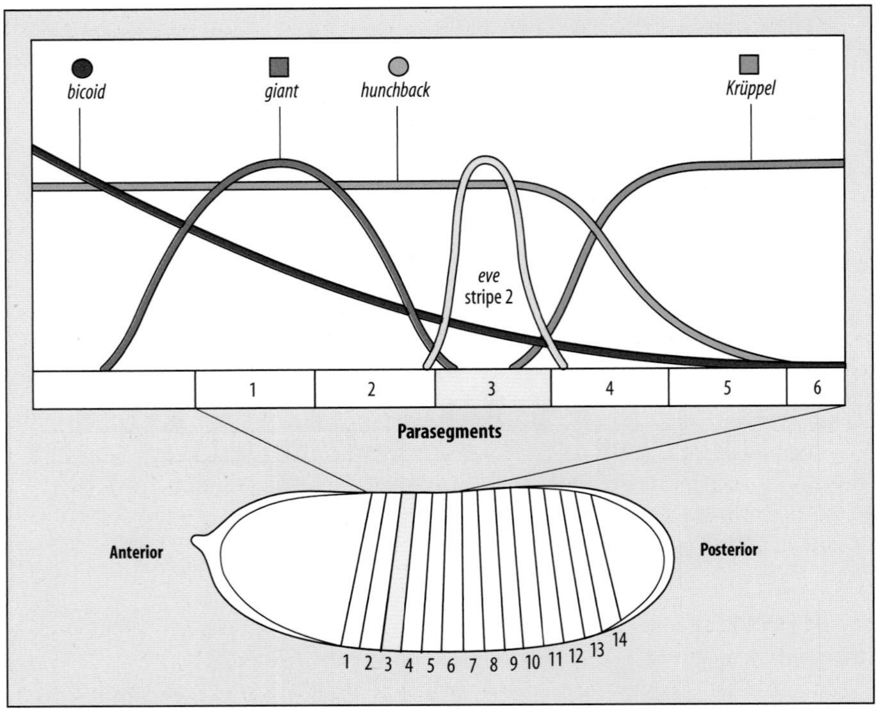

Fig. 2.35 The specification of the second *even-skipped* (*eve*) stripe by gap gene proteins. The different concentrations of transcription factors encoded by the gap genes *hunchback, giant,* and *Krüppel* localize *even-skipped*—expressed in a narrow stripe at a particular point along their gradients—in parasegment 3. Bicoid and Hunchback proteins activate *eve* in a broad domain, and the anterior and posterior borders are formed through repression of the gene by Giant and Krüppel proteins, respectively.

the second *eve* stripe; see Fig. 2.28). Bicoid and Hunchback proteins are required to activate the *eve* gene, but they do not define the boundaries of the stripe. These are defined by Krüppel and Giant proteins, by a mechanism based on repression of *eve*. When concentrations of Krüppel and Giant are above certain threshold levels, *eve* is repressed, even if Bicoid and Hunchback are present. The anterior edge of the stripe is localized at the point of the threshold concentration of Giant protein, whereas the posterior border is similarly specified by the Krüppel protein.

In contrast, the positioning of the third and fourth *eve* stripes depends on regulatory regions that are repressed by the high concentration of Hunchback in the anterior region. The third stripe is expressed about halfway along the embryo, where Hunchback concentration starts to drop sharply, while the fourth *eve* stripe is expressed more posteriorly, where Hunchback is at an even lower level (Fig. 2.36). The posterior boundary of the third *eve* stripe is delimited by repression by the gap protein Knirps.

The independent localization of each of the stripes by the gap-gene transcription factors requires that, in each stripe, the pair-rule gene responds to different concentrations and combinations of the gap-gene transcription factors. The pair-rule genes thus require complex *cis*-regulatory control regions with multiple binding sites for each of the different factors. We have already seen how this can lead to the patterning of distinct regions in the case of the Dorsal protein and the dorso-ventral axis. Examination of the regulatory regions of the *eve* gene reveals five separate modules, each of which controls the localization of one or two stripes. Regulatory regions of around 500 base pairs have been isolated that each determines the expression of a single stripe (Fig. 2.37).

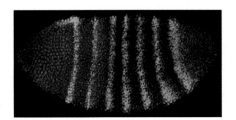

Fig. 2.36 Positioning of the third and fourth *eve* stripes by the Hunchback gradient. Embryo stained with fluorescent antibodies to simultaneously detect the Hb protein (red) and the mRNA of its target gene *eve* (green). The position of the third and fourth *eve* stripes is due directly to lack of repression of *eve* by Hunchback as its gradient drops off sharply.
*From Yu, D., and Small, S.: **Precise registration of gene expression boundaries by a repressive morphogen in Drosophila.*** Curr. Biol, *2008,* **18**: *868-876.*

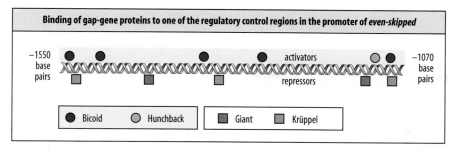

Fig. 2.37 Sites of action of activating and repressing transcription factors in the *eve* control region involved in formation of the second *eve* stripe. A control region of around 500 base pairs, located between 1070 and 1550 base pairs upstream of the transcription start site in the *eve* gene, directs formation of the second *eve* stripe. Gene expression can occur when the Bicoid and Hunchback transcription factors, acting as activators, are present above a given threshold concentration. Even in the presence of Bicoid and Hunchback, however, the gene is repressed when the Giant and Krüppel proteins, acting as repressors, are above threshold levels. They therefore set the boundaries to the stripe (see Fig. 2.35). The repressors may act by preventing binding of activators.

The *eve* gene is an excellent example of modular *cis*-regulatory regions controlling the expression of a gene at different locations (see Sections 1.10 and 1.11).

The presence of control regions that, when activated, lead to gene expression at a specific position in the embryo is a fundamental feature of developmental genes. Other examples of this type of control that we have seen in *Drosophila* are the localized expression of the gap genes, and the expression of genes in distinct regions along the dorso-ventral axis. Each regulatory region of such a gene contains binding sites for different transcription factors, some of which activate the gene, whereas others repress it. In this way, the combinatorial activity of the gap-gene proteins regulates pair-rule gene expression in each parasegment. Some pair-rule genes, such as *fushi tarazu*, may not be regulated by the gap genes directly, but may depend on the prior expression of so-called 'primary' pair-rule genes, such as *eve* and *hairy*. With the initiation of pair-rule gene expression, the future segmentation of the embryo has been initiated. When the patterning along the dorso-ventral axis is also taken into account, the embryo is now divided into a number of unique regions, which each have a different developmental fate. These regions are characterized mainly by the combinations of transcription factors being expressed in each. These include proteins encoded by the gap genes, the pair-rule genes, and the genes expressed along the dorso-ventral axis.

The pair-rule transcription factors set up the spatial framework for the next round of patterning. Their combined activities will result in the expression of segmentation genes encoding transcription factors and signaling molecules on either side of the boundary of each parasegment. The products of these genes will act to delineate and pattern the parasegments, and thus the future segments. Also, together with the gap-gene proteins, the pair rule genes provide regional identities that will create functional specializations. After looking briefly at some different ways of specifying segmentation in insects, we shall consider these regulatory interactions and their consequences in *Drosophila*.

2.23 Some insects use different mechanisms for patterning the body plan

Drosophila belongs to an evolutionarily advanced group of insects, in which all the segments are specified more or less at the same time in development, as shown both by the striped patterns of pair-rule gene expression in the syncytial blastoderm and the appearance of all the segments shortly after gastrulation. This type of development is known as **long-germ development**, as the blastoderm corresponds to the

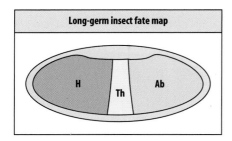

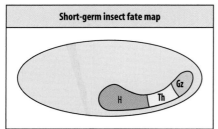

Fig. 2.38 Differences in the development of long-germ and short-germ insects. Top panel: the general fate map of long-germ insects such as *Drosophila* shows that the whole of the body plan—head (H), thorax (Th), and abdomen (Ab)—is present at the time of initial germ-band formation. Bottom panel: in short-germ insects, only the anterior regions of the body plan are present at this embryonic stage. Most of the abdominal segments develop later, after gastrulation, from a posterior growth zone (Gz).

whole of the future embryo. Many other insects, such as the flour beetle *Tribolium*, have a **short-germ development**. In short-germ development the blastoderm is short and forms only the anterior segments. The posterior segments are added by growth after completion of the blastoderm stage and gastrulation (Fig. 2.38). In spite of early differences between them, the mature germ band stages of both long- and short-germ insect embryos look similar. This, therefore, is a common stage through which all insect embryos develop. As noted earlier, *bicoid* is a quite recently evolved gene, found in a restricted group of insects. In the long-germ embryos of the wasp *Nasonia*, for example, gradients in the much more ancient protein Orthodenticle pattern the anterior and posterior regions, in a manner similar to the action of Bicoid in *Drosophila*.

The question clearly arises as to what processes are shared and which are different in the specification of the body plan in long- and short-germ insects. One clear difference is that the patterning of the body plan in the long germ band of *Drosophila* takes place before cell boundaries form, whereas much of the body plan in short-germ insects is laid down at a later stage, when the posterior segments are generated during growth. At that stage the embryo is cellular rather than syncytial. So are the same genes involved?

There is very good evidence that many of the same genes and developmental processes are involved in the patterning of *Tribolium* and *Drosophila*. For example, the gap gene *Krüppel* is expressed at the posterior end of the *Tribolium* embryo at the blastoderm stage and not in the middle region, as in *Drosophila* (Fig. 2.39). It therefore seems to be specifying the same part of the body in the two insects. Similarly, only two repeats of the pair-rule stripes are present at the blastoderm stage in *Tribolium*, together with a posterior cap of pair-rule gene expression, in contrast to the seven repeats in *Drosophila*. The segmentation genes *wingless* and *engrailed*, which we shall meet a little later, are also expressed in a similar relation to that in *Drosophila*.

There is a wide range of short-germ insects with different lengths of the antero-posterior axis specified at blastoderm. *Drosophila* is at the extreme in which all of the segments are specified at once, and this highlights its special status and, perhaps, explains the presence of Bicoid, absent from other insects, which represents a mechanism for the measurement of the length of the organism.

Although not many of the genes involved in segmentation and patterning have been studied in detail in insects other than *Drosophila* and *Tribolium*, at least one, the gene *engrailed*, which is expressed in the posterior region of each segment in *Drosophila*, is also expressed in the posterior region of segments in a variety of insects. The pair-rule gene *eve*, however, (see Section 2.22), although present in the grasshopper (a short-germ insect), may not have a similar role in segmentation. It is, however, involved later in the development of the nervous system in the grasshopper, and is also expressed at the posterior end of the growing germ band.

Fig. 2.39 Gap and pair-rule gene expression in long-germ and short-germ insects at the time of germ-band formation. *Krüppel* (red) is a gap gene and *hairy* (green) is a pair-rule gene. The position of the *Krüppel* stripe in the short-germ embryo of *Tribolium* indicates that posterior regions of the body are not yet present at this stage. Similarly, only three *hairy* stripes, corresponding to the first three *hairy* stripes in a long-germ embryo, are present in *Tribolium*.

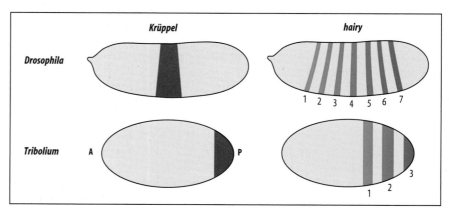

Differences in early development are much more dramatic in some other insects. In certain parasitic wasps, the egg is small and undergoes cleavage to form a ball of cells, which then falls apart. Each of the resulting small clusters of cells—there can be as many as 400—can develop into a separate embryo. This wasp's development apparently does not depend on maternal information to specify the body axes, and in this respect resembles the development of the early mammalian embryo.

SUMMARY

The activation of the pair-rule genes by the gap genes results in the transformation of the embryonic pattern along the antero-posterior axis from a variable regionalization to a periodic one. The pair-rule genes define 14 parasegments. Each parasegment is defined by narrow stripes of pair-rule gene activity. These stripes are uniquely defined by the local concentration of gap-gene transcription factors acting on the regulatory regions of the pair-rule genes. Each pair-rule gene is expressed in alternate parasegments—some in odd-numbered, others in even-numbered. Most pair-rule genes code for transcription factors. In contrast to *Drosophila*, some insects have a short-germ type of development in which posterior segments are added by growth after the cellular blastoderm stage.

SUMMARY: pair-rule genes and parasegments in *Drosophila*

production of local combinations of gap-gene transcription factors

⇩

activation of each pair-rule gene in seven transverse stripes along the antero-posterior axis

⇩

pair-rule gene expression defines 14 parasegments, each pair-rule gene being expressed in alternate parasegments

Segmentation genes and segment patterning

The expression of the pair-rule genes defines the anterior and posterior boundaries of each parasegment but, like the gap genes, their activity is only temporary. How, therefore, are the positions of parasegment boundaries fixed, and how do the boundaries of the segments that are visible in the larval epidermis become established? This is the role of the **segmentation genes**. Segmentation genes are activated in response to pair-rule gene expression. During pair-rule gene expression the blastoderm becomes cellularized, so the segmentation genes are acting in a cellular rather than a syncytial environment. Unlike the gap genes and pair-rule genes, which act in the syncytial environment and all encode transcription factors, the segmentation genes are a diverse group of genes that encode signaling proteins, their receptors, and signaling pathway components, as well as transcription factors, and thus enable intercellular communication.

2.24 Expression of the *engrailed* gene defines the boundary of a parasegment which is also a boundary of cell-lineage restriction

A key gene in establishing parasegment boundaries is the *engrailed* gene, which encodes a transcription factor. Unlike the gap genes and pair-rule genes, whose activity

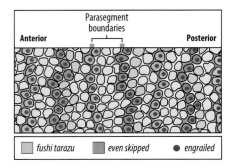

Fig. 2.40 The expression of the pair-rule genes *fushi tarazu* (blue), *even-skipped* (pink), and *engrailed* (purple dots) in parasegments. *engrailed* is expressed at the anterior margin of each stripe and delimits the anterior border of each parasegment. The boundaries of the parasegments become sharper and straighter later on.

After Lawrence, P.: The Making of a Fly. Oxford: Blackwell Scientific Publications, 1992.

is transitory, *engrailed* is expressed throughout the life of the fly. *engrailed* activity first appears at the time of blastoderm cellularization as a series of 14 transverse stripes. It is initially expressed in lines of cells coincident with high levels of expression of the pair-rule genes *fushi tarazu* and *eve*, and delineates the anterior margin of each parasegment, which is only about four cells wide at this time (Fig. 2.40). Figure 2.41 shows *engrailed* expression in relation to the transient parasegment grooves at the later extended germ-band stage, when part of the ventral blastoderm (the germ band) has extended over the dorsal side of the embryo.

Evidence that the pair-rule genes control *engrailed* expression is provided, for example, by embryos carrying mutations in *fushi tarazu*, in which *engrailed* expression is absent only in even-numbered parasegments (in which *fushi tarazu* is normally expressed).

The anterior boundaries of the parasegments delimited by *engrailed* are also boundaries of **cell-lineage restriction**. In other words, cells and their descendants in one parasegment never move into adjacent ones. Such domains of lineage restriction are known as **compartments**. A compartment can be defined as a distinct region that contains all the descendants of the cells present when the compartment is set up, and no others. It therefore has the potential to operate as a cellular unit, in which the cells express some key gene or genes that help to give the compartment an identity and distinguish it from other compartments. Compartments were originally defined in *Drosophila* epidermal structures, but they have since been discovered in the developing limbs and the brain in vertebrates (discussed in Chapters 11 and 12, respectively).

The existence of compartments in *Drosophila* can be detected by cell-lineage studies in which a single cell is marked in such a way that the mark is passed to all its descendants (which form a **clone**) that can be followed into specific structures through many generations. One technique is to inject the egg with a harmless fluorescent compound that is incorporated into all the cells of the embryo. In the early embryo, a fine beam of ultraviolet light directed onto a single cell activates the fluorescent compound. All the descendants of that cell will be fluorescent and can therefore be identified. Examination of these clones, together with an analysis of *engrailed* expression, shows that cells at the anterior margin of the parasegment have no descendants that lie on the other side of that margin. The anterior margin is therefore a boundary of lineage restriction; that is, cells and their descendants that are on one or other side of the boundary when it is formed cannot cross it during subsequent development (Fig. 2.42).

engrailed expression marks the anterior boundary of the compartment. Unlike the pair-rule and gap genes, whose activity is transitory, *engrailed* needs to be expressed continuously throughout the life of the fly to maintain the character of the posterior compartment of the segment. Furthermore, unlike other segmentation genes, it is expressed in the same pattern throughout development and always in association with *hedgehog*. *engrailed* is thus an example of a **selector gene**—a gene whose

Fig. 2.41 The expression of the *engrailed* gene in a late (stage 11) *Drosophila* embryo. The gene is expressed in the anterior region of each parasegment and the transient grooves between parasegments can be seen. At this stage of development, the germ band has temporarily extended and is curved over the back of the embryo. Scale bar = 0.1 mm.

From Lawrence, P.: The Making of a Fly. Oxford: Blackwell Scientific Publications, 1992.

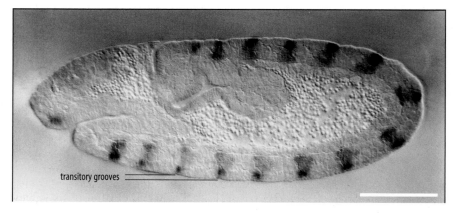

activity is sufficient to cause cells to adopt a particular fate. Selector genes can control the development of a region such as a compartment and, by controlling the activity of other genes, give the region a particular identity.

The compartment boundaries, as well as the expression of *engrailed* and *hedgehog*, are therefore carried over into the imaginal discs that derive from specific embryonic segments (see Fig. 2.6), and into the adult structures that derive from the discs, such as wings and legs (discussed in Chapter 11). As discussed later in this chapter, compartments can be more clearly visualized in these structures, where each compartment is composed of thousands of cells.

The definitive segments of the late embryo and larva are offset from the parasegments, with the anterior boundary of each segment developing immediately posterior to the *engrailed*-expressing cells. This means that each segment is divided into anterior and posterior compartments, with *engrailed* expression defining the posterior compartment (see Fig. 2.33).

2.25 Segmentation genes stabilize parasegment boundaries

Establishment of the parasegment boundary depends on an intercellular signaling circuit being set up between adjacent cells, which delimits the boundary between them, and which is discussed in detail in Section 2.26. As well as *engrailed*, this circuit primarily involves the segmentation genes *wingless* and *hedgehog*, which are also expressed in restricted domains within the parasegment in response to the pair-rule proteins (Fig. 2.43). In contrast to *engrailed*, which encodes a transcription factor, *wingless* and *hedgehog* encode secreted signal proteins, Wingless and Hedgehog, respectively. These act via receptor proteins on cell surfaces to activate intracellular signaling pathways that alter gene expression. *wingless* is named after the effect of its loss-of-function mutation on the adult fly (discussed in Chapter 11); *hedgehog* was so named because the embryo is much shorter than normal and the ventral surface is completely covered in denticles, so that the embryo somewhat resembles a hedgehog. **Wingless** is one of the prototype members of the so-called **Wnt family** of signal proteins (see Fig. 2.19), which are key players in development in vertebrates as well as invertebrates, and are involved in determining cell fate and cell differentiation in many different aspects of development. **Hedgehog** similarly has homologs in other animals, such as Sonic hedgehog in vertebrates, and is involved in similar processes.

The result of the initial activity of the pair-rule genes is the establishment, in the blastoderm, of one-cell-wide stripes of *engrailed* and *hedgehog* expression that will delimit the anterior boundary of each parasegment, and similar stripes of *wingless* expression posterior to them (see Fig. 2.43). Thus, cells expressing *wingless* abut cells expressing *engrailed* and *hedgehog* at the parasegment boundary.

2.26 Signals generated at the parasegment boundary delimit and pattern the future segments

By the late embryo stage, the continuous epithelium of the thorax and abdomen has become divided up into repeating units—the future segments—each composed of anterior and posterior compartments delimited by the original parasegment boundaries (see Fig. 2.43). We shall now discuss how the signals generated on either side of the parasegment boundary delimit and pattern the parasegments, and thus the future segments.

The *engrailed* gene is expressed in cells along the anterior margin of the parasegment. These cells also express the segmentation gene *hedgehog* and secrete the Hedgehog signaling protein. Adjacent cells on either side express Patched, the receptor for Hedgehog, and the *patched* gene is itself a target of Hedgehog signaling (see Fig. 2.43). Hedgehog protein acting on these cells initiates an intracellular signaling

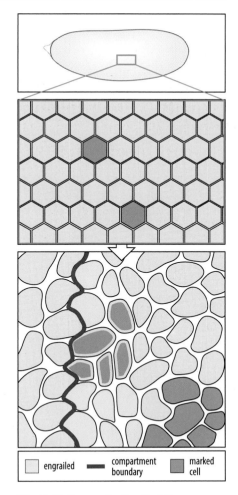

Fig. 2.42 Demonstration that the parasegment boundary is a boundary of lineage restriction. Individual cells are marked (green) in the blastoderm, when all cells are the same and the embryo is unpatterned. The cells divide as the embryo grows and give rise to small clones. When the clones are mapped with regard to the expression of engrailed (yellow), it is possible to visualize the position of the clones with respect to the anterior boundary of engrailed *expression*, corresponding to the parasegment boundary. It was found that no clones crossed this boundary, indicating that the boundary is a boundary of lineage restriction and delimits a compartment.

After Vincent, J.-P., O'Farrell, P.H.: **The state of** engrailed **expression is not clonally transmitted during early** Drosophila **development**. Cell *1992,* **68***: 923-931.*

In the figure legend:
- engrailed
- compartment boundary
- marked cell

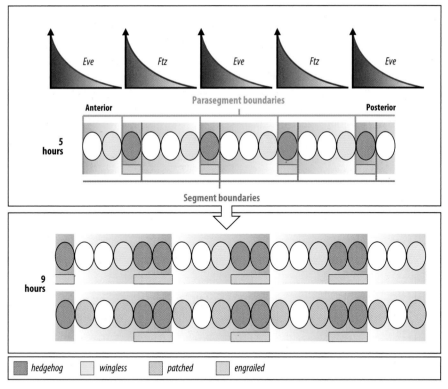

Fig. 2.43 The domains of expression of some key segmentation genes. Upper panel: expression of the pair-rule proteins Ftz and Eve in narrow stripes along the embryo delimit the parasegments. They activate expression of the *engrailed* and *hedgehog* genes in a strip one cell wide at the anterior margin of each parasegment, while the gene *wingless*, which is repressed by Ftz and Eve, is expressed in a strip one cell wide at the posterior of the parasegment as pair-rule gene expression fades away. Lower panel: as the embryo begins to grow, *hedgehog* is being expressed in two cells at the anterior margin of the parasegment, while *wingless* is expressed in an adjacent cell on the other side of the boundary (top row). At this time *patched* (which encodes the receptor for the Hedgehog protein) is induced by the secreted Hedgehog protein and is therefore expressed on either side of the domain of *engrailed*/*hedgehog* expression (bottom row). All the cells express Frizzled, the receptor for Wingless (not shown). The definitive segments of the late embryo and larva (shaded gray in bottom two rows) are offset from the parasegments such that the original parasegment boundary now delimits the posterior compartment of the segment. This posterior compartment continues to be marked by engrailed expression throughout the fly's life.

pathway that results in the activation and maintenance of expression of *wingless* in the cells. The Wingless protein is secreted and feeds back on the *engrailed*-expressing cells to initiate a signaling event that maintains expression of the *engrailed* and *hedgehog* genes. All the cells expressed Frizzled, the receptor for Wingless, but only those close to the source of Wingless signal respond to it. These interactions stabilize and maintain the parasegment boundary (Fig. 2.44), and establish it as a **signaling center** that produces signals (Hedgehog and Wingless) that pattern the segment. As noted earlier, the parasegment boundary is the anterior boundary of the posterior compartment in the future segment. This means that in the segment itself, *engrailed* and *hedgehog* are expressed in the posterior compartment, and *wingless* in cells immediately anterior to this boundary (see Fig. 2.44).

The signaling pathway by which *Drosophila* Hedgehog acts is shown in Box 2F. At parasegment boundaries, and in many of their other developmental functions, Wingless and the vertebrate Wnt proteins act through a conserved signaling pathway illustrated in Box 4B, which is often called the **canonical Wnt/β-catenin pathway**.

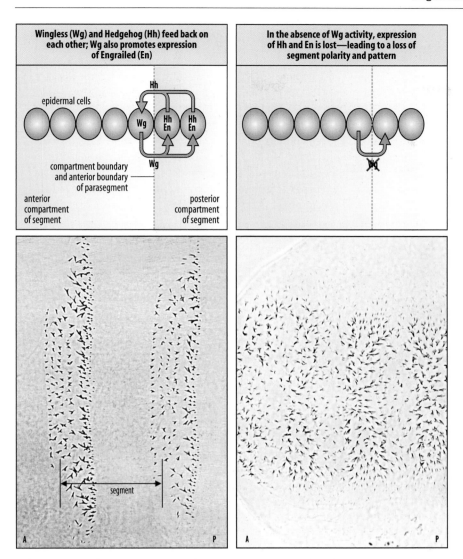

| Wingless (Wg) and Hedgehog (Hh) feed back on each other; Wg also promotes expression of Engrailed (En) | In the absence of Wg activity, expression of Hh and En is lost—leading to a loss of segment polarity and pattern |

Fig. 2.44 Interactions between *hedgehog*, *wingless*, and *engrailed* genes and proteins establish the parasegment boundaries and control denticle pattern. Top left panel: the segmentation gene *hedgehog* is expressed in cells along the anterior margin of the parasegment, which is also delineated by expression of the gene *engrailed*. Hedgehog protein is secreted and its action on adjacent cells results in activation and maintenance of *wingless* gene expression. Secreted Wingless protein signals back to the cells expressing *engrailed* and *hedgehog* and maintains the expression of these genes. These interactions stabilize and maintain the boundary. Top right panel: in mutants where the *wingless* gene is inactivated and the Wingless protein is absent, neither *hedgehog* nor *engrailed* genes are expressed. This leads to loss of the parasegment boundary and conversion of the whole segment into 'anterior' type. Bottom left panel: the denticle bands on the ventral cuticle of the abdominal segments are useful markers for segment pattern. In the wild-type larva, the denticle bands are confined to the anterior part of the segment, while the posterior cuticle is naked, and this pattern is dependent on the activity of the *hedgehog* and *wingless* genes. Bottom right panel: in the *wingless* mutant, the normally well-defined pattern of denticles within each abdominal segment is lost. Denticles are present across the whole ventral surface of the segment in what looks like a mirror-image repeat of the anterior segment pattern.

Illustration after Lawrence, P.: The Making of a Fly. *Oxford: Blackwell Scientific Publications, 1992.*

β-**catenin** is a potential transcriptional co-activator that in the absence of Wingless signaling is degraded in the cytoplasm by a complex of proteins. When Wingless binds to its receptor, Frizzled, this protein complex is inhibited, and β-catenin can then enter the nucleus and switch on the Wingless target genes. Computer modeling of the Wingless–Hedgehog–Engrailed circuitry shows that it is remarkably robust and resistant to variation, such as variations in the levels of expression of the genes governing the circuit's behavior.

About three hours after the parasegment boundary has been established, a deep groove develops at the posterior edge of each *engrailed* stripe and this marks the anterior boundary of each segment. A segmented late embryo is shown in Fig. 2.4 (third panel) and Fig. 2.33.

Once the parasegment boundary is consolidated, the signals localized on either side set up the pattern of each future segment, leading eventually to the differentiation of the epidermal cells to produce the cuticle pattern evident on the late embryo and larva. Each segment in the newly hatched larva has a well-defined antero-posterior pattern, which is most easily seen on the ventral epidermis of the abdomen: the anterior region of each segment bears denticles, whereas in most segments the posterior region is naked. The denticles appear in the late stages of embryonic development (Fig. 2.45), and reflect the prior patterning of the embryonic epidermis by the actions

CELL BIOLOGY BOX 2F The Hedgehog signaling pathway

In the absence of Hedgehog (Figure 1, left panel), the membrane protein Patched, which is the receptor for Hedgehog, inhibits the membrane protein Smoothened. In the absence of Smoothened activity, the transcription factor Cubitus interruptus (Ci) is held in the cytoplasm in two protein complexes—one associated with Smoothened and another with the protein Suppressor of fused (Su(fu)). In the absence of Hedgehog, Ci in the Smoothened complex is also phosphorylated by several protein kinases (glycogen synthase kinase (GSK-3), protein kinase A (PKA) and casein kinase 1 (CK1)), which results in the proteolytic cleavage of Ci and formation of the truncated protein CiRep. This enters the nucleus and acts as a repressor of Hedgehog target genes. When Hedgehog is present, it binds to Patched (Figure 1, right panel) and this lifts the inhibition on Smoothened and blocks production of CiRep. Ci is released from both its complexes in the cytoplasm, enters the cell nucleus, and acts as a gene activator. Genes activated in response to Hedgehog signaling include *wingless* (*wg*), *decapentaplegic* (*dpp*), and *engrailed* (*en*).

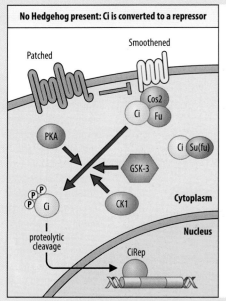
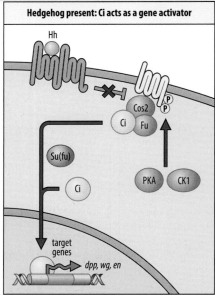

Figure 1

Scan here

Scan this QR code image with your mobile device to see an online animation of the Hedgehog signaling pathway or log on to **http://global.oup.com/uk/orc/biosciences/ devbiol/wolpert5e/qr/qr2b/**

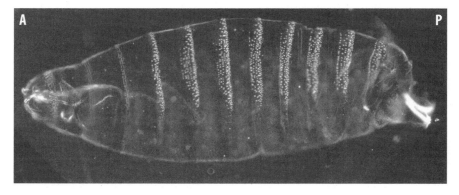

Fig. 2.45 Denticle pattern of a *Drosophila* late embryo just before hatching. Each segment has a characteristic pattern of denticles on its ventral surface. Denticles are confined to the anterior of each segment. This view shows the ventral surface of the embryo.

*From Goodman, R.M., et al.: Sprinter: **a novel transmembrane protein required for Wg secretion and signalling**. Development. 2006, **133**: 4901-4911.*

of the segmentation genes. The rows of denticles make a distinct pattern on each segment and mutations in segmentation genes often alter the pattern in a systematic way—this is how these genes were first discovered (Box 2G).

The pattern of denticle belts is specified in response to the signals provided by Wingless and Hedgehog. At this later stage of embryonic development, Hedgehog and Wingless expression are no longer dependent on each other and can be analyzed separately. Wingless protein moves posteriorly over the parasegment boundary and also moves anteriorly to pattern that part of the segment immediately anterior to the cells in which it is expressed. Wingless moves a shorter distance posteriorly than anteriorly, as it is more rapidly degraded in the posterior part of the segment, and so its effects on patterning extend over a longer range in the anterior region (Fig. 2.46).

Wingless represses genes required for denticle formation, and so the cells that lie immediately anterior to the Wingless-expressing line of cells are specified as epidermal cells that will produce smooth cuticle, while the anterior extent of Wingless signaling delimits the posterior edge of the denticle belts in that segment. In mutants that lack *wingless* function, denticles are produced in the normally smooth region (see Fig. 2.44).

The Hedgehog signals move anteriorly to maintain Wingless function and also more posteriorly to help pattern the anterior region of the next segment. The signals provided by Hedgehog and Wingless lead, by a complex set of interactions, to the expression of a number of genes in non-overlapping narrow stripes in the late embryo that specify the larval segmental pattern of smooth cuticle or specific rows of denticles (see Fig. 2.46).

2.27 Compartment boundaries persist into the adult fly

The segment compartments are carried over into the imaginal discs that derive from specific segments (see Fig. 2.6), and into the adult structures, such as wings and legs, that derive from the imaginal discs (discussed in Chapter 11).

Cell-lineage restriction within compartments is most clearly illustrated by the behavior of cells in the wing of the adult fly. It is easier to distinguish lineage restriction in adult structures than in embryonic and early larval structures, because in adults there has been considerable cell division since the initial event that determined the compartment in the embryo. Techniques such as those described in Box 2H showed that in a normal wing the compartment boundary is remarkably sharp and straight and does not correspond to any structural features in the wing (Fig. 2.47).

Compartments were first defined by making genetically mosaic flies composed of two distinguishable kinds of cells (see Box 2H). Single cells in the embryonic blastoderm or in the larval epidermis are given a distinctive phenotype by X-ray-or laser-induced mitotic recombination. The fates of all the descendants of this marked cell are then followed. Their behavior depends on the stage of development at which the founder nucleus was marked. Descendants of nuclei marked at early stages during cleavage become part of many tissues and organs, but descendants of nuclei marked at the blastoderm stage or later have a more restricted fate. They are found only in the anterior or the posterior part of each segment (or of an appendage such as a wing), never throughout the whole segment.

Cells of imaginal discs divide only about 10 times after the cellular blastoderm state; thus clones of experimentally marked cells are small, even in the adult, and so it is not easy to detect a boundary of lineage restriction (see Fig. 2.47, top panel). Clone size can be increased by the *Minute* technique, which results in the marked cell dividing many more times than the other cells (see Box 2H). A single clone of such cells can almost fill either the anterior or posterior part of the wing, and this makes the boundary that the cells never cross more evident: this boundary separates the anterior and posterior compartments (see Fig. 2.47, middle panel). These experiments also show that the

EXPERIMENTAL BOX 2G Mutants in denticle pattern provided clues to the logic of segment patterning

Mutations that disrupted denticle pattern were discovered long before the signaling pathways underlying segment patterning were worked out. Although in theory there are many ways in which denticle pattern might be disrupted, mutations in denticle pattern all belonged to one of three classes of pattern defects (Figure 1). The existence of these classes meant that there must be a logic to the construction of the epidermal patterning, and that the products of genes with a similar mutant phenotype must be functionally related. This was indeed the case.

One class is exemplified by the loss-of-function mutation in *wingless*, shown in Fig. 2.44. In these mutants, the anterior region of each segment has been duplicated in mirror-image polarity, and the patterning in the posterior region is lost. There are many mutants with this phenotype and all of them encode elements of the Wnt signaling pathway. For example, the *hedgehog* mutation is also a member of this class, which as we know now, reflects the fact that in the absence of Hedgehog signaling, *wingless* expression is lost.

A second class is exemplified by a mutation in a gene called *axin*, which encodes a component of the Wingless intracellular

signaling pathway. This results in the opposite transformation to the *wingless* mutation—the anterior region of every segment is transformed into posterior, and the denticles are substituted by smooth cuticle. The Axin protein is a negative regulator of the intracellular signaling pathway that keeps β-catenin inactive in the absence of Wingless signaling (see Box 4B), and so its loss results in β-catenin being active in cells in which it is normally inactive. Mutations in other regulators of the pathway have the same phenotype, which is also produced when *wingless* is expressed throughout the whole of the embryo.

The third class is exemplified by mutations in *patched*, which encodes a receptor for Hedgehog. Loss-of-function mutations in *patched* therefore have a negative effect on Hedgehog signaling. Mutations in other elements of the Hedgehog signaling pathway have a similar phenotype. These three mutant classes identified Wingless and Hedgehog as the main regulators of the pathway, and were instrumental in identifying the elements of these pathways and the order in which they function.

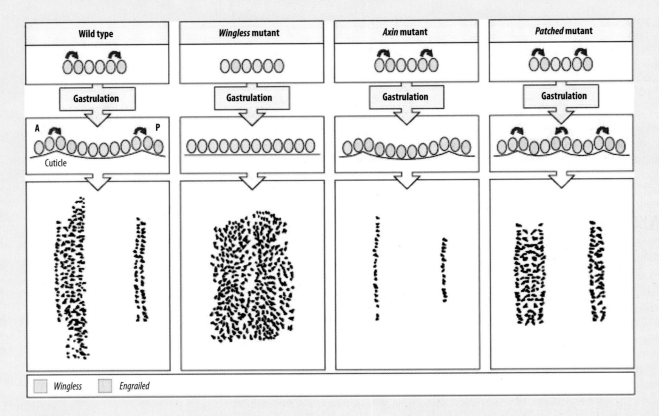

Figure 1

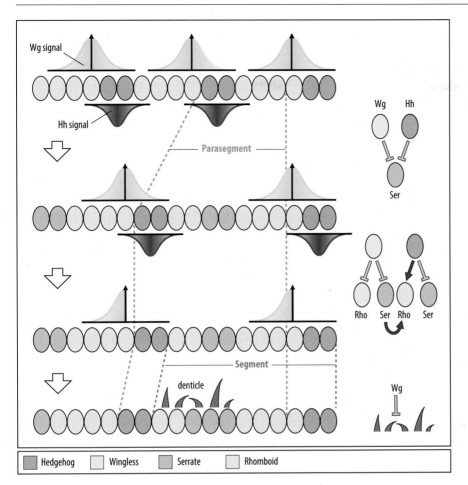

Fig. 2.46 Segment patterning restricts the production of denticles to specific cells.
After gastrulation, the diffusion of Wingless and Hedgehog covers the whole of the segment.
As the embryo grows, domains within each segment lose the influence of Hedgehog and
Wingless and genes that are repressed by these signaling pathways can be expressed. One
of these genes encodes Serrate (Ser), which is a ligand for Notch. Serrate then sends a signal
to adjacent cells that leads to the activation of the gene *rhomboid*, which encodes a protein
involved in regulating epidermal growth factor receptor signaling and which is required for
denticle formation. At the same time, the diffusion of Wingless becomes polarized, and only
cells anterior to its expression domain are exposed to Wingless protein. This sequence of
interactions and gene expression creates a complex landscape, which results in the detailed
patterning of the segment. Wingless suppresses the production of denticles, which results
in the domain of cells that 'see' Wingless becoming naked cuticle, whereas in the rest of
the segment, the fine-grained pattern of gene expression created as the embryo grows and
matures, results in the emergence of different kinds of denticles on each row of cells, as
indicated. The denticle pattern in the second abdominal segment is shown here. Engrailed
continues to be expressed in the Hedgehog-expressing cells (not shown for simplicity).

pattern of the wing is in no way dependent on cell lineage. A single marked embryonic
cell can give rise to about a twentieth of the cells of the adult wing or, using the *Minute*
technique to increase clone size, about half the wing. The lineage of the wing cells in
each case is quite different, yet the wing's pattern is normal. This indicates that what
is being patterned in this case is groups of cells, rather than patterning occurring on
an individual cell-by-cell basis. As we shall see in Chapter 6, this is very different from

EXPERIMENTAL BOX 2H Genetic mosaics and mitotic recombination

Genetic mosaics are embryos derived from a single genome but in which there is a mixture of cells with rearranged or inactivated genes. In flies, genetic mosaics can be generated by inducing rare mitotic recombination events in the somatic cells of the embryo or larva. The original method was to induce chromosome breaks using X-rays just after the chromosomes have replicated to form two chromatids; this results in the exchange of material between homologous chromosomes as the break is repaired. Today, mitotic recombination is induced in strains of transgenic flies that carry the gene for the yeast recombinase FLP and its target sequence, *FRT*, on their chromosomes. Activation of the recombinase will result in recombination involving the *FRT* sequence. A mitotic recombination event can generate a single cell with a unique genetic constitution that will be inherited by all the cell's descendants, which in flies usually form a coherent patch of tissue (Figure 1).

Easily distinguishable mutations like the recessive *multiple wing hairs* can be used to identify the marked clone; if a cell homozygous for this mutation is generated by mitotic recombination in a heterozygous larva, all of its descendant cells will have multiple hairs (Figure 1). Marked epidermal clones made by this method are usually small, as there is little cell proliferation after the recombination event. Larger clones can be made using the *Minute* technique. The cells in flies carrying a mutation in the *Minute* gene grow more slowly than those in the wild type. By using flies heterozygous for the *Minute* mutation, clones can be made in which the mitotic recombination event has generated a marked cell that is normal because it has lost the *Minute* mutation, and so is wild type. This normal cell proliferates faster than the slower-growing background and thus large clones of the marked cells are produced (see Fig. 2.47). The mitotic recombination technique has many applications. If clones of marked cells are generated at different stages of development, one can trace the fate of the altered cells and thus see what structures they are able to contribute to. This can

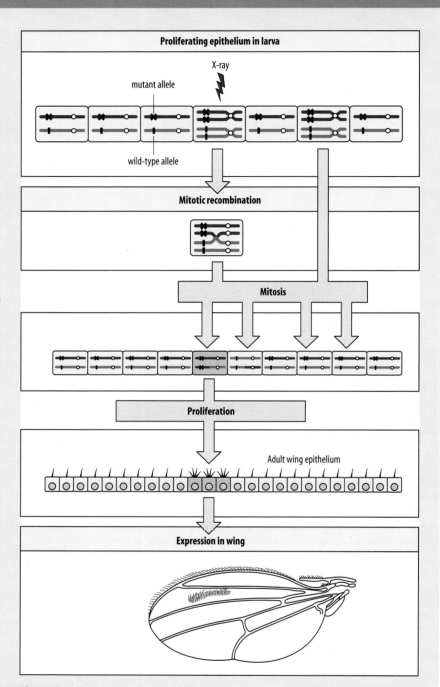

Figure 1

Illustration after Lawrence, P.: The Making of a Fly. Oxford: Blackwell Scientific Publications, 1992.

provide information on their state of determination or specification at different developmental stages. The technique can also be used to study the localized effects of homozygous mutations that are lethal if present in the homozygous state throughout the whole animal.

the situation in the nematode *C. elegans*, where the cell lineages in development are constant and the patterning occurs at the level of the identity of individual cells.

The specification of cells as the posterior compartment of a segment (the anterior of a parasegment) initially occurs when the parasegments are set up, and is due to the *engrailed* gene. Expression of *engrailed* is required both to confer a 'posterior segment' identity on the cells and to change their surface properties so that they cannot mix with the cells adjacent to them, hence setting up the parasegment boundary. Direct evidence for the role of *engrailed* comes from the behavior of clones of wing cells in an *engrailed* mutant (see Fig. 2.47, bottom panel). In the absence of normal *engrailed* expression, clones are not confined to anterior or posterior parts of the segment and there is no compartment boundary. Moreover, in *engrailed* mutants, the posterior compartment is partly transformed so that it comes to resemble the pattern of the anterior part of the wing. For example, bristles normally found only at the anterior margin of the wing are also found at the posterior margin.

2.28 Insect epidermal cells become individually polarized in an antero-posterior direction in the plane of the epithelium

The larval denticle pattern used to study segmental patterning also illustrates another feature of development, which was first investigated in *Drosophila*, but is found in all animals. This is the phenomenon of **planar cell polarity**, in which cells acquire a 'directionality', in the sense that one end is structurally or biochemically different from the other. This type of cell polarity is called planar because the polarization is in the plane of the tissue, in contrast to the apical–basal polarity also found in epithelial cells, which are often polarized across the cell sheet (planar cell polarity in both *Drosophila* and vertebrate development is discussed further in Chapter 9). Planar cell polarity is typically most easily seen in epithelia, in which all the cells in an epithelial sheet (such as the epidermis) are often polarized in the same direction.

In the case of the larval cuticle, planar cell polarity is evident in the fact that the denticles of the rows point in precise directions, reflecting an underlying polarity in the epidermal cells from which the denticles are formed. It is likely that the mirror-image patterns of denticles seen in *wingless* and *patched* mutants (see Figure 2.44 and Box 2G) reflect indirect effects on planar polarity of the loss of function of these genes.

Many different types of cells display planar cell polarity. For example, polarization is clearly seen in cells undergoing chemotaxis in response to a gradient of a chemoattractant, where the advancing end of the cell is quite different from the posterior end. The bristles on the adult Drosophila abdominal epidermis all point in a posterior direction, indicating polarization of the epidermis. Planar cell polarity is also clearly visible in the *Drosophila* wing, which is composed of a thin sheet of epithelium. The epidermal cells on the upper face of the wing bear hairs, and these hairs all point away from the body (see Fig. 2.47). Another example is the insect compound eye, where the individual units (ommatidia) in each half of the eye point away from the midline (discussed in Chapter 11), which is the result of the establishment of planar polarity in the developing photoreceptor cells in the eye imaginal disc. In vertebrates, easily visible examples of planar polarity occur in the stereocilia in the inner ear and in the scales of fish. Planar cell polarity is also important in directing the direction of morphogenetic cell movements, such as the tissue elongation that occurs during gastrulation, notochord extension, and limb development in vertebrates, for example (discussed in Chapters 9 and 11).

Current hypotheses on the molecular mechanisms underlying the establishment of planar cell polarity propose the existence of an informational gradient across the tissue that sets the overall direction of polarity, while local signaling between one end of a cell and the other, and between one cell and its neighbor, are thought to set the polarity of the individual cell (Box 2I). In the case of the adult insect abdominal epidermis, for example, the signaling gradients set up at compartment boundaries

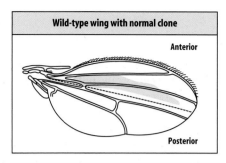

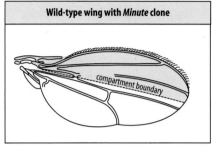

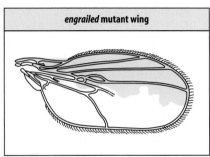

Fig. 2.47 The boundary between anterior and posterior compartments in the *Drosophila* wing can be demonstrated by marked cell clones. Top panel: in the wild-type wing, clones marked by mitotic recombination in the embryo, which gives marked cells a phenotype different from other cells of the wing, are too small to demonstrate the compartment boundary. Middle panel: the use of the *Minute* technique (see Box 2H) produces an increased rate of cell division in the marked cell and gives clones large enough to determine that cells from one compartment do not cross the boundary into an adjoining compartment. Bottom panel: in the wing of an *engrailed* mutant in which the engrailed protein is not produced, there is no posterior compartment or boundary. Clones in the anterior part of the wing cross over into the posterior region and the posterior region is transformed into a more anterior-like structure, bearing anterior-type hairs on its margin. As described earlier, the *engrailed* gene is required for the maintenance of the character of the posterior compartment, and for the formation of the boundary.

BOX 2I Planar cell polarity in *Drosophila*

Cells in an epithelium have a distinct apical–basal polarity, but more relevant to patterning in development is polarity in the plane of the epithelium—**planar cell polarity**. This is visible in the epidermis of the adult fly, where the abdominal epidermis has hairs all pointing in a posterior direction (Figure 1, panel a), and the epidermis of the wing has hairs all pointing in a **distal** direction—away from the proximal base of the wing and towards its tip.

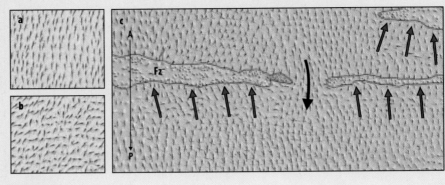

Figure 1

The mechanism for establishing planar polarity is not fully understood, especially how the axis of polarity across the tissue is determined. At the cellular level, however, some key proteins involved have been identified. The cell-surface receptor Frizzled (Fz) is one. In the fly abdominal epidermis, if Fz is absent then tissue polarity is disrupted and the hairs point in random directions (Figure 1, panel b). But if the gene for Fz is turned off in just a small clone of epidermal cells, the unaffected cells immediately posterior to them reverse polarity, showing that polarity is influenced by intercellular—cell to cell—communication (Figure 1, panel c; red arrows indicate the reversed direction; the black arrow shows the normal direction of polarity from anterior (A) to posterior (P)). We met Fz earlier as the receptor for Wingless, but in the case of planar cell polarity it does not act through the canonical Wnt pathway shown in Box 4B, and is unlikely to be stimulated by Wingless or other *Drosophila* Wnts.

Fz is thought to have at least three functions in planar cell polarity. The first is to receive long-range patterning information from upstream cues, which might involve the generation of a gradient of Fz activity across the whole axis of the tissue. Clonal repolarization experiments like those shown in the figure suggest that the direction arrow of polarity points down the Fz activity gradient; cells point their hairs in the direction of a lower level of Fz and away from a higher level. Second, Fz is involved in the cell–cell communication that locally coordinates cell polarity. The third function of Fz is to provide an intracellular cue for the growth of the hairs.

It is now clear that key interactions that determine and maintain the polarity of individual cells take place at the epidermal cell junctions. An asymmetry between the two sides of a junction is established by interactions between proteins in the adjacent cell membranes. In proximal to distal polarity in the wing, for example, Fz becomes associated with an atypical cadherin called Flamingo (Fmi, also known as Starry Night) at the distal end of each epidermal cell, while at the proximal end, Fmi is associated with the membrane protein Van Gogh (Vang, also known as Strabismus). An asymmetric- interaction across the junction between the Fz:Fmi and Fmi:Vang complexes seems to be central

to establishing cell polarity—by some mechanism that is still to be determined. The cytoplasmic signaling proteins Dishevelled (Dsh) and Diego (Dgo) become associated with Fz:Fmi and the cytoplasmic protein Prickled (Pk) associates with Fmi:Vang (Figure 2). These cytoplasmic proteins seem not to be essential for establishing cell polarity *per se* but may act to convert the polarity signal into some of the manifestations of polarity within the cell.

Yet another set of proteins, which include the membrane protein Four-jointed, and the atypical cadherins Dachsous and Fat, have also been implicated in determining planar cell polarity. They form gradients along the axis of polarity in tissues, and one proposed role for these proteins has been as long-range graded signals that set the direction of polarity by interacting in some way with the proteins described above.

However, there is evidence that this system can operate independently in parallel with the Fz-Fmi-Vang system rather than in a strict linear relationship. A clone of cells overexpressing Dachsous or other proteins of that group can repolarize adjacent cells even when those cells lack both Fz and Fmi and so cannot form the junctional complexes. Unraveling planar cell polarity is still a fascinating work in progress.

Establishing cell polarity is an important aspect of development in all animals. We shall see in Chapter 9, for example, how some of these same proteins regulate the changes in cell shape and polarity that help to reshape vertebrate embryos during gastrulation.

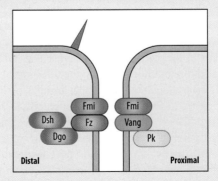

Figure 2

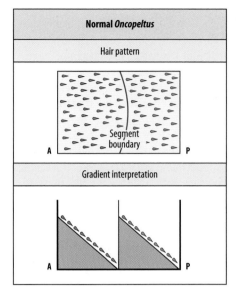

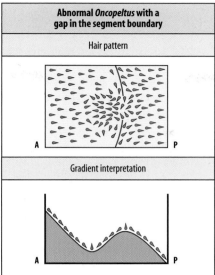

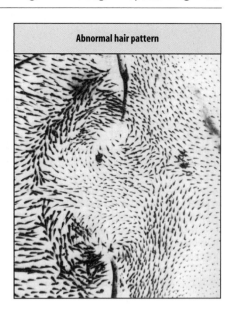

Fig 2.48 Gradients could specify polarity in the segments of
Oncopeltus. The hairs on the cuticle point in a posterior direction
and this may reflect an underlying gradient in a morphogen that is
partly maintained by the segment boundary (left panels). When there
is a gap in the boundary (center panels) the sharp discontinuity in
morphogen concentration smoothes out, with a resulting local reversal
of the gradient and of the direction in which the hairs point. This is
illustrated in the photograph (right panel).

have been proposed to have a role in establishing planar cell polarity, as well as in
patterning the segment.

A natural case where loss of planar cell polarity can be observed is in adults of
the plant-feeding bug *Oncopeltus*, which, like *Drosophila*, have hairs covering each
segment, all pointing in a posterior direction (Fig. 2.48, left panel). In *Oncopeltus*,
some individuals naturally have a gap in the segment boundary; this results in a
rather precise pattern of hairs with altered orientation, many of the hairs near the gap
pointing in the reverse direction (see Fig. 2.48, center and right panels). If the slope
of a gradient within each segment gives the hairs their polarity, a gap in a segment
boundary would result in a local change in the gradient, the sharp change in concentration at the normal boundary being smoothed out and forming a local gradient
running in the opposite direction. This would result in the hairs in this region pointing in the opposite direction. Mutations that cause similar reversals in polarity can be
induced in clones of cells in the *Drosophila* abdominal epidermis, and can be similarly
explained by the disruption of an informational gradient set up within a compartment.
Genes essential for planar cell polarity have been identified in *Drosophila* and other
organisms, together with some of the proteins involved in setting up the long-range
gradient and in the signaling pathways that interpret it to set individual cell polarity
(see Box 2I).

SUMMARY

The segmentation genes are involved in patterning the parasegments within each
of the larval segments. The most important genes in this process are *wingless* and
hedgehog, which encode ligands of major cell–cell signaling pathways and regulate
each other's expression. The activity of the gap and pair-rule genes positions Wingless and Hedgehog protein activity at the posterior (Wingless) and anterior (Hedgehog) sides of the parasegment boundary, which thus serves as a source of patterning

information. Another gene activated at the same time is *engrailed*, which is expressed in the same cells as *hedgehog*. *Drosophila* has a so-called long-germ type of development, in which all the segments are specified at the same developmental stage. In contrast, some insects have a short-germ pattern of development in which posterior segments are added by growth after the cellular blastoderm stage. Regions of cell-lineage restriction, called compartments, were first discovered in *Drosophila*. One compartment boundary is the parasegment boundary, which becomes the compartment boundary between the anterior and posterior parts of a segment in larval and adult tissue. The expression of Engrailed and Hedgehog, but not that of Wingless, is preserved in the adult and these proteins also play a role in the patterning of the adult. The *Drosophila* epidermis also displays planar cell polarity, which gives directionality to a tissue and is a near-universal feature of animal development. Planar cell polarity is most readily seen in the adult, where hairs on the fly body and on the wing point in the same direction.

SUMMARY: gene expression and intercellular signaling defines segment compartments

pair-rule gene expression

segmentation and selector gene *engrailed* activated in anterior of each parasegment, defining anterior compartment of parasegment and posterior compartment of segment

engrailed-expressing cells also express segmentation gene *hedgehog*

cells on other side of compartment boundary express segmentation gene *wingless*

engrailed expression maintained and compartment boundary stabilized by Wingless and Hedgehog proteins

compartment boundary provides signaling center from which segment is patterned

---

Specification of segment identity

Each segment has a unique identity, most easily seen in the larva in the characteristic pattern of denticles on the ventral surface. But since the same segmentation genes are turned on in each segment, what makes the segments different from each other? Their identity is specified by a class of genes known as **homeotic selector genes**, which set the future developmental pathway of each segment. A selector gene is a gene, such as *engrailed*, that controls the activity of other genes and is required throughout development to maintain this pattern of gene expression. Homologs of the *Drosophila* homeotic selector genes that control segment identity have subsequently been discovered in virtually all other animals, where they broadly control identity along the antero-posterior axis, as we shall see in Chapter 5 for vertebrates.

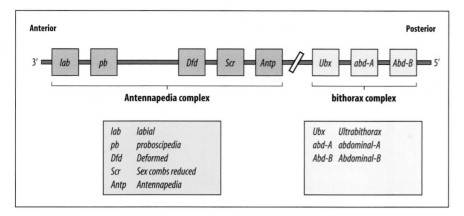

Fig. 2.49 The Antennapedia and bithorax homeotic selector gene complexes. The order of the genes from from 3′ to 5′ in each gene complex reflects both the order of their spatial expression (anterior to posterior) and the timing of expression (3′ earliest).

2.29 Segment identity in *Drosophila* is specified by Hox genes

The first evidence for the existence of genes that specify segment identity came from unusual and striking mutations in *Drosophila* that produced **homeotic transformations**—the conversion of one segment into another. Eventually the genes that produced these mutations were identified and the intricate way in which they specified segment identity was worked out. The discovery of these genes, now called the **Hox genes**, had an enormous impact on developmental biology. Homologous genes were found in other animals, where they acted in a similar way, essentially specifying identity along the antero-posterior axis. Hox genes are now considered as one of the fundamental defining features of animals. In 1995, the American geneticist Edward Lewis shared the Nobel Prize in Physiology or Medicine for his pioneering work on the *Drosophila* homeotic gene complexes and how they function. In *Drosophila*, the Hox genes are organized into two **gene clusters** or **gene complexes**, which together make up the **HOM-C complex** (Fig. 2.49; see also Box 5E). The Hox genes all encode transcription factors and take their name from the distinctive **homeobox** DNA motif that encodes part of the DNA-binding site of these proteins.

The two homeotic gene clusters in *Drosophila* are called the **bithorax complex** and the **Antennapedia complex** and are named after the mutations that first revealed their existence. Flies with the *bithorax* mutation have part of the haltere (the balancing organ on the third thoracic segment) transformed into part of a wing (Fig. 2.50), whereas flies with the dominant *Antennapedia* mutation have their antennae transformed into legs. Genes identified by such mutations are called **homeotic genes** because when mutated they result in **homeosis**—the transformation of a whole

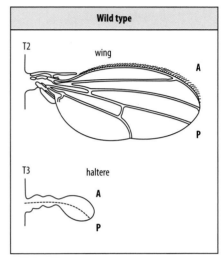

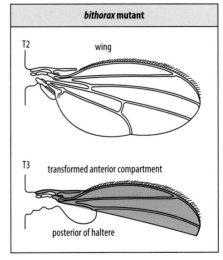

Fig. 2.50 Homeotic transformation of the wing and haltere by mutations in the bithorax complex. Top panel: in the normal adult, both wing and haltere are divided into an anterior (A) and a posterior (P) compartment. Middle panel: the *bithorax* mutation transforms the anterior compartment of the haltere into an anterior wing region. The mutation *postbithorax* acts similarly on the posterior compartment, converting it into posterior wing (illustrated in Fig. 11.44). Bottom panel: if both mutations are present together, the effect is additive and the haltere is transformed into a complete wing, producing a four-winged fly.

*Illustration after Lawrence, P.: The Making of a Fly. Oxford: Blackwell Scientific Publications, 1992. Photograph reproduced with permission from Bender, W., et al.: **Molecular genetics of the bithorax complex in Drosophila melanogaster**. Science 1983, **221**: 23–29 (image on front cover).*

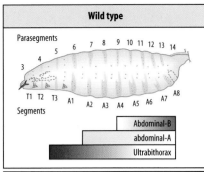

Wild type

Parasegments

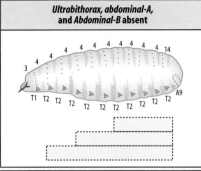

Ultrabithorax, abdominal-A, and Abdominal-B absent

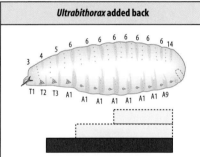

Ultrabithorax added back

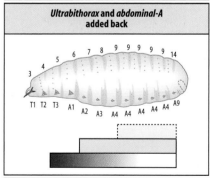

Ultrabithorax and abdominal-A added back

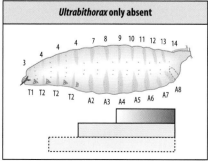

Ultrabithorax only absent

segment or structure into another related one, as in the transformation of antenna to leg. These bizarre transformations arise out of the homeotic selector genes' key role as identity specifiers. They control the activity of other genes in the segments, thus determining, for example, that a particular imaginal disc will develop as wing or haltere. The bithorax complex controls the development of parasegments 5–14, while the Antennapedia complex controls the identity of the more anterior parasegments. The action of the bithorax complex is the best understood and will be discussed first.

2.30 Homeotic selector genes of the bithorax complex are responsible for diversification of the posterior segments

The bithorax complex of *Drosophila* comprises three **homeobox genes**: *Ultrabithorax* (*Ubx*), *abdominal-A* (*abd-A*), and *Abdominal-B* (*Abd-B*). These genes are expressed in the parasegments in a combinatorial manner (Fig. 2.51, top panel). *Ubx* is expressed in all parasegments from 5 to 12, *abd-A* is expressed more posteriorly in parasegments 7–13, and *Abd-B* more posteriorly still from parasegment 10 onwards. Because the genes are active to varying extents in different parasegments, their combined activities define the character of each parasegment. *Abd-B* also suppresses *Ubx*, such that *Ubx* expression is very low by parasegment 14, as *Abd-B* expression increases. The pattern of activity of the bithorax complex genes is determined by gap and pair-rule genes.

The role of the bithorax complex was first indicated by classical genetic experiments. In larvae lacking the whole of the bithorax complex (see Fig. 2.51, second panel), every parasegment from 5 to 13 develops in the same way and resembles parasegment 4. The bithorax complex is therefore essential for the diversification of these segments, whose basic pattern is represented by parasegment 4. This parasegment can be considered as being a type of 'default' state, which is modified in all the parasegments posterior to it by the proteins encoded by the bithorax complex. It is because the genes of the bithorax complex are able to superimpose a new identity on the default state that they are called selector genes.

An indication of the role of each gene of the bithorax complex can be deduced by looking at embryos constructed so that the genes are put one at a time into embryos lacking the whole gene complex (see Fig. 2.51, bottom three panels). If only the *Ubx* gene is present, the resulting larva has one parasegment 4, one parasegment 5, and eight parasegments 6. Clearly, *Ubx* has some effect on all parasegments from 5 backward and can specify parasegments 5 and 6. If *abd-A* and *Ubx* are put into the embryo, then the larva has parasegments 4, 5, 6, 7, and 8, followed by five parasegments 9. So *abd-A* affects parasegments from 7 backward, and in combination, *Ubx* and *abd-A* can specify the character of parasegments 7, 8, and 9. Similar principles apply to *Abd-B*, whose domain of influence extends from parasegment 10 backward, and which is expressed most strongly in parasegment 14. Differences between individual segments may reflect differences in the spatial and temporal pattern of Hox gene expression.

Fig. 2.51 The spatial pattern of expression of genes of the bithorax complex characterizes each parasegment. In the wild-type embryo (top panel) the expression of the genes *Ultrabithorax*, *abdominal-A*, and *Abdominal-B* are required to confer an identity on each parasegment. Mutations in the bithorax complex result in homeotic transformations in the character of the parasegments and the segments derived from them. When the bithorax complex is completely absent (second panel), parasegments 5–13 are converted into nine parasegments 4 (corresponding to segment T2 in the larva), as shown by the denticle and bristle patterns on the cuticle. The three lower panels show the transformations caused by the absence of different combinations of genes. When the *Ultrabithorax*. gene alone is absent (bottom panel), parasegments 5 and 6 are converted into 4. In each case the spatial extent of gene expression is detected by *in situ* hybridization (see Box 1D). Note that the specification of parasegment 14 is relatively unaffected by the bithorax complex.

These results illustrate an important principle, namely that the character of the parasegments is specified by the genes of the bithorax complex acting in a combinatorial manner. Their combinatorial effect can also be seen by taking the genes away one at a time from the wild type. In the absence of *Ubx*, for example, parasegments 5 and 6 become converted to parasegment 4 (see Fig. 2.51, bottom panel). There is a further effect on the cuticle pattern in parasegments 7–14 in that structures characteristic of the thorax are now present in the abdomen, showing that *Ubx* exerts an effect in all these segments. Such abnormalities may result from expression of 'nonsense' combinations of the bithorax genes. For example, in such a mutant, abd-A protein is present in parasegments 7–9 without Ubx protein, and this is a combination never found normally. The site of appendages, such as legs, is determined by the Hox genes. For example, expression of the genes *Antp* and *Ubx* specifies the segments from which the second and third pair of legs, respectively, will emerge. The downstream targets of Hox proteins are beginning to be identified, and there may be a very large number of genes whose expression is affected.

While both the gap and pair-rule proteins initially control the pattern of Hox gene expression, these proteins disappear within about 4 hours. The continued correct expression of the Hox genes involves two groups of genes—the Polycomb and Trithorax groups. The proteins of the Polycomb group maintain transcriptional repression of Hox genes where they are initially off, whereas the Trithorax group maintains expression in those cells where Hox genes have been turned on.

2.31 The Antennapedia complex controls specification of anterior regions

The Antennapedia complex comprises five homeobox genes (see Fig. 2.49), which control the behavior of the parasegments anterior to parasegment 5 in a manner similar to the bithorax complex. Since there are no novel principles involved, we just mention its role briefly. Several genes within the complex are critically involved in specifying particular parasegments. Mutations in the gene *Deformed* affect the ectodermally derived structures of parasegments 0 and 1, those in *Sex combs reduced* affect parasegments 2 and 3, and those in the *Antp* gene affect parasegments 4 and 5, the parasegments that produce the leg imaginal discs. The *Antennapedia* mutation that transforms antennae into legs in the adult fly results from the misexpression of the *Antp* gene in anterior segments in which it is not normally expressed.

2.32 The order of Hox gene expression corresponds to the order of genes along the chromosome

The bithorax and Antennapedia complexes possess some striking features of gene organization. Within both, the order of the genes in the complex is the same as the spatial and temporal order in which they are expressed along the antero-posterior axis during development. *Ubx*, for example, is on the 3′ side of *abd-A* on the chromosome, is more anterior in its pattern of expression, and is activated earlier. As we will see, the related Hox gene complexes of vertebrates, whose ancestors diverged from those of arthropods hundreds of millions of years ago, show the same correspondence between gene order and timing of expression. This highly conserved temporal and spatial **co-linearity** must be related to the mechanisms that control the expression of these genes.

The complex yet subtle control to which the genes of the bithorax complex are subject is seen in experiments in which production of Ultrabithorax protein is forced in all segments. This targeted gene misexpression was achieved by linking the Ultrabithorax protein-coding sequence to a heat-shock promoter—a promoter that is activated at 29°C—and introducing the novel DNA construct into the fly genome using a P element (see Box 2D). When this transgenic embryo was given a heat shock for a few minutes, the extra *Ubx* gene was transcribed and the protein was made at high levels in all cells. This had no effect on posterior parasegments (in which Ubx protein is normally present), with the exception of parasegment 5 which, for unknown reasons, is transformed into parasegment 6 (this may reflect a quantitative effect of the Ubx protein). However, all parasegments anterior to 5 are also transformed into parasegment 6. While this seems a reasonably straightforward and expected result, consider what happens to parasegment 13.

Transcription of *Ubx* is normally suppressed in parasegment 13 in wild-type embryos, yet when the protein itself is produced in the parasegment as a result of heat shock, it has no effect. By some mechanism, the Ultrabithorax protein is rendered inactive in this parasegment. This phenomenon is quite common in the specification of the parasegments and is known as 'phenotypic suppression' or 'posterior dominance': it means that Hox gene products normally expressed in anterior regions are suppressed by more posterior products.

While the roles of the bithorax and Antennapedia complexes in controlling segment identity are well established, we do not yet know very much about how they interact with the genes acting next in the developmental pathway. These are the genes that actually specify the structures that give the segments their unique identities. What, for example, is the pathway that leads to a thoracic rather than an abdominal segment structure? Does each Hox gene activate just a few or many target genes? Do target genes with different functions act in concert to form a particular structure? Answers to these questions will help us understand how changes in a single gene can cause a homeotic transformation like antenna into leg. The role of the Hox genes in the evolution of different body patterns and in the evolution of limbs from fins will be considered in Chapter 14.

2.33 The *Drosophila* head region is specified by genes other than the Hox genes

As mentioned earlier in the chapter, the *Drosophila* larval head anterior to the mandibles is formed by the three most anterior parasegments. The segmental structure is also evident in the embryonic brain, which is composed of three segments (neuromeres). The specification of the anterior head and embryonic nervous system is not, however, under the control of either the pair-rule genes or the Hox genes. Instead, it appears to be determined at the blastoderm stage by the overlapping expression of the genes *orthodenticle*, *empty spiracles*, and *buttonhead*, which all encode gene-regulatory proteins and resemble the gap genes in their effect on phenotype; they are sometimes referred to as the 'head gap genes.' Unlike the gap genes regionalizing the trunk and abdomen, however, they do not regulate each other's activity. Nor is it known whether they work in a combinatorial way like the Hox genes to also confer specific identities on the head segments. *Orthodenticle* and *empty spiracles* have homologs that carry out similar functions in vertebrates, and it is clear that this head-specification system, like the Hox genes, is of ancient evolutionary origin in animals.

The genes *orthodenticle* and *empty spiracles* are examples of **homeodomain** transcription factors that lie outside the Hox gene clusters. The DNA-binding homeodomain is not confined to the proteins encoded by the Hox gene clusters, and similar domains are found in many other transcription factors involved in development in both *Drosophila* and vertebrates.

SUMMARY

Segment identity is conferred by the action of selector or homeotic genes, which direct the development of different parasegments so that each acquires a unique identity. Two clusters of selector genes, known generally as the Hox genes, are involved in specification of segment identity in *Drosophila*: the Antennapedia complex, which controls parasegment identity in the head and first thoracic segment, and the bithorax complex, which acts on the remaining parasegments. Segment identity seems to be determined by the combination of Hox genes active in the segment. The spatial pattern of expression of selector genes along the embryo body is initially determined by the preceding gap gene activity, but selector genes need to be switched on continuously throughout development to maintain the required phenotype. Mutations in genes of the Antennapedia and bithorax complexes can bring about homeotic transformations, in which one segment or structure is changed into a related one, for example an antenna into a leg. The Antennapedia and bithorax complexes are remarkable in that gene order on the chromosome corresponds to the spatial and temporal order of gene expression along the body.

SUMMARY: homeotic genes and segment identity

gap and pair-rule gene expression delimits parasegments

selector genes of the HOM-C complex are expressed along the antero-posterior axis
in an order co-linear with the order of genes on the chromosome

Antennapedia complex
lab, pb, Dfd, Scr, Antp

bithorax complex
Ubx, abd-A, Abd-B

specify segment identity
in head and
first thoracic segment

specify segment identity in second
and third thoracic segments and
abdominal segments

Summary to Chapter 2

- During oogenesis, maternal proteins and mRNAs are deposited in the egg in a particular spatial pattern. This pattern will determine the main body axes and lay down a framework of positional information.
- The earliest stages of development in *Drosophila* take place when the embryo is a multinucleate syncytium. After fertilization, maternally supplied mRNAs are translated and gradients of transcription factors are formed that broadly regionalize the body along the antero-posterior and dorso-ventral axes.
- These transcription factors enter the nuclei and turn on the embryo's own genes, activating a cascade of zygotic gene activity that further patterns the body.
- Along the dorso-ventral axis, zygotic gene expression defines several regions, including the future mesoderm and future neural tissue.
- The first zygotic genes to be activated along the antero-posterior axis are the gap genes, all encoding transcription factors, whose pattern of expression subdivides the embryo into a number of regions.
- The transition to a segmented body organization starts with the activity of the pair-rule genes, whose sites of expression are specified by the gap-gene proteins and which divide the antero-posterior axis into 14 parasegments.
- At the time of pair-rule gene expression, the blastoderm becomes cellularized.
- The pair-rule genes initially define the spatial expression of segmentation genes, which pattern the parasegments. Once the embryo becomes cellular, patterning depends on intercellular communication via signaling proteins and their receptors as well as transcription factors.
- After gastrulation, the definitive final segments of the late embryo and larva are delimited, offset from the original parasegments by about half a segment.

- The parasegment boundaries delimit regions of cell-lineage restriction in the epidermis, known as compartments. These compartment boundaries persist in the segments, where they delimit the anterior and posterior regions of each segment, and in the imaginal discs and the adult structures, such as wings and legs, that derive from the imaginal discs.
- Segment and compartment boundaries are involved in patterning and polarizing segments.
- The identity of each segment is determined by two gene complexes that contain homeotic selector genes, known generally as the Hox genes. The spatial pattern of expression of these genes is largely determined by gap gene activity.
- The *Drosophila* epidermis displays planar cell polarity, which is most easily seen in the adult, in which hairs on the body and on the wing point in the same direction.

■ End of chapter questions

Long answer (concept questions)

1. As we will see in future chapters, cleavage divides the eggs of vertebrates and many other animals into a cluster of separate cells. In contrast, the equivalent stage in the *Drosophila* embryo generates a syncytium. What is a syncytium, and how does this influence the early development of *Drosophila*?

2. Much attention has been paid to the process of segmentation in the *Drosophila* embryo. What are segments, and what is their significance to the strategy used by *Drosophila* to carry out pattern formation? Do humans have segments? Give examples.

3. Briefly summarize the mutation screen devised by Nüsslein–Volhard and Wieschaus to find developmental mutants. What role did denticle patterns play in their search for developmental mutants?

4. Developmental genes act hierarchically during pattern formation, first defining broad regions, which are secondarily refined to form a larger number of smaller regions (see Section 2.6). Summarize how this general principle is illustrated by early *Drosophila* embryogenesis.

5. The anterior end of the embryo is established by a locally high level of activity of the transcription factor Bicoid; the ventral side of the embryo is established by a locally high level of activity of the transcription factor Dorsal. Contrast the mechanisms by which these activities come to be localized in the required places.

6. Bicoid was the first morphogen whose mechanism of action was molecularly characterized; an important outcome of this work was the understanding of how a smoothly declining gradient of Bicoid can lead to a sharply delimited boundary of Hunchback expression. Use the Bicoid–Hunchback system to illustrate how an activation threshold can translate a smooth gradient into a sharp developmental boundary. Be sure to incorporate the role of cooperative binding of Bicoid to the *hunchback* promoter in your answer.

7. The dorsal–ventral axis is patterned by a gradient of nuclear localization of the transcription factor Dorsal. Outline the steps that lead to this gradient of nuclear localization: include the Pipe enzyme, the Spätzle ligand, the Toll receptor, the Cactus protein, and finally Dorsal.

8. The specification of the 14 parasegments along the anterior–posterior axis depends on precise spatial expression of the pair-rule genes. Pair-rule gene expression depends in turn on the gap genes. Using the second stripe of *even-skipped* expression as a model, summarize the interactions that lead to proper expression of Bicoid, Hunchback, Krüppel, and Giant. In particular, describe how the anterior border of Krüppel is set, and how this boundary sets the posterior border of *eve* stripe 2 expression.

9. What is the special role of *engrailed* in the patterning of the *Drosophila* body plan?

10. *Drosophila* has played a central role in our understanding of insect development, but is not necessarily representative of all insects. How does the long-germ development of *Drosophila* differ from the short-germ band type of development seen in beetles, grasshoppers, and wasps?

11. What is meant by a homeotic transformation? Give an example.

12. Experiments demonstrate that the role of the bithorax complex is to drive the identity of the parasegments in the posterior portion of the animal away from the default identity of parasegment 4. Describe these experiments. How do these experiments illustrate the importance of the combinations of gene products present in specifying segment identity?

13. The generation of transgenic *Drosophila* using the P element transposon has played a complementary role to traditional mutation analysis in deciphering the genetic regulation of *Drosophila* development. Answer the following questions: Why is the P element important to this process? Why is the P element injected into the posterior of the egg, before any cellularization has occurred? Why are flies mutant for the *white* gene typically used as the hosts for these experiments? Last, what genetic crosses are necessary before the consequences of the P element injection can be observed?

Multiple choice (factual recall questions)

NB There is only one right answer to each question.

1. The 'spinal cord' of *Drosophila*—its 'nerve cord'

a) consists of two parallel structures that run laterally, along the sides of the animal.

b) is in no way analogous to the spinal cord of vertebrates, since *Drosophila* has no spine!

c) runs along the dorsal side, opposite the mouth as is the case in vertebrates such as humans

d) runs along the ventral surface, on the same side as the mouth

SUMMARY: main genes involved in specifying pattern in the early *Drosophila* embryo

	Gene	Maternal/ zygotic	Nature of protein	Transcription factor (T), receptor (R), or signal protein (S)	Function (where known)
Antero-posterior system	*bicoid*	M	Homeodomain	T	Morphogen, provides positional information along AP axis
	hunchback	M/Z	Zinc fingers	T	Morphogen, provides positional information along AP axis
	nanos	M	RNA-binding protein		Helps to establish AP gradient of hunchback protein
	caudal	M	Homeodomain	T	Involved in specifying posterior region
	gurken	M	Secreted protein of TGF-a family	S	Posterior oocyte-follicle cell signaling
	oskar	M			Pole-cell determination
Terminal system	*torso*	M	Receptor tyrosine kinase	R	Terminal specification
	trunk	M		S	Ligand for torso
Gap genes	*hunchback*	Z	Zinc fingers	T	Localize pair-rule gene expression
	Krüppel	Z	Zinc fingers	T	
	knirps	Z	Zinc fingers	T	
	giant	Z	Leucine fingers	T	
	tailless	Z	Zinc fingers	T	
Pair-rule genes	*even-skipped*	Z	Homeodomain	T	Delimits odd-numbered parasegments
	fushi tarazu	Z	Homeodomain	T	Delimits even-numbered parasegments
				T	
Segmentation and segment patterning	*engrailed*	Z	Homeodomain	T	Defines anterior region of parasegment and posterior region of segment
	hedgehog	Z	Secreted	S	Components of signaling pathways that pattern segments and stabilize compartment boundaries
	wingless	Z	Secreted	S	
	frizzled	Z	7-span receptor for Wingless	R	
	patched	Z	12-span receptor for Hedgehog	R	
	smoothened	Z	7-span receptor	R	
	Notch	Z	Membrane receptor	R	
	Serrate	Z	Membrane-bound ligand for Notch	S	
Selector genes bithorax complex	*Ultrabithorax*	Z	Homeodomain	T	Combinatorial activity confers identity on parasegments 5-13
	abdominal-A	Z	Homeodomain	T	
	Abdominal-B	Z	Homeodomain	T	
Antennapedia complex	*Deformed*	Z	Homeodomain	T	Combinatorial activity confers identity on parasegments anterior to 5
	Sex combs reduced	Z	Homeodomain	T	
	Antennapedia	Z	Homeodomain	T	
	labial	Z	Homeodomain	T	
Maintenance genes	Polycomb group	Z		T	Maintain state of homeotic genes
	Trithorax	Z		T	
Dorso-ventral system					
Maternal genes	*Toll*	M	Membrane	R	Activation results in Dorsal protein entering nucleus
	spätzle	M	Extracellular	S	Ligand for Toll protein
	dorsal	M		T	Morphogen, sets dorso-ventral polarity
	tube	M	Adaptor protein		Components of the Toll signaling pathway leading to Dorsal protein entering nuclei
	pelle	M	Protein kinase		
	cactus	M	Cytoplasmic inhibitor		
	gurken	M	Secreted protein of TGF-α family	S	Specifies oocyte axis
	pipe	M	Sulfotransferase	Enzyme	Part of pathway leading to Spätzle processing
Zygotic genes	*twist*	Z	Helix-loop-helix	T	Define mesoderm
	snail	Z	Zinc finger	T	
	rhomboid	Z	Membrane protein	S	
	zerknüllt	Z	Homeodomain	T	
	decapentaplegic	Z	Secreted protein of TGF-β family	S	Confer regional identity on dorso-ventral axis
	tolloid	Z	BMP-1 family	S	
	short gastrulation	Z		S	

AP, antero-posterior; TGF, transforming growth factor.

2. The process of segmentation will form how many parasegments?

a) 8 in the abdomen
b) 11 to form the thorax and abdomen
c) 14: 3 for the mouth, 3 in the thorax, and 8 in the abdomen
d) 3: head, thorax, and abdomen

3. When do the legs and wings of a fly form?

a) during embryogenesis, after imaginal disc cells are segregated away at the cellular blastoderm stage
b) during pupation, from the imaginal discs
c) during the larval instars
d) legs and wings are present during larval stages, but take their final form only after metamorphosis

4. Which protein (gene) functions to determine an anterior fate in the *Drosophila* embryo?

a) Bicoid
b) Caudal
c) Nanos
d) Torso

5. Maternal determinants are

a) genes that control differentiation as a female or male fruit fly
b) proteins and mRNAs packaged into the egg by the mother
c) proteins and mRNAs that determine the organizing axes of the ovary
d) substances present in the female that determine whether or not she will become a mother

6. Which is the correct hierarchy of gene activity in early *Drosophila* segmentation?

a) gap, segment polarity, pair-rule, maternal
b) maternal, gap, pair-rule, segment polarity
c) maternal, pair-rule, gap, segment-polarity
d) segment polarity, pair-rule, gap, maternal

7. A gap gene mutation could cause which of the following defects in the embryonic body plan?

a) every other segment is missing, resulting in T1, T3, A2, A4, etc., but no T2, A1, A3, and so on
b) patterning within each segment is abnormal, causing denticle belts to form across the entire segment
c) segments A2 through A6 are missing, but the rest of the pattern is essentially normal
d) the identity of one or more segments is transformed to that of a different segment

8. Segments are first positioned at the cell-by-cell level by the

a) gap genes
b) maternal genes
c) pair-rule genes
d) segment polarity genes

9. Engrailed expression defines the

a) anterior compartment of the segment
b) anterior margin of each segment
c) posterior compartment of the segment
d) posterior margin of each parasegment

10. Anterior patterning in the long-germ band wasp *Nasonia* is under the control of

a) Bicoid
b) Engrailed
c) Krüppel
d) Orthodenticle

Multiple choice answer key

1: d, 2: c, 3: b, 4: a, 5: b, 6: b, 7: c, 8: c, 9: c, 10: d.

■ General further reading

Adams, M.D., *et al.*: **The genome sequence of *Drosophila melanogaster***. *Science* 2000, **267**: 2185–2195.

Bate, M., Martinez Arias, A. (Eds): *The Development of* Drosophila melanogaster. Cold Spring Harbor Press: 1993.

Greenspan, R. *Fly Pushing: The Theory and Practice of* Drosophila *Genetics*. Cold Spring Harbor Press: 2004.

Lawrence, P.A.: *The Making of a Fly*. Oxford: Blackwell Scientific Publications, 1992.

Matthews, K.A., Kaufman, T.C., Gelbart, W.M.: **Research resources for *Drosophila*: the expanding universe**. *Nat Rev. Genet.* 2005, **6**: 179–193.

Stolc, V., *et al.*: **A gene expression map for the euchromatic genome of *Drosophila melanogaster***. *Science* 2004, **306**: 655–660.

Perrimon, N., Pitsouli, C., Shilo, B.Z.: **Signaling mechanisms controlling cell fate and embryonic patterning**. *Cold Spring Harb. Perspect. Biol.* 2012, **4**: a005975.

■ Section further reading

2.2 Cellularization is followed by gastrulation and segmentation

Foe, V.E., Alberts, B.M.: **Studies of nuclear and cytoplasmic behaviour during the five mitotic cycles that precede gastrulation in *Drosophila* embryogenesis**. *J. Cell Sci.* 1983, **61**: 31–70.

Leptin, M.: **Gastrulation in Drosophila: the logic and the cellular mechanisms**. *EMBO J.* 1999, **18**: 3187–3192.

Stathopoulos, A., Levine, M.: **Whole-genome analysis of *Drosophila* gastrulation**. *Curr. Opin. Genet. Dev.* 2004, **14**: 477–484.

2.4 Many developmental mutations were identified in *Drosophila* through induced large-scale genetic screening

Nusslein-Volhard. C., Wieschaus, E.: **Mutations affecting segment number and polarity in *Drosophila***. *Nature* 1980, **287**: 795–801.

Box 2A Mutagenesis and genetic screening strategy for identifying developmental mutants in *Drosophila*

Greenspan, R.: *Fly Pushing: The Theory and Practice of Drosophila Genetics*. Cold Spring Harbor Press, 2004

2.7 Three classes of maternal genes specify the antero-posterior axis

St Johnston, D., Nüsslein-Volhard, C.: **The origin of pattern and polarity in the *Drosophila* embryo.** *Cell* 1992, **68**: 201–219.

2.8 Bicoid protein provides an antero-posterior gradient of a morphogen

Driever, W., Nüsslein-Volhard, C.: **The bicoid protein determines position in the *Drosophila* embryo in a concentration dependent manner.** *Cell* 1988, **54**: 95–104.

Ephrussi, A., St Johnston, D.: **Seeing is believing: the bicoid morphogen gradient matures.** *Cell* 2004, **116**: 143–152.

Gibson, M.C.: **Bicoid by the numbers: quantifying a morphogen gradient.** *Cell* 2007, **130**: 14–16.

Morrison, A.H., Scheeler, M., Dubuis, J., Gregor, T.: **Quantifying the bicoid morphogen gadient in living fly embryos.** *Cold Spring Harb. Perspect. Biol.* 2012, doi:10.1101/pdb.top068536

Gregor, T., Wieschaus, E.F., McGregor, A.P., Bialek, W., Tank, D.W.: 2007, **Stability and nuclear dynamics of the bicoid morphogen gradient.** *Cell* **130**: 141–152.

Porcher, A., Dostatni, N.: **The bicoid morphogen system.** *Curr. Biol.* 2010, **20**: R249–R254.

2.9 The posterior pattern is controlled by the gradients of Nanos and Caudal proteins

Irish, V., Lehmann, R., Akam, M.: **The *Drosophila* posterior-group gene *nanos* functions by repressing *hunchback* activity.** *Nature* 1989, **338**: 646–648.

Murafta, Y., Wharton, R.P.: **Binding of pumilio to maternal *hunchback* mRNA is required for posterior patterning in *Drosophila* embryos.** *Cell* 1995, **80**: 747–756.

Rivera-Pomar, R., Lu, X., Perrimon, N., Taubert, H., Jackle, H.: **Activation of posterior gap gene expression in the *Drosophila* blastoderm.** *Nature* 1995, **376**: 253–256.

Struhl, G.: **Differing strategies for organizing anterior and posterior body pattern in *Drosophila* embryos.** *Nature* 1989, **338**: 741–744.

2.10 The anterior and posterior extremities of the embryo are specified by activation of a cell-surface receptor

Casali, A., Casanova, J.: **The spatial control of Torso RTK activation: a C-terminal fragment of the Trunk protein acts as a signal for Torso receptor in the *Drosophila* embryo.** *Development* 2001, **128**: 1709–1715.

Casanova, J., Struhl, G.: **Localized surface activity of torso, a receptor tyrosine kinase, specifies terminal body pattern in *Drosophila*.** *Genes Dev.* 1989, **3**: 2025–2038.

Coppey, M., Boettiger, A.N., Berezhkovskii, A.M., Shvartsman, S.Y.: **Nuclear trapping shapes the terminal gradient in the *Drosophila* embryo.** *Curr. Biol.* 2008, **18**: 915–919.

Furriols, M., Casanova, J.: **In and out of Torso RTK signalling.** *EMBO J.* 2003, **22**: 1947–1952.

Furriols, M., Ventura, G., Casanova, J.: **Two distinct but convergent groups of cells trigger Torso receptor tyrosine kinase activation by independently expressing torso-like.** *Proc. Natl Acad. Sci. USA* 2007, **104**: 11660–11665.

Li, W.X.: **Functions and mechanisms of receptor tyrosine kinase Torso signaling: lessons from *Drosophila* embryonic terminal development.** *Dev. Dyn.* 2005, **232**: 656–672.

2.11 The dorso-ventral polarity of the embryo is specified by localization of maternal proteins in the egg vitelline envelope

Anderson, K.V.: **Pinning down positional information: dorsal-ventral polarity in the *Drosophila* embryo.** *Cell* 1998, **95**: 439–442.

Sen, J., Goltz, J.S., Stevens, L., Stein, D.: **Spatially restricted expression of *pipe* in the *Drosophila* egg chamber defines embryonic dorsal-ventral polarity.** *Cell* 1998, **95**: 471–481.

Shilo, B.Z., Haskel-Ittah, M., Ben-Zvi, D., Schejter, E.D., Barkai, N.: **Creating gradients by morphogen shuttling.** *Trends Genet.* 2013, **29**: 339–347.

2.12 Positional information along the dorso-ventral axis is provided by the Dorsal protein

Roth, S., Stein, D., Nüsslein-Volhard, C.: **A gradient of nuclear localization of the dorsal protein determines dorso-ventral pattern in the *Drosophila* embryo.** *Cell* 1989, **59**: 1189–1202.

Steward, R., Govind, R.: **Dorsal-ventral polarity in the *Drosophila* embryo.** *Curr. Opin. Genet. Dev.* 1993, **3**: 556–561.

Box 2B The Toll signaling pathway: a multifunctional pathway

Belvin, M.P., Anderson, K.V.: **A conserved signaling pathway: the *Drosophila* Toll-dorsal pathway.** *Annu. Rev. Cell Dev. Biol.* 1996, **12**: 393–416.

Ferrandon, D., Imler, J.L., Hoffmann, J.A.: **Sensing infection in *Drosophila*: Toll and beyond.** *Semin. Immunol.* 2004, **16**: 43–53.

Imler, J.L., Hoffmann, J.A.: **Toll receptors in *Drosophila*: a family of molecules regulating development and immunity.** *Curr. Top. Microbiol. Immunol.* 2002, **270**: 63–79.

Valanne, S., Wang, J.H., Rämet, M.: **The *Drosophila* Toll signaling pathway.** *J. Immunol.* 2011, **186**: 649–656.

2.13 The antero-posterior axis of the *Drosophila* egg is specified by signals from the preceding egg chamber and by interactions of the oocyte with follicle cells

González-Reyes, A., St Johnston, D.: **The *Drosophila* AP axis is polarised by the cadherin-mediated positioning of the oocyte.** *Development* 1998, **125**: 3635–3644.

Huynh, J.R., St Johnston, D.: **The origin of asymmetry: early polarization of the *Drosophila* germline cyst and oocyte.** *Curr. Biol.* 2004, **14**: R438–R449.

Riechmann, V., Ephrussi, A.: **Axis formation during *Drosophila* oogenesis.** *Curr. Opin. Genet. Dev.* 2001, **11**: 374–383.

Roth, S., Lynch, J.A.: **Symmetry breaking during *Drosophila* oogenesis.** *Cold Spring Harb. Perspect. Biol.* 2009, **1**: a001891.

Torres, I.L., Lopez-Schier, H., St Johnston, D.: **A Notch/Delta-dependent relay mechanism establishes anterior-posterior polarity in *Drosophila*.** *Dev. Cell* 2005, **5**: 547–558.

Zimyanin, V.L., Belaya, K., Pecreaux, J., Gilchrist, M.J., Clark, A., Davis, I., St Johnston, D.: **In vivo imaging of *oskar* mRNA transport reveals the mechanism of posterior localization.** *Cell* 2008, **134**: 843–853.

Box 2C The JAK-STAT pathway

Arbouzova, N.I., Zeidler, M.P.: **JAK/STAT signalling in Drosophila: insights into conserved regulatory and cellular functions.** *Development* 2006, **133**: 2605–2616.

Hombría, J.C., Sotillos, S.: **JAK-STAT pathway in *Drosophila* morphogenesis: from organ selector to cell behavior regulator.** *JAKSTAT* 2013, **2**: e26089.

2.14 Localization of maternal mRNAs to either end of the egg depends on the reorganization of the oocyte cytoskeleton

Becalska, A.N., Gavis, E.R.: **Lighting up mRNA localization in *Drosophila* oogenesis.** *Development* 2009, **136**: 2493–2503.

Martin, S.G., Leclerc, V., Smith-Litière, K., St Johnston, D.: **The identification of novel genes required for *Drosophila* anteroposterior axis formation in a germline clone screen using GFP-Staufen.** *Development* 2003, **130**: 4201–4215.

St Johnston, D.: **Moving messages: the intracellular localization of mRNAs.** *Nat Rev. Mol. Cell Biol.* 2005, **6**: 363–375.

2.15 The dorso-ventral axis of the egg is specified by movement of the oocyte nucleus followed by signaling between oocyte and follicle cells

González-Reyes, A., Elliott, H., St Johnston, D.: **Polarization of both major body axes in *Drosophila* by gurken-torpedo signaling.** *Nature* 1995, **375**: 654–658.

Jordan, K.C., Clegg, N.J., Blasi, J.A., Morimoto, A.M., Sen, J., Stein, D., McNeill, H., Deng, W.M., Tworoger, M., Ruohola-Baker, H.: **The homeobox gene *mirror* links EGF signalling to embryonic dorso-ventral axis formation through Notch activation.** *Nat. Genet.* 2000, **24**: 429–433.

Roth, S., Neuman-Silberberg, F.S., Barcelo, G., Schupbach, T.: **Cornichon and the EGF receptor signaling process are necessary for both anterior-posterior and dorsal-ventral pattern formation in *Drosophila*.** *Cell* 1995, **81**: 967–978.

van Eeden, F., St Johnston, D.: **The polarisation of the anterior-posterior and dorsal-ventral axes during *Drosophila* oogenesis.** *Curr. Opin. Genet. Dev.* 1999, **9**: 396–404.

2.16 The expression of zygotic genes along the dorso-ventral axis is controlled by Dorsal protein

Cowden, J., Levine, M.: **Ventral dominance governs segmental patterns of gene expression across the dorsal-ventral axis of the neurectoderm in the *Drosophila* embryo.** *Dev. Biol.* 2003, **262**: 335–349.

Harland, R.M.: **A twist on embryonic signalling.** *Nature* 2001, **410**: 423–424.

Markstein, M., Zinzen, R., Markstein, P., Yee, K.P., Erives, A., Stathopoulos, A., Levine, M.: **A regulatory code for neurogenic gene expression in the *Drosophila* embryo.** *Development* 2004, **131**: 2387–2394.

Reeves, G.T., Stathopoulos, A.: **Graded dorsal and differential gene regulation in the Drosophila embryo.** *Cold Spring Harb. Perspect. Biol.* 2009, **1**: a000836.

Reeves, G.T., Trisnadi, N., Truong, T.V., Nahmad, M, Katz, S., Stathopoulos, A.: **Dorsal-ventral gene expression in the *Drosophila* embryo reflects the dynamics and precision of the dorsal nuclear gradient.** *Dev. Cell* 2012, **22**: 544–557.

Rusch, J., Levine, M.: **Threshold responses to the dorsal regulatory gradient and the subdivision of primary tissue territories in the *Drosophila* embryo.** *Curr. Opin. Genet. Dev.* 1996, **6**: 416–423.

Stathopoulos, A., Levine, M.: **Dorsal gradient networks in the *Drosophila* embryo.** *Dev. Biol.* 2002, **246**: 57–67.

2.17 The Decapentaplegic protein acts as a morphogen to pattern the dorsal region

Affolter, M., Basler, K.: **The Decapentaplegic morphogen gradient: from pattern formation to growth regulation.** *Nat. Rev. Genet.* 2007, **8**: 663–674.

Ashe, H.L., Levine, M.: **Local inhibition and long-range enhancement of Dpp signal transduction by Sog.** *Nature* 1999, **398**: 427–431.

Mizutani, C.M., Nie, Q., Wan, F.Y., Zhang, Y.T., Vilmos, P., Sousa-Neves, R., Bier, E., Marsh, J.L., Lander, A.D.: **Formation of the BMP activity gradient in the *Drosophila* embryo.** *Dev. Cell* 2005, **8**: 915–924.

Shimmi, O., Umulis, D., Othmer, H., O'Connor, M.B.: **Facilitated transport of a Dpp/Scw heterodimer by Sog/Tsg leads to robust patterning of the *Drosophila* blastoderm embryo.** *Cell* 2005, **120**: 873–886.

Srinivasan, S., Rashka, K.E., Bier, E.: **Creation of a Sog morphogen gradient in the *Drosophila* embryo.** *Dev. Cell* 2002, **2**: 91–101.

Wang, Y.-C., Ferguson, E.L.: **Spatial bistability of Dpp-receptor interactions driving *Drosophila* dorsal-ventral patterning.** *Nature* 2005, **434**: 229–234.

Wharton, K.A., Ray, R.P., Gelbart, W.M.: **An activity gradient of decapentaplegic is necessary for the specification of dorsal pattern elements in the *Drosophila* embryo.** *Development* 1993, **117**: 807–822.

2.18 The antero-posterior axis is divided up into broad regions by gap-gene expression

Rivera-Pomar, R., Jäckle, H.: **From gradients to stripes in *Drosophila* embryogenesis: filling in the gaps.** *Trends Genet.* 1996, **12**: 478–483.

2.19 Bicoid protein provides a positional signal for the anterior expression of zygotic *hunchback*

Brand, A.H., Perrimon, N.: **Targeted gene expression as a means of altering cell fates and generating dominant phenotypes.** *Development* 1993, **118**: 401–415.

Gibson, M.C.: **Bicoid by the numbers: quantifying a morphogen gradient.** *Cell* 2007, **130**: 14–16.

Gregor, T., Tank, D.W., Wieschaus, E.F., Bialek W.: **Probing the limits to positional information.** *Cell* 2007, **130**: 153–164.

Jaeger, J., Martinez Arias, A.: **Getting the measure of positional information.** *PLoS Biol.* 2009, **7**: e81.

Ochoa-Espinosa, A., Yucel, G., Kaplan, L., Pare, A., Pura, N., Oberstein, A., Papatsenko, D., Small, S.: **The role of binding site cluster strength in Bicoid-dependent patterning in *Drosophila*.** *Proc. Natl Acad. Sci. USA* 2005, **102**: 4960–4965.

Simpson-Brose, M., Treisman, J., Desplan, C.: **Synergy between the Hunchback and Bicoid morphogens is required for anterior patterning in *Drosophila*.** *Cell* 1994, **78**: 855–865.

Struhl, G., Struhl, K., Macdonald, P.M.: **The gradient morphogen Bicoid is a concentration-dependent transcriptional activator.** *Cell* 1989, **57**: 1259–1273.

Box 2D P-element-mediated transformation

Venken, K.J.T., Bellen, H.J.: **Transgenesis upgrades for** *Drosophila* **melanogaster.** *Development* 2007, **134**: 3571–3584.

Box 2E Targeted gene expression and misexpression screening

Duffy, J.B.: **GAL4 system in** *Drosophila*: **a fly geneticist's Swiss knife.** *Genesis* 2002, **34**: 1–15.

Zhong, J., Yedvobnick, B.: **Targeted gain-of-function screening in** *Drosophila* **using GAL4-UAS and random transposon insertions.** *Genet. Res. (Camb)* 2009, **91**: 243–258.

2.20 The gradient in Hunchback protein activates and represses other gap genes

Struhl, G., Johnston, P., Lawrence, P.A.: **Control of** *Drosophila* **body pattern by the hunchback morphogen gradient.** *Cell* 1992, **69**: 237–249.

Wu, X., Vakani, R., Small, S.: **Two distinct mechanisms for differential positioning of gene expression borders involving the** *Drosophila* **gap protein giant.** *Development* 1998, **125**: 3765–3774.

2.21 Parasegments are delimited by expression of pair-rule genes in a periodic pattern

Clyde, D.E., Corado, M.S., Wu, X., Paré, A., Papatsenko, D., Small, S.: **A self-organizing system of repressor gradients establishes segmental complexity in** *Drosophila*. *Nature* 2003, **426**: 849–853.

Hughes, S.C., Krause, H.M.: **Establishment and maintenance of parasegmental compartments.** *Development* 2001, **128**: 1109–1118.

Jaynes, J.B., Fujioka, M.: **Drawing lines in the sand: even skipped et al. and parasegment boundaries.** *Dev. Biol.* 2004, **269**: 609–622.

Yu, D., Small, D.: **Precise registration of gene expression boundaries by a repressive morphogen in** *Drosophila*. *Curr. Biol.* 2008, **18**: 888–876.

2.22 Gap gene activity positions stripes of pair-rule gene expression

Small, S., Levine, M.: **The initiation of pair-rule stripes in the** *Drosophila* **blastoderm.** *Curr. Opin. Genet. Dev.* 1991, **1**: 255–260.

2.23 Some insects use different mechanisms for patterning the body plan

Akam, M., Dawes, R.: **More than one way to slice an egg.** *Curr. Biol.* 1992, **8**: 395–398.

Angelini, D.R., Liu, P.Z., Hughes, C.L., Kaufman, T.C.: **Hox gene function and interaction in the milkweed bug** *Oncopeltus fasciatus* **(Hemiptera).** *Dev. Biol.* 2005, **287**: 440–455.

Brown, S.J., Parrish, J.K., Beeman, R.W., Denell, R.E.: **Molecular characterization and embryonic expression of the** *even-skipped* **ortholog of** *Tribolium castaneum*. *Mech. Dev.* 1997, **61**: 165–173.

French, V.: **Segmentation (and** *eve***) in very odd insect embryos.** *BioEssays* 1996, **18**: 435–438.

French, V.: **Insect segmentation: genes, stripes and segments in 'Hoppers'.** *Curr. Biol.* 2001, **11**: R910–R913.

Lynch, J.A., Brent, A.E., Leaf, D.S., Pultz, M.A., Desplan, C.: **Localized maternal** *orthodenticle* **patterns anterior and posterior in the long germ wasp** *Nasonia*. *Nature* 2006, **439**: 728–732.

Sander, K.: **Pattern formation in the insect embryo.** In *Cell Patterning, Ciba Found. Symp. 29*. London: Ciba Foundation, 1975: 241–263.

Tautz, D., Sommer, R.J.: **Evolution of segmentation genes in insects.** *Trends Genet.* 1995, **11**: 23–27.

2.24 Expression of the *engrailed* **gene defines the boundary of a parasegment which is also a boundary of cell-lineage restriction**

&

2.25 Segmentation genes stabilize parasegment boundaries

&

2.26 Signals generated at the parasegment boundary delimit and pattern the future segments

Alexandre, C., Lecourtois, M., Vincent, J.-P.: **Wingless and hedgehog pattern** *Drosophila* **denticle belts by regulating the production of short-range signals.** *Development* 1999, **126**: 5689–5698.

Chanut-Delalande, H., Fernandes, I., Roch, F., Payre, F., Plaza, S.: *Shavenbaby* **couples patterning to epidermal cell shape control.** *PLoS Biol.* 2006, **4**: e290.

Dahmann, C., Basler, K.: **Compartment boundaries: at the edge of development.** *Trends Genet.* 1999, **15**: 320–326.

Gordon, M., Nusse, R.: **Wnt signaling: multiple pathways: multiple receptors, and multiple transcription factors.** *J. Biol. Chem.* 2006, **281**: 22429–22433.

Hatini, V., DiNardo, S.: **Divide and conquer: pattern formation in** *Drosophila* **embryonic epidermis.** *Trends Genet.* 2001, **17**: 574–579.

Hooper, J.E., Scott, M.P.: **Communicating with Hedgehogs.** *Nat. Rev. Mol. Cell Biol.* 2005, **6**: 306–317.

Larsen, C.W., Hirst, E., Alexandre, C., Vincent, J.-P.: **Segment boundary formation in** *Drosophila* **embryos.** *Development* 2003, **130**: 5625–5635.

Lawrence, P.A., Casal, J., Struhl, G.: **Hedgehog and engrailed: pattern formation and polarity in the Drosophila abdomen.** *Development* 1999, **126**: 2431–2439.

Sanson, B.: **Generating patterns from fields of cells. Examples from** *Drosophila* **segmentation.** *EMBO Rep.* 2001, **2**: 1083–1088.

Tolwinski, N.S., Wieschaus, E.: **Rethinking Wnt signaling.** *Trends Genet.* 2004, **20**: 177–181

von Dassow, G., Meir, E., Munro, E.M., Odell, G.M.: **The segment polarity network is a robust developmental module.** *Nature* 2000, **406**: 188–192.

Vincent, J.P., O'Farrell, P.H.: **The state of** *engrailed* **expression is not clonally transmitted during early** *Drosophila* **development.** *Cell* 1992, **68**: 923–931.

Box 2F The Hedgehog signaling pathway

Briscoe, J., Thérond, P.P.: **The mechanisms of Hedgehog signalling and its roles in development and disease.** *Nat. Rev. Mol. Cell Biol.* 2013, **14**: 416–429.

Box 2G Mutants in denticle pattern provided clues to the logic of segment patterning

Martinez Arias, A., Baker, N.E., Ingham, P.W.: **Role of segment polarity genes in the definition and maintenance of cell states in the *Drosophila* embryo**. *Development* 1988, **103**: 157–170.

Martinez Arias, A.: **A cellular basis for pattern formation in the insect epidermis**. *Trends Genet.* 1989, **5**: 262–267.

2.27 Compartment boundaries persist into the adult fly

&

2.28 Insect epidermal cells become individually polarized in an antero-posterior direction in the plane of the epithelium

&

Box 2I Planar cell polarity in *Drosophila*

Lawrence, P.A., Casal, J., Struhl, G.: **Cell interactions and planar polarity in the abdominal epidermis of *Drosophila***. *Development* 2004, **131**: 4651–4664.

Lawrence, P.A., Struhl, G., Casal, J.: **Planar polarity: one or two pathways**. *Nat. Rev. Genet.* 2008, **8**: 555–363.

Ma, D., Yang, C., McNeill, H., Simon, M.A., Axelrod, J.D.: **Fidelity in planar cell polarity signalling**. *Nature* 2003, **421**: 543–547.

Strutt, D.: **The planar polarity pathway**. *Curr. Biol.* 2008, **16**: R898–R902.

Strutt, D., Strutt, H.: **Differential activities of the core planar polarity proteins during *Drosophila* wing patterning**. *Dev. Biol.* 2007, **302**: 181–194.

Wu, J., Mlodzik, M.: **The Frizzled extracellular domain is a ligand for Van Gogh/Stm during nonautonomous planar cell polarity signaling**. *Dev. Cell* 2008, **15**: 462–469.

2.29 Segment identity in *Drosophila* is specified by Hox genes

Hueber, S.D., Lohmann, I.: **Shaping segments: Hox gene function in the genomic age**. *BioEssays* 2008, **30**: 965–979.

Lewis E.B.: **A gene complex controlling segmentation in *Drosophila***. *Nature* 1978, **276**: 565–570.

Mallo, M., Alonso, C.R.: **The regulation of Hox gene expression during animal development**. *Development* 2013, **140**: 3951–3963.

2.30 Homeotic selector genes of the bithorax complex are responsible for diversification of the posterior segments

&

2.31 The Antennapedia complex controls specification of anterior regions

Castelli-Gair, J., Akam, M.: **How the Hox gene *Ultrabithorax* specifies two different segments: the significance of spatial and temporal regulation within metameres**. *Development* 1995, **121**: 2973–2982.

Lawrence, P.A, Morata, G.: **Homeobox genes: their function in *Drosophila* segmentation and pattern formation**. *Cell* 1994, **78**: 181–189.

Mann, R.S., Morata, G.: **The developmental and molecular biology of genes that subdivide the body of *Drosophila***. *Annu. Rev. Cell Dev. Biol.* 2000, **16**: 143–271.

Mannervik, M.: **Target genes of homeodomain proteins**. *BioEssays* 1999, **4**: 267–270.

Simon, J.: **Locking in stable states of gene expression: transcriptional control during *Drosophila* development**. *Curr. Opin. Cell Biol.* 1995, **7**: 376–385.

2.32 The order of Hox gene expression corresponds to the order of genes along the chromosome

Morata, G.: **Homeotic genes of *Drosophila***. *Curr. Opin. Genet. Dev.* 1993, **3**: 606–614.

2.33 The *Drosophila* head region is specified by genes other than the Hox genes

Rogers B.T., Kaufman, T.C.: **Structure of the insect head in ontogeny and phylogeny: a view from *Drosophila***. *Int. Rev. Cytol.* 1997, **174**: 1–84.

3

Vertebrate development I: life cycles and experimental techniques

- Vertebrate life cycles and outlines of development
- Experimental approaches to studying vertebrate development

In this chapter we examine the similarities and differences in the outlines of development of four vertebrate model animals: amphibians represented by the frog Xenopus, fish by the zebrafish, birds by the chick, and mammals by the mouse. We also briefly consider human development. The second part of the chapter introduces some of the experimental approaches used to investigate vertebrate development using the four models.

We saw in Chapter 2 how early development in insects is largely under the control of maternal factors that interact with each other to specify broadly different regions of the body. This initial plan is then elaborated on by the embryo's own genes. We shall now look at how the same task of establishing the outline of the body plan is achieved in early vertebrate development. All vertebrates, despite their many outward differences, have a similar basic body plan. The defining structures are the segmented backbone or **vertebral column** surrounding the spinal cord, with the brain at the head end enclosed in a bony or cartilaginous skull (Fig. 3.1). These prominent structures mark the **antero-posterior axis**, the main body axis of vertebrates. The head is at the anterior end of this axis, followed by the trunk with its paired appendages—limbs in terrestrial vertebrates (with the exception of snakes and limbless lizards) and fins in fish—and in many vertebrates this axis terminates at the posterior end in a post-anal tail. The vertebrate body also has a distinct **dorso-ventral axis** running from the back to the belly, with the spinal cord running along the dorsal side and the mouth defining the ventral side. The antero-posterior and dorso-ventral axes together define the left and right sides of the animal. Vertebrates have a general **bilateral symmetry** around the dorsal midline so that outwardly the right and left sides are mirror images of each other. Some internal organs, such as lungs, kidneys, and gonads, are present as symmetrically paired structures although their detailed anatomy or position may differ on right and left sides of the body, but single organs such as the heart and liver are arranged asymmetrically with respect to the dorsal midline with the heart on the left and the liver on the right.

Fig. 3.1 The skeleton of a 17.5-day (E17.5) mouse embryo illustrates the vertebrate body plan. The skeletal elements in this embryo have been stained with Alcian blue (which stains cartilage) and Alizarin red (which stains bone). The brain is enclosed in a bony skull. The vertebral column, which develops from blocks of somites, is divided up into cervical (neck), thoracic (chest), lumbar (lower back), and sacral (hip and lower) regions. The paired limbs can also be seen. Scale bar = 1 mm.

Photograph courtesy of M. Maden.

In this and the following two chapters we will look at four vertebrates whose early development has been particularly well studied—*Xenopus*, zebrafish, chick, and mouse. We also touch on the early development of a human embryo. In this chapter, we will first outline the life cycles of these organisms and briefly describe the key features of their early development, providing the context for the later discussions. We will not at this stage consider the mechanisms of development, but focus on the changes in form that occur in early embryonic development to give the well-defined overall **body plan** of the vertebrate embryo, with its characteristic structures such as the notochord and neural tube. We will take the embryo from the earliest cell divisions of the fertilized egg, through the cell movements of gastrulation, the dynamic process in which the embryo is rearranged to form the three germ layers in their correct positions in the body, to the end of **neurulation**, which forms the neural tube—the earliest appearance of the nervous system. The germ layers are the **ectoderm**, which gives rise to the nervous system and epidermis, the **mesoderm**, which gives rise to the skeleton, muscles, heart, blood, and some other internal organs and tissues, and the **endoderm**, which gives rise to the gut and associated glands and organs (see Box 1C). In the second half of the chapter we will introduce some of the experimental techniques that are used to study vertebrate development.

Chapter 4 then looks in detail at the mechanisms that establish the body plan in *Xenopus*. We will consider how the antero-posterior and dorso-ventral axes are determined and how the germ layers are specified and then broadly patterned. We will also look at the setting up of the organizer (see Section 1.4) and how it directs the organization of the body, including the all-important induction of the future nervous system. We then compare the mechanisms involved in establishing the body plan in zebrafish. In Chapter 5, we look at how the same processes are accomplished in chick and mouse and consider how the body plan is completed using chick and mouse as our main models. We will discuss how the further development of the mesoderm is controlled, how the notochord is formed and how the neighboring slabs of mesoderm on either side of the notochord segment into somites, which subsequently develop into muscle and skeletal tissues. We will also look at the origin and development of the neural crest and how this gives rise to a wide range of different tissues, including the peripheral nervous system and connective tissues in the head and finally touch on how left–right asymmetry is established. At the end of all these processes the embryo has become recognizably vertebrate. The development of individual structures and organs, such as limbs, eyes, heart, and the nervous system, is covered in later chapters. We shall take a more detailed look at the mechanics of vertebrate gastrulation and the underlying cell biology that allows such extensive remodeling of tissues in Chapter 9, along with gastrulation in simpler non-vertebrate animals such as the sea urchin.

Vertebrate life cycles and outlines of development

All vertebrate embryos pass through a broadly similar set of developmental stages, which are outlined for each organism in the life-cycle diagrams in this chapter (for example, the one for *Xenopus* in Fig. 3.3). After fertilization, the zygote undergoes **cleavage**. These are rapid cell divisions by which the embryo becomes divided into a number of smaller cells. This is followed by **gastrulation**, a set of cell movements that generates the three germ layers, ectoderm, mesoderm, and endoderm (see Box 1C). By the end of gastrulation, the ectoderm covers the embryo, and the mesoderm and endoderm have moved inside. The endoderm gives rise to the gut, and to its derivatives such as liver and lungs; the mesoderm forms the skeleton, muscles, connective tissues, kidneys, heart, and blood, as well as some other tissues; and the ectoderm gives rise to the epidermis and the nervous system.

Figure 3.2 shows the differences in form of the early embryos from the four model species and of a human embryo at equivalent stages. After gastrulation, all vertebrate embryos pass through a stage at which they all more or less resemble each other and show the specific embryonic features of the chordates, the phylum of animals to which vertebrates belong. This stage is called the **phylotypic stage**. At the phylotypic stage the head is distinct and the **neural tube**, the forerunner of the nervous system, runs along

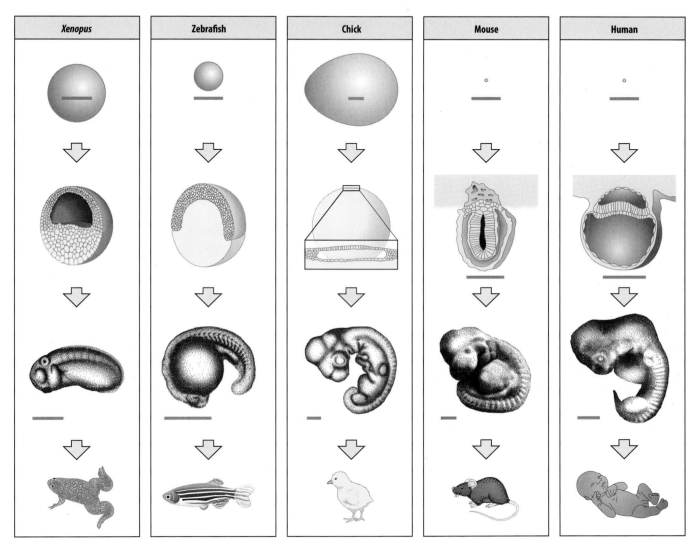

Fig. 3.2 Vertebrate embryos show considerable differences in form before gastrulation but subsequently all go through a stage at which they look similar. The eggs of the frog (*Xenopus*), the zebrafish, the chicken, and the mouse are very different in size, while human eggs are about the same size as mouse eggs (top row). Scale bars in this row all represent 1 mm, except for the chicken egg, which represents 10 mm. Inside the chicken egg, the egg cell contains the 'yolk'. The bulk of the egg is taken up by the egg white. Their early development (second row) is rather different. In this row, the embryos are shown in cross-section at the stage corresponding roughly to the *Xenopus* blastula (left panel) just before gastrulation commences. The main determinant of tissue arrangement is the amount of yolk (yellow) in the egg. The mouse and human embryos (on the right) at this stage have implanted into the uterine wall and have already developed some extra-embryonic tissues required for implantation. The mouse embryo proper is the small cup-shaped structure at the center, seen here in cross-section as a U-shaped layer of epithelium. The human embryo proper is a disc-shaped structure at the center, seen here in cross-section as a flat layer of epithelium. Scale bars for mouse and human embryos = 100 µm. After gastrulation and formation of the neural tube, vertebrate embryos pass through an embryonic stage at which they all look rather similar (third row), and which is known as the phylotypic stage. The body has developed, and neural tube, somites, notochord, and head structures are present. Scale bars = 1 mm. After this stage their development diverges again. Paired appendages, for example, develop into fins in fish, wings and legs in the chick and arms and legs in the human (bottom row).

the dorsal midline of the antero-posterior axis. Immediately under the neural tube runs the **notochord**, a signature structure of chordates, flanked on either side by the **somites**, blocks of tissue from which the muscles and skeleton will derive. In vertebrates, structures that run along this main body axis, such as the notochord, neural tube, somites, and the vertebral column, are known as **axial structures**. Features special to the different vertebrate groups, such as beaks, wings, and fins, appear later.

The rod-shaped notochord is one of the earliest mesodermal structures that can be recognized in vertebrate embryos. This structure does not persist and its cells eventually become incorporated into the vertebral column that forms the spine (see Fig. 3.1). The rest of the vertebral column, the skeleton of the trunk, and the muscles of the trunk and limbs develop from the somites, blocks of cells that form in an antero-posterior sequence from the mesoderm on either side of the notochord. Towards the end of gastrulation, the ectoderm overlying the notochord becomes specified as neural tissue and then rolls up to form the neural tube, from which the brain, spinal cord, and the rest of the nervous system derive (see Fig. 3.7). The overall similarity of the body plan in all vertebrates suggests that the developmental processes that establish it are broadly similar in the different animals. This is largely the case, although there are also considerable differences in development, especially at the earliest stages.

The differences in development in the model organisms particularly relate to how and when the axes are set up, and how the germ layers are established, as we shall see in Chapters 4 and 5, and are mainly due to the different modes of reproduction and the consequent form of the early embryo. Yolk provides all the nutrients for fish, amphibian, reptile, and bird embryonic development, and for the few egg-laying mammals such as the platypus. The eggs of most mammals, by contrast, are small and non-yolky, and the embryo is nourished for the first few days by fluids in the oviduct and uterus. Once implanted in the uterus wall, the mammalian embryo develops specialized **extra-embryonic membranes** which surround and protect the embryo and through which it receives nourishment from the mother via the **placenta**. Avian embryos also develop extra-embryonic membranes for obtaining nutrients from the yolk, for oxygen and carbon dioxide exchange through the permeable eggshell, and for waste disposal. Birds and mammals, both of which form an extra-embryonic membrane called an amnion, are known as **amniotes**, whereas amphibians and fish, which do not form extra-embryonic membranes, are known as **anamniotes**.

To study development, it is necessary to have a reliable way of identifying and referring to a particular stage of development. Simply measuring the time from fertilization is not satisfactory for most species, as there is considerable variation. Amphibians, for example, will develop quite normally over a range of temperatures, but the rate of development is quite different at different temperatures. Developmental biologists, therefore, divide the normal embryonic development of each species into a series of numbered stages, which are identified by their main features rather than by time after fertilization. A stage-10 *Xenopus* embryo, for example, refers to an embryo at a very early stage of gastrulation, whereas stage 54 is the fully developed tadpole with legs. Numbered stages have similarly been characterized for the chick embryo, and these provide a much finer staging system than measuring the time since the egg was laid. Even for mouse embryos, which develop in a much more constant environment inside the mother, there is considerable variation in developmental timing, even among embryos in the same litter. Early mouse embryos are often staged according to days post-conception, with the morning on which the vaginal plug appears called 0.5 dpc, or E0.5 (embryonic day 0.5)—the vaginal plug is formed by hardening of material deposited during mating along with the sperm. The number of days post-conception is however only a guide, and once somites have formed, somite number can be used as a more accurate indication of developmental stage. Links to websites that describe all the stages for our four model vertebrates are given in the General further reading section. The techniques that can be used to study these four vertebrates are discussed in the later part of the chapter.

3.1 The frog *Xenopus laevis* is the model amphibian for studying development of the body plan

Although much of the classical work was carried out on urodele (tailed) amphibians (salamanders), such as newts, the amphibian species now most commonly used for developmental work is the anuran amphibian, the African claw-toed frog *Xenopus laevis*. *Xenopus laevis* is completely aquatic and able to develop normally in tap water but it is a tetraploid species, however, and the diploid species *Xenopus tropicalis* is becoming increasingly used for genetic studies (see Section 1.6). Throughout this book, *Xenopus* will refer to *X. laevis*, unless otherwise specified. A great advantage of *Xenopus* is that its fertilized eggs are easy to obtain; females and males injected with the human hormone chorionic gonadotropin and put together overnight will mate, and the female will lay hundreds of eggs in the water, which are fertilized by sperm released by the male. Eggs can also be fertilized in a dish by adding sperm to eggs released after hormonal stimulation of the female. One advantage of artificial fertilization is that the resulting embryos develop synchronously, so that many embryos, all at the same stage, can be obtained. The eggs are large (1.2–1.4 mm in diameter) and so are quite easily manipulated. The embryos of *Xenopus* are extremely hardy and are highly resistant to infection after microsurgery. It is also easy to culture fragments of early *Xenopus* embryos in a simple, chemically defined solution.

The *Xenopus* life cycle and main developmental stages are summarized in Fig. 3.3. The mature *Xenopus* egg has a distinct polarity, with a dark, pigmented **animal region** and a pale, yolky, and heavier **vegetal region** (Fig. 3.4). The axis running from the animal pole to the vegetal pole is known as the **animal–vegetal axis**. Before fertilization, the egg is enclosed in a protective **vitelline membrane**, which is embedded in a

Fig. 3.3 Life cycle of the African claw-toed frog *Xenopus laevis*. The numbered stages refer to standardized stages of *Xenopus* development. An illustrated list of all stages can be found on the Xenbase website listed in General further reading. The photographs show: an embryo at the blastula stage (top, scale bar = 0.5 mm); a tadpole at stage 41 (middle, scale bar = 1 mm); and an adult frog (bottom, scale bar = 1 cm).

Top photograph courtesy of J. Slack. Middle and bottom photographs courtesy of J. Smith.

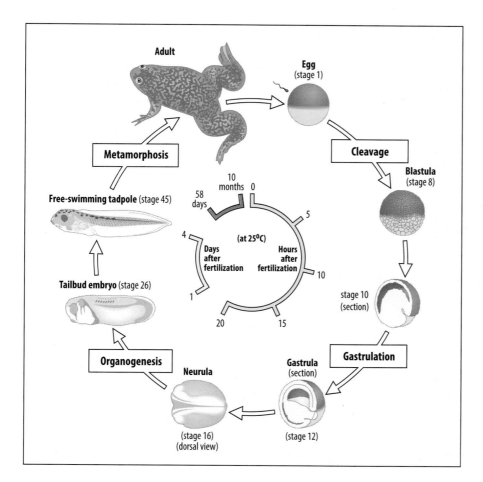

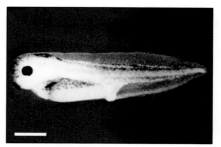

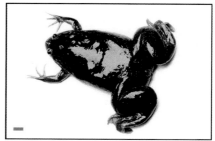

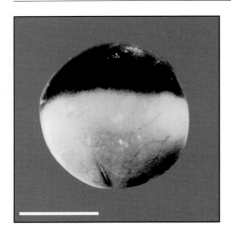

Fig. 3.4 A late-stage *Xenopus* oocyte. The surface of the animal half (top) is pigmented and the paler, vegetal half of the egg is heavy with yolk. Scale bar = 1 mm.

Photograph courtesy of J. Smith.

Fig. 3.5 Cleavage of the *Xenopus* embryo. The upper panels are diagrams showing the fertilized egg and the first three cleavage divisions seen from the side. Polar bodies shown in the fertilized egg and the two-cell stage are attached to the animal pole. The photographs below show the cleaving *Xenopus* egg, but are taken from various angles (see also Fig. 1.14).

Photographs reproduced with permission from Kessel, R.G., Shih, C.Y.: Scanning Electron Microscopy in Biology: A Student's Atlas of Biological Organization. London, Springer-Verlag, 1974. © 1974 Springer-Verlag GmbH & Co.

gelatinous coat. Meiosis is not yet complete: the first meiotic division has resulted in a small cell—a **polar body**—forming at the animal pole, but the second meiotic division is completed only after fertilization, when the second polar body also forms at the animal pole (see Box 9A). At fertilization, one sperm enters the egg in the animal region. The egg completes meiosis and the female pronucleus (from the egg) and the male pronucleus (from the sperm) fuse to form the diploid zygote nucleus.

The first cleavage division of the fertilized egg occurs about 90 minutes after fertilization and is along the plane of the animal–vegetal axis, dividing the embryo into equal left and right halves (Fig. 3.5). Further cleavages follow rapidly at intervals of about 30 minutes. The second cleavage is also along the animal–vegetal axis but at right angles to the first. The third cleavage is equatorial, at right angles to the first two, and divides the embryo into four animal cells and four larger vegetal cells. In *Xenopus* there is no cell growth between cell divisions at this early stage and so continued cleavage results in the formation of smaller and smaller cells; the cells deriving from cleavage divisions in animal embryos are often called **blastomeres**. Cleavage initially occurs synchronously, but because the yolk impedes the cleavage process, division is slowed down in the yolky vegetal half of the embryo, resulting in cells in this region being larger than those in the animal half. Inside this spherical mass of cells a fluid-filled cavity—the **blastocoel**—develops in the animal region, and the embryo is now called a **blastula**.

At the end of blastula formation, the *Xenopus* embryo has gone through about 12 cell divisions and is made up of several thousand cells. Fate maps of this stage (see Section 1.12) show that the **marginal zone**, located around the equator will give rise to the mesoderm, which will develop into internal structures, the vegetal region will give rise to the endoderm, while the animal region will give rise to the ectoderm, which will eventually cover the whole of the embryo (Fig. 3.6, first panel).

The next stage is gastrulation. During this stage there is little cell proliferation; rather, there is extensive movement and rearrangement of the germ layers specified in the blastula so that they become located in their proper positions in the body. Gastrulation involves changes in three dimensions over time and so it can be difficult to visualize. The first external sign of gastrulation is a small slit-like infolding—the **blastopore**—that forms on the surface of the blastula on the future dorsal side (see Fig. 3.6, second panel). The dorsal lip of the blastopore is of particular importance in development, as it is the site of the **embryonic organizer**, known as the **Spemann organizer** in amphibians, without which development of dorsal structures and structures

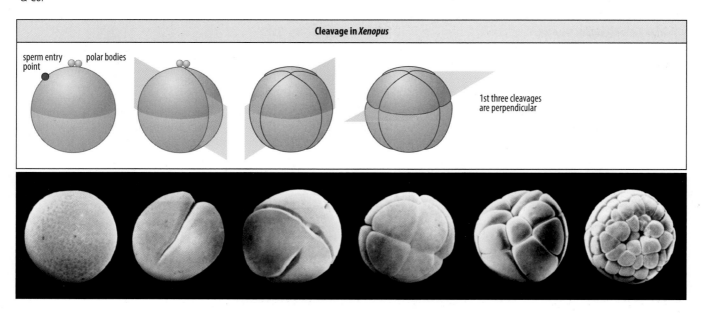

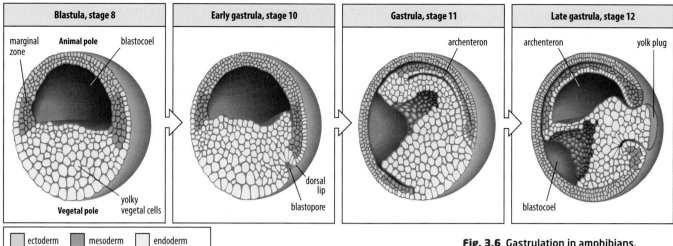

| Blastula, stage 8 | Early gastrula, stage 10 | Gastrula, stage 11 | Late gastrula, stage 12 |

marginal zone   **Animal pole**   blastocoel

yolky vegetal cells

Vegetal pole

dorsal lip

blastopore

archenteron

archenteron   yolk plug

blastocoel

☐ ectoderm   ■ mesoderm   ☐ endoderm

Fig. 3.6 Gastrulation in amphibians.
The blastula (first panel) contains several thousand cells and there is a fluid-filled cavity, the blastocoel, beneath the cells at the animal pole. Gastrulation begins (second panel) at the blastopore, which forms on the dorsal side of the embryo. Future mesoderm and endoderm of the marginal zone move inside at this site through the dorsal lip of the blastopore, the mesoderm ending up sandwiched between the endoderm and ectoderm in the animal region (third panel). The tissue movements create a new internal cavity—the archenteron—that will become the gut. Future endoderm in the ventral region also moves inside through the ventral lip of the blastopore (fourth panel) and will eventually completely line the archenteron. At the end of gastrulation the blastocoel is considerably reduced in size.

Illustration after Balinsky, B.I.: An Introduction to Embryology. *4th edition. Philadelphia, W.B. Saunders, 1975.*

along the antero-posterior axis will not occur. The famous experiment by Hans Spemann and Hilde Mangold that led to the discovery of the organizer is described in Fig. 1.9. Once gastrulation has started, the embryo is known as a **gastrula**. In *Xenopus*, the cells that will give rise to the endoderm and mesoderm in the marginal zone move inside the gastrula through the blastopore by rolling under the lip as coherent sheets of cells. This type of inward movement is known as **involution**. Once inside, the tissues converge towards the midline and extend along the antero-posterior axis beneath the dorsal ectoderm, elongating the embryo along the antero-posterior axis. At the same time, the ectoderm spreads downward to cover the whole embryo by a process known as **epiboly**. The involuting layer of dorsal endoderm is closely applied to the mesoderm, and the space between it and the yolky vegetal cells is known as the **archenteron** (see Fig. 3.6, third panel). This is the precursor of the gut cavity.

The inward movement of endoderm and mesoderm begins dorsally and spreads laterally and then ventrally to form a complete circle of involuting cells around the blastopore (see Fig. 3.6, fourth panel). By the end of gastrulation, the blastopore has almost closed—the remaining slit will form the anus. The dorsal mesoderm now lies beneath the dorsal ectoderm, the lateral and ventral mesoderm is in its definitive position in the body plan, and the ectoderm has spread to cover the entire embryo. There is still a large amount of yolk present, which provides nutrients until the larva—the tadpole—starts feeding. During gastrulation, the mesoderm in the dorsal region starts to develop into the notochord and the somites, while the more lateral mesoderm immediately adjacent to this will later form the **intermediate mesoderm**, from which the kidneys develop, and more laterally still is the **lateral plate mesoderm,** with the anterior lateral plate mesoderm giving rise to the heart. The cell and tissue movements that occur during gastrulation in *Xenopus* and other animals are discussed in more detail in Chapter 9.

Gastrulation is succeeded by **neurulation**—the formation of the neural tube, which is the early embryonic precursor of the central nervous system. At this stage the embryo is called a **neurula**. The earliest visible sign of neurulation is the formation of the **neural folds**, which form on the edges of the **neural plate,** an area of thickened ectoderm overlying the notochord. The folds rise up, fold towards the midline and fuse together to form the neural tube, which sinks beneath the epidermis (Fig. 3.7). **Neural crest cells** detach from the top of the neural tube on either side of the site of fusion and migrate throughout the body to form various structures, as we describe in Chapter 5. The anterior neural tube gives rise to the brain; further back, the neural tube overlying the notochord will develop into the spinal cord.

The embryo now begins to look something like a tadpole and we can recognize the main vertebrate features (Fig. 3.8). At the anterior end, the brain is already divided up into a number of regions, and the eyes and ears have begun to develop. There are also three **branchial arches** on each side, the most anterior of which will form the jaws. More posteriorly, the somites and notochord are well developed. In *Xenopus*, the mouth breaks through at stage 40, about 2.5 days after fertilization. The post-anal

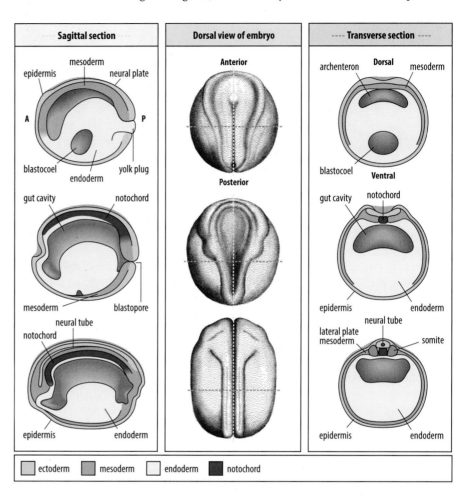

Fig. 3.7 Neurulation in amphibians. Top row: the neural plate develops neural folds just as the notochord begins to form in the midline (see middle row). Middle and bottom rows: the neural folds come together in the midline to form the neural tube, from which the brain and spinal cord will develop. During neurulation, the embryo elongates along the antero-posterior axis. The left panel shows sections through the embryo in the planes indicated by the yellow dashed lines in the center panel. The center panel shows dorsal surface views of the amphibian embryo. The right panel shows sections through the embryo in the planes indicated by the green dashed lines in the center panel. This diagram shows neurulation in a urodele amphibian embryo rather than *Xenopus*, as the neural folds in urodele embryos are more clearly defined.

Fig. 3.8 The early tailbud stage (stage 26) of a *Xenopus* embryo. At the anterior end, in the head region, the future eye is prominent and the ear vesicle (otic vesicle), which will develop into the ear, has formed. The brain is divided into forebrain, midbrain, and hindbrain. Just posterior to the site at which the mouth will form are the branchial arches, the first of which will form the lower jaw. More posteriorly, a succession of somites lies on either side of the notochord (which is stained brown in the photograph). The embryonic kidney (pronephros) is beginning to form from lateral mesoderm. Ventral to these structures is the gut (not visible in this picture). The tailbud will give rise to the tail of the tadpole, forming a continuation of somites, neural tube, and notochord. Scale bar = 1 mm.

Photograph courtesy of B. Herrmann.

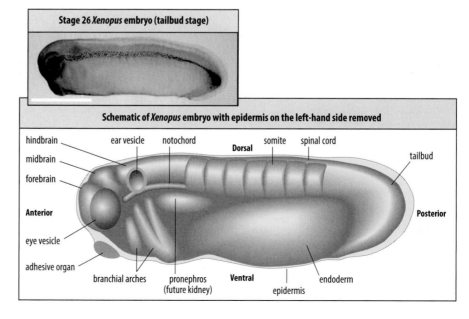

tail of the tadpole is formed last. It develops from the **tailbud**, which gives rise to a continuation of notochord, somites, and neural tube in the tail. Further development gives rise to various organs and tissues such as blood and heart, kidneys, lungs, and liver. After organ formation, or **organogenesis**, is completed, the tadpole hatches out of its jelly covering and begins to swim and feed. Later, the tadpole larva will undergo metamorphosis to produce the adult frog; the tail regresses and the limbs grow (metamorphosis is discussed in Chapter 13).

3.2 The zebrafish embryo develops around a large mass of yolk

The zebrafish has two great advantages as a vertebrate model for development: its short life-cycle of approximately 12 weeks makes genetic analysis, including large-scale genetic screening, relatively easy (see Box 3C), and the transparency of the embryo allows the fate and movements of individual cells during development to be observed. Because of the feasibility of genetic analysis in the zebrafish, it has turned out to be a useful model organism for studying some human diseases caused by genetic defects, in particular certain blood and cardiovascular disorders. The life cycle is shown in Fig. 3.9. The zebrafish egg is about 0.7 mm in diameter, with a clear animal–vegetal axis: the cytoplasm and nucleus at the animal pole sit upon a large mass of yolk. After fertilization, the zygote undergoes cleavage, but cleavage does not extend into the yolk and results in a mound of blastomeres perched above a large yolk cell. The first five cleavages are all vertical, and the first horizontal cleavage gives rise to the 64-cell stage about 2 hours after fertilization (Fig. 3.10).

Further cleavage leads to the **sphere stage**, in which the embryo is now in the form of a **blastoderm** of around 1000 cells lying on top of the yolk cell. The hemispherical

Fig. 3.9 Life cycle of the zebrafish. The zebrafish embryo develops as a mound-shaped blastoderm sitting on top of a large yolk cell. It develops rapidly and by 2 days after fertilization the tiny fish, still attached to the remains of its yolk, hatches out of the egg. The top photograph shows a zebrafish embryo at the sphere stage of development, with the embryo sitting on top of the large yolk cell (scale bar = 0.5 mm). The middle photograph shows an embryo at the 14-somite stage, showing the segmental trunk muscle that has developed. Its transparency is useful for observing cell behavior (scale bar = 0.5 mm). The bottom photograph shows an adult zebrafish (scale bar = 1 cm). An illustrated list of the numbered stages in zebrafish development can be found on the website listed in the General further reading section.

*Photographs courtesy of C. Kimmel (top, from Kimmel, C. B., et al.: **Stages of embryonic development of the zebrafish**. Dev. Dyn. 1995, **203**: 253–310), N. Holder (middle), and M. Westerfield (bottom).*

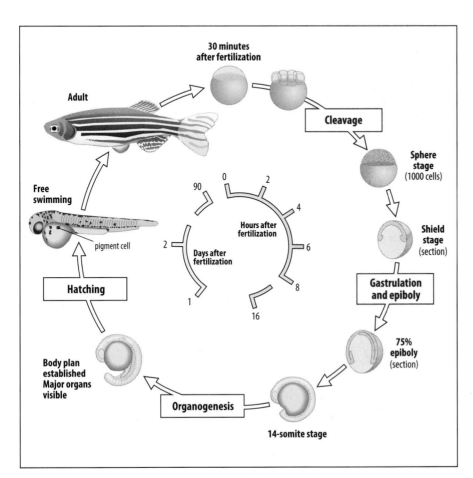

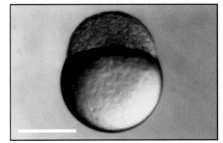

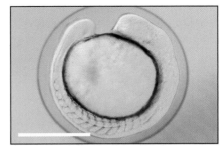

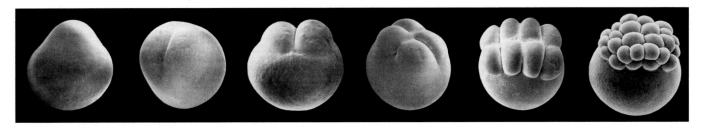

Fig. 3.10 Cleavage of the zebrafish embryo is initially confined to the animal (top) half of the embryo.

Photographs reproduced with permission from Kessel, R.G., Shih, C.Y.: Scanning Electron Microscopy in Biology: A Student's Atlas of Biological Organization. London, Springer-Verlag, 1974. © 1974 Springer-Verlag GmbH & Co.

Fig. 3.11 Epiboly and gastrulation in the zebrafish. At the end of the first stages of cleavage, around 4.3 hours after fertilization, the zebrafish embryo is composed of a mound of blastomeres sitting on top of the yolk, separated from the yolk by a multinucleate layer of cytoplasm called the yolk syncytial layer (first panel). With further cleavage and spreading out of the layers of cells (epiboly), the upper half of the yolk becomes covered by a blastoderm with a thickened edge, the germ ring, and a shield-shaped region is visible in the blastoderm on the dorsal side (second panel). Gastrulation occurs by involution of cells in a ring around the edge of the blastoderm (third panel). The involuting cells converge on the dorsal midline to form the body of the embryo encircling the yolk (fourth panel).

blastoderm has an outer layer of flattened cells, one cell thick, known as the **outer enveloping layer**, and a **deep layer** of more rounded cells from which the embryo develops (Fig. 3.11). During the early blastoderm stage, blastomeres at the margin of the blastoderm collapse into the yolk cell, forming a continuous layer of multinucleated non-yolky cytoplasm underlying the blastoderm that is called the **yolk syncytial layer**. The blastoderm, together with the yolk syncytial layer, spreads in a vegetal direction by epiboly and eventually covers the yolk cell.

Although the fish blastoderm and the amphibian blastula are different in shape, they are corresponding stages in development. In the fish, the endoderm is derived from the deep cells that lie right at the blastoderm margin. This narrow marginal layer is often known as **mesendoderm** because the descendants of a single cell in this region can give rise to either endoderm or mesoderm. Deep cells located between four and six cell diameters away from the margin give rise exclusively to mesoderm. This overlap between prospective endoderm and mesoderm in the zebrafish differs somewhat from the arrangement in *Xenopus* in which the endoderm and mesoderm occupy more distinct locations. The ectoderm in the fish embryo, as in *Xenopus*, comes from cells in the animal region of the blastoderm.

By about 5.5 hours after fertilization, the blastoderm has spread halfway to the vegetal pole, and the deep-layer cells accumulate to form a thickening around the blastoderm edge known as the embryonic **germ ring** (Fig. 3.11, second panel). At the same time, deep-layer cells within the germ ring converge towards the dorsal side of the embryo, eventually forming a compact thickened shield-shaped region in the germ ring on the dorsal side. The shield becomes visible at about 6 hours after fertilization, and this stage is known as the **shield stage**. The shield region is analogous to the Spemann organizer of *Xenopus*. Gastrulation ensues, and the mesendodermal cells and mesoderm cells roll under the blastoderm margin in the process of involution and move into the interior under the ectoderm. The inward movement of cells occurs all around the periphery of the blastoderm at about the same time, with the earliest cells that involute predominantly from the dorsal and lateral margins of the blastoderm

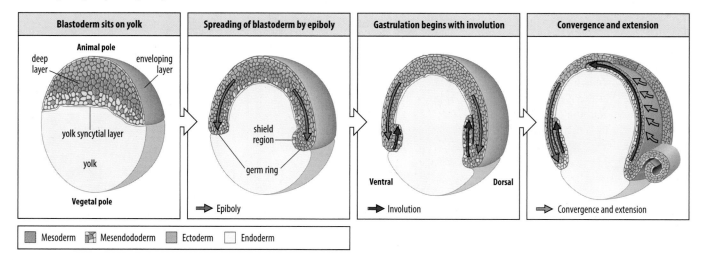

| Blastoderm sits on yolk | Spreading of blastoderm by epiboly | Gastrulation begins with involution | Convergence and extension |

Animal pole — deep layer — enveloping layer — yolk syncytial layer — yolk — Vegetal pole

shield region — germ ring

→ Epiboly

Ventral — Dorsal

→ Involution

→ Convergence and extension

☐ Mesoderm ☐ Mesendododerm ☐ Ectoderm ☐ Endoderm

forming the endoderm. As the mesendoderm is becoming internalized, the ectoderm continues to undergo epiboly, spreading in the vegetal direction until it covers the whole embryo, including the yolk (discussed in more detail in Chapter 9).

Once internalized into the gastrulating embryo, the mesendodermal cells migrate under the ectoderm towards the animal pole, the tissue converging to form the main axis of the embryo, and extending and elongating the embryo in an antero-posterior direction, as in *Xenopus*. The future mesodermal and endodermal cells are now beneath the ectoderm, and by the time that the blastoderm has spread about three-quarters of the way to the vegetal pole, an inner layer of endoderm cells has formed in the embryo next to the yolk, with the more superficial cells becoming mesoderm. By 9 hours, the notochord has become distinct in the dorsal midline of the embryo, and involution of cells around the blastoderm margin is complete by 10 hours. Somite formation, neurulation, and migration of neural crest cells then follow.

Over the next 12 hours the embryo elongates further, and rudiments of the organ systems become recognizable. Somites first appear at the anterior end at about 10.5 hours, and new ones are formed at intervals of initially 20 minutes and then 30 minutes; by 18 hours, 18 somites are present. Formation of the gut in zebrafish differs in several respects from that of *Xenopus*, chick, or mouse. In zebrafish, the gut starts to develop at a relatively late stage in gastrulation—at the 18-somite stage—and is formed by reorganization of cells within the internalized layer of endoderm to form a tubular structure.

Neurulation in zebrafish begins, as in *Xenopus*, towards the end of gastrulation with the formation of the neural plate, an area of columnar ectoderm overlying the notochord. But unlike *Xenopus*, the whole of the neural plate in the zebrafish first forms a solid rod of cells that later becomes hollowed out internally to form the neural tube. The nervous system develops rapidly. Optic vesicles, which give rise to the eyes, can be distinguished at 12 hours as bulges from the brain, and by 18 hours the body starts to twitch. At 48 hours the embryo hatches, and the young fish begins to swim and feed.

Scan here

Scan this QR code image with your mobile device to see an online movie showing epiboly in zebrafish or log on to **http://global.oup.com/uk/orc/biosciences/ devbiol/wolpert5e/qr/qr3a/**

3.3 Birds and mammals resemble each other and differ from *Xenopus* in some important features of early development

Before we describe the course of development in chick and mouse embryos separately, it is worth emphasizing some features of their development that differ from *Xenopus*. The first is the shape of the early embryo. The avian or mammalian structure corresponding to the spherical amphibian blastula just before gastrulation is not a hollow blastula but a layer of epithelium called the **epiblast**. The second difference is that at this stage there are no distinct contiguous regions of the epiblast that correspond to the future ectoderm, endoderm, and mesoderm. As we shall discuss in more detail in later chapters, the timing of specification of the germ layers is somewhat different from *Xenopus*. There is also considerable cell proliferation in the epiblast in both chick and mouse before and during gastrulation, which causes cell mixing at this stage, while there is little cell proliferation during gastrulation in *Xenopus*. The third difference is that the gastrulation process that leads to internalization of endoderm and mesoderm and the organization of the germ layers is rather different in appearance from that in *Xenopus*. Internalization of cells occurs along a straight furrow, the **primitive streak**, rather than through a circular blastopore.

The primitive streak in mouse and chick is equivalent to the amphibian blastopore. The primitive streak is most easily visualized in the flat epiblast of the chick embryo and can be seen in the top photograph in Fig. 3.14. At gastrulation, epiblast cells converge on the primitive streak and pass through it as individual cells, spreading out underneath the surface and forming a bottom layer of endoderm and a middle layer of mesoderm (Fig. 3.12). Cells become specified as endoderm and mesoderm during

Fig. 3.12 Ingression of mesoderm and endoderm during gastrulation in the chick embryo. During gastrulation, future mesodermal and endodermal cells migrate from the epiblast through the primitive streak into the interior of the blastoderm. An aggregation of cells, known as Hensen's node, forms at the anterior end of the streak. As the streak extends, cells of the epiblast move toward the primitive streak (arrows), move through it, and then outward again underneath the surface to give rise internally to the mesoderm and endoderm, the latter displacing a lower layer of cells called the endoblast. Cells that remain in the epiblast form the ectoderm.

Adapted from Balinsky B. I., et al. 1975. An introduction to embryology, 4th edn. W.B. Saunders.

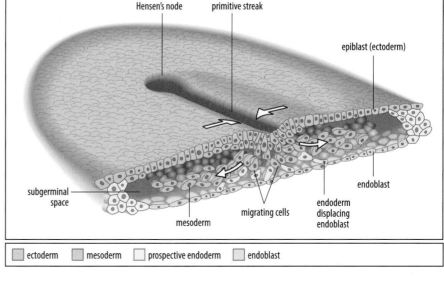

	ectoderm		mesoderm		prospective endoderm		endoblast

their passage through the streak, with the cells remaining on the surface becoming ectoderm. Because of the essentially sheet-like form of the avian or mammalian epiblast at this stage, this initial phase of gastrulation does not directly form a gut cavity in the same way as in the spherical *Xenopus* gastrula. The gut is formed later, by a folding together of the lateral edges of the embryo, which results in a gut cavity entirely surrounded by layers of endoderm, mesoderm, and ectoderm. We shall now return to the chick to look at the course of its development in more detail.

3.4 The early chicken embryo develops as a flat disc of cells overlying a massive yolk

Avian embryos are similar to those of mammals in the morphological complexity of the embryo and the general course of embryonic development, but are easier to obtain and observe. Many observations can be carried out simply by making a window in the egg shell, and the embryo can also be cultured outside the egg. These features make the chick embryo particularly convenient for experimental microsurgical manipulations, tracing cell lineages by injecting marker dyes, and investigations of the effects of introduced genes and other treatments (see Section 3.8). Despite considerable differences in shape between chick and mouse embryos at an early stage of development just prior to gastrulation (see Fig. 3.2), gastrulation and later development are very similar in both, and studies of chick embryos complement those of mouse. The chick's flat early embryo is similar in shape to that of a human embryo (see Fig. 3.2) and the topology of early human embryonic development is therefore more easily compared with that of the chick than with the mouse, which has a cup-shaped embryo.

The large yolky egg cell is fertilized and begins to undergo cleavage while still in the hen's oviduct. Because of the massive yolk, cleavage is confined to a small patch of cytoplasm several millimeters in diameter, which contains the nucleus and lies on top of the yolk. Cleavage in the oviduct results in the formation of a disc of cells called the **blastoderm** or **blastodisc**. During the 20-hour passage down the oviduct, the egg becomes surrounded by extracellular albumen (egg white), the shell membranes, and the shell (Fig. 3.13). At the time of laying, the blastoderm, which is analogous to the early amphibian blastula, is composed of some 32,000–60,000 cells. The chick life cycle is shown in Fig. 3.14.

The early cleavage furrows extend downward from the surface of the cytoplasm but do not completely separate the cells, whose ventral faces initially remain open to the

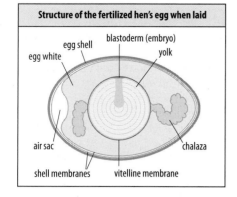

Fig. 3.13 The structure of a hen's egg at the time of laying. Cleavage begins after fertilization while the egg is still in the oviduct. The albumen (egg white) and shell are added during the egg's passage down the oviduct. At the time of laying, the embryo is a disc-shaped cellular blastoderm lying on top of a massive yolk, which is surrounded by the egg white and shell. The twisted chalazae are thought to act as balancers to support the yolk.

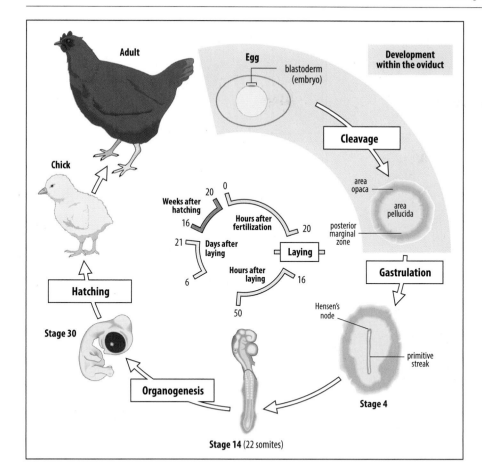

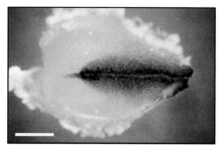

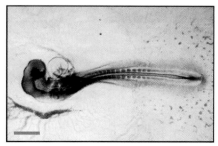

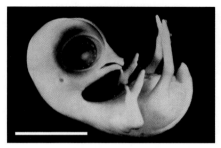

yolk. The central region of the blastoderm, under which a cavity called the **subgerminal space** develops, is translucent and is known as the **area pellucida**, in contrast to the outer region, which is the darker **area opaca** (Fig. 3.15). A layer of cells called the **hypoblast** develops over the yolk to form the floor of the cavity. The hypoblast eventually gives rise to extra-embryonic structures such as those that connect the embryo to its source of nutrients in the yolk. The embryo proper is formed from the remaining upper layer of the blastoderm, known as the epiblast.

The first morphological structure that presages the antero-posterior polarity of the chick embryo is a crescent-shaped ridge of small cells called **Koller's sickle**, located at the boundary between the area opaca and area pellucida at the posterior end of the embryo. Koller's sickle defines the position in which the streak will form, and the region of the epiblast immediately adjacent to the sickle is known as the **posterior marginal zone**. The streak is first visible as a denser region that then gradually extends as a narrow stripe to just over half way across the area pellucida, eventually forming a furrow in the dorsal face of the epiblast (see Fig. 3.12).

Unlike *Xenopus*, cell proliferation and growth of the chick embryo continues throughout gastrulation. Epiblast cells converge on the primitive streak, and as the streak moves forward from the posterior marginal zone, cells in the furrow move inward and spread out anteriorly and laterally beneath the upper layer, forming a layer of loosely connected cells, or **mesenchyme**, in the subgerminal space (see Figs 3.12 and 3.15). The primitive streak is thus similar to the blastopore region of amphibians, but cells move inwards individually, rather than as a coherent sheet. This type of inward movement is called **ingression**. The ingressing cells give rise to mesoderm and endoderm, whereas the cells that remain on the surface of the epiblast give rise to the ectoderm.

Fig. 3.14 Life cycle of the chicken. The egg is fertilized in the hen and by the time it is laid cleavage is complete and a cellular blastoderm lies on the yolk. The start of gastrulation is marked by the appearance of the primitive streak at the posterior marginal zone of the area pellucida. Hensen's node develops at the anterior end of the streak, which then starts regress, accompanied by somite formation. The photographs show: the primitive streak (stained brown by staining with antibody against Brachyury protein) surrounded by the area pellucida (top, scale bar = 1 mm); a stage 14 embryo (50–53 hours after laying) with 22 somites (the head region is well defined and the transparent organ adjacent to it is the ventricular loop of the heart; middle, scale bar = 1 mm); a stage-35 embryo, about 8.5–9 days after laying, with a well-developed eye and beak (bottom, scale bar = 10 mm). A descriptive list of the numbered stages (Hamilton and Hamburger stages) in chick development can be found on the website listed in the General further reading section.

*Top photograph courtesy of B. Herrmann reproduced with permission from Kispert, A., et al.: **The chick brachyury gene: developmental expression pattern and response to axial induction by localized activin.** Dev. Biol. 1995, **168**: 406–415.*

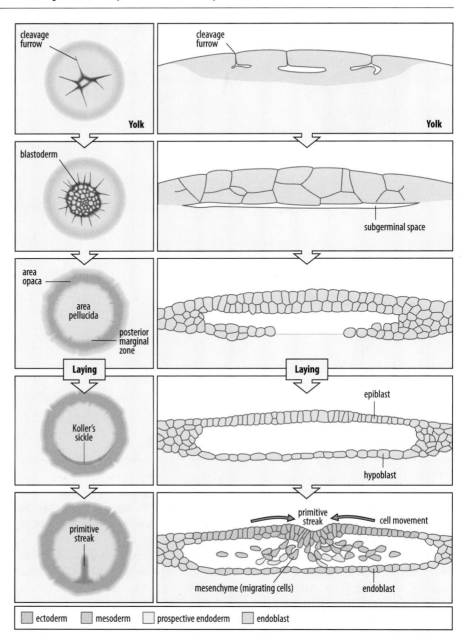

Fig. 3.15 Cleavage and epiblast formation in the chick embryo. By the time the egg is laid, cleavage has divided the small area of egg cytoplasm free from yolk into a disc-shaped cellular blastoderm. The panels on the left show the view of the embryo from above, while the panels on the right show cross-sections through the embryo. The first cleavage furrows extend downward from the surface of the egg cytoplasm and initially do not separate the blastoderm completely from the yolk. In the cellular blastoderm the central area overlying the subgerminal space is called the area pellucida and the marginal region is called the area opaca. A layer of cells develops immediately overlying the yolk and is known as the hypoblast. This will give rise to extra-embryonic structures, while the upper layers of the blastoderm—the epiblast—give rise to the embryo. The chick primitive streak starts to form when the hypoblast becomes displaced forward from the posterior marginal zone by a new layer of cells called the endoblast, which grows out from the zone. The cells of the epiblast move toward the primitive streak (green arrows), move through it, and then outward again underneath the surface to give rise to the mesoderm and endoderm internally, the latter displacing the endoblast.

The primitive streak in the chick embryo is fully extended by 16 hours after laying. At the anterior end of the streak there is a condensation of cells known as **Hensen's node**, where cells are also moving inwards. Hensen's node is the major organizing center for the early chick embryo, equivalent to the Spemann organizer in amphibians.

After the primitive streak has elongated to its full length, some cells from Hensen's node start to migrate forwards along the midline under the epiblast, to give rise to the **prechordal plate mesoderm** and the **head process**. The head process is the emerging rod-like notochord, and the prechordal plate mesoderm is a looser mass of cells anterior to it. After the head process forms, the primitive streak begins to regress, with Hensen's node moving back towards the posterior end of the embryo (Fig. 3.16). As the node regresses, the notochord, continuous with the head process, is laid down in its wake along the dorsal midline of the trunk of the embryo. As the notochord is laid down, the mesoderm immediately on each side of it begins to form the somites. Somite formation and body elongation is discussed in detail in Chapter 5. The first pair of somites is formed at about 24 hours after laying, and new ones are formed

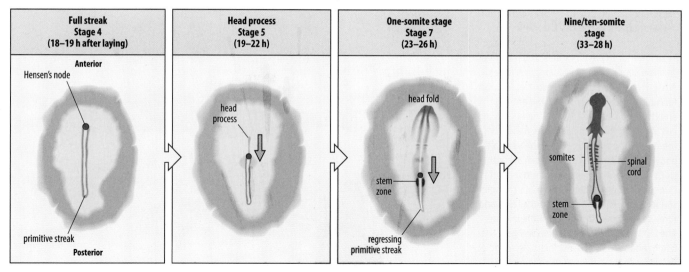

| Full streak
Stage 4
(18–19 h after laying) | Head process
Stage 5
(19–22 h) | One-somite stage
Stage 7
(23–26 h) | Nine/ten-somite stage
(33–28 h) |

Fig. 3.16 Regression of Hensen's node.
After the primitive streak has extended about halfway across the blastoderm, Hensen's node appears at its anterior end. The primitive streak then begins to regress, with Hensen's node moving in a posterior direction as the head fold and neural plate begin to form anterior to it. As the node moves backward, the notochord develops in the area anterior to it. The somites begin to form on either side of the notochord and the spinal cord is progressively laid down in the overlying ectoderm. The medial part of the somites and the spinal cord are formed progressively from a stem zone in the epiblast on either side of the streak immediately posterior to the node.

progressively at intervals of 90 minutes. The mesoderm immediately lateral to the somites is the intermediate mesoderm, which will develop into the kidneys, and the remaining mesoderm, still more laterally—the lateral plate mesoderm—will develop into the vascular system and blood (Fig. 3.17). The medial regions of the somites together with the trunk spinal cord is produced progressively from a region of stem-cell-like cells, the **stem zone**, as the embryo elongates. The stem zone is located in the epiblast on either side of the streak immediately posterior to the node. The remnants of the node and stem zone later contribute to the **tailbud**, which gives rise to the post-anal tail.

As the notochord emerges, neural tissue begins to develop, first as the neural plate, which is first evident as a thickened region of ectoderm above the notochord. The neural plate then folds upwards and inwards so that its sides come together and eventually fuse in the dorsal midline to form the neural tube, initially leaving the anterior and posterior ends open. The fused neural tube becomes covered over by epidermis; Fig. 3.18 shows a section through the chick embryo showing the fused neural tube and the notochord beneath it.

Just after the head process is formed and Hensen's node starts to regress, all three germ layers in the head region start to fold ventrally, to begin to generate the **head fold** (see Fig. 3.16, third image from left). This delimits a pouch lined by endoderm, from which the pharynx and foregut will arise (see top of Fig. 3.17). Eventually, a similar fold appears at the tail end to define the hindgut. In addition, the sides of the embryo fold together to form the rest of the gut, which only remains open in the middle (umbilical region) and subsequently the mesoderm and ectoderm grow over to form the ventral body wall. This key morphogenetic event is known as **ventral closure**. As this folding takes place, the two heart rudiments that start out on either side come together at the midline to form one organ lying ventral to the gut (discussed in Chapter 11). By just over 2 days after laying, the embryo has reached the 22-somite stage, the head is well-developed, optic vesicles and auditory pits are present, and the heart and blood vessels have formed (see Fig. 3.14, middle panel). Blood vessels and blood islands, where the first blood cells are being formed, have developed in the extra-embryonic tissues; the vessels connect up with those of the embryo to provide a circulation with a beating heart. At this stage the embryo starts to turn on its side; the head develops a flexure so that the right eye comes to lie uppermost towards the shell.

By 3–3.5 days after laying, 40 somites have formed, the head is now much more developed with prominent eyes, and the limbs are beginning to develop (Fig. 3.19). At 4 days after laying, extra-embryonic membranes have developed through which the embryo gets its nourishment from the yolk and which also provide protection (Fig. 3.20). The

Fig. 3.17 Development of the neural tube and mesoderm in the chick embryo. Once the notochord has formed, neurulation begins in an anterior to posterior direction. The figure shows a series of sections along the antero-posterior axis of a chick embryo. Neural-tube formation is well advanced at the anterior end (top two sections), where the head fold has already separated the future head from the rest of the blastoderm and the ventral body fold has brought endoderm from both sides of the body together to form the gut. During neurulation, the neural plate changes shape: neural folds rise up on either side and form a tube when they meet in the midline. The mesenchyme in this region will give rise to head structures. Further back (middle sections), in the future trunk region of the embryo, notochord and somites have formed and neurulation is starting. At the posterior end, behind Hensen's node (bottom section), notochord formation, somite formation, and neurulation have not yet begun. The mesoderm internalized through the primitive streak starts to form structures appropriate to its position along the antero-posterior and dorso-ventral axes. For example, in the future trunk region, the intermediate mesoderm will form the mesodermal parts of the kidney, and the anterior splanchnic mesoderm will give rise to the heart. The body fold will continue down the length of the embryo, forming the gut and also bringing paired organ rudiments that initially form on each side of the midline (e.g. those of the heart and dorsal aorta) together to form the final organs lying ventral to the gut. Blood islands, from which the first blood cells are produced, form from the ventral-most part of the lateral mesoderm.

Illustration after Patten, B.M.: Early Embryology of the Chick. New York, Mc Graw-Hill, 1971.

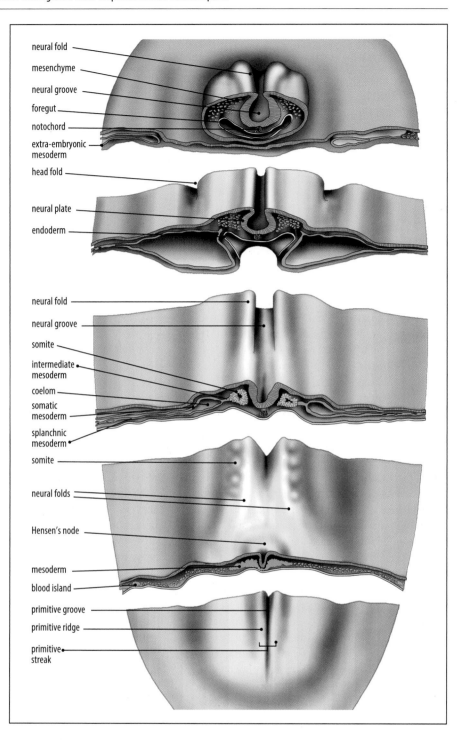

Fig. 3.18 Scanning electron micrograph of chick showing early somites and neural tube. Blocks of somites can be seen adjacent to the neural tube, with the notochord lying beneath it. The lateral plate mesoderm flanks the somites. Scale bar = 0.1 mm.

Photograph courtesy of J. Wilting.

amnion surrounds a fluid-filled amniotic sac in which the embryo lies and which provides mechanical protection; the **chorion** lies outside the amnion just beneath the shell; the **allantois** receives excretory products and provides the site of oxygen and carbon dioxide exchange; and the **yolk sac** surrounds the yolk. By about 9 days after laying, the embryo is very well developed; the wings, legs, and beak are now formed (see Fig 3.14, bottom panel). In the remaining time before hatching, the embryo grows in size, the internal organs become fully developed, and down feathers grow on the wings and body. The chick hatches 21 days after the egg is laid. A picture of a live quail embryo inside the egg obtained by magnetic resonance imaging can be seen in Fig. 3.31.

3.5 The mouse egg has no yolk and early development involves the allocation of cells to form the placenta and extra-embryonic membranes

The mouse has a life cycle of 9 weeks from fertilization to mature adult (Fig. 3.21), which is relatively short for a mammal. This facilitates genetic analysis, and is one reason that the mouse has become the main model organism for mammalian development. The mouse was the first mammal after humans to have its complete genome sequenced. As in all our model organisms, knowledge of the genome sequence enables profiles of all the genes expressed at a particular stage in development to be determined (Box 3B). The mouse is particularly useful to study gene function because it is relatively easy to produce mice with a particular genetic constitution using **transgenic** techniques (see Section 3.10), and mice in which particular genes are rendered completely inactive or 'knocked out'.

Mouse eggs are much smaller than those of either chick or *Xenopus*, about 80–100 μm in diameter, and they contain no yolk. The unfertilized egg is shed from the ovary into the oviduct and is surrounded by a protective external coat, the **zona pellucida**, which is composed of mucopolysaccharides and glycoproteins. Fertilization takes place internally in the oviduct; meiosis is then completed and the second polar body forms (see Box 10A). Cleavage starts while the fertilized egg is still in the oviduct. Early cleavages are very slow compared with *Xenopus* and chick, the first occurring about 24 hours after fertilization, the second about 20 hours later, and subsequent cleavages at about 12-hour intervals. Cleavage produces a solid ball of cells, or blastomeres, called a **morula** (Fig. 3.22). At the eight-cell stage the blastomeres increase the area of cell surface in contact with each other in a process called **compaction**. After compaction, the cells are polarized; their exterior surfaces carry microvilli, whereas their inner surfaces are smooth. Further cleavages are somewhat variable and are both radial and tangential, so that by about the equivalent of the 32-cell stage the morula contains about 10 internal cells and more than 20 outer cells.

In early mammalian embryos, cell-division times vary with developmental time and place. The first two cleavage cycles in the mouse last about 24 hours, and subsequent cycles take about 10 hours each. After implantation, the cells of the epiblast proliferate rapidly; cells in front of the primitive streak have a cycle time of just 3 hours.

A special feature of mammalian development is that the early cleavages give rise to two distinct groups of cells—the **trophectoderm** and the **inner cell mass**. The

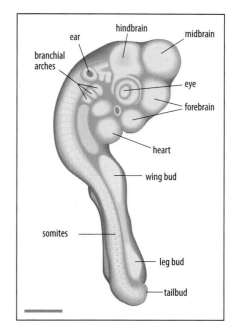

Fig. 3.19 The 40-somite stage of the chick embryo. Development of the head region and the heart are quite well advanced and the wing and leg buds are present as small protrusions.

Fig. 3.20 The extra-embryonic structures and circulation of the chick embryo. A chick embryo at 4 days of incubation is depicted *in situ*. The embryo lies within the fluid-filled amniotic cavity enclosed by the amnion, which provides a protective chamber. The yolk is surrounded by the yolk sac membrane. The vitelline vein takes nutrients from the yolk sac to the embryo and the blood is returned to the yolk sac via the vitelline artery. The umbilical artery takes waste products to the allantois and the umbilical vein brings oxygen to the embryo. The arteries are shown in red and the veins in blue but this does not denote the oxygenation status of the blood. As the embryo grows, the amniotic cavity enlarges; the allantois also increases in size and its outer layer fuses with the chorion, the membrane under the shell, while at the same time the yolk sac shrinks. Note that in this diagram the allantois has been enlarged so that the umbilical vessels can be shown clearly.

After Patten, B.M.: 1951

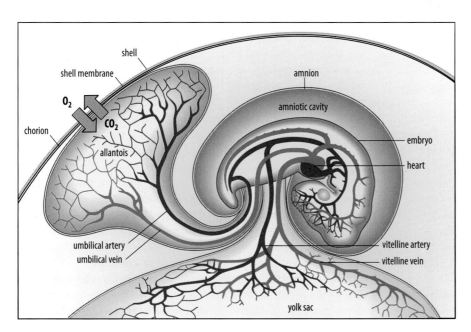

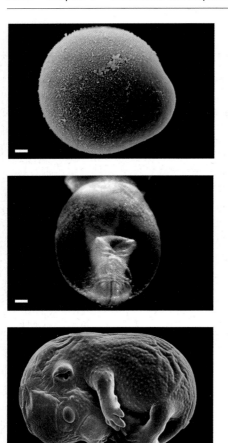

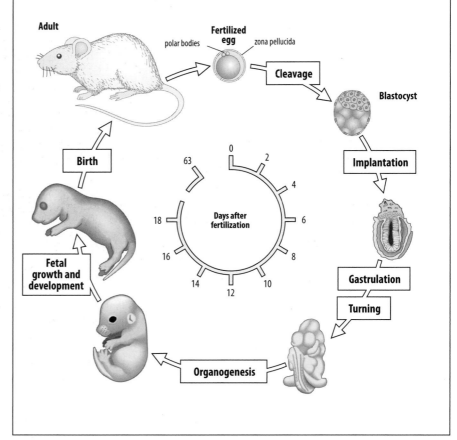

Fig. 3.21 The life cycle of the mouse.
The egg is fertilized in the oviduct, where cleavage also takes place before implantation of the blastocyst in the uterine wall at 5 days after fertilization. Gastrulation and organogenesis then take place over a period of about 7 days and the remaining 6 days before birth are largely a time of overall growth. After gastrulation the mouse embryo undergoes a complicated movement known as turning, in which it becomes surrounded by its extra-embryonic membranes (not shown here). The photographs show (from top): a fertilized mouse egg just before the first cleavage (scale bar = 10 μm); the anterior view of a mouse embryo at E8 (scale bar = 0.1 mm); and a mouse embryo at E14 (scale bar = 1 mm). An illustrated list of the stages in mouse development can be found on the website listed in General further reading.

*Top photograph reproduced with permission from Bloom, T.L.: **The effects of phorbol ester on mouse blastomeres: a role for protein kinase C in compaction?** Development 1989, **106**: 159-171. Middle photograph courtesy of N. Brown and bottom photograph courtesy of J. Wilting.*

trophectoderm arises from the cells on the outside of the embryo and will give rise to extra-embryonic structures, such as the **placenta**, through which the embryo gains nourishment from the mother. The embryo proper develops from a subset of the internal cells that form the inner cell mass. At this stage (E3.5) the mammalian embryo is known as a **blastocyst** (see Fig. 3.21). Fluid is pumped by the trophectoderm into the interior of the blastocyst, which causes the trophectoderm to expand and form a fluid-filled cavity (the blastocoel) containing the inner cell mass at one end.

From E3.5 to E4.5 the inner cell mass becomes divided into two regions. The surface layer in contact with the blastocoel becomes the **primitive endoderm**, and will contribute to extra-embryonic membranes, whereas the remainder of the inner cell mass—will develop into the embryo proper in addition to giving rise to some extra-embryonic components. At this stage, about E4.5, the blastocyst is released from the zona pellucida and implants into the uterine wall.

The course of early post-implantation development of the mouse embryo from around E4.5 to E8.5 appears more complicated than that of the chick, partly because of the need to produce a larger number of extra-embryonic membranes, and partly because the epiblast is distinctly cup-shaped in the early stages. This is a peculiarity of mouse and other rodent embryos. Human and rabbit embryos, for example, are flat blastodiscs, much more resembling that of the chick. Despite the different topology, however, gastrulation and the later stages in development of the mouse embryo are in essence very similar to those of the chick.

The first 2 days of mouse post-implantation development are shown in Fig. 3.23. After the initial adhesion of the blastocyst to the uterine epithelium, the cells of the **mural trophectoderm**—the region not in contact with the inner cell mass—replicate their DNA without cell division, giving rise to **primary trophoblast** giant cells that

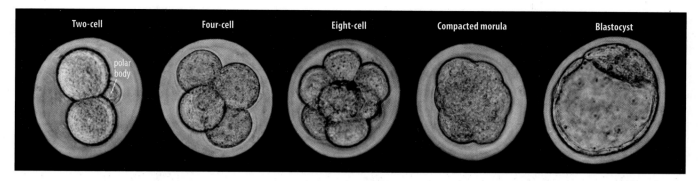

invade the uterus wall and surround most of the conceptus, forming an interface with the maternal tissue. The uterus wall envelops the blastocyst, and the **polar trophectoderm** cells in contact with the inner cell mass continue to divide, forming the ectoplacental cone and the **extra-embryonic ectoderm**, which both contribute to the placenta. The outer cells of the ectoplacental cone differentiate into secondary trophoblast giant cells. Some cells from the primitive endoderm migrate to cover the whole inner surface of the mural trophectoderm. They become the parietal endoderm, which eventually becomes Reichert's membrane, a sticky layer of cells and extracellular matrix that has a barrier function. The remaining primitive endoderm cells form the **visceral endoderm**, which covers the elongating **egg cylinder** containing the epiblast.

By E5.5, an internal cavity has formed inside the epiblast, which then becomes cup-shaped—U-shaped when seen in cross-section (see Fig. 3.23, second and third panels). The epiblast is now a curved single layer of epithelium, which at this stage

Fig. 3.22 Cleavage in the mouse embryo. The photographs show the cleavage of a fertilized mouse egg from the two-cell stage through to the formation of the blastocyst. After the eight-cell stage, compaction occurs, forming a solid ball of cells called the morula, in which individual cell outlines can no longer be discerned. The internal cells of the morula give rise to the inner cell mass, which can be seen as the compact clump at the top of the blastocyst. It is from this that the embryo proper forms. The outer layer of the hollow blastocyst—the trophectoderm—gives rise to extra-embryonic structures.

Photographs courtesy of T. Fleming.

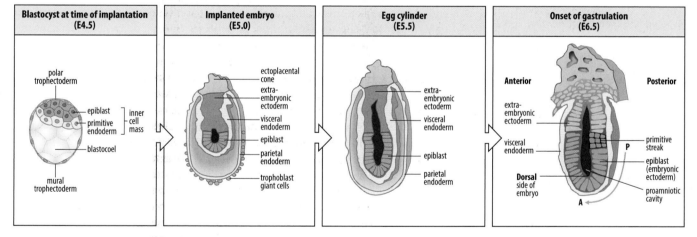

Fig. 3.23 Early post-implantation development of the mouse embryo. First panel: before implantation, the fertilized egg has undergone cleavage to form a hollow blastocyst, in which a small group of cells, the inner cell mass (blue), will give rise to the embryo, while the rest of the blastocyst forms the trophectoderm, which will develop into extra-embryonic structures. At the time of implantation the inner cell mass consists of two regions: the epiblast which will develop into the embryo proper, and the primitive endoderm, which will contribute to extra-embryonic structures. Second panel: the mural trophectoderm gives rise to trophoblast giant cells, which invade the uterine wall, helping to anchor the blastocyst to it. The blastocyst becomes surrounded by the uterine wall. The polar trophectoderm in contact with the epiblast proliferates and forms extra-embryonic tissues—the ectoplacental cone and extra-embryonic ectoderm—which

contribute to the placenta. The epiblast elongates and develops an internal cavity (proamniotic cavity), which gives it a cup-shaped form. Third panel: the cylindrical structure containing both the epiblast and the extra-embryonic tissue derived from the polar trophectoderm is known as the egg cylinder. Fourth panel: the beginning of gastrulation is marked by the appearance of the primitive streak (brown) at the posterior of the epiblast (P). It starts to extend anteriorly (arrow) towards the bottom of the cylinder. For simplicity, the parietal endoderm and trophoblast giant cells are not shown in this panel or in subsequent figures. A, anterior; P, posterior.

Illustration after Hogan, B., et al.: Manipulating the Mouse Embryo: A Laboratory Manual, *2nd edition. New York: Cold Spring Harbor Laboratory Press, 1994.*

Fig. 3.24 Gastrulation in the mouse embryo. Left panel: as in the chick, gastrulation in the mouse embryo begins when epiblast cells converge on the posterior of the epiblast and move under the surface, forming the denser primitive streak (brown) where the cells are becoming internalized. Once inside, the proliferating cells spread out laterally between the epiblast and the visceral endoderm to give a layer of prospective mesoderm (light brown). Some of the internalized cells will eventually replace the visceral endoderm to give definitive endoderm (not shown on these diagrams for simplicity), which will form the gut. Right panel: as gastrulation proceeds, the primitive streak lengthens and reaches the bottom of the cup, with the node at the anterior end. Cells from the node move anteriorly to give rise to the notochord in the head region, which is known as the head process. Part of the visceral endoderm and the mesoderm has been cut away in this diagram to show the node and notochord. Note that, given the topology of the mouse embryo at this stage, the germ layers appear inverted (ectoderm on the inner surface of the cup, endoderm on the outer) if compared with the frog gastrula.

*Illustration adapted, with permission, from McMahon, A.P.: **Mouse development. Winged-helix in axial patterning.** Curr. Biol. 1994, **4**: 903-906.*

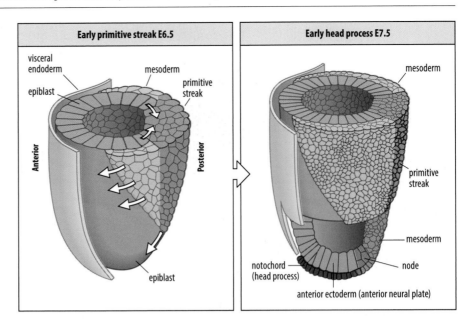

contains about 1000 cells. The embryo develops from this layer. The first easily visible sign that marks the future antero-posterior axis of the embryo is the appearance of the primitive streak at about E6.5. The streak starts as a localized thickening at the edge of the cup on one side; this side will correspond to the future posterior end of the embryo. The initial development of the primitive streak in the mouse is similar to that in the chick. Over the next 12–24 hours it elongates until it reaches the bottom of the cup. Here a condensation of cells—the **node**—becomes distinguishable at the anterior end of the extended streak and corresponds to Hensen's node in the chick embryo. To make primitive streak formation easier to compare with that of the chick, imagine the epiblast cup spread out flat (see Fig 5.12). In mouse gastrulation, as in chick, epiblast cells converge on the primitive streak, and proliferating cells migrate through it to spread out laterally and anteriorly between the ectoderm and the visceral endoderm to form a mesodermal layer (Fig. 3.24).

Some epiblast-derived cells pass through the streak and enter the visceral endoderm, gradually displacing it to form the definitive **embryonic endoderm** on the outside of the cup, which is the future ventral side of the embryo. Development from E7.5 to E10.5 is shown in Fig. 3.25. Cells migrating anteriorly from the node form the prechordal plate mesoderm and the head process. Cell proliferation continues during gastrulation, and the embryo anterior to the node grows rapidly in size and the ectoderm in this region forms the anterior neural plate. As in the chick, as the streak regresses, the trunk notochord is progressively laid down with the spinal cord developing above it and somites forming on either side. The cells that give rise to the spinal cord and the medial parts of the somites are generated from the epiblast on either side of the anterior streak, which acts as a stem zone. The stem zone eventually contributes to the tailbud, which gives rise to the post-anal tail.

Somite formation and organogenesis start at the anterior end of the mouse embryo and proceed posteriorly. The first somite is formed at around E7.5 and new ones are formed at intervals of 120 minutes. At around E8, a head is apparent and neural folds have started to form on its dorsal side (see Fig. 3.25). The embryonic endoderm—initially exposed on the ventral surface of the embryo—becomes internalized to form the foregut and hindgut, the lateral surfaces eventually folding together to internalize the gut completely with subsequent overgrowth of mesoderm and ectoderm to form the ventral body wall in the process of ventral closure. The heart and liver move into their final positions relative to the gut.

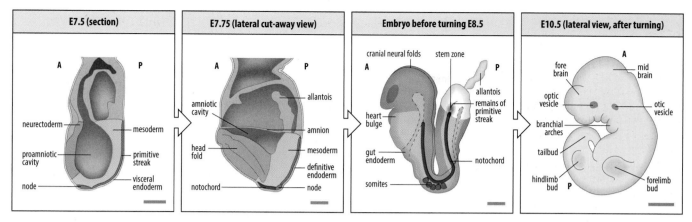

Fig. 3.25 Schematic views of the early development of the mouse embryo to the completion of gastrulation and neurulation. First panel: by E7.5, the primitive streak (brown) has extended to the bottom of the epiblast and the node has formed. The anterior ectoderm (blue) becomes the anterior neural plate, which will give rise to the brain. Mesoderm is shown as light brown. Second panel: the anterior part of the embryo grows in size and the head fold appears. Definitive endoderm (green) replaces visceral endoderm (yellow) to form an outer layer on the ventral surface of the embryo. The notochord (red) begins to be laid down in the trunk. Third panel: by E8.5, there has been further growth of the embryo anterior to the node, the head is distinct, the neural folds have formed, the foregut and hindgut have closed, and somites are beginning to form on either side of the notochord. The embryo is covered by a layer of ectoderm that will form the epidermis, which is not shown on this diagram. Fourth panel: by E10.5, gastrulation and neurulation are complete. The embryo has undergone a complicated turning process around E9 that brings the dorsal and ventral sides into their final positions. Scale bars for the first three panels = 100 μm; fourth panel, scale bar = 75 μm.

By about midway through embryonic development, gastrulation and neurulation are complete and the mouse embryo has a very distinct head, and the forelimb buds are starting to develop. A complicated 'turning' process has occurred around E9 that produces a more recognizable mouse embryo, surrounded by its extra-embryonic membranes (Fig. 3.26). As a result of turning, the initial cup-shaped epiblast has turned inside out so that the dorsal surface is now on the outside; the ventral surface, with the umbilical cord that connects it to the placenta, is facing inwards. (Turning is another developmental quirk peculiar to rodents; human embryos are surrounded by their extra-embryonic membranes from the beginning.) Organogenesis in the mouse proceeds very much as in the chick embryo. From fertilization to birth takes around 18 to 21 days, depending on the strain of mouse.

3.6 The early development of a human embryo is similar to that of the mouse

As we shall see, studying development of the embryos of all of our four main model vertebrates has provided general principles that help us to understand how a human embryo develops. The human life cycle is shown in Fig. 3.27. The length of time from fertilization to birth is 38 weeks, much longer than in any of our model vertebrates. A landmark advance in the 1970s was the devising of methods for fertilizing human

Fig. 3.26 Turning in the mouse embryo. Between E8.5 and E9.5, the mouse embryo becomes entirely enclosed by the protective amnion and lies in the amniotic cavity containing amniotic fluid. The visceral yolk sac, a major source of nutrition, surrounds the amnion and the allantois connects the embryo to the placenta.

Illustration after Kaufman, M.H.: The Atlas of Mouse Development. *London: Academic Press, 1992.*

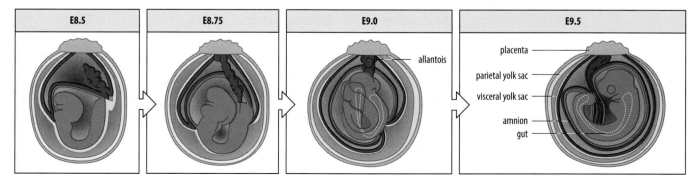

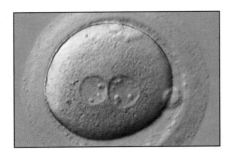

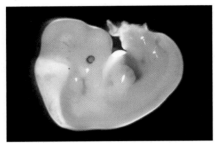

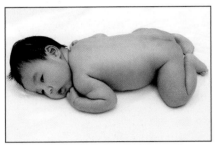

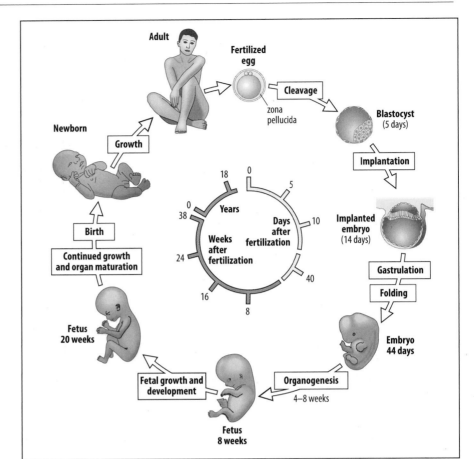

Fig. 3.27 Life cycle of a human being. The egg is fertilized in the oviduct. By the time it reaches the uterus—about 5 days after fertilization—it has developed into a blastocyst, this 'hatches' from the zone pellucida and implants in the uterus wall. During the next week, the embryo proper emerges as the epiblast, the upper layer of epithelium in a bilaminar disc of cells with the amniotic cavity above and the yolk sac cavity below. The other changes that take place at this time are connected with the development of the placenta and the formation of protective membranes. Gastrulation occurs during the third week of development, giving rise to the three germ layers. This is followed by the rapid folding of the flat embryo into a three-dimensional body. Organogenesis takes place between 4 and 8 weeks of development. After this, the embryo is known as a fetus and the next 30 weeks are devoted mainly to growth and the maturation of the tissues and organs. The photographs show: from top, a fertilized human egg just before the first cleavage; a human embryo; a baby.

Top photograph Courtesy Alpesh Doshi. Middle photograph reproduced courtesy of the MRC/Wellcome funded Human Developmental Biology Resource. Baby photograph reproduced via the creative commons attribution 2.0 generic license, ©Tarotastic

eggs *in vitro* and then culturing them as they underwent the earliest stages in development. Not only could the earliest stages in human development be observed for the first time, but there was also a major and lasting impact on the treatment of infertility. It also opened up new clinical treatments, for example, preimplantation genetic diagnosis (Box 3A). Our knowledge of the later stages of human development comes from descriptions of collections of clinical material.

Unfertilized human eggs are about the same size as mouse eggs. Fertilization normally takes place in the oviduct (in humans also known as the Fallopian or uterine tube) and the earliest stages of development then take place while the fertilized egg travels towards the uterus. These stages are similar to those in mice—cleavage divisions, compaction, and blastocyst formation (Fig. 3.28). Implantation usually occurs around day 7 of development, when about 256 cells are present. Cells from the inner cell mass of a human blastocyst can be cultured as embryonic stem cells (see Chapter 8), and these can provide a model system for studying processes in human development such as cell differentiation.

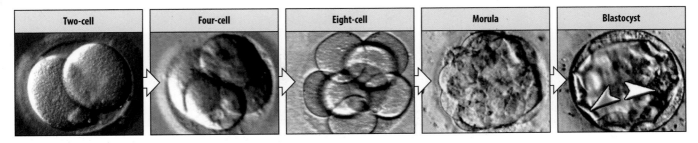

| Two-cell | Four-cell | Eight-cell | Morula | Blastocyst |

Fig. 3.28 Cleavage in a human embryo developing *in vitro*. The photographs show cleavage of a human egg fertilized *in vitro* from the two-cell stage through to the formation of a blastocyst. The stages are very similar to those in the early mouse embryo. In the last panel, the gray arrow points to the trophectoderm, the white arrow to the inner cell mass.

*Reproduced from Niakan, K.K., et al.: **Human pre-implantation embryo development.** Development 2012, **139**: 829–841.*

The processes of implantation and formation of the placenta differ from those in mice, so the mouse is not a good model for these aspects of human development. Furthermore, as remarked earlier, the topology of the human embryo just before gastrulation is more similar to that of the chick embryo rather than the mouse (Fig. 3.29). The embryo proper comes from the epiblast, the upper layer of epithelium of a flat bilaminar disc of cells; the lower layer is the hypoblast. A cavity, the amniotic cavity lies above the epiblast, the yolk sac cavity lies below the hypoblast. The primitive streak becomes apparent in the upper surface of the epiblast at around the beginning of the third week of development. The epiblast cells are thought to ingress through the primitive streak in the same way as the epiblast cells in the chick embryo to give rise to the mesoderm and endoderm. The epiblast cells that do not ingress form the ectoderm.

During gastrulation, the main axes of the body become apparent and the neural plate is induced and folds up to the form the neural tube. As the flat embryo continues to grow and the organs start to form, it undergoes a complex process of folding, as in the chick and the mouse, followed by ventral closure to generate the three-dimensional form of the body. The expansion of the embryo also forces it into a convex shape with both ends of the main body axis curling round and the amniotic cavity, enlarges substantially and comes to surround the embryo while the yolk sac cavity becomes very reduced (Fig. 3.30).

Fluid from the amniotic cavity can later be collected by a procedure known as amniocentesis, which is carried out between 14 and 16 weeks of gestation. Amniocentesis is used for prenatal diagnosis. The levels of metabolic products in the amniotic fluid that have been produced by the fetus can be measured and cells that have been sloughed off the embryo into the fluid can be grown in culture and analyzed, for example, for chromosomal abnormalities.

The early stages in human development take longer than the same stages in the mouse. Blastocyst formation, for example, does not occur until 5–6 days after fertilization whereas in mouse this occurs between 3–4 days, and gastrulation takes place at 14 days in the human embryo whereas in the mouse it takes place at 6 days. The laying down of the body plan and formation of organs is even slower, and it takes about 8 weeks until the structure of the body and organs has been laid in miniature. After 8 weeks, the human embryo is known as a fetus and during the rest of gestation, known as the fetal period, there is substantial growth accompanied by further maturation of the organs (see Fig. 3.27).

Experimental approaches to studying vertebrate development

This part of the chapter introduces some of the main techniques used to study the developmental biology of vertebrates. It is not intended to be comprehensive, but to indicate general experimental approaches and to describe in more detail a few

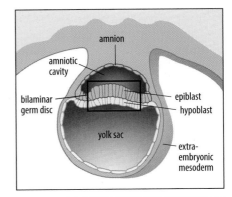

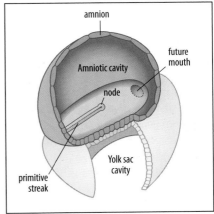

Fig 3.29 The human embryo at the beginning of gastrulation. The primitive streak begins to form in the upper surface of the epiblast, which lies on top of the hypoblast. The cavity above the epiblast is the amniotic cavity and the cavity below the hypoblast is the yolk sac. The embryo is suspended in the chorionic cavity and connected to its wall by a stalk of extraembryonic mesoderm (not shown). The chorionic cavity is lined by extraembryonic mesoderm which gives rise to blood vessels that link with blood vessels in the embryo, establishing a circulation between the embryo and placenta. The depression in the upper surface of the epiblast shown marks the position in which the future mouth will form.

MEDICAL BOX 3A Preimplantation genetic diagnosis

We now take *in vitro* fertilization (IVF) of human eggs as a treatment for infertility almost for granted, even though the first IVF baby, Louise Brown, was born in 1978. The UK scientist Robert Edwards and the clinician Patrick Steptoe were responsible for developing IVF. Edwards received the Nobel prize in 2010, but Steptoe had died some years earlier and so could not share the prize. The successful development of IVF has led to further clinical treatments. It has become possible to determine the genotype of embryos produced by IVF before implantation without harming the embryo. This procedure is called **preimplantation genetic diagnosis** and was developed in the late 1980s for use with fertile couples at risk of passing on a serious genetic disease to their children. The aim was to provide an alternative to prenatal diagnosis and the potential termination of an affected pregnancy.

For pre-implantation diagnosis, one blastomere is removed from an IVF-produced embryo during its early cleavage *in vitro* without affecting subsequent development (Figure 1). The DNA from this blastomere can be amplified *in vitro* and tested for the presence or absence of mutations known to cause disease. Where there is a known high risk that the parents will pass on a particular genetic disease—for example, when the parents are both carriers for the cystic fibrosis gene—preimplantation genetic diagnosis can be used to ensure that an embryo that would develop the disease is not implanted into the mother (Figure 2).

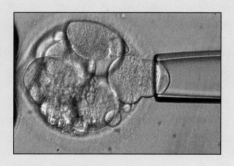

Figure 1

Photograph courtesy of Dr. Malpani, MD, Malpani Infertility Clinic, Mumbai, India. www.drmalpani.com.

The demand for preimplantation genetic diagnosis is likely to increase because it can be used not only to identify embryos with mutations that will inevitably cause a potentially fatal disease in infancy and childhood, but also to identify embryos with mutations in genes that predispose an individual to disease in later life. An example is the gene *BRCA1*, certain mutations in which predispose women to develop breast and ovarian cancer, and account for 80% of these tumors in women with a genetically inherited predisposition (5–10% of all breast and ovarian cancer). In males, mutations in *BRCA1* are linked to an increased susceptibility to prostate cancer. By determining whether an embryo from a high-risk family carries a *BRCA1* disease allele the genetic susceptibility to these cancers could, in principle, be eliminated from a family. At least one baby has been born to such a high-risk couple after IVF and selection by preimplantation genetic diagnosis for the absence of the *BRCA1* disease allele. In the United Kingdom, preimplantation genetic diagnosis has been licensed by the Human Fertilisation and Embryology Authority (HFEA) for more than 60 genetic diseases.

There are practical and ethical questions in relation to preimplantation genetic diagnosis, such as which genetic conditions it should be applied to. In the United Kingdom, the HFEA made a general ruling allowing decisions to be made on a case-by-case basis for conditions not on their list. An example of the kind of ethical question that has arisen is illustrated by parents wishing to select an IVF embryo with the best HLA match to a sibling suffering from a rare blood disease, so that the IVF child could eventually donate stem cells to their sibling. Cases like this have been approved for preimplantation genetic diagnosis in the United Kingdom.

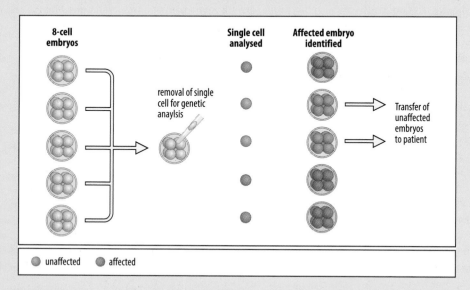

Figure 2

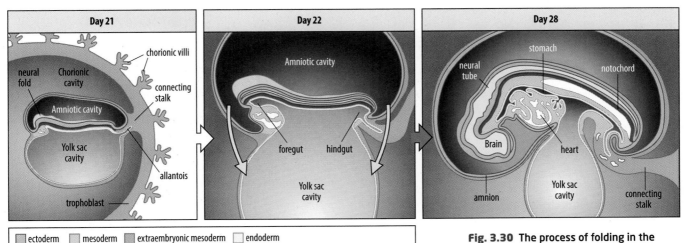

| Day 21 | Day 22 | Day 28 |

neural fold · Chorionic cavity · chorionic villi · connecting stalk · Amniotic cavity · Yolk sac cavity · allantois · trophoblast

Amniotic cavity · foregut · hindgut · Yolk sac cavity

stomach · neural tube · notochord · Brain · heart · amnion · Yolk sac cavity · connecting stalk

ectoderm   mesoderm   extraembryonic mesoderm   endoderm

Fig. 3.30 The process of folding in the human embryo and the expansion of the amniotic cavity. Between day 21 and day 28 of development, the flat embryo becomes converted into the three-dimensional form of the body. Lateral folding brings the edges of the flat embryo together ventrally, while expansion of the embryo also causes both ends of the main body axis to curl round (blue arrows in middle panel). This flexure of the main body axis is accompanied by expansion of the amniotic cavity, which comes to surround the embryo, while the yolk sac becomes very reduced (right panel).

techniques that are very commonly used. Rather than reading it straight through, you may wish to use it as a reference to come back to, to put into context some of the experiments described in the rest of the book. Many of the techniques described here are also applicable to non-vertebrates (see Chapters 2 and 6).

The earliest studies of vertebrate embryonic development involved careful and detailed study of whole and dissected embryos under the microscope, and simply watching, describing, and drawing the development of easily accessible embryos such as those of newts and frogs. From the seventeenth century onwards, observations of animal and human embryos by zoologists and physicians gradually yielded the comprehensive anatomical descriptions of embryonic development that are the foundation of modern developmental biology. Hensen's node and Koller's sickle (see Section 3.4), for example, are named after embryologists working at the end of the nineteenth and in the early twentieth centuries. Careful observation is just as important in developmental biology today, even if the need to draw by hand the images you see in the microscope has been superseded by digital-imaging technology. Another important technique in studying normal development is making fate maps that show what a particular region of the embryo will give rise to, and we will discuss the fate maps of our model vertebrates in Chapters 4 and 5.

Anatomical and cellular descriptions of normal development can now be supplemented by mapping the expression patterns of developmentally important genes. This advance depended on the invention in the middle of the twentieth century of ways of detecting the expression of genes in cells and tissues by the technique of *in situ* **nucleic acid hybridization**, which involves the use of labeled RNA (riboprobes) complementary to the RNA of interest to detect the gene transcripts (see Box 1D). The proteins expressed from the gene transcripts can be detected by specific antibodies labeled with fluorescent dyes (see Box 1D), and these techniques can also be applied to tissue sections and whole-mount specimens (see Fig. 12.1). Large-scale projects are currently under way to produce comprehensive online 'atlases' of the normal expression patterns of hundreds of genes at different developmental stages for the mouse and the chick (see General further reading).

Techniques such as **DNA microarray analysis** and **RNA seq** can also detect and identify the expression of large numbers of genes at the same time and so can provide information about all the genes being expressed in a particular tissue or at a particular stage in development (Box 3B).

Observation on its own cannot unravel the mechanisms underlying developmental processes, however, and the only way to find out more about these is to disturb the developmental process in some specific way and see what happens. Techniques

EXPERIMENTAL BOX 3B Gene-expression profiling by DNA microarrays and RNA seq

In this book we deal with the general principles underlying development so we do not describe every gene known so far to be involved in laying down the body plans of the animals we cover. However, now that the genomes of the model vertebrates have been sequenced, systematic genome-wide approaches are being used to identify all the genes involved in a particular developmental process. Genomic sequence information can also be used to identify the targets of the transcription factors that coordinate gene expression in space and time.

The identification of all the genes expressed in a particular tissue or at a particular stage of development can be accomplished by carrying out genome-wide screens for gene expression using DNA microarrays, often known as DNA chips. This technology enables the levels of RNA transcripts of thousands of genes to be measured simultaneously, and the main use of microarrays in developmental biology is to monitor the changes in gene expression that occur within a tissue or embryo at different stages of development or after experimental manipulation. Microarrays for studying gene expression come in various formats, but the most widely used are flat surfaces studded with a regular array of clusters or 'spots' of DNA fragments of known sequence, each cluster representing a DNA sequence found in a specific protein-coding gene. These DNA fragments are known as the 'probes,' and a microarray can contain thousands to hundreds of thousands of different probes. Microarrays composed of hundreds of thousands of oligonucleotide probes specially synthesized to cover the whole genome sequences of all four model vertebrates are now commercially available.

To determine which genes are being expressed, the total mRNA from the tissue of interest is extracted, then either amplified as RNA or converted into cDNA and amplified by the polymerase chain reaction (PCR). The nucleic acid is then tagged with a fluorescent dye and hybridized to the microarray. A reference sample of mRNA, similarly treated and tagged with a different fluorescent dye, is also hybridized to the same microarray. The microarray is then scanned and the ratios of the signals from both dyes at each spot are recorded and converted into a relative expression level for the test sample.

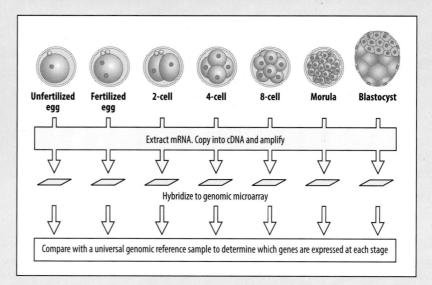

Figure 1

Landmark experiments carried out in the early 2000s used microarrays to find all the genes expressed at particular stages in early mouse development. Genes expressed in the unfertilized egg, fertilized egg, two-cell stage, four-cell stage, eight-cell stage, morula, and blastocyst were identified by extracting mRNA from 500 eggs or embryos at the appropriate stage, then labeling it and hybridizing it to miniaturized microarrays of unique oligonucleotides of 60 bases each, representing all of the mouse genes (see Figure 1). The binding patterns of the labeled RNAs indicate which genes are expressed at the different stages. Similar genome-wide expression analyses using microarrays have since been carried out on the same stages in development of early human embryos.

Another technique that is being increasingly used to determine which genes are expressed in a particular tissue or at a particular time in development involves direct sequencing and is known as **RNA seq**. This approach is now feasible because of the development of rapid next-generation DNA sequencing methods. mRNA is extracted from cells, fragmented, copied into DNA and amplified. The cDNAs are tagged and directly sequenced to generate large numbers of high-quality sequences which are then mapped to a reference genome. The level of expression of a particular gene is measured by the density of sequences that map to that gene. The advantage of RNAseq is that it is more accurate than microarrays, and can provide additional information, for example, about alternative gene transcripts.

for interfering with development can be very broadly divided into two types, and many experiments in developmental biology will use a combination of these two approaches. Classical experimental embryological techniques manipulate the embryo by physical intervention—by removing or adding cells to cleavage-stage embryos or transplanting blocks of cells from one embryo to another, for example. The other

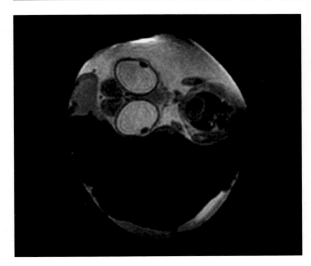

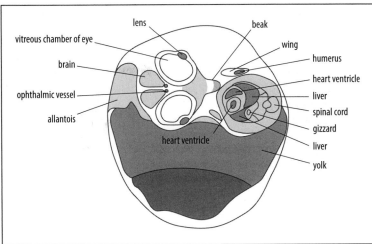

Fig. 3.31 A live nine-day quail embryo inside its egg. The image on the left was obtained by magnetic resonance imaging (MRI).

Photograph courtesy of Suzanne Duce.

class of techniques is based on genetics and is used to disturb the expression of developmentally important genes by mutation, gene silencing, overexpression or misexpression—expressing the gene at a time or place where it is not normally expressed.

We shall start by taking a brief look at fate-mapping techniques in *Xenopus*, then consider experimental manipulation in *Xenopus* and chick embryos, and finally describe some of the techniques from molecular biology, genetics, and genomics that have revolutionized the study of developmental biology over the past few decades. Other technical advances that have benefited developmental biology over the past 20 years are the great improvement in computer-aided microscopic imaging techniques, the development of fluorescent labels in a vast range of colors that allow the imaging of living embryos, and the introduction of new forms of imaging into the field, such as magnetic resonance imaging (MRI) and optical projection tomography (OPT). Images of live avian embryos inside the egg can now be obtained by magnetic resonance imaging (Fig. 3.31).

3.7 Fate mapping and lineage tracing reveal which parts of the body cells in the early embryo give rise to which adult structures

For both biological and technical reasons, fate mapping in the early embryo is easiest in *Xenopus*. Even at the blastula stage, cells that will eventually form the three germ layers are arranged in distinct regions and are accessible from the surface of the embryo (see Fig. 3.6). Single blastomeres can be easily labeled by injecting a non-toxic marker such as fluorescein-linked dextran-amine or by expression of a fluorescent protein such as green fluorescent protein (GFP) from injected RNA. Fluorescein-labeled dextran-amine is a stable high-molecular-weight molecule, which cannot pass through cell membranes and so is restricted to the injected cell and its progeny; as fluorescein fluoresces green when excited by UV light, the fluorescein-labeled dextran can be easily detected under a fluorescence microscope. GFP produces a green light when excited by UV and stable forms are now widely used because the intense fluorescence can be viewed in living embryos and allows long-term lineage tracing. Figure 3.32 shows a fate map for a *Xenopus* embryo made by injecting one of the cells at the 32-cell stage with fluorescein-dextran-amine.

By following the fate of individual cells, or groups of cells in *Xenopus* embryos, we can build up a map on the blastula surface showing the regions that will give rise to, for example, notochord, somites, nervous system, or gut. The fate map shows

Fig. 3.32 Fate mapping in the early *Xenopus* embryo. Left panel: a single cell in the embryo, C3, is labeled by injection of fluorescein-dextran-amine, which fluoresces green under UV light. Right panel: a cross-section of the embryo at the tailbud stage shows that the labeled cell has given rise to mesoderm cells on one side of the embryo. Scale bar = 0.5 mm.

Photograph courtesy of L. Dale

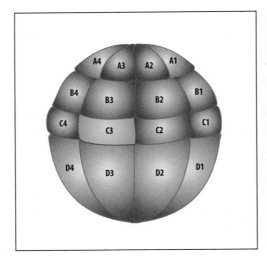
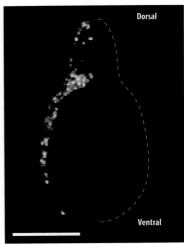

where the tissues of each germ layer normally come from, but it indicates neither the full potential of each region for development nor to what extent its fate is already specified or determined in the blastula. In other words, we know what each of the early cells will give rise to—but the embryo does not. Vertebrate embryos have considerable capacity for regulation when pieces are removed or are transplanted to a different part of the same embryo (see Section 1.12). This implies the existence and persistence of considerable developmental plasticity (see Chapters 4 and 5), and also that the actual fate of cells is heavily dependent on the signals they receive from neighboring cells.

Fate mapping has been carried out in the chick embryo by staining small groups of cells with a lipophilic dye such as DiI and observing where the labeled cells end up. It is more difficult to inject molecules such as fluorescein-dextran-amine into individual cells, as the cells in chick embryos are much smaller, although this technique it has been used successfully to trace the lineage of cells derived from Hensen's node.

Chick–quail chimeras have also been used for cell-lineage tracing. A **chimera** is an organism that is made up of cells of two different genetic constitutions, and a chick–quail chimeric embryo can be made by transplanting tissue or cells from a quail embryo into a chick embryo. We shall come across the use of chimeras again in relation to investigating early mouse development (Chapter 5), and later in this chapter in regard to making transgenic mice. Chick and quail embryos are very similar in their development, and embryos composed of a mixture of chick and quail cells will develop normally. The cells of the two species can be distinguished by the different appearance of their nuclei; quail cells have a prominent nucleolus that stains red with an appropriate stain (Fig. 3.33). Quail cells can also be distinguished from chick cells by immunohistochemical staining with labeled antibodies for quail proteins. For lineage tracing, cells are taken from a particular site in an early quail embryo and transplanted into the same site in a chick embryo of the same age. Later-stage embryos are then stained and sectioned to determine where the quail cells have ended up. Recently, transgenic chicken embryos expressing GFP in all cells have become available (see Section 3.10) and can be used for long-term fate mapping of transplanted cells. We will also come across the use of tissue transplants from transgenic axolotls expressing GFP in studying limb regeneration (Chapter 13).

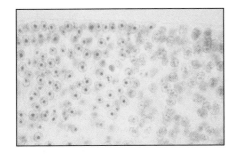

Fig. 3.33 Photograph of quail–chick chimeric tissue. The quail cells are on the left and the chick cells are on the right. Note the heavily staining nuclei of the quail cells.

Photograph courtesy of N. Le Douarin

Defined small populations of cells, or even single cells, can also be conveniently marked in chick embryos in the egg (*in ovo*) or in culture by introducing a DNA construct encoding a fluorescent protein. DNA constructs can be targeted to a relatively small number of cells by the technique of **electroporation**. The DNA is injected into the embryo at the desired site with a fine pipette and pulses of electric current applied using very fine electrodes. The current makes the membranes of nearby cells permeable to the injected DNA.

Cell-lineage tracing and fate mapping by dye injection or electroporation of nucleic acids is technically much more difficult in mammalian embryos than in *Xenopus* or the chick because the embryos are less accessible inside the mother. Nevertheless these same techniques have been used to follow cell fate in mouse embryos between E6.5 and E8.5, stages that can be maintained in culture (see Section 3.8). Researchers have found other ingenious ways around the problem, although some of the techniques require a great deal of skill. For example, to study the migration of neurons in the developing brain, a DNA construct encoding GFP is injected into the lateral ventricles of the brains of mouse embryos *in utero* (at around stage E14), accompanied by electroporation, without harming the embryos. Sections of the brains of later-stage embryos and of mouse pups after birth are examined by fluorescence microscopy to reveal the locations of the GFP-labeled neurons, which originate in the layer of cells immediately under the ventricle surface (see Chapter 12). Lineage tracing in the mouse can now be done even more conveniently using transgenic embryos, as we shall discuss in Section 3.10.

3.8 Not all techniques are equally applicable to all vertebrates

You will soon notice as you read further in this book that experiments involving the microsurgical manipulation of vertebrate embryos largely feature *Xenopus* and the chick. Few spontaneous developmental mutations are known in these animals, and chick and amphibian embryos were being studied long before developmental genes had been identified. *Xenopus laevis* is also unsuitable for conventional genetic analysis because of its long generation time (about a year) and its tetraploid genome. Amphibian and avian embryos are tough, however, and can be readily manipulated surgically at all stages in their development, unlike those of mammals (Fig. 3.34).

Fig. 3.34 The suitability of the different model vertebrates for classical experimental embryological manipulation.

The suitability of different vertebrate embryos for classical experimental embryological manipulation				
	Frog (*Xenopus laevis*)	**Zebrafish**	**Chick**	**Mouse**
Are living embryos easily available and observable at all stages of their development?	Yes. Eggs can be obtained and externally fertilized and embryos develop to swimming tadpole stage in the aquarium. Large batches of eggs can be fertilized at the same time and develop synchronously to give large numbers of embryos at the same developmental stage.	Yes. As for *Xenopus*	For most of it. Eggs are fertilized internally and so the very early stages of development—to blastoderm stage—occur within the mother hen. Laid eggs are easily obtainable, can be incubated in the laboratory, and the development of the embryo observed at any time up to hatching. Early embryos can also be cultured up to about stage 10 out of the egg.	Only very early stages. *In vitro* fertilized eggs can be grown in culture up to the blastocyst stage, after which they have to be replaced in a surrogate mother mouse to continue development. Early isolated embryos can be cultured for a limited time (see Section 3.8).
Is the embryo amenable to experimental manipulation, e.g. blastomere removal or addition, or tissue grafting?	Embryo can be manipulated surgically up to the neurula stage (see Chapter 4). Embryos are comparatively large, and are highly resistant to infection after microsurgery.	Amenable to surgical manipulation up to the neurula stage and to some extent in later stages. The embryo is transparent, which helps in observing developmental changes and the movement of labeled cells.	Can be manipulated surgically up to quite late stages in embryonic development, including limb development (see Chapters 5, 11, and 12).	Only very early embryos up to blastocyst stage can be manipulated if they are to be replaced in a surrogate mother for further development (see Fig. 3.39). Isolated postimplantation embryos can be cultured for a limited time (see Section 3.8).
Is fate mapping and lineage tracing relatively easy?	Yes, in the early embryo. Single cells in the early blastula can be injected with a fluorescent non-toxic dye or RNA encoding fluorescent protein (GFP) (see Fig. 3.32).	Yes, as in *Xenopus*	Relatively easy and cells can be labeled at quite late embryonic stages by dye injection; the use of chick–quail chimeric embryos; grafts from GFP chickens (see below) or electroporation of DNA encoding a reporter protein or fluorescent protein (see Section 3.7).	In the past it was difficult. Now becoming easier owing to transgenic techniques for restricting reporter gene expression to particular cells (see Section 3.10 and Box 3D) and also inducible transgenes.

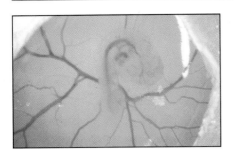

Fig. 3.35 A chick embryo can be observed through a window in the eggshell. The almost transparent 3.5-day embryo (H&H stage 21) can be just seen lying on top of the yolk. After an experimental procedure has been carried out, the window can be sealed with adhesive tape and the embryo left to develop further.

Microsurgical manipulations in amphibians, such as the removal of particular blastomeres, or the grafting of cells from one embryo into another, provided much of the information about the location and function of organizer regions in the blastula and their role in development; several experiments of this type are illustrated for *Xenopus* in Chapter 4. Blastomere removal identifies regions of the embryo that are essential for further development, while transplantation experiments are used to test the developmental potential of a particular region of the embryo or to determine the time at which the fate of cells in a particular location becomes irreversibly determined (see Section 1.12). The cells of early amphibian embryos carry their own store of nutrients in the form of yolk platelets and so explants of embryonic tissue can be cultured in a simple salt medium for several days. This allows experiments that investigate the inducing effects of one tissue on another when placed together in culture, and experiments of this type in *Xenopus* are described in Chapter 4. Similar explant experiments have been carried out with chick embryonic tissue embedded in collagen gels and using cell-culture medium.

The early chick embryo (blastoderm stage) can be removed from the egg and cultured on its vitelline membrane with thin albumen in a watch glass. The embryo will then develop for up to about 36 hours (Hamilton and Hamburger stage 10). Experiments in which pieces of one chick embryo are grafted into another cultured chick embryo at different stages have also provided much of the information on the organizer properties of different regions of the embryo and the generation of the primitive streak (discussed in Chapter 5). Older embryos can be accessed through a window cut in the egg shell and manipulated while remaining in the egg (Fig. 3.35). The window is then re-sealed with adhesive tape and the embryo allowed to continue development so that the effects of the procedures can be assessed. The accessibility of the chick embryo allows many types of experimental procedures: part of a developing structure such as a limb bud can be removed; tissue can be grafted to new sites; and the effects of chemicals such as growth factors on development can be investigated by implanting small inert beads soaked in these agents into a specific location in the embryo (examples of all these types of experiment in the study of chick limb development are described in Chapter 11).

After implantation, mouse development *in vivo* is hidden from view and can only be followed by isolating embryos at different stages. Tissues from very early embryos can be cultured for a limited time. For example, inner cell masses isolated from 3.5-day embryos and cultured *in vitro* for several days will develop certain embryonic and extraembryonic tissues and some structures typical of a normal embryo at the corresponding stage. Mouse embryos between stages E6.5 and E12.5 have been cultured for 24–48 hours in roller tubes, which allows manipulations such as node transplantation. Experimental investigation of living embryonic tissues at later stages requires dissection of embryos and the growth of tissue or organ explants (such as limb buds, embryonic kidney, lung) *in vitro*.

3.9 Developmental genes can be identified by spontaneous mutation and by large-scale mutagenesis screens

One approach to identifying developmental genes, known as forward genetics, first recognizes a mutant organism by its unusual phenotype and then genetic experiments are carried out to find the gene responsible. Rare spontaneous mutations were for a long time the only way that genetic disturbances of development could be studied, and the starting point for a forward genetics approach. Such spontaneous mutations have been identified in mice (see Fig. 1.12) and in chickens. Informative spontaneous mutations are rare, however, even in mice, and as they are only noticed if the animal survives birth, mutations in important developmental genes that lead to the death of a mouse embryo *in utero* (**embryonic-lethal mutations**) are likely to be missed. A lethal developmental mutation in chickens was only discovered because of the reduced hatchability of the eggs.

Both spontaneous and inherited gene mutations lead to congenital abnormalities in human patients, and clinical genetics has provided one route for the discovery of developmentally important genes. Since the human population is very large, exceptionally rare events can be detected, and because the abnormalities in human patients are scrutinized in detail, mutations that produce rather subtle phenotypes have been identified. In this way, human studies have been able to match specific gene mutations with very specific phenotypes. In some cases, mutations that result in the substitution of a single amino acid in a protein with another amino acid can have dramatically different effects depending on which amino acid is changed, and which amino acid replaces it. The understanding of the underlying developmental mechanisms relies however on experiments in model vertebrates.

Many more developmental genes have been identified by inducing many random mutations in large numbers of experimental organisms by chemical treatments or irradiation by X-rays, and then screening for mutants with unusual phenotypes of developmental interest. These can then be analyzed by classical and molecular genetic techniques to discover which gene is responsible for the mutant phenotype. The aim, where possible, is to treat a large enough population so that, in total, a mutation is induced in every gene in the genome. This sort of approach is best used in organisms that breed rapidly and that can conveniently be obtained and treated in very large numbers. Among vertebrates, this approach has been applied to both mice and zebrafish. Zebrafish in particular, are a potentially valuable vertebrate system for large-scale mutagenesis because large numbers can be handled, and the transparency and large size of the embryos makes it easier to identify developmental abnormalities. However, unlike the genetic screening in *Drosophila* described earlier in the book (see Box 2A), there are no genetic means so far of automatically eliminating unaffected individuals in zebrafish. This means that all the progeny of a cross have to be examined visually for any developmental abnormality.

Chemical mutagenesis generates both dominant mutations with varying degrees of gene function, which are particularly relevant to human conditions, and recessive lack-of-function mutations. Dominant mutations can be recognized in offspring in the F_1 generation, as shown for the mouse in Fig. 3.36, but more complicated breeding programs like that described for zebrafish in Box 3C have to be used to reveal

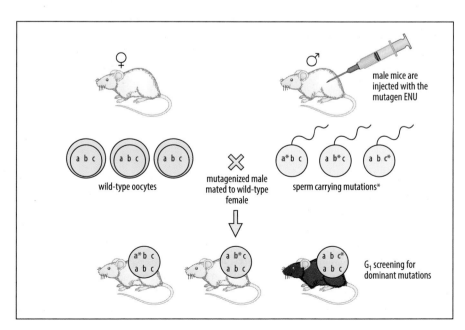

Fig. 3.36 A mutagenesis screen for dominant mutations in the mouse. The male mouse is treated with the mutagen ethylnitrosourea (ENU). After mating with an untreated female, progeny with a dominant mutation, such as the dark-brown coat color shown here, can be detected in the first generation (G_1).

EXPERIMENTAL BOX 3C Large-scale mutagenesis screens for recessive mutations in zebrafish

A screening program for recessive mutations in zebrafish involves breeding for three generations (see Figure 1). Male fish treated with a chemical mutagen are crossed with wild-type (+/+) females to generate F_1 offspring, each of which is likely to be heterozygous for a point mutation (m*) in a different gene. Each F_1 male offspring is crossed again with a wild-type female to generate different F_2 families of fish. If the parent F_1 fish carries a mutation (say, m1), 50% of the fish in the m1 F_2 family will be heterozygous (m1/+) for that mutation, so that when the female and male siblings from this F_2 family are themselves crossed, 25% of these matings will be between two heterozygotes and 25% of their offspring (the F_3 generation, shown here as fertilized eggs) will be homozygous for the mutation. The final step is then to identify the genetic lesion that has led to the mutant phenotype. Zebrafish can also be made to develop as haploids by fertilizing the egg with sperm heavily irradiated with UV light. This allows one to detect early-acting recessive mutations without having to breed the fish to obtain homozygous embryos.

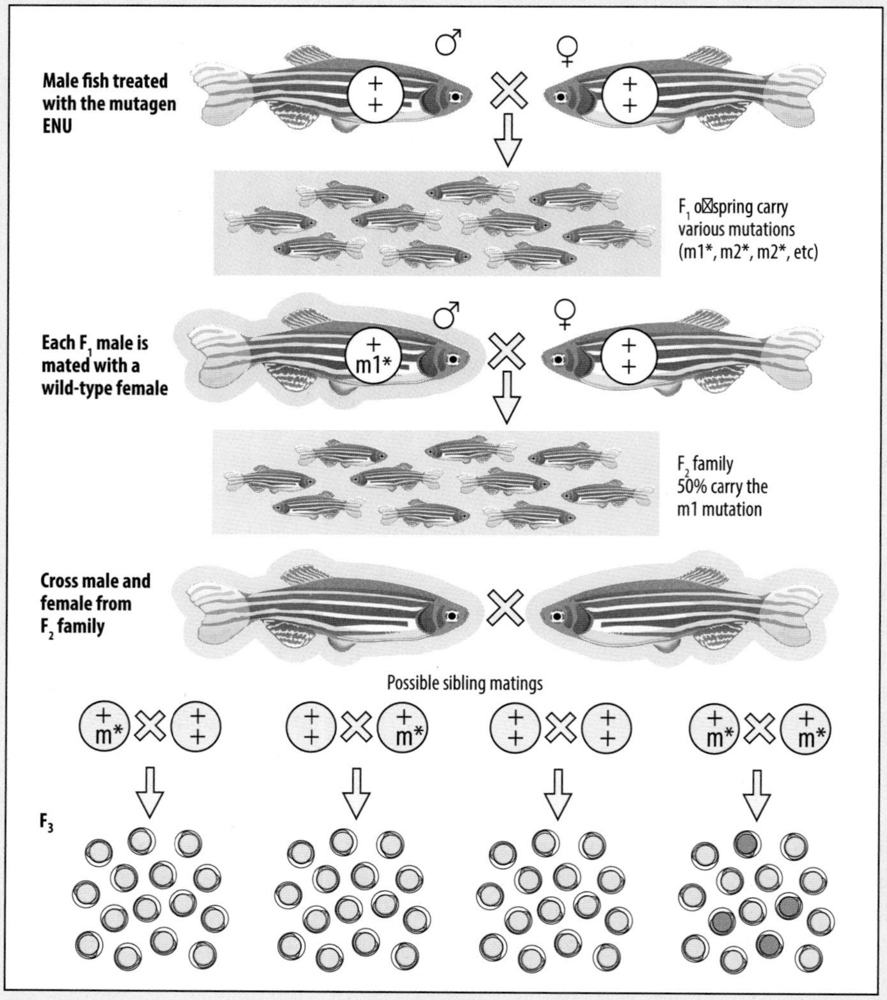

Offspring of male fish treated with a chemical mutagen, which will have mutations in different genes, can be screened to identify an individual fish with mutations in a specific gene by an approach known as **TILLING** (targeting-induced local lesions in genomes). This technique was first developed for plants, but has since been applied to *Drosophila, Caenorhabditis,* and the rat, as well as zebrafish. All that is necessary is knowledge of the DNA sequence of the gene. Male fish treated with ENU are crossed with wild-type females. A sample of tissue from each F_1 male offspring, which will have a point mutation in a different gene, is then collected and, at the same time, its sperm are frozen

Figure 1

and stored. To identify the individual fish that has a mutation in the gene of interest, large amounts of the DNA are made from the relevant genomic region from the tissue samples from all the F_1 male offspring by the polymerase chain reaction (PCR). This DNA is then hybridized with unmutated DNA from the gene in question. Mismatched bases will reveal the site of a mutation and the DNA is then sequenced to identify the precise change. The individual male fish with the desired mutation and/or its frozen sperm is then used to generate families of mutant fish, which will reveal any phenotypic changes associated with the mutant gene.

recessive mutations. Screens for recessive mutations have also been carried out in mice, but because of the smaller number of progeny in the mouse, and the fact that embryos cannot be screened so easily, they are more difficult and costly to set up. Nevertheless, many such screens have been carried out.

3.10 Transgenic techniques enable animals to be produced with mutations in specific genes

Reverse genetics, the other main approach to discovering genes that have important developmental functions, starts with the gene and then analyses the phenotype produced when that particular gene is altered. In the TILLING technique as used for zebrafish (see Box 3C), random mutagenesis is used to generate families of mutant fish. Fish with the desired mutation are identified first by DNA analysis and their phenotype is then analyzed to characterize any changes. It is, however, even more efficient to alter the gene of interest in the animal directly, rather than having to trust to random mutagenesis. This can be done by a variety of techniques that either delete the targeted gene or replace it with another version in which mutations have been made. Animals into which an additional or altered gene has been introduced are known as **transgenic** animals, and strains of transgenic mice with a particular mutant genetic constitution can now be produced relatively routinely. And as long as the mutation is recessive, animals heterozygous for the mutation can be maintained as breeding stock, even for mutations that may be lethal to the embryo when homozygous.

Transgenic zebrafish can be obtained by injecting DNA into the one-cell-stage embryo and the first transgene to be successfully delivered into zebrafish was the delta-crystallin gene of the chicken. Transgenesis has been accomplished in chickens, by the injection of lentiviral vectors containing the gene encoding GFP under the blastoderm of a newly laid egg (Fig. 3.37). And although *X. laevis* is less amenable to germline transgenesis, the transgenic techniques that have been developed for its close relative *X. tropicalis*, which is diploid and has a much shorter generation time of 5–9 months, can also be used. For *Xenopus*, the strategy is to introduce the DNA transgene into sperm nuclei and then transplant these nuclei into unfertilized eggs. The relative suitability of our model vertebrates for genetics-based techniques is summarized in Fig. 3.38.

Transgenic techniques are most highly developed in mice. Two main techniques for producing transgenic mice are currently in use. One is to inject a **transgene** composed of DNA encoding the required gene and any necessary regulatory regions directly into the male pronucleus of a fertilized egg, and then replace the egg in a surrogate mother. If the transgene becomes incorporated into the genome it will be present in all the cells of the embryo, including the germline, and will be expressed according to the promoter and other regulatory regions it contains. In this way, mice that over-express a given gene in particular cells or at particular times during development can be produced, and the effect of misexpressing a gene in a tissue or at a time where it is not usually active can be investigated.

The other technique for producing transgenic mice uses embryonic stem cells (ES cells) that have been mutated *in vitro*, and this is now the most commonly used transgenic technique for producing specific **gene knock-outs**—the permanent deletion of a gene function. ES cells are pluripotent cells derived from the inner cell mass

Fig. 3.37 Chicks transgenic for green fluorescent protein (GFP). Transgenic chickens can be obtained by injecting DNA carried by a self-inactivating lentiviral vector into the subgerminal space beneath the blastodisc in newly laid eggs. The chicks illustrated are second-generation offspring from an original transgenic bird which carried the GFP transgene in the germline. The bird in the center is not transgenic.

*Photograph from McGrew, et al.: **Efficient production of germline transgenic chickens using lentiviral vectors**. EMBO Rep 2004, **5**: 728-733.*

The suitability of the different model vertebrates for genetics-based techniques for studying development				
	Frog (*Xenopus laevis*)	**Zebrafish**	**Chick**	**Mouse**
Genome sequenced	The diploid *X. tropicalis* genome has been sequenced	Yes	Yes	Yes
Spontaneous mutations	No	Yes	Yes	Yes
Induced mutations (mutational screens)	Yes	Yes (see Box 3C)	No	Yes
Gene silencing or gene knock-out in somatic cells	Gene silencing by morpholino antisense RNAs (MOs) (Box 6B)		Gene silencing by MOs and RNA interference (RNAi) (Box 6B)	Gene knock-out by Cre/*loxP* system (requires making transgenic animals first, Box 3D)
Germline gene knock-out (transgenic animals)	No	Yes	Not yet	Yes
Gain-of-function (e.g. misexpression and ectopic expression) of specific genes in the whole embryo	By injection of mRNAs into the fertilized egg or very early embryo (see e.g. Fig. 4.6)		Not routinely possible so far but transgenic chicks have been produced (Fig. 3.37)	Transgenic mice with gain-of-function mutations can be made
Targeting of gene overexpression or misexpression in time and space	Expression can be targeted to some extent by injection of RNAs into specific blastomeres	No	Yes in somatic tissue	Transgenic mice with inducible or cell-type specific expression of a transgene can be made
Embryonic stem cells can be cultured and differentiated *in vivo*	No	No	Yes	Yes. Mutant transgenic mice can also be made using ES cells mutated by homologous recombination (see Section 3.10 and Chapter 8)

Fig. 3.38 The suitability of the different model vertebrates for genetics-based study of development.

of a mouse blastocyst, as described in Chapter 8. They can be maintained in culture indefinitely and grown in large numbers, and are being investigated for their possible use in regenerative medicine.

ES cells injected into the cavity of an early blastocyst become incorporated into the inner cell mass. When the blastocyst is then introduced into a surrogate mother to continue its development, the ES cells that have been incorporated into the inner cell mass can contribute to all the tissues of the mouse that develops, even giving rise to germ cells and gametes. For example, if ES cells from a brown-pigmented mouse are introduced into the inner cell mass of a blastocyst from a white mouse, the mouse that develops from this embryo will be a chimera made up of cells of the ES cell genotype and the blastocyst cell genotype. In the skin, this mosaicism is visible as patches of brown and white hairs (Fig. 3.39).

To generate transgenic mice with a particular mutation, the ES cells are genetically manipulated while growing in culture and these altered cells are then introduced into the inner cell mass of a blastocyst (see Fig. 3.39). Mutations are targeted to a particular gene by the technique of **homologous recombination**, in which specially tailored DNA constructs are introduced into the cell by **transfection**. A transfected DNA molecule will usually insert randomly in the genome, but by including some sequence that is homologous with the desired target gene, it is possible to make it insert at a predetermined site (Fig. 3.40). Homologous recombination between the transfected DNA and the target gene in the ES cell results in an insertion that renders the gene non-functional. Insertional mutagenesis has been the main way of producing gene knock-outs, as homologous recombination works particularly well in mice. The DNA to be introduced must contain enough sequence homology with its target to insert within it in at least a few cells in the ES cell culture. The introduced DNA usually carries a drug-resistance gene so that cells containing the insertion can be selected by adding the drug, which kills all the other, unmodified, cells. When one gene is replaced by another functional gene using these techniques, it is called a **gene knock-in**.

Because the animals produced by ES-cell transfer are chimeras composed of both genetically modified and normal cells, they may show few, if any, effects of the mutation. If ES cells with the mutant gene have entered the germline, however, strains of transgenic mice heterozygous for the altered gene can be intercrossed to produce homozygotes that are viable or fail to develop, depending on the gene involved. We shall see examples of gene knock-outs in mice being used to study the functions of Hox genes in regionalizing the somitic mesoderm and the neural tube in Chapters 5 and 12. Strategies, such as the Cre/*loxP* system, have been developed in order to target gene knock-out to a particular tissue and/or to a particular time in development (Box 3D).

Transgenic mouse embryos expressing a reporter gene linked to a suitable promoter, or in a particular set of cells, under the control of the Cre/*loxP* system provide a convenient way of tracing cell lineage. Techniques such as these are continually being refined to give better resolution, and are particularly useful for tracing cell lineage during later events in embryogenesis, such as organ development, the development of the nervous system, and cell behavior in epithelia in which cells are continually being replaced, such as skin and gut (see Section 8.12). Clonal cell-lineage analysis is now becoming possible using a transgenic mouse in which a drug-inducible Cre recombinase (linked to the reporter gene *lacZ* (see Box 1D) or to *GFP*) is expressed in every cell; the pregnant mouse can then be treated with an appropriate dose of the drug such that a low frequency of recombination occurs and clones of marked cells are generated in the embryos. This strategy led to the discovery of dorso-ventral compartments in the developing mouse limb (see Chapter 11).

Recent novel methods for generating targeted mutations rely on engineering specificity into transgenes that encode nuclease enzymes that cleave DNA (**zinc-finger nucleases (ZFNs)** and **TALENS** (transcription activator-like effector nucleases)) or into RNAs that guide nucleases to cut the DNA (the **CRISPR/Cas9 system**), and using them to cleave and disrupt chosen genes. Cas9 (CRISPR-associated protein) is a nuclease originally derived from bacteria that uses 'guide' RNAs encoded by the CRISPR array of repetitive sequences to target the enzyme to its target gene. The CRISPR array and the *Cas9* gene have been adapted for use as transgenes that can be expressed in mammalian cells, and the array can be engineered to provide guide RNAs that target the Cas9 nuclease to specific mammalian genes.

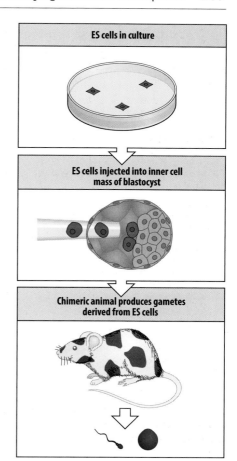

Fig. 3.39 Production of a chimeric mouse by the introduction of ES cells into the blastocyst. ES cells injected into a blastocyst will become part of the inner cell mass, and can give rise to all the cell types in the mouse, including the germline.

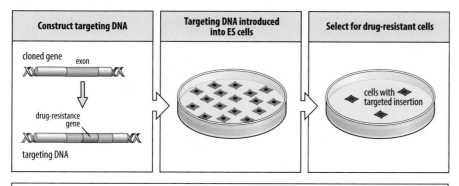

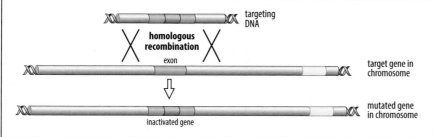

Fig. 3.40 Production of mutant mouse ES cells by homologous recombination. The gene to be targeted for gene knock-out is first cloned and a drug-resistance or other selectable marker is inserted into it, which also renders the gene non-functional. When the modified gene is introduced into cultured mouse ES cells, it will undergo homologous recombination with the corresponding gene in a small number of the ES cells. These can then be selected by their drug resistance and then introduced into a normal mouse blastocyst, where they can contribute to all the tissues in the mouse that develops.

EXPERIMENTAL BOX 3D The Cre/*loxP* system: a strategy for making gene knock-outs in mice

One very powerful and widely used way of targeting a gene knock-out to a specific tissue and/or a particular time in development is provided by the **Cre/*loxP* system** (Figure 1). The target gene is first 'floxed' by inserting a *loxP* sequence of 34 base pairs on either side of the gene (hence 'floxed', for 'flanked by *lox*'). This targeting is carried out using homologous recombination in ES cells in culture (see Section 3.10). Transgenic mice are then produced by introducing the genetically modified ES cells into blastocysts (see Fig. 3.39) and deriving lines of homozygous transgenic mice all carrying the gene. The line of mice with the floxed gene is then crossed with another line of transgenic mice carrying the gene for the recombinase Cre. *loxP* sequences are recognized by Cre, and the resulting recombination reaction excises all the DNA between the two *loxP* sites. If Cre is expressed in all cells in the offspring, then all cells will excise the floxed target, causing a knock-out of the target gene in every cell. However, if the gene for Cre is under the control of a tissue-specific promoter, so that, for example, it is only expressed in heart tissue, the target gene will only be excised in heart tissue (see Figure 1). If the *Cre* gene is linked to an inducible control region, it is possible to induce excision of the target gene at will by exposing the mice to the inducing stimulus (see Figure 1).

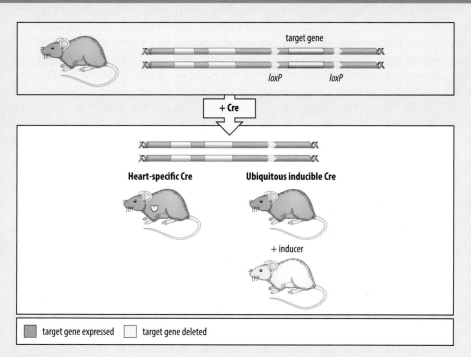

Figure 1

As well as knocking out a coding sequence, the Cre/*loxP* method can be used to target promoters in such a way that the associated gene is permanently turned on. This technique is used in cell-lineage studies in mice to turn on a reporter gene such as *GFP* or *lacZ* in particular cells at a particular stage in development. The progeny of the labeled cells can then be distinguished in later-stage embryos or after birth.

These new methods not only work in mouse ES cells but also can be applied directly to fertilized mouse oocytes. An advantage of the CRISPR/Cas9 system is that it is possible to produce mutations in several genes simultaneously by including the sequences for several guide RNAs in the transgene. This means that it is possible to make a double-mutant mouse in one step. The same approaches also work in zebrafish. Zinc-finger nucleases, for example, have been used to target the No-tail (*ntl*) gene, the zebrafish version of *brachyury*. Like *brachyury* in the mouse (see Fig. 1.12), *No-tail* is required for tail formation. About 30% of zebrafish embryos injected at the one-cell stage with mRNA for *ntl*-specific ZFN lacked tails, showing that *ntl* had been rendered non-functional.

A significant number of knock-outs of a single gene result in mice developing without any obvious abnormality, or with fewer and less severe abnormalities than might be expected from the normal pattern of gene activity. A striking example is that of *myoD*, a key gene in muscle differentiation, which is described in Chapter 8. In *myoD* knock-outs, the mice are anatomically normal, although they do have a reduced survival rate. The most likely explanation for this is that the process of muscle differentiation contains a certain amount of redundancy and that other genes can substitute for some of the functions of *myoD* (see Section 8.5).

However, it is unlikely that any gene is without any value at all to an animal. It is much more likely that there is an altered phenotype in these apparently normal animals, which is too subtle to be detected under the artificial conditions of life in a laboratory. Redundancy is thus probably apparent rather than real. A further complication is the possibility that, under such circumstances, related genes with similar functions may increase their activity to compensate for the mutated gene.

3.11 Gene function can also be tested by transient transgenesis and gene silencing

Germline transgenesis is not the only way of altering gene expression, and producing transgenic mice is still a relatively expensive process. Techniques of **transient transgenesis** or **somatic transgenesis** are easier and cheaper to implement, and are particularly valuable in amphibians and the chick, in which routine germline transgenesis has not been a practical option. The early *Xenopus* embryo has large blastomeres and is very suitable for experiments in which the mRNA of a gene suspected of having a developmental role is injected into a specific blastomere to see the effect of its overexpression. mRNA injection is also used in *Xenopus* and other embryos to see if a specific gene can 'rescue' normal development after a region of the early embryo has been removed or the embryo's own gene has been silenced. Experiments of this type are described in Chapter 4.

Transient transgenesis for both short-term and long-term overexpression or misexpression of genes is also widely used to study gene function in the chick embryo. For short-term expression, genes are delivered into cells by electroporation (see Section 3.7). For longer-term overexpression, the gene is inserted into a replication-competent retroviral vector, which is then injected into the embryo. The infected cells express the gene as part of the viral replication cycle and, as the virus replicates, it spreads the gene to neighboring cells.

The transient counterpart to the gene knock-out is the technique of **gene silencing** or **gene knockdown**, in which a gene is not mutated or removed from the genome, but its expression is blocked by targeting the mRNA and preventing its translation. Techniques for doing this are discussed in Box 6B. Because it is not possible to generate loss-of-function mutations in *Xenopus* and chick embryos, gene silencing by the injection of **morpholino antisense RNA** is particularly useful, and this technique is also widely used in zebrafish. Morpholinos are stable RNAs designed to be complementary to a specific mRNA. When introduced into the cells of an embryo, they hybridize with the target mRNA, thus preventing its translation into protein. Injection of a morpholino into a fertilized *Xenopus* egg will knock down gene expression in all the cells of the early embryo; electroporation of morpholinos into cells of early chick embryos will produce local knockdown. The related technique of **RNA interference** (RNAi), also described in Box 6B, is used in embryology. In this case, 'short interfering RNAs' of defined sequence act as specific guide RNAs to target a nuclease to the relevant mRNA, which is then degraded. RNAi constructs encoding the interfering RNAs have been electroporated into the neural tube of chick embryos to knock down gene expression.

3.12 Gene regulatory networks in embryonic development can be revealed by chromatin immunoprecipitation techniques

One powerful approach to investigating development relies on a combination of bioinformatics and experimental molecular and cell biology to describe the network of interactions between genes and their transcription factors, and so identify **gene**

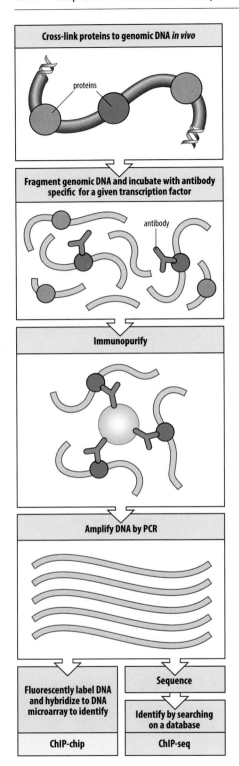

Cross-link proteins to genomic DNA *in vivo*

proteins

Fragment genomic DNA and incubate with antibody specific for a given transcription factor

antibody

Immunopurify

Amplify DNA by PCR

Fluorescently label DNA and hybridize to DNA microarray to identify

ChIP-chip

Sequence

Identify by searching on a database

ChIP-seq

Fig. 3.41 ChIP-chip and ChIP-seq. These two variants of chromatin immunoprecipitation are used to discover the binding sites in DNA for transcription factors and other DNA-binding proteins. After cross-linking any bound proteins to the DNA by chemical treatment, the DNA is fragmented and treated with an antibody specific for the protein of interest. The antibody complex is purified and the DNA extracted. It is then amplified by the polymerase chain reaction (PCR) and either analyzed by DNA microarray (ChIP-chip) or by DNA sequencing (ChIP-seq).

regulatory networks that underlie a particular developmental process. Once the genome sequence of an animal is known, the technique of chromatin immunoprecipitation followed by DNA microarray analysis (**ChIP-chip**), can be used to identify the target genes that a given transcription factor binds to *in vivo* (Fig. 3.41).

Embryos at a given stage are treated so that proteins bound to the *cis*-regulatory elements in the DNA are chemically linked to the DNA. The DNA is then isolated and sheared into small segments and treated with an antibody specific for a given transcription factor. This will precipitate, or 'pull down,' all the DNA segments to which that transcription factor was bound in the embryo's cells. The identity of the DNA can then be determined by digesting away the protein, labeling the DNA segment and hybridizing it to DNA microarrays (see Box 3B) containing genomic sequences lying upstream of known genes, and which are most likely to contain regulatory sequences. Alternatively, the protein-free DNA can be sequenced, when the technique is known as **ChIP-seq**. The gene associated with these regulatory sequences can then be identified by searching the databases of genome sequences. ChIP-seq has been used, for example, to identify the gene targets of the transcription factor Ntl (No-tail), the zebrafish equivalent of Brachyury, in mid-gastrula zebrafish embryos, and thus begin to uncover the gene regulatory network involved in mesoderm specification and patterning in this model animal. Other gene regulatory networks are also being discovered downstream of Gli transcription factors that mediate Shh signaling in the developing limb (Chapter 11) and in the neural tube (Chapter 12).

Summary to Chapter 3

- All vertebrates have a similar basic body plan. The defining vertebrate structures are the vertebral column, which surrounds the spinal cord, and the bony or cartilaginous skull that forms the head and encloses the brain.
- The vertebrate body has three main axes of symmetry—the antero-posterior axis running from head to tail, the dorso-ventral axis running from back to belly, and bilateral symmetry around the dorsal midline, so that outwardly the right and left sides are mirror images of each other. Some internal organs, such as lungs, kidneys, and gonads, are also present as symmetrically paired structures although their position and morphology may differ left versus right, but single organs, such as the heart and liver, are arranged asymmetrically with respect to the dorsal midline.
- Early vertebrate embryos pass through a set of developmental stages—cleavage, gastrulation, and neurulation—to form embryos that broadly resemble each other—a stage known as the phylotypic stage.
- There are considerable differences in the development of the different model organisms before the phylotypic stage, which relate to how and when the axes are set up, and how the germ layers are established. These differences are mainly due to the different modes of reproduction of the animals which affects the amount of yolk in the egg and hence the consequent form of the earliest embryo.
- Structures typical of the different vertebrate groups, such as fins, beaks, wings, and tails, develop after the phylotypic stage.

- The body pattern and the initial organ rudiments are laid down when the embryo is still very small. Growth in size takes place at later stages.
- The mechanisms underlying development can only be studied by disturbing the normal developmental process in specific ways and observing what happens.
- Techniques for interfering with development can be broadly divided into two types: experimental embryological techniques that manipulate the embryo by removing or adding cells to cleavage-stage embryos or transplanting blocks of cells from one embryo to another; and genetics-based techniques that disturb the expression of developmentally important genes by mutation, gene silencing, overexpression or misexpression.

■ End of chapter questions

Long answer (concept questions)

1. All vertebrates share certain features; what are those features? Referring to Fig. 3.2, identify those features in the row of illustrations that shows the phylotypic stage.

2. After fertilization, all embryos undergo cleavage. In the *Xenopus* embryo, what is the relationship of the animal-vegetal axis to the plane of the first three cleavages? Describe how cleavage in the chick embryo differs from cleavage in frogs and mice; why do these differences occur?

3. Compare the terms blastula, blastoderm, and blastocyst. To what organisms do they refer? What features do they share, and how do they differ?

4. Gastrulation in *Xenopus* leads to the establishment of the three germ layers through the process of involution. What are the three germ layers, and what is their relationship to each other in the late gastrula (stage 12)? What is the archenteron, and what will it become?

5. Distinguish between the following regions in the chick embryo: epiblast and hypoblast; posterior marginal zone, Koller's sickle, and Hensen's node; endoblast and endoderm; mesenchyme and mesoderm.

6. Hensen's node of the gastrulating chick embryo is conceptually analogous to the blastopore of the *Xenopus* embryo, but despite this analogy, several differences between the two structures exist. Elaborate on the following differences: (a) the ingression of cells in the chick versus the involution of cells in the frog; (b) the movement of Hensen's node in the chick versus the lack of movement of the blastopore.

7. At the neurulation stage in the chick embryo, the precursors to several adult structures first form. What structures will form from: neural plate; somites; intermediate mesoderm; splanchnic mesoderm; notochord? (See Fig. 3.17.)

8. Distinguish between the inner cell mass and the trophectoderm in the mouse embryo. Outline the events that will lead to the formation of the epiblast and a primitive streak.

9. What are some of the features that would make an animal species a good 'model organism' for the study of development? For example, what features make a chicken an attractive model for development compared to a frog? In contrast, what advantages does the mouse offer over chickens?

10. Briefly summarize the production of transgenic mice and of knock-out mice using embryonic stem cells. How do the two processes differ, both in technique and in outcome?

11. You have discovered that a particular gene encoding a transcription factor is expressed in the developing liver at a certain stage in both mouse and chick embryos. How could you set about finding its function? Describe the experiments you would carry out in both model organisms.

Multiple choice (factual recall questions)

NB There is only one right answer to each question.

1. In which portion of the frog's life cycle would a frog appear most similar to a mammal?

a) blastula
b) phylotypic stage
c) gastrulation
d) fertilized egg

2. The darkly pigmented end of a *Xenopus* egg is called

a) dorsal
b) the animal pole
c) the vegetal pole
d) yolk

3. Which of the following choices puts the early developmental stages of the frog in their correct order?

a) blastula, fertilized egg, gastrula, neurula
b) cleavage, gastrulation, neurulation, organogenesis
c) neurulation, gastrulation, cleavage, organogenesis
d) tadpole, embryo, neurula, gastrula

4. Cleavage in the zebrafish gives rise to the

a) 14-somite stage
b) gastrula
c) sphere stage
d) shield stage

5. The primitive streak of mouse and chick embryos is equivalent to which amphibian structure?

a) archenteron
b) blastocoel
c) blastopore
d) neural folds

6. At the blastocyst stage, the mammalian embryo is composed of

a) a blastocoel, yolk, and animal cap cells
b) an epiblast overlying a hypoblast
c) ectodermal, mesodermal, and endodermal cells
d) trophectoderm and inner cell mass

7. All the genes expressed in a particular tissue or at a particular stage of development can be identified using

a) *in situ* hybridization
b) insertional mutagenesis
c) microarrays
d) zinc-finger nucleases

8. The advantage of using the Cre/*loxP* system for targeted mutagenesis is

a) a gene can be deleted in a particular tissue or at a desired time in development
b) the Cre/*loxP* system allows germline transgenesis in amphibians and chickens
c) the Cre/*loxP* system is more specific than homologous recombination
d) the deletion of a gene by the Cre/*loxP* system is reversible

9. The least favorable model organism for traditional genetic analysis is

a) *Drosophila*
b) mouse
c) *Xenopus*
d) Zebrafish

10. Electroporation is

a) fusion of adjacent cells using an electrical pulse
b) injection of a gene into the nucleus of a cell
c) the introduction of a fluorescent lineage maker into cells
d) use of brief pulses of electricity to cause cells to take up DNA

Multiple choice answer key

1: b, 2: b, 3: b, 4: c, 5: c, 6: d, 7: c, 8: a, 9: c, 10: d.

■ General further reading

Bard, J.B.L.: *Embryos. Color Atlas of Development*. London: Wolfe, 1994.
Carlson, B.M.: *Patten's Foundations of Embryology*. New York: McGraw-Hill, 1996.

Descriptions of developmental stages of frog, zebrafish, chick and mouse embryos

Xenopus: developmental stages: Xenbase – Nieuwkoop and Faber stage series [http://www.xenbase.org/xenbase/original/atlas/NF/NF-all.html] (accessed 21 June 2014); movies of cleavage and gastrulation [http://www.xenbase.org] (accessed 21 June 2014).
Zebrafish: developmental stages: Karlstrom Lab ZFIN Embryonic Developmental Stages [http://zfin.org/zf_info/zfbook/stages/stages.html] (accessed 21 June 2014); developmental movie: Karlstrom, R.O., Kane, D.A.: **A flipbook of zebrafish** embryogenesis. *Development* 1996, **123**: 1–461; Karlstom Lab [http://www.bio.umass.edu/biology/karlstrom/] (accessed 21 June 2014).
Chick: list and description of Hamilton & Hamburger developmental stages: UNSW Embryology – Chicken Developmental Stages [http://embryology.med.unsw.edu.au] (accessed 21 June 2014 2010).
Mouse: Edinburgh Mouse Atlas Project [http://www.emouseatlas.org] (accessed 21 June 2014).
Human: The Multidimensional Human Embryo [http://embryo.soad.umich.edu] (accessed 21 June 2014); UNSW Embryology: Carnegie staging of human embryos [http://embryology.med.unsw.edu.au] (accessed 21 June 2014).

■ Section further reading

3.1 The frog *Xenopus laevis* is the model amphibian for developmental studies

Hausen, P., Riebesell, H.: *The Early Development of* Xenopus laevis. Berlin: Springer-Verlag, 1991.
Nieuwkoop, P.D., Faber, J.: *Normal Tables of* Xenopus laevis. Amsterdam: North Holland, 1967.

3.2 The zebrafish embryo develops around a large mass of yolk

Kimmel, C.B., Ballard, W.W., Kimmel, S.R., Ullmann, B., Schilling, T.F.: **Stages of embryonic development of the zebrafish**. *Dev. Dyn.* 1995, **203**: 253–310.
Warga, R.M., Nusslein-Volhard, C.: **Origin and development of the zebrafish endoderm**. *Development* 1999, **26**: 827–838.
Westerfield, M. (Ed.): *The Zebrafish Book; A Guide for the Laboratory Use of Zebrafish* (Brachydanio rerio). Eugene, Oregon: University of Oregon Press, 1989.

3.4 The early chicken embryo develops as a flat disc of cells overlying a massive yolk

Bellairs, R., Osmond, M.: *An Atlas of Chick Development* (2nd edn). London: Academic Press, 2005.
Chuai, M., Weijer, C.J.: **The mechanisms underlying primitive streak formation in the chick embryo**. *Curr. Top. Dev. Biol.* 2008, **81**: 135–156.
Hamburger, V., Hamilton, H.L.: **A series of normal stages in the development of a chick**. *J. Morph.* 1951, **88**: 49–92.
Lillie, F.R.: *Development of the Chick: An Introduction to Embryology*. New York: Holt, 1952.
Patten, B.M.: *The Early Embryology of the Chick*. New York: McGraw-Hill, 1971.
Stern, C.D.: **Cleavage and gastrulation in avian embryos (version 3.0)**. *Encyclopedia of Life Sciences* 2009, http://www.els.net/.

3.5 The mouse egg has no yolk and early development involves the allocation of cells to form the placenta and extra-embryonic membranes

Cross, J.C., Werb, Z., Fisher, S.J.: **Implantation and the placenta: key pieces of the developmental puzzle**. *Science* 1994, **266**: 1508–1518.

Hu, D., Cross J.C.: **Development and function of trophoblast giant cells in the rodent placenta**. *Int. J. Dev. Biol.* 2010, **54**: 341–354.

Kaufman, M.H.: *The Atlas of Mouse Development* (2nd edition). London: Academic Press, 1992.

Kaufman, M.H., Bard, J.B.L.: *The Anatomical Basis of Mouse Development*. London: Academic Press, 1999.

3.6 The early development of a human embryo is similar to that of the mouse

Drews, U.: *Color Atlas of Embryology*. Stuttgart and New York: Thieme, 1995.

Niakan, K.K., Han, J., Pedersen, R.A., Simon, C., Pera, R.A.R.: **Human pre-implantation embryo development**. *Development* 2012, **139**: 829–841.

Schoenwolf, G.C., Belyl, S.B., Brauer, P.R., Francis-West, P.H.: *Larsen's Human Embryology* (4th edn). London and New York: Churchill Livingstone, 2008.

Zhu, Z., Huangfu, D.: **Human pluripotent stem cells: an emerging model in developmental biology**. *Development* 2013, **140**: 705–717.

Experimental approaches to studying vertebrate development

Dale, L., Slack, J.M.: **Fate map for the 32-cell stage of Xenopus laevis**. *Development* 1987, **99**: 527–551.

Doyon, Y., McCammon, J.M., Miller, J.C., Faraji, F., Ngo, C., Katibah, G.E., Amora, R., Hocking, T.D., Zhang, L., Rebar, E.J., Gregory, P.D., Urnov, F.D., Amacher, S.L.: **Heritable targeted gene disruption in zebrafish using designed zinc-finger nucleases**. *Nat. Biotechnol.* 2008, **26**: 702–708.

Hogan, H., Beddington, R., Costantini, F., Lacy, E.: *Manipulating the Mouse Embryo. A Laboratory Manual* (2nd edn). New York: Cold Spring Harbor Laboratory Press, 1994.

McGrew, M.J., Sherman, A., Ellard, F.M., Lillico, S.G., Gilhooley, H.J., Kingsman, A.J., Mitrophanous, K.A., Sang, H.: **Efficient production of germline transgenic chickens using lentiviral vectors**. *EMBO Rep.* 2004, **5**: 728–733.

Sharpe, P., Mason, I.: *Molecular Embryology Methods and Protocols* (2nd edn). New York: Humana Press; 2008.

Southall, T.D., Brand, A.H.: **Chromatin profiling in model organisms**. *Brief. Funct. Genomics Proteomics* 2007, **6**: 133–140.

Stern, C.D.: **The chick: a great model system just became even greater**. *Dev. Cell* 2004, **8**: 9–17.

Tabata, H., Nakajima, K.: **Efficient *in utero* gene transfer system to the developing mouse brain using electroporation: visualization of neuronal migration in the developing cortex**. *Neuroscience* 2001, **103**: 865–872.

Wang, H., Yang, H., Shivalila, C.S., Dawlaty, M.M., Cheng, A.W., Zhang, F., Jaenisch, R.: **One-step generation of mice carrying mutations in multiple genes by CRISPR/Cas-mediated genome engineering**. *Cell* 2013, **153**: 910–918.

Zernicka-Goetz, M., Pines, J., Ryan, K., Siemering, K.R., Haseloff, J., Evans, M.J., Gurdon, J.B.: **An indelible lineage marker for *Xenopus* using a mutated green fluorescent protein**. *Development* 1996, **122**: 3719–3724.

Box 3A Preimplantation genetic diagnosis

Brezina, P.R., Brezina, D.S., Kearns, W.G.: **Preimplantation genetic testing**. *BMJ* 2012, **345**: e5908.

Box 3B Gene-expression profiling by DNA microarrays and RNA seq

Hamatani, T., Carter, M.G., Sharov, A.A., Ko, M.S.: **Dynamics of global gene expression changes during mouse preimplantation development**. *Dev. Cell* 2004, **6**: 117–131.

Malone, J.H., Oliver, B.: **Microarrays, deep sequencing and the true measure of the transcriptome**. *BMC Biol.* 2011, **9**: 34–43.

Vassena, R., Boué, S., González-Roca, E., Aran, B., Auer, H., Veiga, A., Izpisúa Belmonte, J.C.: **Waves of early transcriptional activation and pluripotency program initiation during human preimplantation development**. *Development* 2011, **138**: 3699–3709.

Box 3C Large-scale mutagenesis in zebrafish

Lawson, N.D., Wolfe, S.A.: **Forward and reverse genetic approaches for analysis of vertebrate development in the zebrafish**. *Dev. Cell* 2011, **21**: 48–64.

Stemple, D.: **Tilling—a high throughput harvest for functional genomics**. *Nat. Rev. Genet.* 2004, **5**: 1–6.

Box 3D The Cre/*loxP* system: a strategy for making gene knock-outs in mice

Menke, D.B.: **Engineering subtle targeted mutations into the mouse genome**. *Genesis* 2013, **51**: 605–618.

Yu, Y., Bradley, A.: **Engineering chromosomal rearrangements in mice**. *Nat. Rev. Genet.* 2001, **2**: 780–790.

4

Vertebrate development II: *Xenopus* and zebrafish

- Setting up the body axes
- The origin and specification of the germ layers

- The Spemann organizer and neural induction

- Development of the body plan in zebrafish

In this chapter we examine the earliest stages of vertebrate development, using Xenopus as our model organism. We will consider how the embryo develops from the fertilized egg to a stage in which it has distinct body axes and the germ layers have been specified. We will first explore the relative roles of pre-existing maternal factors in the egg and the embryo's own developmental program in the formation of the body axes. Another crucial process in early vertebrate development is the development of the three germ layers—ectoderm, mesoderm, and endoderm—and we shall see how this is achieved. We will then consider the formation of the embryonic organizer—the Spemann organizer—and its role in patterning the emerging body plan along the dorso-ventral and antero-posterior axes. This patterning is carried out by a combination of signals from the organizer together with signals from other regions of the embryo that provide positional information which is then interpreted by cells. After the initial patterning, the germ layers move into their appropriate positions in the body plan in the process of gastrulation. A key developmental event that accompanies gastrulation is the induction of the nervous system from the dorsal ectoderm by a complex cascade of signals from adjacent tissues, including the organizer, and we shall consider that here. Having followed the whole process in Xenopus, we will then look briefly at how the body plan is established in another vertebrate, the zebrafish.

Vertebrate development proceeds with respect to well-defined antero-posterior (head to tail) and dorso-ventral (back to belly) axes as outlined in Chapter 3. In this chapter we will look in detail at the key processes in vertebrate development that establish the body plan. The first step is the establishment of these axes, and the second is the specification and early patterning of the germ layers, in particular the induction of the mesodermal germ layer and its patterning along the dorso-ventral axis. We shall then look at patterning of the vertebrate embryo along the antero-posterior axis and the all-important induction of the prospective nervous system from ectoderm, taking

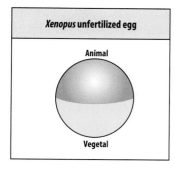
Xenopus unfertilized egg

Animal

Vegetal

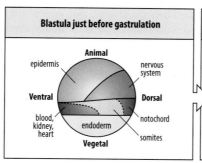
Blastula just before gastrulation

Animal

epidermis · nervous system

Ventral · Dorsal

blood, kidney, heart · endoderm · notochord · somites

Vegetal

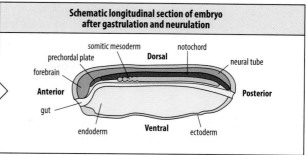
Schematic longitudinal section of embryo after gastrulation and neurulation

somitic mesoderm · notochord · **Dorsal**

prechordal plate · neural tube

forebrain

Anterior · **Posterior**

gut

endoderm · **Ventral** · ectoderm

Xenopus as our model organism. We shall look at *Xenopus* first because many of the major discoveries about how these processes take place were made in amphibian embryos. Once we have seen how the complete process of laying down the body plan is accomplished in *Xenopus*, we will consider how it is accomplished in the embryo of another model vertebrate, the zebrafish, which is quite similar. We will examine how the body plan develops in chick and mouse embryos in Chapter 5.

Some aspects of axis formation and germ-layer development in amphibians are already specified by cytoplasmic factors deposited in the egg by the mother during oogenesis. Maternal factors in the animal region of the *Xenopus* egg specify that region as prospective ectoderm, whereas maternal factors in the vegetal half of the egg specify that region as prospective endoderm. The prospective mesoderm, in contrast, is produced after development has begun by signals from the vegetal-region cells acting on adjacent animal-region cells to specify a band of prospective mesoderm around the equator of the blastula. As the blastula develops, the future mesoderm becomes patterned along the dorso-ventral axis so that different regions give rise to different structures. During gastrulation, the future germ layers—ectoderm, mesoderm, and endoderm—move into the positions in which they will develop into the structures of the larval body. The antero-posterior body axis emerges clearly with a head at one end and the future tail at the other (Fig. 4.1). In addition, the dorsal ectoderm becomes specified as neural tissue. The embryonic organizer (the Spemann organizer), which arises at the early gastrula stage, is a signaling center that has a global organizing function in establishing the body plan, coordinating patterning of both dorso-ventral and antero-posterior axes and inducing neural tissue. Discussion of the mechanisms underlying the movements of cells and tissues during gastrulation in *Xenopus* is deferred to Chapter 9.

Fig. 4.1 Comparison of the axes and fate maps of the *Xenopus* egg and blastula with the structures in the tailbud embryo. The *Xenopus* egg has an initial axis of symmetry running from the animal to the vegetal pole—the animal-vegetal axis (first panel). This symmetry is broken by the specification of one side of the blastula as the dorsal side (second panel). After gastrulation and neurulation, the embryo has elongated and now has a distinct anterior (head) to posterior (tail) axis and a distinct dorso-ventral axis at right angles to it (third panel). In *Xenopus*, the endoderm, which gives rise to the gut and associated organs, and ectoderm, which gives rise to epidermis and the nervous system, are specified by maternal cytoplasmic factors already present in the egg, but the mesoderm, which gives rise to notochord, somites and blood, and other organs, is induced from ectoderm by signals from the vegetal region during the blastula stage.

Setting up the body axes

We first consider how the antero-posterior and dorso-ventral axes are established in *Xenopus*, and to what extent the embryo is already patterned as a result of maternal cytoplasmic factors laid down in the egg during its development in the ovary. A crucial event that occurs after fertilization is the specification of one side of the cleaving egg as the future dorsal side of the embryo, a decision that guides the future development of the embryo.

4.1 The animal-vegetal axis is maternally determined in *Xenopus*

The very earliest stages in development of *Xenopus* are exclusively under the control of maternal factors present in the egg. As we saw in *Drosophila* (see Chapter 2), maternal factors are the mRNA and protein products of genes that are expressed by the mother during oogenesis and laid down in the egg while it is being formed. The *Xenopus* egg possesses a distinct polarity even before it is fertilized, and this polarity influences the pattern of cleavage. One end of the egg, the **animal pole**, which has a heavily pigmented surface, comes to lie uppermost after laying, whereas most of the yolk becomes located toward the opposite, unpigmented end, the **vegetal pole**, under the influence of gravity (see Fig. 3.4). The animal half contains the egg nucleus, which is located close to the animal pole. These differences define the **animal–vegetal axis**.

The pigment itself has no role in development but is a useful marker for real developmental differences that exist between the animal and vegetal halves of the egg. The planes of early cleavages that occur after fertilization are related to the animal–vegetal axis. The first cleavage is parallel with the axis and often defines a plane corresponding to the midline, dividing the embryo into left and right sides. The second cleavage is in the same plane at right angles to the first and divides the egg into four cells, while the third cleavage is at right angles to the axis and divides the embryo into animal and vegetal halves, each composed of four blastomeres (see Fig. 3.5).

Maternal factors are of key importance in the very early development of *Xenopus* because most of the embryo's own genes do not begin to be transcribed until after a stage called the mid-blastula transition, which will be discussed later in this chapter. The egg contains large amounts of maternal mRNAs and there are also large amounts of stored proteins; there is, for example, sufficient histone protein for the assembly of more than 10,000 nuclei, quite enough to see the embryo through the first 12 cleavages until its own genes begin to be expressed. The egg also contains a number of maternal mRNAs that encode proteins with specifically developmental roles. These mRNAs are localized along the animal–vegetal axis of the egg during its development in the mother (oogenesis), with most of the developmentally important maternal products ending up in the vegetal hemisphere of the mature egg. If the vegetal-most cytoplasm is removed early in the first cell cycle leading up to the first cleavage division of the fertilized egg, the embryo never develops beyond a radially symmetric mass of tissue that lacks any indication of characteristic dorsal structures, such as notochord or somites, or a defined antero-posterior axis.

Localization of maternal mRNAs occurs by one of two routes. Early in oogenesis, some mRNAs are concentrated in the cortex at the future vegetal pole via transport involving a cytoplasmic region known as the message transport organizer region (METRO), which is associated with the mitochondrial cloud—a region of the oocyte cytoplasm where mitochondria are being generated. The mechanism underlying the selective trapping and transport of mRNAs by the METRO is not yet clear, but it may involve association of the mRNAs with the cytoskeleton and small vesicles of endoplasmic reticulum. Later in oogenesis, other mRNAs are transported to the vegetal cortex along microtubules by kinesin motor proteins, a similar mechanism to that of mRNA transport in the *Drosophila* oocyte (see Fig. 2.21).

The expression of the developmentally important maternal mRNAs in the *Xenopus* egg is regulated such that in most cases they are only translated after fertilization. The proteins encoded by several of these mRNAs are candidates for signals that specify early polarity and initiate further development. One maternal mRNA encodes the signaling protein Vg-1, which is a member of the large transforming growth factor-β (TGF-β) family (secreted signaling proteins used in vertebrate early development are listed in Box 4A). *vg1* mRNA is synthesized during early oogenesis and becomes localized in the vegetal cortex of mature oocytes (Fig. 4.2), moving into the vegetal cytoplasm before fertilization. It is translated after fertilization and functions as an early signal for some aspects of mesoderm induction. Another vegetally localized mRNA encodes the protein Wnt-11, which provides one of the early signals required for dorso-ventral axis specification. Wnt-11 is a member of the Wnt family of vertebrate signaling proteins related to the Wingless protein in *Drosophila* (the Wingless/Wnt signaling pathway is shown in Box 4B). Other components of the Wnt signaling pathway are also encoded by maternal mRNAs or are laid down as protein.

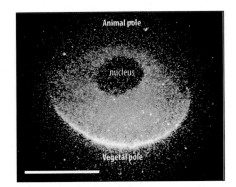

Fig. 4.2 Distribution of mRNA for the growth factor Vg-1 in the unfertilized amphibian egg. *In situ* hybridization with a radioactive probe for maternal *vg1* mRNA in a *Xenopus* late oocyte shows its localization in the cortex of the vegetal region, as indicated by the yellow color (which represents silver grains viewed by special lighting). In this image, made before the use of fluorescent labels to detect nucleic acid hybridization was commonplace, the section through the egg was covered with a radiographic emulsion after hybridization of the probe, and the presence and density of the radioactive probe at any location determines the amount of silver grains deposited. Scale bar = 1 mm.

Photograph courtesy of D. Melton.

CELL BIOLOGY BOX 4A Intercellular protein signals in vertebrate development

Proteins that are known to act as signals between cells during development belong to seven main families. Some of these families, such as the fibroblast growth factors (FGFs), were originally identified because they were essential for the survival and proliferation of mammalian cells in tissue culture. The members of the seven families are either secreted or are signaling proteins spanning the plasma membrane, and they provide intercellular signals at many stages of development. Other signal molecules, such as insulin and the insulin-like growth factors, and the neurotrophins, are used by embryos for particular tasks, but the seven families listed in the table (Figure 1) are the main families involved throughout development in vertebrates. Homologs of these proteins are also important in embryonic development in non-vertebrates (see Chapter 2, Chapter 6, and examples throughout the rest of the book).

The signaling molecules listed in the table all act by binding to receptors on the surface of a target cell. This produces a signal that is passed, or transduced, across the cell membrane by the receptor to the intracellular biochemical signaling pathways. Each type of signal protein has a corresponding receptor or set of receptors, and only cells with the appropriate receptors on their surface can respond.

Common intercellular signaling proteins in vertebrate embryonic development		
Signal molecule family	**Receptors**	**Examples of roles in development**
Fibroblast growth factor (FGF)		
Twenty-five FGF genes have been identified in vertebrates, not all of which are present in all vertebrates	FGF receptors (receptor tyrosine kinases)	Roles at all stages in development. Maintenance of mesoderm formation (this chapter). Induction of spinal cord (see this chapter and Box 4A for signaling pathway). Signal from apical ectodermal ridge in limb bud (Chapter 11)
Epidermal growth factor (EGF)		
Epidermal growth factor	EGF receptors (receptor tyrosine kinases)	Cell proliferation and differentiation. *Drosophila* limb patterning (Chapter 11)
Transforming growth factor-β (TGF-β)		
A large protein family that includes activin, Vg-1, bone morphogenetic proteins (BMPs), Nodal, and Nodal-related proteins	Type I and Type II receptor subunits (receptor serine/threonine kinases) Receptors act as heterodimers	Roles at all stages in development. Signaling pathways in Box4. Mesoderm induction and patterning (this chapter)
Hedgehog		
Sonic hedgehog and other members	Patched	Positional signaling in limb and neural tube (Chapters 11 and 12). Signaling pathway in Box 11D. Involved in determination of left–right asymmetry in chick embryos (Chapter 5)
Wingless (Wnt)		
Nineteen Wnt genes have been identified in vertebrates, not all of which are present in all vertebrates	Frizzled family of receptors (seven-span transmembrane proteins)	Roles at all stages in development. For canonical and non-canonical signaling pathways see Box 4B and Box 9C. Dorso-ventral axis specification in *Xenopus* (this chapter). Limb development (Chapter 11)
Delta and Serrate		
Transmembrane signaling proteins	Notch	Roles at all stages in development. Left–right asymmetry (Chapter 5). Somite development (see Chapter 5, signaling pathway in Box 5D)
Ephrins		
Transmembrane signaling proteins	Eph receptors (receptor tyrosine kinases)	Formation of rhombomeres in embryonic hindbrain (Chapter 12). Guidance of developing blood vessels (Chapter 11). Guidance of axon navigation in the nervous system (Chapter 12)

Figure 1

The intracellular pathways can be complex and involve numerous different proteins. We have already introduced the general principles employed in intracellular signaling pathways in Box 1E and the Wnt/β-catenin signaling pathway is illustrated in Box 4B, the **TGF-β** signaling pathways are illustrated in Box 4C, and the FGF signaling pathway in Box 4E. Figures illustrating the other signaling pathways are cross-referenced in the table. In the developmental context, the important responses of cells are either a change in gene expression, with specific genes being switched on or off, or a change in the cytoskeleton, leading to changes in cell shape and motility.

Some signaling proteins, such as those of the transforming growth factor-β (TGF-β) family, act as dimers; two molecules form a covalently linked complex that binds to and activates the receptor, bringing two receptor proteins together to form a dimer (see Box 4C). In some cases, the active form of the protein signal is a heterodimer, made up of two different members of the same family. Secreted signals can diffuse or be transported short distances through tissue and set up concentration gradients. Some signal molecules, however, remain bound to the cell surface and thus can only interact with receptors on a cell that is in direct contact. One such is the protein Delta, which is membrane-bound and interacts with the receptor protein Notch on an adjacent cell (see Box 5D, for the Notch signaling pathway).

For consistency in this book we have hyphenated the abbreviations for all secreted signaling proteins where the name allows—for example Vg-1, BMP-4, and TGF-β.

CELL BIOLOGY BOX 4B The Wnt/β-catenin signaling pathway

The important developmental signaling pathway initiated by the Wnt family of intercellular signaling proteins (pronounced 'Wints') is named after the proteins encoded by the *wingless* gene in *Drosophila* and the *Int* gene (now *Wnt1*) in mammals. Wnt proteins are involved in specifying cell fate in a multiplicity of developmental contexts in all multicellular animals studied so far. We have already met functions of Wingless in *Drosophila* and will meet many more of the functions of the Wnt family throughout the book. In adult humans, Wnt is required to maintain stem cells in their self-renewing state, and incorrect control of the Wnt signaling pathway is implicated in common cancers, such as colon cancer.

Wnt proteins can activate a number of different intracellular signaling pathways: the one we will illustrate here is involved in specifying cell fates in the early development of all animals. It is known as the **canonical Wnt/β-catenin pathway** as it was the first to be discovered, and it leads to the accumulation of the protein β-catenin and its action as a transcriptional co-activator. (β-Catenin can also act as a protein that links cell-adhesion molecules to the cytoskeleton (see Box 9A); that function is quite distinct from its role as a transcriptional co-activator). Other Wnt pathways will be described when they are encountered in the course of the book. A simplified version of the canonical Wnt/β-catenin pathway is shown in Figure 1. As with many developmental signaling pathways, the initial Wnt signal is converted, or 'transduced', into changes in gene expression in the nucleus. In this case, the Wnt signal prevents the degradation of the protein β-catenin in the cytoplasm, and by doing so enables β-catenin to accumulate and to enter the nucleus, where it binds to and activates transcription factors of the TCF (T-cell-specific factor) family and so activates the expression of selected target genes.

In the absence of a Wnt signal, β-catenin is bound by a 'destruction complex' of proteins in the cytoplasm (left-hand side of Figure 1). This includes protein kinases (CK1γ and GSK-3β) that phosphorylate β-catenin, thus targeting it for ubiquitination and degradation in the proteasome. In the absence of β-catenin

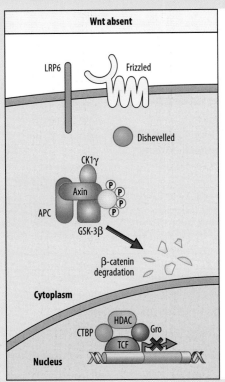
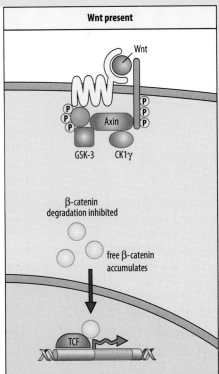

Figure 1

in the nucleus, transcriptional co-repressors bind to the TCF transcription factors and prevent gene expression.

Like most extracellular signals, Wnt acts on its target cells by binding to a specific transmembrane receptor protein (Frizzled) at the cell surface (right-hand side of Figure 1). The signal that Wnt has bound is transmitted across the membrane by Frizzled and an associated co-receptor, LRP, and Wnt itself does not enter the cell. Activation of these receptors causes the same protein kinases that phosphorylated β-catenin in the destruction complex to now become associated with the membrane and phosphorylate the cytoplasmic tail of the activated LRP. Protein phosphorylation, which changes the activity of proteins or their interactions with other proteins, is ubiquitous in signaling pathways as a means of transmitting the signal onwards. The intracellular signaling protein Dishevelled (Dsh) and the protein Axin (which is a component of the destruction complex) are recruited to the tails of LRP and Frizzled at the membrane. This prevents the destruction complex forming and leaves β-catenin free to accumulate and to enter the nucleus. Inside the nucleus, β-catenin binding to TCF displaces the co-repressors, lifts the repression and enables the target genes to be expressed.

The other main class of maternal factors with developmental roles comprises the transcription factors that switch on and regulate zygotic gene expression. Maternal mRNA encoding the T-box transcription factor VegT is localized to the vegetal hemisphere and is translated after fertilization. VegT has a key role in specifying both endoderm and mesoderm, and its localization in the vegetal region is crucial to this.

The animal–vegetal axis of the egg is related to the antero-posterior axis of the tadpole, as the head will form from the animal region, but the axes of the tadpole and the fertilized egg are not directly comparable (see Fig. 4.1). Which side of the animal region will form the head is not determined until the dorsal side of the embryo is specified, which does not happen until after fertilization. The precise position of the embryo's antero-posterior axis thus depends on the specification of the dorso-ventral axis.

Scan here

Scan this QR code image with your mobile device to see an online animation of the Wnt/β-catenin signaling pathway or log on to **http://global.oup.com/uk/orc/biosciences/ devbiol/wolpert5e/qr/qr4a/**

4.2 Local activation of Wnt/β-catenin signaling specifies the future dorsal side of the embryo

The spherical unfertilized egg of *Xenopus* is radially symmetric about the animal–vegetal axis, and this symmetry is broken only when the egg is fertilized. Sperm entry initiates a series of events that defines the dorso-ventral axis of the gastrula (see Section 3.1 for a general outline of *Xenopus* development), with the dorsal side forming more or less opposite the sperm's entry point.

Sperm entry can occur anywhere in the animal hemisphere of the *Xenopus* egg, and it causes the outer layer of cytoplasm, the cortex, to loosen from the inner dense cytoplasm so that it is able to move independently. The cortex is a gel-like layer rich in actin filaments and associated material. During the first cell cycle, the cortex rotates about 30° against the underlying cytoplasm in a direction away from the site of sperm entry (Fig. 4.3). As a result of this **cortical rotation**, an array of sub-cortical microtubules becomes oriented with their plus ends pointing away from the site of sperm entry.

The oriented microtubules act as tracks for the directed movement of some maternal proteins and mRNAs from the vegetal pole to sites on the equator opposite the point of sperm entry, creating asymmetry in the fertilized egg. These relocated factors specify the future dorsal side of the embryo, conferring a second axis of symmetry—the dorso-ventral axis—on the fertilized egg (see Fig. 4.3). Cells that inherit these factors form dorsal structures, and for this reason the factors are often called **dorsalizing factors**. Because the plane of the first cleavage division is usually aligned with the site of sperm entry and essentially divides the embryo into right and left halves, both blastomeres inherit some dorsalizing factors. This explains why, when they are separated, each can give rise to a near-perfect embryo, as discussed in Chapter 1. The second cleavage division is orthogonal to the first and therefore results in two 'dorsal' blastomeres containing dorsalizing factors and two

Fig. 4.3 The future dorsal side of the amphibian embryo develops opposite the site of sperm entry as a result of the redistribution of maternal dorsalizing factors by cortical rotation. After fertilization (first panel), the cortical layer just under the cell membrane rotates by 30° towards the future dorsal region. This results in dorsalizing factors of the Wnt pathway, such as *wnt11* mRNA and Dishevelled (Dsh) protein— indicated by green dots—being relocated from their initial position at the vegetal pole to a position approximately opposite to the site of sperm entry (second panel). The factors are transported further away from the vegetal pole along microtubules that lie at the interface of the cortex and the yolky cytoplasm. The Wnt pathway is activated in cells that inherit these factors during cleavage (third panel, red dots indicate activation of the Wnt signaling pathway), as a result of an accumulation of β-catenin in nuclei on the future dorsal side of the blastula (not shown here). In the late blastula (fourth panel), the Spemann organizer is formed in this dorsal region, and is where gastrulation will start. V, ventral; D, dorsal.

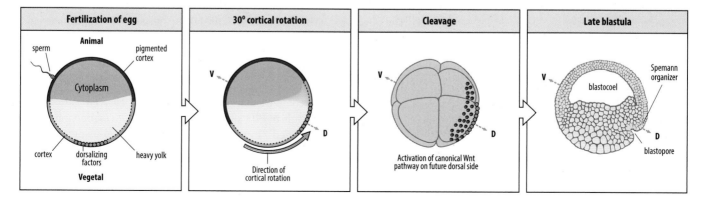

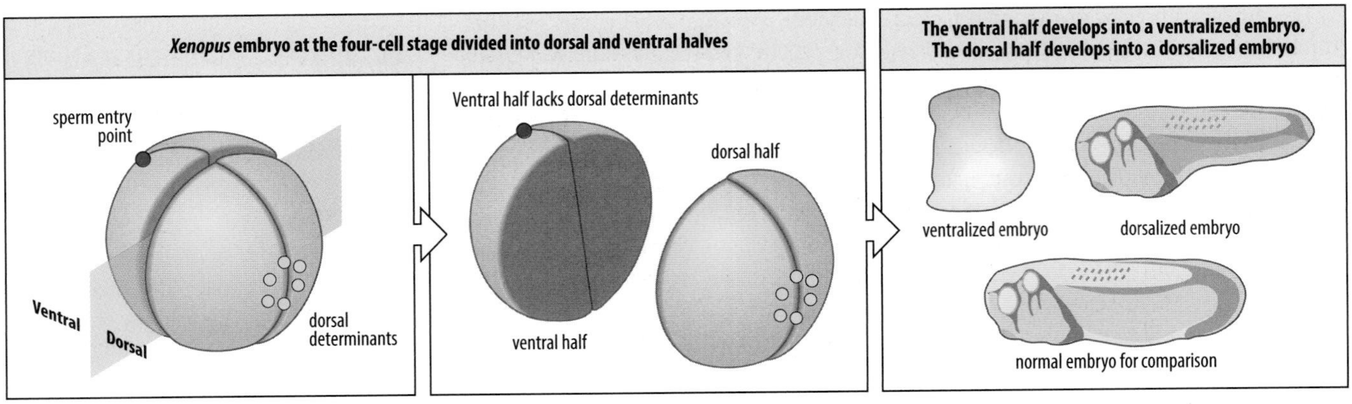

Fig. 4.4 Dorsal determinants are essential for normal development. If a *Xenopus* embryo is divided into a dorsal and a ventral half at the four-cell stage, the dorsal half containing dorsal determinants develops as a dorsalized embryo lacking a gut (compare with normal embryo), while the ventral half, which does not contain any dorsal determinants, is ventralized and lacks both dorsal and anterior structures.

'ventral' blastomeres which contain no dorsalizing factors. When the embryo is divided into two at this four-cell stage, so that one half comprises the two dorsal blastomeres and the other the two ventral blastomeres, the 'dorsal' blastomeres will develop most structures of the embryo, although it lacks a gut. In contrast, the 'ventral' blastomeres give rise to radially symmetric embryos that lack all dorsal and anterior structures (Fig. 4.4). This experiment shows that dorsal and ventral regions are specified by the four-cell stage. The ventralized embryos that develop from the two 'ventral' blastomeres are similar to those produced by removal of the vegetal-most yolk soon after fertilization of the egg, before the relocation of the dorsalizing factors by cortical rotation.

Establishment of dorso-ventral axis is a key event in *Xenopus* development, as it sets up the conditions necessary for the formation of the embryonic organizer, the Spemann organizer, which is located at the dorsal blastopore lip of the gastrula. As we saw in Chapter 1, the Spemann embryonic organizer was originally discovered by grafting the dorsal lip of the blastopore of one newt embryo into the blastocoel of another newt embryo and finding that it could induce the formation of a second embryonic axis. When the dorsal blastopore lip from an early *Xenopus* gastrula is grafted to the opposite side of another early *Xenopus* gastrula, a twinned embryo also develops. A second main body axis is induced, having its own head at the anterior, and a complete dorso-ventral axis (Fig. 4.5).

The importance of cortical rotation for establishing the dorso-ventral axis was shown by experiments in which rotation was blocked. Cortical rotation can be prevented by irradiating the ventral side of the egg with ultraviolet (UV) light, which disrupts the microtubules responsible for transporting the dorsalizing factors. Embryos developing from irradiated eggs are ventralized; that is, they are deficient in structures normally formed on the dorsal side and instead develop excessive amounts of blood-forming mesoderm, a tissue normally only present at the embryo's ventral midline. With very high doses of radiation, both dorsal and anterior structures are lost and the ventralized embryo appears as little more than a small distorted cylinder (see Fig. 4.4). The effects of UV irradiation can, however, be 'rescued' by tipping the egg. Because of gravity, the dense yolk slides against the cortex, thus effectively restoring rotation and the localization of maternal factors.

The maternal dorsalizing factors include *wnt11* mRNA, *wnt5a* mRNA and Dishevelled protein, components of the Wnt signaling pathway. After translation, the Wnt-11 and Wnt-5a proteins form a complex that activates the **Wnt/β-catenin pathway** in dorsal cells. To investigate the role of Wnt signaling in dorsalization, expression of one of the Wnt receptors, Fz7 (Frizzled 7), was inhibited in the embryo using antisense morpholino oligonucleotides (see Box 6B). In the absence of the receptor, the Wnt/β-catenin pathway cannot be activated by an extracellular Wnt signal. Dorsal structures did not develop, showing that Wnt signaling was required for dorsalization.

The Wnt/β-catenin pathway (see Box 4B) is one of several signaling pathways that can be stimulated by Wnt proteins, and we have already seen it in action in *Drosophila* in response to Wingless. As indicated in Box 4B, when the pathway is inactive, the transcriptional co-activator β-catenin is bound to a 'destruction complex' of proteins and is phosphorylated by the protein kinase GSK-3β. Phosphorylation targets β-catenin for proteosomal degradation. In contrast, when the signaling pathway is activated, β-catenin is stabilized, enters the nucleus and helps to regulate gene expression.

Maternal β-catenin is initially present throughout the vegetal region and activation of the canonical Wnt signaling pathway by the dorsalizing factors prevents the degradation of β-catenin on the prospective dorsal side of the embryo. By the two-cell stage, β-catenin has started to accumulate in the cytoplasm on one side of the two cells, and by the second or third cleavages, it has entered the nuclei of the dorsal cells, where it later helps to switch on zygotic genes required for the next stages in development.

The crucial role of activation of the Wnt signaling pathway in specifying dorsal cells is demonstrated by experiments in which β-catenin mRNA was injected into a ventral vegetal cell of a 32-cell stage *Xenopus* embryo. These injections resulted in twinned embryos, which develop due to the dorsalization of the ventral cells, which leads to a series of events that culminates in the formation of a second organizer (Fig. 4.6).

Wnt signaling is also limited to the dorsal region of the blastula by the maternally provided 'antagonist' of Wnt signaling, Dickkopf1 (Dkk1), which is present in the ventral region. Dkk1 is a secreted protein that binds to the Wnt co-receptor LRP6 (see Box 4B) blocking its function and inhibiting activation of the pathway. As we shall see throughout this chapter, inhibition of the activity of secreted signaling proteins is an important developmental strategy for restricting signaling to specific regions. The Wnt pathway could also be activated directly in dorsal cells via pathway components such as the Dishevelled protein, which is moved into the future dorsal region as a result of cortical rotation (see Fig. 4.3). Dishevelled prevents formation of the destruction complex containing the enzyme GSK-3β, which is required for β-catenin degradation (see Box 4B). The key role played by inhibition of GSK-3β activity in establishing the future dorsal region explains the long-established finding that treatment of early *Xenopus* embryos with lithium chloride dorsalizes them. Lithium is known to block GSK-3β activity, and lithium treatment promotes the formation of dorsal and anterior structures at the expense of ventral and posterior structures.

4.3 Signaling centers develop on the dorsal side of the blastula

The Spemann organizer is induced in response to signals from another signaling center that arises in the dorsal region of the early blastula. The combined actions of Wnt-11, Wnt-5a, β-catenin, and other maternal factors present in the dorsal side of the vegetal region lead to the formation of a signaling center known as the **blastula organizer** or **Nieuwkoop center** (Fig. 4.7). The center is named after the Dutch embryologist Pieter Nieuwkoop, who discovered the dorsalizing and mesoderm-inducing properties of this vegetal region through experiments in which explants of animal cap tissue from blastulas were combined with vegetal explants. Before those experiments, in the late 1960s, it had been thought that mesoderm (from which some key dorsal structures such as the notochord are formed) was intrinsically specified by a particular cytoplasmic region in the amphibian egg, as are endoderm and ectoderm. These experiments combining animal cap and vegetal cells suggested instead that mesoderm was induced from the prospective ectoderm by signaling from the prospective endoderm, as we shall see later in the chapter.

The Nieuwkoop center arises in the early blastula and sets the initial dorso-ventral polarity. The Spemann organizer is then induced just above the Nieuwkoop center in the equatorial region of the early gastrula. The signaling activity of the Nieuwkoop center can be demonstrated by grafting the cells from the dorsal–vegetal region of a

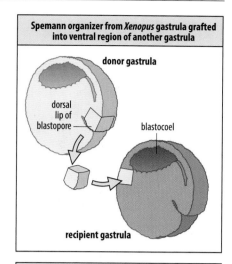

Spemann organizer from *Xenopus* gastrula grafted into ventral region of another gastrula

donor gastrula

dorsal lip of blastopore

blastocoel

recipient gastrula

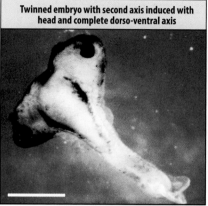

Twinned embryo with second axis induced with head and complete dorso-ventral axis

Fig. 4.5 Transplantation of the Spemann organizer can induce a new axis in *Xenopus*. When the Spemann organizer from a *Xenopus* gastrula is grafted into the ventral region of another gastrula, a partially twinned embryo results, with the second body axis induced by the transplanted organizer. The second axis has a complete dorso-ventral axis and a distinct head. The organizer therefore produces signals that not only result in patterning the mesoderm dorso-ventrally, but also induce neural tissue and anterior structures. Scale bar = 1 mm.

Photograph courtesy of J. Smith

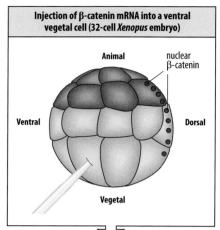

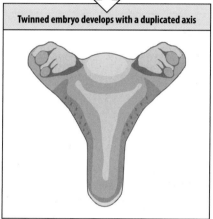

Fig. 4.6 Induction of a new dorsal side by injection of β-catenin mRNA. Injection of mRNA encoding β-catenin into ventral vegetal cells activates Wnt signaling and specifies a new organizer at the site of injection, leading to a twinned embryo.

32-cell *Xenopus* embryo into the ventral side of another embryo. This gives rise to a twinned embryo with two dorsal sides (Fig. 4.8). Cells from the graft itself contribute to endodermal tissues but not to the mesodermal and neural tissues of the new axis. This shows that the grafted cells act as a signaling center and are able to induce these fates in the cells of the host embryo. In contrast, grafting ventral cells to the dorsal side has no effect.

Although β-catenin becomes localized in nuclei in both the animal and vegetal regions, and specifies a large region as dorsal (see Fig 4.7), the Nieuwkoop center arises in the vegetal region. This is because its specification depends not only on nuclear β-catenin but also on the maternal transcription factor VegT, which is present throughout the vegetal region from fertilization onwards. In the blastula, nuclear β-catenin in vegetal cells also expressing VegT helps to activate zygotic genes such as *siamois*, which encodes a transcription factor that defines the Nieuwkoop center and is also involved in the induction of the Spemann organizer. VegT is also required to induce expression of the genes encoding **Nodal-related proteins**, secreted signaling proteins that induce the Spemann organizer in the equatorial region and also induce the mesoderm (see Box 4A). The frog Nodal-related proteins are named after the mouse protein Nodal, with which they are homologous, and members of the Nodal protein family are key factors in mesoderm induction in all vertebrates. They belong to the larger TGF-β family of signaling proteins, and we will discuss them further when we look at the induction and patterning of the mesoderm later in the chapter.

SUMMARY

In the amphibian embryo, maternal factors determine the animal–vegetal axis, which corresponds approximately to the future antero-posterior axis. The dorso-ventral axis is specified by the site of sperm entry and the resulting cortical rotation. This relocates a set of 'dorsalizing' maternal factors to a new location, which is thus specified as the dorsal side. These maternal factors include components of the Wnt signaling pathway and the VegT transcription factor. Their actions lead to the local establishment of a signaling center—the Nieuwkoop center—on the dorsal side of the blastula which, when grafted, can direct the development of a twinned embryo. The Nieuwkoop center induces the formation of the Spemann embryonic organizer above it in the dorsal equatorial region of the early gastrula, as we shall see in the next part of the chapter, when we consider the induction of the mesoderm.

The origin and specification of the germ layers

We have seen in the preceding sections how the animal–vegetal axis of the *Xenopus* egg is maternally determined and how the dorso-ventral axis is established by a series of events triggered by sperm entry. We now focus on the earliest patterning of the Xenopus embryo with respect to these axes: the specification of the three germ layers—ectoderm, mesoderm, and endoderm—and how these germ layers themselves become broadly patterned so that different regions give rise to different structures (see Box 1C).

All the tissues of the body are derived from these three germ layers. The ectoderm becomes patterned in such a way that the dorsal region of the ectoderm gives rise to the nervous system and the rest to the epidermis of the skin. The mesoderm becomes patterned so that different regions along the dorso-ventral axis give rise to notochord, muscle, heart, kidney, and blood-forming tissues, among others. The endoderm gives rise to the gut, and becomes patterned so that different regions of the gut form associated organs such as the lungs, liver, and pancreas. We will first look at the fate map of the early *Xenopus* embryo, which tells us which tissues are generated from particular regions of the embryo. We then consider how the three germ layers are specified and how they are then patterned. Most is known about patterning of the mesoderm and the ectoderm and many of the genes and proteins involved have been identified.

4.4 The fate map of the *Xenopus* blastula makes clear the function of gastrulation

The fate map of the late *Xenopus* blastula (Fig. 4.9) shows that the yolky vegetal region, which occupies the lower third of the spherical blastula, gives rise to most of the endoderm. The yolk, which is present in all cells, provides the nutrition for the developing embryo, and is gradually used up as development proceeds. At the other pole, the animal hemisphere becomes ectoderm. The future mesoderm forms from a belt-like region, the marginal zone, around the equator of the blastula. In *Xenopus*, but not in all amphibians, a thin outer layer of presumptive endoderm overlies the presumptive mesoderm in the marginal zone.

The fate map of the blastula makes clear the function of gastrulation. At the blastula stage, the future endoderm, which will line the gut, is on the outside and so must move inside. Similarly, the future mesoderm, which will form internal tissues and organs, must move inwards. During gastrulation, the marginal zone moves into the interior through the dorsal lip of the blastopore, the site of the Spemann organizer, which has arisen above the Nieuwkoop center. The fate map of the future mesoderm (see Fig. 4.9) shows that different regions along the dorso-ventral axis of the blastula have different fates. Going from dorsal to ventral, the most dorsal mesoderm gives rise to the notochord and the adjacent mesoderm develops into the somites, which give rise to muscle and some skeletal tissue. Then comes the lateral plate, which contains heart and kidney mesoderm, and the blood islands, the tissue where blood formation (hematopoiesis) first occurs in the embryo. There are also important differences between the fate of the dorsal and ventral sides of the animal hemisphere, which will develop as ectoderm. The ventral side of the animal hemisphere will give rise to the epidermis, whereas the dorsal side gives rise to the nervous system. After neural tube formation, the epidermis spreads to cover the whole of the embryo.

The terms dorsal and ventral in relation to the fate map can be somewhat confusing, because the fate map does not correspond exactly to a neat set of axes at right angles to each other. As a result of cell movements during gastrulation, cells from the dorsal side of the blastula give rise to structures at the anterior end of the embryo, such as the head, in addition to dorsal structures. This explains why the dorsalized embryos described in Section 4.2 have well-developed anterior structures. Cells from the ventral side of the blastula not only give rise to ventral structures but will also form the posterior end of the embryo.

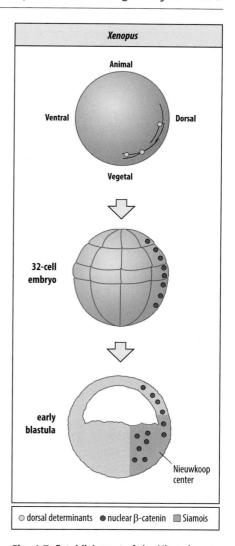

○ dorsal determinants ● nuclear β-catenin ▢ Siamois

Fig. 4.7 Establishment of the Nieuwkoop center. In *Xenopus*, cortical rotation moves dorsal determinants (green) towards the future dorsal side of the embryo, creating a large region where β-catenin moves from the cytoplasm into nuclei (red). Cells in the region of the blastula in which β-catenin is intranuclear, and where VegT is also present, express the transcription factor Siamois (blue shading). The Nieuwkoop center arises in the dorsal vegetal region of the early blastula.

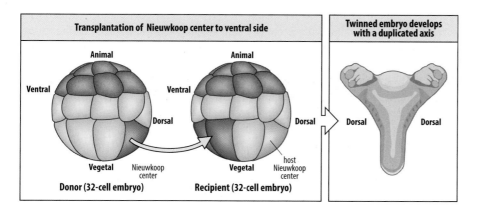

Fig. 4.8 The Nieuwkoop center can specify a new dorsal side. When vegetal cells containing the Nieuwkoop center from the dorsal side of one 32-cell *Xenopus* blastula are grafted to the ventral side of another blastula, a second axis forms and a twinned embryo develops. The grafted cells signal and contribute to endodermal tissues but not to the mesodermal and neural tissues of the second axis.

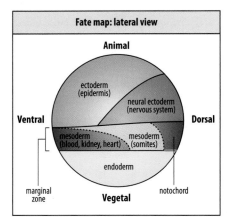

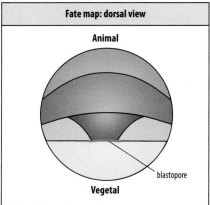

Fig. 4.9 Fate map of a late *Xenopus* blastula. The ectoderm (shades of blue) gives rise to the epidermis and nervous system. Along the dorso-ventral axis, the mesoderm (red and peach) gives rise to notochord, somites, heart, kidneys, and blood. Note that blood can also form in more dorsal regions. The endoderm (yellow) gives rise to the gut. In *Xenopus*, although not in all amphibians, there is also endoderm (not shown here) overlying the mesoderm in the marginal zone.

4.5 Cells of the early *Xenopus* embryo do not yet have their fates determined and regulation is possible

Early vertebrate embryos have considerable powers of regulation (see Section 1.3) when parts of the embryo are removed or rearranged. Fragments of a fertilized *Xenopus* egg that are only one-fourth of the normal volume can develop into small but more-or-less normally proportioned embryos. Similarly, as we discussed earlier in the chapter, each of the blastomeres at the two-cell stage can develop into a normal embryo. There must therefore be a patterning mechanism involving cell–cell interactions that can cope with such differences in size. There are, however, limits to the capacity for regulation. Although the dorsal halves of four-cell embryos regulate to produce reasonably normal embryos apart from not having a gut, the ventral halves do not. As we saw in Section 4.3, essential signaling centers develop in the dorsal half and this why ventral halves, which will not develop such centers, form radially symmetrical embryos.

Remarkably, regulation can even occur when a late *Xenopus* blastula is bisected, as long as each half contains some of the Spemann organizer (Fig. 4.10). This experiment shows that at the blastula stage many cells are not yet specified or determined (see Section 1.12 and Fig. 1.22) and this may also be true for even later stages. That means that the developmental potential of cells at these quite late stages in laying down the body plan is greater than their position on the fate map suggests.

The state of determination of individual cells, or of small regions of an embryo, is studied by transplanting them to a different region of a host embryo and seeing how they develop. If they are already determined, they will develop according to their original position. If they are not yet determined, they will develop in line with their new position. In a series of experiments, single labeled cells from *Xenopus* blastulas and gastrulas were introduced into the blastocoel of gastrula-stage embryos and the fate of their progeny followed during gastrulation. In general, transplanted cells from early blastulas are not yet determined; their progeny will differentiate according to the signals they receive at their new location. So the progeny of single cells from the vegetal pole, which would normally form endoderm, can contribute to a wide range of other tissues such as muscle or nervous system. Similarly, the progeny of single animal pole cells from an early blastula, whose normal fate is epidermis or nervous tissue, can form endoderm or mesoderm. With time, cells gradually become determined, so that similar cells taken from later-stage blastulas and early gastrulas develop according to their fate at the time of transplantation.

4.6 Endoderm and ectoderm are specified by maternal factors, whereas mesoderm is induced from ectoderm by signals from the vegetal region

When explants from different regions of an early *Xenopus* blastula are cultured in a simple medium containing the necessary salts for ion balance, tissue from the region near the animal pole, usually referred to as the **animal cap**, will form a ball of epidermal cells, while explants from the vegetal region will form endoderm (Fig. 4.11). These results are in line with the normal fates of these regions. It is thus generally accepted that in *Xenopus* all the ectoderm and most of the endoderm is specified by maternal factors in the egg. There is no evidence that any signals from other regions of the embryo are necessary for their initial specification.

A strong candidate for a maternal determinant of ectoderm is the mRNA for the E3 ubiquitin-ligase Ectodermin, which is located at the animal pole of the fertilized egg. Ectodermin mRNA is inherited by cells that develop from the animal region and Ectodermin protein becomes localized in the nuclei of cells throughout the animal

Fig. 4.10 Both halves of a bisected *Xenopus* blastula can give rise to a complete tadpole. The bisection of the *Xenopus* blastula is illustrated in the upper panel. The formation of the identical twins from each half (in the lower panel) shows that regulation can take place at quite a late stage of development. Twins are only formed if each half contains some of the Spemann organizer. A normal embryo is shown above for comparison.

*Reproduced with permission from De Robertis, E.M. **Spemann's organizer and self-regulation in amphibian embryos** Nature Rev. Mol. Cell Biol. 2006, **7**: 296–302.*

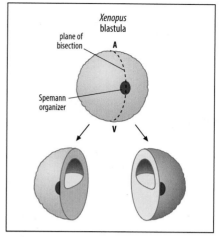

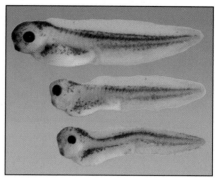

cap (Fig. 4.12). Ectodermin adds ubiquitin to Smad4, a gene-regulatory protein involved in both BMP signaling and Nodal-type TGF-β signaling (Box 4C), targeting it for degradation and thereby reducing signaling through both these pathways. As we will see, these pathways are involved in mesoderm induction and patterning and also in neural specification. The main function of Ectodermin is to limit mesoderm formation to the marginal zone. Another, later, determinant of ectoderm is the zygotic transcription factor Foxl1e, which is expressed in the cells of the animal half of the late blastula. Foxl1e is required to maintain the regional identity of these cells and their development as ectoderm (see Fig 4.12). In the absence of Foxl1e, the animal cells mix with cells of the other germ layers and differentiate according to their new positions.

A key maternal determinant of endoderm is the mRNA for the transcription factor VegT, which is located in the vegetal region of the egg and is inherited by the cells that develop from this region. Injection of *vegt* mRNA into animal cap cells induces expression of endoderm-specific markers, whereas blocking the translation of *vegt* mRNA in the vegetal region, by injecting oligonucleotides that are antisense to the mRNA (see Box 6B), results in a loss of endoderm. In VegT-depleted embryos, ectoderm forms throughout the embryo. As we will see later, VegT is also essential for the production of signals involved in mesoderm formation.

Unlike the ectoderm and endoderm, the mesoderm is not specified by inheritance of maternal determinants in the egg. Instead, it is induced from prospective ectoderm in a band round the equator of the blastula in response to signals from the prospective endoderm. The presence of mesoderm in a late blastula can be demonstrated by culturing explants from the marginal zone where animal and vegetal cells are adjacent. When these explants are cultured in isolation, they not only form ectodermal derivatives but also mesoderm (Fig. 4.13, top panels). Mesoderm can be distinguished by its histology because, after a few days' culture, it can form muscle, notochord, blood, and loose mesenchyme (connective tissue). It can also be identified by the typical proteins that cells of mesodermal origin produce, such as muscle-specific actin.

The key experiment that showed that mesoderm is induced by interactions between animal and vegetal cells in the early blastula involved culturing the animal cap in

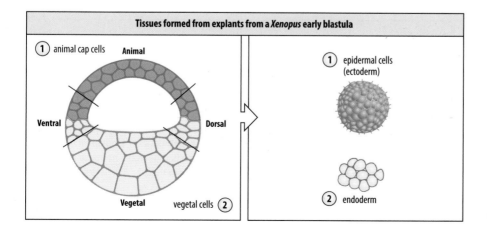

Fig. 4.11 Fate of explants of animal cells and vegetal cells from a *Xenopus* early blastula. Animal cap cells (blue, 1) explanted on their own form epidermis, whereas vegetal cell explants (yellow, 2) form endoderm.

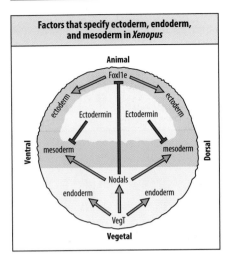

Factors that specify ectoderm, endoderm, and mesoderm in *Xenopus*

Fig. 4.12 Factors for germ-layer specification in *Xenopus*. In the vegetal half of the blastula, the maternal factor VegT is essential for both endoderm specification and, indirectly, for the production of Nodal proteins, which are one of the main signals that induce mesoderm from the prospective ectodermal cells of the animal cap. In the animal half of the blastula, the maternal factor Ectodermin counteracts potential mesoderm-inducing signals, and limits mesoderm formation to the equatorial zone. The zygotic transcription factor Foxl1e is expressed in the animal region and is necessary for the development of the animal cells as ectoderm and for preventing their mixing with more vegetal cells and developing as mesoderm or endoderm. In turn, Nodal signaling inhibits the expression of Foxl1e in the mesoderm.

contact with vegetal cells taken from the region near the vegetal pole (see Fig. 4.13, bottom panels). In such explants from an early blastula, mesodermal tissues are formed and this explant culture system has become a standard system for studying mesoderm induction. Confirmation that it is the animal cap cells, and not the vegetal cells, that are forming mesoderm was obtained by pre-labeling the animal region of the blastula with a cell-lineage marker and showing that the labeled cells form the mesoderm. Clearly, the vegetal region is producing a signal or signals that can induce mesoderm.

If the animal and vegetal explants are separated by a filter with pores too small to allow cell-to-cell contacts to form, mesoderm induction still takes place. This suggests that the mesoderm-inducing signals are in the form of secreted molecules that diffuse across the extracellular space, and do not pass directly from cell to cell via cell junctions (see Fig. 1.25).

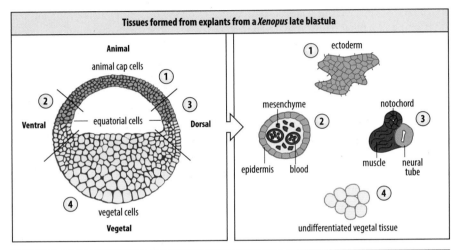

Tissues formed from explants from a *Xenopus* late blastula

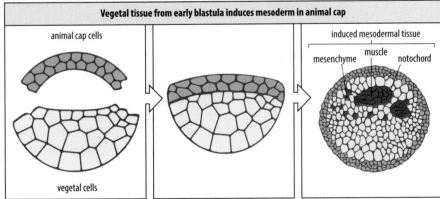

Vegetal tissue from early blastula induces mesoderm in animal cap

Fig. 4.13 Induction of mesoderm by the vegetal region in the *Xenopus* blastula. Top panels: explants of animal cap cells (blue, 1) or vegetal cells (yellow, 2) on their own from a late blastula form only ectoderm or endoderm, respectively. Explants from the equatorial region (3 and 4), where animal and vegetal regions are adjacent, form both ectodermal (epidermis and neural tube, blue) and mesodermal tissues (mesenchyme, blood cells (such as erythrocytes), notochord, and muscle, red), showing that mesoderm induction has taken place. The reason for the differences in the mesodermal tissues formed by ventral and dorsal explants is explained later in the chapter. Bottom panels: the standard experiment for studying mesoderm induction is to take explants of animal cap cells and vegetal cells that would normally make only ectoderm and endoderm, respectively, and culture them together for 3 days. When such explants are taken from an early blastula, in which the mesoderm has not yet been specified, and are combined and cultured, mesodermal tissues including notochord, muscle, blood, and loose mesenchyme (red) are induced from the animal cap tissue.

CELL BIOLOGY BOX 4C Signaling by members of the TGF-β family of growth factors

Signaling by the TGF-β family uses different combinations of receptor subunits and transcriptional regulatory proteins to activate different sets of target genes. Members of this family involved in early vertebrate development, such as Nodal and the Nodal-related proteins, bone morphogenetic proteins (BMPs), and activin, are dimeric ligands that act at cell-surface receptors. These receptors are heterodimers of two different subunits—type I and type II—with intracellular serine/threonine kinase domains. There are several different forms of type I and type II subunits and these combine to form distinctive receptors for different TGF-β family members. Binding of ligand causes subunit II to phosphorylate receptor subunit I, which in turn phosphorylates intracellular signaling proteins called Smads (Figure 1). These are named for the Sma and Mad proteins in *Drosophila*, in which they were discovered. The phosphorylated Smads themselves bind to another type of Smad, sometimes called co-Smads, forming a transcriptional regulatory complex that enters the nucleus to either activate or repress target genes. Different receptors use different Smads, thus leading to the activation of different sets of target genes. The identity of the type I subunit determines which Smads get activated. The biological response depends on the combination of the activated target genes and the particular cellular environment.

Scan here

Scan this QR code image with your mobile device to see an online animation of the TGF-β signaling pathway or log on to **http://global.oup.com/uk/orc/biosciences/ devbiol/wolpert5e/qr/qr4b/**

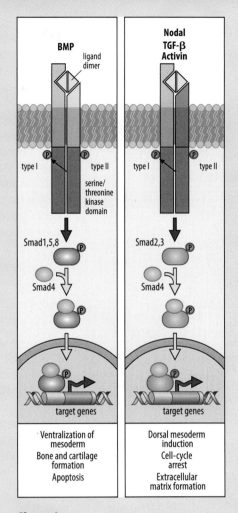

Figure 1

Differentiation of a mesodermal tissue such as muscle in the explant system appears to depend on a **community effect** in the responding cells. A few animal cap cells placed on vegetal tissue will not be induced to express muscle-specific genes. Even when a small number of individual animal cap cells are placed between two groups of vegetal cells, induction does not occur. By contrast, larger aggregates of animal cap cells respond by strongly expressing muscle-specific genes (Fig. 4.14).

4.7 Mesoderm induction occurs during a limited period in the blastula stage

Mesoderm induction is one of the key events in vertebrate development, as the mesoderm itself is a source of essential signals for further development of the embryo. In *Xenopus*, mesoderm induction is largely completed before gastrulation begins. Explant experiments like those described above have shown that in *Xenopus* there is a period of about 7 hours during the blastula stage when animal cells are **competent** to respond—that is, able to respond—to a mesoderm-inducing signal; cells lose their competence to respond about 11 hours after fertilization. An

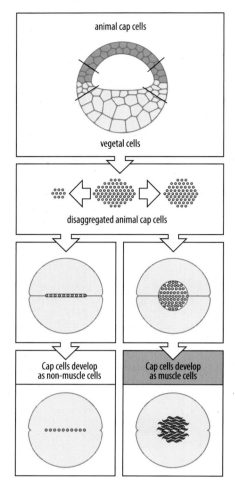

Fig. 4.14 The community effect. A single animal cap cell or a small number of animal cap cells in contact with vegetal tissue are not induced to become mesodermal cells and do not begin to express mesodermal markers such as muscle-specific proteins. A minimum of about 200 animal cap cells must be present for induction of muscle differentiation to occur.

Fig. 4.15 Timing of muscle gene expression is not linked to the time of mesoderm induction. Animal cap cells isolated from a *Xenopus* early blastula are competent to respond to mesoderm-inducing signals only for a period of about 7 hours, between 4 and 11 hours after fertilization. For expression of target genes to occur, exposure to inducer must be for at least 2 hours within this period. Irrespective of when the induction occurs within this competence period, muscle-gene activation occurs at the same time: 16 hours after fertilization.

exposure of about 2 hours to inducing signal is sufficient for the induction of at least some mesoderm.

One might expect that the timing of expression of mesoderm-specific genes, such as those of muscle, would be closely coupled to the time at which the mesoderm is induced—but it is not. Explant experiments have shown that, irrespective of when during the competent period the cells are exposed to the 2-hour induction, muscle-specific gene expression always starts at about 5 hours after the end of the competent period, a time that corresponds to mid-gastrula stage in the embryo (Fig. 4.15). Muscle gene expression can occur as early as 5 hours after induction, if induction occurs late in the period of competence, or as late as 9 hours after induction, if induction occurs early in the competent period. These results suggest that there is an independent timing mechanism by which the cells monitor the time elapsed since fertilization and then, provided that they have received the mesoderm-inducing signal, they express muscle-specific genes.

The extended period of competence gives the embryo latitude as to when mesoderm induction actually takes place, and means that the inductive signal does not have to be occur at a specific time. Given this latitude, an independent timing mechanism for mesodermal gene expression is one way to keep subsequent development coordinated, whatever the actual time of mesoderm induction.

4.8 Zygotic gene expression is turned on at the mid-blastula transition

A global event in the embryo that appears to be under intrinsic temporal control is the beginning of transcription from the embryo's own genes, often referred to as zygotic genes. The *Xenopus* egg contains large amounts of maternal mRNAs, which are laid down during oogenesis. After fertilization, the rate of protein synthesis increases 1.5-fold, and a large number of new proteins are synthesized by translation of the maternal mRNA. There is, in fact, very little new mRNA synthesis until 12 cleavages have taken place and the embryo contains 4096 cells. At this stage, transcription from zygotic genes begins, cell cycles become asynchronous, and various other changes occur. This is known as the **mid-blastula transition**, although it does in fact occur in the late blastula, just before gastrulation starts. A few zygotic genes are expressed earlier, most notably two *nodal-related* genes, which are expressed as early as the 256-cell stage.

How is the mid-blastula transition triggered? Suppression of cleavage but not DNA synthesis does not alter the timing of transcriptional activation, and so the timing is not linked directly to cell division. Nor are cell–cell interactions involved, because dissociated blastomeres undergo the transition at the same time as intact embryos. The key factor in triggering the mid-blastula transition seems to be the ratio of DNA to cytoplasm—the quantity of DNA present per unit mass of cytoplasm.

Direct evidence for this comes from increasing the amount of DNA artificially by allowing more than one sperm to enter the egg or by injecting extra DNA into the egg. In both cases, transcriptional activation occurs prematurely, suggesting that there

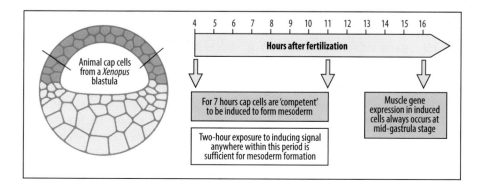

may be some fixed amount of a general repressor of transcription present initially in the egg cytoplasm. As the egg cleaves, the amount of cytoplasm does not increase, but the amount of DNA does. The amount of repressor in relation to DNA gets smaller and smaller until there is insufficient to bind to all the available sites on the DNA and the repression is lifted. Timing of the mid-blastula transition thus seems to fit with an hourglass egg-timer model (Fig. 4.16). In such a model something has to accumulate, in this case DNA, until a threshold is reached. The threshold is determined by the initial concentration of the cytoplasmic factor, which does not increase.

4.9 Mesoderm-inducing and patterning signals are produced by the vegetal region, the organizer, and the ventral mesoderm

Experiments using the explant system just described for studying mesoderm induction established that mesoderm is induced by signals from the vegetal region. But how does the mesoderm become patterned along the dorso-ventral axis? The fate map of the late *Xenopus* late blastula shows that different regions of mesoderm along the dorso-ventral axis give rise to different structures (see Fig. 4.9): the most dorsal region, also the region where the organizer will form, gives rise to the notochord; the next most dorsal to the somites (embryonic tissue that produces the skeletal muscles and the skeleton of the trunk); then comes the mesoderm giving rise to organs such as the heart and the kidney, with the most ventral region of mesoderm giving rise to blood (although significant amounts of blood will also form from more dorsal mesoderm).

Positional information for antero-posterior patterning of the mesoderm is provided by signals from other regions of the blastula and early gastrula. Experiments that combined different parts of the vegetal region with animal cap explants suggested that there are differences in mesoderm-patterning activity along the dorso-ventral axis of the vegetal region of the blastula (Fig. 4.17). For example, dorsal vegetal tissue containing the Nieuwkoop center induces notochord and muscle from animal cap cells, whereas ventral vegetal tissue induces mainly blood-forming tissue and little muscle. As we have already discussed, the Spemann organizer arises in the dorsal region of the marginal zone in the early gastrula stage (see Section 4.3). The organizer also produces mesoderm-patterning signals. This was shown by experiments in which a

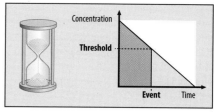

Fig. 4.16 A timing mechanism that could operate in development. A mechanism based on an analogy to an egg-timer could measure time to the mid-blastula transition. The decrease in the active concentration of some molecule, such as a transcriptional repressor, could occur with time, and the transition could occur when the repressor reaches a critically low threshold concentration. This would be equivalent to all the sand running into the bottom of the egg-timer. In the embryo a reduction in repressor activity per nucleus would in fact occur, because repressor concentration in the embryo as a whole remains constant but the number of nuclei increases as a result of cell division. The amount of repressor per nucleus thus gets progressively smaller over time.

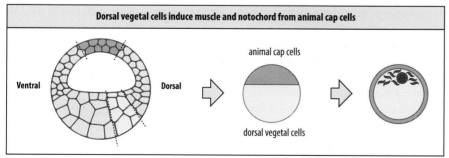

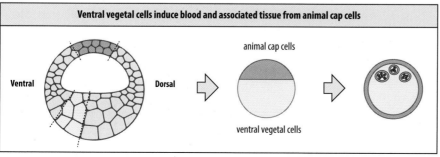

Fig. 4.17 Differences in mesoderm induction by dorsal and ventral vegetal regions. The dorsal vegetal region of the *Xenopus* blastula, which contains the Nieuwkoop center, induces notochord and muscle from animal cap tissues, whereas ventral vegetal cells induce blood and associated tissues.

Fig. 4.18 Signals involved in mesoderm induction and patterning originate from different regions of the embryo. Signals from the vegetal region of the blastula initially induce the mesoderm from the prospective ectoderm of the animal cap, with the organizer region being specified on the dorsal-most side where the inducing signal is strongest and of longest duration. Signals from the ventral side of the embryo pattern the mesoderm ventrally, and the extent of their influence is limited by signals emanating from the Spemann organizer region (O), which counteract them.

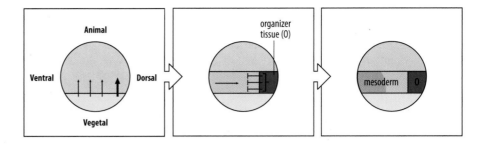

fragment of dorsal marginal zone from a late blastula, which will form the organizer, was combined with a fragment of the ventral presumptive mesoderm. The ventral fragment will now form substantial amounts of muscle, whereas ventral mesoderm isolated from an early blastula and cultured on its own will produce mainly blood-forming tissue and mesenchyme.

These types of experiments, together with other evidence, indicate that the induction and initial patterning of the mesoderm in *Xenopus* is due to the concerted actions of a series of signals emanating from different parts of the developing embryo (Fig. 4.18). The vegetal region of the blastula initially produces signals that induce the mesoderm, with the Nieuwkoop center in the dorsal vegetal region establishing the Spemann organizer on the dorsal-most side, where the inducing signal is strongest and of longest duration. Another set of signals then gives the ventral mesoderm its character. At the same time the organizer emits yet another set of signals that counteract the ventral signals, limiting their range of action and so, for example, enabling dorsal mesoderm to give rise to somites. The outcome of all this activity is that by the early-gastrula stage there is a band of prospective mesoderm around the equator of the embryo with different regions of the mesoderm along the dorso-ventral axis having acquired different regional identities.

4.10 Members of the TGF-β family have been identified as mesoderm inducers

The explant experiments discussed earlier suggested that extracellular signaling proteins secreted from the vegetal region in the blastula induce mesoderm. So the next question was what is the nature of those signals? Two main approaches were used to try and identify these factors. One was to apply candidate proteins directly to isolated animal caps in culture. The other was to inject mRNA encoding suspected inducers into animal cells of the early blastula and then culture the cells. By itself, however, the ability to induce mesoderm in culture does not prove that a particular protein is a natural inducer in the embryo. Rigorous criteria must be met before such a conclusion can be reached. These include the presence of the protein signal and its receptor in the right concentration, place, and time in the embryo; the demonstration that the appropriate cells can respond to it; and the demonstration that blocking the response prevents induction. On all these criteria, growth factors of the TGF-β family (see Box 4C) have been identified as the most likely candidates for mesoderm inducers. Members of this family known to be involved in early *Xenopus* development include the Nodal-related proteins (Nodal-related-1, -2, -4, -5, -6), the protein Derrière, the bone morphogenetic proteins (BMPs), Vg-1, and activin.

The signaling pathways stimulated by TGF-β growth factors are illustrated in Box 4C. Confirmation of the role of TGF-β growth factors in mesoderm induction came initially from experiments that blocked activation of the receptors for these factors (Box 4D). The activin type II receptor is a receptor for several TGF-β family growth factors and is uniformly distributed throughout the early *Xenopus* blastula. When mRNA for

EXPERIMENTAL BOX 4D Investigating receptor function using dominant-negative mutations

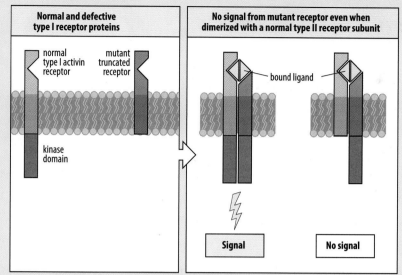

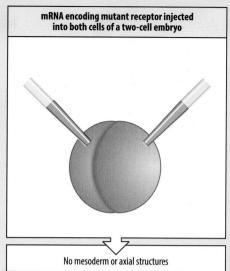

Figure 1

The receptors for proteins of the TGF-β family function as dimers of type I and type II subunits, with the type II subunit participating only in binding and the type I subunit generating the intracellular signal (see Box 4C). One way of investigating the requirement for a particular receptor in a developmental process is to block receptor function (Figure 1). In the case of dimeric receptors, receptor function in a cell can be blocked by introducing mRNA encoding a mutant receptor subunit that lacks most of the cytoplasmic domain, and so cannot function. The mutant receptor subunit can bind ligand and forms heterodimers with normal receptor subunits, but the complex cannot signal. The mutant subunit thus acts in a **dominant-negative** fashion preventing receptor function. When mRNA encoding the mutant activin receptor subunit is injected into cells of the two-cell *Xenopus* embryo, subsequent mesoderm formation is blocked. No mesoderm or axial structures are formed except for the cement gland, the most anterior structure of the embryo.

a mutant receptor subunit is injected into the early embryo and blocks TGF-β family signaling, mesoderm formation is prevented. Because different TGF-β growth factors can act through the same receptor, these experiments were not able to pinpoint the individual proteins responsible.

More recently, it has become possible to study mesoderm induction using more specific inhibitors of Nodal proteins, and also to detect the presence of signaling by using labeled antibodies to detect phosphorylated Smads (the transcription factors that are activated as a result of TGF-β signaling) and to distinguish their type. This latter technique enables the spatial patterns and dynamics of normal growth factor signaling to be visualized directly, and it is now widely used to detect signaling by TGF-β family members. For example, signaling by BMPs can be distinguished from that of Nodal-related proteins or activin by looking for phosphorylated Smad1 or Smad2, respectively (see Box 4C).

4.11 The zygotic expression of mesoderm-inducing and patterning signals is activated by the combined actions of maternal VegT and Wnt signaling

One maternal vegetal factor of great importance in the earliest stage of mesoderm induction is the transcription factor VegT (see Section 4.4). Directly and indirectly, VegT activates the zygotic expression of the genes encoding the Nodal-related proteins and

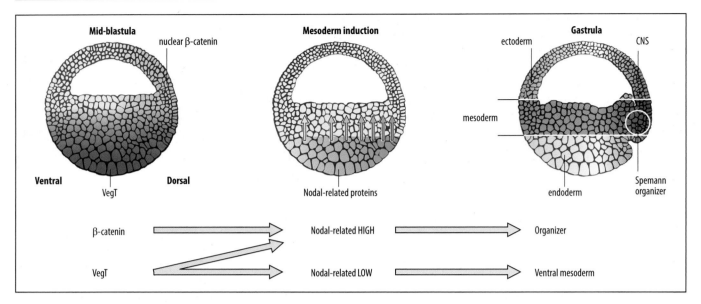

Fig. 4.19 A gradient in Nodal-related proteins may provide the initial signals in mesoderm induction. Maternal VegT in the vegetal region activates the transcription of *nodal-related* genes. The presence of nuclear β-catenin on the dorsal side stimulates *nodal-related* transcription, resulting in a dorsal-to-ventral gradient in the Nodal-related proteins. These induce mesoderm and, at high doses, specify the Spemann organizer on the dorsal side. This is a simplified version because other signals, such as Vg-1 and activin, also have roles. CNS, central nervous system.

Derrière. If VegT is severely depleted, almost no mesoderm develops, and ectoderm forms throughout the embryo but injection of *nodal-related* and *derrière* mRNAs can rescue mesoderm formation. This strongly suggests that the signaling proteins encoded by these mRNAs are most likely to be the direct mesoderm inducers. The rescue of head, trunk, and tail mesoderm by the mRNAs for several of the *nodal-related* genes indicates that they are involved in the general induction of all mesoderm. *derrière* mRNA rescues only trunk and tail, indicating that Derrière protein on its own cannot induce the anterior-most mesoderm. Assays of the activity of the Nodal-related proteins on animal cap cells in culture have, however, shown that they can heterodimerize with each other and with Derrière and so can cooperate to induce mesoderm *in vivo*.

Numerous experiments have now shown that Nodal-related proteins are expressed in the *Xenopus* vegetal region in a graded fashion at the appropriate time for mesoderm induction, and that the difference between a dorsal and a ventral mesoderm fate is likely to be due to expression of a cocktail of different Nodal-related proteins and/or differences in their concentrations along the dorso-ventral axis. Nuclear β-catenin stimulates *nodal-related* gene transcription, and thus the highest concentration of Nodal-related proteins will be produced by the Nieuwkoop center on the dorsal side of the vegetal region where the cells express VegT and also contain nuclear β-catenin (Fig. 4.19). Signaling via Nodal-related proteins stimulates phosphorylation and activation of Smad2, and a wave of Smad2 activation, detected by anti-phospho-Smad2 antibodies, appears to sweep from the dorsal to the ventral side of the embryo from early to mid-gastrulation. All this evidence is consistent with the Spemann organizer being induced in the dorsal-most equatorial region where Nodal signaling is most intense and of longest duration (see Fig. 4.19).

4.12 Threshold responses to gradients of signaling proteins are likely to pattern the mesoderm

The mesoderm-inducing and patterning signals described in previous sections exert their developmental effects by inducing the expression of specific genes that are required for the further development of the mesoderm. The dorsal-most region of the mesoderm, where the Spemann organizer arises, produces the notochord, and more ventral regions form, for example, muscle or blood. Expression of the *brachyury* gene, which encodes a T-box family transcription factor (see Chapter 1), is one of the earliest markers of mesoderm, and is a key transcription factor in mesoderm specification and patterning. *brachyury* is expressed throughout the presumptive mesoderm in the late blastula and early gastrula (Fig. 4.20), later becoming confined to the notochord and also to the tailbud (posterior mesoderm). Various members of the fibroblast growth factor (FGF) family of growth factors are also expressed in the mesoderm in the late blastula and early gastrula (the FGF signaling pathway is shown in Box 4E). One of the functions of FGF signaling in the mesoderm at these stages is to maintain expression of *brachyury*; in turn, Brachyury protein maintains expression of the genes encoding the FGFs. This positive-feedback loop (see Fig 1.21) ensures that the mesoderm continues robustly to express *brachyury*.

One of the first zygotic genes to be expressed specifically in the dorsal-most mesoderm that will become the Spemann organizer is *goosecoid*, which encodes a transcription factor somewhat similar to both the Gooseberry and Bicoid proteins of *Drosophila*—hence the name. In line with its presence in the organizer, microinjection of *goosecoid* mRNA into the ventral region of a *Xenopus* blastula mimics to some extent the transplantation of the Spemann organizer (see Fig. 4.5), resulting in formation of a secondary axis. Genes for other transcription factors are also specifically expressed in the organizer (Fig. 4.21 and see Summary table).

It is still not clear how the secreted signaling proteins present in the mesoderm turn on genes like *goosecoid* and *brachyury* in the right place. For example, *brachyury* can be experimentally switched on in *Xenopus* animal caps by the TGF-β family member activin, but it is more likely to be induced normally in the presumptive mesoderm *in vivo* by Nodal signaling, which also induces expression of FGFs. As we have seen in the previous sections, there are gradients of activity of secreted signaling molecules throughout the mesoderm that could provide positional information for switching on developmental genes.

Experiments with the TGF-β family member activin provide a beautiful example of how a diffusible protein could pattern a tissue by turning on particular genes at specific threshold concentrations. Activin is one of the molecules that plays a part in mesoderm patterning *in vivo* and acts through the same receptor as Nodal. Explanted animal cap cells from a *Xenopus* blastula in culture respond to increasing doses of activin by activation of different mesodermal genes at different threshold concentrations. In this explant system, increasing concentrations of activin specify several distinct cell states that correspond to the different regions along the dorso-ventral axis. At the lowest concentrations of activin, only epidermal genes are expressed and no mesoderm is induced. Then, as the concentration increases and exceeds a particular threshold concentration, *brachyury* is expressed, together with muscle-specific genes such as that for actin. With a further increase in activin concentration, *goosecoid* is expressed, corresponding to the dorsal-most region of the mesoderm—which will form the Spemann organizer (Fig. 4.22). One can see, therefore how graded growth-factor signaling along the dorso-ventral axis could, in principle, activate transcription factors in particular regions and thus pattern the mesoderm. A 1.5-fold increase in activin concentration is sufficient to cause a change from formation of muscle to formation of notochord in explants of animal caps. Just how gradients are set up *in vivo* with the necessary precision is not clear. Although simple diffusion of morphogen

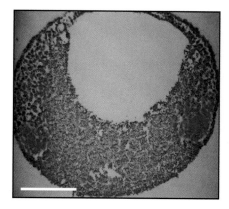

Fig. 4.20 Expression of *brachyury* in a late *Xenopus* blastula. A cross-section through the embryo along the animal-vegetal axis shows that *brachyury* (red) is expressed in the prospective mesoderm. Scale bar = 0.5 mm. *Photograph courtesy of M. Sargent and L. Essex.*

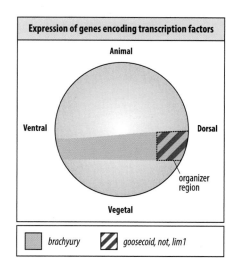

Fig. 4.21 Expression of transcription factor genes in a *Xenopus* late blastula/early gastrula. The expression domains of a number of zygotic genes that code for transcription factors correspond quite well to demarcations on the specification map, as judged by the distribution of their mRNAs. *brachyury* is expressed in a ring around the embryo corresponding quite closely to the future mesoderm (see also Fig. 9.29). A number of transcription factors are specifically expressed in the dorsal-most region of mesoderm that will give rise to the Spemann organizer and are required for its function. Not (notochord homeobox protein) appears to have a role in the specification of notochord; both Goosecoid and Lim1 (LIM homeobox 1 protein) together are required for a transplanted Spemann organizer to be able to induce the formation of a new head.

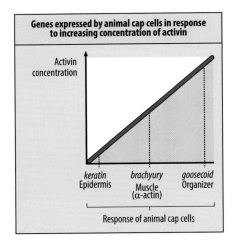

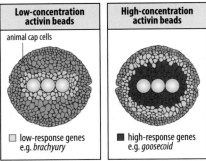

Fig. 4.22 Graded responses of *Xenopus* animal cap cells to increasing concentrations of activin. When animal cap cells are treated with increasing concentrations of activin, particular genes are activated at specific concentrations, as shown in the top panel. As the concentration of activin is increased, *brachyury* is induced, whereas *goosecoid,* which is characteristic of the organizer region, is only induced at high concentrations. If beads releasing a low concentration of activin are placed in the center of a mass of animal cap cells (lower left panel), expression of low-response genes such as *brachyury* is induced immediately around the beads. With a high concentration of activin in the beads (lower right panel), *goosecoid* and other high-response genes are now expressed around the beads and the low-response genes farther away.

molecules may play a part, more complex processes are likely to be involved. Some common cellular mechanisms that could be involved in gradient formation are indicated in Box 11B.

How do cells distinguish between different concentrations of growth factor? Occupation of just 100 activin receptors per cell is required to activate expression of *brachyury*, whereas 300 receptors must be occupied for *goosecoid* to be expressed. The link between the strength of the signal and gene expression may not be so simple, however. There are additional layers of intracellular regulation that act later to refine distinct regions of gene expression. Cells expressing *goosecoid* at high activin concentrations repress *brachyury* expression, and this involves the action of the Goosecoid protein itself, together with other proteins. In turn, Brachyury inhibits *goosecoid* expression, although this is not direct. The mutual repression between Brachyury and Goosecoid enables the graded activin signal to be converted into two discrete domains of gene expression with a sharp boundary between them.

SUMMARY

The endoderm and ectoderm are maternally specified in *Xenopus*, but nevertheless the embryo can still undergo considerable regulation at the blastula stage. This implies that interactions between cells, rather than intrinsic factors, have a central role in early amphibian development. The Spemann organizer region and the rest of the mesoderm are induced from prospective ectoderm at the equator of the blastula by signals from the vegetal region which includes the Nieuwkoop center dorsally. Patterning of the mesoderm along the dorso-ventral axis produces regions that, from ventral to dorsal, give rise to blood and blood vessels, internal organs such as kidneys, the somites, and the organizer region in the most dorsal mesoderm, which produces the notochord.

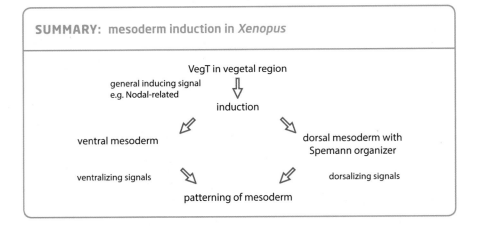

SUMMARY: mesoderm induction in *Xenopus*

The Spemann organizer and neural induction

The Spemann organizer has a global organizing function. As discussed earlier, it can induce a second body axis if transplanted to another embryo at an appropriate stage, resulting in a twinned embryo. In other words, it can organize and coordinate both dorso-ventral and antero-posterior aspects of the body plan, as well as induce neural tissue from ectoderm. We have already alluded to its role in dorso-ventral patterning of the mesoderm in *Xenopus* embryos by producing signals that limit the range of action of signals produced ventrally. We will now consider the function of

CELL BIOLOGY BOX 4E The FGF signaling pathway

The developmentally important growth factors FGF and EGF (epidermal growth factor), along with many other growth and differentiation factors, including neuregulin, signal through transmembrane receptors with intracellular tyrosine kinase domains. Several different signaling pathways lead from these receptors, which control many aspects of cell behavior throughout an animal's life. The signaling pathway shown here is a simplified version of a pathway triggered by FGF that leads to changes in gene expression that promote cell survival, cell growth, cell division, or differentiation, depending on the cells involved and the developmental context. There are multiple different forms of FGF and several different FGF receptors, which are also used in different contexts. For example, in *Xenopus*, FGF-4 is involved in maintaining *brachyury* expression in the mesoderm, and in the chick, FGF-8 is involved in signaling from the organizing centers in the developing brain (discussed in Chapter 12) whereas FGF-10 is crucial for limb development (discussed in Chapter 11).

The intracellular signaling pathway leading from the FGF receptors is often known as the Ras-MAPK pathway because of the key involvement of the small G protein Ras and the activation of a cascade of serine/threonine kinases that ends in the activation of a mitogen-activated protein kinase (MAPK). The name comes from the fact that FGF and other growth factors can act as mitogens, agents that stimulate cell proliferation. In one variation or another, the central Ras-MAPK module occurs in the signaling pathways from many different tyrosine-kinase receptors (Figure 1).

Binding of extracellular FGF to its receptor results in dimerization of two ligand-bound receptor molecules, activating the intracellular tyrosine kinase domains, which then phosphorylate each other. The phosphorylated receptor tails recruit adaptor proteins (Grb and Sos), which in turn recruit and activate Ras at

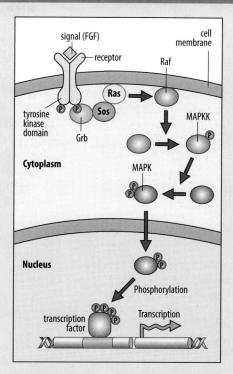

Figure 1

the plasma membrane. This results in the binding and activation of the first serine/threonine kinase in the cascade, which in the mammalian pathway is called Raf. This phosphorylates and activates the next protein kinase, the MAPK kinase (MAPKK), which phosphorylates and activates MAPK. MAPK may then phosphorylate other kinases and can also enter the nucleus and phosphorylate transcription factors, thus activating gene expression. There are several different MAPKs in mammalian cells that are used in different pathways and target different transcription factors.

the organizer in organizing the dorso-ventral axis in more detail, and in particular the nature of the signaling molecules involved. We will also examine its function in organizing the antero-posterior axis of the mesoderm, which emerges during gastrulation.

During gastrulation, the ectoderm in the dorsal region of the embryo becomes specified as neural ectoderm and forms the neural plate, which will give rise to the nervous system (see Fig. 3.7). The nervous system must develop in the correct relationship with other body structures, particularly the mesodermally derived structures that give rise to the skeleto-muscular system. This means that the antero-posterior patterning of the nervous system must be linked to that of the mesoderm, and this is coordinated through the organizer, which is involved in both. We shall therefore consider the role of the organizer in induction of the neural plate and how the neural plate becomes patterned along the antero-posterior axis. The Spemann organizer is located at the dorsal lip of the blastopore, which is the site where gastrulation movements are initiated. These movements are described in Chapter 9.

Scan here

Scan this QR code image with your mobile device to see an online animation of the FGF signaling pathway or log on to **http://global.oup.com/uk/orc/biosciences/devbiol/wolpert5e/qr/qr4c/**

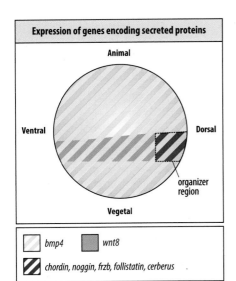

Fig. 4.23 Distribution of mRNAs encoding secreted proteins in the *Xenopus* late blastula/early gastrula. The secreted proteins that are made in the organizer block the action of BMP-4 and Wnt-8, proteins that are made throughout the late blastula/early gastrula and in the prospective mesoderm, respectively, where their genes are expressed. Note that the gene encoding Nodal-related-3, which differs from other members of the *Xenopus* Nodal-related family of proteins in that it has no direct mesoderm-inducing activity, is also expressed in the organizer. Frzb, Frizzled-related protein.

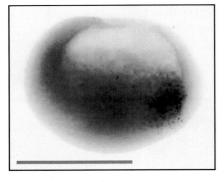

Fig. 4.24 Expression of *noggin* in the *Xenopus* late blastula. Staining for expression of *noggin* mRNA locates it in the region where the Spemann organizer will develop. Scale bar = 1 mm.

Reproduced with permission from Smith W.C., Harland, R.M.: **Expression cloning of noggin, a new dorsalizing factor localized to the Spemann organizer in Xenopus embryos.** Cell *1992,* **70:** *829–840. © 1992 Cell Press.*

4.13 Signals from the organizer pattern the mesoderm dorso-ventrally by antagonizing the effects of ventral signals

One of the main functions of the organizer is to help pattern the mesoderm along the dorso-ventral axis of the embryo. As we discussed earlier, dorso-ventral mesoderm patterning involves not only signals from the vegetal region in the blastula but also dorsalizing signals emanating from the Spemann organizer, which counteract ventralizing signals produced from the ventral side of the embryo (see Section 4.9).

Ventralization of the mesoderm is promoted by gradients of activity of proteins from two families of signaling molecules, BMP and Wnt, which have their high points on the ventral side of the embryo. Initially, BMP-4 protein and other BMP proteins are found throughout the late blastula/early gastrula. Zygotic Wnt-8 protein is made in the prospective mesoderm at these stages (Fig. 4.23). When the action of BMPs is blocked throughout the embryo by introducing a dominant-negative mutant receptor (see Box 4D), the embryo is dorsalized, with ventral cells differentiating as both muscle and notochord. Conversely, overexpression of BMP-4 ventralizes the embryo.

The restriction of BMP and Wnt-8 signaling to the more ventral regions in the gastrula is due to extracellular dorsalizing signals that are secreted by the organizer and inhibit, or antagonize, BMP and Wnt-8 activity. The first antagonists to be discovered were the proteins Noggin and Chordin, both of which inhibit BMP activity. Noggin, for example, was discovered in a screen for factors that could rescue UV-irradiated ventralized *Xenopus* embryos, in which a dorsal region had not been specified by the usual activity of the β-catenin pathway (see Section 4.2) and the *noggin* gene is strongly expressed in the Spemann organizer (Fig. 4.24). Noggin protein cannot induce mesoderm in animal-cap explants, but it can dorsalize explants of ventral marginal zone tissue, making it a good candidate for a signal that patterns the mesoderm along the dorso-ventral axis. The organizer was subsequently found to secrete a mixture of antagonists—the BMP antagonists Noggin, Chordin, and Follistatin, and the Wnt antagonist Frizzled-related protein (Frzb).

The finding that dorsalizing factors are secreted antagonists of signaling molecules was completely unexpected. The secretion of antagonists that restrict signaling to a particular region, or modulate the level of signaling, is a common strategy in embryonic development. We have already come across an example in Section 4.2 where a Wnt antagonist helps to localize activation of Wnt/β-catenin signaling to the dorsal side of the very early embryo, and we shall come across other examples of the role of signal antagonists in Chapter 5.

Noggin, Chordin, and Follistatin proteins interact with BMP proteins and prevent them from binding to their receptors. This sets up a functional gradient of BMP activity across the dorso-ventral axis of the *Xenopus* gastrula with its highest point ventrally and little or no activity in the presumptive dorsal mesoderm. The way in which this gradient is produced is not simply due to BMP antagonists diffusing from the organizer across the dorso-ventral axis but to a shuttling mechanism (see Box 11B). Chordin produced dorsally binds to BMPs in this region, inhibiting their action and forming a complex. Some of the BMPs and BMP-like molecules in the dorsal region of the gastrula are paradoxically produced by the organizer itself. The Chordin–BMP complex then diffuses away from the dorsal region of the embryo until it reaches the ventral region, where an extracellular metalloproteinase is produced that degrades the Chordin portion of the complex, thereby depositing BMP (Fig. 4.25). The metalloproteinase also acts as a clearing agent for Chordin, reducing the extent of its long-range diffusion and helping to maintain a gradient of Chordin dorsalizing activity.

The action of Chordin in creating a gradient of BMPs, including BMP-4, mirrors that of the *Drosophila* Chordin homolog Sog on the BMP-4 homolog Decapentaplegic (Dpp) in the patterning of the dorso-ventral axis in the fly (see Section 2.17). In the fly, however, the dorso-ventral axis is inverted compared with that of vertebrates, and so Dpp specifies a dorsal fate whereas BMP-4 specifies a ventral fate (the inversion of the dorso-ventral axis during evolution is discussed in Section 14.6). Other components of the mechanism that forms the gradient of Dpp activity in *Drosophila* are also conserved in vertebrates, with the *Xenopus* metalloproteinase being related to the fly metalloproteinase Tolloid.

The Wnt antagonist Frzb also generates a ventral-to-dorsal gradient of Wnt activity in the gastrula by binding to Wnt proteins and preventing them acting dorsally. Yet another signaling antagonist produced by the organizer is Cerberus, which inhibits the activity of BMPs, Nodals, and Wnt proteins in prospective anterior tissues. Cerberus is involved in the suppression of mesoderm fate and the induction of anterior structures, in particular the head.

4.14 The antero-posterior axis of the embryo emerges during gastrulation

The Spemann organizer is located in the dorsal lip of the blastopore. If the blastopore dorsal lip from a very early gastrula is grafted to the ventral side of the marginal zone of another early gastrula, it can induce a second embryo with a well-defined head, a central nervous system, a trunk region, and tail (Fig. 4.26, left panels, and see Fig. 4.5). The dorsal blastopore lip of the very early gastrula is therefore often said to function as a 'head organizer'. Various other experimental manipulations, such as grafting dorsal vegetal blastomeres containing the Nieuwkoop center to the ventral side, produce a similar result (see Fig. 4.8). What all these treatments have in common is that, directly or indirectly, they result in the formation of a new Spemann organizer region with head organizer function. Classic experiments also showed that a blastopore dorsal lip taken from a mid-gastrula and grafted to an early gastrula induces a trunk and tail but no head—the dorsal blastopore lip at this stage is often said to function as a 'trunk organizer'. A grafted dorsal lip from a late gastrula induces only a tail (Fig. 4.26, right panels).

These results show that the inductive properties of the organizer change over time as gastrulation proceeds. During gastrulation, dorsal endoderm cells and mesoderm cells move around the dorsal blastopore lip and internalize, moving towards the animal pole and laying down the antero-posterior axis perpendicular to the dorso-ventral axis (see Fig. 3.6). Cells that give rise to anterior structures internalize first and cells that give rise

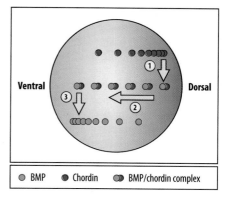

Fig. 4.25 A shuttling mechanism generates a ventral–dorsal gradient of BMP signaling in the gastrula. (1) Chordin (red) secreted from the organizer binds to BMP (green) in the dorsal region of the embryo. Some BMPs and BMP-like signaling molecules are secreted by cells throughout the early gastrula, some are secreted by the organizer itself. (2) The Chordin–BMP complex diffuses ventrally. (3) In the ventral region, Chordin is cleaved by a metalloproteinase which is made in this region and this releases BMP from the complex and deposits it ventrally. This shuttling mechanism generates a gradient of BMP signaling running from high ventrally (green background) to low dorsally (red background) across the embryo.

*After Lewis, J.: **From signals to patterns: space, time, and mathematics in developmental biology.** Science 2008, **322**: 399–403.*

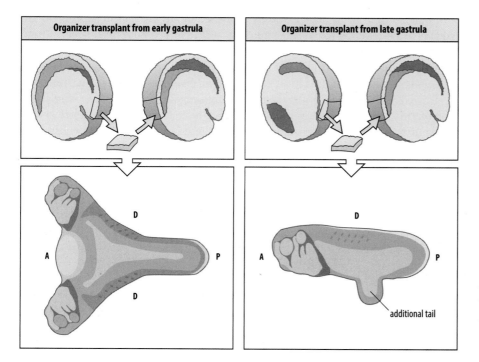

Fig. 4.26 The inductive properties of the organizer change during gastrulation. A graft of the organizer region—the dorsal lip of the blastopore of an early *Xenopus* gastrula—to the ventral side of another early gastrula, results in the development of a twinned embryo with the second embryo induced by the graft having a head (left panels). A graft from the dorsal lip region of a late gastrula to an early gastrula only induces tail structures (right panels). A = anterior; P = posterior; D = dorsal.

Fig. 4.27 Different parts of the *Xenopus* organizer region give rise to different tissues. In the early gastrula (left panel) the organizer is located in the dorsal lip of the blastopore. The cells of the leading edge (orange) are the first to internalize and give rise to anterior endoderm of the neurula stage (right panel). The deep cells (brown) are next to internalize and give rise to the prechordal plate—the mesoderm anterior to the notochord—which forms the ventral head mesoderm. The remainder of the organizer gives rise to notochord (red).

*Adapted from Kiecker, C., Niehrs, C: **A morphogen gradient of Wnt/β-catenin signalling regulates anteroposterior neural patterning in** Xenopus. Development 2001, **128**: 4189-4201.*

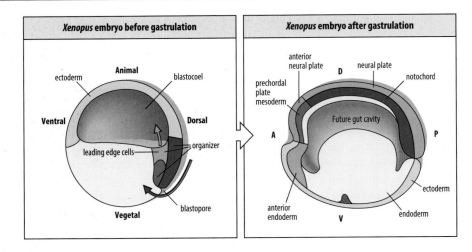

Scan here

Scan this QR code image with your mobile device to see an online movie showing gastrulation in *Xenopus* or log on to http://global.oup.com/uk/orc/biosciences/devbiol/wolpert5e/qr/qr4d/

to more posterior structures later. This means that the dorsal blastopore lip is not a uniform cell population and is composed of different cell populations at different times. This can help to explain the complex patterning and inductive properties of the organizer, and why the inductive properties of grafts of the dorsal lip of the blastopore change over time.

Detailed fate maps of the dorsal lip of the blastopore in the *Xenopus* early gastrula (which has head organizer function) show that it is composed of different regions of cells that give rise during gastrulation to anterior endoderm, prechordal plate mesoderm, and the notochord (Fig. 4.27). The prechordal plate mesoderm ends up anterior to the notochord and forms the ventral head mesoderm. At the late gastrula stage, however, the dorsal lip of the blastopore will just be composed of cells that give rise to the notochord. Experiments investigating the inductive capacity of different parts of the organizer of the early gastrula show that the ability to induce heads—head organizer function—is restricted to the vegetal region (shown in brown in Fig. 4.27). The more dorsal part of the organizer (shown in red in Fig. 4.27) can induce trunk and tail structures but not heads.

A number of proteins are specifically expressed in the *Xenopus* organizer and are known to be required for its function (Fig. 4.28 and see Summary Table). These proteins include a number of transcription factors. We have already discussed how Goosecoid, for example, is an early and reliable marker of the organizer (see Section 4.12). Goosecoid is expressed in all the cells of the organizer of the early gastrula and is required for head formation in the normal course of development. The vegetal portion of the organizer expresses proteins such as the transcription factor Otx2, which is characteristic of anterior structures, and the transcription factor Lim1, whose gene is switched on by VegT. The dorsal part of the organizer expresses the transcription factor Not, which is required for specification of the notochord.

Fig. 4.28 Proteins expressed in the Spemann organizer region of the *Xenopus* gastrula. The expression of some of these proteins, such as Brachyury, is not confined to the organizer.

Proteins expressed in organizer region		
Xenopus gastrula	Transcription factors	Secreted proteins
mesoderm—organizer	Brachyury, Goosecoid	Nodal-related 3
	Not	Chordin, Noggin
	Otx2, Lim1	Cerberus

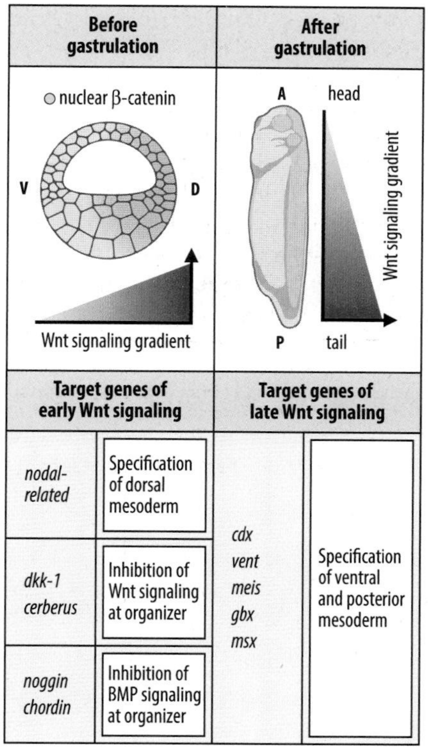

Fig. 4.29 Wnt signaling specifies the dorso-ventral axis in the early *Xenopus* embryo before gastrulation, while a second round of Wnt signaling after gastrulation patterns the antero-posterior axis. Before gastrulation, accumulation of nuclear β-catenin following cortical rotation specifies the future dorsal region of the embryo (shown on the left). At these early stages, Wnt signaling is graded with its high point at the dorsal side of the embryo, where it participates in specifying the Nieuwkoop center and the organizer by activating genes that specify dorsal mesoderm and genes expressed in the organizer, including Wnt antagonists. Zygotic *wnt8* is activated in the mesoderm at the late blastula stage. During gastrulation, Wnt antagonists secreted by the organizer inhibit Wnt signaling in dorsal mesoderm and anterior mesoderm. After gastrulation, Wnt signaling is now graded along the antero-posterior axis of the gastrula with its high point at the posterior of the embryo. Target genes activated include those for transcription factors involved in specifying ventral mesoderm and the posterior mesoderm that later gives rise to the tail of the tadpole (shown on the right). D, dorsal; V, ventral; A, anterior; P, posterior.

*After Hikasa, H., Sokol, S.Y.: **Wnt Signaling in Vertebrate Axis Specification.** Cold Spring Harb. Perspect., 2013, 5:a007955.*

Before gastrulation	After gastrulation
○ nuclear β-catenin	A — head / Wnt signaling gradient / P — tail
Wnt signaling gradient	

Target genes of early Wnt signaling		Target genes of late Wnt signaling	
nodal-related	Specification of dorsal mesoderm	*cdx* *vent* *meis* *gbx* *msx*	Specification of ventral and posterior mesoderm
dkk-1 *cerberus*	Inhibition of Wnt signaling at organizer		
noggin *chordin*	Inhibition of BMP signaling at organizer		

The organizer secretes a cocktail of antagonists of BMP and Wnt signaling—the proteins Chordin, Noggin and Follistatin that antagonize BMPs, and the proteins Frzb and Dickkopf1 that antagonize Wnts. Rather paradoxically, the organizer also secretes signaling molecules—BMP-2 and another BMP-related signaling protein, and in addition a Nodal-related protein that has no direct mesoderm-inducing activity. The protein Cerberus, which antagonizes Wnt, Nodal, and BMP signaling, is produced by the region of the organizer that gives rise to the anterior endoderm. We have already seen the importance of the BMP antagonists produced by the organizer in creating a gradient of BMP signaling that patterns the dorso-ventral axis of the mesoderm in the gastrula (see Sections 4.17 and 4.18). The secretion of the Wnt antagonists by the organizer also produces a gradient of Wnt/β-catenin signaling, so that at the organizer itself, levels of both BMP and Wnt signaling are low.

The gradient of Wnt signaling in the early gastrula provides positional information along the antero-posterior axis. This later function of Wnt signaling contrasts with its earlier function of specifying the dorsal side of the embryo, which we discussed earlier (Fig. 4.29). The essential ingredient for head formation is the inhibition in anterior tissues of Nodal, BMP and Wnt signals. An extra head can be induced in the ventral region of an early gastrula by the simultaneous inhibition of BMP and Wnt signaling, or of Nodal and BMP signaling. The tail forms where both BMP signaling and Wnt signaling are high.

These observations and experimental results have led to a model of early *Xenopus* development that incorporates orthogonal BMP gradients and Wnt gradients into a three-dimensional coordinate system of positional information provided by morphogen gradients (the gradients of BMPs and Wnts) that provides the framework for patterning the embryonic axes in the gastrula (Fig. 4.30). Morphogen gradients at right angles to each other also operate to pattern the *Drosophila* wing (discussed in Chapter 11), and represent an important way of patterning an embryonic tissue.

4.15 The neural plate is induced in the ectoderm

The Spemann organizer-transplant experiment in newts illustrated in Fig. 1.9, and the later experiments in *Xenopus* (see Figs 4.5 and 4.26), showed that neural tissue is induced from the ectoderm. In the second embryonic axis that forms at the site of transplantation of the Spemann organizer, a nervous system develops from host ectoderm that would normally have formed ventral epidermis. This suggested that neural tissue could be induced from as-yet-unspecified ectoderm by signals emanating from the organizer. This hypothesis was confirmed by experiments in which prospective neural plate ectoderm was replaced with prospective epidermis before gastrulation;

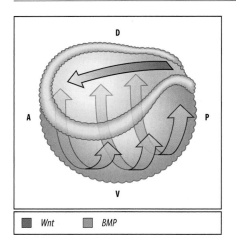

Fig. 4.30 Double-gradient model of embryonic axis formation. The model shows how gradients of Wnts (blue) and BMPs (green) perpendicular to each other regulate antero-posterior and dorso-ventral patterning in *Xenopus*. The color scales of the arrows indicate the signaling gradients; arrows indicate the direction of spreading of the signals. For example, the formation of the tail requires a higher level of Wnt signaling than does the head. Patterning begins at gastrula stages, but for clarity is depicted here in an early neurula.

Adapted from Niehrs C.: **Regionally specific induction by the Spemann-Mangold organiser** *Nat. Rev. Genet. 2004, 5: 425–434.*

Wnt	BMP

the transplanted prospective epidermis developed into neural tissue (Fig. 4.31). This showed that the formation of the nervous system is dependent on an inductive signal.

An enormous amount of effort was devoted in the 1930s and 1940s to trying to identify the signals involved in neural induction in amphibians. Researchers were encouraged by the finding that a dead organizer region could still induce neural tissue. It seemed to be merely a matter of hard work to isolate the chemicals responsible. The search proved fruitless, however, for it appeared that an enormous range of different substances were capable of varying degrees of neural induction. As it turned out, this was because newt ectoderm, the main experimental material used, has a high propensity to develop into neural tissue on its own. This is not the case with *Xenopus* ectoderm, although simply dissociating *Xenopus* ectodermal cells for a few hours can result in their differentiation as neural cells after reaggregation.

A key discovery in the study of neural induction was the finding that the BMP antagonist Noggin, one of the first secreted proteins isolated from the Spemann's organizer, could induce neural differentiation in ectoderm explants from *Xenopus*

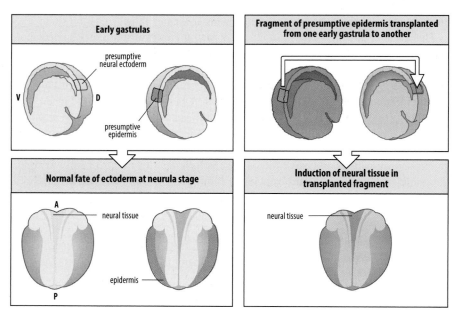

Fig. 4.31 The nervous system of *Xenopus* is induced during gastrulation. The left panels show the normal developmental fate of ectoderm at two different positions in the early gastrula. The right panels show the transplantation of a piece of ventral ectoderm, whose normal fate is to form epidermis, from the ventral side of an early gastrula to the dorsal side of another, where it replaces a piece of dorsal ectoderm whose normal fate is to form neural tissue. In its new location, the transplanted prospective epidermis develops not as epidermis but as neural tissue, and forms part of a normal nervous system. This shows that the ventral tissue has not yet been determined at the time of transplantation, and that neural tissue is induced during gastrulation.

blastulas. In the late *Xenopus* blastula, BMPs are expressed throughout the ectoderm, but their expression is subsequently lost in the neural plate. These results suggested that neural plate could only develop if BMP signaling is absent. This conclusion is supported by the dramatic effect of simultaneously knocking down expression of BMP-4 and the three other BMPs expressed in the late blastula with antisense morpholino oligonucleotides (see Box 6B). In the absence of BMP signaling, the embryo becomes covered in neural tissue (Fig. 4.32). BMP antagonists such as Noggin and Chordin produced by the organizer (see Fig. 4.28) therefore became attractive candidates for inhibiting BMP signaling in the prospective neural plate and allowing neural induction to proceed. As BMP signaling maintains expression of the BMP genes, antagonizing BMP signaling would also shut down BMP expression.

These observations led to the so-called 'default model' for neural induction in *Xenopus*. This proposed that the default state of the dorsal ectoderm is to develop as neural tissue, but that this pathway is blocked by the presence of BMPs, which promote the epidermal fate (Fig. 4.33). The role of the organizer is to lift this block by producing proteins that inhibit BMP activity; the region of ectoderm that comes under the influence of the organizer will then develop as neural ectoderm. According to this model, the antagonists secreted from the organizer act on ectoderm that lies adjacent to the organizer at the beginning of gastrulation. As gastrulation proceeds, internalized cells derived from the organizer continue to secrete these antagonists, which continue to act on the ectoderm that now overlies them (see Fig. 4.27).

Elimination of individual organizer antagonists in amphibians has rather modest effects on neural induction. But when combinations of the BMP antagonists, such as a combination of Chordin, Noggin, and Follistatin or of Cerberus, Chordin, and Noggin, are simultaneously depleted in the organizer by antisense morpholino oligonucleotides, there is a dramatic failure of neural and other dorsal development, and an expansion of ventral and posterior fates. These experiments show that BMP inhibitors, which are produced at the right place and at the right time, are required for neural induction.

The default model for neural induction is not the complete answer, however, as neural development in *Xenopus* also requires the growth factor FGF, even when BMP inhibition is lifted by the presence of Noggin and Chordin. FGF signals are produced by cells in the blastula and are thought to act as early neural inducers. It has also been shown that the apparently spontaneous differentiation of explanted newt ectoderm into neural tissue is caused by activation of the mitogen-activated protein kinase

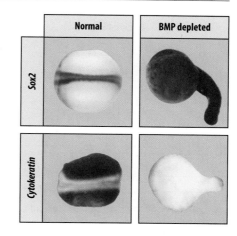

Fig. 4.32 In the absence of BMP signaling, the embryo becomes covered in brain tissue. Left panels: normal neurula-stage *Xenopus* embryos stained blue for expression of *Sox2* mRNA, a neural marker (upper panel) or *cytokeratin* mRNA, an epidermal marker. Right panels: neurula-stage *Xenopus* embryos that have been simultaneously depleted of all four BMP mRNAs by injection of antisense morpholinos at the 2–4-cell stage, and are stained similarly to the embryos on the left. The BMP-depleted embryos are completely covered in neural tissue.

From De Robertis, E.M.: ***Spemann's organizer and the self-regulation of embryonic fields.*** *Mech. Dev. 2009,* ***126*** *925-941.*

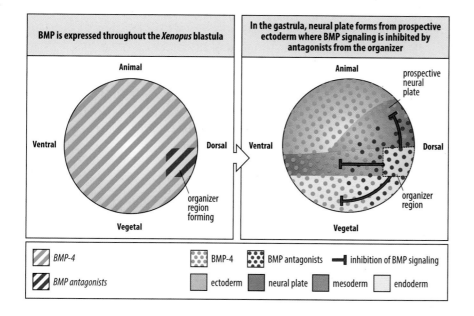

Fig. 4.33 Inhibition of BMP signaling is required for induction of neural plate. BMP (green) is expressed throughout the ectoderm of the blastula. In the early gastrula, BMP antagonists secreted from the organizer inhibit BMP signaling in adjacent regions in all three germ layers. This negative action is not on its own enough to induce formation of neural tissue from the ectoderm; an earlier positive input from FGF signaling, from the cells in the blastula, is also required.

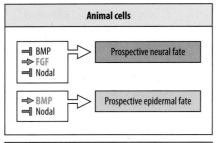

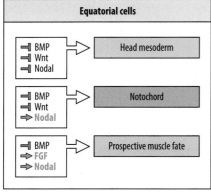

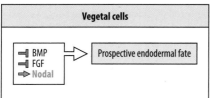

Fig. 4.34 Summary of active and antagonized signals in the different germ layers at the late blastula and gastrula stages. Four main families of secreted signaling proteins are involved in germ-layer specification and patterning in *Xenopus*: Wnts, FGFs, BMPs, and Nodals. Green lettering with a green arrow indicates that the signal is active. Black lettering with a blue barred line indicates that the signal is inhibited.

*Adapted from Heasman, J.: **Patterning the early Xenopus embryo.** Development 2006, 133: 1205-1217.*

(Ras-MAPK) pathway. Activation of this pathway occurs as a result of FGF signaling (see Box 4E) and treatment of explanted newt ectoderm with small-molecule inhibitors of the pathway prevents neural differentiation. The Ras-MAPK signaling pathway is also activated when *Xenopus* ectodermal cells are dissociated, which might explain, as noted earlier, why this dissociation so easily induces a neural fate. Induction of the neural plate therefore involves not only antagonism of BMP signaling but also FGF signaling.

Not all the effects of FGF in neural induction may result from a direct inducing effect. Activated MAPK can interfere with BMP signaling by causing an inhibitory phosphorylation of Smad1, which is part of the intracellular signaling pathway stimulated by BMPs (see Box 4C). Therefore, FGF signaling could also contribute to the inhibition of BMP signaling, and could help to allow neural induction this way.

Although neural induction is a complex process, we can see that the role of the organizer in patterning the ectoderm is similar to its role in patterning the mesoderm. How endoderm is patterned is much less well understood, but BMP antagonists can induce explants of animal cap cells to form endoderm. This suggests that the organizer may pattern all three germ layers by antagonizing BMP signaling in the tissue adjacent to it (see Fig. 4.33). The Summary table lists the main genes involved in setting up the axes, specifying the germ layers and patterning them in the early *Xenopus* embryo and Fig. 4.34 summarizes which signaling pathways are active and which are antagonized in the different germ layers at the late blastula stage. The four main signaling pathways are the TGF-β pathways (Nodals and BMPs), the FGFs, and the Wnts. For all three germ layers, we can see that their specification and patterning involves not only activating particular signaling pathways but also ensuring that other pathways are inactive. Although this diagram is a useful summary, it needs to be borne in mind that early development of the body plan in *Xenopus* does not simply depend on the presence or absence of particular signals in different regions of the embryo but also on graded signaling and the duration of signaling.

4.16 The nervous system is patterned along the antero-posterior axis by signals from the mesoderm

The pattern of neural structures that develop along the antero-posterior axis is governed by signals from the underlying dorsal mesoderm. Classic experiments showed that pieces of mesoderm taken from different positions along the antero-posterior axis of a newt neurula and placed in the blastocoel of an early newt embryo induce neural structures at the site of transplantation that correspond more or less to the original position of the transplanted mesoderm. Pieces of anterior mesoderm form a head with an induced brain, whereas posterior pieces form a trunk with an induced spinal cord (Fig. 4.35).

It was originally suggested from the results of these grafting experiments that the mesoderm underlying the neural plate in different positions along the antero-posterior axis produces different signals, with anterior mesoderm producing a signal that induces the brain and posterior mesoderm a signal that induces the spinal cord. But it turns out that both qualitative and quantitative differences in signaling by the mesoderm are involved in antero-posterior neural patterning.

Qualitatively different inducers are secreted by the mesoderm at different positions along the antero-posterior axis. At the neurula stage, Wnt antagonists such as Cerberus, Dickkopf1, and Frzb are predominantly secreted from anterior endoderm and mesoderm beneath the future head, whereas BMP antagonists are secreted from mesoderm beneath both head and trunk. The combined action of BMP and Wnt antagonists produced by anterior mesoderm could therefore be promoting the development of the brain from the anterior region of the neural plate, whereas BMP antagonists acting alone could be inducing the spinal cord from posterior neural plate. When the Wnt antagonist Dickkopf1 is overexpressed, it is only able to induce extra

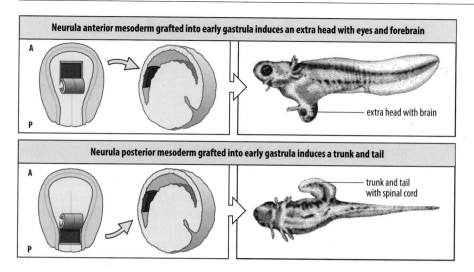

Neurula anterior mesoderm grafted into early gastrula induces an extra head with eyes and forebrain

extra head with brain

Neurula posterior mesoderm grafted into early gastrula induces a trunk and tail

trunk and tail with spinal cord

Fig. 4.35 Induction of the nervous system by the mesoderm is region specific. Mesoderm from different positions along the dorsal antero-posterior axis of early newt neurulas induces neural structures appropriate to that region when transplanted to ventral regions of early gastrulas. Anterior mesoderm forms a head with a brain (top panels), whereas posterior mesoderm forms a posterior trunk with a spinal cord ending in a tail (bottom panels).

*Illustration after Mangold, O.: **Über die induktionsfahigkeit der verschiedenen bezirke der neurula von urodelen.*** Naturwissenschaften *1933, **21**: 761-766.*

heads when in combination with BMP antagonists. This shows that the combined inhibition of BMP and Wnt signaling is indeed required to mimic the action of the anterior mesoderm.

There are also quantitative differences in Wnt/β-catenin signaling along the antero-posterior axis in the neurula. The levels of Wnt/β-catenin signaling progressively increase from anterior to posterior, resulting in a gradient with the highest level of Wnt/β-catenin signaling at the posterior end of the embryo (see Fig. 4.29). This gradient could pattern the neural plate with higher levels of Wnt signaling, conferring more posterior positional values (Fig. 4.36). In animal cap explants taken from embryos injected with mRNA for the BMP antagonist Noggin, and used to represent anterior neuralized tissue, increasing levels of Wnt signaling induce the expression of different posterior marker genes. FGFs also act as posteriorizing factors. Thus, both quantitative and qualitative differences in signaling can account for antero-posterior neural patterning, and several different signals are involved.

4.17 The final body plan emerges by the end of gastrulation and neurulation

Gastrulation transforms the spherical *Xenopus* blastula into an embryo with the three germ layers in their final positions in the body. The future dorsal mesoderm (the organizer region) in the late blastula is internalized at gastrulation to form the prechordal plate mesoderm, which lies anterior to the notochord and gives rise to the ventral head mesoderm. The internalized organizer also forms the notochord, a rigid rod-like structure that runs along the dorsal midline. The notochord is flanked immediately on each side by more ventral mesoderm, the **paraxial mesoderm**, which by the end of gastrulation is beginning to segment to form blocks of somites, starting at the anterior end (Fig. 4.37, left panel). The vertebrate notochord is a transient structure, and its cells eventually become incorporated into the vertebrae and intervening discs of the spinal column. As gastrulation proceeds, the first stage of nervous system formation, or neurulation, begins. The ectoderm overlying the notochord rolls upwards on both sides of the dorsal midline to form the neural folds, which meet over the midline to form a tubular structure, the neural tube (see Fig. 3.7). The internal structure of the *Xenopus* embryo just after the end of neurulation is illustrated in Fig. 4.37, center and right panels. The main structures that can be recognized are the neural tube, the notochord, the somites, the lateral plate mesoderm, which lies ventral to the somites on either side, the endoderm lining the gut, and the ectoderm covering the whole embryo. At this stage, the basic body plan is similar in all vertebrate embryos.

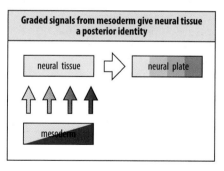

Graded signals from mesoderm give neural tissue a posterior identity

neural tissue → neural plate

mesoderm

Fig. 4.36 Mesoderm patterns the neural plate. Quantitative differences in a signal along the axis could help pattern the neural plate. For example, the levels of Wnt/β-catenin signaling increase from anterior to posterior and the high levels at the posterior end could confer a more posterior positional value on the neural plate.

*Ilustration after Kelly, O.G., Melton, D.A.: **Induction and patterning of the vertebrate nervous system.** Trends Genet. 1995, **11**: 273-278.*

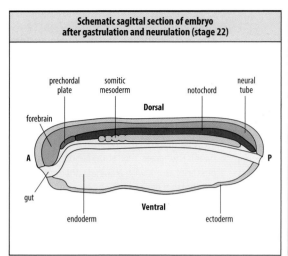

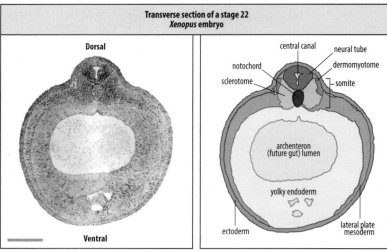

Fig. 4.37 The final body plan of the embryo emerges during gastrulation and neurulation. Left panel, schematic diagram of a sagittal section through a tailbud stage (stage 22) *Xenopus* embryo after gastrulation and neurulation. The germ layers are now all in place for future development and organogenesis. The mesoderm gives rise to the prechordal plate (brown), notochord (red), somites (orange), and lateral plate mesoderm (not shown in this view). The endoderm (yellow) has moved inside to line the gut. The neural tube (dark blue) forms from dorsal ectoderm and the ectoderm that will form the epidermis (light blue) covers the whole embryo. The antero-posterior axis has emerged, with the head at the anterior end. A, anterior; P, posterior. Center and right panels, a cross-section through a stage-22 *Xenopus* embryo. The most dorsal parts of the somites have already begun to differentiate into the dermomyotome, which will give rise to the trunk and limb muscles and the dermis, while the ventral part of the somite, the sclerotome, will give rise to the vertebral column, as described in Chapter 5. Scale bar: center and right panels = 0.2 mm. Left panel not to same scale.

Photograph reproduced with permission from Hausen, P., Riebesell, M.: The Early Development of Xenopus laevis. *Berlin: Springer-Verlag, 1991.*

SUMMARY

Patterning of the early *Xenopus* embryo, along both the antero-posterior and dorso-ventral axes, is closely related to the action of the Spemann organizer and its morphogenesis during gastrulation. The organizer secretes growth-factor antagonists, which inhibit the actions of BMPs and Wnts. This sets up gradients of BMP signaling from ventral to dorsal that pattern the dorso-ventral axis, and a gradient of Wnt signaling from anterior to posterior that patterns the embryo along the antero-posterior axis. Development of the vertebrate nervous system, which forms from the neural plate, depends both on early signals from the organizer and on signals from the mesoderm that comes to lie beneath the prospective neural plate ectoderm during gastrulation. Inhibition of BMP signaling by proteins produced by the organizer, such as Noggin, is required for neural induction. Patterning of the neural plate involves both quantitative and qualitative differences in signaling by the underlying mesoderm, with high levels of Wnt and FGF signaling specifying the more posterior positional values.

Development of the body plan in zebrafish

The way in which the body plan is established in zebrafish is similar to that in *Xenopus* and many of the same proteins are involved. There is, however, much more yolk in the zebrafish egg compared to the *Xenopus* egg and cleavage divisions do not extend into it, and this affects the formation of the endoderm. At the zebrafish sphere stage, which is equivalent to the blastula stage in *Xenopus*, the embryo is in the form of a blastoderm lying on top of the yolk cell (see Fig. 3.11). So what has studying zebrafish added to our understanding of how the vertebrate body plan is laid down? As in *Xenopus*, it is possible to carry out classic embryological experiments, for example transplantation of cells to different positions in the embryo. But in addition, zebrafish are good models for genetic analysis and, for example, mutagenesis screens for mutant fish with defects in the body plan can identify the genes involved in this process. These screens have produced few surprises, and this indicates that, in general, the same families of transcription factors and signaling proteins are involved in laying down the body plan in both *Xenopus* and zebrafish. The other advantage of zebrafish as a model for vertebrate development is the transparency of its embryos, which has allowed the tracking of single cells within the early embryo. Here we will concentrate on general principles, highlighting where the study of zebrafish has brought insights. We will also note any significant differences between *Xenopus* and zebrafish.

4.18 The body axes in zebrafish are established by maternal determinants

The very earliest stages in development of the zebrafish, as in *Xenopus*, are exclusively under the control of maternal factors laid down in the egg, which are then inherited by different blastomeres during cleavage of the fertilized egg and determine their subsequent fate. In the zebrafish egg, maternal factors are distributed along the animal–vegetal axis and key factors for axis formation are present in the vegetal region. The importance of such localized factors is shown by the fact that removal of the vegetal-most yolk early in the first cell cycle of the fertilized egg results in radially symmetric embryos that lack both dorsal and anterior structures. The dorso-ventral axis is also specified in zebrafish by activation of Wnt signaling by β-catenin in cells of the future dorsal side, and this determines where the organizer—known as the **shield** in zebrafish—will develop.

The zebrafish has provided genetic evidence for the importance of the Wnt/β-catenin pathway in dorsal specification. Mutants have been isolated in which expression of pathway components is affected. In the *ichabod* mutant, in which the mutation maps near the zebrafish *beta-catenin2* gene, expression of β-catenin2 protein is downregulated and the mutant embryos are radially symmetric. In zebrafish, the maternal dorsalizing factor is *wnt8a* mRNA rather than the *wnt11* and *wnt5a* mRNAs in *Xenopus*. The *wnt8a* mRNA moves from the vegetal region towards the animal pole on parallel arrays of microtubules that assemble shortly after fertilization. Unlike *Xenopus*, however, where we saw that the site of sperm entry determines the direction of cortical rotation, the way in which the microtubules are oriented in zebrafish is unknown. Wnt-8a activates the β-catenin pathway in the dorsal yolk syncytial layer (see Section 3.2) and dorsal margin blastomeres, and β-catenin accumulates in nuclei there (Fig. 4.38). In zebrafish, maternally provided Wnt antagonists also serve to limit Wnt signaling to the future dorsal side.

The dorsal yolk syncytial layer and the marginal blastomeres overlying it will become the shield region (see Section 3.2). The dorsal yolk syncytial layer can be thought of as being equivalent to the Nieuwkoop center and the overlying marginal blastomeres as being equivalent to the Spemann organizer. Zebrafish embryos, like *Xenopus*, go through a mid-blastula transition at which zygotic transcription begins. It occurs at the 512-cell stage, coincidentally with the formation of the syncytial layer at the margin between the blastoderm and yolk. Shortly after this, β-catenin helps to activate expression of a zygotic gene encoding a transcription factor called Dharma, which is essential for inducing the organizing function of the shield and the development of head and trunk regions (see Fig. 4.38).

4.19 The germ layers are specified in the zebrafish blastoderm by similar signals to those in *Xenopus*

There is extensive cell mixing during the transition from blastula to gastrula in the zebrafish embryo, and so it is not possible to construct a reproducible fate map at

Fig. 4.38 A comparison of the establishment of the dorsal organizer in *Xenopus* and zebrafish embryos. Left panels: in *Xenopus*, cortical rotation moves the dorsal determinants such as *wnt11* and *wnt5a* mRNAs (green) towards the future dorsal side of the embryo, creating a large region where, starting at the 32-cell stage, β-catenin moves from the cytoplasm into nuclei (red). Cells in the region in which β-catenin is intranuclear and where VegT is also present express the transcription factor Siamois, which defines the Nieuwkoop center in the dorsal vegetal region of the blastula (blue shading). The dorsal Spemann organizer will arise in the equatorial region above the Nieuwkoop center in the early gastrula. Right panels: in zebrafish, the dorsal determinant *wnt8a* mRNA is transported towards the future dorsal side of the embryo by oriented parallel arrays of microtubules and enters the blastoderm. At the mid-blastula transition, β-catenin moves into nuclei of the cells of the dorsal yolk syncytial layer where it induces the expression of the transcription factor Dharma (blue shading). The shield region, the zebrafish organizer, will arise in the dorsal region from the overlying marginal blastomeres.

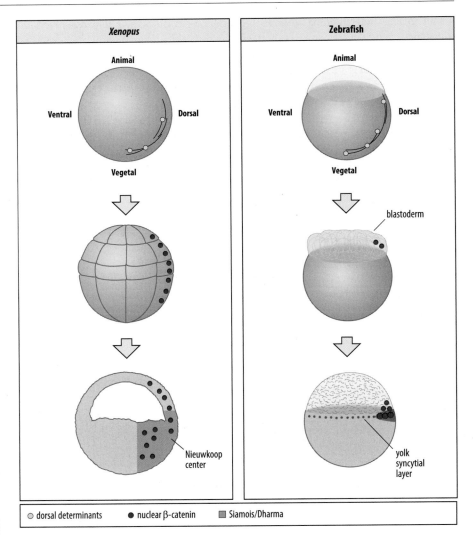

| ○ dorsal determinants | ● nuclear β-catenin | ▪ Siamois/Dharma |

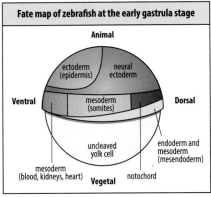

Fig. 4.39 Fate map of zebrafish at the early gastrula stage. The three germ layers come from the blastoderm, which sits on the lower hemisphere composed of an uncleaved yolk cell. The future endoderm comes from the dorsal and lateral margins of the blastoderm, which also give rise to mesoderm (mesendoderm), and some has already moved inside. The cells in the equatorial region give rise to mesoderm and the ectoderm comes from the cells nearest the animal pole.

cleavage stages, but only from the early gastrula stage. The zebrafish early gastrula comprises a blastoderm of deep cells and a thin overlying enveloping layer of cells (Fig. 4.39). The overlying enveloping layer is largely protective and is eventually lost. At the beginning of gastrulation, the fate of cells in the deep layer, from which all the cells of the embryo will come, is correlated with their position in respect to the animal pole. Cells nearest the animal pole give rise to the ectoderm, while cells beneath them slightly further away from the animal pole give rise to mesoderm and cells at the dorsal and lateral margins of the blastoderm form a layer of **mesendoderm** sitting on top of the yolk (see Fig. 4.39). The mesendoderm produces both mesoderm and endoderm, and mesoderm cells and endoderm cells subsequently differentiate in this region and then sort out. In general terms, the fate map of the zebrafish is rather similar to that of *Xenopus*, if one imagines the vegetal region of the *Xenopus* late blastula being replaced by one large yolk cell.

Experiments in which single labeled cells are transplanted from a zebrafish gastrula into a later stage embryo show that many cells are not yet specified or determined (see Section 1.12 and Fig. 1.22) and that their developmental potential is greater than the fate map would suggest. This implies that the actual fate of a cell in the blastoderm is dependent on cell-cell interactions. As in *Xenopus*, the mesodermal germ layer in the zebrafish embryo is not pre-specified and one of the key events during early development that involves cell–cell interactions is the induction of the mesoderm.

Mutagenesis screens have identified genes in zebrafish similar to those involved in mesoderm induction and dorso-ventral patterning in *Xenopus*, and they confirm the general outline of germ-layer specification. The zebrafish Nodal-related signaling protein Squint is expressed in the yolk syncytial layer and immediately overlying cells. Together with another Nodal-related protein, Cyclops, Squint is required for the specification of the marginal blastomeres as future endoderm and mesoderm. Higher levels of Nodal signaling will specify the future endoderm and lower levels the mesoderm. A long-range antagonist of Nodal signaling is also produced and this helps to prevent Nodal signaling in the region of the blastoderm that will form the future ectoderm (Fig. 4.40).

Double mutants of Squint and Cyclops lack both head and trunk mesoderm, but some mesoderm still develops in the tail region. Nodal proteins signal via Smad2 (see Box 4C). Nuclear accumulation of Smad2 and the formation of the complex between Smad2 and its co-Smad, Smad4, have been visualized in living transgenic zebrafish embryos carrying a Smad2 gene fused with a sequence encoding a fluorescent tag. This experiment revealed a graded distribution of Nodal-type signaling, most probably due to Squint, in the blastoderm of these embryos, with highest activity at the margin of the blastoderm and grading off towards the animal pole. This confirms that there is a gradient of Nodal signaling that could provide cells in the blastoderm with positional information. Nodal signaling also appears to be highest and of longest duration in the dorsal marginal region of the blastoderm, suggesting that it may also help to pattern the mesoderm dorso-ventrally. Injection of *squint* mRNA into a single cell of an early zebrafish embryo results in high-threshold Squint target genes being activated in adjacent cells, whereas low-threshold genes are activated in more distant cells. This is just what would be expected if Squint acts as a morphogen molecule, turning on different mesodermal genes at different threshold concentrations (compare with Fig 4.22).

Brachyury is one of the earliest markers of mesoderm. Zebrafish mutants called *no tail*, which lack a tail, have been isolated, and the mutation has been shown to affect the zebrafish *brachyury* gene. Using the techniques of chromatin immunoprecipitation followed by DNA sequencing (ChIP-seq, see Fig. 3.41), *no tail* has been shown to activate a network of genes encoding transcription factors and signaling molecules involved in mesoderm patterning and differentiation.

4.20 The shield in zebrafish is the embryonic organizer like the Spemann organizer in *Xenopus*

The shield in zebrafish has a similar global organizing function to the Spemann organizer of *Xenopus* and can induce a complete body axis if transplanted to another embryo at an appropriate stage. This shows that the shield is able to organize and coordinate both dorso-ventral and antero-posterior aspects of the body plan, as well as induce neural tissue from ectoderm (Fig. 4.41). Many of the proteins involved in dorso-ventral patterning of the germ layers in zebrafish are zebrafish versions of those involved in *Xenopus*, such as Chordin and BMPs.

In *Xenopus*, we saw that the inductive capacity of the Spemann organizer changes over time as different cell populations come to occupy the dorsal blastopore lip as the

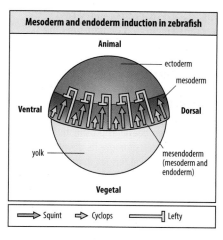

Fig. 4.40 Specification of endoderm and mesoderm in the zebrafish blastoderm. The Nodal-related signaling molecules Squint and Cyclops are secreted by cells at the blastoderm margin and diffuse into the overlying blastoderm to set up a gradient of Nodal signaling. High levels of Nodal signaling near the blastoderm margin specify mesendoderm, from which endoderm and mesoderm will form, while lower levels of Nodal signaling slightly further away specify mesoderm. Squint has long-range activity while Cyclops has short-range activity. An inhibitor of Nodal signaling, Lefty, is also secreted by cells at the blastoderm margin and diffuses into the blastoderm. Lefty has long-range activity and limits the range of Nodal signaling, ensuring that ectoderm cells are not exposed to Nodal signaling.

*After Schier, A.F.: **Nodal morphogens.** Cold Spring Harb. Perspect. Biol., 2009; 1:a003459.*

Fig. 4.41 Transplantation of the shield can induce a new axis in zebrafish. The shield region from the dorsal region of a gastrula-stage embryo in which all the cells had been labeled by fluorescein- linked dextran-amine is transplanted to the ventral region of an embryo at same stage. The transplant induces a complete second axis with head, brain and heart from the host tissue; the transplanted cells give rise to prechordal plate, notochord in the second axis and also to the hatching gland (hg).

*From Saude, L., et al.: **Axis-inducing activities and cell fates of the zebrafish organizer.** Development 2000 **127**, 3407-3417*

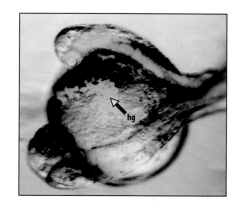

Fig. 4.42 Prospective spinal cord in the zebrafish embryo is distant from the organizer. In the late gastrula, the ectoderm that will form the spinal cord is situated on the other side of the gastrula from the organizer, in the ventral ectoderm, and is too far away to be influenced by signals from the organizer. FGF signals in the ventral region induce this ectoderm as prospective neural tissue and BMPs promote its development as posterior neural tissue (spinal cord). The epidermis is not shown in the 1-day embryo for simplicity. By this stage it covers the whole embryo.

Illustration from Kudoh, T., et. al.: **Combinatorial Fgf and Bmp signalling patterns the gastrula ectoderm into prospective neural and epidermal domains.** Development *2004,* **131:** *3581–3592.*

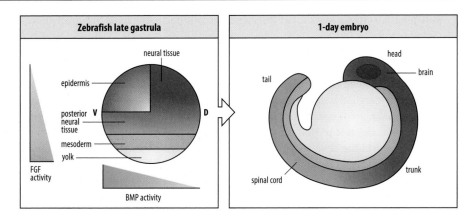

embryo gastrulates. At equivalent stages in the zebrafish, the entire gastrula margin along the dorso-ventral axis has organizing activity, with the dorsal shield region inducing twinned embryos with complete antero-posterior axis (see Fig. 4.41), and progressively more ventral regions inducing additional axes comprised of progressively more posterior structures. Quantitative differences in signaling are involved in specifying the antero-posterior axis, with the head being induced by dorsal shield tissue, where Nodal, BMP and Wnt signaling is low and the tail being induced by ventral tissue of thegastrula where Nodal, BMP and Wnt signaling is high. These gradients in signaling are produced by the secretion of antagonists by the shield region. The requirement for inhibition of Wnt signaling for head formation is confirmed by a mutation in the zebrafish version of a gene for one of the family of T-cell-specific transcription factors, T-cell-specific factor 3 (TCF3). TCF3 represses Wnt target genes when the pathway is inactive (see Box 4B). When TCF3 is mutated, this enhances Wnt activity and the mutant lacks a head; hence the mutation was consequently named *headless*.

In zebrafish, induction of the posterior neural ectoderm, which will give rise to the spinal cord, appears to be induced by FGF signaling independently of the inhibition of BMP signaling. The shield region contributes to the induction of anterior neural tissue by inhibiting BMP signaling, just as the organizer does in *Xenopus*, but the ectoderm that develops into the spinal cord is some distance away from the organizer on the ventral-vegetal side of the zebrafish embryo, where organizer signals do not reach (Fig. 4.42). The initiator of neural development in this ventral vegetal ectoderm is FGF, and BMP signals, which are high in the ventral region, in this case appear to push the neural ectoderm towards a posterior fate.

Summary to Chapter 4

- Both *Xenopus* and zebrafish have the same basic body plan. During early development, the antero-posterior and dorso-ventral axes of this body plan are set up. In frog and zebrafish this involves localized maternal determinants and cell–cell interactions.
- It is possible to construct a fate map in the early embryo for the three germ layers—mesoderm, endoderm, and ectoderm. The fate maps of *Xenopus* and zebrafish have strong similarities, the major differences being due to the amount of yolk in the egg that affects the cleavage divisions and the way in which the endoderm forms in zebrafish.
- At this early stage, the embryos are still capable of considerable regulation and this emphasizes the essential role of cell–cell interactions in development.
- In *Xenopus,* signals emanating from different regions of the blastula are involved in mesoderm induction and early patterning. Good candidates for these signals have been

identified and include members of the TGF-β family, in particular Nodal proteins, and Wnt proteins.

- At particular concentrations, these signals activate mesoderm-specific genes such as *brachyury* and so gradients of these signals could pattern the mesoderm. *brachyury* expression in the mesoderm is maintained by FGF signaling, which also helps to pattern the mesoderm.
- Many of the same proteins are involved in mesoderm induction in *Xenopus* and zebrafish.
- Patterning of all three germ layers along the dorso-ventral axis, together with the establishment of the antero-posterior axis and neural induction during gastrulation, depends on signaling by the Spemann organizer in *Xenopus*, and by the shield region in zebrafish.
- When grafted to the ventral side of an early *Xenopus* gastrula, the Spemann organizer induces a second body axis with a head and a complete dorso-ventral axis, and a twinned embryo develops. The shield can likewise induce a twinned fish embryo with a second body axis.
- The organizers secrete growth-factor antagonists, which set up gradients of TGF-β signaling and Wnt signaling to pattern the embryo.
- A gradient of BMP signaling provides positional information for dorso-ventral patterning of the mesoderm, with low levels of BMP specifying dorsal tissues, such as somites and low levels of BMP signaling specifying ventral tissues such as blood.
- The head forms where Nodal, BMP and Wnt signaling are low; the tail forms where BMP and Wnt signaling are high.
- Specification and patterning of the neural plate, which will form the nervous system, depends on signals from the organizer and on signals from the mesoderm that comes to lie beneath the prospective neural-plate ectoderm during gastrulation.
- Inhibition of BMP signaling by proteins such as Noggin produced by the organizer is required for neural induction.
- Patterning of the neural plate involves both quantitative and qualitative differences in signaling by the mesoderm, with high levels of Wnt and FGF signaling specifying the more posterior structures.
- The way in which the body plan is laid down is very similar in *Xenopus* and zebrafish. This indicates that the developmental mechanisms are likely to be highly conserved between all vertebrates. We will examine this proposition in more detail in Chapter 5, where we will consider how the body plan is laid down in chick and mouse embryos.

■ End of chapter questions

Long answer (concept questions)

1. The organization of the *Xenopus* embryo begins with the establishment of the dorso-ventral axis. A critical component of this event is the accumulation of β-catenin in cells of the dorsal marginal zone. Describe the series of events that lead to this accumulation. Include the maternal deposition of factors into the oocyte, events accompanying sperm entry, cortical rotation, and the Wnt signaling pathway.

2. Give a molecular explanation for the ventralizing effects of: UV irradiation of the egg, and treatment of the embryo with lithium.

3. The Nieuwkoop center was defined by various experiments in *Xenopus* embryos. Describe the results of the following experiments:

(a) The two dorsal cells of a four-cell embryo are isolated from the two ventral cells. What does each half embryo develop into?
(b) Cells from the animal region of a late blastula are cultured together with cells from the dorsal vegetal region.

(c) Cells from the dorsal vegetal region of a 32-cell embryo are grafted into the ventral side of another 32-cell stage embryo.

4. What is the result of the following experiment? The dorsal lip of the blastopore from an early *Xenopus* gastrula is grafted into the ventral marginal zone of another early gastrula (see also Section 1.4). What is the explanation of this result?

5. Review what we have learned about the specification of anterior-most mesoderm in the *Xenopus* gastrula including *goosecoid* activation in response to TGF-β family signaling. What is the role of this mesoderm in subsequent development?

6. The 'default model' for neural induction in *Xenopus* proposes that the development of dorsal ectoderm as neural tissue is the default pathway. What is the evidence for this model and what is the proposed molecular basis? How has the default model been modified in the light of more recent work? Include FGF signaling in your answer.

7. Describe the developmental roles of the VegT protein found in fertilized eggs and early embryos in *Xenopus* embryonic development. Include in your answer what kind of protein it is; whether

SUMMARY: main genes involved in patterning of axes and germ layers in *Xenopus*

Gene	Maternal/ zygotic	Type of protein	Where expressed	Function of protein
Specification of germ layers and dorso-ventral axis				
vegt	M	Transcription factor	Vegetal region	Endoderm specification; activates expression of mesoderm inducers
ectodermin (thim33)	M	Ubiquitin ligase	Animal hemisphere	Ectoderm specification; inhibits mesoderm formation
vg1	M	TGF-β family	Vegetal region; RNA enriched dorsally after fertilization	Mesoderm induction
wnt11	M	Wnt family	Vegetal region; RNA translocated to dorsal side after fertilization	Specification of dorsal structures and organizer formation
dishevelled	M	Wnt pathway signaling protein	Protein associated with vesicles that move to dorsal side of embryo after fertilization	Specification of dorsal structures and organizer formation
β-catenin (ctnnb1)	M	Acts in Wnt pathway; regulates gene expression	Protein enriched in dorsal nuclei in response to Wnt signaling	Specification of dorsal structures and organizer formation
gsk3b	M	Protein kinase	Protein depleted on dorsal side	Suppression of dorsalizing signals
axin	M	Binds β-catenin	RNA throughout zygote; more protein dorsally than ventrally	Suppression of dorsalizing signals
dickkopf1 (dkk1)	M	Secreted protein	Ventral region	Limits Wnt signaling to dorsal region
Foxl1e (foxi1)	Z	Transcription factor	Animal half	Maintains cells as ectoderm
derrière	Z	TGF-β family	Vegetal hemisphere and marginal region	Posterior mesoderm induction
nodel related-1,2,4,5,6	Z	TGF-β family	Vegetal hemisphere and marginal region; higher dorsally	Mesoderm induction
fgf (several types)	Z	Secreted signals	Vegetal hemisphere and marginal region	Posterior mesoderm development
Mesoderm patterning				
brachyury	Z	Transcription factor	Throughout prospective mesoderm	Posterior mesoderm formation
wnt8	Z	Wnt family	Ventral and lateral regions of prospective mesoderm	Mesoderm ventralization
bmp4	Z	TGF-β family	Throughout late blastula; then excluded from organizer	Mesoderm ventralization
activin	Z	TGF-β family	Throughout late blastula/ early gastrula	Mesoderm induction and patterning
Organizer function				
siamois	Z	Transcription factor	Nieuwkoop center	Induction of organizer
noggin	Z	Secreted signal	Spemann organizer	Mesoderm dorsalization by antagonizing BMP-4
chordin	Z	Secreted signal	Spemann organizer	Mesoderm dorsalization by antagonizing BMP-4
frzb	Z	Secreted signal	Spemann organizer	Mesoderm dorsalization by antagonizing Wnt-8
cerberus-like	Z	Secreted signal	Spemann organizer	Promotes head development by inhibiting Wnt, Nodal-related, and BMP signaling
follistatin	Z	Secreted protein	Spemann organizer	BMP antagonist
goosecoid	Z	Transcription factor	Spemann organizer	Organizer function
lim1	Z	Transcription factor	Spemann organizer	Gastrulation and head formation
otx2	Z	Transcription factor	Spemann organizer	Formation of anterior structures
not	Z	Transcription factor	Spemann organizer	Notochord specification

it is maternally supplied, or is transcribed and translated from the embryo's own genes; where it is located in the fertilized egg; and what germ layers it is involved in specifying, whether alone or in combination with other proteins.

8. A simple view of gastrulation in *Xenopus* would be that the blastula is like a soft ball, and the involution of cells during gastrulation is analogous to poking your finger into the ball at the blastopore, until you can touch the ventral surface. Examine the fate map of the *Xenopus* embryo (Fig. 4.9); what is wrong with this simple view? Provide a more accurate description of the process in *Xenopus* (refer to Fig. 3.6 for help). Where would the mesoderm have to be located in the blastula for the simple invagination method to work?

9. Specification, determination, and differentiation describe different states of development of a cell. Define each of these terms, and describe what experiments are used to distinguish these three different states of development (refer to Sections 1.12, and 4.5 in constructing your answer)? Speculate on what sort of changes might have occurred at the molecular level in a cell between specification and determination.

10. Early amphibian development is driven largely by maternally supplied factors until the mid-blastula transition. What is the mid-blastula transition? What is the model that explains how it occurs? What experimental evidence supports this model?

11. 'Dorsal vegetal tissue containing the Nieuwkoop center induces notochord and muscle from animal cap cells, whereas ventral vegetal tissue induces mainly blood-forming tissue and little muscle.' (Section 4.9). What is the significance of these observations? What signaling events are required to explain these observations?

12. An early event in axis formation is the accumulation of the transcription factor β-catenin in the nuclei of cells on one side of the *Xenopus* embryo. Which side of the embryo does this specify? Describe briefly how the actions of β-catenin lead to specification of the Spemann organizer.

13. What is the developmental function of the TGF-β family member BMP-4 in the early gastrula? Since BMP-4 is initially present throughout the embryo, how are its effects restricted to particular regions (a) of the ectoderm, and (b) of the dorsal mesoderm?

14. The TGF-β signaling pathway is an example of the remarkable conservation of signaling pathways during evolution. Compare the activities of BMP-4 in *Xenopus* and Decapentaplegic in *Drosophila*. Describe how a gradient of BMP signaling is established in the *Xenopus* gastrula. To what extent is the mechanism conserved in *Drosophila*?

15. Zebrafish embryos are good models for genetic analysis of development. Which zebrafish mutants have given insights into the laying down of the vertebrate body plan?

16. How does the activity of the shield region of the zebrafish compare with the activity of the Spemann organizer of the amphibian embryo?

Multiple choice (factual recall questions)

NB There is only one right answer to each question.

1. The nervous system, heart, and liver are derived from _____, _____, and _____, respectively.

a) all are derived from mesoderm
b) ectoderm, mesoderm, and endoderm

c) endoderm, mesoderm, and ectoderm
d) mesoderm, ectoderm, and endoderm

2. The 'organizer' in *Xenopus* is responsible for

a) inducing a mesodermal fate in nearby cells
b) helping to pattern the mesoderm along the dorso-ventral axis
c) specifying the dorsal region of the embryo
d) all of these events

3. A key factor that specifies the organizer in *Xenopus* embryos is:

a) GSK-3β
b) Fz7
c) Wnt-11
d) β-catenin

4. If cells from the animal pole of a *Xenopus* blastula (animal cap cells) are placed in direct contact with cells from the vegetal region, what is the result?

a) An embryo forms with only ectodermal and endodermal derivatives.
b) Animal cap cells are induced by the vegetal cells to form mesodermal derivatives.
c) The embryo regulates and forms a normal embryo.
d) Vegetal hemisphere cells are induced by the animal cap cells to form mesodermal derivatives.

5. Maternal factors are

a) the energy-rich yolk present in the eggs of many species
b) the genes contributed to the embryo by the maternal haploid genome
c) the genes expressed by the zygotic genome in the egg
d) developmentally active mRNAs and proteins packaged into the egg by the mother

6. The activity of the Chordin protein is

a) to antagonize BMP signaling and block ventralization of the mesoderm.
b) to signal through the Chordin pathway and establish the organizer.
c) to specify a mesodermal fate.
d) to turn on genes that determine the anterior-most cells in the embryo, which will form the head.

7. Vg-1, Nodals, BMPs and activin are all members of which family of signaling molecules?

a) FGF
b) Hedgehog
c) TGF-β
d) Wnt

8. In the early blastula, Wnt signaling and the subsequent nuclear accumulation of β-catenin specifies a dorsal fate, whereas after gastrulation, Wnt/β-catenin signaling specifies

a) a dorsal fate
b) a posterior fate
c) an anterior fate
d) all of the above

9. The Brachury protein is

a) an early marker of the mesoderm
b) an early marker of the endoderm
c) expressed throughout the ectoderm
d) only expressed after gastrulation

10. Endoderm and mesoderm are specified in the zebrafish blastoderm by

a) signals from the shield region
b) maternal signaling molecules laid down in the egg
c) a gradient of β-catenin signaling
d) a gradient of Nodal signaling

Multiple choice answer key

1: b, 2: b, 3: d, 4: b, 5: d, 6: a, 7: c, 8: b, 9: a, 10: d.

■ Section further reading

4.1 The animal-vegetal axis is maternally determined in *Xenopus*

Heasman, J.: **Patterning the early *Xenopus* embryo.** *Development* 2006, **133**: 1205–1217.

Weaver, C., Kimelman, D.: **Move it or lose it: axis specification in *Xenopus*.** *Development* 2004, **131**: 3491–3499.

Box 4B Wnt signaling pathway

Logan, C.Y., Nusse, R.: The **Wnt signalling pathway in development and disease.** *Annu. Rev. Cell Dev. Biol.* 2004, **20**: 781–801.

Nusse, R., Varmus, H.: **Three decades of Wnts: a personal perspective on how a scientific field develops.** *EMBO J.* 2012, **31**: 2670–2684.

4.2 Local activation of the Wnt/β-catenin signaling pathway specifies the future dorsal side of the embryo

Cha, S.W., Tadjuidje, E., Tao, Q., Wylie, C., Heasman, J.: **Wnt 5a and Wnt11 interact in a maternal Dkk1-regulated fashion to activate both canonical and non-canonical signaling in *Xenopus* axis formation.** *Development* 2008, **135**: 3719–3729.

Cha, S.W., Tadjuidje, E., White, J., Wells, J., Mayhew, C., Wylie, C., Heasman, J.: **Wnt 11/5a complex formation caused by tyrosine sulfation increases canonical signaling activity.** *Proc. Natl Acad. Sci. USA* 2009, **19**: 1573–1580.

Gerhart, J., Danilchik, M., Doniach, T., Roberts, S., Browning, B., Stewart, R.: **Cortical rotation of the *Xenopus* egg: consequences for the antero-posterior pattern of embryonic dorsal development.** *Development* (*Suppl.*) 1989, 37–51.

Heasman, J.: **Maternal determinants of embryonic cell fate.** *Semin. Cell Dev. Biol.* 2006, **17**: 93–98.

4.3 Signaling centers develop on the dorsal side of the blastula

Glimich, R.L., Gerhart, J.C.: **Early cell interactions promote embryonic axis formation in *Xenopus laevis*.** *Dev Biol* 1984, **104**: 117–130.

Kodjabachian, L., Lemaire, P.: **Embryonic induction: is the Nieuwkoop centre a useful concept?** *Curr Biol.* 1998, **8**: R918–921.

Nieuwkoop, P.D.: **The formation of the mesoderm in urodelean amphibians. I. Induction by the endoderm.** *Wilhelm Roux Arc. EntwMech.Org.* 1969, **162**: 341–373.

Vonica, A., Gumbiner, B.M.: **The *Xenopus* Nieuwkoop center and Spemann-Mangold organizer share molecular components and a requirement for maternal Wnt activity.** *Dev. Biol.* 2007, **312**: 90–102.

4.4 A fate map of the *Xenopus* blastula makes clear the function of gastrulation

Dale, L., Slack, J.M.W.: **Fate map for the 32 cell stage of *Xenopus laevis*.** *Development* 1987, **99**: 527–551.

Gerhart J.: **Changing the axis changes the perspective.** *Dev. Dyn.* 2002, **225**: 380–383.

4.5 Cells of the early *Xenopus* embryo do not yet have their fates determined and regulation is possible

Snape, A., Wylie, C.C., Smith, J.C., Heasman, J.: **Changes in states of commitment of single animal pole blastomeres of *Xenopus laevis*.** *Dev. Biol.* 1987, **119**: 503–510.

Wylie, C.C., Snape, A., Heasman, J., Smith, J.C.: **Vegetal pole cells and commitment to form endoderm in *Xenopus laevis*.** *Dev. Biol.* 1987, **119**: 496–502.

Box 4C TGF-β signaling pathway

Massagué, J.: **How cells read TGF-β signals.** *Nat Rev. Mol. Cell Biol.* 2000, **1**: 169–178.

Schier, A.F.: **Nodal signals.** *Cold Spring Harb Perspect Biol.* 2009, **1**: a003459.

Schier, A.F.: **Nodal signaling in vertebrate development.** *Annu. Rev. Cell. Dev. Biol.* 2003, **19**: 589–621.

4.6 Endoderm and ectoderm are specified by maternal factors whereas mesoderm is induced from ectoderm by signals from the vegetal region

&

4.7 Mesoderm induction occurs during a limited period in the blastula stage

Dupont, S., Zacchigna, L., Cordenonsi, M., Soligo, S., Adorno, M., Rugge, M., Piccolo, S.: **Germ-layer specification and control of cell growth by Ectodermin, a Smad4 ubiquitin ligase.** *Cell* 2005, **121**: 87–99.

Gurdon, J.B., Lemaire, P., Kato, K.: **Community effects and related phenomena in development.** *Cell* 1993, **75**: 831–834.

Mir, A., Kofron, M., Zorn, A.M., Bajzer, M., Haque, M., Heasman, J., Wylie, C.C.: **FoxI1e activates ectoderm formation and controls cell position in the *Xenopus* blastula.** *Development* 2007, **134**: 779–788.

White, J.A., Heasman, J.: **Maternal control of pattern formation in *Xenopus laevis*.** *J. Exp. Zool. B* 2008, **310**: 73–84.

Xanthos, J.B., Kofron, M., Wylie, C., Heasman, J.: **Maternal VegT is the initiator of a molecular network specifying endoderm in *Xenopus laevis*.** *Development* 2001, **128**: 167–180.

4.8 Zygotic gene expression is turned on at the mid-blastula transition

Yasuda, G.K., Schübiger, G.: **Temporal regulation in the early embryo: is MBT too good to be true?** *Trends Genet.* 1992, **8**: 124–127.

4.9 Mesoderm-inducing and patterning signals are produced by the vegetal region, the organizer, and the ventral mesoderm

Agius, E., Oelgeschläger, M., Wessely, O., Kemp, C., De Robertis, E.M.: **Endodermal Nodal-related signals and mesoderm induction in *Xenopus*.** *Development* 2000, **127**: 1173–1183.

Heasman, J.: **Patterning the early *Xenopus* embryo.** *Development* 2006, **133**: 1205–1217.

Kimelman, D.: **Mesoderm induction: from caps to chips.** *Nat. Rev. Genet.* 2006, **7**: 360–372.

4.10 Members of the TGF-β family have been identified as mesoderm inducers

Jones, C.M., Kuehn, M.R., Hogan, B.L., Smith, J.C., Wright, C.V.: **Nodal-related signals induce axial mesoderm and dorsalize mesoderm during gastrulation.** *Development* 1995, **121**: 3651–3662.

Osada, S.I., Wright, C.V.: **Xenopus nodal-related signaling is essential for mesendodermal patterning during early embryogenesis.** *Development* 1999, **126** : 3229–3240.

White R.J., Sun, B.I., Sive, H.L., Smith, J.C.: **Direct and indirect regulation of derrière, a Xenopus mesoderm-inducing factor, by VegT.** *Development* 2002, **129**: 4867–4876.

Box 4D Investigating receptor function using dominant-negative mutations

Dyson, S., Gurdon, J.B.: **Activin signalling has a necessary function in *Xenopus* early development.** *Curr Biol.* 1997, **7**: 81–84.

4.11 The zygotic expression of mesoderm-inducing and patterning signals is activated by the combined actions of maternal VegT and Wnt signaling

De Robertis, E.M., Larrain, J., Oelgeschläger, M., Wessely, O.: **The establishment of Spemann's organizer and patterning of the vertebrate embryo.** *Nat Rev. Genet.* 2000, **1**: 171–181.

Kofron, M., Demel, T., Xanthos, J., Lohr, J., Sun, B., Sive, H., Osada, S-I., Wright, C., Wylie, C., Heasman, J.: **Mesoderm induction in *Xenopus* is a zygotic event regulated by maternal VegT via TGFβ growth factors.** *Development* 1999, **126**: 5759–5770.

Niehrs, C.: **Regionally specific induction by the Spemann-Mangold organizer.** *Nat. Rev. Genet.* 2004, **5**: 425–434.

4.12 Threshold responses to gradients of signaling proteins are likely to pattern the mesoderm

Fletcher, R.B., Harland, R.M.: **The role of FGF signaling in the establishment and maintenance of mesodermal gene expression in *Xenopus*.** *Dev. Dyn.* 2008, **237**: 1243–1254.

Green, J.B.A., New, H.V., Smith, J.C.: **Responses of embryonic *Xenopus* cells to activin and FGF are separated by multiple dose thresholds and correspond to distinct axes of the mesoderm.** *Cell* 1992, **71**: 731–739.

Gurdon, J.B., Standley, H., Dyson, S., Butler, K., Langon, T., Ryan, K., Stennard, F., Shimizu, K., Zorn, A.: **Single cells can sense their position in a morphogen gradient.** *Development* 1999, **126**: 5309–5317.

Piepenburg, O., Grimmer, D., Williams, P.H., Smith, J.C.: **Activin redux: specification of mesodermal pattern in *Xenopus* by graded concentrations of endogenous activin B.** *Development* 2004, **131**: 4977–4986.

Saka, Y., Smith, J.C. **A mechanism for the sharp transition of morphogen gradient interpretation in *Xenopus*.** *BMC*, 2007, **7**: 47.

Schulte-Merker, S., Smith, J.C.: **Mesoderm formation in response to *Brachyury* requires FGF signalling.** *Curr. Biol.* 1995, **5**: 62–67.

Box 4E FGF signaling pathway

Dorey, K., Amaya, E; **FGF signalling: diverse roles during early vertebrate embryogenesis.** *Development* 2010, **137**: 3731–3742.

4.13 Signals from the organizer pattern the mesoderm dorso-ventrally by antagonizing the effects of ventral signals

De Robertis, E.M.: **Spemann's organizer and the self-regulation of embryonic fields.** *Mech.Dev.* 2009, **126**: 925–941.

De Robertis, E.M.: **Spemann's organizer and self-regulation in amphibian embryos.** *Nature Mol. Cell Biol. Rev.* 2006, **7**: 296–302.

Piccolo, S., Sasai, Y., Lu, B., De Robertis, E.M.: **Dorsoventral patterning in *Xenopus*: inhibition of ventral signals by direct binding of chordin to BMP-4.** *Cell* 1996, **86**: 589–598.

Zimmerman, L.B., De Jesús-Escobar, J.M., Harland, R.M.: **The Spemann organizer signal noggin binds and inactivates bone morphogenetic protein 4.** *Cell* 1996, **86**: 599–606.

4.14 The antero-posterior axis of the embryo emerges during gastrulation

Glinka, A., Wu, W., Delius, H., Monaghan, P., Blumenstock, C., Niehrs, C.: **Dickkopf-1 is a member of a new family of secreted proteins and functions in head induction.** *Nature* 1998, **391**: 357–362.

Glinka, A., Wu, W., Onichtchouk, D., Blumenstock, C., Niehrs, C.: **Head induction by simultaneous repression of Bmp- and Wnt-signalling in *Xenopus*.** *Nature* 1997, **389**: 517–519.

Hikasa, H., Sokol, S.Y.: **Wnt signaling in vertebrate axis specification.** *Cold Spring Harb. Perspect. Biol.* 2013, **5**: a 007955.

Kiecker, C., Niehrs, C.: **The role of prechordal mesendoderm in neural patterning.** *Curr. Opin. Neurobiol.* 2001, **11**: 27–33.

Niehrs, C.: **Regionally specific induction by the Spemann-Mangold organizer.** *Nat. Rev. Genet.* 2004, **5**: 425–434.

Piccolo, S., Agius, E., Leyns, L., Bhattacharya, S., Grunz, H., Bouwmeester, T., De Robertis, E.M.: **The head inducer Cerberus is a multifunctional antagonist of Nodal, BMP and Wnt signals.** *Nature* 1999, **397**: 707–710.

4.15 The neural plate is induced in the ectoderm

De Robertis, E.M.: **Spemann's organizer and self-regulation in amphibian embryos.** *Nat. Rev. Mol. Cell Biol.* 2006, **4**: 296–302.

De Robertis, E.M., Kuroda, H.: **Dorsal-ventral patterning and neural induction in *Xenopus* embryos.** *Annu. Rev. Dev. Biol.* 2004, **20**: 285–308.

Reversade, B., De Robertis, E.M.: **Regulation of ADMP and BMP2/4/7 at opposite embryonic poles generates a self-regulating embryonic field.** *Cell* 2005, **123**: 1147–1160.

4.16 The nervous system is patterned along the antero-posterior axis by signals from the mesoderm

Kiecker, C., Niehrs, C.: **A morphogen gradient of Wnt/beta-catenin signaling regulates anteroposterior neural patterning in *Xenopus*.** *Development* 2001, **128**: 4189–4201.

Pera, E.M., Ikeda, A., Eivers, E., De Robertis, E.M.: **Integration of IGF, FGF, and anti-BMP signals via Smad1 phosphorylation in neural induction.** *Genes Dev.* 2003, **17**: 3023–3028.

Sasai, Y., De Robertis, E.M.: **Ectodermal patterning in vertebrate embryos.** *Dev. Biol.* 1997, **182**: 5–20.

4.18 The body axes in zebrafish are established by maternal determinants

Dosch, R., Wagner, D.S., Mintzer. K.A., Runke, G., Wiemelt, A.P., Mullins, M.C.: **Maternal control of vertebrate development before the midblastula transition: mutants from the zebrafish I.** *Dev. Cell* 2004, **6**: 771–780.

Kodjabachian, L., Dawid, I.B., Toyama, R.: **Gastrulation in zebrafish: what mutants teach us.** *Dev. Biol.* 1999, **126**: 5309–5317.

Langdon, Y.G., Mullins, M.C.: **Maternal and zygotic control of zebrafish dorsoventral axial patterning.** *Annu. Rev. Genet.* 2011, **45**: 357–377.

Lu, F.I., Thisse, C., Thisse, B.: **Identification and mechanism of regulation of the zebrafish dorsal determinant.** *Proc. Natl Acad. Sci. USA* 2011, **108**: 15876–15880.

Pelegri, F.: **Maternal factors in zebrafish development.** *Dev. Dyn.* 2003, **228**: 535–554.

Schier, A.F., Talbot, W.S.: **Molecular genetics of axis formation in zebrafish.** *Annu. Rev. Genet.* 2005, **39**: 561–613.

Tran, L.D., Hino, H., Quach, H., Lim, S., Shindo, A., Mimori-Kiyosue, Y., Mione, M., Ueno, N., Winkler, C., Hibi, M., Sampath, K.: **Dynamic microtubules at the vegetal cortex predict the embryonic axis in zebrafish.** *Development* 2012, **139**: 3644–3652.

4.19 The germ layers are specified in the zebrafish blastoderm by similar signals to those in *Xenopus*

Agathon, A., Thisse, C., Thisse, B.: **The molecular nature of the zebrafish tail organizer.** *Nature* 2003, **424**: 448–452.

Griffin, K., Patient, R., Holder, N.: **Analysis of FGF function in normal and *no tail* zebrafish embryos reveals separate mechanisms for formation of the trunk and tail.** *Development* 1995, **121**: 2983–2994.

Harvey, S.A., Smith, J.C.: **Visualisation and quantification of morphogen gradient formation in the zebrafish.** *PLoS Biol.* 2009, **7**: e101.

Kimmel, C.B., Warga, R.M., Schilling, T.F.: **Origin and organization of the zebrafish fate map.** *Development* 1990, **108**: 581–594.

Morley, R.H., Lachanib, K., Keefe, D., Gilchrist, M.J., Flicek, P., Smith, J.C., Wardle, F.C.: **A gene regulatory network directed by zebrafish No tail accounts for its roles in mesoderm formation.** *Proc. Natl Acad. Sci. USA* 2009, **106**: 3829–3834.

O'Boyle, S., Bree, R.T., McLoughlin, S., Grealy, M., Byrnes, L.: **Identification of zygotic genes expressed at the midblastula transition in zebrafish.** *Biochem. Biophys. Res. Commun.* 2007, **358**: 462–468.

Schier, A.F.: **Axis formation and patterning in zebrafish.** *Curr. Opin. Genet. Dev.* 2001, **11**: 393–404.

4.20 The shield in zebrafish is the embryonic organizer like the Spemann organizer in *Xenopus*

Fauny, J.D., Thisse, B., Thisse, C.: **The entire zebrafish blastula-gastrula margin acts as an organizer dependent on the ratio of Nodal to BMP activity.** *Development* 2009, **136**: 3811–3819.

Gonzalez, E.M., Fekany-Lee, K., Carmany-Rampey, A., Erter, C., Topczewski, J., Wright, C.V., Solnica-Krezel, L.: **Head and trunk in zebrafish arise via coinhibition of BMP signaling by bozozok and chordino.** *Genes Dev.* 2000, **14**: 3087–3092.

Kudoh, T., Concha, M.L., Houart, C., Dawid, I.B., Wilson, S.W.: **Combinatorial Fgf and Bmp signalling patterns the gastrula ectoderm into prospective neural and epidermal domains.** *Development* 2004, **131**: 3581–3592.

Londin, E.R., Niemiec, J., Sirotkin, H.I.: **Chordin, FGF signaling, and mesodermal factors cooperate in zebrafish neural induction.** *Dev. Biol.* 2005, **279**: 1–19.

Saúde, L., Woolley, K., Martin, P., Driever, W., Stemple, D.L.: **Axis-inducing activities and cell fates of the zebrafish organizer.** *Development* 2000, **127**: 3407–3417.

5

Vertebrate development III: Chick and mouse—completing the body plan

- Development of the body plan in chick and mouse
- Somite formation and antero-posterior patterning
- The origin and patterning of neural crest
- Determination of left-right asymmetry

In this chapter we will examine in more detail some features of vertebrate development involved in completing the vertebrate body plan, such as the development of somites, the regionalization of the antero-posterior axis of the body, the generation and fate of neural crest cells, and the establishment of left-right asymmetry in the body. We will start, however, by describing the early development of chick and mouse embryos and comparing it with that of Xenopus. Much of the process of laying down the body plan in avian and mammalian embryos follows the same principles as in amphibians and fish; fate maps at equivalent stages just before gastrulation show strong similarities, and the genes involved in germ-layer specification are similar. However, there are some striking differences in early development between birds, mammals and amphibians, and we shall focus on these in the first part of the chapter. We shall then use chick and mouse as our model vertebrates to look in detail at somite formation, the regionalization and patterning of the embryo along the antero-posterior axis by the Hox genes, the generation of neural crest and its multifarious fates, and the development of left-right asymmetry.

In this chapter we will examine the mechanisms that control the development of the final vertebrate body plan, using chick and mouse as our main model organisms. As discussed briefly in Chapter 3, chick and mouse development differ in some important ways from that of amphibians and fish—including the shape of the early embryo, the timing of the specification of the germ layers, and gastrulation through a primitive streak rather than a blastopore—and we start the chapter by looking at the early development of mouse and chick starting from the fertilized egg, and comparing it to that of *Xenopus*. We shall also see considerable differences in the early development of the chick compared to the mouse, but that gastrulation and later development in birds and mammals are much more similar. As we approach the phylotypic stage—the embryonic stage just after gastrulation when the final body plan has been established (see Fig. 3.2)—the similarities between all vertebrate embryos become greater.

In the second and third parts of this chapter, we will focus on events around the phylotypic stage. We will consider the formation of the dorsal tissues of the trunk—the notochord, the spinal cord, and the somites—in chick and mouse, and in particular, the formation and patterning of the somites, blocks of tissue formed from mesoderm on either side of the notochord that give rise to the vertebrae of the spine, the skeletal muscles of the trunk, and some of the dermis of the skin.

Both the ectodermally derived nervous system and the mesodermal structures along the antero-posterior axis of the trunk, such as the somites, have a distinct antero-posterior identity. In chick and mouse, an initial antero-posterior patterning of the germ layers occurs during mesoderm induction, when the future mesoderm becomes differentiated from endoderm and ectoderm, but some important aspects of antero-posterior patterning occur later. This later patterning of both the mesoderm and the nervous system is controlled by the Hox genes, which are the vertebrate equivalent of the Hox genes responsible for antero-posterior patterning in *Drosophila* (see Chapter 2). We will discuss the role of the Hox genes in the antero-posterior patterning of the somitic mesoderm, as shown by their effects on the somite-derived vertebrae of the spinal column in the mouse. Hox genes and their role in antero-posterior patterning are common to all vertebrates, but their function has been particularly well studied in the mouse, in which techniques for knocking out specific genes to order are available (see Section 3.10).

A unique feature of vertebrates is the **neural crest**, which arises at the edges of the neural plate. Neural crest cells start to migrate from the crests of the forming neural tube at neurulation. The cells migrate to different locations in the body to give rise to a wide range of tissues including the autonomic and sensory nervous systems, and in the head they give rise to the connective tissues of the face, including skeletal tissue. In this chapter we will look at the generation of the neural crest and how Hox gene expression specifies the regional identity of neural crest cells in the head. The migration of the neural crest is discussed in Chapter 9, and patterning and the development of the brain and of the central nervous system are considered in Chapter 12.

Finally, we will briefly consider how an intriguing aspect of the vertebrate body plan arises—that of left–right asymmetry. Although the vertebrate body is outwardly symmetric about the midline, there is left–right internal asymmetry, which is indicated, for example, by the positions of the heart and liver, and we shall see that this asymmetry originates early in embryonic life.

Development of the body plan in chick and mouse

5.1 The antero-posterior polarity of the chick blastoderm is related to the primitive streak

Relatively little is known about the early stages in chick development, as they take place before the egg is laid and embryos at these stages are difficult to obtain. By the time the chick egg is laid, the embryo consists of a circular blastoderm composed of many thousands of cells—estimates range from 32,000 to as many as 60,000 cells—sitting on the top of a huge yolk cell (see Section 3.3). The central region of the blastoderm, the area pellucida, consists of a single layer of epiblast cells, surrounded by the area opaca, with a thin ring of cells in between forming the marginal zone (see Fig. 3.15). Even at this stage, the blastoderm is already asymmetrical, and very soon this becomes more obvious, when a ridge of cells known as Koller's sickle develops at

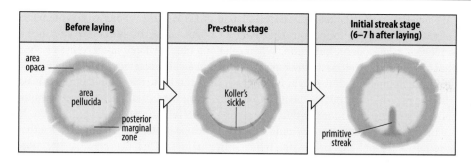

| Before laying | Pre-streak stage | Initial streak stage (6–7 h after laying) |

Fig. 5.1 Chick blastoderm at the beginning of primitive streak formation. Just before laying, the embryo consists of circular blastoderm (right panel). The central translucent region, the area pellucida, which consists of single layer of epiblast cells is surrounded by the darker area opaca. After laying, Koller's sickle develops adjacent to the posterior marginal zone (pre-streak stage; middle panel) and this defines the position in which the primitive streak will form in the epiblast (initial streak stage; right panel). The streak then extends anteriorly about half way across the epiblast with Hensen's node forming at its anterior end.

one point on the circumference of the area pellucida (Fig. 5.1). As we saw in Section 3.3, this sickle of cells lies adjacent to the posterior marginal zone, and marks the site at which the primitive streak will develop in the epiblast at gastrulation. The primitive streak indicates the direction of the antero-posterior axis of the embryo: the start of the streak marks the posterior end of the embryo and the streak elongates in an anterior direction, with the node (often called Hensen's node in the chick, see Fig. 3.16) forming at the anterior end of the streak.

The streak also defines the future dorso-ventral axis of the embryo, with more ventral structures (such as extra-embryonic mesoderm and lateral plate mesoderm) being derived from the more posterior end of the streak, and dorsal structures such as the notochord from the node, at the anterior end of the streak, as discussed later in the chapter. The location of the posterior marginal zone is therefore crucial to setting the body axes.

So how is the radial symmetry of the chick blastoderm broken and the location of the posterior marginal zone determined? We saw in *Xenopus* that the symmetry of the blastula is broken by the asymmetric redistribution of maternal dorsal determinants shortly after fertilization. The chick embryo resembles frog and fish embryos in that the cleavage stages are under the control of maternal factors laid down in the egg— zygotic gene activation occurs at late cleavage stages. It is not clear however whether any maternal factors are inherited asymmetrically by cells during cleavage in the chick and determine the embryonic axis. Instead, the antero-posterior axis seems to be specified later, as a result of rotational forces as the egg passes through the hen's oviduct and uterus.

The egg moves down the uterus pointed end first and rotates slowly around its long axis, each revolution taking about 6 minutes. As the shell rotates, the blastoderm is held in a tilted position at an angle of about 45° to the vertical so that different edges of the blastoderm are exposed to different regions of the yolk which are reoriented under the influence of gravity. The downward-pointing edge will be the future head end of the embryo and the posterior marginal zone develops at the edge of the blastoderm that is uppermost (Fig. 5.2). The molecules in the yolk that might be redistributed by gravity and specify the antero-posterior axis of the blastoderm are completely unknown. Furthermore, the axis is not irreversibly determined by these events. The chick blastoderm is highly regulative. Even at the stage when the blastoderm contains several tens of thousands of cells, it can be cut into many fragments, each of which will develop a complete embryonic axis.

The posterior marginal zone can be thought of as an organizing center analogous in some ways with the Nieuwkoop center in *Xenopus*, because it can induce the formation of a new body axis. If the posterior marginal zone from one blastoderm is grafted to another position on a second blastoderm of the same age it can induce a primitive streak at this new position (Fig. 5.3). Among the earliest signaling proteins involved in streak initiation in the chick is the TGF-β family member Vg-1, which is expressed in the posterior marginal zone, and Wnt-8C, which is present throughout the marginal zone with a high point of concentration at the posterior marginal zone.

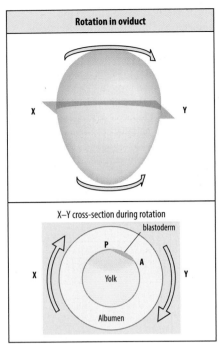

Rotation in oviduct

X–Y cross-section during rotation

Development after laying

Surface view of yolk

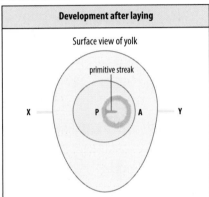

Fig. 5.2 Gravity defines the antero-posterior axis of the chick. Rotation of the egg in the oviduct of the hen results in the blastoderm being tilted in the direction of rotation, although it tends to remain uppermost. Gravitional forces reorient the yolk contents, as indicated by blue shading. The posterior marginal zone (P) develops at that side of the blastoderm that was uppermost, and initiates the primitive streak. A, anterior.

We have already encountered these signaling proteins in *Xenopus*, with Wnts being associated with the formation of signaling centers in the blastula and Vg-1 being involved in mesoderm induction. When cells of a fibroblast cell line expressing Vg-1 are grafted to another part of the marginal zone in a chick blastoderm (essentially the same experiment as that shown in Fig. 5.3), they can induce a complete new streak. This is just one example of the same proteins being used in different vertebrates for slightly different functions within a generally conserved process.

In the chick, however, often only one axis develops in the grafted embryos—either the host's normal axis or the one induced by the graft. This suggests that the more advanced of the two organizing centers inhibits streak formation elsewhere. It has been shown that Vg-1 induces an inhibitory signal that can travel across the embryo, a distance of around 3 mm in about 6 hours, to prevent another streak forming. Vg-1 is thus required for initiating streak formation but at the same time is involved in inhibiting the formation of a second streak.

Soon after laying, a continuous layer of hypoblast cells forms immediately over the yolk in the area pellucida (see Fig. 3.15). Primitive-streak formation in the epiblast only starts when the hypoblast becomes displaced away from the posterior marginal region by another layer of cells known as the **endoblast**, which grows out from the posterior marginal zone. The hypoblast actively inhibits streak formation, as shown by experiments in which its complete removal leads to the formation of multiple streaks in random positions.

The inhibition of streak formation is due to the hypoblast producing the antagonist Cerberus, which is homologous with *Xenopus* Cerberus, and inhibits Nodal signaling by the epiblast cells adjacent to the posterior marginal zone. *Nodal* expression in these cells is induced by Wnt-8C and Vg-1. Nodal signaling by the epiblast, together with FGF signaling from Koller's sickle, is required for streak formation (Fig. 5.4), and so the primitive streak does not form until the hypoblast has been displaced by endoblast. The capacity of the chick embryo to regulate persists right up to the time at which the primitive streak starts to form. When the embryo is cut into pieces, a new site of *Vg-1* expression appears in each piece.

5.2 Early stages in mouse development establish separate cell lineages for the embryo and the extra-embryonic structures

The mammalian egg differs considerably from the eggs of either *Xenopus* or the chick, as it contains no yolk and the early stages in development are concerned with generating extra-embryonic yolk-sac membranes and the placenta connecting the embryo to the mother, so that the embryo can be nourished. Although mouse development takes place inside the mother, early stages can be studied because mouse eggs fertilized *in vitro* undergo cleavage divisions and develop up to the blastocyst stage in culture.

The early cleavage divisions in mouse embryos do not follow a well-ordered pattern. Some are parallel to the egg's surface, so that the morula has distinct outer and inner cell populations. By the 32-cell stage, the morula has developed into a blastocyst, a hollow sphere of epithelium containing the inner cell mass—some 10–15 cells—attached to the epithelium at one end (Fig. 5.5, left panels). The epithelium of the blastocyst will form the **trophectoderm**, which gives rise to extra-embryonic structures associated with implantation in the uterine wall and formation of the **placenta**, a structure unique to mammalian development. The cells of the inner cell mass give rise to the epiblast, from which the embryo will develop, and to the primitive endoderm, from which the yolk sac—an extraembryonic structure—will form. The cells of the epiblast are pluripotent, meaning that they can give rise to all the cell types in the embryo (see Section 1.17), and they remain pluripotent up to stage E4.5.

Cells from mouse and human inner cell masses can be isolated and grown in culture to produce pluripotent embryonic stem cells (ES cells). As described in Chapter 8,

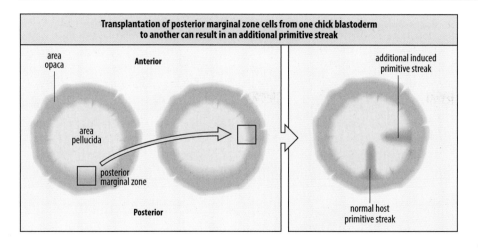

Transplantation of posterior marginal zone cells from one chick blastoderm to another can result in an additional primitive streak

area opaca

Anterior

area pellucida

posterior marginal zone

Posterior

additional induced primitive streak

normal host primitive streak

Fig. 5.3 The posterior marginal zone of the chick blastodisc induces the primitive streak. The posterior marginal zone marks the posterior end of the antero-posterior axis. Grafting posterior marginal zone cells expressing the signaling protein Vg-1 to a lateral site in the marginal zone of another blastoderm can result in the formation of an extra primitive streak at that position. Two streaks often do not develop, however, as whichever forms first tends to inhibit the development of the other.

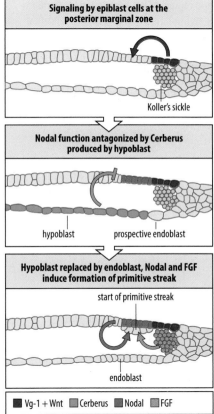

Signaling by epiblast cells at the posterior marginal zone

Koller's sickle

Nodal function antagonized by Cerberus produced by hypoblast

hypoblast   prospective endoblast

Hypoblast replaced by endoblast, Nodal and FGF induce formation of primitive streak

start of primitive streak

endoblast

■ Vg-1 + Wnt   □ Cerberus   ■ Nodal   □ FGF

Fig. 5.4 Signals at the posterior marginal zone of the chick epiblast that initiate primitive streak formation. Epiblast cells in the posterior marginal zone overlying Koller's sickle secrete Vg-1 and Wnts. These signals induce expression of the gene *Nodal* in neighboring epiblast cells, but Nodal protein function is blocked by Cerberus, which is produced by the hypoblast. When the hypoblast has been displaced by endoblast, Nodal signaling in the epiblast and FGF signals from Koller's sickle induce the internalization of epiblast cells and formation of the primitive streak.

mouse ES cells can give rise to all cell types of the embryo, including the germ cells, in culture. ES cells genetically modified in culture are used to produce mice carrying particular mutations or transgenes (see Section 3.10).

The specification of cells as either inner cell mass or trophectoderm in early mouse embryos depends on their position in the cleaving embryo. Determination of their fate occurs after the 32-cell stage, and during earlier stages all the cells seem to be equivalent in their ability to give rise to either tissue. The most direct evidence for the effect of position comes from making chimeric embryos. Individual blastomeres disaggregated from eight-cell embryos are labeled and then combined in different positions with unlabeled blastomeres from other embryos. If the labeled cells are placed on the outside of a group of unlabeled cells they usually give rise to trophectoderm; if they are placed inside, so that they are surrounded by unlabeled cells, they more often give rise to inner cell mass (Fig. 5.5, right panels). Aggregates composed entirely of either 'outside' or 'inside' cells of early embryos can develop into normal blastocysts, showing that there is no specification of these cells, other than by their position, at this stage. To investigate the effect of position, chimeric mice can also be created by combining blastomeres from strains of mice that have different coat colors (see Fig. 1.24). In this case, coat color acts as an indelible label. When cells from an eight-cell embryo from a white strain of mouse are placed on the outside of a group of cells from an eight-cell embryo from a black strain of mouse, the resulting chimeric mouse will have a black coat, showing that it has been formed by the 'inside' cells.

At these early stages in mouse development, there is a remarkable degree of regulation. Even a single cell of a 32-cell mouse embryo can generate an entire embryo when aggregated with tetraploid cells, which will exclusively form the trophectoderm and support development of the embryo (see Box 8B). On the other hand, giant embryos formed by the aggregation of several embryos in early cleavage stages can achieve normal size within about 6 days by reducing cell proliferation.

The specification of the trophectoderm and the pluripotent inner cell mass lineages is the first differentiation event in mammalian embryonic development, and is essential for the formation of the placenta. The trophectoderm forms both the epithelial interface between fetal and maternal tissues and also produces invasive cells that help to remodel uterine blood vessels and come to line maternal blood spaces. The transcription factor Cdx2 is involved in the differentiation of trophectoderm, while the transcription factor Oct4 is involved in keeping the inner cell mass pluripotent. The *Oct4* gene is expressed from the oocyte onwards and by the eight-cell stage all blastomeres express both transcription factors (Fig. 5.6). As the morula forms, levels of Cdx2 become slightly higher in the outer cells. Then, because of reciprocal negative feedback between *Cdx2* and *Oct4*—in which the Cdx2 protein represses *Oct4* gene

Fig. 5.5 The specification of the blastomeres that will form the inner cell mass of a mouse embryo depends on their position. The specification of blastomeres as inner cell mass depends on whether they are on the outside or the inside of the embryo. Left panels: following cleavage, the mouse embryo is a solid ball of cells (a morula). This develops into the hollow blastocyst, which consists of an inner cell mass surrounded by a layer of trophectoderm. Right panels: to investigate whether the position of a blastomere in the morula determines whether it will develop into trophectoderm or inner cell mass, labeled blastomeres (blue) from a four-cell mouse embryo are separated and combined with unlabeled blastomeres (gray) from another embryo. By following the fate of the descendants of the labeled cells, it can be seen that blastomeres on the outside of the aggregate more often give rise to trophectoderm, 97% of the labeled cells ending up in that layer. The reverse is true for the origin of the inner cell mass, which derives predominantly from blastomeres inside the aggregate.

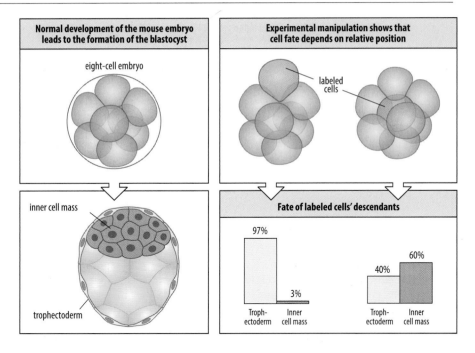

expression and vice versa—*Cdx2* expression levels increase and *Oct4* expression levels decrease in the outer cells, whereas *Oct4* expression increases and *Cdx2* expression decreases in the inner cells. By the blastocyst stage, *Cdx2* is expressed only in the trophectoderm and *Oct4* only in the cells of the inner cell mass. If *Cdx2* function is absent, trophectoderm initially forms in the blastocyst but does not differentiate and is not maintained.

Positive regulation of *Cdx2* expression depends on the transcription factor Tead4, whose activity is controlled by its co-activator Yap. Localization of Yap to the nucleus, where it can interact with transcription factors such as Tead4, is regulated by the Hippo signaling pathway (described in Box 13A). The Hippo pathway is known to act as a sensor of a cell's environment and its interactions with other cells. Although the details are not yet known, one can begin to see how a cell might be informed of its position within the morula by the activity of the Hippo pathway, and how this then determines its fate by controlling the expression of specific transcription factors.

By E4.5, the inner cell mass of the blastocyst has become differentiated into two tissues: a monolayer of **primitive endoderm** (which will form extra-embryonic structures, not the gut) forms its outer, or blastocoelic, surface, and encloses the epiblast, from which the embryo and some extra-embryonic structures will develop (see Fig. 5.6). The specification of cells as either primitive endoderm or epiblast can be traced back to the cells of the morula, which express low levels of two transcription factors, Nanog and Gata6. As the blastocyst matures, cells emerge that express higher levels of Nanog and low levels of Gata6, and vice versa, and, in consequence, two cell populations become distinguished. Nanog-expressing cells give rise to the epiblast, whereas Gata6-expressing cells give rise to the primitive endoderm. FGF signaling is important in this separation. Nanog-expressing cells make FGF, which promotes Gata6 expression in surrounding cells, which express FGF receptors, enhancing the probability that they will become primitive endoderm (see Fig. 5.6).

The orderly arrangement in the inner cell mass of epiblast enclosed by primitive endoderm is produced by a completely different strategy to that which produced the arrangement of inner cell mass enclosed by trophoblast. Instead of cells becoming different because of their position, the two different cell types—epiblast and primitive endoderm—are specified, possibly according to time, in a more-or-less random

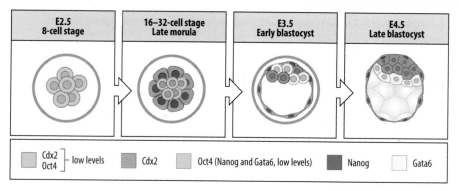

E2.5 8-cell stage	16–32-cell stage Late morula	E3.5 Early blastocyst	E4.5 Late blastocyst

Cdx2 Oct4 — low levels   Cdx2   Oct4 (Nanog and Gata6, low levels)   Nanog   Gata6

Fig. 5.6 **Establishing the three separate cell lineages in the mouse blastocyst.** The first two cell types to be specified are the inner cell mass and the trophectoderm. This involves progressive restriction of expression of the genes that code for transcription factors Oct4 and Cdx2. At the eight-cell stage, all the cells express both genes at low levels, but by the late morula/early blastocyst stage, Cdx2 expression has increased in, and eventually becomes restricted to, the cells on the outside of the embryo, and these will form trophectoderm. Oct4 expression becomes restricted to the inner cell mass cells at early blastula stage. Within the inner cell mass, two separate cell types arise: the epiblast cells, from which the embryo will develop; and the primitive endoderm, which will form extraembryonic structures. This involves a progressive restriction of expression of the genes for the transcription factors Nanog and Gata6. At the early blastula stage, some of the inner cell mass cells apparently at random start to express higher levels of Nanog, whereas the other cells express higher levels of Gata6. By the late blastula, the cells expressing the two different transcription factors have sorted out within the inner cell mass so that cells expressing high levels of Gata6 form an outer layer of primitive endoderm enclosing the epiblast cells expressing high levels of Nanog.

After Stephenson, R.O., et al.: Intercellular interactions, position and polarity in establishing blastocyst cell lineages and embryonic axes. Cold Spring Harb. Perspect. Biol. 2012, *4:* a008235.

arrangement and then sort out to take up their appropriate positions within the inner cell mass. Cell death may also play a role by eliminating cells in the wrong position.

Before the blastocyst stage, the embryo is a symmetric spheroidal ball of cells. The first asymmetry appears in the blastocyst, where the blastocoel develops asymmetrically, leaving the inner cell mass attached to one part of the trophectoderm (Fig. 5.7). The blastocyst now has one distinct axis running from the site where the inner cell mass is attached (the embryonic pole) to the opposite (**abembryonic pole**) site, with the blastocoel occupying most of the abembryonic half. This axis is known as the **embryonic–abembryonic axis**. It corresponds in a geometric sense to the dorso-ventral axis of the future epiblast, but is not related to specification of cell fate.

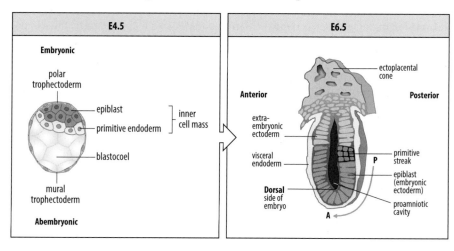

Fig. 5.7 **The axes of the early mouse embryo.** At the late blastocyst stage (left), at about E4.5, the inner cell mass is confined to the embryonic region and this defines an embryonic–abembryonic axis (which relates geometrically, although not in terms of cell fate, to the dorso-ventral axis of the future epiblast). The inner cell mass is oval, and thus also has an axis of bilateral symmetry. At E6.5 (right) the antero-posterior axis in the epiblast becomes visible, with the formation of the primitive streak at the posterior end. The interior of the epiblast cup corresponds to the dorsal side of the future embryo, and the outer side to the future ventral side.

5.3 Movement of the anterior visceral endoderm indicates the definitive antero-posterior axis in the mouse embryo

There is no clear sign of **polarity** in the mouse egg, and any contribution of maternal factors is difficult to determine, as unlike the situation in *Xenopus*, zygotic gene activation starts in the one-cell embryo and is essential for development beyond the two-cell stage. Several claims have been made that the first cleavage of the zygote is related to an axis of the embryo and that the first two blastomeres are biased towards different fates. This is an interesting possibility but is not yet clear, as different workers get different results.

Furthermore, the highly regulative nature of early development in the mouse argues against the importance of maternal determinants. It has been known for a long time that both blastomeres of the two-cell mouse embryo can give rise to normal embryos if separated. In addition, individual blastomeres from mouse embryos up to the eight-cell stage can contribute extensively to both embryonic and placental tissues, and normal development can still occur even after cells are removed or added to a blastocyst before it implants into the wall of the uterus. This extensive regulative ability would not, however, preclude the possibility that, in normal development, individual blastomeres in early embryos have different developmental properties and fates and maternal determinants might be distributed differently between the blastomeres—the debate over this continues.

At around E4.5, the mouse blastocyst implants in the uterine wall, and the trophectoderm at the embryonic pole proliferates to form the ectoplacental cone, producing extra-embryonic ectoderm that pushes the inner cell mass across the blastocoel. A cavity—the proamniotic cavity—is then formed within the epiblast as its cells proliferate. The mouse epiblast is now a monolayer and is formed into a cup shape, U-shaped in section, with a layer of visceral endoderm, which arises from the primitive endoderm, on the outside. By about E5.0–E5.5, the egg cylinder has a proximo-distal polarity in relation to the site of implantation with the ectoplacental cone at the proximal end, and the cup-shaped epiblast at the distal end (see Fig. 3.23). Extra-embryonic ectoderm forms a distinct layer of cells between the ectoplacental cone and the epiblast.

The next key event that takes place around E5.5, about 12–24 hours before gastrulation begins, is that the visceral endoderm becomes regionalized along the proximo-distal axis of the egg cylinder. The first indications of asymmetry can be detected earlier in the primitive endoderm of the blastocyst where cells characterized by the expression of the gene for Lefty-1, a Nodal antagonist, are located on one side (the name 'Lefty' comes from the fact that this antagonist was first identified because it is expressed very strongly on the left side of the embryo at later stages). The *Lefty-1*-expressing cells subsequently come to occupy a small region of visceral endoderm over the distal-most end of the epiblast cup, known as the **distal visceral endoderm** (**DVE**). The DVE expresses genes encoding a range of extracellular antagonists of Nodal, Wnt and BMP signaling, whereas the signaling molecule Nodal is expressed throughout the visceral endoderm and induces *Nodal* gene expression throughout the epiblast. Nodal signaling by the epiblast then induces additional visceral endoderm cells in the vicinity of the DVE to express the same signaling molecules as those expressed by the DVE. These additional cells constitute the **anterior visceral endoderm** (**AVE**). The AVE then moves rapidly to one side of the epiblast cup at random, displacing more proximal visceral endoderm, and this side of the cup will become the anterior region of the embryo (Fig. 5.8). This symmetry-breaking event converts the proximo-distal axis of the egg cylinder into the true antero-posterior axis of the mouse embryo. The AVE continues to express genes for Wnt and Nodal antagonists, including Dickkopf-1, Cerberus-like 1 and Lefty-1. (The various ways in which Nodal signaling can be fine-tuned are shown in Box 5A). As a result, the overlying epiblast becomes anterior ectoderm, from which the forebrain and other anterior structures will eventually develop.

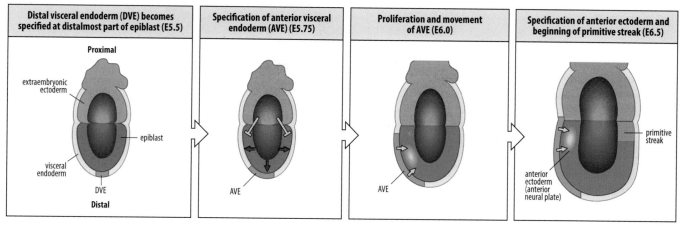

| Distal visceral endoderm (DVE) becomes specified at distalmost part of epiblast (E5.5) | Specification of anterior visceral endoderm (AVE) (E5.75) | Proliferation and movement of AVE (E6.0) | Specification of anterior ectoderm and beginning of primitive streak (E6.5) |

Fig. 5.8 The symmetry-breaking event in the early mouse embryo is the movement of the anterior visceral endoderm to one side of the epiblast cup. At around E5.5, before primitive streak formation, Lefty-1-expressing visceral endoderm (green) becomes located at the distal-most end of the cup and is known as distal visceral endoderm (DVE). The rest of the visceral endoderm is shown in yellow. At around E5.75, visceral endoderm at the distal-most end of the cup in the vicinity of the DVE begins to be specified as anterior visceral endoderm (AVE) by Nodal signals from the epiblast (red arrows). The formation of the AVE is restricted to the distal region of the visceral endoderm by inhibitory signals from the proximal extra-embryonic ectoderm (barred blue lines). The AVE then moves to one side of the cup and as a result of signaling by the AVE (yellow arrows), the overlying epiblast becomes anterior ectoderm (pale green), which includes prospective neural ectoderm. At E6.5, the primitive streak (pale red) has begun to form on the opposite side of the epiblast, thus marking the posterior end of the axis. Wnt signaling, initially in the posterior visceral endoderm, induces Wnt signaling in the epiblast, and this activates Nodal signaling, which is required for streak formation. *Illustration after Rodriguez, T.A., et al.:* **Induction and migration of the anterior visceral endoderm is regulated by the extra-embryonic ectoderm**. Development 2005, **132**: 2513-2520.

Despite the differences in topology, this process is developmentally similar to that described earlier for the chick (see Section 5.1): the AVE is equivalent to the chick hypoblast. Like the hypoblast, the AVE prevents primitive-streak formation. In both cases, this influence is due to these cells producing proteins that counteract Nodal signaling: Cerberus in chick; Lefty-1 and Cerberus-like 1 in mouse (see Box 5A). The movement of the AVE to the anterior side of the cup thus leads directly to the initiation of the primitive streak on the opposite side of the cup, which becomes the posterior end of the embryo. Primitive streak formation starts at around E6.5. *Wnt-3* is initially expressed in the posterior visceral endoderm and then also in the posterior epiblast together with *Nodal*. Thus, the signals that initiate formation of the primitive streak are likely to be Wnt and Nodal—the same signals that are involved in initiating the streak at the posterior marginal zone of the chick epiblast (see Fig. 5.4).

The extra-embryonic ectoderm at the proximal end of the egg cylinder in the E5.75 embryo plays a role in initially restricting the formation of the AVE to the most distal region of the visceral endoderm. Nodal signals are present throughout the epiblast at this stage and could, in principle, induce AVE fate throughout the visceral endoderm. However, the extra-embryonic ectoderm is a source of BMP signals (see Box 4C) that prevent formation of AVE. The proximal rim of the epiblast is exposed to high levels of BMP signaling, whereas the AVE forms distally where BMP signaling is lowest. If the extra-embryonic ectoderm is removed, there is a dramatic expansion of the AVE in the absence of the BMP signal.

5.4 The fate maps of vertebrate embryos are variations on a basic plan

It is difficult to make reproducible fate maps of chick and mouse embryos at the stages that correspond to the blastula stage in *Xenopus*. This is because the cells that will contribute to the different germ layers are mixed up in the epiblast and there are extensive cell movements both before and during primitive streak formation. In the

CELL BIOLOGY BOX 5A Fine-tuning Nodal signaling

Mouse Nodal and the corresponding Nodal-related proteins in other vertebrates are secreted signaling proteins that are essential for the induction and patterning of the mesoderm and the establishment of left-right asymmetry in early vertebrate embryos. Signaling molecules that have such powerful effects on development must be tightly regulated to ensure that the right amounts are present in the right place at the right time, and the strength of Nodal signaling can be modulated and regulated at several different levels as indicated in Figure 1.

One is the production of a functional protein (1). Like many other growth factors, Nodal is synthesized and secreted as an inactive precursor protein that has to be converted to the active form by proteolytic cleavage; this is carried out by enzymes called proprotein convertases. If these are not present, then even if the *Nodal* gene is expressed, no functional protein will be made.

A second level of control is due to extracellular inhibitors that bind to Nodal and/or its receptors (2) and prevent the signal from reaching the target cell, as discussed in this chapter. Like other TGF-β family members, Nodal binds to heterodimeric receptors known as type I/II receptors, and initiates an intracellular signaling cascade in which proteins called Smads are phosphorylated and enter the nucleus to regulate the expression of target genes (see Box 4C). Extracellular antagonists of Nodal signaling include the proteins Cerberus-1 and Lefty-1 and 2.

A third level of control of Nodal signaling is through regulation of transcription of the *Nodal* gene, which is itself a target of Nodal signaling (3). The expression of *Lefty*, *cerberus* and *Nodal* can all be increased in response to Nodal signaling, and this can lead to either negative or positive feedback on the amount of Nodal present. A negative-feedback loop, in which Nodal increases the expression of *Lefty*, fine tunes the level of Nodal signaling in the zebrafish blastoderm and in the proximo-distal patterning of the mouse epiblast. In the mouse epiblast, Nodal induces the anterior visceral endoderm (AVE), which then secretes Cerberus-like 1 and Lefty-1, which in turn downregulate *Nodal* expression in an adjacent area of the epiblast, thus enabling this region to become the anterior end of the antero-posterior axis. A positive-feedback loop, in which Nodal signaling upregulates the expression of *Nodal*, operates in the establishment of left-right asymmetry and leads to high levels of Nodal on the left side of the node in mouse embryos (see Section 5.16). A reaction–diffusion type of mechanism may also operate (see Box 11E). Yet another level of fine tuning of Nodal signaling discovered more recently involves the specific inhibition of translation of *Nodal* and *Lefty* mRNAs by microRNAs (4 on the figure) (see Box 6C for how microRNAs act).

Surprisingly many of the molecules that play key roles in specifying the vertebrate body plan act as extracellular antagonists, or inhibitors of intercellular signaling molecules and block or modulate signaling. Other widely used developmental signals

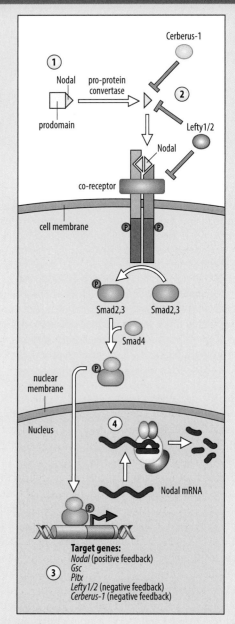

Figure 1

such as Wnts and BMPs have antagonists that counteract their expression or activity at some level (see Chapter 4 summary table), and also make positive- and negative-feedback loops with these antagonists in a similar way to Nodal. In *Xenopus*, for example, the dorsalizing signals produced by the Spemann organizer include the proteins Noggin and Chordin, which block the action of the ventralizing signal BMP-4 (see Section 4.13). In antero-posterior patterning of the epiblast in mouse embryos, the Wnt posteriorizing signal is blocked by the anteriorly produced extracellular antagonist Dickkopf-1.

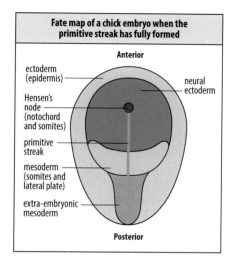

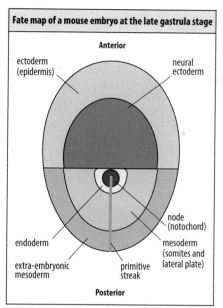

Fig. 5.9 Fate map of a chick embryo when the primitive streak has fully formed. The diagram shows a view of the dorsal surface of the embryo. Almost all the endoderm has already moved through the streak to form a lower layer, so is not represented.

Fig. 5.10 Fate map of a mouse embryo at the late gastrula stage. The embryo is depicted as if the 'cup' has been flattened and is viewed from the dorsal side. At this stage the primitive streak is at its full length.

chick, the epiblast cells adjacent to the posterior marginal zone at the early blastula stage are roughly equivalent to the dorsal cells of the *Xenopus* blastula (future notochord). The equivalent cells in the mouse are the epiblast cells initially adjacent to the posterior visceral endoderm in the egg cylinder. The picture becomes clearer once the primitive streak has formed, and cells have started to move inside and become determined as mesoderm and endoderm.

The fate map shown in Fig. 5.9 is of a chick embryo, which at this stage has a three-layered structure. We are viewing the upper surface, which is the epiblast. Some cells from the epiblast have already ingressed through the primitive streak into the interior and formed mesodermal and endodermal layers—the fate of these layers is not represented. The mesoderm cells that will form the heart have ingressed and migrated in an anterior-lateral direction on each side and will give rise to the bilateral first heart fields (discussed in Chapter 11). Most of the cells in the anterior epiblast are now prospective ectoderm, and will form neural tube and epidermis, but there are still cells in regions of the epiblast that will move through the streak and give rise to mesoderm. Hensen's node and the anterior end of the streak is prospective mesoderm; as the node regresses it leaves cells behind that will form the notochord and will contribute to the somites, while the epiblast lying along the antero-posterior midline not only contributes to the somites but also to the lateral-plate mesoderm (shown as a single region in Fig. 5.9) and organs such as the kidney. The most posterior region of the streak will become incorporated into the tail bud and form extra-embryonic mesoderm.

In the fate map of the mouse embryo at around E7–7.5 shown in Fig. 5.10, the epiblast is also becoming transformed into the three germ layers by gastrulation. Gastrulation in the mouse is essentially very similar to gastrulation in the chick, but the mouse epiblast is cup-shaped at this stage, which makes the process of primitive streak and node formation more difficult to follow. Figure 5.10 shows the embryo as if the cup were opened out and flattened and viewed from the dorsal side, which is the interior of the cup. This shows that the fate map of the mouse at the

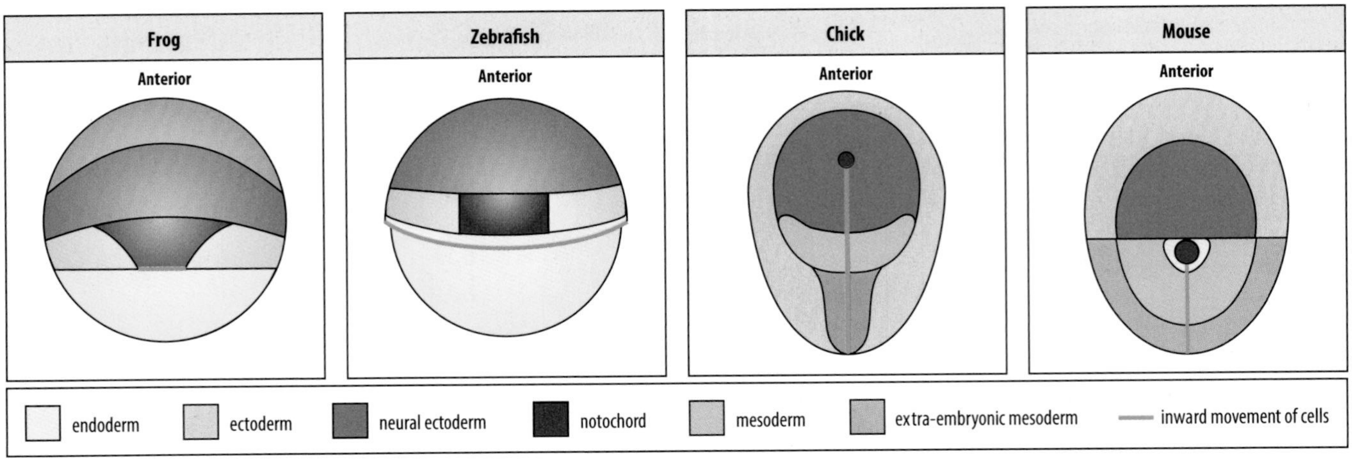

Frog	Zebrafish	Chick	Mouse
Anterior	Anterior	Anterior	Anterior

☐ endoderm   ☐ ectoderm   ◼ neural ectoderm   ◼ notochord   ☐ mesoderm   ☐ extra-embryonic mesoderm   —— inward movement of cells

Fig. 5.11 The fate maps of vertebrate embryos at comparable developmental stages. In spite of all the differences in early development, the fate maps of vertebrate embryos at stages equivalent to a late blastula or early gastrula show strong similarities. All maps are shown in a dorsal view. The future notochord mesoderm occupies a central dorsal position. The prospective neural ectoderm lies adjacent to the notochord, with the rest of the ectoderm anterior to it. The mouse fate map depicts the late gastrula stage. The future epidermal ectoderm of the zebrafish is on its ventral side. The prospective endoderm has already ingressed in the chick.

primitive-streak stage is basically similar to that of the chick. It may be helpful to look back at Figs 3.24 and 3.25, which show the cup-shaped mouse epiblast, to help interpret its fate map.

As in the fate map of the chick epiblast, the anterior region of the mouse epiblast will give rise to the ectoderm and the node will give rise to the notochord. The cells that will form the heart have already ingressed and moved forward of the node, as have some cells that will give rise to the definitive gut endoderm and dorsal mesoderm. The extreme anterior end of the primitive streak will give rise to more endoderm, whereas the middle part of the streak gives rise mainly to somites and lateral-plate mesoderm. The posterior part of the streak will become the tail bud, and also provides the extra-embryonic mesoderm. The mouse embryo retains considerable capacity to regulate until late in gastrulation. Even at this primitive-streak stage, up to 80% of the cells of the epiblast can be destroyed by treatment with the cytotoxic drug mitomycin C, and the embryo can still recover and develop with relatively minor abnormalities. This shows the continuing importance of cell–cell interactions.

The fate maps of the different vertebrates are similar when we look at the relationship between the germ layers and the site of the inward movement of cells at gastrulation (Fig. 5.11). The relative location of cells that will give rise to equivalent tissues in vertebrate embryos is also conserved, despite the very different shapes of the embryos. This is also the case at the pre-gastrulation stage when, as discussed earlier, the mouse AVE is equivalent to the anterior hypoblast in the chick. If one imagines the chick blastoderm rolled up with the epiblast on the outside, one can also see that the vegetal cells in the blastocoel floor of a *Xenopus* blastula and the dorsal yolk syncytial layer in the zebrafish could be considered equivalent to the anterior hypoblast of the chick (Fig. 5.12) This kind of representation also makes it easier to see, at early gastrula stages, the similar spatial relationship of different tissues, such as the node (or Spemann organizer in *Xenopus*) and the prospective germ layers and neural tissue in the different embryos.

5.5 Mesoderm induction and patterning in the chick and mouse occurs during primitive-streak formation

In the chick, most mesoderm induction and patterning occurs at the primitive streak. Chick epiblast isolated before streak formation will form some mesoderm containing blood vessels, blood cells, and some muscle, but no dorsal mesodermal structures, such as notochord. Like the experiments in *Xenopus* with animal cap cells (see Section 4.12), treatment of the isolated epiblast with activin results in the additional appearance of notochord and more muscle. This shows that TGF-β family members

Fig. 5.12 Comparison of the topology of the different regions of pre-gastrula and early gastrula stage embryos of mouse, chick, *Xenopus* and zebrafish. It is possible to see that a simple topological transformation reveals a similarity in the organization of the different embryos. Flattening of the mouse egg cylinder produces the equivalent of the chick blastoderm. Rolling up of the chick blastoderm with the epiblast on the outside, together with a relative increase in the mass of the 'hypoblast', produces the *Xenopus* embryo. Closing off the blastocoel almost completely with a further increase in yolk mass produces an approximation of the zebrafish embryo. The key refers to mouse tissues. A, anterior; P, posterior; D, dorsal; V, ventral.

*From Beddington, R. S. and Robertson, E.J.: **Anterior patterning in mouse.** Trends Genet. 1998, 14: 277–284.*

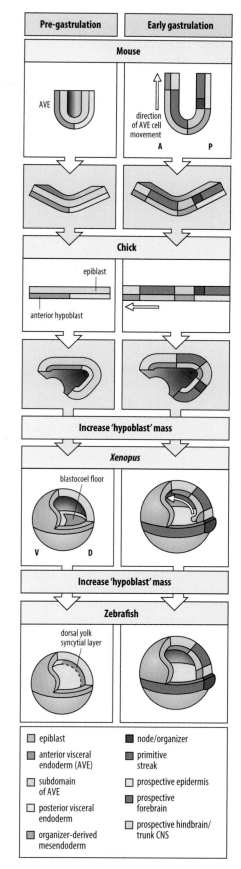

can act as mesoderm-inducing and/or mesoderm-patterning signals in chick embryos as they can in *Xenopus*.

As we saw in Section 5.1, the posterior marginal zone in the chick embryo is crucial for the formation of the primitive streak. The TGF-β family member Nodal, which is produced by the posterior marginal zone and the streak (see Fig. 5.3), together with fibroblast growth factor (FGF), produced by Koller's sickle, induce the internalization of epiblast cells to form mesoderm. The mesoderm expresses the gene *brachyury*. As we have seen, *brachyury* expression is one of the earliest markers of mesoderm in both *Xenopus* and zebrafish and this is also the case in the chick. Dorso-ventral mesodermal patterning occurs in the streak, with different regions of the streak giving rise to different mesodermal tissues along the dorso-ventral axis of the body. The dorso-ventral axis is orthogonal to the antero-posterior axis, but at this stage the body plan is specified in a flat sheet of cells and the descriptions axial or **medial** (at and alongside the midline) and lateral are more useful than dorsal and ventral. The somitic mesoderm, which runs alongside the midline is often specifically referred to as the **paraxial mesoderm**. The anterior region of the streak, including Hensen's node, gives rise to axial and paraxial mesoderm—the notochord and somites, respectively—and also the heart, whereas the posterior streak generates lateral mesoderm, including blood and blood vessels, and extra-embryonic mesoderm (Fig. 5.13). The more posterior cells move progressively forward and outwards to take up their positions more laterally.

In the mouse, mesoderm induction occurs in the primitive streak, which is initiated at around E6.5, starting from one small region at one side of the epiblast (see Fig. 3.24). This region is specified as a result of the determination of the anterior end of the embryo by the movement of the AVE (see Section 5.3). As the anterior ectoderm becomes determined, proximal epiblast cells that express genes characteristic of prospective mesoderm such as *brachyury* (*T*), start to converge on the posterior region and internalize at the point of convergence to initiate the primitive streak (Fig. 5.14, first and second panels). BMP-4 is expressed in the extra-embryonic ectoderm above the proximal rim of the epiblast and activates the expression of mesodermal markers in adjacent proximal epiblast cells. Wnt-3 is initially expressed in the posterior visceral endoderm and then in the posterior region of the epiblast. Wnt-3 activates expression of *Nodal* in the posterior region of the epiblast, which is required for mesoderm formation. In mutants lacking *Nodal* function, mesoderm does not form. Wnt-3 also activates expression of *brachyury*.

As in the chick, dorso-ventral mesodermal patterning occurs in the mouse streak during gastrulation, with different regions of the streak giving rise to different mesodermal tissues along the dorso-ventral axis. Cells at the very anterior end of the streak are the first to ingress, and give rise to the axial prechordal plate together with definitive gut endoderm. These cells ingress together and therefore are known as axial mesendoderm, but how the cells segregate into the different tissues is not understood. Cells that ingress in more posterior regions of the streak form paraxial and lateral plate mesoderm, while cells that ingress in the very posterior end of the streak give rise to extra-embryonic mesoderm (see Fig. 5.14).

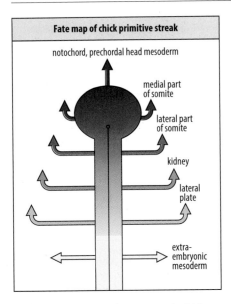

Fate map of chick primitive streak

notochord, prechordal head mesoderm

medial part of somite

lateral part of somite

kidney

lateral plate

extra-embryonic mesoderm

Fig. 5.13 Mesoderm is patterned within the chick primitive streak. Different parts of the primitive streak give rise to mesoderm with different fates. The arrows indicate the direction of movement of the mesodermal cells.

Fig. 5.14 Specification of the primitive streak in the mouse embryo. First panel: by E6 the anterior visceral endoderm (AVE) has promoted anterior character in the overlying epiblast (small arrows) and early markers of the future primitive streak are restricted to the proximal rim of the epiblast. Purple dots represent cells expressing *brachyury*. Second panel: *BMP-4* is transiently expressed in cells of the adjacent extra-embryonic ectoderm (red dots), and the posterior movement (white arrow) of cells expressing primitive-streak markers (purple) results in the primitive streak forming at the end of the embryo opposite to the AVE at stage E6.5. Third panel: by E7, extra-embryonic mesoderm (brown) is produced from the posterior end of the streak (white arrow) while the node (red) forms at the anterior end. The extreme anterior end of the streak gives rise to the tissue known as axial mesendoderm, which will form the prechordal plate and the gut endoderm (white arrow).

The generation of the axial mesendoderm in the mouse requires Nodal signaling. Nodal signaling by the posterior epiblast in the mouse embryo at around E6.5 leads to graded signaling in the streak, with high levels in the anterior streak specifying axial mesendoderm and lower levels of Nodal signal in the intermediate regions of the streak specifying paraxial and lateral-plate mesoderm.

5.6 The node that develops at the anterior end of the streak in chick and mouse embryos is equivalent to the Spemann organizer in *Xenopus*

The avian equivalent of the Spemann organizer is Hensen's node, the region at the anterior end of the primitive streak (see Fig. 3.16). The mouse equivalent is known simply as the node and also can be found at the anterior end of the primitive streak (see Fig. 3.24). Both Hensen's node and the mouse node have organizing activity and can induce a second axis when transplanted to another embryo at the appropriate stage. The second axis is patterned along both dorso-ventral and antero-posterior axes and has its own nervous system. In a classical experiment in the 1930s, the British embryologist Conrad Waddington demonstrated directly that organizers from mammalian and bird embryos are equivalent by grafting the node from a rabbit embryo into a chick embryo. The grafted node induced the formation of a second chick axis. This shows that the signals produced by the node in chick and rabbit are the same, but that the interpretation is governed by the nature of the responding tissue. We will see this same principle at work when we consider other vertebrate signaling centers, for example, the polarizing region in the developing limb (discussed in Chapter 11).

The detailed inducing properties of the avian node have been investigated by transplanting quail nodes to chick embryos. Quail cells can be distinguished from chick cells by their distinctive nuclei, which can be detected in histological sections (see Fig. 3.33), or by a species-specific antibody, and this provides a way of finding out precisely which tissues arise from the graft and which tissues are induced. When, for example, a node from a gastrula-stage quail embryo is grafted beneath the lateral epiblast of a chick embryo at the same stage of development, the transplanted node can induce the formation of a complete additional axis with a head. When, however, the node is taken from an older embryo at the head-process-stage, it only induces a trunk and not a head (Fig. 5.15). Thus, the inductive properties of the node change during gastrulation in the same way as those of the Spemann organizer in *Xenopus* change (see Section 4.14).

In the mouse, grafting the node from a late gastrula-stage E7.5 embryo to the lateral posterior region of the epiblast of an embryo at the same stage induces an additional axis consisting of posterior structures including neural tissue and ectopic somites but lacking a head. The node itself gives rise to the notochord and some endoderm. Grafts of node precursors from earlier streak stages also have inducing ability but never induce a head.

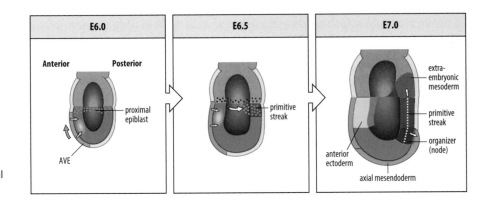

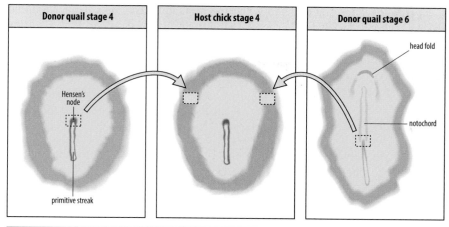

| Donor quail stage 4 | Host chick stage 4 | Donor quail stage 6 |

Fig. 5.15 Hensen's node can induce a new axis in avian embryos. When Hensen's node from a stage-4 quail embryo (left) is grafted to the extra-embryonic area opaca (dark gray), just outside the embryonic area pellucida (light gray), of a host chick embryo at the same stage of development (center), a complete new axis, including brain tissue, forms at the site of transplantation (photo, embryo on the left). The embryo in the center of the photograph has formed from the original primitive streak of the host chick embryo. The two stripes of dark purple staining in the left and center embryos indicate expression of *Krox20*, a gene encoding a transcription factor that serves as a marker for the hindbrain, as detected by *in situ* hybridization. Quail cells are distinguished from chick cells by staining reddish brown with anti-quail antibody. Although some of the tissue of the new axis (embryo on left) is formed from the quail graft, most has been induced from chick tissue that does not normally form an embryo. A node from a quail embryo at the head-fold stage of development (stage 6, right) grafted to a stage-4 chick embryo produces a new axis with only a trunk (photo, right embryo) and most of the new tissue is derived from the quail graft, as indicated by the reddish-brown staining throughout.

*From Stern, C.D.: **Neural induction: old problem, new findings, yet more questions.*** Development, *2005, **312**: 2007-2021.*

The extent of axis duplication induced by grafts of the organizer in different vertebrates depends on its cellular composition when it is transplanted. In the mouse, the activity of the AVE keeps Wnt, Nodal and BMP signaling low in the overlying epiblast and is required to initiate anterior pattern before gastrulation and development of the node. This explains why duplicated axes produced by node grafts in the mouse lack anterior structures. It has been suggested that the vegetal cells in the organizer region in *Xenopus* and the dorsal yolk syncytial cells in the shield region in zebrafish (see Section 4.18) fulfill the same role as the anterior visceral endoderm in the mouse and this is why organizer transplants from *Xenopus* and zebrafish which contain these cells induce full duplications. In the chick, the hypoblast corresponds to the visceral endoderm of the mouse (see Fig. 5.12) and early chick node transplants, which produce full duplications, may contain some hypoblast cells. As would be expected from the ability of a mammalian node to produce a secondary axis in a chick embryo, the signaling molecules produced by vertebrate organizers are common to all vertebrates.

A number of genes for both transcription factors and signaling proteins are specifically expressed in vertebrate organizers and are known to be required for their function. Figure 5.16 lists some of the genes expressed in both the Spemann organizer of *Xenopus* and in the mouse node. These include, for example, the transcription factor Goosecoid, which is an early marker of the organizer in all vertebrates. In addition, several secreted antagonists are expressed that will produce gradients of TGF-β and Wnt signaling in the embryo. Such gradients will provide positional information for dorso-ventral and antero-posterior patterning of the mesoderm and also for the induction of the neural plate. It is important to bear in mind, however, that although many of the same proteins are produced in and around the organizer in different vertebrates, they do not all have precisely the same functions in all our model animals.

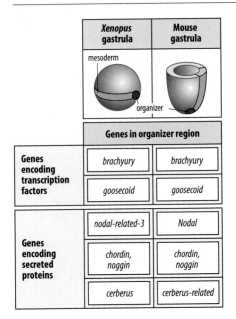

	Xenopus gastrula	Mouse gastrula
	mesoderm	
	organizer	
Genes in organizer region		
Genes encoding transcription factors	*brachyury*	*brachyury*
	goosecoid	*goosecoid*
Genes encoding secreted proteins	*nodal-related-3*	*Nodal*
	chordin, noggin	*chordin, noggin*
	cerberus	*cerberus-related*

Fig. 5.16 Genes expressed in the Spemann organizer region of the *Xenopus* gastrula, and in the node in the mouse gastrula. There is a similar pattern of gene activity in the two animals, with homologous genes being expressed. The expression of some of these genes, such as *brachyury*, is not confined to the organizer.

In both the chick and the mouse, the specification of axial mesoderm (produced by the anterior streak) depends on the production of antagonists of BMP signaling by the node, such as Noggin, Chordin and other antagonists such as Cerberus/Cerberus-related 1. Axial mesoderm, like dorsal mesoderm in *Xenopus* and the zebrafish, is specified where BMP signaling is low. Lateral plate mesoderm (produced by the posterior streak) is specified by high levels of BMP signaling. Likewise, patterning along the antero-posterior axis in the chick and mouse follows the same principles as in *Xenopus* and zebrafish, with high levels of Wnt signaling specifying posterior positional values. The head develops where Nodal, BMP and Wnt signaling are low, and we have seen in the mouse how the specification of anterior structures is permitted in the anterior epiblast through production of antagonists by the underlying AVE (see Fig. 5.8). Mouse embryos mutant for the Wnt antagonist Dickkopf1 (Dkk1) are headless, indicating that Wnt inhibition is essential for head formation. However when Dkk1 is overexpressed in *Xenopus*, it is only able to induce extra heads when in combination with BMP inhibitors. This shows that inhibition of both BMP and Wnt signaling are required to mimic the activity of the AVE and confirms the essential similarity in head specification in mouse and frog.

5.7 Neural induction in chick and mouse is initiated by FGF signaling with inhibition of BMP signaling being required in a later step

As shown in the fate map in Fig. 5.9, the region of the epiblast in the chick embryo that will give rise to the neural plate lies immediately anterior to the node. As mesodermal cells emerge from the anterior tip of the streak and move forward to form the anterior part of the notochord (the head process), the prospective neural plate overlying the head process, and which will form the brain, begins to thicken. The primitive streak then begins to shorten, and the head fold and neural plate form in the epiblast anterior to the furthest extent of the primitive streak. As the streak continues to regress, somites begin to form on either side of the notochord (Fig. 5.17). In the mouse, the neural plate similarly develops from the epiblast anterior to the node.

Signals emitted by the node and by the axial mesendoderm that comes to underlie the anterior epiblast are involved in inducing prospective neural tissue in chick and mouse embryos. Grafts of both Hensen's node and the mouse node induce a new body axis, including neural tissue, when transplanted to another early embryo. This is evidence that in the normal embryo signals from the organizer play a part in inducing anterior neural tissue from ectoderm. In the chick embryo, neural tissue can be induced in the epiblast—in both the area pellucida and the area opaca—by grafts from the primitive-streak mesoderm (see Fig. 5.15). Inducing activity is initially located in the anterior primitive streak and Hensen's node. Later, during regression of the node (see Fig. 3.16), its inducing activity becomes much reduced and by the four-somite stage has completely disappeared. The competence of the ectoderm to respond to neural-inducing signals disappears at the time of head-process formation, suggesting that in normal development neural induction has ended by this time.

FGF is the signal required to initiate neural induction in both chick and mouse embryos. In chick embryos, FGF is initially produced in the blastoderm before gastrulation by Koller's sickle, and at that time is involved primarily in mesoderm induction (see Fig. 5.4). Epiblast cells are thought to be prevented from taking on a neural-plate fate at that time because the chromosomal regions encoding neural-inducing genes are in a 'silenced' state in which genes cannot be switched on and transcribed. The lifting of this inhibition during gastrulation is associated with the removal of repressor proteins from the regulatory regions of such genes and the remodeling of the chromatin into an active state by **chromatin-remodeling complexes** (Box 5B). Epiblast cells that receive FGF signals but have not formed mesoderm by this stage can now form neural plate.

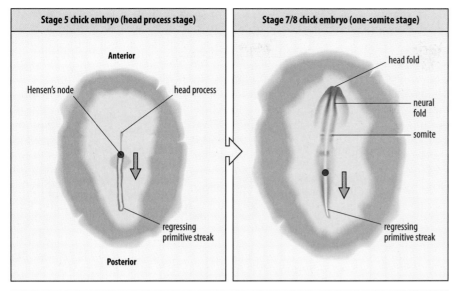

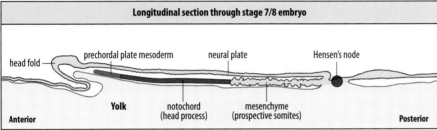

Fig. 5.17 Notochord and head-fold formation in the chick embryo. Upper panels: After extending to its full length, the primitive streak begins to regress posteriorly. As it does so, the notochord (head process) formed by cells that emerged from the anterior tip of the streak and moved forward becomes apparent (left upper panel). The head fold and neural plate form in the epiblast anterior to the furthest extent of the primitive streak. The edges of the neural plate begin to rise up to form the neural folds and the first somite appears (right upper panel). Lower panel: schematic diagram of a sagittal section through the chick embryo around the stage of head-fold formation as Hensen's node starts to regress. The notochord in the head region (the head process) is formed anterior to the node, with the prechordal plate mesoderm immediately anterior to the notochord. The undifferentiated mesenchyme on either side of the notochord will form somites. The anterior somitic mesoderm (in which a somite has already formed, see upper right panel) has been cut away to show the notochord.

One of the key genes activated by FGF signaling in neural induction in chick embryos is *Churchill*, which encodes a zinc finger transcription factor. When *Churchill* expression is downregulated using an antisense morpholino oligonucleotide (see Box 6B), the neural plate does not form. The activation of *Churchill* as a result of FGF signaling leads indirectly to repression of genes characteristic of mesoderm and activation of the gene for the neural-specific transcription factor Sox2, one of the earliest definitive markers of neural tissue.

In both chick and mouse, inhibition of BMP signaling by antagonists produced by the node and cells at the anterior tip of the streak, including Chordin and Noggin, is then required in a later step in the specification of neural cell fate. As in *Xenopus*, antero-posterior patterning of the neural plate in chick and mouse embryos also depends on interactions with the mesoderm that comes to underlie it. Forebrain, midbrain and hindbrain identities are induced. In the mouse, the anterior mesendoderm extends under the anterior epiblast and maintains the forebrain identity that was initially promoted due to the anterior visceral endoderm coming to underlie this region.

Neural induction in vertebrates is therefore a complex multistep process and the very first stages are likely to occur in the epiblast in chick and mouse, or the blastula in amphibians, even before a distinct organizer region becomes detectable. Nevertheless, despite this complexity, there is an essential similarity in the mechanism of neural induction among vertebrates, as Hensen's node from a chick embryo can induce neural gene expression in *Xenopus* ectoderm (Fig. 5.18). This suggests that there has been an evolutionary conservation of inducing signals and confirms the essential similarity of Hensen's node and the Spemann organizer. Moreover, early nodes induce gene expression characteristic of anterior amphibian neural structures, whereas older nodes induce expression typical of posterior structures. These results are in line with the hypothesis that owing to the changing tissue composition of the vertebrate node

CELL BIOLOGY BOX 5B Chromatin-remodeling complexes

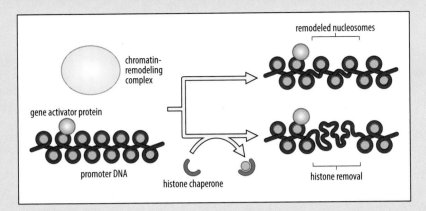

Figure 1

Whether a gene is expressed or not depends not only on whether appropriate gene-specific DNA-binding regulatory proteins are present, but also on the state of the chromatin—the complex of DNA and proteins of which the chromosomes are composed. Chromatin proteins include the histones, which package DNA into structures called nucleosomes, and other proteins that bind directly to DNA or to proteins bound to DNA, and affect chromatin structure and the availability of the DNA for transcription. In highly simplified terms, **euchromatin** is chromatin in which genes are available for transcription whereas in **heterochromatin** they are unavailable—or 'silenced'.

In the case of the chick embryo, the gene *Sox2* is an early marker for neural plate, but is not expressed before gastrulation, even though FGF, which is involved in neural-plate induction, is active early on. Experiments investigating the regulation of *Sox2* in cultured chick embryos have shown that, like many other genes, its expression requires the action of a chromatin-remodeling protein called Brm. Brm has ATPase activity and is an essential com-ponent of numerous multiprotein ATP-dependent chromatin-remodeling complexes that are active in cells. Such complexes can be recruited to control regions in the DNA (see Fig. 1.19) and they are thought to use the energy of ATP hydrolysis to loosen the attachment of DNA to histones and thus enable nucleosomes to be slid along DNA (Figure 1, upper arrow) or the histone core to be removed from the DNA (Figure 1, lower arrow). This frees the control-region DNA to bind other gene regulatory proteins, RNA polymerase, and the rest of the transcriptional machinery, thus enabling the gene to be transcribed.

Chromatin-remodeling complexes work in conjunction with other gene regulatory pro-teins. To determine the effects of candidate regulatory proteins on *Sox2* expression, DNAs encoding active or inactive forms of the proteins, in different combinations, were transfected into the area opaca of early chick embryos, an area outside the embryo that would never normally express *Sox2*, to see whether they could induce *Sox2* expression. From the results of these experiments, a series of steps leading from inactive *Sox2* in the early embryo to active *Sox2* in the neural plate could be

proposed. The early exposure to FGF is proposed to prime *Sox2* for eventual expression by inducing an activator protein that binds to the *Sox2* control region along with Brm, but the gene is kept inactive by the binding of repressive heterochromatin-inducing proteins, which are ubiquitously expressed throughout the embryo. The inhibition is lifted by the specifically timed expression of a protein that disrupts the complex, removing the repressors and enabling the activator protein, in conjunction with the now active chromatin-remodeling complex, to switch on gene expression (Figure 2). This general strategy for preventing the premature expression of a crucial gene in the presence of potential activating signals is likely to be widely deployed in development.

The recruitment of chromatin-modeling complexes and heterochromatin-inducing proteins to specific sites on the chromosome often depends on pre-existing chemical modifications to chromatin such as DNA methylation and histone methylation and acetylation, which are discussed in Box 8A.

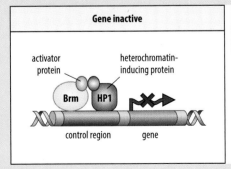
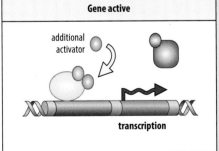

Figure 2

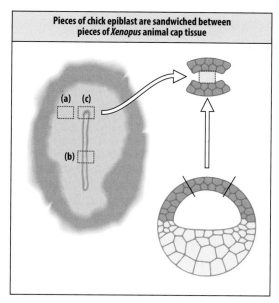

Pieces of chick epiblast are sandwiched between pieces of *Xenopus* animal cap tissue

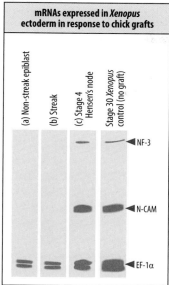

mRNAs expressed in *Xenopus* ectoderm in response to chick grafts

(a) Non-streak epiblast

(b) Streak

(c) Stage 4 Hensen's node

Stage 30 *Xenopus* control (no graft)

◄ NF-3

◄ N-CAM

◄ EF-1α

Fig. 5.18 Hensen's node from a chick embryo can induce gene expression characteristic of neural tissue in *Xenopus* ectoderm. Tissues from different parts of the primitive-streak stage of a chick epiblast are placed between two fragments of animal cap tissue (prospective ectoderm) from a *Xenopus* early blastula. The induction in the *Xenopus* ectoderm of genes that characterize the nervous system is detected by looking for expression of mRNAs for neural cell adhesion molecule (N-CAM) and neurogenic factor-3 (NF-3), which are expressed specifically in neural tissue in stage-30 *Xenopus* embryos. Only transplants from Hensen's node induce the expression of these neural markers in the *Xenopus* ectoderm. (EF-1α is a ubiquitous transcription factor expressed in all cells and is used as a control.)

Illustration after Kintner, C.R., Dodd, J.: **Hensen's node induces neural tissue in *Xenopus* ectoderm. Implications for the action of the organizer in neural induction.** Development 1991, **113**: 1495-1505.

over time (see Section 5.6), it will specify different antero-posterior positional values in the neural plate at different times.

5.8 Axial structures in chick and mouse are generated from self-renewing cell populations

In *Xenopus* and fish embryos, the main antero-posterior body axis is patterned during gastrulation; there is little growth, and convergent–extension movements lead to elongation of the embryo, as discussed in Chapter 9. The most posterior axial tissues are then generated from a tail bud in a process that involves growth. In contrast, in chick and mouse embryos, although early patterning of the embryo is also established during gastrulation—for example, patterning of the brain—most of the body axis is then generated over a period of time as the node and primitive streak gradually regress and the embryo grows and elongates. The axial tissues that give rise to the spinal cord, the skeleton, and the musculature of the body are laid down in an anterior to posterior sequence by cells from the primitive streak and the adjacent epiblast.

The regressing node and adjacent anterior region of the streak give rise to the notochord, and also contribute to the somites, which are formed from the mesoderm on either side of the notochord. The ventral midline of the neural tube that will form the spinal cord is also derived from cells in this region (the neural plate that will form the brain has already been induced, as described earlier). Epiblast cells in an arc of tissue on either side of the primitive streak immediately posterior to the node give rise to most of the neural tube and also contribute to the somites in the trunk (Fig. 5.19). Descendants of the primitive streak and neighboring epiblast later contribute to the tail bud, from which the most posterior region of the body develops. This last phase in axis formation is known as **secondary body formation** and differs in some respects from the earlier processes that form the head and trunk, which are known as **primary body formation**. We will focus here on the generation of axial tissues during primary body formation.

The axial tissues that are laid down anterior to the node as it regresses are formed from cells in the region encompassing the node and the arc of epiblast on either side of the streak immediately posterior to the node. The cells in the node and the region of neighboring epiblast constitute self-renewing populations and have stem-cell-like properties (see Chapter 1). The arc of epiblast has therefore been called the

Fig. 5.19 Location of cell populations that generate axial tissues during body elongation in mouse and chick embryos. In both mouse and chick embryos, the axial tissues—notochord, neural tube (spinal cord), and somites—are laid down progressively along the antero-posterior axis of the trunk from the node and adjacent epiblast as the node regresses. The schematic diagrams show the dorsal views of the posterior regions of mouse and chick embryos at an early stage in axis formation in which 6 somites have already been laid down; another 24 trunk somites remain to be formed in the mouse and 16 in the chick. The neural tube is partly closed but is still open posteriorly. Equivalent regions of the mouse and chick embryo that contain the cell populations that contribute to these axial tissues have been identified. The emerging trunk notochord is generated mainly from a self-renewing population of cells in the node while almost the entire length of the neural tube and the medial parts (those nearest the midline) of the somites in the trunk are generated from populations of self-renewing cells in an arc of epiblast on either side of the streak immediately posterior to the node. This region of epiblast is known as the stem zone (also called the caudal lateral epiblast) and the cells in this region have stem-cell-like properties.

From Wilson, V., et al.: ***Stem cells, signals and vertebrate body axis extension*** Development *2009,* ***136***, *1591-1604*

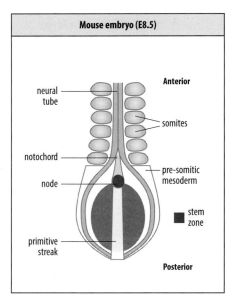

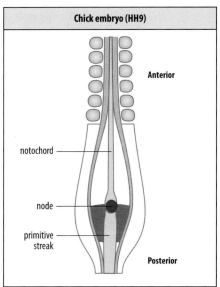

stem zone. When single cells are marked in the stem zone in both chick and mouse embryos, some descendants are found to contribute to the extending axis, whereas other descendants remain in the stem zone. Longer-term fate maps show that descendants of a cell in the stem zone can contribute to axial tissues along almost the entire length of the antero-posterior axis. Furthermore, detailed cell-lineage analyses show that some descendants of cells in the stem zone give rise to neural tissue whereas others give rise to mesoderm, more specifically to the medial part of the somites, the part nearest to the dorsal midline, showing that cells in the stem zone are multipotent.

The other tissues of the extending body axis are not generated from the stem zone. The mesodermal cells that form the lateral parts of the somites and the lateral plate mesoderm come from more posterior regions of the streak, and the gut endoderm is most probably generated by the expansion of endoderm formed by cells that ingressed through the early streak.

Although the way in which trunk axial tissues are generated differs in chick and mouse compared with *Xenopus* and zebrafish, the same signals—Wnt and FGF—are involved. Mutations in mice in genes encoding components of these signaling pathways, for example Wnt-3a and the FGF receptor FGFR1, cause truncation of the axis. As we have seen, the key target for Wnt and FGF signaling is expression of the gene for the transcription factor Brachyury, the earliest marker for mesoderm formation. Spontaneous and induced mutations in the *brachyury* gene have long been known to cause axis truncation in mice (see Chapter 1).

The combination of Wnt and FGF signaling by cells in the posterior region of chick and mouse embryos maintains the pool of undifferentiated proliferating cells in the stem zone. An antero-posterior gradient of FGF and Wnt signaling is set up in the posterior region of the embryo with its high point at the node (Fig. 5.20). The FGF signaling gradient is not formed by diffusion. Instead, cells in the node and primitive streak transcribe the *Fgf8* gene and the *Fgf8* mRNA is then gradually degraded in the cells left behind as the node regresses. The result is that cells at increasing distances from the node produce progressively lower amounts of FGF-8, and so a gradient in that signal is generated.

There is also a gradient of the small secreted signaling molecule **retinoic acid** (Box 5C), which runs in the opposite direction to the FGF gradient, and this has an antagonistic effect on FGF action. Retinoic acid is synthesized by mesoderm cells in the somites and diffuses posteriorly. Posterior cells produce the retinoic-acid-metabolizing enzyme Cyp26a1, and this ensures that retinoic-acid signaling is kept low in cells in the stem zone. Mouse embryos completely lacking *Cyp26a1* gene function have axis truncations, as do mouse embryos treated with excess retinoic acid.

Expression of *Cyp26a1* is regulated by FGF signaling and thus the formation of the opposing gradients is interlinked.

As might be expected from the results of the detailed cell-lineage analysis mentioned earlier, the cell population in the stem zone in chick and mouse embryos expresses genes that are characteristic of both early neural tissue (*Sox2*) and early paraxial mesoderm (*brachyury*). When the cells leave the stem zone, they will be exposed to progressively lower levels of FGF signaling and progressively higher levels of retinoic-acid signaling, and this promotes their differentiation into the neural tissue of the spinal cord (in the case of ectodermal cells) and into the paraxial mesoderm (in the case of mesodermal cells), which segments into somites.

The body axis in mouse and chick is completed by the generation of most posterior structures from the tail bud, which develops when 30 somites and 22 somites, respectively, have formed. At this stage, the final plan of the body has emerged and the chick and mouse embryos look similar to those of frogs and fish (see Fig. 3.2). Establishment of the body plan has also fixed the positions in which various organs will develop, even though there are not yet any overt signs of organ formation. Nevertheless, numerous grafting experiments have shown that the potential to form a given organ is now confined to specific regions of the embryo. Each of these regions has, however, considerable capacity for regulation, so that if part of the region is removed, a normal structure can still form.

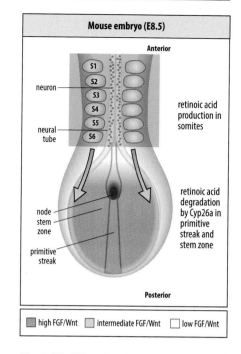

Fig. 5.20 FGF and retinoic-acid gradients control the generation of axial tissues from the stem zone in the mouse embryo. This schematic diagram shows a dorsal view of the posterior half of an E8.5 mouse embryo with its last-formed six pairs of somites. The neural tube has closed anteriorly to form the spinal cord, and neurons have differentiated. There is an antero-posterior gradient of FGF and Wnt signaling in the embryo, with its high point at the node. Both FGF and Wnt signals are produced by cells in the node and stem zone and this combination of signals maintains the pools of self-renewing undifferentiated cells. The FGF gradient is formed by gradual degradation of *Fgf* mRNA in the cells as they leave the stem region anteriorly. This results in a gradient of translated and secreted FGF protein that is continuously moving posteriorly as the embryo elongates. Retinoic acid synthesized and secreted by the newly formed somites forms an opposing gradient (graded shading in the green arrows) that antagonizes the actions of FGF. Similar opposing gradients of FGF/Wnt and retinoic acid signaling are set up in the chick embryo and fulfill the same roles in generating axial tissue.

SUMMARY

The early development of the chick is affected by the large amount of yolk in the egg and the basis for first breaking the symmetry is unknown. Early development in the mouse generates cell populations that will form extra-embryonic tissues, and there is no evidence in the mouse for cytoplasmic localization of maternal factors making cells different. Organized patterns of different cell types in the early mouse embryos are produced by cells being specified by their position and by sorting out. The inner cell mass of the mouse embryo gives rise to all the cells of the body and pluripotent stem cells—embryonic stem cells—can be isolated from it and grown in culture. As in *Xenopus* and zebrafish, localized Wnt signaling specifies the position where the organizer will form in chick embryos and secreted antagonists help to localize signaling.

Specification of the germ layers occurs during gastrulation in chick and mouse, later than in *Xenopus* and zebrafish, and requires cell-cell interactions. In all vertebrate embryos, Nodal is involved in specifying endoderm and mesoderm and providing mesoderm with positional information. Hensen's node in the chick serves a function similar to that of the Spemann organizer and can specify a new antero-posterior axis. The mouse node can specify a new axis apart from a head, for which previous exposure of the epiblast to the anterior visceral endoderm is required.

The organizers in all four vertebrate model embryos produce and secrete antagonists to signaling molecules and this produces gradients of signaling molecules that provide cells in the embryo with positional information. High levels of BMP signaling specify ventral, low levels dorsal. High levels of Wnt signaling specify posterior, low levels anterior. The head forms where levels of BMP, Wnt and Nodal signaling are low. The anterior nervous system, which develops from the neural plate, is induced by FGF signaling in chick and mouse. As in *Xenopus*, inhibition of BMP signaling by proteins such as Chordin produced by the organizer is required for later steps. Patterning of the neural plate involves both quantitative and qualitative differences in signaling by the mesoderm, with high levels of Wnt and FGF signaling specifying the more posterior structures. In the chick and the mouse, the spinal cord and somites of the main body axis are generated from self-renewing cell populations in the node and stem zones, which are maintained by Wnt and FGF signaling.

CELL BIOLOGY BOX 5C Retinoic acid: a small-molecule intercellular signal

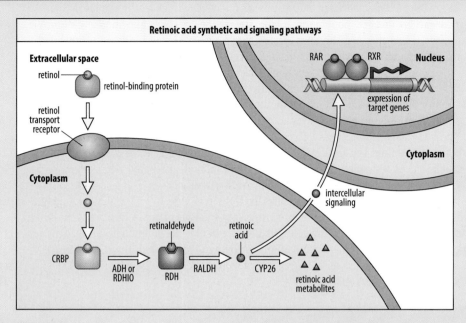

Figure 1

Retinoic acid is a small diffusible molecule derived from vitamin A (retinol). Animals obtain vitamin A through their diet, and it has been known for a long time through work with pigs that vitamin A deficiency can lead to developmental defects. It subsequently emerged that an excess of vitamin A also leads to developmental defects. Sadly, babies with craniofacial malformations have been born to women who had been treated for severe skin diseases with synthetic vitamin A derivatives, not knowing they were pregnant.

Retinol is present at high levels in chicken eggs, and in mammalian embryos is supplied by the maternal circulation. Retinol binds to retinol-binding proteins in the extracellular space and is transferred into cells by the transport receptor Simulated by retinoic acid 6 (STRA6). Once inside the cell, retinol is bound by an intracellular retinol-binding protein (CRBP). Retinoic acid is produced in a two-step reaction (Figure 1), in which retinol is first metabolized to retinaldehyde by retinol dehydrogenases (mainly RDH10) or alcohol dehydrogenase (ADH); retinaldehyde is then metabolized to retinoic acid by retinaldehyde dehydrogenases (RALDH). There are three vertebrate retinaldehyde dehydrogenases, encoded by the genes *Raldh-1, -2,* and *-3*, and expression of these genes is tightly regulated both temporally and spatially in vertebrate embryos to ensure that retinoic acid is produced at very precise locations and specific times during development.

Retinoic acid is broken down by enzymes of the P450 cytochrome system, mainly by the subfamily of CYP26 enzymes which, like the RALDHs, are expressed in very precise patterns in the embryo. Intricate homeostatic buffering mechanisms involving regulation

---

SUMMARY: vertebrate axis determination		
	Dorso-ventral axis	**Antero-posterior axis**
Xenopus	sperm entry point and cortical rotation lead to specification of dorsal side opposite the point of sperm entry. Localization of maternal *wntll* and *wnt5a* mRNAs, and nuclear β-catenin specifies dorsal side and position of organizer	specified in organizer (Spemann organizer)
Zebrafish	initial symmetry-breaking event unknown. Localization of maternal *wnt8a* mRNA and nuclear β-catenin specifies dorsal side and position of organizer	specified in organizer (shield)
Chick	posterior marginal zone specifies ventral end of dorso-ventral axis	gravity determines position of posterior marginal zone and thus the posterior end of the A–P axis
Mouse	interaction between inner cell mass and trophectoderm	specification and movement of anterior visceral endoderm

of expression of the genes encoding retinoic-acid-metabolizing enzymes and end-product feedback serve to maintain appropriate levels of retinoic acid. If retinoic acid levels increase, *Raldh* gene expression is reduced and expression of *Cyp26* genes is increased; when retinoic acid levels fall the reverse happens.

Retinoic acid acting as an intercellular signal is shown in the figure, although it probably also acts as a signal within the cells that produce it. Retinoic acid is lipid-soluble and so can pass through cell membranes. Inside a cell it is bound by proteins that transport it to the nucleus, where it binds to receptors that act directly as transcriptional regulators. Mammals and birds have three retinoic acid receptor (RAR) proteins—α, β and γ—which act as dimers in combination with the retinoid X receptor proteins (RXR; also α, β and γ). In the absence of retinoic acid, the heterodimeric RAR:RXR receptors are bound to regulatory DNA sequences known as retinoic-acid-response elements (RAREs) and repress the associated genes. Binding of retinoic acid to its nuclear receptor leads to activation of the gene associated with the response element. Gene activation by retinoic acid can involve not only the direct initiation of transcription but also the recruitment of co-activator proteins that induce chromatin remodeling, which enables the gene to be expressed (see Box 5B).

Among the genes with a retinoic-acid-response element is the gene for RARβ, and so one direct response to retinoic acid is a switching on of the *RARβ* gene. Expression of *RARβ* can therefore be used as a reporter of retinoic acid signaling, and transgenic mice have been made in which the reporter gene *lacZ* (encoding β-galactosidase, see Box 1D) is hooked up to the *RARβ* retinoic-acid-response element. β-Galactosidase can then be revealed in embryos and tissues by a histochemical reaction that gives a blue color, showing which cells are responding to retinoic acid signaling. Figure 2 shows an E9.5 mouse embryo stained in this way.

Other important developmental signals that act through intracellular receptors of the same superfamily as the RAR and RXR proteins are the steroid hormones (such as estrogen and testosterone, discussed in Chapters 10 and 13) and the thyroid hormones (thyroxine and tri-iodothyroxine; discussed in Chapter 13). Like retinoic acid, these molecules are lipid-soluble; they diffuse through the plasma membrane unaided and bind to intracellular receptor proteins. The ligand-receptor complex is then able to act as a transcriptional regulator, binding directly to specific response elements in the DNA to activate (or in some cases repress) transcription of a number of genes simultaneously.

Figure 2

Photograph courtesy of Pascal Dolle from Ribes, V., et al. **Rescue of cytochrome P450 oxidoreductase (Por) mouse mutants reveals functions in vasculogenesis, brain and limb patterning linked to retinoic acid homeostasis.** Dev. Biol. *2007,* ***303***: *66–81.*

Somite formation and antero-posterior patterning

The paraxial mesoderm on either side of the notochord segments to form blocks of tissue called somites. The somites are formed in a rhythmical fashion in the anterior to posterior direction, and give rise to the bone and cartilage of the body axis, including the spinal column, to the skeletal muscles, and to the dermis of the skin on the dorsal side of the body. In the limb-forming regions, the somites supply the myogenic cells for the muscles of the limbs (see Chapter 11).

As we have seen, mesoderm cells along the **medio-lateral axis** of the somite (the axis running from the part of the somite nearest the midline of the embryo outwards) originate from different sites along the primitive streak and are brought together during gastrulation in the chick embryo (see Fig. 5.13). The cells of the medial part of the somite come from the stem zone, whereas the cells in the lateral part come from a slightly more posterior region of the primitive streak. The level of BMP-4 signaling mediates

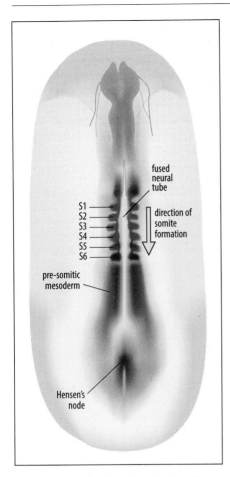

Fig. 5.21 Somites form in pairs from the paraxial mesoderm. Somites are transient blocks of tissue that form from the paraxial mesoderm, which lies immediately on either side of the notochord. They are made in an anterior to posterior direction from unsegmented pre-somitic mesoderm. The medial region of each somite is produced from a pool of stem cells around the node and the lateral region from cells from the more posterior regions of the streak. The first five somites give rise to the posterior part of the skull; the remaining somites produce the muscles and bones of the trunk and tail and the muscles of the limbs.

Scan here

Scan this QR code image with your mobile device to see an online movie showing somitogenesis or log on to **http://global.oup.com/ uk/orc/biosciences/devbiol/wolpert5e/qr/ qr5b/**

medio-lateral patterning of the mesoderm; it is high in the lateral plate mesoderm and lower in pre-somitic mesoderm. The importance of low-level BMP signaling for somite formation is shown by the fact that inhibition of BMP signaling by Noggin is sufficient to convert lateral plate mesoderm into somites.

The segmental nature of the somites imposes a segmental organization on other features of the vertebrate body such as the nervous system—the spinal nerves and ganglia of the trunk (see Chapter 9)—and the vasculature. In addition, the paraxial mesoderm is supplied with positional information by the expression of the Hox genes, and this determines its regional identity along the antero-posterior axis of the body. This regional identity results in the formation of the appropriate structures in the proper places. For example, the vertebrae that form at different positions along the spine have characteristic shapes. In this part of the chapter we will first examine the development of the somites, how their formation is controlled, and what determines the number of somites in different organisms. We will then consider how their different regional identities along the antero-posterior axis are specified by Hox gene expression. We conclude our look at somite formation by discussing how different regions within an individual somite are specified and which structures they give rise to.

5.9 Somites are formed in a well-defined order along the antero-posterior axis

Much of the work on somite formation has been done in the chick embryo because the process can be observed easily in these embryos, and so we will use the chick as our main model organism. We will, however, also discuss work in zebrafish and mouse that makes use of genetic analysis to identify the molecular elements underlying the process.

In the chick embryo, the most anterior paraxial mesoderm does not segment and gives rise to some facial muscles, including muscles associated with the eye. Somite formation occurs on either side of the notochord starting in the paraxial mesoderm adjacent to the posterior region of the hindbrain. The somites are formed anterior to the regressing Hensen's node (Fig. 5.21). The first five somites contribute to the most posterior part of the skull; the rest of the skull is of neural crest origin. These first few somites also contribute to some of the head and neck muscles. The next 12 or so somites (somites 6–19) give rise to the cervical vertebrae, somites 20–26 to the thoracic vertebrae, and the remaining 14 somites to the lumbosacral and caudal vertebrae. The process of somite formation or somitogenesis takes place simultaneously with the growth of the posterior pre-somitic mesoderm, which is very tightly linked to the proliferation of the self-renewing population of multipotent cells of the stem zone described above.

Somite formation proceeds in an anterior to posterior direction in the unsegmented mesoderm—the **pre-somitic mesoderm**—lying between the most recently formed somite and the node. As discussed in Section 5.8, in chick and mouse mesoderm cells are being added to the pre-somitic mesoderm from the stem zone as the node regresses. In this way, the length of the pre-somitic mesoderm remains more or less constant. Somites are formed in pairs, one on either side of the notochord, with the two somites in a pair forming simultaneously. A pair of somites forms every 90 minutes in the chick, every 120 minutes in the mouse, every 45 minutes in *Xenopus*, every 30 minutes in zebrafish, and every 6 hours in humans.

The sculpting of the somites from the slabs of pre-somitic mesoderm involves tissue separation and cell movements, as well as integration of cells at the anterior and posterior somite borders. As is often the case with developmental processes, different organisms achieve this through different means, even if they use the same molecular tools. The formation of the somite unit from the pre-somitic mesoderm requires a **mesenchymal-to-epithelial transition**, in which the relatively loosely connected mesenchymal mesoderm cells adhere tightly to each other to form an epithelial structure.

In *Xenopus*, segmentation is achieved through the formation of a fissure (the intersomitic furrow) in the pre-somitic mesoderm that separates off a block of cells of the correct size which then undergoes a 90° rotation. In zebrafish, the furrow expands in the medial to lateral direction, and in chicken and mouse there is no furrow, and somite formation seems to involve a reshuffling of cells that leads to a 'ball-and-socket' kind of separation (Fig. 5.22). Notch signaling from the most anterior region of the pre-somitic mesoderm seems to be important for the separation event. The mesenchymal-to-epithelial transition involves coordinated changes in the cell's cytoskeleton, in the adherens junctions that bind cells together (discussed in Chapter 9), and in the extracellular matrix, and it is likely that the forces generated in the transition also contribute to the separation.

The precision with which each somite is cut out of the pre-somitic mesoderm suggests that the number of cells in each somite, and thus the size of each segment, is specified in the pre-somitic mesoderm. Experiments in the chick embryo show that this is the case, as the sequence of somite formation in the unsegmented region is unaffected by transverse cuts in the plate of pre-somitic mesoderm. This suggests that somite formation is an autonomous process and that, at this time, no extracellular signal specifying antero-posterior position or timing is involved. Even if a piece of the unsegmented mesoderm is rotated through 180°, each somite still forms at the normal time, but with the sequence of formation running in the opposite direction to normal in the inverted tissue (Fig. 5.23). This shows that, before the somites begin to form, a molecular pattern that specifies the time of formation of each somite has already been laid down in the pre-somitic mesoderm; we shall return to this patterning process later. Given the existence of this pattern, the prospective identity of each somite must also be related to the temporal order in which their cells entered the pre-somitic mesoderm. The commitment to a specific sequence of somite formation occurs at a particular level in the pre-somitic mesoderm. If a piece of mesoderm posterior to that level is rotated, the rotated region will regulate to produce somites in the normal sequence.

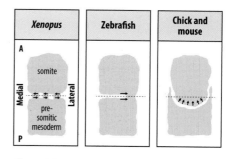

Fig. 5.22 Formation of somites in Xenopus, zebrafish, and chick. The arrows depict the direction of separation of the tissue as the somites form. The dotted line represents the presumptive somite border.

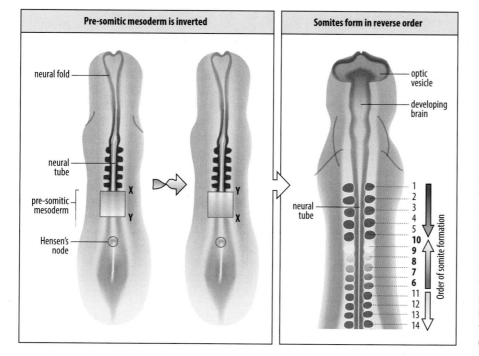

Fig. 5.23 The temporal order of somite formation is specified early in embryonic development. Somite formation in the chick proceeds in an antero-posterior direction. Somites form sequentially in the pre-somitic region between the last-formed somite and Hensen's node, which moves posteriorly. If the antero-posterior axis of the pre-somitic mesoderm is rotated through 180°, as shown by the arrow, the temporal order of somite formation is not altered—somite 6 still develops before somite 10.

Developmental biologists' view of body segmentation in vertebrates has been largely guided by a very successful model, the **clock and wavefront model**. This model was proposed before the advent of molecular biology, to account for the observed periodic and spatially organized appearance of the somites. It is a tissue-level model and proposes that the size of a somite is defined as the distance that a wave of determination (the 'wavefront') travels from anterior to posterior during one oscillation of a 'clock' that operates in the mesoderm (Fig. 5.24). The observation of oscillatory expression of the gene *hairy-1* during somitogenesis in chicken embryos provided the first molecular evidence for a clock. The periodicity of the oscillations equals the time it takes for a pair of somites to form (for example, 90 minutes in the chick), and thus these oscillations in gene expression were suggested to be the molecular clock that drives segmentation. Subsequently, the expression of many other genes in the pre-somitic mesoderm has been found to oscillate.

The clock and wavefront model proposes that when cells in the pre-somitic mesoderm that have been passed by the determination wavefront receive a signal from the clock, in the form of oscillating gene expression, this commits them to form a somite (see Fig. 5.24).

The position of the wavefront is specified by the levels of FGF and Wnt signaling, which are both graded in the pre-somitic mesoderm from posterior to anterior (see Section 5.8 and Fig. 5.24). Somite formation occurs where FGF/Wnt signaling drops below a threshold level. As the FGF/Wnt gradient moves posteriorly along with the growth of the pre-somitic mesoderm and the node, it therefore successively defines regions in the pre-somitic mesoderm where somites will segment. The positions at which these somites will form are specified by the passage of the wavefront some

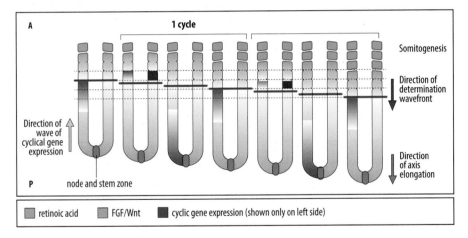

Fig. 5.24 The clock and wavefront model of somite formation. Two cycles of somite specification are illustrated. Retinoic acid (green) produced by the somites forms a gradient extending from anterior (A) to posterior (P) in the mesoderm and meets a gradient of FGF/Wnt signaling (peach) running from posterior to anterior. A wavefront of somite determination (red), specified by a threshold level of FGF/Wnt signaling, advances through the mesoderm from anterior to posterior as the mesoderm elongates. At the same time, cycles of gene expression (purple, the sequential expression of one gene is illustrated) sweep through the mesoderm from posterior to anterior. Cycling gene expression is shown only on the left-hand side of the mesoderm for clarity. During each cycle a pair of new somites is specified (black box, shown on the right-hand side only for clarity) in mesoderm that has experienced both the determination signal and expression of the cycling gene. Gene expression characteristic of a somite 'prepattern' is switched on, and somite segmentation and differentiation (somitogenesis) eventually ensues. Somites specified in previous cycles, but which have not yet differentiated, are indicated in the anterior mesoderm (shown in more detail in Fig. 5.25).

*From Gomez, C., Pourquié, O.: **Developmental control of segment numbers in vertebrates.** J. Exp. Zool. B Mol. Dev. Evol. 2009, **312**: 533–544.*

time before segmentation actually occurs (Fig. 5.25). The gradient of retinoic acid that runs in the opposite direction has an antagonistic effect on FGF action and prevents the pre-somitic region from continually getting longer.

The clock and wavefront model provides an explanation for the results of rotating pieces of presomitic mesoderm from different levels discussed earlier. If the rotated piece comes from pre-somitic mesoderm that the determination wavefront has already passed over, the somite that forms will be rotated relative to the antero-posterior axis (see Fig. 5.23); if it comes from a region that the wavefront has not yet reached, it has a chance to regulate.

There is now a large list of genes whose expression oscillates during somitogenesis; most of them are associated with Wnt, Notch, and FGF signaling. Their pattern of expression is always the same: a wave that sweeps through the tissue from the posterior of the pre-somitic mesoderm towards the anterior, where the somites are being formed (see Fig. 5.24). It is important to appreciate that what travels is the wave of gene expression, not the cells expressing the genes. In other words, in each cycle of gene expression, each cell in the pre-somitic mesoderm expresses a given gene for a given period of time and then stops.

Gene expression is initially cell-autonomous and is synchronized locally by inter-cellular communication in some way, so that the wave of gene expression moves anteriorly and also so that the cells in the specified somite are all in the same state and ready to enter differentiation together. The wave of gene expression meets the determination wavefront and then stops at a boundary where the somites are beginning to form. Gene expression characteristic of a somite 'prepattern' is switched on, and segmentation and differentiation eventually ensue. The cyclical gene expression becomes restricted to the anterior half of each newly forming somite, where it persists for a time before fading away. The cycle then starts again, with a new wave of expression being initiated in the posterior pre-somitic mesoderm.

Apart from the coincidence of timing and the spatially restricted events, the idea that the gene oscillations are linked to the formation of somites is derived from the observation that mutations in oscillating genes, for example *Lunatic fringe* and *Delta-like 1* (both involved in Notch signaling, see Box 5D) in the mouse, and the corresponding zebrafish homologs, result in segmentation defects. Other Notch pathway genes, such as *Notch1* (which encodes the Notch receptor protein itself), are expressed in the pre-somitic mesoderm in a non-cyclical pattern, although the formation of a cleaved Notch intracellular domain, the hallmark of Notch signaling, has been visualized experimentally as a traveling wave, indicating cyclical Notch signaling. Using these observations, an oscillator can be built from a simple autoregulatory loop of gene expression with a time delay that is crucial to generate the oscillation, and it is therefore possible to see how these genes could create the basic oscillators that drive the timing of segmentation. However, where and how the oscillations can be transmitted from single cells to the tissue level at which the process of segmentation occurs is still under investigation.

There is evidence from zebrafish and mice that the all-important step of synchronization of gene oscillations between neighboring cells is dependent on Delta–Notch signaling acting on the oscillating genes. Mutants of the Notch signaling pathway exhibit disruptions of the segmentation process associated with a loss of synchrony, rather than with the loss of oscillations at the level of single cells. In mouse embryos in which Delta–Notch signaling is either made constitutively active or abolished entirely, somites rarely form, and if they do, they vary in size and shape and are different on each side of the body. In addition, experimentally controlled Notch signaling has shown a clear correlation between Notch signaling and the periodic patterning and emergence of somites.

However, which of the clock genes control the periodicity of the clock is unclear; there could be considerable redundancy between different components of the clock,

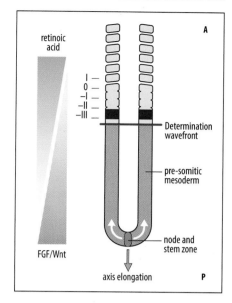

Fig. 5.25 Somites are specified some time before they differentiate and segment from the mesoderm. The most recently specified pair of somites is indicated by the black boxes. I, the most recently formed somite; 0, somite in the process of formation; -I, -II, -III, blocks of cells that have been specified and will form somites during successive cycles. As described in Fig. 5.24, the position of somite formation is specified as a result of threshold levels of of FGF/Wnt signaling (peach), and retinoic acid signaling (green) which form opposing gradients in the mesoderm. A, anterior; P, posterior.

*From Gomez, C., Pourquié, O.: **Developmental control of segment numbers in vertebrates.** J. Exp. Zool. B Mol. Dev. Evol. 2009, **312**: 533-544.*

CELL BIOLOGY BOX 5D The Notch signaling pathway

The family of Notch transmembrane proteins act as receptors and initiate the Notch intracellular signaling pathway. This pathway is involved in numerous developmental decisions that determine cell fate, ranging from neuronal precursors in *Drosophila* and vertebrates to the differentiation of epidermal cells in mammalian skin. The ligands that bind to Notch are also membrane-bound proteins, and so Notch signaling requires direct cell-cell contact. This means that the Notch signaling pathway can be activated in a single cell out of a population of several cells, and that this cell can then have a different fate from surrounding cells. We shall see examples of this when we look at the development of the *Drosophila* eye in Chapter 11 and neuroblast specification in Chapter 12.

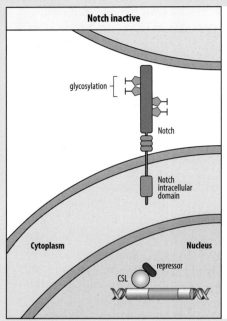
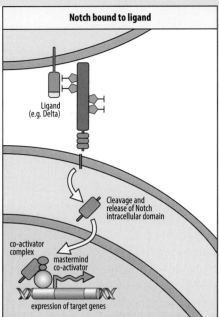

Figure 1

The ligands that bind to Notch are the members of the Delta and Serrate/Jagged1 families. Binding of ligand on one cell to Notch on an adjacent cell initiates intracellular signaling from the receptor. Notch has a large extracellular domain with several glycosylation sites that are specifically modified by glycosyltransferases, including those of the Fringe family, which includes Lunatic Fringe (Figure 1). Glycosylation modulates Notch signaling by affecting the ability of the extracellular domain to bind to different ligands, facilitating its binding to Delta ligands while reducing its ability to bind to Serrate ligands.

Activation of Notch by its ligand causes the enzymatic cleavage of the cytoplasmic tail of the receptor (the Notch intracellular domain) and its translocation to the nucleus to bind and activate a transcription factor of the CSL family (CSL is named after the Notch-activated transcription factors CBF1/RBPkJ in mammals, Suppressor of Hairless (Su(H)) in *Drosophila* and LAG-1 in *C. elegans*). In the absence of Notch signaling, CSL is held in a complex with repressor proteins. Activation of the CSL complex by binding of the Notch intracellular domain involves release of the repressors and recruitment of the co-activator protein Mastermind and other co-activators. The outcome of the Notch signaling pathway is the transcription of specific genes. Notch signaling is very versatile. In different organisms and in different developmental circumstances, activation of the Notch pathway is used to switch a wide range of different genes on or off.

Scan here

Scan this QR code image with your mobile device to see an online animation of the Notch signaling pathway or log on to **http://global.oup.com/uk/orc/biosciences/devbiol/wolpert5e/qr/qr5a/**

and information from different vertebrates is sometimes contradictory. It is also not known how the somite clock is started and how it is stopped.

Notch signalling is also involved in determining a boundary within the somite. Somites are functionally divided into anterior and posterior halves with different fates, and in mice, Notch signaling seems to be required downstream of its possible clock function to establish this anterior–posterior boundary.

An important developmental question is what determines the final number of somites. Somite number is highly variable between different vertebrates: birds and humans have around 50 whereas snakes have up to several hundred. One answer could be that the total number of somites is determined by the length of pre-somitic mesoderm available to form somites. The length of the pre-somitic mesoderm will depend on the extent of axis elongation, and this varies between embryos with small numbers of somites and embryos with large numbers.

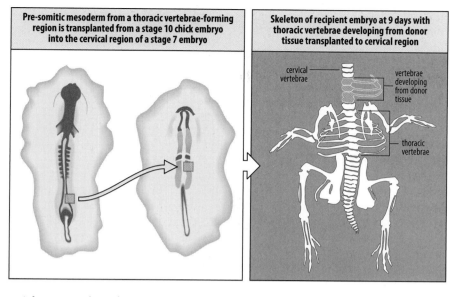

| Pre-somitic mesoderm from a thoracic vertebrae-forming region is transplanted from a stage 10 chick embryo into the cervical region of a stage 7 embryo | Skeleton of recipient embryo at 9 days with thoracic vertebrae developing from donor tissue transplanted to cervical region |

cervical vertebrae

vertebrae developing from donor tissue

thoracic vertebrae

Fig. 5.26 The pre-somitic mesoderm has acquired a positional value before somite formation. Pre-somitic mesoderm that will give rise to thoracic vertebrae is grafted to an anterior region of a younger embryo that would normally develop into cervical vertebrae. The grafted mesoderm develops according to its original position and forms ribs in the cervical region.

A larger number of somites could also be generated from the same length of pre-somitic mesoderm if each of the somites was smaller. This could be accomplished by making the clock go faster so that more, and smaller, segments form in a given time period from a length of pre-somitic mesoderm. Alternatively, the pace of the clock could be kept the same and the advance of the determination wave slowed down, so that more oscillations of the clock take place in a length of pre-somitic mesoderm before the cells become committed to form a segment.

Experiments in chick embryos have shown that transiently increasing FGF signaling in the pre-somitic mesoderm, which would transiently change the shape of the FGF gradient and decelerate the speed of advance of the determination wavefront, does indeed lead to the formation of smaller somites. Corn snakes have 260 somites, which are each a third of the size of chick or mouse somites, and this seems to be due to acceleration of the pace of the clock and thus a shortening of the time allowed for one somite to form.

Somites differentiate into particular axial structures depending on their position along the antero-posterior axis. The easiest somite-derived structures to detect are the vertebrae. The anterior-most somites contribute to the most posterior region of the skull, those posterior to them form cervical vertebrae, and those more posterior still develop as thoracic vertebrae with ribs. The most posterior somites develop into caudal vertebrae. Not only does somite number vary between different vertebrate embryos but also the number of somites with a particular regional identity. A goose, for example, has a larger number of cervical vertebrae than a chicken. Positional values that lead to acquisition of regional identity are specified before somite formation itself begins; for example, if unsegmented somitic mesoderm from the presumptive thoracic region of the chick embryo is grafted in place the presumptive mesoderm of the neck region, it will still form thoracic vertebrae with ribs (Fig. 5.26). How then is the pre-somitic mesoderm patterned so that somites acquire their identity and form particular vertebrae?

5.10 Identity of somites along the antero-posterior axis is specified by Hox gene expression

The antero-posterior patterning of the mesoderm is most clearly seen in the differences in the vertebrae, each individual vertebra having well defined anatomical characteristics depending on its location along the axis. The most anterior vertebrae are specialized for attachment and articulation of the skull, while the cervical vertebrae of the neck are followed by the rib-bearing thoracic vertebrae and then those of the lumbar region, which do not bear ribs, and finally, those of the sacral and caudal regions. Patterning of the skeleton along the body axis is based on the mesodermal cells acquiring a positional value that reflects their position along the axis and so determines their subsequent development. Mesodermal cells that will form thoracic vertebrae, for example, have different positional values from those that will form cervical vertebrae. The concept of positional value has important implications for developmental strategy; it implies that a cell or a group of cells in the embryo acquires a unique state related to its position at a given time. This state is then interpreted to specify regional identity and this then determines later development (see Section 1.15).

Patterning along the antero-posterior axis in all animals involves the expression of genes that specify regional identity along the axis. Homeotic genes were originally identified in *Drosophila* (see Chapter 2) where they specify segment identity along the antero-posterior axis, and it was later discovered that the related Hox genes are involved in patterning the antero-posterior axis in vertebrates. The Hox genes are members of the large family of homeobox genes that are involved in many aspects of development (Box 5E). As we shall see later, patterning of vertebrate embryos along the antero-posterior axis by Hox genes and other homeobox genes is not confined to mesodermal structures; in the hindbrain, for example, there are distinct regions that have characteristic patterns of Hox gene expression.

The Hox genes are the most striking example of a widespread conservation of developmental genes in animals. Although it was widely believed that there are common mechanisms underlying the development of all animals, it seemed at first unlikely that genes involved in specifying the identity of the segments of *Drosophila* would specify identity of the somites in vertebrates. However, the strategy of comparing genes by sequence homology proved extremely successful in identifying the related vertebrate genes. Numerous other genes first identified in *Drosophila*, in which the genetic basis of development is better understood than in any other animal, have also proved to have counterparts involved in development in vertebrates.

Most vertebrates have four separate clusters of Hox genes—the **Hox clusters** Hoxa, Hoxb, Hoxc, and Hoxd—that are thought to have arisen by duplications of the gene clusters themselves during vertebrate evolution. Some of the Hox genes within each cluster may also have arisen by duplication (see Box 5E). The zebrafish is unusual in having seven Hox gene clusters, as a result of further duplication. A particular feature of Hox gene expression in both insects and vertebrates is that the genes in the clusters are expressed in a temporal and spatial order that reflects their order on the chromosome. The Hox clusters are the only known case where a spatial arrangement of genes on a chromosome corresponds to a spatial pattern of expression in the embryo.

A simple idealized model illustrates the key features by which a Hox gene cluster could record regional identity by a combinatorial mechanism. Consider four genes—I, II, III, and IV—arranged along a chromosome in that order (Fig. 5.27). The genes are expressed in a corresponding order along the antero-posterior axis of the body. Thus, gene I is expressed throughout the body with its anterior boundary at the head end. Gene II has its anterior boundary in a more posterior position and expression continues posteriorly. The same principles apply to the two other genes. This pattern of expression defines four distinct regions, each expressing a different combination of genes. If the amount of gene product is varied within each expression domain, for example by interactions between the genes, many more regions can be specified.

The function of the Hox genes in vertebrate axial patterning has been best studied in the mouse, because it is possible to either knock out particular Hox genes or alter their expression in transgenic mice (see Section 3.10 and Box 3D). As in all vertebrates, the Hox genes start to be expressed in mesoderm cells of early mouse embryos during gastrulation when these cells begin to move internally away from the primitive streak

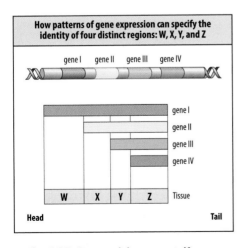

Fig. 5.27 Gene activity can specify regional identity. The model shows how the pattern of gene expression along the body can specify the distinct regions W, X, Y, and Z. For example, only gene I is expressed in region W, but all four genes are expressed in region Z.

BOX 5E The Hox genes

The vertebrate Hox genes encode proteins belonging to a large group of gene-regulatory proteins that all contain a similar DNA-binding region of around 60 amino acids. This region, known as the homeodomain, contains a helix-turn-helix DNA-binding motif that is characteristic of many DNA-binding proteins. The homeodomain is encoded by a DNA motif of around 180 base pairs termed the homeobox, a name that came originally from the fact that this gene family was discovered through mutations in *Drosophila*, that produce a homeotic transformation—a mutation in which one structure replaces another. For example, in one homeotic mutant in *Drosophila*, a segment in the fly's body that does not normally bear wings is transformed to resemble the adjacent segment that does bear wings, resulting in a fly with four wings instead of two (see Fig. 2.50).

The Hox genes in *Drosophila* are organized into two distinct gene clusters or gene complexes—the Antennapedia complex and the bithorax complex (known collectively as the HOM-C complex). In vertebrates, including humans, the Hox genes are organized into four clusters, known

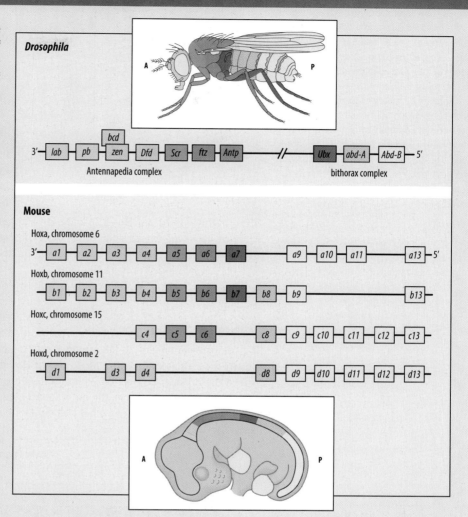

Figure 1

as the Hox clusters or the Hox complexes. (Zebrafish have seven Hox clusters rather than four as a result of a partial duplication of the whole genome in their evolutionary history). The homeoboxes of the genes in each Hox cluster are related in DNA sequence to the homeoboxes of genes of the HOM-C complex in *Drosophila*. In each Hox cluster, the order of the genes from 3' to 5' in the DNA is the order in which they are expressed along the antero-posterior axis and specify regional identity. In the mouse, the four Hox clusters, designated Hoxa, Hoxb, Hoxc, and Hoxd (originally called Hox1, Hox2, Hox3, and Hox4), are located on chromosomes 6, 11, 15, and 2, respectively (see Figure 1; in the human, Hoxa, Hoxb, Hoxc and Hoxd clusters are located on chromosomes 7, 17, 12, and 2). All Hox genes resemble each other to some extent; the homology is most marked within the homeobox and less marked in sequences outside it. The four Hox gene clusters are thought to have arisen by duplication of an ancestral gene cluster. The corresponding genes in the different

clusters (for example, *Hoxa4, Hoxb4, Hoxc4, Hoxd4*) are known as **paralogous genes**, and collectively as a **paralogous gene subgroup**. In the mouse and human there are 13 paralogous gene subgroups.

The Hox genes and their role in development are of ancient origin. The mouse and frog genes are similar to each other and to those of *Drosophila*, both in their coding sequences and in their order on the chromosome. In both *Drosophila* and vertebrates, these homeotic genes are involved in specifying regional identity along the antero-posterior axis. Many other genes contain a homeobox but they do not, however, belong to a homeotic complex, nor are they involved in homeotic transformations. Other subfamilies of homeobox genes in vertebrates include the **Pax genes**, which contain a homeobox typical of the *Drosophila* gene *paired*. All these genes encode transcription factors with various functions in development and cell differentiation.

Fig. 5.28 Hox gene expression in the mouse embryo after neurulation. The three panels show lateral views of 9.5 dpc mouse embryos stained for the product of a *lacZ* gene reporter (blue) linked to promoters for the *Hoxb1*, *Hoxb4*, and *Hoxb9* genes. These genes are expressed in the neural tube and the mesoderm. The arrowheads indicate the anterior boundary of expression of each gene within the neural tube. Note that the anterior boundary of expression in the mesoderm is not the same as in the neural tube. The position of the three genes within the Hoxb gene cluster is indicated (inset). Scale bar = 0.5 mm.

Photographs courtesy of A. Gould.

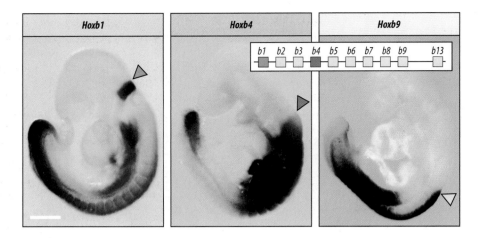

and towards the anterior. The 'anterior' Hox genes are expressed first. More 'posterior' Hox genes are expressed as gastrulation proceeds and clearly defined patterns of Hox gene expression are most easily seen in the mesoderm after somite formation and in the neural tube after neurulation.

Figure 5.28 shows the expression patterns of three Hox genes in mouse embryos at day E9.5. Typically, the expression of each gene is characterized by a sharp anterior border and extends posteriorly. Although there is considerable overlap in the spatial pattern of expression of Hox genes, particularly of neighboring genes in a cluster, almost every region in the mesoderm along the antero-posterior axis is characterized by a specific combination of expressed Hox genes, as shown by a summary picture of the anterior borders of expression of Hox genes in mouse embryonic mesoderm (Fig. 5.29). For example, the most anterior somites are characterized by expression of *Hoxa1* and *Hoxb1* and no other Hox genes are expressed in this region. By contrast, all

Fig. 5.29 A summary of Hox gene expression along the antero-posterior axis of the mouse mesoderm. The anterior border of expression of each gene is shown by the red blocks. Expression usually extends backwards some distance but the posterior margin of expression may be poorly defined. The pattern of Hox gene expression specifies the regional identity of the tissues at different positions. This figure represents an overall picture of Hox gene expression and is not a snapshot of gene expression at any particular time.

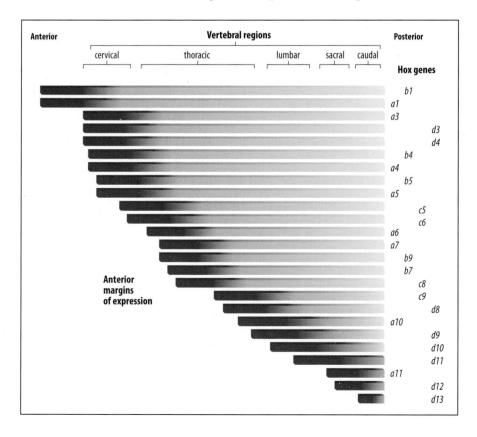

Hox genes are expressed in the more posterior regions. The Hox genes can therefore provide a code for regional identity. The spatial and temporal order of expression is similar in the mesoderm and the ectodermally derived nervous system, but the positions of the boundaries between regions of gene expression in the two germ layers do not always correspond (see Fig. 5.28). The most anterior region of the body in which Hox genes are expressed is the hindbrain; other homeobox genes such as *Emx* and *Otx*, but not Hox genes, are expressed in more anterior regions of the vertebrate body—midbrain, forebrain, and the anterior part of the head (see Chapter 12).

If we focus on just the Hoxa cluster of genes, we see that the most anterior border of expression in the mesoderm is that of *Hoxa1* in the posterior head mesoderm, while *Hoxa11*, a very posterior gene in the Hoxa cluster, has its anterior border of expression in the sacral region (see Fig. 5.29). This exceptional correspondence, or **co-linearity**, between the order of the genes on the chromosome and their order of spatial and temporal expression along the antero-posterior axis is typical of all the Hox clusters. The genes of each Hox cluster are expressed in an orderly sequence, with the gene lying most 3′ in the cluster being expressed the earliest and in the most anterior position. The correct expression of the Hox genes is dependent on their position within the cluster, and 'anterior' genes (3′) are expressed before more 'posterior' (5′) genes. This typical sequential activation of Hox genes in the order they occur along the chromosome is likely to be due, at least in part, to changes in chromatin status (see Box 8A).

Support for the idea that the Hox genes are involved in specifying regional identity comes from comparing their patterns of expression in mouse and chick with the well-defined anatomical regions—cervical, thoracic, sacral, and lumbar (Fig. 5.30). Hox gene expression corresponds well with the different regions. For example, even though the number of cervical vertebrae in birds (14) is twice that of mammals, the anterior boundaries of *Hoxc5* and *Hoxc6* gene expression in the somites in both chick

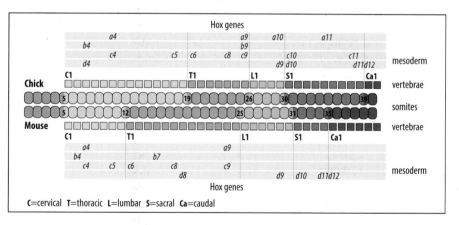

Fig. 5.30 Patterns of Hox gene expression in the mesoderm of chick and mouse embryos, and their relation to regionalization. The pattern of Hox gene expression varies along the antero-posterior axis. The vertebrae (squares) of the spinal column are derived from the embryonic somites (circles), 40 of which are shown here. The vertebrae have characteristic shapes in each of the five regions along the antero-posterior axis: cervical (C), thoracic (T), lumbar (L), sacral (S), and caudal (Ca). Which somites form which vertebrae differs in chick and mouse. For example, thoracic vertebrae start at somite 19 in the chick, but at somite 12 in the mouse. Despite this difference, the transition from one region to another corresponds to a similar pattern of *Hox* gene expression in chick and mouse, such that the anterior boundaries of expression of *Hoxc5* and *Hoxc6* lie on either side of the transition from cervical to thoracic vertebrae (light blue to mid blue) in both species. Similarly, the anterior boundaries of *Hoxd9* and *Hoxd10* are at the transition between the lumbar and sacral regions (green to dark blue) in both chick and mouse.

*Illustration after Burke, A.C., et al.: **Hox genes and the evolution of vertebrate axial morphology.** Development 1995, **121**: 333–346.*

and mouse lie on either side of the cervical/thoracic boundary. A correspondence between Hox gene expression and regional identity is also similarly conserved among vertebrates at other anatomical boundaries.

It must be emphasized that the summary picture of Hox gene expression given in Fig. 5.29 does not represent a 'snapshot' of expression at a particular time but rather an integrated picture of the overall pattern of expression. Some genes are switched on early and expression is then reduced, whereas others are switched on considerably later; the most 'posterior' Hox genes, such as *Hoxd12* and *Hoxd13*, are expressed in the tail. Moreover, this summary picture reflects only the general expression of the genes in any particular region; not all Hox genes expressed in a region are expressed in all the cells. Nevertheless, the overall pattern suggests that the combination of Hox genes expressed specifies regional identity. In the cervical region, for example, each somite, and thus each vertebra, could be specified by a unique pattern of Hox gene expression or **Hox code**.

As we saw in Fig. 5.26, grafting experiments show that the character of the somites is already determined in the pre-somitic mesoderm, and that pre-somitic tissue transplanted to other levels along the axis retains its original identity. As would be expected if the Hox code specifies vertebral identity, the transplanted pre-somitic mesoderm also retains its original pattern of Hox gene expression.

5.11 Deletion or overexpression of Hox genes causes changes in axial patterning

If the Hox genes specify regional identity that determines a region's subsequent development, then one would expect morphological changes if their pattern of expression is altered. This is indeed the case. To investigate the function of Hox genes, their expression can be abolished by mutation, or they can be expressed in abnormal positions. Hox gene expression can be eliminated from the developing mouse embryo by gene knock-out (see Section 3.10 and Box 3D). Experiments along these lines indicate that the absence of given Hox genes affects patterning in a way that agrees with the idea that they are normally providing cells with regional identity. In some cases, however, knocking out an individual Hox gene produces little effect, or affects only one tissue, and more substantial changes in patterning are only seen when two or more genes, particularly **paralogous genes** from different Hox clusters (see Box 5E), are knocked out together.

Mice in which *Hoxd3* is deleted show structural defects in the first and second cervical vertebra, where this gene is normally strongly expressed (see Fig. 5.29), and in the sternum (breastbone). More posterior structures, where the inactivated gene is also normally expressed together with other Hox genes, show no evident defects, however. This observation illustrates a general principle of Hox gene expression, which is that more posteriorly expressed Hox genes tend to over-ride the function of more anterior Hox genes whenever they are co-expressed. This phenomenon is known as **posterior dominance** or **posterior prevalence**. It has been proposed that the action of microRNAs (miRNAs) encoded by genes within Hox clusters might explain posterior prevalence. The encoded miRNAs in the Hox clusters are likely to be involved in post-transcriptional regulation of Hox protein expression (see Box 6C, for how miRNAs work). More of the predicted target genes of a given miRNA tend to lie on the 3' side of the miRNA gene rather than the 5' side. Given this arrangement, the Hox miRNAs might help to repress the function of more anterior Hox genes and thus explain the general principle of posterior prevalence.

Another general point illustrated by the *Hoxd3* knock-out is that the effects of a Hox gene knock-out can be tissue specific, in that only certain tissues in which the Hox gene is normally expressed are affected, whereas other tissues at the same position along the antero-posterior axis appear normal. For example, although *Hoxd3* is expressed in neural tube, branchial arches, and paraxial mesoderm at the same antero-posterior level in the body, only the vertebrae are defective in a *Hoxd3* knock-out.

The apparent absence of an effect in such cases may be due to redundancy, with paralogous genes from another cluster being able to compensate. *Hoxa3* is expressed strongly in the same tissues as *Hoxd3* in the cervical region, but when *Hoxa3* alone is knocked out there are no defects in cervical vertebrae; instead other defects are observed, including reductions in cartilage elements derived from the second branchial arch. When both *Hoxd3* and *Hoxa3* are knocked out together, much more severe defects are seen in both vertebrae and branchial arch derivatives, suggesting that the two genes normally function together in these tissues.

Loss of Hox gene function often results in **homeotic transformation**—the conversion of one body part into another, as seen most spectacularly in *Drosophila* (see Fig. 2.50). This is the case in the *Hoxd3* knock-out mouse we have just considered. Detailed analysis of the cervical region in these mice revealed that the atlas vertebra (the first cervical vertebra) appears to be transformed into the basal occipital bone at the base of the skull, and that the axis vertebra (the second cervical vertebra) has acquired morphological features resembling the atlas. Thus, in the absence of *Hoxd3* expression, cells acquire a more anterior regional identity and develop into more anterior structures. The double knock-out of *Hoxa3* and *Hoxd3* results in a deletion of the atlas. The complete absence of this bone in these double knock-outs suggests that one target of Hox gene action is the cell proliferation required to build such a structure from the somite cells. Unfortunately, rather few direct gene targets transcriptionally regulated by Hox proteins have yet been identified.

Another example of homeotic transformation is seen in *Hoxc8*-knock-out mice. In normal embryos, *Hoxc8* is expressed in the thoracic and more posterior regions of the embryo from late gastrulation onward (see Fig. 5.29). Mice homozygous for deletion of *Hoxc8* die within a few days of birth, and have abnormalities in patterning between the seventh thoracic vertebra and the first lumbar vertebra. The most obvious homeotic transformations are the attachment of an 8th pair of ribs to the sternum and the development of a 14th pair of ribs on the first lumbar vertebra (Fig. 5.31). Thus, as with the knock-outs discussed above, the absence of *Hoxc8* has led to some of the cells that would normally express it acquiring a more anterior regional identity, and developing accordingly.

Some homeotic transformations only occur if all the paralogous genes are knocked out together. In mice, in the absence of all the Hox10 paralogous subgroup genes, there are no lumbar vertebrae and there are ribs on all posterior vertebrae; in the absence of all the Hox11 paralogous subgroup genes several sacral vertebrae become lumbar. These homeotic transformations do not occur if only some members of the paralogous subgroup are mutated, again suggesting redundancy of Hox gene function.

In contrast to the effects of knocking out Hox genes, the ectopic expression of Hox genes in anterior regions that normally do not express them can result in transformations of anterior structures into structures that normally develop in a more posterior position. For example, when *Hoxa7*, whose normal anterior border of expression is in the thoracic region (see Fig. 5.29), is expressed throughout the whole antero-posterior axis, the basal occipital bone of the skull is transformed into a pro-atlas structure, normally the next more posterior skeletal structure.

5.12 Hox gene expression is activated in an anterior to posterior pattern

In all vertebrates, the Hox genes begin to be expressed at an early stage of gastrulation, when mesodermal cells begin their gastrulation movements. The genes expressed most anteriorly, which correspond to those at the 3′ end of a cluster, are expressed first. If an 'early' *Hoxd* gene is relocated to the 5′ end of the *Hoxd* cluster, for example, its spatial and temporal expression pattern then resembles that of the neighboring *Hoxd13* (see Box 5E). This shows that the structure of the Hox cluster is crucial in determining the pattern of Hox gene expression.

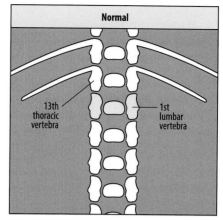

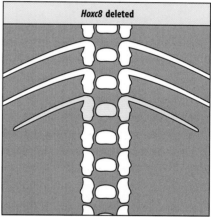

Fig. 5.31 Homeotic transformation of vertebrae due to deletion of Hoxc8 in the mouse. In loss-of-function homozygous mutants of *Hoxc8*, the first lumbar vertebra (yellow) is transformed into a rib-bearing 'thoracic' vertebra. The mutation has resulted in the transformation of the lumbar vertebra into a more anterior structure.

Unlike the situation in *Drosophila*, where the activation of Hox genes clearly depends on factors unequally distributed along the antero-posterior axis (see Chapter 2), the mechanism of Hox gene activation in vertebrates is more complex and less well understood. One way the antero-posterior pattern of Hox gene expression might be established in the somitic mesoderm is through linking gene activation to the time at which cells undergo gastrulation movements. In the chick, Hox gene expression is activated in epiblast cells that will form the lateral region of the somite before they ingress through the posterior streak and follows a strict temporal order, the 3′ genes being activated before the 5′ ones. Expression patterns of Hox genes are maintained when the cells leave to form the pre-somitic mesoderm, and in this way the temporal pattern of Hox gene activation can be converted into regional identity; the cells expressing different Hox genes are generated progressively along the antero-posterior axis as structures are laid down in an antero-posterior sequence. This hypothesis is similar to one proposed for specifying position along the proximo-distal axis of the vertebrate limb bud, in which Hox gene expression is also involved (see Chapter 11).

Retinoic acid signaling may also be involved in activating expression of Hox genes in the early embryo. The gene for the retinoic-acid metabolizing enzyme, Cyp26A1, is expressed in the anterior half of the mouse embryo during gastrulation stages while the gene for the retinoic-acid generating enzyme Ralhd2 is expressed in the posterior half. This leads to graded retinoic signaling from low levels anteriorly to high levels posteriorly. Later on in development, as already discussed in Section 5.8, the gene for Cyp26A1 is expressed in the posterior region of the elongating body axis as the somites are being generated, and the retinoic acid gradient then runs from anterior to posterior.

In mice, retinoic acid has been shown experimentally to alter the expression of Hox genes. More recently, genes encoding retinoic acid receptors and other proteins of the retinoid signaling pathway have been knocked out in mouse embryos, resulting in homeotic transformations of the axial skeleton. Mouse embryos lacking the gene *Cyp26a1* not only have a truncated axis, but also have posterior homeotic transformation, with each vertebra taking on the identity of its neighboring more posterior vertebra—for example, the last cervical vertebra is transformed into a vertebra carrying ribs. These transformations occur in regions where retinoic acid signaling is higher than normal and where anterior expansion of *Hoxb4* expression has also been seen. A role for retinoic acid in Hox gene activation is supported by studies in zebrafish. The zebrafish mutation *giraffe*, which affects the *Cyp26* gene, has anteriorly extended Hox gene expression and vertebral transformations.

5.13 The fate of somite cells is determined by signals from the adjacent tissues

We shall now return to the individual somite and consider how it is patterned along its antero-posterior, dorso-ventral, and medio-lateral axes. The differences between the anterior and posterior halves of the somite are later used by neural crest cells and motor nerves as guidance cues to generate the periodic arrangement of spinal nerves and ganglia (see Chapter 9). The subdivision is also important for vertebra formation, as each vertebra derives from the anterior half of one somite and the posterior half of the preceding one, a process known as 'resegmentation'.

The antero-posterior patterning process within an individual somite is quite independent of the antero-posterior patterning of the whole pre-somitic mesoderm by Hox gene expression. Individual somites are subdivided into anterior and posterior halves with different properties, with this subdivision being linked to the segmentation clock. In the mouse, the transcription factor Mesp2 is expressed in a somite-wide stripe in the pre-somitic mesoderm that will give rise to the next somite (S0), and its expression then shrinks to just the anterior part of the somite (Fig. 5.32). Mesp2 activates a cascade of gene expression that leads to its expression being maintained

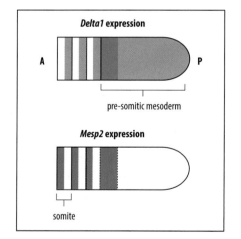

Fig. 5.32 Division of somites into anterior and posterior halves. Positive and negative feedback loops between Delta1 and Mesp2 (lower panel) are essential for dividing the somite into an anterior half expressing Mesp2 and a posterior half expressing Delta1 (upper panel), shown in dorsal views of the tail regions of E9.5 mouse embryos. Mesp2 is initially expressed in a somite-wide stripe in pre-somitic mesoderm and then as the somite forms, Mesp expression is maintained in the anterior half.

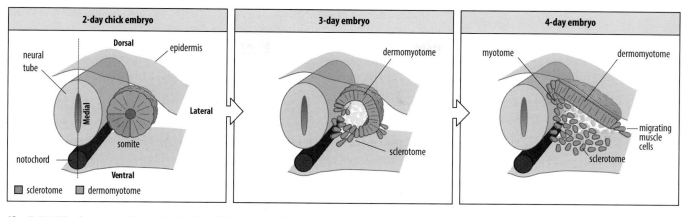

Fig. 5.33 The fate map of a somite in the chick embryo. The ventral medial quadrant (blue) gives rise to the sclerotome cells, which migrate to form the cartilage of the vertebrae. The rest of the somite—the dermomyotome—forms the myotome, which gives rise to all the trunk muscles, while the epithelial dermomyotome region gives rise to the dermis of the skin. The dermomyotome also gives rise to muscle cells that migrate into the limb buds.

in the anterior part of the somite. The restricted expression of *Mesp2* is thought to be a consequence of Notch–Delta signaling, which as already discussed, is essential for establishing the differences between anterior and posterior somite halves.

Somites are also patterned along the dorso-ventral axis and medio-lateral axis. Cells located in the dorsal and lateral regions of a newly formed somite make up the **dermomyotome**, which expresses the gene *Pax3*, a homeobox-containing gene of the *paired* family. The dermomyotome gives rise to the **myotome**, from which muscle cells originate, and it later forms the epithelial sheet dorsal to the myotome from which the dermis originates. The dermomyotome also contains cells that contribute to the vasculature. Cells from the medial region of the myotome form mainly axial and back (epaxial) muscles, whereas lateral myotome cells migrate to give rise to abdominal and limb (hypaxial) muscles. All committed muscle precursors express muscle-specific transcription factors of the MyoD family. The ventral part of the medial somite contains **sclerotome** cells that express the *Pax1* gene and migrate ventrally to surround the notochord and develop into vertebrae and ribs (Fig. 5.33).

Which cells will form cartilage, muscle, or dermis is not yet determined at the time of somite formation. This is clearly shown by experiments in which the dorso-ventral orientation of newly formed somites is inverted but normal development nevertheless ensues. Specification of the fate of somite tissue requires signals from adjacent tissues. In the chick, determination of myotome occurs within hours of somite formation, whereas the future sclerotome is only determined later. Both the neural tube and the notochord produce signals that pattern the somite and are required for its future development. If the notochord and neural tube are removed, the cells in the somites undergo apoptosis; neither vertebrae nor axial muscles develop, although limb musculature still does.

The role of the notochord in specifying the fate of somitic cells has been shown by experiments in the chick, in which an extra notochord is implanted to one side of the neural tube, adjacent to the somite. This has a dramatic effect on somite differentiation, provided the operation is carried out on unsegmented pre-somitic mesoderm. When the somite develops, there is an almost complete conversion to cartilage precursors (Fig. 5.34), suggesting that the notochord is an inducer of cartilage. In normal development, cartilage-inducing signals come from the notochord and the ventral region of the neural tube, the floor plate. Other signals come from the dorsal neural tube and the overlying ectoderm to specify the medial part of the dermomyotome, while signals from the lateral plate mesoderm are involved in specifying the lateral dermomyotome (Fig. 5.35).

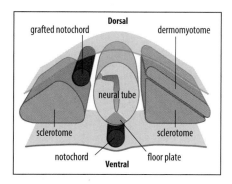

Fig. 5.34 A signal from the notochord induces sclerotome formation. A graft of an additional notochord to the dorsal region of a somite in a 10-somite chick embryo suppresses the formation of the dermomyotome from the dorsal portion of the somite, and instead induces the formation of sclerotome, which develops into cartilage. The graft also affects the shape of the neural tube and its dorso-ventral pattern.

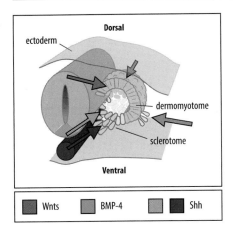

Fig. 5.35 The somite is patterned by signals secreted by adjacent tissues. The sclerotome is specified by a diffusible signal, Sonic hedgehog protein (Shh) (red and orange arrows), produced by the notochord (red) and the floor plate (orange) of the neural tube. Signals from the dorsal neural tube and ectoderm such as Wnts (blue arrows) specify the dermomyotome, together with lateral signals such as BMP-4 (green arrow) from the lateral plate mesoderm.

*Illustration after Johnson, R.L., et al.: **Ectopic expression of Sonic hedgehog alters dorsal-ventral patterning of somites.** Cell 1994, **79**: 1165-1173.*

Some of these signals have been identified. In the chick, both the notochord and the floor plate of the neural tube express Sonic hedgehog (Shh), a secreted protein that is a key molecule for positional signaling in many developmental situations. We encountered Shh in Chapter 1 as a signal involved in craniofacial development (see Box 1F) and will meet it later in this chapter with respect to the asymmetry of structures about the midline. We shall meet it again in Chapter 11 (see Box 11D, for the Shh signaling pathway) in connection with limb development, and in Chapter 12 in the dorso-ventral patterning of the neural tube. In somite patterning, high levels of Shh signaling specify the ventral region of the somite and are required for sclerotome development. Low levels of Shh signaling, together with signals from the dorsal neural tube and from the overlying non-neural ectoderm, specify the dorsal region of the somite as dermo-myotome (see Fig. 5.35). Wnts provide dorsal signals. BMP-4 is the signal from the lateral plate mesoderm, where it is present at a higher concentration than in the somites. Tendons arise from cells that come from the dorso-lateral domain of the sclerotome and specifically express the transcription factor Scleraxis. This region, which gives rise to tendons, is induced at the boundary of sclerotome and myotome.

Regulation of expression of the *Pax* homeobox genes in the somite by signals from the notochord and neural tube is important in determining cell fate. *Pax3* is initially expressed in all cells that will form somites. Its expression is then modulated by signals from BMP-4 and Wnt proteins so that it becomes confined to muscle precursors. It is then further downregulated in cells that differentiate as the muscles of the back, but remains switched on in the migrating presumptive muscle cells that populate the limbs. Mice that lack a functional *Pax3* gene—*Splotch* mutants—lack limb muscles. In the chick, *Pax1* has been implicated in the formation of the scapula, a key element in the shoulder girdle, part of which is contributed by somites. All the scapula-forming cells express *Pax1*. But unlike the *Pax1*-expressing cells of the vertebrae, which are of sclerotomal origin, the blade of the scapula is formed from dermomyotome cells of chick somites 17–24, whereas the head of the scapula is derived from lateral plate mesoderm.

SUMMARY

Somites are blocks of mesodermal tissue that are formed after gastrulation. They form sequentially in pairs on either side of the notochord, starting adjacent to the posterior region of the hindbrain. The somites give rise to the vertebrae, to the muscles of the trunk and limbs, and to most of the dermis of the skin. The pre-somitic mesoderm is patterned with respect to the antero-posterior axis during gastrulation stages. The manifestation of this pattern is the expression of the Hox genes in the pre-somitic mesoderm. The regional identity of the somites is specified by the combinatorial expression of genes of the Hox clusters along the antero-posterior axis, from next to the hindbrain to the posterior end of the embryo, with the order of expression of these genes along the axis corresponding to their order along the chromosome. Mutation or overexpression of a Hox gene results, in general, in localized defects in the anterior parts of the regions in which the gene is expressed, and can cause homeotic transformations. We can think of Hox genes as encoding positional information that specifies the identity of a region and its later development. They act on downstream gene targets about which we know relatively little.

Patterning of the mesoderm with respect to the medio-lateral axis and the formation of somites in the paraxial mesoderm depends on the inhibition of signals that would otherwise specify lateral plate mesoderm. Individual somites are patterned by signals from the notochord, neural tube, and ectoderm, which induce particular regions of each somite to give rise to muscle, cartilage, or dermis.

The origin and patterning of neural crest

In this part of the chapter we will consider the origin, and an example of patterning, of the neural crest, which is a population of cells that is specific to vertebrates. Neural crest cells arise at the borders of the neural plate, which at neurulation rise up to form the neural folds and come together to form the dorsal portion of the neural tube. From there, they migrate to give rise to an enormous range of different tissues and cell types. The migration of neural crest cells is described in more detail in Chapter 9. The migration of neural crest from the hindbrain to the branchial arches, from which some of the structures and tissues of the vertebrate head are derived, is touched on briefly in this chapter in relation to the antero-posterior patterning of the neural crest.

5.14 Neural crest cells arise from the borders of the neural plate and migrate to give rise to a wide range of different cell types

The neural crest is induced at the lateral borders of the neural plate (Fig. 5.36). Induction of the neural crest occurs during gastrulation, at the same time as the neural plate is induced, and the first markers of neural crest are detected at the 5–7-somite stage in the chick and at the headfold stage in the mouse. The current view of neural crest induction is that crest cells form in a band in the ectoderm along each lateral border of the neural plate, at a position where BMP signaling is just above the level that blocks neural plate formation (see Fig. 4.33). As might be expected from their role in neural plate induction, Wnt and FGF signals also appear to be involved in neural crest specification. Induction of neural crest is marked by expression of the genes for the transcription factors Sox9 and Sox10, which in turn activate the gene *snail*, an early marker of neural crest. This gene is the vertebrate version of the *Drosophila* gene *snail* (see Section 2.16). As we shall see in Chapter 9, the Snail protein is involved in enabling the morphological transition that an epithelial cell undergoes to become a migratory mesenchymal cell. Such epithelial-to-mesenchymal transitions occur in many situations during animal development, and neural crest migration is just one example. *Sox10* expression provides a marker for neural crest and the gene continues to be expressed in migrating neural crest cells (Fig. 5.37).

Once neural crest cells have detached from the neural tube they migrate to many sites, where they differentiate into a remarkable number of different cell types (Fig. 5.38). Neural crest cells are, for example, the origin of the neurons and glia of the autonomic nervous system, the glial Schwann cells of the peripheral nervous system, the pigment cells (melanocytes) of the skin, and the adrenaline (epinephrine)-producing cells of the adrenal glands. Surprisingly, much of the bone, cartilage, and connective tissue of the face—tissue types that are more commonly of mesodermal origin—are of neural crest origin, and the head mesenchyme derived from neural crest is called **ectomesenchyme**.

The tremendous differentiation potential of the neural crest was originally discovered by removing the neural folds from amphibian embryos and noting which structures failed to develop. This was later followed by an extensive series of experiments using quail neural crest transplanted into chick embryos (Fig. 5.39). As we have already discussed, it is possible to trace the transplanted quail cells in the chick embryos and see which tissues they contribute to and what cell types they differentiate into. To construct a fate map of the neural crest, small regions of quail dorsal neural tube were grafted to equivalent positions in chick embryos of the same age, at a stage before neural crest migration. The fate map shows a general correspondence between the position of crest cells along the antero-posterior axis and the axial position of the cells and tissues they will give rise to (Fig. 5.40). For example, the neural crest cells that give rise to tissues in the facial and pharyngeal regions are derived from crest anterior to somite 5, whereas ganglion cells of the autonomic sympathetic system come from crest posterior to somite 5.

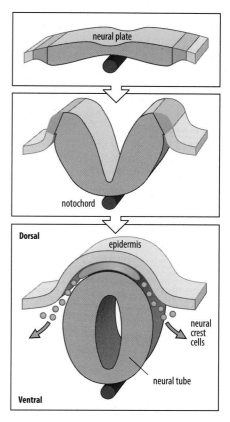

Fig. 5.36 Neural crest cells are specified at the lateral borders of the neural plate. Top panel: prospective neural crest cells (green) are specified at the sides of the neural plate (blue) where levels of BMP are just high enough to prevent neural plate specification. Middle panel: formation of the neural folds carries these cells to the dorsal crests of the neural tube, from which they soon migrate away (bottom panel).

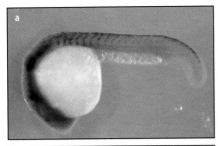

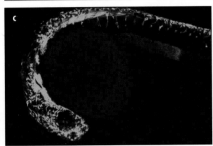

Fig. 5.37 Migrating neural crest cells in zebrafish. a, Zebrafish embryo (25-somite stage, around 19 hours post-fertilization) in which the expression of RNA for the neural crest marker *crestin* (blue) in migrating neural crest cells is detected by *in situ* nucleic acid hybridization in a fixed embryo. b, The expression of the neural crest gene *Sox10* mRNA (blue) in neural crest cells detected by *in situ* hybridization in a fixed embryo (24 hours post-fertilization). c, A live transgenic zebrafish embryo at the same stage of development that carries a reporter transgene in which a GFP coding sequence is attached to the promoter region of *Sox10*. GFP is therefore expressed in the neural crest cells that express *Sox10*. Note the contribution of the neural crest to the head.

*Photographs (a) courtesy of Roberto Mayor (UCL); (b, c) Robert Kelsh. (b) reproduced with permission from Kelsh: **Sorting out Sox10 functions in neural crest development.** 2006, BioEssays **28**: 788-789.*

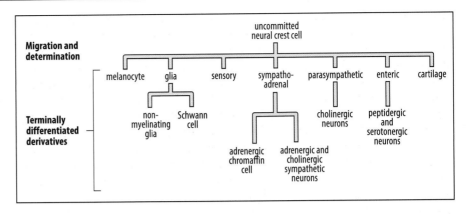

Fig. 5.38 Derivatives of the neural crest. Neural crest cells give rise to a wide range of cell types that include melanocytes, cartilage, glia, and various types of neurons distinguished by their functional specializations and the neurotransmitters they produce. Cholinergic neurons use acetylcholine as their neurotransmitter, adrenergic neurons principally use norepinephrine (noradrenaline), and peptidergic and serotonergic neurons produce peptide neurotransmitters and serotonin, respectively.

To what extent is the fate of neural crest cells fixed before migration? Some crest cells are unquestionably multipotent. Single neural crest cells injected with a tracer shortly after they have left the neural tube can be seen to give rise to a number of different cell types, both neuronal and non-neuronal. Also, by changing the position of neural crest before cells start to migrate, it has been shown that the developmental potential of these cells is greater than their normal fate would suggest. Multipotent stem cells have been isolated from neural crest that can give rise to neurons, glia, and smooth muscle. Similar multipotent cells have even been isolated from the peripheral nerves of mammalian embryos days after neural crest migration.

Almost all the cell types to which the neural crest can give rise will differentiate in tissue culture, and the developmental potential of single neural crest cells has been studied in this way. Most of the clones derived from single neural crest cells taken at the time of migration contain more than one cell type, showing that the cell was multipotent at the time it was isolated. As neural crest cells migrate, however, their potential progressively decreases: the size of clones derived from these cells gets smaller and fewer cell types are produced. Thus, shortly after the beginning of migration, the neural crest is a mixed population of multipotent cells and cells whose potential is already restricted.

5.15 Neural crest cells migrate from the hindbrain to populate the branchial arches

Neurons and neural crest cells obtain their regional identity from the Hox genes they express as part of the neural tube and, in the case of the neural crest cells, they retain this Hox gene expression when they migrate. This is illustrated here by the neural crest from the hindbrain, which migrates to populate the branchial arches where it will contribute to mesenchymal tissues (bones, cartilage, and connective tissues) of the face and skull. In the chick embryo this migration of ectomesenchyme cells is accomplished in about 10 hours between 1 and 2 days after laying.

The cranial neural crest that migrates from the dorsal region of the hindbrain has a segmental arrangement, correlating closely with the segmentation of this region of the brain into units called rhombomeres. We will discuss the way in which this organization develops in the brain in Chapter 12. The migration pathways have been followed by marking chick neural crest cells *in vivo*. Crest cells from rhombomeres 2, 4, and 6 populate the first, second, and third branchial arches, respectively (Fig. 5.41).

The crest cells have already acquired a regional identity before they begin to migrate. When crest cells of rhombomere 4 are replaced by cells from rhombomere 2 taken from another embryo, these cells enter the second branchial arch, but develop into structures characteristic of the first arch, to which they would normally have migrated. This can result in the development of an additional lower jaw in the chick embryo. However, neural crest cells have some developmental plasticity and their ultimate differentiation depends on signals from the tissues into which they migrate.

The involvement of Hox genes in patterning the neural crest of the hindbrain region has also been demonstrated by gene knock-outs in mice. The results are not always easy to interpret, as knock-out of a particular Hox gene can affect different populations of neural crest cells in the same animal, such as those that will form neurons and those that will form ectomesenchyme. Knock-out of the *Hoxa2* gene, for example, results in skeletal defects in the region of the head corresponding to the normal domain of expression of the gene, which extends from rhombomere 2 backwards. Skeletal elements in the second branchial arch, all of which come from neural crest cells derived from rhombomere 4, are abnormal and some of the skeletal elements normally formed by the first branchial arch develop in the second arch. Thus, suppression of *Hoxa2* causes a partial homeotic transformation of one arch into another. This shows that Hox genes encode positional information that specifies the regional identity of the segmental cranial neural crest cells in the same way as the Hox genes do in the somites.

SUMMARY

Neural crest cells arise at the lateral margins of the neural plate and migrate away from the forming neural tube to differentiate into a wide range of cell types throughout the body, including the sympathetic, parasympathetic, and enteric nervous systems, the melanocytes of the skin, and some of the bones of the face. Before they leave the crest, some crest cells still have a broad developmental potential whereas others are already restricted in their fate. A Hox gene code provides a regional code for the neural tube and the neural crest cells. Neural crest cells that migrate from the hindbrain populate the branchial arches in a position-dependent fashion.

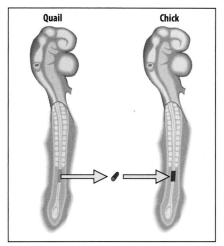

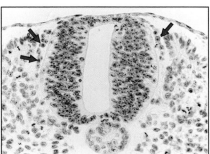

Fig. 5.39 Tracing cell-migration pathways after grafting a piece of quail neural tube into a chick host. A piece of neural tube from a quail embryo is grafted to a similar position in a chick host. The photograph shows migration of the quail neural crest cells (red arrows). Their migration can be traced because quail cells have a nuclear marker that distinguishes them from chick cells.

Photograph courtesy of N. Le Douarin.

SUMMARY: patterning of the axial body plan

gastrulation and organizer activity

⇩

the Hox gene complexes are expressed along the antero-posterior axis

⇩

Hox gene expression specifies regional identity for mesoderm, endoderm, and ectoderm

⇙　　　　⇘

mesoderm develops into notochord, somites, and lateral plate mesoderm　　　　early signals and mesoderm induce neural plate from ectoderm

⇩　　　　⇩

somites receive signals from notochord, neural tube, and ectoderm　　　　neural crest cells are specified at edges of neural plate

⇩　　　　⇩

somite develops into sclerotome and dermomyotome　　　　neural crest in the hindbrain is characterized by regional patterns of Hox gene expression

Fig. 5.40 Fate and developmental potential of neural crest cells. The fate map of the neural crest (bottom panel) shows that there is a general correspondence between the position of the neural crest cells along the antero-posterior axis and the position of the structure to which these cells give rise (top panel). For example, the crest that gives rise to the ectomesenchyme of the head comes from the anterior region. However, the developmental potential of neural crest cells, with the exception of the presumptive ectomesenchyme, is very much greater than their presumptive fate.

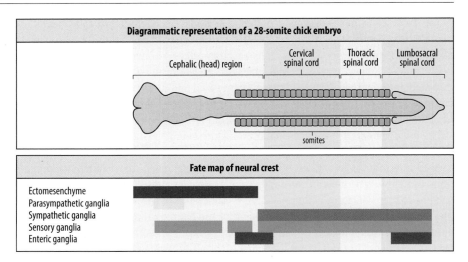

Determination of left-right asymmetry

We end our consideration of how the body plan is specified in vertebrate embryos by looking at how the internal asymmetry between the right and left sides of the body arises. Vertebrates are bilaterally symmetric about the midline of the body for many structures, such as eyes, ears, and limbs. But whereas the vertebrate body is outwardly symmetric, most internal organs such as the heart and liver are in fact asymmetric with respect to left and right. This is known as **left–right asymmetry**.

5.16 The bilateral symmetry of the early embryo is broken to produce left-right asymmetry of internal organs

Specification of left and right is fundamentally different from specifying the other axes of the embryo, as left and right have meaning only after the antero-posterior and dorso-ventral axes have been established. If one of these axes were reversed, then so too would be the left–right axis (this is the reason that handedness is reversed when you look in a mirror: your dorso-ventral axis is reversed, and so left becomes right and vice versa). One suggestion to explain the development of left–right asymmetry is that an initial asymmetry at the molecular level is converted to an asymmetry at the cellular and multicellular

Fig. 5.41 Expression of Hox genes in the branchial region of the head. The expression of Hox genes in the hindbrain (rhombomeres r1 to r8), neural crest, branchial arches (b1 to b4), and surface ectoderm is shown. The arrows indicate the migration of neural crest cells into the branchial arches. There is no significant neural crest migration from r3 and r5.

*Illustration after Krumlauf, R.: **Hox genes and pattern formation in the branchial region of the vertebrate head.** Trends Genet. 1993, **9**: 106-112.*

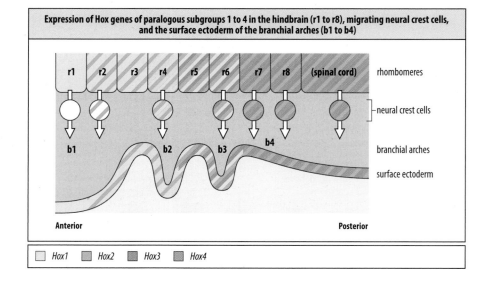

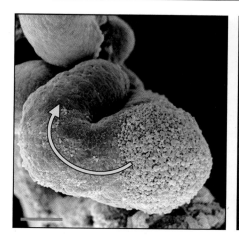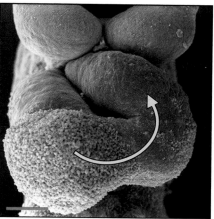

Fig. 5.42 **Left–right asymmetry of the mouse heart is under genetic control.** Each photograph shows a mouse heart viewed frontally after the loops have formed. The normal asymmetry of the heart results in it looping to the right, as indicated by the arrow (left panel). 50% of mice that are homozygous for the mutation in the iv gene have hearts that loop to the left (right panel). Scale bar = 0.1 mm.

Photographs courtesy of N. Brown.

level. If that were so, the asymmetric molecules or molecular structure would need to be oriented with respect to both the antero-posterior and dorso-ventral axes.

The breaking of left–right asymmetry occurs early in development. Although the initial symmetry-breaking mechanism is still not known, the subsequent cascade of events that lead to organ asymmetry is better understood. These give rise to left–right differences in gene expression in the lateral plate mesoderm in the early embryo around the time the somites are forming, and these differences in gene expression are then translated into asymmetric positioning and shape of organs. In mice and humans, for example the heart is on the left side, the right lung has more lobes than the left, the stomach and spleen lie towards the left, and the bulk of the liver towards the right. This handedness of organs is remarkably consistent, but there are rare individuals, about one in 10,000 in humans, who have the condition known as **situs inversus**, a complete mirror-image reversal of handedness. Such people are generally asymptomatic, even though all their organs are reversed. A similar condition can be produced in mice that carry the *iv* mutation, in which organ asymmetry is randomized (Fig. 5.42). In this context, 'randomized' means that some *iv* mutant mice will have the normal organ asymmetry and some the reverse. In another mouse mutant, *inversion of turning* (*inv*), laterality is consistently reversed.

The *iv* mutation in mice affects the gene that encodes left–right dynein, a protein that acts as a directional microtubular motor that is required for ciliary motion and intracellular transport. Patients with Kartagener's syndrome characterized by situs inversus have abnormal cilia and a large number of other human syndromes in which ciliary development is abnormal—known as **ciliopathies**—are also characterized by left–right asymmetry defects. This suggested that in mice and humans the normal functioning of cilia is required for left–right patterning.

Although it is still not known whether it is the activity of cilia that initiates left–right symmetry breaking, ciliary activity is critical in inducing the asymmetric expression of genes involved in establishing left versus right in mouse, zebrafish and *Xenopus* embryos. In these embryos at around the gastrula stage, a 'leftward' flow of extracellular fluid is stimulated across the embryonic midline by transient populations of ciliated cells (Fig. 5.43). In the mouse, ciliated cells are found on the ventral surface of the node and create fluid flow in a pit formed below the node where there is no underlying endoderm. The directed flow depends on the alignment of the basal bodies of the cilia. There are two types of cilia on the cells lining the pit: central motile cilia that generate the flow, and peripheral immotile cilia. Coordinated rotational beating of the central cilia generates a leftward flow of extracellular fluid. One hypothesis is that the immotile cilia act as mechanosensors for the direction of flow and their stimulation results in release of intracellular Ca^{2+} in the cells on the left side of the node. Propagation of the Ca^{2+} signal to nearby cells then leads to the upregulation of Nodal

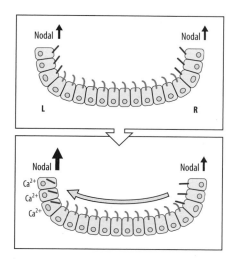

Fig. 5.43 **In the mouse embryo, cilia-directed leftward flow of extracellular fluid generates left–right asymmetry across the midline.** Fluid flow occurs by the beating of cilia on cells lining a pit on the ventral side of the node. There are two types of cilia: central motile cilia (blue) that generate the flow and peripheral immotile cilia (red). Coordinated rotational beating of the central cilia generates a leftward flow of extracellular fluid. One hypothesis is that the immotile cilia sense the direction of flow and their mechanical stimulation results in release of intracellular Ca^{2+} in the cells on the left side of the node, as illustrated here. Propagation of the Ca^{2+} signal to nearby cells then leads to the upregulation of Nodal expression on the left side (bottom panel, larger arrow).

expression on the left side. Other hypotheses propose that the leftward flow leads to the accumulation of morphogen molecules on the left-hand side of the node, and that this morphogen directly or indirectly causes the upregulation of Nodal.

In the zebrafish, the ciliated epithelium lines a fluid-filled vesicle known as Kupffer's vesicle, which develops at the midline in the zebrafish embryo's tailbud at the end of gastrulation, while in *Xenopus*, a triangular region of ciliated mesoderm, the amphibian gastrocoel roof plate, straddles the midline in the roof of the archenteron. Disrupting the action of the cilia in either the Kupffer's vesicle of zebrafish or the gastrocoel roof plate of *Xenopus*, by, for example, using antisense morpholinos (see Box 6B) to knock down expression of ciliary dynein genes, leads to a breakdown in right–left asymmetry. Reversal of the leftward flow in the ventral node by an imposed fluid movement also reverses left–right asymmetry in mouse embryos. In chick embryos, however, motile cilia do not appear to be present on the ventral surface of the node and instead, passive leftward movement of cells appears to be involved in establishing asymmetrical gene expression.

One of the key proteins in establishing 'leftness,' is Nodal, and the gene for Nodal comes to be expressed at high levels on the left side of the embryo. *Nodal* is initially expressed symmetrically on both sides of the node in both mouse and chick embryos, and adjacent to Kupffer's vesicle in zebrafish, and the gastrocoel roof plate in *Xenopus*. Initial small differences in *Nodal* expression on right and left sides induced by leftward flow of fluid across the midline in mouse, zebrafish and *Xenopus* then become amplified as a result of a positive-feedback loop—as one effect of Nodal signaling is to activate the *Nodal* gene, and so produce more Nodal protein (see Box 5A).

Asymmetric *Nodal* expression on the left side is followed by widespread and stable activation of *Nodal* expression in the left lateral-plate mesoderm, together with expression of the gene for the transcription factor Pitx2, another key determinant of leftness. This left-sided pattern of gene expression in the lateral plate mesoderm that gives rise to internal organs such as the heart is highly conserved, being found in the mouse, zebrafish, *Xenopus*, and chick. It has been shown in chick embryos that if the expression of either *Nodal* or *Pitx2* in the lateral plate mesoderm is made symmetric, then organ asymmetry is randomized.

5.17 Left–right symmetry breaking may be initiated within cells of the early embryo

An early indication of left–right asymmetry in both *Xenopus* and chick embryos is the asymmetric activity of a proton–potassium pump (H^+/K^+-ATPase). This ATPase can be detected as early as the third cleavage division—the eight-cell stage—in *Xenopus*, and treatment of early embryos with drugs that inhibit the pump cause randomization of asymmetry. How the asymmetry of the H^+/K^+-ATPase itself originates is not known, but one possibility is that the mRNA encoding the ion transporter and/or the ion transporter protein itself might be moved by motor proteins along oriented cytoskeletal elements in the fertilized egg. In *Xenopus*, disruption of cortical actin at the first cleavage division can affect left–right asymmetry. This left–right asymmetry established during early cleavage stages could lead to the subsequent asymmetrical alignment of the basal bodies of the cilia on the cells in the gastrocoel roof plate.

In the chick, a detailed mechanism for left–right asymmetry based on the asymmetric activity of an H^+/K^+-ATPase in Hensen's node has been proposed (Fig. 5.44). It has been found that the activity of the pump is reduced on the left side of the node, leading to membrane depolarization and release of calcium from cells to the extracellular space on the left-hand side. The release of calcium in turn leads to the expression of the Notch ligands, Delta-like and Serrate, in a spatial pattern that could activate the transmembrane signaling protein Notch on the left side of the node only (the Notch signaling pathway is illustrated in Box 5D).

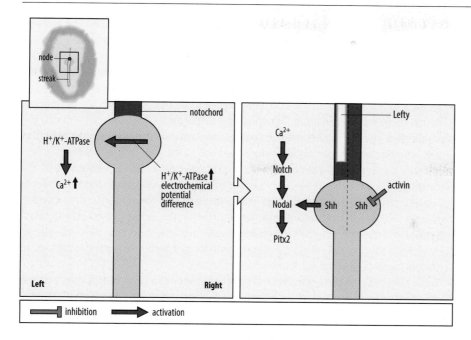

Fig. 5.44 Determination of left-right asymmetry in the chick. The activity of a proton-potassium pump (H⁺/K⁺-ATPase) in Hensen's node is reduced on the left side of the node, leading to a membrane-potential difference across the node and an increased release of calcium (Ca²⁺) into the extracellular space on the left side of the node (left panel). This leads to a greater activation of Notch signaling on the left side of the node, which in turn activates expression of the gene *Nodal* in cells on the left side of the node. Nodal signaling together with Sonic hedgehog (Shh) then switches on *Nodal* expression in the lateral plate mesoderm on the left side, which leads to expression of the transcription factor Pitx2, an important determinant of leftness (right panel). Shh activity is inhibited by activin on the right-hand side. The Nodal antagonist Lefty, expressed in the left half of the notochord and the floorplate of the neural tube, provides a midline barrier that prevents Nodal signaling occurring on the right-hand side. This is a simplified version of a much more complex set of interactions.

Notch activity, in conjunction with signaling by the secreted protein Sonic hedgehog (Shh) (see Box 11D for the Shh signaling pathway), which is expressed at higher levels on the left side of the node, then leads to the left-sided expression of *Nodal* in the lateral-plate mesoderm. Left-sided expression of *Shh* may be due to passive leftward movement of *Shh*-expressing cells; *Shh* expression on the right side is repressed by signaling by the TGF-β family member activin. Left-sided Nodal signaling induced by combined Notch and Shh signaling activates the expression of the gene for the transcription factor Pitx2 on the left side. If a pellet of cells secreting Shh is placed on the right side, then organ asymmetry is randomized.

The right half of the embryo is protected from Nodal signaling by the protein Lefty, which is secreted by the notochord and the floor of the neural tube. Lefty antagonizes Nodal signaling by binding to the Nodal protein and so prevents Nodal binding to and activating its receptor. Lefty might also interact with the receptor itself and so compete with Nodal for binding to the receptor (see Box 5A).

SUMMARY

The generation of the consistent left-right organ asymmetry found in vertebrates is marked by the expression of the secreted protein Nodal and the transcription factor Pitx2 on the left side only. This asymmetry appears to be generated by various mechanisms, involving a proton-potassium pump asymmetry (in *Xenopus* and chick), the leftward movement of cilia (in mouse, humans, zebrafish and *Xenopus*), and asymmetrically located calcium signaling, Notch pathway activation, Sonic hedgehog signaling, and Nodal signaling.

Summary to Chapter 5

- The early development of the chick and the mouse is related to the way in which the embryo is nourished; the chick egg contains a large amount of yolk, whereas an early event in mouse embryo development is the generation of cells to make the placenta.

- There are many similarities in the way in which the germ layers are specified and patterned in all vertebrates despite the different topologies of the embryos.
- The organizer has equivalent functions in all four model vertebrate embryos. During gastrulation, the germ layers specified during the earlier stages of embryonic development become further patterned along the antero-posterior and dorso-ventral axes.
- At the end of gastrulation, the basic body plan has been laid down and the prospective neural tissue has been induced from the dorsal ectoderm.
- In the chick and the mouse, the notochord, spinal cord and somites are generated from self-renewing populations of cells around the node as the body axis elongates.
- Somites are formed as a series of blocks of mesodermal tissue on either side of the notochord.
- Specific regions of each somite give rise to cartilage (the forerunner of the skeleton), muscle, and dermis, and these regions are specified by signals from the notochord, neural tube, lateral plate mesoderm, and ectoderm.
- Regional identity of cells along the antero-posterior axis is encoded by the expression of combinations of genes of the four Hox gene clusters.
- There is both spatial and temporal co-linearity between the order of Hox genes on the chromosomes and the order in which they are expressed along the axis from the level of the hindbrain to the posterior end of the embryo.
- Inactivation or overexpression of Hox genes can lead both to localized abnormalities and to homeotic transformations of one 'segment' of the axis into another, indicating that these genes are crucial in specifying regional identity.
- The neural crest arises at the lateral borders of the neural plate and is carried to the dorsal crests of the neural tube, from where the cells migrate to give rise to a wide range of different cell types, including cells of the peripheral nervous system, pigment cells and head connective tissue cells.
- Left-right asymmetry involves an initial symmetry-breaking event leading to asymmetric gene expression on the left and right sides of the embryo, which is translated into asymmetric positioning and morphogenesis of organs.

■ End of chapter questions

Long answer (concept questions)

1. What experiments define the organizer in vertebrate embryos? To what extent are the organizers of different vertebrate embryos equivalent? Include discussion of *Xenopus*, zebrafish, chick and mouse.

2. How does the way the embryo is nourished affect early development? Compare all four vertebrate model vertebrates and include discussion of the relative amounts of yolk in the eggs and the importance of extra-embryonic tissues.

3. Distinguish between the notochord, the spinal cord, and the vertebral column and describe the relationships between the following structures: neural plate, neural folds, neural tube, and neural crest.

4. Somite formation in chick embryos proceeds from anterior to posterior in response to a gradient of FGF-8. Describe this gradient: Where does the FGF-8 originate? Why does it form a gradient? How does the gradient trigger somite formation? What other gradients are present in the embryo during this period and what are their functions?

5. The Hox gene clusters in mice (Box 5E) illustrate several mechanisms that operate in genome evolution: genes can become duplicated to give rise to paralogs, the entire cluster can become duplicated to give rise to paralogous groups of genes, and within any given cluster, genes can be lost. Give an example of each of these three mechanisms.

6. The Hox genes are expressed in the mesoderm of vertebrate embryos both temporally and spatially in order, from one end of the Hox gene cluster to the other. Refer to Fig. 5.29 and compare Hox gene expression in the mouse posterior cervical region to that in the anterior thoracic region. How do the gene expression patterns illustrate the temporal gradient of expression along the Hox clusters? How do the gene expression patterns illustrate how different combinations of gene expression can produce spatial differences in regional identity along an axis?

7. Compare the consequences of *Hoxc8* gene deletion in the mouse (Fig. 5.31) with deletion of the *Ubx* gene in *Drosophila* (Fig. 2.51). Are the transformations seen toward anterior or posterior identities? What do these transformations say about the roles of Hox genes in identity specification? Do these transformations in mice qualify as homeotic transformations?

8. How does the zebrafish *giraffe* (*gir*) mutation support the hypothesis that a gradient of retinoic acid may be involved in the co-linearity of the vertebrate axis and Hox gene expression? Recalling that genes are typically named for the mutant phenotype produced, speculate on the phenotype of the *gir* mutation. If a deletion (a loss-of-function) of the *Cyp26a1* gene in mice causes cervical vertebrae to become more posterior rib-bearing vertebrae, would the *gir* mutation similarly be a loss-of-function, or, conversely, might it instead be a mutation that causes increased expression of the gene (gain of function)? Explain.

9. Describe the effects of the experiment shown in Fig. 5.34, in which a notochord from a donor embryo is grafted adjacent to the dorsal region of the neural tube in a host embryo. What signaling molecule is proposed to be responsible for the observed induction? What kind of proteins are the Pax proteins? What is the role of Pax3 in the somite?

10. What is the posterior marginal zone of the chick embryo? What is its significance to chick development? Describe the external forces that lead to its formation and the signaling molecules associated with it.

11. The establishment of left–right asymmetry is associated with higher levels of the TGF-β family signaling molecule Nodal on the left side of the embryo, relative to the right. What are the influences that led to this asymmetry of Nodal concentration? Include positive feedback, Notch signaling, Sonic hedgehog signaling, and the Lefty protein in your answer.

12. How have chick–quail chimeras been used in the study of the neural crest? How have experiments on neural crest cells in culture added to our knowledge of the development of neural crest derivatives *in vivo*?

Multiple choice (factual recall questions)

NB There is only one right answer to each question.
1. Somites will give rise to

(a) heart and blood
(b) internal organs of the abdomen
(c) the vertebral column and skeletal muscles
(d) the spinal cord.

2. The positioning of the dorso-ventral axis in Xenopus is determined by _____, while the positioning of the antero-posterior axis in chicks is determined by _____.

(a) gravity; point of sperm entry
(b) maternal factors in both cases
(c) position in the egg chamber; gravity
(d) the point of sperm entry; gravity.

3. How would you produce a chimeric mouse?

(a) Divide a morula into two separate clusters of blastomeres and let each cluster form an embryo.
(b) Introduce cells from the inner cell mass of one blastocyst into the inner cell mass of another blastocyst of the same genetic constitution.
(c) Divide a fertilized mouse egg into two halves and let each develop separately.
(d) Introduce cells from the inner cell mass of one blastocyst into the inner cell mass of another blastocyst of a different genetic constitution.

4. The equivalents of the *Xenopus* organizer in chick and zebrafish are

(a) Hensen's node and the embryonic shield

(b) the hypoblast and the yolk syncytial layer
(c) the notochord in both cases
(d) the primitive streak and the mesendoderm.

5. Somite formation in chicks proceeds

(a) from anterior to posterior, as Hensen's node retreats
(b) from posterior to anterior, as the primitive streak elongates
(c) from the dorsal midline outward, into lateral plate
(d) simultaneously along the axis.

6. The sclerotome of the somite gives rise to _____, whereas the myotome gives rise to _____

(a) muscles, vertebrae
(b) peripheral nervous system, muscles
(c) spinal cord, limb muscles
(d) vertebrae, muscles of the back.

7. Hox genes that are expressed more anteriorly comprise Hox genes

(a) with all possible numbers and letters, since there is no systematic relationship between the gene designations and the expression patterns of the Hox genes
(b) with higher numbers, such as *Hoxa13*
(c) with letters, so that anterior genes are *Hoxa1–13*, and posterior genes are *Hoxp1–13*
(d) with lower numbers, such as *Hoxa1*.

8. Neural crest cells will contribute to which tissues or structures?

(a) melanocytes in the skin
(b) mesoderm of the face and skull
(c) peripheral nervous system
(d) all of these.

9. Lefty

(a) acts intracellularly to phosphorylate Smads
(b) acts as an extracellular antagonist of Nodal signalling
(c) is a transcription factor that regulates *Nodal* gene expression
(d) is a membrane-bound receptor for Nodal.

10. Oct4

(a) is expressed in the trophectoderm of the mouse blastocyst
(b) is expressed in the inner cell mass of the mouse blastocyst
(c) is a signaling molecule that induces pluripotency
(d) is a mouse-specific gene.

Multiple choice answer key

1: c, 2: d, 3: d, 4: a, 5: a, 6: d, 7: d, 8: d, 9: b, 10: b.

■ Section further reading

5.1 The antero-posterior polarity of the chick blastoderm is related to the primitive streak

Bertocchini, F., Skromne, I., Wolpert, L., Stern, C.D.:
Determination of embryonic polarity in a regulative system: evidence for endogenous inhibitors acting sequentially during primitive streak formation in the chick embryo. *Development* 2004, **131**: 3381–3390.

Khaner, O., Eyal-Giladi, H.: **The chick's marginal zone and primitive streak formation. I. Coordinative effect of induction and inhibition.** *Dev. Biol.* 1989, **134**: 206–214.

Kochav, S., Eyal-Giladi, H.: **Bilateral symmetry in chick embryo determination by gravity.** *Science* 1971, **171**: 1027–1029.

Seleiro, E.A.P., Connolly, D.J., Cooke, J.: **Early developmental expression and experimental axis determination by the chicken Vg-1 gene.** *Curr. Biol.* 1996, **11**: 1476–1486.

Sheng, G.: **Day-1 chick development.** *Dev. Dyn.* 2014, **243**: 357–367.

Stern, C.D.: **Cleavage and gastrulation in avian embryos (version 3.0).** *Encyclopedia of Life Sciences* 2009 http://www.els.net.

5.2 Early stages in mouse development establish separate cell lineages for the embryo and extra-embryonic structures

Arnold, S.J., Robertson, E.J.: **Making a commitment: cell lineage allocation and axis patterning in the early mouse embryo.** *Nature Mol. Cell Biol. Rev.* 2009, **10**: 91–103.

Clavería, C., Giovinazzo, G., Sierra, R., Torres, M.: **Myc-driven endogenous cell competition in the early mammalian embryo.** *Nature* 2013, **500**: 39–44.

Deb, K., Sivaguru, M., Yul Yong, H., Roberts, M.: **Cdx2 gene expression and trophectoderm lineage specification in mouse embryos.** *Science* 2006, **311**: 992–996.

Hillman, N., Sherman, M.I., Graham, C.: **The effect of spatial arrangement on cell determination during mouse development.** *J. Embryol. Exp. Morph.* 1972, **28**: 263–278.

Manzanares, M., Rodriguez, T.A.: **Hippo signaling turns the embryo inside out.** *Curr. Biol.* 2013, **23**: R559–R561.

Rossant, J., Tam, P.P.L.: **Blastocyst lineage formation, early embryonic asymmetries and axis patterning in the mouse.** *Development* 2009, **136**: 701–713.

Stephenson, R.O., Rossant, J., Tam, P.P.: **Intercellular interactions, position and polarity in establishing blastocyst cell lineages and embryonic axes.** *Cold Spring Harb. Perspect. Biol.* 2012, **4**: a008235.

5.3 Movement of the anterior visceral endoderm indicates the definitive antero-posterior axis in the mouse embryo

Bischoff, M., Parfitt, D.E., Zernicka-Goetz, M.: **Formation of the embryonic-abembryonic axis of the mouse blastocyst: relationships between orientation of early cleavage divisions and pattern of symmetric/asymmetric divisions.** *Development* 2008, **135**: 953–962.

Rodriguez, T.A., Srinivas, S., Clements, M.P., Smith, J.C., Beddington, R.S.: **Induction and migration of the anterior visceral endoderm is regulated by the extra-embryonic ectoderm.** *Development* 2005, **132**: 2513–2520.

Srinivas, S., Rodriguez, T., Clements, M., Smith, J.C., Beddington, R.S.P.: **Active cell migration drives the unilateral movements of the anterior visceral endoderm.** *Development* 2004, **131**: 1157–1164.

Takaoka, K., Yamamoto, M., Hamada, H.: **Origin and role of distal visceral endoderm.** *Nature Cell Biol.* 2011, **13**: 743–752.

5.4 The fate maps of vertebrate embryos are variations on a basic plan

Beddington, R.S., Robertson, E.J.: **Anterior patterning in mouse.** *Trends Genet.* 1998, **14**: 277–284.

5.5 Mesoderm induction and patterning in the chick and mouse occurs during primitive-streak formation

Chapman, S.C., Matsumoto, K., Cai, Q., Schoenwolf, G.C.: **Specification of germ layer identity in the chick gastrula.** *BMC Dev. Biol.* 2007, **7**: 91–107.

Shen, M.M.: **Nodal signaling—developmental roles and regulation.** *Development* 2007, **134**: 1023–1034.

5.6 The node that develops at the anterior end of the streak in chick and mouse embryos is equivalent to the Spemann organizer in *Xenopus*

Beddington, R.S.: **Induction of a second neural axis by the mouse node.** *Development* 1994, **120**: 613–620.

Joubin, K., Stern, C.D.: **Molecular interactions continuously define the organizer during the cell movements of gastrulation.** *Cell* 1999, **98**: 559–571.

Kinder, S.J., Tsang, T.E., Wakamiya, M., Sasaki, H., Behringer, R.R., Nagy, A., Tam, P.P.: **The organizer of the mouse gastrula is composed of a dynamic population of progenitor cells for the axial mesoderm.** *Development* 2001, **128**: 3623–3634.

Stern, C.D.: **Neural induction: old problem, new findings, yet more questions.** *Development* 2005, **132**: 2007–2021.

5.7 Neural induction in chick and mouse is initiated by FGF with inhibition of BMP being required a later step

Bachiller, D., Klingensmith, J., Kemp, C., Belo, J.A., Anderson, R.M., May, S.R., MacMahon, J.A., McMahon, A.P., Harland, R.M., Rossant, J., De Robertis, E.M.: **The organizer factors Chordin and Noggin are required for mouse forebrain development.** *Nature* 2000, **403**: 658–661.

Mukhopadhyay, M., Shtrom, S., Rodriguez-Esteban, C., Chen, L., Tsuku, T., Gomer, L., Dorward, D.W., Glinka, A., Grinberg, A., Huang, S.-P., Niehrs, C., Izpisúa Belmonte, J.-C., Westphal, H.: **Dickkopf1 is required for embryonic head induction and limb morphogenesis in the mouse.** *Dev. Cell* 2001, **1**: 423–434.

Ozair, M.Z., Kintner, C., Brivanlou, A.H.: **Neural induction and patterning in vertebrates.** *WIREs Dev. Biol.* 2013, **2**: 479–498.

Sheng, G., dos Reis, M., Stern, C.D.: **Churchill, a zinc finger transcriptional activator, regulates the transition between gastrulation and neurulation.** *Cell* 2003, **115**: 603–613.

Stern, C.: **Neural induction: old problems, new findings, yet more questions.** *Development* 2005, **132**: 2007–2021.

Storey, K., Crossley, J.M., De Robertis, E.M., Norris, W.E., Stern, C.D.: **Neural induction and regionalization in the chick embryo.** *Development* 1992, **114**: 729–741.

Streit, A., Berliner, A.J., Papanayotou, C., Sirulnik, A., Stern, C.D.: **Initiation of neural induction by FGF signalling before gastrulation.** *Nature* 2000, **406**: 74–78.

Box 5B Chromatin-remodeling complexes

Ho, L., Crabtree, G.R.: **Chromatin remodelling during development.** *Nature* 2010, **463**: 474–484.

5.8 Axial trunk structures in chick and mouse are generated from self-renewing cell populations

Delfino-Machin, M., Lunn, J.S., Breitkreuz, D.N., Akai, J., Storey, K.G.: **Specification and maintenance of the spinal cord stem zone.** *Development* 2005, **132**: 4273–4283.

Diez del Corral, R., Olivera-Martinez, I., Goriely, A., Gale, E., Maden, M., Storey, K.: **Opposing FGF and retinoid pathways control ventral neural pattern, neuronal differentiation, and segmentation during body axis extension.** *Neuron* 2003, **40**: 65–79.

Dubrulle, J., Pourquié, O.: *fgf8* **mRNA decay establishes a gradient that couples axial elongation to patterning in the vertebrate embryo.** *Nature* 2004, **427**: 419–422.

Neijts, R., Simmini, S., Giuliani, F., van Rooijen, C., Deschamps, J.: **Region-specific regulation of posterior axial elongation during vertebrate embryogenesis.** *Dev. Dyn.* 2014, **243**: 88–98.

Sakai, Y., Meno, C., Fujii, H., Nishino, J., Shiratori, H., Saijoh, Y., Rossant, J., Hamada, H.: **The retinoic acid-inactivating enzyme CYP26 is essential for establishing an uneven distribution of retinoic acid along the anteroposterior axis within the mouse embryo.** *Genes Dev.* 2006, **15**: 213–225.

Wilson, V., Olivera-Martinez, I., Storey, K.G.: **Stem cells, signals and vertebrate body axis extension.** *Development* 2009, **136**: 1591–1604.

Box 5C Retinoic acid: a small-molecule intercellular signal

Niederreither, K., Dolle, P.: **Retinoic acid in development: towards an integrated view.** *Nat. Rev. Genet.* 2008, **9**: 541–553.

Rossant, J., Zirngibl, R., Cado, D., Shago, M., Giguère, V.: **Expression of a retinoic acid response element-hsplacZ transgene defines specific domains of transcriptional activity during mouse embryogenesis.** *Genes Dev.* 1991, **5**: 1333–1344.

Box 5D The Notch signaling pathway

Bray, S.J.: **Notch signalling: a simple pathway becomes complex.** *Nat. Rev. Mol. Cell Biol.* 2006, **7**: 678–689.

Hori, Y., Sen, A., Artavanis-Tsakonas, S.: **Notch signaling at a glance.** *J. Cell Sci.* 2013, **126**: 2135–2140.

Lai, E.: **Notch signaling: control of cell communication and cell fate.** *Development* 2004, **131**: 965–973.

5.9 Somites are formed in a well-defined order along the antero-posterior axis

Dequeant, M.L., Pourquié, O.: **Segmental patterning of the vertebrate embryonic axis.** *Nat. Rev. Genet.* 2008, **9**: 370–382.

Gomez, C., Pourquié, O.: **Developmental control of segment numbers in vertebrates.** *J. Exp. Zool. B Mol. Dev. Evol.* 2009, **312**: 533–544.

Kieny, M., Mauger, A., Sengel, P.: **Early regionalization of somitic mesoderm as studied by the development of the axial skeleton of the chick embryo.** *Dev. Biol.* 1972, **28**: 142–161.

Kulesa, P.M., Schnell, S., Rudloff, S., Baker, R.E., Maini, P.K.: **From segment to somite: segmentation to epithelialization analyzed within quantitative frameworks.** *Dev Dyn.* 2007, **236**: 1392–402.

Lewis, J.: **Autoinhibition with transcriptional delay: a simple mechanism for the zebrafish somitogenesis oscillator.** *Curr. Biol.* 2003, **13**: 1398–408.

Maroto, M., Bone, R.A., Dale, J.K.: **Somitogenesis.** *Development* 2012, **139**: 2453–2456.

Oates, A.C., Morelli, L.G., Ares, S.: **Patterning embryos with oscillations: structure, function and dynamics of the vertebrate segmentation clock.** *Development* 2012, **139**: 625–639.

Stern, C.D., Charité, J., Deschamps, J., Duboule, D., Durston, A.J., Kmita, M., Nicolas, J.-F., Palmeirim, I., Smith, J.C., Wolpert, L.: **Head–tail patterning of the vertebrate embryo: one, two or many unsolved problems?** *Int. J. Dev Biol.* 2006, **50**: 3–15.

5.10 Identity of somites along the antero-posterior axis is specified by Hox gene expression

Burke, A.C., Nelson, C.E., Morgan, B.A., Tabin, C.: **Hox genes and the evolution of vertebrate axial morphology.** *Development* 1995, **121**: 333–346.

Nowicki, J.L., Burke, A.C.: **Hox genes and morphological identity: axial versus lateral patterning in the vertebrate mesoderm.** *Development* 2000, **127**: 4265–4275.

Wellik, D.M.: **Hox genes and vertebrate axial pattern.** *Curr. Top. Dev. Biol.* 2009, **88**: 257–278.

Wellik, D.M.: **Hox patterning of the vertebrate axial skeleton.** *Dev. Dyn.* 2007, **236**: 2454–2463.

Box 5E The Hox genes

Pearson, J.C., Lemons, D., McGinnis, W.: **Modulating Hox gene functions during animal body patterning.** *Nat. Rev. Genet.* 2005, **6**: 893–904.

5.11 Deletion or overexpression of Hox genes causes changes in axial patterning

Condie, B.G., Capecchi, M.R.: **Mice with targeted disruptions in the paralogous genes Hoxa3 and Hoxd3 reveal synergistic interactions.** *Nature* 1994, **370**: 304–307.

Favier, B., Le Meur, M., Chambon, P., Dollé, P.: **Axial skeleton homeosis and forelimb malformations in Hoxd11 mutant mice.** *Proc. Natl Acad. Sci. USA* 1995, **92**: 310–314.

Kessel, M., Gruss, P.: **Homeotic transformations of murine vertebrae and concomitant alteration of the codes induced by retinoic acid.** *Cell* 1991, **67**: 89–104.

Mallo, M., Wellik, D.M., Deschamps, J.: **Hox genes and regional patterning of the vertebrate body plan.** *Dev. Biol.* 2010, **344**: 7–15.

Wellik, D.M., Capecchi, M.R.: **Hox10 and Hox11 genes are required to globally pattern the mammalian skeleton.** *Science* 2003, **301**: 363–367.

Yekta, S., Tabin, C., Bartel, D.P.: **MicroRNAs in the Hox network: an apparent link to posterior prevalence.** *Nat. Rev. Genet.* 2008, **9**: 789–796.

5.12 Hox gene expression is activated in an anterior to posterior pattern

Abu-Abed, S., Dollé, P., Metzger, D., Beckett, B., Chambon, P., Petkovich, M.: **The retinoic acid-metabolizing enzyme CYP26A1 is essential for normal hindbrain patterning, vertebral identity and development of posterior structures.** *Genes Dev.* 2001, **15**: 226–240.

Emoto Y., Wada, H., Okamoto, H., Kudo, A., Imai, Y.: **Retinoic acid-metabolizing enzyme Cyp26a1 is essential for determining territories of hindbrain and spinal cord in zebrafish.** *Dev. Biol.* 2005, **278**: 415–427.

5.13 The fate of somite cells is determined by signals from the adjacent tissues

Brand-Saberi, B., Christ, B.: **Evolution and development of distinct cell lineages derived from somites.** *Curr. Topics Dev. Biol.* 2000, **48**: 1–42.

Brent, A.E., Braun, T., Tabin, C.J.: **Genetic analysis of interactions between the somitic muscle, cartilage and tendon cell lineages during mouse development.** *Development* 2005, **132**: 515–528.

Olivera-Martinez, I., Coltey, M., Dhouailly, D., Pourquié, O.: **Mediolateral somitic origin of ribs and dermis determined by quail-chick chimeras.** *Development* 2000, **127**: 4611–4617.

Pourquié, O., Fan, C.-M., Coltey, M., Hirsinger, E., Watanabe, Y., Bréant, C., Francis-West, P., Brickell, P., Tessier-Lavigne, M., Le Douarin, N.M.: **Lateral and axial signals involved in avian somite patterning: a role for BMP-4.** *Cell* 1996, **84**: 461–471.

Shearman, R.M., Tulenko, F.J., Burke, A.C.: **3D reconstructions of quail-chick chimeras provide a new fate map of the avian scapula.** *Dev. Biol.* 2011, **355**: 1–11.

Tonegawa, A., Takahashi, Y.: **Somitogenesis controlled by Noggin.** *Dev. Biol.* 1998, **202**: 172–182.

5.14 Neural crest cells arise from the borders of the neural plate and migrate to give rise to a wide range of different cell types

Huang, X., Saint-Jeannet, J-P.: **Induction of the neural crest and the opportunities of life on the edge.** *Dev. Biol* 2004, **275**: 1–11.

Kelsh, R.N., Erickson, C.A.: **Neural crest: origin, migration and differentiation.** *Encyclopedia of Life Sciences*, 2013, http://www.els.net/

Le Douarin, N., Kalcheim, C.: *The Neural Crest.* 2nd edn. Cambridge: Cambridge University Press, 1999.

5.15 Neural crest cells from the hindbrain migrate to populate the branchial arches

Grammatopoulos, G.A., Bell, E., Toole, L., Lumsden, A., Tucker, A.S.: **Homeotic transformation of branchial arch identity after Hoxa2 overexpression.** *Development* 2000, **127**: 5355–5365.

Krumlauf, R.: **Hox genes and pattern formation in the branchial region of the vertebrate head.** *Trends Genet.* 1993, **9**: 106–112.

Le Douarin, N.M., Creuzet, S., Couly, G., Dupin, E.: **Neural crest plasticity and its limits.** *Development* 2004, **131**: 4637–4650.

Rijli, F.M., Mark, M., Lakkaraju, S., Dierich, A., Dolle, P., Chambon, P.: **A homeotic transformation is generated in the rostral branchial region of the head by disruption of Hoxa2, which acts as a selector gene.** *Cell* 1993, **75**: 1333–1349.

5.16 The bilateral symmetry of the early embryo is broken to produce left-right asymmetry of internal organs

Brennan, J., Norris, D.P., Robertson, E.J.: **Nodal activity in the node governs left-right asymmetry.** *Genes Dev.* 2002, **16**: 2339–2344.

Brown, N.A., Wolpert, L.: **The development of handedness in left/right asymmetry.** *Development* 1990, **109**: 1–9.

McGrath, J., Somlo, S., Makova, S., Tian, X., Brueckner, M.: **Two populations of node monocilia initiate left-right asymmetry in the mouse.** *Cell* 2003, **114**: 61–73.

Rana, A.A., Barbera, J.P., Rodriguez, T.A., Lynch, D., Hirst, E., Smith, J.C., Beddington, R.S.P.: **Targeted deletion of the novel cytoplasmic dynein mD2LIC disrupts the embryonic organiser, formation of body axes and specification of ventral cell fates.** *Development* 2004, **131**: 4999–5007.

Raya, A., Izpisua Belmonte, J.C.: **Unveiling the establishment of left-right asymmetry in the chick embryo.** *Mech. Dev.* 2004, **121**: 1043–1054.

Schlueter, J., Brand, T.: **Left-right axis development: examples of similar and divergent strategies to generate asymmetric morphogenesis in chick and mouse embryos.** *Cytogenet. Genome Res.* 2007, **117**: 256–267.

Shen, M.M.: **Nodal signaling: developmental roles and regulation.** *Development* 2007, **134**: 1023–1034.

5.17 Left-right symmetry breaking may be initiated within cells of the early embryo

Blum, M., Beyer, T., Weber, T., Vivk, P., Andre, P., Bitzer, E., Schweickert, A.: ***Xenopus,* an ideal model system to study vertebrate left-right asymmetry.** *Dev. Dyn.* 2009, **238**: 1215–1225

Gros, J., Feistel. K., Viebahn, C., Blum, M., Tabin, C.J.: **Cell movements at Hensen's node establish left/right asymmetric gene expression in the chick.** *Science* 2009, **324**: 941–944.

Raya, A., Izpisua Belmonte, J.C.: **Insights into the establishment of left-right asymmetries in vertebrates.** *Birth Defects Res C Embryo Today* 2008, **84**: 81–94.

Vandenberg, L.N., Levin, M.: **A unified model for left-right asymmetry? Comparison and synthesis of molecular models of embryonic laterality.** *Dev Biol.* 2013, **379**: 1–15.

Yoshiba, S., Hamada, H.: **Roles of cilia, fluid flow, and Ca^{2+} signaling in breaking of left-right symmetry.** *Trends Genet.* 2014, **30**: 10–17.

6

Development of nematodes and sea urchins

- ● Nematodes
- ● Echinoderms

Having looked at early embryonic development in Drosophila and vertebrates, we will now examine some aspects of the early development of two other model invertebrate organisms that introduce some different developmental mechanisms. The nematode Caenorhabditis elegans is a very important model for the specification of fate associated with asymmetric cell division, as much of its early development involves patterning on a cell-by-cell basis rather than by morphogens affecting groups of cells. Sea urchins, which represent the echinoderms, are models for highly regulative development, and their development illustrates in a simpler system some principles common to the development of vertebrates, which is also highly regulative. Discussion of a third group of invertebrates, the ascidians, can be found online.

This chapter considers aspects of body-plan development in two model invertebrate organisms—nematodes and sea urchins (representing the echinoderms). The evolutionary relationships between the organisms discussed in this chapter are shown in Fig. 1.11. All conform to the general plan of animal development: cleavage leads to a blastula (or blastula-equivalent stage in *Caenorhabditis*), which then undergoes gastrulation with the emergence of a body plan.

There is an old, and now less fashionable, distinction sometimes made between so-called regulative and mosaic development—the difference being that in regulative development decisions about what cells become are made by interactions between groups of cells, whereas in mosaic development, the decisions are taken by individual cells autonomously (see Section 1.12). Nematodes show many instances of mosaic-like development, whereas sea-urchin development is highly regulative. There are, however, elements of both types of development in most embryos.

A feature of nematodes is that cell fate is often specified cell by cell, the characteristic of mosaic development, and in general does not rely on positional information established by gradients of morphogens. This contrasts with the specification of cell fate in groups of cells in *Drosophila*, vertebrates, and the sea urchin. The early embryos of many invertebrates contain far fewer cells than those of flies and vertebrates, with each cell acquiring a unique identity at an early stage of development. In

Scan here

Scan this QR code image with your mobile device to see the online supplementary material on Ascidians or log on to **http://global.oup.com/uk/orc/biosciences/devbiol/wolpert5e/qr/qr1a/**

the nematode, for example, there are only 28 cells when gastrulation starts, compared with thousands in *Drosophila*.

Specification on a cell-by-cell basis often makes use of **asymmetric cell division** and the unequal distribution of cytoplasmic factors (see Section 1.17). Daughter cells resulting from asymmetric cell division sometimes adopt different fates autonomously, as a result of the unequal distribution of some factor between them. However, asymmetric cell division in the early stages of development does not mean that cell–cell interactions are absent or unimportant in these organisms.

We begin with the nematode *Caenorhabditis elegans*, which has been studied intensively and in which many key developmental genes and signaling pathways have been identified. We then consider sea urchins, which develop much more like vertebrates, in that their embryos rely heavily on intercellular interactions and are highly regulative, and patterning involves groups of cells.

Fig. 6.1 Life cycle of the nematode
Caenorhabditis elegans. After cleavage and embryogenesis there are four larval stages (L1–L4) before the sexually mature adult develops. Adults of *C. elegans* are usually hermaphrodite, although males can develop. The photographs show: the two-cell stage (top, scale bar = 10 μm); an embryo after gastrulation with the future larva curled up (middle, scale bar = 10 μm); and the four larval stages and adult (bottom, scale bar = 0.5 mm).

Photographs courtesy of J. Ahringer.

Nematodes

The free-living soil nematode *C. elegans*, whose life cycle is shown in Fig. 6.1, is an important model organism in developmental biology. Its advantages are its suitability for genetic analysis, the small number of cells and their invariant lineage, and the transparency of the embryo, which allows the formation of each cell to be observed. This led Sydney Brenner to choose *C. elegans* as a model organism for studying the genetic basis of development. Brenner and his co-workers, Robert Horvitz and John Sulston, were awarded the Nobel Prize in Physiology or Medicine in 2002 for

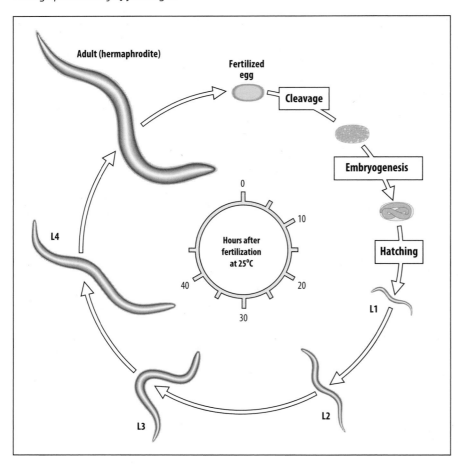

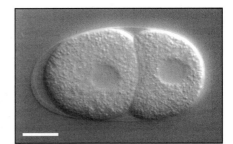

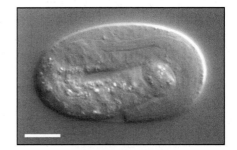

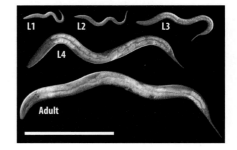

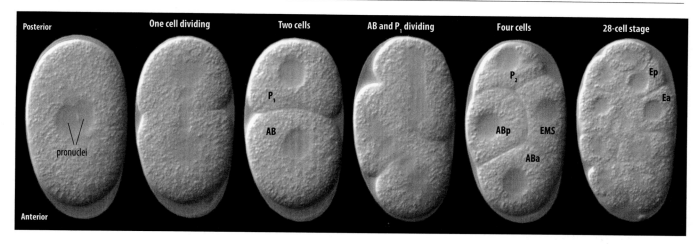

discoveries concerning the genetic control of organ development and apoptosis—cell suicide—in their studies on the worm.

C. elegans has a simple anatomy and the adults are about 1 mm long and just 70 μm in diameter. Nematodes can be grown on agar plates in large numbers and early larval stages can be stored frozen and later resuscitated. *C. elegans* adults are mainly **hermaphrodite**; in nematodes, these are females that make their own sperm for a short time and then switch to making oocytes. They are capable of self-fertilization. Small numbers of males exist and can be used in mating experiments. Embryonic development is rapid, the larva hatching after 15 hours at 20°C, although maturation through larval stages to adulthood takes about 50 hours.

The nematode egg is small, ovoid and only 50 μm long. Polar bodies are formed after fertilization. Before the male and female nuclei fuse, there is what appears to be an abortive cleavage, but after fusion of the nuclei, true cleavage begins (Fig. 6.2). The first cleavage is asymmetric and generates an anterior AB cell and a smaller posterior P_1 cell. At the second cleavage, AB divides to give ABa anteriorly and ABp posteriorly, while P_1 divides to give P_2 and EMS. At this stage the main axes can already be identified, as P_2 is posterior and ABp dorsal.

The AB cells and the P_2 cell will also divide in a well-defined pattern to give rise to the other tissues of the worm. Gastrulation starts at the 28-cell stage, when the descendants of the E cell (itself produced by division of the EMS cell) that will form the gut move inside. Not all cells that are formed during embryonic development survive. **Programmed cell death**, or **apoptosis**, of specific cells is an integral feature of nematode development (Box 6A).

The newly hatched larva (Fig. 6.3), while similar in overall organization to the mature adult, is sexually immature and lacks a gonad and its associated structures,

Fig. 6.2 Cleavage of the *Caenorhabditis elegans* embryo. After fertilization, the pronuclei of the sperm and egg fuse. The egg then divides into a large anterior AB cell and a smaller, posterior P_1 cell. At the next cell division, AB divides into ABa and ABp, while P_1 divides into P_2 and EMS. Each of the cells will continue to divide in a well-defined pattern within these groups to give rise to particular cell types and tissues. Gastrulation begins at the 28-cell stage. By this stage, further division of the EMS cell has produced the E cells, Ea and Ep, which will give rise to the gut, and MS cells (not labeled here) which give rise to a variety of other cell types.

Photographs courtesy of J. Ahringer.

Fig. 6.3 *Caenorhabditis elegans* larva at the L1 stage (20 hours after fertilization). The vulva will develop from the gonad primordium.

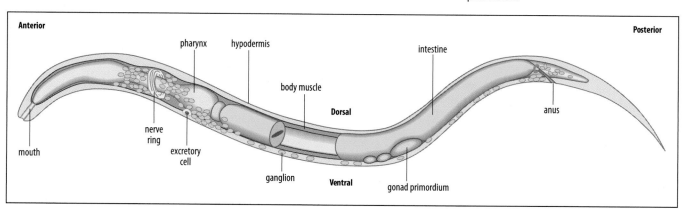

CELL BIOLOGY BOX 6A Apoptotic pathways in nematodes, *Drosophila* and mammals

Programmed cell death, or apoptosis, is an essential part of animal development. In the nematode, with its invariant cell lineage, particular cell lineages or particular cells within specific lineages are destined to end in programed cell death (see Fig. 6.10), and it was this feature that allowed the identification of the process and the analysis of its genetic control. Apoptosis also occurs in the developing vertebrate limb, where cell death separates the digits (see Chapter 11), and in the developing vertebrate nervous system, in which large numbers of neurons die during the formation of neural circuits (see Chapter 12). Apoptosis also plays an important role in homeostasis in adult tissues, and has a key role in controlling growth and preventing cancer (see Chapter 13). Apoptosis is also triggered by specific stimuli, which can be extracellular signals or endogenous factors produced by responses to stress, or by the lack of survival signals from neighboring cells.

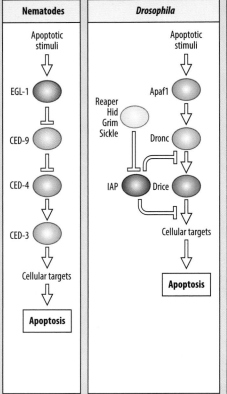

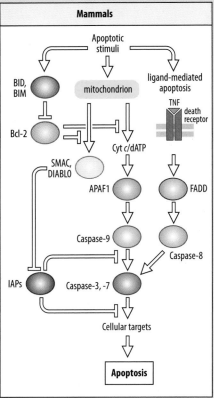

Figure 1

In apoptosis, endonucleases fragment the chromatin into short pieces. The cytoplasm shrinks and breaks up, but the cell contents remain enclosed by cell membrane, thus avoiding damage to neighboring cells by the release of potentially cytotoxic molecules normally sequestered within organelles. Changes in the cell membrane facilitate the disposal of the dying cells by scavenger cells such as macrophages. These features distinguish apoptosis from cell death by **necrosis**, in which the dying cell bursts open, spilling its contents to the detriment of adjacent cells.

The cell-death pathway in the nematode is centered on the actions of CED-3, a type of protease called a **caspase**. Activated CED-3 triggers cellular changes that lead to cell death. The nematode caspase-activation pathway and the equivalent pathways in *Drosophila* and mammals are shown in Figure 1 (homologous proteins are color-coded). In nematodes with mutations that inactivate either the gene for CED-3 or for the adaptor protein CED-4 that activates CED-3, none of the 131 cells that normally die do so, but appear instead to differentiate in the same way as their sister cells. Somewhat surprisingly, such worms appear to survive well and die at the usual age of several weeks.

The nematode's cell-death program is controlled by the protein CED-9, which acts as a brake by constitutively binding CED-4 and

preventing it from activating CED-3. Apoptosis occurs when the protein EGL-1 is produced in response to pro-apoptotic signals and releases CED-4 from CED-9. If the *ced-9* gene is inactivated by mutation, many cells that normally do not die will die, leading to death of the embryo. In *Drosophila* (see Figure 1, middle panel) the function of nematode CED-9 is fulfilled by an inhibitor of apoptosis protein (IAP), which inhibits caspase action. Inhibition by IAP is lifted by Reaper and other proteins.

In mammals, apoptosis is inhibited by Bcl-2 and other related proteins. Some apoptotic stimuli cause the release of cytochrome *c* from mitochondria, which can trigger the caspase-activation pathway (see Figure 1, right panel). The apoptotic stimulus also induces production of the EGL-1 homologs Bim and Bid, which inhibit the action of Bcl-2, thus enabling apoptosis to proceed. As in *Drosophila*, the apoptotic caspases can be inhibited by IAPs. Inhibition is lifted by the pro-apoptotic protein DIABLO, which is also released from mitochondria. Mammalian caspase-9 initiates the proteolytic cascade leading to cell death.

In mammals, and in *Drosophila* (not shown), there is an additional input called **ligand-mediated apoptosis**. Pro-apoptotic ligands such as tumor necrosis factor (TNF) and Fas ligand interact with 'death receptors' (TNFR and Fas, respectively). These signals activate caspase-8 and caspase-3.

such as the vulva of hermaphrodites, which are required for reproduction. Post-embryonic development takes place during a series of four larval stages separated by molts. The additional cells in the adult are derived largely from precursor blast cells (P cells) that are distributed along the body axis. Each of these blast cells founds an invariant lineage involving between one and eight cell divisions. The vulva, for example, is derived from blast cells P5, P6, and P7. One can think of post-embryonic development in the nematode as the addition of adult structures to the basic larval plan.

6.1 The cell lineage of *Caenorhabditis elegans* is largely invariant

It is a triumph of direct observation that, with the aid of Nomarski interference microscopy, the complete lineage of every cell in the nematode *C. elegans* has been worked out (see Further reading). The pattern of cell division is largely invariant—it is virtually the same in every embryo. The larva, when it hatches, is made up of 558 cells, and after four further molts this number has increased to 959, excluding the germ cells, which vary in number. This is not the total number of cells derived from the egg, as 131 cells die during development as a result of apoptosis, mainly during the development of the nervous system. As the fate of every cell at each stage is known, a fate map can be accurately drawn at any stage and has a precision not found in any vertebrate. As with any fate map, however, even where there is an invariant cell lineage, this precision in no way implies that the lineage must determine the fate or that the fate of the cells cannot be altered. As we shall see, interactions between cells have a major role in determining cell fate in the nematode.

The complete genome of *C. elegans* has been sequenced, and contains nearly 20,000 predicted genes. *C. elegans* does not have extensive alternative RNA splicing, so that most genes code for just one protein, which may be part of the reason that the anatomically simple worm apparently needs more genes than the more complex fruit fly. Around 1700 genes have been identified as affecting development, two-thirds of which were found using the technique of RNA interference (RNAi; Box 6B). Many nematode developmental genes are related to genes that control development in *Drosophila* and other animals; they include the Hox genes (see Box 5E), and genes for signaling proteins of the TGF-β, Wnt, and Notch families (see Box 4A) and their associated intracellular pathways. One exception is the Hedgehog signaling pathway (see Box 2F), which is not present in *Caenorhabditis*, although it does have genes for proteins that can be considered to be related to Hedgehog. Although zygotic gene expression begins at the four-cell stage, maternal components control almost all of the development up to gastrulation at the 28-cell stage.

6.2 The antero-posterior axis in *Caenorhabditis elegans* is determined by asymmetric cell division

The development of *C. elegans* relies on an invariant, determinate pattern of cell divisions in which the fates of the cells are linked to the division. The first cleavage of the nematode egg is unequal, dividing the egg into a large anterior AB cell and a smaller posterior P_1 cell. This asymmetry defines the antero-posterior axis and is determined at fertilization.

The P_1 cell behaves rather like a stem cell; at each further division it produces one P-type cell and one daughter cell, anterior in position, that will embark on another developmental pathway. For the first three divisions, the anterior P-cell daughters give rise to precursors of body cells, but after the fourth cleavage they only give rise to germ cells (Fig. 6.4).

Division of the AB cell gives rise to anterior and posterior AB daughter cells. The anterior daughter, ABa, gives rise to typically ectodermal tissues, such as the epidermis (called the **hypodermis** in nematodes) and nervous system, but also to a small portion

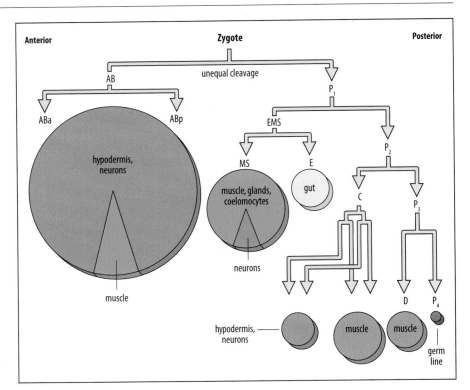

Fig. 6.4 Cell lineage and cell fate in the early *Caenorhabditis elegans* embryo. The cleavage pattern is invariant. The first cleavage divides the fertilized egg into a large anterior AB cell and a smaller posterior P_1 cell and the descendants of these cells have constant lineages and fates. AB divides into ABa, which produces neurons, hypodermis, and muscle (of the pharynx), and ABp, which produces neurons, hypodermis, and some specialized cells. P_1 divides to give EMS and P_2. The EMS cell then divides to give cells MS and E. MS gives rise to muscle, glands, and coelomocytes (free-floating spherical cells), and E produces the gut. Further divisions of the P lineage are rather like stem-cell divisions, with one daughter of each division (C and D) giving rise to a variety of tissues while the other (P_2 and P_3) continues to act as a stem cell. Eventually, P_4 gives rise to the germ cells.

of the mesoderm of the pharynx. The posterior daughter, ABp, also makes neurons and hypodermis as well as some specialized cells. At second cleavage, P_1 divides asymmetrically to give P_2 and EMS, which will subsequently divide into MS and E. MS gives rise to many of the body's muscles and the posterior half of the pharynx, while E is the precursor of the 20 cells of the endoderm of the mid-gut. The C cell, which is derived from the P_2 cell at the third cleavage, forms hypodermis and body-wall muscle; D from the fourth cleavage of the P cell produces only muscle. All the cells undergo further invariant divisions, and at about 100 minutes after fertilization gastrulation begins.

Before fertilization, there is no evidence of any asymmetry in the nematode egg, but sperm entry sets up an antero-posterior polarity in the fertilized egg that determines the position of the first cleavage division and the future embryonic antero-posterior axis. This cleavage is both unequal and asymmetric. The existence of polarity in the fertilized egg becomes evident before the first cleavage. A cap of actin microfilaments forms at the future anterior end, and a set of cytoplasmic granules—the so-called **P granules**, which contain maternal mRNAs and proteins required for development of the germline cells—become localized at the future posterior end. (The P granules are not the determinants of polarization; their redistribution simply reflects it.) P granules remain in the P daughters of cell division, eventually becoming localized to the P_4 cell, which gives rise to the germline (Fig. 6.5).

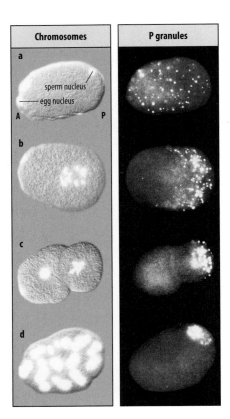

Fig. 6.5 Localization of P granules after fertilization. The movement of P granules is shown during the development of a fertilized egg of *Caenorhabditis elegans* from sperm entry up to the 26-cell stage. In the left panel, the embryo is stained for DNA, to visualize the chromosomes, and in the right panel for P granules. (a) Fertilized egg with egg nucleus at anterior end and sperm nucleus at posterior end. At this stage the P granules are distributed throughout the egg. (b) Fusion of sperm and egg nucleus. The P granules have already moved to the posterior end. (c) Two-cell stage. The P granules are localized in the posterior cell. (d) 26-cell stage. All the P granules are in the P_4 cell, which will give rise to the germline only.

*Photograph reproduced with permission from Strome, S., Wood, W.B.: **Generation of asymmetry and segregation of germ-line granules in early C. elegans embryos**. Cell 1983, 35: 15–25. © 1983 Cell Press.*

EXPERIMENTAL BOX 6B Gene silencing by antisense RNA and RNA interference

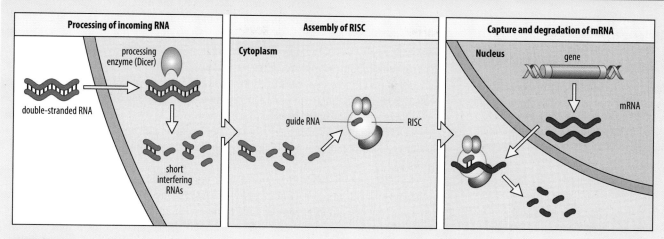

Figure 1

Gene knock-outs remove the function of a gene permanently by disrupting its DNA. Another set of techniques for suppressing gene function rely on destroying or inhibiting the mRNA. Because the gene itself is left untouched, and thus suppression is, in principle, reversible, this type of loss of function is generally known as **gene silencing**. All methods of gene silencing rely on the introduction into embryos or cultured cells of an RNA with a sequence complementary to that of the specific mRNA being targeted. Depending on the particular technique, the introduced RNA either binds to the mRNA and prevents it from being translated, and/or targets a nuclease to bind to the mRNA and degrade it.

The first technique of this type to be developed uses short synthetic **antisense RNAs**, which are usually chemically modified (for example, morpholino RNAs) to increase their stability within the cell. Morpholino oligonucleotides have been used extensively to block specific gene expression, especially in sea urchins, *Xenopus*, and zebrafish.

A more recent addition to the gene-silencing armoury is the technique of **RNA interference (RNAi)**, which recruits a natural RNA-degrading mechanism that is apparently ubiquitous in multicellular eukaryotes from plants to mammals. The phenomenon was first uncovered in plants during transgenic experiments, when it was found that introduction of a gene very similar to one of the plant's own genes blocked the expression of both the introduced gene and the endogenous gene. It turned out that suppression was occurring at the level of the mRNAs, which appeared to be rapidly degraded. RNAi is now known to be caused by the degradation of double-stranded RNAs by a cellular enzyme called Dicer, which chops them into short lengths of around 21–23 nucleotides. These so-called **short interfering RNAs (siRNAs)** are unwound into single-stranded RNA, which becomes incorporated into a nuclease-containing protein complex known as RISC (for RNA-induced silencing complex). The RNA-binding component in RISC is a ribonuclease called Argonaute, also called Slicer in mammalian cells. The siRNA acts a 'guide RNA' to target RISC to any mRNA containing an exactly complementary sequence. The mRNA is then degraded by the nuclease in the RISC (see Figure 1). RNAi is particularly effective and easy to carry out in *C. elegans*. Adult worms injected with double-stranded RNA will show suppression of the corresponding gene in the embryos they produce. Worms can also be soaked in the appropriate RNA or fed on bacteria expressing the required double-stranded RNA. The natural role of the cellular RNAi machinery is thought to be defense against transposons and viruses, many of which produce double-stranded RNAs in the course of replication.

In plants, fungi, nematodes, *Drosophila*, and most non-mammalian cells, RNAi is carried out by introducing an appropriate double-stranded RNA, which is then processed into siRNAs within the cells. In mammalian cells, however, double-stranded RNAs trigger another defensive response that interferes with the gene-silencing effect, and RNAi is instead achieved by introducing the siRNAs themselves, or by expressing artificial DNA constructs from which siRNAs are transcribed.

The initial polarization of the fertilized egg seems to be due to the centrosome that the sperm nucleus brings into the egg. The centrosome, or cell center, is a small structure composed of microtubules that organizes the microtubule cytoskeleton of the cell; it duplicates to form the two centrosomes that nucleate each pole of the microtubule spindle that organizes chromosome distribution in mitosis and meiosis (the mitotic cell cycle is discussed in Box 1B, and the role of the centrosome in determining

Scan here

Scan this QR code image with your mobile device to see an online movie showing early polarisation of the *C. elegans* embryo or log on to **http://global.oup.com/uk/orc/biosciences/devbiol/wolpert5e/qr/qr6b/**

Fig. 6.6 Spindle movement in the nematode zygote leads to an unequal first division. Before the first division, the posterior pole of the mitotic spindle moves towards the posterior end of the cell, leading to an asymmetric spindle in which the mid-zone has moved from the center of the cell. Division at the site of the midzone leads to daughter cells (AB and P_1) of unequal sizes.

Adapted from: The Cell Cycle, Morgan, D.O. 2007 and information from Nance & Zallen. **Elaborating polarity: PAR proteins and the cytoskeleton.** Development *2011,* **138:** *799-809.*

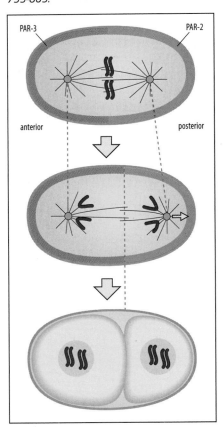

the plane of cell division is discussed further in Chapter 9. Microtubules, actin, and other components of the cytoskeleton are described in Box 9B). The egg cortex, the layer of cytoplasm immediately under the plasma membrane, contains a dense network of actin filaments associated with the motor protein myosin, forming contractile macromolecular assemblies of actomyosin. Interaction of the centrosome with the actin network causes the network to contract asymmetrically away from the point of sperm entry about 30 minutes after fertilization. This leads to a general flow of cortical components towards the future anterior end of the zygote, and cytoplasmic flow towards the future posterior end. Among the cortical proteins that are relocated by this movement are a group of maternal proteins known as **PAR (partitioning) proteins**. These are initially uniformly distributed, but become concentrated in anterior or posterior cortex after cortical movement and control the asymmetric divisions in the early embryo.

The PAR proteins were discovered in a very ingenious way. It was thought that mutations affecting the asymmetry of the first cleavage would be very rare, and that vast numbers of embryos would have to be screened to find them. The process was streamlined by using a mutant—*egl*—in which the fertilized eggs are not released and the larvae hatch out inside the parent, devouring it from the inside and killing it. Mutations that halt development early on spare the parent, and this is how the maternal *par* genes were discovered.

Partitioning-defective embryos have defects in the asymmetric divisions that divide the zygote into blastomeres. If the normal pattern of division is not precisely maintained, future development is affected. The *par* genes encode several different types of proteins, including protein kinases and other proteins involved in intracellular signaling pathways. PAR proteins similar to those in *C. elegans* are involved in establishing and maintaining cell polarity in many different situations, and are present in animals ranging from nematodes to mammals. In *C. elegans*, the correct localization of PAR proteins is required for the correct positioning and orientation of mitotic spindles in the early cleavages to ensure division at the required place in the cell and in the required plane.

The cortical flow resulting from sperm entry and actomyosin network contraction leads to the anterior localization of PAR-3 and PAR-6, and the posterior localization of PAR-1 and PAR-2. This asymmetric distribution then results, by mechanisms that are not yet completely understood, in the positioning of the first mitotic spindle posterior to the center of the cell, and thus to a first cleavage that is both unequal (producing cells of different sizes) and functionally asymmetric (producing cells containing different developmental determinants) (Fig. 6.6). The PAR proteins are thought to be involved in regulating the forces that pull on each pole of the spindle to position it in the cell. The mechanisms that control the first two cleavage divisions are extremely complex, as a large-scale RNAi experiment that individually suppressed around 98% of the nematode's genes showed that more than 600 gene products were involved in these two stages.

6.3 The dorso-ventral axis in *Caenorhabditis elegans* is determined by cell–cell interactions

Despite the highly determinate cell lineage in the nematode, cell–cell interactions are involved in many cell-fate specification events. Specification of the dorso-ventral axis is one example. At the time of the second cleavage, if the future anterior ABa cell is pushed and rotated with a glass needle, not only is the antero-posterior order of the daughter AB cells reversed, but the cleavage of P_1 is also affected, so that the position of the P_1 daughter cell EMS relative to the AB cells is inverted (Fig. 6.7). The manipulated embryo develops completely normally but 'upside down', with its dorso-ventral axis inverted within the eggshell. ABp develops as the anterior cell and ABa

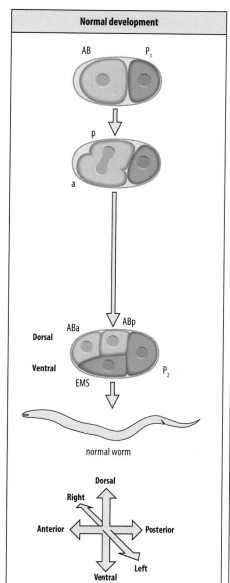

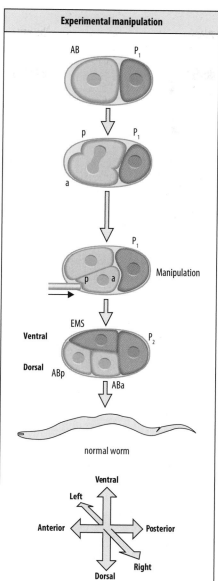

Fig. 6.7 Reversal of dorso-ventral polarity at the four-cell stage of the nematode. In normal embryos, the AB cell rotates at cleavage so that ABa is anterior. If the AB cell is mechanically rotated in the opposite direction at second cleavage, the ABp cell is now anterior. This manipulation also displaces the P_1 cell so that when it divides, the position of the EMS daughter cell is reversed with respect to the AB cells. Development is still normal, but the dorso-ventral axis is inverted (the worm develops 'upside down' in the egg case) a and p show the anterior/posterior orientation of the AB cell before and after manipulation.

Illustration after Sulston, J.E., et al.: **The embryonic cell lineage of the nematode C. elegans.** Dev. Biol. *1983, 100: 69-119.*

as the posterior cell of the pair, and the polarity of P_2 has been reversed compared to its polarity in the unmanipulated embryo. This shows that the dorso-ventral polarity of the embryo cannot already be fixed at this stage and that the usual developmental fate of these cells can be changed: this implies that their fate is specified by cell–cell interactions. The normal development after this manipulation also implies that the left–right axis of symmetry, which is defined in relation to the antero-posterior and dorso-ventral axes, is not yet determined. Indeed, the left–right axis can be reversed by manipulation at a slightly later stage, as we see next.

Adult worms have a well defined left–right asymmetry in their internal structures, and the pattern of development on the left and right sides of the embryo is strikingly different, more so than in many other animals. Not only do cell lineages on the left and right differ, but some cells move from one side to the other during embryonic development. The concept of left and right only has meaning once the antero-posterior and dorso-ventral axes are defined (see Section 5.16). In *C. elegans*, specification of left and right normally occurs at the third cleavage, and handedness can be reversed by experimental manipulation at this stage. At third cleavage, ABa and ABp

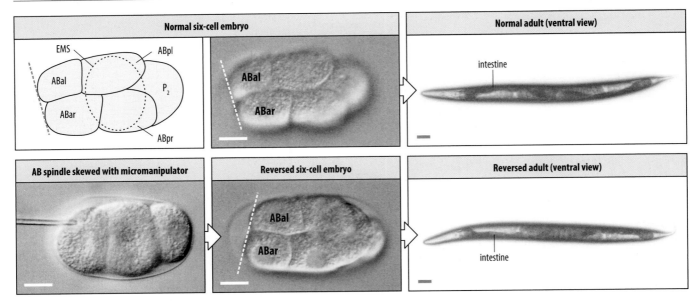

Fig. 6.8 Reversal of handedness in *Caenorhabditis elegans*. At the six-cell stage in a normal embryo, the AB cell on the left side (Abal) is slightly anterior to that on the right (top left panels, scale bar = 10 μm). Manipulation that makes the AB cell on the right side (ABar) more anterior results in an animal with reversed handedness (bottom left panels, scale bars = 10 μm). The right panels show the resulting normal and reversed adults; scale bars = 50 μm.

*Photographs reproduced with permission from Wood W.B: **Evidence from reversal of handedness in C. elegans embryos for early cell interactions determining cell fates**. Nature 1991, **349**: 536-538. © 1991 Macmillan Magazines Ltd.*

each divides to produce laterally disposed right and left daughter cells (for example, ABa produces ABal and ABar). The plane of cleavage is, however, slightly asymmetric, so that the left daughter cell lies just a little anterior to its right-hand sister. If the cells are manipulated with a glass rod during this cleavage, their positioning can be reversed so that the right-hand cell lies slightly anterior (Fig. 6.8). This small manipulation is sufficient to reverse the handedness of the animal.

While the molecular mechanisms specifying handedness are not known, mutation of the maternal gene *gpa-16* results in near-randomization of spindle orientation. Using temperature-sensitive mutant alleles of *gpa-16*, and raising the embryos at a temperature at which the first two cleavages are often normal, it proved possible to pinpoint the third cleavage as the one at which handedness is determined. In mutant worms that survived to maturity, handedness had been reversed from right to left in around 50%, and this could be traced back to a skewing of the mitotic spindle axis through 90° at third cleavage. Embryos that died had complete misorientation of spindle axes at the third cleavage. The protein GPA-16 is a G-protein α subunit which, together with other proteins, is involved in the positioning of centrosomes, and thus of the mitotic spindle.

There are ways of generating functional asymmetry other than unequal cell division: for example, at a much later stage in nematode development a functional difference is generated between two taste receptors, one on the left and one on the right, so that each expresses genes for different chemoreceptors. This asymmetry in gene expression is controlled by a microRNA, lsy-6.

6.4 Both asymmetric divisions and cell–cell interactions specify cell fate in the early nematode embryo

Although cell lineage in the nematode is invariant, experimental evidence, such as that described above, shows that cell–cell interactions are crucial in specifying cell fate at very early stages. Otherwise, the reversal of ABa and ABp by micromanipulation just as they are being formed (see Fig. 6.7) would not give rise to a normal worm. This experiment indicates that ABa and ABp must initially be equivalent, and their fate must be specified by interactions with adjacent cells. Evidence for such an interaction comes from removing P_1 at the first cleavage: the result is that the pharyngeal cells normally produced by ABa are not made. These experiments and observations also demonstrate that in lineage-driven organisms such as *C. elegans*, the lineage positions the cells that need to interact in the right positions realtive to each other. This explains the apparently paradoxical importance of cell–cell interactions in an otherwise lineage-driven organism.

What are the interactions that specify the non-equivalence of the two AB descendants? If ABp is prevented from contacting P_2 it develops as an ABa cell. Thus, the P_2 blastomere is responsible for specifying ABp. The induction of an ABp fate by P_2 involves the proteins GLP-1 in the ABp cell and APX-1 in P_2. These proteins are the nematode counterparts of the receptor protein Notch and its ligand Delta (see Box 5D), respectively and, like Notch and Delta, are associated with the cell membrane. As we have seen in previous chapters, interactions between Notch and its various ligands set cell fate in many developmental situations throughout the animal kingdom.

GLP-1 is one of the earliest proteins to be spatially localized during nematode embryogenesis. The maternal *glp-1* mRNA is uniformly present throughout the embryo but its translation is repressed in the P-cell lineage; at the two-cell stage the GLP-1 protein is thus only present in the AB cell. This strategy is highly reminiscent of the localization of maternal Hunchback protein in *Drosophila* (see Section 2.9), and, as in that case, in *C. elegans* other maternal proteins binding to the 3′ untranslated region of *glp-1* mRNA suppress its translation.

After second cleavage the ABa and ABp cells both contain GLP-1. The two cells are then directed to become different by a local inductive signal, sent to the ABp cell from the P_2 cell at the four-cell stage (Fig. 6.9). This signal depends on cell contact: it is delivered by APX-1 (produced by the P_2 cell), which acts as an activating ligand for GLP-1 in the ABp cell membrane. As a result of this induction, the descendants of ABa and ABp respond differently to later signals from descendants of the EMS lineage.

The EMS cell gives rise to daughter cells (MS and E) that give rise to mesoderm and endoderm, respectively (see Fig. 6.4). Specification of the EMS cell as a mesendodermal precursor also involves localized maternal determinants in the EMS cell and inductive signals from the adjacent P_2 cell (see Fig. 6.9). One maternal determinant involved in giving EMS its mesendodermal fate is the transcription factor SKN-1, the product of the *skin excess* (*skn-1*) gene. *skn-1* mRNA is uniformly distributed at the two-cell stage, but there are much higher levels of SKN-1 protein in the nucleus of P_1 than in AB. Maternal-effect mutations that abolish the function of *skn-1* result in the EMS blastomere adopting a mesectodermal fate; that is, it develops much like its sister cell P_2 (see Fig. 6.4), giving rise to mesoderm (body-wall muscle) and ectoderm (hypodermis—hence the name 'skin excess' for the *skn-1* mutant), but no pharyngeal cells and, in most mutants, no endoderm. SKN-1 acts by switching on the zygotic genes *med-1* and *med-2*, which encode transcription factors that are necessary for MS fate and which contribute to the E fate in parallel with other maternal inputs.

Formation of endoderm from the EMS cell also requires an inductive signal and is therefore another example of cell interactions in a lineage-based system. If removed from the influence of its neighbors towards the end of the four-cell stage, an EMS cell can develop in isolation to produce gut structures, but if isolated at the beginning of that stage, it cannot. Removal of P_2 at the early four-cell stage results in no gut being formed, indicating that P_2 is the cell that delivers the inductive signal. This was confirmed by recombining isolated EMS and P_2 cells, which restores development of endoderm; recombining EMS with other cells from the four-cell embryo has no effect. The P_2 cell signal to the EMS cell is a Wnt protein (MOM-2), which interacts with a Frizzled receptor (MOM-5) on the EMS cell surface (see Fig. 6.9), the point of contact

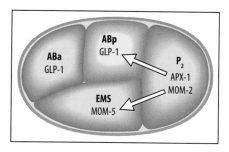

Fig. 6.9 The P_2 cell is the source of inductive signals that determine anteroposterior polarity and cell fate. ABp becomes different from ABa as a result of signaling from the adjacent P_2 cell by the Delta-like protein APX-1, which interacts with the Notch-like receptor GLP-1 on the ABp cell. P_2 also sends a Wnt-like inductive signal, MOM-2, to the EMS cell that interacts with a Frizzled-like receptor, MOM-5, on EMS to set up a posterior difference in the cell that determines the different fates of anterior and posterior daughter cells. These interactions depend on cell contact.

*Illustration after Mello, C.C., et al.: **The maternal genes apx-1 and glp-1 and establishment of dorsal-ventral polarity in the early** C. elegans **embryo**. Cell 1994, 77: 95-106.*

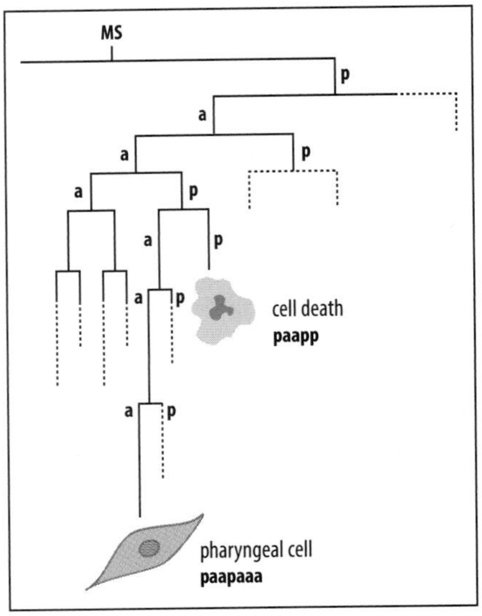

Fig. 6.10 Cell fate is linked to the pattern of cell divisions. The figure illustrates part of the cell lineages generated by the MS blastomere, which give rise to body muscle and the mesodermal cells of the pharynx. Each division produces an anterior (a) and a posterior (p) cell. The lineage paapp, for example, always gives rise to a cell that undergoes programmed cell death and dies by apoptosis (see Box 6A). The lineage paapaaa, on the other hand, always results in a particular pharyngeal cell.

specifying the future posterior cell, which will be E. Signaling by P$_2$ thus serves to break two symmetries in the embryo: P$_2$ makes ABp different from ABa, and it also polarizes the EMS cell so that its division produces a posterior E cell and an anterior MS cell.

So we begin to see how a combination of cell–cell interactions and localized cytoplasmic determinants specify cell fate in the early nematode embryo. The result of killing individual cells by laser ablation at the 32-cell stage, at around the time gastrulation begins, suggests that many of the cell lineages are now determined, because when a cell is destroyed at this stage it cannot be replaced and its normal descendants are absent. Cell–cell interactions are, however, required again later to effect cells' final differentiation.

6.5 Cell differentiation in the nematode is closely linked to the pattern of cell division

Each blastomere in the early *C. elegans* embryo undergoes a unique and nearly invariant series of cleavages that successively divide cells into anterior and posterior daughter cells. Cell fate appears to be specified by whether the final differentiated cell is descended through the anterior (a) or posterior (p) cell at each division. In the lineage generated from the MS blastomere, for example, the cell resulting from the sequence p-a-a-p-p undergoes apoptosis (Fig. 6.10), whereas that resulting from p-a-a-p-a-a gives rise to a particular pharyngeal cell. What is the causal relation between the pattern of division and cell differentiation?

One answer appears to lie in the anterior daughter cell having a higher intranuclear level of the maternal protein POP-1, a transcriptional regulatory protein, compared with its posterior sister. In the case of the EMS blastomere, for example, the anterior daughter cell, the MS blastomere, has a higher level of intranuclear POP-1 than the posterior daughter, the E blastomere, as a result of an asymmetric division of the EMS cell. In a maternal-effect *pop-1* mutant that does not have any POP-1 protein, MS adopts an E-like fate because the antero-posterior distinction is lost.

POP-1 is the *C. elegans* version of the vertebrate transcription factor TCF. In the canonical Wnt signaling pathway, β-catenin enters the nucleus as a result of Wnt signaling, interacts with TCF and converts it from a transcriptional repressor to a transcriptional activator (see Box 4B). The differential distribution of POP-1 in the EMS daughter cells is the ultimate result of the P$_2$ Wnt signal (MOM-2) acting on the receptor Frizzled (MOM-5) on the EMS cell (see Fig. 6.9). This Wnt/POP-1/β-catenin 'asymmetry pathway' has not been found in *Drosophila* or vertebrates. This pathway has two branches that independently control the actions of two nematode-specific β-catenins, WRM-1 and SYS-1. The MOM-2 Wnt signal received at the posterior end of the EMS cell polarizes the cell, resulting in the differential localization of cytoplasmic factors controlling the pathway in the anterior (MS) and posterior (E) daughter cells after cell division. As a result of the localization, a complex of WRM-1 and a protein kinase (LIT-1) accumulates in the E-cell nucleus, which leads to the phosphorylation of POP-1 and its export from the nucleus (Fig. 6.11). The other β-catenin, SYS-1, also accumulates in the anterior nucleus, and is able to act as co-activator to the reduced amounts of POP-1 and help to switch on endoderm-specifying genes. In the MS cell, on the other hand, WRM-1–LIT-1 is exported from the nucleus, thus keeping POP-1 levels high there and suppressing the endodermal fate.

The Wnt/β-catenin asymmetry pathway is thought to be a general mechanism of regulating anterior/posterior cell fate throughout nematode development, switching on different target genes in different situations. This Wnt/POP-1 pathway is iteratively used in many fate assignments linked to cell division in *C. elegans*.

At the 80-cell gastrula stage a fate map can be made for the nematode embryo (Fig. 6.12). At this stage, cells from different lineages that will contribute to the same organ, such as the pharynx, cluster together. The gut is made clonally from E derivatives only. Genes that appear to act as organ-identity genes may now be expressed by

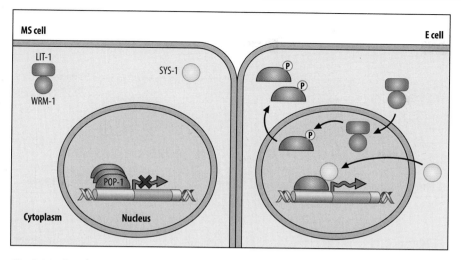

Fig. 6.11 The *Caenorhabditis* Wnt/β-catenin asymmetry pathway determines anterior and posterior cell fates in the MS and E cells. The Wnt (MOM-2) signal originally received by Frizzled receptors (MOM-5) at the posterior end of the EMS cell (not shown) sets up anterior-posterior differences in the EMS cell that become evident in its daughter MS (anterior) and E (posterior) cells. POP-1 is the *C. elegans* versions of the transcription factor TCF. SYS-1 and WRM-1 are nematode-specific β-catenins. In the E cell, SYS-1 is stabilized and accumulates in the nucleus. In addition, a complex of WRM-1 and the protein kinase LIT-1 also accumulates in the nucleus and phosphorylates POP-1. This leads to the export of POP-1 from the E-cell nucleus, reducing the amount to a level at which SYS-1 can act as co-activator for POP-1 and switch on endoderm-specifying target genes. In the MS cell, in contrast, neither SYS-1 nor the WRM-1-LIT-1 complex accumulate in the nucleus and so levels of POP-1 remain high and target genes remain suppressed.

*Illustration after Mello, C.C., et al.: **The maternal genes apx-1 and glp-1 and establishment of dorsal-ventral polarity in the early C. elegans embryo**. Cell 1994, **77**: 95-106.*

the cells in the cluster. The *pha-4* gene, for example, seems to be an organ-identity gene for the pharynx. Mutations in *pha-4* result in loss of the pharynx, whereas ectopic expression of normal *pha-4* in all the cells of the embryo leads to all cells expressing pharyngeal cell markers.

6.6 Hox genes specify positional identity along the antero-posterior axis in *Caenorhabditis elegans*

The nematode body plan is very different from that of vertebrates or *Drosophila*, and there is no segmental pattern along the antero-posterior axis; nevertheless, as in other organisms, Hox genes are involved in specifying positional identity along this axis. Like other organisms the nematode contains a large number of homeobox genes, of which only six are orthologous with those in the Hox gene clusters of *Drosophila* or vertebrates. Four of these genes—*lin-39*, *ceh-13*, *mab-5*, and *egl-5*—are relatively closely linked, with the remaining two, *php-3* and *nob-1*, some distance away on the same chromosome. Comparisons with other nematode species indicate loss of Hox genes and disruption of the ancestral Hox gene cluster in the lineage leading to *C. elegans* and its close relatives.

Of the *C. elegans* Hox genes only *ceh-13*, which is required for anterior organization, is essential for embryonic development, and it seems that the other Hox genes carry out their function of regional specification at the larval stage. With the exception of *ceh-13*, the order of the genes along the chromosome is reflected in their spatial pattern of expression, as shown for three Hox genes in Fig. 6.13. *lin-39* appears to control the fate of mid-body cells and is known to be involved in regulating the development of the vulva in hermaphrodites (discussed later in this chapter), whereas *mab-5* controls the development of a region slightly more posterior, and *egl-5* provides positional

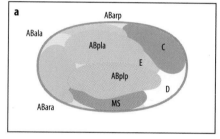

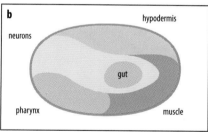

Fig. 6.12 Fate map of the 80-cell gastrula of *Caenorhabditis elegans*. In panel (a), the regions of the embryo are color-coded according to their blastomere origin. In panel (b) they are colored according to the organs or tissues they will ultimately give rise to. At this stage in development, cells from different lineages that contribute to the same tissue or organ have been brought together in the embryo. For clarity, the domains occupied by ABpra and ABprp are not shown in panel a. Anterior is to the left, ventral down.

*After Labouesse, M. and Mango, S. E.: **Patterning the C. elegans embryo: moving beyond the cell lineage**. Trends Genet. 1999, **15**: (8)8, 307-313.*

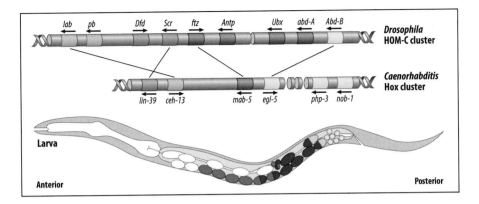

Fig. 6.13 The *Caenorhabditis elegans* Hox gene cluster and its relation to the HOM-C gene cluster of *Drosophila*. The nematode contains a cluster of six Hox genes, four of which show homologies with Hox genes of the fly. The pattern of expression of three of the genes in the larva is shown.

Illustration after Bürglin, T.R., Ruvkun, G.: **The** *Caenorhabditis elegans* **homeobox gene cluster.** Curr. Opin. Gen. Dev. *1993,* **3:** *615–620.*

information for posterior structures. *php-3* and *nob-1* are also posterior group genes. As in other organisms, mutations in Hox genes can cause cells in one region of the body to adopt fates characteristic of other body regions. For example, a mutation in *lin-39* can result in larval mid-body cells expressing fates characteristic of more anterior or more posterior body regions.

Despite the fact that nematode Hox gene expression occurs in a regional pattern, the pattern may not be lineage dependent. For example, in the larva, cells expressing the Hox gene *mab-5* all occur in the same region (see Fig. 6.13) but are quite unrelated by lineage. The expression of *mab-5* is due to extracellular positional signals.

6.7 The timing of events in nematode development is under genetic control that involves microRNAs

Embryonic development gives rise to a larva of 558 cells and there are then four larval stages, which produce the adult. Because each cell in the developing nematode can be identified by its lineage and position, genes that control the fates of individual cells at specific times in development can also be identified. This enables the genetic control of timing during development to be studied particularly easily in *C. elegans*. The order of developmental processes is of central importance, as well as the time at which they occur. Genes must be expressed both in the right place and at the right time. One well-studied example of timing in nematode development is the generation of different patterns of cell division and differentiation in the four larval stages of *C. elegans*, which are easily distinguished in the developing cuticle.

Mutations in a small set of genes in *C. elegans* change the timing of cell divisions in many tissues and cell types. Mutations that alter the timing of developmental events are called **heterochronic**. The first two heterochronic genes to be discovered in *C. elegans* were *lin-4* and *lin-14*, and we shall use them to illustrate both the phenomenon of heterochrony and the control of this process by microRNAs. Different mutations in either of these genes can produce either 'retarded' or 'precocious' development.

Changes in developmental timing resulting from mutations in *lin-14* can be illustrated by what happens to the T-cell lineage (Fig. 6.14). In wild-type individuals, the T.ap cell gives rise to hypodermal cells, neurons, and their support cells by a pattern of stereotyped divisions in larval stages L1 and L2. During larval stages L3 and L4, some of the T-cell descendants divide further to give rise to other adult structures. Loss-of-function mutations in *lin-14* result in a precocious phenotype—for example, the pattern of cell divisions seen in L1 is lost and post-embryonic development starts with the cell divisions normally seen in L2. In contrast, gain-of-function mutations in *lin-14* result in retarded development. Post-embryonic development begins normally, but the cell-division patterns of the first or second larval stages are repeated.

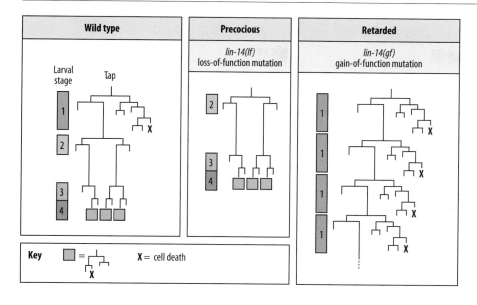

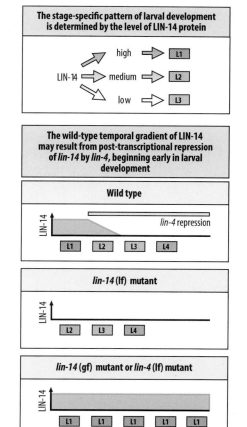

Fig. 6.14 Cell-lineage patterns in wild-type and heterochronic mutants of *Caenorhabditis elegans.* The lineage of the T-blast cell (T.ap) continues through four larval stages (left panel). Mutants in the gene *lin-14* show disturbance in the timing of cell division, resulting in changes in the patterns of cell lineage. Loss-of-function mutations result in a precocious lineage pattern, with the pattern of development of early stages being lost (center panel). Gain-of-function mutations result in retarded lineage patterns, with the patterns of the early larval stage being repeated (right panel).

It has been suggested that the genes that control timing of developmental events may do so by controlling the concentration of some substance, causing it to decrease with time (Fig. 6.15). This temporal gradient could control development in much the same way that a spatial gradient can control patterning. This type of timing mechanism seems to operate in *C. elegans*, because the concentration of LIN-14 protein drops tenfold between the first and later larval stages. Differences in the concentration of LIN-14 at different stages of development may specify the fates of cells, with high concentrations specifying early fates and low concentrations later fates. Thus, the decrease of LIN-14 during development could provide the basis of a precisely ordered temporal sequence of cell activities. Dominant gain-of-function mutations of *lin-14* keep LIN-14 protein levels high, and so in these mutants the cells continue to behave as if they are at an earlier larval stage. In contrast, loss-of-function mutations result in abnormally low concentrations of LIN-14, and the larvae therefore behave as if they were at a later larval stage.

Developmental timing in *C. elegans* was one of the first processes discovered in which gene expression was found to be controlled by microRNAs (Box 6C). The expression of *lin-14* is controlled post-transcriptionally by *lin-4*, which encodes a microRNA that represses the translation of *lin-14* mRNA and downregulates the concentration of LIN-14 protein. LIN-14 itself is a transcription factor, but no target genes have yet been identified that can explain its heterochronic loss-of-function and gain-of-function phenotypes. Increasing synthesis of *lin-4* RNA during the later larval stages generates the temporal gradient in LIN-14, as indicated by the fact that loss-of-function mutations in *lin-4* have the same effect as gain-of-function mutations in *lin-14*. Expression of another gene involved in timing, *lin-41*, is also regulated post-transcriptionally by a microRNA, *let-7*, which represses its translation.

Fig. 6.15 A model for the control of the temporal pattern of *Caenorhabditis elegans* larval development. Top panel: the stage-specific pattern of larval development is determined by a temporal gradient of the protein LIN-14, which decreases during larval development. A high concentration in the early stages specifies the pattern of normal early-stage development as shown in Fig. 6.14 (left panel). The reduction in LIN-14 at later stages is due to the inhibition of *lin-14* mRNA translation by *lin-4* RNA. Bottom panels: loss-of-function (lf) mutations in *lin-14* result in the absence of the first larval stage (L1) lineage, whereas gain-of-function (gf) mutations, which maintain a high level of LIN-14 throughout development, keep the lineage in an L1 phase. Loss-of-function mutations in *lin-4* result in a lifting of repression of *lin-14*, a continued high activity of LIN-14, and a repetition of the L1 lineage.

BOX 6C Gene silencing by microRNAs

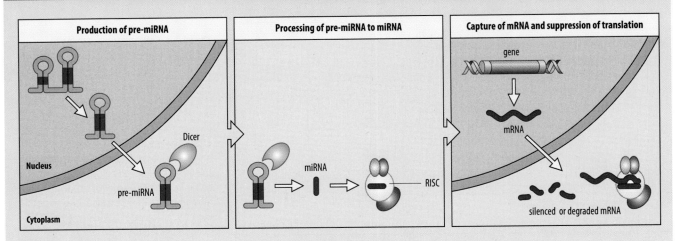

Figure 1

A new category of genes involved in development has been recognized quite recently, which codes not for proteins but for specialized short RNAs, called **microRNAs (miRNAs)**. These regulate gene expression by preventing specific mRNA transcripts from being translated. This natural method of gene silencing has similarities with RNA interference by small interfering RNAs (siRNAs) described in Box 6B, but seems to be a distinct phenomenon, although there is some overlap in the processing machinery. MicroRNAs were first identified in *Caenorhabditis elegans* as the products of the genes *let-7* and *lin-4*, which control the timing of development (see text). Developmental roles are known for a few other miRNAs in nematodes, and for the miRNA *bantam* in *Drosophila*, which prevents cell death. There is now evidence that many human genes code for miRNAs and regulatory roles for human miRNAs are being revealed in cancer.

The primary miRNA transcript is several hundred nucleotides long and contains an inverted repeat of the mature miRNA sequence. After initial processing in the nucleus, the transcript is folded back on itself to form a double-stranded RNA hairpin, the pre-miRNA, which is exported to the cytoplasm (see Figure 1). Here it is cleaved by the enzyme Dicer to produce a short single-stranded miRNA about 22 nucleotides long. As in RNA interference, miRNAs become incorporated as guide RNAs in an RNA-induced silencing protein complex (RISC; see Box 6B) and the complex is specifically targeted to an mRNA. Unlike siRNAs, animal miRNAs are typically not perfectly complementary to their target mRNA and have a few mismatched bases. Once bound, the protein complex renders the mRNA inactive and suppresses translation, and the mRNA may be degraded. miRNA can also repress transcription of genes.

Other non-coding RNAs have a role in regulating gene expression but work in a different way by physically binding to the chromosomes to silence expression. *Xist* RNA, which is involved in **X-chromosome inactivation** in mammals, is one example and is discussed in Chapter 10.

6.8 Vulval development is initiated by the induction of a small number of cells by short-range signals from a single inducing cell

The vulva forms the external genitalia of the adult hermaphrodite worm and connects with the uterus. It is an important example of a structure that is initially specified as a small number of cells—just four, one inducing and three responding. Unlike the other aspects of *Caenorhabditis* development discussed so far, the vulva is an adult structure that develops in the last larval stage. The mature vulva contains 22 cells, with a number of different cell types, and more than 40 genes are involved in its development. It is derived from ectodermal cells originating from the AB blastomere. Six of these precursor blast cells (P cells, not to be confused with the P (posterior) cells of the early embryo) persist in the larva in a row aligned antero-posteriorly on the ventral side of the larva, and posterior daughters of three of these—P5p, P6p, and P7p—give rise to the vulva, each having a well defined lineage. One of the functions of the Hox gene

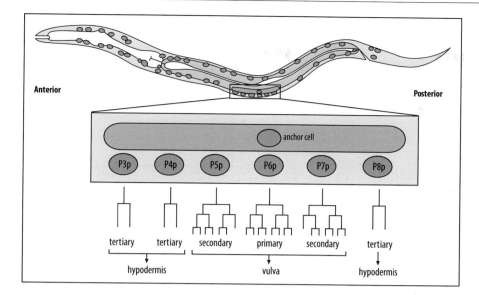

Fig. 6.16 Development of the nematode vulva. The vulva develops from three cells, P5p, P6p, and P7p, in the post-embryonic stages of nematode development. Under the influence of a fourth cell—the anchor cell—P6p undergoes a primary pathway of differentiation that gives rise to eight vulval cells. P6p is flanked by P5p and P7p, which each undergo a secondary pathway of differentiation that gives rise to seven cells of different vulval cell type. Three other P cells nearby adopt a tertiary fate and give rise to hypodermis.

lin-39 is to prevent the fusion of the six P cells to form hypodermis in the early larval stages, which is the fate of the rest of the ventral ectodermal cells of this lineage.

In discussing the development of the vulva, three distinct cell fates are conventionally distinguished—primary (1°), secondary (2°), and tertiary (3°). The primary and secondary fates refer to different cell types in the vulva, whereas the tertiary fate is non-vulval. P6p normally gives rise to the primary lineage, whereas P5p and P7p give rise to secondary lineages. The other three P cells give rise to the tertiary lineage, divide once and then fuse to form parts of the hypodermis (Fig. 6.16). Initially, however, all six P cells are equivalent in their ability to develop as vulval cells. One of the questions we address here is how the final three cells become selected as the vulval precursor cells.

The fates of the three vulval precursor cells are specified by an inductive signal from a fourth cell, the gonadal anchor cell, which confers a primary fate on P6p, the cell nearest to it, and a secondary fate on P5p and P7p, the cells lying just beyond it (see Fig. 6.16). In addition, once induced, the primary cell inhibits its immediate neighbors from expressing a primary fate. P cells that do not receive the anchor cell signal adopt a tertiary fate.

After induction, the lineage of these P cells is fixed. If one of the daughter cells is destroyed, the other does not change its fate. There is no evidence for cell interactions in the further development of these three P-cell lineages, and cell fate is thus probably specified by asymmetric cell divisions. The vulva is formed from the 22 cells derived from the three lineages by a precise pattern of cell division, movement and cell fusion that forms a structure of seven concentric rings stacked on each other to form a conical structure (Fig. 6.17).

How are just three Pp cells specified, and how is the primary fate of the central cell made different from the secondary fate of its neighbors? The six Pp cells are initially equivalent, in that any of them can give rise to vulval tissue. The key determining signal is that provided by the anchor cell. The vital role of the anchor cell is shown by cell-ablation experiments; when the anchor cell is destroyed with a laser microbeam, the vulva does not develop.

Familiar signaling pathways are involved in vulval induction. The anchor cell's signal is the secreted product of the *lin-3* gene, which is the counterpart of the growth factor EGF in other animals. Mutations in *lin-3* result in the same abnormal development that follows removal of the anchor cell: no vulva is formed. The receptor for the inductive signal is a transmembrane tyrosine kinase of the EGF receptor (EGFR) family, encoded by the *let-23* gene (Fig. 6.18). The inductive signal from the anchor

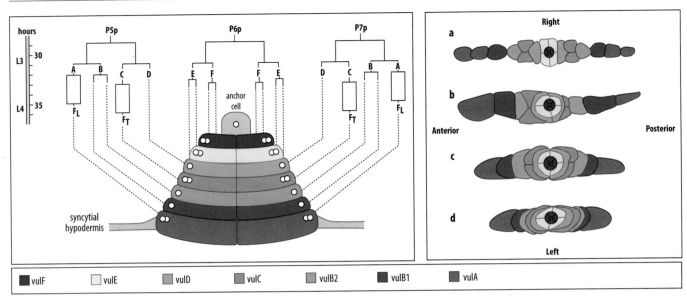

vulF ☐ vulE ☐ vulD ☐ vulC ☐ vulB2 ☐ vulB1 ☐ vulA

Fig. 6.17 Formation of the vulva by migration and fusion of specified precursors. The left-hand diagram shows a lateral representation of the seven rings of the vulva at around 39 hours and the cell lineages that give rise to them. The daughter cells of the A cell fuse, as do the daughter cells of the C cell. The open circles represent nuclei. At a later stage there will be further fusion of cells in the rings. The blue bar on the left shows the time in hours and larval stages. The right panel shows a ventral view of the developing vulva. The sequence (a) to (d) shows morphogenesis of the vulva over a period of 5 hours, from 34 hours. Changes in the shape of the cells and cell fusion give rise to the conical structure seen in the left panel.

*After Sharma-Kishore R. et al.: **Formation of the vulva in Caenorhabditis elegans: a paradigm for organogenesis.*** Development *1999,* **126**: 691–699.

cell activates the EGFR intracellular signaling pathway most strongly in P6p, but also detectably in P5p and P7p. The strongest activation of this pathway leads to the P6p cell adopting a primary cell fate, and the lower level of signaling results in P5p and P7p. P6p then sends a lateral signal to its two neighbors that both prevents them from adopting a primary fate and promotes a secondary fate. This signal is composed of three proteins of the Delta family, which interact with the transmembrane receptor LIN-12, a member of the Notch family, on P5p and P7p. The canonical Wnt signaling pathway (see Box 4B), involving a third *C. elegans* β-catenin (BAR-1), is involved in the adoption of cell fate in response to this signal.

Vulval development varies in many ways, both minor and major, among nematode species as a result of evolutionary divergence and convergence within this large group of organisms. Although the vulva has the same form in all species examined and is derived from much the same small set of P cells, some species have a vulva located posteriorly, whereas others, like *C. elegans*, have a centrally located vulva. The

Fig. 6.18 Intercellular interactions in vulval development. The anchor cell produces a diffusible signal LIN-3, which induces a primary fate in the precursor cell closest to it by binding to the receptor LET-23. It also induces a secondary fate in the two P cells slightly further away, where the concentration of the signal is lower. The cell adopting a primary fate inhibits adjacent cells from adopting the same fate by a mechanism of lateral inhibition involving LIN-12, and also induces a secondary fate in these cells. A constitutive signal from the hypodermis inhibits the development of both primary and secondary fates, but is overruled by the initial inductive signal from the anchor cell.

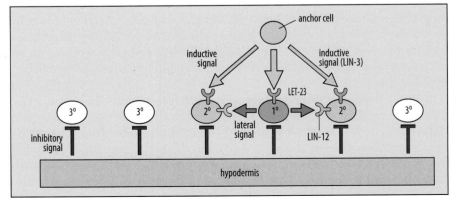

distinctive *C. elegans* mechanism of vulval induction, in which general vulval fate and a primary cell fate are induced at the same time via a signal from a single gondal anchor cell, is very recently evolved and probably arose within the genus *Caenorhabditis*. Other species of nematode have a variety of vulval induction mechanisms that differ from that of *C. elegans* to a greater or lesser extent.

SUMMARY

Specification of cell fate in the nematode embryo is intimately linked to the pattern of cleavage, and provides an excellent example of the subtle relationships between maternally specified cytoplasmic differences and very local and immediate cell–cell interactions. The antero-posterior axis is specified at the first cleavage by the site of sperm entry and specification of the dorso-ventral and left-right axes involves local cell–cell interactions. Gene products are asymmetrically distributed during the early cleavage stages, but specification of cell fates in the early embryo is crucially dependent on local cell–cell interactions, which are mediated by Wnt and Delta-Notch signaling. The development of the gut, which is derived from a single cell, requires an inductive signal by an adjacent cell. A small cluster of homeobox genes encodes positional information along the antero-posterior axis of the larva. The timing of developmental events in the larva is controlled by changes in the levels of various LIN proteins that decrease over time as a result of post-transcriptional repression by microRNAs. Formation of the adult nematode vulva involves both a well-defined cell lineage and inductive cell interactions involving the EGF and Notch signaling pathways. The detailed mechanism of vulval development varies among nematode species.

SUMMARY: early nematode development: axis specification

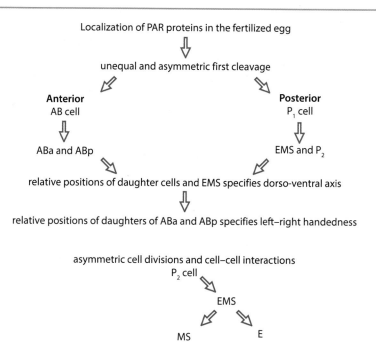

Echinoderms

Echinoderms include the sea urchins and starfish. Because of their transparency and ease of handling, sea-urchin embryos have long been used as a model developmental system. Another useful feature is that echinoderms are more closely related to vertebrates than are the other main model invertebrate organisms, *Drosophila* and *Caenorhabditis*. Echinoderms are **deuterostomes**, like vertebrates, and so these two groups have certain basic developmental features in common. Deuterostomes are coelomate animals that have radial cleavage of the egg, and in which the primary invagination of the gut at gastrulation forms the anus, with the mouth developing independently. Arthropods (which include *Drosophila*) and nematodes, on the other hand, belong to the **protostomes**, in which cleavage of the zygote is not radial, and in which gastrulation primarily forms the mouth.

Sea urchin embryos are the experimental system in which Hans Driesch performed his seminal experiments that revealed regulative behavior in development (see Section 1.3), but for some years they have been somewhat neglected as a developmental model as they were not amenable to investigation by conventional genetics or by the transgenic techniques that have proved so successful in other organisms. With the advent of new genetic techniques and the ability to identify genes by homology with genes from other organisms, the genetic and molecular basis of sea urchin development can now be studied. The rewards from this effort are large, as there are a number of detailed and informative experiments that now can find a molecular underpinning. The genome sequence of the purple sea urchin *Strongylocentrotus purpuratus* has been completed, and the complete gene-regulatory network that governs early development is being worked out.

6.9 The sea-urchin embryo develops into a free-swimming larva

The fertilized sea-urchin egg is surrounded by a double membrane, inside which the embryo develops to blastula stage. The blastula hatches from the membrane, undergoes gastrulation, and develops into a bilaterally symmetrical free-swimming larva called a **pluteus**. This eventually undergoes metamorphosis into the radially symmetrical adult (Fig. 6.19). Developmental studies are confined to the development of the embryo into the larva, as metamorphosis is a complex and poorly understood process.

The sea-urchin egg divides by radial cleavage. The first three cleavages are symmetric but the fourth is asymmetric, producing four small cells at one pole of the egg, the vegetal pole, thus defining the animal–vegetal axis of the egg. The first two cleavages are at right angles to each other and divide the egg in the plane of the animal–vegetal axis (Fig. 6.20). The third cleavage, in contrast, is equatorial and divides the embryo into animal and vegetal halves. At the next cleavage the animal cells again divide in a plane parallel to the animal–vegetal axis, but the vegetal cells divide asymmetrically to produce four **macromeres**, which contain about 95% of the cytoplasm, and four much smaller **micromeres**. At fifth cleavage, the micromeres divide asymmetrically again, so that there are four small micromeres at the vegetal pole with four larger micromeres above them. Continued cleavage results in a hollow spherical blastula composed of about 1000 ciliated cells that form an epithelial sheet enclosing the blastocoel.

Gastrulation in sea-urchin embryos, the mechanism of which we consider in detail in Chapter 9, starts about 10 hours after fertilization, with the mesoderm and endoderm moving inside from the vegetal region. In *S. purpuratus*, the first event is the entry into the blastocoel of about 32 primary mesenchyme (mesodermal) cells at the vegetal pole (see Fig. 6.20). They migrate along the inner face of the blastula wall to form a ring in the vegetal region and lay down calcareous rods that will form the internal skeleton of the pluteus larva. The endoderm, together with the mesodermal

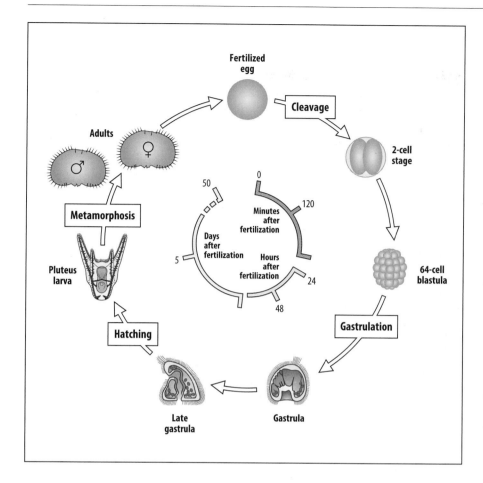

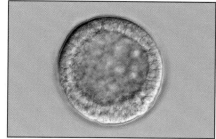

Fig. 6.19 Life cycle of the sea urchin *Strongylocentrotus purpuratus.* Eggs released by the female are fertilized externally by male sperm and develop into a blastula. After cleavage and gastrulation, the embryo hatches as a pluteus larva, which then undergoes metamorphosis into the mature radially symmetrical adult. The photographs show: the blastula (top); the pluteus larva (middle); and the adult (bottom).

Top and middle photographs reproduced with permission from J. Coffman. Lower photograph Copyright © 2007 D. Monniaux reproduced under the Creative Commons Attribution-Share Alike 3.0 Unported license.

secondary mesenchyme, then starts to undergo **invagination** at the vegetal pole. Invagination involves an inward movement of cells in a sheet, rather like pushing a finger into a balloon. In the embryo, the invagination is driven by local cellular forces, which are discussed in Chapter 9.

The invagination eventually stretches right across the blastocoel, where it fuses with a small invagination in the future mouth region on the ventral side. Thus the mouth, gut, and anus are formed. Before the invaginating gut fuses with the mouth, secondary mesenchyme cells at the tip of the invagination migrate out as single cells and give rise to mesoderm, such as muscle and pigment cells. The embryo is then a feeding pluteus larva. We will focus here on the initial patterning of the early sea-urchin embryo along its two main axes. We shall see that, despite the great differences in final form between sea urchins and vertebrates, some of the same basic developmental mechanisms and gene circuits are used in both.

6.10 The sea-urchin egg is polarized along the animal–vegetal axis

The sea-urchin embryo is conventionally defined in respect to two perpendicular axes: the animal–vegetal axis and an **oral–aboral** axis (see Fig. 6.20). The oral–aboral axis is defined by the position of the mouth (the oral end of the axis) in the larva and in turn defines the larva's plane of bilateral symmetry. An animal–vegetal asymmetry is already present in the unfertilized egg, but the future oral–aboral axis is only established after fertilization and early cleavage. The sea-urchin egg has a well-defined animal–vegetal polarity that appears to be related to the site of attachment of the egg in the ovary. Polarity is marked in some species by a fine canal at the animal pole, whereas in other species there is a band of pigment granules in the vegetal region. Early development is

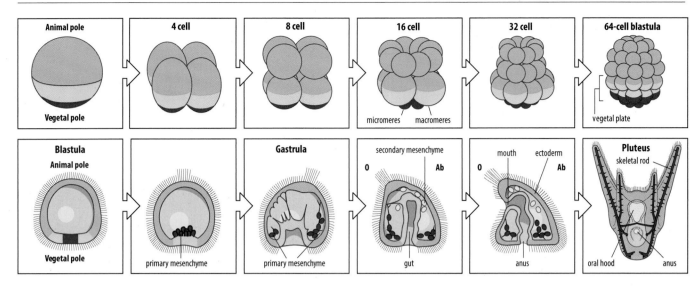

Fig. 6.20 Development of the sea-urchin embryo. Top panels: external view of cleavage to the 64-cell stage. The first two cleavages divide the egg along the animal–vegetal axis. Prospective ectoderm is in blue; endoderm in yellow, and mesoderm in red. Blastomeres that are colored in more than one color give rise to cells of all the respective germ layers. The third cleavage divides the embryo into animal and vegetal halves. At the fourth cleavage, which is unequal, four small micromeres (red) are formed at the vegetal pole, only two of which can be seen here. At fifth cleavage, the micromeres divide asymmetrically again, giving rise to four small micromeres and four large micromeres above them, not all of which can be seen here. Further cleavage results in a hollow blastula. Bottom panels: gastrulation and development of the pluteus larva. Sections of the developing embryo through the plane of the animal–vegetal axis are shown, starting from a hollow blastula. Gastrulation begins at the vegetal pole, with the entry of about 40 primary mesenchyme cells into the interior of the blastula. The gut invaginates from this site and fuses with the mouth, which invaginates from the opposite side of the embryo. The mouth defines one end of an oral (O)–aboral (Ab) axis, which will be the main axis of symmetry of the pluteus larva. The secondary mesenchyme comes from the tip of the invagination. During further development, growth of skeletal rods laid down by the primary mesenchyme results in the extension of the four 'arms' of the pluteus larva. The view of the pluteus is from the oral side. The oral hood narrows into the mouth and channels food into it.

intimately linked to this egg axis. The first two planes of cleavage are always parallel to the animal–vegetal axis, and at the fourth cleavage, which is unequal (see Fig. 6.20), the micromeres are formed at the vegetal pole. The micromeres give rise to the primary mesenchyme, and both the micromeres and the primary mesenchyme are initially specified by cytoplasmic factors localized at the vegetal pole of the egg.

The animal–vegetal axis is stable and cannot be altered by centrifugation, which redistributes larger organelles such as mitochondria and yolk platelets. Isolated fragments of egg also retain their original polarity. Blastomeres isolated at the two-cell and four-cell stages, each of which contains a complete animal–vegetal axis, give rise to mostly normal but small pluteus larvae (Fig. 6.21, left panel). Similarly, eggs that are fused together with their animal–vegetal axes parallel to each other form giant, but otherwise normal, larvae.

There is, in contrast, a striking difference in the development of animal and vegetal halves if they are isolated at the eight-cell stage. An isolated animal half merely forms a hollow sphere of ciliated ectoderm, whereas a vegetal half develops into a larva that is variable in form but is usually vegetalized; that is, it has a large gut and skeletal rods, and a reduced ectoderm lacking the mouth region (see Fig. 6.21, right panel). On occasion, however, vegetal halves from the eight-cell stage can form normal pluteus larvae if the third cleavage is slightly displaced toward the animal pole.

Together, these observations show that there are maternally determined differences along the animal–vegetal axis that are necessary to specify either animal or vegetal fates. Despite this, it is clear that the sea-urchin embryo has considerable capacity for regulation, implying the occurrence of cell–cell interactions.

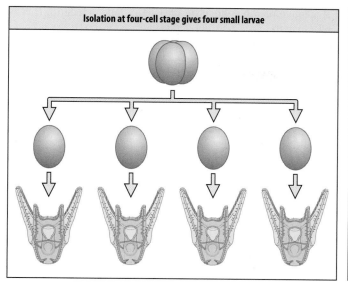

Isolation at four-cell stage gives four small larvae

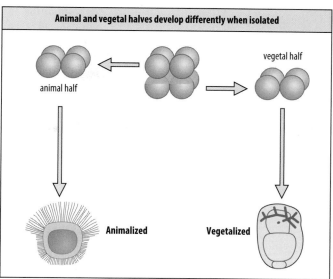

Animal and vegetal halves develop differently when isolated

Fig. 6.21 Development of isolated sea-urchin blastomeres. Left panel: if isolated at the four-cell stage, each blastomere develops into a small but normal larva. Right panel: an isolated animal half from the eight-cell stage forms a hollow sphere of ciliated ectoderm, whereas an isolated vegetal half usually develops into a highly abnormal embryo with a large gut, and some skeletal structures, but with a reduced ectoderm.

6.11 The sea-urchin fate map is finely specified, yet considerable regulation is possible

At the 60-cell stage a simplified fate map can be constructed (Fig. 6.22). The embryo is composed of three bands of cells along the animal–vegetal axis: the small and large micromeres at the vegetal pole, which will give rise to mesoderm (the primary mesenchyme that forms the skeleton and some adult structures); the vegetal plate, which comprises the next two tiers of cells (Veg1 and Veg2) and gives rise to endoderm, mesoderm (the secondary mesenchyme that forms muscle and connective tissues), and some ectoderm; and the remainder of the embryo, which gives rise to ectoderm and is divided into future oral and aboral regions. The ectoderm gives rise to the outer epithelium of the embryo and to neurogenic cells.

Using vital dyes as lineage tracers, the pattern of cleavage and of cell fates have both been mapped. The pattern of cleavage has been shown to be invariant. Unlike the nematode, however, this stereotypical cleavage pattern does not determine the fate of the individual cells. Compressing an early embryo by squashing it with a cover slip, and thus altering the pattern of cleavage, still results in a normal embryo after further development. Thus the embryo is highly regulative: lineage and cell fate are to a large extent separable. The regulative capacity is not uniform, however. The micromeres, for example, have a fixed fate once they are formed and this fate cannot be changed. They give rise only to primary mesenchyme, and no other fate has been observed, even if they are grafted to another site on the embryo. The fate of the vegetal macromeres is not fixed when they are formed. Vegetal macromeres

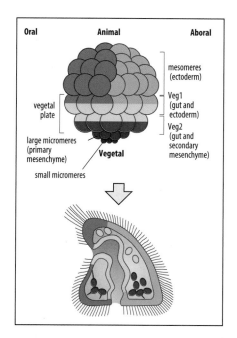

Fig. 6.22 Fate map of the sea-urchin embryo. Lineage analysis has shown four main regions at the 60-cell stage. The embryo is divided along the animal-vegetal axis into three bands: the micromeres, which give rise to the primary mesenchyme; the vegetal plate, which gives rise to the endoderm, secondary mesenchyme, and some ectoderm; and the mesomeres, which give rise to ectoderm. The ectoderm is divided into oral and aboral regions. Prospective mesoderm is colored in red; prospective endoderm in yellow; prospective neural tissue in dark blue; and prospective epidermis in light blue.

After Logan, C. Y., and McClay, D. R.: ***The allocation of early blastomeres to the ectoderm and endoderm is variable in the sea urchin embryo.*** Development *1997,* ***124****: 2213-2223.*

labeled at the 16-cell stage give rise to endoderm, mesoderm, and ectoderm. If cells are labeled at the 60-cell stage, the Veg2 cells derived from these macromeres become either mesoderm (secondary mesenchyme) or endoderm, but never ectoderm, while the Veg1 cells give rise to endoderm and ectoderm but not mesoderm. This shows that **endomesoderm** (also called mesendoderm) specification has occurred by this stage, but that the boundary between endoderm and ectoderm has not yet been set. This occurs slightly later. Single Veg1 cell progeny labeled after the eighth cleavage give rise to either endodermal or ectodermal tissues, but not both.

When isolated and cultured, some regions of the sea-urchin embryo develop more or less in line with their normal fate. Micromeres isolated at the 16-cell stage will form primary mesenchyme cells and may even form skeletal rods. An isolated animal half of presumptive ectoderm forms an animalized ciliated epithelial ball, but there is no indication of mouth formation. Although these experiments suggest that localized cytoplasmic factors are involved in specifying cell fate, this does not tell the full story, as they do not reveal the role of inductive cell-cell interactions in the normal process of development. The ability of a vegetal half to form an almost complete embryo has already been noted. Another dramatic example of regulation is seen when the animal and vegetal blastomeres from half of an eight-cell embryo are combined with the animal blastomeres from another; this unlikely combination can develop into a completely normal embryo. How does this occur? The answer, as we see next, is characteristic of many regulative embryos, and relies on the ability of the vegetal region to act as an organizer.

6.12 The vegetal region of the sea-urchin embryo acts as an organizer

The regulative capacities of the early sea-urchin embryo are due in large part to an organizing center that develops in the vegetal region and which, like the amphibian Spemann organizer (see Fig. 4.5), has the potential ability to induce the formation of an almost complete body axis. This center is initially set up by the activities of maternal factors and becomes located in the micromeres formed at the fourth cleavage division. It produces signals that induce the adjacent vegetal macromeres to adopt an endomesodermal fate. The endomesoderm gives rise to the sea-urchin spicules (skeletal rods), muscle, and gut. The signals from the organizer set in train a relay of events that subsequently specifies the endomesoderm more finely as mesoderm or endoderm, and helps set the boundary between endoderm and ectoderm. Ablation of the micromeres at the fourth cleavage results in abnormal development, but if they are removed at the sixth cleavage, after 2–3 hours of contact, gastrulation is delayed but normal larvae can develop.

Striking evidence for an organizer came from experiments that combined micromeres isolated from a 16-cell embryo with an isolated animal half from a 32-cell embryo. This combination can give rise to an almost normal larva (Fig. 6.23). The micromeres clearly induce some of the presumptive ectodermal cells in the animal half to form a gut, and the ectoderm becomes correctly patterned. Further evidence for the organizer-like properties of the micromeres comes from grafting them to the side of an intact embryo. Here, they induce endoderm in the presumptive ectoderm, which then invaginates to form a second gut. There is some evidence that the closer the graft is to the vegetal region, the greater the size of the invagination, implying a graded ability along the animal–vegetal axis to respond to a signal from micromeres.

The regulative capacity of the very early embryo can even compensate for loss of the micromeres. If micromeres are removed as soon as they are formed at the fourth cleavage (see Fig. 6.20), regulation occurs and a normal larva will still develop. The most vegetal region assumes the property of micromeres, giving rise to skeleton-forming cells and acquiring organizing properties.

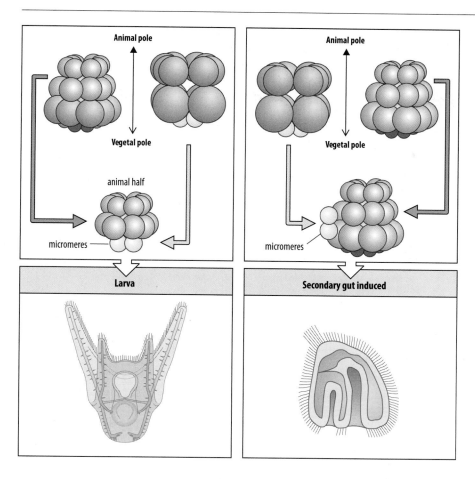

Animal pole

Vegetal pole

animal half

micromeres

Larva

Animal pole

Vegetal pole

micromeres

Secondary gut induced

Fig. 6.23 The inductive action of micromeres. Left panels: combining the four micromeres from a 16-cell sea-urchin embryo with an animal half from a 32-cell embryo results in a normal larva. An animal half cultured on its own merely forms ectoderm (not shown). Right panels: micromeres implanted into the side of another 32-cell embryo induce the formation of another gut at the implantation site.

6.13 The sea-urchin vegetal region is demarcated by the nuclear accumulation of β-catenin

In the sea urchin, the early blastomeres accumulate maternal β-catenin in their nuclei as a result of activation of part of the canonical Wnt pathway (see Box 4B). This accumulation is not uniform, however; β-catenin concentrations are high in the micromere nuclei whereas it is almost absent from nuclei in the most animal region of the future ectoderm. This differential stabilization of intranuclear β-catenin along the animal–vegetal axis is at least partly controlled by the local activation in the vegetal region of maternal Dishevelled protein, which is an intracellular activator of the pathway by which β-catenin is stabilized. Within the nucleus, β-catenin acts as a co-activator of the transcription factor TCF to activate the expression of zygotic genes required to specify a vegetal fate.

Treatments that inhibit β-catenin degradation, such as lithium chloride (see Section 4.2) cause vegetalization, resulting in sea-urchin embryos with a reduced ectoderm and an enlarged gut. The main effect of lithium on whole embryos is to expand the area of intranuclear β-catenin accumulation beyond the normal vegetal region (the region receiving a Wnt signal), shifting the border between endoderm and ectoderm toward the animal pole. Similarly, overexpression of β-catenin also causes vegetalization and can induce endoderm in animal caps. In contrast, preventing the nuclear accumulation of β-catenin in vegetal blastomeres completely inhibits the development of both endoderm and mesoderm. Micromeres depleted of β-catenin, for example, are unable to induce endoderm when transplanted to animal regions.

Maternal β-catenin and the transcription factor Otx activate the gene *pmar1* in the micromeres at the fourth cleavage. *pmar1* encodes a transcription factor that is

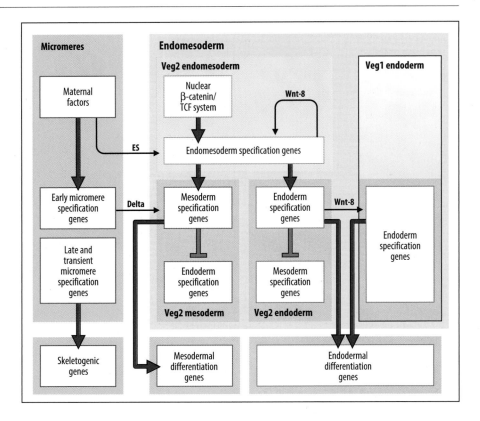

Fig. 6.24 An outline of endomesoderm specification in the sea urchin. The white boxes represent networks of genes that are reponsible for the developmental functions indicated. The background shading represents the embryonic territories involved. Peach, the large micromere precursors that will give rise to skeletal tissue; yellow, the early endomesoderm, which segregates at late cleavage into mesoderm (violet) and endoderm (blue). The red arrows and blue barred lines refer to gene interactions; the thin black arrows represent intercellular signals. ES, early signal.

Illustration from Oliveri, P., Davidson, E.H.: ***Gene regulatory network controlling embryonic specification in the sea urchin.*** Curr. Opin. Genet. Dev. *2004, **14**: 351–360.*

produced exclusively in micromeres during cleavage and early blastula stages and is required for the micromeres to express their organizer function and also for their development as primary skeletogenic mesenchyme. The outline of endomesoderm specification is illustrated in Fig. 6.24. Soon after they are formed, the micromeres produce a signal, called the early signal (ES), which induces the endomesoderm state in the Veg2 cells. A second signal, produced after the seventh cleavage, is the ligand Delta, which acts on the Notch receptor in adjacent Veg2 cells to specify them as pigment and secondary mesenchyme cells. At later stages Notch–Delta signaling is also involved in endoderm specification and in delimiting the endoderm–ectoderm boundary.

The secreted signaling protein Wnt-8, first expressed by micromeres after the fourth cleavage, acts back on the micromeres that produce it to maintain the stabilization of β-catenin and thus maintain micromere function. Wnt-8 acting later in the Veg2 endoderm is involved in directing the specification of endoderm in adjacent Veg1 cells. In this way, the sea urchin mesoderm and endoderm is specified by a series of short-range signals that result in changes in gene expression.

6.14 The animal–vegetal axis and the oral–aboral axis can be considered to correspond to the antero-posterior and dorso-ventral axes of other deuterostomes

Because of its body form, the sea-urchin pluteus larva does not have obvious antero-posterior and dorso-ventral axes like vertebrate embryos. However, on the basis of the location of common developmental signals (BMP, Wnt, Nodal), we can see that the basic coordinate system of the sea-urchin blastula shares many features with that of other deuterostomes.

As we have seen, classically the sea-urchin embryo is defined in respect to two perpendicular axes: the animal–vegetal axis and an oral–aboral axis. Analysis of the

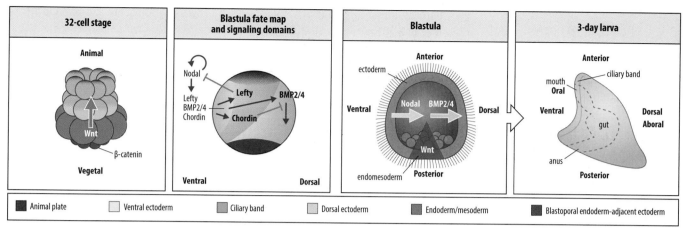

| 32-cell stage | Blastula fate map and signaling domains | Blastula | 3-day larva |

| ■ Animal plate | ☐ Ventral ectoderm | ■ Ciliary band | ☐ Dorsal ectoderm | ■ Endoderm/mesoderm | ■ Blastoporal endoderm-adjacent ectoderm |

Fig. 6.25 A possible correspondence between the sea-urchin animal–vegetal and oral–aboral axes with the antero-posterior and dorso-ventral axes of amphibians. First panel: a wave of Wnt/β-catenin signaling from the most vegetal cells sweeps through the embryo to specify endomesoderm (blue) in adjacent tiers of cells; the most distant tiers are specified as neuroectoderm (red). This event defines an antero-posterior axis along the animal–vegetal axis. Second and third panels: a dorso-ventral axis is specified perpendicular to the antero-posterior axis by the expression of Nodal on one side, which leads to the expression of BMP and of positive (red arrows) and negative (blue barred lines) regulators of these signaling pathways. The result of signaling is a blastula with the two main axes specified. Fourth panel: a sequence of morphogenetic events leads to a distorsion of these axes into the final morphology of the early sea urchin larva.

*Adapted from Angerer, L.M. et al.: **The evolution of nervous system patterning: insights from sea urchin development.*** Development *2011,* **138***: 3613-3623.*

patterns of expression of signaling molecules and comparison with other embryos reveals that the sea urchin oral–aboral axis is likely to correspond to the dorso-ventral axis in frogs and fish, and the sea urchin animal–vegetal axis to the antero-posterior axis (Fig. 6.25).

The oral region expresses members of the Nodal, BMP, and Chordin families, which in vertebrate embryos are distributed along the dorso-ventral axis. Functional analysis reveals that whereas Nodal signaling is restricted to the oral half, BMP signaling operates in the aboral half, and so suggests that this axis corresponds to the dorso-ventral axis.

As we have seen in Section 6.13, the vegetal side of the sea urchin expresses β-catenin as a result of Wnt signaling and marks the site where gastrulation is initiated. Although in *Xenopus*, β-catenin is associated with the dorsal side of the embryo, it also marks the position of the blastopore (and thus the future anus), and thus, by equivalence with other deuterostomes, the posterior part of the embryo. On this basis we could say that β-catenin marks the posterior pole of the sea-urchin embryo and that the animal–vegetal axis is likely to correspond to the antero-posterior axis.

It is clear that the sea-urchin embryo is a derived structure, as it does not have proper dorsal and ventral structures in the sense that an amphibian, a chick or a mouse does. However, these comparisons do suggest an underlying conserved organization in the molecular coordinate systems.

6.15 The pluteus skeleton develops from the primary mesenchyme

The pluteus larva has an internal skeleton of rods composed of calcium magnesium salts and matrix proteins, many of which are acidic glycoproteins. The skeletal material is secreted by skeleton-forming (skeletogenic) primary mesenchyme cells, which are derived from the micromeres. The gene-regulatory network underlying the development of the skeletogenic cells, from the events occurring in the early blastula to activation of genes for matrix proteins, has been worked out in some detail. The genes in this developmental pathway, and the interactions between them, were identified

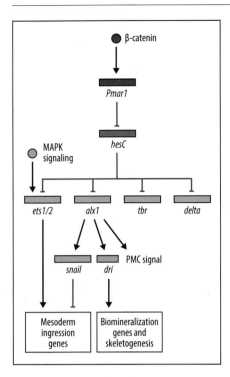

Fig. 6.26 Skeleton-forming gene regulatory network activity in micromeres. The localized activation of β-catenin in the most vegetal cells of the sea urchin early embryo (see Fig. 6.25) leads to the expression of the protein Pmar1, which represses the gene *hesC*. This results in the expression of a suite of genes, normally repressed by the HesC protein, which control the expression of proteins that endow cells with skeletogenic characteristics, and also give them an identity that is manifest in their ability to signal to influence the fate of other cells (for example, the signal protein Delta and the primary mesenchyme cell (PMC) signal). The effect of β-catenin signaling is reinforced in some cells by MAPK activity. Note that although the snail protein is a transcriptional repressor, it stimulates ingression.

*Adapted from Ettensohn, C.A. et al.: **Gene regulatory networks and developmental plasticity in the early sea urchin embryo: alternative deployment of the skeletogenic gene regulatory network.** Development 2007, **134**: 3077-3087.*

both by measuring gene activity directly and by using antisense morpholino RNAs (see Box 6B) to silence specific gene expression. Every known regulatory gene that is expressed in the skeletogenic lineage has been incorporated into a genetic regulatory network; here we focus on the activation of genes that specify the primary mesenchyme as skeleton-forming tissue.

As described in Section 6.13, the zygotic gene *pmar1* is activated exclusively in the micromeres, and this is the first step in specifying the micromeres as future skeleton-forming tissue (Fig. 6.26). If *pmar1* mRNA is made to be present in all the cells, the whole embryo turns into skeleton-forming mesenchyme. *pmar1* encodes a transcriptional repressor for the gene *hesC*, which itself encodes a repressor. The HesC protein is repressing genes essential for the micromeres to develop as primary mesenchyme, and so the spatially localized repression of *hesC* transcription by Pmar1 allows the activation of these genes (for example, *alx1*, *ets1* and *tbr*) in the micromeres only (see Fig. 6.26). This is another example of a common developmental strategy by which many cells of an embryo have a potential that is repressed (by *hesC* in this case) and is revealed only in some cells (the primary mesenchyme cells) as the result of appropriate signals (the activation of *pmar1* expression by β-catenin). This leads to the expression of a second tier of fate-determining genes—*alx1* and *ets1*—that encode transcription factors, which activate the differentiation genes. These are genes that encode matrix proteins, proteins involved in the biomineralization of the skeleton, and signal proteins, such as Delta, that coordinate the development of adjacent cells. This transcriptional cascade is aided by other signaling pathways, such as a MAPK pathway that is necessary for the expression of the *ets* genes and thus reinforces the effects of *pmar1* (see Fig. 6.26).

The network of interactions leading to skeletogenic mesoderm is just a small part of the whole network that specifies the endomesoderm and endoderm (Box 6D). The complexity of these networks is hardly surprising given how complex the regulatory regions of developmental genes can be. As we saw in Chapter 2, the regulatory region of a *Drosophila* pair-rule gene, such as *even-skipped*, contains many binding sites for transcription factors that can both activate and repress gene activity (see Fig. 2.37). These target sites are grouped into discrete subregions—regulatory modules—which each control expression of the gene at a particular site in the embryo. Many sea urchin genes similarly have *cis*-regulatory control regions composed of a number of modules that govern gene expression at different times and in different parts of the embryo.

Although all the genes controlling sea-urchin development have yet to be identified, many whose expression varies with both space and time have been analyzed in detail. Their regulatory regions prove to be complex and modular, as illustrated by the regulatory region of the sea-urchin *Endo-16* gene, a gene coding for a secreted glycoprotein of unknown function. *Endo-16* is initially expressed in the vegetal region of the blastula, in the presumptive endoderm (Fig. 6.27). After gastrulation, *Endo-16* expression increases and becomes confined to the middle region of the gut. The regulatory region controlling *Endo-16* expression is about 2200 base pairs long and contains at least 30 target sites, to which 13 different transcription factors can bind. These regulatory sites seem to fall into several subregions, or modules. The function of each module has been determined by attaching it to a reporter gene and injecting the recombinant DNA construct into a sea-urchin egg. In such experiments, module A, for example, is found to promote expression of its attached reporter gene in the presumptive endoderm, whereas modules D and C prevent expression in the primary mesenchyme. Midgut expression is under control of module B. Modules D, C, E, and F are activated by the action of added lithium and are involved in the vegetalizing action of *Endo-16*.

The modular nature of the *Endo-16* control region is similar to that of the pair-rule genes in *Drosophila*, where different modules are responsible for expression

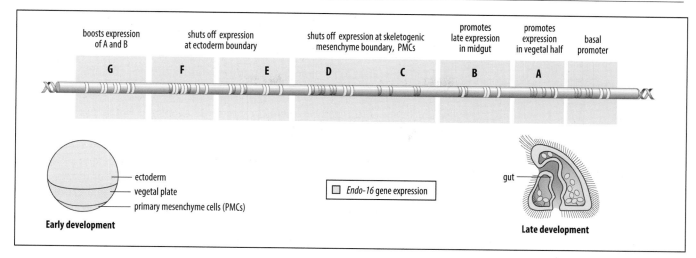

Fig. 6.27 Modular organization of the *endo-16* gene regulatory region. The modular subregions are indicated A to G. Thirteen different transcription factors bind to the more than 50 individual binding sites in these modules. The spatial domains of the embryo in which each module regulates gene expression are indicated, together with other regulatory modules. Early in development, regulation of gene expression restricts *Endo*-16 activity to the vegetal plate and the endoderm of the future gut; later in development it is restricted to the midgut.

in each of the seven stripes. The sea-urchin example illustrates yet again the importance of gene-control regions in integrating and interpreting developmental signals.

6.16 The oral–aboral axis in sea urchins is related to the plane of the first cleavage

The other main axis in the sea urchin-embryo is the oral–aboral axis, which defines the main axis of the pluteus larva and also defines the plane of larval bilateral symmetry. The oral ectoderm gives rise to the cells of the stomodeum (the mouth), the epithelium that forms the larval outer layer in the oral region, and neurogenic structures in the larva. The aboral ectoderm gives rise only to epithelium, which will cover most of the larval body. There are clear differences in skeletal patterning in the oral and aboral regions, and the oral side can be recognized well before invagination forms the mouth, because the skeleton-forming mesenchyme migrates to form two columns of cells on the future oral face. The migration of these cells is considered in Chapter 9.

Unlike the animal–vegetal axis, the oral–aboral axis cannot be identified in the egg and appears to be potentially labile up to as late as the 16-cell stage. This is why Driesch could get a normal complete larva from one isolated cell of the two-cell embryo (see Fig. 1.8). In normal development, however, the axis is predicted by the plane of the first cleavage. In the sea urchin *S. purpuratus*, a good correlation was found between the oral–aboral axis and the plane of first cleavage: the future axis lies 45° clockwise from the first cleavage plane as viewed from the animal pole (Fig. 6.28). In other sea urchin species, however, the relationship between the oral–aboral axis and plane of cleavage is different; the axis may coincide with the plane of cleavage or be at right angles to it.

One clear indication of the establishment of the oral–aboral axis is the expression of the TGF-β family member Nodal exclusively in the presumptive oral ectoderm in the early blastula, at around the 60- to 120-cell stage (see Fig. 6.25). The initial event that breaks the radial symmetry of the fertilized egg and sets the oral end of the oral–aboral axis by initiating Nodal expression may be an underlying mitochondrial gradient. Oral–aboral polarity correlates with a mitochondrial gradient (highest in the oral region), which produces an intracellular gradient in hydrogen peroxide (H_2O_2) and reactive oxygen species. Experimental destruction of H_2O_2 in the two-cell embryo suppresses the initial expression of Nodal and pushes oral–aboral polarity towards

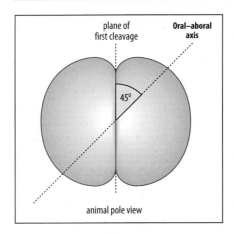

Fig. 6.28 Position of the oral-aboral axis of the sea urchin, *Strongylocentrotus purpuratus,* **embryo in relation to first cleavage.** When viewed from the animal pole, the oral-aboral axis is usually 45° clockwise from the plane of first cleavage. Note that at this stage the polarity of the axis is not determined, only the positions of the two ends.

aboral. It is likely that the stress-activated kinase p38, which is upregulated by H_2O_2, is an intermediary in this process.

6.17 The oral ectoderm acts as an organizing region for the oral-aboral axis

Nodal is one of the earliest zygotic genes to be expressed in the sea urchin, and Nodal signaling seems to be instrumental in organizing and patterning the oral–aboral axis (see Fig. 6.25). If Nodal activity is blocked, the embryo remains radially symmetrical and differentiation of ectoderm into oral and aboral ectoderm is blocked. Overexpression of Nodal also abolishes the oral–aboral axis, but in this case all the ectoderm is converted to an oral fate. But if Nodal expression is completely blocked up to the eight-cell stage and Nodal is then expressed in a single blastomere, even in one that would not normally express it, development can be 'rescued.' The treated blastulas develop oral–aboral polarity, with the Nodal-expressing cells becoming oral ectoderm, and will frequently even develop into normal pluteus larvae.

These experiments suggest that Nodal activity sets up a signaling center with organizing properties along the oral–aboral axis. High levels of Nodal signaling in the oral ectoderm result in the expression of the transcription factor Goosecoid (a characteristic protein of the amphibian Spemann's organizer) and the secreted signaling protein BMP-2/4. In the sea urchin, Goosecoid is thought to suppress aboral fate in the oral ectoderm, whereas BMP-2/4 is thought to diffuse from its source in the oral ectoderm and induce an aboral fate in the rest of the ectoderm. Another protein expressed in the prospective oral ectoderm as a consequence of Nodal activity is the TGF-β family member Lefty, which is an antagonist of Nodal signaling and restricts Nodal activity to the oral ectoderm. If the function of Lefty is blocked, most of the ectoderm develops into oral ectoderm. This situation is reminiscent of the interactions involved in mesoderm induction and patterning in vertebrates, showing how the same developmental gene circuits can be deployed for different purposes in different species. Indeed, as indicated above, the spatial organization of signaling along this axis suggests that it correlates with a deuterostome dorso-ventral axis, but in the early sea-urchin embryo, Nodal appears to have no role in specifying and patterning the mesoderm, which is one of its primary roles in vertebrates (see Chapter 4).

For correct overall development, patterning along one axis must be spatially and temporally linked to what is happening along the other. In the sea urchin, the oral–aboral axis is linked to the animal–vegetal axis by an indirect effect of the intranuclear localization of β-catenin in the vegetal region (described in Section 6.13). In the early blastula, the transcription factor FoxQ2 is initially produced throughout the animal hemisphere and suppresses expression of the *Nodal* gene there. As the wave of intranuclear β-catenin localization proceeds from the vegetal pole towards the animal hemisphere, the signals it generates restrict FoxQ2 expression to a small region at the animal pole known as the animal plate (which eventually gives rise to neurogenic ectoderm). This removal of FoxQ2 from the rest of the blastula allows expression of *Nodal* in the prospective oral ectoderm (see Fig. 6.25).

There is a second role for Nodal signaling in the sea urchin—the establishment of internal left–right asymmetry (see Section 5.16). Inside the sea-urchin larva, a structure called the adult rudiment, from which the adult tissues will be produced during morphogenesis, is formed only on the left side. As in vertebrates, a regulatory circuit comprising *Nodal*, *Lefty,* and *Pitx2* regulates the generation of this asymmetry, but the expression of these genes is reversed in the sea urchin compared with vertebrates. Nodal signals produced in the ectoderm on the right side of the larva inhibit rudiment formation on the right side.

EXPERIMENTAL BOX 6D The gene regulatory network for sea-urchin endomesoderm specification

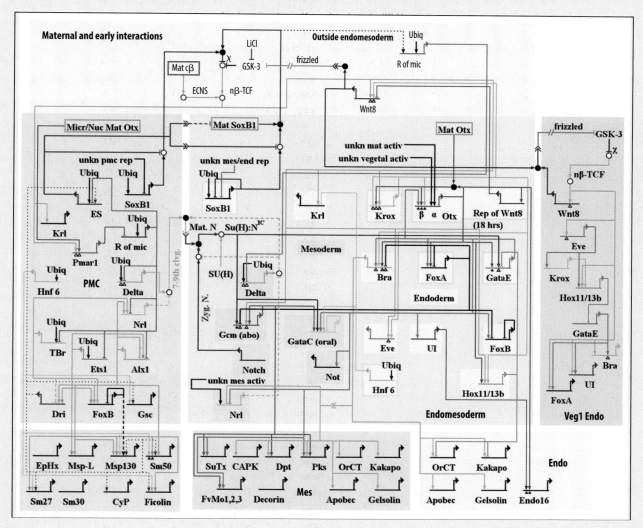

Figure 1

*Illustration from Oliveri, P., Davidson, E.H.: **Gene regulatory network controlling embryonic specification in the sea urchin**. Curr. Opin. Genet. Dev. 2004, **14**: 351–360.*

The complexity of the interactions between developmental regulatory genes can be visualized in 'wiring diagrams' representing gene-regulatory networks, in which genes encoding transcription factors are linked into a network showing which genes any given transcription factor acts on. The first developmental gene-regulatory network worked out in considerable detail was the one that specifies the various fates of sea-urchin endomesoderm. Figure 1 shows a relatively early version. The part of the network shaded in gray at the top corresponds to the nuclear localization of β-catenin under the control of maternal factors. The rest of the network refers to zygotic gene expression. PMC, primary mesenchyme (micromeres); Mes, secondary

mesenchyme; Veg1 Endo, Veg1-derived endoderm; Endo, rest of endoderm. The network is being continually updated and verified (see URL in Further reading for the latest interactive version, which shows the timing of interactions). The interactions of each regulatory gene were originally uncovered by systematic gene silencing and computational biology accompanied by laborious 'wet biology' to verify each interaction. Similar gene-regulatory networks are now being produced for other model organisms, using techniques such as chromatin immunoprecipitation followed by DNA sequencing (ChIP-seq) to identify genes bound by specific transcription factors (see Fig. 3.41).

SUMMARY

Sea-urchin embryos have a remarkable capacity for regulation even though they also have an invariant early cell lineage. The sea-urchin egg has a stable animal–vegetal axis before fertilization, which appears to be specified by the localization of maternal cytoplasmic factors in the egg. After fertilization, maternal factors specify the formation of an organizer region in the most vegetal region of the egg, where the micromeres are formed. This organizer, like the Spemann organizer in amphibians, is able to specify an almost complete body axis. Localized activation of the Wnt pathway and nuclear accumulation of β-catenin specifies the vegetal region and the organizing properties of the micromeres. The oral–aboral axis is determined after fertilization. Signals from the micromeres induce and pattern the endomesoderm along the animal–vegetal axis, whereas the oral–aboral axis is specified and patterned by the activation of the Nodal signaling pathway in the future oral region. The site and time of expression of developmental genes are controlled by complex gene-regulatory networks, which have been worked out in some detail.

SUMMARY: sea urchin early development

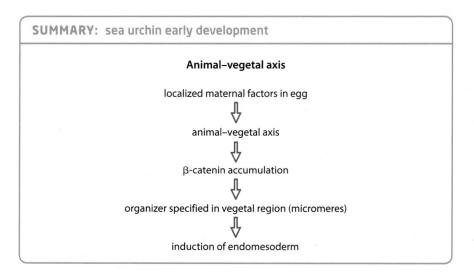

Animal–vegetal axis

localized maternal factors in egg

⇓

animal–vegetal axis

⇓

β-catenin accumulation

⇓

organizer specified in vegetal region (micromeres)

⇓

induction of endomesoderm

Summary to Chapter 6

- In many invertebrates, such as the nematode, cytoplasmic localization and cell lineage, together with local intercellular interactions, seem to be very important in specifying cell fate.
- There is very little evidence in these animals for key signaling regions corresponding to the vertebrate organizer, or for morphogen gradients as in *Drosophila*.
- This may reflect the predominance of mechanisms that specify cell fate on a cell-by-cell basis, rather than by groups of cells as in vertebrates and insects.
- Although genetic evidence of the kind available in *Drosophila* is lacking in other invertebrates, except in the case of the nematode *Caenorhabditis elegans*, it is clear that maternal genes are crucial in specifying early patterning.
- There are numerous examples of the first cleavage in the embryo being related to the establishment of the embryo's axes and evidence that this plane of cleavage is maternally determined.
- Asymmetric cleavage associated with the unequal distribution of cytoplasmic determinants seems to be very important at early stages.

- Cytoplasmic movements that are involved in this localization of determinants often occur after fertilization, and in this respect they may resemble the cortical rotation of the amphibian egg.
- There are cell interactions, even at early stages, in all of the organisms in this chapter; an example is the specification of left–right asymmetry in nematodes, where a small alteration in cell relationships can reverse their handedness.
- In most of these embryos the interactions are highly local, only affecting an adjacent cell. This is unlike vertebrates, in which interactions may occur over several cell diameters and involve groups of cells.
- Sea-urchin development, by contrast, is highly regulative, and cell interactions play a major role; thus echinoderms resemble the vertebrates in this respect.
- The micromeres at the vegetal pole have properties similar to those of the vertebrate organizer, and cells are specified in groups rather than on a cell-by-cell basis.

■ End of chapter questions

Long answer (concept questions)

1. What is meant by asymmetric cell division: does it necessarily mean that one daughter is bigger than the other? What other possibility qualifies a cell division event as being asymmetric? How are both types of asymmetric cell division illustrated by the first division in *Caenorhabditis elegans*?

2. Compare and contrast RNAi and miRNA gene silencing. Incorporate these hints into your answer: experimental versus natural; ~22 nucleotides; RISC; and degradation versus translation.

3. How is the antero-posterior axis of *C. elegans* determined? What is the role of the PAR proteins in axis determination?

4. During studies of cell lineage in *C. elegans*, a convention arose that allowed the unambiguous identification of each of its 558 cells (at hatching). Key to this system is the designation of the anterior daughter of each cell division with an 'a', and the posterior daughter with a 'p.' Refer to Fig. 6.10 and describe why the indicated pharyngeal cell is called 'MS paapaaa.' Similarly, why is the apoptotic cell called 'MS paapp'?

5. To what class of signaling proteins do the *C. elegans* proteins GLP-1 and APX-1 belong? What is the role of the GLP-1/APX-1 system in dorso-ventral cell fate specification? 'APX' stands for 'anterior pharynx in excess;' why was the mutant gene named 'APX' (refer to Section 6.4 and Fig. 6.9)?

6. What are heterochronic genes? What feature of *C. elegans* development makes this a favorable organism for the identification of such genes?

7. *lin-4* encodes a miRNA that represses translation of *lin-14* mRNA. Since *lin-14* maintains the Tap cell in a first larval (L1) state, this repression is required for the cell to transition to a L2 identity. Based on this mechanism, what is the phenotype of a *lin-14* loss-of-function mutant? What is the phenotype of a *lin-4* loss-of-function mutant? What would be the phenotype of a worm with loss-of-function mutations in both genes? Explain your answer.

8. The anchor cell of *C. elegans* is required for the induction of vulval development. To trace this induction, indicate which cells express the EGF, EGF receptor, Delta, and Notch proteins, and the consequences of these proteins in those cells.

9. What are the similarities between gastrulation in *Strongylocentrotus* and *Xenopus* that make them both deuterostomes?

10. Compare the first two cleavages in the frog and sea urchin. In what ways are they similar or different morphologically? Are the embryos similar or different molecularly; that is, have the determinative events that occur in *Xenopus* by the four-cell stage also occurred in sea urchins? Describe the differences in terms of 'regulation' during development.

11. The micromeres of sea urchins act as an organizer. Describe two experiments that support this statement (see Fig. 6.23).

12. Sea-urchin larvae have two overt axes, animal–vegetal and oral–aboral, and a more subtle internal left–right asymmetry. Compare the molecular bases of these three axes with the three axes of frogs (this may be most easily done in the form of a table, with the six types of axes on the side and the signaling molecules and transcription factors along the top, or in some other form of your own choosing).

Multiple choice (factual recall questions)

NB There is only one right answer to each question.

1. *C. elegans* is best known for

a) its formation of an inductive signaling center analogous to the organizer of *Xenopus*
b) its fully defined cell lineage during development
c) the large size and easy manipulation of its blastomeres
d) its mechanisms of metamorphosis

2. P granules of nematodes are

a) determinants of the antero-posterior axis
b) determinants of the germ line
c) precursors of TGF-β family signaling proteins
d) remnants of the sperm pronucleus

3. In flies, frogs, and chicks, gradients of morphogens determine the future antero-posterior and dorso-ventral axes of the developing embryo. How is the antero-posterior axis determined in *C. elegans*?

a) Bicoid protein is translated in the anterior of the fertilized egg, leading to a gradient that determines the antero-posterior axis.
b) Opposing gradients of chordin and BMP-4 establish the antero-posterior axis.

c) Sperm entry leads to a reorganization of the cytoskeleton and redistribution of maternally packaged PAR proteins, which in turn determine the antero-posterior axis.

d) β-catenin becomes localized to the nucleus in the future anterior cells after fertilization.

4. What are the *lin-4* and *lin-14* genes of *C. elegans*?

a) *lin-4* and *lin-14* are *C. elegans* versions of the Hox genes.

b) *lin-4* and *lin-14* are genes that are named for their control of the first division, and hence the lineage of the AB and P1 cells.

c) *lin-4* controls the timing of transcription of *lin-14*, and so is referred to a 'heterochronic' gene.

d) *lin-4* encodes a miRNA that represses *lin-14* translation, which in turn regulates timing during larval development.

5. In *C. elegans*, the only Hox gene utilized during embryogenesis is

a) *ceh-13*, the *C. elegans* ortholog of the *Drosophila labial* gene

b) *egl-5*, the *C. elegans* ortholog of the *Drosophila Abd-B* and the vertebrate *Hox9-13* genes

c) *lin-39*, a *C. elegans* gene related to the *Drosophila Deformed* and *Sex combs reduced* family of homeodomain proteins

d) *mab-5*, a *C. elegans* gene related to the *Drosophila Antennapedia*, *Ultrabithorax*, and *abdominal-A* family of homeodomain proteins

6. Which of these model organisms is closest evolutionarily to humans?

a) Fruit flies, because of their sophisticated nervous system and the presence of legs.

b) Plants, because their early divergence from animal lineages means they are closely related to animals, especially humans.

c) Sea urchins, despite their radial symmetry and lack of limbs.

d) Worms, despite their few cells, small genomes, and abbreviated complement of Hox genes.

7. Which of the following statements about regulation in sea-urchin embryos is true?

a) Any blastomere at the eight-cell stage can still regulate to form a full larva, but by the 60-cell stage, blastomeres will develop according to their presumptive fate.

b) Blastomeres at the four-cell stage each still possesses a portion of the original animal–vegetal axis, and if isolated will go on the form a fully formed, yet smaller, larva.

c) Blastomeres from the animal or vegetal poles of the embryo can be isolated and will go on to form an embryo, as long as either an animal or vegetal pole is included.

d) Up until the hatching of the embryo as a pluteus larva, any blastomere can be separated from the others and will regulate to go on and develop into a fully formed and full-sized larva.

8. In which ways is the organizer region of the sea urchin similar to that of frogs?

a) The organizer accumulates β-catenin, which in turn activates the expression of cell–cell signaling molecules.

b) The organizer forms near the dorsal lip of the blastopore.

c) The organizer induces the dorso-ventral axis.

d) The organizer will express high levels of Nodal.

9. In the sea urchin embryo, the function of BMP and Nodal is

a) to establish the anteroposterior axis

b) to specify mesoderm

c) to control the early patterns of cell division

d) to organize the dorso-ventral axis

10. A classical view of development sees embryos as either 'mosaic' or 'regulative'. Which of the following statements is true?

a) *C. elegans* is a mosaic embryo with cell interactions whereas sea urchin embryos are regulative

b) *C. elegans* and sea urchin embryos are regulative

c) *C. elegans* is a regulative embryo whereas sea urchin is mosaic

d) *C. elegans* is a strictly mosaic embryo while sea urchin is regulative

Multiple choice answer key

1: b, 2: b, 3: c, 4: d, 5: a, 6: c, 7: b, 8: a, 9: d, 10: a.

■ Section further reading

Nematodes

Sommer, R.J.: **Evolution of development in nematodes related to *C. elegans*** (December 14, 2005), *WormBook*, ed. The *C. elegans* Research Community, WormBook, doi/10.1895/wormbook.1.46.1, http://www.wormbook.org (date accessed 13 May 2010).

Sulston, J.E., Scherienberg, E., White, J.G., Thompson, J.N.: **The embryonic cell lineage of the nematode *Caenorhabditis elegans*.** *Dev. Biol.* 1983, **100**: 64–119. A view of the complete lineage is available at http://www.wormatlas.org/celllineages.html (accessed 23 May 2014)

6.1 The cell lineage of *Caenorhabditis elegans* is largely invariant

Sulston, J.: **Cell lineage.** In *The Nematode* Caenorhabditis elegans. Edited by Wood, WB. New York: Cold Spring Harbor Laboratory Press, 1988: 123–156.

Wood, W.B.: **Embryology.** In *The Nematode Caenorhabditis elegans*. Edited by Wood, WB. New York: Cold Spring Harbor Laboratory Press, 1988: 215–242.

6.2 The antero-posterior axis in *Caenorhabditis elegans* is determined by asymmetric cell division

Cheeks, R.J., Canman, J.C., Gabriel, W.N., Meyer, N., Strome, S., Goldstein, B.: ***C. elegans* PAR proteins function by mobilizing and stabilizing asymmetrically localized protein complexes.** *Curr. Biol* 2004, **14**: 851–862.

Cowan, C.R., Hyman, A.A.: **Acto-myosin reorganization and PAR polarity in *C. elegans*.** *Development* 2007, **134**: 1035–1043.

Goldstein, B., Macara, I.G.: **The PAR proteins: fundamental players in animal cell polarization.** *Dev. Cell* 2007, **13**: 609–622.

Munro, E., Nance, J., Priess, J.R.: **Cortical flows powered by asymmetrical contraction transport PAR proteins to establish and maintain anterior-posterior polarity in the early *C. elegans* embryo.** *Dev. Cell* 2004, **7**: 413–424.

Box 6A Apoptotic pathways in nematodes, *Drosophila* and mammals

Meier, P., Finch, A., Evan, G.: **Apoptosis in development**. *Nature* 2000, **407**: 796–801.

Jacobson, M.D., Weil, M., Raff, M.C.: **Programmed cell death in animal development**. *Cell* 1997, **88**: 347–354.

Box 6B Gene silencing by antisense RNA and RNA interference

Brennecke, J., Malone, C.D., Aravin, A.A., Sachidanandam, R., Stark, A., Hannon, G.J.: **An epigenetic role for maternally inherited piRNAs in transposon silencing**. *Science* 2008, **322**: 1387–1392.

DasGupta, R., Nybakken, K., Booker, M., Mathey-Prevot, B., Gonsalves, F., Changkakoty, B., Perrimon, N.: **A case study of the reproducibility of transcriptional reporter cell-based RNAi screens in *Drosophila***. *Genome Biol.* 2007, **8**: R203.

Dykxhoorn, D.M., Novind, C.D., Sharp, P.A.: **Killing the messenger: short RNAs that silence gene expression**. *Nat. Rev. Mol. Cell. Biol* 2003, **4**: 457–467.

Eisen, J.S., Smith, J.C.: **Controlling morpholino experiments: don't stop making antisense**. *Development* 2008, **135**: 1735–1743.

Kamath, R.S., Fraser, A.G., Dong, Y., Poulin, G., Durbin, R., Gotta, M., Kanapin, A., Le Bot, N., Moreno, S., Sohrmann, M., Welchman, D.P., Zipperlen, P., Ahringer, J.: **Systematic functional analysis of the *Caenorhabditis elegans* genome using RNAi**. *Nature* 2003, **421**: 231–237.

Nature Insight: **RNA interference**. *Nature* 2004, **431**: 337–370.

Sonnichsen, B., *et al.*: **Full-genome RNAi profiling of early embryogenesis in *Caenorhabditis elegans***. *Nature* 2005, **434**: 462–469.

6.3 The dorso-ventral axis in *Caenorhabditis elegans* is determined by cell-cell interactions

Bergmann, D.C., Lee, M., Robertson, B., Tsou, M.F., Rose, L.S., Wood, W.B.: **Embryonic handedness choice in *C. elegans* involves the Gα protein GPA-16**. *Development* 2003, **130**: 5731–5740.

Wood, W.B.: **Evidence from reversal of handedness in *C. elegans* embryos for early cell interactions determining cell fates**. *Nature* 1991, **349**: 536–538.

6.4 Both asymmetric divisions and cell-cell interactions specify cell fate in the early nematode embryo

Eisenmann, D.M.: **Wnt signaling** (June 25, 2005), *WormBook*, ed. The *C. elegans* Research Community, WormBook, doi/10.1895/wormbook.1.7.1, http://www.wormbook.org (date accessed 13 May 2010).

Evans, T.C., Crittenden, S.L., Kodoyianni, V., Kimble, J.: **Translational control of maternal *glp-1* mRNA establishes an asymmetry in the *C. elegans* embryo**. *Cell* 1994, **77**: 183–194.

Kidd, A.R.3rd, Miskowski, J.A., Siegfried, K.R., Sawa, H., Kimble, J.: **A beta-catenin identified by functional rather than sequence criteria and its role in Wnt/MAPK signaling**. *Cell* 2005, **121**: 761–772.

Lyczak, R., Gomes, J.E., Bowerman, B.: **Heads or tails: cell polarity and axis formation in the early *Caenorhabditis elegans* embryo**. *Dev. Cell* 2002, **3**: 157–166.

Mello, G.C., Draper, B.W., Priess, J.R.: **The maternal genes *apx-1* and *glp-1* and the establishment of dorsal-ventral polarity in the early *C. elegans* embryo**. *Cell* 1994, **77**: 95–106.

6.5 Cell differentiation in the nematode is closely linked to the pattern of cell division

Goldstein, B., Macara, I.G.: **The PAR proteins: fundamental players in animal cell polarization**. *Dev Cell.* 2007, **13**: 609–622.

Maduro, M.F.: **Structure and evolution of the *C. elegans* embryonic endomesoderm network**. *Biochim. Biophys. Acta* 2008, **1789**: 250–260.

Mizumoto, K., Sawa, H.: **Two βs or not two βs: regulation of asymmetric division by β-catenin**. *Trends Cell Biol.* 2007, **17**: 465–473.

Mizumoto, K., Sawa, H.: **Cortical β-catenin and APC regulate asymmetric nuclear β-catenin localization during asymmetric cell division in *C. elegans***. *Dev. Cell* 2007, **12**: 287–299.

Nance, J., Zallen, J.A.: **Elaborating polarity: PAR proteins and the cytoskeleton**. *Development* 2011, **138**: 799–809.

Park, F.D., Priess, J.R.: **Establishment of POP-1 asymmetry in early *C. elegans* embryos**. *Development* 2003, **130**: 3547–3556.

Park, F.D., Tenlen, J.R., Priess, J.R.: ***C. elegans* MOM-5/frizzled functions in MOM-2/Wnt-independent cell polarity and is localized asymmetrically prior to cell division**. *Curr. Biol.* 2004, **14**: 2252–2258.

Phillips, B.T., Kidd, A.R.IIIKing, R., Hardin, J., Kimble, J.: **Reciprocal asymmetry of SYS-1/β-catenin and POP-1/TCF controls asymmetric divisions in *Caenorhabditis elegans***. *Proc. Natl Acad. Sci. USA* 2007, **104**: 3231–3236.

6.6 Hox genes specify positional identity along the antero-posterior axis in *Caenorhabditis elegans*

Aboobaker, A.A., Blaxter, M.L.: **Hox gene loss during dynamic evolution of the nematode cluster**. *Curr. Biol.* 2003, **13**: 37–40.

Clark, S.G, Chisholm, A.D., Horvitz, H.R.: **Control of cell fates in the central body region of *C. elegans* by the homeobox gene *lin-39***. *Cell* 1993, **74**: 43–55.

Cowing, D., Kenyon, C.: **Correct Hox gene expression established independently of position in *Caenorhabditis elegans***. *Nature* 1996, **382**: 353–356.

Salser, S.J., Kenyon, C.: **A *C. elegans* Hox gene switches on, off, on and off again to regulate proliferation, differentiation and morphogenesis**. *Development* 1996, **122**: 1651–1661.

Van Auken, K., Weaver, D.C., Edgar, L.G, Wood, W.B.: ***Caenorhabditis elegans* embryonic axial patterning requires two recently discovered posterior-group Hox genes**. *Proc. Natl Acad. Sci. USA* 2000, **97**: 4499–4503.

6.7 The timing of events in nematode development is under genetic control that involves microRNAs

Ambros, V.: **Control of developmental timing in *Caenorhabditis elegans***. *Curr. Opin. Genet. Dev.* 2000, **10**: 428–433.

Austin, J., Kenyon, C.: **Developmental timekeeping: marking time with antisense.** *Curr. Biol.* 1994, **4**: 366–396.

Grishok, A., Pasquinelli, A.E., Conte, D., Li, N, Parrish, S., Ha, I., Baillie, D.L., Fire, A., Ruvkun, G., Mello, C.C.: **Genes and mechanisms related to RNA interference regulate expression of the small temporal RNAs that control *C. elegans* developmental timing.** *Cell* 2001, **106**: 23–34.

Box 6C Gene silencing by microRNAs

Bartel, D.P.: **MicroRNAs: genomics, biogenesis, mechanism and function.** *Cell* 2004, **116**: 281–297.

He, L., Hannon, G.J.: **MicroRNAs: small RNAs with a big role in gene regulation.** *Nat Rev. Genet.* 2004, **5**: 522–531.

Liu, J.: **Control of protein synthesis and mRNA degradation by microRNAs.** *Curr. Opin. Cell Biol.* 2008, **20**: 214–221.

Murchison, E.P., Hannon, G.J.: **miRNAs on the move: miRNA biogenesis and the RNAi machinery.** *Curr. Opin. Cell Biol.* 2004, **16**: 223–229.

6.8 Vulval development is initiated by the induction of a small number of cells by short-range signals from a single inducing cell

Kenyon, C.: **A perfect vulva every time: gradients and signaling cascades in *C. elegans*.** *Cell* 1995, **82**: 171–174.

Sharma-Kishore, R., White, J.G., Southgate, E., Podbilewicz, B.: **Formation of the vulva in *Caenorhabditis* elegans: a paradigm for organogenesis.** *Development* 1999, **126**: 691–699.

Sundaram, M., Han, M.: **Control and integration of cell signaling pathways during *C. elegans* vulval development.** *BioEssays* 1996, **18**: 473–480.

Yoo, A.S., Bais, C., Greenwald, I.: **Crosstalk between the EGFR and LIN-12/Notch pathways in *C. elegans* vulval development.** *Science* 2004, **303**: 663–636.

Echinoderms

Horstadius, S.: *Experimental Embryology of Echinoderms.* Oxford: Clarendon Press, 1973.

6.9 The sea-urchin embryo develops into a free-swimming larva

Angerer, L.M., Yaguchi, S., Angerer, R.C., Burke, R.D.: **The evolution of nervous system patterning: insights from sea urchin development.** *Development* 2011, **138**: 3613–3623.

Raff, R.A., Raff, E.C.: **Tinkering: new embryos from old— rapidly and cheaply.** *Novartis Found. Symp.* 2007, **284**: 35–45; discussion 45–54, 110–115.

Sea Urchin Genome Sequencing Consortium: **The genome of the sea urchin *Strongylocentrotus purpuratus*.** *Science* 2006, **314**: 941–952. Erratum in: *Science* 2007, **315**: 766.

Sea urchin interactive online poster: *Science* 2006, www. sciencemag.org/sciext/seaurchin (date accessed 13 May 2010).

Wilt, F.H.: **Determination and morphogenesis in the sea-urchin embryo.** *Development* 1987, **100**: 559–576.

6.10 The sea-urchin egg is polarized along the animal-vegetal axis

Davidson, E.H., Cameron, R.S., Ransick, A.: **Specification of cell fate in the sea-urchin embryo: summary and some proposed mechanisms.** *Development* 1998, **125**: 3269–3290.

Henry, J.J.: **The development of dorsoventral and bilateral axial properties in sea-urchin embryos.** *Semin. Cell Dev. Biol.* 1998, **9**: 43–52.

Logan, C.Y., McClay, D.R.: **The allocation of early blastomeres to the ectoderm and endoderm is variable in the sea-urchin embryo.** *Development* 1997, **124**: 2213–2223.

6.11 The sea urchin fate map is finely specified, yet considerable regulation is possible

Angerer, L.M., Angerer, R.C.: **Patterning the sea-urchin embryo: gene regulatory networks, signaling pathways, and cellular interactions.** *Curr. Top. Dev. Biol.* 2003, **53**: 159–198.

Angerer, L.M., Oleksyn, D.W., Logan, C.Y., McClay, D.R., Dale, L., Angerer, R.C.: **A BMP pathway regulates cell fate allocation along the sea urchin animal-vegetal embryonic axis.** *Development* 2000, **127**: 1105–1114.

Sweet, H.C., Hodor, P.G., Ettensohn, C.A.: **The role of micromere signaling in Notch activation and mesoderm specification during sea-urchin embryogenesis.** *Development* 1999, **126**: 5255–5265.

Wessel, G.M., Wikramanayake, A.: **How to grow a gut: ontogeny of the endoderm in the sea-urchin embryo.** *BioEssays* 1999, **21**: 459–471.

6.12 The vegetal region of the sea-urchin embryo acts as an organizer

Ransick, A., Davidson, E.H.: **A complete second gut induced by transplanted micromeres in the sea-urchin embryo.** *Science* 1993, **259**: 1134–1138.

6.13 The sea-urchin vegetal region is specified by the nuclear accumulation of β-catenin

Ettensohn, C.A.: **The emergence of pattern in embryogenesis: regulation of beta-catenin localization during early sea urchin development.** *Sci STKE* 2006, **14**: pe48.

Weitzel, H.E., Illies, M.R., Byrum, C.A., Xu, R., Wikramanayake, A.H., Ettensohn, C.A.: **Differential stability of beta-catenin along the animal-vegetal axis of the sea-urchin embryo mediated by dishevelled.** *Development* 2004, **131**: 2947–2955.

6.14 The animal-vegetal axis and the oral-aboral axis can be considered to correspond to the antero-posterior and dorso-ventral axes of other deuterostomes

Angerer, L.M., Yaguchi, S., Angerer, R.C., Burke, R.D.: **The evolution of nervous system patterning: insights from sea urchin development.** *Development* 2011, **138**: 3613–3623.

Ettensohn, C.A., Kitazawa, C., Cheers, M.S., Leonard, J.D., Sharma, T.: **Gene regulatory networks and developmental plasticity in the early sea urchin embryo: alternative deployment of the skeletogenic gene regulatory network.** *Development* 2007, **134**: 3077–3087.

6.15 The pluteus skeleton develops from the primary mesenchyme

Ettensohn, C.A.: **Lessons from a gene regulatory network: echinoderm skeletogenesis provides insights into evolution, plasticity and morphogenesis.** *Development* 2009, **136**: 11–21.

Kirchnamer, C.V., Yuh, C.V., Davidson, E.H.: **Modular *cis*-regulatory organization of developmentally expressed genes: two genes transcribed territorially in the sea-urchin embryo, and additional examples**. *Proc. Natl Acad. Sci. USA* 1996, **93**: 9322–9328.

6.16 The oral-aboral axis in sea urchins is related to the plane of the first cleavage

Cameron, R.A., Fraser, S.E., Britten, R.J., Davidson, E.H.: **The oral-aboral axis of a sea-urchin embryo is specified by first cleavage**. *Development* 1989, **106**: 641–647.

6.17 The oral ectoderm acts as an organizing region for the oral-aboral axis

Coffman, J.A., Coluccio, A., Planchart, A., Robertson, A.J.: **Oral-aboral axis specification in the sea-urchin embryo III. Role of mitochondrial redox signaling via H_2O_2**. *Dev. Biol.* 2009, **330**: 123–130.

Duboc, V., Lepage, T.: **A conserved role for the nodal signaling pathway in the establishment of dorso-ventral and left-right axes in deuterostomes**. *J. Exp. Zool. (Mol. Dev. Evol.)* 2008, **310B**: 41–53.

Duboc, V., Lapraz, F., Besnardeau, L., Lepage, T.: **Lefty acts as an essential modulator of Nodal activity during sea urchin oral-aboral axis formation**. *Dev. Biol.* 2008, **320**: 49–59.

Duboc, V., Rottinger, E., Besnardeau, L., Lepage, T.: **Nodal and BMP2/4 signaling organizes the oral-aboral axis of the sea-urchin embryo**. *Dev. Cell* 2004, **6**: 397–410.

Yaguchi, S., Yaguchi, J., Angerer, R.C., Angerer, L.M.: **A Wnt-FoxQ2-nodal pathway links primary and secondary axis specification in sea-urchin embryos**. *Dev. Cell* 2008, **14**: 97–107.

Box 6D The gene regulatory network for sea-urchin endomesoderm specification

The Davidson Lab: Endomesoderm & Ectoderm Models http://sugp.caltech.edu/endomes/ (accessed 27 May 2014)

Davidson, E.H.: **A view from the genome: spatial control of transcription in sea urchin development**. *Curr. Opin. Genet. Dev.* 1999, **9**: 530–541.

Davidson, E.H. *et al.*: **A genomic regulatory network for development**. *Science* 2002, **295**: 1669–1678.

Oliveri, P., Davidson, E.H.: **Gene regulatory network controlling embryonic specification in the sea urchin**. *Curr. Opin. Genet. Dev.* 2004, **14**: 351–360.

Oliveri, P., Tu, Q., Davidson, E.H.: **Global regulatory logic for specification of an embryonic cell lineage**. *Proc. Natl Acad. Sci. USA* 2008, **105**: 5955–5962.

7

Plant development

- **Embryonic development**
- **Meristems**
- **Flower development and control of flowering**

Because of plant cells' rigid cell walls and lack of cell movement within tissues, a plant's architecture is very much the result of patterns of oriented cell divisions. Despite this apparent invariance, however, cell fate in development is largely determined by similar means as in animals—by a combination of positional signals and intercellular communication. As well as communicating by extracellular signals and cell-surface interactions, plant cells are interconnected by cytoplasmic channels, which allow movement of proteins such as transcription factors directly from cell to cell.

The plant kingdom is very large, ranging from the algae, many of which are unicellular, to the multicellular land plants, which exist in a prodigious variety of forms. Plants and animals probably evolved the process of multicellular development independently, their last common ancestor being a unicellular eukaryote some 1.6 billion years ago. Plant development is, therefore, of interest not simply for its agricultural importance, but also for its own sake. By looking at the similarities and differences between plant and animal development, a study of plant development can shed light on the way that developmental mechanisms in two groups of multicellular organisms have evolved independently and under different sets of developmental constraints.

Do plants and animals use the same developmental mechanisms? As we shall see in this chapter, the logic behind the spatial layouts of gene expression that pattern a developing flower is similar to that of Hox gene action in patterning the body axis in animals, but the genes involved are completely different. Such similarities between plant and animal development are due to the fact that the basic means of regulating gene expression are the same in both, and thus similar general mechanisms for patterning gene expression in a multicellular tissue are bound to arise. As we shall see, many of the general control mechanisms we have encountered in animal development, such as asymmetric cell division, the response to positional signals, lateral inhibition, and changes in gene expression in response to extracellular signals, are all present in plants. Differences between plants and animals in the way development is controlled arise from some different ways in which plant cells can communicate with each other compared with animal cells, from the existence of rigid cell walls and the lack of large-scale cell movements, and to the fact that environment has a much greater impact on plant development than on that of animals.

One general difference between plant and animal development is that most of the development occurs not in the embryo but in the growing plant. The mature plant embryo inside a seed is not simply a smaller version of the organism it will become. All the 'adult' structures of the plant—shoots, roots, stalks, leaves, and flowers—are produced after germination from localized groups of undifferentiated cells known as **meristems**. Two meristems are established in the embryo: one at the tip of the root and the other at the tip of the shoot. These persist in the adult plant, and almost all the other meristems, such as those in developing leaves and flower shoots, are derived from them. Cells within meristems can divide repeatedly and can potentially give rise to all plant tissues and organs. This means that developmental patterning within meristems to produce organs such as leaves and flowers continues throughout a plant's life.

Plant and animal cells share many common internal features and much basic biochemistry, but there are some crucial differences that have a bearing on plant development. One of the most important is that plant cells are surrounded by a framework of relatively rigid cell walls. There is therefore virtually no cell migration in plants, and major changes in the shape of the developing plant cannot be achieved by the movement of sheets of cells. In plant development, form is largely generated by differences in rates of cell division and by division in different planes, followed by directed enlargement of the cells. When a plant cell divides, it is not pinched in half by a contractile actomyosin ring as in an animal cell. Instead, a new cell membrane and cell wall is formed *in situ* midway between the two poles of the mitotic spindle. In plants, the plane of division seems to be determined before mitosis begins, as a circumferential band of microtubules and actin filaments transiently forms in the cortex at the position of the future cell division.

As in animal development, one of the main questions in plant development is how cell fate is determined. Many structures in plants normally develop by stereotyped patterns of cell division, but despite this observation, which implies the importance of lineage, cell fate is, in many cases, also known to be determined by factors such as position in the meristem and cell–cell signaling. The cell wall would seem to impose a barrier to the passage of large signaling molecules, such as proteins, although it is very thin in some regions, such as the meristems; but most of the known plant extracellular signaling molecules—such as ethylene, auxins, gibberellins, cytokinins, steroids and peptides—are small molecules that penetrate cell walls. Steroids and peptides have transmembrane receptors of the protein kinase type, which are very numerous in plants. The receptors for other types of plant signals are mainly intracellular. Plant cells also communicate with each other through fine cytoplasmic channels known as **plasmodesmata**, which link neighboring plant cells through the cell wall, and they may be the channels by which some developmentally important gene-regulatory proteins, and even mRNAs, move directly between cells. The size of the channels varies, with those in shoot cells having the largest diameter.

Another important difference between plants and most animals is that a complete, fertile plant can develop from a single differentiated somatic cell and not just from a fertilized egg. This suggests that, unlike the differentiated cells of adult animals, some differentiated cells of the adult plant may remain **totipotent**. Perhaps they do not become fully determined in the sense that adult animal cells do, or perhaps they are able to escape from the determined state, although how this could be achieved is as yet unknown. In any case, this difference between plants and animals illustrates the dangers of the wholesale application to plant development of concepts derived from animal development. Nevertheless, the genetic analysis of development in plants is turning up instances of genetic strategies for developmental patterning rather similar to those of animals.

The small cress-like weed *Arabidopsis thaliana* has become the model plant for genetic and developmental studies, and will provide many of the examples in this

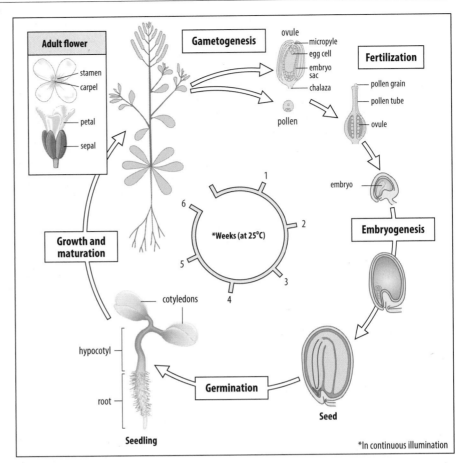

Fig. 7.1 Life cycle of *Arabidopsis*. In flowering plants, egg cells are contained separately in ovules inside the carpels. Fertilization of an egg cell by a male nucleus from a pollen grain takes place inside the ovary. The egg then develops into an embryo contained within the ovule coat, forming a seed. *Arabidopsis* is a dicotyledon, and the mature embryo has two wing-like cotyledons (storage organs) at the apical (shoot) end of the main axis—the hypocotyl—which contains a shoot meristem at one end and a root meristem at the other. Following germination, the seedling develops into a plant with roots, a stem, leaves, and flowers. The photograph shows five mature *Arabidopsis* plants.

chapter. We will begin by describing the main features of its morphology, life cycle, and reproduction.

7.1 The model plant *Arabidopsis thaliana* has a short life cycle and a small diploid genome

The equivalent of *Drosophila* in the study of plant development is the small crucifer *Arabidopsis thaliana*, commonly called thale cress, which is well suited to genetic and developmental studies. It is a diploid (unlike many plants, which are polyploid) and has a relatively small, compact genome that has been sequenced and which contains about 27,000 protein-coding genes. It is an annual, flowering in the first year of growth, and develops as a small ground-hugging rosette of leaves, from which a branched flowering stem is produced with a flowerhead, or **inflorescence**, at the end of each branch. It develops rapidly, with a total life cycle in laboratory conditions of some 6–8 weeks, and like all flowering plants, mutant strains and lines can easily be stored in large quantities in the form of seeds. The life cycle of *Arabidopsis* is shown in Fig. 7.1.

Each *Arabidopsis* flower (Fig. 7.2) consists of four sepals surrounding four white petals; inside the petals are six stamens, the male sex organs, which produce pollen containing the male gametes. Petals, sepals, and the other floral organs are thought to derive evolutionarily from modified leaves. at the center of the flower are the female sex organs, which consist of an ovary of two carpels, which contain the **ovules**. Each ovule contains an egg cell. Fertilization of an egg cell occurs when a pollen grain deposited on the carpel surface grows a tube that penetrates the carpel and delivers two haploid pollen nuclei to an ovule. One nucleus fertilizes the egg cell, while the other fuses with two other nuclei in the ovule. This forms a triploid cell that will develop into a specialized nutritive tissue—the **endosperm**—that surrounds the fertilized egg cells and provides the food source for embryonic development.

Following fertilization, the embryo develops inside the ovule, taking about 2 weeks to form a mature seed, which is shed from the plant. The seed will remain dormant until suitable external conditions trigger germination. The early stages of germination and seedling growth rely on food supplies stored in the **cotyledons** (seed leaves), which are storage organs developed by the embryo. *Arabidopsis* embryos have two cotyledons, and thus belong to the large group of plants known as dicotyledons (dicots for short). The other large group of flowering plants is the monocotyledons (monocots), which have embryos with one cotyledon; monocots typically have long, narrow leaves and include many important staple crops, such as wheat, rice, and maize. Agriculturally important dicots include soy, potato, tomato, sugar beet, and most other vegetables.

At germination, the shoot and root elongate and emerge from the seed. Once the shoot emerges above ground it starts to photosynthesize and forms the first true leaves at the shoot apex. About 4 days after germination the seedling is a self-supporting plant. Flower buds are usually visible on the young plant 3–4 weeks after germination and will open within a week. Under ideal conditions, the complete *Arabidopsis* life cycle thus takes about 6–8 weeks (see Fig. 7.1).

Fig. 7.2 An individual flower of *Arabidopsis*. Scale bar = 1 mm.

*From Meyerowitz, E. et al.: **A genetic and molecular model for flower development in** Arabidopsis thaliana. Development 1991, 113: 157-167.*

Embryonic development

Formation of the embryo, or embryogenesis, occurs inside the ovule, and the end result is a mature, dormant embryo enclosed in a seed awaiting germination. During embryogenesis, the shoot–root polarity of the plant body, which is known as the **apical–basal axis**, is established, and the shoot and root meristems are formed. Plants also possess **radial symmetry**, as seen in the concentric arrangement of the different tissues in a plant stem, and this **radial axis** is also set up in the embryo. Development of the *Arabidopsis* embryo involves a rather invariant pattern of cell division (which is not the case in all plant embryos) and this enables structures in the *Arabidopsis* seedling to be traced back to groups of cells in the early embryo to provide a fate map.

7.2 Plant embryos develop through several distinct stages

Arabidopsis belongs to the angiosperms, or flowering plants, one of the two major groups of seed-bearing plants; the other is the gymnosperms, or conifers. The typical course of embryonic development in angiosperms is outlined in Box 7A. Like an animal zygote, the fertilized plant egg cell undergoes repeated cell divisions, cell growth, and differentiation to form a multicellular embryo. The first division of the zygote is at right angles to the long axis, dividing it into an apical cell and a basal cell, and establishing an initial polarity that is carried over into the apical–basal polarity of the embryo and into the apical–basal axis of the plant. In many species, the first zygotic division is unequal, with the basal cell larger than the apical cell. The basal cell divides to give rise to the **suspensor**, which may be several cells long (see Box 7A). This attaches the embryo to maternal tissue and is a source of nutrients. The apical cell divides vertically to form a two-celled **proembryo**, which will give rise to the rest of the embryo. In some species, the basal cell contributes little to the further development of the embryo, but in others, such as *Arabidopsis*, the topmost suspensor cell is recruited into the embryo, where it is known as the **hypophysis**, and contributes to the embryonic root meristem and root cap.

The next two divisions produce an eight-cell **octant-stage** embryo, which develops into a **globular-stage** embryo of around 32 cells (Fig. 7.3). The embryo elongates and

BOX 7A Angiosperm embryogenesis

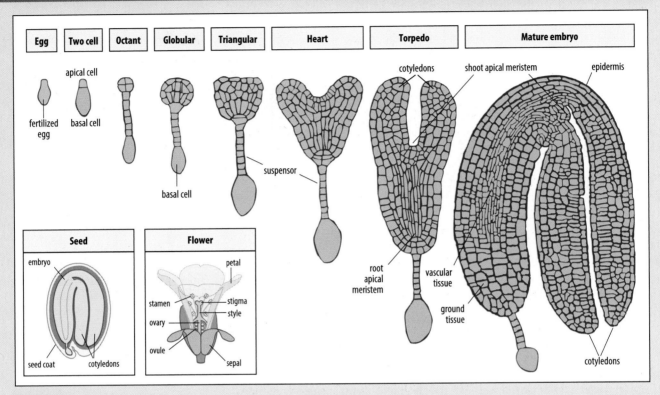

Figure 1

In flowering plants (angiosperms), the egg cell is contained within an ovule inside the ovary in the flower (right inset in Figure 1). at fertilization, a pollen grain deposited on the surface of the stigma puts out a pollen tube, down which two male gametes migrate into the ovule. One male gamete fertilizes the egg cell while the other combines with another cell inside the ovule to form a specialized nutritive tissue, the endosperm, which surrounds, and provides the food source for, the developing embryo.

The small annual weed *Capsella bursa-pastoris* (shepherd's purse) is a typical dicotyledon. The first, asymmetric, cleavage divides the zygote transversely into an apical and a basal cell (Figure 1). The basal cell then divides several times to form a single row of cells—the suspensor—which, in many angiosperm embryos, takes no further part in embryonic development, but may have an absorptive function; in *Capsella*, however, it contributes to the root meristem. Most of the embryo is derived from the apical cell. This undergoes a series of stereotyped divisions, in which a precise pattern of cleavages in different planes gives rise to the heart-shaped embryonic stage typical of dicotyledons. This develops into a mature embryo that consists of a cylindrical body with a meristem at either end, and two cotyledons.

The early embryo becomes differentiated along the radial axis into three main tissues—the outer epidermis, the prospective vascular tissue, which runs through the center of the main axis and cotyledons, and the ground tissue (prospective cortex) that surrounds it.

The ovule containing the embryo matures into a seed (left inset in Figure 1), which remains dormant until suitable external conditions trigger germination and growth of the seedling. A typical dicotyledon seedling (Figure 2) comprises the shoot apical meristem, two cotyledons, the trunk of the seedling (the hypocotyl), and the root apical meristem. The seedling may be thought of as the phylotypic stage of flowering plants. The seedling body plan is simple. One axis—the apical-basal axis—defines the main polarity of the plant. The shoot forms at the apical pole and the root at the basal pole. The shoot meristem is, therefore, referred to in the seedling and adult plant simply as the apical meristem. A plant stem also has a radial axis, which is evident in the radial symmetry in the hypocotyl and is continued in the root and shoot. In the center is the vascular tissue, which is surrounded by cortex, and an outer covering of epidermis. at later stages, structures such as leaves and other organs have a dorso-ventral axis running from the upper surface to the lower surface.

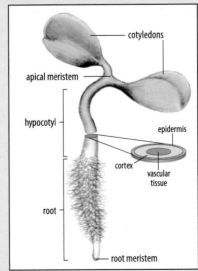

Figure 2

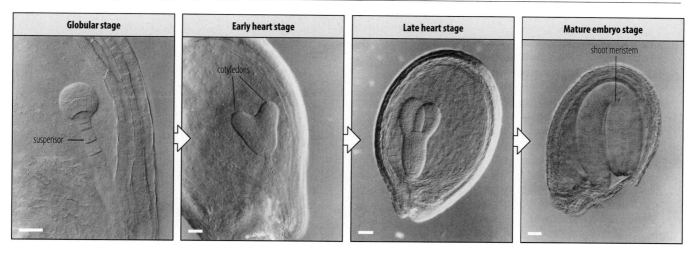

Globular stage | **Early heart stage** | **Late heart stage** | **Mature embryo stage**

suspensor (Globular stage)
cotyledons (Early heart stage)
shoot meristem (Mature embryo stage)

the cotyledons start to develop as wing-like structures at one end, while an embryonic root forms at the other. This stage is known as the **heart stage**. Two groups of undifferentiated cells capable of continued division are located at each end of this axis; these are known as the **apical meristems**. The meristem lying between the cotyledons gives rise to the shoot, while the one at the opposite end of the axis, towards the end of the embryonic root, will drive root growth at germination. The region in between the embryonic root and the future shoot will become the seedling stem or **hypocotyl**. Almost all above ground adult plant structures are derived from the apical meristems. The main exception is radial growth in the stem, which is most evident in woody plants, and which is produced by the **cambium**, a ring of secondary meristematic tissue in the stem. After the embryo is mature, the apical meristems remain quiescent until germination.

Seedling structures can be traced back to groups of cells in the early embryo to provide a fate map (Fig. 7.4). In *Arabidopsis*, patterns of cleavage up to the 16-cell

Fig. 7.3 *Arabidopsis* **embryonic development.** Light micrographs (Nomarski optics) of cleared wild-type seeds of *A. thaliana*. The cotyledons can already be seen at the heart stage. The embryo proper is attached to the seed coat through a filamentous suspensor. Scale bar = 20 μm.

*Photographs courtesy of D. Meinke, from Meinke, D.W.: **Seed development in** Arabidopsis thaliana. In Arabidopsis, CSHLP: 1994.*

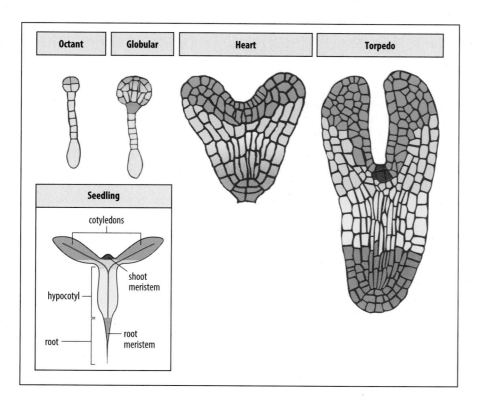

Fig. 7.4 Fate map of the *Arabidopsis* embryo. The stereotyped pattern of cell division in dicotyledon embryos means that at the globular stage it is already possible to map the three regions that will give rise to the cotyledons (dark green) and shoot meristem (red), the hypocotyl (yellow), and the root meristem (purple) in the seedling.

*Illustration after Scheres, B., et al.: **Embryonic origin of the** Arabidopsis primary root and root meristem initials. Development 1994, **120**: 2475-2487.*

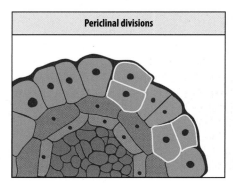

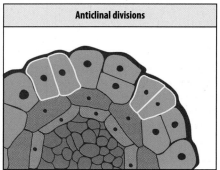

Fig. 7.5 Periclinal and anticlinal divisions. Periclinal cell divisions are parallel to the organ surface, whereas anticlinal divisions are at right angles to the surface.

*Ilustration after Friml, J., et al.: **Efflux-dependent auxin gradients establish the apical-basal axis of Arabidopsis**. Nature 2003, **426**: 147-153.*

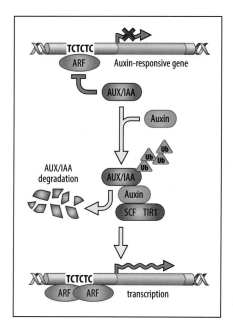

Fig. 7.6 Auxin signaling pathway. In the absence of auxin, AUX/IAA proteins bind to and repress the activity of transcription factors in the AUXIN RESPONSE FACTOR (ARF) family, which bind TGTCTC-containing DNA sequence elements in the promoters of auxin-responsive genes. Auxin targets the ubiquitin ligase complex SCF/TIR1 to the AUX/IAA protein, causing it to be ubiquitinated (Ub). This modification targets AUX/IAA for degradation and thus lifts the repression of the auxin-responsive gene, which can then be transcribed.

*Adapted from Chapman, E.J., Estelle, M.: **Cytokinin and auxin intersection in root meristems**. Genome Biol. 2009, **10**: 210. doi:10.1186/gb-2009-10-2-210*

stage are highly reproducible, and even at the octant stage it is possible to make a fate map for the major regions of the seedling along the apical–basal axis. The upper tier of cells gives rise to the cotyledons and the shoot meristem, the next tier is the origin of the hypocotyl, and the bottom tier together with the region of the suspensor where it joins the embryo will give rise to the root. At the heart stage, the fate map is clear.

The radial pattern in the embryo comprises three concentric rings of tissue: the outer **epidermis**, the ground tissue (**cortex** and **endodermis**), and the vascular tissue at the center. This radial axis appears first at the octant stage, when adaxial (central) and abaxial (peripheral) regions become established. Subsequently, **periclinal** divisions, in which the plane of division is parallel to the outer surface, give rise to the different rings of tissue, and **anticlinal** divisions, in which the plane of cell division is at right angles to the outer surface, increase the number of cells in each ring of tissue (Fig. 7.5). at the 16-cell stage the epidermal layer, or dermatogen, is established.

It is not known to what extent the fate of cells is determined at this stage or whether their fate is dependent on the early pattern of asymmetric cell divisions. It seems that cell lineage is not crucial, as mutations have been discovered that show that cell division can be uncoupled from pattern formation. The *fass* mutation in *Arabidopsis* disrupts the regular pattern of cell division by causing cells to divide in random orientation, and produces a much fatter and shorter seedling than normal. But although the *fass* seedling is misshapen, it has roots, shoots, and even eventually flowers in the correct places and the correct pattern of tissues along the radial axis is maintained.

7.3 Gradients of the signal molecule auxin establish the embryonic apical-basal axis

The small organic molecule **auxin** (indole-3-acetic acid or IAA) is one of the most important and ubiquitous chemical signals in plant development and plant growth. It causes changes in gene expression by promoting the ubiquitination and degradation of transcriptional repressors known as **Aux/IAA proteins**. In the absence of auxin, these bind to proteins called **auxin-response factors (ARFs)** to block the transcription of so-called **auxin-responsive genes**. Auxin-stimulated degradation of Aux/IAA proteins frees the auxin-response factors, which can then activate, or in some cases repress, these genes (Fig. 7.6). Auxin-responsive genes include genes involved in the regulation of cell division and cell expansion, as well as genes involved in specifying cell fate. In some cases auxin appears to be acting as a classical morphogen, forming a concentration gradient and specifying different fates according to a cell's position along the gradient.

The earliest known function of auxin in *Arabidopsis* is in the very first stage of embryogenesis, where it establishes the apical–basal axis. Immediately after the first division of the zygote, auxin is actively transported from the basal cell into the apical cell, where

Fig. 7.7 The role of auxin in patterning the early embryo. Left panel: auxin produced in the original basal cell accumulates in the two-celled proembryo (green) through transport in the basal to apical direction mediated by the protein PIN7, which is located in the apical membranes of the basal and suspensor cells (purple arrows). Another PIN protein, PIN1, transports auxin between the two cells of the proembryo (red arrows). Cell division distributes auxin into the developing embryo. Right panel: at the globular stage, free auxin starts to be produced at the apical pole and the direction of auxin transport is reversed. PIN7 becomes localized in basal membranes of the suspensor cells and the proteins PIN1 and PIN4 (orange arrows) transport auxin from the apical region into the most basal cell of the embryo, known as the hypophysis, where it accumulates.

From Friml, J., et al: **Efflux-dependent auxin gradients establish the apical-basal axis of Arabidopsis**. Nature 2003, **426**: 147-153.

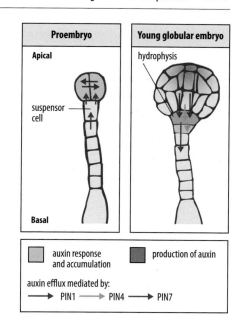

it accumulates. It is transported out of the basal cell by the auxin-efflux protein PIN7, which is localized in the apical plasma membrane of the basal cell. The auxin is required to specify the apical cell, which gives rise to all the apical embryonic structures such as shoot apical meristem and cotyledons. Through the subsequent cell divisions, transport of auxin continues up through the suspensor cells and into the cell at the base of the developing embryo until the globular-stage embryo of about 32 cells. The apical cells of the embryo then start to produce auxin, and the direction of auxin transport is suddenly reversed. PIN7 proteins in the suspensor cells move to the basal faces of the cells. Other *PIN* genes are activated, and the concerted actions of the PIN proteins cause auxin from the apical region to be transported into the basal region of the globular embryo, from which will develop the hypocotyl, root meristem, and embryonic root (Fig. 7.7).

Expression of the homeobox genes *WOX2* and *WOX8* in the asymmetrically dividing zygote seems to be essential for the formation of the embryonic auxin gradient. Both genes are expressed in the zygote, but after division, *WOX2* expression is restricted to the apical cell and its descendants whereas *WOX8* is expressed only in the basal cell lineage. Expression of *WOX8* appears to be required for the continued expression of *WOX2* and for the establishment of an auxin gradient, most probably by effects of the WOX proteins on expression of the PIN proteins that direct auxin transport.

The importance of auxin in specifying the basal region of the embryo is illustrated by the effects of mutations in the cellular machinery for auxin transport or auxin response. The gene *MONOPTEROS* (*MP*), for example, encodes one of the auxin-response factors, ARF5. Embryos mutant for *MP* lack the hypocotyl and the root meristem, and abnormal cell divisions are observed in the regions of the octant-stage embryo that would normally give rise to these structures.

An early step in axis formation in the embryo is the specification of cells as potential shoot or root. The mutation *topless-1* causes a dramatic switching of cell fate, transforming the whole of the apical region into a second root region and abolishing development of cotyledons and shoot meristem. The normal function of the TOPLESS protein is to repress expression of root-promoting genes in the top half of the early embryo. It does this by acting as a transcriptional co-repressor, forming a complex together with Aux/IAA proteins and the auxin-response factors they regulate.

The *topless-1* mutation is temperature-sensitive, and so can be made to act at different times during development by simply changing the temperature at which the embryos are grown. Such manipulations show that the apical region can be respecified as root between the globular and heart stage, even after it has begun to show molecular signs of developing cotyledons and shoot meristem. This suggests that even though apical cell fate is normally specified by this stage of embryonic development, the decision is not irreversible.

Mutations in the gene *SHOOT MERISTEMLESS* (*STM*) completely block the formation of the shoot apical meristem, but have no effect on the root meristem or other parts of the embryo. *STM* encodes a transcription factor that is also required to maintain cells in the pluripotent state in the adult shoot meristem. The pattern

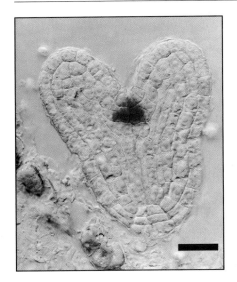

Fig. 7.8 Section through a late heart-stage *Arabidopsis* embryo showing expression of *SHOOT MERISTEMLESS* (*STM*). At this stage, the *STM* RNA (stained red) is expressed in cells located between the cotyledons. Scale bar = 25 μm.

*Photograph courtesy of K. Barton. From Long, J.A., Barton, K.B.: **The development of apical embryonic pattern in** Arabidopsis. Development 1999, **125**: 3027-3035.*

of *STM* expression develops gradually, which is also typical of several other genes that characterize the shoot apical meristem. Expression is first detected in the globular stage in one or two cells and only later in the central region between the two cotyledons (Fig. 7.8).

7.4 Plant somatic cells can give rise to embryos and seedlings

As gardeners well know, plants have amazing powers of regeneration. A new complete plant can develop from a small piece of stem or root, or even from the cut edge of a leaf. This reflects an important difference between the developmental potential of plant and animal cells. In animals, with few exceptions, cell determination and differentiation are irreversible. By contrast, many somatic plant cells remain totipotent. Cells from roots, leaves, stems, and even, for some species, a single, isolated protoplast—a cell from which the cell wall has been removed by enzymatic treatment—can be grown in culture and induced by treatment with the appropriate growth hormones to give rise to a new plant (Fig. 7.9). Plant cells proliferating in culture can give rise to cell clusters that pass through a stage resembling normal embryonic development, although the pattern of cell divisions are not the same. These 'embryoids' can then develop into seedlings. Plant cells can also form callus, an apparently disorganized mass of cells, and the callus can form new shoot or root meristems, and therefore new shoots and roots. The ability of plants to regenerate from somatic cells means that transgenic plants can be easily generated using the plant pathogenic bacterium *Agrobacterium tumefaciens* as the carrier of the gene to be transferred (Box 7B).

The ability of single somatic cells to give rise to whole plants has two important implications for plant development. The first is that maternal determinants may be of little or no importance in plant embryogenesis, as it is unlikely that many somatic cells would still be carrying such determinants. Second, it suggests that many cells in the adult plant body are not fully determined with respect to their fate, but remain totipotent. Of course, this totipotency seems only to be expressed under special conditions, but it is, nevertheless, quite unlike the behavior of animal cells. It is as if such plant cells have no long-term developmental memory, or that such memory is easily reset.

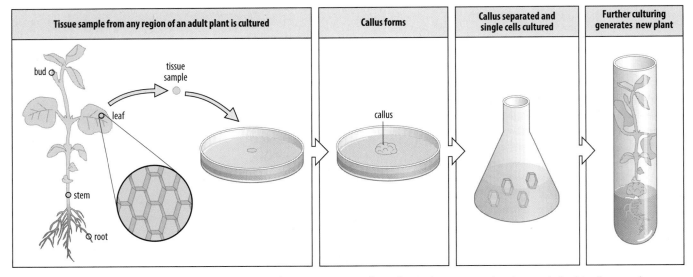

| Tissue sample from any region of an adult plant is cultured | Callus forms | Callus separated and single cells cultured | Further culturing generates new plant |

Fig. 7.9 Cultured somatic cells from a mature plant can form an embryo and regenerate a new plant. The illustration shows the generation of a plant from single cells. If a small piece of tissue from a plant stem or leaf is placed on a solid agar medium containing the appropriate nutrients and growth hormones, the cells will start to divide to form a disorganized mass of cells known as a callus. The callus cells are then separated and grown in liquid culture, again containing the appropriate growth hormones. In suspension culture, some of the callus cells divide to form small cell clusters. These cell clusters can resemble the globular stage of a dicotyledon embryo, and with further culture on solid medium, develop through the heart-shaped and later stages to regenerate a complete new plant.

EXPERIMENTAL BOX 7B Transgenic plants

One of the most common ways of generating transgenic plants containing new and modified genes is through infection of plant tissue in culture with the bacterium *Agrobacterium tumefaciens*, the causal agent of crown gall tumors. *Agrobacterium* is a natural genetic engineer. It contains a plasmid—the Ti plasmid—that contains the genes required for the transformation and proliferation of infected cells to form a callus. During infection, a portion of this plasmid—the T-DNA (shown in red in Figure 1)—is transferred into the genome of the plant cell, where it becomes stably integrated. Genes experimentally inserted into the T-DNA will therefore also be transferred into the plant cell chromosomes. Ti plasmids, modified so that they do not cause tumors but still retain the ability to transfer T-DNA, are widely used as vectors for gene transfer in dicotyledonous plants. The genetically modified plant cells of the callus can then be grown into a complete new transgenic plant that carries the introduced gene in all its cells and can transmit it to the next generation.

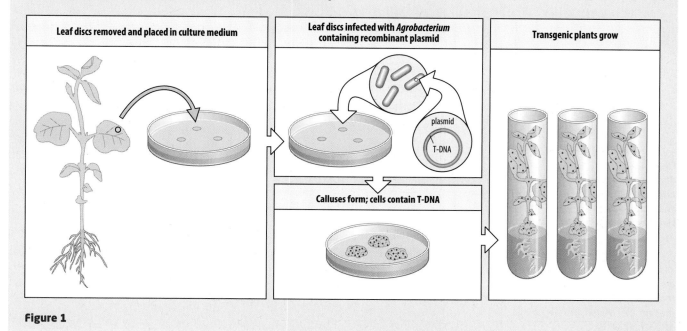

Leaf discs removed and placed in culture medium

Leaf discs infected with *Agrobacterium* containing recombinant plasmid

Transgenic plants grow

plasmid

T-DNA

Calluses form; cells contain T-DNA

Figure 1

7.5 Cell enlargement is a major process in plant growth and morphogenesis

Plants grow in size by a combination of cell division and cell expansion. Cell expansion can provide up to a 50-fold increase in the volume of a growing plant tissue, and much of the growth in size of plants and their organs is due to cell enlargment. The driving force for expansion is the hydrostatic pressure—turgor pressure—exerted on the cell wall as the protoplast swells as a result of the entry of water into cell vacuoles by osmosis (Fig. 7.10). Plant-cell expansion involves synthesis and deposition of new cell-wall material, and is an example of a process called **directed dilation**, in which an increase in internal hydrostatic pressure causes a distinct change in shape in a particular direction. Directed dilation is also seen in animal development, in the development of the vertebrate notochord, and elongation of the nematode body (discussed in Chapter 9). In plants, the direction of cell growth is determined by the orientation of the cellulose fibrils in the cell wall. Enlargement occurs primarily in a direction at right angles to the fibrils, where the wall is weakest. The orientation of cellulose fibrils in the cell wall is determined by the microtubules of the cell's cytoskeleton, which are responsible for positioning the enzyme assemblies that synthesize cellulose at the cell wall. Plant growth hormones, such as ethylene and gibberellins alter the orientation in which the fibrils

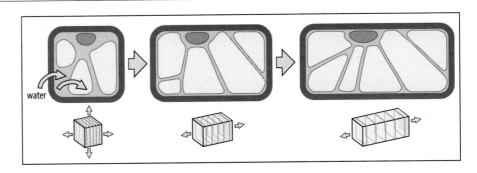

Fig. 7.10 Enlargement of a plant cell.
Plant cells expand as water enters the cell vacuoles and thus causes an increase in intracellular hydrostatic pressure. The cell elongates in a direction perpendicular to the orientation of the cellulose fibrils in its cell wall.

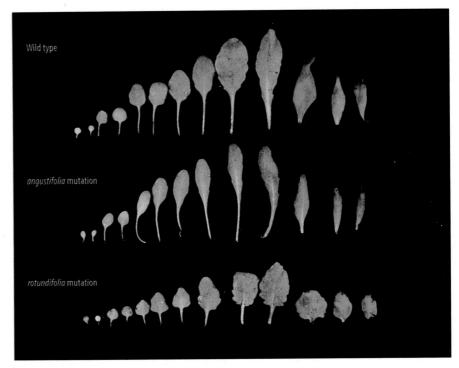

Fig. 7.11 The shape of the leaves of *Arabidopsis* is affected by mutations affecting cell elongation. The *angustifolia* mutation results in cells with narrow leaves, while *rotundifolia* mutations cause short, fat leaves to develop.

*Photograph courtesy of H. Tsukaya, from Tsuge, T., et al.: **Two independent and polarized processes of cell elongation regulate blade expansion in Arabidopsis thaliana (L.) Heynh**. Development 1996, 122: 1589-1600.*

are laid down and so can alter the direction of expansion. Auxin aids expansion by loosening the structure of the cell wall.

The development of a leaf in a particular shape involves a complex pattern of cell division and cell elongation along the proximo-distal and medio-lateral axes of the outgrowing leaf, which is thought to be organized by a region at the base of the leaf. Cell elongation plays a central part in the expansion of the leaf blade in the later stages of leaf outgrowth. Two mutations that affect the shape of the blade by affecting the direction of cell elongation have been identified. Leaves of the *Arabidopsis* mutant *angustifolia* are similar in length to the wild type but are much narrower (Fig. 7.11). In contrast, *rotundifolia* mutations reduce the length of the leaf relative to its width. Neither of these mutations affects the number of cells in the leaf. Examination of the cells in the developing leaf shows that these mutations are affecting the direction of elongation of the enlarging cells.

SUMMARY

Early embryonic development in many plants is characterized by asymmetric cell division of the zygote, which specifies apical and basal regions. Embryonic development in flowering plants establishes the shoot and root meristems from which the adult plant develops. Auxin gradients are involved in specifying the apical–basal axis of the *Arabidopsis* embryo. One major difference between plants and animals is that, in culture, a single somatic plant cell can develop through an embryo-like stage and regenerate a complete new plant, indicating that some differentiated plant cells retain totipotency.

SUMMARY: early development in flowering plants

first asymmetric cell division and auxin signaling in embryo establishes apical–basal axis

⇩

embryonic cell fate is determined by position

⇩

shoot and root meristems of seedling give rise to all adult plant structures

---

Meristems

In plants, most of the adult structures are derived from just two regions of the embryo, the embryonic shoot and root meristems, which are maintained after germination. The embryonic shoot meristem, for example, becomes the shoot apical meristem of the growing plant, giving rise to all the stems, leaves, and flowers. As the shoot grows, lateral outgrowths from the meristem give rise to leaves and to side shoots. In flowering shoots, the vegetative meristem becomes converted into one capable of producing **floral meristems** that make flowers, not leaves. In *Arabidopsis*, for example, the shoot apical meristem changes from a vegetative meristem that makes leaves around it in a spiral pattern to an **inflorescence meristem** that then produces floral meristems, and thus flowers, around it in a spiral pattern. The first stages of future organs are known as **primordia** (singular **primordium**). Each primordium consists of a small number of **founder cells** that produce the new structure by cell division and cell enlargement, accompanied by differentiation.

There is usually a time delay between the initiation of two successive leaves in a shoot apical meristem, and this results in a plant shoot being composed of repeated modules. Each module consists of an **internode** (the cells produced by the meristem between successive leaf initiations), a **node** and its associated leaf, and an axillary bud (Fig. 7.12). The axillary bud itself contains a meristem, known as the **lateral shoot meristem**, which forms at the base of the leaf. The growth of lateral buds immediately below the apical bud is suppressed, the phenomenon known as **apical dominance.** Apical dominance is caused by the hormone auxin, which is produced by the shoot apex and diffuses back to inhibit development of the lateral meristem. The existence of a diffusible inhibitor was demonstrated by putting an excised apex in contact with an agar block, which could then, on its own, inhibit lateral growth. Auxin produced by the apical bud is transported down the stem, and suppresses the outgrowth of buds that fall within its sphere of influence. If the apical bud is removed, apical dominance is also removed, and lateral buds start to grow (Fig. 7.13). Application of auxin to the cut tip replaces the suppressive effect of the apical bud.

Root growth is not so obviously modular, but similar considerations apply, as new lateral meristems initiated behind the root apical meristem give rise to lateral roots.

Shoot apical meristems and root apical meristems operate on the same principles, but there are some significant differences between them. We will use the shoot apical meristem to illustrate the basic principles of meristem structure and properties, and then discuss roots.

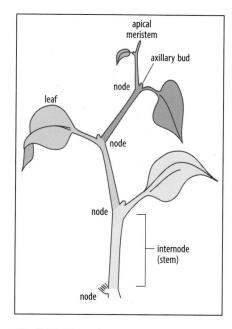

Fig. 7.12 Plant shoots grow in a modular fashion. The shoot apical meristem produces a repeated basic structural module. The vegetative shoot module typically consists of internode, node, leaf, and axillary bud (from which a side branch may develop). Successive modules are shown here in different shades of green. As the plant grows, the internodes behind the meristem lengthen and the leaves expand.

Illustration after Alberts, B., et al.: Molecular Biology of the Cell, *2nd edition. New York: Garland Publishing, 1989.*

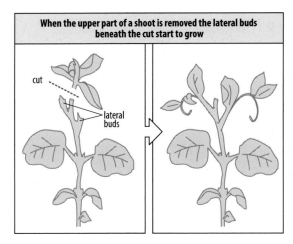

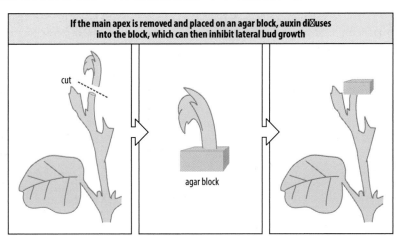

Fig. 7.13 Apical dominance in plants.
Left panels: the growth of lateral buds is
inhibited by the apical meristem above them.
If the upper part of the stem is removed,
lateral buds start growing. Right panels: an
experiment to show that apical dominance
is due to inhibition of lateral bud outgrowth
by a substance secreted by the apical region.
This substance is the plant hormone auxin.

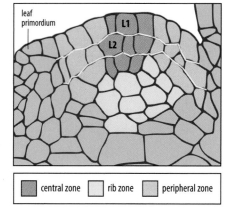

**Fig. 7.14 Organization of the *Arabidopsis*
shoot meristem.** A longitudinal section is
shown. The meristem has three main layers,
L1, L2, and an inner layer, as indicated by
the yellow lines, and is divided functionally
into a central zone (red), a rib zone (yellow),
and a peripheral zone (blue). The stem cells
or initials lie in the central zone, while the
peripheral zone contains proliferating cells
that will give rise to leaves and side shoots.
The rib zone gives rise to the central tissues
of the plant stem.

7.6 A meristem contains a small, central zone of self-renewing stem cells

Shoot apical meristems are rarely more than 250 μm in diameter in angiosperms and contain a few hundred relatively small, undifferentiated cells that are capable of cell division. Most of the cell divisions in normal plant development occur within meristems, or soon after a cell leaves a meristem, and much of a plant's growth in size is due to cell elongation and enlargement. Cells leave the periphery of the meristem to form organs such as leaves or flowers, and are replaced from a small central zone of slowly dividing, self-renewing stem cells or **initials** at the tip of the meristem (Fig. 7.14). In *Arabidopsis* this zone comprises around 12 to 20 cells. Initials behave in the same way as animal stem cells (see Chapter 8). They can divide to give one daughter that remains a stem cell and one that loses the stem-cell property. This daughter cell continues to divide and its descendants are displaced towards the peripheral zone of the meristem, where they become founder cells for a new organ or internode, leave the meristem, and differentiate. A very small number of long-term stem cells at the meristem center may persist for the whole life of the plant.

Meristem stem cells are maintained in the self-renewing state by cells underlying the central zone that form the **organizing center**. As we shall see, it is the microenvironment maintained by the organizing center that gives stem cells their identity.

The undetermined state of meristem stem cells is confirmed by the fact that meristems are capable of regulation. If, for example, a seedling shoot meristem is divided into two or four parts by vertical incision, each part becomes reorganized into a complete meristem, which gives rise to a normal shoot. Provided that some subpopulation of organizing center cells plus overlying stem cells is present, a normal meristem will regenerate. If a shoot apical meristem is completely removed, no new apical meristem forms, but the incipient meristem at the base of the leaf is now able to develop and form a new side shoot. In the presence of the original meristem, this prospective meristem remains inactive, as active meristems inhibit the development of other meristems nearby—as a result of auxin transport from the shoot apex, among other factors. This regulative behavior is in line with cell–cell interactions being a major determinant of cell fate in the meristem.

7.7 The size of the stem-cell area in the meristem is kept constant by a feedback loop to the organizing center

Numerous genes control the behavior of the cells in the meristem. The gene *STM*, which is involved in specifying the shoot meristem in the development of the *Arabidopsis* embryo (see Section 7.3) is, for example, expressed throughout adult shoot meristems but is suppressed as soon as cells become part of an organ primordium.

Fig. 7.15 Regulation of the stem-cell population in a shoot meristem. Intercellular signals control the position and size of the stem-cell population in the *Arabidopsis* shoot meristem. Top panel: the organizing center (mauve) expresses the transcription factor WUS, which is thought to act as an intercellular signal (red arrow) that helps specify and maintain the overlying cells as stem cells (orange). The stem cells express and secrete the signal protein CLAVATA3 (CLV3; orange dots), which moves laterally and downwards, and indirectly represses transcription of the *WUS* gene in the surrounding cells, acting through its cell-surface receptor proteins CLV1 and CLV2. CLV3 thus limits the extent of the area specified as stem cells (blue barred lines). The descendants of the stem cells are continuously displaced into the peripheral zone of the meristem (pale yellow), where they are recruited into leaf primordia. Bottom panel: the negative-feedback loop whereby WUS expression is restricted by CLV3. WUS proteins produced in the organizing center cells move into the overlying cells, where they induce the expression of CLV3. CLV3 in turn signals through CLV1/2 to suppress the expression of *WUS*.

Adapted from Brand U., et al.: **Dependence of stem cell fate in Arabidopsis on a feedback loop regulated by CLV3 activity.** *Science 2000, 289: 635–644.*

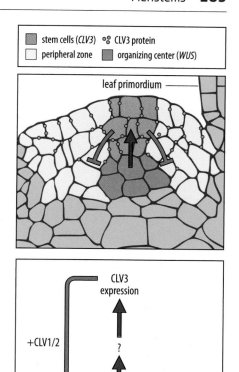

Its role seems to be to maintain meristematic cells in an undifferentiated state, as loss of *STM* function results in all the meristem cells being incorporated into organ primordia. In contrast, mutations in the *Arabidopsis CLAVATA* (*CLV*) genes increase the size of the meristem as a result of an increase in the number of stem cells. Normally, despite the continual exit of cells from the stem-cell pool, the number of stem cells in a meristem is kept roughly the same throughout a plant's life, by division of the remaining stem cells. The role of the *CLV* genes in regulating stem-cell numbers is understood in some detail, and involves feedback between the stem cells and the organizing center that underlies them.

In *Arabidopsis*, cells of the organizing center express a homeobox transcription factor, WUSCHEL (WUS). This is required to produce a signal that gives the overlying cells their stem-cell identity. At least part of this signal seems to be due to the WUS protein itself moving from the cells of the organizing center into the overlying cells via plasmodesmata. Mutations in *WUS* result in termination of the shoot meristem and cessation of growth as a result of the loss of stem cells, while its overexpression increases stem-cell numbers. The stem cells express *CLAVATA3* (*CLV3*), which encodes a secreted protein that acts indirectly to repress *WUS*. This feedback loop could control *WUS* activity in the organizing center and suppress *WUS* activation in neighboring cells, thus limiting the extent of *WUS* expression (Fig. 7.15). In turn, this would regulate the extent of the stem-cell zone above the organizing center. If stem-cell numbers temporarily fall, for example, less CLV3 is produced and *WUS* activity increases, with a consequent increase in stem-cell numbers. More CLV3 is then produced and limits the extent of *WUS* activity. Other CLAVATA proteins are involved in the feedback loop (see Fig. 7.15).

Meristems can be induced elsewhere in the plant by the misexpression of genes involved in specifying stem-cell identity, yet another indication that stem-cell identity is conferred by cell–cell interactions and not by an embryonically specified cell lineage.

7.8 The fate of cells from different meristem layers can be changed by changing their position

Other evidence that the fate of a meristematic cell is determined by its position in the meristem, and thus the intercellular signals it is exposed to, comes from observing the fates of cells in the different meristem layers. As well as being organized into central and peripheral zones, the apical meristem of a dicotyledon, such as *Arabidopsis*, is composed of three distinct layers of cells (Fig. 7.16). The outermost layer, L1, is just one cell thick. Layer L2, just beneath L1, is also one cell thick. In both L1 and L2, cell divisions are anticlinal—that is, the new wall is in a plane perpendicular to the

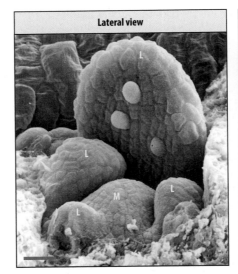
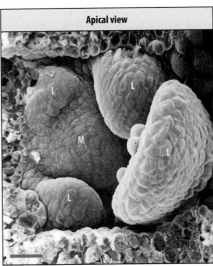

Lateral view	Apical view

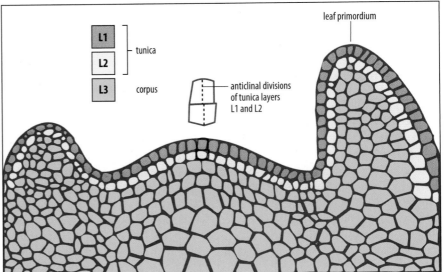

Fig. 7.16 Apical meristem of *Arabidopsis*.
Top panels: scanning electron micrographs showing the organization of the meristem at the young vegetative apex of *Arabidopsis*. The plant is a *clavata1* mutant, which has a broadened apex that allows for a clearer visualization of the leaf primordia (L) and the meristem (M). Scale bar = 10 μm. Bottom panel: diagram of a vertical section through the apex of a shoot. The three-layered structure of the meristem is apparent in the most apical region. In layer 1 (L1) and layer 2 (L2), the plane of cell division is anticlinal; that is, at right angles to the surface of the shoot. Cells in layer 3 (L3, the corpus) can divide in any plane. A leaf primordium is shown forming at one side of the meristem.

Reprinted with permission from Bowman J. ed. Arabidopsis An Atlas of Morphology and Development. Springer-Verlag, 1994.

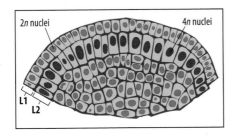

Fig. 7.17 A periclinal meristem chimera composed of cells of two different genotypes. In L1 the cells are diploid whereas the cells of L2 are tetraploid—that is, they have double the normal chromosome number—and are larger and easily recognized.
Illustration after Steeves, T.A., Sussex, I.M.: Patterning in Plant Development. Cambridge: Cambridge University Press, 1989.

layer—thus maintaining the two-layer organization. The innermost layer is L3, in which the cells can divide in any plane. L1 and L2 are often known as the tunica, and L3 as the corpus.

To find which tissues each layer can give rise to, the fates of cells in the different layers can be followed by marking one layer with a distinguishable mutation, such as a change in pigmentation or a polyploid nucleus. When a complete layer is genetically different from the others in this way, the organism is known as a **periclinal chimera** (Fig. 7.17) and the fate of cells from this layer can be traced. As we saw in relation to animal embryos, chimeras are organisms composed of cells of two different genotypes (see Section 3.9). Plant chimeras can be made by inducing mutation in the apical meristem of a seed or shoot tip by X-irradiation, or by treatment with chemicals such as colchicine that induce polyploidy.

Layer L1 gives rise to the epidermis that covers all structures produced by the shoot, while L2 and L3 both contribute to cortex and vascular structures. Leaves and flowers are produced mostly from L2; L3 contributes mainly to the stem. Although the three layers maintain their identity in the central region of the meristem over long periods of growth, cells in either L1 or L2 occasionally divide periclinally; new cell walls are formed parallel to the surface of the meristem, and thus one of the new cells invades an adjacent layer. This migrant cell now develops according to its new position, showing that cell fate is not necessarily determined by the meristem layer in which the cell originated, and that intercellular signaling is involved in specifying or changing its fate in its new position. The anticlinal pattern of cell division in L2 becomes disrupted when leaf primordia start to form, when the cells divide periclinally, as well as anticlinally.

The transcription factor Knotted-1 in maize is homologous to *Arabidopsis* STM and, like STM, it is expressed throughout the shoot meristem to keep cells in an undifferentiated state. It is one example of a transcription factor that moves directly from cell to cell. The *KNOTTED-1* gene is expressed in all layers except L1, but the protein is also found in L1, suggesting that it can move between cells, perhaps via plasmodesmata. In *Knotted-1* gain-of-function mutants, in which the gene is misexpressed in leaves, Knotted-1 protein fused to green fluorescent protein has also been observed to move from the inner layers of the leaf to the epidermis, but not in the opposite direction.

7.9 A fate map for the embryonic shoot meristem can be deduced using clonal analysis

Much of our knowledge about the general properties of meristems comes from studies some decades ago that determined how the embryonic shoot apical meristem is related to the development of a plant over its whole lifetime. What was the fate of individual 'embryonic initials', as the stem cells were known at the time? Did particular regions in the embryonic meristem give rise to particular parts of the adult plant? Individual initials can be marked by mutagenesis of seeds by X-irradiation or transposon activation, to give cells with a different color from the rest of the plant, for example. If the marked meristem cell and its immediate progeny populate only part of one layer of the meristem (in contrast to a periclinal chimera), then this area will give rise to visible sectors of marked cells in stems and organs as the plant grows; this type of chimera is known as a **mericlinal chimera** (Fig. 7.18). The fate of individually marked initials in mericlinal chimeras can be determined using clonal analysis in a manner similar to its use in *Drosophila* (see Box 2H).

In maize, the marked sectors usually start at the base of an internode and extend apically, terminating within a leaf. Some sectors include just a single internode and leaf, representing the progeny of an embryonic initial that is lost from the meristem before the generation of the next leaf primordium. Others, on the other hand, extend through numerous internodes, showing that some embryonic initials remain in the meristem for a long time, contributing to a succession of nodes and internodes. In sunflowers, marked clones have been observed to extend through several internodes up into the flower, showing that a single initial can contribute to both leaves and flowers.

From the analysis of hundreds of mericlinal chimeras, fate maps of the embryonic shoot meristems of several species were constructed, which shed light on the properties of the shoot meristem and how it behaves during normal development. These fate maps are probabilistic because it was not possible to know the location of the marked cell in the embryonic meristem, which is inaccessible inside the seed.

The probabilistic fate map for the maize embryonic shoot apical meristem indicates that the three most apical cells in L1 give rise to the male inflorescence (the tassel and spike; Fig. 7.19). The remainder of the maize meristem can be divided into five tiers of cells that produce internodes and leaves, and which form overlapping concentric domains on the fate map. The outermost domain contributes to the earliest internode–leaf modules, while the inner domains give rise to internodes and leaves successively higher up the stem. A fate map has been similarly constructed for the embryonic shoot apical meristem of *Arabidopsis* (Fig. 7.20). Most of the *Arabidopsis* embryonic meristem gives rise to the first six leaves, whereas the remainder of the shoot, including all the flowerheads, is derived from a very small number of embryonic cells at

Fig. 7.18 Tobacco plant mericlinal chimera. This plant has grown from an embryonic shoot meristem in which an albino mutation has occurred in a cell of the L2 layer. The affected area occupies about a third of the total circumference of the shoot, suggesting that there are three apical initial cells in the embryonic shoot meristem.

Photograph courtesy of S. Poethig.

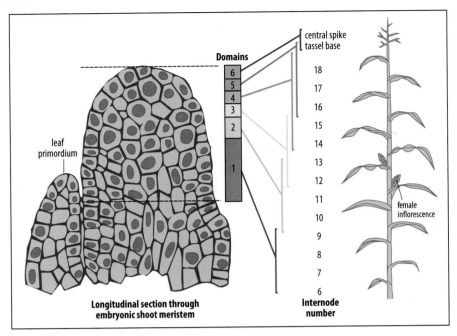

Fig. 7.19 Probabilistic fate map of the shoot apical meristem in the mature maize embryo.
In maize, the primordia of the first six leaves are already present in the mature embryo and are excluded from the analysis. at the time the cells were marked, the meristem contained about 335 cells, which will give rise to 12 more leaves, the female inflorescences, and the terminal male inflorescence (the tassel and spike). A longitudinal section through the embryonic apical dome—the shoot meristem—is shown on the left. Clonal analysis shows that it can be divided into six vertically stacked domains, each comprising a set of initials that can give rise to a particular part of the plant. The number of initials in layers L1 and L2 of each domain at the embryonic stage can be estimated from the final extent of the corresponding marked sectors in the mature plant (see text). The fate of each domain, as estimated from the collective results of clonal analysis of many different plants, is shown on the right. Domain 6, comprising the three most apical L1 cells of the meristem, will give rise to the terminal male inflorescence. The fate of cells in the other domains is less circumscribed. Domain 5, for example, which consists of around eight L1 cells surrounding domain 6 and the underlying four or so L2 cells, can contribute to nodes 16–18; domain 4 to nodes 14–18; and domain 3 to nodes 12–15. The female inflorescences, which develop in the leaf axils, are derived from the corresponding domains.
*Illustration after McDaniel, C.N., Poethig, R.S.: **Cell lineage patterns in the shoot apical meristem of the germinating maize embryo**. Planta 1988, **175**: 13–22.*

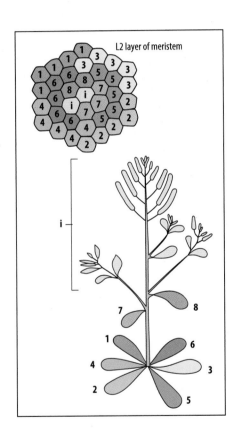

Fig. 7.20 Probabilistic fate map of the embryonic shoot meristem of *Arabidopsis*.
The L2 layer of the meristem is depicted as if flattened out and viewed from above. The numbers indicate the leaf, as shown on the plant below, to which each group of meristem cells contributes, and indicate the sequence in which the leaves are formed. The inflorescence shoot (i) is derived from a small number of cells in the center of the layer.
*After Irish, V. E. **Cell lineage in plant development**. Curr Opin. Genet. Dev. 1991, **1**: 169–173.*

the center of the meristem. Unlike maize, the number of leaves in *Arabidopsis* is not fixed—growth is said to be **indeterminate**. There is no relation between particular cell lineages and particular structures, which indicates that position in the meristem is crucial in determining cell fate. One exception is that germ cells always arise from L2. The L2 layer is a clone, and so there is, in the case of germ cells, a relation between fate and a particular cell lineage.

The conclusions from the clonal analysis experiments are that the initials that contribute to a particular structure are simply those that happen to be in the appropriate region of the meristem at the time; they have not been prespecified in the embryo as, say, flower or leaf.

7.10 Meristem development is dependent on signals from other parts of the plant

To what extent does the behavior of a meristem depend on other parts of the plant? It seems to have some autonomy, because if a meristem is isolated from adjacent tissues

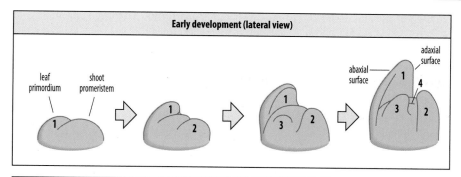

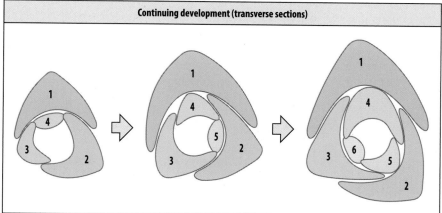

Fig. 7.21 Leaf phyllotaxis. In shoots where single leaves are arranged spirally up the stem, the leaf primordia arise sequentially in a mathematically regular pattern in the meristem. Leaf primordia arise around the sides of the apical dome, just outside the **promeristem** region. A new leaf primordium is formed slightly above and at a fixed radial angle from the previous leaf, often generating a helical arrangement of primordia visible at the apex. Top panel: lateral views of the shoot apex. Bottom panel: view looking down on cross-sections through the apex near the tip, at successive stages from the top panel.

*Top panel, illustration after Poethig, R.S., Sussex, I.M.: **The cellular parameters of leaf development in tobacco: a clonal analysis**. Planta 1985, **165**: 170-184. Bottom panel, illustration after Sachs, T.:* Pattern Formation in Plant Tissues. *Cambridge: Cambridge University Press, 1994.*

by excision it will continue to develop, although often at a much slower rate. Excised shoot apical meristems of a variety of plants can be grown in culture, where they will develop into shoots complete with leaves if the growth hormones auxin and cytokinin are added. The behavior of the meristem *in situ*, however, is influenced in more subtle ways by interactions with the rest of the plant.

As we saw earlier, the apical meristem of maize gives rise to a succession of nodes, and terminates in the male flower. The number of nodes before flowering is usually between 16 and 22. This number is not controlled by the meristem alone, however. Evidence comes from culturing shoot tips consisting of the apical meristem and one or two leaf primordia. Meristems taken from plants that have already formed as many as 10 nodes still develop into normal maize plants with the full number of nodes. The isolated maize meristem has no memory of the number of nodes it has already formed, and repeats the process from the beginning. The meristem itself is therefore not determined in the embryo with respect to the number of nodes it will form. In the plant, control over the number of nodes that are formed must therefore involve signals from the rest of the plant to the meristem.

7.11 Gene activity patterns the proximo-distal and adaxial–abaxial axes of leaves developing from the shoot meristem

Leaves develop from groups of founder cells within the peripheral zone of the shoot apical meristem. The first indication of leaf initiation in the meristem is usually a swelling of a region to the side of the apex, which forms the **leaf primordium** (Fig. 7.21). This small protrusion is the result of increased localized cell multiplication and altered patterns of cell division. It also reflects changes in polarized cell expansion.

Two new axes that relate to the future leaf are established in a leaf primordium. These are the **proximo-distal axis** (leaf base to leaf tip) and the **adaxial–abaxial axis** (upper surface to lower surface, sometimes called dorsal to ventral). The latter is termed adaxial–abaxial as it is related to the radial axis of the shoot. The upper surface of the leaf derives from cells near the center of this axis (adaxial), while the lower surface derives from more peripheral cells (abaxial). The two leaf surfaces carry out different functions and have different structures, with the top surface being specialized for light capture and photosynthesis. In *Arabidopsis*, flattening of the leaf along the adaxial–abaxial axis occurs after leaf primordia begin developing, but in monocots like maize the leaf is flattened as it emerges. The establishment of adaxial–abaxial polarity is likely to make use of positional information along the radial axis of the meristem.

Arabidopsis leaf primordia emerge from the shoot meristem with distinct programs of development in the adaxial and abaxial halves. This asymmetry can be seen from the beginning, as the leaf primordium has a crescent shape in cross-section, with a convex outer (abaxial) side and a concave inner (adaxial) surface (see Fig. 7.21). Different genes are expressed in the future adaxial and abaxial sides. For example, the *Arabidopsis* gene *FILAMENTOUS FLOWER* (*FIL*) is normally expressed in the abaxial side of the leaf primordium, and specifies an abaxial cell fate. Its ectopic expression throughout the leaf primordium can cause all cells to adopt an abaxial cell fate, and the leaf develops as an arrested cylindrical structure.

The specification of adaxial cell fate in *Arabidopsis* involves the genes *PHABULOSA* (*PHAB*), *PHAVOLUTA* (*PHAV*), and *REVOLUTA* (*REV*). These encode transcription factors and are initially expressed in the shoot meristem and in the adaxial side of the primordium. Loss-of-function mutations in these genes result in radially symmetrical leaves with only abaxial cell types characteristic of the underside of the leaf. A microRNA (see Box 6C) is involved in the restriction of *PHAB*, *PHAV*, and *REV* expression to the adaxial side, targeting and destroying their mRNAs on the abaxial side and thus limiting their activity to the adaxial side.

It has been suggested that, in normal plants, the interaction between adaxial and abaxial initial cells at the boundary between them initiates lateral growth, resulting in the formation of the leaf blade and the flattening of the leaf. The importance of boundaries in controlling pattern and form has already been seen in the parasegments of *Drosophila* (see Section 2.24) and further examples will be found in Chapter 11.

Development along the leaf proximo-distal axis also appears to be under genetic control. Like other grasses, a maize leaf primordium is composed of prospective leaf-sheath tissue proximal to the stem and prospective leaf-blade tissue distally. Mutations in certain genes result in distal cells taking on more proximal identities; for example, making sheath in place of blade. Similar proximo-distal shifts in pattern occur in *Arabidopsis* as a result of mutation. Regional identity along the proximo-distal axis may reflect the developmental age of the cells, distal cells adopting a different fate from proximal cells because they mature later.

7.12 The regular arrangement of leaves on a stem is generated by regulated auxin transport

As the shoot grows, leaves are generated within the meristem at regular intervals and with a particular spacing. Leaves are arranged along a shoot in a variety of ways in different plants, and the particular arrangement, known as phyllotaxy or **phyllotaxis**, is reflected in the arrangement of leaf primordia in the meristem. Leaves can occur singly at each node, in pairs, or in whorls of three or more. A common arrangement is the positioning of single leaves spirally up the stem, which can sometimes form a striking helical pattern in the shoot apex.

In plants in which leaves are borne spirally, a new leaf primordium forms at the center of the first available space outside the central region of the meristem and above the previous primordium (see Fig. 7.21). This pattern suggests a mechanism for leaf arrangement based on lateral inhibition (see Section 1.16), in which each leaf primordium inhibits the formation of a new leaf within a given distance. In this model,

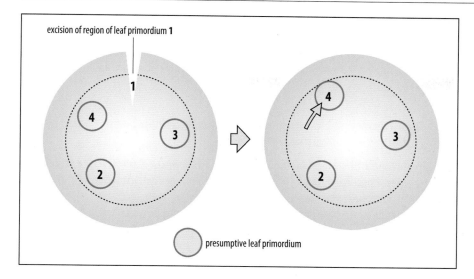

excision of region of leaf primordium **1**

presumptive leaf primordium

Fig. 7.22 Leaf primordia may be positioned by lateral inhibition or by competition. Leaf primordia on a fern shoot tip form in a regular order in positions 1 to 4. Primordia appear to form as far as possible from existing primordia, so 2 forms almost opposite 1. Normally, 4 will develop between 1 and 2, but if 1 is excised, 4 forms much further away from 2. This result can be interpreted either by the removal of lateral inhibition by primordium 1 or by the removal of the competitive effect of 1 for some primordium-inducing factor such as auxin.

inhibitory signals emanating from recently initiated primordia prevent leaves from forming close to each other. There is some experimental evidence for this. In ferns, leaf primordia are widely spaced, allowing experimental microsurgical interference. Destruction of the site of the next primordium to be formed results in a shift toward that site by the future primordium whose position is closest to it (Fig. 7.22). Recent studies indicate, however, that competition, as well as inhibition, is responsible for the spacing.

As we have seen in the determination of apical–basal polarity in the embryo, auxin is transported out of cells with the help of proteins such as PIN1 (see Section 7.3). In the shoot, auxin produced in the shoot tip below the meristem is transported upwards into the meristem, through the epidermis and the outermost meristem layer. The direction of auxin flow in the shoot apex is controlled by PIN1 and follows a simple rule: the side of the cell on which PIN1 is found is the side nearest the neighbor cell with the highest auxin concentration. Thus auxin transport is always towards a region of higher concentration. A high concentration of auxin is a primordium activator, and initially, auxin is pumped towards a new primordium that is developing at a site of high auxin concentration (Fig. 7.23). This, however, depletes a zone of cells around the primordium of auxin, such that cells nearer the center of the meristem now have more auxin than the cells adaxial to the new primordium. This activates a feedback mechanism

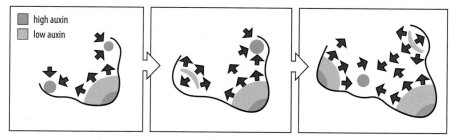

high auxin
low auxin

Fig. 7.23 Auxin-dependent mechanism of phyllotaxis in *Arabidopsis*. Auxin is transported through PIN proteins towards areas of high auxin concentration (dark green), at which an organ primordium will form. Cells around the developing primordium become depleted of auxin (light green), which causes the polarity of the PIN proteins to reverse so that auxin then flows away from the primordium. The red arrows indicate the direction of PIN polarity. It is proposed that this pattern of auxin circulation could set up the regular pattern of leaf and flower formation from meristems.

*Adapted from Heisler, M.G., et al.: **Patterns of auxin transport and gene expression during primordium development revealed by live imaging of the Arabadopsis inflorescence meristem.*** Curr. Biol. *2005, **15**: 1899-1911.*

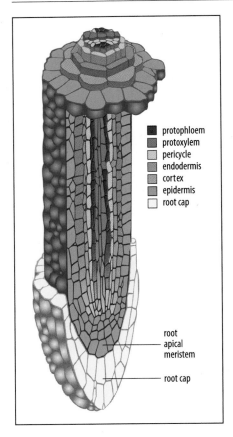

Fig. 7.24 The structure of the root tip in *Arabidopsis*. Roots have a radial organization. In the center of the growing root tip is the future vascular tissue (protoxylem and protophloem). This is surrounded by further tissue layers.

that causes PIN1 to move to the other side of these cells, and auxin now flows out of the new primordium towards the meristem, creating a new spot of high auxin concentration in the meristem farthest away from any new primordium (see Fig. 7.23). This leads to auxin peaks occurring sequentially, at the regular positions later occupied by new leaves.

7.13 Root tissues are produced from *Arabidopsis* root apical meristems by a highly stereotyped pattern of cell divisions

The organization of tissues in the *Arabidopsis* root tip is shown in Fig. 7.24. The radial pattern comprises single layers of epidermal, cortical, endodermal, and pericycle cells, with vascular tissue in the center (protophloem and protoxylem). Root apical meristems resemble shoot apical meristems in many ways and give rise to the root in a similar manner to shoot generation. But there are some important differences between the root and shoot meristems. The shoot meristem is at the extreme tip of the shoot, whereas the root meristem is covered by a root cap (which is itself derived from one of the layers of the meristem); also, there is no obvious segmental arrangement at the root tip resembling the node–internode–leaf module.

The root is set up early (see Section 7.2) and a well-organized embryonic root can be identified in the late heart-stage embryo (Fig. 7.25). An antagonistic interaction between auxin and cytokinin controls the establishment of the root stem-cell niche. Clonal analysis has shown that the seedling root meristem can be traced back to a set of embryonic initials that arise from a single tier of cells in the heart-stage embryo.

As in the shoot meristem, a root meristem is composed of an organizing center, called the **quiescent center** in roots, in which the cells divide only very rarely, and which is surrounded by stem-cell-like initials that give rise to the root tissue (see Fig. 7.25). The quiescent center is essential for meristem function. When parts of the meristem are removed by microsurgery, it can regenerate, but regeneration is always preceded by the formation of a new quiescent center. Laser destruction of individual quiescent-center cells shows that, as in the shoot meristem, a key function of the quiescent center is to maintain the immediately adjacent initials in the stem-cell state and prevent them from differentiating.

Each initial undergoes a stereotyped pattern of cell divisions to give rise to a number of columns, or **files**, of cells in the growing root (see Fig. 7.24); each file of cells in the root thus has its origin in a single initial. Some initials give rise to both endodermis and cortex, whereas others give rise to both epidermis and the root cap. Before it leaves the meristem, therefore, the undifferentiated progeny of an endodermis/cortex

Fig. 7.25 Fate map of root regions in the heart-stage *Arabidopsis* embryo. The root grows by the division of a set of initial cells. The root meristem comes from a small number of cells in the heart-shaped embryo. Each tissue in the root is derived from the division of a particular initial cell. At the center of the root meristem is a quiescent center, in which cells rarely divide.

*Illustration after Scheres, B., et al.: **Embryonic origin of the Arabidopsis primary root and root meristem initials**. Development 1994, **120**: 2475-2487.*

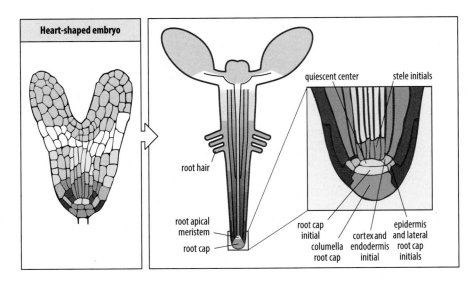

initial, for example, will divide asymmetrically to give one daughter that produces cortex and one that produces endodermis. The gene *SCARECROW* is necessary to confer this asymmetry on the dividing cell, and mutations in this gene give roots with no distinct endodermis or cortex but with a tissue layer with characteristics of both.

The normal pattern of cell divisions is not obligatory, however. As discussed earlier, *fass* mutants, which have disrupted cell divisions, still have relatively normal patterning in the root. In addition, laser ablation of individual meristem cells does not lead to an abnormal root. The remaining initials undergo new patterns of cell division that replace the progeny of the cells that have been destroyed. Such observations show that, as in the shoot meristem, the fate of cells in the developing root meristem depends on their recognition of positional signals and not on their lineage.

As we saw in Section 7.3, auxin gradients play a major role in patterning the embryo and specifying the root region, and mutations that affect auxin localization lead to root defects. at the globular stage of embryonic development, the auxin-transport protein PIN1 is localized in cells in the future root region and the highest level of auxin is found adjacent to where the quiescent center will develop.

The role of auxin in root development continues into the adult plant. Auxin in the root is transported out of cells via the PIN proteins, and enters adjacent cells. Cells with raised auxin levels transport auxin better, probably due to an increase in the number of PIN proteins in the membrane as auxin prevents their endocytosis and recycling, and so there is a positive feedback loop that raises the local concentration.

Auxin plays a key role in patterning the growing root. Modeling of auxin movement, using the known distribution of PIN proteins in the membranes of different cell types, has shown that PIN-directed flow can explain the formation and maintenance of a stable auxin maximum at the quiescent center. Auxin is transported down the central vascular tissue of the root tip, forming a concentration maximum at the quiescent center, and then outwards and upwards through the outer layers of the root tip, forming gradients along both the basal-apical axis and laterally in the root. This modeling successfully simulates the effects of such auxin gradients on pattern formation, cellular differentiation and root growth in real time.

An idea of how the auxin gradients might be translated into effects on cell fate and cell behavior is provided by the graded distribution of the four PLETHORA-family transcription factors in the root. These transcription factors are required for proper root development and are expressed in a graded manner along the apical–basal axis, with expression maxima for all at the quiescent center—the region of maximum auxin concentration. Here they are essential for stem-cell maintenance and function. Lower concentrations of PLETHORA proteins correspond to the meristem region, where cells are proliferating, while even lower concentrations appear to be necessary for exit from the meristem and cell differentiation in the elongation zone. Although not yet proven experimentally, *PLETHORA* gene expression could provide a graded read-out of the auxin gradient that helps direct root patterning.

Other influences on root patterning are the interplay between auxin and the hormone cytokinin. Cytokinin helps regulate root meristem size and set the position of the transition zone—where cells stop dividing and start elongating and differentiating—by repressing both auxin transport and cellular responses to auxin in this region.

Auxin is also involved in the ability of plants to regenerate from a small piece of stem. In general, roots form from the end of the stem that was originally closest to the root, whereas shoots tend to develop from dormant buds at the end that was nearest to the shoot. This polarized regeneration is related to vascular differentiation and to the polarized transport of auxin. Transport of auxin from its source in the shoot tip toward the root leads to an accumulation of auxin at the 'root' end of the stem cutting, where it induces the formation of roots. One hypothesis suggests that polarity is both induced and expressed by the oriented flow of auxin.

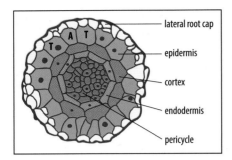

Fig. 7.26 **Organization of cell types in the root epidermis.** The epidermis is composed of two types of cells: trichoblasts (T), which will form root hairs, and atrichoblasts (A), which will not. Trichoblasts overlie the junction between two cortical cells and atrichoblasts are located over the outer tangential wall of cortical cells.

After Dolon, L., Scheres, B: **Root pattern: Shooting in the dark?** Semin. Cell Dev. Biol. *1998,* **9***: 201–206.*

One of the best examples of developmentally important transcription factor movement from one cell to another is found in roots. As noted earlier, expression of the gene *SCARECROW* is required for root cells to adopt an endodermal fate. This expression requires the transcriptional activator SHORT-ROOT (SHR). SHR is, however, not synthesized in the prospective endodermal cells, but in the adjacent cells on the inner side. SHR protein is transported from these cells outwards into the prospective endodermis, and this movement appears to be regulated and not simply due to diffusion.

7.14 Root hairs are specified by a combination of positional information and lateral inhibition

Root hairs are formed from epidermal cells at regular intervals around the root, and this regularity is thought to be achieved by a combination of responses to positional information and lateral inhibition by the movement of transcription factors between cells. Files of cells that will make root hairs alternate with files of non-hair-producing cells on the surface of the developing root. The importance of position is shown by the fact that if an epidermal cell overlies a junction between two cortical cells it forms a root hair, whereas if it contacts just one cortical cell it does not (Fig. 7.26). And if a cell changes its position in relation to the cortex, its fate will also change, from a potential hair-forming cell to a non-hair-forming cell and vice versa. Most cell divisions in the future epidermis are horizontal, increasing the number of cells per file, but occasionally a vertical anticlinal division occurs, pushing one of the daughter cells into an adjacent file. The daughter cell then assumes a fate corresponding to its new position in relation to the adjacent cortical cells.

The positional cues, as yet unknown, are thought to be detected by the epidermal cells through the SCRAMBLED protein, which is a receptor-like protein kinase. SCRAMBLED activity influences the activity of a network of transcription factors that control cell fate. Two groups of transcription factors have been identified by mutation experiments, one group promoting a root-hair fate and one suppressing it. A key transcription factor that seems to be regulated by SCRAMBLED activity is WEREWOLF, whose expression is suppressed in presumptive hair-forming cells, presumably in response to the positional signal. As well as promoting an atrichoblast fate, however, WEREWOLF is also required for the expression of transcription factors (CAPRICE, TRYPTYCHON, and ENHANCER OF TRYPTYCHON) that are needed to specify hair cells. After positional signaling, these proteins will only be produced in the presumptive atrichoblasts, but they are thought to move laterally into the adjacent epidermal cells and promote these cells' differentiation as trichoblasts by inhibiting genes that would otherwise give an atrichoblast fate.

SUMMARY

Meristems are the growing points of a plant. The apical meristems, found at the tips of shoots and roots, give rise to all the plant organs—roots, stem, leaves, and flowers. They consist of small groups of a few hundred undifferentiated cells that are capable of repeated division. The center of the meristem is occupied by self-renewing stem cells, which replace the cells that are lost from the meristem when organs are formed. The fate of a cell in the shoot meristem depends upon its position in the meristem and interactions with its neighbors, as when a cell is displaced from one layer to another it adopts the fate of its new layer. Meristems can also regulate when parts are removed, in line with cell–cell interactions determining cell fate. Fate maps of embryonic shoot meristems show that they can be divided into domains, each of which normally contributes to the tissues of a particular region of the plant, but the fate of the embryonic initials is not fixed. The shoot meristem gives rise to leaves in species-specific patterns—phyllotaxy—which seem best accounted for by regulated transport of auxin. Lateral inhibition is involved in the regular spacing of hairs on root and leaf surfaces. In the root meristem, the cells are organized rather differently from those in the shoot meristem, and there is a much more stereotyped pattern of cell division. A set of initial cells maintains root structure by dividing along different planes.

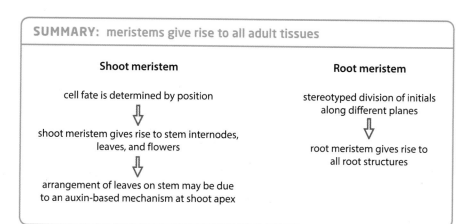

SUMMARY: meristems give rise to all adult tissues

Shoot meristem	**Root meristem**
cell fate is determined by position	stereotyped division of initials along different planes
⇓	⇓
shoot meristem gives rise to stem internodes, leaves, and flowers	root meristem gives rise to all root structures
⇓	
arrangement of leaves on stem may be due to an auxin-based mechanism at shoot apex	

Flower development and control of flowering

Flowers contain the reproductive cells of higher plants and develop from the shoot meristem. In most plants, the transition from a vegetative shoot meristem to a floral meristem that produces a flower is largely, or absolutely, under environmental control, with daylength and temperature being important determining factors. In a plant such as *Arabidopsis*, in which each flowering shoot produces multiple flowers, the vegetative shoot meristem first becomes converted into an inflorescence meristem, which then forms floral meristems, each of which develops completely into a single flower (Fig. 7.27). Floral meristems are thus determinate, unlike the indeterminate shoot apical meristem. Flowers, with their arrangement of floral organs (sepals, petals, stamens, and carpels), are rather complex structures, and it is a major challenge to understand how they arise from the floral meristem.

The conversion of a vegetative shoot meristem into one that makes flowers involves the induction of so-called **meristem identity genes**. A key regulator of floral induction in *Arabidopsis* is the meristem identity gene *LEAFY* (*LFY*); a related gene in *Antirrhinum* is *FLORICAULA* (*FLO*). How environmental signals, such as daylength,

Fig. 7.27 Scanning electron micrograph of an *Arabidopsis* inflorescence meristem. The central inflorescence meristem (shoot apical meristem, SAM) is surrounded by a series of floral meristems (FM) of varying developmental ages. The inflorescence meristem grows indeterminately, with cell divisions providing new cells for the stem below, and new floral meristems on its flanks. The floral meristems (or floral primordia) arise one at a time in a spiral pattern. The most mature of the developing flowers is on the right (FM1), showing the initiation of sepal primordia surrounding a still-undifferentiated floral meristem. Eventually such a floral meristem will also form petal, stamen, and carpel primordia.

*Photograph reproduced with permission from Meyerowitz, E.M., et al.: **A genetic and molecular model for flower development in Arabidopsis thaliana.** Development Suppl. 1991, 157-167. Published by permission of The Company of Biologists Ltd.*

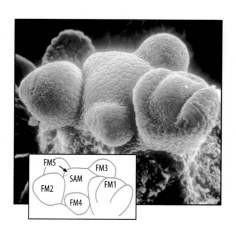

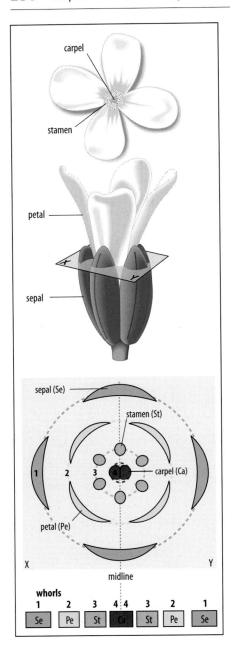

Fig. 7.28 Structure of an Arabidopsis flower. *Arabidopsis* flowers are radially symmetrical and have an outer ring of four identical green sepals, enclosing four identical white petals, within which is a ring of six stamens, with two carpels in the center. Bottom: floral diagram of the *Arabidopsis* flower representing a cross-section taken in the plane indicated in the top diagram. This is a conventional representation of the arrangement of the parts of the flower, showing the number of flower parts in each whorl and their arrangement relative to each other.

Illustration after Coen, E.S., Meyerowitz, E.M.: **The war of the whorls: genetic interactions controlling flower development**. *Nature 1991,* **353**: *31–37.*

influence floral induction is discussed later. We will first consider the mechanisms that pattern the flower, in particular those that specify the identity of the floral organs.

7.15 Homeotic genes control organ identity in the flower

The individual parts of a flower each develop from a **floral organ primordium** produced by the floral meristem. Unlike leaf primordia, which are all identical, the floral organ primordia must each be given a correct identity and be patterned according to it. An *Arabidopsis* flower has four concentric whorls of structures (Fig. 7.28), which reflect the arrangement of the floral organ primordia in the meristem. The sepals (whorl 1) arise from the outermost ring of meristem tissue, and the petals (whorl 2) from a ring of tissue lying immediately inside it. An inner ring of tissue gives rise to the male reproductive organs—the stamens (whorl 3). The female reproductive organs—the carpels (whorl 4)—develop from the center of the meristem. In a floral meristem of *Arabidopsis*, there are 16 separate primordia, giving rise to a flower with four sepals, four petals, six stamens and a pistil made up of two carpels (see Fig. 7.28).

The primordia arise at specific positions within the meristem, where they develop into their characteristic structures. After the emergence of the primordia in *Antirrhinum*, cell lineages become restricted to particular whorls, rather like the lineage restriction to compartments in *Drosophila* (see Section 2.26). Lineage restriction occurs at the time when the pentagonal symmetry of the flower becomes visible and genes that give the different floral organs their identity are expressed. The lineage compartments within the floral meristem appear to be delineated by narrow bands of non-dividing cells.

Like the homeotic selector genes that specify segment identity in *Drosophila*, mutations in floral identity genes cause homeotic mutations in which one type of flower part is replaced by another. In the *Arabidopsis* mutant *apetala2*, for example, the sepals are replaced by carpels and the petals by stamens; in the *pistillata* mutant, petals are replaced by sepals and stamens by carpels. These mutations identified the floral organ identity genes, and have enabled their mode of action to be determined.

Homeotic floral mutations in *Arabidopsis* fall into three classes, each of which affects the organs of two adjacent whorls (Fig. 7.29). The first class of mutations, of which *apetala2* is an example, affect whorls 1 and 2, giving carpels instead of sepals in whorl 1, and stamens instead of petals in whorl 2. The phenotype of the flower, going from the outside to the center, is therefore carpel, stamen, stamen, carpel. The second class of homeotic floral mutations affects whorls 2 and 3. In this class, *apetala3* and *pistillata* give sepals instead of petals in whorl 2 and carpels instead of stamens in whorl 3, with a phenotype sepal, sepal, carpel, carpel. The third class of mutations affects whorls 3 and 4, and gives petals instead of stamens in whorl 3 and sepals or variable structures in whorl 4. The mutant *agamous*, which belongs to this class, has an extra set of sepals and petals in the center instead of the reproductive organs.

These mutant phenotypes can be accounted for by an elegant model in which overlapping patterns of gene activity specify floral organ identity (Fig. 7.30) in a manner highly reminiscent of the way in which *Drosophila* homeotic genes specify segment

| *apetala2* mutant | *apetala3* mutant | *agamous* mutant |
| Ca | St | St | Ca | Ca | St | St | Ca | | Se | Se | Ca | Ca | Ca | Ca | Se | Se | | Se | Pe | Pe | Se | Se | Pe | Pe | Se |

Fig. 7.29 Homeotic floral mutations in *Arabidopsis*. Left panel: an *apetala2* mutant has whorls of carpels and stamens in place of sepals and petals. Center panel: an *apetala3* mutant has two whorls of sepals and two of carpels. Right panel: *agamous* mutants have a whorl of petals and sepals in place of stamens and carpels. Transformations of whorls are shown inset, and can be compared to the wild-type arrangement, as shown in Fig. 7.28.

*Photographs reproduced with permission from Meyerowitz, E.M., et al.: **A genetic and molecular model for flower development in Arabidopsis thaliana.** Development Suppl. 1991, 157–167.*

*Center panel from Bowman, J.L., et al.: **Genes directing flower development in Arabidopsis.** Plant Cell 1989, **1**: 37–52. Published by permission of The American Society of Plant Physiologists.*

identity along the insect's body. In detail, however, there are many differences, and quite different genes are involved. In this instance, plants and animals have, perhaps not surprisingly, independently evolved a similar approach to patterning a multicellular structure, but have recruited different proteins to carry it out.

In essence, the floral meristem is divided by the expression patterns of the homeotic genes into three concentric overlapping regions, A, B, and C, which partition the meristem into four non-overlapping regions corresponding to the four whorls. Each of the A, B, and C regions corresponds to the zone of action of one class of homeotic genes and the particular combinations of A, B, and C functions give each whorl a unique identity and so specify organ identity. Of the genes mentioned in Fig. 7.29, *APETALA1* (*AP1*) and *APETALA2* (*AP2*) are A-function genes, *APETALA3* (*AP3*) and *PISTILLATA* (*P1*) are B-function genes, and *AGAMOUS* (*AG*) is a C-function gene. The expression of *AP3* and *AG* in the developing flower is shown in Fig. 7.31. All the homeotic genes, also known as **floral organ identity genes**, encode transcription factors, and the B- and C-function proteins such as AP3 and AG contain a conserved DNA-binding sequence known as the MADS box. MADS-box genes are present in animals and yeast, but a role in development is known mainly in plants—although a MADS-box transcription factor, MEF2, is involved in muscle differentiation in animals. The original simple model for specifying floral organ identity is presented in more detail in Box 7C. Since the model was first proposed, more has become known about the activities and functions of the genes identified by the homeotic mutations, more genes controlling flower development have been discovered, and more 'functions' added.

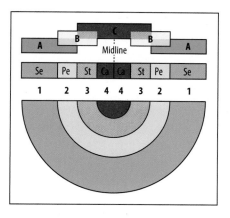

Fig. 7.30 The three overlapping regions of the *Arabidopsis* floral meristem that have been identified by the homeotic floral identity mutations. Region A corresponds to whorls 1 and 2, B to whorls 2 and 3, and C to whorls 3 and 4.

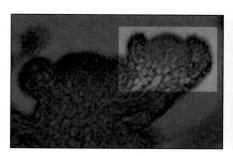

Fig. 7.31 Expression of *APETALA3* and *AGAMOUS* during flower development. *In situ* hybridization shows that *AGAMOUS* is expressed in the central whorls (left panel), whereas *APETALA3* is expressed in the outer whorls that give rise to petals and stamens (right panel).

Reprinted with permission from Meyerowitz. ***The Genetics of Flower Development.*** Scientific American *1994.*

BOX 7C The basic model for the patterning of the *Arabidopsis* flower

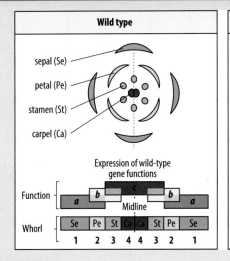

Wild type

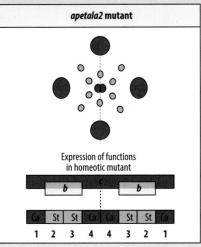

apetala2 mutant

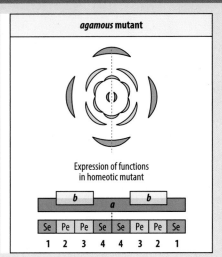

agamous mutant

Figure 1

The floral meristem is divided into three overlapping regions, A, B, and C, each region corresponding to a class of homeotic mutations, as shown in Fig. 7.29 (see text). Three regulatory functions—*a*, *b*, and *c*—operate in regions A, B, and C, respectively, as shown in Figure 1. In the wild-type flower (Figure 1, left panel), it is assumed that *a* is expressed in whorls 1 and 2, *b* in 2 and 3, and *c* in whorls 3 and 4. In addition, *a* function inhibits *c* function in whorls 1 and 2 and *c* function inhibits *a* function in whorls 3 and 4—that is, *a* and *c* functions are mutually exclusive. *a* alone specifies sepals, *a* and *b* together specify petals, *b* and *c* stamens, and *c* alone carpels.

The homeotic mutations eliminate the functions of *a*, *b*, or *c*, and alter the regions within the meristem where the various functions are expressed. Mutations in *a*, such as *apetala2* (see Figure 1, center panel), result in an absence of function *a*, and *c* spreads throughout the meristem, resulting in the half-flower pattern of carpel, stamen, stamen, carpel. Mutations in *b*, such as *apetala3* (see Fig. 7.29), result in only *a* functioning in whorls 1 and 2, and *c* in whorls 3 and 4, giving sepal, sepal, carpel, carpel. Mutations in *c* genes (such as *agamous*), result in *a* activity in all whorls, giving the phenotype sepal, petal, petal, sepal (see Figure 1, right panel).

All the floral homeotic mutants discovered so far in *Arabidopsis* can be quite satisfactorily accounted for by this model (although there are small variations in gene numbers and expression patterns in other species that allow mutant phenotypes not seen in *Arabidopsis*), and particular genes can be assigned to each controlling function. Function *a* corresponds to the activity of genes such as *APETALA2*, *b* to *APETALA3* and *PISTILLATA*, and *c* to *AGAMOUS*. The model also accounts for the phenotype of double mutants, such as *apetala2* with *apetala3*, and *apetala3* with *pistillata*, as shown in Figure 2.

This system emphasizes the similarity in function between the homeotic genes in animals and those controlling organ identity in flowers, although the genes themselves are completely

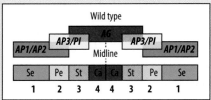

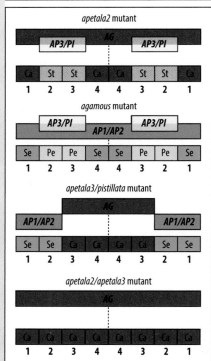

Figure 2

different. The functional similarity with the HOM-C complex of *Drosophila* is further illustrated by the role of the *CURLY LEAF* gene *of Arabidopsis*, which is necessary for the stable repression of a floral homeotic gene. *CURLY LEAF* is related to the *Polycomb* family of animal genes, which are similarly required for stable repression of homeotic genes in the fly and in mammals.

One question that arose when the original floral organ identity model was investigated experimentally was why the ABC genes only showed their homeotic properties in the floral meristem and did not convert leaves into floral organs when artificially overexpressed in vegetative meristems, as might be expected for homeotic genes of this type. The answer came with the discovery of the *SEPALLATA* (*SEP*) genes, which also encode MADS-box proteins. These genes are required for the B and C functions and are only active in floral meristems. The SEP proteins are thought to combine with *B* and *C* gene products to form active gene-regulatory complexes. A current view of the mechanism of specifying floral organ identity is shown in Fig. 7.32.

There is a better understanding of the functions and patterning of the floral homeotic genes than when the ABC model was first proposed. The MADS-box homeotic A-class gene *AP1* has been found to have a dual role: it acts early with other genes to specify general floral meristem identity and only later contributes to A function. It is induced by the meristem identity gene *LFY*, which is expressed throughout the meristem, and *AP1* is actively inhibited in the central regions of floral meristems by *AGAMOUS*. The expression of the A-function gene *APETALA2* (*AP2*) is translationally repressed by a microRNA, keeping the AP2 protein at a low level. *APETALA3* and *PISTILLATA1* are thought to be activated as a result of a meristem identity gene called *UNUSUAL FLORAL ORGANS* (*UFO*), which is expressed in the meristem in a pattern similar to that of B-function genes (see Fig. 7.32). *UFO* encodes a component of ubiquitin ligase, and is thought to exert its effects on flower development by targeting specific proteins for degradation. As we saw in animals, in relation to the control of β-catenin degradation (see Chapters 4 and 6), regulated degradation of proteins can be a powerful developmental mechanism. In the center of the floral meristem, the expression of *AGAMOUS* is partly controlled by *WUS*, which as we have seen earlier, is expressed in the organizing center of the vegetative shoot meristem and continues to be expressed in floral meristems.

Another group of genes that help pattern the floral organ primordia are genes that control cell division. The gene *SUPERMAN* is one example, controlling cell proliferation in stamen and carpel primordia, and in ovules. Plants with a mutation in this gene have stamens instead of carpels in the fourth whorl. *SUPERMAN* is expressed in the third whorl, and maintains the boundary between the third and fourth whorls.

Despite the enormous variation in the flowers of different species, the mechanisms underlying flower development seem to be very similar. For example, there are striking similarities between the genes controlling flower development of *Arabidopsis* and *Antirrhinum*, despite the quite different final morphology of the snapdragon flower. In developing *Arabidopsis* flowers, the patterns of activity of the corresponding genes fit well with the cell-lineage restriction to whorls seen in *Antirrhinum*.

7.16 The *Antirrhinum* flower is patterned dorso-ventrally as well as radially

Like *Arabidopsis* flowers, those of *Antirrhinum* consist of four whorls, but unlike *Arabidopsis*, they have five sepals, five petals, four stamens, and two united carpels (Fig. 7.33, left). Floral homeotic mutations similar to those in *Arabidopsis* occur in *Antirrhinum*, and floral organ identity is specified in the same way. Several of the *Antirrhinum* homeotic genes have extensive homology with those of *Arabidopsis*, the MADS box in particular being well conserved.

An extra element of patterning is required in the *Antirrhinum* flower, which has a bilateral symmetry imposed on the basic radial pattern common to all flowers. In whorl 2, the upper two petal lobes have a shape quite distinct from the lower three, giving the flower its characteristic snapdragon appearance. In whorl 3, the uppermost stamen is absent, as its development is aborted early on. *The Antirrhinum flower*

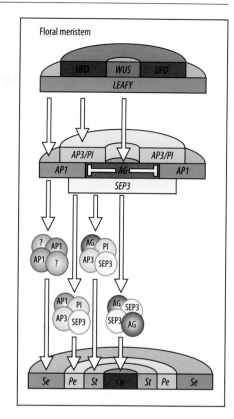

Fig. 7.32 Current status of the ABC model of floral organ identity. The regulatory genes *LEAFY*, *WUSCHEL* (*WUS*), and *UNUSUAL FLORAL ORGANS* (*UFO*) are expressed in specific domains in the floral meristem, which, together with repression of *APETALA1* by *AGAMOUS*, results in the pattern of ABC functions. ABC proteins and the co-factor SEP proteins assemble into complexes that specify the different organ identities.

Adapted from Lohmann, J.U., Weigel, D.: **Building beauty: the genetic control of floral patterning.** *Dev. Cell 2002, 2: 135–142.*

Fig. 7.33 Mutations in *CYCLOIDEA* make the *Antirrhinum* flower symmetrical. In the wild-type flower (left) the petal pattern is different along the dorso-ventral axis. In the mutant (right) the flower is symmetrical. All the petals are like the most ventral one in the wild type and are folded back.

*Photograph reproduced with permission from Coen, E.S., Meyerowitz, E.M.: **The war of the whorls: genetic interactions controlling flower development**. Nature 1991, **353**: 31–37. © 1991 Macmillan Magazines Ltd.*

therefore has a distinct dorso-ventral axis. Another group of homeotic genes, different from those that govern floral organ identity, appear to act in this dorso-ventral patterning. For example, mutations in the gene *CYCLOIDEA*, which is expressed in the dorsal region, abolish dorso-ventral polarity and produce flowers that are more radially symmetrical (Fig. 7.33, right).

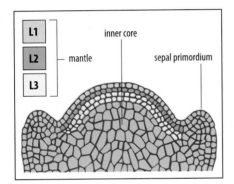

Fig. 7.34 Floral meristem. The meristem is composed of layers L1, L2, and L3. The inner core cells are derived from L3. The sepal primordia are just beginning to develop.

*Illustration after Drews, G.N., Goldberg, R.B.: **Genetic control of flower development**. Trends Genet. 1989, **5**: 256–261.*

7.17 The internal meristem layer can specify floral meristem patterning

Although all three layers of a floral meristem (Fig. 7.34) are involved in organogenesis, the contribution of cells from each layer to a particular structure may be variable. Cells from one layer can become part of another layer without disrupting normal morphology, suggesting that a cell's position in the meristem is the main determinant of its future behavior. Some insight into positional signaling and patterning in the floral meristem can be obtained by making periclinal chimeras (see Section 7.8) from cells that have different genotypes and that give rise to different types of flower. From such chimeras, one can find out whether the cells develop autonomously according to their own genotype, or whether their behavior is controlled by signals from other cells.

As well as being produced by mutation, chimeras can also be generated by grafting between two plants of different genotypes. A new shoot meristem can form at the junction of the graft, and sometimes contains cells from both genotypes. Such chimeras can be made between wild-type tomato plants and tomato plants carrying the mutation *fasciated*, in which the flower has an increased number of floral organs per whorl. This phenotype is also found in chimeras in which only layer L3 contains *fasciated* cells (Fig. 7.35). The increased number of floral organs is associated with an overall increase in the size of the floral meristem, and in the chimeric plants this cannot be achieved unless the *fasciated* cells of layer L3 induce the wild-type L1 cells to divide more frequently than normal. The mechanism of intercellular signaling between L3 and L1 is not yet known. In *Antirrhinum*, the abnormal expression of *FLORICAULA* in only one meristem layer can result in flower development. These results illustrate the importance of signaling between layers in flower development.

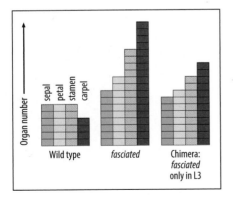

Fig. 7.35 Floral organ number in chimeras of wild-type and *fasciated* tomato plants. In the *fasciated* mutant there are more organs in the flower than in wild-type plants. In chimeras in which only layer L3 of the floral meristem contains *fasciated* mutant cells, the number of organs per flower is still increased, showing that L3 can control cell behavior in the outer layers of the meristem.

7.18 The transition of a shoot meristem to a floral meristem is under environmental and genetic control

Flowering plants first grow vegetatively, during which time the apical meristem generates leaves. Then, triggered by environmental signals such as increasing daylength,

Fig. 7.36 Flowering can be controlled by daylength and LEAFY expression. As shown in the top panels, when wild-type *Arabidopsis* is grown under long-day conditions (left), few lateral shoots are formed before the apical shoot meristem begins to form floral meristems. When grown under short-day conditions, flowering is delayed and there are in consequence more lateral shoots. The gene *LEAFY* is normally expressed only in inflorescence and floral meristems, but if it is expressed throughout the plant (bottom panels), all shoot meristems produced are converted to floral meristems in both daylengths.

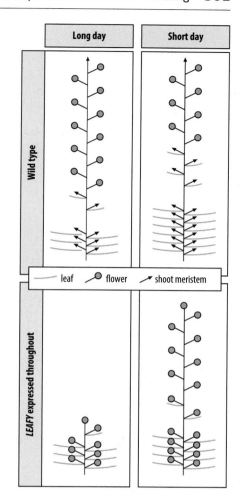

the plant switches to a reproductive phase and from then on the apical meristem gives rise only to flowers. There are two types of transition from vegetative growth to flowering. In the determinate type, the inflorescence meristem becomes a terminal flower, whereas in the indeterminate type the inflorescence meristem gives rise to a number of floral meristems. *Arabidopsis* is of the indeterminate type (Fig. 7.36). A primary response to floral inductive signals in *Arabidopsis* is the expression of floral meristem identity genes such as *LEAFY* and the dual-function *AP1* (see Section 7.15), which are necessary and sufficient for this transition. *LEAFY* potentially activates *AP1* throughout the meristem while also activating *AGAMOUS* in the center of the flower. *AGAMOUS* then represses the expression of *AP1* in the center, helping to restrict its floral organ identity function to region A (see Fig. 7.32). Mutations in floral meristem identity genes partly transform flowers into shoots. In a *leafy* mutant, which lacks LEAFY function, the flowers are transformed into spirally arranged sepal-like organs along the stem, whereas expression of *LEAFY* throughout a plant is sufficient to confer a floral fate on lateral shoot meristems and they develop as flowers (Fig. 7.36, bottom panels).

In *Arabidopsis*, flowering is promoted by increasing daylength, which predicts the end of winter and the onset of spring and summer (see Fig. 7.36). This behavior is called **photoperiodism**. In some strains, flowering is also accelerated after the plant has been exposed to a long period of cold temperature, a cue that winter has passed. This phenomenon is known as **vernalization**. Grafting experiments have shown that daylength is sensed not by the shoot meristem itself, but by the leaves. When the period of continuous light reaches a certain length, a diffusible flower-inducing signal is produced that is transmitted through the phloem to the shoot meristem. The pathway that triggers flowering involves the plant's **circadian clock**, the internal 24-hour timer that causes many metabolic and physiological processes, including the expression of some genes, to vary throughout the day. One of the genes regulated by the circadian clock is *CONSTANS* (*CO*), which is a key gene in controlling the onset of flowering and provides the link between the plant's day-length-sensing mechanism and production of the flowering signal. The expression of *CO* oscillates on a 24-hour cycle under the control of the circadian clock, and its timing is such that the peak *CO* expression occurs towards the end of the afternoon. This means that in longer days, peak expression occurs in the light, whereas in short days, it will already be dark at this time. In the dark, the CO protein is degraded and so the circadian control ensures that CO only accumulates to high enough levels to trigger the flowering pathway when light conditions are favorable (Fig. 7.37).

CO is a transcription factor that activates a gene known as *FLOWERING LOCUS T* (*FT*), producing the FT protein, which appears to act as the flowering signal. The FT protein is thought to travel from the leaf through the phloem to the shoot apical

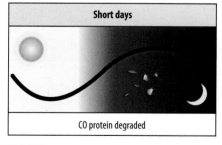

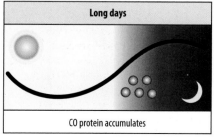

Fig. 7.37 The initiation of flowering is under the dual control of daylength and the circadian clock. The transcription factor CONSTANS (CO) is required for production of the flowering signal and is expressed in leaves under the control of the circadian clock. In short days, expression of the *CO* gene peaks in the dark and the protein is rapidly degraded. In long days, peak expression occurs in the light, and the CO protein accumulates.

Fig. 7.38 Signals that initiate flowering in Arabidopsis. When the daylength becomes longer after winter, the transcription factor CO accumulates in the leaf, which in turn activates the transcription of *FT* in leaf phloem cells. The FT protein travels through the phloem to the shoot apex. It interacts there with the transcription factor FD to form a complex, which acts together with the transcription factor LEAFY (whose own expression is upregulated in the shoot apex by FT) to activate key floral meristem identity genes such as *AP1*, which convert the vegetative meristem into one that will produce floral meristems.

*Illustration from Blazquez, M.A.: **The right time and place for making flowers**. Science 2005, **309**: 1024–1025.*

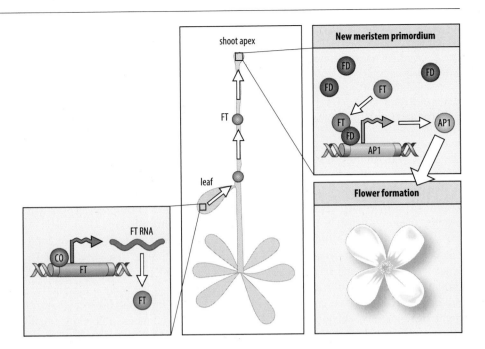

meristem, where it acts in a complex with the transcription factor FLOWERING LOCUS D (FD), which is expressed in the meristem, to turn on the expression of genes such as *AP1* that promote flowering (Fig. 7.38). If *FT* is activated in a single leaf, this is sufficient to induce flowering. Induction of flowering also requires the downregulation of a set of floral repressor genes such as *FLOWERING LOCUS C* (*FLC*). These suppress the transition from a vegetative to the flowering state until the positive signals to flower are received. FLC encodes a protein that binds to FT and suppresses its activity. After cold exposure, for example, FLC activity is low, and the repression of FT can be released.

7.19 Most flowering plants are hermaphrodites, but some produce unisexual flowers

Unlike animals, plants do not set aside germ cells in the embryo, and germ cells are only specified when a flower develops. Any meristem cell can, in principle, give rise to a germ cell of either sex, and there are no sex chromosomes. The great majority of flowering plants (angiosperms) give rise to flowers that contain both male and female sexual organs, in which meiosis occurs. The male sexual organs are the stamens; they produce pollen, which contains the male gamete nuclei corresponding to the sperm of animals. The female reproductive structures are the carpels, which are either free or are fused to form a compound ovary. Carpels are the site of ovule formation, and each ovule produces an egg cell.

Angiosperms, such as *Arabidopsis*, undergo so-called 'double fertilization.' Each pollen grain contains two sperm nuclei, which are delivered into the ovule by the growth of a vegetative pollen tube. One nucleus fuses with the haploid egg cell, forming the zygote that develops into the embryo. The second nucleus fuses with the diploid 'central cell' in the ovule, forming a triploid cell that proliferates and develops into a storage tissue—the endosperm. In some plants, such as *Arabidopsis*, the endosperm nourishes the growth of the embryo, while in others it is broken down at germination to produce nutrients for the seedling.

Most flowering plants are hermaphrodite, bearing flowers with functional male and female sexual organs, as in *Arabidopsis*. In about 10% of flowering plant species, flowers of just one sex are produced. Flowers of different sexes may occur on the same plant, or be confined to different plants. The development of male or female flowers usually involves the selective resorption of either the stamens or pistil after they have been specified and have started to grow. In maize, for example, male and female flowers develop at particular sites on the shoot. The tassel at the tip of the main stem only bears flowers with stamens; the 'ears' at the ends of the lateral branches bear female flowers containing pistils. Sex determination becomes visible when the flower is still small, with the stamen primordia being larger in males and the pistil longer in females. The smaller organs eventually degenerate. The plant hormone gibberellic acid may be involved in sex determination, as differences in gibberellin concentration are associated with the different sexual organs. In the maize tassel, gibberellin concentration is 100-fold lower than in the developing ears. If the concentration of gibberellic acid is increased in the tassel, pistils can develop.

Genomic imprinting (see Section 10.8) occurs in flowering plants, but has only been detected in the endosperm and not in the embryo. The imprinting events take place in the central-cell genome and the sperm genome. In plants, as in mammals, imprinting involves gene silencing due to DNA methylation, histone modifications, Polycomb proteins and non-coding RNAs. Unlike mammals, however, the imprinting does not have to be removed to produce a new generation, as the endosperm is a temporary tissue that does not contribute cells to the embryo. As well as the silencing of paternal genes in sperm by DNA methylation and histone modification, a feature of angiosperm imprinting is the DNA demethylation of previously silenced maternal genes in the central cell, which then enables their expression. The activated genes include members of the Polycomb group, which then maintain the repression of incoming silenced paternal alleles and silence the maternal alleles of certain other genes. Imprinting in angiosperms could have evolved together with double fertilization as a means of preventing endosperm proliferation in the absence of fertilization.

SUMMARY

Before flowering, which is triggered by environmental conditions such as daylength, the vegetative shoot apical meristem becomes converted into an inflorescence meristem, which either then becomes a flower or produces a series of floral meristems, each of which develops into a single flower. Genes involved in the initiation of flowering and patterning of the flower have been identified in both *Arabidopsis* and *Antirrhinum*. Flowering is induced by daylength acting together with the plant's natural circadian rhythms of gene expression to turn on a gene in the leaves that produces a flowering signal that is transported to the shoot meristem. This signal turns on the expression of meristem identity genes that are required for the transformation of the vegetative shoot meristem to an inflorescence meristem and the formation of floral meristems from the inflorescence meristem. Homeotic floral organ identity genes, which specify the organ types found in the flowers, have been identified from mutations that transform one flower part into another. On the basis of these mutations, a model has been proposed in which the floral meristem is divided into three concentric overlapping regions, in each of which certain floral identity genes act in a combinatorial manner to specify the organ type appropriate to each whorl. Studies with chimeric plants have shown that different meristem layers communicate with each other during flower development and that transcription factors can move between cells. Unlike animals, plants do not set aside germ cells in the embryo, and germ cells are only specified when a flower develops. Genomic imprinting occurs in flowering plants, but has only been detected in the endosperm and not in the embryo.

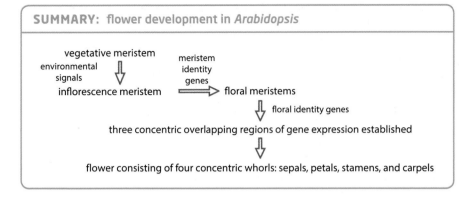

SUMMARY: flower development in *Arabidopsis*

vegetative meristem

environmental signals ⟱

inflorescence meristem ⟹ meristem identity genes ⟹ floral meristems

⟱ floral identity genes

three concentric overlapping regions of gene expression established

⟱

flower consisting of four concentric whorls: sepals, petals, stamens, and carpels

Summary to Chapter 7

- A distinctive feature of plant development is the presence of relatively rigid walls and the absence of any cell migration.
- Another distinctive feature compared with animals is that a single, isolated somatic cell from a plant can regenerate into a complete new plant.
- Early embryonic development is characterized by asymmetric cell division of the fertilized egg, which specifies the future apical and basal regions.
- During early development of flowering plants, both asymmetric cell division and cell–cell interactions are involved in patterning the body plan. During this process, the shoot and root meristems are specified and these meristems give rise to all the organs of the plant—stems, leaves, flowers, and roots.
- The shoot meristem gives rise to leaves in well-defined positions, a process involving regulated transport of a morphogen, auxin.
- The shoot meristem eventually becomes converted to an inflorescence meristem, which either becomes a floral meristem (in determinate inflorescences) or gives rise to a series of floral meristems, retaining its shoot meristem identity indefinitely (in indeterminate inflorescences).
- In floral meristems, each of which develops into a flower, homeotic floral organ identity genes act in combination to specify the floral organ types.
- Increasing daylength induces the synthesis of a flowering signal in the leaves that is transported to the shoot meristem where it induces flower formation.
- Unlike animals, plants do not set aside germ cells in the embryo, and germ cells are only specified when a flower develops. Genomic imprinting occurs in flowering plants, but has only been detected in the endosperm and not in the embryo.

▉ End of chapter questions

Long answer (concept questions)

1. What features led to the adoption of *Arabidopsis thaliana* as the predominant model for plant development?

2. Distinguish between the following parts of a plant: shoot, root, node, leaf, meristem, sepal, petal, stamen, carpel.

3. What is the role of auxin before the 32-cell stage in *Arabidopsis* embryogenesis? What is the mechanism by which a differential in auxin concentrations is generated? What is the mechanism by which auxins influence gene expression?

4. Describe the process by which transgenic plants are produced. Include the role of the Ti plasmid, and of *Agrobacterium tumefaciens*.

5. What is a meristem? Describe the structure of the shoot meristem of *Arabidopsis*.

6. Contrast the roles of the homeobox genes *WUSCHEL* (*WUS*) and *SHOOT MERISTEMLESS* (*STM*) in formation and maintenance of shoot meristems.

7. What has the study of mericlinal chimeras revealed about cell specification during plant embryogenesis? Are cells specified in the embryo to form leaves and flowers, in a way analogous, for example, to the specification of cells as dorsal mesoderm in animal embryos?

8. What are the terms applied to the top and bottom surfaces of a leaf, in reference to the radial axis of the shoot? How are the transcription factors PHAB, PHAV, and REV restricted to the top surface?

9. How does auxin control the positioning of leaf primordia? Include the role of the PIN transporter and the concept of lateral inhibition in your answer.

10. What is SHORT-ROOT? How does it come to be present in endodermal cells? Might plasmodesmata be involved? Is there an analogous process involving the SHOOT MERISTEMLESS ortholog KNOTTED-1 of maize?

11. What is the nature of the homeotic mutations that can occur in *Arabidopsis*? What genes are involved?

12. What is the ABC model for flower development? How does it illustrate combinatorial control of cellular identity?

13. The MADS-box is named for the prototypic proteins in which it was found: MCM1 (*Saccharomyces*), AGAMOUS (*Arabidopsis*), DEFICIENS (*Antirrhinum*), and SRF (*Homo*). What is the function of the MADS-box in a protein? (Note that the MADS-box proteins are not related to SMADS; see Fig. 4.33.)

14. What modification of the ABC model derived from studies of *Arabidopsis* is required to explain flower development in *Antirrhinum*?

15. Through what mechanism is the photoperiod interpreted to trigger flower development in *Arabidopsis*?

Multiple choice (factual recall questions)

NB There is only one right answer to each question
1. How many genes are present in humans, *Drosophila melanogaster*, *Caenorhabditis elegans*, and *Arabidopsis*, respectively. (Although the number for humans has not yet been presented in the text, the number for the other organisms has been described.)

a) 19,000 − 27,000 − 14,000 − 19,000
b) 21,000 − 14,000 − 19,000 − 27,000
c) 27,000 − 21,000 − 19,000 − 14,000
d) 14,000 − 19,000 − 21,000 − 27,000

2. Embryogenesis in plants occurs

a) in the ovule, after the seed is fertilized and shed by the plant
b) in the seed after it germinates
c) in the seed, after the seed is fertilized
d) inside the ovule, before the seed is shed by the plant

3. Which statement is true about the totipotency of cells?

a) All animal and plant cells are totipotent.
b) In plants, many cells are totipotent, whereas in animals, only the fertilized egg is totipotent.
c) Mammalian embryonic stem cells are totipotent.
d) Only the stem cells of animals, and the meristem cells of plants are totipotent.

4. One of the earliest events in *Arabidopsis* development is formation of the _____ axis, in response to a gradient of _____.

a) adaxial–abaxial, cytokinins
b) apical–basal, auxin

c) apical–basal, Pin proteins
d) dorsal–ventral, miRNA

5. The fate map of the *Arabidopsis* embryo at the heart stage indicates that

a) although none of the adult structures have formed, the regions that will give rise to the meristems, which will in turn give rise to adult structures, can be identified
b) development in plants is so indeterminate that a true fate map cannot be drawn
c) the primordia of the leaves, stems, and roots have already formed
d) the three germ layers that will give rise to roots, stems, and leaves have formed

6. The *agamous* mutation causes the formation of

a) flowers with only petals and sepals
b) flowers with only sepals and carpels
c) flowers with only stamens and carpels
d) plants completely lacking flowers

7. Maintenance of the shoot meristems in adult *Arabidopsis* plants relies on which of the following mechanisms?

a) A homeobox transcription factor encoded by the WUSCHEL gene is expressed in the organizing center and initiates a signal to the overlying cells to behave as stem cells.
b) A transcription factor encoded by the SHOOT MERISTEMLESS gene is expressed in shoot meristem cells and maintains them in their undifferentiated state.
c) Shoot meristems cells secrete proteins encoded by the CLAVATA family that antagonize WUSCHEL expression, thereby restricting the size of the shoot meristems.
d) All of these are involved in specification and maintenance of shoot meristems.

8. What is meant by the word 'whorl' in discussing floral meristems?

a) Flowers consist of four different types of organs, which occur in concentric rings called 'whorls.'
b) The floral meristem as to rotate during flower formation, giving the process the name 'whorl.'
c) The flowers of *Arabidopsis* appear as the stem elongates in a pattern called a 'whorl.'
d) The six stamens in a dicot flower like that of *Arabidopsis* form a ring that is called the flower's 'whorl.'

9. In what way are the homeotic genes of flowering plants similar to those of *Drosophila* and other animals?

a) All homeotic genes encode transcription factors of the homeobox class.
b) Homeotic genes in both plants and animals encode transcription factors of the MADS-box type.
c) Mutations in the homeotic genes of flowers cause transformation of one organ into another.
d) The homeotic genes of flowers are derived during evolution from the same primordial genes used in animals.

10. How is the ABC model for floral identity in *Arabidopsis* reminiscent of the models for homeotic gene function derived from studies in *Drosophila*?

a) In both organisms, each homeobox gene specifies the identity of a different region of the adult.

b) In both organisms, one homeotic gene is expressed at the two ends, a second expressed in domains more central to that of the first, and a third expressed most centrally, thus contributing unambiguous identities to all regions of the organism.

c) In both organisms, it is often the combination of genes present that is critical in unambiguously specifying structures in the adult.

d) The key responsibility of the homeotic genes in both organisms is the patterning of antero-posterior identity.

Multiple choice answer key

1: b, 2: d, 3: b, 4: b, 5: a, 6: a, 7: d, 8: a, 9: c, 10: c.

■ General further reading

Meyerowitz, E.M.: *Arabidopsis—a useful weed. Cell* 1989, **56**: 263–264.

Meyerowitz, E.M.: **Plants compared to animals: the broader comparative view of development.** *Science* 2002, **295**: 1482–1485.

■ Section further reading

7.1 The model plant *Arabidopsis thaliana* has a short life cycle and a small diploid genome & 7.2 Plant embryos develop through several distinct stages

Lloyd, C.: **Plant morphogenesis: life on a different plane.** *Curr. Biol.* 1995, **5**: 1085–1087.

Mayer, U., Jürgens, G.: **Pattern formation in plant embryogenesis: a reassessment.** *Semin. Cell Dev. Biol.* 1998, **9**: 187–193.

Meyerowitz, E.M.: **Genetic control of cell division patterns in developing plants.** *Cell* 1997, **88**: 299–308.

Torres-Ruiz, R.A., Jürgens, G.: **Mutations in the *FASS* gene uncouple pattern formation and morphogenesis in Arabidopsis development.** *Development* 1994, **120**: 2967–2978.

7.3 Gradients of the signal molecule auxin establish the embryonic apical-basal axis

Breuninger, H., Rikirsch, E., Hermann, M., Ueda, M., Laux, T.: **Differential expression of *WOX* genes mediates apical-basal axis formation in the *Arabidopsis* embryo.** *Dev. Cell* 2008, **14**: 867–876.

Friml, J., Vieten, A., Sauer, M., Weijers, D., Schwarz, H., Hamann, T., Offringa, R., Jürgens, G.: **Efflux-dependent auxin gradients establish the apical-basal axis of *Arabidopsis*.** *Nature* 2003, **426**: 147–153.

Jenik, P.D., Barton, M.K.: **Surge and destroy: the role of auxin in plant embryogenesis.** *Development* 2005, **132**: 3577–3585.

Jürgens, G.: **Axis formation in plant embryogenesis: cues and clues.** *Cell* 1995, **81**: 467–470.

Long, J.A., Moan, E.I., Medford, J.I., Barton, M.K.: **A member of the knotted class of homeodomain proteins encoded by the *STM* gene of *Arabidopsis*.** *Nature* 1995, **379**: 66–69.

Szemenyei, H., Hannon, M., Long, J.A.: **TOPLESS mediates auxin-dependent transcriptional repression during *Arabidopsis* embryogenesis.** *Science* 2008, **319**: 1384–1386.

7.4 Plant somatic cells can give rise to embryos and seedlings

Zimmerman, J.L.: **Somatic embryogenesis: a model for early development in higher plants.** *Plant Cell* 1993, **5**: 1411–1423.

7.5 Cell enlargement is a major process in plant growth and morphogenesis

Backues, S.K., Konopka, C.A., McMichael, C.M., Bednarek, S.Y.: **Bridging the divide between cytokinesis and cell expansion.** *Curr. Opin. Plant. Biol.* 2007, **10**: 607–615.

Kuchen, E.K., Fox, S., Barbier de Reuille, P., Kennaway, R., Bensmihen, S., Avondo, J., Calder, G.M., Southam, P., Robinson, S., Bangham, A., Coen, E.: **Generation of leaf shape through early patterns of growth and tissue polarity.** *Science* 2012, **335**: 1092–1096.

Tsuge, T., Tsukaya, H., Uchimiya, H.: **Two independent and polarized processes of cell elongation regulate leaf blade expansion in *Arabidopsis thaliana* (L.) Heynh.** *Development* 1996, **122**: 1589–1600.

7.6 A meristem contains a small central zone of self-renewing stem cells

Byrne, M.E., Kidner, C.A., Martienssen, R.A.: **Plant stem cells: divergent pathways and common themes in shoots and roots.** *Curr. Opin. Genet. Dev.* 2003, **13**: 551–557.

Großhardt, R., Laux, T.: **Stem cell regulation in the shoot meristem.** *J. Cell Sci.* 2003, **116**: 1659–1666.

Ma, H.: **Gene regulation: Better late than never?** *Curr. Biol.* 2000, **10**: R365–R368.

7.7 The size of the stem-cell area in the meristem is kept constant by a feedback loop to the organizing center

Brand, U., Fletcher, J.C., Hobe, M., Meyerowitz, E.M., Simon, R.: **Dependence of stem cell fate in *Arabidopsis* on a feedback loop regulated by *CLV3* activity.** *Science* 2000, **289**: 635–644.

Carles, C.C., Fletcher, J.C.: **Shoot apical meristem maintenance: the art of dynamical balance.** *Trends Plant Sci.* 2003, **8**: 394–401.

Clark, S.E.: **Cell signalling at the shoot meristem.** *Nat Rev. Mol. Cell Biol.* 2001, **2**: 277–284.

Lenhard, M., Laux, T.: **Stem cell homeostasis in the *Arabidopsis* shoot meristem is regulated by intercellular movement of CLAVATA3 and its sequestration by CLAVATA1.** *Development* 2003, **130**: 3163–3173.

Reddy, G.V., Meyerowitz, E.M.: **Stem-cell homeostasis and growth dynamics can be uncoupled in the *Arabidopsis* shoot apex.** *Science* 2005, **310**: 663–667.

Schoof, H., Lenhard, M., Haecker, A., Mayer, K.F.X., Jürgens, G., Laux, T.: **The stem cell population of *Arabidopsis* shoot meristems is maintained by a regulatory loop between the *CLAVATA* and *WUSCHEL* genes.** *Cell* 2000, **100**: 635–644.

Vernoux, T., Benfey, P.N.: **Signals that regulate stem cell activity during plant development.** *Curr. Opin. Genet. Dev.* 2005, **15**: 388–394.

7.8 The fate of cells from different meristem layers can be changed by changing their position

Castellano, M.M., Sablowski, R.: **Intercellular signalling in the transition from stem cells to organogenesis in meristems.** *Curr. Opin. Plant Biol.* 2005, **8**: 26–31.

Gallagher, K.L., Benfey, P.N.: **Not just another hole in the wall: understanding intercellular protein trafficking.** *Genes Dev.* 2005, **19**: 189–195.

Laux, T., Mayer, K.F.X.: **Cell fate regulation in the shoot meristem.** *Semin. Cell Dev. Biol.* 1998, **9**: 195–200.

Sinha, N.R., Williams, R.E., Hake, S.: **Overexpression of the maize homeobox gene *knotted-1*, causes a switch from determinate to indeterminate cell fates.** *Genes Dev.* 1993, **7**: 787–795.

Turner, I.J., Pumfrey, J.E.: **Cell fate in the shoot apical meristem of *Arabidopsis thaliana*.** *Development* 1992, **115**: 755–764.

Waites, R., Simon, R.: **Signaling cell fate in plant meristems: three clubs on one tousle.** *Cell* 2000, **103**: 835–838.

7.9 A fate map for the embryonic shoot meristem can be deduced using clonal analysis

Irish, V.F.: **Cell lineage in plant development.** *Curr. Opin. Genet. Dev.* 1991, **1**: 169–173.

7.10 Meristem development is dependent on signals from other parts of the plant

Doerner, P.: **Shoot meristems: intercellular signals keep the balance.** *Curr. Biol.* 1999, **9**: R377–R380.

Irish, E.E., Nelson, T.M.: **Development of maize plants from cultured shoot apices.** *Planta* 1988, **175**: 9–12.

Sachs, T.: *Pattern Formation in Plant Tissues.* Cambridge: Cambridge University Press, 1994.

7.11 Gene activity patterns the proximo-distal and adaxial-abaxial axes of leaves developing from the shoot meristem

Bowman, J.L.: **Axial patterning in leaves and other lateral organs.** *Curr. Opin. Genet. Dev.* 2000, **10**: 399–404.

Kidner, C.A., Martienssen, R.A.: **Spatially restricted microRNA directs leaf polarity through *ARGONAUTE1*.** *Nature* 2004, **428**: 81–84.

Waites, R., Selvadurai, H.R., Oliver, I.R., Hudson, A.: **The *PHANTASTICA* gene encodes a MYB transcription factor involved in growth and dorsoventrality of lateral organs *in Antirrhinum*.** *Cell* 1998, **93**: 779–789.

7.12 The regular arrangement of leaves on a stem is generated by regulated auxin transport

Berleth, T., Scarpella, E., Prusinkiewicz, P.: **Towards the systems biology of auxin-transport-mediated patterning.** *Trends Plant Sci.* 2007, **12**: 151–159.

Heisler, M.G., Ohno, C., Das, P., Sieber, P., Reddy, G.V., Long, J.A., Meyerowitz, E.M.: **Patterns of auxin transport and gene expression during primordium development revealed by live imaging of the *Arabidopsis* inflorescence meristem.** *Curr. Biol.* 2005, **15**: 1899–1911.

Jönsson, H., Heisler, M.G., Shapiro, B.E., Meyerowitz, E.M., Mjolsness, E.: **An auxin-driven polarized transport model for phyllotaxis.** *Proc. Natl. Acad. Sci. USA* 2006, **103**: 1633–1638.

Mitchison, G.J.: **Phyllotaxis and the Fibonacci series.** *Science* 1977, **196**: 270–275.

Reinhardt, D.: **Regulation of phyllotaxis.** *Int. J. Dev. Biol.* 2005, **49**: 539–546.

Scheres, B.: **Non-linear signaling for pattern formation.** *Curr. Opin. Plant Biol.* 2000, **3**: 412–417.

Schiefelbein, J.: **Cell-fate specification in the epidermis: a common patterning mechanism in the root and shoot.** *Curr. Opin. Plant Biol.* 2003, **6**: 74–78.

Smith, L.G., Hake, S.: **The initiation and determination of leaves.** *Plant Cell* 1992, **4**: 1017–1027.

7.13 Root tissues are produced from *Arabidopsis* root apical meristems by a highly stereotyped pattern of cell divisions

Costa, S., Dolan, L.: **Development of the root pole and patterning in *Arabidopsis* roots.** *Curr. Opin. Genet. Dev.* 2000, **10**: 405–409.

Dello Ioio, R., Nakamura, K., Moubayidin, L., Perilli, S., Taniguchi, M., Morita, M.T., Aoyama, T., Costantino, P., Sabatini, S.: **A genetic framework for the control of cell division and differentiation in the root meristem.** *Science* 2008, **322**: 1380–1384.

Dolan, L.: **Positional information and mobile transcriptional regulators determine cell pattern in *Arabidopsis* root epidermis.** *J. Exp. Bot.* 2006, **57**: 51–54.

Grieneisen, V.A., Xu, J., Marée, A.F.M., Hogeweg P., Scheres, B.: **Auxin transport is sufficient to generate a maximum and gradient guiding root growth.** *Nature* 2007, **449**: 1008–1013.

Sabatini, S., Beis, D., Wolkenfeldt, H., Murfett, J., Guilfoyle, T., Malamy, J., Benfey, P., Leyser, O., Bechtold, N., Weisbeek, P., Scheres, B.: **An auxin-dependent distal organizer of pattern and polarity in the *Arabidopsis* root.** *Cell* 1999, **99**: 463–472.

Scheres, B., McKhann, H.I., van den Berg, C.: **Roots redefined: anatomical and genetic analysis of root development.** *Plant Physiol.* 1996, **111**: 959–964.

van den Berg, C., Willemsen, V., Hendriks, G., Weisbeek, P., Scheres, B.: **Short-range control of cell differentiation in the *Arabidopsis* root meristem.** *Nature* 1997, **390**: 287–289.

Veit, B.: **Plumbing the pattern of roots.** *Nature* 2007, **449**: 991–992.

7.14 Root hairs are specified by a combination of positional information and lateral inhibition

Kwak, S.-H., Schiefelbein, J.: **The role of the SCRAMBLED receptor-like kinase in patterning the *Arabidopsis* root epidermis.** *Dev. Biol.* 2007, **302**: 118–131.

Schiefelbein, J., Kwak, S.-H., Wieckowski, Y., Barron, C., Bruex, A.: **The gene regulatory network for root epidermal cell-type pattern formation in *Arabidopsis*.** *J. Exp. Bot.* 2009, **60**: 1515–1521.

7.15 Homeotic genes control organ identity in the flower

Bowman, J.L., Sakai, H., Jack, T., Weigel, D., Mayer, U., Meyerowitz, E.M.: **SUPERMAN, a regulator of floral homeotic genes in *Arabidopsis*.** *Development* 1992, **114**: 599–615.

Breuil-Broyer, S., Morel, P., de Almeida-Engler, J., Coustham, V., Negrutiu, I., Trehin, C.: **High-resolution boundary analysis during *Arabidopsis thaliana* flower development.** *Plant J.* 2004, **38**: 182–192.

Coen, E.S., Meyerowitz, E.M.: **The war of the whorls: genetic interactions controlling flower development.** *Nature* 1991, **353**: 31–37.

Irish, V.F.: **Patterning the flower.** *Dev. Biol.* 1999, **209**: 211–220.

Krizek, B.A., Meyerowitz, E.M.: **The *Arabidopsis* homeotic genes**

APETALA3 and *PISTILLATA* are sufficient to provide the B class organ identity function. *Development* 1996, **122**: 11–22.

Krizek, B.A., Fletcher, J.C.: **Molecular mechanisms of flower development: an armchair guide**. *Nat Rev. Genet.* 2005, **6**: 688–698.

Lohmann, J.U., Weigel, D.: **Building beauty: the genetic control of floral patterning**. *Dev. Cell* 2002, **2**: 135–142.

Ma, H., dePamphilis, C.: **The ABCs of floral evolution**. *Cell* 2000, **101**: 5–8.

Meyerowitz, E.M., Bowman, J.L., Brockman, L.L., Drews, G.M., Jack, T., Sieburth, L.E., Weigel, D.: **A genetic and molecular model for flower development in *Arabidopsis thaliana***. *Development Suppl.* 1991, **1**: 157–167.

Meyerowitz, E.M.: **The genetics of flower development**. *Sci. Am.* 1994, **271**: 40–47.

Sakai, H., Medrano, L.J., Meyerowitz, E.M.: **Role of *SUPERMAN* in maintaining *Arabidopsis* floral whorl boundaries**. *Nature* 1994, **378**: 199–203.

Vincent, C.A., Carpenter, R., Coen, E.S.: **Cell lineage patterns and homeotic gene activity during Antirrhinum flower development**. *Curr. Biol.* 1995, **5**: 1449–1458.

Wagner, D., Sablowski, R.W.M., Meyerowitz, E.M.: **Transcriptional activation of *APETALA 1***. *Science* 1999, **285**: 582–584.

7.16 The *Antirrhinum* flower is patterned dorso-ventrally as well as radially

Coen, E.S.: **Floral symmetry**. *EMBO J.* 1996, **15**: 6777–6788.

Luo, D., Carpenter, R., Vincent, C., Copsey, L., Coen, E.: **Origin of floral asymmetry in *Antirrhinum***. *Nature* 1996, **383**: 794–799.

7.17 The internal meristem layer can specify floral meristem patterning

Szymkowiak, E.J., Sussex, I.M.: **The internal meristem layer (L3) determines floral meristem size and carpel number in tomato periclinal chimeras**. *Plant Cell* 1992, **4**: 1089–1100.

7.18 The transition of a shoot meristem to a floral meristem is under environmental and genetic control

An, H., Roussot, C., Suarez-Lopez, P., Corbesier, L., Vincent, C., Pineiro, M., Hepworth, S., Mouradov, A., Justin, S., Turnbull, C., Coupland, G.: **CONSTANS acts in the phloem to regulate a systemic signal that induces photoperiodic flowering of *Arabidopsis***. *Development* 2004, **131**: 3615–3626.

Becroft, P.W.: **Intercellular induction of homeotic gene expression in flower development**. *Trends Genet.* 1995, **11**: 253–255.

Blázquez, M.A.: **The right time and place for making flowers**. *Science* 2005, **309**: 1024–1025.

Corbesier, L., Vincent, C., Jang, S., Fornara, F., Fan, Q., Searle, I., Giakountis, A., Farrona, S., Gissot, L., Turnbull, C., Coupland, G.: **FT protein movement contributes to long-distance signaling in floral induction of *Arabidopsis***. *Science* 2007, **316**: 1030–1033.

Hake, S.: **Transcription factors on the move**. *Trends Genet.* 2001, **17**: 2–3.

Jaeger, K.E., Wigge, P.A.: **FT protein acts as a long-range signal in *Arabidopsis***. *Curr. Biol.* 2007, **17**: 1050–1054.

Kobayashi, Y., Weigel, D.: **Move on up, it's time for change - mobile signals controlling photoperiod-dependent flowering**. *Genes Dev.* 2007, **21**: 2371–2384.

Lohmann, J.U., Hong, R.L., Hobe, M., Busch, M.A., Parcy, F., Simon, R., Weigel, D.: **A molecular link between stem cell regulation and floral patterning in *Arabidopsis***. *Cell* 2001, **105**: 793–803.

Putterill, J., Laurie, R., Macknight, R.: **It's time to flower: the genetic control of flowering time**. *BioEssays* 2004, **26**: 363–373.

Sheldon, C.C., Hills, M.J., Lister, C., Dean, C., Dennis, E.S., Peacock, W.J.: **Resetting of *FLOWERING LOCUS C* expression after epigenetic repression by vernalization**. *Proc. Natl. Acad. Sci. USA* 2008, **105**: 2214–2219.

Valverde, F., Mouradov, A., Soppe, W., Ravenscroft, D., Samach, A., Coupland, G.: **Photoreceptor regulation of CONSTANS protein in photoperiodic flowering**. *Science* 2004, **303**: 1003–1006.

7.19 Most flowering plants are hermaphrodites, but some produce unisexual flowers

Irisa, E.N.: **Regulation of sex determination in maize**. *BioEssays* 1996, **18**: 363–369.

Huh, J.H., Bauer, M.J., Hsieh, T.-F., Fischer, R.L.: **Cellular programming of plant gene imprinting**. *Cell* 2008, **132**: 735–744.

8

Cell differentiation and stem cells

- ● The control of gene expression
- ● Models of cell differentiation and stem cells
- ● The plasticity of the differentiated state

The differentiation of unspecialized cells into many different cell types occurs first in the developing embryo and continues after birth and throughout adulthood. The character of specialized cells, such as blood, muscle or nerve cells, is the result of a particular pattern of gene activity, which determines which proteins are synthesized. Thus, how this particular pattern of gene expression develops and is maintained are central questions in cell differentiation, and are discussed in this chapter in relation to some well-studied model systems. Gene expression is under a complex set of controls that include the actions of transcription factors, chemical modification of DNA, and post-translational modifications to chromatin proteins. External signals play a key role in differentiation by triggering intracellular signaling pathways that affect gene expression. We shall also see how, after birth, many tissues are renewed throughout life from their own adult stem cells, which are both self-renewing and capable of differentiating into the range of cell types needed to renew that tissue. And we will consider the properties of mammalian embryonic stem cells, cells derived from the inner cell mass of the blastocyst (see Chapter 4) which are pluripotent: that is, they can give rise to all the cell types present in the embryo. Another intriguing question in development is the degree of plasticity of the differentiated state: can a differentiated cell change into another cell type? We shall see that such changes are possible, although rare under normal circumstances. And it is now well established that a nucleus from a differentiated cell can be reprogrammed when transferred into an enucleated egg, and become able to direct embryonic development—the process that underlies animal cloning. More recently, differentiated cells themselves have been reprogrammed genetically to return to a stem-cell-like pluripotent state, from which they can be induced to differentiate into a range of different cell types. These induced stem cells, along with embryonic stem cells and adult stem cells have potential uses in regenerative medicine.

Embryonic cells start out morphologically similar to each other but eventually become different, acquiring distinct identities and specialized functions. As we have seen in previous chapters, many developmental processes, such as the early specification of the germ layers, involve transient changes in cell form, patterns of gene activity, and proteins synthesized. **Cell differentiation**, in contrast, involves the gradual emergence of cell types that have a clear-cut identity in the adult, such as nerve cells, red blood cells, and fat cells (Fig. 8.1). Initially, embryonic cells fated to become different cell types can be distinguished from each other only by small differences in patterns of gene expression and thus in the proteins produced. Differentiation is a progressive restriction in fates that occurs over successive cell generations as cells gradually acquire special structural features associated with their specialized functions.

As with earlier processes in development, the central feature of cell differentiation is a progressive change in the set of genes expressed by the cell. All cells continue to produce the housekeeping proteins concerned with general metabolism, but as cells differentiate, they also begin to express proteins tailored to the specialized function of the differentiated cell. Red blood cells produce the oxygen-carrying protein hemoglobin; skin epidermal cells produce the fibrous protein keratin, which helps to give skin its resilience; neurons produce voltage-gated ion channels that enable them to produce an electrical signal; and muscle cells produce muscle-specific actins and myosins that enable contraction. It is important to remember that out of the several thousand genes active in any cell in the embryo at any given time, only a small number may be involved in specifying cell fate or differentiation. The genes that are being expressed in a particular tissue or at a particular stage of development can be detected by DNA microarray techniques or by RNAseq (see Box 3B).

In contrast to the transitory differences in gene activity characteristic of earlier stages in development, fully differentiated cells have achieved a stable state in which

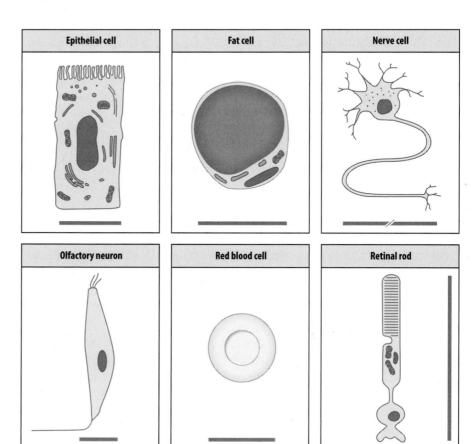

Fig. 8.1 Differentiated cell types. Mammalian cell types come in various shapes and sizes. Scale bars: epithelial cell, 15 μm; fat cell, 100 μm; nerve cell, 100 μm–1 m; olfactory neuron, 8 μm; red blood cell, 8 μm; retinal rod, 20 μm.

they undergo no further changes and do not change into another cell type. In this state they are called **terminally differentiated**. A fully differentiated cell has often undergone considerable structural change. For example, mature mammalian red blood cells lose their nucleus and become biconcave discs stuffed with hemoglobin, whereas neutrophils, a type of white blood cell, develop a multi-lobed nucleus and a cytoplasm filled with secretory granules containing proteins involved in protecting the body against infection. Yet both cells derive from the same early lineage of blood-forming (hematopoietic) cells.

The first step in differentiation is for a cell to become committed, or determined, to becoming a particular cell type (see Section 1.12). The earliest determined precursors of different cell types have no obvious structural differences from each other. The precursors of cartilage and muscle cells, for example, look the same and might be described as undifferentiated, but will differentiate as cartilage and muscle, respectively. There are however, subtle differences in the sets of genes expressed by these precursor cells, and thus in the proteins produced, which control their future development. Determination is a key stage preceding differentiation. Cells that have become determined with respect to their eventual fate will form only the appropriate cell types, even if grafted to a different site in the embryo: that is, they retain their identity (see Fig. 1.23). Once determined for a particular fate, cells pass on that determined state to all their progeny.

In some cases, expression of a single gene can determine how a cell will differentiate. The gene *myoD*, for example, encodes a transcription factor that is essential for muscle differentiation, and if it is introduced experimentally into fibroblasts, a cell type that does not normally express *myoD*, the fibroblasts will develop into muscle cells. MyoD is therefore often called a **master regulatory gene** for muscle development. Master genes are genes, usually encoding transcription factors, whose expression is both necessary and sufficient to trigger activation of many other genes in a specific program leading to the development of a particular tissue or organ. And as we shall see later in this chapter, a very dramatic effect is obtained when genes encoding four of the transcription factors expressed in mammalian embryonic stem cells (ES cells) are introduced into fibroblasts. The fibroblasts are converted into **pluripotent** cells, which means that they lose all their specialized features and, like ES cells, which are derived from the inner cell mass and give rise to all the cells of the embryo (see Section 5.2), can then be induced to differentiate into cells belonging to tissues derived from all three germ layers.

The initiation and progress of cell differentiation is under strict control by extracellular signals produced by other cells, which range from cell-surface proteins to secreted signaling molecules, such as growth factors, and molecules of the extracellular matrix. It is important to remember that, while the extracellular signals that stimulate differentiation are often referred to as being 'instructive,' they are in general 'selective,' in the sense that, in practice, the number of developmental options open to a cell at any given time is limited. These options are set by the cell's internal state, which in turn reflects its developmental history. Extracellular signals cannot, for example, convert an endodermal cell into a muscle or nerve cell. These restrictions mean that the signals we have seen acting in early vertebrate development, such as Wnts, FGFs, Notch, and members of the TGF-β family (see Box 4A), can continue to be used in determination and differentiation with different effects.

One feature of fully differentiated cells is that they cease to divide altogether or divide far less frequently than undifferentiated embryonic cells. Embryonic cells continue to proliferate after they have been determined, but cessation of cell division is necessary for full cell differentiation to occur. Some cells, such as skeletal muscle and nerve cells, do not divide at all after they have become fully differentiated. In cells that can divide after differentiation, such as fibroblasts and liver cells, the differentiated state, like the determined state, is passed on through subsequent cell divisions.

As the pattern of gene activity is the key feature in cell differentiation, this raises the question of how a particular pattern of gene activity is first established, and then how it is passed on to daughter cells.

We therefore start this chapter with a consideration of the general mechanisms by which a pattern of gene activity can be established, maintained, and inherited at cell division. We then look at the molecular basis of cell differentiation in specific instances, with muscle cells and blood cells providing our main examples. We shall also discuss the special properties of mammalian stem cells, both the pluripotent ES cells derived from the inner cell mass, and the 'adult' stem cells of more limited potential from which tissues such as blood and skin are continually renewed throughout life.

Finally, we look at the reversibility and plasticity of the differentiated state, particularly the reprogramming in oocytes of nuclei from differentiated cells, so that the egg gives rise to an embryo with the genome of the differentiated nucleus. This makes it possible to clone some animals: that is, to make an animal that is genetically identical to the animal from which the nucleus came. We shall also discuss the exciting discovery that fully differentiated cells taken from adults can be reprogrammed genetically back to a stem-cell-like pluripotent state. This chapter concentrates on differentiation in animal cells. As we have seen in Chapter 7, differentiated plant cells do not have a permanently determined state, and a small piece of tissue, or even a single somatic cell, can readily give rise to a whole plant.

The control of gene expression

Every nucleus in the body of a multicellular organism is derived from the single zygotic nucleus in the fertilized egg and thus carries the same genome, which in humans comprises around 20,000 distinct genes. But the subsets of these genes that are active in differentiated cells vary enormously from one cell type to another. The egg itself has a pattern of gene activity that is different from that of the embryonic cells to which it gives rise. So how do distinct and unique combinations of genes come to be expressed in each differentiated cell type? Although there are only a few hundred morphologically recognizable basic mammalian cell types, there are in fact millions of functionally distinct cells in the mammalian body, each with its particular pattern of gene activity: think, for example, of the number of functionally different neurons required to enable the nervous system to work (see Chapter 12). There are fewer genes in the genome than types of differentiated cell that need to be specified, and therefore each distinct cell type must be defined by the expression of a combination of genes rather than by a single gene (Fig. 8.2). This raises the question of how the particular pattern of gene activity in a differentiated cell is specified and how it is inherited.

To understand the molecular basis of cell differentiation, we first need to know how a gene can be expressed in a cell-specific manner—why a given gene gets switched on in one cell and not in another. A second question is how the particular combination of genes expressed by a differentiated cell comes to be selected, and how their expression is coordinated. We focus here on the regulation of **transcription**, which is the first (and generally the most significant) step in the expression of a gene and the subsequent synthesis of a protein (see Section 1.10). **Translation** of the mRNA to produce a protein, however, is also subject to several layers of control. We have seen an example of translational regulation in early *Drosophila* development, where the Nanos protein controls the production of Hunchback protein by preventing translation of maternal *hunchback* RNA (see Section 2.9). But for the most part, it is the transcriptional regulation of gene expression that underlies the process of differentiation.

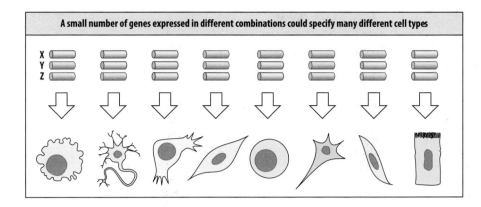

A small number of genes expressed in different combinations could specify many different cell types

Fig. 8.2 Different cell types are the result of the expression of different combinations of genes. To make the most efficient use of the relatively modest numbers of genes in the genome, different functional cell types are specified by the expression of different combinations of genes. This enables many more cell types to be specified than if each gene was only able to uniquely specify one cell type. This hypothetical example shows how a set of three genes (X, Y, and Z) could specify eight different cell types by the unique combinations of those genes that are active (indicated by the red bars) and inactive (gray bars) in each cell type.

8.1 Control of transcription involves both general and tissue-specific transcriptional regulators

Most of the key genes in development are initially in an inactive state and require activating transcription factors, or **activators**, to turn them on. These activators bind to sites in the *cis*-regulatory regions in the DNA surrounding the gene (the '*cis*' simply refers to the fact that the regulatory region is on the same DNA molecule as the gene it controls). The site to which an individual transcription factor binds is a short stretch of nucleotides with a sequence specifically recognized by the protein. Sites to which transcription factors bind to switch on a gene are often clustered within regulatory regions called **enhancers**. The importance of regulatory regions in tissue-specific gene expression was clearly demonstrated by experiments in which the control region of one tissue-specific gene was replaced by the control region of another (Fig. 8.3). We have already seen how such replacements are routinely used in experimental developmental biology to force the expression or misexpression of a gene in a particular tissue, and to mark cells expressing a particular gene in order to investigate its developmental function (see Box 3D and Box 2E).

In eukaryotic cells, most regulated protein-coding genes are transcribed by RNA polymerase II. The polymerase binds to a control region in the gene called the **promoter**, where it is positioned so that it can start transcription in the correct place. The binding of RNA polymerase II to the promoter requires the cooperation of a set of **general transcription factors**, so called because they are required by all genes transcribed by RNA polymerase II. They form a **transcription initiation complex** with

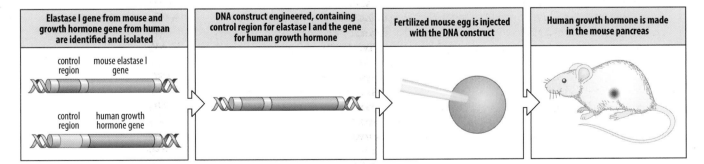

| Elastase I gene from mouse and growth hormone gene from human are identified and isolated | DNA construct engineered, containing control region for elastase I and the gene for human growth hormone | Fertilized mouse egg is injected with the DNA construct | Human growth hormone is made in the mouse pancreas |

control region   mouse elastase I gene

control region   human growth hormone gene

Fig. 8.3 Tissue-specific gene expression is controlled by regulatory regions. Growth hormone is normally made in the pituitary and the enzyme elastase I only in the pancreas. To demonstrate that the regulatory regions of a gene determine its tissue-specific expression, the control region of the mouse elastase I gene was joined to a DNA sequence coding for human growth hormone. This DNA construct was injected into the nucleus of a fertilized mouse egg, where it becomes integrated into the genome. When the mouse develops, human growth hormone is made in the pancreas. A 213-base-pair stretch of DNA that includes the elastase I promoter and other regulatory regions is sufficient to drive growth hormone expression in the pancreas.

the polymerase at the promoter. For highly regulated genes, such as those involved in development, this complex cannot be formed and activated without additional help. This is provided by the binding of regulatory proteins to specific sites, defined by their nucleotide sequence, in the regulatory regions of the gene, which help to attract the general transcription factors and polymerase and position them correctly on the promoter (Fig. 8.4). Some regulatory sites, including some enhancers, are alternatively bound by **repressors**, gene-regulatory proteins that prevent a gene being expressed.

For any given gene, when and where it is expressed at any given stage of development is determined by particular combinations of gene-regulatory proteins binding to individual sites in its control regions (Fig. 8.5). These sites are typically 7–10 nucleotides long. At least 1000 different transcription factors are encoded in the genomes of *Drosophila* and *C. elegans*, and as many as 3000 in the human genome. On average, around five different transcription factors act together at the control region of a gene, and in some cases considerably more. The efficient use of the transcription factor repertoire is maximized by the fact that many transcription factors act on the control regions of more than one gene. This means that a single transcription factor can help to coordinate the activity of these genes, activating some and repressing others (Fig. 8.6).

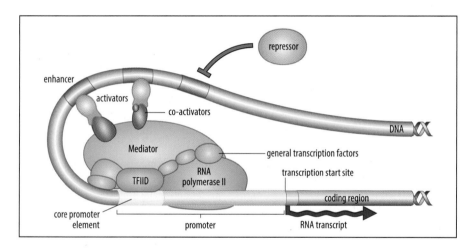

Fig. 8.4 Gene expression is regulated by the coordinated action of gene-regulatory proteins that bind to regulatory regions in DNA. Most protein-coding genes in eukaryotic cells, including all highly regulated developmental genes, are transcribed by RNA polymerase II. The transcriptional machinery itself, composed of the polymerase and its associated general transcription factors (the core transcriptional machinery; for example, TFIID and the Mediator complex), recognizes the promoter region by binding to specific sequences within it that are common to all genes transcribed by RNA polymerase II. These core promoter sequences are platforms for the assembly of the core transcriptional complex, and determine the nucleotide at which transcription will start (the transcription start site, TSS). Other control sites are gene specific, and can be occupied by specific activator or repressor proteins together with their associated co-factors. These proteins interact with the polymerase and associated proteins to either initiate or prevent transcription. Thus, the particular combination of activators and repressors occupying the control sites determines whether a gene is expressed or not in a particular cell. These control sites can be located adjacent to the promoter, or further upstream, or sometimes downstream, of the coding region of the gene. For some genes, control sites can be many kilobases away from the start point of transcription. The proteins bound to these distant sites are brought into contact with the transcription machinery by looping of the DNA. The regulation of gene expression requires the connection of activators and repressors with the RNA polymerase II complex, and this is effected by a large multi-protein complex called Mediator. The fully operational initiation complex activates the RNA polymerase and releases it to start transcription.

*Illustration after Tjian, R.: **Molecular machines that control genes**. Sci. Am. 1995, **272**: 54-61.*

The combinatorial action of gene-regulatory proteins is a key principle in the control of gene expression and is fundamental to the exquisite control and complexity of the gene expression that drives development. Complex control regions that regulate the temporal and spatial expression of a gene during embryonic development are shown in Fig. 2.37 for another part of the *eve* gene. The spatially delimited expression of *eve* in the second of seven transverse stripes in the embryo involves four different gene-regulatory proteins, which include activators and repressors, acting at 11 distinct sites. The control region of the sea urchin *Endo-16* gene, which encodes a calcium-binding extracellular matrix protein, is diagrammed in Fig. 6.27. The correct expression of *Endo-16* throughout sea-urchin embryo development requires the participation of 13 gene-regulatory proteins acting at a potential 56 regulatory motifs within a control region that spans 2.3 kilobases.

Sites within the promoter region determine the start site of transcription, and bind the general transcription factors and the RNA polymerase (see Fig. 8.4). Tissue-specific or developmental-stage expression is controlled by sites outside the immediate promoter region, such as those at which tissue-specific gene-regulatory proteins act. These sites vary greatly in type and position from gene to gene, and may be thousands of base pairs away from the start point of transcription. These distant sites are able to control gene activity because DNA can form loops, thus bringing the sites and their bound proteins into close proximity to the promoter. Proteins bound at distant sites can thus make contact with proteins at the promoter, forming a fully active **transcription initiation complex** (see Fig. 8.4).

Another important class of gene-regulatory proteins comprises the **co-activators** and **co-repressors**, which do not bind DNA themselves but link the transcriptional machinery to the DNA-binding activator and repressor proteins (see Fig. 8.4). One co-activator we have already encountered is β-catenin (whose activity is the end product of the Wnt signaling pathway), which binds transcription factors of the TCF/LEF family (see Box 4B). In the absence of Wnt signaling, TCFs bound to control elements in genes that are targets of Wnt signaling are bound by a co-repressor, called Groucho in *Drosophila* and TLE in humans, and transcription is repressed. β-Catenin accumulating in the nucleus as the result of Wnt signaling displaces Groucho/TLE and binds to TCF, turning it into a transcriptional activator.

Transcriptional regulators fall into two main groups: those that are required for the transcription of a wide range of genes, and which are found in many different cell types; and those that are required for a particular gene or set of genes whose expression is tissue restricted, or **tissue-specific**, and which are only found in one or a very few cell types. We shall encounter both types of transcription factor in the following sections, when we look at the regulation of muscle-specific genes in muscle precursor

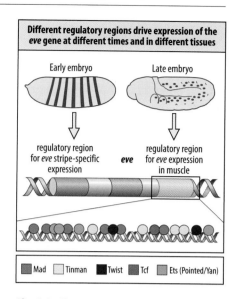

Fig. 8.5 Changes in the expression of the *eve* gene throughout development are due to the use of different regulatory regions. The *even-skipped (eve)* gene in *Drosophila* is expressed in a pattern of transverse stripes in the early embryo, and this expression is controlled by regulatory modules (one for each stripe) located on the 5′ side of the gene (described in Section 2.22, not shown in detail here). In the late embryo, *eve* is expressed in muscle precursors, under the control of a different regulatory region on the 3′ side of the gene. Illustrated here is the combination of transcription factors that binds to that region to regulate gene expression in a particular muscle group in each segment. Five different transcription factors bind at 17 different sites in this region to switch on *eve* expression. Some of the transcription factors are effectors of signaling pathways: Mad (BMP, TGF-β), Tcf (Wnt) and Ets (FGF). Different combinations of factors control *eve* expression in the other muscle groups.

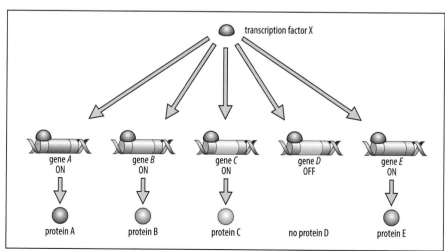

Fig. 8.6 A single transcription factor can regulate the expression of different genes in a positive or negative fashion. Transcription factors can activate or repress the activity of genes by binding to their control regions. In the illustration, the production of transcription factor X in the cell eventually leads to the production of four new proteins (A, B, C, and E), and the repression of protein D production.

*Illustration after Alberts B., et al.: **Molecular Biology of the Cell**, 2nd edition. New York: Garland Publishing, 1989.*

cells and the differentiation of red blood cell precursors. In general, it can be assumed that activation of each gene involves a unique combination of transcription factors.

As noted earlier, developmental genes generally have complex control regions, containing binding sites for a range of activating or inhibitory transcription factors, so whether or not a gene is activated depends on the precise combinations, and levels, of these transcription factors, and their associated co-activators and co-repressors, in the cell. The particular combination of gene-regulatory proteins present in a given embryonic cell at a given time is the result of its developmental history so far, and determines the next step in its development.

8.2 Gene expression is also controlled by chemical and structural modifications to DNA and histone proteins that alter chromatin structure

A central feature of development in general, and cell differentiation in particular, is that in any given cell, some genes are maintained in an active state, while others are repressed and inactive. As development and differentiation proceed, genes that were previously inactive become active in particular cells, while previously active genes become permanently shut down. Transcriptional control by transcription factors is not the only means available to the cell for controlling gene expression. That process operates against a larger-scale mechanism by which genes can be either made accessible for transcription or kept permanently or semi-permanently inactive. This mechanism involves the modification of **chromatin**, the complex of DNA, histones, and other proteins of which the chromosomes are made. Genes that are not required are packed away, or **silenced**, in the relevant cells by changes in chromatin that render the genes inaccessible to transcription. In this way, large numbers of genes can be shut down over the long term by a general mechanism, without requiring the continued production of a unique set of regulatory proteins for each gene. Chromatin that is packaged into a relatively compact structure that cannot be transcribed is called **heterochromatin**. For other genes, the chromatin is in a different, more open, state, in which it is accessible to the transcriptional machinery and to control by gene-regulatory proteins such as those discussed in Section 8.1.

Whether a stretch of chromosome can be transcribed or not therefore depends in the first instance on its chromatin structure. The compaction of chromatin that occurs at mitosis, for example, when the condensed chromosomes become visible under the light microscope, renders the chromosomes transcriptionally inactive. More localized structural alterations of this sort that persist throughout the cell cycle, and are reinstated after cell division, provide a means of shutting down genes in the long term. Localized changes in chromatin structure and protein composition provide a general mechanism for long-term gene inactivation. Underlying these changes in the transcriptional properties of chromatin are covalent chemical modifications to the DNA itself and to the associated histone proteins (Box 8A). Unlike mutation, these modifications do not change the DNA sequence and are known as **epigenetic** modifications. They are carried out by dedicated sets of enzymes, some of which add the chemical tags and some of which remove them. Although some epigenetic modifications are normally permanent for the lifetimes of a somatic cell and its progeny, they can be reversed given the right environment, as shown by the reprogramming of a somatic cell nucleus when transplanted into an enucleated oocyte (see Fig. 1.20).

An important epigenetic modification that keeps genes silenced in many organisms, especially in mammals and in plants, is **DNA methylation**. Cytosines occurring in cytosine–guanine (CpG) sequences at certain sites in the DNA are methylated by the enzyme **DNA methylase**. DNA methylation preferentially occurs in promoters and enhancers, and is correlated with the absence of transcription in the neighboring genes. DNA methylation is one way in which certain genes are rendered inactive in

CELL BIOLOGY BOX 8A Epigenetic control of gene expression by chromatin modification

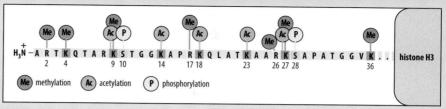

Figure 1

In chromatin, DNA is typically packaged together with four different histone proteins into structures called nucleosomes, in which the DNA is wrapped around the outside of a core composed of eight histone molecules. The histones in chromatin can undergo post-translational covalent modifications, such as methylation, acetylation, or phosphorylation, on specific amino-acid residues, particularly lysine (K), and these modifications are closely linked to the regulation of gene activity (Figure 1). Methylation and acetylation have been the most studied and their patterns within the histones associated with a gene are thought to create a code that determines whether the gene is ON or OFF: that is, it is becoming possible to gauge the pattern of activity of a gene in a cell during development from the pattern of modifications of its associated histones. Histone acetylation, for example, tends to be associated with regions of chromatin that are being transcribed, whereas the effect of methylation depends on the residue that is methylated and the number of methyl groups covalently attached to the lysine residue. Up to three methyl groups can be added. On some lysine residues, for example K4 in histone H3, monomethylation is associated with gene expression, but on the K9 and K27 residues of the same histone, the effect depends on the degree of methylation. Monomethylation is associated with expression, whereas trimethylation is associated with repression, promoting the formation of transcriptionally silent heterochromatin.

Changes to chromatin, such as DNA methylation, histone acetylation, and histone methylation, exert their effects on chromatin structure and gene expression by the recruitment of proteins that recognize the altered sites and maintain the transcriptional status of the chromatin. Histone methylation can recruit proteins that either initiate heterochromatin formation or promote transcription, depending on the residue methylated, whereas histone acetylation recruits proteins that keep the DNA available for transcription. These in turn can recruit other protein complexes that will, for example, specifically affect gene expression. We have seen an example of this in *Drosophila*, where proteins of the Polycomb group maintain the repression of Hox genes in cells in which they have been turned off, whereas proteins of the Trithorax group maintain the expression of Hox genes in cells in which they have been turned on (see Section 2.30).

Changes in chromatin status are often self-propagating, because histone modifications, such as acetylation/deacetylation or methylation/demethylation, recruit the corresponding chromatin-modifying enzymes to the chromatin, causing a 'domino effect' that automatically propagates the change in chromatin structure along the chromosome.

This type of effect is likely to underlie the typical sequential activation of Hox genes in the order they occur in the gene cluster (see Box 5E and Fig. 5.29). The relationship between histone modification and expression of the genes of the Hoxd cluster has been followed in excised mouse tailbuds at successive stages of development using chromatin immunoprecipitation (ChIP) to detect the modified histone proteins and ChIP-chip to detect the target genes (see Section 3.12). At stage E8.5, the genes *Hoxd1–9* are already expressed and their chromatin is heavily acetylated, but there are few acetylation marks on the chromatin associated with *Hoxd10–13*, which are not being expressed at this stage, although acetylation is starting to appear on *Hoxd10* (Figure 2). Twelve hours later, at stage E9.0, *Hoxd10* chromatin is more heavily acetylated and *Hoxd10* is expressed. At stage E9.5, histone acetylation has spread over the remaining Hoxd genes, which are all now starting to be expressed. The opposite is seen when the progress of repressive methylation on H3 K27 is tracked, with less methylation on histones associated with active genes and strong histone methylation associated with genes not yet expressed.

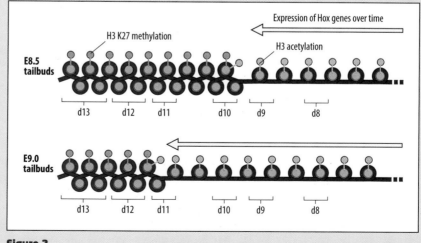

Figure 2

*Data for Fig. 2 is from Soshnikova, N., Duboule, D.: **Epigenetic temporal control of mouse Hox genes in vivo**. Science 2009, **324**: 1320-1323.*

the maternal or paternal genome in the phenomenon of genomic imprinting (discussed in Section 10.8). The inactive X chromosome in the cells of female mammals also has a pattern of DNA methylation that differs from that of the active X, and this is likely to play a part in keeping it inactive (discussed in Section 10.17). The pattern of methylation can be faithfully inherited when the DNA replicates, thus providing a mechanism for passing on a pattern of gene activity to daughter cells.

Genes can also be silenced, or be made potentially active, by modifications to the protein constituents of chromosomes. In chromatin, the DNA is wrapped around complexes of **histone proteins** to form structures called **nucleosomes**. Histones can undergo various post-translational chemical modifications—such as methylation, acetylation, and phosphorylation—on specific amino acids, which affect whether the DNA they package can be transcribed or not (see Box 8A). These modifications are carried out by specific enzymes, for example, histone acetyltransferases and histone methyltransferases, and can be reversed by other enzymes—histone deacetylases and histone demethylases—which remove the chemical tag. Like DNA methylation, the pattern of histone modification can be passed on to daughter cells, thus creating a lasting memory of the transcriptional status of a region of DNA. However, patterns of histone modification tend to be more dynamic than those of DNA methylation.

8.3 Patterns of gene activity can be inherited by persistence of gene-regulatory proteins or by maintenance of chromatin modifications

Once committed to a particular differentiation pathway, embryonic cells continue to proliferate, usually only ceasing when they enter the final stages of differentiation. And some cell types also continue dividing after differentiation. In both cases, a cell's determined or differentiated state is transmitted unchanged to their daughters over many cell generations. The question then is: how can a particular pattern of gene activity be inherited intact after DNA replication and the disruption of cell division?

One way of maintaining gene activity is to ensure the continued production of the required gene-regulatory proteins in the differentiated cell and its progeny. If the gene product itself acts as the positive regulatory protein, all that is required for gene activity to be maintained is the initial event that first activates the gene; once switched on the gene remains active. The requisite regulatory protein will be present in the cell after division and so can restart gene expression immediately (Fig. 8.7). We shall see this type of positive-feedback control of gene expression occurring in muscle-cell differentiation, where the MyoD protein acts as an activator of its own gene, *myoD*. In *Drosophila*, the selector gene *engrailed* remains active in the posterior compartment of segments throughout embryonic and larval development and into the adult (see Chapter 2). Its expression is at least partly maintained by a similar positive feedback of the Engrailed transcription factor on its own gene, long after the pair-rule transcription factors that initiated expression of *engrailed* in the embryo have disappeared.

Another way of maintaining a particular pattern of active and inactive genes and transmitting it to daughter cells is through the maintenance of chromatin structure. Evidence that changes in the packing state of chromatin can be inherited through many cell divisions and can keep genes inactive over a long period was first provided by the phenomenon of X-chromosome inactivation in female mammals. One of the two X chromosomes in each somatic cell in females becomes highly condensed and inactive early in embryogenesis and is maintained in this state throughout the individual's lifetime. In this case the silencing mechanism primarily involves the production of a non-coding RNA that coats the chromosome (described in Section 10.17). The inactive and active X chromosomes also have different DNA methylation patterns. Inactivation is not irreversible, however, as the inactive X is reactivated in the

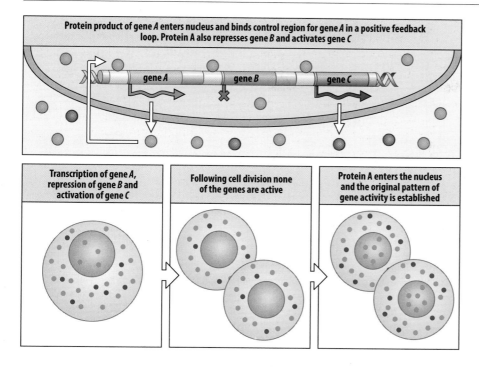

Transcription of gene A, repression of gene B and activation of gene C

Following cell division none of the genes are active

Protein A enters the nucleus and the original pattern of gene activity is established

Fig. 8.7 Continued expression of gene regulatory proteins can maintain a stable pattern of gene activity in a differentiated cell and its progeny. Top panel: transcription factor A is produced by gene A, and acts as a positive gene-regulatory protein for its own control region. Once activated, therefore, gene A remains switched on and the cell always contains A. Transcription factor A also acts on the control regions of genes B and C to repress and activate them, respectively, setting up a cell-specific pattern of gene expression. Bottom panels: after cell division, the cytoplasm of both daughter cells contains sufficient amounts of protein A to reactivate gene A, and thus maintain the pattern of expression of genes B and C.

germline during the formation of egg cells, which each end up containing a single X chromosome.

Both the pattern of DNA methylation and the pattern of histone modifications on chromatin can be transmitted to daughter cells. In the case of DNA methylation, DNA replication produces a double-stranded DNA with the original methylation pattern on the parental strand. DNA methylase then recognizes the methylated cytosines in CpG sequences and methylates the corresponding cytosine on the newly synthesized strand to restore the pattern on that strand (Fig. 8.8).

Patterns of histone modification are thought to be inherited according to a similar principle. When the DNA is replicated, nucleosomes are displaced, and new nucleosomes then reassemble on the two new DNA molecules. Some of the 'parental' marked histones are retained at the DNA during replication and are distributed at random to the two daughter DNA molecules. They reassemble with newly synthesized unmarked histones to form new nucleosomes. As noted in Section 8.2, chromatin states can be self-propagating because modified histones recruit the corresponding chromatin-modifying enzymes to the chromatin. This is thought to be the way that the original state of a chromatin region is reconstructed after chromosome replication.

8.4 Changes in patterns of gene activity during differentiation can be triggered by extracellular signals

As we have seen in previous chapters, the effect of developmentally important cell–cell signaling is almost always to induce a change in gene expression in the responding cell. We have already encountered extracellular signalling proteins such as Wnt, Hedgehog, TGF-β, FGF, and Notch, which bind to receptors in the cell membrane; the signal is then relayed to the cell nucleus by intracellular signaling pathways (see Box 4B (Wnt), Box 2F (Hedgehog), Box 4C (TGF-β), Box 4E (FGF), and Box 5D (Notch)). All these signaling pathways end in the activation or inactivation of a gene-regulatory protein (Fig. 8.9).

Some signal molecules do not act at cell-surface receptors. Retinoic acid (see Box 5C) and the steroid hormones testosterone and estrogen all act as important

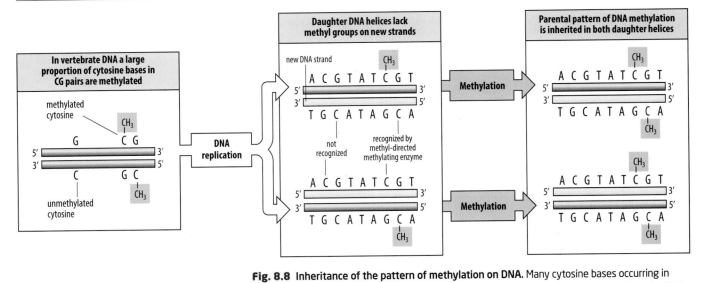

Fig. 8.8 Inheritance of the pattern of methylation on DNA. Many cytosine bases occurring in cytosine-guanine (CG) pairs in vertebrate DNA are methylated (a CH$_3$ group is added). When the DNA is replicated, the new complementary strand lacks methylation but the pattern can be reinstated on the new strand by methylase enzymes that recognize the methylated cytosine of the CG pair on the old DNA strand and methylate the cytosine of the corresponding CG pair on the new strand.

Illustration after Alberts, B., et al.: Molecular Biology of the Cell, 2nd edition. New York: Garland Publishing, 1989.

Fig. 8.9 Extracellular signals trigger changes in gene expression. Extracellular signals act through various signal effectors—the receptors and intracellular signaling proteins that transduce the signal within the cell to eventually affect gene expression. These effectors can influence various inputs to gene expression, including the expression or activation of transcription factors, effects on the transcription initiation complex, and effects on the epigenetic factors controlling gene expression, such as DNA methylases and the enzymes that add or remove chemical markers to histone proteins (see Box 8A).

signals for differentiation. Retinoic acid is involved, for example, in the differentiation of neurons in the embryonic central nervous system (discussed in Chapter 12), whereas testosterone, produced in the testis, is responsible for the secondary sexual characteristics that make male mammals different from females (discussed in Chapter 10). Non-protein, lipid-soluble signal molecules such as these cross the plasma membrane unaided. Once inside the cell, they bind to receptor proteins, and the complex of signal protein and receptor is then able to act as a transcriptional regulator, binding directly to control sites in the DNA, known as response elements, to activate (or in some cases repress) transcription. In insects, the steroid hormone ecdysone is responsible for metamorphosis (discussed in Chapter 13); it acts through similar intracellular receptors and induces the differentiation of a wide range of cell types. In all these cases, the hormone coordinates the expression of a whole set of genes, turning some on and some off, by acting, via its intracellular receptors, at a control site that is present in all these genes.

Estrogen is involved in a classic example of tissue-specific gene expression, the induction of chicken oviduct cells to produce the protein ovalbumin, a major component of egg white. The continued presence of estrogen is required for transcription of the ovalbumin gene; when estrogen is withdrawn, *ovalbumin* mRNA and protein disappear. The circulating estrogen activates expression of the *ovalbumin* gene only in the cells of the oviduct, whereas the *ovalbumin* gene in other cells, such as those of the liver, is unaffected. Indeed, the expression of the *ovalbumin* gene is among the most restricted known, but the precise mechanisms that lead to such a tissue-specific response are still elusive after decades of research. There is evidence that estrogen affects the chromatin structure of the chromosomal region containing the *ovalbumin* gene in oviduct cells but not in others, thus allowing the hormone–receptor complexes access to the regulatory regions of the gene only in oviduct cells. This tissue-specific effect of estrogen implies the existence of other tissue-specific activators, repressors, and co-regulators that allow *ovalbumin* gene expression in the oviduct when stimulated with estrogen and prevent its expression in all other tissues.

SUMMARY

Transcriptional control is a key feature of cell differentiation. Tissue-specific expression of a eukaryotic gene depends on control sites located in the regulatory regions flanking the gene. These sites comprise both the promoter sequences adjacent to the start site of transcription, which bind RNA polymerase, and more distant sites that control tissue-specific or developmental-stage-specific gene expression. The combination of different regulatory proteins bound to sites in the regulatory regions determines whether a gene is active or not, and so even small differences between two genes in the number and type of control sites in their regulatory regions can produce quite different patterns of expression. In some cases, cell-specific expression is due to the correct combination of regulatory proteins only being present in that cell type; in other cases, gene expression may be prevented by the gene being packed away in a state in which it is inaccessible to transcription factors and RNA polymerase. Gene-regulatory proteins interact not only with DNA but also with each other, forming multi-component transcription complexes that are responsible for initiating transcription by RNA polymerase. Extracellular signals, such as protein growth factors, do not themselves enter the cell, but can specifically influence gene expression by acting at cell-surface receptors to stimulate intracellular signal transduction pathways. In contrast, signal molecules that can diffuse through cell membranes, such as retinoic acid or steroid hormones, form a complex with an intracellular receptor protein that then acts as a transcription factor to influence gene expression.

Once established, the pattern of gene activity in a differentiated cell can be maintained over long periods of time and is transmitted to the cell's progeny. Mechanisms that maintain the pattern include the continued action and production of gene regulatory proteins, and long-term localized changes in the structure and properties of chromatin that are due to chemical modifications to DNA and to histone proteins.

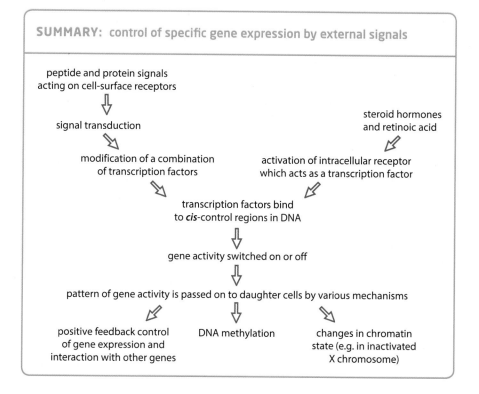

SUMMARY: control of specific gene expression by external signals

peptide and protein signals
acting on cell-surface receptors

signal transduction

steroid hormones
and retinoic acid

modification of a combination
of transcription factors

activation of intracellular receptor
which acts as a transcription factor

transcription factors bind
to *cis*-control regions in DNA

gene activity switched on or off

pattern of gene activity is passed on to daughter cells by various mechanisms

positive feedback control
of gene expression and
interaction with other genes

DNA methylation

changes in chromatin
state (e.g. in inactivated
X chromosome)

Models of cell differentiation and stem cells

In this part of the chapter we will first look at some particularly well-studied examples of mammalian cell differentiation, which between them illustrate many of the principles underlying differentiation. We start with the molecular basis of the differentiation of a single cell type—mammalian striated skeletal muscle—from undifferentiated embryonic mesodermal cells in the somites (see Section 5.13). But cellular differentiation is not confined to the embryo. Most differentiated tissues do not last forever, and most can be renewed from undifferentiated stem cells, specific to each tissue. Among the best-studied examples of cell differentiation and the properties of adult stem cells are the blood, skin, and the gut epithelial lining, which are subject to continual wear and tear and cell loss, and are continually being renewed throughout life. We will then return to skeletal muscle to see how it is repaired when necessary in adulthood from a muscle-specific stem-cell population set aside during embryonic development. We shall also see that even the central nervous system has a limited capacity for self-renewal via stem cells. Finally, we consider embryonic stem cells, and what makes them pluripotent.

8.5 Muscle differentiation is determined by the MyoD family of transcription factors

In Chapter 5 we considered the signals involved in specifying cells in the somites of vertebrate embryos to become muscle or cartilage. Here we focus on the differentiation of those cells that will become skeletal muscle; differentiation of cardiac muscle from cells of the lateral plate mesoderm involves a different set of transcription factors. Differentiation of vertebrate skeletal striated muscle can be studied in cell culture and provides a valuable model system for cell differentiation.

Skeletal muscle cells derive from the myotome of the somites. Somitic cells become committed to form muscle by the activity of the homeodomain transcription factors Pax3 and Pax7. These committed, but as yet undifferentiated, cells are called **myoblasts**, and can be isolated from chick or mouse embryos and grown in culture. Mouse myoblasts can be cloned to produce cell cultures derived from a single cell. Myoblasts will continue to proliferate until growth factors are removed from the culture medium, whereupon proliferation ceases and overt differentiation into muscle cells begins. This behavior illustrates the key importance of withdrawal from the cell cycle in promoting the final stages of differentiation. The cells then begin to synthesize muscle-specific proteins such as actin, myosin II, and tropomyosin, which are part of the contractile apparatus, and muscle-specific enzymes such as creatine phosphate kinase.

Myoblasts also undergo structural change during differentiation. They first become bipolar in shape as a result of reorganization of the microtubule cytoskeleton, and then fuse to form multinucleate **myotubes** (Fig. 8.10). Fusion *in vivo* requires the

Fig. 8.10 Differentiation of striated muscle in culture. Myoblasts are cells that are committed to becoming muscle but have yet to show signs of differentiation. In the presence of growth factors they continue to multiply but do not differentiate. When growth factors are removed the myoblasts stop dividing; they align and fuse into multinucleate myotubes, which develop into muscle fibers that contract spontaneously.

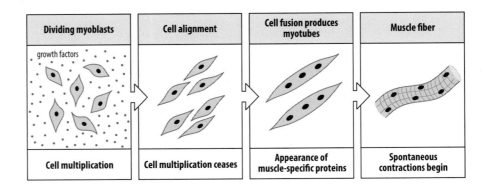

Dividing myoblasts	Cell alignment	Cell fusion produces myotubes	Muscle fiber
growth factors			
Cell multiplication	**Cell multiplication ceases**	**Appearance of muscle-specific proteins**	**Spontaneous contractions begin**

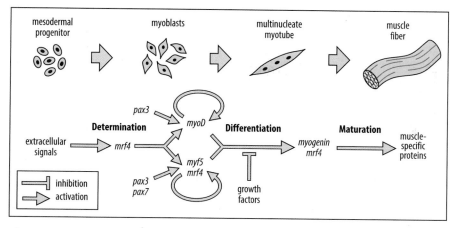

Fig. 8.11 Key features of the differentiation of vertebrate skeletal muscle. Extracellular signals lead to activation of the genes *mrf4*, *myf5*, and *myoD* and the initiation of muscle differentiation. One of these genes tends to be expressed preferentially (depending on the species) and their activity is self-sustaining. The transcription factor products of these genes, Mrf4, Myf5, and MyoD, activate muscle-differentiation genes such as *myogenin*, which in turn activate expression of muscle-specific proteins, such as the proteins that make up the contractile apparatus.

presence of integrin β_1 on the cell membrane. Within about 20 hours of withdrawal of growth factors, typical striated **muscle fibers** can be seen.

MyoD is a member of a family of basic helix-loop-helix transcription factors that are expressed only in muscle precursors and muscle cells. The proteins in this family can be considered as key controlling transcription factors for muscle differentiation, as they switch on muscle-specific genes and lead to differentiation. As well as the *myoD* gene, the family includes three other genes—*mrf4*, *myf5*, and *myogenin*. If transfected into fibroblasts and some other non-muscle cells, which do not normally express either these genes or muscle-specific structural proteins, they can all induce differentiation into muscle cells. *mrf4*, *myf5*, and *myoD* are the first genes to be switched on in muscle precursors in mammals and they act as muscle-determination genes (Fig. 8.11). They are also activated by the transcription factor Pax3, whose expression becomes restricted to muscle-cell precursors (see Section 5.13). Once turned on, these factors maintain their own expression by means of positive-feedback loops of the type illustrated in Fig. 8.11. The Mrf4, Myf5, and MyoD proteins can all activate the *myogenin* gene, which is involved in muscle structural differentiation and functional maturation. Mrf4, MyoD and Myf5 are expressed in proliferating, undifferentiated myogenic cells, whereas myogenin is only expressed in non-dividing, differentiating cells; Mrf4 also has a second role in muscle fiber maturation.

Despite the demonstrated muscle-determining effects of MyoD, gene knock-out experiments in transgenic mice (see Section 3.10 and Box 3D) show that mice lacking *myoD* make apparently normal skeletal muscle if *myf5* is active, suggesting redundancy in the system: that is, one protein can functionally cover for the absence of another. Apparent redundancy is not uncommon in development, and has probably evolved to serve as a back-up mechanism to ensure the continuation of development in the face of fluctuations in gene expression (see Section 1.19). Mice lacking *myf5* alone also make skeletal muscle, although they have other defects; the most obvious abnormality is a shortening of the ribs. Mice in which both *myf5* and *myoD* are lacking, however, die before birth, although if such embryos still express *mrf4*, they make some muscle.

Other types of experiments have confirmed the muscle-determining effects of these transcription factors. Individual prospective muscle cells can be identified experimentally by making mice carrying mutant versions of *myf5* and *myoD* that become activated as normal (so that the mRNA can be detected) but which do not make functional Myf5 or MyoD protein. These cells do not become myoblasts; they do not end up in their normal location, and may be integrated into other differentiation pathways, such as those that end in cartilage and bone.

In contrast to what happens when the *myoD* gene is knocked-out, mice in which the *myogenin* gene has been knocked out form myoblasts, but most skeletal muscle is absent. Myogenin is thus considered to be a myogenic **differentiation factor** rather than a protein essential for muscle determination.

The MyoD family of transcription factors activates transcription of muscle-specific genes by binding to a nucleotide sequence called the E-box, which is present in the regulatory regions of these genes. MyoD forms a heterodimer with products of the E2.2 gene such as E_{12}, and this binds to the E-box. As the activity of MyoD and its relatives is essential for muscle-cell differentiation, we now look at how the activity of MyoD could be regulated.

8.6 The differentiation of muscle cells involves withdrawal from the cell cycle, but is reversible

The proliferation and differentiation of myoblasts are mutually exclusive activities. Skeletal myoblasts that are multiplying in culture do not differentiate. Only when proliferation ceases does differentiation begin. In the presence of protein growth factors, myoblasts that are expressing both MyoD and Myf5 proteins continue to proliferate and do not differentiate into muscle. This means that the presence of MyoD and Myf5 is not in itself sufficient for muscle differentiation, and that an additional signal or signals are required. In culture, this stimulus to differentiate can be supplied by the removal of growth factors from the medium. The myoblasts then withdraw from the cell cycle, cell fusion occurs, and differentiation takes place.

There is an intimate relationship between the proteins that control progression through the cell cycle and the muscle determination and differentiation factors. MyoD and Myf5 are phosphorylated by cyclin-dependent protein kinases, which become activated at key points in the cell cycle (see Fig. 13.4). Phosphorylation affects the activity of MyoD and Myf5 by making them more likely to be degraded. Thus, in actively proliferating cells, the levels of the myogenic factors will be kept low. A number of other proteins also interact with myogenic factors or with their active partners such as E_{12}, and inhibit their activity. One interacting protein is the transcription factor Id, which is present at high concentrations in proliferating cells, and which probably helps to prevent premature activation of muscle-specific genes.

The myogenic factors themselves actively interfere with the cell cycle to cause cell-cycle arrest. Myogenin and MyoD activate transcription of members of the p21 family of proteins, which block cell-cycle progression, causing the cell to withdraw from the cell cycle and to differentiate. Another key protein in preventing progression through the cell cycle is the retinoblastoma protein RB. RB is phosphorylated in proliferating cells by the cyclin-dependent kinase 4 (Cdk4), and dephosphorylation of RB is essential both for withdrawal from the cell cycle and for muscle differentiation. MyoD can interact strongly and specifically with Cdk4 to inhibit RB phosphorylation, which would also help to cause cell-cycle arrest.

A more general mechanism for shutting down genes involved in cell proliferation has also been found to operate in muscle cells. Rather unexpectedly, this involves changing components of the core transcriptional machinery. One of the general

Fig. 8.12 The cessation of myoblast proliferation and the onset of muscle-cell differentiation is accompanied by a change in the core transcription complex. In myoblasts, as in most other cells studied, the core transcription complex contains the general transcription factor TFIID, which recognizes the core promoter element in gene promoters. In differentiating myotubes, TFIID is dismantled and is thought to be replaced by a complex containing the proteins TRF3 and TAF3. Along with specific activators, this complex recognizes the *myogenin* gene promoter and initiates transcription.

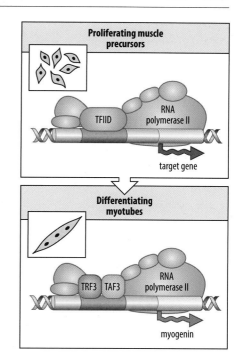

transcription factors in the basic promoter–recognition complex described in Fig. 8.4 is TFIID, which is a multisubunit protein that recognizes the core promoter element. TFIID is used in the promoter-recognition complex that controls the expression of genes involved in progression through the cell cycle and cell division. When the myoblasts undergo terminal differentiation, however, TFIID is dismantled and is replaced by a different protein complex, which controls the expression of differentiation genes such as *myogenin* (Fig. 8.12). This switch selectively turns on one transcriptional program in the cell while effectively shutting down another, and could be a general mechanism for shutting down cell proliferation and initiating differentiation.

The opposition between cell proliferation and terminal differentiation makes developmental sense. In those tissues where fully differentiated cells do not divide further, it is essential that sufficient cells are produced to make a functional structure, such as a muscle, before differentiation begins. The reliance on external signals to induce differentiation, or to permit a cell already poised for differentiation to start differentiating, is a way of ensuring that differentiation only occurs in the appropriate conditions.

Despite the apparently terminal nature of muscle-cell differentiation, muscle cells provide an example of a differentiated cell that can undergo **dedifferentiation**—the loss of differentiated characteristics—re-enter the cell cycle, and even give rise to other cell types when manipulated in culture, a phenomenon called **transdifferentiation**. The gene *Msx1* encodes a homeodomain-containing transcriptional repressor, which is expressed in determined but undifferentiated myoblasts during limb development and in other undifferentiated cells at the tip of the limb bud. When mouse myotubes in culture were transfected with *Msx1*, 5–10% of the myotubes dedifferentiated, cleaving to form both smaller multinucleate myotubes and, more significantly, mononuclear cells that could undergo cell division. When the dividing mononuclear cells were cultured in the appropriate conditions, some progeny of a single myotube expressed markers for other cell types, such as cartilage or fat cells. It was thought that a similar transdifferentiation of muscle cells to produce other cell types might occur in the regeneration of amphibian limbs, but this has been shown not to be the case in the axolotl (see Chapter 13).

Once formed, skeletal muscle cells can enlarge by cell growth but do not divide. They can, however, be replaced, as we shall see later, from specialized muscle stem cells.

8.7 All blood cells are derived from multipotent stem cells

Hematopoiesis, or blood formation, is a particularly well-studied example of differentiation, partly because hematopoietic cells at all stages of differentiation are relatively accessible in adult animals and can be grown in culture, and partly because of its medical importance. The process is continuous: in order to keep roughly 25 trillion (25×10^{12}) blood cells circulating in our bodies, most of which are red blood cells, we make 2 million new blood cells per second. In hematopoiesis, we are looking at the differentiation of a **multipotent** stem cell—a stem cell that can give rise to a large but limited range of differentiated cell types (see Section 1.17). We first look at the general process of hematopoiesis and at the external signals that influence it, and then consider the control of gene expression in one cell type—the developing red blood cell.

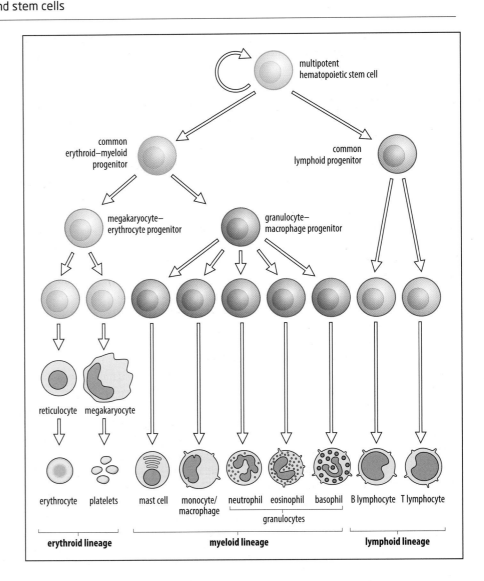

Fig. 8.13 Blood cells derive from multipotent hematopoietic stem cells. Multipotent stem cells (the hematopoietic stem cells) in the bone marrow give rise to all cells that circulate in the blood, as well as to some cells that reside in tissues. As well as renewing themselves, the hematopoietic stem cells give rise to progenitor cells with more restricted potential, which in turn give rise to the committed precursors of the three blood-cell lineages. The erythroid lineage produces erythrocytes and megakaryocytes. The myeloid lineage produces the monocytes and granulocytes that circulate in the blood, as well as the macrophages that differentiate from monocytes in tissues, tissue mast cells, the dendritic cells of the immune system, and osteoclasts (cells that remodel bone) (not shown on this simplified diagram). The lymphoid lineage produces the T lymphocytes and B lymphocytes and the natural killer cells (not shown here).

All the blood cells in an adult mammal originate from **hematopoietic stem cells** located in the bone marrow. Hematopoietic stem cells originate from mesoderm in the yolk-sac blood islands of the early embryo; later, self-renewing hematopoietic stem cells are found in a region of the aorta and then in a number of definitive blood-forming sites, including the fetal liver and adult bone marrow. Hematopoietic stem cells are both self-renewing and give rise to progenitor cells that act as small amplifying compartments and become committed to one of the hematopoietic lineages. Thus, hematopoiesis is, in effect, a complete developmental system in miniature, in which a single cell—the multipotent stem cell—gives rise to numerous different cell types. There is a continual turnover of blood cells, so that hematopoiesis must continue throughout life. As a measure of its complexity, hematopoietic stem cells from fetal liver have been found to express at least 200 transcription factors, a similar number of membrane-associated proteins, and around 150 signaling molecules.

The cells circulating in the mammalian bloodstream comprise numerous fully differentiated cell types, together with immature cells at various stages of differentiation (Fig. 8.13). The cell types are derived from three main lineages—the **erythroid**, **lymphoid** and **myeloid**, lineages. The first yields the red blood cells, or erythrocytes, and the megakaryocytes, which give rise to blood platelets.

The lymphoid lineage produces the lymphocytes, which include the two antigen-specific cell types of the immune system, the B and T lymphocytes. In mammals, B lymphocytes develop in the bone marrow, whereas T lymphocytes develop in the thymus, from precursors that originate from the multipotent stem cells in the bone marrow and migrate from the bloodstream into the thymus. Both B and T lymphocytes undergo a further terminal differentiation after they encounter antigen. Terminally differentiated B lymphocytes become antibody-secreting plasma cells, whereas T cells undergo terminal differentiation into at least six functionally distinct effector T cells. The B and T lymphocytes of the immune system are the only known examples in vertebrates in which the DNA in the differentiated cell becomes irreversibly altered, in a somatic recombination process that potentially enables billions of different antigen receptors to be generated.

The myeloid lineage gives rise to the rest of the white blood cells, or leukocytes. These include the eosinophils, neutrophils, and basophils (collectively known as granulocytes or polymorphonuclear leukocytes), mast cells, and monocytes. Monocytes differentiate into macrophages after they have left the bone marrow and entered tissues. Mast cells also reside in tissues.

Within the bone marrow, the different types of blood cells and their precursors are intimately mixed up with connective-tissue cells—the **bone marrow stromal cells**. The ability of stem cells both to self-renew and to differentiate into different cell types is tightly regulated by the cells and molecules in their immediate environment—the so-called 'niche' that they inhabit and on which they depend for their survival and function. In the case of hematopoiesis, this **stem-cell niche** is provided by the bone marrow stromal cells. Secreted signal proteins of the Wnt and BMP families, cell-surface Notch ligands (which stimulate the Notch pathway in responding cells on cell–cell contact), retinoic acid, and prostaglandin E_2 have been implicated in maintaining proliferation and self-renewal of hematopoietic stem cells in the stem-cell niche of the bone marrow.

The multipotent hematopoietic stem cell has not been directly observed, but its existence can be inferred from the ability of bone marrow to reconstitute a complete blood and immune system when transplanted into individuals whose own bone marrow has been destroyed. Experiments have shown that this ability resides in single cells and it is exploited therapeutically in the use of bone-marrow transplantation to treat diseases of the blood and the immune system. The key experiment that showed this capacity, some 60 years ago, was the transfusion of a suspension of bone marrow cells into a mouse that had received a potentially lethal dose of X-irradiation. The mouse would normally have died as a result of lack of blood cells, which, like other proliferating cells, are particularly sensitive to radiation, but transfusion of bone marrow cells reconstituted the hematopoietic system and the mouse recovered. Additional sources of hematopoietic stem cells such as umbilical cord blood are now also routinely used in transplantation.

The multipotent stem cell generates uncommitted progenitor cells that become irreversibly committed to one or other of the three main lineages, after which they undergo further rounds of proliferation and further commitment. There is early commitment to either the myeloid or the lymphoid lineage, followed by later commitment to individual lineages and overt differentiation into recognizable, functionally distinct cell types. The hematopoietic system can thus be considered as a hierarchical system with the multipotent hematopoietic stem cell at the top (see Fig. 8.13). All stages of the differentiation process are regulated by extracellular signals in the form of hematopoietic growth factors. For those blood cells differentiating in the bone marrow, most of the signal molecules are produced by the bone marrow stromal cells. In this way, the numbers of the different types of blood cells can be regulated according to the individual's physiological need. Loss of blood results in an increase in red cell production, for example, whereas infection leads to an increase in the production of lymphocytes and other white blood cells. Several types of leukemia are caused by white blood cells continuing to proliferate instead of differentiating. These cells

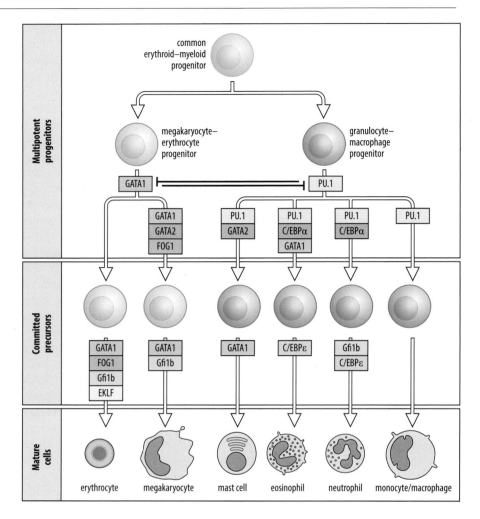

Fig. 8.14 Hierarchical requirements for the lineage-specific differentiation of cell types within the hematopoietic lineages. At the top of the hierarchy, the common erythroid-myeloid progenitor cells choose between the megakaryocyte-erythrocyte and the granulocyte-macrophage lineages, and they do this by the levels of GATA1 and PU.1 that arise in individual cells. Progenitor cells express low levels of these factors, which repress each other at the same time as they reinforce their own expression, and this leads to the stabilization of expression of just one of them in different cells. GATA1 and PU.1 then drive sequential programmes of differentiation as indicated, through combinatorial transcriptional codes. Note that the same transcription factor can contribute to different cell types through being a member of different transcription factor combinations.

Adapted from Orkin, S.H., Zon L.I.: **Hematopoiesis An evolving paradigm for stem cell biology.** *Cell, 2008, **132**: 631-644.*

become stuck at a particular immature stage in their normal development, a stage that can be identified by the molecules expressed on their cell surfaces.

A central question in considering the behavior of stem cells is how a single stem cell can divide to produce two daughter cells one of which remains a stem cell while the other gives rise to a lineage of differentiating cells. One possibility is that there is an intrinsic difference between the two daughter cells because the stem-cell division is an asymmetric division that results in the two cells acquiring a different complement of proteins. A clear example of this type of division is neuroblast development in *Drosophila*, which is considered in Chapter 12. The second possibility is that external signals make the daughter cells different: a daughter cell that remains in the stem-cell niche continues to renew itself, whereas one that ends up outside the niche differentiates. In the case of hematopoiesis, both mechanisms may operate. On the one hand, four proteins have been found to segregate asymmetrically at cell division when human hematopoietic cells are cultured without the bone marrow stroma that provides the stem-cell niche *in vivo*. On the other hand, factors produced by bone marrow stromal cells that control stem-cell renewal have also been identified.

8.8 Intrinsic and extrinsic changes control differentiation of the hematopoietic lineages

Hematopoiesis can be characterized by the activities of a hierarchy of transcription factors, reflecting the hierarchical organization of the different cell types and

Some hematopoietic growth factors and their target cells	
Type of growth factor	Responding hematopoietic cells
Erythropoietin (EPO)	Erythroid progenitors
Granulocyte colony-stimulating factor (G-CSF)	Granulocytes, neutrophils
Granulocyte-macrophage colony-stimulating factor (GM-CSF)	Granulocytes, macrophages
Interleukin-3	Multipotent precursor cells
Stem cell factor (SCF)	Stem cells
Macrophage colony-stimulating factor (M-CSF)	Macrophages, granulocytes

Fig. 8.15 Hematopoietic growth factors and their target cells.

whose overlapping patterns of expression specify the various cell lineages. Some are only expressed in immature cells and are not lineage specific. Somewhat surprisingly, hematopoietic precursors express low levels of some of the genes that will be expressed later only in the myeloid or erythroid lineages. Thus, gene repression, as well as activation, must occur when a cell becomes committed to a particular lineage. Some of the transcription-factor combinations that are required for differentiation of particular cell types in the erythroid and myeloid lineages are illustrated in Fig. 8.14.

In blood-cell precursors, the transcription factor GATA1, for example, is necessary for the differentiation of the erythroid lineage, whereas another key transcription factor, PU.1, drives differentiation of the myeloid lineage. The relative concentration of the two proteins determines cell fate, because GATA1 and PU.1 bind to each other, thus inhibiting each other's ability to regulate the transcription of lineage-specific genes. When the concentration of GATA1 is low, PU.1 will activate the myeloid program, and when the concentration of PU.1 is low, GATA1 will activate the erythroid program. As with all cell types, it is the combination of transcription factors, and not any particular one, that is responsible for the specific pattern of gene expression that leads to differentiation into a given cell type.

How then is the activity of all these factors controlled? Signals delivered by extracellular protein growth factors and differentiation factors have a key role. Studies on blood-cell differentiation in culture have identified at least 20 extracellular proteins, known generally as **colony-stimulating factors** or **hematopoietic growth factors**, that together with some of the more general signaling molecules, can affect cell proliferation and cell differentiation at various points in hematopoiesis (Fig. 8.15). Both stimulators and inhibitors of differentiation have been identified. Not all the factors controlling blood-cell production are made by blood cells or stromal cells. The protein erythropoietin, for example, which induces the differentiation of committed precursor red blood cells, is mainly produced by the kidney in response to physiological signals that indicate red cell depletion.

Despite the well-established role of these factors in blood-cell proliferation and differentiation, it is possible that the commitment of a cell to one or other pathway in the myeloid lineage might be a chance event, with the growth factors simply promoting the survival of particular cell lineages. Evidence for this comes from experiments where the two daughters of a single early progenitor cell (which can give rise to a range of different blood-cell types) are cultured separately but under the same conditions. Both daughter cells usually give rise to the same combination of cell types, but in about 20% of cases they give rise to dissimilar combinations.

Among the plethora of growth factors, it is difficult to distinguish those that might be exerting specific effects on differentiation from those that may be required for the survival and proliferation of a lineage. However, the contrasting functions of three growth factors—**granulocyte-macrophage colony-stimulating factor (GM-CSF)**, **macrophage colony-stimulating factor (M-CSF)**, and **granulocyte colony-stimulating factor (G-CSF)**—are well established. GM-CSF is required for the development of most myeloid cells from the earliest progenitors that can be identified. In combination with G-CSF,

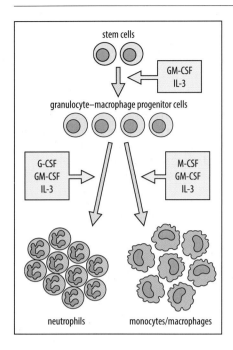

Fig. 8.16 Colony-stimulating factors can direct the differentiation of neutrophils and macrophages. Neutrophils and macrophages are derived from a common granulocyte-macrophage progenitor cell. The choice of differentiation pathway can be determined by, for example, the growth factors G-CSF and M-CSF. The growth factors GM-CSF and IL-3 are also required, in combination, to promote the survival and proliferation of cells of the myeloid lineage generally.

Illustration after Metcalf, D.: **Control of granulocytes and macrophages: molecular, cellular, and clinical aspects**. Science *1991,* ***254:*** *529–533.*

however, it tends to stimulate only granulocyte (principally neutrophil) formation from the common granulocyte–macrophage progenitor. This is in contrast to M-CSF, which, in combination with GM-CSF, tends to stimulate differentiation of monocytes (which will ultimately become macrophages) from the same progenitors (Fig. 8.16).

These growth and differentiation factors have no strict specificity of activity, in the sense of each factor exerting a specific effect on one type of target cell only; rather, they act in different combinations on target cells to produce different results, the outcome also depending on the developmental history of their target cell, which determines which proteins are already bound to DNA and which chromatin modifications are present.

8.9 Developmentally regulated globin gene expression is controlled by regulatory sequences far distant from the coding regions

We now focus on one type of blood cell, the red blood cell or erythrocyte, to consider in more detail how the production of one cell-specific protein, the oxygen-carrying hemoglobin, is regulated within a differentiating lineage. The main feature of red blood cell differentiation is the synthesis of large amounts of hemoglobin. All the hemoglobin contained in a fully differentiated red blood cell is produced before its terminal differentiation. Terminally differentiated mammalian red blood cells lose their nuclei, whereas those of birds retain the nucleus but it becomes transcriptionally inactive.

Vertebrate hemoglobin is a tetramer of two identical α-type and two identical β-type globin chains, and so its synthesis involves the coordinated regulation of two different sets of globin genes. The genes encoding α-globin and β-globin belong to different multigene families. Each family consists of a cluster of genes, and the two families are located on different chromosomes. In mammals, different members of each family are expressed at various stages of development so that distinct hemoglobins are produced during embryonic, fetal, and adult life. Hemoglobin switching is a mammalian adaptation to differing requirements for oxygen transport at different stages of life; human fetal hemoglobin, for example, has a higher affinity for oxygen than adult hemoglobin, and so is able to efficiently take up the oxygen that has been released by maternal hemoglobin in the placenta. In turn, the oxygen affinity of adult hemoglobin is sufficient for it to pick up oxygen in the relatively oxygen-rich environment of the lungs. Here, we look at the control of expression of the β-globin genes, as examples of genes that are not only uniquely cell specific, but are also developmentally regulated.

The human β-globin gene cluster contains five genes—ε, γ_G, γ_A, δ, and β (Fig. 8.17). These are expressed at different times during development: ε is expressed in the early embryo in the embryonic yolk sac; the two γ genes, which encode proteins differing by only one amino acid, are expressed in the fetal liver; and the δ and β genes are expressed in erythroid precursors in adult bone marrow. The protein products of all these genes combine with globins encoded by the α-globin complex to form physiologically different hemoglobins at each of these three stages of development (see Fig. 8.17).

Mutations in the globin genes are the cause of several relatively common inherited blood diseases. One of these, sickle-cell anemia, is caused by a point mutation in the

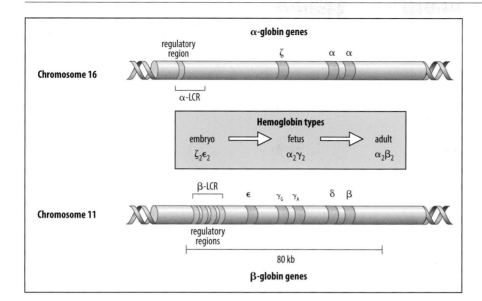

Fig. 8.17 The human globin genes. Human hemoglobin is made up of two identical α-type globin subunits and two identical β-type globin subunits, which are encoded by α-family and β-family genes on chromosomes 16 and 11, respectively. The composition of human hemoglobin changes during development. In the embryo it is made up of ζ (α-type) and ε (β-type) subunits; in the fetal liver, of α and γ subunits; and in the adult most of the hemoglobin is made up of α and β subunits, but a small proportion is made up of α and δ subunits. The regulatory regions α-LCR and β-LCR are locus control regions involved in switching on the hemoglobin genes at different stages.

gene coding for β-globin. In people homozygous for the mutation, the abnormal hemo-globin molecules aggregate into fibers, forcing the cells into the characteristic sickle shape. These cells cannot pass easily through fine blood vessels and tend to block them, causing many of the symptoms of the disease. They also have a much shorter lifetime than normal red cells. Together these effects cause anemia—a lack of functional red blood cells. Sickle-cell anemia is one of the few genetic diseases where the link between a mutation and the subsequent developmental effects on health is fully understood.

The control regions that regulate gene expression are often very complex and exten-sive, and we highlight this here with the example of the β-globin gene cluster. Each gene in the cluster has a promoter and control sites immediately upstream (to the 5′ side) of the transcription start point, and there is also an enhancer downstream (to the 3′ side) of the β-globin gene, which is the last gene in the cluster (Fig. 8.18). β-globin expression is primarily driven by the transcription factor GATA1, which has numerous binding sites in the control regions of the gene. But these local control sequences, which contain binding sites for transcription factors specific for erythroid cells, as well as for other more widespread transcriptional activators, are not sufficient to provide properly regulated expression of the β-globin genes. As in other instances, there is a need for a combination of inputs.

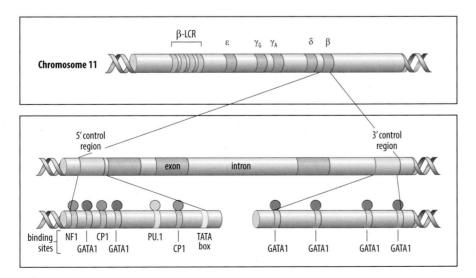

Fig. 8.18 The control regions of the β-globin gene. The β-globin gene is part of a complex of other β-type globin genes and is only expressed in the adult. Binding sites for the relatively erythroid-specific transcription factor GATA1, as well as for other non-tissue-specific transcription factors, such as NF1 and CP1, are located in the control regions. Upstream of the whole β-globin cluster, in the locus control region (LCR), are additional control regions required for high-level expression and full developmental regulation of the β-globin genes.

Fig. 8.19 A possible mechanism for the successive activation of β-family globin genes by the LCR during development. The LCR is thought to make contact with the promoter of each gene in succession, at different stages of development, thus controlling their temporal expression.

Illustration after Crossley, M., Orkin, S.H.: ***Regulation of the β-globin locus.*** Curr. Opin. Genet. Dev. *1993,* **3***: 232-237.*

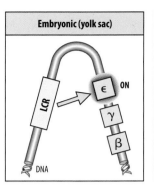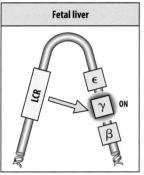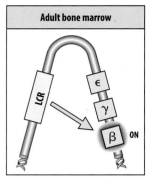

The successive switching of different globin genes on and off during development is an intriguing feature of globin gene expression. This developmentally regulated expression of the β-globin gene cluster depends on a region a considerable distance upstream from the ε gene. This is the **locus control region** (LCR; see Fig. 8.18), which stretches over some 10,000 base pairs at around 5000 base pairs from the 5′ end of the ε gene. The LCR confers high levels of expression on any β-family gene linked to it, and is also involved in directing the developmentally correct sequence of expression of the whole β-globin gene cluster, even though the β-globin gene itself, for example, is around 50,000 base pairs away from the extreme 5′ end of the LCR. A similar control region is located upstream of the α-globin gene cluster (see Fig. 8.17).

An attractive model for the control of globin-gene switching envisages an interaction of LCR-bound proteins with proteins bound to the promoters of successive globin genes (Fig. 8.19). The DNA between the LCR region and the globin genes is thought to loop in such a way that proteins binding to the LCR can physically interact with proteins bound to the globin gene promoters. Thus, in erythrocyte precursors in the embryonic yolk sac, the LCR would interact with the ε promoter; in fetal liver it would interact with the two γ promoters; and in the adult bone marrow with the β-gene promoter. One proposed mechanism is that the β-globin LCR enhances transcription by delivering preassembled core transcription initiation complexes to the appropriate gene, which are then activated at the gene and start elongation of the RNA transcript.

8.10 The epidermis of adult mammalian skin is continually being replaced by derivatives of stem cells

The epidermis of the skin is in constant contact with the harsh external environment and is subject to the continual loss of large numbers of cells from its surface. Like the blood, it needs to be constantly replaced. This task is carried out by tissue-specific stem cells, which are embedded at the base of the epidermis and provide a constant supply of new cells.

Mammalian skin is composed of two cellular layers—the **dermis**, which mainly contains fibroblasts, and the protective outer **epidermis**, which contains mainly keratin-filled cells called **keratinocytes**. We are concerned here with the epidermis, which is renewed from stem cells. Damage to the dermis is repaired by proliferation of fibroblasts and regrowth of damaged blood vessels. A **basal lamina** or **basement membrane** composed of extracellular matrix separates the epithelial epidermis from the dermis. The epidermis comprises a multilayered epithelium of keratinocytes that provides a tough, watertight and resilient barrier, protecting internal tissues from the outside environment (Fig. 8.20). Associated with this epithelium are evenly spaced hair follicles and sebaceous and sweat glands. The epidermis between the hairs is known as the **interfollicular epidermis**. Cells are being

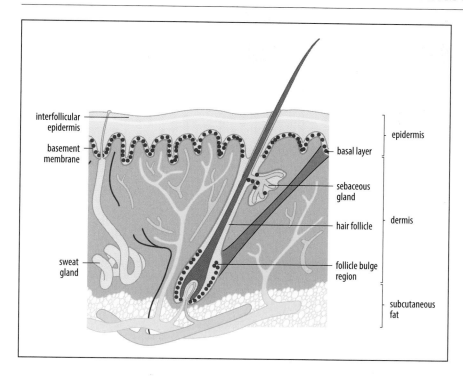

Fig. 8.20 Section through human skin showing epidermal structures. The skin is composed of an outer epidermis (pale orange) separated from the underlying dermis by a basement membrane. Structures that are epidermal in origin are shown in color here; dermal structures are in gray. The locations of stem cells are shown in red. In the case of injury, the stem cells at the base of the hair follicles and in the sebaceous gland can migrate to contribute to a new basal layer.

continuously lost from outer surface of the epidermis and must be replaced—every four weeks we have a brand new epidermis. New cells of the interfollicular epidermis are produced throughout life by the differentiation of stem cells that reside in the **basal layer** of the epidermis in contact with the basement membrane. Each hair has its own system of support in the form of stem cells at the base of each follicle (see Fig. 8.20).

Basal-layer cells depend on stimuli received from fibroblasts in the dermis and the extracellular molecules of the basement membrane to remain proliferative. The stem-cell niche for epidermal stem cells therefore comprises the basement membrane and dermis. Once a cell leaves this stem-cell niche, it becomes committed to differentiate. After becoming committed, the epidermal cells differentiate further after leaving the basal layer. One terminal differentiation pathway culminates in the production of keratinocytes, which form the outermost, cornified epidermal layers that provide the skin's protective covering (Fig. 8.21). In the interfollicular epidermis, differentiating keratinocytes move through the skin as they mature, until by the time they reach the outermost layer they are completely filled with the fibrous protein keratin and their membranes are strengthened with the protein involucrin. The dead cells eventually flake off the surface (see Fig. 8.21). The terminally differentiated cells of the sebaceous glands, by contrast, are lipid-filled sebocytes that eventually burst, releasing their contents onto the surface of the skin. Differentiation of a hair follicle is even more complex, comprising eight different cell lineages.

In living epidermal cells, keratins are an essential cytoskeletal component, forming the intermediate filaments through which epidermal cells are attached to each other, and giving skin its resilience and elasticity. Basal layer cells express keratin types 5 and 14, and mutations in the genes encoding these keratins are found in patients with one form of epidermolysis bullosa, a disease in which the epidermis is not anchored properly to the basement membrane and so is very fragile and easily blisters. Spinous cells expressing keratins 1 and 10 are located in the suprabasal layer of the epidermis and the switch from expression of keratin types 5 and 14 to expression of keratins 1 and 10 is a reliable indicator that cells have become committed to terminal differentiation (see Fig. 8.21).

Fig. 8.21 Differentiation of keratinocytes in human epidermis. Descendants of basal-layer cells, which will become the keratinocytes of the epidermis, detach from the basement membrane, divide several times, and leave the basal layer before beginning to differentiate. Basal-layer cells express the intermediate filament proteins keratin 14 and keratin 5, and also the integrin $\alpha_6\beta_4$, required for hemidesmosome formation and adhesion to the basement membrane. Differentiation into keratinocytes involves a loss of integrin expression, a change in the keratin genes expressed and the production of large amounts of keratin 1 and keratin 10. In the intermediate layers, the cells are still large and metabolically active, whereas in the outer layers of the epidermis, the cells lose their nuclei, become filled with keratin filaments, and their membranes become insoluble owing to deposition of the protein involucrin. The dead cells are eventually shed from the skin surface.

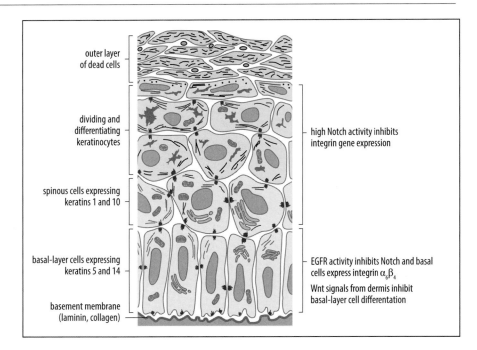

outer layer of dead cells

dividing and differentiating keratinocytes

spinous cells expressing keratins 1 and 10

basal-layer cells expressing keratins 5 and 14

basement membrane (laminin, collagen)

high Notch activity inhibits integrin gene expression

EGFR activity inhibits Notch and basal cells express integrin $\alpha_6\beta_4$

Wnt signals from dermis inhibit basal-layer cell differentation

Several signals are now known to regulate this switch. Activation of Notch signaling in basal-layer cells promotes differentiation, and is inhibited by various mechanisms, including signaling through the epidermal growth factor receptor (EGFR), which, as a consequence, contributes to maintaining basal-cell fate. Wnt produced by the dermis and acting on the basal-layer cells also contributes to inhibiting differentiation (see Fig. 8.21). Cell-adhesion molecules also have a key role. The basal-cell layer is attached to the basement membrane by specialized cell junctions that include integrin-containing hemidesmosomes and focal contacts (see Box 9A). Basal cells express the integrin $\alpha_6\beta_4$, which is required for hemidesmosome formation and interacts with collagen and laminin in the extracellular matrix. Mutations in the collagen associated with epidermal hemidesmosomes have been found in a subset of patients with epidermolysis bullosa. They also express high levels of integrin $\alpha_3\beta_1$, which is a receptor for laminin. Detachment of a cell from the basement membrane is accompanied by a decrease in the amount of integrins on the cell surface, which is a result of activation of Notch.

The interfollicular epidermis can be also generated from the hair follicle, where stem cells have been located in a region known as the bulge (see Fig. 8.20). Lineage-tracing experiments in transgenic mice in which, for example, green fluorescent protein (GFP) is expressed under the control of the promoter for the keratin 5 gene, showed that the cells that give rise to new hair follicles originate in the bulge, and under certain conditions, usually as a result of injury, bulge cells can also contribute to sebaceous gland and epidermis. The descendants of single follicle-bulge cells from mouse epidermis cultured in the laboratory can give rise to hair follicles, sebaceous glands, and epidermis when combined with dermal fibroblasts and used as skin grafts. These experiments show that within mouse epidermis there are different populations of stem cells that can be brought into play when required. They have also highlighted several modes of stem-cell division, which we discuss next.

8.11 Stem cells use different modes of division to maintain tissues

The precise identification of the stem cells in the basal layer of the epidermis, and whether all the cells within the basal layer can, under certain circumstances such

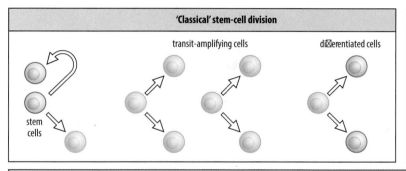

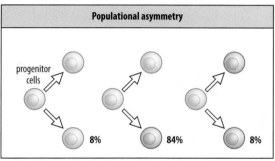

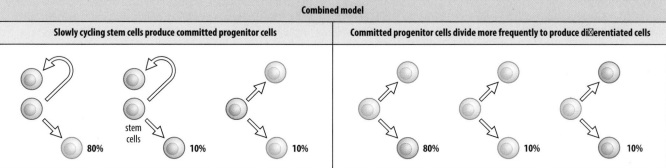

Fig. 8.22 Different modes of stem-cell division contribute to tissue homeostasis. In the 'classical' model of stem-cell division, the stem cell (red) always divides asymmetrically to give rise to another stem cell and a 'transit-amplifying cell' cell (orange) that is committed to divide several times and then differentiate (blue). In the populational asymmetry model, a progenitor cell (green) can divide in the three modes as indicated. In the combined model, very slowly dividing stem cells (undergoing only a few divisions per year) produce progenitor cells which then divide much more rapidly following the rules of populational asymmetry to renew the epidermis. To achieve tissue homeostasis, the proportions of each type of division must be correctly balanced; in particular, the symmetric divisions must occur in equal proportions. Thus, if a stem cell (S) can divide in any of three ways to give S + S, D + D or S + D, where D is a cell that will go on to differentiate, then as long as the number of S + S divisions is equal to the number of D + D divisions, the system will remain at a constant size. The percentages shown on the figure represent the likelihood that this particular division will occur within the starting population of cells.

*Source: Mascré, G.,et al.: **Distinct contribution of stem and progenitor cells to epidermal maintenance.** Nature 2012, **489**: 257-262.*

as injury, act as stem cells, is still not resolved. Basal epidermal cells are heterogeneous, with some cells proliferating rapidly and others dividing far less frequently. It has been suggested that the slowly cycling cells are the stem cells and that the surrounding proliferating basal cells, known as **transit-amplifying cells** or sometimes as progenitor cells, are committed to terminal differentiation. These observations fit the classical model of stem-cell division, which is one of invariant asymmetric cell division. Each stem-cell division results in one stem cell and one cell committed to differentiate, which then undergoes several rounds of amplifying divisions before differentiating (Fig. 8.22).

However, other investigations tracking the fate of individual marked basal cells observed continuous proliferation of these cells in certain parts of the epidermis, which suggested that all basal cells could be equivalent in their developmental potential—that is, that they were all acting as progenitor cells rather than classical stem cells. These observations led to additional models of basal cell division in which the cells can divide symmetrically or asymmetrically, and the proportions of the different types of divisions keeps the balance between stem/progenitor and differentiated cells and maintains tissue homeostasis (see Fig. 8.22). According to the **populational asymmetry** model, basal-layer cells are progenitor cells that can divide in any of three ways: symmetrically to give rise to two cells that remain progenitor cells and do not differentiate; symmetrically to give rise to two cells that go on to differentiate; or asymmetrically to give one of each type (as in the classical model) (see

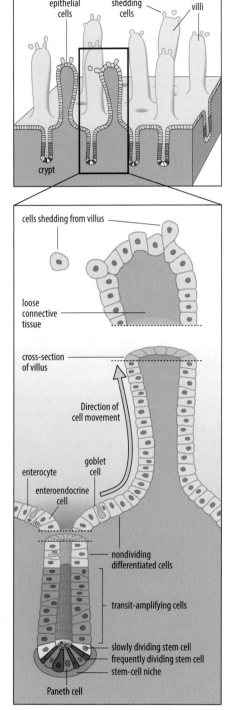

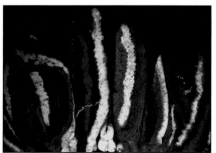

Fig. 8.23 Epithelial cells lining the mammalian gut are continually replaced. The upper panel shows the villi, which are epithelial extensions covering the wall of the small intestine. The lower panel is a detail of a crypt and villus. The stem cells from which the intestinal epithelium is renewed are found in the base of the crypts. One type of stem cell (yellow) is characterized by a slow rate of division, and is found at some distance from the bottom of the crypt, while a second type of stem cell (red) is interspersed between the Paneth cells in the base of the crypt. These stem cells divide more frequently than the first type and are the ones mostly involved in routinely renewing the epithelium. The stem cells give rise to cells that proliferate and move upward. By the time they have left the crypt they have differentiated into epithelial cell types. They continue to move upwards on the projecting villus, and are finally shed from the tip. The total time of transit is about 4 days. The photograph shows a section through mouse small intestine. Individual stem cells at the base of each crypt were initially labeled in different colors. As the crypts were renewed, each was, at first, a mixture of clones of all the stem cells (not shown). But over time, competition between the different clones (see Fig. 8.22) results in each crypt gradually becoming populated by the descendants of a single stem cell, as shown here.

From Snippert, H.J., et al.: **Intestinal crypt homeostasis results from neutral competition between symmetrically dividing Lgr5 stem cells.** Cell *2010,* **143***: 134–144.*

Fig. 8.22). In this mode of tissue renewal there is no need for an amplifying phase, as the progenitor cells are dividing continuously. And so long as there is a correct balance between the three kinds of divisions, the tissue is maintained. The populational asymmetry model, however, does not account for the presence of the slowly cycling cells in the epidermal basal layer. Analysis of the abundant experimental and observational data on epidermal renewal suggests that a model in which classical stem cells (the slowly cycling cells) produce progenitor cells that then proliferate and differentiate according to the rules of populational asymmetry best fits the observed dynamics of epidermal renewal over time.

8.12 The lining of the gut is another epithelial tissue that requires continuous renewal

Like the outer layer of the epidermis, the epithelium lining the gut is exposed to a relatively harsh environment—in this instance the lumen of the gut and its contents—and undergoes continuous cell loss. It is also replaced from stem cells, and there are considerable similarities to epidermal renewal, especially in the signaling molecules used to regulate stem-cell differentiation. The stem cells in the gut have also been observed to divide by all the modes of division outlined in Section 8.11.

In the small intestine, the endoderm-derived epithelial cells form a single-layered epithelium that is folded to form villi that project into the gut lumen and crypts that penetrate the underlying mesoderm-derived connective tissue. Cells are continuously shed from the tips of the villi, while the stem cells that produce their replacements lie in the base of the crypts. The new cells generated by the stem cells move upward toward the tips of the villi. *En route,* they multiply in the lower half of the crypt and constitute a transit-amplifying population (Fig. 8.23). The turnover of cells in the mammalian small intestine is about 4 days.

A single crypt in the small intestine of the mouse contains about 250 cells comprising four main differentiated cell types. Three are secretory cells: Paneth cells secrete antimicrobial peptides, goblet cells secrete mucus, and enteroendrocine cells secrete various peptide hormones. The fourth cell type, the enterocyte, is the predominant cell in the epithelium of the small intestine and absorbs nutrients from the gut contents. About 150 of the crypt cells are proliferative, dividing about twice a day so that the crypt produces 300 new cells each day; about 12 cells move from the crypt into the villus each hour.

The base of each crypt has a very stereotyped structure and contains two types of stem cell. One type, originally identified by its slow rate of division, is located a short distance above the bottom of the crypt, at the edge of the stem-cell niche. The stem-cell niche is composed of epithelial cells in the crypt and the mesenchymal

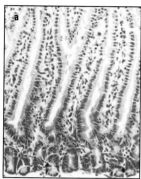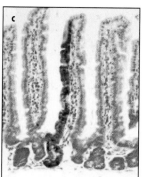

Fig. 8.24 Cell lineage tracing of *Lgr5*-expressing cells in the mouse small intestine. Transgenic mice which expressed LacZ from the *Lgr5* promoter were used to trace the origin and movement of *Lgr*-expressing cells. The photos show histological analysis of LacZ activity (blue-staining cells) in small intestine: (a) 1 day after induction; (b) 5 days after induction; and (c) 60 days after induction.

*From Barker, N., et al.: **Identification of stem cells in small intestine and colon by marker gene Lgr5.** Nature 2007, 449: 1003–1007.*

cells surrounding the base of the crypt. The other stem cells, which are considerably more proliferative, are slender columnar cells—between four to six cells per crypt, that are interspersed between Paneth cells at the bottom of the crypt (see Fig. 8.23). Both sets of stem cells have distinctive patterns of gene expression. The slowly dividing cells express high levels of Bmi1, a component of the Polycomb complex of chromatin-binding proteins, and long-term lineage tracing of cells expressing *Bmi1* in transgenic mice using LacZ as a marker resulted in the labeling of entire crypts and villi after 12 months, showing that the *Bmi1*-expressing cells are multipotent stem cells. The other population of stem cells expresses a Wnt target, the gene encoding Lgr5, a G-protein-coupled receptor whose ligand is still unknown. Experiments in which *Lgr5*-expressing cells and their progeny were marked irreversibly by expression of LacZ showed that a single one of these columnar cells could give rise to entire crypts and to ribbons of cells on adjacent villi after 2 months (Fig. 8.24). Thus, like the slowly cycling cells, the *Lgr5*-expressing columnar cells are also multipotent stem cells. It is not yet clear whether these two stem-cell populations serve different functions, nor how they are related in the maintenance of the lining of the gut. Evidence suggests that one is involved in normal tissue homeostasis while the other provides a reserve of stem cells for regeneration, but it has also been shown that they are interconvertible—that is, each can serve the other function if necessary. If all *Lgr5*-positive cells are eliminated, the *Bmi1*-expressing cells regenerate this population and contribute to the maintenance of the crypt.

The multipotency of the *Lgr5*-expressing stem cells was confirmed by isolating them and growing them in culture. When cultured under the right conditions, a single *Lgr5*-expressing cell can give rise to organized intestinal villus-crypt tissue containing all four differentiated cell types.

Notch signaling stimulated by cells of the stem-cell niche in the crypt of the small intestine is involved in specifying the fate of the proliferating cells as either absorptive or secretory cells. Under conditions in which Notch signaling is reduced, proliferative crypt cells are completely converted to goblet cells, whereas when Notch signaling is activated, there is increased proliferation of stem cells and a severe reduction in differentiated secretory cells.

The maintenance of stem cells in both the epidermis and the small intestine is under the control of the same signals. In both instances, Wnt signaling through β-catenin promotes stem-cell renewal whereas BMP inhibits it. Populations of stem cells at particular sites in the tissue may be kept in a quiescent state by the action of BMP and inhibition of Wnt signaling, while other populations are stimulated to divide by Wnt signaling in the presence of BMP antagonists (Fig. 8.25). The differentiation of the stem cells in both systems requires Notch signaling. This interplay between Wnt, BMP, and Notch signaling pathways in the maintenance and differentiation of stem-cell pools is common to other types of adult stem cells as well. Their importance in the maintenance and differentiation of stem-cell populations is

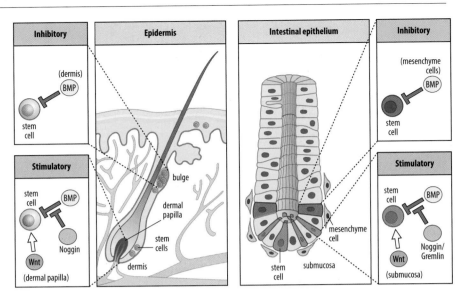

Fig. 8.25 Signaling in the stem-cell niche. Some signaling pathways perform conserved functions in different stem-cell populations. For example, Wnt promotes stem-cell division and BMP inhibits it in both the skin and the intestine. Stem-cell populations in different locations within the tissue are kept quiescent by the action of BMP and the inhibition of Wnt signaling, or are activated by the action of Wnt and the inhibition of BMP signaling by the BMP antagonists Noggin or Gremlin.

From Li, L., Clevers, H.: **Coexistence of quiescent and active adult stem cells in mammals.** Science 2010, **327**: 542–545.

highlighted by the incidence of mutations in elements of these pathways in several types of cancers.

8.13 Skeletal muscle and neural cells can be renewed from stem cells in adults

Muscles are also subject to injury and are often in need of repair, but these structurally highly differentiated cells do not divide. Instead, mammals have a reserve of muscle stem cells, called **satellite cells**, that are associated with mature skeletal muscle fibers, and that become activated in response to muscle injury. When satellite cells are activated, they proliferate and develop into myoblasts, which then differentiate into new muscle fibers (see Section 8.5).

Satellite cells emerge during the differentiation of skeletal muscle in the embryo and generate a potential reservoir of muscle tissue. Set aside from the myoblast pool before the final stages of differentiation, they are small mononuclear cells that lie between the basement membrane and the plasma membrane of mature muscle fibers, and can be recognized by their expression of characteristic proteins, such as the cell-surface protein CD34 and the transcription factor Pax7 (Fig. 8.26). Under favorable conditions, the satellite cells associated with one muscle fiber (about 7 satellite cells), grafted into an irradiated muscle in a mouse, can generate more than 100 new fibers and will also undergo self-renewal. The irradiation kills the mouse's own satellite cells, thus making it impossible that the muscle could have regenerated on its own. Even a single muscle stem cell transplanted into irradiated mouse muscle can produce both muscle fibers and more satellite cells.

The function of Pax7 seems to overlap with that of the closely related Pax3. As discussed earlier, Pax3 is essential for early determination of muscle-cell precursors. In contrast, Pax7 is required for the maintenance of the satellite-cell population at the time of birth and is expressed in adult muscle. If, however, the *pax7* gene is deleted in mice 2–3 weeks after birth, limb skeletal muscle growth and regeneration is, surprisingly, unaffected and the satellite cells still make new muscle even though they no longer express the gene.

It is becoming clear that the existence of stem cells to support the normal life of an adult tissue, as in the skin and intestine, or as a reservoir in case of injury, as in muscle, is very widespread, and that just about every tissue and organ in the body has a population of cells which, upon need, can be activated to repair damage. This is also true, although to a more limited extent, of the nervous system. We will see in

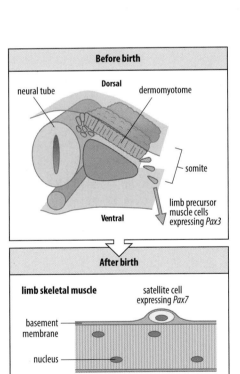

Fig. 8.26 Muscle satellite cells. During embryonic development, *Pax3*-expressing muscle cell precursors leave the dermomyotome and migrate into the limbs. In the limb muscle after birth, muscle stem cells remain associated with the muscle as satellite cells but now express *Pax7*.

Fig. 8.27 Sites of neurogenesis in the adult mammalian brain. Neurogenesis occurs mainly in the subgranular layer of the dentate gyrus in the hippocampus and the anterior part of the subventricular zone, which is shown in more detail in this figure. Radial glial cells act as neural stem cells, giving rise to rapidly dividing transit-amplifying cells, which differentiate into neurons. New neuronal cells generated in the dentate gyrus migrate to the granular layer, whereas new neuronal cells generated in the subventricular zone migrate to the olfactory bulb and differentiate there into interneurons. Endothelial cells of nearby blood vessels also provide signals for stem-cell differentiation (not shown).

Illustration after Taupin, P.: **Adult neurogenesis in the mammalian central nervous system: functionality and potential clinical interest.** *Med. Sci. Monit. 2005,* **11***: 247-252.*

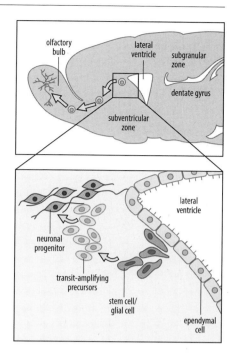

Chapter 12 how neurons and other nervous-system cells develop from embryonic neural stem cells. Once fully differentiated, however, neurons in the mammalian brain do not divide, and for many years it was thought that no new brain neurons at all were produced in the adult mammal, although adult neurogenesis was known to occur in the brains of songbirds, reptiles, amphibians and fish. Relatively recently, however, the production of new neurons, or **neurogenesis**, has been demonstrated as a normal occurrence in the adult mammalian brain, and neural stem cells have been identified in adult mammals that can generate neurons, astrocytes, and oligodendrocytes in culture.

These adult neural stem cells are a population of slowly dividing radial glial cells, like those in the embryonic cortex, that produce neurons—at least in the adult rodent. Neurogenesis in adult mammals occurs in just two regions: the subventricular zone of the lateral ventricle, and the subgranular zone of the dentate gyrus in the hippocampus. New neurons generated in the subventricular zone migrate to give rise to **interneurons** in the olfactory bulb, whereas the stem cells in the hippocampus give rise to new neurons of the dentate granule layer of the hippocampus. Slowly proliferating radial glial cells that express astrocyte-specific genes act as the adult neural stem cells. In the subventricular zone of the lateral ventricle, these stem cells give rise to proliferating transit-amplifying cells, which then mature into neural progenitors (Fig. 8.27), whereas in the hippocampus, radial glial cells can give rise to neurons directly. Other adjacent glial cells in these regions provide a stem-cell niche, producing signals that control neural stem-cell behavior. Wnt signaling through the β-catenin pathway in neural stem cells has been shown to play an important role in regulating adult neurogenesis. Notch signaling determines the number of stem cells and regulates Sonic hedgehog signaling, which promotes their survival. Other sources of signals in the adult brain include the ependymal cells lining the ventricle, the blood vessels, and blood components.

Evidence as to whether new neurons are born also in the adult human brain has remained sparse and contentious. However, confirmation that new neurons are indeed produced in the human hippocampus (in large numbers—possibly as many as 1400 a day—and throughout life) has been provided by an ingenious study that exploited the fact that the testing of atomic bombs during the 1950s and early 60s caused a spike in atmospheric ^{14}C. Starting with uptake and fixation by plants, the carbon radioisotope was incorporated into newly born neurons through the food chain, providing a 'natural' form of pulse labeling that has provided a reasonably accurate marker of neuronal birth date. The hippocampus is a part of the cortex that is crucial to the acquisition and storage of new memory, and it is tempting to ascribe a related functional significance to the persistence of neurogenesis in this brain region.

8.14 Embryonic stem cells can proliferate and differentiate into many cell types in culture and contribute to normal development *in vivo*

The multipotent stem cells that are responsible for tissue renewal in adults are limited in the range of different cell types they give rise to, and require a specific stem-cell niche to maintain their stem-cell state, such as the bone-marrow stroma that supports hematopoietic stem cells (see Section 8.7) or the basement membrane and the underlying dermis in the case of epidermal stem cells (see Fig. 8.21). By contrast, pluripotent early embryonic cells can differentiate into cell types of all three germ layers and do not require a special niche.

The pluripotent stem cells in mammalian embryos are the **embryonic stem cells (ES cells)**, which can be obtained from the inner cell mass of the blastocyst. Mouse ES cells have been studied intensively in culture. They are clonally derived and can be maintained in culture for long periods, apparently indefinitely, but if injected into a blastocyst that is then returned to the uterus, they can contribute to all the types of cells in that embryo, but not to extraembryonic structures. Such experiments have shown that just two or three ES cells can make a whole embryo.

A useful property of mouse ES cells is that they can be grown in culture to provide large numbers of pluripotent stem cells of known genetic constitution for further experiments. For example, ES cells derived from mouse models of human diseases can be differentiated into the cell types affected in the disease and the cells studied and experimented on.

The maintenance of mouse ES cells in culture requires serum as well as leukemia inhibitory factor (LIF), which inhibits differentiation (Box 8B). A search for the specific component in serum that complemented LIF in the maintenance of pluripotency yielded BMP-4. Studies of the requirements for pluripotency revealed a small set of transcription factors acting downstream of LIF and BMP-4, which are essential for maintaining cultured mouse ES cells in a pluripotent state. These factors are Nanog, Oct 3/4, Sox2, Esrrb and Kfl4. The expression of some of them, in particular Nanog, is restricted to pluripotent stem cells. ES cells grown in LIF and BMP-4 appear to be poised for differentiation, and this is highlighted by the existence of both repressive and activating chromatin marks (see Box 8A) associated with the regulatory regions of many of the genes expressed in ES cells, which become resolved into cell-type-specific sets of marks as differentiation signals lead to the activation or inactivation of these genes.

The main function of the transcription factors that mediate pluripotency is to prevent differentiation, and to ensure this they maintain each other's expression and suppress the expression of differentiation genes. For this reason, loss of function of any of them does not result in the death of the cells, but in their differentiation into specific cell types. The expression of pluripotency factors fluctuates in cells in culture and perhaps for this reason, ES cell cultures are very heterogeneous in terms of gene expression and always contain a fraction of differentiating cells. These features reflect the origin of ES cells in the embryo, which also expresses the pluripotency transcription factors, but where pluripotency is a transient state. In the so-called 2i culture conditions, which suppress differentiation (see Box 8B), these heterogeneities are eliminated.

Apart from the inner mass cells, the other two lineages of the preimplantation mouse embryo, the primitive endoderm and the trophectoderm, which normally give rise to extraembryonic tissues (see Fig. 5.6), can also be converted into self-renewing stem cells in culture. They only give rise to their usual cell types on differentiation and are thus known as 'lineage-specific stem cells.'

The first human ES cells were produced in 1998. They express the same transcription factors as do mouse ES cells, but the signaling pathways that support self-renewal in culture are different. FGF and activin/Nodal signaling, but not LIF, are required for the maintenance of human ES cells, whereas BMP signaling promotes differentiation, the opposite effect to that in mouse ES cells. These differences have been resolved to some extent by the discovery that in mice, pluripotent stem cells can also be derived from the 5-day epiblast, a more advanced developmental stage than the blastocyst (see Fig. 3.23), which have the same properties as human ES cells. These cells from

EXPERIMENTAL BOX 8B The derivation and culture of mouse embryonic stem cells (ES cells)

Mouse embryonic stem (ES) cells can be obtained by culturing mouse blastocysts in a dish and selecting for cells that can grow in the presence of serum and the cytokine leukemia-inhibiting factor (LIF). After a few days, colonies emerge, derived from single cells, which can be propagated without any sign of differentiation. A more recent, and very efficient, derivation procedure replaces the serum with inhibitors of the protein kinases MAPK and GSK3 (known as the 2i condition), because signaling by MAPK and GSK3 pro-

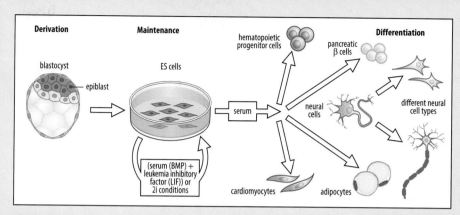

Figure 1

motes cell differentiation (Figure 1). These conditions can be used to derive ES cells from rat as well as mouse blastocysts, and the 2i condition may turn out to be a good general approach to deriving pluripotent ES cells in many animals. On removal of LIF and the signaling inhibitors, ES cells can be differentiated into many different cell types and tissues in culture by the application of different cocktails of growth factors and culture times tailored to produce specific cell fates (see Section 8.14).

To determine whether cultured ES cells are truly pluripotent and can support compete development they are injected into

host blastocysts of a genetically different mouse strain to produce a chimera. The chimeric blastocyst is replaced in a surrogate mother and its development into a fertile adult is taken as proof of pluripotency of the introduced ES cells. Another assay for pluripotency is the tetraploid chimeric assay in which the host blastocyst is tetraploid (formed from the fusion of the blastomeres of a two-cell embryo in culture). The tetraploid cells are only able to contribute to the placenta and so the embryo must derive from the ES cells only.

the 5-day mouse epiblast, called **epiblast stem cells (epiSCs)**, have many of the properties of mouse ES cells in terms of their ability to differentiate into a wide range of tissues, and the transcription factors they express, but they cannot make chimeras, which ES cells can do when injected into preimplantation blastocysts (Fig. 8.28).

Interestingly, mouse epiSCs have the same culture requirements as human ES cells, which has led to the suggestion that human ES cells more closely resemble mouse epiSCs than mouse ES cells. Several hundred human ES cell lines have now been established around the world and it would be interesting to find out if cells with all the properties shown by mouse ES cells can be derived from these cell lines, or whether the existing human ES cell lines have the same limitations as mouse epiSCs. A chimera assay cannot, of course, be performed with human ES cells.

Cultured mouse and human ES cells can be made to differentiate into particular cell types. Their differentiation follows the pattern of embryonic development and, as they exit the pluripotent state, ES cells face a choice of either a neuroectodermal or mesendodermal fate. Which is chosen depends on the growth factors in the culture medium, which can be adjusted to mimic the growth factors present in the embryo (see Chapter 4). When cultured in the absence of serum, ES cells differentiate into neural cells, and this is facilitated by the suppression of signals, such as activin/Nodal, BMP and Wnt, which promote differentiation into endoderm and into mesodermal derivatives such as heart and somatic muscle, blood islands, fat cells, macrophages, and even germ cells (see Box 8B). Of particular medical interest, neural stem cells obtained from ES cells and maintained in culture can be differentiated into particular types of neurons with therapeutic aims in mind.

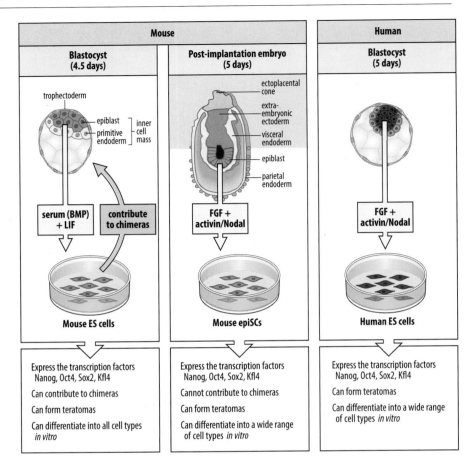

Fig. 8.28 A comparison of mouse ES cells, mouse epiblast stem cells (epiSCs) and human ES cells. Different kinds of embryonic stem cells and their properties. Human ES cells resemble more the mouse EpiSCs, and this suggests that they are in a more advanced stage of development than the mouse ES cells.

While the initial choice (neuroectoderm versus mesendoderm) is clear, differentiation into specific cell types requires precise changes in the composition of the growth medium over time. For example, aggregates of ES cells successively treated with particular combinations of FGF-2, EGF, and platelet-derived growth factor continue to proliferate as long as these growth factors are present. But when they are withdrawn, the cells differentiate into either of two glial cell types—astrocytes or oligodendrocytes.

SUMMARY

Cell differentiation is controlled by complex combinations of transcription factors, whose expression and activity are influenced and temporally guided by external signals. Some transcription factors are common to many cell types, but others have a very restricted pattern of expression. Combinations of transcription factors define individual differentiated cell types. Muscle differentiation, for example, can be brought about by the expression of transcription factors specific to the muscle lineage, such as MyoD, which initiate the differentiation program and bind to the control regions of muscle-specific genes.

Several adult tissues, such as the blood, the epidermis and the lining of the gut, are continuously being replaced by the differentiation of new cells from stem cells. Hematopoiesis illustrates how a single progenitor cell, the multipotent stem cell, can be both self-renewing and generate a range of different cell lineages. The differentiation of the various lineages of blood cells seems to depend on external signals, although there is some evidence for an intrinsic tendency to diversification. External factors are, however, essential for the survival and proliferation of the specific cell types. The derivation of blood cells from the multipotent stem cell involves a successive restriction of developmental potential as cells become committed to the different lineages.

In all these instances the adult stem cells are multipotent, not pluripotent, because they give rise only to the specialized cell types present in a given tissue. Embryonic stem cells are derived from mammalian blastocysts and have the potential to differentiate into all the cell types of an organism—that is, they are pluripotent. Their differentiation in culture can be steered in a particular way by controlling the signals that they are exposed to, and they therefore hold promise for regenerative medicine.

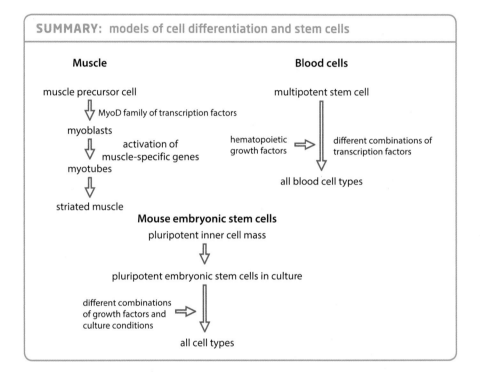

SUMMARY: models of cell differentiation and stem cells

Muscle

muscle precursor cell

⇩ MyoD family of transcription factors

myoblasts

⇩ activation of muscle-specific genes

myotubes

⇩

striated muscle

Blood cells

multipotent stem cell

hematopoietic growth factors ⇒ different combinations of transcription factors

all blood cell types

Mouse embryonic stem cells

pluripotent inner cell mass

⇩

pluripotent embryonic stem cells in culture

different combinations of growth factors and culture conditions ⇒

all cell types

The plasticity of the differentiated state

In earlier parts of this chapter we have looked at examples of genes being turned on and off during development and have considered some of the mechanisms involved. These chiefly involve the binding of different combinations of gene-regulatory proteins to the control regions of genes, and the epigenetic modifications to chromatin that silence genes in the long term. How reversible are these changes, and thus how reversible is cell differentiation? We have seen several examples of naturally multipotent adult stem cells that both self-renew and give rise to a range of different cell types. If such stem cells could be produced reliably and in sufficient numbers, it might be possible to use them to replace cells that have been damaged or lost by disease or injury. This is one of the main aims of the field of **regenerative medicine**. The therapeutic use of stem cells will depend on understanding precisely how gene activity can be controlled in stem cells to give the desired cell type, and just how plastic stem cells are. Could, for example, blood stem cells give rise to neurons under appropriate conditions? How easy is it to alter the pattern of gene activity? Is it possible to make pluripotent stem cells from differentiated cells?

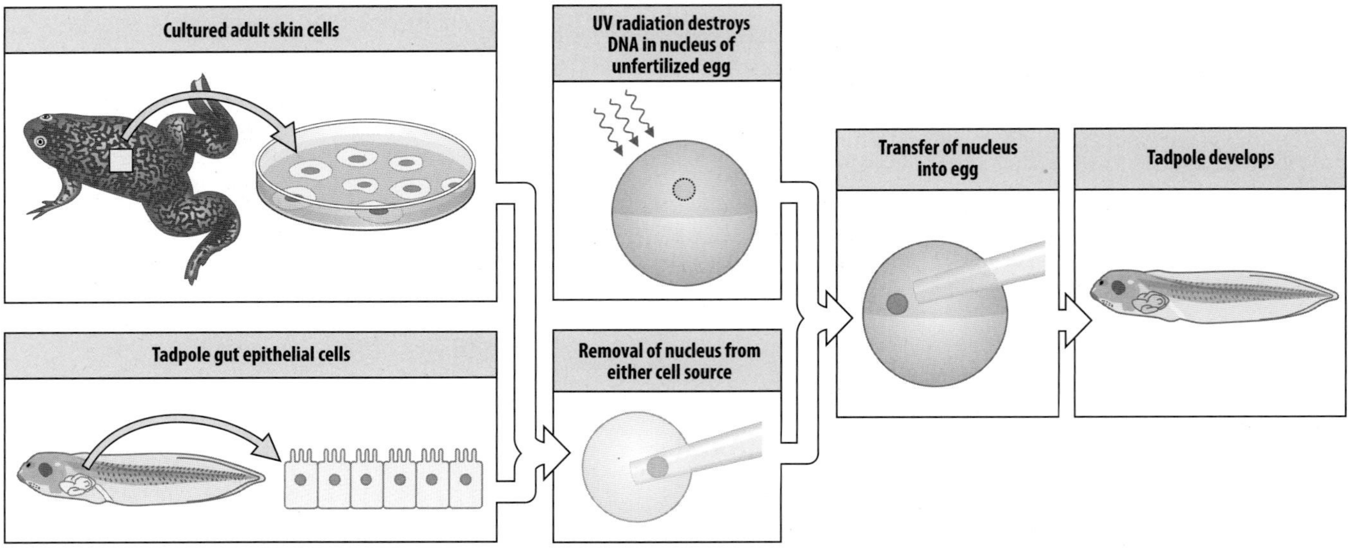

Fig. 8.29 Nuclear transplantation. The haploid nucleus of an unfertilized *Xenopus* egg is rendered non-functional by treatment with ultraviolet radiation. A diploid nucleus taken from an epithelial cell lining the tadpole gut or from a cultured adult skin cell is transferred into the enucleated egg, and can support the development of an embryo until at least the tadpole stage.

We will start the discussion of cell plasticity by examining the extent to which the pattern of gene activity in differentiated cells can revert to that present in the fertilized egg. One way of finding out whether it can be reversed in practice is to place the nucleus of a differentiated cell in a different cytoplasmic environment, one that contains a different set of gene-regulatory proteins.

8.15 Nuclei of differentiated cells can support development

The most dramatic experiments addressing the reversibility of differentiation investigated the ability of diploid nuclei from cells at different stages of development to replace the nucleus of an egg and to support normal development. If they can do this, it would indicate that no irreversible changes have occurred to the genome during differentiation. It would also show that a particular pattern of nuclear gene activity is determined by whatever transcription factors and other regulatory proteins are being synthesized in the cytoplasm of the cell. Such experiments were first carried out using the eggs of amphibians, which are large and particularly robust, and therefore very suitable for experimental manipulation.

In the unfertilized eggs of *Xenopus*, the nucleus lies directly below the surface at the animal pole. A dose of ultraviolet radiation directed at the animal pole destroys the DNA within the nucleus, and thus effectively removes all nuclear function. These enucleated eggs can then be injected with a diploid nucleus taken from a cell at a later stage of development to see whether it can function in place of the inactivated nucleus. The results are striking: nuclei from early embryos and from some types of differentiated cells of larvae and adults, such as gut and skin epithelial cells, can replace the egg nucleus and support the development of an embryo up to the tadpole stage (Fig. 8.29), and in a small number of cases even into an adult. The organisms that result are a **clone** of the animal from which the somatic cell nucleus was taken, as they have the identical genetic constitution. The process by which they are produced is called **cloning**. In 2012, the British biologist John Gurdon was awarded a share in the Nobel Prize for Physiology or Medicine for this work.

Nuclei from adult skin, kidney, heart, and lung cells, as well as from the intestinal cells and cells in the myotome of *Xenopus* tadpoles, can support development when transplanted into enucleated eggs. This technique is known as **somatic cell nuclear transfer**. However, the success rate with nuclei from somatic cells of adults is very low, with only a small percentage of nuclear transplants developing past the cleavage stage. In general, the later the developmental stage the nuclei come from, the less likely they are to be able to support development. Transplantation of nuclei taken from blastula cells is much more successful; by transplanting nuclei from the same blastula into several enucleated eggs, clones of genetically identical frogs can be obtained (Fig. 8.30). An even greater success rate is achieved if the process is repeated with nuclei from the blastula stage of a cloned embryo. These results show that, at least in those differentiated cell types tested, the genes required for development are not irreversibly altered. More importantly, when exposed to the egg's cytoplasmic factors, they behave as the genes in the zygotic nucleus of a fertilized egg would. So in this sense at least, many of the nuclei of the embryo and the adult are equivalent, and their behavior is determined entirely by factors present in the cell.

What about organisms other than *Xenopus*? The genetic equivalence of nuclei at different developmental stages has also been demonstrated in ascidians. In plants, single somatic cells can give rise to fertile adult plants, demonstrating complete reversibility of the differentiated state at the cellular level (see Fig. 7.9). The first mammal to be cloned by somatic cell nuclear transfer was a sheep, the famous Dolly. In this case, the nucleus was taken from a cell line derived from the udder. Mice have been similarly cloned using, for example, nuclei from the somatic cumulus cells that surround the recently ovulated egg (see Chapter 10), and several generations of these mice have been produced by repeating the procedure.

In general, the success rate of cloning by somatic cell nuclear transfer in mammals is extremely low, and the reasons for this are not yet well understood. Most cloned mammals derived from nuclear transplantation die before birth, and those that survive are usually abnormal in some way. The most likely cause of failure is incomplete reprogramming of the donor nucleus to remove all the epigenetic modifications, such as DNA methylation and histone modification, that are involved in determining and maintaining the differentiated cell state (see Section 8.2). In such a case the DNA would not be restored to a state resembling that of a newly fertilized oocyte.

A related cause of abnormality may be that the reprogrammed genes have not gone through the normal **imprinting** process that occurs during germ-cell development, where different genes are silenced in the male and female parents (see Section 10.8). The abnormalities in adults that do develop from cloned embryos include early death, limb deformities and hypertension in cattle, and immune impairment in mice, and all these defects are thought to be due to abnormalities of gene expression that arise from the cloning process. Dolly the sheep developed arthritis at 5 years old and was humanely put down at the age of 6 years (relatively young for a sheep) after developing a progressive lung disease, although there is no evidence that cloning was a factor in Dolly contracting the disease. Studies have shown that some 5% of the genes in cloned mice are not correctly expressed and that almost half of the imprinted genes are incorrectly expressed.

The generation of cloned blastocysts following the transfer of an adult somatic cell nucleus has more recently been achieved for primates and humans by modifying the protocols used for nuclear transfer in other animals. In both cases, nuclei from adult skin cells were transferred into enucleated oocytes, many of which then underwent cleavage, with a small number eventually forming blastocysts. The production of a cloned adult primate by nuclear transfer has not yet been accomplished, and cloning a human being by these means, even if it could be achieved, is banned in most countries on both practical and ethical grounds. Despite reports in the media that humans have been cloned, none of these reports has been verified.

One of the reasons for producing cloned blastocysts of mice and other mammals is to generate ES cells of known genetic constitution from the cells of the inner cell mass, for example, for studies of disease (see Section 8.18). ES cells have been isolated from cloned monkey blastocysts and, very recently, from human blastocysts produced by somatic nuclear transfer.

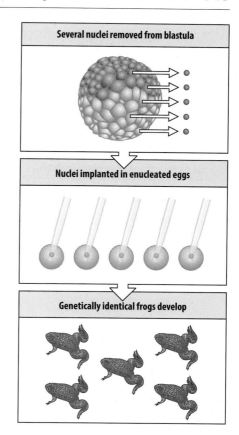

Fig. 8.30 Cloning by nuclear transfer.
Nuclei from cells of the same blastula are transferred into enucleated, unfertilized *Xenopus* eggs. The frogs that develop are a clone, as they all have the same genetic constitution.

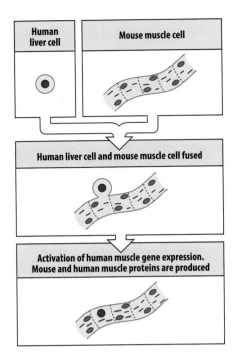

Fig. 8.31 Cell fusion shows the reversibility of gene inactivation during differentiation. A human liver cell is fused with a mouse muscle cell. The exposure of the human nucleus (red) to the mouse muscle cytoplasm results in the activation of muscle-specific genes and repression of liver-specific genes in the human nucleus. Human muscle proteins are made together with mouse muscle proteins. This shows that the inactivation of muscle-specific genes in human liver cells is not irreversible.

8.16 Patterns of gene activity in differentiated cells can be changed by cell fusion

Nuclear transplantation into eggs, particularly frog eggs, is facilitated by their size and the large amount of cytoplasm. With other cell types, particularly differentiated cells, injection of nuclei into foreign cytoplasm is not possible. It is possible, however, to expose the nucleus of one cell type to the cytoplasm of another by fusing the two cells. In this process, which can be brought about by treatment with certain chemicals or viruses, the plasma membranes fuse and nuclei from the different cells come to share a common cytoplasm. A cell-division inhibitor is then used to ensure that the two nuclei remain separate.

A striking example of the reversibility of gene activity after cell fusion is provided by the fusion of chick red blood cells with human cancer cells in culture. Unlike mammalian red blood cells, mature chick red blood cells have a nucleus. Gene activity in the nucleus is, however, completely turned off and it produces no mRNA. When a chick red blood cell is fused with a cell from a human cancer cell line, gene expression in the red blood cell nucleus is reactivated, and chick-specific proteins are expressed. This chick-specific expression not only shows that the chick genes can be reactivated, but that human cells contain cytoplasmic factors capable of doing this, yet another striking example of the conservation of crucial developmental genes within the vertebrates.

Fusion of differentiated cells with striated muscle cells from a different species provides further evidence for reversibility of gene expression in differentiated cells (Fig. 8.31). Multinucleate striated muscle cells are good partners for cell-fusion studies as they are large, and muscle-specific proteins can easily be identified. Differentiated human cells representative of each of the three germ layers have been fused with mouse multinucleate muscle cells, so that the human cell nuclei are exposed to the mouse muscle cytoplasm. This results in muscle-specific gene expression being switched on in the human nuclei. For example, human liver cell nuclei in mouse muscle cytoplasm no longer express liver-specific genes; instead, their muscle-specific genes are activated and human muscle proteins are made. Furthermore, expression of genes such as *MyoD*, which can initiate muscle differentiation, is activated, thus showing that reprogramming of the liver-cell nuclei in mouse muscle cytoplasm appears to proceed through the same steps as those taken when a muscle cell differentiates.

These results clearly show that patterns of gene expression in differentiated cells can be changed, and that gene expression can be controlled by factors present in the cytoplasm. At least some of these factors are transcription factors, as we will see below, and this leads to the conclusion that the differentiated state may be at least partly maintained by the continuous action of transcription factors on their target genes (see Section 8.3).

8.17 The differentiated state of a cell can change by transdifferentiation

A fully differentiated cell is generally stable, which is essential if it is to serve a particular function in the mature animal. If it has retained the ability to divide, the cell passes its differentiated state on to all its descendants. In some long-lived cells, such as neurons, which do not divide once they are differentiated, the differentiated state must remain stable for many years. Plant cells maintain their differentiated state while in the plant, but do not maintain it when placed in cell culture (see Chapter 7). But under certain conditions, differentiated animal cells are not stable and this provides yet another demonstration of the potential reversibility of patterns of gene activity. The change of one differentiated cell type into another is known as **transdifferentiation**. The allied phenomenon in which there is a switch of

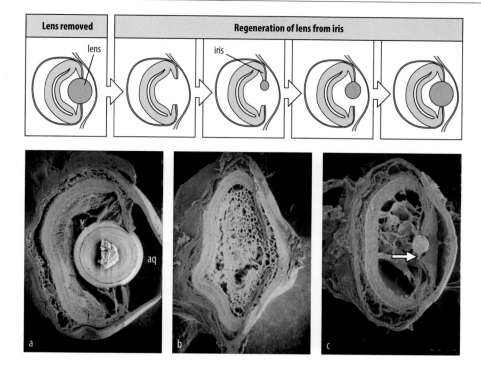

Lens removed | **Regeneration of lens from iris**

Fig. 8.32 Lens regeneration in adult newt. Top panels: Removal of the lens from the eye of a newt results in regeneration of a new lens from the dorsal pigmented epithelium of the iris. Bottom panels: sections through an intact and a regenerating newt eye viewed by scanning electron microscopy. (a) A section through an intact eye. Note that the aqueous anterior chamber (aq) is clear, whereas the vitreous chamber (v) is dominated by compacted layers of extracellular matrix (pink). Lens is gray, cornea is shown as light blue, the retina as purple and the ligaments and matrix associated with the attachment to the lens are colored brown. (b) A section through an eye 5 days post-lentectomy. The aqueous chamber has been invaded by extracellular matrix and the iris has become thickened. (c) A section through an eye 25 days post-lentectomy. The aqueous chamber is gradually clearing, the lens, which developed from the dorsal rim of the iris, is now attached to the ventral iris (arrow), and extracellular matrix in the vitreous chamber is becoming reorganized. Magnification × 50.

*Micrographs from Tsonis, P.A., et al. 2004 reproduced with permission from Int. J. Dev. Biol. 2004, **48**: 975–980.*

committed but not yet differentiated progenitor cells into a different lineage is known as **transdetermination**. The best-studied examples of transdetermination occur in regenerating *Drosophila* imaginal discs, where rare transdetermination events result in the homeotic transformation of one type of adult structure into another.

Transdifferentiation occurs very rarely as part of normal development, and in vertebrates is more commonly seen in regeneration (discussed in Chapter 13) and certain pathological conditions. There are some examples of transdifferentiation in normal invertebrate development, however. In *Caenorhabitis elegans*, in which the fate of every cell can be followed, a few cells appear to undergo transdifferentiation. An epithelial cell (known as Y), which forms part of the rectum, is generated in the embryo and has the morphological and molecular hallmarks of a differentiated epithelial cell. In the second larval stage, this cell leaves the rectum epithelium and migrates to form a **motor neuron** (known as PDA), which now possesses an axonal process and makes synaptic connections. Thus a fully differentiated epithelial cell with no detectable neuronal features undergoes transdifferentiation into a neuron with no residual epithelial characteristics. It is not yet clear whether the Y cell dedifferentiates and then redifferentiates as the PDA cell or whether dedifferentiation and redifferentiation occur at the same time.

A role for the extracellular matrix in determining the state of cell differentiation is suggested by a particularly interesting example of transdifferentiation in jellyfish striated muscle. This can be made to transdifferentiate into two different types in succession. When a small patch of striated muscle and its associated extracellular matrix (mesogloea) is cultured, the striated muscle state is maintained. However, if the cultured tissue is treated with enzymes that degrade the extracellular matrix, the cells form an aggregate, and within 1–2 days some have transdifferentiated into smooth muscle cells, which have a different cellular morphology. This is followed by the appearance of a second cell type—nerve cells. This example indicates a role for the extracellular matrix in maintaining the differentiated state of the striated muscle.

The classic example of transdifferentiation in vertebrates is lens regeneration in the adult newt eye (Fig. 8.32). When the lens is completely removed by surgery (lentectomy), a new lens vesicle starts to form around 10 days post-lentectomy from the dorsal region of the pigmented epithelium of the iris. The iris cells first dedifferentiate—they lose their pigmentation and alter their shape from a flat to a columnar epithelium—and start to proliferate. They then redifferentiate to form a new lens. Remarkably, it has been found that the ability of the iris in a newt eye to undergo transdifferentiation and regenerate the lens does not diminish with age. In a long-term project, the lens was removed 18 times from the same animals, which were over 30 years old at the time of the last operation.

The lens of the adult mammalian or avian eye cannot regenerate *in vivo*, but the pigmented epithelial cells of the embryonic chick retina can be induced to transdifferentiate in culture. A single pigmented cell from the embryonic retina can be grown in culture to produce a monolayer of pigmented cells. On further culture in the presence of hyaluronidase, serum, and phenylthiourea, the cells lose their pigment and retinal cell characteristics. If cultured at a high density with ascorbic acid, the cells then start to take on the structural characteristics of lens cells, and to produce the lens-specific protein crystallin. It should be noted that, in both the newt and the chick, transdifferentiation occurs to a developmentally related cell type. Retinal and iris pigment cells and lens cells are all derived from the ectoderm (vertebrate eye development is discussed in Chapter 11).

Transdifferentiation could also underlie well-documented examples in humans, usually associated with tissue damage and repair, in which rare differentiated cells of one type appear in an inappropriate location in another tissue. Such examples include the appearance of hepatic cells in the pancreas and vice versa, and gastric-type epithelium in the esophagus in place of stratified squamous epithelium. In the latter example, known as Barrett's metaplasia, the change predisposes the patient to developing esophageal cancer, and illustrates the clinical importance of transdifferentiation. More optimistically, it might be possible to harness the process of transdifferentiation to produce selected differentiated cell types *in situ* for regenerative medicine (discussed in Section 8.18).

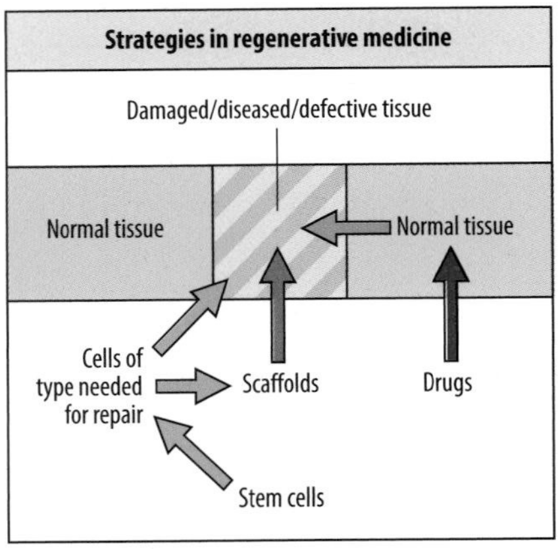

Fig. 8.33 Strategies in regenerative medicine. Damaged, diseased, or defective tissue could be repaired by cell-replacement therapies in which stem cells are the source for producing cells for transplantation, either alone or on scaffolds (blue arrows). ES cells, adult stem cells, or iPS cells could be used. An alternative strategy is to develop drugs that could be applied to nearby undamaged tissue (purple arrow) to stimulate endogenous adult stem cells and/or induce transdifferentiation in order to generate cells to replace the damaged tissue (green arrow).

8.18 Stem cells could be a key to regenerative medicine

The goal of regenerative medicine is to restore the structure and function of damaged or diseased tissues. One possible strategy employs **cell-replacement therapies**, the restoration of tissue function by the introduction of new healthy cells. As stem cells can proliferate and differentiate into a wide range of cell types, they have tremendous potential for providing the appropriate cells for making repairs. This could involve transplanting the replacement cells, either alone or seeded on a scaffold (a three-dimensional support made out of an inert material; Box 8C). An alternative strategy is to apply drugs to stimulate endogenous adult stem cells and/or to induce transdifferentiation of nearby undamaged tissue to generate replacement cells (Fig. 8.33). These types of therapy might eventually offer an alternative to conventional organ transplantation from a donor, with its attendant problems of rejection and shortage of organs, and might also be able to restore function to tissues such as brain and nerves.

Both ES cells and adult stem cells have been studied with cell-replacement therapy in mind. Some cell-therapy treatments are already being tested in patients for their safety and efficacy. Clinical trials of a potential cell-based therapy for certain types of macular degeneration, a degenerative disease of the eye that eventually leads to blindness, have been carried out recently. The main purpose was to assess the safety of the procedure. Retinal pigment epithelial cells derived from human ES cells were injected into the eyes of the patients. Transplantation of the cells did not cause any ill effects and a slight improvement in vision was even detected.

MEDICAL BOX 8C Tissue engineering using stem cells

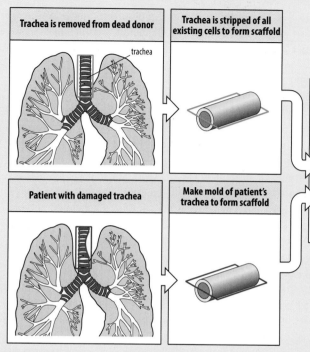

Figure 1

Regenerative medicine is the branch of medicine that aims to repair or replace damaged cells and tissues to restore normal function (see Figure 8.33). Current and prospective developments in **cell-replacement therapy** using healthy cells derived from various types of stem cells are discussed in Sections 8.17 and 8.18. Stem cells are also starting to be used to build replacement tissues and body parts for transplantation using the techniques of **tissue engineering**.

One, still highly experimental, example, is the transplantation of tissue-engineered windpipes (tracheas) to replace diseased tracheas. Although it is not yet clear how successful such operations can be at restoring completely normal tracheal function over the long term, they show how tissue engineering requires multidisciplinary teams to pool their knowledge and expertise. In the first operation of this type, to replace a trachea severely damaged by tuberculosis, an inert scaffold was made from the trachea of a person who had recently died. All cells were stripped off to leave a structure made of extracellular matrix components. More recently, a patient with an inoperable tumor of the trachea was similarly treated using a completely artificial scaffold. This was made out of a polymer molded to the size and shape of the patient's own trachea using information from magnetic resonance imaging. The scaffolds in both cases were then seeded on the inside with epithelial cells from the lining of the patient's trachea, and covered on the outside with chondrocytes that had been differentiated in culture from **mesenchymal stem cells** derived from the patient's bone marrow. The scaffolds with the attached cells were then placed in a bioreactor—a specialized culture vessel—to promote cell proliferation and differentiation. In the first procedure, the new tissue-engineered trachea was ready to be surgically implanted to replace the patient's damaged trachea after about 4 days in the bioreactor (Figure 1). In refinements to the protocol used for the second patient, signaling molecules were added to the bioreactor and treatment with signaling molecules continued after the artificial trachea was transplanted, in order to stimulate regeneration and repair from the patient's own tissues.

In both cases, function of the trachea seemed to be restored, although it is not clear exactly how the tracheal lining was re-established. A follow-up report five years after the transplant of the tissue-engineered cadaveric trachea suggests that the procedure can be effective, but that there are still many questions to be answered before such operations could become routine. It also needs to be borne in mind that the trachea has a very simple structure and it will be much more challenging to replace more complicated organs.

Induced transdifferentiation is also emerging as an alternative route to generating selected cell types, but the development of **induced pluripotent stem cells (iPS cells)** directly from differentiated adult cells offers the most exciting new opportunities, as these cells can, in principle, circumvent some of the problems associated with ES cells and adult stem cells (Box 8D).

EXPERIMENTAL BOX 8D Induced pluripotent stem cells (iPS cells)

The most dramatic example of the reprogramming of adult differentiated cells is the production of ES-cell-like cells, known as **induced pluripotent stem cells (iPS cells)** by the introduction of the genes for four transcription factors associated with pluripotency into the differentiated cell. It had previously been shown that when adult differentiated somatic cells are fused with ES cells they are transformed into pluripotent cells, suggesting that cytoplasmic factors in the ES cells might be able to confer ES-cell-like properties on differentiated cells. Nevertheless, it was surprising that so few factors were needed to induce pluripotency.

iPS cells were first derived from mice in 2006. These first iPS cells were produced from skin fibroblasts isolated from transgenic mice that carried a drug-selectable marker linked to expression of a gene associated with pluripotency, either *Oct4* or *Nanog*. The genes encoding four transcription factors associated with pluripotency in ES cells—*Oct4*, *Sox2*, *Kfl4*, and *c-Myc*—were introduced into the adult fibroblasts using retroviral vectors (Figure 1). The transfected cells were then treated with the appropriate drugs to select for the rare iPS cells. Most transfected cells did not express either *Oct4* or *Nanog* and therefore had no drug resistance and died. The remaining colonies of cells that survived resembled ES cells and could be isolated and expanded in culture. The inserted genes in the iPS cells shut themselves off once the reprogramming process was completed, and the cells reactivated their own pluripotency genes to maintain the pluripotent state. Even though *Nanog* is not part of the iPS cocktail, its expression in the final stages of the reprogramming is essential for obtaining true iPS cells. Only cells that expressed *Nanog* are truly pluripotent by the criteria of self-renewal, being able to differentiate in culture and, most significantly, to contribute to fertile chimeras (see Box 8B). The same set of transcription factor genes was very soon found to be able reprogram adult human fibroblasts, but the conditions needed to expand human iPS cell populations are different, with mouse cells requiring LIF and human cells FGF. Human iPS cells were also derived by other groups using a slightly different mix of transcription factors.

The pluripotency of mouse iPS cells was tested by injecting them into normal mouse blastocysts, which were then replaced in a surrogate mother. The iPS cells contributed to all cell types in the resulting embryos, including the germline. When these iPS chimeric mice were crossed with normal mice, iPS-derived gametes successfully participated in fertilization and live mice were born. More recently, fertile adult mice generated entirely from iPS cells have been derived from tetraploid blastocysts into which fibroblast-derived iPS cells were injected.

Advances have been made over the past few years in inducing pluripotency without the need to introduce a persisting viral vector. This will be a prerequisite for any medical use for iPS cells in cell-replacement therapy (see Section 8.18). Transposon-based vectors have been produced that snip themselves and the genes they carry cleanly out of the genome after the transformation to pluripotency is complete. Other approaches involve searching for small-molecule drugs that will act as co-activators for a cell's own pluripotency genes.

An equal effort is going into reducing the number of genes that need to be introduced. Adult neural stem cells have been reprogrammed by introducing and expressing the gene for just one of the transcription factors, Oct4. Adult neural stem cells already express high levels of Sox2 and c-Myc, however, and this may explain why introduction of Oct4 alone is sufficient to induce pluripotency.

ES cells have the advantage over adult stem cells in that they can differentiate into a wide range of different cell types and could, in theory, be used to repair any tissue. But stem cells from an unrelated donor embryo will still cause immune rejection reactions, and because of the proliferative capacity and pluripotency of ES cells, the risk of embryonal tumors—tumors derived from embryonic cells—is high. Cultured ES cells introduced into another embryo will develop normally, but when mouse ES cells are introduced under the skin of a genetically identical adult mouse they give rise to a tumor known as a **teratocarcinoma** (Fig. 8.34). These unusual tumors contain a mixture of differentiated cells.

To generate ES cells of the patient's own tissue type, it would be a great advantage to be able to make use of **therapeutic cloning** by somatic cell nuclear transfer (see Section 8.15). A nucleus from a somatic cell from the patient would be introduced into an enucleated human oocyte which would then be allowed to develop to the blastocyst stage. The blastocyst could then be used as a source of ES cells that could be maintained in

Some methods for deriving induced pluripotent stem cells (iPS cells)			
Method of delivery or activation of pluripotency genes	**Integration into genome**	**Advantages**	**Disadvantages**
Viral vectors			
Retrovirus-derived vector. Original method of Takahashi and Yamanaka (2006). Multiple viruses each carrying one transcription factor gene	Yes. Transgenes usually become silenced in the iPS cells after reprogramming and integration	Efficient and stable reprogramming	Vector and transgenes remain in genome and might be reactivated in differentiated cells
Lentivirus-derived vector; vectors often encode multiple transcription factors	Yes. Transgenes may remain active after reprogramming and integration, but can be made excisable using a Cre/lox system (see Box 3D) or silenced using a doxycyclin-inducible promoter and withdrawal of the drug		The long terminal repeats of the vector remain in the genome and might activate oncogenes if they integrate near by
Retroviral and lentiviral vectors are widely used to generate iPS cells for research purposes, but because of the persistence of the vector and transgenes in the genome they cannot be used to generate iPS cells for potential clinical use			
DNA plasmids and transposons			
Modified *piggybac* transposon + plasmid carrying PB transposase	Vector integrates but is excised by the transposase	Efficient reprogramming. Differentiated cells transgene-free and vector free	Some cells might still carry small pieces of integrated DNA
Linear DNA plasmid with *loxP* sites	Vector integrates but is excised in the presence of Cre recombinase	Differentiated cells transgene-free and vector free	Less efficient than viral vectors
Non-integrative plasmids	No. Plasmids diluted out as iPS cells divide		
Non-DNA methods			
Transfection with synthetic RNAs (RNA-induced iPS cells (RiPS cells))	Differentiated cells free of any exogenous DNA		Repeated transfections required
Transcription factor proteins			Slow and less efficient than viral vectors
Activation of endogenous pluripotency genes by treatment with small-molecule drugs (chemically induced iPS cells (CiPS cells))			Relatively new method: complete reprogramming using only chemical inducers first reported in 2013. Slow

Figure 1

culture, and which would not cause immune rejection when implanted into the patient. Human blastocysts have now been made by somatic cell nuclear transfer and ES cells have been derived from them, but the success rate of the procedure is extremely low.

There are claimed to be ethical issues associated with the use of ES cells and therapeutic cloning. To make human ES cells, the blastocyst must be destroyed, and there are those who believe that this is a destruction of a human life. There is good evidence that the blastocyst does not necessarily represent an individual at this very early stage, because it is still capable of giving rise to twins at a later stage. And in practice, many early embryos are lost during assisted reproduction involving *in vitro* fertilization (IVF), a widely accepted medical intervention. Acceptance of IVF and rejection of the use of ES cells could be seen as a contradiction. New alternative approaches, such as transdifferentiation and the use of iPS cells, could produce cells of a patient's own tissue type but avoid the ethical issues associated with therapeutic cloning.

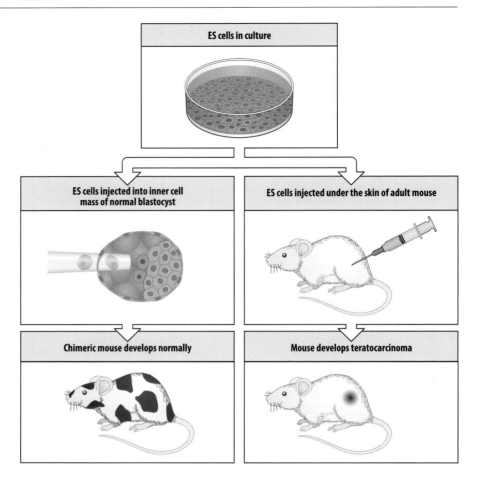

Fig. 8.34 **Embryonic stem cells (ES cells)** **can develop normally or into a tumor,** **depending on the environmental signals** **they receive.** Cultured ES cells originally obtained from an inner cell mass of a mouse contribute to a healthy chimeric mouse when injected into an early embryo (bottom left panels), but if injected under the skin of an adult mouse, the same cells will develop into a teratocarcinoma (bottom right panels).

Adult stem cells have also been considered for use in cell-replacement therapy. Adult hematopoietic stem-cell transplants in the form of bone marrow transplants from donors have been successfully used for many years as therapy for certain immunological diseases and cancers, despite associated problems such as graft-versus-host reactions. The advantage of adult stem cells in certain other conditions is that they could be taken from the patient's own undamaged tissues and re-implanted where needed, without any problems of tissue rejection. This is already done in the case of skin grafts, which use the patient's own skin. But the range of cell types and tissues that could be repaired using a patient's own stem cells is limited. The isolation of stem cells from the brain, for example, would not be feasible. There have been reports over the years from experiments in mice that adult hematopoietic stem cells will differentiate into cell types quite different from their normal range when transplanted into another type of tissue. However, it was subsequently shown that hematopoietic stem cells had not been able to give rise to cells outside their normal repertoire, but instead had fused with existing cells.

8.19 Various approaches can be used to generate differentiated cells for cell-replacement therapies

One of the challenges facing cell-replacement therapy is devising differentiation protocols that will generate the particular cell types needed for repair, whether from ES cells, adult stem cells, iPS cells, or by transdifferentiation of pre-existing cell types. The generation of insulin-producing pancreatic β cells to replace those destroyed in type 1 diabetes is a prime medical target, and we shall use it to illustrate the experimental approaches that are being developed for cell-replacement therapy generally.

In type 1 diabetes, the insulin-producing β cells of the endocrine pancreas (the pancreatic islets) are destroyed by an autoimmune reaction, resulting in non-production of insulin and the potentially lethal build-up of sugar in the blood. Unlike blood or muscle, the pancreas does not contain dedicated stem cells from which it can regenerate, and people with type 1 diabetes face a lifetime of dependence on insulin injections. Pancreas or islet-cell transplants from cadaver donors have had some success, but tissue for transplantation is scarce and treatment with powerful immunosuppressant drugs is necessary to prevent rejection. Transplantation of insulin-producing pancreatic cells derived from the patient's own cells, along with the induction of immune tolerance to these cells, might eventually provide one long-term solution.

Treatments that direct the differentiation of ES cells towards making endodermal derivatives such as pancreatic cells have been particularly difficult to find. And to be useful for cell replacement, the insulin-producing cells generated must also be responsive to the signals, such as glucose, that switch on insulin production as required. Nevertheless, using knowledge of the signals that induce endoderm and pancreas development in mouse embryos, progress has been made in devising methods for differentiating human ES cells into pancreatic progenitor cells. Human ES cells put through a 9-day laboratory protocol involving stepwise treatments with different signal molecules differentiate into endoderm, expressing the endoderm-specific transcription factor FoxA2, but only a few of the cells in these cultures (around 5%) express Pdx1, a transcription factor essential for pancreas development. The next step was therefore to find ways of boosting the differentiation of pancreatic progenitor cells. A library of 5000 organic compounds was screened and one called indolactam V was found to stimulate the differentiation of Pdx1-positive cells. The percentage of Pdx1-expressing cells increased to 45% when endoderm cultures derived from the human ES cells were treated with indolactam V together with FGF-10, one of the signals involved in embryonic pancreatic development. When the *Pdx1*-expressing cells were transplanted under the kidney capsule in a mouse, a site that allows continued differentiation, they differentiated into insulin-secreting cells (Fig. 8.35).

iPS cells offer the greatest opportunities for generating pluripotent cells of a patient's own tissue type for cell replacement. As described in Box 8D, iPS cells were first derived from mouse fibroblasts by introducing and expressing genes for four transcription factors, Oct 3/4, Sox2, c-Myc and Kfl4, which are associated with pluripotency in ES cells. In regard to diabetes, iPS cells have been derived from fibroblasts taken from patients with type 1 diabetes. The fibroblasts were reprogrammed using the human genes *OCT4*, *SOX2* and *KLF4*, and were subsequently redifferentiated into insulin-producing cells by the route shown in Fig. 8.35. In the immediate future, such cells will be used to investigate the cellular and molecular defects underlying type 1 diabetes, the details of which are still largely unknown.

Induced transdifferentiation may be yet another way of generating replacement cells, although this approach is likely to be limited to developmentally related cell types. The feasibility of this approach for generating insulin-producing endocrine cells from liver cells and from pancreatic exocrine cells has been demonstrated experimentally. The liver and the pancreas both derive from endoderm, and arise from adjacent regions of endoderm in the embryo, but the transcription factor Pdx1, which is essential for pancreatic development, is not expressed in the liver. When an activated form of Pdx1 is expressed transiently in the liver in transgenic *Xenopus* tadpoles under the control of a liver-specific promoter, however, part or all of the liver is converted to both exocrine and endocrine pancreatic tissue, while proteins characteristic of differentiated liver cells disappear from the cells in the converted regions. In another

Fig. 8.35 Three experimental routes for making insulin-producing cells by nuclear reprogramming. Two routes start with a skin fibroblast. In somatic cell nuclear transfer, the nucleus of the fibroblast is introduced into an unfertilized egg and embryonic stem cells (ES cells) are isolated from the resulting blastocyst. Alternatively, cultured fibroblasts can be transfected with the genes for the pluripotency transcription factors Oct4, Sox2, and Kfl4 to produce induced pluripotent cells (iPS cells). The ES cells and iPS cells can then be differentiated into insulin-producing cells *in vitro*. A third possible route is the transdifferentiation of adult exocrine pancreatic cells or liver cells into insulin-producing cells *in vivo* by the targeted transfection of these organs with genes (including that for the transcription factor Pdx1) required for differentiation of endocrine pancreatic cells. Steps in red have been accomplished for human cells although the success rate of deriving ES cells from human blastocysts made by somatic cell nuclear transfer is extremely low. This means that this route for making human insulin-producing cells is not feasible at present. Insulin-producing glucose-responsive cells have been made starting with human ES cells derived from a normal blastocyst and differentiated *in vitro*.

Adapted from Gurdon J. B., Melton, D. A.: **Nuclear reprogramming in cells.** Science *2008,* **322:** *1811-1815.*

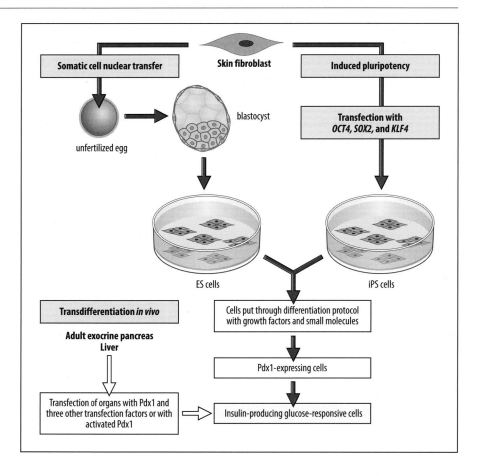

example of transdifferentiation, progenitor liver cells in adult mice have also been induced to develop into insulin-secreting cells resembling pancreatic islet cells by infecting the liver with viral vectors expressing the transcription factor neurogenin, which is another pancreatic transcription factor.

Transdifferentiation of pancreatic exocrine cells into endocrine β cells has also been achieved in adult mice. A combination of three transcription factors required for endocrine pancreatic cell differentiation, including Pdx1, were delivered into the pancreas using genetically modified adenoviruses. Endocrine cells were induced and, although they were not found in islets, these induced cells closely resembled their normal counterparts and had the ability to secrete insulin. Transdifferentiation was efficient (around 20%) and relatively fast, with the first insulin-positive cells appearing on day 3 after treatment.

There are, however, many obstacles to overcome before the potential therapeutic benefits of technologies based on ES cells, iPS cells, or transdifferentiation can be realized. The most effective current procedures for making iPS cells or causing transdifferentiation use viral vectors to introduce the pluripotency genes. Viral vectors can integrate randomly into the genome and might activate genes that cause cancer, and so alternative ways of inducing expression of the necessary transcription factors will need to be found before medical applications are possible. Fortunately, it is possible to remove the vectors along with the introduced genes after successful conversion to pluripotency and still maintain a stable iPS cell phenotype. This is because once pluripotency is induced by the introduced genes, the cell's own pluripotency genes become activated. In addition, there has been some success in obtaining iPS cells by introducing the four pluripotency proteins into mouse embryonic fibroblasts rather than the genes.

But even if non-viral methods of inducing pluripotency are perfected, it will not eliminate the risk of tumor induction in patients undergoing cell-replacement therapy with ES cells or iPS cells. If undifferentiated pluripotent cells remain after the differentiation process and are introduced into the patient they could cause tumors (see Fig. 8.34). Only stringent selection procedures that ensure no undifferentiated cells are present in the transplanted cell population will overcome this problem. And it is not yet clear how stable differentiated ES cells and iPS cells will be in the long term. A final, but not insignificant, obstacle to be overcome is finding ways of generating enough cells for an effective treatment.

We have focused here on just one medical target, the generation of pancreatic insulin-producing cells to treat type 1 diabetes. Similar strategies could also be used in other diseases. The neurodegenerative Parkinson's disease is another medical target and iPS cells have been made successfully from people with this disease. In the near future, iPS cells derived from patients will probably be most useful for investigating disease processes in the context of human cells, as animal models do not always mimic human disease completely. For example, iPS cells have been derived from skin fibroblasts from a child with genetically determined spinal muscular atrophy, and motor neurons have been produced in culture from these cells. Neuron production in the child's iPS cell cultures differed from neuron production in cultures of iPS cells made from the child's healthy mother, with the number and size of motor neurons produced decreasing over time in cultures derived from the child. Such cultures could be used to test the effects of drugs, which could lead to new clinical treatments. More generally, human iPS cells could be useful for screening compounds for toxicity and for teratogenicity in drug development.

SUMMARY

Transplantation of nuclei from differentiated animal cells into fertilized eggs, and cell-fusion studies, show that the pattern of gene expression in the nucleus of a differentiated cell can often be reversed, implying that it is determined by factors supplied by the cytoplasm and that no genetic material has been lost. Although the differentiated state of an animal cell *in vivo* is usually extremely stable, some cases of differentiation are reversible. Transdifferentiation of one differentiated cell type into another has been shown to occur during regeneration in some cases and in cultured cells. Both amphibians and some mammals can be cloned by nuclear transplantation from an adult somatic cell into an enucleated egg. Stem cells are self-renewing and can differentiate into a wide range of cell types. They are being investigated with a view to being used in cell-replacement therapy. It has been found that the introduction of four transcription factors expressed in embryonic stem cells into adult differentiated cells induces these cells to become pluripotent. These induced pluripotent stem cells and induced transdifferentiation provide alternative approaches for producing cells for cell-replacement therapy.

Summary to Chapter 8

- Cell differentiation leads to distinguishable cell types, such as blood cells, nerve cells and muscle cells, whose specialized functions and properties are determined by their pattern of gene activity and thus by the specialized proteins they produce.
- The first step in differentiation is the commitment, or determination, of the cell to that particular fate. Subsequent differentiation occurs gradually, usually over several cell generations.

- Differentiation reflects the selection of a particular subset of genes to be expressed and the silencing of other genes.
- The pattern of gene activity in a differentiated cell can be maintained over long periods of time and can be transmitted to the cell's progeny. Maintenance is likely to involve a combination of several mechanisms, including the continued action of gene-regulatory proteins and epigenetic chemical modifications to chromatin, such as DNA methylation or the acetylation or methylation of histone proteins.
- The differentiated state of an animal cell *in vivo* is usually extremely stable, but in some cases transdifferentiation of one differentiated cell type into another can occur.
- Many adult tissues, such as blood, epidermis and gut lining, are continually being renewed from undifferentiated, self-renewing cells called stem cells. In others, such as skeletal muscle, stem cells are present and can be mobilized to produce new tissue after injury.
- Pluripotent embryonic stem cells can be isolated from the inner cell mass of the mammalian blastocyst and grown in culture. They can be induced to differentiate into a great range of cell types in culture and mouse embryonic stem cells can produce all the cell types in the animal.
- The cloning of amphibians and some mammals by somatic cell nuclear transfer into enucleated eggs shows that the nuclei of differentiated cells can be reprogrammed and restored to a state in which they are able to direct embryonic development.
- Adult differentiated somatic cells can also be induced to develop into pluripotent cells, called induced pluripotent stem cells, by expression of four transcription factors associated with pluripotency in embryonic stem cells.
- Pluripotent cells that can proliferate and differentiate into cell types from all three germ layers could be the key to regenerative medicine.

■ End of chapter questions

Long answer (concept questions)

1. Use the Glossary to review the terms 'fate,' 'specification,' and 'determination.' How do these cellular states differ from the topic of this chapter, differentiation?

2. Contrast the concept of 'housekeeping' genes (and proteins) from the cell-specific genes (and proteins) that characterize, and define, a differentiated cell type. Propose a microarray or RNAseq experiment that would distinguish the tissue-specific gene expression from housekeeping gene expression in muscle cells and neurons.

3. Gene function is often studied by misexpression of the protein in an abnormal site in the embryo or abnormal tissue in the adult body to see what effects the protein has. What are the essential features of the transcriptional apparatus that make misexpression experimentally possible (include the role of both *cis*-regulatory DNA sequences and transcription factors)?

4. Create two lists: one of peptide signaling molecules, and one of lipid-soluble signaling molecules. How do their mechanisms of action differ? What generalization(s) can you make regarding the way they activate a transcription factor?

5. Cell division entails the separation of the two strands of the DNA during DNA replication, which disrupts patterns of gene expression that have been previously established in the cell. Nonetheless, the two daughter cells must somehow 'remember' the differentiated state of the parental cell. Discuss how the differentiated state of a cell is remembered, using transcription factors, chromatin structure, and DNA methylation.

6. Hematopoiesis involves the development of lymphocytes (B and T cells), red blood cells, macrophages, and neutrophils (among other blood-cell types) from cells in the bone marrow. Outline the general stages that lead to each of the cell types named, giving the name of the cell from which they all derive.

7. What is meant by 'stem-cell niche' and what is its role? For one of the stem-cell niches described in the text give examples of the signaling molecules that enable it to carry out this role.

8. Differentiation of the erythroid lineage requires signaling by erythropoietin (Epo), a peptide signaling molecule made by the kidney in response to red blood cell deficiency. One target of Epo is the GATA-1 transcription factor, and Epo is itself a target of GATA-1. β-globin is also a target of GATA-1. Using these facts, propose a pathway from a physiological need for more red blood cells, through the multipotent stem cell, to the globin-expressing red blood cell.

9. Describe the choice between differentiation as a neutrophil and differentiation as a macrophage at the level of the transcription factors involved. Describe the same choice at the level of the signaling molecules involved.

10. Why are different globins required at different stages of development? How is the switching between globin types thought to be controlled?

11. Describe the process by which epithelial cells of the intestine are replaced from a stem-cell population. What is the role of Wnt signaling in this process?

12. Despite its ability to act as a master regulator of muscle differentiation, mouse knockouts of *myoD* are viable. Refer to Fig. 8.11, and propose a reason that *myoD* is not essential. In contrast, knock-out of both *myoD* and *myf5* is lethal; knock-out of *mrf4* is also lethal; how would you interpret this result?

13. The differentiation of muscle cells requires withdrawal from the cell cycle. Elaborate on the following points regarding growth versus differentiation of myoblasts: (1) Removal of growth factors causes muscle precursor cells to differentiate in culture. (2) MyoD activates expression of p21. (3) Loss of Rb will allow myoblasts to re-enter the cell cycle.

14. Describe the process referred to as cloning by nuclear transplantation. How does the success of this process illustrate the importance of the complement of transcription factors present in any given cell? In contrast, how does the difficulty of the process illustrate the importance of epigenetic mechanisms in development?

15. ES cells and iPS cells are both pluripotent stem cells but are obtained in different ways. Describe how each is obtained.

16. What are the potential advantages of iPS cells over ES cells for therapeutic purposes? What are the disadvantages?

Multiple choice (factual recall questions)

NB There is only one correct answer to each question.

1. The many different cell types in the blood of an adult mammal are derived from

a) a single type of multipotent stem cell found in the bone marrow
b) differentiated cells that first developed in the fetal liver, but now reside in the bone marrow
c) division of differentiated cell types while circulating in the blood
d) stem cells that became committed to each specific blood cell type during embryonic development

2. What might be the result of producing a mouse knock-out for G-CSF?

a) The mouse would be unable to make granulocytes.
b) The mouse would be unable to make macrophages.
c) The mouse would lack red blood cells.
d) The mouse would lose its hematopoietic stem cell population completely.

3. The stem cells responsible for renewal of the keratinocytes of the skin are found in which layer of the epidermis?

a) basal lamina
b) basal layer
c) dermis
d) keratinocyte layer

4. The introduction of *myoD* into fibroblasts will cause

a) apoptosis
b) commitment to the myeloid lineage
c) differentiation into muscle cells
d) formation of red blood cells

5. Differentiation, as opposed to determination, of muscle cells is dependent on

a) Mrf4
b) Myf5
c) MyoD
d) myogenin

6. What are satellite cells?

a) Satellite cells are circulating cells 'orbiting' in the blood.
b) Satellite cells are muscle stem cells.
c) Satellite cells are muscle-associated support cells, much as glial cells are neuron-associated support cells.
d) Satellite cells are neuronal stem cells.

7. If the lens of the eye is removed in the adult, which of the following organisms can produce a new lens by transdifferentiation of iris cells?

a) chicken
b) human
c) mouse
d) newt

8. ES cells are

a) cells isolated from teratocarcinomas
b) embryonic stem cells derived from the inner cell mass of the mammalian embryo
c) epidermal stem cells found in the basal layer of the epidermis
d) human stem cells generated by treatment of mature cell types with a specific combination of transcription factors

9. EpiSCs are

a) Epidermal stem cells derived from the embryonic epidermis
b) Stem cells derived from reprogramming which are different from embryonic stem cells
c) Embryonic stem cells derived from the embryonic epiblast
d) Epidermal cells committed for tissue repair

10. Which of the following cocktails of transcription factors, aka Yamanaka factors, can be used to generate iPS cells?

a) Oct4, Nanog, Smad and Cdx2
b) Gata4, Nanog, Gli and Cdx2
c) Oct4, Nanog, Sox2 and Myc
d) Oct4, Sox2, Gata4 and Rb

Multiple choice answer key

1: a, 2: a, 3: b, 4: c, 5: d, 6: b, 7: d, 8: b, 9: c, 10: c

■ Section further reading

8.1 Control of transcription involves both general and tissue-specific transcriptional regulators

Buecker, C., Wysocka, J.: **Enhancers as information integration hubs in development: lessons from genomics.** *Trends Genet.* 2012, **28**: 276–284.

Levine, M., Tjian, R.: **Transcription regulation and animal diversity.** *Nature* 2003, **424**: 147–151.

Mannervik, M., Nibur, Y., Zhang, H., Levine, M.: **Transcriptional coregulators in development.** *Science* 1999, **284**: 606–609.

Spitz, F., Furlong, E.E.: **Transcription factors: from enhancer binding to developmental control.** *Nat. Rev. Genet.* 2012, **13**: 613–626.

8.2 Gene expression is also controlled by chemical and structural modifications to DNA and histone proteins that alter chromatin structure

&

8.3 Patterns of gene activity can be inherited by persistence of gene-regulatory proteins or by maintenance of chromatin modifications

Cheung, W.L., Briggs, S.D., Allis, C.D.: **Acetylation and chromosomal functions.** *Curr. Opin. Cell Biol.* 2000, **12**: 328–333.

de Laat, W., Grosveld, F.: **Spatial organization of gene expression: the active chromatin hub.** *Chromosome Res.* 2003, **11**: 447–459.

Ho, L., Crabtree, G.R.: **Chromatin remodelling during development.** *Nature* 2010, **463**: 474–484.

Box 8A Epigenetic control of gene expression by chromatin modification

Soshnikova, N., Duboule, D.: **Epigenetic temporal control of mouse Hox genes *in vivo*.** *Science* 2009, **324**: 1320–1323.

8.4 Changes in patterns of gene activity during differentiation can be triggered by extracellular signals

Barolo, S., Posakony, J. **Three habits of highly effective signaling pathways: principle of transcriptional control by developmental cell signalling.** *Genes Dev.* 2002, **16**: 1167–1181.

Brivanlou, A.H., Darnell, J.E.: **Signal transduction and control of gene expression.** *Science* 2002, **295**: 813–818.

Dougherty, D.C., Park, H.M., Sanders, M.M.: **Interferon regulatory factors (IRFs) repress transcription of the chicken ovalbumin gene.** *Gene* 2009, **439**: 63–70.

8.5 Muscle differentiation is determined by the MyoD family of transcription factors

Kassar-Duchossoy, L., Gayraud-Morel, B., Gomes, D., Rocancourt, D., Buckingham, M., Shinin, V., Tajbakhsh, S.: **Mrf4 determines skeletal muscle identity in Myf5: MyoD double-mutant mice.** *Nature* 2004, **431**: 466–471.

Tapscott, S.J.: **The circuitry of a master switch: MyoD and the regulation of skeletal muscle transcription.** *Development* 2005, **132**: 2685–2695.

8.6 The differentiation of muscle cells involves withdrawal from the cell cycle, but is reversible

Deato, M.D., Tjian, R.: **An unexpected role of TAFs and TRFs in skeletal muscle differentiation; switching of core promoter complexes.** *Cold Spring Harb. Symp.* 2008, **73**: 217–225.

Novitch, B.G., Mulligan, G.J., Jacks, T., Lassar, A.B.: **Skeletal muscle cells lacking the retinoblastoma protein display defects in muscle gene expression and accumulate in S and G_2 phases of the cell cycle.** *J. Cell Biol.* 1996, **135**: 441–456.

Odelberg, S.J., Kolhoff, A., Keating, M.: **Dedifferentiation of mammalian myotubes induced by *msx1*.** *Cell* 2000, **103**: 1099–1109.

8.7 All blood cells are derived from multipotent stem cells

Ema, H., Nakauchi, H.: **Self-renewal and lineage restriction of hematopoietic stem cells.** *Curr. Opin. Genet. Dev.* 2003, **13**: 508–512.

Phillips, R.L., Ernst, R.E., Brunk, B., Ivanova, N., Mahan, M.A., Deanehan, J.K., Moore, K.A., Overton, G.C., Lemischka, I.R.: **The genetic program of hematopoietic stem cells.** *Science* 2000, **288**: 1635–1640.

Zon, L.I.: **Intrinsic and extrinsic control of haematopoietic stem-cell renewal.** *Nature* 2008, **453**: 306–313.

8.8 Intrinsic and extrinsic changes control differentiation of the hematopoietic lineages

Anguita, E., Hughes, J., Heyworth, C., Blobel, G.A., Wood, W.G., Higgs, D.R.: **Globin gene activation during haemopoiesis is driven by protein complexes nucleated by GATA-1 and GATA-2.** *EMBO J.* 2004, **23**: 2841–2852.

Kluger, Y., Lian, Z., Zhang, X., Newburger, P.E., Weissman, S.M.: **A panorama of lineage-specific transcription in hematopoiesis.** *BioEssays* 2004, **26**: 1276–1287.

Metcalf, D.: **Control of granulocytes and macrophages: molecular, cellular, and clinical aspects.** *Science* 1991, **254**: 529–533.

Orkin, S.H.: **Diversification of haematopoietic stem cells to specific lineages.** *Nat Rev. Genet.* 2000, **1**: 57–64.

8.9 Developmentally regulated globin gene expression is controlled by regulatory sequences far distant from the coding regions

Engel, J.D., Tanimoto, K.: **Looping, linking, and chromatin activity: new insights into β-globin locus regulation.** *Cell* 2000, **100**: 499–502.

Tolhuis, B., Palstra, R.J., Splinter, E., Grosveld, F., de Laat, W.: **Looping and interaction between hypersensitive sites in the active beta-globin locus.** *Mol. Cell* 2002, **10**: 1453–1465.

8.10 The epidermis of adult mammalian skin is continually being replaced by derivatives of stem cells

Clayton, E. *et al.*: **A single type of progenitor cell maintains normal epidermis.** *Nature* 2007, **446**: 185–189.

Coulombe, P.A., Kerns, M. L., Fuchs, E.: **Epidermolysis bullosa simplex: a paradigm for disorders of tissue fragility.** *J. Clin. Invest.* 2009, **119**: 1784–1793.

Fuchs, E.: **The tortoise and the hair: slow-cycling cells in the stem cell race.** *Cell* 2009, **137**: 811–819.

Fuchs, E.: **Finding one's niche in the skin.** *Cell Stem Cell* 2009, **4**: 499–502.

Fuchs, E., Nowak, J.A.: **Building epithelial tissues from skin stem cells.** *Cold Spring Harb. Symp. Quant. Biol.* 2008, **73**: 333–350.

Jones, P.H., Simons, B.D., Watt, F.M.: **Sic transit gloria: farewell to the epidermal transit amplifying cell?** *Cell Stem Cell* 2007, **1**: 371–381.

Mascré, G. *et al.*: **Distinct contribution of stem and progenitor cells to epidermal maintenance.** *Nature* 2012, **489**: 257–262.

Schepeler, T., Page, M.E., Jensen, K.B.: **Heterogeneity and plasticity of epidermal stem cells.** *Development* 2014, **141**: 2559–2567.

8.11 Stem cells use different modes of division to maintain tissues

Li, X., Upadhyay, A.K., Bullock, A.J., Dicolandrea, T., Xu, J., Binder, R.L., Robinson, M.K., Finlay, D.R., Mills, K.J., Bascom, C.C., Kelling, C.K., Isfort, R.J., Haycock, J.W., MacNeil, S., Smallwood, R.H.: **Skin stem cell hypotheses and long term clone survival explored using agent-based modeling.** *Sci. Rep.* 2013, **3**: 1904.

Mascré, G., Dekoninck, S., Benjamin Drogat, B., Youssef, K.K., Brohée, S., Sotiropoulou, P.A., Simons, B.D., Blanpain, C.: **Distinct contribution of stem and progenitor cells to epidermal maintenance.** *Nature* 2012, **489**: 257–262.

Simons, B.D., Clevers, H.: **Strategies for homeostatic stem cell self-renewal in adult tissues.** *Cell* 2011, **145**: 851–862.

8.12 The lining of the gut is another epithelial tissue that requires continuous renewal

Barker, N., van Es, J.H., Kuipers, J., Kujala, P., van den Born, M., Cozijnsen, M., Haegebarth, Andrea, Korving, J., Begthel, H., Peters, P.J., Clevers, H.: **Identification of stem cells in small intestine and colon by marker gene *Lgr5*.** *Nature* 2007, **449**: 1003–1008.

Sato, T., Vries, R.G., Snippert, H.J., van de Wetering, M., Barker, N., Stange, D.E., van Es, J.H., Abo, A., Kujala, P., Peters, P.J., Clevers, H.: **Single *Lgr5* stem cells build crypt villus structures *in vitro* without a mesenchymal cell niche.** *Nature* 2009, **459**: 262–265.

Van der Flier, L.G., Clevers, H.: **Stem cells, self-renewal and differentiation in the intestinal epithelium.** *Annu. Rev. Physiol.* 2009, **71**: 241–260.

8.13 Skeletal muscle and neural cells can be renewed from stem cells in adults

Collins, C.A., Partridge T.A.: **Self-renewal of the adult skeletal muscle satellite cell.** *Cell Cycle* 2005, **4**: 1338–1341.

Dhawan, J., Rando, T.A.: **Stem cells in postnatal myogenesis: molecular mechanisms of satellite cell quiescence, activation and replenishment.** *Trends Cell Biol.* 2005, **15**: 663–673.

Kemperman, G.: **New neurons for 'survival of the fittest'.** *Nat. Rev. Neurosci.* 2012, **13**: 727–736.

Lepper, C., Conway, S.J., Fan, C.M.: **Adult satellite cells and embryonic muscle progenitors have distinct genetic requirements.** *Nature* 2009, **460**: 627–631.

Li, G., Pleasure, S.J.: **Ongoing interplay between the neural network and neurogenesis in the adult hippocampus.** *Curr. Opin. Neurobiol.* 2010, **20**: 126–133.

Lie, D-C., Colamarino, S.A., Song, H-J., Desire, L., Mira, H., Consiglio, A., Lein, E.S., Jessberger, S., Lansford, H., Dearier, A.R., Gage, F.H.: **Wnt signalling regulates adult hippocampal neurogenesis.** *Nature* 2005, **437**: 1370–1375.

Ninkovic, J., Gotz, M.: **Signaling in adult neurogenesis: from stem cell niche to neuronal networks.** *Curr. Opin. Neurobiol.* 2007, **17**: 338–344.

Relaix, F., Rocancourt, D., Mansouri, A., Buckingham, M.A.: ***pax3/pax7* dependent population of skeletal muscle progenitor cells.** *Nature* 2005, **435**: 948–953.

Sacco, A., Doyonnas, R., Kraft, P., Vitorovic, S., Blau, H.M.: **Self-renewal and expansion of single transplanted muscle stem cells.** *Nature* 2008, **456**: 502–506.

Spalding, K.L., Bergmann, O., Alkass, K., Bernard, S., Salehpour, M., Huttner, H.B., Boström, E., Westerlund, I., Vial, C., Buchholz, B.A., Possnert, G., Mash, D.C., Druid, H., Frisén J.: **Dynamics of hippocampal neurogenesis in adult humans.** *Cell* 2013, **153**: 1219–1227.

8.14 Embryonic stem cells can proliferate and differentiate into many cell types in culture and contribute to normal development *in vivo*

Chambers, I., Silva, J., Colby, D., Nichols, J., Nijmeijer, B., Robertson, M., Vrana, J., Jones, K., Grotewold, L., Smith, A.: **Nanog safeguards pluripotency and mediates germline development.** *Nature* 2007, **450**: 1230–1234.

Kojima, Y., Kaufman-Francis, K., Studdert, J.B., Steiner, K.A., Power, M.D., Loebel, D.A., Jones, V., Hor, A., de Alencastro, G., Logan, G.J., Teber, E.T., Tam, O.H., Stutz, M.D., Alexander, I.E., Pickett, H.A., Tam, P.P.: **The transcriptional and functional properties of mouse epiblast stem cells resemble the anterior primitive streak.** *Cell Stem Cell* 2013, **14**: 107–120.

Nichols, J., Smith, A.: **The origin and identity of embryonic stem cells.** *Development* 2011, **138**: 3–8.

Silva, J., Chambers, I., Pollard, S., Smith, A. **Nanog promotes transfer of pluripotency after cell fusion.** *Nature* 2006, **441**: 997–1001.

West, J.A., Daley, G.Q.: *In vitro* **gametogenesis from embryonic stem cells.** *Curr. Opin. Cell Biol.* 2004, **16**: 688–692.

Ying, Q.L., Wray, J., Nichols, J., Batlle-Morera, L., Doble, B., Woodgett, J., Cohen, P., Smith, A.: **The ground state of embryonic stem cell self-renewal.** *Nature* 2008, **453**: 519–523.

Young, R. **Control of the embryonic stem cell state.** *Cell* 2011, **144**: 940–954.

Box 8B The derivation and culture of mouse embryonic stem cells (ES cells)

Wray, J., Kalkan, T., Gomez-Lopez, S., Eckardt, D., Cook, A., Kemler, R., Smith, A.: **Inhibition of glycogen synthase kinase-3 alleviates Tcf3 repression of the pluripotency network and increases embryonic stem cell resistance to differentiation.** *Nat. Cell Biol.* 2011, **13**: 838–845.

8.15 Nuclei of differentiated cells can support development

Eggan, K., Baldwin, K., Tackett, M., Osborne, J., Gogos, J., Chess, A., Axel, R., Jaenisch, R.: **Mice cloned from olfactory sensory neurons.** *Nature* 2004, **428**: 44–49.

Gurdon, J.B.: **Nuclear transplantation in eggs and oocytes.** *J. Cell Sci. Suppl.* 1986, **4**: 287–318.

Gurdon, J.B., Melton, D.A.: **Nuclear reprogramming in cells.** *Science* 2008, **322**: 1811–1815.

Humpherys, D., Eggan, K., Akutsu, H., Friedman, A., Hochedlinger, K., Yanagimachi, R., Lander, E.S., Golub, T.R., Jaenisch, R.: **Abnormal gene expression in cloned mice derived from embryonic stem cell and cumulus cell nuclei.** *Proc. Natl Acad. Sci. USA* 2002, **99**: 12889–12894.

Rhind, S.M., Taylor, J.E., De Sousa, P.A., King, T.J., McGarry, M., Wilmut, I.: **Human cloning: can it be made safe?** *Nat. Rev. Genet.* 2003, **4**: 855–864.

Wilmut, I., Taylor, J.: **Primates join the club.** *Nature* 2007, **450**: 485–486.

8.16 Patterns of gene activity in differentiated cells can be changed by cell fusion

Blau, H.M.: **How fixed is the differentiated state? Lessons from heterokaryons.** *Trends Genet.* 1989, **5**: 268–272.

Blau, H.M., Baltimore, D.: **Differentiation requires continuous regulation.** *J. Cell Biol.* 1991, **112**: 781–783.

Blau, H.M., Blakely, B.T.: **Plasticity of cell fates: insights from heterokaryons.** *Cell Dev. Biol.* 1999, **10**: 267–272.

Pomerantz, J.H., Mukherjee, S., Palermo, A.T., Blau, H.M.: **Reprogramming to a muscle fate by fusion recapitulates differentiation.** *J. Cell Sci.* 2009, **122**: 1045–1053.

8.17 The differentiated state of a cell can change by transdifferentiation

Eguchi, G., Eguchi, Y., Nakamura, K., Yadav, M.C., Millán, J.L., Tsonis, P.A.: **Regenerative capacity in newts is not altered by repeated regeneration and ageing.** *Nat. Commun.* 2011, **2**: 384.

Horb, M.E., Shen, C.N., Tosh, D., Slack, J.M.: **Experimental conversion of liver to pancreas.** *Curr. Biol.* 2003, **13**: 105–115.

Jarriault, S., Shwab, Y., Greenwald, I.A.: *Caenorhabitis elegans* **model for epithelial-neuronal transdifferentiation.** *Proc. Natl Acad. Sci. USA* 2008, **105**: 3790–3795.

Slack, J.M.W.: **Metaplasia and transdifferentiation from pure biology to the clinic.** *Nat. Rev. Mol. Cell Biol.* 2007, **8**: 369–378.

Tsonis, P.A., Madhavan, M., Tancous, E.E., Del Rio-Tsonis, K.: **A newt's eye view of lens regeneration.** *Int. J. Dev. Biol.* 2004, **48**: 975–980.

8.18 Stem cells could be a key to regenerative medicine

Jaenisch, R.: **Human cloning—the science and ethics of nuclear transplantation.** *N. Engl. J. Med.* 2004, **351**: 2787–2792.

McClaren, A.: **Ethical and social considerations of stem cell research.** *Nature* 2001, **414**: 129–131.

Pera, M.F., Trounson, A.O.: **Human embryonic stem cells: prospects for development.** *Development* 2004, **131**: 5515–5525.

Pomerantz, J., Blau, H.M.: **Nuclear reprogramming: a key to stem cell function in regenerative medicine.** *Nat. Cell Biol.* 2004, **6**: 810–816.

Rolletschek, A., Wobus, A.M.: **Induced human pluripotent stem cells: promises and open questions.** *Biol. Chem.* 2009, **390**: 845–849.

Schwartz, S.D., Hubschman, J.P., Heilwell, G., Franco-Cardenas, V., Pan, C.K., Ostrick, R.M., Mickunas, E., Gay, R., Klimanskaya, I., Lanza, R.: **Embryonic stem cell trials for macular degeneration: a preliminary report.** *Lancet* 2012, **379**: 713–720.

Tachibana, M., Amato, P., Sparman, M., Marti Gutierrez, N., Tippner-Hedges, R., Ma, H., Kang, E., Fulati, A., Lee, H.-S., Sritanaudomchai, H., Masterson, K., Larson, J., Eaton, D., Sadler-Fredd, K., Battaglia, D., Lee, D., Wu, D., Jensen, J., Patton, P., Gokhale, S., Stouffer, R.L., Wolf, D., Shoukhrat Mitalipov, S.: **Human embryonic stem cells derived by somatic cell nuclear transfer.** *Cell* 2013, **153**: 1228–1238.

Wurmer, A.E., Palmer, T.D., Gage, F.H.: **Cellular interactions in the stem cell niche.** *Science* 2004, **304**: 1253–1255.

Yechoor, V., Liu, V., Espiritu, C., Paul, A, Oka, K., Kojima, H., Chan, L.: **Neurogenin3 is sufficient for transdetermination of hepatic progenitor cells into neo-islets** *in vivo* **but not transdifferentiation of hepatocytes.** *Dev. Cell* 2009, **16**: 358–373.

Box 8C Tissue engineering using stem cells

Gonfiotti, A., Jaus, M.O., Barale, D., Baiguera, S., Comin, C., Lavorini, F., Fontana, G., Sibila, O., Rombolà, G., Jungebluth, P., Macchiarini, P.: **The first tissue-engineered airway transplantation: 5-year follow-up results.** *Lancet* 2014, **383**: 238–244.

Jungebluth, P. *et al.*: **Tracheobronchial transplantation with a stem-cell-seeded bioartificial nanocomposite: a proof-of-concept study.** *Lancet* 2011, **378**: 1997–2004.

Macchiarini, P., Jungebluth, P., Go, T., Asnaghi, M.A., Rees, L.E., Cogan, T.A., Dodson, A., Martorell, J., Bellini, S., Parnigotto, P.P., Dickinson, S.C., Hollander, A.P., Mantero, S., Conconi, M.T., Birchall, M.A.: **Clinical transplantation of a tissue-engineered airway.** *Lancet* 2008, **372**: 2023–2030.

Box 8D Induced pluripotent stem cells (iPS cells)

Okita, I., Ichisaka, T., Yamanka, S.: **Generation of germline-competent induced pluripotent stem cells.** *Nature* 2007, **448**: 313–317.

Rolletschek, A., Wobus, A.M.: **Induced human pluripotent stem cells: promises and open questions.** *Biol. Chem.* 2009, **390**: 845–849.

Takahashi, K., Yamanaka, S. **Induction of pluripotent stem cells from mouse embryonic and adult fibroblast cultures by defined factors.** *Cell* 2006, **126**: 663–676.

Woltjen, K., Michael, I.P., Mohseni, P., Desai, R., Mileikovsky, M., Hämäläinen, R., Cowling, R., Wang, W., Liu, P., Gertsenstein, M., Kaji, K., Sung, H.K., Nagy, A.: **piggyBac transposition reprograms fibroblasts to induced pluripotent stem cells.** *Nature* 2009, **458**: 766–770.

Yusa, K., Rad, R., Takeda, J., Bradley, A.: **Generation of transgene-free induced pluripotent mouse stem cells by the piggyBac transposon.** *Nat. Meth.* 2009, **6**: 363–369.

8.19 Various approaches can be used to generate differentiated cells for cell-replacement therapies

Chen, S., Borowiak, M., Fox, J.L., Maehr, R., Osafune, K., Davidow, L., Lam, K., Peng, L.F., Schreiber, S.L., Rubin, L.L., Melton, D.: **A small molecule that directs differentiation of human ESCs into the pancreatic lineage.** *Nat. Chem. Biol.* 2009, **5**: 258–265.

Maehr, R., Chen, S., Snitow, M., Ludwig, T., Yagasaki, L., Goland, R., Leibel, R.L., Melton, D.A.: **Generation of pluripotent stem cells from patients with type 1 diabetes.** *Proc. Natl Acad. Sci. USA* 2009, **106**: 15768–15773.

Zhou, Q., Brown, J., Kanarek, A., Rajagopal, J., Melton, D.A.: *In vivo* **reprogramming of adult pancreatic exocrine cells to beta-cells.** *Nature* 2008, **455**: 627–632.

9

Morphogenesis: change in form in the early embryo

- Cell adhesion
- Cleavage and formation of the blastula
- Gastrulation movements
- Neural tube formation
- Cell migration
- Directed dilation

Changes in form in animal embryos are brought about by cellular forces that are generated in various ways, including cell division, change in cell shape, rearrangement of cells within tissues, and the migration of individual cells and groups of cells from one part of the embryo to another. These cellular forces are most frequently generated within epithelial sheets by contraction of the cells' internal cytoskeleton; resisting these forces are the cells' own structure and their adhesive interactions with other cells and with the extracellular material, which hold them together in tissues. Cell division, cell migration, and the generation of hydrostatic pressure within a structure also play a part in morphogenetic processes. The major morphogenetic event in early animal embryos is gastrulation, when a wholesale reorganization of the cells in the tissues occurs. Other examples of morphogenesis discussed in this chapter include notochord and neural tube formation, and the migration of neural crest cells to specific locations.

So far, we have discussed early development mainly from the viewpoint of the assignment of cell fate, resulting from different genes being expressed in particular cell populations. In this chapter, we look at embryonic development from a different perspective—the generation of form, or **morphogenesis**. All animal embryos undergo a dramatic change in shape during their early development. This occurs principally during gastrulation, the process that transforms a uniform two-dimensional sheet of cells into an arrangement of layers of cells with different identities arranged in three dimensions. Gastrulation involves extensive rearrangements and directed movement of cells from one location to another, and in tribloblastic animals yields a universal arrangement of cells into the three germ layers—ectoderm, mesoderm, and endoderm—with an antero-posterior organization (see Fig. 4.37).

If pattern formation can be likened to painting, morphogenesis is more akin to modeling a formless lump of clay into a recognizable shape. An important general principle is that isolated cells on their own cannot drive morphogenesis. It is their organization into coherent tissues that enables cells to coordinate their behavior and generate forces that model the embryo into a different shape. Change in form is largely a problem of how the behavior of individual cells affects the tissues of which they are a part, and at its heart is a problem in the mechanics of cells and tissues. Understanding morphogenesis requires an appreciation of the forces involved in bringing about changes in cell shape and cell movement, the molecular machinery that carries them out, and how this can be controlled during development.

The generation of animal embryonic form involves three key cell properties. The first is **cell adhesiveness**, the tendency of cells to stick to one another. In tissues, animal cells adhere to one another, and to the extracellular matrix, through interactions involving various cell-surface proteins known collectively as **adhesion molecules**. Changes in the adhesion molecules at the cell surface can therefore alter the strength of cell–cell adhesion and its specificity, with far-reaching effects on the tissue of which the cells are a part.

The second property is **cell shape**; cells can actively change shape by means of internal contractions and constrictions. Changes in cell shape, which are caused by cytoskeletal rearrangements, are crucial in many developmental processes. For example, folding or rolling of a cell sheet—a very common feature in embryonic development—is caused by coordinated changes in the shape of selected cells within the sheet.

The third is **cell migration**, the ability of many types of cells to travel as individuals or groups from one location to another. The ability of cells to move is also crucial to development, as many developmental processes involve the guided coordinated migration of individual cells or groups of cells from their site of origin to their final location.

The organization of cells into tissues amplifies the potential of these key cell properties, enabling them to operate over larger dimensions. A cell shrinking or expanding within a tissue will not only affect its neighbors but will have an effect several cell diameters away. Cell behavior can also be controlled by extracellular signals, both chemical and mechanical, received from neighboring cells and the extracellular matrix.

Changes in the adhesiveness of cells is a central component in both cell-shape changes and cell migration. Not surprisingly, there are many points of interaction between the molecular machinery of cell adhesion and that controlling the cytoskeleton. Some cell-adhesion molecules we will discuss in this chapter also act as signaling receptors, transmitting signals from the extracellular environment into the cell to effect changes in the cytoskeleton. Cell adhesion, cell-shape changes, and cell migration can therefore be controlled in a coordinated fashion both by the previous developmental history of the cell, which determines, for example, which adhesion molecules it expresses, and by extracellular signals.

An additional force that operates during morphogenesis, more particularly in plants but also in some aspects of animal embryogenesis, is hydrostatic pressure, which is generated by osmosis and fluid accumulation. In plants there is no cell movement during growth, and changes in form are generated by oriented cell division and cell expansion, as we saw in Chapter 7.

Changes in embryonic form are therefore the final consequence of the precise spatio-temporal expression of proteins that control cell adhesion, **cell motility**, oriented cell division, and the generation of hydrostatic pressure. Gastrulation, for example, can be thought of as the animation of a detailed map of instructions that has been progressively laid down on the canvas that is the early embryo. The earlier patterning process determines which cells will express those proteins that are required to generate and harness the appropriate forces.

In this chapter we shall discuss some examples of the morphological changes that occur during the development of the animal body plan. First, we will look at how cleavage of the zygote gives rise to the simple shape of the early embryo, of which the spherical blastocyst of the mouse and the blastulas of sea urchin and amphibian are good examples. We then consider the movements that occur during gastrulation and during neurulation—the formation of the neural tube in vertebrates—which involve folding of cell sheets and rearrangement of cell layers. In vertebrates, migration of cells from the neural crest after neurulation generates various structures in the trunk and head, and we consider how these cells are guided to their correct sites. Finally, we look at directed dilation, in which hydrostatic pressure is the force that drives changes in shape. Other morphogenetic mechanisms, such as cell growth, cell proliferation, and cell death are considered in relation to the development of particular organs and the growth of an organism as a whole in Chapters 11, 12, and 13. Some aspects of growth in plants are considered in Chapter 7.

We begin by considering how cells adhere to each other, and how differences in adhesiveness and specificity of adhesion are involved in maintaining boundaries between tissues.

Cell adhesion

Adhesive interactions between cells, and between cells and the extracellular matrix, are an important element in cell behavior in morphogenesis, as they promote the assembly of groups of cells that will behave coordinately as well as setting-up differences in cell adhesiveness that help maintain the boundaries between different tissues and structures.

Cells stick to each other by means of **cell-adhesion molecules**, which are proteins on the cell surface that can bind strongly to proteins on other cell surfaces or in the extracellular matrix (Box 9A). The particular adhesion molecules expressed by a cell determine which cells it can adhere to, and changes in the adhesion molecules expressed are involved in many developmental phenomena. In epithelial tissues, with which we shall be most concerned here, adjacent cells are joined together by specialized structures called **adherens junctions** that incorporate cell-adhesion molecules.

All cells have a surface tension at their membrane, where they encounter the extracellular environment. It is surface tension that results in isolated cells adopting a spherical shape, just as water forms drops in air, because this minimizes the energy of interaction with their surroundings. Surface tension is also what drives cells to cohere, again as a way to reduce their overall surface tension by reducing the amount of surface exposed. Adhesive interactions affect the surface tension at the cell membrane, reflected by flattening of the cell at point of contact with another cell or with the substratum. Different surface tensions are what keep two immiscible liquids such as water and oil separate, and embryologists had long noted the 'liquid-like' behavior of the tissues in gastrulating frog embryos, observing how the separate germ layers formed and remained distinct from each other, slid over each other, spread out, and underwent internal rearrangement to shape the embryo. The **differential adhesion hypothesis** explains these liquid-like tissue dynamics as the consequences of surface tensions and interfacial tensions generated by cells that tend to stick to each other but are also able to change shape and to move.

9.1 Sorting out of dissociated cells demonstrates differences in cell adhesiveness in different tissues

Differences in cell adhesiveness can be illustrated by experiments in which different tissues are confronted with one another in an artificial setting. Two pieces of early endoderm from an amphibian blastula will fuse to form a smooth sphere if placed in contact. In contrast, when a piece of early endoderm and a piece of early ectoderm are combined they initially fuse, but in time the pieces of endoderm and

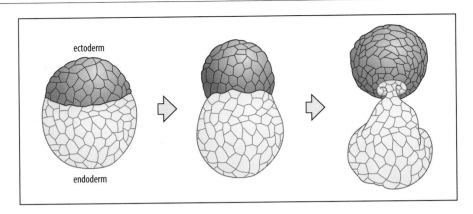

Fig. 9.1 Separation of embryonic tissues with different adhesive properties. When pieces of early ectoderm (blue) and early endoderm (yellow) from an amphibian blastula are placed together, they initially fuse, but then separate until only a narrow strip of tissue joins the two tissues.

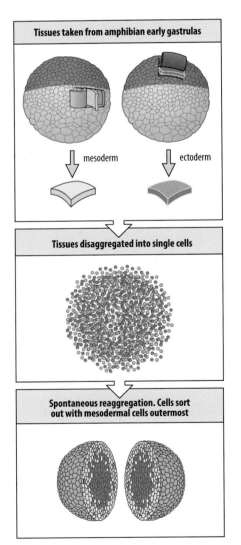

Fig. 9.2 Sorting out of different cell types. Ectoderm (gray) and mesoderm (blue) from amphibian blastulas are disaggregated into single cells by treatment with an alkaline solution. The cells, when mixed together, sort out with the mesoderm cells on the outside.

ectoderm do not remain stuck together but separate, until only a narrow bridge connects the two types of tissue (Fig. 9.1). A different result is obtained when a piece of early ectoderm is placed in contact with a piece of early mesoderm. Ectoderm and mesoderm normally adhere to each other within the blastula, and the tissues do not separate from each other like ectoderm and endoderm. Instead, the pieces of tissue remain in contact and the mesoderm spreads over the ectoderm and eventually envelops it.

Remarkably, when the different tissues are disaggregated into single cells, and the cells mixed together and allowed to reaggregate, cells from the different tissues sort out, so that within the aggregate, like cells associate preferentially with like cells. For example, cells disaggregated from the ectoderm and mesoderm of amphibian blastulae or gastrulae will sort themselves out into ectoderm and mesoderm. Furthermore, ectoderm cells are on the inside of the aggregate and mesoderm on the outside (Fig. 9.2), exactly the same arrangement as adopted by the fused pieces of tissue. Similar experiments with zebrafish embryos have shown that the ectoderm ends up on the inside of the aggregate and mesendoderm on the outside. In a similar way, when cells of amphibian presumptive epidermis and presumptive neural plate are dissociated, mixed, and left to reaggregate, they sort out to reform the two different tissues. In this case, the epidermal cells are eventually found on the outer face of the aggregate, surrounding a mass of neural cells.

The sorting out is the combined result of differences in adhesiveness and surface tension between cells of the different types, so that the overall energy of interfacial interaction within the cell mass is eventually minimized. Cells of the same type tend to adhere preferentially to each other. Initially, cells in the mixed aggregate start to exchange weaker for stronger adhesions. Over time, the adhesive interactions between cells produce different degrees of surface tension that are sufficient to generate the sorting-out behavior, just as two immiscible liquids, such as oil and water, separate out when mixed. As the constituent cells form stable contacts with each other, the tissues become reorganized and the strength of intercellular binding in the system as a whole is maximized. In general, if the adhesion between unlike cells is weaker than the average of the adhesions between like cells, the cells will segregate according to type, with the more adhesive cell type on the inside of the aggregate. In terms of the surface tension, those cells with the higher surface tension become surrounded by those with lower surface tension.

It is important to note that experiments on isolated cells and tissues *in vitro* like those described above simply indicate the different adhesiveness and different surface tensions of the tissues in isolation; they do not define which tissue will be on the inside or the outside in the embryo itself. This is because in a real embryo, tissues are under pre-existing developmental constraints that define their location and the specialization of their adhesive properties and surface tension, and they may also adhere to other tissues. Nevertheless, these *in vitro* experiments show how differential cell

CELL BIOLOGY BOX 9A Cell-adhesion molecules and cell junctions

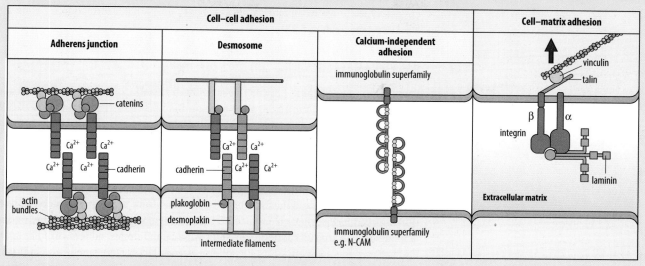

Figure 1

Three classes of adhesion molecules are particularly important in development (Figure 1). The **cadherins** are transmembrane proteins that, in the presence of calcium ions (Ca^{2+}), adhere to cadherins on the surface of another cell. Calcium-independent cell-cell adhesion involves a different structural class of proteins members of the large **immunoglobulin superfamily**. The neural cell-adhesion molecule (N-CAM), which was first isolated from neural tissue, is a typical member of this family. Some immunoglobulin superfamily members, such as N-CAM, bind to similar molecules on other cells; others bind to a different class of adhesion molecule, the **integrins**. Integrins also act as receptors for molecules of the extracellular matrix, mediating adhesion between a cell and its substratum.

About 30 different types of cadherins have been identified in vertebrates. Cadherins bind to each other through one or more binding sites located within the extracellular amino-terminal 100 amino acids. In general, a cadherin binds only to another cadherin of the same type, but they can also bind to some other molecules. Cadherins are the adhesive components in both **adherens junctions** (adhesive cell-cell junctions that are present in many tissues) and **desmosomes** (cell-cell junctions present mainly in epithelia). **Hemidesmosomes** are an integrin-based type of junction that rivets epithelial cells firmly to the extracellular matrix of the basement membrane.

As cells approach and touch one another, the cadherins cluster at the site of contact. They interact with the intracellular cytoskeleton through the connection of their cytoplasmic tails with **catenins** and other proteins, and thus can be involved in transmitting signals to the cytoskeleton. This adhesive and signaling role of α-, β-, and γ-catenins is separate from the role of β-catenin as a gene-regulatory protein (see, for example, Section 4.2). The interaction with the cytoskeleton is required for normal strong cell-cell

adhesion. In adherens junctions the connection is to cytoskeletal actin filaments, whereas in desmosomes and hemi-desmosomes it is to intermediate filaments such as keratin. Nectin, a member of the immunoglobulin superfamily of adhesion molecules, also clusters to adherens junctions in mammalian tissues and connects to the actin cytoskeleton.

Adhesion of a cell to the extracellular matrix, which contains proteins such as collagen, fibronectin, laminin, and tenascin, as well as proteoglycans, is by the binding of integrins to these matrix molecules. An integrin molecule is composed of two different subunits, an α subunit and a β subunit. Twenty-four different integrins are known so far in vertebrates, made up from eight β subunits and 19 α subunits. Many extracellular matrix molecules are recognized by more than one integrin.

Integrins not only bind to other molecules through their extracellular face, but they also associate with the actin filaments or intermediate filaments of the cell's cytoskeleton through complexes of proteins that are in contact with the integrin's cytoplasmic region. This association enables integrins to transmit information about the extracellular environment, such as extracellular matrix composition or the type of intercellular contact, to the cell. Integrins can, therefore, transmit signals from the matrix that affect cell shape, motility, metabolism, and cell differentiation. Integrins are also involved in adhesion between cells, binding either to adhesion molecules of the immunoglobulin superfamily or via a shared ligand to integrins on another cell surface.

A third type of cell junction present in some, but not all, epithelia is the **tight junction**, which forms a seal that prevents water and other molecules passing between the epithelial cells. Multiple lines of transmembrane proteins called claudins and occludins in adjoining cell membranes form a continuous seal around the apical end of the cell (see Fig. 9.13).

adhesion and different surface tensions can, in principle, create and stabilize boundaries between tissues. Furthermore, there are a few examples where sorting out does occur during development, as in the early mouse embryo, where sorting out of different cell types provides an alternative strategy to positional information for generating a pattern (see Section 5.2).

9.2 Cadherins can provide adhesive specificity

Differential adhesiveness between cells is the result of differences in the types and numbers of adhesion molecules they have on their surfaces. The main types of adhesion molecules we will be concerned with in this chapter are outlined in Box 9A. One class comprises the **cadherins**, which depend on calcium for their function and are components of a type of adhesive junction called the **adherens junction**, which is the junction that holds cells together in all epithelial tissues. There are many kinds of cadherins and, in many instances, they are cell-type specific. Evidence that the cadherins can provide adhesive specificity comes from studies in which cells with different cadherins on their surface are mixed together. Fibroblasts of the mouse L cell line do not express cadherins on their surface and do not adhere strongly to each other. But if the gene encoding E-cadherin is transfected into L cells and expressed, the cells produce that cadherin on their surface and stick together, forming a structure resembling a compact epithelium. The adherence is both calcium-dependent, indicating that it is due to the cadherin, and specific, as the transfected cells do not adhere to untreated L cells lacking surface cadherins.

When populations of L cells are transfected with different types of cadherin and mixed together in suspension, only those cells expressing the same cadherin adhere strongly to each other: cells expressing E-cadherin adhere strongly to other cells expressing E-cadherin, but only weakly to cells expressing P- or N-cadherin. The amount of cadherin on the cell surface can also have an effect on adhesion. When cells expressing different amounts of the same cadherin are mixed together, those cells with more cadherin on their surface end up on the inside of the reaggregate, surrounded by the cells with less surface cadherin (Fig. 9.3). Thus, quantitative differences in cell-adhesion molecules could maintain differential cell adhesion.

Cadherin molecules bind to each other through their extracellular domains, whose structure is strictly dependent on calcium binding, but adhesion is not exclusively controlled by these domains. The cytoplasmic domain of a cadherin associates with the actin filaments of the cytoskeleton by means of a protein complex containing α- and β-catenins (see Box 9A), and failure to make this association results in weak adhesion. That is, just as there is information flowing from outside the cell to the inside via the cadherin–catenin complex, there is also information flowing from the inside of the cell outwards.

In early *Xenopus* blastulas, for example, E-cadherin is expressed in the ectoderm just before gastrulation and N-cadherin appears in the prospective neural plate. If E-cadherin lacking an extracellular domain is produced in the blastulas from injected

Fig. 9.3 Sorting out of cells carrying different amounts of adhesion molecules on their surfaces. When two cell lines with different amounts of N-cadherin on their surface are mixed, they re-assort with the cells containing the most N-cadherin (green stained cells) ending up on the inside.

*Photographs courtesy of M. Steinberg, from Foty, R.A., and Steinberg, M.S.: **The differential adhesion hypothesis: a direct evaluation.** Developmental Biology 2005, 278: 255–263.*

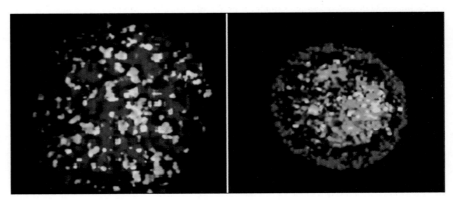

mutated mRNA, the mutant cadherin will compete with the embryo's own intact cadherin molecules for the cytoskeletal association sites. The defective cadherin cannot influence cell–cell adhesion because it has no extracellular domain, but it blocks the intact cadherin's access to the cytoskeleton and affects the adhesion between cells. The result is disruption of the ectoderm during gastrulation, indicating that a cadherin molecule must bind both to its partner on an adjacent cell and to the cytoskeleton of its own cell to create a stable adhesion. The initial binding of the extracellular cadherin domains transmits a signal to the cytoskeleton, which then stabilizes the interaction. The elements of the cytoskeleton are outlined in Box 9B.

9.3 Transitions of tissues from an epithelial to a mesenchymal state, and vice versa, involve changes in adhesive junctions

Morphogenesis is driven by interactions within and between two kinds of supracellular architectures—epithelial and mesenchymal. In an epithelium, adherens junctions hold the cells tightly together to form a continuous sheet. Mesenchyme has a much looser organization. Associations between individual mesenchymal cells are weaker, and in some cases the cells lack mature adhesive junctions. They are often partly surrounded by extracellular matrix and tend to contact each other through very thin processes, and become connected by **gap junctions**. These are protein pores in the apposed cell membranes of two adjacent cells through which ions and some small molecules can pass directly from the cytoplasm of one cell to the other. Epithelial cells can also be connected by gap junctions.

As well as serving familiar barrier functions in differentiated tissues, epithelia are present within early embryos as dynamic and transient structures, as we have seen in previous chapters. The first morphogenetic event in the embryo is the formation of an epithelium. Epithelia serve as the starting tissues from which different cell types emerge, and they can be sculpted into different forms by the coordinated actions of the cells within them.

A frequent event in early development is the conversion of an epithelium into a more loosely connected mass of mesenchyme, or even into individual mesenchymal cells that can migrate. This change, which involves the dissolution of the adherens junctions between the epithelial cells, is known as an **epithelial-to-mesenchymal transition (EMT)** (Fig. 9.4). We have already seen such transitions at work in gastrulation in *Drosophila* (see Section 2.2) and the sea urchin (see Section 6.9), in which mesoderm cells detach from the ectoderm once inside the embryo, and in chick and mouse, where cells of the epiblast detach from the epithelium and move inside the embryo through the primitive streak (see Chapter 3). In this chapter we shall look more closely at the mechanisms underlying the epithelial-to-mesenchyme transition in these situations.

The reverse process, a **mesenchymal-to-epithelial (MTE)** transition, is also found in embryogenesis. For example, mesenchymal mesoderm condenses into blocks of epithelia to form the somites (see Chapter 5) and into tubules of epithelium to form blood vessels and kidney tubules. This type of transition involves the formation of

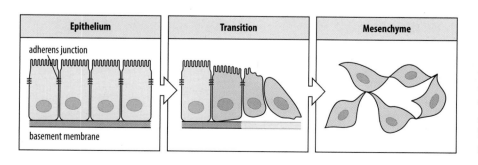

Fig. 9.4 Epithelial-to-mesenchymal transition. A common feature of embryogenesis is the conversion of an epithelium into more loosely organized mesenchyme. This occurs by dissolution of the adherens junctions between adjoining epithelial cells, the detachment of cells from the basement membrane, and their conversion through intermediate phenotypes into individual or loosely adhering mesenchymal cells.

CELL BIOLOGY BOX 9B The cytoskeleton, cell-shape change and cell movement

Cells actively undergo major changes in shape during development. The two main changes are associated with cell migration and with the infolding of epithelial sheets. Changes in shape are generated by the cytoskeleton, an intracellular protein framework that also controls changes in cell movement. There are three principal types of protein polymers in the cytoskeleton—**actin filaments (microfilaments)**, **microtubules**, and **intermediate filaments**—as well as many other proteins that interact with them. Actin filaments and microtubules are dynamic structures, polymerizing and depolymerizing according to the cell's requirements. Intermediate filaments are more stable, forming rope-like structures that transmit mechanical forces, spread mechanical stress, and provide mechanical stability to the cell. Microtubules play an important part in maintaining cell asymmetry and polarity, and provide tracks along which motor proteins convey other molecules and even organelles. They also form the spindles that segregate chromosomes at mitosis and meiosis. Polymerization and depolymerization of actin filaments at the leading edges of cells are involved in cell shape change and cell movements, and actin filaments also associate with the motor protein myosin to form contractile bundles of **actomyosin**, which are primarily responsible for force-generating contractions within cells.

Actin filaments are fine threads of protein about 7 nm in diameter and are polymers of the globular protein actin. They are organized into bundles and three-dimensional networks, which, in most cells, lie mostly just beneath the plasma membrane, forming a gel-like cell cortex. Numerous actin-binding proteins are associated with actin filaments, and are involved in bundling them together, forming networks, and aiding the polymerization and depolymerization of the actin subunits. Actin filaments can form rapidly by polymerization of actin subunits and can be equally rapidly depolymerized. This provides the cell with a highly versatile system for assembling actin filaments in a variety of different ways and in different locations, as required. The fungal drug cytochalasin D causes the disruption of actin filaments and also prevents actin polymerization and is a useful agent for investigating the role of actin networks. Actin filaments can also assemble with myosin into contractile structures, which act as miniature muscles. For example, contraction of the **contractile ring**, bundles of actomyosin arranged in a ring in the cell cortex, pinches animal cells in two at cell division (see Figure 1, left panel). The contraction of a similar ring of actomyosin around the apical end of a cell leads to apical constriction and elongation of the cell (see Figure 1, center panel).

Many embryonic cells can migrate over a solid substratum, such as the extracellular matrix. They move by extending a thin sheet-like layer of cytoplasm known as a lamellipodium (see Figure 1, right panel), or long, fine cytoplasmic processes called filopodia. Both these temporary structures are pushed outwards from the cell by the assembly of actin filaments. Contraction of the actomyosin network at the front of the cell then draws the cell forward. To do this, the contractile system must be able to exert a force on the substratum and this occurs at focal contacts, points at which the advancing filopodia or lamellipodium are anchored to the surface over which the cell is moving. At focal contacts, integrins (see Box 9A) both adhere to extracellular matrix molecules through their extracellular domains, and provide an anchor point for actin filaments through their cytoplasmic domains. Integrins transduce signals from the extracellular matrix across the plasma membrane at focal contacts, enabling the cell to sense the environment over which it is traveling and to adjust its movement accordingly.

The small GTP-binding proteins known as the Rho-family GTPases, which include Rho, Rac, and Cdc42, have a key role in regulating the actin cytoskeleton. Rac is required for extension of lamellipodia, for example, and Cdc42 is necessary to maintain cell polarity. Rho is involved in the planar cell polarity signaling pathway (see Box 9C). Signals from the environment relayed to the cytoskeleton via the Rho-family proteins are thus able to control cell movement.

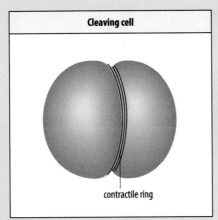

Cleaving cell

contractile ring

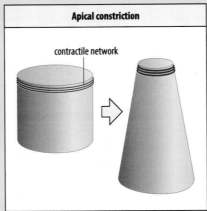

Apical constriction

contractile network

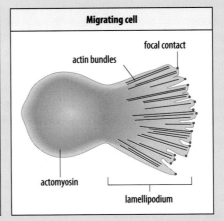

Migrating cell

focal contact

actin bundles

actomyosin

lamellipodium

Figure 1

adherens junctions between the cells. Later in their development, cells from the somites will undergo epithelial-to-mesenchymal transitions as they give rise to prospective muscle cells.

SUMMARY

The adhesion of cells to each other and to the extracellular matrix maintains the integrity of tissues and the boundaries between them. The associations of cells with each other are determined by the cell-adhesion molecules they express on their surface: cells bearing different adhesion molecules, or different quantities of the same molecule, sort out into separate tissues. This is due to the effects of intercellular adhesiveness on the cortical cytoskeleton and thus on cell-surface tension and interfacial tension between tissues.

Cell-cell adhesion is due mainly to two classes of surface proteins: the cadherins, which bind in a calcium-dependent manner to identical cadherins on another cell surface; and members of the immunoglobulin superfamily, some of which bind to similar molecules on other cells, while others bind to different molecules, such as integrins. The binding of this second class of surface proteins is calcium independent. Adhesion to the extracellular matrix is mediated by a third class of adhesion molecules—the integrins. Stable cell-cell adhesion through cadherins involves both physical contact between the external domains of cadherins on adjacent cells, and connection of the cytoplasmic domain of the cadherin molecule to the cell cytoskeleton via a protein complex containing catenins. Because adhesion molecules are connected to the cytoskeleton, they can transmit signals from the extracellular environment to the cytoskeleton to cause, for example, changes in cell shape. An important feature of morphogenesis is the transition of cells from an epithelial organization to form the more loosely organized mesenchyme, a change known as an epithelial-to-mesenchymal transition. This involves the dissolution of adhesive junctions. The opposite transition is the condensation of mesenchymal cells into epithelial structures, called a mesenchymal-to-epithelial transition.

Cleavage and formation of the blastula

The first step in animal embryonic development is the division of the fertilized egg by cleavage into a number of smaller cells (blastomeres), followed in many animals by the formation of a hollow sphere of cells—the blastula (see Chapter 3). In early embryos, cleavage involves short mitotic cell cycles, in which cell division and mitosis succeed each other repeatedly without intervening periods of cell growth. During cleavage, therefore, the mass of the embryo does not increase.

Early cleavage patterns vary widely between different groups of animals (Fig. 9.5). In **radial cleavage**, the divisions occur at right angles to the egg surface and the first few cleavages produce tiers of blastomeres that sit directly over each other. This type of cleavage is characteristic of the deuterostomes, such as sea urchins (see Chapter 6) and vertebrates. The eggs of molluscs and annelids, which are protostomes, illustrate another cleavage pattern, called **spiral cleavage**, in which successive divisions are at planes at slight angles to each other, producing a spiral arrangement of cells. In both radial and spiral cleavage, some divisions may be unequal. The first three cleavages in the sea urchin, for example, give rise to equal-sized blastomeres, whereas the later cleavage that forms the micromeres is unequal, with one daughter cell being smaller than the other (see Fig. 6.20). In nematodes, the very first cleavage of the egg is unequal, producing two cells of different sizes (see Fig. 9.5). In the early *Drosophila* embryo, the nuclei undergo repeated divisions without cell division, forming a syncytium, and separation into individual cells only occurs later, with the growth of cell membranes between the nuclei (see Section 2.1).

Radial cleavage—sea urchin	Spiral cleavage—annelids and molluscs	Unequal cleavage—nematode

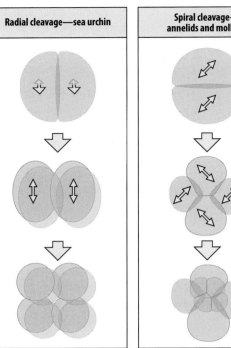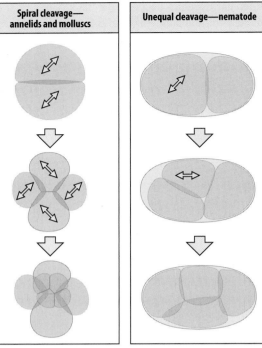

Fig. 9.5 Different patterns of early cleavage are found in different animal groups. Radial cleavage (for example, sea urchin) gives tiers of blastomeres sitting directly above each other. In spiral cleavage (for example, molluscs and annelid worms), the mitotic apparatus is oriented at a slight angle to the long axis of the cell, resulting in a spiral arrangement of blastomeres. Unequal cleavage (for example, nematode) results in one daughter cell being larger than the other. The blue arrows indicate the orientation of the mitotic spindle. Cleavage occurs at right angles to this axis.

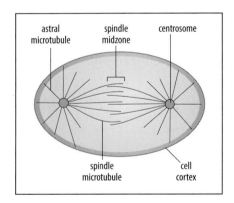

Fig. 9.6 The mitotic spindle. The mitotic spindle is the structure constructed by the cell that will segregate the duplicated chromosomes into two identical sets. It is composed of microtubules that emanate from two centrosomes that become positioned opposite to each other in the cell and which form the two poles of the spindle. One set of microtubules, the spindle microtubules, grows out from each centrosome to meet and overlap with the spindle microtubules from the opposite pole at the spindle midzone. The spindle midzone marks the point at which the cell will divide. A second set of microtubules, the astral microtubules, anchors each centrosome to the cell cortex. They are so called because they form a star-shaped structure, called an aster, at each pole. Chromosomes are not shown for simplicity.

The amount of yolk in the egg can influence the pattern of cleavage. In yolky eggs undergoing otherwise equal cleavages, a cleavage furrow develops in the least yolky region and gradually spreads across the egg, but its progress is slowed or even halted by the presence of the yolk. In these embryos, cleavage may thus be incomplete for some time. This effect is most pronounced in the heavily yolked eggs of birds and zebrafish, where complete divisions are restricted to a region at one end of the egg, and the embryo is formed as a cap of cells sitting on top of the yolk (see Figs. 3.13 and 3.15). Even in moderately yolky eggs, such as those of amphibians, the presence of yolk can influence cleavage patterns. In frog eggs, for example, the later cleavages are unequal and asynchronous, resulting in an animal half composed of a mass of small cells and a yolky vegetal region composed of fewer and larger cells.

Two key questions arise in relation to the mechanics of early cleavages. How are the positions of the cleavage planes determined? And how does cleavage lead to a hollow blastula (or its equivalent), which has a clear inside–outside polarity (that is, the outer surface of the blastula epithelium is markedly different from the inner surface)?

9.4 The orientation of the mitotic spindle determines the plane of cleavage at cell division

Mitosis and cell division depend on the formation of a **mitotic spindle**, a structure formed by microtubules that grow out from two **centrosomes** that are positioned opposite each other in the cell and form the two poles of the spindle. The microtubules from each pole overlap where they meet to form the **spindle midzone**, which is in general equidistant from each spindle pole. Other microtubules radiate away from the centrosome with their ends anchored to motor proteins in the actomyosin-rich cortex of the cell, forming an arrangement called an **aster** at each pole (Fig. 9.6). The duplicated chromosomes become attached to spindle microtubules at the midzone, and the chromatids of each pair are subsequently pulled back to opposite ends of the cell by shortening of these microtubules. Two new nuclei are formed, and the cell divides to form two new cells by cleaving at right angles to the spindle axis at the position of the spindle midzone. In animal cells, cleavage is achieved by the constriction of an actomyosin-based **contractile ring**, and the position of cleavage becomes visible on the cell surface as the cleavage furrow (see Box 9B).

The orientation and position of the plane of cleavage at cell division is of great importance in embryonic development. They not only determine the spatial arrangement of daughter cells relative to each other, and whether the cells will be equal or unequal in size, but the position of the plane of cleavage can also result in the apportionment of different cytoplasmic determinants to different daughter cells, thus giving them different fates, as for example in the first cleavage division in nematodes (see Section 6.1).

The orientation of cleavage in general follows an old rule, the Hertwig rule, which says that a cell orients the mitotic spindle along the longer axis of the cell before division. Cleavage then occurs perpendicular to this axis. All cells, even epithelial cells, round up during division, which would erase any cue of asymmetry and orientation. However, it seems that the actin cytoskeleton maintains a memory of the original shape of the cell by leaving fine, actin-based retraction fibers attached to the cell's original points of attachment to the extracellular matrix. The guide for the spindle to orient itself, therefore, is a footprint of the shape of the cell before the division in the form of actin fibers, and this can position the spindle accordingly. Spindle orientation is particularly important in the cells of an epithelium, because to maintain a continuous single layer the plane of cleavage must be at right angles to the epithelial surface. The rule of the long axis can be overridden by signals delivered through the actin cytoskeleton that target the spindle and orient it in another direction.

A common form of early animal embryo is the hollow spherical blastula, composed of an epithelial sheet enclosing a fluid-filled interior. The development of such a structure from a fertilized egg depends both on particular patterns of cleavage and on changes in the way cells pack together, as shown in schematic form in Fig. 9.7. If the plane of cell division is always at right angles to the surface, the cells will remain in a single layer, as shown in Fig. 9.7. As cleavage proceeds, individual cells get smaller, but the surface area of the cell sheet increases and a space—the blastocoel—forms in the interior, which increases in volume at each round of cell division. This is essentially how the sea urchin blastula is formed.

The orientation of the spindle and the resulting plane of cleavage are determined by the behavior of the centrosomes. Before mitosis, the single centrosome in the cell becomes duplicated and the daughter centrosomes move to opposite sides of the nucleus and form the asters, whose interactions with cortical microtubules will anchor and help to orient the mitotic spindle. The pattern of centrosome duplication and movement usually results in successive planes of cell division being at right angles to each other. An example of this is the division of the nematode AB cell, one of the pair of cells formed by the first cleavage of the zygote, where the mitotic spindle in the dividing AB cell is oriented roughly at right angles to that of the zygote (Fig. 9.8). However, in the other product of

Fig. 9.7 Specifically oriented cell divisions convert a small ball of cells into a hollow spherical blastula. Left panel: when adjacent cells in a prospective epithelium, such as the early amphibian blastula, make contact with each other over large areas of cell surface, the overall volume of the cell aggregate is relatively small as there is little space at its center. Middle panel: with a decrease in the area of intercellular contact, the size of the internal space (the blastocoel), and thus the overall volume of the blastula, is greatly increased without any corresponding increase in cell number or total cell volume. Right panel: if the number of cells is increased by radial cleavage, and the packing remains the same, the blastocoel volume will increase further, again without any increase in total cell volume.

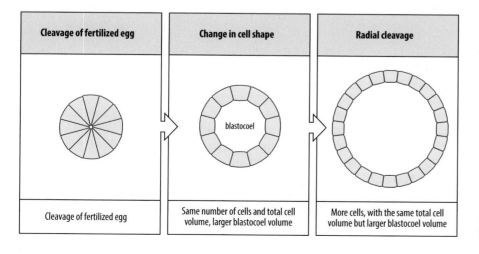

Cleavage of fertilized egg	Change in cell shape	Radial cleavage
Cleavage of fertilized egg	Same number of cells and total cell volume, larger blastocoel volume	More cells, with the same total cell volume but larger blastocoel volume

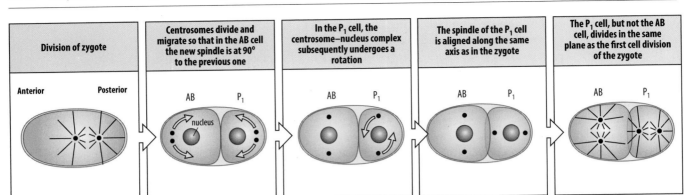

Fig. 9.8 Different planes of cleavage in different cells are determined by the behavior of the centrosomes. The first cleavage of the nematode zygote divides it into an anterior AB cell and a posterior P_1 cell. Before the next division, the duplicated centrosomes of the AB cell move apart in such a way that the plane of cleavage will be at right angles to the first division. The duplicated centrosomes in the P_1 cell initially move into a similar alignment, but before mitosis, the nucleus and its associated centrosomes rotate 90° so that cleavage is in the same plane as the first cleavage.

Illustration after Strome, S.: **Determination of cleavage planes.** *Cell 1993, 72: 3–6.*

the first cleavage, the P_1 cell, the centrosomes initially migrate to the same positions as in the AB cell but the nucleus and its associated centrosomes then rotates 90° towards the anterior of the cell, so that the eventual plane of cleavage of P_1 is the same as that of the zygote. The different planes of cleavage in the AB and P_1 cells are likely to be under the ultimate control of cytoplasmic factors, such as the PAR proteins, that are differentially distributed at the first cleavage and have effects on spindle positioning (see Section 6.2).

The poles of the mitotic spindle are positioned by a balance between forces that tend to pull the centrosomes apart and those that tend to pull them towards each other. Microtubule-associated motor proteins anchor the astral microtubules in the cortex of the cell and tend to pull the spindle poles towards the cortex. Similar motor proteins cross-link the overlapping spindle microtubules and either tend to pull them inwards, pulling the spindle poles towards each other, or push them outwards, pushing the spindle poles away from each other. The actomyosin-rich cell cortex, by undergoing contraction or relaxation, can also exert force on the centrosomes and the subsequent spindle poles through the astral microtubules, and thus influence the position of the spindle in the cell.

9.5 The positioning of the spindle within the cell also determines whether daughter cells will be the same or different sizes

The position of the cleavage site in relation to the axis of the cell determines whether the daughter cells are equal or unequal in size. When the mitotic spindle is positioned centrally within the cell, and each arm of the spindle is of equal length, the cleavage plane at the spindle midzone will be equidistant from the two ends of the cell, and two cells of equal size will be produced at division. If the two cells have the same contents and the same developmental potential, such a division is called a **symmetric division**. Most cell proliferation is of this type. However, depending on the location of a cytoplasmic determinant in the cell, geometrically equal divisions can sometimes lead to cells with different content, and thus with different fates. In the developmental sense, these are considered to be **asymmetric divisions**, because the result is two cells with different content and often with different fates.

If the spindle moves away from the center of the cell, and/or one of its arms becomes extended so that the structure is no longer symmetrical, the spindle midzone will no longer correspond to the center of the cell's long axis and cleavage will produce daughter cells of unequal sizes. Such divisions are generally, but not necessarily, asymmetric and lead to cells with different developmental fates as a result of the differential distribution of cytoplasmic determinants. The position of the cleavage plane depends on the forces acting on the poles of the spindle. In an equal division, the forces are equal on both poles, and the spindle is symmetrical, whereas in an unequal division, signals from the cortex at one pole lead to a change in the position of the pole, resulting in an asymmetrically positioned spindle (see Fig. 6.6).

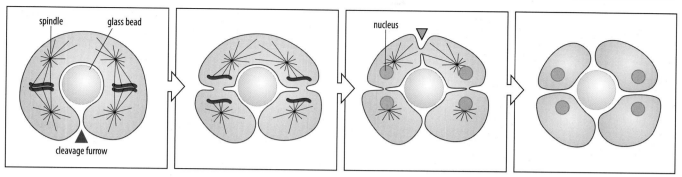

Early experiments investigating the role of the spindle in cell division used ingenious ways of manipulating dividing echinoderm eggs so that spindles were moved out of their normal position or placed in unusual configurations. In one such experiment, a glass bead on the end of a needle was inserted into a fertilized sand-dollar egg so that it interrupted the first cleavage furrow. This created a horseshoe-shaped cell in which the two nuclei generated by the first mitotic division were segregated into separate arms of the horseshoe (Fig. 9.9). At the next mitosis, chromosome-laden spindles formed normally in each arm of the horseshoe and each was bisected by a cleavage furrow, but a third furrow also formed between the two adjacent asters at the top of the horseshoe.

This observation was originally interpreted as showing that the asters were able to direct the formation of a cleavage furrow independently of the spindle by delivering a signal of some kind to the cortex. However, the two centrosomes involved do have some microtubules extending in the cytoplasm towards the plane of the cleavage furrow, where they meet (see Fig. 9.9), and these could also be exerting an influence.

Subsequently, other experiments also indicated that the position of cleavage can be determined by signals from the spindle midzone—directing the formation of the contractile ring in the cortex nearest to the midzone. In one set of experiments, a physical barrier placed between the spindle midzone and the cortex in both sand-dollar eggs and cultured mammalian cells was found to block cleavage furrow formation.

The asters might also influence contractile-ring positioning indirectly. In preparation for mitosis, tension increases in the cell cortex overall as a result of cytoskeletal contraction, which causes the cell to round up. It has been suggested that the interaction of astral microtubules with the cortex at each spindle pole relaxes the cortex equally in the area around each pole, leaving a region of greater contractility midway between the two poles—in which the contractile ring will form. It is likely, therefore, that the position of the contractile ring is determined by a combination of signals from various sources acting on the cortex.

Fig. 9.9 The role of asters and spindles in determining the position of the cleavage furrow. If the mitotic apparatus of a fertilized sea-urchin egg is displaced at the first cleavage by a glass bead, the cleavage furrow forms only on the side of the egg to which the mitotic apparatus has been moved. At the next cleavage, furrows bisect each mitotic spindle as expected (only a single chromosome is shown on the spindle for simplicity), but an additional furrow, indicated by the blue triangle, forms in the absence of chromosomes between the two adjacent asters of different mitotic spindles. This was originally interpreted as the asters being able to direct the formation of a cleavage furrow independently of the presence of a spindle. However, microtubules emanating from each of these asters do meet at the cleavage position, and could therefore have a role in defining the position of the cleavage furrow.

9.6 Cells become polarized in the sea-urchin blastula and the mouse morula

The emergence of polarity in the cells of early embryos is a most important event in development, as in some cases it is instrumental in assigning differences in cell fate by the polarized localization of cytoplasmic determinants and their consequent unequal distribution at cell division. One type of polarity found in epithelia is **apico-basal polarity**, which describes the differences between the two faces of an epithelium.

The sea-urchin egg cleaves in such a way that the surface of the egg, which is covered in non-motile actin-based protrusions called microvilli, becomes the outer surface of the blastula. A layer of extracellular matrix called the hyaline layer, which contains the proteins hyalin and echinonectin, is secreted onto the external surface and the cells—the blastomeres—adhere to this sheet. Adhesive cell–cell junctions develop between adjacent blastomeres, forming them into an epithelium and the cell contents become polarized in an apical–basal direction, with the Golgi apparatus

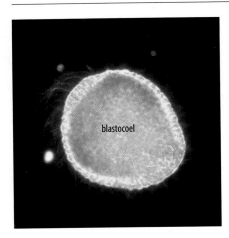

Fig. 9.10 Blastula of a sea-urchin embryo, *Lytechinus pictus*. A single-layered epithelium surrounds the hollow blastocoel. The epithelial cells bear cilia on their outer surface. Scale bar = 10 μm.

From Raff, E.C., et al.: ***Experimental taphonomy shows the feasibility of fossil embryos.*** *PNAS 2006, **103**: 5846-5851. Copyright 2006 National Academy of Sciences, USA.*

oriented toward the apical surface, which in this case is the outer surface of the blastula. In the sea-urchin blastula the cells are held together laterally by cadherin-based adherens junctions and by **septate junctions**, a type of cell junction found in invertebrates but not in chordates. The septate junction is thought to provide the same type of permeability barrier as the vertebrate **tight junction**—sealing the epithelium and so preventing water, small molecules, and ions from permeating freely between the cells. The epithelial cells can then control, through their cell membranes, what enters and exits the blastula. On the inner (basal) surface of the epithelium, a basement membrane (an organized layer of extracellular matrix) is laid down. As the blastula develops, cilia are produced on the outer surface (Fig. 9.10).

In the mouse embryo, the earliest sign of structural differentiation is at the eight-cell stage, when the morula undergoes a process known as **compaction** (Fig. 9.11). Until then, cells divide equally without a preferred orientation and the blastomeres form a loosely packed ball, with each cell surface uniformly covered by microvilli. At compaction, blastomeres flatten against each other, maximizing cell–cell contact, and cells become polarized, with an apico-basal polarity, defined by the presence of E-cadherin-based adherens junctions on the lateral faces and by microvilli confined to the apical surface (Fig. 9.12). At the next round of division, each cell divides along one of two possible planes: radial cleavages produce two polarized cells, but some cleavages now occur tangentially (parallel to the surface), each producing one outer polarized cell and one inner non-polarized cell. This results in two cell populations with different developmental fates: the non-polarized cells become the inner cell mass, express the pluripotency markers *Nanog* and *Oct4* (see Chapter 8), and will give rise to the embryo proper and the yolk sac. The outer cells develop adherens and tight junctions (Fig. 9.13), express a different set of genes, for example, *Cdx2*, and form the trophectoderm, the outer epithelium of the blastocyst that will give rise to the placenta (see Fig. 5.6, first panel).

The formation of cadherin-based adherens junctions is a major feature of compaction. At the two-cell and four-cell stages, E-cadherin is uniformly distributed over blastomere surfaces and contact between cells is not extensive. At the eight-cell stage, however, E-cadherin becomes restricted to regions of intercellular contact, where it now acts for the first time as an adhesion molecule, forming adherens junctions. The adhesive properties of E-cadherin result from its becoming linked to cytoskeletal elements in the cortex via β-catenin (see Box 9A). Formation of the outer epithelial layer is essential for further development; in embryos mutant for E-cadherin it does not form, resulting in a disorganized mass of cells with abnormal ratios of cells expressing *Oct4* (prospective inner cell mass) or *Cdx2* (prospective trophectoderm) (Fig. 9.14).

Compaction is associated with a radical remodeling of the cell cortex. Microvilli disappear from the regions of intercellular contact at the same time that actin, the motor

Fig. 9.11 Compaction of the mouse embryo. At the eight-cell stage the cells have relatively smooth surfaces and microvilli are distributed uniformly over the surface. At compaction, microvilli are confined to the outer surface and cells increase their area of contact with one another. Scale bar = 10 μm.

Photograph courtesy of T. Bloom reproduced with permission from Bloom T.L.: ***The effects of phorbol ester on mouse blastomeres: a role for protein kinase C in compaction?*** *Development 1989, **106**: 159-171.*

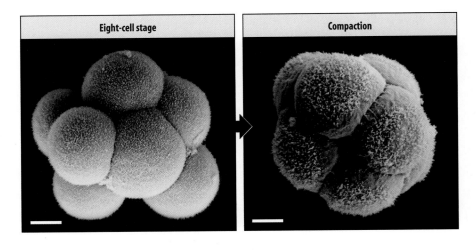

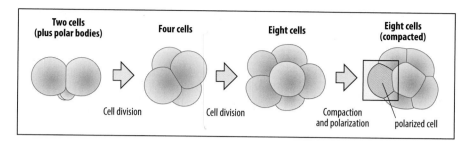

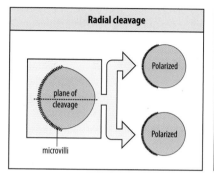 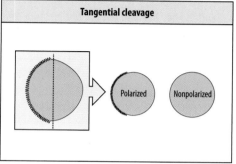

Fig. 9.12 Polarization of cells during cleavage of the mouse embryo. At the eight-cell stage, compaction takes place, with the cells forming extensive contacts with each other (top panel). The cells also become polarized. For example, microvilli, which were initially uniformly distributed over the cell surface, become confined to the outward-facing cell surface. Future cleavages can thus divide the cell either into two polarized cells (by a radial cleavage, bottom left panel) or into a polarized and a nonpolarized cell (by a tangential cleavage, bottom right panel). The tangential cleavages give rise to the inner cell mass.

protein myosin, and another cytoskeletal protein, spectrin, are cleared from these regions to become concentrated in a band around the apical face of the polarized cell. Conversely, E-cadherin becomes localized within the regions of contact.

9.7 Fluid accumulation as a result of tight-junction formation and ion transport forms the blastocoel of the mammalian blastocyst

Accumulation of fluid in the interior space, or blastocoel, of a blastocyst or blastula exerts an outward pressure on the blastula wall, and this hydrostatic pressure is one of the forces that forms and maintains the spherical shape. The blastocoel is formed in the preimplantation mammalian embryo by a cavitation process in which fluid accumulates in the interior of the developing blastocyst. The inflow of water is linked to the active transport of sodium ions and other electrolytes into the extracellular spaces (Fig. 9.15).

At the eight-cell stage, tight junctions start to be formed between the polarized cells of the outer layer of the mammalian morula. By the 32-cell stage this barrier is fully formed. At the same time, the Na^+/K^+-ATPase sodium pump and other membrane transport proteins become active in the basolateral membranes of the outer cells, transporting sodium and other electrolytes into the extracellular space, which is continuous with the blastocoel. As the ion concentration in the blastocoel fluid increases, water is drawn into the blastocoel by osmosis through aquaporin water channels in the cell membranes, and the increase in hydrostatic pressure caused by the accumulating fluid stretches the surrounding epithelial layer. A similar mechanism based on ion transport seems to operate in drawing fluid into the blastocoel in the *Xenopus* blastula. In the case of *Xenopus*, a small blastocoel can be observed by electron microscopy as early as the two-cell stage.

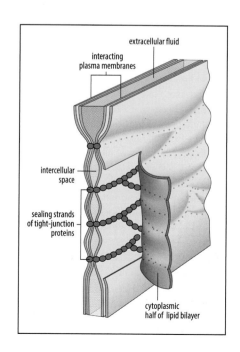

Fig. 9.13 The structure of the tight junction. The apposing membranes of adjacent epithelial cells are held closely together by the tight junction, which is located towards the apex of the cell and forms a tight seal between the extracellular and intracellular environment, preventing water, small molecules, and ions from permeating freely between the cells. The membranes are held together by horizontal rows (sealing strands) of membrane proteins (occludins and claudins), which bind tightly to each other. The pink shading represents the space outside the epithelium; the yellow shading represents the spaces between cells inside the epithelium.
From Alberts, B., et al.: Molecular Biology of the Cell, Fifth Edition, Garland Science, New York, 2008.

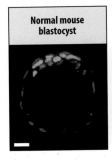

| Normal mouse blastocyst | Mutant mouse embryo lacking E-cadherin |

Fig. 9.14 Mouse embryos mutant for E-cadherin do not undergo compaction and do not develop into a blastocyst. The first panel shows a normal mouse blastocyst. The inner cell mass (green, cells expressing the pluripotency factor Oct4) lies at one end of a fluid-filled space surrounded by the trophectoderm (red, cells expressing Cdx2). The second panel shows the disorganized mass of cells that develops from cleavage of a mutant zygote lacking E-cadherin. Most of the cells express the trophectodermal marker Cdx2, with many fewer cells expressing Oct4. Scale bars, 20 μm.

*From Stephenson, R. O. et al.: **Disorganized epithelial polarity and excess trophectoderm cell fate in preimplantation embryos lacking E-cadherin**. Development 2010, **137**: 3383-3391.*

Fig. 9.15 The blastocoel of the mammalian blastocyst is formed by the inflow of water, creating an internal fluid-filled cavity. Left panel: active transport of sodium ions (Na⁺) across the basolateral membranes of trophectoderm cells into the extracellular space, which is continuous with the blastocoel cavity of the mouse blastocyst (yellow arrow), results in an increase in ion concentration in the extracellular space, which causes water to flow in from the cytoplasm by osmosis. The sodium in the trophectoderm cells is replaced by sodium from the isotonic fluid surrounding the blastocyst (white arrow) and water also enters the cell by osmosis. The blastocoel is sealed off from the outside by tight junctions between the trophectoderm cells at their outer edge, and so the inflow of water creates hydrostatic pressure that enlarges the blastocoel. Right panel: a surface view of a mouse blastocyst shows the tight junctions between epithelial cells.

*From Eckert, J.J., and Fleming, T.P.: **Tight junction biogenesis during early development**. Biochim. Biophys. Acta 2008, **1778**: 717-728.*

SUMMARY

In many animals, the fertilized egg undergoes a cleavage stage that divides the egg into a number of small cells (blastomeres) and eventually gives rise to a hollow blastula. Various patterns of cleavage are found in different animal groups. In some animals, such as nematodes and molluscs, the plane of cleavage is of importance in determining the position of particular blastomeres in the embryo, and the distribution of cytoplasmic determinants. The plane of cleavage in animal cells is determined by the orientation of the mitotic spindle, which is in turn determined by the final positioning of the asters. At the end of the cleavage period, the blastula essentially consists of a polarized epithelium surrounding a fluid-filled blastocoel. Fluid accumulation in the blastocoel may be partly due to active ion transport into the blastocoel, drawing water in after it by osmosis.

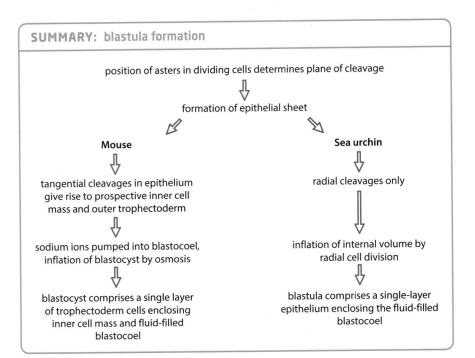

SUMMARY: blastula formation

position of asters in dividing cells determines plane of cleavage
⇩
formation of epithelial sheet

Mouse
⇩
tangential cleavages in epithelium give rise to prospective inner cell mass and outer trophectoderm
⇩
sodium ions pumped into blastocoel, inflation of blastocyst by osmosis
⇩
blastocyst comprises a single layer of trophectoderm cells enclosing inner cell mass and fluid-filled blastocoel

Sea urchin
⇩
radial cleavages only
⇩
inflation of internal volume by radial cell division
⇩
blastula comprises a single-layer epithelium enclosing the fluid-filled blastocoel

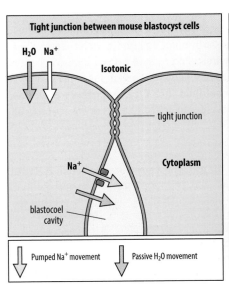

Tight junction between mouse blastocyst cells

H₂O   Na⁺
Isotonic
tight junction
Na⁺
Cytoplasm
blastocoel cavity

Pumped Na⁺ movement   Passive H₂O movement

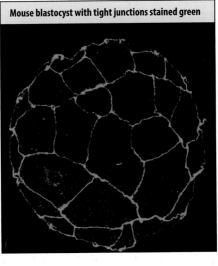

Mouse blastocyst with tight junctions stained green

Gastrulation movements

Movements of individual cells and cell sheets at gastrulation bring most of the tissues of the blastula (or its equivalent stage) into their appropriate position in relation to the body plan. The necessity for gastrulation is clear from fate maps of the blastula in, for example, sea urchins and amphibians, where the presumptive endoderm, mesoderm, and ectoderm can be identified as adjacent regions on the outside of the spherical blastula in the continuous cell sheet that encloses the blastocoel; the endoderm and mesoderm must therefore move from the surface to inside the early embryo (see Chapters 4, 5, and 6). After gastrulation, these tissues become completely rearranged in relation to each other. For example, the endoderm develops into an internal gut and is separated from the outer ectoderm by a layer of mesoderm.

Gastrulation thus involves dramatic changes in the overall structure of the embryo, converting it into a complex three-dimensional structure. During gastrulation, a program of cell activity involving changes in cell shape and adhesiveness remodels the embryo, so that the endoderm and mesoderm move inside and only ectoderm remains on the outside. The primary force for gastrulation is provided by changes in cell shape, as well as the ability of cells to migrate (see Box 9B). In some embryos, all this remodeling occurs with little or no accompanying increase in cell number or total cell mass.

In this part of the chapter, we first consider the relatively simple process of gastrulation in sea urchins and insects. *Xenopus* serves as our main model for the more complex gastrulation process in vertebrates. Because of the shape of the very early embryo, gastrulation in chick and mouse has some notable differences from that in *Xenopus*, even though the eventual body plan is very similar. We will finish this part of the chapter by briefly comparing and contrasting the types of gastrulation movements in chick and mouse with those in *Xenopus*.

9.8 Gastrulation in the sea urchin involves an epithelial-to-mesenchymal transition, cell migration, and invagination of the blastula wall

Just before gastrulation begins, the sea-urchin late blastula consists of a single-layered ciliated epithelium surrounding a fluid-filled blastocoel. The future mesoderm occupies the most vegetal region, with the future endoderm adjacent to it (Fig. 9.16). The rest of the embryo gives rise to ectoderm. The epithelium is polarized in an apico–basal direction: on its apical surface is the hyaline layer, whereas the basal surface facing the blastocoel is lined with a basement membrane, or basal lamina. The epithelial cells are attached laterally to each other by septate junctions and adherens junctions. Gastrulation involves various cell movements and cell behaviors that make use of the adhesive and motile properties of embryonic cells discussed in previous sections.

Gastrulation begins with an epithelial-to-mesenchymal transition, in which the most vegetal mesodermal cells leave the blastula epithelium and become motile and mesenchymal in form (Fig. 9.17). They become detached from each other and from the hyaline layer, and migrate into the blastocoel as single cells that have lost both their epithelial apico–basal polarity and their cuboidal shape. They are now known as **primary mesenchyme** (see Section 6.9). The transition to primary mesenchyme cells and entry into the blastocoel is foreshadowed by intense pulsatory activity on the inner face of these cells, while still part of the epithelium, and on occasion a small transitory infolding or invagination is seen in the surface of the blastula before migration begins properly. Cell internalization requires loss of cell–cell adhesion and is associated with repression of cadherin expression, the removal of cadherin from the cell surface by endocytosis, and the disappearance of the α- and β-catenins that link cadherins to the cytoskeleton (see Box 9A).

The epithelial-to-mesenchymal transition is regulated in part by the sea-urchin version of the gene *snail*, which encodes a transcription factor and, like many other

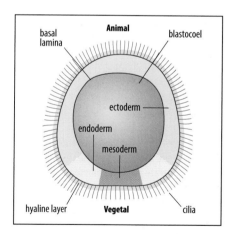

Fig. 9.16 The sea-urchin blastula before gastrulation. The prospective endoderm (yellow) and mesoderm (orange) are at the vegetal pole. The rest of the blastula is prospective ectoderm (gray). There is an extracellular hyaline layer and a basal lamina lines the blastocoel. The blastula surface is covered with cilia.

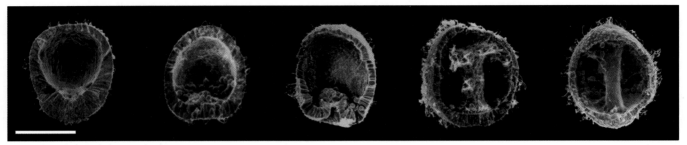

Fig. 9.17 Sea-urchin gastrulation. Cells of the vegetal mesoderm undergo the transition to primary mesenchyme cells and enter the blastocoel at the vegetal pole. This is followed by an invagination of the endoderm, which extends inside the blastocoel toward the animal pole, forming the embryonic gut, or archenteron. Filopodia extending from the secondary mesenchyme cells at the tip of the invaginating endoderm contact the blastocoel wall and draw the invagination to the site of the future mouth, with which it fuses, forming the gut. Scale bar = 50 μm.

Photographs courtesy of J. Morrill.

developmental genes, was first identified in *Drosophila* (see Section 2.16). Experiments using antisense morpholinos (see Box 6B) to knock down *snail* expression in early sea-urchin embryos show that it is required for the repression of cadherin gene expression and for cadherin endocytosis in the epithelial-to-mesenchymal transition. The role of the Snail family of transcription factors in this type of cellular change is highly conserved across the animal kingdom, as we shall see in this chapter. The actions of Snail and a related transcription factor, Slug, are also of medical interest. A similar epithelial-to-mesenchymal transition takes place in cancer cells, enabling them to undergo **metastasis** and migrate from the original site of the tumor to set up secondary cancers elsewhere in the body. In these metastatic cells, the normal low expression of *snail* and *slug* has increased.

In the sea urchin, the primary mesenchyme cells migrate within the blastocoel to form a characteristic pattern on the inner surface of the blastocoel wall. They first become arranged in a ring around the gut in the vegetal region at the ectoderm–endoderm border. Some then migrate to form two extensions towards the animal pole on the ventral (oral) side (Fig. 9.18). The migration path of individual cells varies considerably from embryo to embryo, but their final pattern of distribution is fairly constant. The primary mesenchyme cells later lay down the skeletal rods of the sea urchin endoskeleton by secretion of matrix proteins (see Section 6.15).

Fig. 9.18 Migration of primary mesenchyme in early sea-urchin development. The primary mesenchyme cells enter the blastocoel at the vegetal pole and migrate over the blastocoel wall by filopodial extension and contraction. Within a few hours they take up a well-defined ring-like pattern in the vegetal region with extensions along the ventral side.

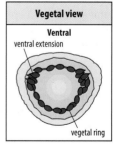

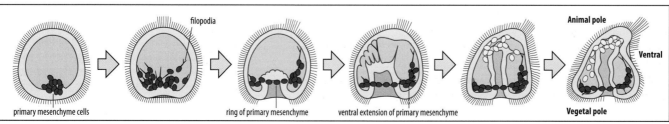

Primary mesenchymal cells move over the inner surface of the blastocoel wall by means of fine filopodia, which can be up to 40 μm long and can extend in several directions. Filopodia contain cross-linked bundles of actin filaments and extend forward by the rapid assembly of actin filaments that push out the leading edge of the filopodium (see Box 9B). At any one time, each cell has on average six filopodia, most of which are branched. When filopodia make contact with, and adhere to, the basal lamina lining the blastocoel wall, they retract, drawing the cell body toward the point of contact. As each cell extends several filopodia (Fig. 9.19), some or all of which may retract on contact with the wall, there seems to be competition between the filopodia, the cell being drawn towards that region of the wall where the filopodia make the most stable contact. The movement of the primary mesenchymal cells therefore resembles a random search for the most stable attachment. As the cells migrate, their filopodia fuse with each other, forming cable-like extensions.

If primary mesenchyme cells are artificially introduced by injection at the animal pole, they move in a directed manner to their normal positions in the vegetal region. This suggests that guidance cues, possibly graded, are distributed in the basal lamina. Even cells that have already migrated will migrate again to form a similar pattern when introduced into a younger embryo. The stability of contacts between filopodia and the blastocoel wall is one factor determining the pattern of cell migration. Analysis of movies of migrating cells suggests that the most stable contacts are made in the regions where the cells finally accumulate, namely the vegetal ring and the two ventro-lateral clusters (see Fig. 9.18).

The signaling molecules fibroblast growth factor A (FGF-A) and vascular epithelial growth factor (VEGF) have been implicated as guidance cues. In the blastula, FGF-A is expressed in ectodermal regions to which the primary mesenchyme will migrate and the FGF receptor is expressed on the primary mesenchyme cells. If signaling by FGF-A is inhibited, then so is cell migration and skeleton formation. VEGF is expressed by the prospective ectoderm cells at the most ventral sites to which the mesenchyme cells migrate, and the VEGF receptor is again expressed exclusively in the prospective primary mesenchyme. As with FGF, if VEGF or its receptor are lacking, the cells do not reach their correct positions and the skeleton does not form.

The entry of the primary mesenchyme is followed by the invagination and extension of the endoderm to form the embryonic gut (the archenteron). The endoderm invaginates as a continuous sheet of cells (see Fig. 9.17). Formation of the gut occurs in two phases. During the initial phase, the endoderm invaginates to form a short, squat cylinder extending up to halfway across the blastocoel. There is then a short pause, after which extension continues. In this second phase, the cells at the tip of the invaginating gut, which will later detach as the secondary mesenchyme, form long filopodia, which make contact with the blastocoel wall. Filopodial extension and contraction pull the elongating gut across the blastocoel until it comes in contact with and fuses with the mouth region, which forms a small invagination on the ventral side of the embryo (see Fig. 9.17). During this process the number of invaginating cells doubles, and this is in part due to cells around the site of invagination contributing to the hindgut.

How is invagination of the endoderm initiated? The simplest explanation is that a change in shape of the endodermal cells causes, and initially maintains, the change in curvature of the cell sheet (Fig. 9.20). At the site of invagination, the cuboid cells adopt a more elongated, wedge-shaped form, which is narrower at the outer (apical) face. This change in cell shape is the result of constriction of the cell at its apex by the contraction of actomyosin in the cortex (see Box 9B). Such a shape change is sufficient to pull the outer surface of the cell sheet inward and to maintain invagination, as shown by a computer simulation in which apical constriction, modeled to spread over the vegetal pole, results in an invagination (Fig. 9.21).

The primary invagination only takes the gut about a third of the way to its final destination. The second phase of gastrulation involves two different mechanisms:

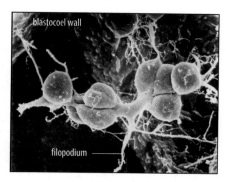

Fig. 9.19 Filopodia of sea-urchin mesenchyme cells. The scanning electron micrograph shows a group of primary mesenchyme cells moving over the blastocoel wall by means of their numerous filopodia, which can extend and contract. As illustrated here, the filopodia of adjacent cells fuse together.

Photograph courtesy of J. Morrill, from Morrill, J.B., Santos, L.L.: **A scanning electron micrographical overview of cellular and extracellular patterns during blastulation and gastrulation in the sea urchin, Lytechinus variegatus.** *In* The Cellular and Molecular Biology of Invertebrate Development. *Edited by Sawyer, R.H. and Showman, R.M. University of South Carolina Press, 1985; pp. 3-33.*

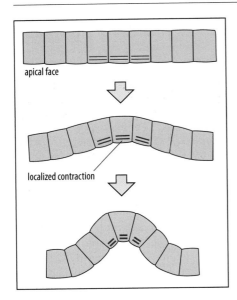

Fig. 9.20 Change in cell shape in a small number of cells can cause invagination of endodermal cells. Bundles of contractile filaments composed of actin and myosin contract at the outer edge of a small number of adjacent cells, making them wedge-shaped. As long as the cells remain mechanically linked to each other and to adjacent cells in the sheet, this local change in cell shape draws the sheet of cells inward at that point.

Fig. 9.21 Computer simulation of the role of apical constriction in invagination. Computer simulation of the spreading of apical constriction over a region of a cell sheet shows how this can lead to an invagination.

*Illustration after Odell, G.M., et al.: **The mechanical basis of morphogenesis. I. Epithelial folding and invagination**. Dev. Biol. 1981, **85**: 446–462.*

contraction of filopodia in contact with the future mouth region, as described above, and a filopodia-independent process. Treatments that interfere with filopodial attachment to the blastocoel wall result in failure of the gut to elongate completely, but it still reaches about two-thirds of its complete length. Filopodia-independent extension is due to active rearrangement of cells within the endodermal sheet. If a sector of cells in the vegetal mesoderm is labeled with a fluorescent dye before gastrulation, it is seen to become a long, narrow strip as the gut extends (Fig. 9.22). The sheet of cells has elongated in one direction while becoming narrower in width. We will look at this type of cell rearrangement—known as 'convergent extension'—in more detail later in this chapter, in relation to similar phenomena in *Drosophila*, amphibian and zebrafish gastrulation (see Box 9C).

What guides the tip of the gut to the future mouth region? As the long filopodia at the tip of the gut initially explore the blastocoel wall, they make more stable contacts at the animal pole and then in the region where the future mouth will form. Filopodia making contact there remain attached for 20–50 times longer than when attached to other sites on the blastocoel wall. Gastrulation in the sea urchin clearly shows how changes in cell shape, changes in cell adhesiveness, and cell migration all work together to cause a major change in embryonic form.

9.9 Mesoderm invagination in *Drosophila* is due to changes in cell shape controlled by genes that pattern the dorso-ventral axis

At the beginning of gastrulation, the *Drosophila* embryo consists of a blastoderm of about 6000 cells, which forms a superficial single-celled layer (see Section 2.1). Gastrulation begins with the invagination of a longitudinal strip of future mesodermal cells (9–10 cells wide) on the ventral side of the embryo, to form a ventral furrow and then a transient tube inside the body. The tube breaks up into individual cells, which spread out to form a single layer of mesoderm on the interior face of the ectoderm (Fig. 9.23). The gut develops slightly later, by invaginations of prospective endoderm near the anterior and posterior ends of the embryo (see Fig. 2.3), which we shall not consider here.

Invagination of the mesoderm occurs in two phases. First, the central strip of cells develops flattened and smaller apical surfaces, and their nuclei move away from the apical surface. These changes in cell shape result in formation of the furrow and its folding into the interior of the embryo to form a tube, the central cells forming the tube proper while the peripheral cells form a 'stem' (see Fig. 9.23, third panel). This first phase takes about 30 minutes. During the second phase, which takes about an hour, the tube dissociates into individual cells, which proliferate and spread out laterally. The entry of cells into cell division is delayed during invagination and this delay is essential for the cell-shape changes to occur.

Mutant *Drosophila* embryos that have been dorsalized or ventralized (see Section 2.12) show that this behavior of the mesodermal cells is autonomous, as in the sea urchin (see Section 9.8), and is not affected by adjacent tissues. In dorsalized embryos, which have no mesoderm, no cells undergo nuclear migration or apical contraction. In ventralized embryos, in which most of the cells are mesoderm, these changes occur throughout the whole of the dorso-ventral axis.

Gastrulation in *Drosophila* provided the first direct link between the action of patterning genes and morphogenesis, in this case through the involvement of such genes

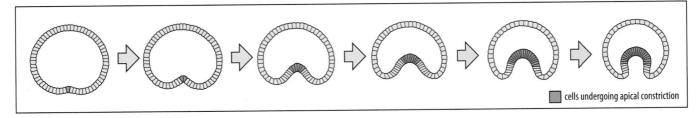

cells undergoing apical constriction

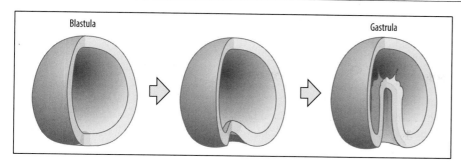

Blastula

Gastrula

Fig. 9.22 Gut extension during sea-urchin gastrulation. Labeling of cells in the vegetal region of the blastula (green) shows that gut extension involves cell rearrangement within the endoderm, causing the labeled cells to become distributed in a long narrow strip.

in causing changes in cell shape and cell adhesiveness. In *Drosophila*, the transcription factors Twist and Snail are expressed in the prospective mesoderm before gastrulation as part of the patterning of the dorso-ventral axis (see Section 2.19). It was then shown that mesoderm invagination is affected by mutations in these patterning genes and that they are associated with epithelial-to-mesenchymal transitions.

When the outlines of ventral cells in normal embryos are visualized by labeling cell membranes with green fluorescent protein, apical constriction of the prospective mesoderm cells can be seen to occur. Labeling of myosin and actin showed that the constriction is due to a succession of pulsed contractions of an actomyosin network that forms under the apical surface of the cell. No contractile network forms at all in loss-of-function *snail* mutants, whereas in *twist* mutants, pulsed contractions still occur but they do not produce apical constriction. Further investigations confirmed that while Snail was required to initiate the whole process, Twist was required to stabilize the state of contraction between pulses. In the absence of Twist function, the cell surface relaxes back to its original area after each contraction.

Twist and Snail also control the next stage of mesoderm internalization. When the central part of the mesoderm is fully internalized, its cells make contact with the ectoderm. Adherens junctions that had been established in the blastoderm are now lost—one of the results of Snail activity—and the mesoderm undergoes an epithelial-to-mesenchymal transition, dissociating into individual cells that spread out to cover the inner side of the ectoderm as a single-cell layer. Like the situation in the sea urchin, FGF signals from the ectoderm acting on an FGF receptor on the mesodermal cells are needed for the invaginating cells to spread. Failure to activate the FGF signaling pathway results in the mesodermal cells remaining near the site of invagination. Expression of the FGF receptor in the mesoderm is induced by Twist and Snail, while Snail also represses expression of FGF in the mesoderm, thus restricting FGF to the ectoderm.

Mesodermal cell spreading involves the cells switching from expression of ectodermal E-cadherin to N-cadherin, so that the mesoderm cells no longer adhere to the ectodermal cells. The change from E- to N-cadherin is also under the control of Snail and Twist. The actions of Snail suppress the production of E-cadherin in the mesoderm, while expression of Twist induces that of N-cadherin.

Fig. 9.23 *Drosophila* gastrulation. The first panel shows a transverse section through a *Drosophila* embryo in the early stages of gastrulation, showing the apical constriction of the mesodermal cells highlighted by the high levels of myosin II (red) and β-catenin (green). Nuclei are stained blue. The other three panels show the internalization of the mesoderm. The mesodermal cells (stained brown) are present in a longitudinal strip in the epithelium on the ventral side of the embryo. A change in their shape from cuboidal to wedge-shaped causes an invagination, which develops into a ventral furrow (left panel). Further apical contraction of the cells results in the mesoderm forming a tube on the inside of the embryo (center panel). The mesodermal cells then start to migrate individually to different sites (right panel). Scale bars = 50 μm.

*Photographs reproduced with permission from Francois Schweisguth and from Leptin, M., et al.: **Mechanisms of early Drosophila mesoderm formation**. Development Suppl. 1992, 23-31. Published by permission of The Company of Biologists Ltd.*

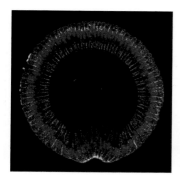

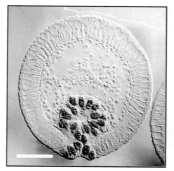

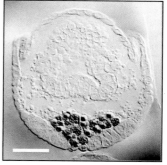

Scan here

Scan this QR code image with your mobile device to see an online movie showing germ band extension or log on to **http://global.oup.com/uk/orc/biosciences/ devbiol/wolpert5e/qr/qr9a/**

9.10 Germ-band extension in *Drosophila* involves myosin-dependent remodeling of cell junctions and cell intercalation

Another dramatic change in shape that occurs in the *Drosophila* embryo at the time of gastrulation is the extension of the germ band (see Fig. 2.4), which leads to an almost doubling in the length of the thorax and abdomen of the embryo. The germ band comprises the mesoderm, ventral ectoderm, and dorsal epidermis but not the amnioserosa. This extension is driven not by cell division or changes in cell shape, but by convergent extension of the ventral part of the epithelial layers. Adjacent cells intercalate between each other laterally—that is, towards the midline—so that the tissue narrows, causing it to extend in the antero-posterior direction. This remodeling occurs over the whole of the ventral epithelium and is due to regulated changes in the cadherin-containing adherens junctions that normally hold the cells tightly together.

At the beginning of germ-band extension, the epidermal cells are packed in a regular hexagonal pattern, with their boundaries either parallel to the dorso-ventral axis or at 60° to it, as shown in Fig. 9.24. When extension starts, the adherens junctions on the faces parallel to the dorso-ventral axis shrink and disappear and the cells become diamond-shaped. New junctions parallel to the antero-posterior axis appear, making the boundaries hexagonal again and intercalating the cells along the dorso-ventral axis—which makes the tissue extend along the antero-posterior axis.

This ability to undergo convergent extension along the antero-posterior axis reflects a prior antero-posterior patterning of cells in the epidermal sheet, which gives each cell an individual polarity, reflected in differences between opposite ends of the cell. The polarity can be seen in the distribution of some surface proteins, which are expressed only along the 'antero-posterior' axis of the cell, forming parallel stripes in the tissue when visualized. How the cell determines which axis is antero-posterior is not yet known. This type of polarity (which can reflect any axis, not just the

Fig. 9.24 Junctional remodeling enables cells to undergo intercalation in *Drosophila* germ-band extension. The disassembly and reassembly of intercellular contacts in the germ-band epithelium drives the tissue to elongate along the antero-posterior axis and to narrow along the dorso-ventral axis. The left-hand diagrams show how four cells in contact with each other change their intercellular contacts during junction remodeling. The right-hand diagram indicates schematically how this remodeling elongates the tissue.

*Adapted from Bertet, C., et al.: **Myosin-dependent junction remodelling controls planar cell intercalation and axis elongation**. Nature 2004, **429**: 667–671.*

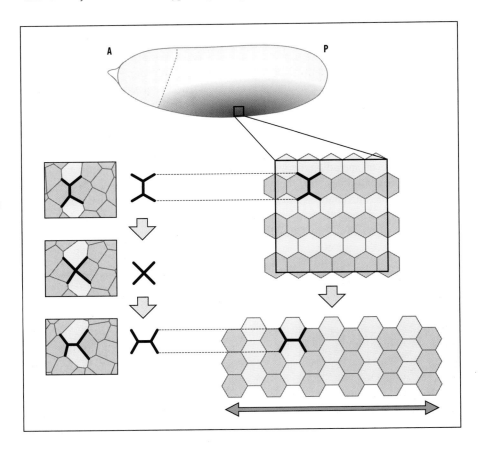

Fig. 9.25 Tissue movements during gastrulation of *Xenopus*. The late blastula basically consists of a continuous sheet of cells surrounding the blastocoel cavity. The future mesoderm (red) is in the marginal zone, overlaid by the future endoderm (yellow). The future ectoderm is shown in blue. Gastrulation is initiated by the formation of bottle cells in the blastopore region, which is followed by the involution of mesoderm over the dorsal lip of the blastopore. Marginal zone endoderm and mesoderm move inside over the dorsal lip of the blastopore. The marginal zone endoderm, which was on the surface of the blastula, now lies ventral to the mesoderm and forms the roof of the archenteron or future gut. At the same time, the ectoderm of the animal cap spreads downward. The mesoderm converges and extends along the antero-posterior axis. The region of involution spreads ventrally to include more endoderm, and forms a circle around a plug of yolky vegetal cells. The ectoderm spreads by epiboly.
Illustration after Balinsky, B.I.: An Introduction to Embryology, 4th edition. Philadelphia, W.B. Saunders, 1975.

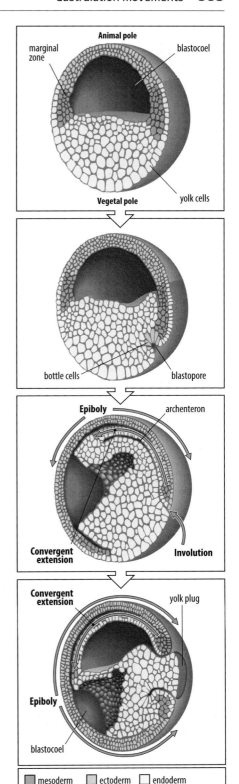

antero-posterior one) is called **planar cell polarity**, because cells are polarized in the plane of the tissue, and is a feature of most epithelia and organized mesenchyme in embryos. It is quite distinct from the apico-basal polarity of epithelia discussed earlier (see Section 9.6). We have already seen planar cell polarity at work in determining the direction in which the bristles point in the *Drosophila* adult epidermis (see Box 2I, which describes some of the proteins involved in determining planar cell polarity in adult *Drosophila* epidermis). In the next section, and in Box 9C, we shall consider the importance of planar cell polarity in the extension of the vertebrate body.

The mechanism of intercalation in the germ band involves the regulated localization and activity of myosin, which co-localizes with β-catenin–E-cadherin–actin complexes at the adherens junctions and is enriched in shrinking junctions. The proposed mechanism is that regulated myosin contraction at the junctions prevents E-cadherin holding the cells together, enabling new contacts to be made that result in cell intercalation.

9.11 Gastrulation in amphibians and fish involves involution, epiboly, and convergent extension

Gastrulation in vertebrates involves a much more dramatic and complex rearrangement of tissues than in sea urchins, because of the need to produce a more complex body plan. In amphibians, fish, and birds, there is also the added complication of the presence of large amounts of yolk. But the outcome is the same: the transformation of a two-dimensional sheet of cells into a three-dimensional embryo, with ectoderm, mesoderm, and endoderm in the correct positions for further development of body structure. We shall first consider gastrulation in amphibians and fish. The main movements of gastrulation here are **involution**, which is the rolling-in of the coherent sheet of endoderm and mesoderm at the blastopore, as occurs in *Xenopus* (Fig. 9.25); **epiboly**, which is the spreading and thinning of the ectodermal sheet as the endoderm and mesoderm move inside and which can be observed in gastrulating frogs and zebrafish; and **convergent extension**, which elongates the body axis. We have already seen a simple process of convergent extension in the cell intercalation that elongates the germ band in *Drosophila* (see Section 9.10).

To undergo convergent extension, cells must have acquired planar cell polarity. This is a polarization of the cell in the plane of the tissue such that one end of the cell differs molecularly or structurally from the other, for example in the formation of filopodia or lamellipodia, the distribution of the cytoskeleton, or the location of surface receptors. Planar polarity is a near-universal characteristic of embryonic epithelia and can also be seen in some mesenchyme. The polarity of epithelia is most readily visible in surface structures such as cilia, which all point in one direction (see Fig. 9.10). This reflects an underlying polarized organization of the cytoskeleton and other organelles. In general, all the cells in a particular epithelium, or region of an epithelium, will have

Fig. 9.26 Formation of the blastopore and gastrulation in *Xenopus*. This figure shows details of tissue movements around the blastopore in a gastrulating Xenopus embryo as illustrated in the first three panels of Fig. 9.25. First panel: before gastrulation, in the marginal zone, presumptive endoderm (yellow) overlies the presumptive mesoderm (red). Second panel: bottle cells at the site of the blastopore undergo apical constriction and elongate, causing the involution of surrounding cells and the formation of a groove which defines the dorsal lip of the blastopore. Third panel: as gastrulation proceeds, the presumptive endoderm and mesoderm involute and move anteriorly under the ectoderm (blue). Fourth panel: the archenteron—the future gut lined by endoderm—starts to form and the mesoderm converges and extends along the antero-posterior axis. The ectoderm continues to move down by epiboly to cover the whole embryo.

*After Hardin, J.D. and Keller, R.: **The behaviour and function of bottle cells during gastrulation of** Xenopus laevis. Development 1988, **103**: 211-230.*

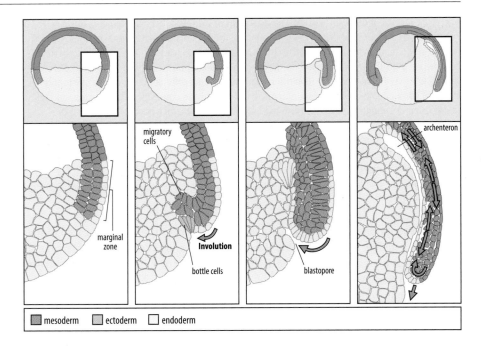

the same polarity, thus giving a directionality to the epithelial sheet, which can influence its further development.

In this chapter we shall see planar cell polarity at work in notochord extension and in neurulation and body-axis extension; we shall discuss its involvement in limb-bud outgrowth in Chapter 11. Planar polarity provides information to the cell on its alignment relative to the body axes, and helps to ensure that convergent extension or other cell movements occur in the correct direction. Convergent extension and planar cell polarity are discussed in more detail in Box 9C.

In the late *Xenopus* blastula, the presumptive endoderm extends from the most vegetal region to cover the presumptive mesoderm. During gastrulation, the presumptive endoderm moves inside through the blastopore to line the gut, while all the cells in the equatorial band of mesoderm move inside to form a layer of mesoderm underlying the ectoderm, extending antero-posteriorly along the dorsal midline of the embryo (Fig. 9.26).

Gastrulation in *Xenopus* starts at a site on the dorsal side of the vegetal half of the blastula. The first visible sign is the conversion of some of the presumptive mesodermal cells into bottle-shaped cells as a result of apical constriction (see Box 9B). As in the sea urchin and *Drosophila*, this change in cell shape forms a groove in the blastula surface. In *Xenopus*, this starts as a small groove—the blastopore—whose dorsal lip is the site of the Spemann organizer. The coherent sheet of mesoderm and endoderm starts to involute around the blastopore, rolling into the interior of the blastula against its own undersurface (see Fig. 9.26). As gastrulation proceeds, the region of involution spreads laterally and ventrally to involve the vegetal endoderm, and the blastopore eventually forms a circle around a plug of yolky cells. In time, the blastopore contracts, forcing the yolky cells into the interior where they form the floor of the gut.

The first mesodermal cells to involute migrate as individual cells over the blastocoel roof to give rise to the anterior-most mesodermal structures in the head. Behind them is mesoderm that enters, together with the overlying endoderm, as a single multilayered sheet. For this cell sheet, going through the narrow blastopore is rather like going through a funnel, and the cells become rearranged by convergent extension, which is a major feature of gastrulation.

BOX 9C Convergent extension

Convergent extension plays a key role in gastrulation and other morphogenetic processes. It is a mechanism for elongating a sheet of cells in one direction while narrowing its width, and occurs by rearrangement of cells within the sheet, rather than by cell migration or cell division. It occurs, for example, in the elongation of the antero-posterior axis in amphibian embryos (see Fig. 9.28). For convergent extension to take place, the axes along which the cells will intercalate and extend must already have been defined. The cells first become elongated in a direction at right angles to the antero-posterior axis—the medio-lateral direction. They also become aligned parallel to one another in a direction perpendicular to the direction of tissue extension (see Figure 1). Active movement is largely confined to each end of these elongated bipolar cells, which form filopodia, enabling them to exert traction on neighboring cells on each side and on the underlying substratum and to shuffle in between each other—or intercalate—always along the medio-lateral axis, with some cells moving medially and some laterally. This rearrangement is known as **medio-lateral intercalation**. At a tissue boundary, one end of the cell is captured and movement ceases, but cells can still exert traction on each other to draw the boundaries closer together.

The boundary of a tissue undergoing convergent extension is maintained by the behavior of cells at the boundary. When an active cell tip is at the boundary of the tissue, its movement ceases and the cell becomes monopolar (Figure 1), with only the tip facing into the tissue still active. It is as if the cell tip that has reached the boundary has become fixed there. Precisely what defines a boundary is not clear, but boundaries between mesoderm, ectoderm, and endoderm seem to be specified before gastrulation commences.

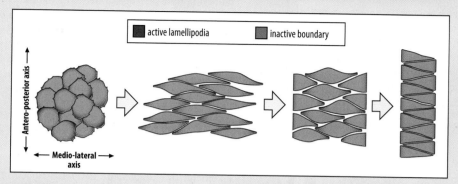

Figure 1

active lamellipodia   inactive boundary

Antero-posterior axis

Medio-lateral axis

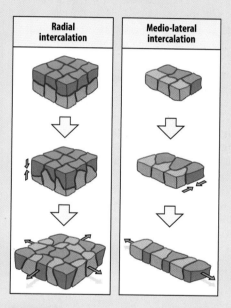

Figure 2

Radial intercalation   Medio-lateral intercalation

A mechanically similar process called **radial intercalation** (Figure 2) causes the thinning of a multi-cell layer into a thinner sheet and its consequent extension around the edges, as seen during epiboly of the ectoderm during frog and zebrafish gastrulation (Fig. 9.30). Radial intercalation occurs in the multilayered ectoderm of the animal cap, in which cells intercalate in a direction perpendicular to the surface, moving from one layer into that immediately above. This leads to an increase in the surface area of the cell sheet, as well as its thinning, and is in part the cause of epiboly.

The molecular mechanisms controlling convergent extension in vertebrates are related to those controlling planar cell polarity in *Drosophila* (see Box 2I), and involve signaling by Wnts through Frizzled via the non-canonical Wnt planar cell polarity pathway (Figure 3). The much-simplified pathway shown here is a general compilation from different organisms. Signaling by Wnt through its receptor Frizzled (Fz) via Dishevelled activates RhoA and the Rho kinase Rok2, leading to changes in the cell's actin cytoskeleton. The transmembrane protein Van Gogh (Vang) and other proteins identified from *Drosophila* are also components of the vertebrate pathway. Vang is thought to interact with Dishevelled to stimulate the activation of another protein kinase, the Jun N-terminal kinase (JNK), which can activate the transcription factor AP1.

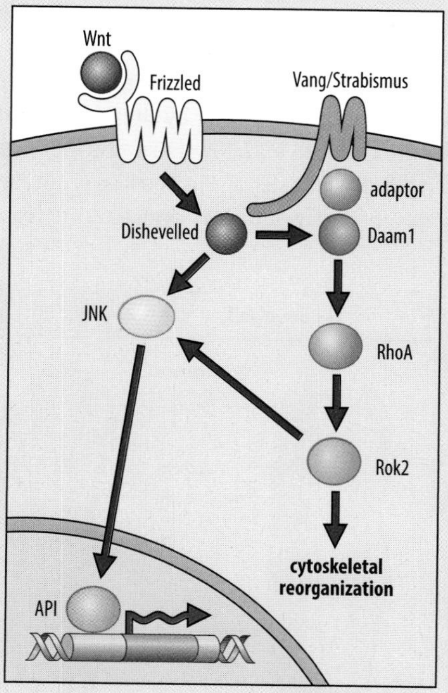

Figure 3

Wnt

Frizzled

Vang/Strabismus

Dishevelled

adaptor

Daam1

JNK

RhoA

Rok2

AP1

cytoskeletal reorganization

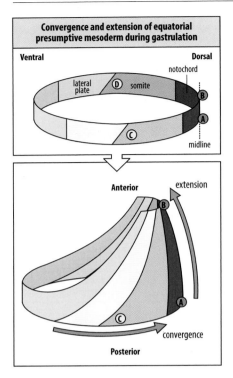

Fig. 9.27 Convergent extension of the mesoderm. The mesoderm (red, orange and yellow) is initially located in an equatorial ring, but during gastrulation it converges and extends along the antero-posterior axis. A–D are reference points used to illustrate the extent of movement during convergent extension. In the bottom panel, D is hidden behind C.

Fig. 9.28 The rearrangement of mesoderm and endoderm during gastrulation in Xenopus. The presumptive mesoderm (red and orange) is present in a ring around the blastula underlying the endoderm (yellow), which is shown peeled back. During gastrulation, both these tissues move inside the embryo through the blastopore and completely change their shape, converging and extending along the antero-posterior axis, so that points A and B move apart. Thus, the two points C and D, originally on opposite sides of the blastula, come to lie very close to each other. In the neural ectoderm (blue) there is also convergent extension, as shown by the movement of points E and F. Note that little cell division or cell growth has occurred in these stages, and all these changes have occurred by rearrangement of cells within the tissues.

The mesoderm is initially in the form of an equatorial ring, but during gastrulation it converges and extends along the antero-posterior axis (Fig. 9.27). Thus, cells initially on opposite sides of the embryo come to lie very close to each other (Fig. 9.28), whereas others become separated from each other along the long axis of the embryo. The migrating mesoderm cells are polarized in the direction of the animal pole and their migration depends on interaction with the fibronectin fibrils in the extracellular matrix lining the blastocoel roof. The assembly of fibronectin fibrils in the matrix is due to tension at the surface of the blastocoel roof epithelium, caused by cell–cell adhesion, being transmitted via integrins in the cell membrane that interact with fibronectin.

In *Xenopus*, convergent extension occurs in both the mesoderm and endoderm as they involute, and in the future neural ectoderm that gives rise to the spinal cord. Together with the elongation of the notochord, all these processes elongate the embryo in an antero-posterior direction. One can appreciate the dramatic nature of convergent extension by viewing the early gastrula from the blastopore. The future mesoderm can be identified by the expression of the gene *brachyury*, and can be seen as a narrow ring around the blastopore (Fig. 9.29). As the *brachyury*-expressing ring of tissue enters the gastrula it converges into a narrow band along the dorsal midline of the embryo, extending in the antero-posterior direction, and eventually *brachyury* expression is restricted to the prospective notochordal mesoderm along the midline itself. Genetic analysis has shown that *brachyury* activates a network of genes including not only genes involved in mesoderm specification but also genes encoding molecules involved in convergent extension, thus linking patterning and the behavior of the cells during gastrulation and notochord extension.

As gastrulation proceeds, the mesoderm comes to lie immediately beneath the ectoderm, while the endoderm immediately below it lines the roof of the archenteron, the future gut (see Fig. 9.25). Even though they are in contact, the mesoderm and ectoderm move independently of each other. As the mesoderm and endoderm move inside, the ectoderm of the animal cap region spreads down over the vegetal region by epiboly and will eventually cover the whole embryo. Epiboly involves both stretching of the cells themselves and cell rearrangements, in which cells lower down in the ectodermal layers intercalate between the cells immediately above them—a process called **radial intercalation** (see Box 9C). Together, these processes lead to thinning of the ectoderm layer and an increase in its surface area.

Gastrulation in the zebrafish has both similarities and differences to that of *Xenopus* (see Fig. 3.11 for an outline of zebrafish gastrulation). It begins by the blastoderm at the animal pole starting to spread out over the yolk cell and down towards the vegetal pole. As in *Xenopus*, this epibolic spreading is due to cells of the deep layer undergoing radial intercalation, leading to the layer becoming thinner and thus increasing

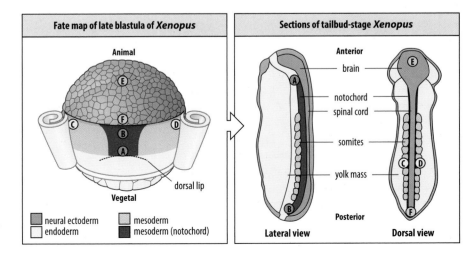

its surface area (Fig. 9.30, left panel). By the time the blastoderm margin reaches about half-way down the yolk cell, the deep-layer cells have accumulated around the blastoderm edge to form a thickening known as the embryonic germ ring, and have formed two cell layers—prospective ectoderm overlying mesendoderm—under the superficial enveloping layer. The mesendodermal cells now move internally at the edge of the germ ring and then move upwards under the ectoderm towards the future anterior end of the embryo (Fig. 9.30, center panel). The ectoderm continues to undergo epiboly and move downwards. The next step is convergent extension of the mesendodermal and ectodermal layers towards the dorsal midline and along the future antero-posterior axis, which elongates the embryo (Fig. 9.30, right panel).

In amphibian and zebrafish gastrulation, the cells in tissues undergoing convergent extension along the antero-posterior axis are polarized in the medio-lateral direction (see Box 9C). The contractile forces generated at the tips of the elongated cells both move the cells towards the midline and extend the tissue. One can visualize the process of convergent extension in terms of lines of cells pulling on one another in a direction at right angles to the direction of extension. Because the cells at the boundary are trapped at one end, this tensile force causes the tissue to narrow and hence extend anteriorly.

Figure 9.31 compares the process of convergent extension in the elongation of the notochord mesoderm in *Xenopus* and zebrafish during axis elongation. In fish, notochord elongation also involves an initial directed migration of individual mesodermal cells from lateral regions towards the midline, a movement known as **dorsal convergence**, followed by medio-lateral intercalation as mesodermal cells become incorporated into the future notochord at the midline.

9.12 *Xenopus* notochord development illustrates the dependence of medio-lateral cell polarity on a pre-existing antero-posterior polarity

In *Xenopus* the presumptive notochord mesoderm is among the first of the tissues to internalize during gastrulation (see Fig. 4.27), and the notochord is the first of the axial structures to differentiate. It will eventually form a solid rod of cells running along the dorsal midline of the embryo, where it has a structural supporting function and also provides signals for somite patterning (see Section 5.13). During gastrulation, the notochord mesoderm elongates along with the body axis by the process of medio-lateral intercalation of cells, leading to convergent extension along the antero-posterior axis, as shown in Fig 9.32. This intercalation depends on the pre-existing antero-posterior polarity set up by previous patterning processes in the notochord mesoderm.

Initially the notochord can be distinguished from the adjacent somitic mesoderm by the closer packing of its cells, and by the slight gap that forms the boundary between the two tissues, possibly reflecting differences in cell adhesiveness. After an initial elongation by medio-lateral intercalation and convergent extension, the *Xenopus* notochord undergoes a further dramatic narrowing in width accompanied by an increase in height (Fig. 9.31). The cells become elongated at right angles to the main antero-posterior axis, foreshadowing a later 'pizza-slice' arrangement of cells (see Fig. 9.32, bottom). The cells intercalate between their neighbors again, thus causing further convergent extension. A later stage in notochord development, which stiffens and further elongates the rod of mesoderm, involves a process called directed dilation and is discussed later in this chapter.

The medio-lateral polarization of the individual notochord cells requires the vertebrate version of the Wnt-mediated planar cell-polarity pathway (see Box 9C). When this pathway is rendered non-functional, *Xenopus* mesodermal cells fail to elongate or to develop polarized lamellipodia. But another polarization system is also at work in the convergent extension of the vertebrate body axis, which maintains the correct direction of extension in this population of continually rearranging and moving cells. The prospective notochord mesoderm originates from the organizer at the dorsal lip of

Scan here

Scan this QR code image with your mobile device to see an online movie showing convergent extension in zebrafish or log on to **http://global.oup.com/uk/orc/biosciences/devbiol/wolpert5e/qr/qr9d/**

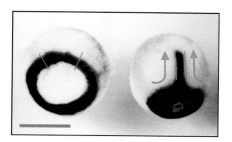

Fig. 9.29 Expression of *brachyury* during *Xenopus* gastrulation illustrates convergent extension. Left: before gastrulation, the expression of *brachyury* (dark stain) marks the future mesoderm, which is present as an equatorial ring when viewed from the vegetal pole. Right: as gastrulation proceeds, the mesoderm that will form the notochord (delimited by blue lines) converges and extends along the midline. Scale bar = 1 mm.

Photograph reproduced with permission from Smith, J.C., et al.: **Xenopus Brachyury**. *Semin. Dev. Biol. 1995, 6: 405–410. © 1995 by permission of the publisher, Elsevier.*

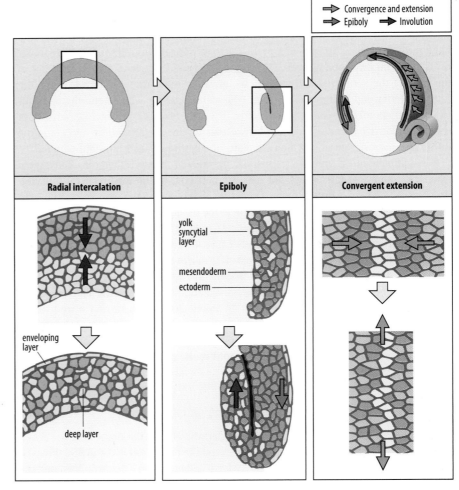

Fig. 9.30 Zebrafish gastrulation movements. Left panel: the first sign of gastrulation in the zebrafish embryo is spreading of the blastoderm over the surface of the yolk cell. This epiboly is caused by a thinning of the deep layer by a radial intercalation process. Center panel: when the blastoderm has reached about half the way down the yolk cell, separate layers of prospective mesendoderm (brown and yellow cells) and ectoderm (blue) have formed and the mesendoderm starts to internalize by involution (blue arrow). Ectoderm does not internalize and will continue to move down and spread out over the yolk cell by epiboly (blue arrow). It will eventually cover the whole embryo. Right panel: convergent extension (green arrows) of the mesendoderm towards the dorsal midline results in elongation of the embryo.

*Illustration after Montero, J.A., Heisenberg, C.P.: **Gastrulation dynamics: cells move into focus**. Trends Cell Biol. 2004,**14**: 620–627.*

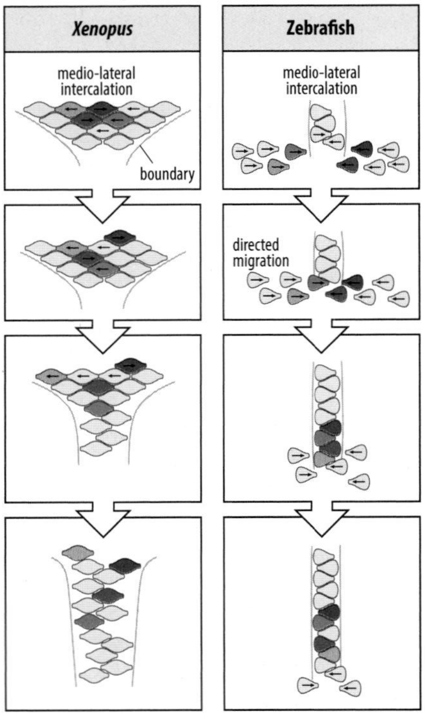

the blastopore (Fig. 4.27) and is patterned along the antero-posterior axis, with the cells nearest the dorsal lip, which will form the anterior notochord involuting before cells that will form posterior notochord. After gastrulation, the notochord helps to pattern the overlying neural ectoderm by antagonizing the activity of BMPs (see Section 4.16).

Antero-posterior patterning of the mesoderm seems to be a prerequisite for cell intercalation and convergent extension. This can be demonstrated by disaggregating explants of 'anterior' and 'posterior' prospective notochord from the organizer of a *Xenopus* early gastrula and then mixing the cells together. The cells recognize each other by their different antero-posterior positional identities and sort out. It was found that only combined explants of anterior and posterior tissue could undergo intercalation and convergent extension; combinations of 'all anterior' or 'all posterior' cells did not. These experiments suggest that the patterning along the antero-posterior axis is required to set up the medio-lateral planar polarity in individual notochord cells that

Fig. 9.31 Convergent extension during notochord formation in *Xenopus* and zebrafish. The left panels show convergent extension of mesoderm in *Xenopus* during the formation of the notochord, which occurs by medio-lateral intercalation of cells in a coherent tissue sheet. The right panels show formation of the notochord by convergent extension in the zebrafish, which occurs by directed movement of loosely packed mesenchymal mesodermal cells from the lateral regions towards the midline, insertion into the future notochord and medio-lateral intercalation within the notochord boundary.

*Illustration after Wallingford, J.B., et al.: **Convergent extension: the molecular control of polarized cell movement during embryonic development**. Dev. Cell 2002, **2**: 695–706.*

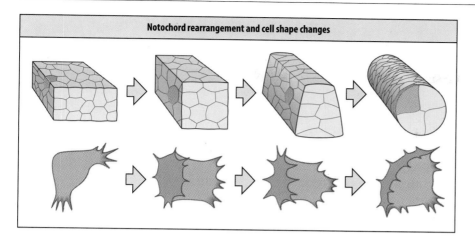

Notochord rearrangement and cell shape changes

Fig. 9.32 Changes in cell arrangement and cell shape during notochord development in *Xenopus*. During convergent extension of the notochord, its height increases. At the same time, the cells become elongated. The bottom of the figure shows the eventual arrangement of cells, and their pizza-slice shape.

*After Keller, R., et al.: **Cell intercalation during notochord development in** Xenopus laevis.* J. Exp. Zool. *1989, **251**: 134–154.*

enables them to intercalate. How cells integrate the cues they receive from the antero-posterior polarization system and from the Wnt-mediated planar polarity system has still to be worked out.

9.13 Gastrulation in chick and mouse embryos involves the delamination of cells from the epiblast and their ingression through the primitive streak

The events associated with gastrulation in frogs and fishes repeat themselves in amniotes (birds and mammals) and the endpoint is remarkably similar in all cases (see Fig. 3.2). But the different topologies and organization of chick and mouse embryos compared with those of frogs and fishes impose mechanical constraints that have led to adaptation of the movements.

Gastrulation in mammals and birds involves the formation of the structure known as the primitive streak, through which epiblast cells move into the interior of the embryo, becoming specified as mesoderm and endoderm as they become internalized (described in Chapters 3 and 5). The primitive streak can be likened functionally to the blastopore of amphibian embryos, but in the essentially flat embryos of chick and mouse, the internalization movements differ from those in the spherical, hollow *Xenopus* blastula (see Section 3.3). In mouse and chick, instead of the movement of mesoderm and endoderm as continuous sheets of cells, epiblast cells undergo a complete epithelial-to-mesenchymal transition and separate from the epiblast epithelium, a process called **delamination**. They move through the streak and into the interior as individual cells, a process called **ingression** (Fig. 9.33).

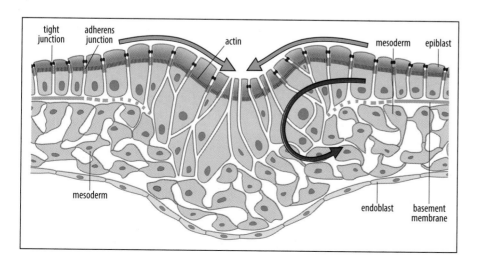

Fig. 9.33 Delamination of cells and movement through the primitive streak involves apical constriction and an epithelial-to-mesenchymal transition. The primitive streak is a groove in the epithelial epiblast created by apical constriction of the cells in this region. The cells become wedge-shaped as a result of the contraction of actin microfilaments which run along the apical surfaces of the epiblast cells. Cells in the streak undergo an epithelial-to-mesenchymal transition involving loss of cell junctions, including adherens junctions, and the basement membrane in this region also breaks down. This allows cells to separate from the epiblast and then migrate as individuals beneath it (red arrow). The cells that ingress in this way will form the mesoderm and the endoderm. The cells that remain in the epiblast will form the ectoderm. As cells ingress through the streak, the epiblast on either side is drawn towards the streak (green arrows).

*From Nayaka, Y., Sheng., G.: **An amicable separation: Chick's way of doing EMT**.* Cell Adhesion Migration *2009, **3**: 160–163.*

Fig. 9.34 Polonaise movements in the chick epiblast during primitive-streak formation. Left panel: cell movements in the epiblast before the egg is laid start to bring cells towards the site at which the primitive streak will form. Second panel: around 6-7 hours after laying (stage 2 embryo) cells in the epiblast undergo characteristic extensive 'polonaise' movements outwards from the midline and back towards the start of the primitive streak; the region forming the streak starts to converge and extend by medio-lateral intercalation (see Box 9C). Third panel: at around 12-13 hours after laying (stage 3) the streak has extended to about half its full length; as it extends, the epiblast cells in front of it move forward.

Adapted from Voiculescu, O., et al.: **The amniote primitive streak is defined by epithelial cell intercalation before gastrulation.** Nature *2007,* **449***: 1049-1052.*

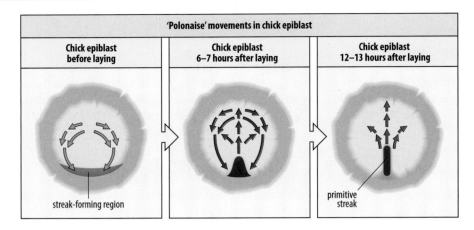

'Polonaise' movements in chick epiblast

| Chick epiblast before laying | Chick epiblast 6–7 hours after laying | Chick epiblast 12–13 hours after laying |

streak-forming region

primitive streak

In the chick embryo, the primitive streak is initiated from a point in the posterior region of the blastodisc as a dynamic population of cells, and this process of streak formation progresses forward , while epiblast cells ingress behind it to give rise to the endoderm and the mesoderm (see Figs 3.15, 3.16, 3.17 and Fig. 4.11). The epiblast cells remaining on the surface become the ectoderm. Once internalized, cells migrate internally to form the mesodermal and endodermal layers, and the embryo elongates by a combination of cell proliferation and convergent extension.

The movements and changes in behavior of epiblast cells during streak formation are less well understood than the involution movements in *Xenopus*, but streak formation involves many of the cell behaviors and morphogenetic movements we have looked at already. Delamination and movement of epiblast cells into the streak is achieved by a combination of processes. Coordinated epithelial-to-mesenchymal transitions (a combination of apical constriction, loss of adherens junctions and breakdown of the basement membrane) start at the posterior end of the blastoderm and progress forward. At the same time there is convergent extension in the epiblast. *brachyury* is expressed in the streak and, as in *Xenopus* (see Fig. 9.29), identifies the ingressing cells.

One view of the mechanics of streak extension is that convergence of cells at the posterior end of the streak provides much of the force for its extension, and that the contraction and delamination of cells at the streak draw lateral epiblast cells towards and into the streak. In the chick epiblast, primitive-streak formation is preceded by extensive bilaterally symmetric movements of anterior epiblast cells outwards from the midline and backwards towards the posterior edge where the streak will begin (Fig. 9.34). These movements are known as 'polonaise' movements, as they resemble a formal eighteenth-century dance of that name. High-resolution time-lapse videos of the chick epiblast show that convergent extension is the most likely mechanism generating the cell movements. The intercalation of cells close to the streak generates forces in the epithelium that cause the movement of cells elsewhere in the epiblast. As with axis elongation in amphibians and zebrafish, this example of convergent extension in the chick involves the Wnt planar cell polarity pathway (see Box 9C), and inhibition of this pathway blocks streak formation.

Under the primitive streak, cells form a loose mass of mesenchyme that spreads out under the ectoderm of the epiblast to populate the whole of the embryo. Cells that enter the streak give rise to endoderm and mesoderm. Precursors for the various tissues come out of the streak in a definite order. Endoderm is produced first, then cardiac mesoderm, and then notochord. As in sea urchin and *Drosophila* embryos, the migration of the mesoderm cells is directed by FGFs. FGF-4 produced by the forming head process acts as a chemoattractant for mesoderm cells from the anterior part of the streak while FGF-8 produced by the streak itself repels them. The balance between attractive and repulsive cues determines where mesoderm cells will go. Remarkably FGF signaling as a mechanism for controlling mesoderm cell migration appears to have been highly conserved during evolution.

The process of gastrulation in the mouse is similar to that in chick, but the fact that the mouse epiblast is a hollow cylinder as opposed to a disc (see Fig. 3.23), results in a different initial topology, which has consequences for the progress of gastrulation. The primitive streak starts at a point on the upper rim of the cup, which identifies the posterior region of the embryo, and cells enter it from the inside of the cup; this means that the endoderm—the fate of the first cells to enter the primitive streak—ends up as the outer layer of the early embryo, rather than the inner (see Fig. 3.24). Then, much as in the chick, the streak advances anteriorly, with cells forming different mesodermal derivatives in the wake of the streak as they migrate out under the epiblast, which will form the ectoderm. The 'endoderm outside' configuration is resolved later in development when the mouse embryo turns inside out (see Fig. 3.26). This turning process is exclusive to rodents. In humans, where the flat blastoderm is topologically more similar to that of the chick, gastrulation resembles that of the chick, with the primitive streak forming in the upper surface of the epiblast.

Nevertheless, like the chick, gastrulation in mice is associated with the expression of *brachyury* in the streak and epithelial-to-mesenchyme transitions during delamination. FGF signaling is also involved in the movement of cells away from the streak once they have ingressed and Wnt signaling is also involved. In recent live-imaging studies of gastrulating mouse embryos, however, no polonaise movements in the posterior epiblast or convergent extension before streak formation could be detected. Mutational screens in the mouse have confirmed the involvement of epithelial-to-mesenchymal transitions and cell movement in gastrulation, because mutations in genes affecting the dynamics of the cytoskeleton abort the process of gastrulation.

SUMMARY

At the end of cleavage the animal embryo is essentially a continuous sheet of cells, which is often in the form of a sphere enclosing a fluid-filled interior. Gastrulation converts this sheet into a solid three-dimensional embryonic animal body. During gastrulation, cells move from the exterior into the interior of the embryo, and the endodermal and mesodermal germ layers take up their appropriate positions inside the embryo. Gastrulation results from a well-defined spatio-temporal pattern of changes in cell shape, cell movements, and changes in cell adhesiveness, and the main forces underlying the morphological changes are generated by changes in cell motility and localized cellular contractions. In the sea urchin, gastrulation occurs in two phases. In the first, changes in shape and adhesion of prospective mesoderm cells in a small region of the blastula wall result in an epithelial-to-mesenchymal transition and migration of these cells into the interior and along the inner face of the blastula wall, followed by invagination of the adjacent part of the blastula epithelium, the prospective endoderm, to form the gut. Filopodia on the migrating primary mesenchyme cells explore the environment, drawing the cells to those regions of the wall where the filopodia make the most stable attachment; in this way guidance cues in the extracellular matrix lead the migrating cells to their correct locations. In the second phase, extension of the gut to reach the prospective oral (mouth) region on the opposite side of the embryo occurs by cell rearrangement within the endoderm, which causes the tissue to narrow and lengthen (convergent extension), and by traction of filopodia extending from the gut tip against the blastocoel wall. Invagination of the mesoderm in *Drosophila* occurs by a similar mechanism to that in sea urchin, whereas germ-band extension is due to myosin-driven remodeling of cell junctions and intercalation of cells, leading to convergent extension. Gastrulation in vertebrates involves more complex movements of cells and cell sheets and also results in elongation of the embryo along the antero-posterior axis. In *Xenopus* it involves three processes: involution, in which a double-layered sheet of endoderm and mesoderm rolls into the interior over the lip of the blastopore; convergent extension of endoderm and mesoderm in the antero-posterior direction to form the roof of the gut, and the notochord and somitic mesoderm, respectively; and epiboly—the spreading of the ectoderm from the animal cap region to cover the whole outer surface of the embryo. Both convergent extension and epiboly are due to cell intercalation, the cells becoming rearranged with respect to their neighbors. Convergent extension of the antero-posterior axis at gastrulation involves the vertebrate version of the Wnt planar cell polarity signaling pathway and requires pre-existing antero-posterior patterning of the mesoderm. Gastrulation in the chick and the mouse involves delamination of individual cells from the epiblast and their ingression through the primitive streak into the interior of the embryo, where they form mesoderm and endoderm.

SUMMARY: gastrulation in sea urchin, *Xenopus*, chick, and mouse

Sea urchin

formation of blastopore: mesodermal cells migrate individually into the interior; endoderm moves inside by invagination of the epithelial sheet

convergent extension of endoderm and traction via filopodia completes gut extension

Xenopus and zebrafish

formation of blastopore: mesoderm and endoderm move inside by involution or internalization

convergent extension of mesoderm along antero-posterior axis

extension of ectoderm over whole surface by epiboly

Chick and mouse

formation of primitive streak: epiblast cells migrate individually into the interior (ingression) through the streak and become specified as mesoderm and ectoderm. Remaining epiblast becomes ectoderm

mesoderm cells migrate laterally and anteriorly under the epiblast

gut is formed by fusion of ventral edges of embryo

Neural tube formation

Neurulation in vertebrates results in the formation of the **neural tube**—a tube of epithelium derived from the dorsal ectoderm—which develops into the brain and spinal cord. Following induction by the mesoderm during gastrulation (see Chapter 5), the ectoderm that will give rise to the neural tube initially appears as a thickened plate of epithelium—the **neural plate**—in which the cells have become more columnar than the cells of the adjacent ectoderm. This plate will roll up into a tube from which the nervous system will emerge. In this chapter we will look at the formation of the neural tube itself. The further development of the nervous system from the cells of the neural tube is considered in Chapter 12.

The vertebrate neural tube is formed by two different mechanisms in different regions of the body. The anterior neural tube forms the brain and the central nervous system in the trunk, and is essentially formed by the folding of the neural plate into a tube, a process called **primary neurulation** (although in birds and mammals a stem-cell-like population of cells makes an important contribution to the nervous system of the trunk, see Section 5.8). The posterior neural tube beyond the lumbo-sacral region, in contrast, develops from stem-like cells in the tailbud, which form a solid rod of mesenchymal cells that then undergoes a mesenchymal-to-epithelial transition and develops an interior cavity or lumen that connects with that of the anterior neural tube. This process is known as **secondary neurulation**. There are, however, variations in different vertebrates; in zebrafish and other fishes, for example, the whole of the neural plate first forms a solid rod, which later becomes hollow in a similar manner. Here we shall concentrate on primary neurulation mainly in relation to the chick embryo.

9.14 Neural tube formation is driven by changes in cell shape and convergent extension

The chick embryo serves as a good model for mammalian neurulation and is more easily studied. Neurulation in the cranial region of the chick embryo is shown in Fig. 9.35. It starts with formation of a **neural furrow** formed by the infolding of the neural plate along the midline. The edges of the neural plate become raised above the plane of the epithelium, forming two parallel **neural folds** with a depression—the **neural groove**—between them. The neural folds eventually come together along the dorsal midline of the embryo and fuse at their edges to form the neural tube, which then separates from the adjacent ectoderm. In amphibians, the neural tube closes along its whole length almost simultaneously; in the chick, closure starts at the midbrain region and proceeds both anteriorly and posteriorly, while in mammals closure occurs at different regions at different times.

The surface layer of ectoderm becomes epidermis, and the cells that were at the boundary of the neural plate with the prospective epidermis give rise to the **neural crest**. This is a population of cells that leaves the dorsal surface of the neural tube on both sides just before and after closure, and migrates to form a wide range of different tissues. Neural crest cells give rise to the peripheral nervous system, the melanocytes of the skin, and even to some of the tissues of the face; neural crest and its migration are discussed in the next part of this chapter.

The process of neurulation is driven by the activity of small, spatially defined sub-populations of cells located along the midline of the neural plate, along its boundaries with the prospective epidermis, and in the sides of the neural folds. The cells at the boundary of the neural plate are under mechanical constraint from the neighboring ectoderm and undergo apical constriction. Cells along the midline of the neural plate also undergo apical constriction, resulting in folding of the neural plate at the so-called **median hinge point** to form the neural furrow (Fig. 9.36). Together, the forces generated by the cell-shape changes in this constrained situation lead to the elevation of the edges of the neural plate to form the neural folds. This is a very important step

Fig. 9.35 Neural tube formation results from the bending of the neural plate and fusion of the neural folds. Bending of the neural plate inwards at both sides creates the neural groove, with the neural folds at the edges. The neural folds come together and fuse to form a tube of epithelium—the neural tube. This detaches from the rest of the ectoderm, which becomes the epidermis. Neural crest cells detach from the dorsal neural tube and migrate away from it.

*Adapted from Yamaguchi, Y., Miura, M.: **How to form and close the brain: insight into the mechanism of cranial neural tube closure in mammals**. Cell. Mol. Life Sci. 2013, 70: 3071-3186.*

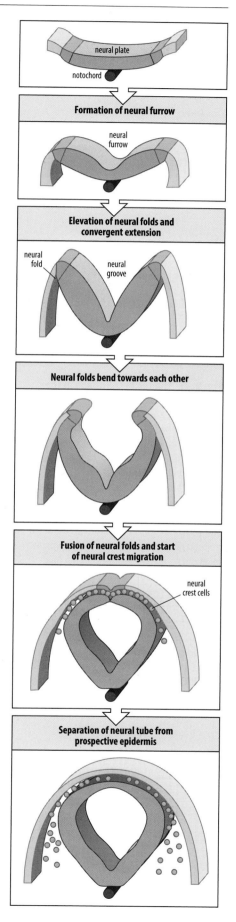

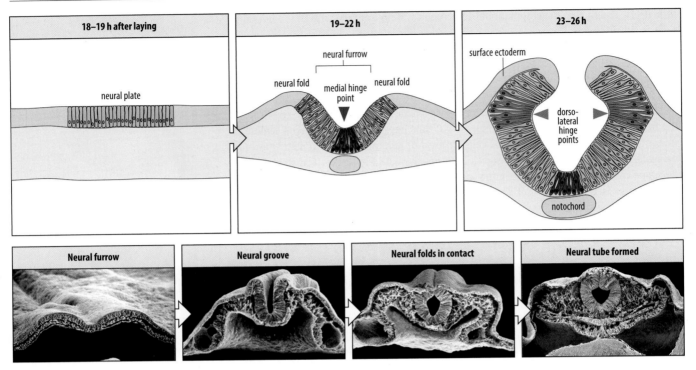

Fig. 9.36 Change in cell shape in the neural plate during chick neurulation.
Cells in the center of the neural plate become wedge-shaped as a result of apical constriction, defining a median 'hinge point' at which the neural plate bends. Additional dorso-lateral hinge points with wedge-shaped cells also form on the sides of the furrow and help to bring the neural folds together. The electron micrographs show successive stages in neurulation in the chick embryo in sections taken at the approxiate level of the future mesencephalon (midbrain).

*Illustration after Schoenwolf, G.C., Smith, J.L.: **Mechanisms of neurulation: traditional viewpoint and recent advances**.* Development *1990,* **109***: 243–270. Photographs courtesy of G. Schoenwolf from Colas, J.-F., Schoenwolf, G.C.: **Towards a cellular and molecular understanding of neurulation**.* Dev. Dyn. *2001,* **221***: 117–145.*

in neural tube formation, and failure to elevate the folds leads to a complete failure of neurulation.

The neural folds are now brought together through two coordinated movements within the neural plate. The first, and most important, is a process of convergent extension similar to those that take place during gastrulation. This reduces the width of the neural plate, thus bringing the folds closer to each other, and also elongates the forming neural tube along the antero-posterior axis as the rest of the body elongates. The cells of the neural plate seem to become oriented for convergent extension in response to signals from midline structures such as the underlying notochord. The second process changes the shape of the tips of the neural folds so that they are brought together. Cells on the dorso-lateral faces of the folds become wedge-shaped, in part through apical constriction, forming two more hinge points, known as the **dorso-lateral hinge points** (see Fig. 9.36). These cause the neural folds to buckle inwards, so that the tips are now pointing towards each other.

As in many other situations, apical constriction in the cells at the midline and the dorso-lateral hinge points is thought to occur by the purse-string mechanism of actomyosin contraction illustrated in Box 9B.

The last stage of neural tube formation is **neural tube closure**. The cells at the tips of the folds develop lamellipodia and filopodia that interdigitate when the folds meet. The edges of the neural folds then fuse with each other and the tube separates from the rest of the ectoderm, which closes above it (see Fig. 9.35). The folds zip together so closely that the lumen almost disappears; it only opens up again after the folds fuse.

The neural tube, which is initially part of the ectoderm, separates from the presumptive epidermis after its formation as a result of changes in cell adhesiveness. In the chick, neural plate cells, like the rest of the ectoderm, initially have the adhesion molecule L-CAM (see Box 9A) on their surface. However, as the neural folds develop, the neural plate ectoderm begins to express both N-cadherin and N-CAM, whereas the adjacent ectoderm expresses E-cadherin. These differences may help the neural tube to separate from the surrounding ectoderm and allow it to sink beneath the surface, the rest of the ectoderm reforming over it in a continuous layer.

CELL BIOLOGY BOX 9D Eph receptors and their ephrin ligands

Eph receptors and their ephrin ligands are transmembrane proteins that can generate both repulsive and adhesive interactions between cells. Because they are both attached to a cell surface, they can only act through direct cell–cell contact, like Notch and Delta. They are best known as contact-dependent guidance molecules for growing axons in the nervous system, especially in the guidance of axons of retinal neurons to their correct locations in the brain's visual-processing centers; their role in axon guidance is discussed in Chapter 12. Among other roles, they act as guidance molecules for migrating neural crest cells, and appear to be involved in the final stages of somite formation—the separation of the new somite from the pre-somitic mesoderm. And according to work in mice, Ephs and ephrins are required to correctly set up the local neuronal circuits in the spinal cord that enable you to walk with alternate legs, rather than have to hop like a rabbit.

The 14 different Eph receptors in mammals fall into two structural classes: EphAs (EphA1–EphA8 and EphA10) and EphBs (EphB1–Eph B4 and EphB6). Ephrins similarly can be classified into ephrin As (ephrins A1–A5) and ephrin Bs (ephrins B1–B3). The A-type receptors typically bind to ephrin As, and the B-type receptors to ephrin Bs, but EphA4 is an exception, being able to bind both ephrin A and B ligands. Ephrins binding to their Eph receptors can generate a bidirectional signal, in which both the Eph and the ephrin send a signal to their own cell (Figure 1). Unlike most of the signaling receptors we have encountered so far, the main target of Eph-ephrin signaling is the actin cytoskeleton, whose rearrangement causes the changes in cell morphology and behavior that underlie repulsion.

Ephs are receptor tyrosine kinases, with an intracellular kinase domain, whereas ephrins signal by associating with and activating a cytoplasmic kinase. Ephrin–Eph contact is best known for producing repulsion between the interacting cells, as occurs at the interface between different rhombomeres (see Section 12.4), but Eph-ephrin interactions can also result in attraction and adhesion. Whether a particular Eph-ephrin pairing repels or attracts is not straightforward, and seems to depend on a range of different factors, including the cell types involved, the numbers of receptors present, and their degree of clustering.

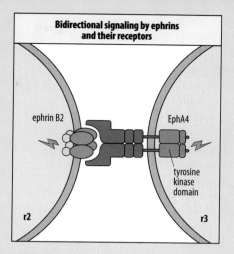

Figure 1

The repulsion commonly mediated by Eph-ephrin pairs is something of a paradox, given their inherent affinity for each other. One solution to the paradox was the discovery that repulsion by ephrin As involves cleavage of their ectodomain, which enables cell disengagement. Internalization of Eph-ephrin complexes by endocytosis is another possible solution. Repulsive interactions seem generally to require a relatively high level of signaling from the Eph tyrosine kinase. If this is lacking or at a low level, then the binding of Eph and ephrin will tend to promote cell adhesion.

An example of this has been found in the closure of the neural tube in mice, in which a form of EphA7 lacking its kinase domain is produced by the cells of the neural folds as a result of alternative RNA splicing. The neural fold cells also express ephrin A5, and the adhesive interaction between EphA7 and ephrin A5 on opposing folds is essential for neural tube closure. Mouse mutants that lack ephrin A5 have neural-tube defects resembling the human condition of anencephaly, in which the cranial neural tube fails to close, the forebrain does not develop, and the fetus is usually stillborn. In tissues other than the neural tube, in which EphA7 is produced with an intact tyrosine kinase domain, its interaction with ephrin A5 causes repulsion between the interacting cells.

The closing of the neural tube is a complex process that requires integration of the activity of many different kinds of proteins—adhesion and signaling proteins such as the cadherins, and the ephrins and their receptors (Box 9D), cytoskeletal components, and regulators of cytoskeletal activity—in addition to changes in gene expression driven by transcription factors.

Because so many cell biological processes have to function perfectly for neural-fold elevation and closure to succeed, it is perhaps not surprising that the process goes wrong relatively frequently, resulting in a failure of neural tube closure. Neurulation defects, or neural tube defects (NTDs) are among the most common human birth

MEDICAL BOX 9E Neural tube defects

Neurulation defects have been studied extensively in the mouse, in which not only are spontaneous mutations known that affect neural tube development, but also large-scale mutagenesis screens can be carried out (see Section 3.9). More than 60 different genes have been found that, when mutated, cause neural tube defects similar to those found in humans. In mice and humans, neural fold elevation occurs in distinct regions of the tube at different times, and neural tube closure is initiated at particular sites within these regions. The first panel of the figure 1 shows the sites of initiation of closure in a mouse embryo. Closure starts at E8.5 at a site (1 on Figure 1) at the hindbrain-cervical boundary, from which it proceeds in both directions. Closure progresses posteriorly along the length of the spinal cord, with completion at the posterior neuropore, which marks the end of the trunk neural tube (the neural tube in the tail region is formed by secondary neurulation). Closure is also initiated a little later at a second site, at the forebrain-midbrain boundary (site 2). A third wave of neural tube closure starts at the anterior-most end of the neural tube (site 3) and proceeds posteriorly to meet closure from the second site at the anterior neuropore.

Failure in neural fold elevation or closure causes a range of different defects including anencephaly, rachischisis, craniorachischisis and spina bifida, depending on the regions involved, as shown in the second and third panels of Figure 1. Anencephaly (called exencephaly in the mouse) is a lethal defect that results from a failure of proper neural tube formation in the fore- and midbrain regions. Rachischisis represents the failure of the spinal cord to close, and craniorachischisis results from the neural tube remaining open from the mid-brain to the lower spine. Spina bifida results

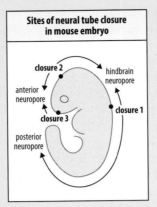
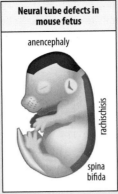
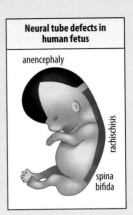

Figure 1

First panel From Harding, B. N. & Copp, A. J. in *Greenfield's Neuropathology* 357–483 (Arnold, London, 2002). Second and third panels from Juriloff and Harris. **Mouse models for neural tube closure defects**. Human Molecular Genetics 2000, **9**: 993–1000.

from a failure to complete neural tube closure at the posterior neuropore, and the consequent failure of the spinal column to form correctly at that point. In humans, the clinically most severe form of this condition is known as myelomeningocele, in which the spinal cord and meninges protrude through the fault in the spine. This form of spina bifida is, in general, seriously disabling even after corrective treatment, as some spinal cord function has been lost. In humans, neural tube defects originate at 4 weeks gestation, the time at which the neural tube is being formed.

Mouse mutants with defects in neural tube formation indicate that regulators of actomyosin assembly and function, as well as elements of the Wnt planar cell polarity pathway that impinge on the cytoskeleton (see Box 9C), are essential for the process. For example, mouse embryos lacking the planar cell polarity pathway components Dishevelled, Flamingo, or the vertebrate Vang proteins, Vang-like 1 and 2, display craniorachischisis.

defects (around 1:1500 births), and in their most severe forms are lethal, resulting in stillbirth or death within a few days. The type of **neural tube defect** depends on the region of the neural tube involved (Box 9E).

SUMMARY

In vertebrates, induction of neural tissue by the mesoderm during gastrulation is followed by neurulation—the formation of the neural tube, which will eventually form the brain and spinal cord. Neurulation in mammals, birds, and amphibians involves both the folding of the neural plate into a tube (in the head and trunk), and the formation of a central cavity in a solid rod of cells (in the tail). The formation of the neural folds and their coming together in the midline to form the neural tube is driven by changes in cell shape and convergent extension within the tube itself, as well as by forces generated by adjacent tissues. Separation of the neural tube from the overlying ectoderm involves changes in cell adhesiveness. The complexity of the process of neural tube closure means that defects in closure are not uncommon, giving rise to conditions such as spina bifida.

Cell migration

Cell migration is a major feature of animal morphogenesis, with cells sometimes moving over relatively long distances from one site to another to integrate into new territories. In this chapter, we shall describe three examples of cell migration—**neural crest cells** in vertebrates, **lateral line cells** in fishes, and an example of motile cells drawing an epithelial sheet together, in the phenomenon of **dorsal closure** in *Drosophila*. Neural crest cells are a hallmark of vertebrates and produce an enormously diverse range of different cell types.

Other examples of cell migration can be found elsewhere in the book. The migration of germ cells from their site of formation to their final location in the gonads is described for zebrafish and mouse in Chapter 10, the migration of cells from the somites into the developing chick limb to form muscle cells is discussed in Chapter 11, and the migration of immature neurons within the mammalian central nervous system in Chapter 12.

9.15 Embryonic neural crest gives rise to a wide range of different cell types

The neural crest cells of vertebrates have their origin at the edges of the neural plate and first become recognizable during neurulation (see Section 5.14). They arise from the dorsal neural tube just before and after fusion of the neural folds, and are recognizable as individual cells on their emergence from the neural epithelium. Once neural crest cells have detached from the neural tube they migrate to many sites, where they differentiate into a remarkable number of different cell types.

Neural crest cells give rise to the neurons and glia of the peripheral nervous system (which includes the sensory and autonomic nervous systems), endocrine cells, such as the chromaffin cells of the adrenal medulla, some structures in the heart (see Section 11.32) and melanocytes, which are pigmented cells found in the skin (see Fig. 5.38). In mammals and birds, the neural crest also gives rise in the head to bone, cartilage, and dermis, tissues that are formed elsewhere in the body by mesoderm.

9.16 Neural crest migration is controlled by environmental cues

Neural crest cells of vertebrates have their origin in the epithelium at the edges of the neural plate and first become recognizable during neurulation through the expression of transcription factors of the Snail and Slug families. At the time of closure of the neural tube in vertebrate embryos, neural crest cells can be seen as an emerging loose mesenchyme on each side of the midline. The expression of Snail and Slug triggers an epithelial-to-mesenchymal transition (see Fig. 9.4). The individual neural cells then migrate away from the neural tube on both sides under the guidance of chemical cues. Figure 9.37 shows neural crest cells migrating in culture.

Migrating chick neural crest cells can be tracked by labeling them with a fluorescent dye such as DiI, or by expressing a photoactivatable green fluorescent protein (pa-GFP) in the neural tube. pa-GFP can then be made to fluoresce when required by a precisely focused beam of light, and the migration patterns of a small group of neural crest cells, or even single cells, can be followed. Migrating neural crest cells in a zebrafish embryo, carrying a reporter transgene in which a GFP coding sequence is linked to a promoter for a gene expressed in neural crest cells (*Sox10*), can be seen in Fig. 5.37.

 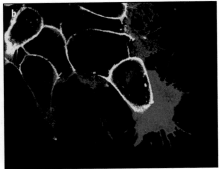

Fig. 9.37 *Xenopus* **neural crest cells migrating in culture.** Lamellipodia at the leading edges of the cells are stained red, cell membranes yellow, and nuclei blue. The photograph on the right shows a higher-magnification view.

*Images courtesy of Roberto Mayor, (a) originally published in Kuriyama, S. et al.: Journal of Cell Biology, 2014, **206**: 113–127.*

There are two main migratory pathways for neural crest cells in the trunk of the chick embryo (Fig. 9.38). One goes dorso-laterally under the ectoderm and over the somites; cells that migrate this way mainly give rise to pigment cells, which populate the skin and feathers. The other pathway is more ventral, primarily giving rise to sympathetic and sensory ganglion cells. Some crest cells move into the somites to form dorsal root ganglia; others migrate through the somites to form sympathetic ganglia and adrenal medulla, but appear to avoid the region around the notochord. Trunk neural crest selectively migrates through the anterior half of the somite and not through the posterior half. Within each somite, neural crest cells are found only in the anterior half, even when they originate in neural crest adjacent to the posterior half of the somite. This behavior is unlike that of the neural crest cells taking the dorsal pathway, which migrate over the whole dorso-lateral surface of the somite.

Neural crest migration to the anterior of each somite results in the distinct segmental arrangement of spinal ganglia in vertebrates, with one pair of ganglia for each pair of somites—one segment—in the embryo (Fig. 9.39). The segmental pattern of

Fig. 9.38 Neural crest migration in the trunk of the chick embryo. One group of cells (1) migrates under the ectoderm to give rise to pigment cells (shown in dotted outline). The other group of cells (2) migrates over the neural tube and then through the anterior half of the somite; the cells do not migrate through the posterior half of the somite. Those that migrate along this pathway give rise to dorsal root ganglia, sympathetic ganglia, and cells of the adrenal cortex, and their future sites are also shown in dotted outline. Those cells opposite the posterior regions of a somite migrate in both directions along the neural tube until they come to the anterior region of a somite. This results in a segmental pattern of migration and is responsible for the segmental arrangement of ganglia.

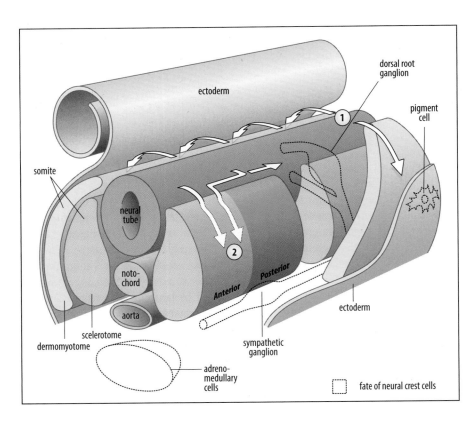

migration is due to different guidance cues on the two halves of each somite. If the somites are rotated through 190° so that their antero-posterior axis is reversed, the crest cells still migrate through the original anterior halves only.

Interactions between ephrins and their Eph receptors (see Box 9D) are likely to contribute to the expulsion of neural crest cells from the posterior halves of the somites. In the chick, ephrin B1 is localized to the posterior halves of somites and acts as a repulsive guidance cue, interacting with EphB3 on the neural crest cells, whereas the anterior halves of the somites, through which the neural crest cells do travel, express EphB3. Membrane proteins called **semaphorins** are also involved in repelling neural crest cells from the posterior half of the somite. Interactions such as these provide a molecular basis for the segmental arrangement of the spinal ganglia.

Both the neural tube and notochord also influence neural crest migration. If the early neural tube is inverted through 190°, before neural crest migration starts, so that the dorsal surface is now the ventral surface, one might think that the cells that normally migrate ventrally, now being nearer to their destination, would move ventrally. But this is not the case, and many of the crest cells move upward through the sclerotome in a ventral-to-dorsal direction, staying confined to the anterior half of each somite. This suggests that the neural tube somehow influences the direction of migration of neural crest cells. The notochord also exerts an influence, inhibiting neural crest cell migration over a distance of about 50 µm, and thus preventing the cells from approaching it.

The direction of neural crest migration does involve some chemotaxis towards specific chemical cues. The migration of streams of neural crest cells in particular directions mainly results from the cells becoming polarized in the direction of migration, probably the result of the actions of the Wnt-mediated planar cell polarity pathway (see Box 9C). Inhibition of components of this pathway, such as Wnts and Frizzled, inhibit neural crest migration.

Neural crest cells are guided by interactions with the extracellular matrix over which they are moving, as well as by cell–cell interactions. Neural crest cells can interact with extracellular matrix molecules by means of their cell-surface integrins (see Box 9A). Avian neural crest cells cultured *in vitro* adhere to, and migrate efficiently on, fibronectin, laminin, and various collagens. Blocking adhesion of neural crest cells to fibronectin or laminin by blocking the integrin β_1 subunit *in vivo* causes severe defects in the head region but not in the trunk, suggesting that the crest cells in these two regions adhere by different mechanisms, probably involving other integrins. It is striking how neural crest cells in culture will preferentially migrate along a track of fibronectin, although the role of this extracellular matrix molecule in guiding the cells in the embryo is still unclear.

9.17 The formation of the lateral-line primordium in fishes is an example of collective cell migration

In certain situations, a group of cells undergoes migration together as a collective entity, remaining linked to each other by adhesive junctions. A good example of this is the **lateral-line primordium** in fishes, which gives rise to a series of sensory organs along the length of the body—the lateral line—that will form a major sensory apparatus. The primordium is a group of more than a hundred cells, looking rather like a garden slug (Fig. 9.40), which moves as a collective along the flank of the embryo under the skin. It derives from the cranial ectoderm and moves from anterior to posterior along a track of the chemoattractant protein stromal-derived factor 1 (SDF-1 or CXCL12) laid down along the myoseptum, the membrane connecting the underlying muscle cells. SDF-1 is sensed by its receptor CXCR4, which is expressed in all cells of the primordium but is only active in those at the leading edge, which drive the movement. In the primordium, cells are attached to each other by junctional complexes

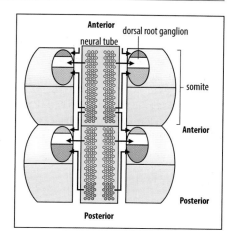

Fig. 9.39 Segmental arrangement of dorsal root ganglia is due to the migration of crest cells through the anterior half of the somite only. Neural crest cells cannot migrate through the posterior (gray) half of a somite but can migrate in either direction along the neural tube and through the anterior half of somites (yellow). The dorsal root ganglion in a given segment is thus made up of crest cells from the posterior region of the segment anterior to it, neural crest cells immediately adjacent to it (white), and crest cells from the posterior of its own segment.

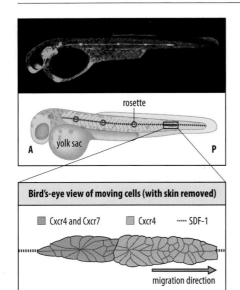

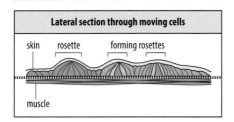

Scan here

Scan this QR code image with your mobile device to see an online movie showing zebrafish lateral-line formation or log on to **http://global.oup.com/uk/orc/biosciences/ devbiol/wolpert5e/qr/qr9b/**

Fig. 9.40 Collective cell migration in zebrafish lateral-line formation. Top panel: photograph and explanatory drawing of zebrafish embryo showing the location of the developing lateral line. A, anterior; P, posterior. The transgenic embryo carries a glycosyl phosphatidylinositol-green fluorescent protein that labels membranes and enables visualization of the developing lateral line in the transparent embryo. As the lateral-line primordium moves along the route from the head, guided by a pre-patterned line of chemoattractant SDF-1, rosettes of cells that will form the lateral-line sensory organs are left behind at regular intervals. Center panel: a bird's-eye view of the migrating cells (the overlying skin has been removed). Cells at the tip of the migrating primordium express CXCR4, the receptor for SDF-1 (dotted pink line). Trailing cells also express the SDF-1 receptor CXCR7. Bottom panel: a section through the lateral-line primordium showing the beginning of rosette formation.

Friedl, P., Gilmour, D.: **Collective cell migration in morphogenesis, regeneration and cancer.** Nat. Rev. Mol. Cell Biol. *2009,* **10***: 445–457.*

that are looser and more dynamic at the leading edge, where the cells have undergone a partial epithelial-to-mesenchymal transition. The structure is dynamic because as it moves along, the cells within it are dividing, and at regular intervals a small cluster of cells is left behind to form a sensory organ.

Similar collective cell migrations occur during the development of the highly branched system of tracheal tubules in *Drosophila*, which conveys oxygen into the tissues (discussed in Chapter 11). Leading cells at the tip of each branch follow a trail of signals, dragging the extending tracheal system behind them.

9.18 Dorsal closure in *Drosophila* and ventral closure in *Caenorhabditis elegans* are brought about by the action of filopodia

Some morphogenetic events involve the movement of an intact epithelial sheet. The process of dorsal closure in the *Drosophila* embryo provides a good example of these movements. Half-way through *Drosophila* embryogenesis, when cell proliferation ceases and the germ band retracts, there is a gap in the dorsal epidermis. This gap is covered by a squamous epithelium, the amnioserosa, which is continuous with the epidermis. The gap is closed by the process of dorsal closure, which involves the coordinated movement of the two halves of the epidermis towards each other, accompanied by internalization of the amnioserosa in the embryo where it undergoes apoptosis (Fig. 9.41).

The process relies on the specific structure and function of the cytoskeleton in the cells of the epidermis and the amnioserosa. In the epidermis, the cells in contact with the amnioserosa remodel their junctions in the plane of the epithelium and form a large supracellular actinomyosin cable whose contractions drive the coordinated expansion of the cells towards the midline and hence the movement of the two sides of the epidermis towards each other. The process is aided by contraction of the cells of the amnioserosa, which provide a complementary force by pulling on the epidermal sheets. As the amnioserosa contracts, its cells fall inside the embryo and eventually undergo apoptosis and disappear.

The leading edge of the epidermis develops filopodia and lamellipodia, extending up to 10 μm from the edge, that have cadherin molecules on their surface. These structures do not provide force for the process but play a central role in the zippering of the two sides when they meet each other at the midline. This general mechanism is also seen in neural tube closure.

The small GTPase Rac has a key role in organizing actin filaments at the leading edges of the cells, and is also involved through the Jun N-terminal kinase (JNK) intracellular signaling pathway in controlling changes in cell shape and in cell fusion. Dorsal closure also requires the segment patterning on the two sides to be in precise register with each other, and this is mediated by these extensions. For example,

Fig. 9.41 Dorsal closure in *Drosophila*. a, Embryo at the beginning of dorsal closure. Expression of the gene *puckered* (brown), which encodes a phosphatase that regulates Jun N-terminal kinase signaling, defines cells at the leading edge of the epidermis. Once these cells have been defined they trigger an elongation of the epidermis (blue arrows) and a contraction of the cells of the amnioserosa (red arrows) which drives the organized closure of the epidermal gap. b, c, During the final phases, the epidermal cells zip up the gap from both ends by means of filopodia developed on the leading-edge cells. d, Visualization of expression of the gene *engrailed* (green) and *patched* (red) shows the precise alignment of cells at the join, with cells expressing *engrailed* fusing only with each other and *patched*-expressing cells fusing only with each other, thus preserving segment patterning across the join.

*a-c, courtesy of Alfonso Martinez Arias; d from Millard, T.H., Martin, P.: **Dynamic analysis of filopodial interactions during the zippering phase of** Drosophila **dorsal closure**. Development 2008, 135: 621-626.*

the gene *engrailed* is expressed in a series of transverse stripes along the length of the embryo, corresponding to the posterior compartment of each segment (see Section 2.24). Cells expressing *engrailed* on one edge interact only with cells expressing *engrailed* on the other edge, and cells expressing the gene *patched*, which form a different set of stripes, fuse only with *patched*-expressing cells on the other edge, thus neatly matching up the stripes across the seam (see Fig. 9.41).

The *C. elegans* embryo undergoes a process similar to dorsal closure, but on the ventral surface. At the end of gastrulation, the epidermis only covers the dorsal region and the ventral region is bare. The epidermis then spreads around the embryo until its edges finally meet along the ventral midline. Time-lapse studies show that the initial ventral migration is led by four cells, two on each side, which extend filopodia towards the ventral midline. Blocking filopodial activity by laser ablation or by cytochalasin D, which disrupts actin filaments and inhibits actin polymerization, prevents ventral closure, showing that the filopodia provide the driving force. Closure probably occurs by a zipper mechanism similar to that in the fly.

Dorsal closure in *Drosophila*

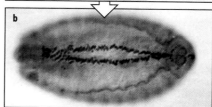

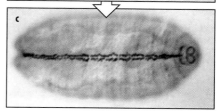

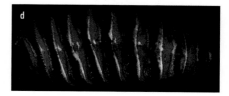

SUMMARY

Cell migration is an important force in morphogenesis and can be observed in individual cells as well as in groups and even in epithelia. It allows the redistribution of cells or the reorganization of tissues. In the case of the neural crest, streams of cells migrate from both sides of the dorsal neural tube to form a wide range of different tissues in different parts of the body. Their migration is guided by signals from other cells and by interactions with the extracellular matrix. In neural crest cells migrating in the trunk, repulsive interactions between ephrins in the posterior halves of the somites and Eph receptors on the neural crest cells prevent their migration through the posterior halves. Thus, neural crest cells that will give rise to the dorsal root ganglia of the peripheral nervous system accumulate adjacent to the anterior half of each somite, resulting in a segmental arrangement of paired ganglia along the antero-posterior axis. In contrast to the migration of the neural crest as individual cells, the development of the lateral line in the fish is an example of collective migration of a group of cells that moves as a slug and sows sensory organs in the epidermis of the fish embryos. In both cases, the cells originate from an epithelium and undergo epithelial-to-mesenchymal transitions, complete in the case of the neural crest, partial in the case of the lateral line. In some instances, of which dorsal closure is one, what moves is a whole epithelium.

Scan here

Scan this QR code image with your mobile device to see an online movie showing dorsal closure in *Drosophila* or log on to **http://global.oup.com/uk/orc/biosciences/devbiol/wolpert5e/qr/qr9c/**

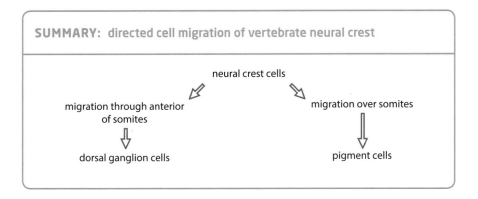

SUMMARY: directed cell migration of vertebrate neural crest

neural crest cells

migration through anterior of somites   migration over somites

dorsal ganglion cells   pigment cells

Directed dilation

Hydrostatic pressure can provide the force for morphogenesis in various situations. We have already seen how an increase in hydrostatic pressure inside the mammalian blastocyst causes an increase in volume and maintains the blastocyst in a roughly spherical shape (see Section 9.7). Here, we consider examples of **directed dilation**, where an increase in internal hydrostatic pressure causes a distinct asymmetric change in shape as a result of the particular properties of the structure. For example, if the circumferential resistance of a stretchable tube to internal pressure is much greater than the resistance lengthways, then an increase in the internal pressure will cause an increase in length (Fig. 9.42).

9.19 Later extension and stiffening of the notochord occurs by directed dilation

After the *Xenopus* notochord has formed, its volume increases threefold, and there is considerable further lengthening as it straightens and becomes stiffer. At this stage, the notochord has become surrounded by a sheath of extracellular material, which because of its molecular structure restricts circumferential expansion but does allow expansion in the antero-posterior direction. The flat cells within the notochord develop fluid-filled vacuoles and the rod of cells expands in volume in an antero-posterior direction. Cells are initially flat and are restricted in the direction in which they swell by junctional connections, with the result that cell expansion is exerting force lengthways along the notochord. The sheath contains helically wound collagen fibrils, which make it resistant to outward expansion but make it elastic as well, and so it can lengthen. The hydrostatic pressure exerted by the cells, combined with the composition of the sheath enables the notochord to undergo directed dilation. Circumferential expansion of the notochord is prevented by the resistance of the sheath, and this ensures that the increase in volume (dilation) is directed along the notochord's long axis, enabling the notochord to lengthen, straighten and stiffen without buckling.

The vacuoles in the notochord cells are filled with glycosaminoglycans which, because of their high carbohydrate content, tend to attract water into the vacuoles by osmosis. This produces the hydrostatic pressure that causes the increase in cell volume, and the consequent stiffening and straightening of the notochord. Changes in the structure of the sheath during the period of notochord elongation fit well with the proposed hydrostatic mechanism. The sheath contains both glycosaminoglycans, which have little tensile strength, and the fibrous protein collagen, which has a high tensile strength; during notochord dilation the number of collagen fibers increases, providing resistance to circumferential expansion. The crucial role of the sheath in

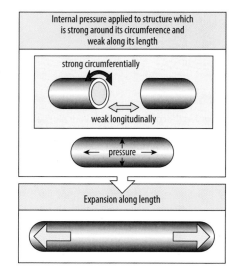

Internal pressure applied to structure which is strong around its circumference and weak along its length

strong circumferentially

weak longitudinally

pressure

Expansion along length

Fig. 9.42 Directed dilation. Hydrostatic pressure inside a constraining sheath or membrane can lead to elongation of the structure. If the circumferential resistance is much greater than the longitudinal resistance, as it is in the notochord sheath, the rod of cells inside the sheath lengthens.

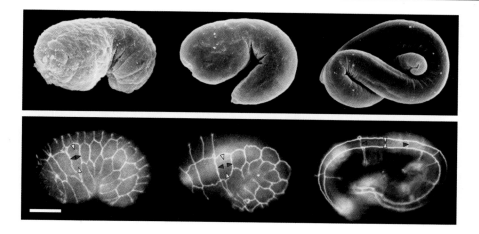

Fig. 9.43 Increase in nematode body length by directed dilation. The change in body shape over 2 hours is illustrated in the top panel. The increase in length is due to circumferential contraction of the hypodermal cells, as shown in the bottom panel. The change in shape of a single cell can be seen in the cell marked with arrows. Scale bar = 10 μm.

Photographs reproduced with permission from Priess, J.R., Hirsh, D.I.: **Caenorhabditis elegans** *morphogenesis: the role of the cytoskeleton in elongation of the embryo.* Dev. Biol. *1986, **117**: 156–173. © 1986 Academic Press.*

dilation and elongation is shown by the fact that if it is digested away, the notochord buckles and folds and the notochord cells, instead of being flat, become rounded.

9.20 Circumferential contraction of hypodermal cells elongates the nematode embryo

During the early development of the nematode there is little change in body shape from the ovoid form of the fertilized egg, even during gastrulation. After gastrulation, about 5 hours after fertilization, the embryo begins to elongate rapidly along its antero-posterior axis. Elongation takes about 2 hours, during which time the embryo decreases in circumference about threefold and undergoes a fourfold increase in length.

This elongation is brought about by a change in shape of the hypodermal (epidermal) cells that make up the outermost layer of the embryo; their destruction by laser ablation prevents elongation. During embryo elongation, these cells change shape so that, instead of being elongated in the circumferential direction, they become elongated along the antero-posterior axis (Fig. 9.43). Throughout this elongation the hypodermal cells remain attached to each other by cell junctions. The junctions are also linked within the cells by actin-containing fibers that run circumferentially, and these fibers appear to shorten as the cells elongate. The disruption of actin filaments and inhibition of actin polymerization by cytochalasin D treatment blocks elongation, and so it is very likely that the contraction of the actin-containing fibers brings about the change in cell shape. Circumferential contraction of the hypodermal cells causes an increase in hydrostatic pressure within the embryo, forcing an extension in an antero-posterior direction. Circumferentially oriented microtubules may also have a mechanical role in constraining the expansion, in the same way as the sheath of the *Xenopus* notochord described earlier. Increase in nematode body length is thus another example of directed dilation.

SUMMARY

Directed dilation results from an increase in hydrostatic pressure inside a structure, and unequal peripheral resistance to this pressure. Extension of the notochord is brought about by directed dilation. The notochord interior increases in volume while its circumferential expansion is constrained by the notochord sheath, forcing it to elongate. Similarly, the nematode embryo elongates after gastrulation as a result of a circumferential contraction of the outer hypodermal cells that generates pressure on the internal cells, forcing the embryo to extend in an antero-posterior direction.

Summary to Chapter 9

- Changes in the shape of animal embryos as they develop are mainly due to forces generated by differences in cell adhesiveness, changes in cell shape, and cell migration.
- Epithelia and mesenchyme are the two main types of cellular organization in embryos, and epithelial-to-mesenchymal transitions are a common feature of embryonic development. Mesenchymal-to-epithelial transitions also occur.
- Formation of a hollow, spherical blastula or blastocyst is the result of particular patterns of cell division, cell packing, and cell polarization.
- The spherical shape of a blastula or blastocyst is maintained by flow of fluid into the interior and the hydrostatic pressure generated.
- Gastrulation involves major movements of cells, such that the future endoderm and mesoderm move inside the embryo to their appropriate positions in relation to the main body plan.
- Convergent extension has a key role in many developmental contexts, such as elongation of the antero-posterior body axis during vertebrate gastrulation, elongation of the notochord, lengthening and closure of the neural tube, and germ-band extension in *Drosophila*.
- Formation of the vertebrate neural tube from the neural plate involves cell-shape changes that cause the edges of the plate to curve upwards to form the neural folds, and convergent extension that narrows the neural plate, enabling fusion of the neural folds to occur.
- Neural crest cells leave the dorsal neural tube and migrate as individual cells to give rise to a wide range of different tissues throughout the body.
- The formation of the lateral line in fishes involves collective cell migration.
- Dorsal closure in *Drosophila* embryos and ventral closure in *Caenorhabditis* embryos is driven by changes in the cytoskeleton of the epidermal cells that brings two edges of epidermis together.
- The phenomenon of directed dilation, involving hydrostatic pressure, is responsible for the later elongation of the notochord and the elongation of the nematode embryo.

■ End of chapter questions

Long answer (concept questions)

1. Name the four types of adhesive interactions between cells and between cells and the extracellular matrix. What are the transmembrane proteins that characterize each of these adhesive interactions? What are the particular features of an adherens junction: what cell adhesion molecule is used, what cytoskeletal component is involved, and what links the adhesion molecule to the cytoskeleton?

2. Describe the results of an experiment in which epidermis and neural plate from a neurula stage amphibian embryo are first disaggregated into single cells, then mixed together and reaggregated. How is this experiment interpreted in terms of the relative strength of 'self' adhesiveness in these two tissues? How does this result reflect the natural organization of these tissues in the embryo?

3. Define 'radial cleavage.' In the radial cleavage of an embryo such as the sea urchin, how are the centrosomes positioned at the first cleavage, with respect to the animal–vegetal axis? The second cleavage? The third cleavage?

4. What is compaction of the mouse embryo? How is compaction related to two important subsequent events in mouse embryogenesis: the formation of the blastocoel, and the formation of the inner cell mass?

5. Contrast the properties of an epithelial tissue with those of a mesenchymal tissue. How are cadherins related to the transition of epithelium to mesenchyme; how is this transition regulated, or triggered (think: 'snail')?

6. Refer to the fate maps of sea urchins (Fig. 6.20) and *Xenopus* (Fig. 4.9): in sea urchins, mesoderm is most vegetal and endoderm lies between mesoderm and ectoderm; in *Xenopus*, endoderm is most vegetal and mesoderm lies between the endoderm and the ectoderm. Can you briefly describe the differences in gastrulation in these two organisms that explain these differing arrangements?

7. Contrast epiboly and convergent extension. How are these two processes used during *Xenopus* gastrulation?

8. If cells in a flat sheet constrict one surface (the apical surface), the sheet will develop a depression in that region. Give examples of how this mechanism is used to initiate gastrulation in sea urchins, *Drosophila*, and *Xenopus*.

9. How do changes in cell shape and changes in cell adhesiveness collaborate in the formation of the neural tube?

10. Describe the migration route of the neural crest cells that will form the dorsal root ganglia. What role do Ephs and ephrins play in this migratory pathway?

11. What is the influence of directed dilation on formation of the notochord? What role do sugars (glycosaminoglycans) play in this process?

12. In your own words, describe the cell movements that occur during gastrulation, from the first appearance of the blastopore, to the first appearance of the neural folds. Do this first for *Xenopus*, then describe the major differences seen in chicks and zebrafish.

Multiple choice (factual recall questions)

NB There is only one right answer to each question.

1. Adherens junctions play in important role in development; they are composed of

a) cadherins linked to other cadherins extracellularly, and to actin filaments intracellularly
b) cadherins linked to other cadherins extracellularly, and to intermediate filaments intracellularly
c) immunoglobulin superfamily (IgSF) molecules linked to other IgSF molecules extracellularly, and to the cytoskeleton intracellularly
d) integrins linked to extracellular matrix molecules extracellularly, and to the cytoskeleton intracellularly

2. If embryos are disaggregated with chemicals or proteases, and the individual cells are mixed together in culture, what will happen?

a) The cells will associate with one another randomly.
b) The cells will often sort themselves so that like cells are together.
c) The cells will reaggregate to form a normal embryo capable of continuing development.
d) The cells will regulate to form one or more normal embryos.

3. What type of cleavage is used in frog embryos?

a) radial
b) spiral
c) superficial
d) unequal

4. How do the endodermal cells of the sea urchin embryo carry out gut formation during gastrulation?

a) Changes in cell shape initiate the invagination; convergent extension extends the sheet of cells into the blastocoel, and finally, filopodia make contact with the future mouth region and pull the tip of the gut to that point.
b) They crawl inside the blastocoel and form a solid rod, which is then hollowed out to form the tube of the gut.
c) They move inside the blastocoel, and form the skeletal structures that will support the adult animal.
d) They move into the blastocoel as mesenchyme and crawl ventrally, to mark the location of the future mouth.

5. During gastrulation in the frog, the very first cells to move into the interior of the embryo through the blastopore come from the surface layer of cells in the marginal zone; they will become:

a) ectoderm
b) endoderm
c) mesoderm
d) yolk

6. Gastrulation in sea urchins, *Drosophila*, and *Xenopus* all begin with a change in cell shape, in which the apical surface of an epithelial sheet contracts. This is called

a) convergent extension
b) epiboly
c) invagination
d) involution

7. In *Xenopus*, the mesoderm moves in through the blastopore by rolling around the dorsal lip in a process called

a) convergent extension
b) epiboly
c) invagination
d) involution

8. In *Xenopus*, the elongation of the mesoderm toward the anterior results from the intercalation of cells during a process called

a) convergent extension
b) epiboly
c) invagination
d) involution

9. In chicks, the formation of the neural tube relies on what cell biological processes?

a) Changes in adhesiveness between the prospective neural tube cells and the prospective epidermis in the ectoderm are the only changes required for neural tube formation.
b) Changes in cell shape in the neural plate and changes in adhesion molecule expression in the neural tube are required during formation of the neural tube.
c) Changes in cell shape in the neural plate are the only process required for neural tube formation.
d) Neural tube formation is a passive by-product of convergent extension in the notochord.

10. Neural crest cells taking the dorsal-lateral route will become

a) adrenal medulla
b) dorsal root ganglia
c) melanocytes
d) sympathetic ganglia

Multiple choice answer key

1: a, 2: b, 3: a, 4: a, 5: b, 6: c, 7: d, 8: a, 9: b, 10: c.

■ General further reading

Alberts, B. *et al.*: *Molecular Biology of the Cell* (5th edn). New York: Garland Science, 2009.

Aman, A., Piotrowski, T.: **Cell migration during morphogenesis**. *Dev. Biol.* 2010, **341**: 20–33.

Engler, A.J., Humbert, P.O., Wehrle-Haller, B., Weaver, V.M.: **Multiscale modeling of form and function**. *Science* 2009, **324**: 209–212.

Mattila, P.K., Lappalainen, P.: **Filopodia: molecular architecture and cellular functions**. *Nat. Rev. Mol. Cell Biol.* 2009, **9**: 446–454.

Stern, C. *et al.*: *Gastrulation: From Cells to Embryos*. New York: Cold Spring Harbor Laboratory Press; 2004.

■ Section further reading

9.1 Sorting out of dissociated cells demonstrates differences in cell adhesiveness in different tissues

Davis, G.S., Phillips, H.M., Steinberg, M.S.: **Germ-layer surface tensions and 'tissue affinities' in** *Rana pipiens* **gastrulae: quantitative measurements.** *Dev. Biol.* 1997, **192**: 630–644.

Krieg, M., Arboleda-Estudillo, Y., Puech, P.-H., Käfer, J., Graner, F., Müller, D.J., Heisenberg, C.-P.: **Tensile forces govern germ-layer organization in zebrafish.** *Nat. Cell Biol.* 2008, **10**: 429–436.

Lecuit, T., Lenne, P.-F.: **Cell surface mechanics and the control of cell shape, tissue patterns and morphogenesis.** *Nat. Rev. Mol. Cell Biol.* 2007, **8**: 633–644.

Steinberg, M.S.: **Differential adhesion in morphogenesis: a modern view.** *Curr. Opin. Genet. Dev.* 2007, **17**: 291–296.

Box 9A Cell-adhesion molecules and cell junctions

Hynes, R.O: **Integrins: bidirectional allosteric signaling mechanisms.** *Cell* 2002, **110**: 673–699.

Tepass, U., Truong, K., Godt, D., Ikura, M., Peifer, M.: **Cadherins in embryonic and neural morphogenesis.** *Nat Rev. Mol. Cell Biol.* 2000, **1**: 91–100.

Thiery, J.P.: **Cell adhesion in development: a complex signaling network.** *Curr. Opin. Genet. Dev.* 2003, **13**: 365–371.

9.2 Cadherins can provide adhesive specificity

Duguay, D., Foty, R.A., Steinberg, M.S.: **Cadherin-mediated cell adhesion and tissue segregation: qualitative and quantitative determinants.** *Dev. Biol.* 2003, **253**: 309–323.

Levine, E., Lee, C.H., Kintner, C., Gumbiner, B.M.: **Selective disruption of E-cadherin function in early** *Xenopus* **embryos by a dominant negative mutant.** *Development* 1994, **120**: 901–909.

Takeichi, M., Nakagawa, S., Aono, S., Usui, T., Uemura, T.: **Patterning of cell assemblies regulated by adhesion receptors of the cadherin superfamily.** *Proc. Roy. Soc. Lond. B* 2000, **355**: 995–996.

9.3 Transitions of tissues from an epithelial to a mesenchymal state, and vice versa, involve changes in adhesive junctions

Ferrer-Vaquer, A., Viotti, M., Hadjantonakis, A.K.: **Transitions between epithelial and mesenchymal states and the morphogenesis of the early mouse embryo.** *Cell Adh. Migr.* 2010, **4**: 447–457.

Lim, J., Thiery, J.P.: **Epithelial–mesenchymal transitions: insights from development.** *Development* 2012, **139**: 3471–3486.

Nakaya, Y., Sheng, G.: **Epithelial to mesenchymal transition during gastrulation: an embryological view.** *Dev. Growth Differ.* 2008, **50**: 755–766.

Nieto, M.A.: **The ins and outs of the epithelial to mesenchymal transition in health and disease.** *Annu. Rev. Cell Dev. Biol.* 2011, **27**: 347–376.

Yang, J., Weinberg, R.A.: **Epithelial–mesenchymal transition: at the crossroads of development and tumor metastasis.** *Dev. Cell* 2008, **14**: 818–829.

9.4 The orientation of the mitotic spindle determines the plane of cleavage at cell division

Glotzer, M.: **Cleavage furrow positioning.** *J. Cell Biol.* 2004, **164**: 347–351.

Théry, M., Racine, V., Pépin, A., Piel, M., Chen, Y., Sibarita, J.-B., Bornens, M.: **The extracellular matrix guides the orientation of the cell division axis.** *Nat. Cell Biol.* 2005, **7**: 947–953.

9.5 The positioning of the spindle within the cell also determines whether daughter cells will be the same or different sizes

Galli, M., van den Heuvel, S.: **Determination of the cleavage plane in early** *C. elegans* **embryos.** *Annu. Rev. Genet.* 2009, **42**: 399–411.

Gönczy, P., Rose, L.S.: **Asymmetric cell division and axis formation in the embryo.** In *WormBook* (ed. The C. elegans Research Community) (October 15, 2005). doi:10.1995/wormbook.1.30.1, http://www.wormbook.org (date accessed 4 February 2014).

9.6 Cells become polarized in the sea-urchin blastula and the mouse morula

Pauken, C.M., Capco, D.G.: **Regulation of cell adhesion during embryonic compaction of mammalian embryos: roles for PKC and beta-catenin.** *Mol. Reprod. Dev.* 1999, **54**: 135–144.

Sutherland, A.E., Speed, T.P., Calarco, P.G.: **Inner cell allocation in the mouse morula: the role of oriented division during fourth cleavage.** *Dev. Biol.* 1990, **137**: 13–25.

9.7 Fluid accumulation as a result of tight-junction formation and ion transport forms the blastocoel of the mammalian blastocyst

Barcroft, L.C., Offenberg, H., Thomsen, P., Watson, A.J.: **Aquaporin proteins in murine trophectoderm mediate transepithelial water movements during cavitation.** *Dev. Biol.* 2003, **256**: 342–354.

Eckert, J.J., Fleming, T.P.: **Tight junction biogenesis during early development.** *Biochim. Biophys. Acta* 2008, **1778**: 717–728.

Fleming, T.P., Sheth, B., Fesenko, I.: **Cell adhesion in the preimplantation mammalian embryo and its role in trophectoderm differentiation and blastocyst morphogenesis.** *Front. Biosci.* 2001, **6**: 1000–1007.

Watson, A.J., Natale, D.R., Barcroft, L.C.: **Molecular regulation of blastocyst formation.** *Anim. Reprod. Sci.* 2004, **92–93**: 593–592.

9.8 Gastrulation in the sea urchin involves an epithelial-to-mesenchymal transition, cell migration, and invagination of the blastula wall

Davidson, L.A., Oster, G.F., Keller, R.E., Koehl, M.A.R.: **Measurements of mechanical properties of the blastula wall reveal which hypothesized mechanisms of primary invagination are physically plausible in the sea urchin** *Stronglyocentrotus purpuratus.* *Dev. Biol.* 1999, **204**: 235–250.

Duloquin, L., Lhomond, G., Gache, C.: **Localized VEGF signaling from ectoderm to mesenchyme cells controls morphogenesis of the sea urchin embryo skeleton.** *Development* 2007, **134**: 2293–2302.

Ettensohn, C.A.: **Cell movements in the sea urchin embryo.** *Curr. Opin. Genet. Dev.* 1999, **9**: 461–465.

Gustafson, T., Wolpert, L.: **Studies on the cellular basis of morphogenesis in the sea urchin embryo. Directed movements of primary mesenchyme cells in normal and vegetalized larvae.** *Exp. Cell Res.* 1999, **253**: 299–295.

Mattila, P.K., Lappalainen, P.: **Filopodia: molecular architecture and cellular functions.** *Nat. Rev. Mol. Cell Biol.* 2009, **9**: 446–454.

Miller, J.R., McClay, D.R.: **Changes in the pattern of adherens junction-associated β-catenin accompany morphogenesis in the sea urchin embryo.** *Dev. Biol.* 1997, **192**: 310–322.

Odell, G.M., Oster, G., Alberch, P., Burnside, B.: **The mechanical basis of morphogenesis. I. Epithelial folding and invagination.** *Dev. Biol.* 1991, **95**: 446–462.

Raftopoulou, M., Hall, A.: **Cell migration: Rho GTPases lead the way.** *Dev Biol.* 2004, **265**: 23–32.

Röttinger, E., Saudemont, A., Duboc, V., Besnardeau, L., McClay, D., Lepage, T.: **FGF signals guide migration of mesenchymal cells, control skeletal morphogenesis and regulate gastrulation during sea urchin development.** *Development* 2009, **135**: 353–365.

9.9 Mesoderm invagination in *Drosophila* is due to changes in cell shape that are controlled by genes that pattern the dorso-ventral axis

Martin, A.C., Kaschube, M., Wieschaus, E.F.: **Pulsed contractions of an actin–myosin network drives apical contraction.** *Nature* 2009, **457**: 495–499.

9.10 Germ-band extension in *Drosophila* involves myosin-dependent remodeling of cell junctions and cell intercalation

Bertet, C., Sulak, L., Lecuit, T.: **Myosin-dependent junction remodelling controls planar cell intercalation and axis elongation.** *Nature* 2004, **429**: 667–671.

9.11 Gastrulation in amphibians and fish embryos involves involution, epiboly, and convergent extension

Dzamba, B.J., Jakab, K.R., Marsden, M., Schwartz, M.A., DeSimone, D.W.: **Cadherin adhesion, tissue tension, and noncanonical Wnt signaling regulate fibronectin matrix organization.** *Dev. Cell.* 2009, **16**: 421–432.

Goto, T., Davidson, L., Asashima, M., Keller, R.: **Planar cell polarity genes regulate polarized extracellular matrix deposition during frog gastrulation.** *Curr. Biol.* 2005, **15**: 797–793.

Heisenberg, C.P., Solnica-Krezel, L.: **Back and forth between cell fate specification and movement during vertebrate gastrulation.** *Curr. Opin. Genet. Dev.* 2008, **18**: 311–316.

Keller, R.: **Cell migration during gastrulation.** *Curr. Opin. Cell Biol.* 2005, **17**: 533–541.

Keller, R., Davidson, L.A., Shook, D.R.: **How we are shaped: the biomechanics of gastrulation.** *Differentiation* 2003, **71**: 171–205.

Montero, J.A., Carvalho, L., Wilsch-Brauninger, M., Kilian, B., Mustafa, C., Heisenberg, C.P.: **Shield formation at the onset of zebrafish gastrulation.** *Development* 2005, **132**: 1197–1199.

Montero, J.A., Heisenberg, C.P.: **Gastrulation dynamics: cells move into focus.** *Trends Cell Biol.* 2004, **14**: 620–627.

Ninomiya, H., Elinson, R.P., Winklbauer, R.: **Antero-posterior tissue polarity links mesoderm convergent extension to axial patterning.** *Nature* 2004, **430**: 364–367.

Ninomiya, H., Winklbauer, R.: **Epithelial coating controls mesenchymal shape change through tissue-positioning effects and reduction of surface-minimizing tension.** *Nat. Cell Biol.* 2009, **10**: 61–69.

Shih, J., Keller, R.: **Gastrulation in *Xenopus laevis*: involution—a current view.** *Dev. Biol.* 1994, **5**: 95–90.

Wacker, S., Grimm, K., Joos, T., Winklbauer, R.: **Development and control of tissue separation at gastrulation in *Xenopus*.** *Dev. Biol.* 2000, **224**: 429–439.

Box 9C Convergent extension

Keller, R., Davidson, L., Edlund, A., Elul, T., Ezin, M., Shook, D., Skoglund, P.: **Mechanisms of convergence and extension by cell intercalation.** *Proc R. Soc. Lond. B* 2000, **355**: 997–922.

Simons, M., Mlodzik, M.: **Planar cell polarity signaling: from fly development to human disease.** *Annu. Rev. Genet.* 2009, **42**: 517–540.

Torban, E., Kor, C., Gros, P.: **Van Gogh-like2 (Strabismus) and its role in planar cell polarity and convergent extension in vertebrates.** *Trends Genet.* 2004, **20**: 570–577.

Wallingford, J.B., Fraser, S.E., Harland, R.M.: **Convergent extension: the molecular control of polarized cell movement during embryonic development.** *Dev. Cell* 2002, **2**: 695–706.

Zallen, J.A.: **Planar polarity and tissue morphogenesis.** *Cell* 2007, **129**: 1051–1063.

9.12 *Xenopus* notochord development illustrates the dependence of medio-lateral cell polarity on a pre-existing antero-posterior polarity

Keller, R., Cooper, M.S., D'Anilchik, M., Tibbetts, P., Wilson, P.A.: **Cell intercalation during notochord development in *Xenopus laevis*.** *J. Exp. Zool.* 1999, **251**: 134–154.

9.13 Gastrulation in chick and mouse embryos involves the delamination of cells from the epiblast and their ingression through the primitive streak

Arkell, R.M., Fossat, N., Tam, P.P.: **Wnt signalling in mouse gastrulation and anterior development: new players in the pathway and signal output.** *Curr Opin Genet Dev.* 2013, **23**: 454–460.

Chuai, M., Hughes, D., Weijer, C.J.: **Collective epithelial and mesenchymal cell migration during gastrulation.** *Curr. Genomics* 2012, **13**: 267–277.

Sun, X., Meyers, E.N., Lewandoski, M., Martin, G.R.: **Targeted disruption of Fgf8 causes failure of cell migration in the gastrulating mouse embryo.** *Genes Dev.* 1999, **13**: 1834–46

Voiculescu, O., Bertocchini, F., Wolpert, L., Keller, R.E., Stern, C.D.: **The amniote primitive streak is defined by epithelial cell intercalation before gastrulation.** *Nature* 2007, **449**: 1049–1052.

Williams, M., Burdsal, C., Periasamy, A., Lewandoski, M., Sutherland, A.: **Mouse primitive streak forms in situ by initiation of epithelial to mesenchymal transition without migration of a cell population.** *Dev. Dyn.* 2012, **241**: 270–283.

9.14 Neural tube formation is driven by changes in cell shape and convergent extension

Colas, J.F., Schoenwolf, G.C.: **Towards a cellular and molecular understanding of neurulation**. *Dev. Dyn.* 2001, **221**: 117–145.

Copp, A.J., Greene, N.D.E., Murdoch, J.N.: **The genetic basis of mammalian neurulation**. *Nat. Rev. Genet.* 2003, **4**: 784–793.

Davidson, L.A., Keller, R.E.: **Neural tube closure in *Xenopus laevis* involves medial migration, directed protrusive activity, cell intercalation and convergent extension**. *Development* 1999, **126**: 4547–4556.

Haigo, S.L., Hildebrand, J.D., Harland, R.M., Wallingford, J.B.: **Shroom induces apical constriction and is required for hingepoint formation during neural tube closure**. *Curr. Biol.* 2003, **13**: 2125–2137.

Ray, H.J., Niswander, L.: **Mechanisms of tissue fusion during development**. *Development* 2012, **139**: 1701–1711.

Rolo, A., Skoglund, P., Keller, R.: **Morphogenetic movements during neural tube closure in *Xenopus* require myosin IIB**. *Dev. Biol.* 2009, **327**: 327–339.

Wallingford, J.B., Harland, R.M.: **Neural tube closure requires Dishevelled-dependent convergent extension of the midline**. *Development* 2002, **129**: 5915–5925.

Box 9E Neural tube defects

Greene, N.D.E., Stanier, P., Copp, A.J.: **Genetics of human neural tube defects**. *Hum. Mol. Genet.* 2009, **19**: R113–R129.

Juriloff, D.M., Harris, M.J.: **Mouse models for neural tube closure defects**. *Hum. Mol. Genet.* 2000, **9**: 993–1000.

Kibar, Z., Capra, V., Gros, P.: **Toward understanding the genetic basis of neural tube defects**. *Clin. Genet.* 2007, **71**: 295–310.

Torban, E., Patenaude, A.M., Leclerc, S., Rakowiecki, S., Gauthier, S., Andelfinger, G., Epstein, D.J., Gros, P.: **Genetic interaction between members of the Vang1 family causes neural tube defects in mice**. *Proc. Natl. Acad. Sci. USA* 2009, **105**: 3449–3454.

Wallingford, J.B., Niswander, L.A., Shaw, G.M., Finnell, R.H.: **The continuing challenge of understanding, preventing, and treating neural tube defects**. *Science* 2013, **339**: 1222002.

9.15 Embryonic neural crest gives rise to a wide range of different cell types

&

9.16 Neural crest migration is controlled by environmental cues

Bronner-Fraser, M.: **Mechanisms of neural crest migration**. *BioEssays* 1993, **15**: 221–230.

Holder, N., Klein, R.: **Eph receptors and ephrins: effectors of morphogenesis**. *Development* 1999, **126**: 2033–2044.

Kuan, C.Y., Tannahill, D., Cook, G.M., Keynes, R.J.: **Somite polarity and segmental patterning of the peripheral nervous system**. *Mech. Dev.* 2004, **121**: 1055–1069.

Kuriyama, S., Mayor, R.: **Molecular analysis of neural crest migration**. *Philos. Trans. R. Soc. Lond. B Biol. Sci.* 2009, **363**: 1349–1362.

Nagawa, S., Takeichi, M.: **Neural crest emigration from the neural tube depends on regulated cadherin expression**. *Development* 1999, **125**: 2963–2971.

Poliakoff, A., Cotrina, M., Wilkinson, D.G.: **Diverse roles of Eph receptors and ephrins in the regulation of cell migration and tissue assembly**. *Dev. Cell* 2004, **7**: 465–490.

Stark, D.A., Kulesa, P.M.: **An *in vivo* comparison of photoactivatable fluorescent proteins in an avian embryo model**. *Dev. Dyn.* 2007, **236**: 1583–1594.

Thevenean, E., Marchant, L., Kuriyama, S., Gull, M., Moeppo, B., Parsons, M., Mayor, R.: **Collective chemotaxis requires contact-dependent cell polarity**. *Dev. Cell* 2010, **19**: 39–53.

Tucker, R.P.: **Neural crest cells: a model for invasive behavior**. *Int. J. Biochem. Cell Biol.* 2004, **36**: 173–177.

Xu, Q., Mellitzer, G., Wilkinson, D.G.: **Roles of Eph receptors and ephrins in segmental patterning**. *Proc. Roy. Soc. Lond. B* 2000, **353**: 993–1002.

9.17 The formation of the lateral-line primordium in fishes is an example of collective cell migration

Friedl, P., Gilmour, D.: **Collective cell migration in morphogenesis, regeneration and cancer**. *Nat. Rev. Mol. Cell Biol.* 2009, **10**: 445–457.

Haas, P., Gilmour, D.: **Chemokine signaling mediates self-organizing tissue migration in the zebrafish lateral line**. *Dev. Cell* 2006, **10**: 673–680.

9.18 Dorsal closure in *Drosophila* and ventral closure in *Caenorhabditis elegans* are brought about by the action of filopodia

Chin-Sang, I.D., Chisholm, A.D.: **Form of the worm: genetics of epidermal morphogenesis in *C. elegans***. *Trends Genet.* 2000, **16**: 544–551.

Jacinto, A., Woolner, S., Martin, P.: **Dynamic analysis of dorsal closure in *Drosophila*: from genetics to cell biology**. *Dev. Cell* 2002, **3**: 9–19.

Millard, T. H., Martin, P.: **Dynamic analysis of filopodial interactions during the zippering phase of Drosophila dorsal closure**. *Development* 2008, **135**: 621–626.

Peralta, X.G., Toyama, Y., Hutson, M.S., Montague, R., Venakides, S., Kiehart, D.P., Edwards, G.S.: **Upregulation of forces and morphogenic asymmetries in dorsal closure during *Drosophila* development**. *Biophys. J.* 2007, **92**: 2583–2596.

Woolner, S., Jacinto, A., Martin, P.: **The small GTPase Rac plays multiple roles in epithelial sheet fusion—dynamic studies of *Drosophila* dorsal closure**. *Dev. Biol.* 2005, **292**: 163–173.

9.19 Later extension and stiffening of the notochord occurs by directed dilation

Adams, D.S., Keller, R., Koehl, M.A.: **The mechanics of notochord elongation, straightening and stiffening in the embryo of *Xenopus laevis***. *Development* 1990, **110**: 115–130.

9.20 Circumferential contraction of hypodermal cells elongates the nematode embryo

Priess, J.R., Hirsh, D.I.: **Caenorhabditis elegans morphogenesis: the role of the cytoskeleton in elongation of the embryo**. *Dev. Biol.* 1996, **117**: 156–173.

10

Germ cells, fertilization, and sex

- The development of germ cells
- Fertilization
- Determination of the sexual phenotype

Animal embryos develop from a single cell, the fertilized egg or zygote, which is the product of the fusion of an egg and a sperm. In previous chapters we have considered how the basic body plan of animal embryos is generated by the development of the somatic cells and their assignment to the three germ layers. The zygote also gives rise to the germline cells—or germ cells—that develop into eggs and sperm, and through which an organism's genes are passed on to the next generation. As we shall discuss in this chapter, in animals, germline cells are specified and set apart in the early embryo, although functional mature eggs and sperm are only produced by the adult. An important property of germ cells is that they remain totipotent—able to give rise to all the different types of cells in the body. Nevertheless, eggs and sperm in mammals have certain genes differentially silenced during germ-cell development by a process known as genomic imprinting, and we will look at the implications of this for development. Embryonic development is initiated by the process of fertilization, and we shall discuss mechanisms in sea urchins and mammals that ensure that only one sperm enters the egg. In the animals we cover in this chapter, sex is determined by the chromosomal constitution, and we shall see that mechanisms of sex determination and of compensating for the unequal complement of sex chromosomes between males and females differ considerably in different species.

In sexually reproducing organisms, there is a fundamental distinction between the **germline cells (germ cells)** and the somatic or body cells (see Section 1.2): the former give rise to the **gametes**—eggs and sperm in animals—whereas somatic cells make no genetic contribution to the next generation. Germ cells have three key functions: the preservation of the genetic integrity of the germline, such as the prevention of aging; the generation of genetic diversity; and the transmission of genetic information to the next generation.

So far in this book, we have looked at the somatic development of our model organisms. Not surprisingly, a great deal of the biology of animals and plants is devoted to reproduction and sex, and in this chapter, we look at germ-cell formation, fertilization of the egg by the sperm, and sex determination, mainly in the mouse, *Drosophila*, and *Caenorhabditis elegans*. As germ cells are the cells that will give rise to the next generation, their development is crucial, and in animals, germ cells are usually specified and set aside early in embryonic development. Plants, although reproducing sexually, differ from most animals in that their germ cells are not specified early in embryonic development, but during the development of the flowers (see Chapter 7). It is worth noting that some simple animals, such as the coelenterate *Hydra*, can reproduce asexually, by budding, and that even in some vertebrates, such as turtles, the eggs can develop without being fertilized.

We will begin this chapter by considering how the germ cells are specified in the early animal embryo and how they differentiate into eggs and sperm. Then, we will look at the **fertilization** and activation of the egg by the sperm—the vital step that initiates development. Finally, we turn to **sex determination**—why males and females are different from each other. In all the animals we look at here, sex is determined genetically, by the number and type of specialized chromosomes known as the **sex chromosomes**. Male and female embryos initially look the same, with sexual differences only emerging as a result of the activity of sex-determining genes located on the sex chromosomes.

The development of germ cells

In all but the simplest animals, the cells of the germline are the only cells that can give rise to a new organism. So, unlike somatic cells, which eventually all die, germ cells in a sense outlive the bodies that produced them. They are, therefore, very special cells. The outcome of germ-cell development is either a male gamete (the sperm in animals) or a female gamete (the egg). The egg is a particularly remarkable cell, as it ultimately gives rise to all the cells in an organism. In species whose embryos receive no nutrition from the mother after fertilization, the egg must also provide everything necessary for development, as the sperm contributes virtually nothing to the organism other than its chromosomes.

Animal germ cells typically differ from somatic cells by dividing less often during early embryogenesis. Later, they are the only cells to undergo the type of nuclear division called **meiosis**, by which gametes with half the number of chromosomes of the germ-cell precursor are produced. The halving of chromosome number at meiosis means that when two gametes come together at fertilization to form the zygote, the diploid number of chromosomes is restored. If there was no reduction division in germ-cell formation, the number of chromosomes in the somatic cells would double in each generation.

Germ cells undergo meiosis and differentiate into eggs and sperm within specialized reproductive organs called **gonads**—the **ovary** in females and the **testis** in males. But in many animals, including *Drosophila* and vertebrates, the germ cells are first specified in a region quite distant from where the gonads will eventually form, and migrate from their site of origin into the forming gonads. From the time they are first specified as germline cells until the time that they enter the gonads, precursor germ cells are known as **primordial germ cells**. In many animals, primordial germ cells are formed in locations that seem to protect them from the inductive signals that determine the fate of somatic cells, and instead shut down the somatic developmental program. In *Caenorhabditis*, for example, there is evidence of a general repression of transcription in the primordial germ cells. In some germlines, other mechanisms that help suppress mutation and maintain genetic integrity are present. In *Drosophila*, for example, the movement of transposons in the genome, which can cause insertion mutations, is prevented by an RNA silencing pathway involving the Piwi-interacting small RNAs (piRNAs).

Germ cells are specified very early in some animals, although not in mammals, by cytoplasmic determinants present in the egg. We therefore start our discussion of germ-cell development by looking at the specification of primordial germ cells by special cytoplasm—the **germplasm**—which is already present in these eggs.

10.1 Germ-cell fate is specified in some embryos by a distinct germplasm in the egg

In flies, nematodes, fish, and frogs, molecules localized in specialized cytoplasm in the egg are involved in specifying the germ cells. The clearest example of this is in *Drosophila*, where primordial germ cells known as **pole cells** become distinct at the posterior pole of the egg about 90 minutes after fertilization, more than an hour before the cellularization of the rest of the embryo (see Fig. 2.2). The cytoplasm at the posterior pole is called the **pole plasm** and is distinguished by large organelles, the polar granules, which contain both proteins and RNAs. Two key experiments demonstrate that there is something special about the posterior cytoplasm. First, if the posterior end of the egg is irradiated with ultraviolet light, which destroys the pole plasm activity, no germ cells develop, although the somatic cells in that region do develop. Second, if pole plasm from an egg is transferred to the anterior pole of another embryo, the nuclei that become surrounded by the pole plasm are specified as germ-cell nuclei (Fig. 10.1). If these anterior cells containing pole plasm are transplanted into the posterior pole region of a third embryo, they develop as functional germ cells (see Fig. 10.1).

In Chapter 2, we saw how the main axes of the *Drosophila* egg are specified by the follicle cells in the ovary, and how the mRNAs for proteins such as Bicoid and Nanos become localized in the egg (see Section 2.14). The pole plasm also becomes localized at the posterior end of the egg under the influence of the follicle cells. Several maternal genes are involved in pole-plasm formation in *Drosophila*. Mutations in any of at least eight genes result in the affected homozygous individual being 'grandchildless.' Its offspring lack a proper pole plasm, and although they may develop normally in other ways, they lack germ cells and are therefore sterile. One of these eight genes is *oskar*, which plays a central role in the organization and assembly of the pole plasm; of the genes involved in pole-plasm formation, *oskar* is the only one to have its mRNA localized at the posterior pole. The signal for localization is contained in the 3' untranslated region of the mRNA. Staufen protein is required for the localization of *oskar* mRNA and may act by linking the mRNA to the microtubule system that is polarized along the antero-posterior axis (see Section 2.14). If the region of the *oskar* gene that codes for the 3' localization signal is replaced by the *bicoid* 3' localization signal, flies made transgenic with this DNA construct have *oskar* mRNA localized at the anterior end of the egg (Fig. 10.2).

In the nematode, a germ-cell lineage is set up at the end of the fourth cleavage division, with all the germ cells being derived from the P_4 blastomere (see Section 6.2). The P4 cell is derived from three stem-cell-like divisions of the P_1 cell. At each of these divisions, one daughter produces somatic cells, whereas the other divides again to produce a somatic cell progenitor and a P cell. The egg contains P granules in its cytoplasm that become asymmetrically distributed before the first cleavage division, and are subsequently confined to the P-cell lineage (Fig. 10.3). The association of germ-cell formation with the P granules suggested that they might have a role in germ-cell specification, and at least one P-granule component, the product of the *pgl-1* gene, has been shown to be necessary for germ-cell development. PGL-1 protein may specify germ cells by regulating some aspect of mRNA metabolism.

In both the fly and the nematode, repression of transcription is necessary to specify a germ-cell fate. In the nematode, the gene *pie-1* is involved in maintaining the stem-cell property of the P blastomeres. It encodes a nuclear protein that is expressed maternally and is not a component of P granules; the PIE-1 protein is present only in

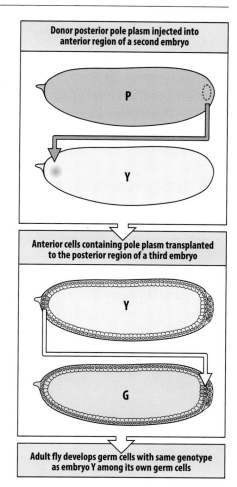

Fig. 10.1 Transplanted pole plasm can induce germ-cell formation in *Drosophila*. Pole plasm from a fertilized egg of genotype P (pink) is transferred to the anterior end of an early cleavage-stage embryo of genotype Y (yellow). After cellularization, cells containing pole plasm induced at the anterior end of embryo Y are transferred to the posterior end (a site from which germ cells can migrate into the gonad) of another embryo, of genotype G (green). The adult fly that develops from embryo G contains germ cells of genotype Y as well as those of G.

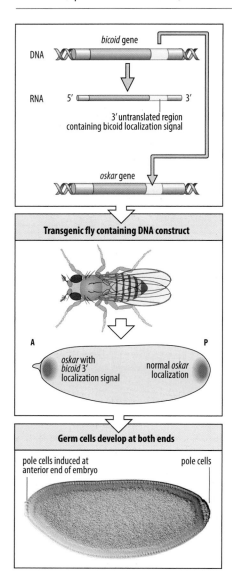

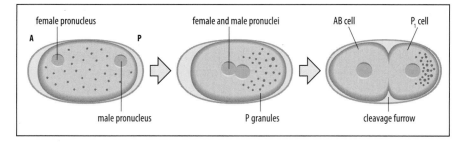

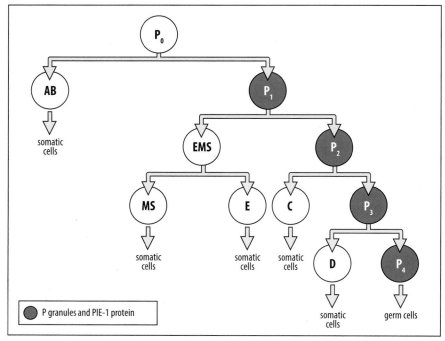

Fig. 10.3 P granules and PIE-1 protein become asymmetrically distributed to germline cells during cleavage of the nematode egg. Before fertilization, P granules (purple) are distributed throughout the egg. After fertilization, P granules become localized at the posterior end of the egg. At the first cleavage, they are only included in the P1 cell (top panel), and thus become confined to the P cell lineage. The PIE-1 protein is only present in P cells. All germ cells are derived from P4, which is formed at the fourth cleavage.

Fig. 10.2 The gene *oskar* is involved in specifying the germplasm in *Drosophila*. In normal eggs, *oskar* mRNA is localized at the posterior end of the embryo, whereas *bicoid* mRNA is at the anterior end. The localization signals for both *bicoid* and *oskar* mRNA are in their 3′ untranslated regions. By manipulating the Drosophila DNA, the localization signal of *oskar* can be replaced by that of bicoid (top panel). A transgenic fly is made containing the modified DNA. In its egg, *oskar* becomes localized at the anterior end (middle panel). The egg therefore has *oskar* mRNA at both ends, and germ cells develop at both ends of the embryo, as shown in the photograph (bottom panel). Thus, *oskar* alone is sufficient to initiate the specification of germ cells. *Photograph courtesy of R. Lehmann.*

the germline blastomeres, and represses new transcription of zygotic genes in these blastomeres until PIE-1 disappears at around the 100-cell stage. This general repression may protect the germ cells from the actions of transcription factors that promote development into somatic cells. In the absence of zygotic transcription, maternal RNAs encode the germ-cell components.

Xenopus and zebrafish also have distinct germplasm. In *Xenopus*, distinct yolk-free patches of cytoplasm aggregate at the yolky vegetal pole after fertilization. When the blastomeres at the vegetal pole cleave, this cytoplasm is distributed asymmetrically, so that it is retained only in the most vegetal daughter cells, from which the germ cells are derived. Ultraviolet irradiation of the vegetal cytoplasm abolishes the formation of germ cells, and transplantation of fresh vegetal cytoplasm into an irradiated egg restores germ-cell formation.

At gastrulation, the germplasm is located in cells in the floor of the blastocoel cavity, among the cells that give rise to the endoderm. Cells containing the germplasm are, however, not yet determined as germ cells and can contribute to all three germ layers if transplanted to other sites. At the end of gastrulation, the primordial germ

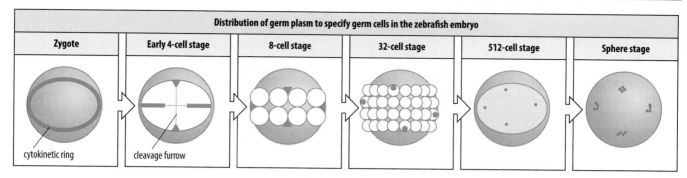

Fig. 10.4 Distribution of germplasm in zebrafish embryo during cleavage. In this bird's-eye view of the animal pole of the egg, maternal *vasa* mRNA (blue) is a marker for the germplasm. The *vasa* mRNA becomes localized to the cleavage furrows at first and second divisions. As a furrow completes division the *vasa* mRNA forms tight clumps at the distal ends, at or just outside the cell boundary but under the yolk cell membrane. At the 32-cell stage, the aggregates ingress into four cells and segregate asymmetrically at subsequent divisions, ending up in four widely separated cells at the 512-cell stage. At the sphere stage (cleavage cycle 13), the *vasa* mRNA becomes evenly distributed throughout the four cells, and subsequent cell divisions produce a total of around 30 primordial germ cells in the four locations.

*Adapted from Pelegri, F.: **Maternal factors in zebrafish development.** Dev. Dyn. 2003, 228: 535-554.*

cells are determined, and migrate out of the presumptive endoderm and into the **genital ridge**, which develops from mesoderm lining the abdominal cavity and will form the gonads. There are, however, quite big differences in germ-cell origin in different groups of amphibians. In the Urodeles—the tailed amphibians that include the salamanders and newts—there is no evidence for cytoplasmic localization of germplasm, and the germ cells arise from lateral plate mesoderm. The evolutionary significance of this difference is not known.

In the zebrafish, germplasm containing a number of maternal mRNAs is present in several locations in the fertilized egg, and becomes localized to the distal ends of the cleavage furrows in the first few cleavages. Its fate can be followed during early development by tracking one of its components, such as maternal *vasa* mRNA (Fig. 10.4). By the 32-cell stage, germplasm has been segregated into four cells. These cells then continue to divide asymmetrically, with the germplasm segregating to one daughter cell at each division, so that even by the 512-cell stage there are still only four prospective germline cells in the embryo. At the sphere stage of the embryo (about 3.8 hours post-fertilization), the germplasm expands to fill the cells, and equal divisions begin, eventually producing a total of around 30 primordial germ cells, which will migrate to the prospective ovaries or testes.

10.2 In mammals germ cells are induced by cell–cell interactions during development

There is no evidence for germplasm in the mouse or other mammals or in the chick. Indeed, it seems that, although germplasm is present in several of our developmental model organisms, it is the less prevalent mode of germline specification in animals generally. Germ-cell specification in the mouse involves cell–cell interactions, as cultured embryonic stem cells can, when injected into the inner cell mass, give rise to both germ cells and somatic cells (see Fig. 3.10). Mammalian germ cells were previously identified mainly by their high concentration of the enzyme alkaline phosphatase, which provides a convenient means of detecting them histochemically. Now that genes involved in germline specification have been identified, it is possible to identify primordial germ cells at a much earlier stage by detecting the expression of such genes.

The earliest detectable primordial germ cells can be identified in the mouse proximal epiblast just before the beginning of gastrulation. They are specified by BMP signaling by neighboring extraembryonic tissues and form a cluster of six to eight cells expressing the gene for the transcriptional repressor protein Blimp1. The *Blimp1*-expressing cells are first found in the proximal epiblast at E6.25, in a region of prospective extraembryonic mesoderm that will, after cell movement, be adjacent to the beginning of the primitive streak (Fig. 10.5). The Blimp1-positive cells proliferate and by E7.5 there are around 40 primordial germ cells in the primitive streak. These represent the full complement of primordial germ cells that will eventually migrate to the mouse

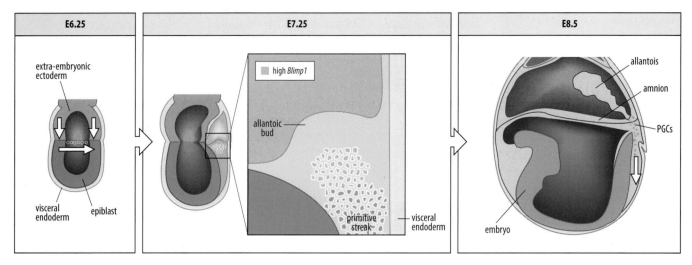

Fig. 10.5 Germ-cell formation in the mouse. First panel: a small number of primordial germ cells (PGCs) (white) expressing *Blimp1* are first detectable in the proximal epiblast at 6.25 days after fertilization (E6.25). Second panel: during gastrulation, these cells and the surrounding prospective extra-embryonic mesoderm cells move to the posterior end of the embryo above the primitive streak. By E7.25, around 40 PGCs (orange) are present in the primitive streak. Third panel: at around E8.5, PGCs start to migrate to the gonads.

*Adapted from Hogan, B.: **Decisions, decisions.** Nature 2002, **418**: 282, and Saitou, M., et al.: **A molecular programme for the specification of germ cell fate in mice.** Nature, 2002, **418**: 293-300.*

gonads (Fig. 10.5, third panel). Blimp1 and two other transcription factors that come to be expressed by primordial cells act in combination to repress expression of somatic genes and to activate expression of genes characteristic of germ cells.

Unlike the somatic cells at this stage, the Hox genes and other typical somatic developmental genes are repressed in the primordial germ cells, which may explain their escape from a somatic cell fate. Instead, the primordial germ cells express genes such as *nanog*, which encodes a transcription factor associated with the maintenance of pluripotency in stem cells (see Section 8.14), and a large number of genes involved in cell adhesion and cell migration. The earliest events in establishing human germ cells remain unknown, but migratory germ cells have been reported at the fourth week of human embryonic development.

10.3 Germ cells migrate from their site of origin to the gonad

In many animals, the primordial germ cells develop at some distance from the gonads, and only later migrate to them, where they differentiate into eggs or sperm. The reason for this separation of site of origin from final destination is not known, but it may be a mechanism for excluding germ cells from the general developmental upheaval involved in laying down the body plan, or a mechanism for selecting the healthiest germ cells, namely those that survive migration. The migration path of primordial germ cells is controlled by their environment; in *Xenopus*, for example, primordial germ cells transplanted to the wrong place in the blastula do not end up in the gonad.

The vertebrate gonad develops from mesoderm lining the abdominal cavity, which is known as the genital ridge. In *Xenopus*, primordial germ cells originating in the floor of the blastocoel migrate to the future gonad along a cell sheet that joins the gut to the genital ridge. Only a small number of cells start this journey, dividing about three times before arrival, so that about 30 germ cells colonize the gonad—about the same number as in the zebrafish. In contrast, the number of primordial germ cells that arrive at the genital ridge of the mouse is about 8000, starting from a population of around 40 at the posterior end of the primitive streak (see Fig. 10.5). In the mouse, the primordial germ cells enter the hindgut endoderm, and migrate to the dorsal mesentery, the cell sheet that suspends the gut in the body cavity, and then across it to reach the genital ridges (Fig. 10.6).

In chick embryos, the pattern of migration is different again: the germ cells originate in the epiblast and then migrate to the head end of the embryo. Most arrive at their final destination in the gonads by circulating in the blood, leaving the bloodstream at the hindgut and then migrating along an epithelial sheet to the gonad.

In zebrafish, small clusters of primordial germ cells are specified in four separate positions relative to the embryonic axis (see Fig. 10.4). These cells migrate to form bilateral lines of cells in the trunk at the level of somites 1–3 at around 6 hours post-fertilization, and then reach their final position at the level of somites 8–10 at around 12 hours, where they and the surrounding somatic cells form the gonads (Fig. 10.7). Zebrafish primordial germ cells face a challenging journey, as they originate in four separate locations, and it may be no coincidence that the journey seems to be broken up into a number of distinct stages. This also means that germ cells taken from the margin of a mid-blastula zebrafish embryo may even find their way to the gonad after transplantation to the head region (the animal pole) of a similar-stage embryo.

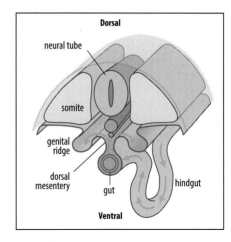

Fig. 10.6 Pathway of primordial germ-cell migration in the mouse embryo. In the final stage of migration, the cells move from the gut tube into the genital ridge, via the dorsal mesentery.

*Illustration after Wylie, C.C., Heasman, J.: **Migration, proliferation, and potency of primordial germ cells**. Semin. Dev. Biol. 1993, **4**: 161-170.*

10.4 Germ cells are guided to their final destination by chemical signals

All migrating germ cells are continuously receiving signals for guidance, survival, and proliferation from the tissues through which they migrate. Some of these signals have been identified; if they are lacking, germ cells are absent from the gonads or their numbers are greatly reduced. The main guidance cue in both zebrafish and in mice seems to be the chemoattractant protein SDF1 (stromal-cell derived factor 1). We have come across this chemoattractant in relation to the collective migration of the cells that give rise to the lateral line in zebrafish (see Section 9.17). SDF1 is also secreted into the extracellular matrix by tissue cells in the regions through which the germ cells migrate in zebrafish, and is thought to be the long-range guidance cue for the clusters of primordial germ cells to reach their final positions from their different starting points (see Fig. 10.7). Experiments that change the spatial expression of the gene for SDF1 change the germ-cell migration patterns correspondingly. Also, knockdown by antisense RNA of the expression of either SDF1 or its receptor, CXCR4, which is found on the surface of primordial germ cells, disrupts migration.

In the mouse, the primordial germ cells in the primitive streak migrate anteriorly and become incorporated into the hindgut endoderm, which they eventually leave to migrate into the genital ridge (see Fig. 10.6). The mouse version of SDF1 seems to act as a guidance cue for this final part of the journey at least, as in animals lacking either SDF1 or CXCR4, most germ cells get no further than the hindgut. SDF1 and CXCR4 are not only of interest to embryos; they are also expressed in some invasive human tumors, where they contribute to the ability of the tumor cells to spread to other sites in the body.

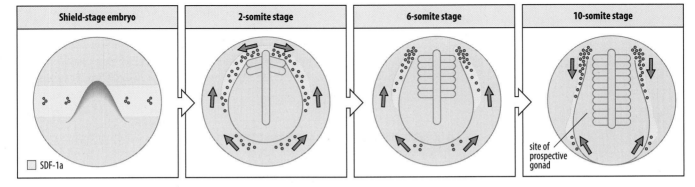

Fig. 10.7 Primordial germ cells in zebrafish migrate to the gonads in two stages. The diagrams show a dorsal view of the embryo. The chemoattractant SDF1 (yellow) is expressed initially in a broad domain that includes the sites at which the primordial germ cells (PGCs, blue) are formed (see Fig. 10.4). PGCs migrate towards the region of highest SDF1, which is initially at the level of the first somite. Alterations in the pattern of SDF1 expression then cause the PGCs to migrate posteriorly to the level of the eighth and tenth somites, where the gonads develop.

*Adapted from Weidinger, G., et al.: **Regulation of zebrafish primordial germ cell migration by attraction towards an intermediate target**. Development, 2002, **129**: 25-36.*

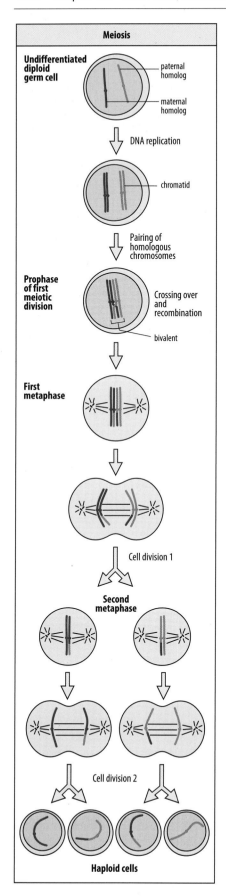

Fig. 10.8 Meiosis produces haploid cells. Meiosis reduces the number of chromosomes from the diploid to the haploid number. Only one pair of homologous chromosomes is shown here for simplicity. Before the first meiotic division the DNA replicates, so that each chromosome entering meiosis is composed of two identical chromatids. The paired homologous chromosomes (known as a bivalent) undergo crossing over and recombination, and align on the meiotic spindle at the metaphase of the first meiotic division. The homologous chromosomes separate, and each is segregated into a different daughter cell at the first cell division. There is no DNA replication before the second meiotic division. The daughter chromatids of each chromosome separate and segregate at the second cell division. The chromosome number of the resulting daughter cells is thus halved.

In *Drosophila*, primordial germ cells are formed next to the posterior midgut, and during gastrulation they are carried along the dorsal side of the embryo with the midgut primordium. They then migrate across the dorsal midgut epithelium to the adjacent mesoderm to join with gonadal precursor cells in forming the gonad. Screening for mutations that affect the direction of germ-cell migration has identified genes that encode proteins keep the cells on track. Two related proteins, Wunen (WUN) and WUN2, are involved in repelling germ cells from the rest of the gut, and so preventing them from dispersing before they reach the gonadal mesoderm, whereas expression of the enzyme HMGCoA reductase in the prospective gonad is needed for attracting the germ cells. This enzyme is involved in regulating production of a chemoattractant. A candidate for this attractant is Hedgehog, which is also produced by the prospective gonad, but is more familiar as the extracellular signaling molecule providing positional information (see Chapter 2).

10.5 Germ-cell differentiation involves a halving of chromosome number by meiosis

The sperm and egg are **haploid**—that is, they contain just one copy of each chromosome in contrast to the two copies present in the **diploid** somatic cells. At fertilization, therefore, the diploid chromosome number is reinstated. The primordial germ cells described in the previous sections are diploid, and reduction from the diploid to the haploid state during germ-cell development occurs during the specialized nuclear division known as meiosis, which does not occur until after germ cells have reached the gonads (Fig. 10.8). Meiosis comprises two cell divisions, in which the chromosomes are replicated before the first division, but not before the second, so that their number is reduced by half. The cell divisions during meiosis can be unequal and give rise to a small structure called a **polar body** (Box 10A).

During prophase of the first meiotic division, replicated homologous chromosomes pair up and undergo **recombination**, in which corresponding DNA sequences are exchanged between the homologs (see Fig. 10.8). Homologous chromosomes usually each carry different versions, or alleles, of many of the genes, and so meiotic recombination generates chromosomes with new combinations of alleles. Meiosis, therefore, results in gametes whose chromosomes carry different combinations of alleles compared with the parent. This means that when two gametes come together at fertilization, the resulting animal will differ in genetic constitution from either of its parents. So although we might resemble our parents, we never look exactly like them. Meiotic recombination is the main source of the genetic diversity present within populations of sexually reproducing organisms, including humans and other vertebrates.

Development of eggs and sperm follow different courses, even though they both involve meiosis, and the course is determined by the sex of the embryo. The development of the egg is known as **oogenesis**, and the main stages of mammalian oogenesis are shown in Fig. 10.9 (left panel). In mammals, germ cells undergo a small

BOX 10A   Polar bodies

Polar bodies are small cells formed by meiosis during the development of an oocyte into an egg. In this highly schematic illustration (Figure 1), the segregation of only one pair of chromosomes is shown for simplicity. There are two cell divisions associated with meiosis, and one daughter from each division is almost always very small compared with the other, which becomes the egg—hence the term polar bodies for these smaller cells.

The timing of meiosis in relation to the development of the oocyte varies in different animals, and in some species meiosis is completed and the second polar body formed only after fertilization (Figure 1). In general, polar-body formation is of little

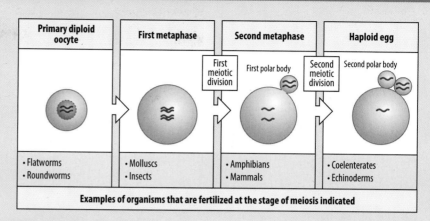

Figure 1

importance for later development, but in some animals the site of formation is a useful marker for the embryonic axes.

number of proliferative mitotic cell divisions as they migrate to the gonad. In the case of developing eggs, the diploid germline cells, now known as **oogonia**, continue to divide mitotically for a short time within the ovary. After entry into meiosis, they are called **primary oocytes**. Within the ovary, individual oocytes are each enclosed in a sheath of somatic ovarian cells called the **follicle**. Mammalian primary oocytes enter meiosis in the embryo but become arrested in the prophase of the first meiotic division, the stage at which homologous chromosomes pair up and recombination occurs (see Fig. 10.8). Arrest depends on high levels of cyclic AMP within the oocyte, which are generated as a result of signaling from G-protein-coupled receptors. The first meiotic division will not be completed until after ovulation in the adult, and the second meiotic division not until after fertilization.

Oocytes never proliferate again after entry into meiosis; thus, the number of oocytes at this embryonic stage is generally considered to be the maximum number of eggs a female mammal can ever have. In humans, most oocytes degenerate before puberty, leaving about 400,000 out of an original 6–7 million to last a lifetime. This number declines with age, with the decline becoming steeper after the mid-30s until menopause, typically in the 50s (Fig. 10.10). In mammals and many other vertebrates, oocyte development is held in suspension after birth for months (mice) or years (humans). When the female becomes sexually mature, the oocytes start to undergo further development, or maturation, as the result of hormonal stimuli. During maturation they grow up to 10-fold in size in mammals, and very much larger in some animals, such as frogs. In some vertebrates, such as frogs and fish, large numbers of oocytes become mature and are released, or ovulated, simultaneously at the end point of each reproductive cycle. In mammals, only a relatively small number of oocytes mature during each cycle and are released under the influence of luteinizing hormone. In mammals, meiosis resumes at the time of ovulation; the released oocyte completes the first meiotic division, with the production of one polar body, and proceeds as far as the metaphase of the second meiotic division. Here meiosis is arrested again, and is only completed after fertilization, with the production of the second polar body. The time of fertilization in relation to the stage of oocyte development differs in different animal groups (see Box 10A).

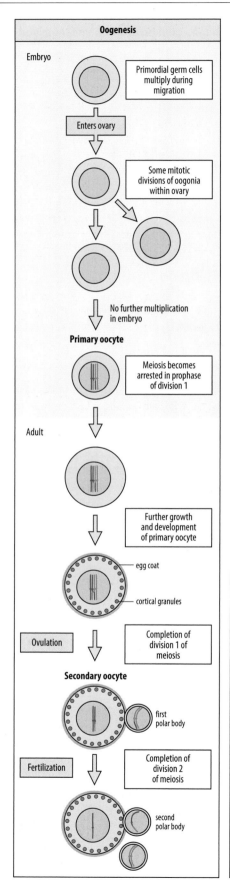

Fig. 10.9 Oogenesis and spermatogenesis in mammals. Left panel: after germ cells enter the embryonic ovary, where they will form oocytes, they divide mitotically a few times and then enter the prophase of the first meiotic division. No further cell multiplication occurs. Further development occurs in the sexually mature adult female. This includes a 10-fold increase in mass (not shown here), the formation of external cell coats, and the development of a layer of cortical granules located under the oocyte plasma membrane. In each reproductive cycle, a group of follicles starts to grow, oocyte growth and maturation follows; a few eggs are ovulated but most degenerate. Eggs continue to mature in the ovary under hormonal influences, but become blocked in the second metaphase of meiosis, which is only completed after fertilization. Polar bodies are formed at meiosis (see Box 10A). Right panel: germ cells enter the embryonic testis, where they will become sperm, and become arrested at the G1 stage of the cell cycle. After birth, they begin to divide mitotically again, forming a population of stem cells (spermatogonia). These give off cells, spermatocytes, that then undergo meiosis and differentiate into sperm. Sperm can therefore be produced indefinitely.

A major error in human oocyte meiosis results in the oocyte having an extra chromosome 21, known as a **trisomy** of chromosome 21, as the result of non-disjunction of homologous chromosomes 21 at the first meiotic division. This trisomy is the cause of Down syndrome, and is one of the most common genetic causes of congenital malformations and learning disability. Trisomies and other errors in chromosome segregation are relatively common in human eggs, far more so than in human sperm, and their incidence rises exponentially as oocytes age. The majority of human embryos with too many or too few chromosomes are not viable, because of the imbalance in the expression of large numbers of genes. They undergo developmental arrest at the preimplantation stage, or fail to implant, or are spontaneously aborted during gestation. Chromosomal abnormalities such as those in Down syndrome can be detected prenatally by techniques such as amniocentesis (see Section 3.6) or chorionic villus sampling, but because of the chance of inducing a miscarriage, these invasive techniques are only used if the risk of a fetus with Down syndrome is high. Routine blood tests for the mother can now estimate a risk of Down syndrome, but the condition can only be confirmed by sampling the fetal cells directly.

The strategy for **spermatogenesis**—the production of sperm—is quite different from oogenesis. Diploid germ cells that give rise to sperm do not enter meiosis in the embryo, but become arrested at an early stage of the mitotic cell cycle in the embryonic testis. They resume mitotic proliferation after birth. Later, in the sexually mature animal, spermatogonial stem cells give rise to differentiating spermatocytes, which undergo meiosis, each forming four haploid spermatids that mature into sperm (see Fig. 10.9, right panel). Thus, unlike the fixed number of oocytes in female mammals, sperm continue to be produced throughout the life of the organism.

In *Drosophila*, there is a continuous production of both eggs and sperm from a population of stem cells in females and males, respectively. Oogenesis begins with the division of a stem cell (see Fig. 2.16), and so there is no intrinsic limitation to the number of eggs that a female fly can produce.

10.6 Oocyte development can involve gene amplification and contributions from other cells

Eggs vary enormously in size among different animals, but they are always larger than the somatic cells. A typical mammalian egg cell is about 0.1 mm in diameter, a frog's egg about 1 mm, and a hen's egg about 3 cm (i.e. the yolk; the white of the egg is extracellular material) (see Fig. 3.13). To achieve growth to such a large size, various mechanisms have evolved, some organisms using several of them together. One strategy is to increase the overall number of gene copies in the developing oocyte, as this proportionately increases the amount of mRNA that can be transcribed, and thus the amount of protein that can be synthesized. Vertebrate oocytes, for example, are arrested in prophase of the first meiotic division and so have double the normal diploid number of genes. Transcription and oocyte growth continue while meiosis is halted.

In addition, insects and amphibians produce many extra copies of selected genes whose products are needed in very large quantities in the egg. During amphibian oocyte development, the rRNA genes are amplified from hundreds of copies to millions; their subsequent translation produces enough ribosomal proteins to ensure sufficient ribosomes for protein synthesis during oocyte growth and early embryonic development. In insects, the genes that encode the proteins of the egg membrane, the chorion, become amplified in the surrounding follicle cells, which produce the egg membrane.

Yet another strategy is for the oocyte to rely on the synthetic activities of other cells. In insects, the nurse cells adjacent to the oocyte make many mRNAs and proteins and deliver them to the oocyte (see Fig. 2.16). Yolk proteins in birds and amphibians are made by liver cells, and are carried by the blood to the ovary, where the proteins enter the oocyte by endocytosis and become packaged into yolk platelets. In *C. elegans*, yolk proteins are made in intestinal cells, and in *Drosophila* in the fat body, from where

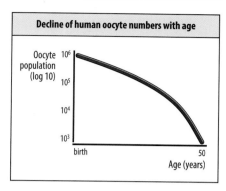

Fig. 10.10 Decline in human oocyte numbers with age. The graph shows the changes in numbers of oocytes in the human ovary with age, peaking at about 6–7 million (in early fetal development) and then decreasing by apoptotic cell death. At puberty there remain about 400,000 oocytes and only a very small number of these, between 400 and 500, are released throughout a woman's lifetime.

they are transported to the oocytes. In amphibian eggs, the oocyte is polarized from an early stage onwards, and the yolk platelets accumulate at the vegetal pole. In animals that depend on maternal determinants for their early development, these mRNAs are produced by the mother and delivered into the egg, where they become localized to the appropriate position by microtubule-based transport.

10.7 Factors in the cytoplasm maintain the totipotency of the egg

Like all other cells in the body, mature gametes are specialized cells that have undergone a program of differentiation. As we have considered in detail in Chapter 8, except in rare instances, cell differentiation does not involve any alteration in the sequence or amount of DNA, but instead involves **epigenetic** changes. These are chemical modifications to DNA and to the chromosomal proteins associated with it that change the structure of the chromosome locally so that some genes are selectively silenced while others can still be expressed. Some somatic cell types, such as muscle cells and nerve cells, do not divide further once differentiation is complete. In cell types that retain the ability to divide, such as liver cells (hepatocytes) or connective tissue fibroblasts, the epigenetic changes associated with that cell type are passed on to the daughter cells, so that they are also hepatocytes or fibroblasts, respectively. Once differentiation is under way, none of these somatic cell types can in normal circumstances give rise to a completely different type of differentiated cell, let alone to the whole range of cell types needed to build a new organism.

Unlike differentiated somatic cells, mature gametes must retain totipotency, so that after fertilization their combined genomes can direct the development of a complete individual. In the egg, factors present in the cytoplasm are responsible for maintaining the egg genome in a totipotent state. This is demonstrated most dramatically by the ability of the cytoplasm of an unfertilized egg to 'reprogram' the nucleus from a differentiated somatic cell so that it reverts to a totipotent state and is able to direct the development of a new organism. This forms the basis of animal cloning by somatic cell nuclear transfer, which has been achieved in amphibians and some species of mammals, including mice, and is described in Chapter 8.

10.8 In mammals some genes controlling embryonic growth are 'imprinted'

Cloning mammals turned out to be much more difficult to achieve than some other animals, and this is most probably due to the phenomenon of **genomic imprinting**— by which certain genes are switched off in either the egg or the sperm during their development and remain silenced in the genome of the early embryo. Evidence for imprinting in mammals came initially from the demonstration that the maternal and paternal genomes make different contributions to embryonic development.

Mouse eggs can be manipulated by nuclear transplantation to have either two paternal genomes or two maternal genomes, and can be reimplanted into a mouse for further development. The embryos that result are known as **androgenetic** and **gynogenetic** embryos, respectively. Although both kinds of embryo have a diploid number of chromosomes, their development is abnormal. The embryos with two paternal genomes have well-developed extra-embryonic tissues, but the embryo itself is abnormal, and does not proceed beyond a stage at which several somites are present. By contrast, the embryos with diploid maternal genomes have relatively well-developed embryos, but the extra-embryonic tissues—placenta and yolk sac— are poorly developed (Fig. 10.11). These results clearly show that both the maternal and the paternal genome are necessary for normal mammalian development: the two parental genomes make different contributions, and both are required for the normal development of the embryo and the placenta. This could explain why mammals cannot be naturally produced **parthenogenetically**, by activation of an unfertilized egg.

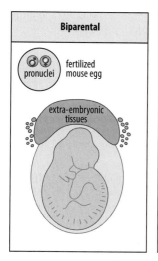
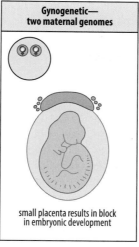
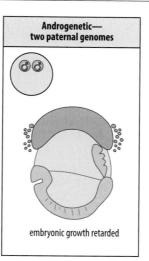

Biparental	Gynogenetic— two maternal genomes	Androgenetic— two paternal genomes

fertilized mouse egg

pronuclei

extra-embryonic tissues

small placenta results in block in embryonic development

embryonic growth retarded

Fig. 10.11 Paternal and maternal genomes are both required for normal mouse development. A normal biparental embryo has both paternal and maternal genomes (contained in pronuclei) in the zygote after fertilization (left panel). Using nuclear transplantation, an egg can be constructed with two paternal or two maternal genomes from an inbred strain. Embryos that develop from an egg with two maternal genomes—gynogenetic embryos (center panel)—have underdeveloped extra-embryonic structures. This results in development being blocked, although the embryo itself is relatively normal and well developed. Embryos that develop from eggs with two paternal genomes—androgenetic embryos (right panel)—have normal extra-embryonic structures, but the embryo itself only develops to a stage where a few somites have formed.

Such observations suggested that the paternal and maternal genomes must be epigenetically modified during germ-cell differentiation. The paternal and maternal genomes contain the same set of genes, but the imprinting process turns on or off certain genes in either the sperm or the egg, so that they are, or are not, expressed during development. Imprinting implies that the affected genes carry a 'memory' of being in a sperm or an egg.

Imprinting is a reversible process. The reversibility is important, because in the next generation any of the chromosomes could end up in male or female germ cells. Inherited imprinting is probably erased during early germ-cell development, and imprinting is later established afresh during germ-cell differentiation (Fig. 10.12). When mammals are cloned using donor nuclei from differentiated somatic cells, these nuclei will not have gone through the normal reprogramming and imprinting

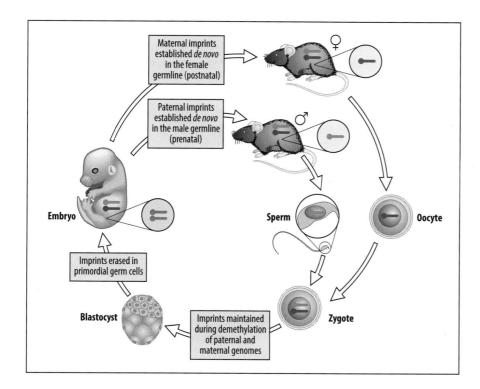

Maternal imprints established *de novo* in the female germline (postnatal)

Paternal imprints established *de novo* in the male germline (prenatal)

Embryo

Imprints erased in primordial germ cells

Blastocyst

Imprints maintained during demethylation of paternal and maternal genomes

Zygote

Sperm

Oocyte

Fig. 10.12 Establishment, maintenance and erasure of genomic imprints during mouse life cycle. Imprints are acquired in sex-specific manner in mature germline (pale green) with male imprints (illustrated by a single brown chromosome for clarity) being established prenatally and female imprints (red chromosome) established postnatally. The imprints are maintained in the embryo despite global changes in de-methylation of male and female genomes that occurs after fertilization and are also maintained in somatic cells. Imprints are erased in primordial germ cells (grey chromosomes) and then re-set for next generation.

*Adapted from Plasschaert, R.N. and Bartolomei, M.S.: **Genomic imprinting in development, growth, behavior and stem cells.** Development 2014, **141**: 1805-1813.*

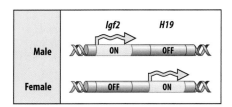

Fig. 10.13 Imprinting of genes controlling embryonic growth. In mouse embryos, the paternal gene for insulin-like growth factor 2 (*Igf2*) is on, but the gene on the maternal chromosome is off. In contrast, the closely linked gene *H19* is switched on in the maternal genome but silenced in the paternal genome.

process that occurs during germ-cell formation, and this could account for the large number of failures and abnormalities in the cloned embryos.

The imprinted genes affect not only early development but also the later growth of the embryo. Further evidence that imprinted genes are involved in embryonic growth comes from studies on chimeras made between normal embryos and androgenetic or gynogenetic embryos. When inner cell mass cells from gynogenetic embryos are injected into normal embryos, growth is retarded by as much as 50%. But when androgenetic inner cell mass cells are introduced into normal embryos, the chimera's growth is increased by up to 50%. The imprinted genes on the male genome thus significantly increase the growth of the embryo.

At least 80 imprinted genes have been identified in mammals, some of which encode **non-coding RNAs** (ncRNAs), RNAs with gene-regulatory roles such as short microRNAs (see Box 6C) and also long non-coding RNAs. Some imprinted genes are involved in growth control. The insulin-like growth factor IGF-2 is required for embryonic growth; its gene, *Igf2*, is imprinted in the maternal genome—that is, it is turned off, so that only the paternal gene is active (Fig. 10.13). Direct evidence for the *Igf2* gene being imprinted comes from the observation that when sperm carrying a mutated, and thus defective, *Igf2* gene fertilize a normal egg, small offspring result. This is because only very low levels of IGF-2 are produced from the imprinted maternal gene and this is not enough to make up for the loss of IGF-2 expression from the paternal genome. In contrast, if the non-functional mutant gene is carried by the egg's genome, development is normal, with the required IGF-2 activity being provided from the normal paternal gene.

Closely linked to the *Igf2* gene on mouse chromosome 7 is the gene *H19*, which encodes a ncRNA that generates two microRNAs. *H19* is imprinted in the opposite direction to the *Igf2* gene and is expressed from the maternal chromosome but not from the paternal one (see Fig. 10.13). By manipulating the chromosomes of an egg it was possible to make a diploid egg with two maternal genomes but with one genome having the normal female imprinting of *Igf2* and *H19* while in the other *Igf2* was not imprinted, as in the male germ cell (*H19* was absent in this genome). This egg could develop normally, unlike an embryo with two female genomes.

A possible evolutionary explanation for the reciprocal imprinting of genes that control growth invokes parental-conflict theory: the theory that the reproductive strategies of the father and mother are different. Paternal imprinting promotes embryonic growth, whereas maternal imprinting reduces it, for example. The father wants to have maximal growth for his own offspring, so that his genes have a good chance of surviving and being carried on. This can be achieved by having a large placenta, as a result of producing growth hormone, whose production is stimulated by IGF-2. The mother, who may mate with different males, benefits more by spreading her resources over all her offspring, and so wishes to prevent too much growth in any one embryo. Thus, a gene that promotes embryonic growth is turned off in the mother. Paternal genes expressed in the offspring could be selected to extract more resources from mothers, because an offspring's paternal genes are less likely to be present in the mother's other children. There are, however, many effects of imprinted genes other than on growth.

Imprinting occurs during germ-cell differentiation, and so a mechanism is required both for maintaining the imprinted condition throughout development and for wiping it out during the next cycle of germ-cell development. One mechanism for maintaining imprinting is DNA methylation, an epigenetic modification in which methyl groups are attached to cytosines in DNA. As we have seen in Chapter 8, DNA methylation is associated with gene silencing. Evidence that DNA methylation is required for imprinting comes from mice in which the methylation process is aberrant. In these mice, the *Igf2* gene is no longer imprinted and is expressed from the maternal chromosome as well as the paternal one. In addition to DNA methylation, other factors that have been implicated in imprinting are ncRNAs, repressor proteins of the Polycomb group, and the chemical modification of histone proteins, the proteins that help package DNA in the chromosomes (see Box 8A). Some of these ncRNAs are involved in imprinting genes adjacent to the genes that encode them. The long non-coding RNA *Airn (antisense Igfr2 RNA)*, for example, is encoded in the *Igf2r* cluster of imprinted genes. *Igfr2* (which encodes an IGF receptor) and the other two genes in the cluster are expressed from the maternal chromosome. *Airn* itself, however, is expressed from the paternal chromosome and represses expression of the three surrounding genes; it is therefore responsible for the paternal imprinting of the cluster.

A number of developmental disorders in humans are associated with imprinted genes. One is Prader–Willi syndrome, which is linked to a loss of expression of the paternal copy of a gene on chromosome 15, usually as a result of a deletion of a small region of the chromosome containing that gene. Infants fail to thrive and later can become extremely obese; they also show mental retardation and mental disturbances such as obsessional-compulsive behavior. Angelman syndrome results from the loss of the same region of the maternal chromosome 15. The effect in this case is severe motor and mental retardation. Beckwith–Wiedemann syndrome is due to a generalized disruption of imprinting on a region of chromosome 7. There is excessive fetal overgrowth and an increased predisposition to cancer.

SUMMARY

In many animals, the germline cells are specified by localized cytoplasmic determinants in the egg, whose localization is controlled in the mother by cells surrounding the oocyte. In contrast, germ cells in mammals are not specified by maternal determinants, but by intercellular interactions in the embryo. Once the germ cells are determined, they migrate from their site of origin to the gonads, where further development and differentiation takes place. In the gonads, the diploid germ-cell precursors undergo meiosis, eventually producing the haploid germ cells—eggs and sperm. The number of oocytes in a female mammal is fixed before birth, whereas sperm production in male mammals is continuous throughout adult life. Eggs are always larger than somatic cells, and in some animal groups are very large indeed. In order to achieve their increased size, specialized somatic cells surrounding the developing oocyte may provide some of the constituents, such as yolk; in addition, some genes producing materials required in large amounts may be amplified in the oocyte.

Both maternal and paternal genomes are necessary for normal mammalian development. Embryos with diploid maternal or paternal genomes develop abnormally. Certain genes in eggs and sperm are imprinted, so that the activity of the same gene is different depending on whether it is of maternal or paternal origin. Several imprinted genes are involved in growth control of the embryo. Inappropriate imprinting can lead to developmental abnormalities in humans.

SUMMARY: specification of germ cells

germ cells specified by germplasm

↗ *Drosophila*
germplasm specified by *oskar* at posterior end of egg

⇒ *C. elegans*
germplasm is characterized by polar granules and PIE-1 protein, and segregates into P4 during cleavage

↘ *Xenopus*
germplasm localized in vegetal region of egg

↘ **Zebrafish**
germplasm in cleavage furrows

Mammals have no germplasm

Fertilization

Fertilization—involving the fusion of egg and sperm—initiates development. In many marine organisms, such as sea urchins, sperm released into the water by the males are attracted to eggs by chemotaxis—the sperm swim up a gradient of chemical released by the egg. The membranes of the egg and sperm fuse, and the contents of the sperm enter the egg cytoplasm, with the sperm nucleus becoming the male **pronucleus**. In mammals and many other animals, fertilization triggers the completion of meiosis in the egg. One set of maternal chromosomes is retained in the egg and becomes the female pronucleus, while the other set is expelled in the second polar body (see Box 10A). The male and female pronuclei (see Fig. 1.1) fuse to form the zygotic nucleus, and the egg starts dividing and embarks on its developmental program. Fertilization can be either external, as in frogs and sea urchins, or internal, as in *Drosophila*, mammals, and birds. Of all the sperm released by the male, only one fertilizes each egg. In many animals, including mammals, sperm penetration activates a blocking mechanism in the egg that prevents any further sperm entering. This is necessary because if more than one sperm nucleus enters the egg there will be additional sets of chromosomes, resulting in abnormal development. In humans, embryos with such abnormalities fail to develop. As we shall see in this chapter, there are multiple overlapping mechanisms for ensuring that only one sperm nucleus contributes to the zygote. In all mammals, except rodents, the sperm centrosome enters the egg and becomes the active zygotic centrosome. Centrosomal defects are one cause of sperm-based infertility in humans.

Both eggs and sperm are structurally specialized for fertilization. The specializations of the egg are directed to preventing fertilization by more than one sperm, and then to start development, while those of the sperm are directed to penetrating the egg. The unfertilized egg is usually surrounded by several protective layers outside the plasma membrane, and the eggs of many organisms have a layer of **cortical granules** just beneath the plasma membrane, whose contents are released on fertilization and help to block the entry of more sperm. These barriers together serve to make the egg impenetrable to more than a small number of sperm.

10.9 Fertilization involves cell-surface interactions between egg and sperm

Sperm are motile cells, typically designed for activating the egg and at the same time delivering their nucleus into the egg cytoplasm. They essentially consist of a nucleus, mitochondria to provide an energy source, a centrosome, and a flagellum for movement (Fig. 10.14). The anterior end is highly specialized to aid penetration. The sperm of *C. elegans* and some other invertebrates are unusual as they more closely resemble tissue cells and move by ameboid motion.

After sperm have been deposited in the mammalian female reproductive tract, they undergo a process known as **capacitation**, which facilitates fertilization and involves membrane remodeling and removal of certain inhibitory factors. There are very few mature eggs—usually one or two in humans and about 10 in mice—waiting to be fertilized, and fewer than a hundred of the millions of sperms deposited actually reach these eggs.

Fig. 10.14 A human sperm. The acrosome at the anterior end of the sperm contains enzymes that are used to digest the protective coats around the egg. The plasma membrane on the head of the sperm contains various specialized proteins that bind to the egg coats and facilitate entry. The sperm moves by its single flagellum, which is powered by mitochondria. The overall length from head to tail is about 60 μm.

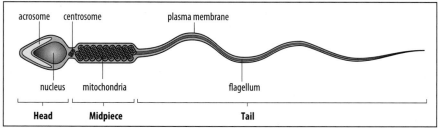

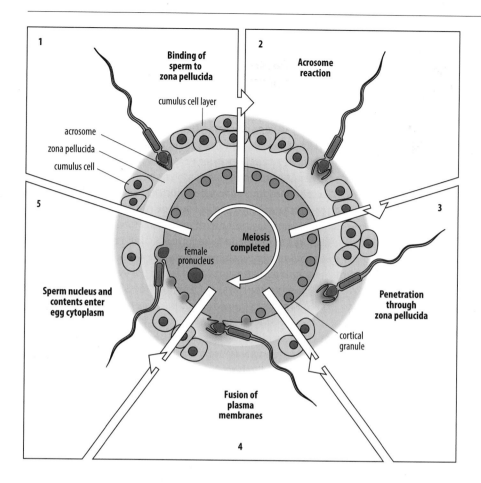

Fig. 10.15 Fertilization of a mammalian egg. After penetrating the follicle-derived cumulus layer, the sperm binds to the zona pellucida (1). This triggers the acrosome reaction (2), in which enzymes are released from the acrosome and break down the zona pellucida. This enables the sperm to penetrate the zona pellucida (3), and bind to the egg plasma membrane. The plasma membrane of the sperm head fuses with the egg plasma membrane and within minutes cortical granules are released from the egg (4). The contents of the sperm including the nucleus then enter the egg (5). By this time meiosis has been completed in the egg and a female pronucleus has been formed.

Illustration after Alberts, B., et al.: Molecular Biology of the Cell, 2nd edition. New York: Garland Publishing, 1989.

The sperm has to penetrate several physical barriers to enter the egg (Fig. 10.15). In mammalian eggs, the first is a sticky layer of hyaluronic acid and embedded somatic follicle cells, which are shed with the egg when it is released. These cells are called **cumulus cells**, as in low-power microscopic images of a released egg it appears to be surrounded by a 'cloud' of material. The first cloned mouse was called Cumulina as she was produced using the nucleus of a cumulus cell (see Chapter 8). Hyaluronidase activity on the surface of the sperm head helps it to penetrate this cumulus layer. The sperm next encounters the **zona pellucida**, a layer of fibrous glycoproteins secreted by the oocyte. This also acts as a physical barrier, but sperm are helped to penetrate it by the **acrosomal reaction**—the release of enzymes contained in the **acrosome,** a Golgi-derived vesicle located in the sperm head (see Fig. 10.14).

Mammalian sperm carry various cell-surface associated proteins that are involved in binding to, and penetrating, the zona pellucida. One of these is the adhesive protein SED1. As they enter the epididymis from the testis, sperm become coated with this protein, which is secreted by the epididymal epithelium. *SED1*$^{-/-}$ male mice, in which both the genes for the protein are defective, are sub-fertile. After initial adhesion due to SED1 and other proteins, proteins in the plasma membrane of the sperm head bind to the zona pellucida glycoprotein ZP2. At some point during the penetration of the layers surrounding the egg, a signaling pathway is triggered in the sperm that results in release of the contents of the acrosome by exocytosis. The acrosome enzymes break down the oligosaccharide side chains on the zona pellucida glycoproteins, making a hole in the zona pellucida that enables the sperm to approach the egg plasma membrane. In the sperm of many invertebrates, such as the sea urchin, the acrosome reaction also results in the extension of a rod-like acrosome process. This forms by polymerization of actin in the sperm cytoplasm and facilitates contact with the egg plasma membrane.

The acrosome reaction also exposes proteins on the sperm surface that can bind to the egg membrane and are involved in the sperm and the egg recognizing each other and allowing fusion of sperm and egg membranes. One of these proteins, called Izumo, after a Japanese marriage shrine, has been shown to be essential for sperm–egg fusion in mice. The ligand for Izumo on the egg plasma membrane is a folate receptor, renamed Juno, after the Roman goddess of fertility and marriage. Female transgenic mice whose eggs lack Juno and male transgenic mice whose sperm lack Izumo are both infertile. The same pair of proteins is required for human sperm-egg fusion. Human and other mammalian eggs can be fertilized in culture and the very early pre-blastocyst embryo transferred to the mother's womb, where it implants and develops normally. This procedure of *in vitro* fertilization (IVF) has been of great help to couples who have, for various reasons, difficulty in conceiving. A human egg can even be fertilized by injecting a single intact sperm directly into the egg in culture, a technique known as intracytoplasmic sperm injection (ICSI), which is useful when infertility is due to the sperm being unable to penetrate the egg.

10.10 Changes in the egg plasma membrane and enveloping layers at fertilization block polyspermy

Although many sperm attach to the layers around the egg, and even reach the plasma membrane, it is important that only one sperm nucleus fuses with the egg nucleus as the unbalanced chromosomal constitution resulting from the fusion of more than one sperm nucleus will be ultimately lethal to the embryo. Different organisms have different ways of ensuring fertilization by only one sperm. In birds, for example, many sperm penetrate the egg but only one sperm nucleus fuses with the egg nucleus; the other sperm nuclei are destroyed in the cytoplasm. It is likely in this case that the position at which the sperm enters the egg is important, with the sperm that enters immediately into the germinal vesicle—the region containing the egg chromosomes—acting as the fertilizing sperm. The DNA of sperm entering the cytoplasm outside this region is probably degraded by cytoplasmic DNases, whose presence has been detected in quail eggs. In sea urchins and mammals, on the other hand, the entry of more than one sperm is prevented. This general strategy is called the **block to polyspermy**.

The eggs of sea urchins and *Xenopus* have a two-stage block to **polyspermy**, an initial rapid reaction followed by a slower one. They are both externally fertilized and so are likely to be bombarded simultaneously by large numbers of sperm, whereas in mammals far fewer sperm eventually reach the oviducts (Fallopian tubes) where fertilization takes place. An immediate means of signaling that one sperm has fused with the egg membrane is therefore desirable. The **rapid block to polyspermy** in sea urchin is triggered within seconds by a transient depolarization of the egg plasma membrane that occurs on sperm–egg fusion. The electrical membrane potential across the plasma membrane goes from −70 mV to +20 mV within a few seconds of sperm entry (Fig. 10.16). The membrane potential then slowly returns to its original level. If depolarization is prevented, polyspermy occurs, but how depolarization blocks polyspermy is not yet clear. A similar rapid electrical block to polyspermy occurs in *Xenopus*.

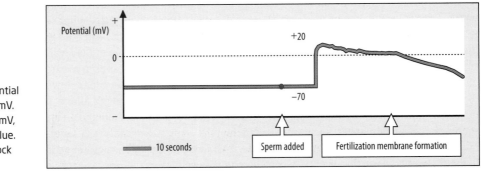

Fig. 10.16 Depolarization of the sea-urchin egg plasma membrane at fertilization. The resting membrane potential of the unfertilized sea-urchin egg is −70 mV. At fertilization, it changes rapidly to +20 mV, and then slowly returns to the original value. This depolarization may provide a fast block to polyspermy.

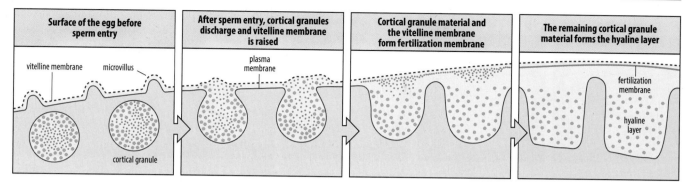

| Surface of the egg before sperm entry | After sperm entry, cortical granules discharge and vitelline membrane is raised | Cortical granule material and the vitelline membrane form fertilization membrane | The remaining cortical granule material forms the hyaline layer |

Fig. 10.17 The cortical reaction at fertilization in the sea urchin. The egg is surrounded by a vitelline membrane, which lies outside the plasma membrane. Membrane-bound cortical granules lie just beneath the egg plasma membrane. At fertilization, the cortical granules fuse with the plasma membrane, and some of the contents are extruded by exocytosis. These join with the vitelline membrane to form a tough fertilization membrane, which then lifts off the egg surface and prevents further sperm entry. Other cortical granule constituents give rise to a hyaline layer, which surrounds the egg under the fertilization membrane.

As the plasma membrane of the sea urchin egg repolarizes, an impenetrable membrane called the **fertilization membrane** is formed around the egg. This is the result of the **slow block to polyspermy** and is triggered by a wave of calcium release in the egg due to sperm entry and leads to the cortical granules releasing their contents to the outside of the plasma membrane by exocytosis. The wave of calcium also activates development, as we shall see in the next section. The unfertilized sea-urchin egg is surrounded by a **vitelline membrane**, which corresponds to the zona pellucida of mammalian oocytes. Release of the granule contents into the space between the egg plasma membrane and the vitelline membrane causes the vitelline membrane to lift off the plasma membrane. The cortical granule contents cross-link molecules in the vitelline membrane to produce a 'hardened' **fertilization membrane**, and also provide an additional jelly-like hyaline layer between it and the egg plasma membrane (Fig. 10.17). Some of the released enzymes also cleave sperm-binding proteins. Together, these changes prevent additional sperm from gaining access to the egg. The sea-urchin fertilization membrane dissolves at the blastula stage, and the blastula 'hatches' out.

In mammals, including humans, the passage of sperm through the reproductive tract is regulated and relatively few sperm—a few thousand out of the millions of sperm deposited during coitus—reach the Fallopian tubes and are in a position to fertilize the egg. Polyspermy is prevented in mammals by changes that occur to both the zona pellucida and egg plasma membrane once the first sperm has entered the egg. There is no rapid electrical block to polyspermy in mammalian eggs, but there is a slow block very similar to that in the sea urchin and *Xenopus*, which develops within 30 minutes to an hour after fertilization. After the first sperm fuses with the plasma membrane, the granules in the egg cortex are released by exocytosis and their contents form a layer immediately outside the egg plasma membrane. One component of the cortical granules is the protease, ovastacin, and this cuts the ZP2 protein in the zona pellucida so that sperm can longer bind to it.

Mammalian eggs also have a membrane block to polyspermy that develops over a similar time scale to the zona pellucida reaction. The membrane block has been demonstrated in mouse eggs whose zona pellucida has been removed before fertilization *in vitro*. Challenge with a second dose of sperm after the initial fertilization results in very few, if any, of these sperm entering the egg. Juno, the receptor that we noted earlier is essential for sperm–egg recognition, disappears from the egg-cell plasma membrane about 40 minutes after fertilization and is the basis for the membrane block to polyspermy in mammalian eggs.

10.11 Sperm–egg fusion causes a calcium wave that results in egg activation

At fertilization, the egg becomes activated and a series of events is initiated that result in the start of development. In the sea-urchin egg, for example, there is a several-fold increase in protein synthesis, and there are often changes in egg structure, such as the cortical rotation that occurs in amphibian eggs (see Fig 4.3). Amphibian and

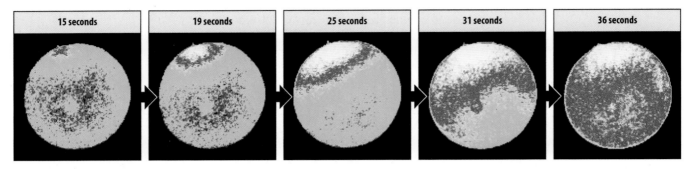

| 15 seconds | 19 seconds | 25 seconds | 31 seconds | 36 seconds |

Fig. 10.18 Calcium wave at fertilization.
A series of images showing an intracellular calcium wave at fertilization in a sea-urchin egg. The fertilizing sperm has fused just to the left of the top of the egg, and triggered the wave. Calcium ion concentration is monitored with a calcium-sensitive fluorescent dye, using confocal fluorescence microscopy. Calcium concentration is shown in false color: red is the highest concentration, then yellow, green, and blue. Times shown are seconds after sperm entry.

Photographs courtesy of M. Whitaker.

mammalian eggs, which are arrested in second meiotic metaphase before fertilization (see Box 10A), now complete meiosis, after which the egg and sperm pronuclei fuse to form the diploid zygotic genome, and the fertilized egg enters mitosis. The pronuclear membranes disappear before the pronuclei come together in mice and humans. In mammals, the sperm mitochondria are destroyed, and so all mitochondria in the fertilized egg are of maternal origin. The mitochondrial DNA genome is, therefore, inherited unchanged down the female line apart from the occasional change in DNA sequence as a result of mutation. This makes mitochondrial DNA an excellent material for studies that use the rare DNA sequence changes accumulated in the mitochondrial genomes of present-day humans to trace the movements of our earliest human ancestors out of Africa and their spread to different parts of the world.

Fertilization and egg activation are associated with an explosive release of free calcium ions (Ca^{2+}) from stores within the egg, producing a wave of Ca^{2+} that travels across it (Fig. 10.18). The calcium release is triggered by sperm entry. In sea-urchin eggs, the wave starts at the point of sperm entry and crosses the egg at a speed of 5–10 µm per second. In all mammals, oscillations in calcium concentration occur for several hours after fertilization until female pronucleus formation. Ca^{2+} release at fertilization is triggered by the activity of a sperm-specific enzyme, phospholipase C zeta. This initiates a signaling pathway that leads to production of the second messenger inositol 1,4,5-trisphosphate, which acts on receptors in intracellular membranes to release calcium stored in the endoplasmic reticulum.

The sharp increase in free Ca^{2+} is the natural trigger for egg activation although mouse eggs can be activated without any change in calcium levels. The eggs of many animals can be activated if the Ca^{2+} concentration in the egg cytosol is artificially increased, for example, by direct injection of Ca^{2+} Conversely, preventing calcium increase by injecting agents that bind it, such as the calcium chelator EGTA, blocks activation. It has been known for many years that *Xenopus* eggs can be activated simply by prodding them with a glass needle; this is due to a local influx in calcium at the site of prodding that triggers a calcium wave.

Calcium initiates the completion of meiosis in the fertilized egg by acting on proteins that control the cell cycle. The unfertilized *Xenopus* egg is maintained in the metaphase of the second meiotic division by the presence of high levels of a protein complex called maturation-promoting factor (MPF), which is a complex of a cyclin-dependent kinase (Cdk) and its partner cyclin B. The effects of MPF are due to phosphorylation of a variety of protein targets by the kinase. For the egg to complete meiosis, the level of MPF activity must be reduced (Fig. 10.19). Similar Cdk–cyclin B complexes control the mitotic cell cycle (see Section 13.2). The calcium wave results in the activation of the enzyme calmodulin-dependent protein kinase II. The activity of this kinase indirectly results in the degradation of the cyclin component of MPF, which allows the egg to complete meiosis. Following this, sperm and egg genomes undergo global changes in demethylation and male and female pronuclei are formed. The pronuclei then fuse, and the zygote moves on to the next stage of its development, which is entry into the mitotic cell cycles of cleavage.

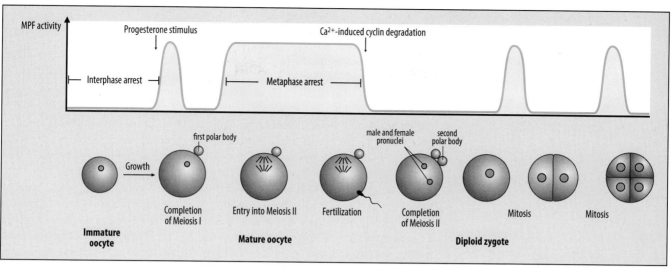

Fig. 10.19 Profile of maturation-promoting factor (MPF) activity in early Xenopus development. The immature *Xenopus* oocyte cell cycle is arrested. On receipt of a hormonal progesterone stimulus it enters, and completes, the first meiotic division, with the formation of the first polar body. It enters the second meiotic division, but becomes arrested again in metaphase. The egg is laid at this point. At fertilization, the calcium wave leads to completion of meiosis, and the second polar body is formed. The zygote starts to cleave rapidly by mitotic divisions. Maturation-promoting factor (MPF) rises sharply just before each division of the meiotic and mitotic cell cycles, remains high during mitosis, and then decreases abruptly and remains low between successive mitoses.

SUMMARY

The fusion of sperm and egg at fertilization stimulates the egg to start dividing and developing. Both sperm and egg have specialized structures relating to fertilization. The initial binding of sperm to the mammalian egg is mediated by molecules in the zona pellucida (in mammals) or the corresponding vitelline membrane (in sea urchins). Sperm binding leads to the release of the contents of the sperm acrosome, which facilitates the penetration of the sperm through the layers surrounding the egg and allows it to reach the egg plasma membrane. A block to polyspermy allows only one sperm to fuse with the egg and deliver its nucleus into the egg cytoplasm. In mammals, one block results from the release of egg cortical granule contents to the exterior which modify the zona pellucida and the second involves changes in the egg plasma membrane. A key role in egg activation after fertilization is played by the release of free calcium ions into the cytosol, which spread in a wave from the site of sperm fusion. In mammals, and most other vertebrates, fertilization triggers the completion of the second meiotic division; the male and female haploid pronuclei give rise to the zygote nucleus, and the fertilized egg divides.

SUMMARY: fertilization in mammals

hyaluronidase activity on surface of mammalian sperm

⇩

sperm penetrates cumulus layer ⇨ binding to zona pellucida elicits
acrosomal reaction in sperm

⇩

enzymes released from acrosome on sperm head
enable sperm to penetrate zona pellucida

⇩

sperm plasma membrane fuses with
egg plasma membrane

⇩

calcium waves

cortical reaction provides block to polyspermy ⇩

sperm nucleus delivered into cytoplasm

⇩

egg completes meiosis and development is initiated

Determination of the sexual phenotype

In organisms that produce two phenotypically different sexes, sexual development involves the modification of a basic developmental program. Early development is similar in both male and female embryos, with sexual differences only appearing at later stages. In the organisms considered here, somatic sexual phenotype—that is, the development of the individual as either male or female—is genetically fixed at fertilization by the chromosomal content of the gametes that fuse to form the fertilized egg. In mammals, for example, sex is determined by the Y chromosome: males are XY, females are XX.

Even among vertebrates, however, sex is not always determined by which chromosomes are present; in alligators, it is determined by the environmental temperature during incubation of the embryo, and some fish can switch sex as adults in response to environmental conditions. In insects, there is a wide range of different sex-determining mechanisms. Intriguing though these are, we focus here on those organisms in which the genetic and molecular basis of sex determination is best understood—mammals, *Drosophila*, and *C. elegans*—in all of which sex is determined by chromosomal content, although by quite different mechanisms.

We first consider the determination of the somatic sexual phenotype. We then deal with the determination of the sex of the germ cells—whether they become eggs or sperm—and finally consider how the embryo compensates for the difference in chromosomal composition between males and females.

10.12 The primary sex-determining gene in mammals is on the Y chromosome

The genetic sex of a mammal is established at the moment of conception, when the sperm introduces either an X or a Y chromosome into the egg (Fig. 10.20). Eggs contain one X chromosome; if the sperm introduces another X the embryo will be female, if a Y it will be male. The presence of a Y chromosome causes testes to develop, and the hormones they produce switch the development of all somatic tissues to a characteristic male pathway and suppress female development. In the absence of a Y chromosome, the development of somatic tissues is along the female pathway. Specification of a gonad as a testis is controlled by a single gene on the Y chromosome, the **sex-determining region of the Y chromosome** (*SRY* in humans and *Sry* in mice), whose product was formerly known as testis-determining factor.

Evidence that a region on the Y chromosome actively determines maleness first came from two unusual human syndromes: Klinefelter syndrome, in which individuals have two X chromosomes and one Y (XXY), but are still males; and Turner syndrome, in which individuals have just one X chromosome (XO) and are female. Both these types of individual have some abnormalities; those with Klinefelter syndrome are infertile males with small testes, whereas females with Turner syndrome do not produce eggs. There are also rare cases of XY individuals who are female, and XX individuals who are phenotypically male. This is due to part of the Y chromosome being lost (in XY females) or to part of the Y chromosome being transferred to the X chromosome (in XX males). This can happen during meiosis in the male germ cells as the X and Y chromosomes are able to pair up, and crossing over can occur between them. Very rarely, this crossing over transfers the *SRY* gene from the Y chromosome onto the X (Fig. 10.21), thus leading to sex reversal.

The sex-determining region alone is sufficient to specify maleness, as shown by an experiment in which the mouse equivalent of the *SRY* gene (*Sry*) was introduced into the eggs of XX mice. These transgenic embryos developed as males, even though they lacked all the other genes on the Y chromosome. The presence of the *Sry* gene, which

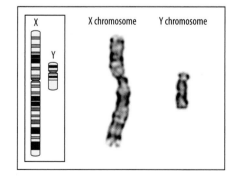

Fig. 10.20 The sex chromosomes in humans. If two X chromosomes (XX) are present, a female develops, whereas the presence of a Y chromosome (XY) leads to development of a male. The inset shows a diagrammatic representation of the banding in the chromosomes, which represent regions of increased chromatin condensation.

encodes a transcription factor, resulted in these XX embryos developing testes instead of ovaries. In these embryos, as in normal males, *Sry* was expressed in the developing gonad just before it started to differentiate and drove testis development by triggering the differentiation of precursor cells into testis-specific Sertoli cells rather than ovarian follicle cells. However, these XX + *Sry* males were not completely normal but were infertile, as other genes on the Y chromosome are necessary for the development of the sperm.

10.13 Mammalian sexual phenotype is regulated by gonadal hormones

All mammals, whatever their genetic sex, start off as embryos along a sexually neutral developmental pathway. The sex of the gonads is genetically determined, as the presence of a Y chromosome results in the somatic cells of the embryo's gonads developing into testes rather than into ovaries. The testes secrete **Müllerian-inhibiting substance**, which suppresses female development by causing the embryonic precursor of the female reproductive organs to regress. It also induces cells to become Leydig cells, which secrete the male hormone testosterone, which stimulates the development of male reproductive organs. In XX individuals, the absence of a Y means that ovaries develop and all subsequent development is female.

The role of hormones in mammalian sexual development means that although the sex of the gonads is genetically determined, all the other cells in the mammalian body are neutral, irrespective of their chromosomal sex. It does not matter if they are XX or XY, as any future sex-specific development they undergo is controlled by hormones. The primary role of the testis in directing male development was originally demonstrated by removing the prospective gonadal tissue from early rabbit embryos. All the embryos developed as females, irrespective of their chromosomal constitution. To develop as a male, therefore, a testis has to be present. The testis exerts its effect on sexual differentiation of somatic tissues primarily by secreting the hormone testosterone.

The gonads in mammals develop in close association with the **mesonephros**; this is an embryonic kidney that contributes to both the male and female reproductive organs. Associated with the mesonephros on each side of the body are the **Wolffian ducts**, which run down the body to the cloaca, an undifferentiated opening. Another pair of ducts, the **Müllerian ducts**, run parallel to the Wolffian ducts and also open into the cloaca. In early mammalian development, before gonadal differentiation, both sets of ducts are present (Fig. 10.22). In females, in the absence of the testes, the Müllerian ducts develop into the **oviducts**, also called the **Fallopian tubes** in mammals, which transport eggs from the ovaries to the uterus, while the Wolffian ducts degenerate.

In males, the expression of Sry results in the differentiation of Sertoli cells, which are the somatic cells of the testis and are essential for testis formation and for spermatogenesis, as they retain the germ cells that migrate into the gonad. Sertoli cell development involves upregulation of production of the transcription factor Sox9, which induces production and secretion of **Müllerian-inhibiting substance**. This protein hormone in turn induces regression of the Müllerian duct, largely by apoptosis. FGF-9 is also required for Sertoli cell differentiation and male mice lacking the gene *Fgf9* develop as females. FGF represses expression of *Wnt4*, which would otherwise promote ovary development by repressing *Sox9* expression. Therefore, development of a testis needs active repression of the female developmental pathway (Fig. 10.23).

The interstitial cell lineage in the testis then differentiates into Leydig cells, which produce testosterone. Under the influence of testosterone, the Wolffian duct develops into the **vas deferens**, the duct that carries sperm to the penis. The main secondary sexual characters that distinguish males and females are the reduced size of mammary glands in males, and the development of a penis and a scrotum in males instead of the clitoris and labia of females (Fig. 10.24). This is due to the action of the hormone

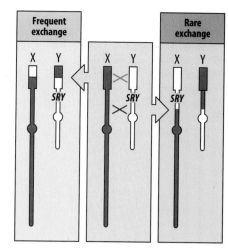

Fig. 10.21 Sex reversal in humans due to chromosomal exchange. At meiosis in male germ cells, the X and Y chromosomes pair up (center panel). Crossing over of the distal region (blue cross) does not affect sexual development (left panel). On rare occasions, crossing over can involve a larger segment that includes the *SRY* gene (red cross), so that the X chromosome now carries this male-determining gene (right panel).
Illustration after Goodfellow, P.N., Lovell-Badge, R.: **SRY and sex determination in mammals.** Ann. Rev. Genet. *1993, **27**: 71-92.*

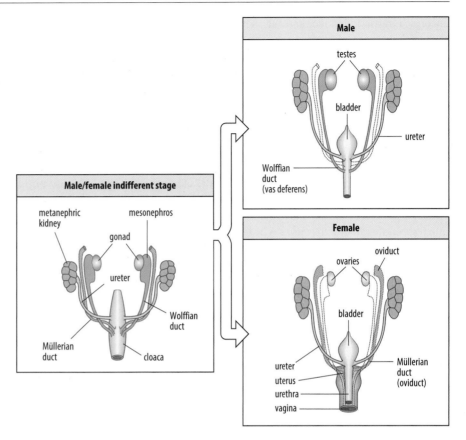

Fig. 10.22 Development of the gonads and related structures in mammals. Left panel: early in development, there is no difference between males and females in the structures that give rise to the gonads and related organs. The future gonads lie adjacent to the mesonephros, which are embryonic kidneys that are not functional in adult mammals. (The true kidney develops from the metanephros, from which the ureter carries urine to the bladder.) Two sets of ducts are present; the Wolffian ducts, which are associated with the mesonephros, and the Müllerian ducts. Both ducts enter the cloaca. Top right panel: after testes develop in the male, their secretion of Müllerian-inhibiting substance results in degeneration of the Müllerian duct by programmed cell death, whereas the Wolffian duct becomes the vas deferens, carrying sperm from the testis. Bottom right panel: in females, the Wolffian duct disappears, also by programmed cell death, and the Müllerian duct becomes the oviduct. The uterus forms at the end of the Müllerian ducts.

*Illustration after Higgins, S.J., et al.: **Induction of functional cytodifferentiation in the epithelium of tissue recombinants II. Instructive induction of Wolffian duct epithelia by neonatal seminal vesicle mesenchyme.** Development 1989, **106**: 235–250.*

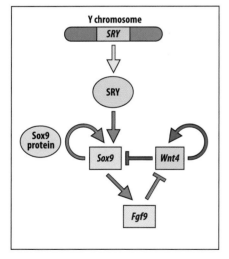

Fig. 10.23 Genetic interactions that determine the sex of the gonads in mammals. In males, SRY protein upregulates *Sox9* expression in cells in the developing testis and *Sox9* expression becomes self-regulating. Sox9 protein upregulates expression of *Fgf9*, and FGF-9 signaling represses *Wnt4* expression. Wnt-4 signaling represses *Sox9* expression. If Wnt-4 signaling is not repressed, male development does not occur.

*Adapted from Jameson, S.A., et al.:**Testis development requires the repression of Wnt4 by Fgf signaling**. Dev. Biol. 2012, **370**: 24–32.*

testosterone. At early stages of embryonic development, the genital regions of males and females are indistinguishable. Differences only arise after gonad development, as a result of the action of testosterone in males. For example, in humans, the phallus gives rise to the clitoris in females and the end of the penis in males.

The role of hormones in sexual development is illustrated by rare cases of abnormal sexual development. Certain XY males develop as phenotypic females in external appearance, even though they have testes and secrete testosterone. They have a mutation that renders them insensitive to testosterone because they lack the testosterone receptor, which is present throughout the body. Conversely, genetic females with a completely normal XX constitution can develop as phenotypic males in external appearance, if they are exposed to male hormones during their embryonic development.

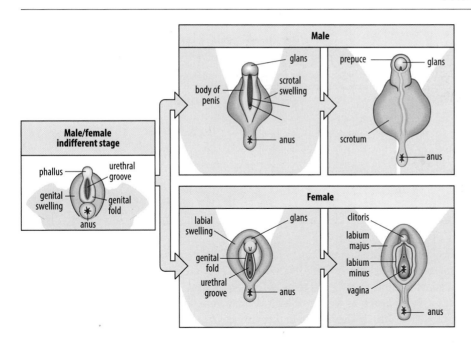

Fig. 10.24 Development of the genitalia in humans. At an early embryonic stage, the genitalia are the same in males and females (left panel). After testis formation in males, the phallus and the genital fold give rise to the penis (top right panels), whereas in females they give rise to the clitoris and the labia minus (bottom right panels). The genital swelling forms the scrotum in males and the labia majus in females.

Sex-specific behavior is also affected by the hormonal environment as a result of the effects of hormones on the brain. For example, male rats castrated after birth develop the sexual behavioral characteristics of genetic females.

10.14 The primary sex-determining signal in *Drosophila* is the number of X chromosomes and is cell autonomous

The external sexual differences between *Drosophila* males and females are mainly in the genital structures, although there are some differences in bristle patterns and pigmentation, and male flies have a sex comb on the first pair of legs. In flies, sex determination of the somatic cells is cell autonomous—that is, it is specified on a cell-by-cell basis—and there is no process resembling the control of somatic sexual differentiation by hormones. Somatic sexual development is the result of a series of gene interactions that are initiated by the primary sex signal, and which act on a binary male/female genetic switch. The end result is the expression of just a few effector genes, whose activity controls the subsequent male or female differentiation of the somatic cells.

Like mammals, fruit flies have two unequally sized sex chromosomes, X and Y, and males are XY and females XX. But these similarities are misleading. In flies, sex is not determined by the presence of a Y chromosome, but by the number of X chromosomes. Thus, XXY flies are female and X flies are male. The chromosomal composition of each somatic cell determines its sexual development. This is beautifully illustrated by the creation of genetic mosaics in which the left side of the animal is XX and the right side X: the two halves develop as female and male, respectively (Fig. 10.25).

In flies, the presence of two X chromosomes results in the production of the protein Sex-lethal (Sxl), whose gene is located on the X chromosome. This leads to female development through a cascade of gene activation that first determines the sexual state and then produces the sexual phenotype. At the end of the sex-determination pathway is the *transformer* (*tra*) gene, which determines how the mRNA of the *doublesex* (*dsx*) gene is spliced; *dsx* encodes a transcription factor whose activity ultimately produces most aspects of somatic sex. The *dsx* gene is active in both males

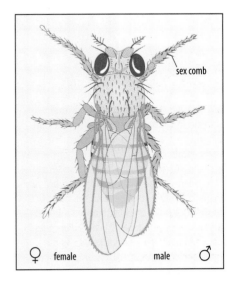

Fig. 10.25 A *Drosophila* female/male genetic mosaic. The left side of the fly is composed of XX cells and develops as a female, whereas the right side is composed of X cells and develops as a male. The male fly has smaller wings, a special structure, the sex comb, on its first pair of legs, and different genitalia at the end of the abdomen (not shown).

Fig. 10.26 Outline of the sex-determination pathway in *Drosophila*. The number of X chromosomes is the primary sex-determining signal, and in females the presence of two X chromosomes activates the gene *Sex-lethal* (*Sxl*). This produces Sex-lethal protein, whereas no Sex-lethal protein is made in males, which have only one X chromosome. The activity of *Sex-lethal* is transduced via the *transformer* gene (*tra*) and causes sex-specific splicing of *doublesex* RNA (*dsx^f*), such that the cells follow a female developmental pathway. In the absence of Sex-lethal protein, the splicing of *doublesex* RNA to give *dsx^m* RNA leads to male development.

Sex determination in *Drosophila*			
Primary signal	**Stable binary genetic switch**	**Transducers of the sexual state**	**Effectors of sexual phenotype**
X chromosomes	*Sxl* ⟹	e.g. *tra* ⟹	splicing *dsx* → *dsx^f* Female / → *dsx^m* Male
XX	ON	ON	*dsx^f* ♀
X	OFF	OFF	*dsx^m* ♂

and females, but different protein products are produced in the two sexes as a result of the sex-specific RNA splicing (Fig. 10.26). Males and females thus express similar but distinct Doublesex (Dsx) proteins, which act in somatic cells to induce expression of sex-specific genes, as well as to repress characteristics of the opposite sex. Production of the male form of the protein in the absence of Sxl results in development as a male, but in the presence of Sxl, *tra* mRNA undergoes RNA splicing, and this, together with the actions of the Transformer-2 protein, leads to the female form of the Dsx protein being made, resulting in development as a female.

Once the *Sex-lethal* (*Sxl*) gene is activated in females, it remains activated through an autoregulatory mechanism, which results in the Sxl protein being synthesized throughout female development. Early expression of *Sxl* in females occurs through activation of a promoter, P_e, at about the time of syncytial blastoderm formation. Sxl protein is synthesized and accumulates in the blastoderm of female embryos. At the cellular blastoderm stage, another promoter for *Sxl*, P_m, becomes active in both males and females and P_e is shut off, but the sex is already determined. Splicing of the RNA transcribed from P_m into functional *Sxl* mRNA needs some Sxl protein to be present already, and so can only happen in females (Fig. 10.27).

How does the number of X chromosomes control these key sex-determining genes? In *Drosophila*, the mechanism involves interactions between the products of so-called 'numerator' genes on the X chromosome and of genes on the autosomes, as well as

Fig. 10.27 Production of Sex-lethal protein in *Drosophila* sex determination. When two X chromosomes are present, the early establishment promoter (P_e) of the *Sex-lethal* (*Sxl*) gene is activated at the syncytial blastoderm stage in future females, but not in males. This results in the production of Sxl protein. Later, at the blastoderm stage, the maintenance promoter (P_m) of *Sxl* becomes active in both females and males, and P_e is turned off. The *Sxl* RNA is only correctly spliced if Sxl protein is already present, which is only in females. A positive feedback loop for Sxl protein production is thus established in females. The continued presence of Sxl protein initiates a cascade of gene activity leading to female development. If no Sxl protein is present, male development ensues.
*Illustration after Cline, T.W.: **The Drosophila sex determination signal: how do flies count to two?** Trends Genet. 1993, **9**: 385–390.*

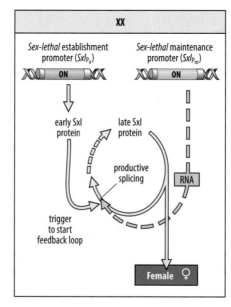
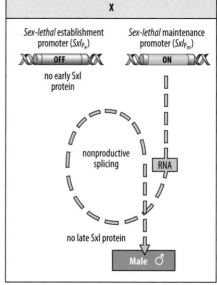

maternally specified factors. Essentially, in females, the double dose of numerator proteins activates *Sxl* by binding to sites in the P_e promoter (see Fig. 10.27), overcoming the repression of Sxl by autosomally encoded proteins that would occur in males.

The pathway outlined in Fig. 10.26 is an oversimplification, as dsx does not control all aspects of somatic sexual differentiation in *Drosophila*. There is an additional branch of the sex-differentiation pathway downstream of *tra*, which controls sexually dimorphic aspects of the nervous system and sexual behavior. This branch includes the gene *fruitless*, whose activity has been shown to be necessary for male sexual behavior.

In other dipteran insects, the same general strategy for sex determination is used, but there are marked differences at the molecular level. Only *dsx* has been found in dipterans distantly related to *Drosophila*; *Sxl* has been found in other dipterans but is not involved in sex determination there.

10.15 Somatic sexual development in *Caenorhabditis* is determined by the number of X chromosomes

In the nematode *C. elegans*, the two sexes are self-fertilizing hermaphrodite (essentially a modified female) and male (Fig. 10.28), although in other nematodes they are male and female. Hermaphrodites produce a limited amount of sperm early in development, with the remainder of the germ cells developing into oocytes. Sex in *C. elegans* (and other nematodes) is determined by the number of X chromosomes: the hermaphrodite (XX) has two X chromosomes, whereas the presence of just one X chromosome leads to development as a male (XO). One of the primary sex signals for hermaphrodite development is the SEX-1 protein, which is encoded on the X chromosome and is a key element in 'counting' the number of X chromosomes present. SEX-1 is a nuclear hormone receptor that represses the sex-determining gene *XO lethal* (*xol-1*), also present on the X chromosome. In the presence of a double dose of SEX-1 produced from the two X chromosomes, xol-1 expression is inhibited, resulting in the development of a hermaphrodite. When only one X chromosome is present, *xol-1* is expressed at a high level and the embryo develops as a male.

A cascade of gene activity converts the level of *xol-1* expression into the somatic sexual phenotype (Fig. 10.29). Unlike *Drosophila*, sex determination in *C. elegans* requires intercellular interactions, as at least one of the genes involved encodes a secreted protein. At the end of the cascade is the gene *transformer-1* (*tra-1*), which encodes a transcription factor. Expression of TRA-1 protein is both necessary and sufficient to direct all aspects of hermaphrodite (XX) somatic cell development, as a gain-of-function mutation in *tra-1* leads to hermaphrodite development in an XO animal, irrespective of the state of any of the regulatory genes that normally control its activity. Mutations that inactivate *tra-1* lead to complete masculinization of XX hermaphrodites.

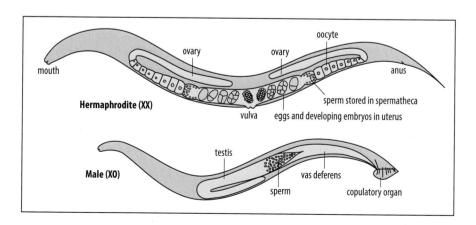

Fig. 10.28 Hermaphrodite and male *Caenorhabditis elegans*. The hermaphrodite has a 'two-armed' gonad and initially makes sperm, which is then stored in the spermotheca. It then switches to make eggs. The eggs are fertilized internally. The male makes sperm only.

Fig. 10.29 Outline of the somatic sex determination pathway in *Caenorhabditis elegans*. The primary signal for sex determination is set by the number of X chromosomes. When two X chromosomes are present, the expression of the gene *XO lethal* (*xol-1*) is low, leading to hermaphrodite development, whereas *xol-1* is expressed at a high level in males. There is a cascade of gene expression starting from *xol-1* that leads to the gene *transformer-1* (*tra-1*), which codes for a transcription factor. If *tra-1* is active, development as a hermaphrodite occurs, but if it is expressed at low levels, males develop. The product of the *hermaphrodite-1* (*her-1*) gene is a secreted protein, which probably binds to a receptor encoded by *transformer-2* (*tra-2*), inhibiting its function.

Sex determination in *Caenorhabditis elegans*			
Primary signal	**Binary genetic switch**	**Transducers of the sexual state**	**Effectors of sexual phenotype**
X chromosomes	*xol-1* —(inhibition)—	*sdc-1* *sdc-2* *sdc-3* ⊣ *her-1* ⊣ *tra-2* *tra-3* ⊣	*fem-1* *fem-2* *fem-3* ⊣ *tra-1* → Hermaphrodite / Male
XX	Low	**High** Low **High**	Low **High** ♂
X	High	Low **High** Low	**High** Low ♂

10.16 Determination of germ-cell sex depends on both genetic constitution and intercellular signals

Determination of the sex of animal germ cells—that is, whether they will develop into eggs or sperm—is strongly influenced by the signals they receive when they become part of a gonad. In the mouse, for example, their future development is determined largely by the sex of the gonad in which they reside, and not by their own chromosomal constitution. In reality, chromosomal constitution and gonadal signals will almost always coincide, but it has been shown that germ cells from male mouse embryos can develop into oocytes rather than sperm if grafted into female embryonic gonads and vice versa.

There is a distinct difference in the timing of meiosis in male and female mammals. In male mouse embryos, the diploid germ cells stop dividing when they enter the gonad, becoming arrested in the G_1 phase of the mitotic cell cycle. They start dividing mitotically again after birth and enter meiosis some 7 to 8 days after birth. In female mouse embryos, the primordial diploid germ cells continue to proliferate for a few days after entering the genital ridge (see Sections 10.4 and 10.5). Diploid germ cells in the gonad first undergo a few rounds of mitotic division and then enter prophase of the first meiotic division; they then arrest at this primary oocyte stage until the mouse becomes a sexually mature female, about 9 weeks after birth, when at each reproductive cycle, selected oocytes complete the first meiotic division and start the second meiotic division (Fig. 10.30). Meiosis is only completed after fertilization.

Although the development of germ cells is normally heavily influenced by the environment in which they find themselves, if the usual external cues are absent, germ cells seem to follow an intrinsic pathway of differentiation. All mouse germ cells that enter meiosis before birth develop as eggs, whereas those not entering meiosis until after birth develop as sperm. Germ cells, whether XX or XY, that fail to enter the genital ridge and instead end up in adjacent tissues such as the embryonic adrenal gland or mesonephros, enter meiosis and begin developing as oocytes in both male and female embryos; thus, the default germ-cell sex appears to be female. XX/XY chimeric mouse embryos can be made by combining male

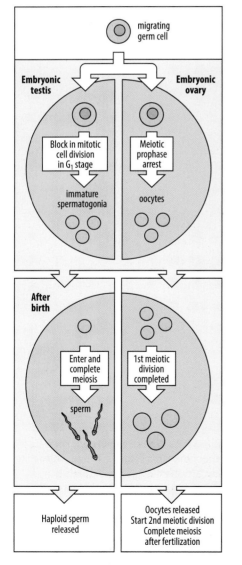

Fig. 10.30 The timing of meiosis in the germ cells differs considerably between males and females in mammals. Top panel: Migrating germ cells, whether XX or XY, enter meiotic prophase and start developing as oocytes unless they enter a testis. In the testis, the germ cells receive an inhibitory signal that blocks mitotic division and prevents them entering meiotic prophase. Bottom panel: In male mice after birth, immature diploid spermatogonia in the testis enter and complete meiosis in the testis to produce haploid germ cells which eventually mature into sperm (left). In female mice after birth, oocytes complete their first meiotic division in the ovary, but do not enter the second meiotic division meiosis until after they are released from the ovary. The egg only completes meiosis after fertilization.

and female four-cell embryos and, because of the presence of one Y chromosome, can develop testes. In these XX/XY embryos, XX germ cells that are surrounded by testis cells start to develop along the spermatogenesis pathway, but the later development of the gametes is abnormal. In a particular strain of Y/XX mice that develops ovaries rather than testes, XY germ cells appear to develop quite normally as oocytes but fail to develop further after fertilization as the spindle does not assembly properly for the second meiotic division. Experiments transferring XY oocyte nuclei into normal XX oocytes produced healthy offpring, indicating that the fault lies in the XY cytoplasm.

In *Drosophila*, the difference in the behavior of XY and XX germ cells depends initially on the number of X chromosomes, as in somatic cells; the *Sxl* gene again plays an important role, although most other elements in the sex-determination pathway may differ from those in somatic cells. Both chromosomal constitution and cell interactions are involved in the development of germ-cell sexual phenotype. Transplantation of genetically marked pole cells (see Section 10.1) into a *Drosophila* embryo of the opposite sex shows that male XY germ cells in a female XX embryo become integrated into the ovary and begin to develop as sperm; that is, their behavior is autonomous with respect to their genetic constitution. By contrast, XX germ cells in a testis attempt to develop as sperm, showing a role for environmental signals. In neither case, however, are functional sperm produced.

The hermaphrodite of *C. elegans* provides a particularly interesting example of germ-cell differentiation, as both sperm and eggs develop within the same gonad (Fig. 10.31). Unlike the somatic cells of the adult nematode, which have a fixed lineage and number (see Section 6.1), the number of germ cells is indeterminate, with about 1000 germ cells in each 'arm' of the gonad. At hatching of the first-stage larva, there are just two founder germ cells, which proliferate to produce the germ cells. The germ cells are flanked on each side by cells called distal tip cells, and their proliferation is controlled by a signal from the distal tip cells. This signal is the protein LAG-2, which is homologous to the Notch ligand Delta. The receptor for LAG-2 on the germ cells is GLP-1, which is similar both to nematode LIN-12, which is involved in vulva formation (see Section 6.8) and to Notch. We have already met GLP-1 acting to determine cell fate in the early embryo (see Section 6.4).

In *C. elegans*, entry of germ cells into meiosis from the third larval stage onward is controlled by the distal tip signal. In the presence of this signal, the cells proliferate, but as they move away from it, they enter meiosis and develop as sperm (see Fig. 10.31, top). In the hermaphrodite gonad, all the cells that are initially outside the range of the distal tip signal develop as sperm, but cells that later leave the proliferative zone and enter meiosis develop as oocytes (see Fig. 10.31, bottom). The eggs are

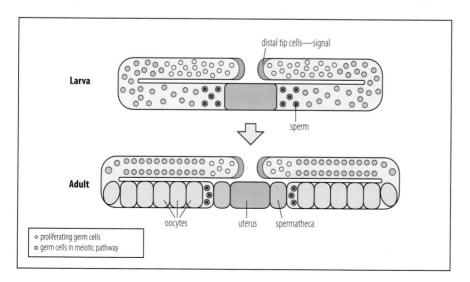

Fig. 10.31 Determination of germ-cell sex in the hermaphrodite *C. elegans* gonad. Top: during the larval stage, germ cells in a zone close to the distal tips of the gonad multiply; when they leave this zone in the larval stage they enter meiosis and develop into sperm. Bottom: in the adult, cells that leave the proliferative zone develop into oocytes. The eggs are fertilized as they pass into the uterus.

*Illustration after Clifford, R., et al.: **Somatic control of germ cell development**.* Semin. Dev. Biol *1994, **5**: 21–30.*

fertilized by stored sperm as they pass into the uterus. The male gonad has similar proliferative meiotic regions, but all the germ cells develop as sperm.

Sex determination of the nematode germ cells is somewhat similar to that of the somatic cells, in that the chromosomal complement is the primary sex-determining factor and many of the same genes are involved in the subsequent cascades of gene expression. The terminal regulator genes required for spermatogenesis are called *fem* and *fog*. In hermaphrodites, there must be a mechanism for activating the *fem* genes in some of the XX germ cells, so allowing them to develop as sperm.

10.17 Various strategies are used for dosage compensation of X-linked genes

In all the animals we have considered in this chapter there is an imbalance of X-linked genes between the sexes. One sex has two X chromosomes, whereas the other has one. This imbalance has to be corrected to ensure that the level of expression of genes carried on the X chromosome is the same in both sexes. The mechanism by which the imbalance in X-linked genes is dealt with is known as **dosage compensation**. Failure to correct the imbalance leads to abnormalities and arrested development. Different animals deal with the problem of dosage compensation in different ways (Fig. 10.32).

Mammals, such as mice and humans, achieve dosage compensation in females by inactivating one X chromosome in each cell after the blastocyst has implanted in the uterine wall. Once an X chromosome has been inactivated in an embryonic cell, this chromosome is maintained in the inactive state in all the resulting somatic cells, and inactivation persists throughout the life of the organism (Fig. 10.33). The inactive X chromosome is replicated at each cell division but remains transcriptionally inactive and

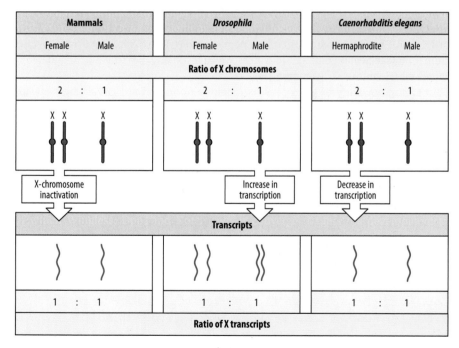

Fig. 10.32 Mechanisms of dosage compensation. In mammals, *Drosophila*, and *Caenorhabditis elegans* there are two X chromosomes in one sex and only one in the other. Mammals inactivate one of the X chromosomes in females; in *Drosophila* males there is an increase in transcription from the single X chromosome; and in *C. elegans* there is a decrease in transcription from the X chromosomes in hermaphrodites. The result of these different dosage compensation mechanisms is that the level of X chromosome transcripts is approximately the same in males and females.

in a different physical state from the other chromosomes during the rest of the cell cycle. At mitosis all chromosomes become highly condensed. During the interphase of the cell cycle—the period between successive mitoses—all the other chromosomes decondense into extended threads of **chromatin**, the complex of DNA and proteins of which chromosomes are made, and are no longer visible under the light microscope. The inactive X chromosome, however, remains condensed and is visible in human cells as the Barr body (Fig. 10.34). Which X chromosome in a cell is inactivated seems to be random, and so female mammals are a mosaic of cells with different X chromosomes inactivated.

The mosaic effect of X inactivation is sometimes visible in the coats of female mammals. Female mice heterozygous for a coat pigment gene carried on the X chromosome have patches of color on their coat, produced by clones of epidermal cells that express the X chromosome carrying a functional pigment gene. The rest of the epidermis is composed of cells in which that X chromosome has been inactivated. The coat pattern of tortoiseshell cats is also due to X-linked mosaicism. They carry two co-dominant alleles, *orange* (X^O) and *black* (X^B), of the main coat color gene, which is located on the X chromosome. These alleles produce orange and black pigments, respectively. Pigment genes are expressed in melanocytes that derive from neural crest cells that migrate to the skin. Cells in which the chromosome carrying X^O is inactivated express the X^B allele and vice versa. In bi-colored tortoiseshell cats (Fig. 10.35), the two cell types are intermingled, producing the characteristic brindled appearance. In calico cats, which have distinct patches of white, orange, and black fur, there has been less intermingling and larger clones of cells of one color develop, along with white patches with no pigment.

How cells count and choose chromosomes for inactivation is not yet fully understood. But the finding that tetraploid cells have two active X chromosomes has led to the following proposal, which depends on the levels of a signal produced by the X chromosomes themselves and invokes a negative-feedback mechanism between autosomes and the X chromosomes. Before inactivation, the X chromosomes are producing a signal that binds to the other chromosomes in the cell. When all the binding sites are occupied, the autosomal chromosomes in turn produce a signal that can bind to and inactivate X chromosomes. The first X chromosome to acquire a critical mass of bound signal initiates its own inactivation. The output of the X-chromosome signal is therefore reduced, the autosomes stop producing the X-inactivating signal, with the result that just one X chromosome is inactivated per diploid genome. This type of mechanism can also accommodate the long-standing observation that in diploid female cells with an abnormal sex-chromosome constitution, such as XXY or XXXY, there is only one active X chromosome per somatic cell, with all the other X chromosomes inactivated.

X inactivation is dependent on a small region of the X chromosome, the X-inactivation center. The center contains many genes, including the gene for the long non-coding RNA, *Xist*, which is the major regulator of X-inactivation and another long non-coding RNA, *Tsix*, which negatively regulates *Xist*. Before inactivation, both RNAs are transcribed at low levels from both X chromosomes. *Xist* expression then dramatically increases on one or other of the X chromosomes and ceases on the other. What causes this to happen is still unknown, but a candidate might be the still-hypothetical inactivating signal from the autosomes noted above. An ubiquitin ligase encoded by a gene on the X chromosome initiates an increase in *Xist* expression. This so-called competence factor could be responsible for sensing the number of X chromosomes present in the cell and ensuring that only one X chromosome is inactivated in a female cell. The *Xist* transcripts spread from their site of synthesis to coat the whole X chromosome on which they are made, blocking transcription, including that of *Tsix*, and inactivating the chromosome. On the chromosome that is not inactivated,

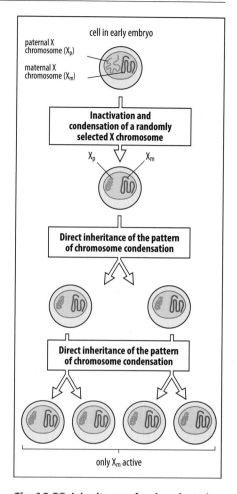

Fig. 10.33 Inheritance of an inactivated X chromosome. In early mammalian female embryos one of the two X chromosomes, either the paternal X (X_p) or the maternal X (X_m), is randomly inactivated. In the figure, X_p is inactivated and this inactivation is maintained through many cell divisions. The inactivated chromosome becomes highly condensed.

Illustration after Alberts, B., et al.: Molecular Biology of the Cell, *2nd edition. New York: Garland Publishing, 1989.*

Fig. 10.34 Inactivated X chromosome (the Barr body). The photograph shows the Barr body (arrow) in the interphase nucleus of a female human buccal cell.

Photograph courtesy of J. Delhanty.

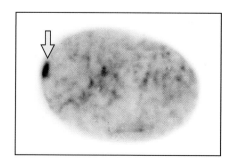

Fig. 10.35 The coloring of tortoiseshell cats is due to mosaicism of X-linked alleles due to X inactivation.

Foxfire, photograph courtesy of Bruce Goatly.

Tsix remains active for a little time but soon after X inactivation is complete, both *Tsix* and Xist are switched off on the active chromosome. *Xist* RNA continues to keep the inactivated chromosome inactive through subsequent cell divisions. The inactive X chromosome has a pattern of DNA methylation that differs from that of the active X, and this methylation is likely to help keep it inactive. DNA methylation is one of the mechanisms used in mammals for long-term silencing of genes, and we have already seen how it can contribute to the differential imprinting of maternal and paternal genes in the germline (see Section 10.8). DNA methylation and other epigenetic mechanisms for controlling gene expression are discussed in more detail in Box 8A.

Dosage compensation in *Drosophila* works in a different way from that in mice and humans (see Fig. 10.32). Instead of repression of the 'extra' X activity in females, transcription of the X chromosome in males is increased nearly twofold. A set of male-specific genes, the MSL complex, controls most dosage compensation, and these are repressed in females by Sxl protein, thereby preventing excessive transcription of the X chromosome. The increased activity in males is regulated by the primary sex-determining signal, which results in the dosage compensation mechanism operating when *Sxl* is 'off'. In females, where *Sxl* is 'on', it turns off the dosage-compensation mechanism. As in the mouse, this regulatory mechanism involves non-coding RNAs.

In *C. elegans*, dosage compensation is achieved by reducing the level of X chromosome expression in XX animals to that of the single X chromosome in XO males (see Fig. 10.32). The number of X chromosomes is communicated by a set of X-linked genes that can repress the master gene *xol-1*. A key event in initiating nematode dosage compensation is the expression of the protein SDC-2, which occurs only in hermaphrodites. It forms a complex specifically with the X chromosome and triggers assembly of a protein complex called the dosage compensation complex, which binds to a specific region on the X chromosome and reduces transcription.

SUMMARY

The development of early embryos of both sexes is very similar. A primary sex-determining signal sets off development toward one or the other sex, and in mammals, *Drosophila*, and *C. elegans*, this signal is determined by the chromosomal complement of the fertilized egg. In mammals, the *Sry* gene on the Y chromosome is responsible for the embryonic gonad developing into a testis and producing hormones that determine male sexual characteristics. The sexual phenotype of the somatic cells is determined by the gonadal hormones. In *C. elegans* and *Drosophila*, the primary sex-determining signal is the number of X chromosomes. In *Drosophila*, the gene *Sex-lethal* is turned on in females but not in males, in response to this signal. In both cases, this results in further gene activity in which sex-specific RNA splicing is involved. In *C. elegans*, the gene *XO lethal* is low in hermaphrodites and high in males, eventually leading to sex-specific expression of the gene *transformer-1*, which determines the sexual phenotype. Somatic sexual differentiation in *Drosophila* is cell autonomous and is controlled by the number of X chromosomes; in *C. elegans*, cell–cell interactions are also involved. In mammals, signals from the gonads determine whether the germ cells develop into oocytes or sperm. Male germ cells in *Drosophila* develop along the sperm pathway even in an ovary, but female germ cells develop along the sperm pathway when placed in a testis. Most *C. elegans* adults are hermaphrodites, and produce both sperm and eggs from the same gonad.

Various strategies of dosage compensation are used to correct the imbalance of X chromosomes between males and females. In female mammals, one of the X chromosomes is inactivated; in *Drosophila* males the activity of the single X chromosome is upregulated; and in *C. elegans* the activity of the X chromosomes in XX hermaphrodites is downregulated to match that from the single X chromosome in males.

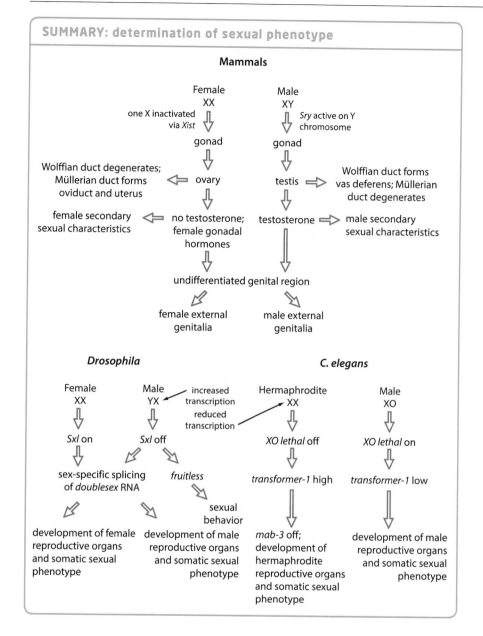

SUMMARY: determination of sexual phenotype

Mammals

Drosophila   **C. elegans**

Summary to Chapter 10

- In many animals, the future germ cells are specified by localized cytoplasmic determinants in the egg. Germ cells in mammals are unusual, as they are specified entirely by cell–cell interactions.

- In animals, the development of germ cells into sperm or egg depends both on chromosomal constitution and interactions with the cells of the gonad.

- At fertilization, fusion of sperm and egg initiates development, and there are mechanisms to ensure that only one sperm enters an egg.

- Both maternal and paternal genomes are required for normal mammalian development, as some genes are imprinted; for such genes, whether they are expressed or not during development depends on whether they are derived from the sperm or the egg.

- In many animals, the chromosomal constitution of the embryo determines which sex will develop. In mammals, the Y chromosome is male-determining; it specifies the development of a testis, and the hormones produced by the testis cause the development of male sexual characteristics. In the absence of a Y chromosome, the embryo develops as a female.

- In *Drosophila* and *C. elegans*, sexual development is initially determined by the number of X chromosomes, which sets in train a cascade of gene activity. In flies, somatic sexual differentiation is cell autonomous; in nematodes, the somatic sexual phenotype is determined by cell–cell interactions.
- Animals use various strategies of dosage compensation to correct the imbalance in the number of X chromosomes in males and females.

■ End of chapter questions

Long answer (concept questions)

1. Distinguish between the germline, gametes, and somatic cells. What key functions are germ cells responsible for?

2. Summarize the evidence for the existence of a special germplasm in *Drosophila* and *Xenopus*. Is the presence of germplasm a universal feature of animal development?

3. Draw a schematic of two generations of *Drosophila*: start with a male and a female, each of which is heterozygous for a loss-of-function recessive *oskar* mutation. Mate them together to get homozygous *oskar* males and females. Mate these *oskar* mutants to normal (wild-type) flies. How do these matings illustrate why such mutations are referred to as 'grandchildless'? Does it matter whether the homozygous *oskar* mutant is a male or a female? Explain.

4. What is the role of SDF-1/CXCR4 in germ-cell migration? To what extent is this conserved between vertebrates?

5. Discuss the following aspects of meiosis: (1) Why is a 'reduction division' (meiosis I) important? (2) What is the significance of recombination in sexual reproduction? (3)Why do four sperm cells result from each cell undergoing meiosis during spermatogenesis, yet only one egg cell results during oogenesis—where do the other 'eggs' go?

6. When in human development do primary oocytes form? What happens to a primary oocyte between this time and its maturation into an egg ready to be fertilized? At what point in meiosis is the oocyte when fertilization occurs?

7. What is meant by 'genomic imprinting'? What are the mechanisms that maintain imprinting?

8. Describe the process by which the nucleus and contents of the sperm enters the mammalian egg. Include SED1, ZP2, and acrosome enzymes in your answer.

9. Fertilization of a sea-urchin egg leads to a wave of calcium that sweeps across the egg. Is this calcium intracellular or extracellular in origin? What are two consequences of this increase in calcium?

10. What is MPF? Summarize the role of MPF-like complexes in the control of the various stages of the cell cycle.

11. Explain how chromosomal abnormalities can be used as evidence for the following statements. (1) Determination of the male sex in humans is controlled by the presence of the Y chromosome, not two X chromosomes. (2) Determination of female sex in *Drosophila* is controlled by the presence of two X chromosomes and not the presence of a Y chromosome. (3) One X chromosome in humans is sufficient to determine the female sex in humans. (4) One X chromosome is sufficient to determine the male sex in *Drosophila*. (5) *SRY* is the gene responsible for male sex determination in humans.

12. How does *SRY* trigger male development in mammals? Include: Sox9, Sertoli cells, Leydig cells, Müllerian inhibiting substance and testosterone in your answer.

13. Summarize the development of the gonads and related structures in mammals, focusing on the Wolffian and Müllerian ducts.

14. Summarize the information presented in Figs 10.26 and 10.27 on how cell-autonomous sexual differentiation occurs in *Drosophila*. Do not just copy the two figures, but instead incorporate the information into one comprehensive schematic.

15. What is different about the sexes in *C. elegans* compared with the sexes in our other model organisms? How does this affect its method of reproduction?

16. Briefly compare and contrast the sex-determination pathways in *Drosophila* and *C. elegans*. Are there similarities in the strategies used? Are there differences in the mechanisms involved?

17. Say how X-chromosome inactivation will achieve dosage compensation in humans with the chromosomal compositions: XY, XX, XXY, or XXX?

18. How does *Xist* control X-chromosome inactivation in mammals? Speculate why the other gene involved in this process, *Tsix*, might have been given this name.

Multiple choice (factual recall questions)

NB There is only one correct answer to each question.

1. The germplasm in *Drosophila* is specified by

a) *bicoid*
b) Hox genes
c) *oskar*
d) the point of sperm entry

2. In mice, germ cells can be first identified

a) in prospective mesoderm of the proximal epiblast, as *Blimp1*-expressing cells
b) in the mesoderm that will form gonadal tissue
c) in the posterior tip of the embryo, as cells containing pole plasm
d) in the genital ridge, as *Oct4*-expressing cells

3. Which feature of meiosis, as compared with mitosis, is most important for the production of gametes?

a) Homologous chromosomes separate in meiosis I, whereas sister chromatids separate in mitosis.
b) Meiosis leads to the production of haploid products, whereas mitosis leads to diploid daughter cells.
c) Meiosis provides an opportunity for recombination, whereas mitosis does not.
d) Typically, three of the four products do not become oocytes during meiosis in female organisms, whereas mitosis will produce four functional products after two rounds of cell division.

4. In what stage of which type of cell division, is the mammalian oocyte at the birth of the animal?

a) The oocyte is in G1 of the mitotic cell cycle.
b) The oocyte is in metaphase of meiosis II.
c) The oocyte is in prophase of meiosis I.
d) The oocyte is in prophase of mitosis.

5. What is the zona pellucida of the mammalian egg?

a) a hardened membrane that forms a physical block to polyspermy, formed from the vitelline membrane and the contents of the cortical granules
b) a layer of follicle-derived cells called cumulus cells
c) an extracellular layer of glycoproteins
d) the plasma membrane of the egg

6. The mammalian oviduct will form from the

a) mesonephros
b) Müllerian duct
c) ureter
d) Wolffian duct

7. The molecular activity of the Sxl protein of *Drosophila* is

a) as a 'numerator', counting the number of X chromosomes
b) as a transcription factor
c) to control RNA splicing
d) to signal to the Tra receptor

8. The human congenital disorder Angelman syndrome is due to the inheritance of a maternal chromosome 15 that has a small deletion of a specific region. Why does this deletion not behave as a recessive allele; that is, why is its loss not made up for by the intact region on the unmutated paternal chromosome 15?

a) Chromosomes with deletions do not go through mitosis correctly, so cell divisions in the embryo result in cells with abnormal numbers of chromosomes, and these cells do not contribute properly to the development of the organism.
b) The father's copy of chromosome 15 has genes in the region of the deletion that are imprinted, and thus inactive; in the absence of any active copies of these genes, development cannot proceed normally.
c) The genes in this portion of chromosome 15 are special in that they are required in two copies for normal development, and so the loss of one set does not allow normal development.
d) Two copies of every gene in the genome are required for development, so loss of one of the copies from this region perturbs development.

9. What is the cortical reaction in sea urchin eggs, and why is it important?

a) The cortical reaction is the depolarization of the plasma membrane after sperm entry, which helps to block polyspermy.
b) The cortical reaction is the entry of Ca^{2+} ions into the egg via the cortex, which initiates development.
c) The cortical reaction is the fusion of the egg cortex with the egg plasma membrane, which allows the sperm to enter.
d) The cortical reaction is the release of the cortical granules after sperm entry, which converts the vitelline membrane into the fertilization membrane, which blocks polyspermy.

10. The *C. elegans* GLP-1 protein is similar to which mammalian signaling protein?

a) BMPs
b) Delta
c) Notch
d) testosterone receptor

Multiple choice answer key

1: c, 2: a, 3: b, 4: c, 5: c, 6: b, 7: c, 8: b, 9: d, 10: c.

■ General further reading

Chadwick, D., Goode, J.: *The Genetics and Biology of Sex Determination 2002*. Novartis Foundation Symposium 244. New York: John Wiley, 2002.
Cinalli, R.M., Rangan, P., Lehman, R.: **Germ cells are forever**. *Cell* 2008, **132**: 559–562.
Crews, D.: **Animal sexuality**. *Sci. Am.* 1994, **270**: 109–114.
Zarkower, D.: **Establishing sexual dimorphism: conservation amidst diversity?** *Nat. Rev. Genet.* 2001, **2**: 175–185.

■ Section further reading

10.1 Germ-cell fate is specified in some embryos by a distinct germplasm in the egg

Extavour, C.G., Akam, M.: **Mechanisms of germ cell specification across the metazoans**: epigenesis and preformation. *Development* 2003, **130**: 5869–5884.
Matova, N., Cooley, L.: **Comparative aspects of animal oogenesis**. *Dev. Biol.* 2001, **231**: 291–320.
Mello, C.C., Schubert, C., Draper, B., Zhang, W., Lobel, R., Priess, J.R.: **The PIE-1 protein and germline specification in *C. elegans* embryos**. *Nature* 1996, **382**: 710–712.
Micklem, D.R., Adams, J., Grunert, S., St. Johnston, D.: **Distinct roles of two conserved Staufen domains in *oskar* mRNA localisation and translation**. *EMBO J.* 2000, **19**: 1366–1377.
Ray, E.: **Primordial germ-cell development: the zebrafish perspective**. *Nat. Rev. Genet.* 2003, **4**: 690–700.
Williamson, A., Lehmann, R.: **Germ cell development in *Drosophila***. *Annu. Rev. Cell Dev. Biol.* 1996, **12**: 365–391.

10.2 In mammals germ cells are induced by cell–cell interactions during development

Magnusdottir, E., Surani, M.A.: **How to make a primordial germ cell**. *Development* 2014, **141**: 245–252.

McLaren, A.: **Primordial germ cells in the mouse**. *Dev. Biol.* 2003, **262**: 1–15.

Ohinata, Y., Payer, B., O'Carroll, D., Ancelin, K., Ono, Y., Sano, M., Barton, S.C., Obukhanych, T., Nussenzweig, M., Tarakhovsky, A., Saitou, M., Surani, M.A.: **Blimp1 is a critical determinant of the germ cell lineage in mice**. *Nature* 2005, **436**: 207–213.

10.3 Germ cells migrate from their site of origin to the gonad & 10.4 Germ cells are guided to their final destination by chemical signals

Deshpande, G., Zhou, K., Wan, J.Y., Friedrich, J., Jourjine, N., Smith, D., Schedl, P.: **The *hedgehog* pathway gene *shifted* functions together with the *hmgcr*-dependent isoprenoid biosynthetic pathway to orchestrate germ cell migration**. *PLoS Genet.* 2013, **9**: e1003720.

Doitsidou, M., Reichman-Fried, M., Stebler, J., Koprunner, M., Dorries, J., Meyer, D., Esguerra, C.V., Leung, T., Raz, E.: **Guidance of primordial germ cell migration by the chemokine SDF-1**. *Cell* 2002, **111**: 647–659.

Knaut, H., Werz, C., Geisler, R., Nusslein-Volhard, C., Tubingen 2000 Screen Consortium: **A zebrafish homologue of the chemokine receptor Cxcr4 is a germ-cell guidance receptor**. *Nature* 2003, **421**: 279–282.

Richardson, B.E., Lehmann, R.: **Mechanisms guiding primordial germ cell migration: strategies from different organisms**. *Nat. Rev. Mol. Cell Biol.* 2010, **11**: 37–49.

Weidinger, G., Wolke, U., Köprunner, M., Thisse, C., Thisse, B., Raz, E.: **Regulation of zebrafish primordial germ cell migration by attraction towards an intermediate target**. *Development* 2002, **129**: 25–36.

10.5 Germ-cell differentiation involves a halving of chromosome number by meiosis

De Rooij, D.G., Grootegoed, J.A.: **Spermatogonial stem cells**. *Curr. Opin. Cell Biol.* 1998, **10**: 694–701.

Hultén, M.A., Patel, S.D., Tankimanova, M., Westgren, M., Papadogiannakis, N., Jonsson, A.M., Iwarsson, E.: **The origins of trisomy 21 Down syndrome**. *Mol. Cytogenet.* 2008, **1**: 21–31.

Mehlmann, L.M.: **Stops and starts in mammalian oocytes: recent advances in understanding the regulation of meiotic arrest and oocyte maturation**. *Reproduction* 2005, **130**: 791–799.

Pacchierottia, F., Adler, I.-D., Eichenlaub-Ritter, U., Mailhes, J.B.: **Gender effects on the incidence of aneuploidy in mammalian germ cells**. *Environ. Res.* 2007, **104**: 46–69.

Vogta, E., Kirsch-Volders, M., Parry, J., Eichenlaub-Rittera, U.: **Spindle formation, chromosome segregation and the spindle checkpoint in mammalian oocytes and susceptibility to meiotic error**. *Mutat. Res.* 2008, **651**: 14–29.

10.6 Oocyte development can involve gene amplification and contributions from other cells

Browder, L.W.: ***Oogenesis***. New York: Plenum Press, 1985.

Choo, S., Heinrich, B., Betley, J.N., Chen, A., Deshler, J.O.: **Evidence for common machinery utilized by the early and late RNA localization pathways in *Xenopus* oocytes**. *Dev. Biol.* 2004, **278**: 103–117.

10.7 Factors in the cytoplasm maintain the totipotency of the egg

Gurdon, J.B.: **Nuclear transplantation in eggs and oocytes**. *J. Cell Sci. Suppl.* 1986, **4**: 287–318.

10.8 In mammals some genes controlling embryonic growth are 'imprinted'

Bartolomei, M.S: **Genomic imprinting: employing and avoiding epigenetic processes**. *Genes Dev.* 2009, **23**: 2124–2133.

Plasschaert, R.N., Bartolomei, M.S.: **Genomic imprinting in development, growth, behavior and stem cells**. *Development* 2014, **141**: 1805–1813.

Reik, W., Walter, J.: **Genomic imprinting: parental influence on the genome**. *Nat. Rev. Genet.* 2001, **2**: 21–32.

Wood, A.J., Oakey, R.J.: **Genomic imprinting in mammals: emerging themes and established theories**. *PLoS Genet.* 2006, **2**: e147.

10.9 Fertilization involves cell-surface interactions between egg and sperm

Avella, M.A., Xiong, B., Dean, J.: **The molecular basis of gamete recognition in mice and humans**. *Mol Hum Reprod.* 2013; **19**: 279–289.

Baibakov, B., Boggs, N.A., Yauger, B., Baibakov, G., Dean, J.: **Human sperm bind to the N-terminal domain of ZP2 in humanized zonae pellucidae in transgenic mice**. *J Cell Biol.* 2012, **197**: 897–905.

Bianchi, E., Doe, B., Goulding, D., Wright, G.J.: **Juno is the egg Izumo receptor and is essential for mammalian fertilization**. *Nature* 2014, **508**: 483–487.

Inoue, N., Satouh, Y., Ikawa, M., Okabe, M., Yanagimachi, R.: **Acrosome-reacted mouse spermatozoa recovered from the perivitelline space can fertilize other eggs**. *Proc. Natl Acad. Sci. USA* 2011, **108**: 20008–20011.

10.10 Changes in the egg plasma membrane and enveloping layers at fertilization block polyspermy,

Bianchi, E., Doe, B., Goulding, D., Wright, G.J.: **Juno is the egg Izumo receptor and is essential for mammalian fertilization**. *Nature* 2014, **508**: 483–487.

Burkart A.D., Xiong, B., Baibakov, B., Jiménez-Movilla, M., Dean, J.: **Ovastacin, a cortical granule protease, cleaves ZP2 in the zona pellucida to prevent polyspermy**. *J Cell Biol.* 2012, **197**: 37–44.

Wong, J.L., Wessel, G.M.: **Defending the zygote: search for the ancestral animal block to polyspermy**. *Curr. Top. Dev. Biol.* 2006, **72**: 1–151.

10.11 Sperm–egg fusion causes a calcium wave that results in egg activation

Kashir, J., Nomikos, M., Lai, F. A., Swann, K.: **Sperm-induced Ca²⁺ release during egg activation in mammals.** *Biochem. Biophys. Res. Commun.* 2014, pii: S0006-291X(14)00729-3.

Suzuki, T., Yoshida, N., Suzuki, E., Okuda, E., Perry, A.C.: **Full-term mouse development by abolishing Zn²⁺-dependent metaphase II arrest without Ca²⁺ release.** *Development* 2010, **137**: 2659–2669.

Whitaker, M.: **Calcium signaling in early embryos.** *Philos. Trans. R. Soc. Lond. B Biol. Sci.* 2008, **363**: 1401–1418.

10.12 The primary sex-determining gene in mammals is on the Y chromosome

Capel, B.: **The battle of the sexes.** *Mech. Dev.* 2000, **92**: 89–103.

Koopman, P.: **The genetics and biology of vertebrate sex determination.** *Cell* 2001, **105**: 843–847.

Schafer, A.J., Goodfellow, P.N.: **Sex determination in humans.** *BioEssays* 1996, **18**: 955–963.

10.13 Mammalian sexual phenotype is regulated by gonadal hormones

Jameson, S.A., Lin, Y.T., Capel, B.: **Testis development requires the repression of Wnt4 signaling by Fgf9 signaling.** *Dev Biol.* 2012, **370**: 24–32.

Svingen, T., Koopman, P.: **Building the mammalian testis: origins, differentiation and assembly of the component cell populations.** *Genes Dev.* 2013, **27**: 2409–2426.

Swain, A., Lovell-Badge, R.: **Mammalian sex determination: a molecular drama.** *Genes Dev.* 1999, **13**: 755–767.

Vainio, S., Heikkila, M., Kispert, A., Chin, N., McMahon, A.P.: **Female development in mammals is regulated by Wnt-4 signalling.** *Nature* 1999, **397**: 405–409.

10.14 The primary sex-determining signal in *Drosophila* is the number of X chromosomes, and is cell autonomous

Brennan, J., Capel, B.: **One tissue, two fates: molecular genetic events that underlie testis versus ovary development.** *Nat. Rev. Genet.* 2004, **5**: 509–520.

Hodgkin, J.: **Sex determination compared in *Drosophila* and *Caenorhabditis*.** *Nature* 1990, **344**: 721–728.

MacLaughlin, D.T., Donahoe, M.D.: **Sex determination and differentiation.** *New Engl. J. Med.* 2004, **350**: 367–378.

10.15 Somatic sexual development in *Caenorhabditis* is determined by the number of X chromosomes

Carmi, I., Meyer, B.J.: **The primary sex determination signal of *Caenorhabditis elegans*.** *Genetics* **152**: 999–1015.

Meyer, B. J.: **X-chromosome dosage compensation.** In WormBook (June 25, 2005) (edited by The C. elegans Research Community). doi/10.1895/wormbook.1.8.1, http://www.wormbook.org (date accessed 21 May 2010).

Raymond, C.S., Shamu, C.E., Shen, M.M., Seifert, K.J., Hirsch, B., Hodgkin, J., Zarkower, D.: **Evidence for evolutionary conservation of sex-determining genes.** *Nature* 1998, **391**: 691–695.

10.16 Determination of germ-cell sex depends on both genetic constitution and intercellular signals

Childs, A.J., Saunders, P.T.K., Anderson, R.A.: **Modelling germ cell development *in vitro*.** *Mol. Hum. Reprod.* 2008, **14**: 501–511.

McLaren, A.: **Signaling for germ cells.** *Genes Dev.* 1999, **13**: 373–376.

Obata, Y., Villemure, M., Kono, T., Taketo, T.: **Transmission of Y chromosomes from XY female mice was made possible by the replacement of cytoplasm during oocyte maturation.** *Proc. Natl Acad. Sci. USA* 2008, **105**: 13918–13923.

Seydoux, G., Strome, S.: **Launching the germline in *Caenorhabditis* elegans: regulation of gene expression in early germ cells.** *Development* 1999, **126**: 3275–3283.

10.17 Various strategies are used for dosage compensation of X-linked genes

Avner, P., Heard, E.: **X-chromosomes inactivation: counting, choice and initiation.** *Nat. Rev. Genet.* 2001, **2**: 59–67.

Csankovszki, G., McDonel, P., Meyer, B.J.: **Recruitment and spreading of the *C. elegans* dosage compensation complex along X chromosomes.** *Science* 2004, **303**: 1182–1185.

Navarro, P., Avner, P.: **An embryonic story: analysis of the gene regulative network controlling Xist expression in mouse embryonic stem cells.** *Bioessays*, 2010, **32**: 581–588

Panning, B.: **X-chromosome inactivation: the molecular basis of silencing.** *J. Biol.* 2008, **7**: 30.

Pollex, T., Heard, E.: **Recent advances in X-chromosome inactivation research.** *Curr. Opin. Cell Biol.* 2012, **24**: 825–832.

Starmer, J., Magnuson, T.: **New model for random X chromosome inactivation.** *Development* 2009, **136**: 1–10.

11

Organogenesis

- The vertebrate limb
- Insect wings and legs

- Vertebrate and insect eyes
- Vertebrate lungs and insect tracheal system

- Vertebrate blood vessels and heart
- Teeth

Once the basic animal body plan has been laid down, the development of organs as varied as insect wings and vertebrate eyes begins. The positions in which these organs will develop have already been specified as part of the process of laying down the body plan, but the subsequent development of the organ rudiments is essentially autonomous. Organogenesis depends on the same basic mechanisms as those used in early development, but is more complex and involves large numbers of genes, as an organ is composed of different types of tissues and their development has to be integrated. Nevertheless, many of the mechanisms used, such as positional information, and cell activities such as migration, are the same as those used in earlier development, and certain signals are used again and again.

So far, we have concentrated almost entirely on the aspects of development involved in laying down the basic body plan in various organisms, and on early morphogenesis and cell differentiation. We now turn to the development of specific organs and structures—**organogenesis**—which is a crucial phase of development that will eventually lead to the embryo becoming a fully functioning organism, capable of independent survival.

The development of certain organs has been studied in great detail and they provide excellent models for looking at developmental processes such as pattern formation, the specification of positional information, induction, change in form, and cellular differentiation. In this chapter, we first consider the development of some classical model systems—the vertebrate limb, the legs and wings of *Drosophila*, and vertebrate and insect eyes. We then consider some internal organs. The structure of many internal organs is based on tubes—for example, the lungs and the vascular system—and we look at the different ways in which tubes can be formed. We also look briefly at the initial patterning and regionalization of the heart, which develops from a tube of mesoderm. The kidney also has a tubular structure and its development can be found in the online material associated with this chapter.

We touch finally on the induction of teeth and patterning of the dentition, which are based on similar principles to the induction and patterning of vertebrate limbs. The development of other major organs, such as the gut, liver, and pancreas, involves no new developmental principles and we will not consider them here. The liver is

Scan here

Scan this QR code image with your mobile device to see the online supplementary material on the kidney or log on to **http://global.oup.com/uk/orc/biosciences/ devbiol/wolpert5e/qr/qr11a/**

interesting as an example of a mammalian organ that can regenerate, and some aspects of liver growth and regeneration are discussed in Chapter 13.

The cellular mechanisms involved in organogenesis are essentially similar to those encountered in earlier stages of development; they are merely employed in different spatial and temporal patterns. Many of the genes and signaling molecules involved will be familiar from earlier chapters. We shall see, however, that the mechanisms involved are much more complex and that, while one can identify some general principles, such as the use of positional information, there is much additional detail for which there are no unifying principles—it just works that way in that particular organ.

The vertebrate limb

The vertebrate embryonic limb is a particularly good system in which to study general developmental principles. Development of the limb involves some simple changes in shape, and the basic pattern of limb structures is initially quite simple and can be easily recognized. The limb is also a good model for studying cellular interactions within a structure containing a large number of cells, and for elucidating the role of cell–cell signaling in development. Mice are used to study some aspects of limb development, mainly through spontaneous mutants and artificial gene knock-outs, but the basic principles of limb morphogenesis and pattern formation have been most extensively investigated in chick embryos, because here the developing limbs themselves are easily accessible for microsurgical manipulation. A window can be made in the eggshell and the limb buds manipulated while the embryo remains in the egg (see Fig. 3.35). After surgery, the window can be re-sealed with adhesive tape and the embryo allowed to continue development, so that the effects of the manipulations on limb development can be assessed. The early development of zebrafish pectoral fin buds is very similar to that of limb buds, and zebrafish are increasingly being used to study these early stages.

In chick embryos, the first signs of limb development can be seen around the third day after the egg is laid, when the structures of the main body axis are already well established. Small protrusions—the **limb buds**—arise from the body wall of the embryo (Fig. 11.1). By 10 days, the main features of the limbs are well developed. Figure 11.2 shows the pattern of the skeletal elements; they are first formed as cartilage and are later replaced by bone (how bones grow is described in Section 13.9). Feather buds have formed and the limb at this stage also has muscles and tendons. The limb has three developmental axes: the **proximo-distal axis** runs from the base of the limb to the tip; the **antero-posterior axis** runs parallel with the body axis (in the human hand it goes from the thumb (anterior) to the little finger (posterior), and in the chick wing from digit 2 to digit 4); the **dorso-ventral axis** is the third axis—in the human hand it runs from the back of the hand to the palm.

11.1 The vertebrate limb develops from a limb bud

The early limb bud has two major components—a core of loose mesenchymal cells and an outer layer of epithelial cells (Fig. 11.3). The mesenchymal cells are of two separate lineages—cells from the lateral plate mesoderm, which give rise to the skeletal elements and other connective tissues of the limb, and cells derived from the somites, which migrate into the limb bud and give rise to the myogenic cells of the muscles (see Section 5.13). The limb vasculature is, at least in part, also formed by cells derived from the somites. The epithelial cells are derived from the ectoderm and give rise to the epidermis of the skin (see Section 8.10).

At the very tip of the limb bud is a thickening in the ectoderm—the **apical ectodermal ridge** or **apical ridge**, which runs along the boundary between dorsal

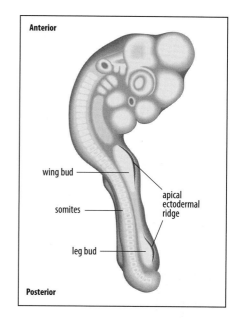

Fig. 11.1 The limb buds of the chick embryo. Limb buds appear on the flanks of the embryo at three days of incubation after the egg has been laid (only the limb buds on the right side are shown here). They are composed of mesoderm, with an outer covering of ectoderm. Along the tip of each runs a thickened ridge of ectoderm, the apical ectodermal ridge.

Fig. 11.2 The embryonic chick wing.
The photograph shows a whole mount of the wing of a chick embryo at 10 days of incubation after the egg has been laid. The wing has been stained to show the pattern of differentiated cartilage tissue. By this time, the main skeletal elements (e.g. humerus, radius, and ulna) have been laid down in cartilage. They later become ossified to form bone. The muscles and tendons are also well developed at this stage but cannot be seen in this type of preparation. Feather buds can be seen, particularly along the posterior margin of the limb. The three developmental axes of the limb are proximo-distal, antero-posterior, and dorso-ventral, as shown in the top panel. Note that the chick wing has only three digits, which have traditionally been called 2, 3, and 4 (in relation to the five-digit tetrapod limb) although evolutionary studies now identify them as 1, 2, 3, as discussed in Chapter 14. We retain the traditional numbering in this chapter as it is the numbering used in the literature until very recently. Scale bar = 1 mm.

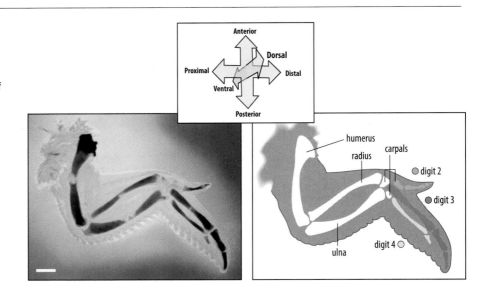

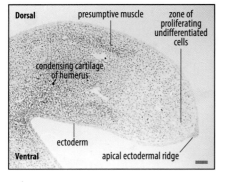

Fig. 11.3 Cross-section through an embryonic chick wing bud. The thickened apical ectodermal ridge is at the tip. Beneath the apical ridge is a region of undifferentiated cells. Proximal to this region, mesenchyme cells condense and differentiate into cartilage. Presumptive muscle cells migrate into the limb from the adjacent somites and form dorsal and ventral muscle masses. Scale bar = 0.1 mm.

and ventral ectoderm (Fig. 11.4). Directly beneath the apical ectodermal ridge lies a region composed of undifferentiated mesenchymal cells, and it is only when cells leave this zone that they begin to differentiate. As the bud grows out, the cells left behind start to differentiate and the cartilaginous elements begin to appear in the mesenchyme. The proximal part of the limb bud—that is, the part nearest to the body—is the first to differentiate, and differentiation proceeds distally as the limb bud grows out, with the 'hand' region differentiating last. Of the structures found in the developing limb, the patterning of the cartilage has been the best studied, as cartilage can be stained and seen easily in whole mounts of the embryonic limb (see Fig. 11.2). The disposition of muscles and tendons is more intricate, and although this can be studied in whole mounts by staining with antibodies for tissue-specific proteins, histological examination of serial sections through the limb may also be needed.

The first sign of cartilage differentiation to form a skeletal element is the increased local packing of the mesenchyme cells, a process known as **condensation**. The cartilage elements are laid down in a proximo-distal sequence in the chick wing—the humerus, followed by radius and ulna, then the wrist elements (carpals), and finally three easily distinguishable digits, 2, 3, and 4 (see Fig. 11.2). Figure 11.5 compares this sequence in the development of a chick wing with the similar sequence in the development of a mouse forelimb. In addition, this figure shows how the shape of the limb changes over these stages of development. An indentation appears in the anterior margin of the elongated bud marking the 'elbow', and later the distal region of the limb bud broadens and forms the flattened **hand-plate (digital-plate)** in which the digits arise.

The chick limb bud at 3 days is about 1 mm wide by 1 mm long, but by 10 days it has grown around 10-fold, mostly in length. The basic pattern has been laid down well before then, but even at 10 days, the limb is still very small compared with the size of the limb when the chick hatches. The apical ridge disappears as soon as all the basic elements of the limb are in place, and growth occupies most of the subsequent development of the limb, both before and after hatching. During the growth phase, the cartilaginous elements are largely replaced by bone. Nerves only enter the limb after the cartilage has been laid down, at around 4.5 days of incubation, and we shall discuss this in Chapter 12. The fundamental change in shape that occurs during early limb development is elongation of the bud, and the general mechanisms of how this elongation is accomplished are thought to be similar in other regions of the embryo that undergo outgrowth—for example, the posterior part of the main body axis. The problem of pattern formation in the limb is to understand how the basic pattern of cartilage, muscle, and tendons is formed in the right places, and how they make the right connections with each other. The first consideration, however, is how limb buds develop at the appropriate positions on the body.

11.2 Genes expressed in the lateral plate mesoderm are involved in specifying the position and type of limb

The forelimbs and hindlimbs of vertebrates arise at precise positions along the antero-posterior axis of the body. Transplantation experiments in chick embryos have shown that the lateral plate mesoderm becomes determined to form limbs in these positions, long before limb buds are visible. As in the pre-somitic mesoderm (see Chapter 5), Hox gene expression becomes regionalized in the lateral plate mesoderm of the trunk along the antero-posterior axis, with 3′ genes being expressed more anteriorly than 5′ Hox genes. Hox genes in the paralogous subgroups 4 and 5, for example, are expressed in the region of the lateral plate mesoderm that will form the wings/forelimbs, and

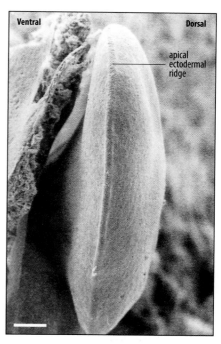

Fig. 11.4 Scanning electron micrograph of a limb bud of a chick embryo at 4.5 days incubation after laying, showing the apical ectodermal ridge. Scale bar = 0.1 mm.

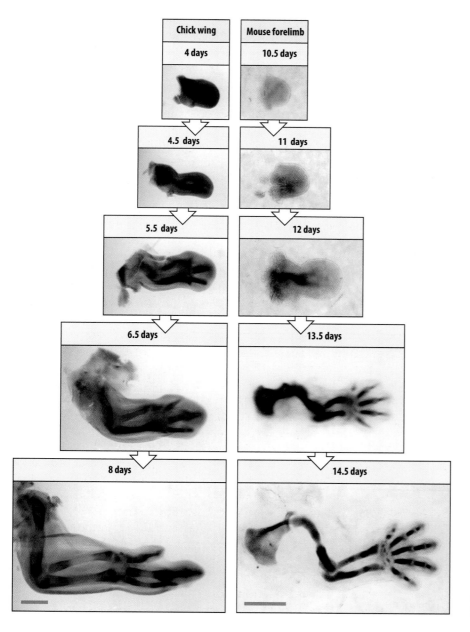

Fig. 11.5 The development of the chick wing and mouse forelimb is similar. The skeletal elements are laid down, as cartilage, in a proximo-distal sequence as the limb bud grows outward. The cartilage of the humerus is laid down first, followed by the radius and ulna, wrist elements, and digits. Scale bar = 1 mm.

Hox genes in the paralogous subgroups 8, 9 and 10 are expressed more posteriorly. A combinatorial Hox code determines the position in which limbs will develop by controlling the expression of the genes for two T-box transcription factors, Tbx5 and Tbx4, which are required for initiation of forelimb and hindlimb development, respectively.

Tbx4 and Tbx5, which are related to the mesodermal marker Brachyury (see Section 4.12), are expressed specifically in the limb-forming regions, with the *Tbx4* gene being expressed in the prospective wing/forelimb and *Tbx5* in the prospective leg/hindlimb (Fig. 11.6). In mouse embryos, *Tbx4* expression becomes restricted to the region where the forelimb will develop as a result of its activation by Hox proteins encoded by the 4 and 5 paralogous group genes combined with repression by more 'posterior' Hox proteins. The gene for the homeodomain transcription factor Pitx1 is also expressed in the prospective leg/hindlimb region, and has a key role in determining the difference between hindlimb and forelimb, but it is not yet known whether its expression is regulated by Hox proteins. Mutations in both *TBX5* and *PITX1* have been associated with limb malformations in human patients. Mutations in *TBX5* are responsible for Holt–Oram syndrome, characterized by defects in the upper limbs and heart, whereas mutations in *PITX1* affect the lower limbs.

Tbx4 and Tbx5 have equivalent roles and are essential for the development of their respective limbs because they control local production of FGFs that initiate limb development.

Local application of FGF to the interlimb region of the flank of an early chick embryo, the region between the prospective wing and prospective leg buds, induces an ectopic limb bud. Wing buds develop when FGF is applied to the anterior interlimb region, whereas leg buds develop from application to the posterior interlimb region (Fig. 11.7). FGF-10 is the ligand that initiates normal limb development and its expression in the limb-forming regions of the lateral plate mesoderm is controlled by the Tbx factors (see Fig. 11.6). In the mouse embryo, in which, unlike chick, transgenic knock-outs are possible, knock-out of either *Fgf10* or the gene encoding its receptor results in embryos lacking limb buds. In chick embryos, Wnt proteins produced and secreted by the mesoderm play a key role in determining where FGFs are expressed and maintained.

FGF is involved in establishing and maintaining the two main organizing regions of a limb. These are the **apical ectodermal ridge**, which is essential for limb bud outgrowth and for correct patterning along the proximo-distal axis of the limb, and a mesodermal **polarizing region**, which is situated on the posterior side of the limb bud

Fig. 11.6 Schematic diagram of limb initiation in chick and mouse embryos. A combinatorial Hox code in the lateral plate mesoderm determines where limbs develop by activating expression of *Tbx5* in the region where the wing/forelimb will develop and *Tbx4* in the region where the leg/hindlimb will develop. Pitx1, which is also expressed in the hindlimb region of the lateral plate mesoderm, is required for *Tbx4* expression and specifies the identity of the limb as leg/hindlimb. Tbx4 and Tbx5 activate *Fgf10* expression in the limb-forming regions and FGF-10 signaling induces expression of *Fgf8* expression in the overlying ectoderm that will form the apical ectodermal ridge. FGF-8 signaling by the apical ridge then maintains *Fgf10* expression in the mesoderm. This establishes a positive feedback loop of FGF signaling, which is common to both wings/forelimbs and legs/hindlimbs. White arrows in the first panel represent signals from somites and paraxial mesoderm which may be involved at the earliest stages. *From Duboc, V., Logan, M.P.: **Regulation of limb bud initiation and limb-type morphology.** Dev. Dyn. 2011, **240**: 1017-1027.*

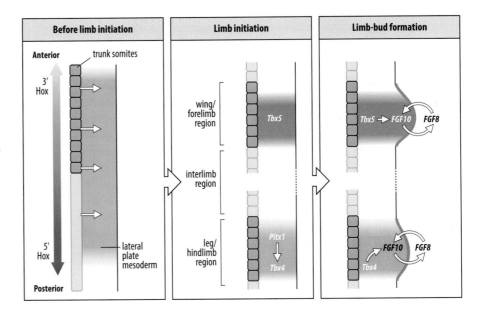

Fig. 11.7 Result of local application of FGF-4 to the interlimb region of a chick embryo.
A bead soaked in FGF-4 was implanted in the flank close to the prospective leg bud region, where it induces outgrowth of an additional leg bud. The dark regions indicate *Sonic hedgehog* transcripts. *Sonic hedgehog* is expressed at the posterior sides of the normal wing and leg buds but at the anterior of the additional leg bud, and the leg that subsequently develops has an opposite polarity to that of the normal limbs. It is not known why *Sonic hedgehog* is expressed in this way in the additional leg bud. Scale bar = 1 mm.

*Photograph reproduced with permission from Cohn, M.J., et al.: **Fibroblast growth factors induce additional limb development from the flank of chick embryos**. Cell 1995, 80: 739-746. © 1995 Cell Press.*

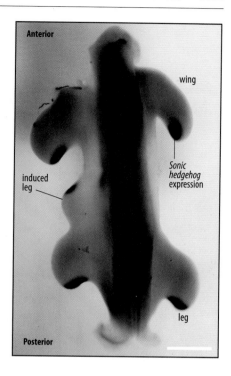

and is crucial for determining pattern along the antero-posterior limb axis. The apical ectodermal ridge arises at a boundary between dorsal and ventral compartments of cell-lineage restriction (see Section 2.24 for an explanation of compartments), which are specified in the trunk ectoderm of early embryos before any buds are visible. The cells that will form the apical ectodermal ridge are initially scattered throughout the ectoderm and then migrate to the compartment boundary. The apical ectodermal ridge begins to express *Fgf8* in response to FGF-10 signaling by the mesenchyme and a positive-feedback loop is established in which FGF-8 from the apical ectodermal ridge maintains *Fgf10* expression in the mesenchyme. Cells of the polarizing region express the signaling protein Sonic hedgehog (Shh), and *in situ* hybridization for *Shh* mRNA has been used to reveal the polarizing regions in the normal wing and leg buds, and the ectopic leg bud in the embryo shown in Fig. 11.7. Once a limb bud has formed equipped with its own signaling regions, it can develop autonomously—that is, on its own without reference to neighboring parts of the embryo.

11.3 The apical ectodermal ridge is required for limb outgrowth and the formation of structures along the proximo-distal axis of the limb

We shall next look at how the limb buds start to grow and develop. One crucial signaling region in the limb bud is the apical ectodermal ridge, which consists of closely packed columnar epithelial cells, directly connected to each other by gap junctions, through which ions and small molecules can flow directly between cells. The apical ectodermal ridge is essential for outgrowth of the limb bud and the progressive formation of skeletal elements along the proximodistal axis of the limb. When the apical ectodermal ridge is surgically removed from a chick wing bud, outgrowth is significantly reduced, and the wing that develops is anatomically truncated, with distal parts missing. The proximo-distal level at which the wing is truncated depends on the time at which the ridge is removed (Fig. 11.8). The earlier the ridge is removed, the greater the effect; removal at a late stage results only in loss of the ends of the digits, whereas when the ridge is removed at an early stage most of the wing is missing. Following apical-ridge removal, proliferation of cells at the tip of the wing bud is greatly reduced and there is also cell death in this region. The importance of signaling by the apical ectodermal ridge in promoting outgrowth has also been shown by grafting an isolated ridge to the dorsal surface of an early chick wing bud. The result is an outgrowth from the dorsal surface, which develops cartilaginous elements and digits.

The key signal from the ridge is provided by FGFs. The gene for FGF-8 is expressed throughout the ridge and genes for FGF-4 and two other FGFs are expressed in the posterior region of the ridge. Experiments in chick wing buds show that FGF-8 (or FGF-4) can act as a functional substitute for an apical ridge. If the ridge is removed and beads soaked in FGF are placed at the tip of the wing bud in its place, more-or-less normal outgrowth of the wing bud continues. If sufficient FGF is provided to the outgrowing cells, a fairly normal wing develops (Fig. 11.9). Experiments in transgenic mice, in which *Fgf* genes are knocked-out specifically in the apical ectodermal ridge, also show that FGF signaling is required for limb-bud outgrowth and normal

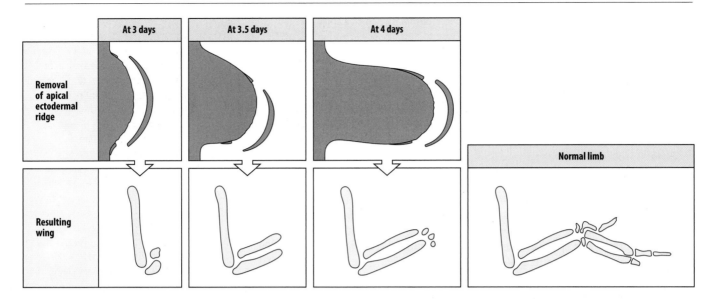

Fig. 11.8 The apical ectodermal ridge is required for proximo-distal development. Limbs develop in a proximo-distal sequence. Removal of the apical ridge from a developing wing bud leads to truncation of the wing; the later the apical ridge is removed, the more complete the resulting wing.

development. Of the four different *Fgf* genes that are expressed in the apical ectodermal ridge, *Fgf8* alone is sufficient for normal limb development. Complete limb truncations are produced when *Fgf8* and two of the other *Fgf* genes are simultaneously knocked out. The simplest explanation is that the individual *Fgf* genes expressed in the apical ridge are to a large extent functionally equivalent, and that having several FGFs involved ensures the robustness of apical ridge signaling.

11.4 Outgrowth of the limb bud involves oriented cell behavior

We have just seen how the apical ridge controls outgrowth of the limb, but how does the bud form in the first place and then acquire its elongated shape? In the first 24–30 hours of chick wing bud development, when a well-defined apical ridge is present, its length along the proximo-distal axis increases about threefold, while the dimensions of its antero-posterior and dorso-ventral axes are almost unchanged.

Perhaps surprisingly, the establishment of the early wing bud in chick embryos does not reflect a local increase in cell proliferation in the limb-bud region, but rather a decrease from a previously high rate of cell proliferation along the rest of the flank of the embryo. Cells are also recruited to the bud through an epithelial-to-mesenchymal transition involving cells from the lateral-plate-derived coelomic epithelium specifically in the limb-forming regions.

Cell proliferation occurs throughout the bud as it elongates and, although most studies have shown that cell proliferation is higher distally and near the ectoderm, computer modeling of mouse limb buds reveals that these local differences are not sufficient to explain bud outgrowth. Instead, cell polarization and oriented cell movements, and cell division in the direction of the apical ridge make an important contribution to formation of the bud and its elongation. Live imaging of mouse embryos has revealed dramatic changes in cell orientation as limb buds form. Just before limb-bud formation, the cells in the body wall are elongated parallel to the main head-to-tail body axis, with the plane of cell division at right angles. Individual cells point towards the tail of the embryo, as indicated by the position of the Golgi apparatus, which is always in front of the nucleus in moving cells, with movement oriented in this direction (see Fig. 11.10, first panel). But as the bud begins to form, the cells lateral to the forming bud become reoriented perpendicular to the body axis and become aligned with the proximo-distal axis of the emerging limb bud. Cell division also becomes reoriented in this direction, as does cell movement. At the same time, cells move from positions in the body wall anterior to the limb bud into the distal region of the

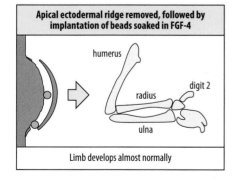

Fig. 11.9 The growth factor FGF-4 can substitute for the apical ectodermal ridge. After the ridge is removed, placing heparin beads soaked in FGF-4 at the tip into the chick wing bud results in almost normal development.

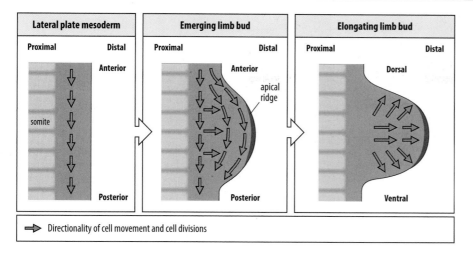

Lateral plate mesoderm	Emerging limb bud	Elongating limb bud

➡ Directionality of cell movement and cell divisions

Fig. 11.10 Cell orientation during formation and outgrowth of the limb bud. First panel: before limb-bud formation, the mesenchyme cells in the lateral plate mesoderm are oriented parallel to the antero-posterior axis of the body, point posteriorly (as indicated by the direction of the arrow) and move in this direction. Cell divisions are also oriented in the same direction. Second panel: the early limb bud forms as a result of anterior cells of the lateral plate mesoderm moving posteriorly, and lateral cells becoming oriented perpendicular to the antero-posterior main body axis and moving in the direction of budding. Third panel: in the elongating bud, cells in the central region and at the tip are oriented parallel to the proximo-distal axis and move distally towards the apical ectodermal ridge with cell divisions oriented in the same direction. Cells in dorsal and ventral regions are oriented towards the ectoderm. The two panels on the left are based on observations in developing mouse embryos, the right panel on the developing chick limb.

*First and second panels: based on Wyngaarden, L.A. et al.: **Oriented cell motility and division underlie early limb bud morphogenesis.** Development 2010, **137**: 2551-2558; third panel: Gros, J., et al.: **Wnt5a/Jnk and FGF/Mapk pathways regulate the cellular events shaping the vertebrate limb.** Curr. Biol. 2010, **20**: 1993-2002.*

bud and across to the posterior side (see Fig. 11.10, middle panel), with the speed of movement increasing as they enter the distal region. Successive cell divisions and the inward movement of cells from the flank push the limb bud outwards.

In buds that have begun to elongate, the orientation of the cells becomes more complex. In both mouse and chick limb buds, the central cells are still oriented along the proximo-distal axis, but cells at the periphery of the bud become oriented almost perpendicularly to the ectoderm. Cells in the central and distal regions generally show oriented cell divisions and movement towards the tip of the limb bud, with distal cells moving faster than proximal cells, and so continuing to push the limb bud outwards (see Fig. 11.10, third panel).

The oriented cell behavior in early limb buds is regulated by the combined activity of FGF and the Wnt planar cell polarity (PCP) signaling pathway (see the Wnt/PCP signaling figure in Box 9C). Wnt-5a, one of the Wnt ligands that signals via the Wnt/PCP pathway, is expressed in a graded fashion in limb-bud mesenchyme, with the highest level of transcripts at the tip. Mutant mice lacking *Wnt5a* have short limbs and the distal parts of the digits are missing, and while the proximo-distal axis of the mutant limb buds is reduced in length, the width of the bud along the dorso-ventral axis is increased. Live imaging of limb buds from this mutant revealed that the mesenchymal cells are not elongated, and that cell movement and cell division are only weakly oriented. This suggests that in the normal limb bud, the perpendicularly oriented cells adjacent to the ectoderm in the dorso-ventral regions undergo convergent extension under the influence of Wnt-5a (see Box 9C), thus contributing to bud elongation and to maintaining the correct dorso-ventral thickness of the bud.

Other experiments show that Wnt-5a may act as a chemoattractant in early limb-bud formation and polarize the migration of mesenchymal cells towards the forming bud. *Wnt5a* is expressed at the tips of all regions of the embryo that undergo outgrowth, not only the limb buds, but also the buds that give rise to the jaws, the bud that gives rise to the genital tubercle, and the posterior region of the main body axis. So the orientation and polarization of cells by Wnt-5a is likely to be a general mechanism for elongation.

FGF-4, which is produced by the apical ridge, has also been shown to act as a chemoattractant for limb mesenchymal cells. But another FGF produced by the ridge, FGF-8, controls the velocity of cell movement, explaining why distal mesenchymal cells move faster than proximal mesenchymal cells in both mouse and chick limb buds. Interestingly, FGF signaling has been shown to control a gradient of cell motility in the extending pre-somitic mesoderm during elongation of the antero-posterior axis in chick embryos. In this case, however, the entire tissue is being deformed and the cells are in fact moving in random directions, with the direction of elongation being imposed by constraining neighboring denser tissues.

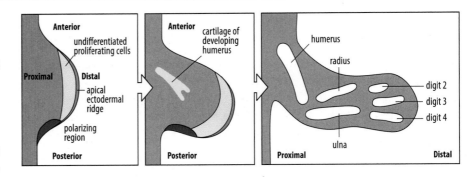

Fig. 11.11 Cells acquire positional values along the antero-posterior and proximo-distal axes. In the early chick wing bud there are two main signaling regions—the polarizing region at the posterior margin (red), and the apical ectodermal ridge (blue). The polarizing region specifies position along the antero-posterior axis. How position is specified along the proximo-distal axis is still under debate (see Section 11.6). Cartilage begins to differentiate first in the most proximal region, and the cartilaginous elements are laid down in a proximo-distal sequence (carpals not shown).

11.5 Patterning of the limb bud involves positional information

We have seen how the apical ridge controls the outgrowth of the limb, but how are the skeletal and other elements of the limb laid down in the correct positions? Some aspects of vertebrate limb development fit very well with a model of pattern formation based on positional information. The developing chick limb bud behaves as if its future development is determined by the positions of its cells with respect to the main limb axes while the cells are in the undifferentiated region at the tip of the limb bud (Fig. 11.11).

Central to the idea of positional information is the distinction between positional specification and interpretation. Cells acquire positional information first and then interpret these positional values according to their developmental history. It is the difference in developmental history that makes wings and legs in chick embryos different, for positional information is signaled in wing and leg buds in the same way. An attractive hypothesis for limb patterning is that a single three-dimensional positional field controls the development of the cells that give rise to all the mesodermal limb elements—cartilage, muscle, and tendons. The specification of the three axes—proximo-distal, antero-posterior, and dorso-ventral—is, as we shall see, linked by molecular signals emanating from different parts of the bud.

11.6 How position along the proximo-distal axis of the limb bud is specified is still a matter of debate

We will start by discussing patterning along the limb's all-important proximo-distal axis, which runs from the shoulder to the tips of the digits. The skeletal elements are laid down in a proximal to distal sequence, with the proximal elements (humerus or femur) beginning to differentiate first. Wnt signals from the ectoderm together with FGFs produced by the apical ridge maintain the region of undifferentiated mesenchymal cells at the tip of the limb bud.

The mechanisms that pattern the proximo-distal axis are, however, still a matter of some debate, with two main classes of models having been proposed. In one, specification of position along the proximo-distal axis is based on timing. This long-standing model derives from work on chick embryo limbs and proposes that proximo-distal patterning is specified by the length of time cells spend in the zone of undifferentiated cells at the tip of the limb. This region has therefore been called the **progress zone**. The second class of model derives from more recent experiments in chick and mouse and proposes that diffusible signals specify proximo-distal values in the early limb bud, with the elements then differentiating progressively in a proximo-distal sequence as the limb grows out. There is evidence in support of both classes of model, but neither is entirely consistent with all the known evidence, so it is clear that there is still more to be learned about how the limb is patterned in the proximo-distal direction.

As the limb bud grows, cells are continually leaving the zone of undifferentiated cells at the distal tip. The earliest cells leaving the zone develop into the humerus/

femur and those leaving last form the tips of the digits. The timing model proposes that, if cells could measure the time they spend in this zone, by, for example, counting cell divisions, this would give them their positional value along the proximo-distal axis (see Fig. 11.12, upper panel). The experimental observation that removal of the apical ridge results in a distally truncated limb (see Fig. 11.8) would be explained by the fact that the progress zone is no longer maintained, and so the more distal structures will not form. A timing mechanism has been suggested to provide positional information to the paraxial mesoderm that will form the somites (see Section 5.8).

Another line of evidence put forward in support of a timing mechanism is that killing cells in the progress zone of a chick wing bud at an early stage (by irradiation with X-rays, for example) results in the absence of proximal structures, whereas distal ones are present and can be almost normal. A similar condition seen in human limbs is called **phocomelia**. Because many cells in the irradiated progress zone do not divide, fewer cells than usual will leave the zone during each unit of time. As the bud grows out, the progress zone becomes repopulated by the proliferation of surviving cells, which would spend a longer time than usual in the progress zone and therefore acquire a more distal positional value and form normal distal structures, such as digits. But other work suggests that proximal structures are most affected simply because

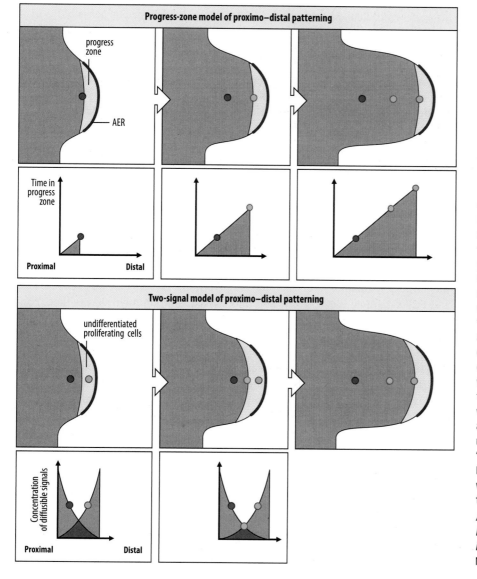

Fig. 11.12 A cell's proximo-distal positional value might depend on the time it spends in the 'progress zone' or might be specified by opposing signals in the early limb bud. The upper two rows show the 'timing' model. Cells are continually leaving the progress zone. If the cells could measure how long they spend in the zone, this could specify their position along the proximo-distal axis. Cells that leave the zone early (red) form proximal structures whereas cells that leave it last (blue) form the tips of the digits. The lower panels show the 'two-signal' model. Cells in the early limb bud acquire positional values from the presence of two opposing signals: one due to a source of retinoic acid in the most proximal region, which specifies proximal (red); and one due to FGF from the apical ectodermal ridge, which specifies distal (blue). One idea is that as the limb bud grows, cells that are out of range of both signaling sources acquire an 'intermediate' positional value (green). As the limb grows, the different domains expand, with the proximal domain expanding first and then the successively more distal domains.

*Adapted from Zeller R., et al.: **Vertebrate limb bud development: moving towards integrative analysis of organogenesis**. Nat. Rev. Genet. 2009, **10**: 845–858.*

although the few cells that are left immediately after irradiation will be molecularly specified as proximal, there will not be enough of them to form a cartilage element. Similar arguments to either of these could be made to account for the absence of proximal limb structures in babies born in the late 1950s and early 1960s to mothers who took the drug thalidomide to ease morning sickness (Box 11A).

We have already seen that FGFs produced by the apical ridge are key signals in limb development, but what specific part do they play in proximo-distal patterning? According to the timing model, the function of FGF signaling is to maintain the progress zone, and cells might measure time by the length of time they are exposed to FGF signals. However, in the other class of model for proximo-distal patterning, FGFs diffusing from the apical ectodermal ridge into the underlying mesenchyme specify distal positional values. One specific version of this model, known as the two-signal model (see Fig. 11.12, lower panel) proposes that FGF signaling distally is opposed by retinoic acid signaling proximally, and that proximal positional values are specified by retinoic acid diffusing from the adjacent body wall into the limb bud.

Such an interplay between FGF and retinoic acid signaling would not be unique to the developing limb. Opposing gradients of retinoic acid and FGF have been proposed to operate during growth and antero-posterior patterning of the main body axis, and to control the formation of the somites (see Sections 5.8 and 5.9) and differentiation of the spinal cord. Retinoic acid targets in the developing limb include genes of the *Meis* family, which encode homeodomain transcription factors. *Meis* genes are expressed in the proximal region of the limb bud and misexpression of *Meis* genes in the chick wing leads to distal abnormalities or truncations. Recent work on chick wing buds has emphasized the role of proximal signals in specifying the proximal part of the wing, and supports the idea that retinoic acid is the signal involved, although this is still under debate. The way in which the more distal parts of the limb are specified is not so clear.

Genetic data from the mouse have been interpreted in terms of the two-signal model. When FGF signaling in the apical ridge of mouse limb buds is incrementally reduced by genetically deleting *Fgf* genes in the apical ridge one by one, a series of progressively reduced limbs develop. Among the limb phenotypes observed when FGF signaling is severely reduced is the unusual phenotype in which both proximal and distal skeletal elements—that is, humerus and digits respectively—are present, but the intermediate elements—radius and ulna—are absent. This limb phenotype is difficult to reconcile with the timing model, in which structures along the proximo-distal axis are specified sequentially. Instead, the proposal is that proximal and distal components of the limb pattern, that is, the humerus (proximal) and the fingers (distal), are first specified by retinoic acid and FGF signaling, respectively, and that subsequent outgrowth of the limb, controlled by FGFs produced by the apical ridge, leads to the emergence of a region of cells outside the range of both distal and proximal signals. As a consequence, this region adopts an 'intermediate' fate—and forms the radius and ulna (see Fig. 11.12). It is not clear whether this suggested mechanism for generating intermediate fates could apply to the chick limb. When the tip of an old chick wing bud, which will form distal limb components, is grafted on to a stump of a young wing bud, which will form proximal limb components, subsequent outgrowth and development generates a wing that lacks radius and ulna.

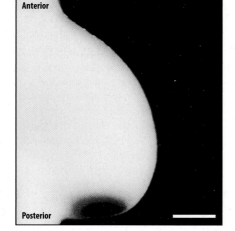

Fig. 11.13 The polarizing region or zone of polarizing activity (ZPA) of a chick limb bud is located at the posterior margin of the bud. The polarizing region expresses *Sonic hedgehog (Shh)* RNA (blue stain). Scale bar = 0.1 mm.

Photograph courtesy of C. Tabin.

11.7 The polarizing region specifies position along the limb's antero-posterior axis

The specification of positional information along the antero-posterior axis of the limb bud is most clearly seen with respect to the digits, as it gives them their identities. In this and the next two sections we shall look at how the pattern of three digits is specified in the chick wing, the signaling molecules that operate in antero-posterior patterning,

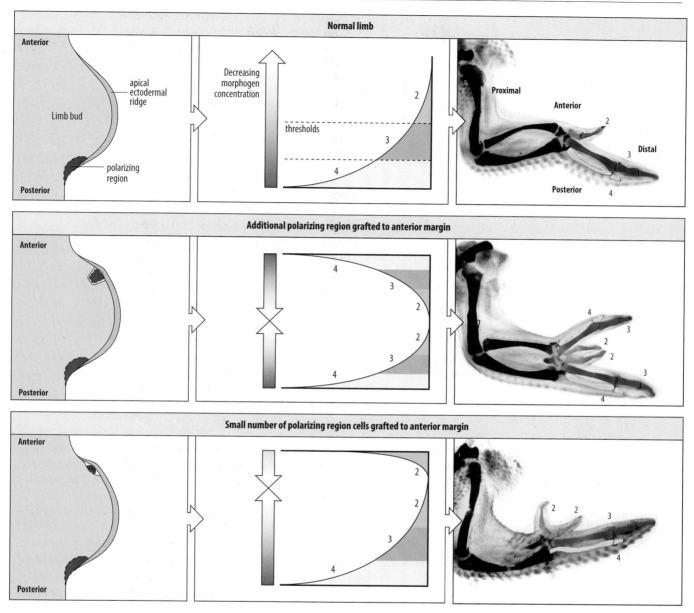

Fig. 11.14 The polarizing region can specify positional information along the antero-posterior axis. If the polarizing region is the source of a graded morphogen, the different digits could be specified at different threshold concentrations of signal, as shown for a normal chick wing bud in the top panels, with digit 4 developing where morphogen concentration is high and digit 2 where it is low. Threshold concentrations of morphogen that give rise to the different digits are indicated in the center panel of each row and the digits in the right-hand panels are color-coded to match. Grafting an additional polarizing region to the anterior margin of a wing bud (center row of panels) would result in a mirror-image gradient of signal, and thus the observed mirror-image duplication of digits. A small number of polarizing region cells grafted to the anterior margin of the wing bud produces only a weak signal and so only an extra digit 2 is produced (bottom row of panels).

and, finally, how each digit is given its unique identity. The organizing region for the antero-posterior axis is a mesodermal region at the posterior margin of the limb bud, known as the **polarizing region** or the **zone of polarizing activity** (ZPA) (Fig. 11.13).

The polarizing region of a vertebrate limb bud has organizing properties almost as striking as those of the Spemann organizer in amphibians. When the polarizing region from an early chick wing bud is grafted to the anterior margin of another early chick wing bud, a wing with a mirror-image pattern develops: instead of the normal pattern of digits—2 3 4—the pattern 4 3 2 2 3 4 develops (Fig. 11.14,

MEDICAL BOX 11A Teratogens and the consequences of damage to the developing embryo

A **teratogen** is an external agent, such as a chemical or an infection, that interferes with embryonic development, leading to damaging effects that are seen when the baby is born. Damage to an embryo can also be due to heritable gene mutations, and it is sometimes difficult to distinguish between these two causes.

One of the best-documented teratogens is the drug thalidomide. This was widely prescribed for morning sickness during early pregnancy in the late 1950s and early 1960s, first in Germany and then throughout Europe and other countries. It is estimated that 10,000 severely affected babies were born worldwide as the result of thalidomide (of which around 40% did not survive beyond their first birthday), with the most obvious effects being seen in the limbs, although other organs may be affected including the heart, eyes and ears. Typically, the arms are either completely absent or have what are known as transverse deficiencies, in which a region along the proximo-distal axis fails to form. One such deficiency seen in those affected by thalidomide is **phocomelia**, in which proximal structures are affected, in this case the lower arm, but distal structures such as the fingers still form (Figure 1). The period during which a teratogen can interfere with development of specific organs is known as the **critical period**. The critical period for thalidomide damage is during days 20–36 of embryonic development, when structures such as the upper limb and heart are forming in the embryo.

So how did such a potent teratogen get approved for clinical use? All drugs have to go through safety tests before they are approved, but the company that developed thalidomide has admitted that it did not test for effects in pregnant animals. Safety tests have become much more stringent since the 1950s, in part at least because of the catastrophic effects of thalidomide. But ironically, mice and rats, which are commonly used to screen drugs for teratogenic effects, are not very sensitive to thalidomide.

Once it was realized that thalidomide was a teratogen, it was withdrawn from sale, and research to classify thalidomide injuries and to find out how they are caused began immediately.

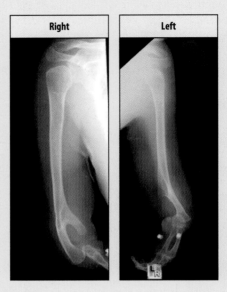

Figure 1

Published with permission from LearningRadiology.com

But even today, there is no consensus on how thalidomide interferes with limb development. One long-standing theory is that the drug targets the embryonic nervous system and thereby blocks the outgrowth of embryonic limb buds. But although thalidomide is known to affect nerves in adults, there is no experimental evidence that blocking nerves affects the outgrowth of embryonic limbs. Another theory is based on the ability of thalidomide to cause cell death. As discussed in Section 11.6, phocomelia can occur when early wing buds in chick embryos are X-irradiated. The irradiation causes extensive cell death, which is thought to be the underlying cause of the phocomelia; a similar explanation has been advanced for thalidomide-caused phocomelia in humans.

compare the first and second rows). The pattern of muscles and tendons in the wing shows similar mirror-image changes. A polarizing-region graft can also specify another ulna anteriorly, showing that the polarizing region influences antero-posterior positional values throughout the region of the wing distal to the elbow.

The additional digits come from the host wing bud and not from the graft, showing that the grafted polarizing region has altered the developmental fate of the host cells in the anterior region of the wing bud. The wing bud widens in response to the polarizing graft, which enables the additional digits to be accommodated. Widening of the wing bud is associated with an increase in cell proliferation and maintenance of the apical ectodermal ridge in the anterior region of the bud.

One way that the polarizing region could specify position along the antero-posterior axis is by producing a morphogen that forms a posterior to anterior gradient (Box 11B). The local concentration of morphogen provides a positional value with

Thalidomide is also anti-angiogenic—that is, it blocks the growth of blood vessels—a property that has led to the reintroduction of thalidomide and its analogs as anticancer drugs, in particular for multiple myeloma, with appropriate precautions. An analog of thalidomide with anti-angiogenic properties has been shown to produce abnormalities in chick wing buds similar to those seen when the wing bud is irradiated with X-rays. The chemical specifically interferes with the development of the delicate blood capillaries in the early wing bud, and this interruption in the blood supply leads to cell death in the mesenchymal core of the bud. Another recent discovery is that thalidomide binds to the protein cereblon, part of a protein complex that promotes the ubiquitination of proteins, making them susceptible to degradation. Bound thalidomide reduces the activity of the complex, but it is not clear how this could cause limb defects.

There have been reports that babies born to parents who were affected by thalidomide also have damaged limbs, which suggested that thalidomide is not only a teratogen but might also be a mutagen, causing a permanent and heritable change in a gene. These reports are now thought to be due to misdiagnosis of patients with unsuspected genetic conditions. One of the parents originally diagnosed as affected by thalidomide was instead found to have a mutation in the *SALL4* gene. *SALL4* is expressed in developing limbs, and mutations in *SALL4* have been found in patients with Okihiro syndrome, which is characterized by limb defects similar to those caused by thalidomide. Another genetic disease that leads to limb and heart defects that could be confused with the effects of thalidomide is Holt-Oram syndrome, in which patients have mutations in the gene *TBX5* (see Section 11.2).

Limb deficiency defects occur in around 5 per 10,000 live births (precise figures vary) but only a small percentage can be confidently ascribed to teratogens, whereas more than 50% are estimated to be due to genetic causes. In over 25% of cases, the cause is unknown. Many of these cases are likely to be multifactorial—the result of a combination of genetic and environmental factors. Just as the genotype of a patient determines her reaction to drugs used to treat disease—the basis of personalized medicine—her genotype similarly could determine her sensitivity to environmental agents that can act as teratogens. This could explain why some pregnant women exposed to environmental factors have affected babies whereas others exposed to the same factors do not.

It is still important to know how thalidomide damages the developing embryo as it has been also reintroduced to treat leprosy. Although thalidomide is not now recommended for use by the World Health Organization (WHO) in patients with leprosy because safer drugs are available, a few babies affected by thalidomide continue to be born every year.

Thalidomide is probably the best-known example of a chemical teratogen. The list of proven chemical teratogens is fairly short, and includes anticonvulsants (trimethadione and valproic acid), folic-acid antagonists, retinoids (derivatives of vitamin A), ethanol and dioxins. One important issue in regard to all chemical teratogens is to understand how they are metabolized, as the products can be more, or less, harmful. Either a deficiency or an excess of vitamin A is teratogenic in animals and, despite the thalidomide catastrophe, affected babies were born to women in the United States who were being treated with retinoid-based drugs for severe skin conditions (see Box 5C). Ethanol can also damage developing embryos when consumed in large amounts by the mother, and this may be due to its being metabolized by the same pathway that produces retinoic acid.

Infectious agents can be teratogenic. The best-known example in humans is the rubella virus, commonly known as German measles. If a woman catches German measles in the first 16 weeks of pregnancy, the critical period for this agent, damage to the embryo can affect ears, eyes, heart and the brain, as these organs are developing at this time. How the virus causes this spectrum of effects, known as congenital rubella syndrome, is not understood. Vaccination against rubella has largely eliminated the syndrome in developed countries, but elsewhere it is still a problem, with the WHO estimating that about 110,000 babies worldwide are born every year with the syndrome. Finally, exposure to physical agents such as high doses of ionizing radiation can also cause congenital abnormalities.

respect to the polarizing region located at the posterior margin of the limb. Cells could then interpret their positional values by developing specific structures at particular threshold concentrations of morphogen. Digit 4, for example, would develop at a high concentration, digit 3 at a lower one, and digit 2 at an even lower one (see Fig. 11.14, top row). According to this model, a graft of an additional polarizing region to the anterior margin would set up a mirror-image gradient of morphogen, which would result in the 4 3 2 2 3 4 pattern of digits that is observed (see Fig. 11.14, center row).

If the action of the polarizing region in specifying the identity of a digit is due to the level of the signal, then when the signal is weakened, the pattern of digits should be altered in a predictable manner. Grafting small numbers of polarizing region cells

BOX 11B Positional information and morphogen gradients

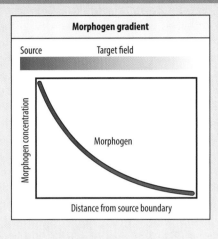

Morphogen gradient

Source   Target field

Morphogen concentration

Morphogen

Distance from source boundary

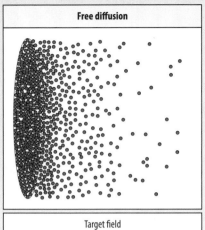

Free diffusion

Target field

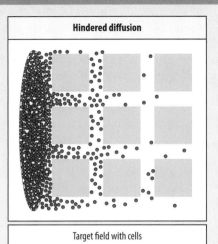

Hindered diffusion

Target field with cells

Pattern formation in development can be specified by positional information, which in turn can be defined by a gradient of some molecular property. The basic idea, related to the French Flag problem (see Section 1.15), is that a specialized group of cells at a boundary of an area to be patterned, for example a field of cells, secretes a molecule whose concentration then decreases with distance from this source, thus forming a gradient (see Figure 1, first panel). A cell at any point along the gradient then 'reads' the local concentration and interprets it to respond in a manner appropriate to its position. Molecules that can specify cell fate in this way are known as **morphogens** and are said to form 'gradients of positional information'. These gradients have some important properties: in particular, they provide a measure for the size of a tissue in terms of the slope and the decay length of the gradient (a measure of how far the gradient extends), and also harbor the possibility of scaling—a most important property of biological systems according to which the proportions of the pattern elements within a tissue are independent of its size. Morphogens are used for various tasks during development: regionalization of the antero-posterior and dorso-ventral axes and patterning of segments and imaginal discs in insects; mesoderm patterning in vertebrates; vertebrate limb patterning; and patterning along the dorso-ventral axis of the vertebrate neural tube, among others.

Two main modes of morphogen transport have been envisaged—extracellular diffusion of the morphogen or its movement along cellular extensions. Morphogens could travel by simple diffusion through extracellular spaces and be degraded and/or permanently trapped by the responding cells so that a gradient forms ('diffusion' in the figure). It is unlikely, however, that simple diffusion operates in tissues because of the obstacles presented by the cells themselves, and the binding of morphogen molecules to receptors and extracellular matrix components. Interactions between a diffusing morphogen and other extracellular proteins ('hindered diffusion' in the figure) can even generate a gradient in activity from an initially fairly uniform distribution of morphogen. In 'positive diffusion' or 'shuttling', a regulator binds to the morphogen and actively transports it to form a gradient (see, for example, the formation of the ventral-dorsal BMP gradient in the *Xenopus* embryo, Fig. 4.25).

Despite years of work on the topic, there is still much discussion about whether mechanisms based on secreted diffusible morphogens will be sufficiently reliable for pattern formation. One of the alternative modes of morphogen transport is from cell to cell via long, specialized filopodia in vertebrates, or by similar actin-based extensions called cytonemes in insects (see Figure 1).

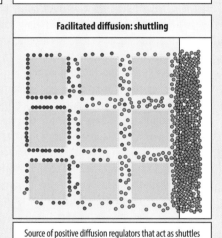

Facilitated diffusion: shuttling

Source of positive diffusion regulators that act as shuttles

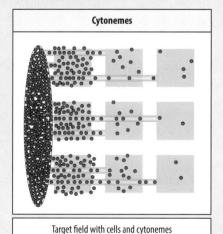

Cytonemes

Target field with cells and cytonemes

Figure 1

*Adapted from Müller, P., et al.: **Morphogen transport.** Development 2013, **140**: 1621-1638.*

The Decapentaplegic protein in *Drosophila* is a good example of a morphogen. It acts in dorso-ventral patterning in the early enbryo (see Section 2.17), and also later in the antero-posterior patterning of the wing (see Section 11.20). In dorso-ventral patterning, the sharp gradient of Dpp in the early blastoderm is formed by the interaction of the diffusing morphogen with other proteins, both secreted extracellular proteins and cell-surface receptors. In the wing, the formation of an extracellular Dpp gradient has been visualized directly using a fusion of GFP to the Dpp protein and expressing it in its normal pattern, but whether or not this gradient forms by simple diffusion is unclear; there is evidence, for example, that binding of Dpp to extracellular molecules alters its distribution.

Cytonemes have also been seen extending from responding cells in the wing disc to the Dpp source. In a later round of Dpp signaling in wing development, Dpp has been shown to move along cytonemes towards recipient cells in the wing disc and there are indications that the range of Dpp signaling is related to the number and length of cytonemes. Similar findings have been reported with respect to the transport of Hh protein in the wing disc. Filopodia have also been seen extending from both polarizing-region cells and responding cells in the chick limb, and have been shown to transport Shh. It is not clear how this transport of Shh contributes to gradient formation in the limb.

It is most likely that a cell measures a particular morphogen concentration according to how many receptors for the morphogen are occupied, and thus by the strength of the signal that is transmitted from the receptors. In the wing, Dpp binds to the receptor protein Thick veins, leading to the phosphorylation of a Smad protein called Mad, which can be followed with a specific anti-P antibody (P-Mad). Superimposition of the distribution of P-Mad and the Dpp gradient shows a linear correlation between the two.

to a wing bud results in the development of only an additional digit 2 (see Fig. 11.14, bottom row). The same result can be obtained by leaving the polarizing-region graft in place for a short time and then removing it. When the graft is left for 15 hours, an additional digit 2 develops, whereas the graft has to be in place for up to 24 hours for a digit 3 to develop.

The limb buds of a number of other vertebrates, including mouse, pig, ferret, and turtles, and even human limb buds, have been shown to have a polarizing region. When the posterior margin of a limb bud from an embryo of these species is cut out and grafted to the anterior margin of a chick wing bud, additional digits are produced. The additional digits induced are chick wing digits, showing that, although the polarizing-region signal is conserved among vertebrates, its interpretation depends on the responding cells. This is the same principle we saw with respect to inter-species grafts of the node between different vertebrate embryos (see Section 5.6).

11.8 Sonic hedgehog is the polarizing region morphogen

The secreted protein Sonic hedgehog (Shh) is the morphogen that patterns the antero-posterior axis of vertebrate limbs. The *Sonic hedgehog* gene (*Shh*) is expressed in the polarizing region of the limb buds of chick (see Fig. 11.13) and mouse embryos and in the limb buds of other vertebrate embryos that have been studied, including fish fin buds. Shh protein is involved in numerous patterning processes elsewhere in vertebrate development; for example, in somite patterning (see Section 5.13), in the establishment of internal left–right asymmetry in chick embryos (see Section 5.17), and in the patterning of the neural tube (discussed in Chapter 12). The related *Drosophila* Hedgehog protein is a key signaling molecule in the patterning of the segments in the *Drosophila* embryo (see Section 2.26), in addition to *Drosophila* wings and legs, as we shall see later in this chapter.

Sonic hedgehog has all the properties expected of the polarizing-region morphogen. The Shh protein is present in a graded distribution across the posterior region of the limb bud (the region of the limb bud where the digits will form) at a distance from the cells in the polarizing region that produce it. Cultured chick fibroblasts transfected with a retrovirus containing the *Shh* gene acquire the properties of a polarizing region; when grafted to the anterior margin of a chick wing bud they cause the development of a mirror-image wing. A bead soaked in Shh protein will give the same result. The bead must be left in place for 16 to 24 hours for extra digits to be specified, and the pattern of digits depends on Shh concentration. Shh also has a direct effect on mesenchymal cell proliferation in chick wing buds by controlling expression of genes encoding cell-cycle regulators, thus revealing a mechanism that contributes to the widening of the wing bud following a polarizing graft.

The signaling molecule retinoic acid (see Box 5C) was the first defined chemical that was found to mimic the signaling of the polarizing region of the chick wing, and it can induce mirror-image duplications of the digits when applied to the anterior margin of a chick wing bud (see online practical). But it was later shown that retinoic acid induces *Shh* expression in anterior tissue, and this explains the digit duplications. We have already discussed the possibility that retinoic acid signaling patterns the proximal part of the limb, but how this would be related to its ability to induce *Shh* expression is not clear.

Evidence that Shh has a crucial role in specifying the pattern of digits in mammalian limbs comes from studies of mutations in mice that cause **preaxial polydactyly**. 'Polydactyly' denotes limbs with extra digits and 'preaxial' that the extra digits form anteriorly. These mutations affect the polarizing region and *Shh* transcripts are present at both anterior and posterior margins of the mutant mouse limb buds. Similar mutations have been found in patients with preaxial polydactyly (Box 11C). In contrast, in *Shh*-deficient mouse embryos, the limbs are truncated and no digits form in the forelimb and only a single digit in the hindlimb.

A concentration gradient of Shh protein provides a good explanation for the way in which the positional values for the three digits of the chick wing are specified, but the specification of the positional values for the five digits of the mouse hand is more complicated, as the two posterior digits have been shown to be derived from the polarizing region itself. It has been proposed, therefore, that although Shh concentration may specify the three anterior mouse digits, the two posterior mouse digits are specified by a timing mechanism based on the duration of Shh signaling. It is still not clear, however, exactly how the Shh gradient across the limb bud is established.

The next question is how limb bud cells measure Shh concentration or duration of Shh signaling. The intracellular signaling pathway that is activated in response to Shh involves the Gli transcription factors Gli1, Gli2, and Gli3, which are analogous to Ci in the Hedgehog signaling pathway in *Drosophila* (see Box 2F). The vertebrate Shh signaling pathway is shown in Box 11D. In the presence of Shh, the Gli proteins function as transcriptional activators, while in its absence, Gli2 and Gli3 are processed to short forms and function as transcriptional repressors. Therefore the ratio of Gli3 activator to Gli3 repressor in a cell reflects the level of Shh signaling. In *Shh*-deficient mouse embryos, high levels of Gli3 repressor are present in cells throughout the bud but when *Gli3* is knocked out in mice, many digits all of the same unpatterned type develop. The function of Shh signaling in the limb bud is, therefore, over time to suppress repressor Gli3 function in a graded fashion in cells in the posterior region of the limb bud, which will form the digits. The differences in the ratios of Gli3 activator to Gli3 repressor in cells across the anterior-posterior axis in this region will determine the digit pattern.

Some mouse mutants with reduced levels of repressor Gli and polydactyly do not have defects in the genes encoding Shh or components of the signaling pathway, but instead have mutations in genes essential for the formation of a cell structure called the **primary cilium**. This immotile microtubule-based structure is essential for Shh signaling and the processing of Gli3 in vertebrate cells, as described in Box 11D.

11.9 How digit identity is encoded is not yet known

A major question still to be answered is how antero-posterior positional information is encoded and then interpreted so that the appropriate type of digit develops. Several genes encoding transcription factors, including Sal1, Tbx2 and Tbx3, together with

MEDICAL BOX 11C Too many fingers: mutations that affect antero-posterior patterning can cause polydactyly

Mutations have been discovered in mice that affect the function of the polarizing region and cause **preaxial polydactyly**. 'Polydactyly' denotes limbs with extra digits and 'preaxial' the fact that the extra digits form anteriorly. These mouse mutants, such as the Sasquatch mutant, provide more evidence for the crucial role of Shh in specifying the pattern of digits, as it has been shown that formation of the extra anterior digit in Sasquatch is due to additional anterior expression of the *Shh* gene in the limb buds. The Sasquatch mutation maps to a long-range *cis*-regulatory region in the mouse genome, which has become known as the ZRS (zone of polarizing activity regulatory sequence). The ZRS governs limb-specific *Shh* expression even though it is located 1 Mb away from the *Shh* gene. This example of long-range control is similar to the developmental control of β-globin expression in red blood cells by the locus control region (discussed in Section 8.9).

The discovery of inherited human conditions based on mutations and/or duplications of the ZRS provides evidence of the importance of Shh signaling in human limb development. Human patients with pre-axial polydactyly have mutations in the ZRS upstream of the human *SHH* gene, and several other polydactylous conditions in human patients are associated with microduplications within this regulatory sequence. Mutations in the ZRS have also been found in polydactylous cats, with extra anterior digits, including the celebrated tribe of cats that inhabit the novelist Ernest Hemingway's old home in Key West, Florida (Figure 1, and look out for one of these cats in the film *The Man with the Golden Gun*). The tribe is descended from a polydactylous cat given to Hemingway by a sea captain, and all the polydactylous individuals have the same single-nucleotide substitution in the ZRS. ZRS mutations have now been found in several other polydactylous animals, including the ancient Silkie breed of chickens, which has an extra toe.

In contrast to the extra digits that develop from limb buds with additional anterior *Shh* gene expression, limbs of mouse embryos in which the *Shh* gene has been functionally deleted lack distal structures, although proximal structures are still present. At best, one very small digit develops in the hindlimb. *Shh*-deficient mouse embryos also have other defects, including craniofacial defects such as holoprosencephaly (see Box 1F) and patterning defects in the neural tube. Complete deletion of the ZRS in mouse embryos also leads to loss of posterior *Shh* expression in the limb buds and results in limb truncations resembling those in *Shh*-deficient embryos. In embryos with the ZRS deletion, however, no other structures are affected because Shh function is deleted just in the limb bud.

Loss of distal limb structures is also seen in the very rare inherited human condition, **acheiropodia**, in which structures below the elbow or knee fail to develop but no other defects are present. Acheiropodia has been shown to be due to a chromosomal deletion near the ZRS, suggesting that there may be a *cis*-regulatory region in this position in the genome that is required for *Shh* expression in the polarizing region.

Figure 1

Photograph courtesy of Lettice, L. A., et al.: **Point mutations in a distant Shh cis-regulator generate a variable regulatory output responsible for preaxial polydactyly.** Hum. Mol. Genet. *2008, 17: 978–985.*

---

the 5' Hoxd genes *Hoxd13* and *Hoxd11*, are among the Gli3 targets in the developing limb and could contribute to encoding digit identity. These targets were found by combining two types of experimental data. In one experiment, a Gli3 repressor tagged with a Flag label (which can be detected by specific antibodies) was expressed in the developing limbs of transgenic mice, and the genomic sites occupied by the tagged repressor protein were identified by chromatin immunoprecipitation (see Section 3.12). The other data came from microarray analysis of genes differentially expressed in anterior and posterior regions of normal developing mouse limbs and in limbs from mouse mutants with defective Shh. From these data, the gene-regulatory network that controls digit patterning in the developing limb is beginning to be constructed (for other examples of gene-regulatory networks see Section 6.13).

Tbx2 and *Tbx3* are not only expressed in the early limb bud but are also expressed later in the digital-plate. Condensations of mesenchymal cells arise at specific positions in the digital-plate and will develop into the digits. The tissue between each condensation is known as the **interdigital region**. In the chick leg, *Tbx2* and *Tbx3* are expressed in an overlapping pattern in the interdigital regions associated with the condensations for the two posterior digits. Misexpression of either *Tbx2* or *Tbx3* appears to change one type of digit into the other. It seems likely that the combined activity of these transcription factors and other transcription factors in the gene regulatory network encode antero-posterior positional information. The acquisition of identity also depends on cell–cell interactions between the developing digit and the interdigital region. Another component of the gene regulatory network is *Bmp2*, which encodes the secreted signaling protein BMP-2. *Bmp2* is expressed in the posterior of the early limb bud and later with other BMPs in the interdigital regions. Each forming digit is characterized by a specific level of activity of Smad transcription factors, which provides the read-out of the levels of BMP signaling (see Box 4C). Removal of interdigital tissue or implanting beads soaked in the BMP antagonist Noggin (see Section 4.13) into the interdigital regions, anteriorizes the adjacent digit. Therefore it appears that BMP signaling in the forming digit activates genes involved in the late stages of morphogenesis that realize digit identity.

The transcription factor genes just discussed are associated with human **syndromes** which include limb defects. Mutations in *TBX3* are associated with Ulnar- mammary syndrome and *SALL1* and *SALL4* with Townes–Brockes syndrome and Okihiro/ Duane–radial ray syndrome, respectively, whereas mutations in *HOXD13* lead to digit fusions and some types of polydactyly. Okihiro/Duane-radial ray syndrome, together with Holt–Oram syndrome (*TBX5* defect) already discussed (see Section 11.2), could be confused with the effects of thalidomide (see Box 11A)).

11.10 The dorso-ventral axis of the limb is controlled by the ectoderm

The chick wing has a well-defined pattern along the dorso-ventral axis: large feathers are only present on the dorsal surface, and muscles and tendons have a complex dorso-ventral organization. Flexor muscles develop on the ventral side, extensors on the dorsal.

The development of pattern along the dorso-ventral axis in the mesoderm has been studied by recombining ectoderm taken from left limb buds with mesoderm from right limb buds in such a way that the dorso-ventral axis of the ectoderm is reversed with respect to the underlying mesoderm, but the antero-posterior axis is unchanged. The right and left wing buds are removed from a chick embryo and, after treatment with cold trypsin, the ectoderm can be pulled off the mesodermal core just like taking off a glove. Ectoderm from a left bud is then recombined with mesoderm from a right bud so that the dorso-ventral axis of the ectoderm runs the opposite way to that of the mesoderm— the equivalent of putting a left-hand glove on the right hand (Fig. 11.15). The 'recombinant' limb buds are grafted to the flank of a host embryo and allowed to develop. In

Fig. 11.15 Reversing the dorso-ventral axis of limb bud ectoderm. The only way a left-hand glove (representing the ectoderm from a left-hand limb bud) can fit on a right hand (the mesoderm of a right hand) is by turning it over so that the former ventral side (patterned) is now uppermost. The dorso-ventral polarity of the glove is now reversed in relation to the dorso-ventral axis of the hand. The antero-posterior relationship remains the same.

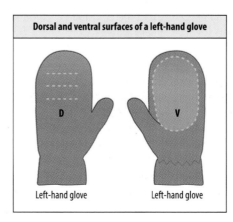

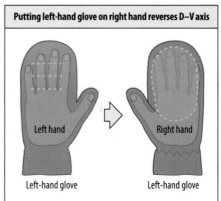

general, the proximal regions of the limbs that develop have dorso-ventral polarity corresponding to that of the mesoderm, whereas the distal regions, in particular the 'hands', have a reversed dorso-ventral axis, with the pattern of muscles and tendons reversed and corresponding to the dorso-ventral polarity of the ectoderm. The ectoderm covering the sides of the limb bud can therefore specify dorso-ventral pattern in the limb bud.

Genes controlling the dorso-ventral axis in vertebrate limbs have been identified from mutations in mice, as the dorso-ventral pattern can be distinguished by the fact that the ventral surface of the paw normally has no fur, whereas the dorsal surface does. The dorsal surfaces of the digits also have claws (nails in humans). Mutations that inactivate the gene *Wnt7a*, for example, result in limbs in which many of the dorsal tissues adopt ventral fates to give a double ventral limb, the two halves being mirror images. The *Wnt7a* gene is expressed in the dorsal ectoderm (Fig. 11.16), which suggests that the ventral pattern may be the ground state and is modified dorsally by the dorsal ectoderm, with the secreted Wnt-7a protein diffusing into, and playing a key role in, patterning the dorsal mesoderm. The ventral ectoderm is characterized by expression of the gene *Engrailed1* (*En1*), which encodes a homeodomain transcription factor first described in *Drosophila* (see Section 2.24). Mutations that destroy *En1* function result in *Wnt7a* being expressed ventrally as well as dorsally, giving a double dorsal limb. *En1* expression is induced by BMP signaling in the ventral ectoderm.

One function of Wnt-7a is to induce expression of the gene encoding the LIM homeobox transcription factor Lmx1b in the mesenchyme underlying the dorsal ectoderm (see Fig. 11.16). Clonal analysis of marked cells in mouse limb buds has revealed dorso-ventral compartments in the mesenchyme, with the progeny of dorsal cells being restricted to the dorsal half of the limb bud and the progeny of ventral cells to the ventral half. This is the only instance in which cell-lineage restrictions have been demonstrated in mesenchymal tissue; all other compartments discovered so far in both vertebrates and invertebrates are in epithelia—for example, the *Drosophila* epidermis (see Chapter 2) and the rhombomeres of the vertebrate hindbrain (see Chapter 12). The compartmentalization of the mesenchyme may serve to control gene expression; *Lmx1b*, for example, is expressed exactly in the dorsal compartment. In chick embryos, Lmx1b has been shown to specify a dorsal pattern in the mesoderm. Ectopic expression of Lmx1b in the ventral mesoderm of the chick limb results in the cells adopting a dorsal fate, leading to a mirror-image dorsal limb. Loss-of-function mutations in human *LMX1B* cause an inherited disease called nail–patella syndrome, in which these dorsal structures (nails and kneecaps) are malformed or absent.

11.11 Development of the limb is integrated by interactions between signaling centers

It is crucial that patterning along all three limb axes is integrated so that the correct anatomy of the limb is generated. Patterning is integrated by interactions between the signals Wnt-7a from dorsal ectoderm, FGFs from the apical ectodermal ridge, and Shh from the polarizing region. In mice mutant for loss of *Wnt7a* function, posterior digits are frequently missing, suggesting that Wnt7a is also required for normal antero-posterior patterning, and expression of *Shh* in the polarizing region is also reduced in these mice. *Shh* expression is similarly reduced when the dorsal ectoderm of the wing bud is removed in chick embryos.

A positive-feedback loop has been discovered between the polarizing region and the apical ectodermal ridge that links development along the proximo-distal and antero-posterior axes. Shh signaling maintains the level of *Fgf* gene expression in the ridge (Fig. 11.17), while FGF signaling in turn maintains *Shh* gene expression in the polarizing region. This feedback loop involves the BMP growth factor BMP-4. A minimum level

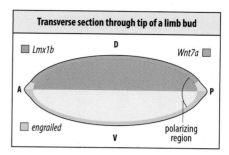

Fig. 11.16 **The ectoderm controls dorso-ventral pattern in the developing limb bud.** The gene for the secreted signal Wnt-7a is expressed in the dorsal ectoderm and the gene *Engrailed-1* is expressed in the ventral ectoderm. Expression of the gene for the transcription factor Lmx1b is induced in the dorsal mesoderm compartment by Wnt-7a signaling and Lmx1b is involved in specifying dorsal structures.

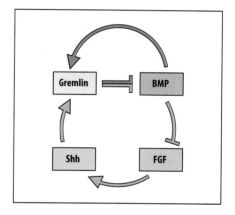

Fig. 11.17 **Integration of limb development signals by feedback control.** Gremlin is a BMP antagonist that keeps BMP activity low in the limb bud mesoderm, allowing a positive feedback loop to be established. FGF signaling from the apical ectodermal ridge maintains *Shh* expression in the polarizing region and, in turn, Shh signaling from the polarizing region maintains *Fgf* expression in the apical ectodermal ridge. Arrows indicate positive control in which a signaling molecule promotes gene expression, while barred lines indicate either negative control of signaling by an antagonist or prevention of gene expression. Blue indicates the initiation loop with negative feedback. Green indicates the propagation loop with positive feedback.

*After Bénazet J., Zeller R.: **Vertebrate limb development: moving from classical morphogen gradients to an integrated 4-dimensional patterning system.** Cold Spring Harb. Perspect. Biol. 2009, **1**: a001339.*

CELL BIOLOGY BOX 11D Sonic hedgehog signaling and the primary cilium

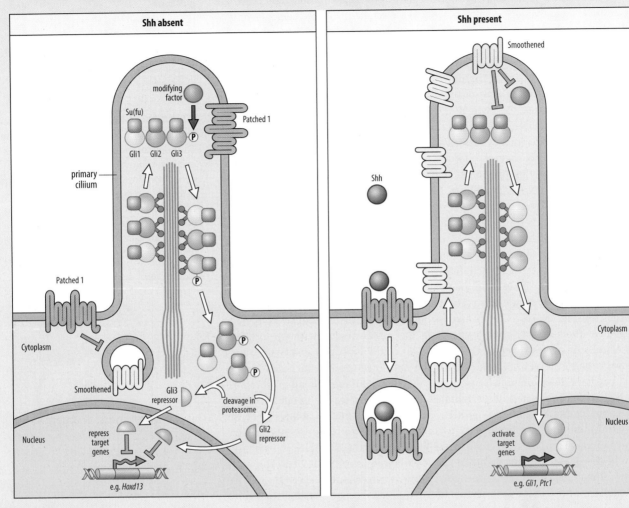

Figure 1

The intercellular signaling protein Sonic hedgehog is a vertebrate homolog of the Hedgehog protein that plays a major role in the patterning of the *Drosophila* embryo (see Chapter 2). Mice and humans have three genes for hedgehog proteins—*Sonic hedgehog* (*Shh*), *Indian hedgehog* (*Ihh*), and *Desert hedgehog* (*Dhh*); zebrafish have a fourth hedgehog gene, *Tiggywinkle*, as a result

of BMP signaling is required to maintain outgrowth of the limb bud, but if the level rises too high, the apical ridge regresses.

The level of BMP signaling in the early limb bud is controlled by the BMP antagonist Gremlin. *Gremlin* expression is first induced in the limb bud mesoderm by BMP-4 signaling, and a negative-feedback loop is rapidly set up in which Gremlin protein antagonizes BMP activity, keeping it low. *Fgf* expression is consequently maintained in the apical ridge and, in turn, maintains *Shh* expression in the polarizing region in the earliest stages of limb-bud development. Shh signaling subsequently maintains *Gremlin* expression, and thus low levels of BMP activity in the limb-bud mesoderm, reinforcing the positive-feedback loop between apical-ridge FGF signaling and polarizing-region Shh signaling (see Fig. 11.17). The epithelial–mesenchymal positive-feedback loop takes about 12 hours to complete, and is linked to the rapid negative-feedback loop between BMP-4 and Gremlin. This self-regulating of genome duplication. The signaling pathway stimulated by Shh and the other vertebrate hedgehog proteins (Figure 1) has many similarities to the Hedgehog signaling pathway in *Drosophila* (see Box 2F). There are vertebrate homologs of the membrane proteins Patched and Smoothened, which sense the presence or absence of a hedgehog signal, and the vertebrate counterparts of the *Drosophila* transcription factor, Cubitus interruptus (Ci), are the Gli transcription factors Gli1, Gli2, and Gli3. The name Gli comes from the initial identification of one of these genes as being amplified and highly expressed in a human glioma. Like Ci, Gli3 and Gli2 can be processed so that they act as transcriptional activators in the presence of a hedgehog signal and transcriptional repressors in its absence. Proteins homologous with Cos2 and Su(fu) in *Drosophila* (see Box 2F) form complexes with the Gli proteins and keep them in an inactive state.

A major difference between a vertebrate hedgehog signaling pathway and Hedgehog signaling in *Drosophila* is that in vertebrates Gli processing takes place at the tip of a cell structure called the primary cilium. Most vertebrate cells have a primary cilium, which, like all cilia, is made of microtubules and extends from a basal body derived from a centrosome. Unlike most other types of cilia, the primary cilium is non-motile.

In the absence of a hedgehog ligand, such as Shh, Patched represses Smoothened activity and moves into the cilium membrane whereas Smoothened does not. The Gli3 and Gli2 proteins are phosphorylated, which targets them for cleavage when they encounter the proteasome at the base of the cilium. Cleavage produces short forms that act as transcriptional repressors. The repressor Gli3 and Gli2 then enter the nucleus and prevent the transcription of Shh target genes. When Shh is present, it binds to Patched and activates Smoothened. Hedgehog ligand bound to Patched is internalized in membrane vesicles and plays no further part in signaling. Membrane vesicles containing the activated Smoothened are targeted to the cilium membrane, fuse with it, and deliver Smoothened into the membrane. At the tip of the cilium, Smoothened antagonizes the inhibition of Gli proteins by their binding proteins and prevents phosphorylation of Gli3 and Gli2. The activated Gli proteins are then transported along microtubules down the cilium to the cell body, where they enter the nucleus and activate transcription of target genes.

The first clues to the importance of the primary cilium in Shh signaling came from a genetic screen in mice for embryonic patterning mutants following mutagenesis with ethylnitrosourea (see Section 3.9). Two mutants were identified with defects in Shh signaling in the neural tube, one of which survived long enough for limb polydactyly to be observed. Most unexpectedly, a search against the databases revealed that the genes mutated in these embryos encoded proteins similar to those that transport material within a flagellum (a longer version of a cilium), and which were known to be essential for the growth and maintenance of flagella in the unicellular alga *Chlamydomonas*. Subsequently, mutations in genes encoding subunits of intraflagellar dynein and kinesin motor proteins were also found to result in defective Shh signaling, providing confirmation of the essential role of cilia in vertebrate hedgehog signaling.

All known mutations that prevent formation of the primary cilium cause patterning defects in the limbs and the neural tube. In the absence of the primary cilium, both Gli repressor and Gli activator functions fail. Because of the importance of Gli repressor function in antero-posterior limb patterning, this leads to polydactyly, whereas the neural tube is dorsalized, which is due to the importance of Gli activator function in its dorso-ventral patterning (see Section 12.7).

Defects in the formation of cilia and basal bodies have been linked to several dozen human congenital disorders, collectively known as **ciliopathies**. For example, patients with Bardet–Beidl syndrome (BBS), in which the primary cilia are affected, have a range of defects that can include polydactyly, polycystic kidney disease, and hearing loss. Mutations in any of 11 genes can cause BBS, and some of these genes encode proteins that are part of the basal body and/or the centrosome, thus providing an explanation for defects in ciliogenesis.

system is an excellent example of the feedback-control mechanisms that make development so robust and reliable (see Section 1.19).

11.12 Different interpretations of the same positional signals give different limbs

The signals controlling limb-bud outgrowth and patterning, such as FGF and Shh, are the same in chick wing and leg, but they are interpreted differently. A polarizing region from a wing bud, for example, will specify additional mirror-image digits if grafted into the anterior margin of a leg bud. However, these digits develop as toes, not wing digits, as the signal is being interpreted by leg cells. In a similar manner, signals are conserved between different vertebrates; for example, a mouse apical ectodermal ridge grafted in place of a chick apical ectodermal ridge can provide the appropriate signal for early chick limb development. Signals are, however, interpreted according to the origin of the responding cells: as we have discussed, a

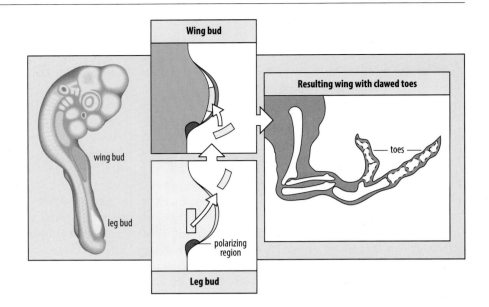

Fig. 11.18 Chick proximal leg bud cells grafted to a distal position in a wing bud acquire distal positional values. Proximal tissue from a leg bud that would normally develop into a thigh is grafted to the tip of a wing bud underneath the apical ectodermal ridge. It acquires more distal positional values and interprets these in terms of leg structures, forming clawed toes at the tip of the wing.

polarizing region from a mouse limb bud grafted into the anterior margin of a chick wing bud, specifies additional wing structures, not mouse structures (see Section 11.7). Thus, the signals from the different polarizing regions are the same, as are the signals from the different apical ridges; the difference in the structures formed by the limbs is a consequence of how the signals are interpreted and depends on the genetic constitution and developmental history of the responding limb bud cells.

A further demonstration of the interchangeability of positional signals comes from grafting limb bud tissue to different positions along the proximo-distal axis of the bud. If tissue that would normally give rise to the thigh is grafted from the proximal part of an early chick leg bud to the tip of an early wing bud, it develops into toes with claws (Fig. 11.18). The tissue has acquired a more distal positional value after transplantation, but interprets this value according to its own developmental program, which is to make leg structures. As we discussed earlier, we now know that the transcription factor Pitx1 is involved in the developmental 'leg' program and its ectopic expression in the chick wing results in the development of leg-like distal structures, such as toes and claws.

11.13 Hox genes have multiple inputs into the patterning of the limbs

Hox genes specify position along the antero-posterior axis of the vertebrate body (see Section 5.10) and these genes are then later employed in the development of the limb. Hox gene expression is involved in establishing the polarizing region, and position-dependent Hox gene expression in the developing limb is associated with the formation of structures along its proximo-distal axis, including the digits. At least 23 different Hox genes are expressed during mouse and chick limb development. Attention has been largely focused on the genes of the Hoxa and Hoxd gene clusters (see Box 5E), because genes in the 5′ regions of these clusters are expressed at high levels in both forelimbs and hindlimbs.

The expression of the 5′ Hoxa and Hoxd genes during limb development changes as the limb develops. Early on, *Hoxd9* and *Hoxd10* genes are expressed in the lateral plate mesoderm as it begins to thicken to become a limb bud. Expression of *Hoxd11*, *Hoxd12*, and *Hoxd13* genes is then initiated in rapid succession in progressively more restricted domains centered on the posterior distal region of both forelimbs and hindlimbs. These domains then extend along the limb bud as it grows out under the influence of apical ridge and polarizing region signaling, as shown for the chick wing bud in Fig. 11.19 (upper panel). Expression of *Hoxa9–13* is initiated slightly later in the early limb bud and then gives rise to similar overlapping domains of expression along the proximo-distal axis, but without a posterior bias, and these domains also extend along the limb bud as it grows out (see Fig. 11.19, lower panel). *Hoxa9* is expressed throughout the limb bud and successive genes are expressed in more and more distal regions, with *Hoxa11* expression becoming stronger in the intermediate region of the limb where *Hoxd11* is also strongly expressed; *Hoxd13* and *Hoxa13* are expressed in the distal region of the limb.

As the digits begin to form, a second phase of *Hoxd* gene expression is initiated in the distal part of the limb with *Hoxd13* expression being initiated first through-out the digital plate, with progressively smaller domains of *Hoxd12–10* expression then being initiated successively within this region except in the most anterior part. These Hoxd genes, together with *Hoxa13*, are involved in digit formation. The gap in Hoxd gene expression between the domain of strong expression of genes like *Hoxd10* in the intermediate region of the limb and the domain of strong Hoxd expression in the digital plate corresponds to the region that will form the wrist/ankle.

Sophisticated genetic experiments in mice, together with analyses of post-transla-tional histone modifications associated with gene activity and chromosomal confor-mation (see Box 8A), have succeeded in identifying remote *cis*-regulatory regions in so-called 'gene deserts' extending for about 1 MB on either side of the Hoxd and Hoxa gene clusters. These regulatory regions act as enhancers and drive the two phases of Hox gene expression in the limb. These gene deserts are conserved in vertebrate genomes, and it is now clear that such regions here and elsewhere in the genome, even though they contain no genes, have important long-range regulatory functions and are not just 'junk DNA.'

The same regulatory logic that governs transcription of the Hox genes in the devel-oping limb in both clusters and we will consider this in more detail by reference to the Hoxd cluster. Two enhancers in the gene desert on the side of the Hoxd cluster nearest the telomere (3′ to the cluster) drive the nested pattern of Hoxd gene expres-sion in early limb buds, while later on, multiple enhancer regulatory regions lying on the side nearest the centromere (5′ to the cluster), drive Hoxd gene expression in the tissue that will form digits (Fig. 11.20).

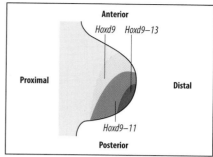

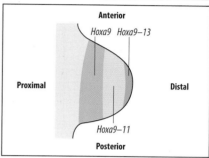

Fig. 11.19 Pattern of Hox gene expression in the early chick wing bud. The *Hoxd* genes (upper panel) are expressed in a nested pattern centered on the posterior tip of the wing bud, *Hoxd13* being expressed most posteriorly and distally. The *Hoxa* genes (lower panel) come to be similarly expressed in a nested pattern along the proximo-distal axis, *Hoxa13* being expressed most distally.

Fig. 11.20 A switch between telomeric and centromeric enhancers controls Hoxd gene expression in the developing mouse limb. Upper panel: in cells in the early limb bud, enhancers (orange diamonds) in the 'gene desert' on the telomeric side of the Hoxd cluster are brought into contact with genes in the cluster by chromosomal looping and activate them (green) and control the first phase of Hox gene expression as illustrated for *Hoxd10* (top right). Lower panel: in cells in the digital-plate, enhancers (purple) in the gene desert on the centromeric side of the cluster are brought into contact with genes towards the 5′ end of the cluster and control the second phase of Hox gene expression as illustrated for *Hoxd10* (bottom right). Expression of *Hoxd10*, which is not on the periphery of the cluster, is controlled by both sets of enhancers but the patterns of expression are different.

*Andrey, G., et al.: **A switch between topological domains underlies HoxD genes collinearity in mouse limbs**. Science 2013, **340**: 1234167.*

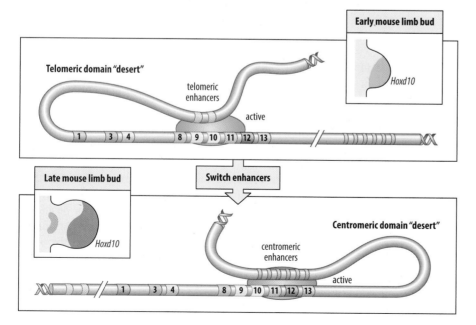

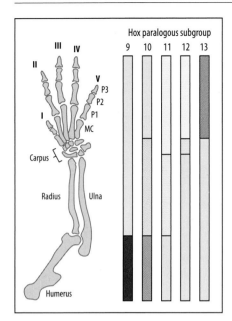

Fig. 11.21 Functional domains of Hox gene expression in the patterning of the mouse forelimb. Colored regions denote regions of *Hox* gene influence on proximo-distal limb pattern as determined from gene knock-out experiments. Hox9 (red) and Hox10 (orange) paralogs function together for the upper forelimb (humerus), Hox10 paralogs also affect the middle forelimb (radius and ulna), as represented by the lighter orange shading. Hox11 paralogs (yellow) are mostly involved in formation of this middle region, with some influence on the wrist and 'hand' region (lighter yellow shading). According to this analysis, Hox12 paralogs (green) predominantly pattern the wrist although Hoxd gene activity is low here, while Hox13 paralogs (purple) predominantly pattern the hand. MC, metacarpal; P, phalanx.

Adapted from Wellik, D.M., Capecchi, M.R.: **Hox10 and Hox11 genes are required to globally pattern the mammalian skeleton.** Science *2003,* **301***: 363-367.*

Thus, during limb development, there is a switch from control of gene expression via the telomeric enhancers to control via the centromeric enhancers, so that genes like *Hoxd10* that are not on the periphery of the cluster are expressed in two distinct domains, in the intermediate part of the limb early in development and in the distal part of the limb later (see Fig. 11.19). Experiments deleting each of the centromeric regulatory regions in turn show that only deletion of the entire regulatory region completely abolishes Hoxd gene expression in the digital-plate, which highlights the complexity of gene regulation.

The initiation of Hox gene expression in the limb is involved in establishing the polarizing region. Genetic deletion of all Hoxa and Hoxd genes in the early mouse forelimb bud results in the absence of *Shh* expression and severe limb truncations. Conversely, genetic manipulations leading to inversions of the Hoxd cluster or deletions of large parts of the downstream control region result in Hoxd genes in the 5′ part of the cluster being expressed in the patterns of 3′ Hoxd genes, so that *Hoxd13* is expressed throughout the early bud instead of being restricted to the posterior. The unrestricted expression of *Hoxd13* results in *Shh* expression both anteriorly and posteriorly, and the development of duplicated limb patterns. *Hoxb8*, one of the few Hox genes outside the Hoxa and Hoxd clusters that is expressed in the forelimb, also seems to be involved in establishing *Shh* expression.

As outlined above, the proximo-distal sequence of expression of Hoxa and Hoxd genes eventually comes to correspond generally with the three main proximo-distal regions of the limb. *Hoxa9* and *Hoxd9* are expressed in the upper part of the fore-limb, where the humerus forms. *Hoxa11* and *Hoxd11* are expressed in the lower limb, where the radius and ulna develop, and *Hoxa13* and *Hoxd13* (together with *Hoxd10–Hoxd12*) are expressed in the digit-forming regions. The patterns of expression of these Hox genes are generally similar in the forelimb and hindlimb, although there are a few differences.

Gene function can only be inferred from expression patterns, and mutational experiments are needed to prove function. We need to find out whether altering Hox gene expression in the limb leads to homeotic changes similar to those seen in vertebrae after Hox gene knock-out (see Section 5.7). The results of Hox gene knock-outs in mice are quite difficult to interpret in relation to the limb, but they do show that Hox genes clearly influence the formation of the different regions of the mouse limb in which they are expressed along the proximo-distal axis (Fig. 11.21). This conclusion is based on the phenotypes of mice mutant for genes in the same paralogous subgroup. For example, paralogous subgroup Hox10 controls the development of proximal elements, whereas Hox11 controls formation of the radius and ulna in the forelimb (tibia and fibula in the hindlimb). Inactivation of genes of the Hox13 subgroup disrupts digit development, as one might expect from the expression of *Hoxd13* and *Hoxa13* in the distal region of the limb.

Mutations in the human HOX genes are known to underlie conditions in which limbs are abnormal. Mutations in *HOXD13* lead to fusion of digits and some types of polydactyly, as already mentioned, while mutations in *HOXA13* lead to hand-foot-genital syndrome, in which the thumb and big toe are short. In both conditions, it is the digits that are affected.

Recent work using mouse genetics has revealed a previously unsuspected function of the Hox genes expressed in the hand-plate in regulating digit spacing. In the broadened limb buds of mouse *Gli3* mutants, which lack antero-posterior positional information, many unpatterned digits develop, and Hox genes are expressed throughout the digit-forming region of the limb. Unexpectedly, however, when the Hox genes expressed in this region (*Hoxa13* and *Hoxd11–Hoxd13*) are progressively deleted in the *Gli3* mutants, the number of digits progressively increases, although the size of the hand-plate remains the same. In *Gli3*-mutant mice in which *Hoxd11–13* are completely deleted and only one copy of *Hoxa13* remains, the limb has around 12–14 digits crammed into the hand-plate. These findings suggest that digit formation is based on a self-organizing system that relies on a **reaction–diffusion** mechanism (Box 11E), and that in normal development Hox genes regulate this mechanism, thereby ensuring that the correct number of digits are formed.

11.14 Self-organization may be involved in the development of the limb bud

The capacity for self-organization in the developing mouse limb indicated by the genetic experiments just outlined in Section 11.13 is also revealed by experimental manipulations in chick limb buds. Cells from a disaggregated chick limb bud can be reaggregated and will develop digits, even in the absence of a polarizing region. Mesodermal cells disaggregated from early chick limb buds are thoroughly mixed to disperse the polarizing region, or the mesodermal cells are taken from just the anterior half of a limb bud, which will not contain the polarizing region. The disaggregated cells are placed in an ectodermal jacket (prepared as described in Section 11.10), and grafted to a site where they can acquire a blood supply, but which is 'neutral' in terms of providing polarizing signals, such as the dorsal surface of an older limb. Limb-like structures develop from the grafted reaggregated limb-bud cells, even though there is no polarizing region. Several long cartilaginous elements may form in the more proximal regions of these 'recombinant' limbs, although none of the proximal elements can be easily identified with normal structures. More distally, however, reaggregated hindlimb-bud cells develop identifiable toes (Fig. 11.22). The fact that well-formed cartilaginous elements can develop at all in the absence of a discrete polarizing region shows that the bud has considerable capacity for self-organization. 'Recombinant' limbs that produce some limb-like structures can even be formed by reaggregating limb-bud mesodermal cells that have been grown in culture for several days and stuffing these into an ectodermal jacket.

These results suggest that the limb bud might be able to generate a basic **prepattern** consisting of equivalent cartilaginous elements. A prepattern is a basic organization that is generated autonomously in the embryo and can be identified in advance of the later development of a similar pattern of structures. The pattern that eventually develops does not have to follow the prepattern exactly but may be a modified form of it. The equivalent cartilaginous elements of the limb prepattern could be given their identities and further refined in response to positional information involving signals such as Shh and the activation of Hox genes and their downstream targets (see Sections 11.8 and 11.13).

The mechanism for generating the prepattern of a series of repeated structures could be based on a reaction–diffusion mechanism. In this model, polydactyly could simply result from a chance widening of the limb bud. If a reaction–diffusion mechanism generates a periodic pattern of cartilage elements across the limb as the digits are forming, merely widening the limb bud by some small developmental accident without altering the periodicity would enable a further digit to develop (see Box 11E).

Fig. 11.22 Reaggregated limb bud cells form digits in the absence of a localized polarizing region. Mesodermal cells from a chick leg bud are disaggregated, mixed to disperse all the cells, including the polarizing region, and then reaggregated, placed in an ectodermal jacket, and grafted to a neutral site. Well-formed toes develop distally.

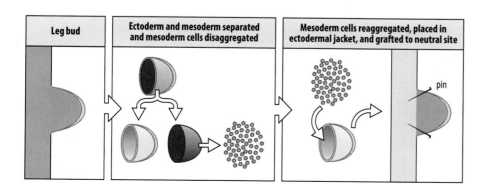

Leg bud	Ectoderm and mesoderm separated and mesoderm cells disaggregated	Mesoderm cells reaggregated, placed in ectodermal jacket, and grafted to neutral site

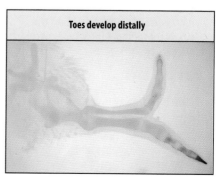

Toes develop distally

BOX 11E Reaction-diffusion mechanisms

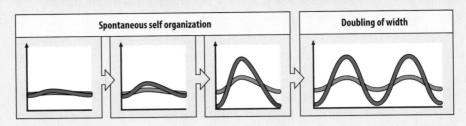

Figure 1

Figure 2

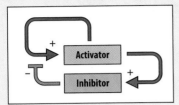

Some chemical systems, such as the famous Belousov–Zhabotinsky reaction, are self-organizing and spontaneously generate spatial patterns of concentration of their components. In two dimensions, the initial distribution of the molecules is uniform, but over time the system forms expanding wave-like patterns. The essential features of such self-organizing systems can be mathematically modeled by chemical reactions between two or more molecules with different rates of diffusion. Such systems are called **reaction–diffusion** systems or **Turing mechanisms** (after the mathematician Alan Turing, who was the first to prove that simple repetitive patterns could arise in this way).

For example, consider a closed two-component system of activator and inhibitor molecules in which the activator molecule stimulates both its own synthesis and that of the inhibitor, which in turn inhibits synthesis of the activator (Figure 1), and the inhibitor diffuses faster than the activator. A type of lateral inhibition will occur such that synthesis of activator becomes confined to one region, forming a peak of activator concentration with a given wavelength. If the wavelength stays the same when the size of the system is increased, two peaks will eventually develop, then three, and so on, as the system grows in size (Figure 2). In two dimensions, a pattern of peaks of high concentration of activator can develop like that in Figure 3. Conversely, if the system stays the same size, a shortening of the wavelength is needed to generate more peaks.

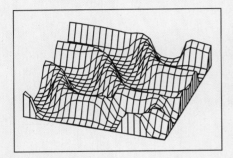

Figure 3

*From Meinhardt, H.: **Turing's theory of morphogenesis of 1952 and the subsequent discovery of the crucial role of local self-enhancement and long-range inhibition.** Interface Focus 2012, **2**: 407–416.*

Reaction–diffusion has interested biologists as, given the appropriate conditions and components, such a mechanism could, in theory, generate periodic patterns such as the spots and stripes on animals (see online Box 11E with additional material) and patterns of repeated structures such as the sepals and petals of flowers.

The limb has considerable capacity for self-organization, and it has been suggested that reaction–diffusion mechanisms could generate the periodic pattern of skeletal elements, such as the series of digits. In the hand-plate of normal mice (Figure 4, left panel), five *Sox9*-expressing stripes across the antero-posterior axis fan out from the proximal wrist region towards the wider distal tip of the hand-plate and prefigure the five digits. In a reaction–diffusion system, the stripes would represent high concentrations of activator, and the fan-like pattern would be accomplished by changing the wavelength of stripe spacing across the antero-posterior axis so that is longer distally than proximally, thus preventing the stripes splitting into two at the wider end of the hand-plate.

In *Gli3*−/− mouse limb buds that lack antero-posterior positional information (see Section 11.8), the hand-plate is broader and, as predicted in a reaction–diffusion system, an increased number of Sox9-expressing stripes are formed, which also fan out

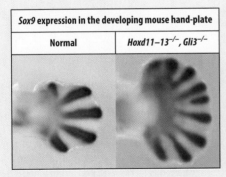

Sox9 expression in the developing mouse hand-plate	
Normal	**Hoxd11–13$^{-/-}$, Gli3$^{-/-}$**

Figure 4

*From Sheth, R., et al.: **Hox genes regulate digit patterning by controlling the wavelength of a Turing-type mechanism.** Science 2012, **338**: 1476–1480.*

from the wrist as in the normal limb. When the dose of Hox genes expressed in the hand-plate of *Gli3*−/− mouse embryos is reduced, many *Sox9*-expressing stripes are produced all crammed into the same-sized hand-plate, and these may bifurcate distally (Figure 4, right panel). This pattern of stripes matches the activator pattern predicted computationally for a reaction-diffusion system in which Hox genes modulate inhibitor production and thus decrease the wavelength of the spacing of the peaks.

11.15 Limb muscle is patterned by the connective tissue

Limb muscle cells have a different origin from that of the limb connective tissue cells, including muscle-associated connective tissue cells. If quail somites are grafted into a chick embryo at a site opposite where the wing bud will develop, the wing that forms will have muscle cells of quail origin, but all the connective tissue cells will be of chick origin. The cells that give rise to limb muscles migrate into the limb bud from the somites as myoblasts at a very early stage (see Section 5.13). After migration, the myoblasts multiply and initially form dorsal and ventral blocks of presumptive muscle (Fig. 11.23). These blocks undergo a series of divisions to give the individual muscles. The initial migration of myoblasts into the limb is restricted so that they enter either the dorsal or the ventral region; cells do not cross the dorsal–ventral compartment boundary, which is characterized by the dorsal expression of the transcription factor Lmx1b (see Fig. 11.16). Presumptive muscle cells, at least initially, do not acquire positional values in the same way as do cartilage and connective tissue cells, and are all equivalent. However, recent work suggests that Shh may directly control myogenic differentiation in the ventral muscle mass and in the absence of Shh signaling in the developing limbs of mouse embryos, ventral paw muscles do not develop.

Evidence for the early equivalence of myoblasts comes from experiments where somites were grafted from the future neck region of the early chick embryo to replace the normal wing somites. The myoblasts that enter the wing therefore come from neck somites, but a normal pattern of limb muscles still develops. This shows that the muscle pattern is determined by the connective tissue into which the myoblasts migrate, rather than by the myoblasts themselves. The prospective muscle-associated connective tissue may have surface or adhesive properties that the myoblasts recognize and are attracted to. If the pattern of adhesiveness in the connective tissue changed over time, this could result in associated changes in the migration of presumptive muscle cells, and this could account for the splitting of the muscle masses. Hox gene expression may be induced in the myoblasts by the surrounding limb mesenchyme. *Hoxa11* is expressed in myoblasts entering the limb, whereas cells in the posterior region of the dorsal and ventral muscle masses express *Hoxa13*. These patterns of Hox gene expression in the myoblasts differ from those in the surrounding mesenchyme, and could signify that the myoblasts have acquired their own positional values.

The myoblasts proliferating in the limb express *Pax3*. When *Pax3* is downregulated, cell proliferation stops and the cells differentiate (see Section 8.5). Signals from the overlying ectoderm, including BMP-4, prevent premature differentiation. The development of tendons is marked by the expression of the transcription factor Scleraxis.

11.16 The initial development of cartilage, muscles, and tendons is autonomous

A graft of the polarizing region causes the development of a mirror-image pattern of all the elements of the limb, not only the skeletal elements but also muscles and tendons (see Fig. 11.23). This indicates that the patterns of cartilage, tendon, and muscle

Scan here

Scan this QR code image with your mobile device to see an extended version of Box 11E or log on to **http://global.oup.com/uk/orc/biosciences/devbiol/wolpert5e/qr/qr11c/**

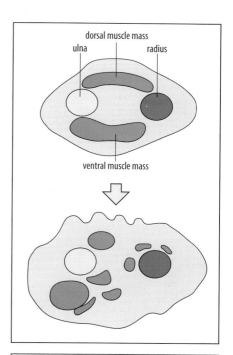
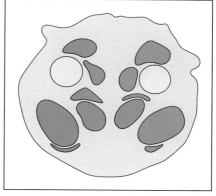

Fig. 11.23 Development of muscle in the chick wing. A cross-section through the chick wing in the region of the radius and ulna shortly after cartilage formation shows presumptive muscle cells present as two blocks—the dorsal muscle mass and the ventral muscle mass (upper panel). These blocks then undergo a series of divisions to give rise to individual muscles. A polarizing region graft to the anterior margin of a chick wing bud causes the formation of a mirror-image pattern of muscles (lower panel).

*From Shellswell, G.B., Wolpert, L.: **The pattern of muscle and tendon development in the chick wing.** In Limb and Somite Morphogenesis. Eds Ede, D.A., Hinchliffe, J.R., Balls, M. Cambridge: Cambridge University Press, 1977.*

in the limb are all specified by the same set of signals and there is little or no communication between them—that is, the development of each element is autonomous. If, for example, just the tip of an early chick wing bud is removed and grafted to the flank of a host embryo, it initially develops into normal distal structures with a wrist and three digits. The long tendon that normally runs along the ventral surface of digit 3 starts to develop, even though both its proximal end and the muscle to which it attaches are absent. The tendon does not continue to develop, however, because it does not make the necessary connection to a muscle, and so is not put under tension. The mechanism whereby the correct connections between tendons, muscles, and cartilage are established has still to be determined. It is clear, however, that there is little or no specificity involved in making such connections; if the tip of a developing limb is inverted dorso-ventrally, dorsal and ventral tendons can join up with inappropriate muscles and tendons. They simply make connections with those muscles and tendons nearest to their free ends—they are promiscuous.

11.17 Joint formation involves secreted signals and mechanical stimuli

The first event in joint formation is the formation of an 'interzone' at the site of the prospective joint, in which the cartilage-producing cells become fibroblast-like and cease producing the cartilage matrix, thus separating one cartilage element from another (Fig. 11.24). At this time, genes for Wnt-9A (previously known as Wnt-14) and the BMP-related protein GDF-5 are expressed in the presumptive joint region. Ectopic expression of *Wnt-9A* can induce the early steps in joint formation, and mice lacking *GDF-5* have some of their joints missing. A little later, the musculature begins to function and muscle contraction occurs. At this point, high levels of hyaluronan, a component of the proteoglycans that lubricate joint surfaces, are made by the cells of the interzone; hyaluronan is secreted in vesicles that coalesce to form the joint cavity, which separates the articulating surfaces of the joint from each other.

In chick embryos in which muscle activity has been inhibited by drugs, the joints fuse, indicating that joint formation depends on the mechanical stimulation that results from muscle contractions in the embryo. Muscle contraction promotes the synthesis of hyaluronan. Cells isolated from the interzone region and grown in culture respond to mechanical stimulation by increased secretion of hyaluronan when

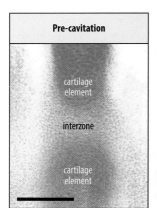
Pre-cavitation

cartilage element

interzone

cartilage element

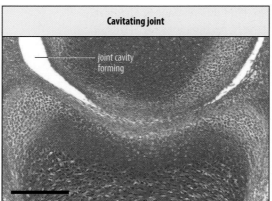
Cavitating joint

joint cavity forming

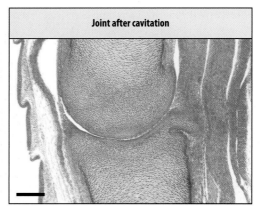
Joint after cavitation

Fig. 11.24 Formation of a joint. Sections through a developing proximal interphalangeal joint in the limb of a chick embryo stained with toluidine blue showing different stages in development. The appearance of the interzone region of fibroblast-like cells between two cartilaginous skeletal elements is the first morphological sign of joint formation (left panel) and cells in this region express *Wnt-9* and *GDF-5*. The cavity of the joint (middle panel) begins to form as interzone cells secrete hyaluronan. The completely cavitated joint (right panel) has a continuous cavity between the articulating surfaces of the two cartilage elements. Scale bar = 50 μm.

Photographs courtesy of A.S. Pollard and A.A. Pitsillides.

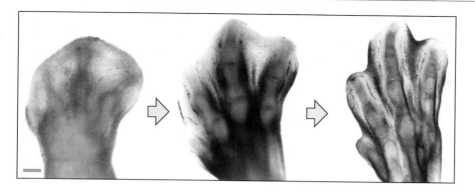

Fig. 11.25 Cell death during leg development in the chick. Programmed cell death in the interdigital region results in separation of the toes. Scale bar = 1 mm.

*Photograph reproduced with permission from Garcia-Martinez, V., et al.: **Internucleosomal DNA fragmentation and programmed cell death (apoptosis) in the interdigital tissue of embryonic chick leg bud**. J. Cell Sci. 1993, **106**: 201–208. Published by permission of The Company of Biologists Ltd.*

the substratum to which they are attached is stretched. In mouse mutants with muscles that are unable to contract or in mutants that lack limb muscles, some, but not all, of the limb joints are fused. In the affected joints, Wnt signaling activity is reduced during the earliest stages in their formation and the joint cells differentiate into cartilage instead of undergoing their normal programme of differentiation to form joint tissues.

11.18 Separation of the digits is the result of programmed cell death

Programmed cell death by apoptosis (see Box 6A) has a key role in molding the form of chick and mammalian limbs, especially the digits. The hand-plate in which the digits form is flattened along the dorso-ventral axis. The cartilaginous elements of the digits develop from condensations of mesenchyme cells at specific positions within this plate at which the transcription factor Sox9 is expressed (see Box 11E). Sox9 controls cartilage cell differentiation. Separation of the digits then depends on the death of the cells in the interdigital regions between these cartilaginous elements (Fig. 11.25). BMPs are involved in signaling this cell death: if the functioning of BMP receptors is blocked in the developing chick leg, cell death does not occur and the digits are webbed. Perhaps not surprisingly, other signals are also involved. For example, FGF-8 from the overlying ectoderm functions as a survival signal for interdigital cells, and it is the balance between cell death and survival signals that ensures the appropriate extent and location of cell death.

Programmed cell death is a normal part of development. The fact that ducks and other waterfowl have webbed feet, whereas chickens do not, is simply the result of less cell death between the digits of waterfowl. When chick-limb mesoderm is recombined with duck-limb ectoderm, cell death between the digits is reduced and the limb develops 'webbed' feet (Fig. 11.26). It is the mesoderm that determines the patterns of cell death both within the mesoderm and in the overlying ectoderm. In amphibians, digit separation is not due to cell death, but results from the digits growing more than the interdigital regions.

Programmed cell death also occurs in other regions of the developing limb, such as the anterior margin of the limb bud, and between the radius and ulna. In the chick wing bud, there is substantial cell death in the polarizing region. Indeed, it was the investigation of cell death in this region of a chick wing bud by transplanting it to the anterior margin of another chick wing bud that led to the discovery that it acts as a signaling center. One suggestion for the role of cell death in the polarizing region is that it could control the number of cells expressing Shh.

Fig. 11.26 The mesoderm determines the pattern of cell death. In the developing webbed feet of ducks and other water birds, there is less cell death between the digits than in birds without webbed feet. When the mesoderm and ectoderm of embryonic chick and duck limb buds are exchanged, webbing develops only when duck mesoderm is present.

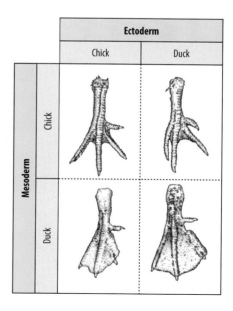

SUMMARY

Positioning and patterning of the vertebrate limb is largely carried out by cell-cell interactions that provide the cells with positional information. The position in which the limb buds form along the antero-posterior axis of the body is determined by a combinatorial Hox gene code in the lateral plate mesoderm. There are two key signaling regions within the limb bud. One is the apical ectodermal ridge, which is required for limb bud outgrowth; the second is the polarizing region at the posterior margin of the limb, which specifies positional information along the antero-posterior axis. The signals from the apical ridge are fibroblast growth factors (FGFs) and are essential for limb outgrowth, which involves oriented cell movements and cell divisions. Shh protein is produced by the polarizing region and provides a graded positional signal across the antero-posterior axis of the limb bud that together with the duration of Shh signaling determines the identities of the digits. Pattern across the dorso-ventral axis is specified by the ectoderm.

Signaling along each of the three limb axes is integrated into a network that involves both negative- and positive-feedback loops. Hox genes are expressed in complex spatiotemporal patterns within the limb bud and have several different functions, including establishing the polarizing region and specifying the identity of structures along the proximo-distal axis. Hox genes later regulate digit spacing in the digital plate. The limb muscle cells are not generated within the limb but migrate in from the somites and are patterned by the limb connective tissue. Formation of the periodic pattern of cartilaginous elements may involve a self-organizing mechanism. The development of joints involves local production of signaling molecules and requires mechanical stimuli. In birds and mammals, separation of the digits is achieved by programmed cell death.

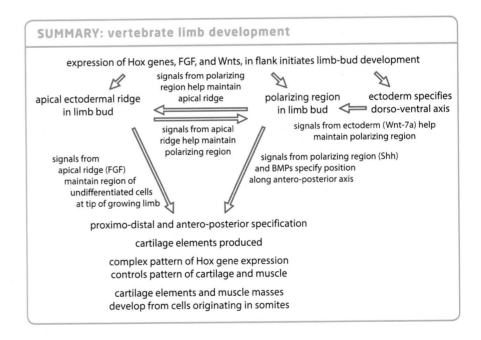

SUMMARY: vertebrate limb development

expression of Hox genes, FGF, and Wnts, in flank initiates limb-bud development

apical ectodermal ridge in limb bud

signals from polarizing region help maintain apical ridge

signals from apical ridge help maintain polarizing region

polarizing region in limb bud

ectoderm specifies dorso-ventral axis

signals from ectoderm (Wnt-7a) help maintain polarizing region

signals from apical ridge (FGF) maintain region of undifferentiated cells at tip of growing limb

signals from polarizing region (Shh) and BMPs specify position along antero-posterior axis

proximo-distal and antero-posterior specification

cartilage elements produced

complex pattern of Hox gene expression controls pattern of cartilage and muscle

cartilage elements and muscle masses develop from cells originating in somites

Insect wings and legs

The adult organs and appendages of *Drosophila*, such as the wings, legs, eyes, and antennae, develop from imaginal discs, which are excellent experimental systems for analyzing pattern formation (see Fig. 2.6). The discs invaginate from the embryonic ectoderm as simple pouches of epithelium during embryonic development and remain as such until metamorphosis (see Fig. 2.1). Although all imaginal discs superficially look rather similar, they develop differently, according to the segment in which they are located. The imaginal discs that give rise to the wings and legs are located in the thoracic segments. They are specified as wing or leg within the embryonic epithelium, at the time when the segments are being patterned and given their identity. In *Drosophila*, wings develop on thoracic segment 2, halteres—appendages that balance the fly during flight—on segment 3, and legs on each of the three thoracic segments. The wing and leg discs are specified in the embryo as clusters of 20–40 cells, and during larval development they grow about 1000-fold.

A feature of imaginal discs is that if some of their cells are removed or destroyed, the disc will regrow to the correct size and develop normally, and the same is true if some cells are given a growth advantage, as with the *Minute* technique (see Section 2.27). The mechanism underlying this is not yet known.

Although insect wings and legs differ so greatly in appearance, they are partially homologous structures, and the strategy of their patterning is similar. Moreover, the mechanisms by which they are patterned, and even the genes involved, show some similarities to the patterning of the vertebrate limb, although this is almost certainly due to parallel evolution adapting the same sets of genes to build functionally similar structures, rather than to evolutionary homology between insect and vertebrate limbs. We will start by discussing the development of the adult wing, and then compare the development of the leg.

11.19 The adult wing emerges at metamorphosis after folding and evagination of the wing imaginal disc

The adult insect wing is a largely epidermal structure in which two epithelial layers—the dorsal and ventral surfaces—are close together. The wing is joined to the dorsal part of the thoracic wall, the notum, by a 'hinge' region. The hinge and part of the notum are also specified within the wing imaginal disc. At metamorphosis, by which time patterning of the imaginal discs is largely complete, the wing disc undergoes a series of profound morphological changes. Essentially, the epithelial pouch is turned inside out, as its cells differentiate and change shape. The epithelial pouch extends and folds so that one half comes to lie beneath the other to form the double-layered wing structure (Fig. 11.27). The zinc-finger transcription factors Elbows and No-ocelli act in both the wing and leg imaginal discs to repress genes specifying the body wall, and thus permit the appendages to develop.

In the embryo, the epidermis of each thoracic segment, and of the wing and leg discs that derive from them, is divided into an anterior and a posterior compartment—regions of cell-lineage restriction (see Section 2.24) that abut at a compartment boundary. In the wing disc epithelium, a second compartment boundary between the dorsal and ventral regions develops during the second larval instar (Fig. 11.28). The

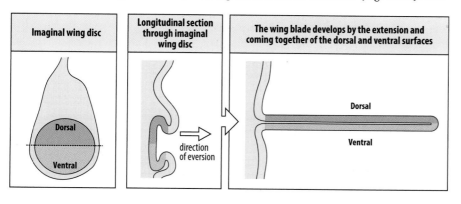

Fig. 11.27 Schematic representation of the emergence of the wing blade from the imaginal disc. Initially, the future dorsal and ventral surfaces of the wing are in the same plane within the imaginal disc (first panel). At metamorphosis, the pouch turns inside out and extends outwards (second and third panels); the dorsal and ventral surfaces of the wing come together.

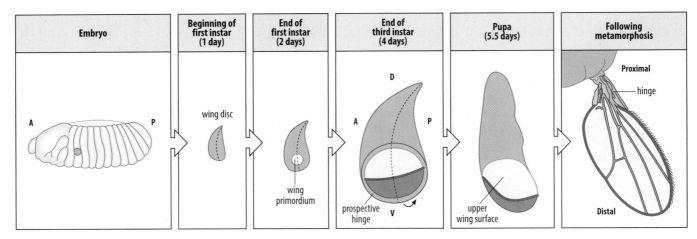

| Embryo | Beginning of first instar (1 day) | End of first instar (2 days) | End of third instar (4 days) | Pupa (5.5 days) | Following metamorphosis |

Fig. 11.28 Development and fate map of the wing imaginal disc of *Drosophila*. When the wing imaginal disc is first specified in the embryonic epidermis there is no particular group of cells that corresponds to a wing primordium. This emerges in the early stages of larval growth as a small cluster of 20–30 cells in one half of the imaginal disc, which proliferate rapidly to form the ovoid wing primordium. The wing disc is bisected by the interface between the anterior (A) and posterior (P) compartments (see Section 2.24), shown by the dashed line. By the end of the third larval instar, the future wing is an ovoid epithelial sheet taking up around half of the imaginal disc, and a second compartment boundary (red) has developed between the future dorsal (D, yellow) and ventral (V, blue) surfaces of the wing. At metamorphosis, the ventral surface folds under the dorsal surface (arrow) in the manner outlined in Fig 11.27. The wing disc also contains the precursors to the hinge region (gray). The notum (brown) develops from the proximal part of the imaginal disc.

After figure 12.18 in Martinez Arias, A., Stewart, A.: Molecular Principles of Animal Development. Oxford University Press, 2002.

proximal parts of the wing and leg imaginal discs give rise to parts of the thoracic body wall.

When first specified, the leg imaginal disc already contains cells that will give rise to parts of the adult leg. The wing imaginal disc, on the other hand, initially contains cells that will form the hinge and the notum, but not the wing itself. The wing primordium emerges within the imaginal disc in the first larval instar.

The development and growth of the *Drosophila* wing is completely dependent on the function of the gene *vestigial*, which encodes a transcriptional co-activator that is expressed exclusively in the wing and which elicits wing tissue if expressed in other imaginal discs. *vestigial* is expressed at a low level throughout the early wing disc and acts at several different stages and in different spatial patterns throughout wing development, being controlled through different *cis*-regulatory modules at different stages. The initiation of the wing primordium is correlated with an increase in *vestigial* expression in a small group of about 30 cells in the middle of the wing imaginal disc in response to Notch and Wingless signaling (Fig. 11.29). These cells will become the wing margin (see Fig. 11.27) and will seed the outgrowth of the wing. Wingless is a member of the large Wnt family of secreted signal proteins, and we have already seen it in action in *Drosophila* embryonic development (see Section 2.26). Wingless is expressed in the wing disc from an early stage. Reduction of *wingless* function results in a failure to specify the wing primordium and therefore the fly has no wings; the gene is named after this adult phenotype.

11.20 A signaling center at the boundary between anterior and posterior compartments patterns the *Drosophila* wing along the antero-posterior axis

The compartment boundaries act as signaling centers, and their activity has provided a number of important clues about how signals pattern a group of cells (see Section 2.24). We shall look first at the antero-posterior patterning that occurs in the early stages of larval life and is mediated from an organizing center located at the anterior–posterior compartment boundary. The *engrailed* gene is expressed in the posterior compartment of the imaginal disc (Fig. 11.30), a pattern of expression inherited from the embryonic parasegment from which the disc derives, and the cells that express *engrailed* also express the gene *hedgehog* (see Section 2.26).

Fig. 11.29 Initiation of the wing primordium by *vestigial* expression in response to Wingless and Notch signaling. The signaling pathways are used in the development and patterning of all imaginal discs. In the wing disc, their activity is channeled by the transcriptional co-activator Vestigial, which confers wing-specific activity on the transcription factor to which it binds, and ensures that, in the wing imaginal disc, the transcriptional effectors of the signaling pathways are directed specifically to the genes involved in making a wing.

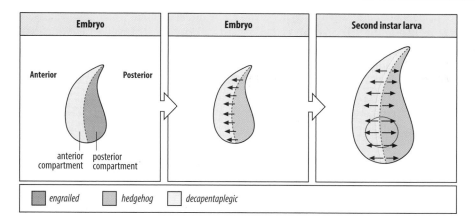

| engrailed | hedgehog | decapentaplegic |

Fig. 11.30 Establishment of a signaling center in the wing imaginal disc at the anterior–posterior compartment boundary. The gene *engrailed* is expressed in the posterior compartment, where the cells also express the gene *hedgehog* and secrete the Hedgehog protein. Where Hedgehog protein signals to anterior compartment cells, the gene *decapentaplegic* is activated, and Decapentaplegic protein is secreted into both compartments, as indicated by the black arrows.

As in the embryo, the secreted Hedgehog protein acts through its receptor Patched (see Box 2F) to activate the transcription factor Ci (the counterpart of the Gli transcription factors in the Shh pathway). The gradient of Hedgehog protein reaches about 10 cell diameters, about half the width of the disc at this early stage. The Hedgehog protein is heavily modified by lipids, and this is likely to be why it does not diffuse very far. The Hedgehog transcriptional effector Ci is only expressed in the anterior compartment, and this ensures that only anterior cells respond to Hedgehog. Signaling generates graded activation of Ci in cells at different distances from the compartment boundary and this leads to different responses in adjacent cells along the gradient. Hedgehog is therefore acting as a morphogen, patterning the future wing tissue along the antero-posterior axis.

A low-threshold target of Ci is *decapentaplegic* (*dpp*), a gene that encodes a member of the BMP family of signaling proteins (see Box 4C). In response to Hedgehog signaling, *dpp* is expressed in a narrow stripe along the anterior side of the anterior–posterior compartment boundary (see Fig. 11.30). Decapentaplegic protein (Dpp) is secreted from the antero-posterior compartment boundary and forms symmetric concentration gradients in the anterior and posterior compartments. It is the positional signal for patterning both compartments along the antero-posterior axis.

The gradient of Dpp across the wing disc provides a long-range signal controlling the localized expression of the genes for the transcription factors Spalt (Sal), Optomotorblind (Omb) and Brinker in a concentration-dependent manner (Fig. 11.31). It is these transcription factors that implement the patterning functions of Dpp and refine the initial coarse patterning of the wing. Thus, Dpp acts as a morphogen (Fig. 11.32): low levels

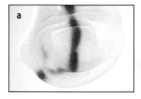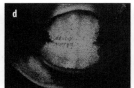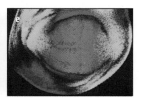

Fig. 11.31 Domains of gene expression initiated by Dpp signaling pattern the *Drosophila* wing disc. Panel (a) shows the expression of *decapentaplegic* (blue) at the antero-posterior compartment border of the wing blade. Dpp signaling represses expression of the gene *brinker* (green, panel (b)), which is therefore expressed only in two regions on either side of the zone of *dpp* expression. *brinker* represses the expression of *spalt* (red, panel (c)) and *omb* (blue, panel (d)), which are therefore expressed in a wide stripe over the antero-posterior border,

where *brinker* is not expressed. Panel (e) shows a composite overlay of (b), (c), and (d).

*(a) photograph reproduced with permission from Nellen, D., et al.: **Direct and long-range action of a dpp morphogen gradient.** Cell 1996, **85**: 357–368. © 1996, Cell Press.*

*(b–e) photographs reproduced with permission from Moser, M. and Campbell, G.: **Generating and interpreting the Brinker gradient in the Drosophila wing.** Dev. Biol. 2005, **286**: 647–658.*

Scan here

Scan this QR code image with your mobile device to see an online movie showing formation of the Dpp gradient or log on to **http://global.oup.com/uk/orc/biosciences/ devbiol/wolpert5e/qr/qr11d/**

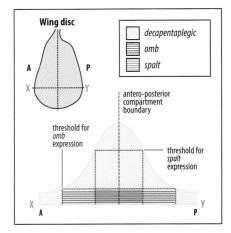

Fig. 11.32 A model for patterning the antero-posterior axis of the wing disc. The Decapentaplegic protein is produced at the compartment boundary. The schematic cross-section of the wing disc (X–Y) along the antero-posterior axis shows how the presumed gradient in Decapentaplegic in both anterior and posterior compartments activates the genes *spalt* and *omb* at their threshold concentrations.

of Dpp induce *omb*, whereas higher levels are required to induce *spalt*. *omb* and *spalt* are related to the *Tbx* and *Sall* genes, respectively, which are involved in vertebrate limb patterning (see Section 11.9). *omb* and *spalt* expression in the wing disc determine the position of various veins in the adult wing. Brinker is a repressor of *spalt* and *omb* that is antagonized by the Dpp signal, and thus provides fine-tuned control of their expression.

The evidence that Dpp is a morphogen in the wing disc comes from several types of experiments. First, clones of cells unable to respond to a Dpp signal do not express *spalt* or *omb*. Second, ectopic expression of the *dpp* gene in a region that does not normally express *spalt* and *omb* leads to the localized activation of these genes around the *dpp*-expressing cells. Formation of the long-range Dpp gradient in the wing disc is a complex process and is not yet fully understood. The glypican Dally, present on the cell surface, is involved in both shaping the Dpp gradient and influencing Dpp signaling via the Dpp receptor Thick veins.

The pattern of veins on the adult wing blade is one of the ultimate outcomes of the antero-posterior patterning activity of Hedgehog and Dpp in the wing disc (Fig. 11.33). This was shown by ectopically expressing the *hedgehog* gene in genetically marked cell clones generated at random in wing discs. When such clones form in the posterior compartment they have little effect (as *hedgehog* is normally expressed throughout this compartment in the disc) and development of the wing is more or less normal. With *hedgehog*-expressing clones in the anterior compartment, however, a wing with a mirror-symmetric pattern is produced (see Fig. 11.33). Ectopic expression of *hedgehog* has set up new sites of *dpp* expression in the anterior compartment of the wing disc, which results in new gradients of Dpp protein being formed.

There is no single or simple process by which the morphogens Hedgehog and Dpp specify the vein pattern: rather, the position of each vein is likely to be specified by a unique combination of factors, including the level of Dpp and Hedgehog signaling, which compartment the cells are in, and the induction or repression of particular transcription factors. For example, the lateral vein L5 in the posterior compartment develops at the border between *omb* and *brinker* expression domains. It seems, therefore, that Dpp and Hedgehog do not directly pattern the veins, but set up a series of cell–cell interactions that specify positional information across the wing disc, perhaps even to the level of single cells. It is striking how precise the pattern of veins is if one compares the wings on both sides of a fly.

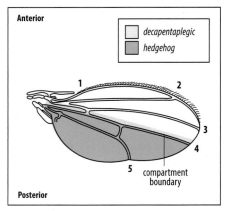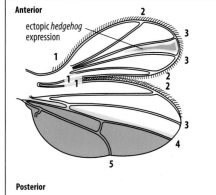

Fig. 11.33 Alteration of wing patterning due to ectopic expression of *hedgehog* and *decapentaplegic*. When *hedgehog* is expressed in a clone of cells in the anterior compartment, a new source of Decapentaplegic protein (Dpp) is set up. Left panel: in the normal wing, Dpp is made at the compartment boundary, where Hedgehog is expressed. Right panel: the ectopic expression of *hedgehog* in the anterior compartment results in a new source of Dpp and a new wing pattern develops in relation to this new source.

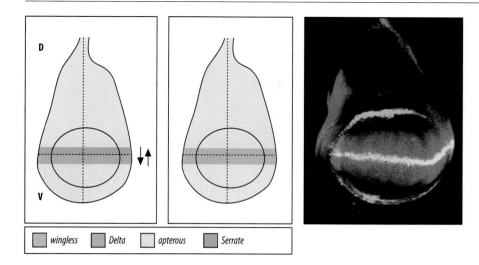

▨ *wingless*	▨ *Delta*	▨ *apterous*	▨ *Serrate*

11.21 A signaling center at the boundary between dorsal and ventral compartments patterns the *Drosophila* wing along the dorso-ventral axis

Fig. 11.34 Wingless is the signaling protein at the boundary between dorsal and ventral compartments. The dorsal-ventral compartment boundary is established late in the second instar larva, when a dorsal compartment is delineated by expression of the gene for the transcription factor Apterous. Apterous induces Notch signaling by first inducing *Serrate* expression in its own domain, which leads to *Delta* expression on the adjacent ventral cells (Delta and Serrate are ligands for Notch). Serrate and Delta enhance each other's expression and lead to the expression of *wingless* in a stripe at the boundary. The combination of these signals increases expression of the gene *vestigial*, which encodes a key regulator of wing development, in a broad stripe at the dorsal-ventral boundary. The photograph shows *wingless* (green) and *vestigial* (red) expression in a wing disc from a third-instar larva. Both *wingless* and *vestigial* are expressed at the dorso-ventral boundary, as indicated by the yellow stripe.

*Photograph reproduced with permission from Zecca, M., et al.: **Direct and long-range action of a wingless morphogen gradient**. Cell 1996, 87: 833–844. © 1996 Cell Press.*

In the second larval instar, the wing disc epithelium also becomes subdivided into dorsal and ventral compartments, which correspond to the dorsal and ventral surfaces of the adult wing (Fig. 11.34). The dorsal and ventral compartments were originally identified by cell-lineage studies, but they are also defined by expression of the selector gene *apterous*, which is confined to the dorsal compartment and defines the dorsal state. The Apterous protein is related to Lmx1b from the mouse, which specifies a dorsal pattern in the mesoderm of the mouse limb bud (see Section 11.10).

Like the anterior–posterior compartment boundary, the boundary between dorsal and ventral compartments acts as an organizing center. Wingless is the signaling molecule at this boundary (see Fig. 11.34). Apterous begins to be expressed in the dorsal compartment shortly after the wing primordium is established. At the dorsal–ventral compartment boundary Apterous induces the expression of the gene encoding Serrate, a Notch ligand, in dorsal cells, and restricts the expression of the gene for the other Notch ligand, Delta, to ventral cells. Notch signaling (see Box 5D) at the compartment boundary induces expression of *wingless* along the boundary. Wingless in turn induces expression of *Delta* and *Serrate* at the boundary, and progressively refines its own expression as well as that of Notch signaling. As the patterns of Wingless and Notch signaling change, they focus expression of *vestigial* in a broad stripe centered on the dorsal–ventral boundary (see Fig. 11.34). Genetic analysis of Wingless function suggests that it may not be acting as a long-range morphogen, but may rather be collaborating with other genes with local effects, particularly genes encoding gene-regulatory proteins, most significantly *vestigial*.

11.22 Vestigial is a key regulator of wing development that acts to specify wing identity and control wing growth

The gene *vestigial* was initially identified by genetic analysis as the key gene controlling the establishment, growth, and patterning of the wing. Loss- and gain-of-function experiments showed that in the complete absence of *vestigial* expression the wing does not develop (see Fig. 1.12), while *vestigial* mutants with partial function lead to wing defects. On the other hand, the ectopic expression of *vestigial* in other imaginal discs (eye, leg, or genital) turns these tissues into wings.

The expression of *vestigial* mirrors the development of the wing tissue as it responds to and channels the various molecular inputs that govern the emergence of this tissue. *vestigial* encodes a transcriptional co-factor and it integrates these signals through position-specific enhancer sites in its control regions that respond to different inputs.

Fig. 11.35 *vestigial* **integrates different signals via different enhancers in its control regions.** The *vestigial* gene contains two different enhancers, the 'boundary enhancer' (BE) and the 'quadrant enhancer' (QE), that respond to different inputs and result in an increase in *vestigial* expression at different times and in different areas of the wing disc. The enhancers are located in introns within the gene. BE responds to Notch signaling and Wingless at the dorso-ventral boundary to increase *vestigial* expression at this boundary. QE is activated later, at the beginning of the third instar, in response to Wingless signaling at the dorso-ventral boundary and Dpp signaling at the antero-posterior boundary, and drives *vestigial* expression in the four quadrants of the wing disc. It is this latter expression that controls the growth and further development of the wing blade.

*Data from Kim, J., et al.: **Integration of positional signals and regulation of wing formation and identity by** Drosophila **vestigial gene**. Nature 1996, **382**: 133-138.*

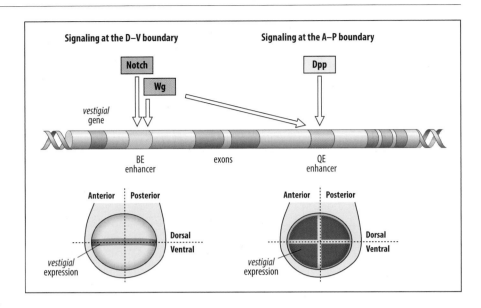

As we have seen (Section 11.19) the initiation of wing development depends on the expression of *vestigial* in a small group of about 30 cells in the middle of the wing disc in the first-instar larva. This expression is under the control of an enhancer (BE) that responds to Notch and Wingless signaling in the disc at this stage. As Wingless expression within the wing disc develops into a particular spatial pattern, *vestigial* expression becomes concentrated at the dorso-ventral boundary of the wing primordium at the beginning of the second instar larva. These cells will become the wing margin. Later, at the beginning of the third instar, a combination of Wingless signaling from the dorso-ventral boundary and Dpp signaling from the antero-posterior boundary switches on *vestigial* expression throughout the remainder of the wing primordium via another enhancer (QE). The function of Vestigial in promoting wing-cell proliferation and wing outgrowth during the later larval instars is due to the combination of these patterns of expression (Fig. 11.35). If a *lacZ* reporter gene is put under the control of this second enhancer, it shows very accurately the growth of the wing in the imaginal disc during the larval instars.

Vestigial is a gene-regulatory protein that is located in the nucleus. It does not have a DNA-binding domain and its ability to regulate gene expression is mediated by its interactions with Scalloped, a widely expressed transcription factor that is also essential for wing development. Although few of the individual targets of the Vestigial–Scalloped complex are known, the effects of mutations in either gene show that the complex must regulate a large battery of genes that ensure that the growing cells make wing structures.

The expression of *vestigial* under the control of the QE enhancer is entirely dependent on its previous expression within the disc, and on the presence of Vestigial; it is therefore an example of keeping a gene active by autoregulation and ensuring that it remains active in daughter cells (see Section 8.3). The QE enhancer also integrates antero-posterior and dorso-ventral positional information by being able to respond to inputs from both Wingless and Dpp signaling. It is likely that other discs in *Drosophila*, and organs in other organisms, develop under the influence of proteins functionally similar to Vestigial—that is, disc- or organ-specific regulatory proteins that serve to channel the activity of widely expressed signaling molecules and transcription factors towards the genes that produce the final structure and pattern particular to that disc or organ. In some instances a combination of proteins is more likely than a single protein. In the *Drosophila* leg, for example, there is no 'leg-specific' gene corresponding to *vestigial*.

The patterning processes described in the previous sections divide the wing disc into overlapping domains of gene expression. These domains specify the position of pattern elements, such as veins, and they also specify the size of the various parts of the wing, although how they do this is still a total mystery. While the development and positioning of these domains of gene expression can be related to the actions of the antero-posterior and dorso-ventral signaling centers, the patterns of growth within the wing, although dependent to a certain degree on Dpp and Wingless, are not obviously related to the gradients of these signals nor to the patterns of expression of their target genes. The wing disc grows in size by cell division, which occurs throughout the disc and then ceases uniformly before metamorphosis when the correct size is reached. The adult wing is produced at metamorphosis by changes in cell shape without further cell division.

It is becoming clear that in addition to the biochemical inputs that cells receive and that stimulate, or inhibit, their proliferation, they are also subject to physical forces and mechanical signals associated with the density of cells and the geometrical constraints of the tissue. The boundary of the circular wing disc, for example, is demarcated by deep furrows that create physical forces (strains and stresses) in particular spatial patterns. These stresses also affect growth rates and directions of tissue outgrowth, and can thus shape the tissue (discussed in Chapter 9).

Genetic analysis has identified the Hippo signaling pathway (see Box 13A) as an important element in sensing changes in cell shape, adhesion, and density and relating them to cell growth. Both Wingless and Dpp can interact with this pathway, and this might be one way of channeling their inputs into directing growth in the wing. A possible clue to the control of wing growth is that Scalloped, the transcription factor that interacts with Vestigial, is a homolog of the TEAD transcription factor, whose activity is controlled by the Hippo pathway. The interaction between Scalloped and Vestigial provides an interesting basis to think about how Hippo signaling could integrate patterning and growth.

11.23 How the proximo-distal axis of the *Drosophila* wing is patterned is not yet clear

The adult wing, like any other appendage, also has a proximo-distal axis. This is less easy to visualize in the flat imaginal disc than the antero-posterior and dorso-ventral axes. As illustrated in Fig. 11.27, the prospective wing tissue can be considered the most distal element of the wing imaginal disc. It is surrounded by cells that will give rise to the hinge structures by which the wing is attached to the body, or to the notum, which is represented in the most proximal part of the imaginal disc. How the proximo-distal axis of the wing blade is patterned remains unclear in *Drosophila*, but there is evidence for both short- and long-range activation of genes in the wing blade by Wingless. We shall see later in the chapter how the colorful pattern on butterfly wings is a manifestation of proximo-distal patterning.

As well as the general asymmetry of wing structure, which is due to patterning along the various axes, the individual cells in the adult *Drosophila* wing epidermis become polarized in a proximo-distal direction; this is reflected by the pattern of hairs (trichomes) on the wing surface. Each of the approximately 30,000 cells in the wing epidermis produces a single hair, which points distally (see Fig. 11.28, which shows only those hairs that would be visible on the edge of the wing). This is an example of planar cell polarity, and what is known of the underlying mechanism is outlined in Box 2I.

11.24 The leg disc is patterned in a similar manner to the wing disc, except for the proximo-distal axis

Insect legs are essentially jointed tubes of epidermis and thus have a quite different structure from those of vertebrates. The epidermal cells secrete the hard outer cuticle that forms the exoskeleton. Internally, there are muscles, nerves, and connective

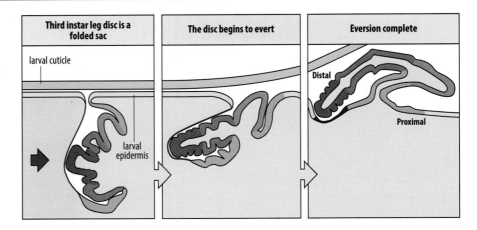

Fig. 11.36 *Drosophila* **leg disc extension at metamorphosis.** The disc epithelium, which is an extension of the epithelium of the body wall, is initially folded internally. At metamorphosis it extends outward, as if pulled out from the center. The red arrow in the first panel is the viewpoint that produces the concentric rings shown in Fig. 11.37.

tissues. A change in the shape of the epithelial cells of the leg imaginal disc is responsible for the outward extension of the leg at metamorphosis. The process is rather like pulling the disc out from the center, with the result that the center of the disc ends up as the distal end, or tip, of the leg (Fig. 11.36).

The easiest way of relating the leg imaginal disc to the adult leg is to think of it as a collapsed cone. Looking down on the disc, one can imagine it as a series of concentric rings, each of which will form a proximo-distal segment of leg (Fig. 11.37). The outermost ring gives rise to the base of the leg, which is attached to the body, and the rings nearer the center give rise to the more distal structures.

The leg disc contains some 30 cells at its initial formation in the embryo but grows to contain more than 10,000 cells by the third instar. Unlike the wing disc, it contains some cells specified to become leg cells at its formation. The first steps in the patterning of the leg disc along its antero-posterior axis are the same as in the wing. The *engrailed* gene is expressed in the posterior compartment and induces expression of *hedgehog*. The Hedgehog protein induces a signaling region at the antero-posterior compartment border. In the dorsal region of the leg disc, the expression of *dpp* is induced in the anterior compartment, as it is in the wing. In the ventral region, however, Hedgehog induces expression of *wingless* instead of *dpp* in the anterior compartment along the border, and Wingless protein acts, as in the wing, coordinating the sequence of short-range signaling events associated with the growth of the leg (Fig. 11.38).

The patterning of the proximo-distal axis is far better understood in the leg disc than in the wing. It involves interactions between Wingless and Dpp that are very likely to mirror the events in the wing, but with different downstream gene targets. The distal end of the proximo-distal axis is the point at which Wingless and Dpp expression meet, and is marked by expression of the homeodomain transcription factor Distal-less, which marks what will be the distal tip of the leg. Although *Distal-less* is also expressed in *Drosophila* wing development, it appears to have no function there, as if the gene is removed, there is no effect on wing development. It is, however, crucial for correct leg development.

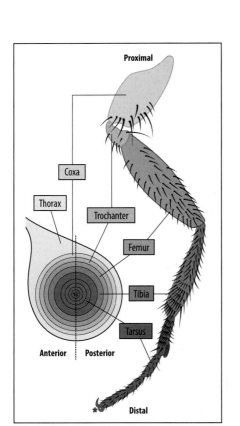

Fig. 11.37 Fate map of the leg imaginal disc of *Drosophila*. The disc is a roughly circular epithelial sheet, which becomes transformed into a tubular leg at metamorphosis. The center of the disc becomes the distal tip of the leg and the circumference gives rise to the base of the leg—this defines the proximo-distal axis. The tarsus is divided into five tarsal segments. The presumptive regions of the adult leg, such as the tibia, are thus arranged as a series of circles with the future tip at the center. A compartment boundary divides the disc into anterior and posterior regions. There is no division of the leg disc into dorsal and ventral compartments as there is in the wing.

Illustration after Bryant, P.J.: **The polar coordinate model goes molecular**. Science *1993,* **259***: 471–472.*

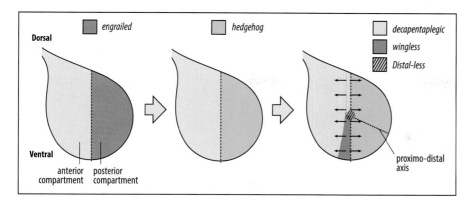

Fig. 11.38 Establishment of signaling centers in the antero-posterior compartment of the leg disc, and the specification of the distal tip. The gene *engrailed* is expressed in the posterior compartment and induces expression of *hedgehog*. Where Hedgehog protein meets and signals to anterior compartment cells, the gene *decapentaplegic* is expressed in dorsal regions and the gene *wingless* is expressed in ventral regions. Both of these genes encode secreted proteins. Expression of the gene *Distal-less*, which specifies the proximo-distal axis, is activated where the Wingless and Decapentaplegic proteins meet.

Patterning along the proximo-distal axis takes place sequentially, and produces a two-dimensional pattern in the leg disc that will be converted at metamorphosis into the three-dimensional tubular leg. Early expression of *Distal-less* is governed by a particular module in the gene's control region and lasts for just a few hours. Later expression of *Distal-less* in the future distal region is directed by other *cis*-regulatory modules. The gene for another homeodomain transcription factor, *homothorax*, is expressed in the peripheral region surrounding *Distal-less*, which is the future proximal region (Fig. 11.39). Their actions lead to the expression of the gene *dachshund*, which also encodes a transcription factor, in a ring between *Distal-less* and *homothorax*, which eventually leads to some overlap of the expression of *dachshund* with *Distal-less* and *homothorax*. Each of these genes is required for the formation of particular leg regions, but the expression domains do not correspond precisely with leg segments. Expression of *dachshund*, for example, corresponds to femur, tibia, and proximal tarsus. There is no evidence for a proximo-distal compartment boundary; cells expressing *homothorax* and *Distal-less* are, however, prevented from mixing at the interface between their two territories.

Additional patterning of the leg involves a gradient of EGF receptor activity from distal to proximal. In the third instar larva, the EGF receptor ligand Vein and the signaling pathway component Rhomboid are expressed at the central point of the disc. Genes coding for transcription factors Bric-a-brac and Bar are expressed at different levels in the tarsal segments and may determine their identity. The level of activity of these genes may be determined by the gradient of EGF receptor activity. Leg joint formation requires Notch signaling. Delta and Serrate, ligands for Notch, are expressed as a ring in each leg segment and their activation of Notch results in specification of the joint-forming cells.

11.25 Butterfly wing markings are organized by additional positional fields

The variety of color markings on butterfly wings is remarkable: more than 17,000 species can be distinguished. Many of these patterns are variations on a basic 'ground

Fig. 11.39 Regional subdivision of the *Drosophila* leg along the proximo-distal axis. The pattern of gene expression in the leg is shown on the right. The stages of gene expression are shown in the two columns on the left, and are viewed as if looking down on the disc. Because of the way in which the leg extends from the disc, the center corresponds to the future tip of the leg, and the more proximal regions to successive rings around it. *decapentaplegic* (*dpp*) and *wingless* (*wg*) are initially expressed in a graded manner along the antero-posterior compartment boundary and together induce *Distal-less* (*Dll*) in the center and repress *homothorax* (*hth*), which is expressed in the outer region. They then induce *dachshund* (*dac*) in a ring between *Dll* and *hth*. Further signaling leads to these domains overlapping. While *hth* expression corresponds to proximal regions and *Dll* to distal regions, there is no simple relation between the other genes and the leg segments. *After Milan, M., Cohen, S. M.: **Subdividing cell populations in the developing limbs of Drosophila: Do wing veins and leg segments define units of growth control?** Dev. Biol. 2000, **217**: 1-9.*

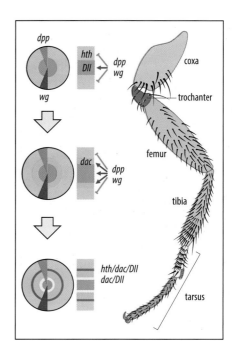

Fig. 11.40 Butterfly wing pattern. Ventral view of a female African butterfly *Bicyclus anynana,* showing wing color pattern and prominent eyespots. Scale bar = 5 mm.

Photograph courtesy of V. French and P. Brakefield.

plan' consisting of bands and concentric eyespots (Fig. 11.40). The wings are covered with overlapping cuticular scales, which are colored by pigment synthesized and deposited by the underlying epidermal cells. How are these patterns specified? Butterfly wings develop from imaginal discs in the caterpillar in a similar way to *Drosophila* wings. Surgical manipulation has shown that the eyespot is specified at a late stage in the development of the wing disc, and that the pattern is dependent on a signal emanating from the center of the spot. A number of the genes that control wing development in *Drosophila*, such as *apterous*, are expressed in the butterfly in a spatial and temporal pattern similar to that in the fly. Thus, as in *Drosophila*, the shape and structure of the butterfly wing is patterned by a field of positional information. The patterning of the pigmentation, however, involves the establishment of additional fields of positional information.

Unlike *Drosophila* wings, development of butterfly wings does involve Distal-less function, and it is this that provides the extra dimension of pattern. The expression pattern of *Distal-less* in butterfly wing discs suggests that the mechanism used to delineate the color pattern on the butterfly wing is similar to the mechanism that specifies positional information along the proximo-distal axis of the insect leg. In the butterfly wing, *Distal-less* is expressed in the center of the eyespot, whereas in the *Drosophila* leg disc it is expressed in the central region corresponding to the future tip of the limb. Thus, it is possible that eyespot development and distal leg patterning involve similar mechanisms, although in the butterfly wing *Notch* expression precedes that of *Distal-less*.

The eyespot may be thought of as a proximo-distal pattern superimposed on the two-dimensional wing surface. The center of the eyespot represents the distal-most positional value, with the surrounding rings representing progressively more proximal positions, as in the leg. The eyespot is likely to be positioned with reference to the primary wing pattern (that is, the anterior, posterior, dorsal, and ventral compartments); a secondary coordinate system centered on the eyespot would then be established, with expression of *Distal-less* defining the central focus.

11.26 Different imaginal discs can have the same positional values

The patterning of legs and wings involves similar signals, such as Dpp and Wingless, yet the actual pattern that develops is very different. This implies that the wing and leg discs interpret positional signals in different ways. This interpretation is under two distinct types of control, both involving transcription factors. The first centers on the wing-specific transcriptional co-factor Vestigial. As discussed earlier (see Fig. 11.29), Vestigial channels the positional information provided by the compartment boundary signaling centers specifically towards genes involved in building wing-specific structures. Indeed, forcing the ectopic expression of Vestigial at places at which Wingless, Dpp, and Notch signals converge triggers the development of wing tissue.

The second type of control relates to the segments from which the appendages derive. It answers the question of why *Drosophila* wings develop in only one particular thoracic segment, legs in all three, and antennae on the head. This form of control is associated with the Hox genes and can be illustrated in regard to leg and antenna. If the Hox gene *Antennapedia*, which is normally expressed in parasegments 4 and 5 (see Section 2.31), and specifies the imaginal discs for the second pair of legs, is expressed in the head region, the antennae develop as legs (Fig. 11.41).

Expression of *Antennapedia* in the head region, where it is not normally expressed, converts the whole antennal imaginal disc to a leg disc and leads to the development of an ectopic leg, which is a second thoracic (T2) leg. But using the technique of mitotic recombination (see Box 2H), it is possible to generate a clone of

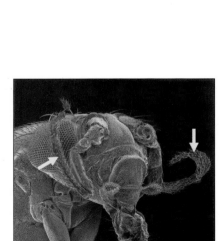

Fig. 11.41 Scanning electron micrograph of *Drosophila* carrying the *Antennapedia* mutation. Flies with this mutation have the antennae converted into legs (arrows). Scale bar = 0.1 mm.

Photograph by D. Scharfe, from Science Photo Library.

Antennapedia-expressing cells within a normal antennal disc. These cells develop as leg cells, and exactly which type of leg cell depends on their position along the proximo-distal axis in the disc. If, for example, they are at the tip, they form a claw. It is as if the positional values of the cells in the antenna and leg are similar, and the difference between the two structures lies in the interpretation of these values, which is governed by the expression or non-expression of the *Antennapedia* gene (Fig. 11.42). We therefore see that cells develop according to both their position and their developmental history, which determines which genes they are expressing at any given time. *Antennapedia* is normally expressed in T2 and T3, allowing legs to develop on those segments, while development of legs on T1 is due to expression of the Hox gene *Sex combs reduced*.

This principle applies also to wing and haltere imaginal discs, which develop in adjacent thoracic segments. They are distinguished by the expression of the Hox gene *Ubx* in the third thoracic segment, which produces the haltere, but not in the second, which produces the wing.

Thus, we find a similarity in developmental strategy between insects and vertebrates; both use the same positional information in appendages such as legs and wings, and interpret it differently. Still to be resolved is the question of how the expression of a single transcription factor like Antennapedia can transform an antenna into a leg. This requires an understanding of its downstream targets. So far, there is evidence that Antennapedia acts as a repressor of antennal identity in the leg by, for example, preventing the co-expression of *homothorax* and *Distal-less* in the femur region. These two genes are expressed together in the corresponding region in the antennal disc, but in the leg disc they are expressed in adjacent and non-overlapping domains (see Fig. 11.39). In evolutionary terms, this indicates that the antennal state may be the ground state for limb development.

The character of a disc and how positional information is interpreted is therefore strongly influenced by the pattern of Hox gene expression (or the absence of Hox expression in the case of the antennal disc). The segmental distribution of wings and legs illustrates this very clearly. Insect legs develop only on the three thoracic segments and not on the abdominal segments and, in *Drosophila*, wings develop only on the second thoracic segment. These adult structures are segment-specific because the particular type of imaginal disc that gives rise to them is only formed by certain parasegments (Fig. 11.43). There are no appendages on abdominal segments in *Drosophila* because the genes required for formation of leg and wing discs are suppressed in the abdomen. The type of disc formed in a particular thoracic segment is typically specified by the action of one of the Hox genes expressed in the segment. For example, expression of the genes *Antennapedia* and *Ultrabithorax* specify the second and third pair of legs, respectively.

The leg imaginal discs arise from small clusters of ectodermal cells in parasegments 3–6, which contribute to the future thoracic segments of the embryo (see Fig. 11.43). The discs arise at the parasegment boundaries, with anterior and posterior compartments of adjacent parasegments contributing to each disc. In the future second thoracic segment, the leg disc splits off a second disc early in its development, which becomes the wing disc. Similarly, an additional disc, which develops into the haltere, a balancing organ, is formed in the future third thoracic segment.

Because imaginal discs are formed across parasegment boundaries, mutations in Hox genes in *Drosophila* embryos can cause compartment-specific homeotic transformations of halteres into wings. In normal *Drosophila* adults, the wing is on the second thoracic segment and the haltere on the third thoracic segment; they arise from imaginal discs originating at the boundaries of parasegments 4/5 and 5/6, respectively (see Fig. 11.43). In the normal embryo, the *Ultrabithorax* gene, one of the genes of the bithorax complex (see Fig. 2.49), is expressed in parasegments 5 and 6 and is involved

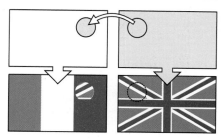

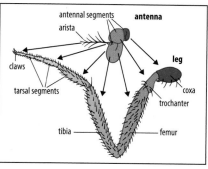

Fig. 11.42 Cells interpret their position according to their developmental history and genetic make-up. If two flags used the same positional information to produce different patterns, then a graft from one to the other would result in the graft developing according to its new position, but with its original pattern (upper panels). Imaginal discs similarly use the same positional information to produce differently patterned appendages (lower panel).

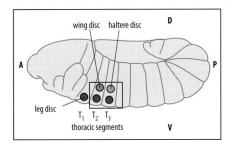

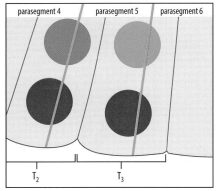

Fig. 11.43 The position in the late *Drosophila* embryo of the imaginal discs that give rise to the adult thoracic appendages. The imaginal discs for the legs, wings, and halteres lie across the parasegment boundaries in the thoracic segments, as shown for the wing and haltere discs in the lower panel.

in specifying their identity. The *bithorax* mutation (*bx*), which causes the *Ultrabithorax* gene to be misexpressed, can transform the anterior compartment of the third thoracic segment, and thus of the haltere, into the corresponding anterior compartment of the second segment: the anterior half of the haltere thus becomes a wing (see Fig. 2.50). The *postbithorax* mutation (*pbx*), which affects a regulatory region of the *Ultrabithorax* gene, transforms the posterior compartment of the haltere into a wing (Fig. 11.44). If both mutations are present in the same fly, the effect is additive and the result is a fly that has four wings but cannot fly (see Fig. 2.50). Another mutation, *Haltere mimic*, causes a homeotic transformation in the opposite direction: the wing is transformed into a haltere.

A most dramatic example of these transformations can be observed if one deletes the whole bithorax complex, responsible for the patterning of all segments posterior to the second thoracic segment (T2). These mutants die as embryos and all their abdominal segments develop like T2 (see Fig. 2.51, second panel). In each segment it is possible to detect a wing and a leg disc, which cannot develop fully because the embryo cannot hatch.

As with the antenna and leg, it is possible to generate a mosaic haltere with a small clone of cells containing an *Ultrabithorax* mutation (such as *bithorax*) in the haltere imaginal disc; the cells in the clone make wing structures, which correspond exactly to those that would form in a similar position in a wing. It is as if the positional values in haltere and wing discs are identical, and all that has been altered in the mutant is how this positional information is interpreted. In fact, other imaginal discs seem to have similar positional fields.

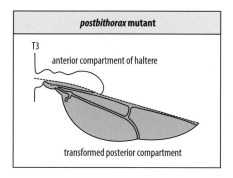

Fig. 11.44 Effects of mutations in the bithorax complex on the identity of the imaginal discs. The mutation *postbithorax* acts on the posterior compartment of a haltere, converting it into the posterior half of a wing. Deletion of all of the bithorax complex genes leads to an embryo with a wing and a leg disc in every segment (not shown).

SUMMARY

The legs and wings of *Drosophila* develop from epithelial sheets—imaginal discs—that are set aside in the embryo. Hox genes acting in the parasegments specify which sort of appendage will form. The leg and wing imaginal discs are divided at an early stage into anterior and posterior compartments. The boundary between the compartments is an organizing center and a source of signals that pattern the disc. In the wing disc, expression of *decapentaplegic* is activated by the Hedgehog protein at the antero-posterior compartment boundary and the Decapentaplegic and Hedgehog proteins act as antero-posterior patterning morphogens. The wing disc also has dorsal and ventral compartments, with Wingless produced at the dorso-ventral boundary acting as a patterning signal. The essential wing-specific regulatory gene *vestigial* integrates this positional information through its control regions and channels it towards genes that build wing-specific structures and that control wing growth. In the leg disc, the antero-posterior compartment boundary is established in a very similar way, except that Hedgehog activates *wingless* instead of *decapentaplegic* in the ventral region and the Wingless protein acts as the patterning signal in this region. There is no evidence of dorsal and ventral compartments in the leg disc. The proximo-distal axis of the leg is specified by the interaction between Decapentaplegic and Wingless proteins, which activates genes such as *dachshund* and *homothorax* in concentric domains corresponding to regions along the leg. The colorful eyespots on butterfly wings may be patterned by a mechanism similar to that used to organize the proximo-distal axis of the insect leg.

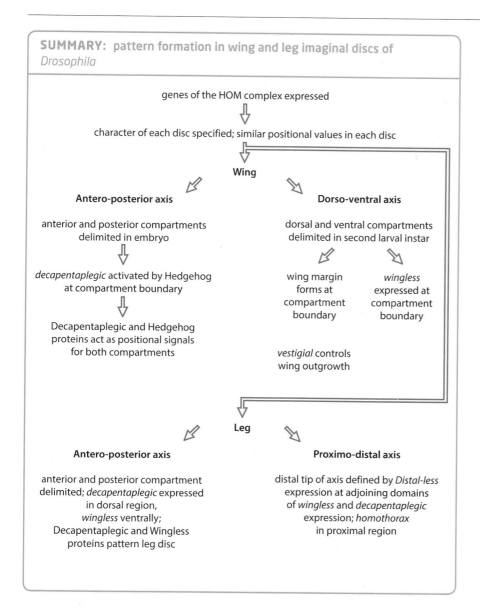

SUMMARY: pattern formation in wing and leg imaginal discs of *Drosophila*

genes of the HOM complex expressed

character of each disc specified; similar positional values in each disc

Wing

Antero-posterior axis

anterior and posterior compartments delimited in embryo

decapentaplegic activated by Hedgehog at compartment boundary

Decapentaplegic and Hedgehog proteins act as positional signals for both compartments

Dorso-ventral axis

dorsal and ventral compartments delimited in second larval instar

wing margin forms at compartment boundary

wingless expressed at compartment boundary

vestigial controls wing outgrowth

Leg

Antero-posterior axis

anterior and posterior compartment delimited; *decapentaplegic* expressed in dorsal region, *wingless* ventrally; Decapentaplegic and Wingless proteins pattern leg disc

Proximo-distal axis

distal tip of axis defined by *Distal-less* expression at adjoining domains of *wingless* and *decapentaplegic* expression; *homothorax* in proximal region

Vertebrate and insect eyes

Structures as complex as the compound eyes of insects and the 'camera' eyes of vertebrates are a remarkable achievement of evolution (Fig. 11.45). Camera eyes and compound eyes share some basic similarities. They all contain a lens to focus the light, a retina composed of light-sensing photoreceptor cells, and a pigmented layer that absorbs stray light and prevents it interfering with photoreceptor signaling. And despite the great difference in the anatomy of the final eye, some of the same transcription factors specify eye formation in insects and vertebrates. We shall begin here with the development of the vertebrate camera eye, which is essentially an offshoot of the forebrain. In the vertebrate eye, light enters through the pupil at the front of the eye and passes through a convex transparent lens, which focuses the light on the photosensitive retina lining the back of the eyeball (see Fig. 11.45). Photoreceptor cells in the retina—the rods and cones—register the incoming photons and pass signals on to nerve cells, which transmit them via the optic nerve to the brain, where they are decoded.

A long-standing puzzle about the vertebrate eye is that the photoreceptors occupy the innermost layer of the retina rather than the surface adjacent to the vitreous

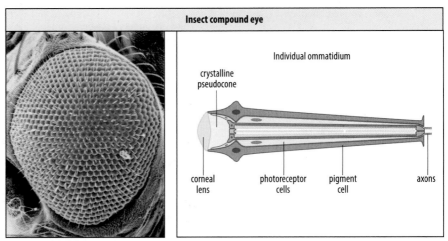

Fig. 11.45 Vertebrate and insect eyes. Vertebrate eyes (upper panel) are 'camera' eyes with a single lens that focuses light on photoreceptor cells in the neural retina lining the back of the eyeball. The space between the lens and the neural retina is filled with vitreous humor. The photoreceptor cells connect to retinal neurons, whose axons form the optic nerve, which connects the eye to the brain. The different cell layers of the neural retina are indicated. A, anterior; P, posterior. The *Drosophila* compound eye (lower panel) is composed of numerous individual light-sensing structures called ommatidia; a section through one is shown on the right. Each ommatidium has a lens, through which light passes to activate the photoreceptor cells, which are surrounded by pigment cells, and whose axons pass through the optic stalk to the brain.

humor, where they would get more light and produce a sharper image, as light would not be dispersed and diffracted by the overlying layers of cells (see Fig. 11.45). A potential solution to the puzzle is the demonstration that radial glia called Müller cells, which span the whole thickness of the retina, act as fibre-optic light guides, efficiently transmitting the image pixel by pixel from the surface to the photoreceptors.

11.27 The vertebrate eye develops mainly from the neural tube and the ectoderm of the head

In developmental terms, the vertebrate eye is essentially an extension of the forebrain, together with a contribution from the overlying ectoderm and surrounding mesenchyme, mainly migrating neural crest cells. The development of an eye starts at E8.5 in the mouse and around 22 days in human embryos with the formation of a bulge, or **evagination**, in the epithelial wall of the posterior forebrain, in the region called the diencephalon. This evagination is called the **optic vesicle**; an optic vesicle is formed on either side of the head and extends to meet the surface ectoderm (Fig. 11.46). The optic vesicle interacts with the ectoderm to induce formation of the lens **placode**, which is a thickened region of ectoderm from which the lens will develop. The lens placode is part of a larger region of head ectoderm that gives rise to the epithelial placodes of some other sensory organs, including the placode that gives rise to the semi-circular canals, cochlea, and endolymphatic duct of the ear, and the placode that gives rise to the olfactory epithelium in the nose.

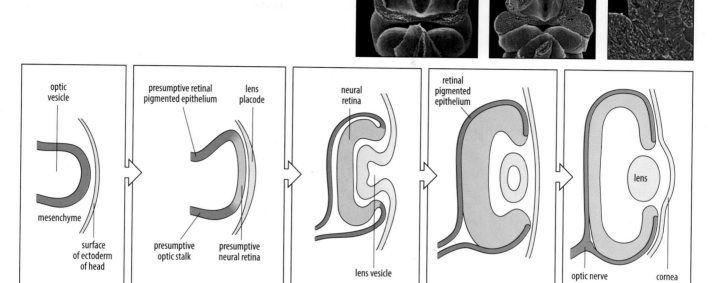

Fig. 11.46 The main stages in the development of the vertebrate eye. The schematic diagrams of the development of the optic cup and lens show that the optic vesicle (blue) develops as an outgrowth of the forebrain and induces the lens placode in the surface ectoderm (yellow) of the head. The optic vesicle invaginates to form a two-layered cup, the optic cup, around the developing lens (the lens vesicle). The inner layer of the optic cup forms the neural retina (pale blue) and the outer layer the pigmented epithelium underlying the retina (dark blue). The lens vesicle detaches from the surface ectoderm to form the lens and the remaining overlying ectoderm forms the cornea. The iris (not shown) develops from the rim of the optic cup. The fluid-filled space between the iris and the cornea is the anterior chamber of the eye. The scanning electron micrographs show frontal sections through the head of a mouse embryo showing (a) formation of the optic vesicle (E9–9.5); (b) formation of the lens vesicle and optic cup (E10.5); and (c) high-magnification view of the optic cup and lens vesicle in (b).

*Diagram adapted from Adler, R., Valeria Canto-Soler, M.: **Molecular mechanisms of optic vesicle development: Complexities, ambiguities, and controversies**. Dev. Biol., 2007, **305**: 1–13.*

*Electron micrographs from Heavner, W., Pevny, L.: **Eye development and retinogenesis**. Cold Spring Harb. Perspect. Biol. 2012, **4**: a008391.*

After induction of the lens placode, the tip of the optic vesicle, adjacent to the placode, invaginates to form a two-layered cup, the optic cup. The inner epithelial layer of the optic cup will form the neural retina, while the outer layer will form the retinal pigment epithelium. The invagination of the lens placode is coordinated with the invagination of the optic vesicle and the placode then detaches from the surface ectoderm to form a small hollow sphere of epithelium, the lens vesicle, that will develop into the lens, while the remaining surface ectoderm fuses to give rise to the cornea.

The lens is formed by proliferation of cells of the epithelium on the anterior side of the lens vesicle—that is, the side nearer the cornea—with the new cells moving into the center of the lens, where they start to manufacture crystallin proteins. The cells eventually lose their nuclei, mitochondria, and internal membranes to become completely transparent lens fibers filled with crystallin. Renewal of lens fibers continues after birth, but proceeds much more slowly than in the embryo. In the adult chicken, the transformation of an epithelial cell into a crystallin-filled fiber takes two years.

The cornea is a transparent epithelium that seals the front of the eye. It is composed of inner and outer layers with different developmental origins. The inner layer is formed by mesenchymal neural crest cells that migrate into the anterior eye chamber to form a thin layer initially overlying the lens. Neural crest cells also contribute to other structures in the anterior part of the eye. The outer layer of the cornea is derived from the surface ectoderm adjacent to the eye. Most of the iris develops from the rim of the optic cup—also

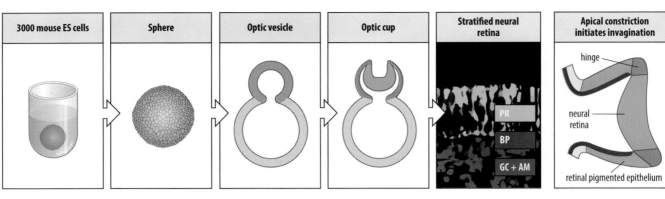

| 3000 mouse ES cells | Sphere | Optic vesicle | Optic cup | Stratified neural retina | Apical constriction initiates invagination |

PR, BP, GC + AM

hinge, neural retina, retinal pigmented epithelium

Fig. 11.47 The generation of an optic cup in culture. Left panels: floating aggregates of mouse ES cells are grown in serum-free medium containing extracellular matrix components including laminin. They form spheres. A retinal epithelial hemisphere subsequently evaginates from the sphere to give an optic-vesicle-like structure. The distal region of the vesicle then undergoes invagination to form an optic cup. The photomicrograph shows a cross-section of the stratified neural retina of the self-organized optic cup with various recognizable cell types. PR, photoreceptors; BP, bipolar cells; GC + CM, ganglion cells + amacrine cells. Right panel: the hinge region (blue) at the junction between the region of epithelium that will give rise to the pigmented retina (pink and red) and the neural retina (green). Apical constriction of cells at the hinge initiates invagination.

*Photomicrograph from Eiraku, et al.: **Self-organizing optic-cup morphogenesis in three-dimensional culture.** Nature 2011, 472, 51-56.*

known as the ciliary margin—with neural crest contributing to the anterior iris. In the adult eye, potential stem cells have been identified in the ciliary margin but whether they are able to self-renew and to give rise to all the cell types in the eye is as yet unclear.

The neural retina in the vertebrate eye develops three distinct layers of cells, with the photoreceptor cells forming the innermost layer, underneath layers of ganglion and bipolar cells (see Fig. 11.45). This is in contrast to the camera eyes of octopus and squid in which the photoreceptor cells are on the surface of the neural retina. Visual signals are transmitted from the eye to the brain via the optic nerve, which is formed by the axons of the neural retina ganglion cells. In Chapter 12 we will see how the neurons of the optic nerve connect with precise positions in the visual-processing centers in the brain in order to produce a 'map' of the retina.

The morphogenesis of the optic cup involves outgrowth and coordinated folding of epithelial cell sheets. Although the optic vesicle and the lens placode are tightly apposed when invagination begins, the optic vesicle can still invaginate in the absence of the lens placode. This conclusion came from classical work carried out in the 1930s on eye development in amphibian embryos. When an optic vesicle is transplanted to the trunk, the optic vesicle can develop into an optic cup, even though trunk ectoderm is unable to form a lens. The capacity of the optic vesicle for self-organization has recently been revealed even more dramatically in three-dimensional aggregates of mouse ES cells, which had been induced to differentiate into retinal epithelial cells and cultured with basement membrane components (Fig. 11.47). After about a week in culture, the ES cell aggregates had formed hollow epithelial spheres, and hemispherical bulges of epithelium had begun to evaginate from the main body of the spheres, thus mimicking the evagination of the optic vesicle from the diencephalon at the start of eye development in the embryo. Even more remarkably, over the next 2 days of culture the hemispherical bulges underwent further dynamic changes in shape to give rise to two-walled cup-like structures mimicking the invagination of the optic vesicle to give rise to the optic cup in the embryo. Cells in the inner wall were shown to express genes characteristic of neural retina and cells in the outer wall those characteristic of pigmented retina. The cells in the inner wall subsequently differentiated into all the main neuronal cell types in the retina, including photoreceptors, arranged in appropriate layers, whereas the cells in the outer wall became pigmented.

The entire process of generation of the optic cup can be followed in some detail by imaging the living three-dimensional cell aggregates in which cells in the 'eye field' (the region across the center of the anterior neural plate from which the eyes will form) and later in the neural retina express a green fluorescent protein reporter gene. The movies revealed, for example, that curvature of the epithelium is associated with changes in the shape of the epithelial cells. In places where the epithelium becomes folded at an acute angle, hinge points with wedge-shaped cells like those seen during neurulation (see Section 9.14) are observed. Self-formation of a two-walled cup and development of a stratified neural retina can similarly be produced from aggregates of human ES cells. The ability of human ES cells to form 3D structures given the appropriate conditions is encouraging in the context of engineering tissues for regenerative medicine (see Box 8C). The two-walled cups can also be frozen and stored, and provide useful material for studying degenerative diseases of the human eye.

Although evagination of the optic vesicles does not occur until neural tube closure is almost completed, the specification of cells as prospective eye cells occurs much earlier, in the neural plate. Genes encoding key conserved eye-specifying transcription factors, such as Pax6, Six3, and Otx2, are expressed in anterior neural plate at late gastrula stages as part of the initial antero-posterior patterning of the neural plate. They continue to be expressed in the optic vesicle epithelium and the lens placode, as well as in other precursors of sensory organs such as the olfactory placode. Retinal precursor cells are first specified as a single eye field across the center of the anterior neural plate. The eye field eventually becomes separated into two lateral regions of cells that will give rise to the optic vesicles after neural tube formation. Separation is achieved by downregulating expression of Pax6 and the other eye-specifying transcription factors along the midline. Failure of this separation could be one of the causes of cyclopia in human embryos, in which a single abnormal central eye is produced. The normal downregulation of the expression of Pax6 and other transcription factors in the forebrain midline is likely to be mediated by Shh signaling, which is known to be required for the correct specification of ventral midline structures in the brain. As we discussed in Chapter 1 (see Box 1F), a failure of Shh signaling can result in the condition of holoprosencephaly, which in its most severe form results in failure of the forebrain to divide into right and left hemispheres, and loss of midline facial structures, resulting in cyclopia.

Later on, dorso-ventral patterning of the optic cup involves Shh signaling ventrally and BMP-4 signalling dorsally, as it does throughout the entire central nervous system. Other signals, including Notch, various BMPs, Wnts, FGFs, TGF-β and retinoic acid, some of them produced by the neural crest surrounding the optic cup, are important in inducing and maintaining differentiation of the cells in different regions of the optic cup. Thus, for example, *Fgf9* is expressed in the region of the optic cup destined to be neural retina. Expression of ectopic *Fgf9* leads to a duplicate neural retina; on the other hand, when *Fgf9* is lacking, the pigmented retinal epithelium extends into the region that would normally form neural retina.

The same set of transcription factors is essential for eye formation throughout the animal kingdom. The classic example of a gene with a conserved basic function is *Pax6*, which is required for the development of light-sensing structures in all bilaterian animals (animals with bilateral symmetry)—from the simple light-sensing organs of planarians to the compound eyes of insects and the camera eyes of vertebrates and cephalopods.

Pax6 was initially identified from the genetic analysis of mutations causing abnormal eye development in mice and humans, and is homologous with the *Drosophila* gene *eyeless*. Mouse embryos with defective *Pax6* function have smaller eyes than normal or no eyes at all. People heterozygous for mutations in *PAX6* have various eye malformations collectively known as **aniridia**, because of the partial or complete absence of the iris; these patients also have cognitive defects, as *PAX6* has a number of roles in brain development other than eye formation. In the very rare cases of homozygous mutations in *PAX6*, eyes are absent, and this may be associated with other severe malformations that can be incompatible with life.

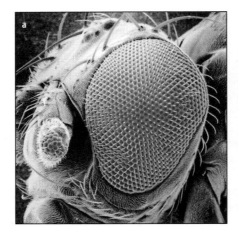

Fig. 11.48 *Pax6* **is a master gene for eye development.** a, The ectopic expression of mouse *Pax6* in a *Drosophila* antennal disc results in compound eye structures developing on the antenna. b, a high-power view of the eye-like structure.

Photographs reproduced with permission from Gehring, W.J.: **New perspectives on eye development and the evolution of eyes and photoreceptors**. J. Hered. *2005,* **96(3):** *171-184.*

In *Xenopus*, injection of *Pax6* mRNA into an animal pole blastomere at the 16-cell stage leads to formation of ectopic eye-like structures, with well-formed lenses and epithelial optic cups, in the tadpole head. Even more amazingly, the ectopic expression of *Pax6* from mouse, *Xenopus*, ascidians, or squid in *Drosophila* imaginal discs causes the development of *Drosophila*-type compound eye ommatidia on adult structures such as antennae (Fig. 11.48).

Another transcription factor, Six3, is also involved in eye development in both vertebrates and *Drosophila* (the *Drosophila* transcription factor is encoded by the *sine oculis* gene), and can induce eye structures when overexpressed in fish embryos. Thus there appears to be a conserved network of 'eye' transcription factors that govern development of this sense organ.

Formation of the lens is a crucial step in development of anterior eye structures such as the cornea and the iris. Among much other evidence, this is nicely demonstrated by experiments in the teleost fish *Astyanax mexicanus*, which has both surface-dwelling sighted forms and cave-dwelling blind forms—called cavefish (Fig. 11.49). A small optic primordium with an optic cup and a lens is formed in cavefish embryos but the lens then undergoes massive apoptosis. An expansion of Shh signaling along the embryonic midline of the brain in the cavefish has been implicated in lens apoptosis. Because the lens does not develop, the cornea, iris, and other eye structures at the anterior of the eye also fail to form. A lens from the surface-dwelling *A. mexicanus* transplanted into a cavefish optic cup during embryogenesis continues to grow and differentiate and also dramatically restores eye development, indicating that the absence of a lens is at least partly responsible for the non-development of the cavefish eye. The effect of expanded Shh signaling in eye development was investigated by increasing Shh expression in embryos of surface-dwelling *Astyanax*, which normally develop into sighted adults. When *Shh* mRNA was injected into one side of a cleavage-stage embryo, expression of the eye gene *Pax6* was downregulated unilaterally in the corresponding developing eye region, and the larvae that developed lacked an eye on that side of the head.

11.28 Patterning of the *Drosophila* eye involves cell–cell interactions

The adult *Drosophila* compound eyes develop from imaginal discs at the anterior end of the embryo (see Fig. 2.6). The *Drosophila* compound eye is quite different in structure from the vertebrate eye. It is composed of about 800 identical photoreceptor organs called **ommatidia** (singular **ommatidium**) arranged in a hexagonal array of crystalline regularity (Fig. 11.50). Each fully developed ommatidium is made up 20 cells: eight photoreceptor neurons (R1–R8), together with four overlying cone cells (which secrete the lens), and eight additional pigment cells (not shown in the inset in Fig. 11.50). The genetic analysis of ommatidium development has provided one of the

Fig. 11.49 The two different forms of *Astyanax mexicanus.* Surface-living fish have eyes and are pigmented (upper photo). In cave-dwelling fish (cavefish) the eyes do not develop and pigmentation has been lost.

Photographs courtesy of A. Strickle, Y. Yamamoto, and W. Jeffery.

Fig. 11.50 Scanning electron micrograph of the compound eye of adult *Drosophila*. Each unit is an ommatidium. At the third larval instar, the ommatidial cluster (inset) is made up of eight photoreceptor neurons (R1-R8) and four cone cells (which will secrete the lens of the adult eye (see Fig. 11.45)). The inset shows the orientation of an ommatidial cluster in the dorsal half of the eye. Ommatidia in the ventral half are oriented in the opposite direction, such that the eye has mirror-image symmetry around the equator (red line). Dorsal is to the top of the photograph and ventral is to the bottom. Scale bar = 50 μm.

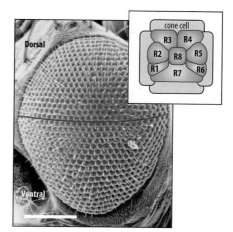

best model systems for studying the patterning of a small group of cells. An important early finding from lineage analysis was that the pattern of each ommatidium is specified by cell–cell interactions and is not based on cell lineages.

As we saw in Section 11.27, the gene *Pax6* is the master gene for eye development throughout the animal kingdom. The *Drosophila* version of Pax6 is called *eyeless*, as mutations in this gene result in the reduction or complete absence of the compound eye. *eyeless* is expressed in the region of the eye disc anterior to the morphogenetic furrow (see Fig. 11.51). Ectopic expression of the *eyeless* gene in other imaginal discs results in the development of ectopic eyes, and these have been induced in this way in wings, legs, antennae, and halteres. The fine structure of these ectopic eyes is remarkably normal, and distinct ommatidia are present, although they are not connected to the nervous system. It is estimated that some 2000 genes are eventually activated as a result of *eyeless* activity, and all of them are required for eye morphogenesis. The function of *eyeless* in the eye disc may be similar to that of *vestigial* in the wing disc, as it appears to change the interpretation of positional information in discs in which it is ectopically expressed.

The eye develops from the single-layered epithelial sheet of the eye imaginal disc, located at the anterior end of the larva. Specification and patterning of the cells of the ommatidia begins in the middle of the third larval instar. Patterning starts at the posterior of the eye disc and progresses anteriorly, taking about 2 days, during which time the disc grows eight times larger.

One of the earliest events in eye differentiation is the formation of a groove, the **morphogenetic furrow**, which sweeps across the disc from posterior to anterior in response to a wave of signals that initiate development of ommatidia from the eye disc cells. As the furrow moves across the epithelium from posterior to anterior, clusters of cells that will give rise to the ommatidia, appear behind it, spaced in a hexagonal array (Fig. 11.51). The furrow moves slowly across the disc, at a rate of 2 hours per row of ommatidial clusters. As the morphogenetic furrow moves forward, the cells behind it start to differentiate to form regularly spaced ommatidia. The ommatidia are arranged in rows, with each row half an ommatidium out of register with the previous one. This gives the characteristic hexagonal packing arrangement. The first cells to differentiate are the R8 photoreceptor neurons. These appear as regularly spaced cells in each row, separated from each other by about eight cells. This separation sets the spacing pattern for the ommatidia.

The passage of the morphogenetic furrow is essential for differentiation of the ommatidia, as mutations that block its progress also block the differentiation of new rows of ommatidia, resulting in a fly with abnormally small eyes. Although there are no anterior and posterior compartments in the eye discs, the cells just behind the furrow can be regarded as resembling posterior cells, similar to the situation in the wing-disc posterior compartment (see Section 11.20). They secrete Hedgehog protein, which triggers the expression of *decapentaplegic*, which in turn causes the cells to become competent to form neural tissue. The gene *wingless* also plays a part in eye disc patterning. It is expressed at the lateral edges of the eye disc, and prevents the furrow starting in these regions. We thus see that, even though imaginal discs give rise to very diverse structures, the key signals involved in patterning the leg, wing, and eye are similar, although they have different roles in each case.

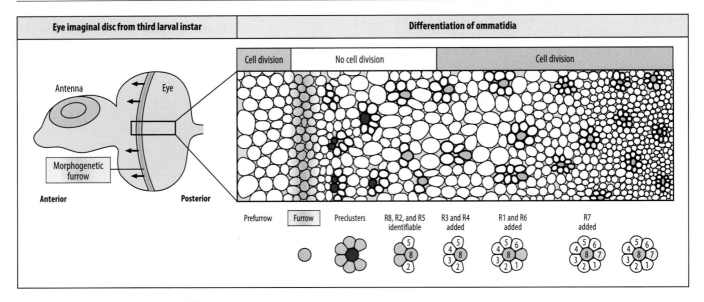

Fig. 11.51 Development of ommatidia in the *Drosophila* compound eye. The compound eye develops from the eye imaginal disc, which is part of a larger disc that also gives rise to an antenna. During the third larval instar, the morphogenetic furrow develops in the eye disc and moves across it in a posterior to anterior direction. Ommatidia develop behind the furrow, the photoreceptor neurons being specified in the order shown, R8 developing first and R7 last. The individual ommatidia are regularly spaced in a hexagonal grid pattern.

Illustration after Lawrence, P.: The Making of a Fly. Oxford: Blackwell Scientific Publications, 1992.

Once specified, each R8 cell initiates a cascade of signals that eventually recruits a cluster of 20 cells that form the mature ommatidium. The first cells to be recruited are the prospective photoreceptor cells, which are **sensory neurons**. R2 and R5 differentiate on either side of R8, to form two functionally identical neurons. R3 and R4, which are a slightly different type of photoreceptor, are specified next. All these cells become arranged in a semi-circle with R8 at the center. R1 and R6 differentiate next and almost complete the circle, which is finally closed by the differentiation of R7 adjacent to R8 (see Fig. 11.51).

The clusters then rotate 90°, so that R7 comes to be closest to the equator of the disc and R3 furthest away; rotation is in the opposite direction in the dorsal and ventral halves of the eye. This means that the dorsal and ventral regions of the eye have distinct and different polarities, with ommatidia on the dorsal and ventral sides of the equator having mirror-image symmetry (see Fig. 11.50). The polarization of the ommatidia is yet another example of planar cell polarity (see Box 2I) and involves a higher level of Frizzled signaling in R3 than in R4.

The determination of the equator of the eye results from the specification of the dorsal region of the eye disc by members of the *Iroquois* gene complex, which is followed by specification of the equator itself, involving the actions of Notch and its ligands Serrate and Delta, in a similar way to the specification of the dorso-ventral compartment boundary in the wing (see Section 11.21).

The regular spacing of the ommatidia within the eye involves a **lateral inhibition** mechanism that spaces the R8 cells. All cells in the eye disc initially have the capacity to differentiate as R8 cells, and as the morphogenetic furrow passes they start to do so. But some inevitably gain a lead and are thus able to inhibit the differentiation of another R8 cell over a range of around three cell diameters. Cells that will give rise to R8 express the gene *atonal*. Inhibitors of *atonal* that space the R8 cells are the secreted Scabrous protein and Notch. In the eye, cell fate is specified and determined cell by cell, not in groups of cells.

Two proteins crucial in the patterning of an individual ommatidium are the *Drosophila* EGF receptor DER, and its ligand Spitz, a membrane-tethered EGF-like molecule. One model for patterning the ommatidium is based on both EGF receptor activation and the age of the cells (Fig. 11.52). Spitz is produced by the three earliest-specified and most centrally located cells—R8, R2, and R5. It activates the EGF receptor on neighboring cells, and this recruits R3, R4, R1, R6, and R7 to a photoreceptor fate. The responding cells also secrete the protein Argos. This diffuses away and inhibits more distant cells from being activated by Spitz, so that no more cells in the prospective ommatidial cluster develop as photoreceptors. The actual character of each photoreceptor may be determined by the age of the cell, with cells passing through a series of 'states', each representing a potential fate. Other signals are also involved, however. The difference between R3 and R4, for example, involves Notch signaling; the cell with the high level of Notch activity is inhibited from becoming R3, and becomes R4. After the photoreceptors have differentiated, the four lens-producing cone cells develop, and finally the surrounding ring of accessory cells.

The specification of R7 as a photoreceptor cell is one of the best-understood cases of the specification of cell fate on an individual cell basis. It requires expression of the gene *seopenless* in the prospective R7 and *bride-of-sevenless* (*boss*) in R8. When either gene is inactivated, the phenotype is the same: R7 does not develop and an extra cone cell is formed. The Sevenless protein is a transmembrane receptor tyrosine kinase, and Boss is its ligand. Sevenless protein is produced not only by R7 but also by other cells in the ommatidium, including lens cells. Thus, expression of Sevenless is a necessary, but not sufficient, condition for R7 specification. Using genetic mosaics it can be shown that for R7 to develop, only R8 need express Boss protein and that this is the signal by which R8 induces R7. Boss is also an integral membrane protein; it is present on the apical surface of the R8 cell, where it makes contact with R7. A second signal for R7 is provided by the R1/R6 pair, which must activate Notch in the R7 cell.

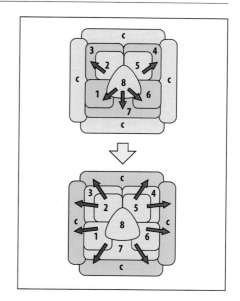

Fig. 11.52 Sequential recruitment of photoreceptors and cone cells during ommatidial development. The three earliest-specified and most centrally located photoreceptor cells—R8, R2, and R5—produce the protein Spitz. It activates the *Drosophila* EGF receptor, DER, on neighboring cells, which recruits R3, R4, R1, R6, and R7 to a photoreceptor fate (top). These then also produce Spitz, which interacts with DER on the prospective cone cells (c) to recruit them to the ommatidium. Once cells are determined they secrete the protein Argos. This diffuses away and inhibits more distant cells from being activated by Spitz, so that no more cells in the prospective ommatidial cluster develop as photoreceptors.

Adapted from Freeman, M.: ***Cell determination strategies in the Drosophila eye.*** Development *1997,* **124***: 261–270.*

SUMMARY

The vertebrate eye develops as an extension of the forebrain. Evagination of the forebrain lateral wall produces the optic vesicle, which in turn induces formation of the lens from the surface ectoderm. Invagination of the tip of the optic vesicle produces a two-layered optic cup, which surrounds the lens and develops into the eyeball, with the inner epithelial layer forming the neural retina, and the outer epithelium developing into the pigmented epithelium at the back of the retina. The optic vesicle has a remarkable capacity for self-organization but the lens is needed for the eye to develop further. The transcription factor Pax6 is essential for eye development in animals as diverse as vertebrates and *Drosophila*. The *Drosophila* compound eye contains around 800 individual ommatidia, arranged in a regular hexagonal pattern, and develops from an imaginal disc. The regular spacing of the ommatidia is achieved by lateral inhibition by the photoreceptor neuron R8. The patterning of the eight photoreceptor neurons in each ommatidium is due to local cell–cell interactions, in which the photoreceptor cells are specified and differentiate in a strict order, initiated by R8.

SUMMARY: development of the ommatidia of *Drosophila* eye

Drosophila

eyeless gene expression required for eye development. Morphogenetic furrow moves across eye disc, associated with hedgehog and decapentaplegic signals

the eight photoreceptors of the future ommatidia begin to develop behind the furrow, R8 developing first and R7 last.
Ommatidia spaced by lateral inhibition

photoreceptors specified by DER activation, other signals, and time;
R7 specification depends on a signal from R8

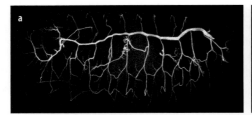

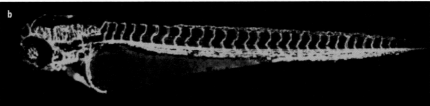

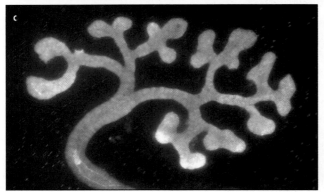

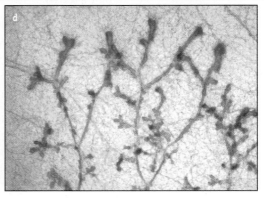

Fig. 11.53 Examples of branching morphogenesis. a, *Drosophila* embryonic tracheal system. b, Zebrafish embryonic vascular system. c, Mouse embryonic kidney. d, Rat virgin mammary gland. The branching epithelia are visualized by antibody staining (panels a, c) or a histological stain (d) or by a fluorescent reporter gene (b).

*(a) reproduced from Luschnig, S., et al.: **Serpentine and vermiform encode matrix proteins with chitin binding and deacetylation domains that limit tracheal tube length in Drosophila**. Curr Biol. 2006, 24, **16**: 186-194.*

*(b) Courtesy Ochoa-Espinosa , A., Affolter, M.: from **Branching Morphogenesis: From cells to organs and back**. Cold Spring Harb. Perspect. Biol. 2012, 4: a008243.*

(c) Image courtesy of J. Davies.

*(d) From Schedin, P., et al.: **Microenvironment of the involuting mammary gland mediates mammary cancer progression**. J Mammary Gland Biol Neoplasia 2007, **12**: 71-82.*

Vertebrate lungs and insect tracheal system

All the structures discussed so far are external, and this has made them relatively easy to study. We now consider some internal organs. A particular feature of their development, which we have not yet encountered, is the formation and branching of tubular epithelia. Epithelia are the most common type of tissue organization in animals, and many organs, such as the lungs, kidneys, and mammary glands in mammals are predominantly composed of functionally specialized epithelia. Epithelial cells adhere tightly together to form a cell sheet, which can be single layered, as in mammalian lung alveoli, or multilayered, as in skin. A common feature of the organs we shall discuss first is that the epithelium forms branching tubes. This phenomenon is known as **branching morphogenesis**. The development of the complex network of interconnecting blood vessels in the vascular system of vertebrates, which are lined by a single layer of epithelium, the **endothelium**, and the network of tubules in the tracheal system in insects also involves extensive branching (Fig. 11.53).

Internal organs in vertebrates, such as the lung, the pancreas, and the mammary glands, all start development as epithelial buds. The position in which these buds develop is determined by the mesoderm as part of the process of the laying down of the body plan. Thus, as in the developing limb, it is the mesoderm that carries the positional information which determines where organs develop. Reciprocal interactions between the epithelial bud and its associated mesoderm are required for outgrowth of the bud and determine its branching morphogenesis.

The crucial role of the mesoderm was first shown in a classical experiment in which the epithelial bud of a developing mammary bud was recombined with the mesoderm associated with an early salivary gland. The branching patterns of the epithelium in the two glands are quite different and mammary gland epithelium combined with salivary gland mesoderm branched like a salivary gland. But despite this abnormal morphogenesis, the epithelial cells still differentiated as mammary gland cells, and produced milk when stimulated. We shall now consider how branching morphogenesis occurs in more detail using the lung as our example.

11.29 The vertebrate lung develops by branching of epithelial tubes

The paired lungs of vertebrates develop from two buds of endoderm (Fig. 11.54). They arise on each side of the end of the embryonic trachea (the windpipe) and grow out to form the bronchi. The lung is one of the internal organs that has left–right asymmetry (see Section 5.16) and this becomes evident very early in its development. Each bud grows out as a tubular epithelium, which will become a bronchus, and in humans it then makes one further branch on the right and three on the left to form the bronchioles. These in turn make finer and finer branches, which terminate in thinwalled sacs of epithelium, the alveoli, which are intimately associated with a highly

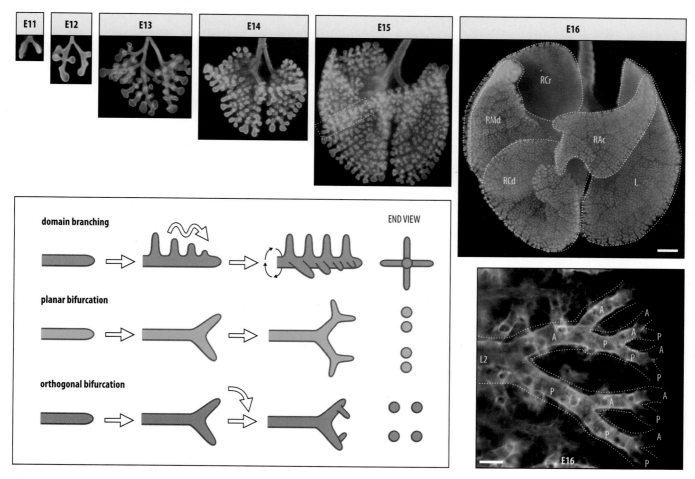

Fig. 11.54 Morphogenesis of the mouse lung involves outgrowth and branching of the bronchial buds. The lung airway epithelium develops by successive branching of the bronchial buds in response to signals from the surrounding mesenchyme. The upper panels show whole mounts of lungs (ventral view) at different embryonic stages immunostained for E-cadherin to show airway epithelium. L, left lobe; RAc, accessory lobe, RCr, right cranial lobe; RCd, right caudal lobe; RMd, right middle lobe. Scale bar = 500 μm. Three basic modes of branching (left lower panel) are used repeatedly and in combination. The simplest of these is planar bifurcation, shown here in the lung of an E16 mouse embryo (right lower panel). L2, lineage from the initial left branch. Sequential branching occurs in the same plane. A, anterior branches; P, posterior branches. Scale bar = 100 μm.

*Images kindly provided by R. Metzger from Metzger, R.J., et al.: **The branching programme of lung development**. Nature 2008, **453**: 745-750.*

branched system of blood capillaries that develops at the same time. Gas exchange occurs in the alveoli—oxygen is taken up by the blood and carbon dioxide released.

The outgrowth and branching of the bronchial tubules in the vertebrate lung is the result of cell proliferation, which is greater towards the tip of the advancing tube. Wnt-5a, which signals via the planar polarity pathway (see Box 8C), is expressed at the tips of the lung tubules and controls outgrowth of the tubule. Mice lacking Wnt-5a have a truncated trachea which branches excessively. We saw that Wnt-5a controls outgrowth of the limb by controlling oriented cell behavior (see Section 11.4) and it may control lung tubule outgrowth in a similar way, although this has yet to be shown.

The patterns of tubule branching in the developing mouse lung have been analyzed in some detail. Just three modes of branching, used in different combinations and at different times, can account for the complex tree-like system of tubules that develops (see Fig. 11.54). The initial sprouting and outgrowth of tubules from the main bronchial tube depends on interaction of the tubular epithelium with signals from the surrounding mesenchyme. The main driver of this sprouting is localized secretion of FGF-10 by the mesoderm cells associated with the lung epithelium; FGF-10 interacts with the receptor FGFR2b expressed by the lung epithelial cells. Therefore, the positions in which mesoderm cells adjacent to the lung epithelium express *Fgf10* will determine the branching pattern. Activation of FGFR2b induces expression of the *Sprouty* gene in the lung epithelial cells. *Sprouty* was first identified by a mutation in *Drosophila* that causes many more branches to form in the tracheal system. Sprouty protein is controlled by FGF signaling and antagonizes its activity, and in the lung it acts as negative feedback to prevent the cells in the main tube away from the tip from forming branches.

Another signaling protein essential for lung development is Shh, which is expressed in endodermal cells at the tips of the extending tubes. Shh inhibits expression of *Fgf10* in the mesoderm, which in turn inhibits *Shh* expression in the endoderm. A reaction–diffusion model (see Box 11E) based on these interactions can mimic the bifurcation mode of lung branching. BMPs, Wnts, Notch, and retinoic acid are also components of the complex network of interacting signaling molecules at the tip of each extending tube that organizes outgrowth and branching.

11.30 The *Drosophila* tracheal system is a prime example of branching morphogenesis

The development of the *Drosophila* **tracheal system**, which delivers oxygen to the tissues, provides an excellent model for branching morphogenesis. Genes such as *Sprouty* were first discovered through *Drosophila* mutants in which branching of the tracheal system is abnormal and were subsequently discovered to control branching in the vertebrate lung.

Air enters the tracheal system of the *Drosophila* larva through openings in the body wall called spiracles, and oxygen is delivered to the tissues by some 10,000 or so fine tubules, which develop during embryogenesis from 20 ectodermal placodes (10 on each side). Each placode produces the tracheal system for one lateral half of a complete segment in the larva. At the start of germ-band retraction, towards the end of embryonic development (see Section 2.2), the placode ectoderm invaginates to form a hollow sac of around 80 cells, which gives rise through successive branching to hundreds of fine terminal branches. Remarkably, the extension of the sacs to form branched tubes does not involve any further cell proliferation but is all achieved by directed cell migration, cell rearrangement by intercalation, and changes in cell shape. As development proceeds, branches from different placodes fuse to form a body-wide network of interconnected tubes (see Fig. 11.54).

The initial invagination of a tracheal placode involves constriction of apical cell surfaces and changes in cell shape, very similar to the internalization of the *Drosophila*

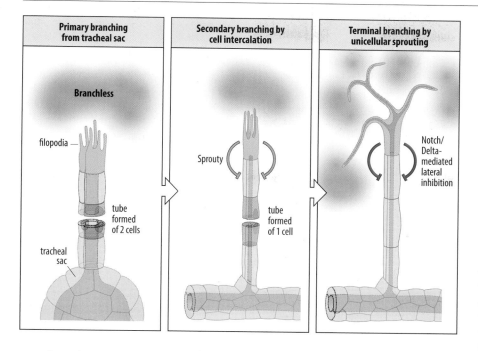

Primary branching from tracheal sac	Secondary branching by cell intercalation	Terminal branching by unicellular sprouting

Fig. 11.55 Branching in the *Drosophila* tracheal system is solely the result of cell rearrangement and remodeling. In the first stage of tracheal development, localized secretion of Branchless protein (blue) causes cells of the tracheal sac epithelium to form filopodia and move towards the source of Branchless, creating a primary branch with walls formed of two cells edge to edge (left panel). A secondary branch is formed by cells intercalating between each other and wrapping round themselves to form a tube of single cells placed end to end, with a lumen running through the center (center panel). At the tip of the secondary branch, expression of the gene *sprouty* is induced, and the Sprouty protein (green) inhibits branching further away from the source of Branchless. Tip cells that will undergo terminal branching are then specified by Notch–Delta signaling. A tip cell puts out fine cytoplasmic extensions that develop a lumen and branch extensively (right panel). Cells behind the tip are inhibited by the Notch signaling (red barred lines) from adopting a tip-cell fate.

mesoderm (see Fig. 9.23). The first tubules are formed when some cells in the wall of the sac develop filopodia, which enable them to migrate toward a source of chemoattractant, drawing an elongating tube of cells behind them (Fig. 11.55). These tubes then branch by a combination of cell intercalation and remodeling of intercellular junctions to produce secondary tubules composed of single cells joined end to end, with each cell wrapped around itself to form the tubule lumen. The chemoattractant guiding tube extension is the protein Branchless, which is the *Drosophila* version of mammalian FGF. It is expressed in clusters of overlying epidermal cells in a pattern determined by the previous antero-posterior and dorso-ventral patterning of the embryo. Branchless acts via the *Drosophila* version of an FGF receptor called Breathless, which is expressed in the tracheal cells and which is also involved in initiating branching. Genes such as *breathless* and *branchless* were identified by mutations that disrupted tracheal morphology—hence their names. Another such gene is *sprouty*, which is expressed by the tracheal cells. Sprouty protein prevents excess branching by antagonizing Breathless signaling, which prevents the more proximal cells from forming branches. In the absence of Sprouty, many more secondary branches form than normal.

Thus, remarkably, the molecules involved in regulating the early branching of the *Drosophila* larval respiratory system are the same as those involved in regulating branching of the embryonic mammalian lung. It has been estimated from screens for mutations that affect tracheal tube growth and branching that more than 200 patterning and morphogenesis genes are required to build the *Drosophila* tracheal system. These screens may identify new genes that are involved not only in branching morphogenesis in the *Drosophila* tracheal system but also in mammalian lung development.

As the tubules extend into the larval body, low oxygen levels lead to local expression of *branchless* in the epidermis. Breathless signaling in response induces cells at the tips of the secondary tubules to sprout to form the fine terminal branches of the tracheal system. Each tip cell forms a much-branched unicellular sprout, with the lumen of the tubules formed within the cell itself (see Fig. 11.55). Excessive terminal branching is prevented by Notch–Delta signaling, which assigns tip-cell fate and prevents the cells away from the tip becoming tip cells and being able to form terminal branches.

SUMMARY

Both the vertebrate lung and the insect tracheal system are examples of the branching morphogenesis of tubular epithelia. In both systems, the branching pattern depends on local cues from the surrounding mesodermal cells and negative feedback from the tip of the epithelial sprouts prevents excessive branching. The signaling proteins that regulate branching are the same in the vertebrate and insect systems.

Vertebrate blood vessels and heart

The vascular system, including blood vessels and blood cells, is among the first organ systems to develop in vertebrate embryos. Its early development is not surprising, because oxygen and nutrients need to be delivered to the rapidly developing tissues and also subsequently to developing organs. The development of the vascular system involves not only extensive branching morphogenesis but also the formation of tubes by a mesenchymal-to-epithelial transition (see Section 9.3), with the heart arising from the tube established ventral to the foregut. There are interesting parallels between the development of the vertebrate vascular system and the tracheal system in the insect, which has the same function of conveying oxygen to the tissues, but a very different developmental origin.

11.31 The vascular system develops by vasculogenesis followed by sprouting angiogenesis

The defining cell type of the vascular system is the **endothelial cell**, which forms the lining of the entire circulatory system, including the heart, veins, and arteries. The development of blood vessels starts with mesodermal cells called **angioblasts**; these are the precursors of endothelial cells. Angioblasts undergo a mesenchymal-to-epithelial transition (see Section 9.3) and migrate to form chains of cells that assemble into a primitive network of vessels, in a process called **vasculogenesis** (Fig. 11.56). A more dramatic example of a mesenchymal-to-epithelial transition is seen in the development of the kidney (see additional material online). In blood-vessel development, the initial vessels formed by vasculogenesis are then elaborated into a vascular system ramifying throughout the body by the process of **angiogenesis**, in which vessels grow and branch to form extensive networks of capillaries of venules and arterioles that can go on to form veins and arteries. The vessels also fuse with each other and undergo anastomosis (the bringing together and joining of two previously separated branches) to create complex interconnected networks of blood vessels.

Angioblast differentiation into endothelial cells requires the growth factor VEGF (vascular endothelial growth factor) and its receptors. VEGF is also a potent mitogen for endothelial cells, stimulating their proliferation. The primary blood vessels, such as the dorsal aorta and the main veins, arise from angioblasts located in the lateral plate mesoderm. VEGF is secreted by axial structures, such as somites, and its expression is driven by Shh signaling in the notochord. Expression of the *Vegf* gene is induced by lack of oxygen or hypoxia, and thus an active organ using up oxygen promotes its own vascularization.

New blood capillaries are formed by sprouting from pre-existing blood vessels. The process is very similar to tracheal branching, although here the chemoattractant is VEGF rather than Branchless, and cell proliferation is involved in growth and branching of the capillaries, whereas in the tracheal system growth is due to cell rearrangements (Fig. 11.57). Cells at the tip of a sprout extend filopodia-like processes that guide

Scan here

Scan this QR code image with your mobile device to see the online supplementary material on the kidney or log on to http:// global.oup.com/uk/orc/biosciences/devbiol/ wolpert5e/qr/qr11a/

Fig. 11.56 Vasculogenesis and angiogenesis. First panel: angioblasts migrating in chains, the first step in vasculogenesis that results in the formation of simple tubes. Individual angioblasts at time point 1 are indicated in red and at time point 2 in green to show movement, with arrows indicating direction. Ringed cells are immotile. Scale bar = 200 μm. The other photographs show sprouting during angiogenesis in model systems used to study the process: (a) mouse retina; (b) formation of a zebrafish intersomitic vessel; (c) formation of blood vessels in culture from an aggregate of mouse ES cells. *Top panel from Sato, Y., et al.: **Dynamic analysis of vascular morphogenesis using transgenic quail embryos.** PLoS One 2010, **5**: e12674. a-c, from Guedens, I., Gerhardt, H.: **Coordinating blood vessel formation during blood vessel formation.** Development 2011, **138**: 4569-4583.*

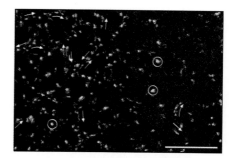

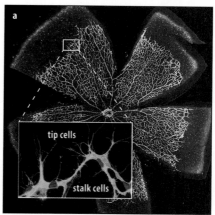

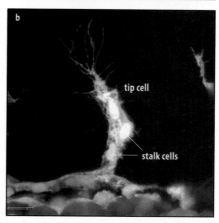

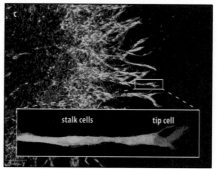

and extend the sprout. The response of the tip cells to VEGF is to express the Notch ligand Delta-like 4, which then activates Notch in adjacent proximal cells. Notch signaling blocks expression of the VEGF receptor. This is one of the mechanisms that confines outgrowth to the tip of the tube and is reminiscent of the Delta–Notch feedback mechanism deployed in *Drosophila* tracheal development (see Section 11.30).

During their development, blood vessels navigate along specific paths towards their targets, with the filopodia at the leading edge responding to both attractant and repellent cues on other cells and in the extracellular matrix. Cues in the extracellular environment are provided by proteins of the netrin and semaphorin families, which can act on the tips of growing blood vessels to block filopodial activity. Netrins and semaphorins are also involved in guidance of axons (discussed in Chapter 12), and there is a striking similarity in the signals and mechanisms that guide developing blood vessels and neurons. Blood-vessel morphogenesis also requires modulation of adhesive interactions between endothelial cells and the extracellular matrix and between the endothelial cells themselves.

Changes in integrin-mediated cell adhesion to the extracellular matrix through focal contacts are particularly important in this context. VEGF stimulates endothelial cells to degrade the surrounding basement membrane matrix and to migrate and proliferate. In the skin of mice, embryonic nerves form a template that directs the growth of arteries. The nerves secrete VEGF, which can both attract blood vessels and specify them as arteries. In the lung, the branching pattern of the developing vasculature is coordinated with the branching pattern of the bronchial epithelium, although the main vessels can develop in the absence of the lungs. Many solid tumors produce VEGF and other growth factors that stimulate angiogenesis, and blocking new vessel formation is a means of reducing tumor growth. Eight drugs that target VEGF have been approved or are pending approval (as of 2012) for clinical use in various solid tumors. Other drugs that target molecules involved in blood-vessel stability are also being developed.

The first tubular structures of the vasculature are formed by endothelial cells, and these delicate, thin-walled vessels are then covered by connective-tissue cells called pericytes and by smooth-muscle cells. Arteries and veins are defined by the direction of blood flow, as well as by structural and functional differences. Evidence from lineage tracing shows that angioblasts are specified as arterial or venous before they form blood vessels; their identities however are still labile. Arterial and venous capillaries join each other in capillary beds, which form the interchange sites between the arterial and venous blood systems. The cell-sorting and guidance molecule ephrin B2 (see Box 9D) is expressed in arterial blood vessels, whereas its receptor EphB4 is expressed in venous vessels. Interaction between them in primitive capillary networks may be required for endothelial cells sorting out into distinct arterial and venous vessels. Lymphatic vessels are also developmentally part of the vascular system and originate by budding from veins. The expression of the homeobox gene *Prox1* marks an early stage of commitment to the lymphatic lineage.

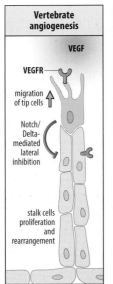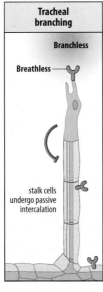

Vertebrate angiogenesis	Tracheal branching

Fig. 11.57 Outgrowth and branching of new vessels in angiogenesis. Left panel: In the first stage of angiogenesis, localized secretion of VEGF (blue) causes nearby endothelial cells of pre-existing blood vessels to form filopodia and move towards the source of VEGF. VEGF also stimulates proliferation and rearrangement of the endothelial cells, so forming a new branch. Delta production is enhanced in the tip cells (green) of the sprout, leading to activation of Notch in neighboring cells, which prevents them from becoming tip cells. These cells become stalk cells and a lumen forms. Right panel: This mechanism of branching is very similar to the branching of the *Drosophila* tracheal system apart from the fact that cell proliferation is involved.

*From Ochoa-Spinosa, A., Affloter, M.: **Branching Morphogenesis: From Cells to Organs and Back**. Cold Spring Harb Perspect Biol 2012; doi: 10.1101/cshperspect.a008243.*

Angiogenesis is not just a property of the embryo. It can occur throughout life and, properly regulated, is the means of repairing damaged blood vessels. Excessive or abnormal angiogenesis is, however, the hallmark of many human diseases, including cancer, obesity, psoriasis, atherosclerosis, and arthritis. Human disorders relating to the nervous system that involve abnormalities in the vasculature include vascular dementia, where blockage of the vasculature is a major cause of disease symptoms, and motor neuron disease.

11.32 The development of the vertebrate heart involves morphogenesis and patterning of a mesodermal tube

The heart is the first organ to form in the embryo. It is mainly of mesodermal origin and is first established by ventral midline fusion of the two arms of the **cardiogenic crescent** or **cardiac crescent**, regions of cardiogenic precursors located in the splanchnic lateral plate mesoderm on either side of the body underlying the head folds (Fig. 11.58). The resulting single tube consists of two layers of cells—the inner **endocardium**, which is a sheet of endothelium, and the outer **myocardium**, which will become the contractile cardiac muscle cells—separated by a layer of extracellular matrix, known as cardiac jelly, produced by the myocardium. Once formed, the primitive cardiac tube soon begins to pump blood. During its subsequent development, this tube elongates and becomes divided transversely into two chambers, the **atrial chamber** and the **ventricular** chamber. A third layer of cells, called the **epicardium**, develops as an epithelial sheet covering the myocardium. A two-chambered heart is the basic adult form in fish, but in higher vertebrates, such as birds and mammals, further partitioning of atrium and ventricle gives rise to the four-chambered heart. In humans, about 1 in 100 live-born infants have some congenital heart

Fig. 11.58 Schematic of human heart development. By day 15 of human embryonic development, cardiogenic precursors have formed a crescent, as shown in the first panel. The two arms of the crescent fuse along the midline to give a linear heart tube, which elongates and becomes patterned along the anteroposterior axis with the regions and chambers of the mature heart (second panel). After looping (third panel), these regions are disposed approximately in their eventual positions. Later development results in further patterning (fourth panel), and the formation of valves between, for example, the atria and ventricles. AVV, atrioventricular valve region.

After Srivastava, D. and Olson, E.N.: 2000.

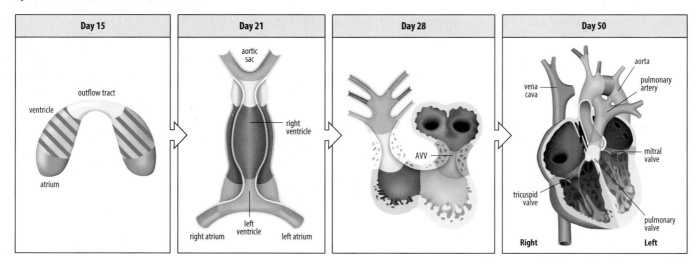

Fig. 11.59 Several different populations of cells contribute to making a heart. In the earliest stage of heart development (E7.5), illustrated here in the mouse, different populations of cardiogenic precursors form the cardiac crescent (red) and the second heart field (green), which extend across the ventral midline on either side. The two arms of the crescent fuse in the midline below the forming foregut and form a simple tube that then begins to loop (E8). Cells of the second heart field extend the length of the heart tube by contributing to both arterial (dark green) and venous (purple) poles (E8.5). In addition, cardiac neural crest cells (yellow) migrate from the pharyngeal arches to the arterial pole. The proepicardial organ (blue) forms near the venous pole (E9.5) and migrates to form the epicardium that covers the entire heart. The contributions of the various cell populations to the chambers of the looped heart tube are indicated (E10.5). By E14.5, the heart chambers have formed and the heart is mature. AA, aortic arch; Ao, aorta; EC, endocardial cushion; IVC, inferior caval vein; IVS, interventricular septum; LA, left atrium; LV, left ventricle; OFT, outflow tract; PT, pulmonary trunk; PV, pulmonary vein; RA, right atrium; RV, right ventricle; SVC, superior caval vein.

*From Vincent, S.D. and Buckingham, M.: **How to make a heart: the origin and regulation of cardiac progenitor cells**. Curr Top Dev Biol 2010, 90: 1–41.*

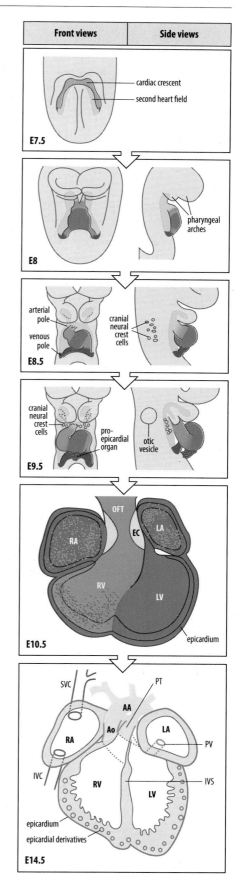

malformation; *in utero*, heart malformation leads to death of the embryo in between 5 and 10% of conceptions (the different numbers reflect different studies and it could even be as high as 30%).

As we saw in Chapter 4, heart mesoderm in *Xenopus* is initially specified by signals from the organizer during the general process of mesoderm induction and dorso-ventral patterning. In chick and mouse embryos, cells that will eventually become heart cells ingress through the primitive streak and become part of the lateral plate mesoderm. A few hours after ingression, the cells have become committed to be heart cells and will differentiate into cardiac muscle cells if isolated. In mice, experiments that traced the lineage of cells in the early heart region revealed the existence of two distinct lineages of heart precursor cells that are derived from a common precursor very early in development at or before gastrulation. One lineage of cells, which is the first to differentiate into cardiac muscle, is the exclusive source of the myocardial cells of the primitive left ventricle, whereas the other is the exclusive source of the myocardial cells of the outflow tract; all other heart regions are colonized by both lineages. The first differentiating myocardial cells are found in the cardiac crescent, which is therefore known as the **first heart field**, whereas the second lineage of myocardial cells lies medially and posteriorly to the crescent and is known as the **second heart field** (Fig. 11.59).

In all vertebrates the cardiac crescent extends across the midline and into the lateral plate mesoderm on both sides of the embryo. This region then undergoes a complex morphogenesis, and the cells move towards the midline and fuse to form the heart tube (see Fig. 11.59). Mutations in zebrafish have been found that disrupt this process and result in two laterally positioned hearts—a condition known as **cardia bifida**. One of the mutated genes is called *miles apart*, and codes for a receptor that binds lysosphingolipids; sphingosine 1-phosphate is the likely ligand in this case. *Miles apart* is not expressed in the migrating heart cells themselves, but in cells on either side of the midline, and thus may be involved in directing the migration of the pre-sumptive heart cells. Zebrafish have proved to be very useful in mutant screens to uncover the genetics of heart development, because their embryos and larvae do not require a beating heart to survive.

The second heart field makes a major contribution. As a result of morphogenetic movements, it comes to lie behind the primitive heart tube in the pharyngeal meso-derm extending both anteriorly and posteriorly. The elongation of the early heart tube depends on contributions from the second heart field, and cardiac precursors move from a proliferative zone within it to contribute to both the anterior end of the tube (arterial/outflow end) and the posterior end (venous/inflow end), called the **arterial** and **venous poles** respectively. The precursor cells for the different regions of the heart seem to acquire specific patterns of gene expression according to their position along the antero-posterior axis, and there is evidence that this may be accomplished by graded retinoic acid signaling, which establishes domains of Hox gene expression. Retinoic acid signaling also defines the posterior limit of the second heart field.

The addition of cells from the second heart field to the heart tube accompanies looping of the tube and also the recruitment of cardiac neural crest cells to the outflow region of the heart. The asymmetric looping of the heart tube is not well understood but is related to the left–right asymmetry of the internal organs of the embryo (see Section 5.16). The extracellular matrix molecule flectin is expressed earlier on the left side compared with the right, and experiments in chick embryos suggest that this may be the result of the expression of *Pitx2* on the left. The cells of neural crest origin contribute to the outflow tracts; they are essential, for example, for the formation of the pulmonary artery and the aorta from the single embryonic outflow tract (see Fig. 11.59). There may also be important interactions between the neural crest cells and the second heart field cells that contribute to this region of heart, because when the cardiac neural crest is ablated there is overproliferation of the cells of the second heart field. Defects in the embryonic outflow tract account for 30% of congenital heart defects in humans, some of which are the result of developmental perturbations in neural-crest cell specification and migration.

Just after heart looping, the heart tube acquires its outer epithelial layer, the epicardium, whose importance has only recently been recognized. The epicardium develops from a separate population of mesodermal cells outside the heart tube near its posterior end known as the proepicardial organ (see Fig. 11.59) which grow over the myocardium. Some epicardial cells later undergo an epithelial-to-mesenchymal transition and invade the myocardium to give rise to the cardiac connective tissue—fibroblasts, interstitial cells, and cells that support the coronary arteries. The endothelial cells of the coronary vessels migrate from the endocardium of the inflow region into the ventricular region after the formation of the epicardium. Some cells stay just beneath the epicardium on the surface of the myocardium and form the coronary veins while others invade the myocardium and form the coronary arteries. The epicardium is essential for the normal development and growth of the heart, with reciprocal interactions between the epicardium and the myocardium involving signaling molecules including FGFs, Shh, Wnt, and Notch. The cardiac fibroblasts are of great medical importance, because they give rise to the massive scarring that takes place as a consequence of a heart attack.

The heart tube then forms the separate chambers—four in the case of the mammalian heart. One of the critical processes in generating these chambers is establishing continuous structures such as the septa that separate atrial and ventricular spaces. These septa form from localized bulges in precise locations on opposite sides of the inner wall of the heart known as **endocardial cushions**. In the cushions, endocardial cells undergo an epithelial to mesenchymal transition and migrate into the cardiac jelly so that when two opposing cushions come into contact and break down, the mesenchyme cells can form a bridge that stabilizes the fusion of the two cushions.

In contrast, the atrial and ventricular septa that separate the right and left chambers of the heart develop from growing spurs of myocardium with mesenchyme cells at their leading edges that eventually fuse with mesenchyme cells of the endocardial cushions. We have already come across tissue fusion in neural tube morphogenesis (see Section 9.14) and in the developing eye during the formation of the optic cup (see Section 11.27). Probably the best-known example of tissue fusion takes place in the development of the face to form the upper lip and palate. When this fails, cleft lip and/or palate results. These conditions represent a very common class of birth defect seen in live births.

An increasing number of transcription factors are known to be involved in specifying development of the heart and the differentiation of cardiac muscle cells. The homeodomain transcription factor Nkx2.5, for example, is required for heart development. It is one of the first markers of early heart cells and is expressed in the cardiac crescent. Mutations in the *Nkx2.5* gene in mice and humans result in heart abnormalities. The T-box gene *Tbx5*, which we came across in relation to limb development (see Section 11.2), is also a key gene in heart development, and this is why patients with mutations in *Tbx5* have both limb and heart defects. Tbx5 and another transcription factor, GATA-4, in the presence of a chromatin-remodeling component, can induce beating myocardial tissue when ectopically expressed in mesoderm.

There are remarkable similarities—although we should no longer be surprised—between the genes involved in heart development in *Drosophila* and in vertebrates. The homeobox gene *tinman* is required for heart formation in *Drosophila*, and is a homolog of vertebrate *Nkx2.5*. When vertebrate *tinman*-like genes are expressed in *Drosophila* they can substitute for *Drosophila tinman* and rescue some of the abnormalities caused by lack of normal *tinman* functions. In *Drosophila*, signaling by Decapentaplegic maintains *tinman* expression in the dorsal mesoderm, whereas in vertebrates, BMPs that are homologs of Decapentaplegic have been shown to induce *Nkx2.5*. Homologs of other key transcription factors in vertebrate heart development are also involved in *Drosophila* heart formation—Dorsocross (a Tbx5/6 homolog) and Pannier (a GATA-4/6 homolog).

Teeth

Teeth develop from a well-defined series of reciprocal interactions between the oral epithelium, which in mammals is of ectodermal origin, and the mesenchyme of the facial primordia, which is of neural crest origin. The first sign of tooth development is a horseshoe-shaped band of thickened oral epithelium called the **dental lamina**. The development of individual teeth is then initiated within specific regions of the dental lamina known as **dental placodes**. The position of the placode determines the type of tooth, for example, incisor or molar. The dental epithelium of the placode invaginates and the surrounding mesenchyme condenses, forming a tooth primordium or tooth germ (Fig. 11.60). In each **tooth primordium**, a specialized group of cells—the enamel knot—acts as a signaling center for later development of the crown of the tooth. In multi-cusped teeth, such as molars, the enamel knot determines the positions of secondary enamel knots, which mark the tips of the future cusps. The mesenchyme forms the dental papilla, which gives rise to the pulp and dentin, whereas the enamel is secreted by the epithelial cells. The mouse is used as the model for human tooth development although it has a simpler dentition with just two types of teeth, incisors and molars, and only a single set of teeth.

The pattern of the permanent human dentition in each half of the jaw (from distal to proximal) consists of two incisors, one canine, two premolars and three molars (Fig. 11.60, upper left photograph). The pattern of the mouse dentition consists of one incisor and three molars separated by a region in which no teeth form (upper right photograph). The lower panels of Fig. 11.60 show stages in initiation of tooth development, formation of the tooth germ and its growth and morphogenesis, formation of the enamel knot, differentiation of epithelial and mesenchymal cells to produce enamel and dentin, respectively, and the final stages of eruption.

11.33 Tooth development involves epithelial-mesenchymal interactions and a homeobox gene code specifies tooth identity

The basic pattern of the dentition is set up before any external signs of tooth formation. The oral epithelium produces signals that establish a spatial pattern of homeobox

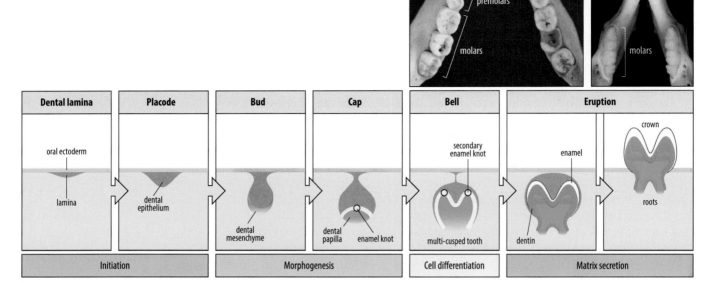

Fig. 11.60 The human and mouse dentition and the stages of tooth development. The photographs show (a) the upper dentition of an adult human and (b) the upper dentition of the mouse. Premolars and canines are absent in the mouse. The diagram outlines the stages of tooth development. The first sign of tooth formation is a thickening of the oral epithelium in the tooth-forming region to form the dental lamina. The oral epithelium invaginates into the dental mesenchyme, which condenses around the epithelium to form a bud. The epithelium then extends further into the mesenchymal tissue and starts to enclose the condensing mesenchyme , the dental papilla, to form the cap; the primary enamel knot is formed. During the bell stage, cusp patterns emerge and the crown of the tooth is formed. In multicusped mammalian teeth, secondary enamel knots form at the places of future cusps. Final growth and matrix secretion follows, during which time the inner enamel epithelium differentiates into ameloblasts, which produce enamel, and the underlying mesenchymal cells differentiate into odontoblasts that secrete dentin. The tooth erupts through the surface of the jaw. Roots continue to develop during eruption.

Reproduced with permission from Jussila and Thesleff: **Signaling networks regulating tooth organogenesis and regeneration, and the specification of dental mesenchymal and epithelial cell lineages.** Cold Spring Harb. Perspect. Biol. *2012.*

gene expression in the facial mesenchyme that provides a code for regional identity in the same way as the spatial pattern of Hox gene expression provides a code for regional identity along the antero-posterior body axis.

We will consider the development of the teeth in the lower jaw, which arises from the fused mandibular primordia, as our example. *Fg8* is expressed in the oral epithelium in the lateral region of each mandibular primordium where the molars will form and FGF signaling positively regulates expression of genes for the homeodomain transcription factors, Barx1 and Dlx2 in the underlying mesenchyme.

Bmp4 is expressed in the oral epithelium in the central region of the mandibular primordia where the incisors will form and BMP4 signaling positively regulates expression of the genes encoding the related homeodomain transcription factors Msx1 and Msx2 in the underlying mesenchyme, while at the same time negatively regulating *Barx1* expression. The gene for the homeodomain transcription factor Islet1, which we shall meet again in neural development, is expressed exclusively in the oral epithelium in the central region and has a positive reciprocal interaction with *Bmp4*. The genes for the Lim homeodomain transcription factors Lhx6 and Lhx7 are also expressed throughout the oral half of the mesenchyme of the mandibular primordia and their expression marks the region where teeth will form (Fig 11.61). *Dlx2, Barx1, Lhx6,* and *Lhx7* are expressed in mesenchyme cells that will form molars, while *Msx1, Msx2, Lhx6,* and *Lhx7* are expressed in cells that will

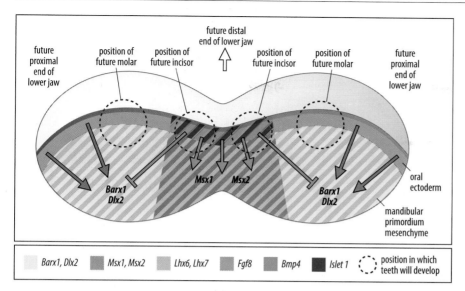

Fig. 11.61 The expression domains of homeobox genes in the lower jaw before initiation of tooth primordia. Schematic diagram of a cut-away section of the fused mandibular primordia of a mouse embryo viewed from the back of the oral cavity. The lateral regions of the mandibular primordia will give rise to the proximal regions of the lower jaws, while the central region will give rise to the distal region. *Barx1* and *Dlx2* expression are induced in the lateral mesenchyme by FGF signaling by overlying epithelium and confined to this region by inhibitory BMP signaling by distal oral ectoderm. *Msx1* and Msx2 expression in central mesenchyme are induced by BMP signaling from overlying epithelium. *Lhx6* and *Lhx7* are expressed throughout the mesenchyme of the oral half of the mandibular primordium. *Islet1* is expressed in the central oral epithelium. Expression of *Islet1* and *Bmp4* is maintained by mutual positive interactions. Dotted circles indicate positions in which incisors and molars will develop. Homeobox gene code for incisor: *Msx1, Msx2, Lhx6, Lhx7*. Code for molar: *Barx1, Dlx2, Lhx6, Lhx7*. Arrows indicate positive signaling; barred lines indicate inhibition.

form incisors. The importance of these transcription factors in determining tooth identity was shown by experiments in which beads soaked in the BMP antagonist Noggin were placed in cultured explants of the presumptive incisor region of a mouse mandibular primordium. This resulted in loss of *Msx1* expression and promotion of *Barx1* expression and upon further culture a molar tooth developed rather than an incisor. In mouse mutants lacking both *Dlx2* and *Dlx1* (a similar gene that can substitute for *Dlx2* in development and is also expressed in the facial mesenchyme), molars do not develop in the upper jaw, whereas the incisors are unaffected.

Early tooth development is controlled by reciprocal interactions between the oral epithelium and the underlying mesenchyme. The same families of signaling molecules—Wnt, TGF-β, FGF and Hh—are involved as in the limb. *Shh* is expressed in the dental lamina and later becomes restricted to the placodes where the Shh protein acts as a mitogen to induce local proliferation of the epithelium. Many Wnt ligands are also expressed in the oral epithelium and Wnt signaling is essential for placode formation. Dental placodes fail to form in mutant mice overexpressing the gene for the Wnt antagonist Dkk1, whereas activation of Wnt signaling can induce additional placodes in the dental lamina which then develop into supernumerary teeth. BMP-4 and FGF-8 signaling by the oral epithelium not only induces position-dependent patterns of expression of transcription factors in the facial mesenchyme, but also induces expression of the genes for the signaling molecules including BMP-4, activin, FGFs and Wnts. At this stage, the direction of signaling changes so that the signals secreted by the mesenchyme direct the development of the epithelium and lead to the formation of the enamel knot which controls tooth morphogenesis. The enamel knot produces signals including FGFs, Wnts, BMPs and FGFs. FGFs stimulate the neighboring epithelium to proliferate while the enamel knot cells themselves do not proliferate, and this shapes the developing tooth.

SUMMARY

Many internal organs, such as the lungs and blood vessels, are formed of tubular epithelial structures. Tubular structures can be formed *de novo* by two basic mechanisms: invagination and budding of a pre-existing epithelium, as in the *Drosophila* trachea and the vertebrate lung, or by the aggregation of mesenchymal cells to form a tubular epithelium, as in the initial formation of blood vessels. Branching of the *Drosophila* tracheal system and the airways of the vertebrate lung are controlled by conserved signals produced by the surrounding mesenchyme. The vascular system of vertebrates develops from mesenchymal endothelial cells that assemble during vasculogenesis into tubular epithelial structures that are the precursors of the large blood vessels. The tubules then grow and branch into finer vessels in the process of sprouting angiogenesis. The vertebrate heart initially forms as a linear tube made up of an inner endocardium and an outer myocardium. Its development is controlled by genes similar to those that regulate heart formation in *Drosophila*. Teeth develop from a bud of oral epithelium and the underlying neural crest mesenchyme. A homeobox gene code in the mesenchyme specifies tooth identity and their further development is controlled by reciprocal signaling between epithelium and mesenchyme.

Summary to Chapter 11

- In vertebrate embryos, the position at which organs develop is specified as part of the process of laying down the body plan. For example, a combinatorial Hox code in the lateral plate mesoderm determines limb position; a combinatorial homeobox gene code in the neural crest cells of the facial primordia determines tooth position.

- In *Drosophila*, adult appendages develop from imaginal discs formed in the embryo, and the segment in which an imaginal disc arises helps to determine appendage identity. Segment identity is defined by a combinatorial Hox code, and within a segment, discs corresponding to different structures (such as leg and wing discs in thoracic segment 2) are distinguished by the expression of different sets of transcription factors.

- The identity of some organs is encoded by 'master' or 'selector' genes and/or specific combinations of genes for transcription factors and other proteins that activate the gene-regulatory network required for subsequent development of that particular organ. *vestigial*, for example, which encodes a gene-regulatory co-activator, is the master gene for *Drosophila* wing development, and *Pax6*, which encodes a transcription factor, is the master gene for eye development in both vertebrates and *Drosophila*.

- There is a remarkable degree of conservation in the transcription factors involved in the development of particular organs in vertebrates and *Drosophila*. The transcription factors encoded by the gene *Nkx2.5* in vertebrates and its homolog *tin-man* in *Drosophila* are both required for heart formation.

- The main processes involved in organogenesis are the same as those used to generate the body plan, but they operate within the organ primordium or imaginal disc in a self-contained fashion. Once established, the organ rudiments (for example, limb buds, wing and leg imaginal discs, tooth germs, and optic cup) develop more-or-less autonomously, without reference to the rest of the embryo.

- Organ development in vertebrates requires integrating cells of different origins. For example, limbs are composed of lateral plate mesoderm and epidermal ectoderm, with contributions from the somites; the eye must integrate epidermal ectoderm, neural ectoderm, and contributions from cranial neural crest; the lung is composed of endoderm and mesoderm; the heart is composed of different cell populations of lateral plate mesoderm and neural crest; and a tooth derives from ectoderm and neural crest.

- In insects, structures also integrate cells of different origins, but it is done in a more modular manner. For example, an imaginal disc is a self-contained unit, with cells that will be the precursors of the muscle cells and sensory organs being part of the disc from very early in development and then developing within the disc.

- Organs also develop a vascular system (a tracheal system in *Drosophila*) and become innervated.

- Cell-cell interactions between different components, mainly mesenchymal and epithelial components, play a key role in organ development. In vertebrates, the limb-bud ectoderm imposes dorso-ventral pattern on the mesoderm, and lung mesenchyme influences the branching of lung epithelium, for example. In the eye, interactions between different epithelia induce the formation of the lens.

- Pattern formation in organs involves the interpretation of positional information. Organ primordia become equipped with signaling regions which produce morphogens that specify positional values. Examples of these signaling regions are the polarizing regions in vertebrate limbs and the compartment boundaries in wing and leg imaginal discs in *Drosophila*.

- The same families of signaling molecules act as morphogens in different organs, the differences in interpretation being due to the different developmental histories of the responding cells. For example, Shh signaling is involved in the development of vertebrate limbs, eyes, lungs, and teeth. Hh signaling acts in wing and eye development in *Drosophila*.

- Cell differentiation generates specialized cells for specific organ functions. The same differentiated cell types can be generated from cells with different developmental histories (for example, bone in the limb is of mesodermal origin while bones in the face derive from neural crest) or a unique cell type can differentiate in a particular organ (photoreceptors in the eye).

- Organ morphogenesis involves many of the same mechanisms used during the laying down of the body plan—for example, oriented cell movements and cell divisions (in the limb) and folding of cell sheets (in the eye).

- An additional process in organogenesis is branching morphogenesis, illustrated in this chapter by the lung, the vascular system, and the tracheal system in *Drosophila*. The same signals control branching morphogenesis in the vertebrate lung and the *Drosophila* tracheal system.

- Organs have a remarkable capacity for self-organization (for example, limb digits, optic vesicle.)

- Reaction-diffusion mechanisms may be involved in generating periodic patterns (digits in the limb, epithelial branches in the lung).

- Mutations in developmentally important genes involved in organogenesis can lead to human congenital abnormalities. Abnormalities can also be caused by exposure to environmental agents that act as teratogens (for example, thalidomide) at critical periods during organogenesis.

■ End of chapter questions

Long answer (concept questions)

1. Describe an experiment that showed that FGFs play a critical role in the initiation of limb bud formation. What causes the expression of FGFs? Propose a linear gene expression pathway from Hox gene expression to limb bud formation based on these facts.

2. Describe the location and extent of the apical ectodermal ridge with respect to the three main axes of the limb.

3. What is the consequence of removing the ridge at different times during limb formation? How do you interpret these results? Cite evidence for your interpretation.

4. Describe the proposed relationship between FGF and retinoic acid signaling in specifying identity along the proximo-distal axis in the developing limb.

5. Where in the limb bud is the polarizing region (also known as the zone of polarizing activity or ZPA)? What properties of this region make it an organizing region (see Fig. 11.14)? Through which signaling molecule does it exert its effects? What is the evidence for the importance of this molecule?

6. What is Gli3? What is its role in specifying digit identity, and through what downstream genes does it appear to exert this role?

7. What is the consequence of grafting the ectoderm from a left limb bud onto a right mesodermal core, such that the dorso-ventral axis is reversed without altering the antero-posterior axis? What conclusion is drawn from this experiment? How are Wnt and Engrailed involved in this patterning, and how do we know?

8. What are the signals from the dorsal ectoderm and the apical ectodermal ridge that are required to maintain *Shh* expression in the polarizing region? In turn, how does Shh maintain signaling from the apical ectodermal ridge?

9. What is the origin of the muscle cells in a limb? What is the evidence that the limb mesoderm controls the pattern of muscles that forms, so that the musculature is appropriate to the limb?

10. How is the identity of the different *Drosophila* imaginal discs established in the embryo—for example, how is it determined whether a disc will form a wing, a leg, or some other structure? How does misexpression of *Antennapedia* in the head region illustrate this?

11. The gene *engrailed* is expressed in the posterior compartment of the wing disc in the larva and in the posterior compartment of segments in the embryo. Compare and contrast the signals downstream of posterior compartment *engrailed* expression in the wing disc and in segmentation (refer back to Chapter 2 for information about segmentation).

12. The *Drosophila* genes *wingless*, *apterous* (Pteron is Greek for wing), and *vestigial* can all reduce or eliminate the formation of the wing when mutated. Based on the roles of these genes in the wing imaginal disc, propose mechanisms by which these mutations cause the wing defects.

13. What type of protein does the gene *Pax6* encode and discuss its role in eye development? What human conditions result from defects in the normal function or expression of *Pax6*?

14. Describe the developmental events leading to formation of the lens of the eye in vertebrate embryos. What have experiments in the teleost genus *Astyanax* contributed to our understanding of the importance of the lens in eye development?

15. What is the role of signaling through the EGF receptor, DER, in development of the *Drosophila* ommatidial cell cluster?

16. Summarize the general processes involved in blood-vessel formation in vertebrates. What is the role of VEGF in blood-vessel formation?

17. How are the three cell layers of the heart formed? Include endocardium, myocardium and epicardium. What is the importance of cell-cell interactions between the epicardium and the myocardium?

18. Teeth form a repeated pattern. What ensures that the appropriate type of tooth forms in its correct position within the jaw? Take as your example the formation of incisors and molars in the mouse.

Multiple choice (factual recall questions)

NB There is only one correct answer to each question.

1. Cartilage (bone-forming tissue) and connective tissue of the vertebrate limb originate from

a) mesodermal cells that migrate into the limb bud from the somites
b) mesodermal mesenchyme of the limb bud
c) the apical ectodermal ridge
d) the progress zone

2. Removal of the apical ectodermal ridge of a limb bud leads to

a) continued development of the proximal structures, but no formation of new distal structures
b) degeneration of the limb bud
c) regeneration of a new apical ridge from adjacent epidermal tissue
d) regeneration of the entire limb bud from underlying mesoderm

3. Grafting of a second polarizing region into the anterior region of a limb bud results in

a) formation of additional digits in mirror symmetry with the normal digits
b) formation of a second limb bud at the site of the graft
c) formation of new digits that lack any specific identity
d) formation of new digits, but only those of anterior identity

4. The programmed cell death that separates the digits is dependent upon which signaling pathway?

a) BMP
b) FGF
c) Shh
d) Wnt

5. The *Drosophila apterous* gene is related to the mouse *Lmx1b* gene, and both are involved in

a) controlling segmentation along the antero-posterior axis
b) determining whether or not wings are made
c) specifying dorsal identity in their respective appendages
d) specifying what structure is made from a given segment

6. The formation of the eyespot on butterfly wings is dependent on the expression of _____ in the disc, which is not expressed in the *Drosophila* wing disc.

a) *Distal-less*
b) *Engrailed*
c) *Hedgehog*
d) *Wingless*

7. The engineered expression of the mouse *Pax6* gene in the *Drosophila* leg disc

a) has no effect, since mouse genes cannot work in insects
b) can cause formation of mouse-type eye structures on the leg
c) will transform the leg disc into a wing disc
d) can cause formation of *Drosophila* eye structures on the leg

8. Drugs that block VEGF are used for

a) correcting birth defects
b) preventing limb development in mice
c) studying eye development in *Drosophila*
d) treating cancer

9. The heart develops from

a) foregut endoderm
b) invaginations of the ectodermal epithelium
c) lateral plate mesoderm
d) mesodermal angioblasts

10. The type of tooth that will form in different regions of the jaw is controlled by

a) BMP expression
b) homeobox gene expression
c) lateral inhibition
d) previous patterning in the mesoderm

Multiple choice answer key

1: b, 2: a, 3: a, 4: a, 5: c, 6: a, 7: d, 8: d, 9: c, 10: b.

■ Section further reading

11.1 The vertebrate limb develops from a limb bud

Delaurier, A., Burton, N., Bennett, M., Baldock, R., Davidson, D., Mohun, T.J., Logan, M.P.: **The Mouse Limb Anatomy Atlas: an interactive 3D tool for studying embryonic limb patterning.** *BMC Dev. Biol.* 2008, **8**: 83–89.

Tickle, C.: **The contribution of chicken embryology to the understanding of vertebrate limb development.** *Mech. Dev.* 2004, **121**: 1019–1029.

11.2 Genes expressed in the lateral plate mesoderm are involved in specifying the position and type of limb

Altabef, M., Clarke, J.D.W., Tickle, C.: **Dorso-ventral ectodermal compartments and origin of apical ectodermal ridge in developing chick limb.** *Development* 1997, **124**: 4547–4556.

DeLaurier, A., Schweitzer, R., Logan M.: **Pitx1 determines the morphology of muscle, tendon, and bones of the hindlimb.** *Dev Biol* 2006, **299**: 22–34.

Duboc, V., Logan, M.P.: **Regulation of limb bud initiation and limb-type morphology.** *Dev Dyn* 2011, **240**: 1017–1027.

Kawakami, Y., Capdevila, J., Buscher, D., Itoh, T., Rodriguez Esteban, C., Izpisua Belmonte, J.C.: **WNT signals control FGF-dependent limb initiation and AER induction in the chick embryo.** *Cell* 2001, **104**: 891–900.

Nishimoto, S., Minguillon, C., Wood, S., Logan, M.P.: **A combination of activation and repression by a colinear Hox code controls forelimb-restricted expression of Tbx5 and reveals Hox protein specificity.** *PLoS Genet.* 2014, **10**: e1004245.

Minguillon, C., Buono, J.D., Logan, M.P.: ***Tbx5* and *Tbx4* are not sufficient to determine limb-specific morphologies but have common roles in initiating limb outgrowth.** *Dev. Cell* 2005, **8**: 75–84.

Minguillon, C., Nishimoto, S., Wood, S., Vendrell, E., Gibson-Brown, J.J., Logan, M.P.: **Hox genes regulate onset of Tbx expression in the forelimb region.** *Development* 2012, **139**: 3180–3189.

11.3 The apical ectodermal ridge is required for limb outgrowth and the formation of structures along the proximo-distal axis of the limb

Fernandez-Teran, M., Ros, M.A.: **The apical ectodermal ridge: morphological aspects and signaling pathways.** *Int. J. Dev. Biol.* 2008, **52**: 857–871.

Mariani, F.V., Ahn, C.P., Martin, G.R.: **Genetic evidence that FGFs have an instructive role in limb proximal-distal patterning.** *Nature* 2008, **453**: 401–405.

Niswander, L., Tickle, C., Vogel, A., Booth, I., Martin, G.R.: **FGF-4 replaces the apical ectodermal ridge and directs outgrowth and patterning of the limb.** *Cell* 1993, **75**: 579–587.

11.4 Outgrowth of the limb bud involves oriented cell behavior

Bénazéraf, B., Francois, P., Baker, R.E., Denans, N., Little, C.D., Pourquié, O.: **A random cell motility gradient downstream of FGF controls elongation of an amniote embryo.** *Nature* 2010, **466**: 248–252.

Boehm, B., Westerberg H., Lesnicar-Pucko, G., Raja, S., Rautschka, M., Cotterell, J., Swoger, J., Sharpe, J.: **The role of spatially controlled cell proliferation in limb bud morphogenesis.** *PLoS Biol* 2010, **8**: 8e1000420.

Gros, J., Hu, J.K., Vinegoni, C., Feruglio, P.F., Weissleder, R., Tabin, C.J.: **Wnt5a/Jnk and FGF/Mapk pathways regulate the cellular events shaping the vertebrate limb.** *Curr. Biol.* 2010, **20**: 1993–2002.

Gros, J., Tabin C.J.: **Vertebrate limb bud formation is initiated by localized epithelial-to-mesenchymal transition.** *Science* 2014, **343**: 1253–1256.

Wyngaarden, L.A., Vogeli, K.M., Ciruna, B.G., Wells, M., Hadjantonakis, A.K., Hopyan, S.: **Oriented cell motility and division underlie early limb bud morphogenesis.** *Development* 2010, **137**: 2551–2558.

11.5 Patterning of the limb bud involves positional information

Niswander, L.: **Pattern formation: old models out on a limb.** *Nat. Rev. Genet.* 2003, **4**: 133–143.

Towers, M., Tickle, C.: **Growing models of vertebrate limb development**. *Development* 2009, **136**: 179–190.

11.6 How position along the proximo-distal axis of the limb bud is specified is still a matter of debate

Cooper, K.L., Hu, J.K., ten Berge, D., Fernandez-Teran, M., Ros, M.A., Tabin, C.J.: **Initiation of proximo-distal patterning in the vertebrate limb by signals and growth**. *Science* 2011, **332**: 1083–1086.

Galloway, J.L., Delgado, I., Ros, M.A., Tabin, C.J.: **A reevaluation of X-irradiation-induced phocomelia and proximodistal limb patterning**. *Nature* 2009, **460**: 400–404.

Rosello-Diez A, Ros MA, Torres M.: **Diffusible signals not autonomous mechanisms determine the main proximodistal limb subdivision**. *Science* 2011, **332**: 1086–1088.

Summerbell, D., Lewis, J.H., Wolpert, L.: **Positional information in chick limb morphogenesis**. *Nature* 1973, **244**: 492–496.

Tabin, C., Wolpert, L.: **Rethinking the proximodistal axis of the vertebrate limb in the molecular era**. *Genes Dev.* 2007, **21**: 1433–1442.

Wolpert, L., Tickle, C., Sampford, M.: **The effect of cell killing by X-irradiation on pattern formation in the chick limb**. *J. Embryol. Exp. Morph.* 1979, **50**: 175–193.

Zeller, R., Lopez-Rios, J., Zuniga, A.: **Vertebrate limb bud development: moving towards integrative analysis of organogenesis**. *Nat. Rev. Genet.* 2009, **10**: 845–858.

Box 11A Teratogens and the consequences of damage to the developing embryo

Brown, N.: **Chemical teratogens hazards, tools and clues**. In *Embryos, Genes and Birth Defects* 2nd edition (eds Ferretti, P., Copp, A., Tickle, C., Moore, G.) Wiley: 2006.

Ferretti, P., Tickle, C. **The limbs**. In *Embryos, Genes and Birth Defects* 2nd edition (eds Ferretti, P., Copp, A., Tickle, C., Moore, G.) Wiley: 2006.

Ito, T., Ando, H., Suzuki, T., Ogura, T., Hotta, K., Imamura, Y., Yamaguchi, Y., Handa, H.: **Identification of a primary target of thalidomide teratogenicity**. *Science* 2010, **327**: 1345–1350.

Tabin, C.J.: **A developmental model for thalidomide defects**. *Nature* 1998, **396**: 322–323.

Therapontos, C., Erskine, L., Gardner, E.R., Figg, W.D., Vargesson, N.: **Thalidomide induces limb defects by preventing angiogenic outgrowth during early limb formation**. *Proc. Natl Acad. Sci. USA* 2009, **106**: 8573–8578.

Vargesson, N.: **Thalidomide embryopathy: an enigmatic challenge**. *ISRN Dev. Biol.* 2013, Article ID 241016.

11.7 The polarizing region specifies position along the limb's antero-posterior axis

Litingtung, Y., Dahn, R.D., Li, Y., Fallon, J.F., Chiang, C.: **Shh and Gli3 are dispensable for limb skeleton formation but regulate digit number and identity**. *Nature* 2002, **418**: 979–983.

Riddle, R.D., Johnson, R.L., Laufer, E., Tabin, C.: *Sonic hedgehog* **mediates polarizing activity of the ZPA**. *Cell* 1993, **75**: 1401–1416.

te Welscher, P., Zuniga, A., Kuijper, S., Drenth, T., Goedemans, H.J., Meijlink, F., Zeller, R.: **Progression of vertebrate limb development through SHH-mediated counteraction of GLI3**. *Science* 2002, **298**: 827–830.

Tickle, C.: **Making digit patterns in the vertebrate limb**. *Nat. Rev. Mol. Cell Biol.* 2006, **7**: 1–9.

Towers, M., Mahood, R., Yin, Y., Tickle, C.: **Integration of growth and specification in chick wing digit-patterning**. *Nature* 2008, **452**: 882–886.

Box 11B Positional information and morphogen gradients

Ashe, H.L., Briscoe, J.: **The interpretation of morphogen gradients**. *Development* 2006, **133**: 385–394.

Kerszberg, M., Wolpert, L.: **Specifying positional information in the embryo: looking beyond morphogens**. *Cell* 2007, **130**: 205–209.

Kicheva, A., González-Gaitán, M.: **The decapentaplegic morphogen gradient: a precise definition**. *Curr. Opin. Cell Biol.* 2008, **20**: 137–143.

Müller, P., Rogers, K.W., Yu, S.R., Brand, M., Schier, A.F.: **Morphogen transport**. *Development* 2013, **140**: 1621–1638.

Roy, S., Huang, H., Liu, S., Kornberg, T.B.: **Cytoneme-mediated contact-dependent transport of the *Drosophila* decapentaplegic signaling protein**. *Science* 2014, **343**: 1244624.

Sanders, T.A., Llagostera, E., Barna, M.: **Specialized filopodia direct long-range transport of SHH during vertebrate tissue patterning**. *Nature* 2013, **497**: 628–632.

11.8 Sonic hedgehog is the polarizing region morphogen

Riddle, R.D., Johnson, R.L., Laufer, E., Tabin, C.: **Sonic hedgehog mediates the polarizing activity of the ZPA**. *Cell* 1993, **75**: 1401–1416.

Yang, Y., Drossopoulou, G., Chuang, P.T., Duprez, D., Marti, E., Bumcrot, D., Vargesson, N., Clarke, J., Niswander, L., McMahon, A., Tickle, C.: **Relationship between dose, distance and time in Sonic Hedgehog-mediated regulation of anteroposterior polarity in the chick limb**. *Development* 1997, **124**: 4393–4404.

Box 11C Too many fingers: mutations that affect antero-posterior patterning can cause polydactyly

Ahn, S., Joyner, A.L.: **Dynamic changes in the response of cells to positive hedgehog signaling during mouse limb patterning**. *Cell* 2004, **118**: 505–516.

Harfe, B.D., Scherz, P.J., Nissim, S., Tian, H., McMahon, A.P., Tabin, C.J.: **Evidence for an expansion-based temporal Shh gradient in specifying vertebrate digit identities**. *Cell* 2004, **118**: 517–528.

Hill, R.E., Lettice, L.A.: **Alterations to the remote control of Shh gene expression cause congenital abnormalities**. *Trans. R. Soc. B* 2013, **368**: 20120357.

Lettice, L.A., Hill, A.E., Devenney, P.S., Hill, R.E.: **Point mutations in a distant Shh cis-regulator generate a variable regulatory output responsible for preaxial polydactyly**. *Hum. Mol. Genet.* 2008, **17**: 978–985.

Box 11D Sonic hedgehog signaling and the primary cilium

Goetz S.C., Anderson, K.V.: **The primary cilium: a signaling centre during vertebrate development**. *Nat. Rev. Genet.* 2010, **11**: 331–344.

Huangfu, D., Anderson, K.V.: **Cilia and Hedgehog responsiveness in the mouse**. *Proc. Natl. Acad. Sci. USA* 2005, **102**: 11325–11330.

Huangfu, D., Liu, A., Rakeman, A.S., Murcia, N.S., Niswander, L., Anderson, K.V.: **Hedgehog signalling in the mouse requires intraflagellar transport proteins**. *Nature* 2003, **426**: 83–87.

11.9 How digit identity is encoded is not yet known

Dahn, R.D., Fallon, J.F.: **Interdigital regulation of digit identity and homeotic transformation by modulated BMP signaling**. *Science* 2000, **289**: 438–441.

Suzuki, T., Takeuchi, J., Koshiba-Takeuchi, K., Ogura, T.: *Tbx* **genes specify posterior digit identity through Shh and BMP signaling**. *Dev. Cell* 2004, **6**: 43–53.

Suzuki, T., Hasso, S.M., Fallon, J.F.: **Unique SMAD1/5/8 activity at the phalanx-forming region determines digit identity**. *Proc. Natl. Acad. Sci. USA* 2008, **105**: 4185–4190.

Vokes, S.A., Ji, H., Wong, W.H., McMahon, A.P.: **A genome-scale analysis of the *cis*-regulatory circuitry underlying Shh-mediated patterning of the mammalian limb**. *Genes Dev.* 2008, **22**: 2651–2663.

11.10 The dorso-ventral axis of the limb bud is controlled by the ectoderm

Arques, C.G., Doohan, R., Sharpe, J., Torres, M.: **Cell tracing reveals a dorsoventral lineage restriction plane in the mouse limb bud mesenchyme**. *Development* 2007, **134**: 3713–3722.

Geduspan, J.S., MacCabe, J.A.: **The ectodermal control of mesodermal patterns of differentiation in the developing chick wing**. *Dev. Biol.* 1987, **124**: 398–408.

Riddle, R.D., Ensini, M., Nelson, C., Tsuchida, T., Jessell, T.M., Tabin, C.: **Induction of the LIM homeobox gene *Lmx1* by Wnt-7a establishes dorsoventral pattern in the vertebrate limb**. *Cell* 1995, **83**: 631–640.

11.11 Development of the limb is integrated by interactions between signaling centers

Bénazet, J.D., Zeller, R.: **Vertebrate limb development: moving from classical morphogen gradients to an integrated 4-dimensional patterning system**. *Cold Spring Harb. Perspect.* 2009, **1**: a001339.

Bénazet, J.D., Bischofberger, M., Tiecke, E., Gonçalves, A., Martin, J.F., Zuniga, A., Naef, F., Zeller, R.: **A self-regulatory system of interlinked signaling feedback loops controls mouse limb patterning**. *Science* 2009, **323**: 1050–1053.

Niswander, L., Jeffrey, S., Martin, G.R., Tickle, C.: **A positive feedback loop coordinates growth and patterning in the vertebrate limb**. *Nature* 1994, **371**: 609–612.

Pizette, S., Niswander, L.: **BMPs negatively regulate structure and function of the limb apical ectodermal ridge**. *Development* 1999, **126**: 883–894.

Zeller, R., Lopez-Rios, J., Zuniga, A.: **Vertebrate limb development: moving towards integrative analysis of organogenesis**. *Nature Rev. Genet.* 2009, **10**: 845–855.

11.12 Different interpretations of the same positional signals give different limbs

Logan, M.: **Finger or toe: the molecular basis of limb identity**. *Development* 2003, **130**: 6401–6410.

Logan, M., Tabin, C.J.: **Role of Pitx1 upstream of Tbx4 in specification of hindlimb identity**. *Science* 1999, **283**: 1736–1739.

Saunders, J.W., Gasseling, M.T., Cairns, J.M.: **The differentiation of prospective thigh mesoderm grafted beneath the apical ectodermal ridge of the wing bud in the chick embryo**. *Dev. Biol.* 1959, **1**: 281–301.

11.13 Hox genes have multiple inputs into the patterning of the limbs

Andrey, G., Montavon, T., Mascrez, B., Gonzalez, F., Noordermeer, D., Leleu, M., Trono, D., Spitz, F., Duboule D.: **A switch between topological domains underlies HoxD genes collinearity in mouse limbs**. *Science* 2013, **340**: 1234167.

Charité, J., De Graaff, W., Shen, S., Deschamps, J.: **Ectopic expression of *Hoxb-8* causes duplication of the ZPA in the forelimb and homeotic transformation of axial structures**. *Cell* 1994, **78**: 589–601.

Goodman, F.R.: **Limb malformations and the human HOX genes**. *Am. J. Med. Genet.* 2002, **112**: 256–265.

Kmita, M., Tarchini, B., Zàkàny, J., Logan, M., Tabin, C.J., Duboule, D.: **Early developmental arrest of mammalian limbs lacking *HoxA/HoxD* gene function**. *Nature* 2005, **435**: 1113–1116.

Montavon, T., Duboule, D.: **Chromatin organization and global regulation of Hox gene clusters**. *Phil. Trans R. Soc.* 2013, **368**: 20120367.

Nelson, C.E., Morgan, B.A., Burke, A.C., Laufer, E., DiMambro, E., Muytaugh, L.C., Gonzales, E., Tessarollo, L., Parada, L.F., Tabin, C.: **Analysis of Hox gene expression in the chick limb bud**. *Development* 1996, **122**: 1449–1466.

Wellik, D.M., Capecchi, M.R.: **Hox10 and Hox11 genes are required to globally pattern the mammalian skeleton**. *Science* 2003, **301**: 363–367.

Zakany, J., Kmita, M., Duboule, D.: **A dual role for Hox genes in limb anterior-posterior asymmetry**. *Science* 2004, **304**: 1669–1672.

Zakany, J., Duboule, D.: **The role of Hox genes during vertebrate limb development**. *Curr. Opin. Genet. Dev.* 2007, **17**: 359–366.

11.14 Self-organization may be involved in the development of the limb bud

Hardy, A., Richardson, M.K., Francis-West, P.N., Rodriguez, C., Izpisúa-Belmonte, J.C., Duprez, D., Wolpert, L.: **Gene expression, polarising activity and skeletal patterning in reaggregated hind limb mesenchyme**. *Development* 1995, **121**: 4329–4337.

Box 11E Reaction-diffusion mechanisms

Kondo, S., Miura, T.: **Reaction-diffusion model as a framework for understanding biological pattern formation.** *Science* 2010, **329**: 1616–1620.

Marcon, L., Sharpe, J.: **Turing patterns in development: what about the horse part?** *Curr. Opin. Genet. Dev.* 2012, **22**: 578–584.

Meinhardt, H.: **Turing's theory of morphogenesis of 1952 and the subsequent discovery of the crucial role of local self-enhancement and long-range inhibition.** *Interface Focus* 2012, **2**: 407–416.

Meinhardt, H., Gierer, A.: **Pattern formation by local self-activation and lateral inhibition.** *BioEssays* 2000, **22**: 753–760.

Miura, T., Shiota, K., Morriss-Kay, G., Maini, P.K.: **Mixed-mode pattern in Doublefoot mutant mouse limb—Turing reaction-diffusion model on a growing domain during limb development.** *J. Theor. Biol.* 2006, **240**: 562–573.

Sheth, R., Marcon, L., Bastida, F.M., Junco, M., Quintana, L., Dahn, R., Kmita, M., Sharpe, J., Ros, M.A.: **Hox genes regulate digit patterning by controlling the wavelength of a Turing-type mechanism.** *Science* 2012, **338**: 1476–1480.

11.15 Limb muscle is patterned by the connective tissue

Amthor, H., Christ, B., Patel, K.: **A molecular mechanism enabling continuous embryonic muscle growth—a balance between proliferation and differentiation.** *Development* 1999, **126**: 1041–1053.

Anderson, C., Williams, V.C., Moyon, B., Daubas, P., Tajbakhsh, S., Buckingham, M.E., Shiroishi, T., Hughes, S.M., Borycki, A.G.: **Sonic hedgehog acts cell-autonomously on muscle precursor cells to generate limb muscle diversity.** *Genes Dev.* 2012, **26**: 2103–2117.

Hashimoto, K., Yokouchi, Y., Yamamoto, M., Kuroiwa, A.: **Distinct signaling molecules control *Hoxa-11* and *Hoxa-13* expression in the muscle precursor and mesenchyme of the chick limb bud.** *Development* 1999, **126**: 2771–2783.

Hu, J.K., McGlinn, E., Harfe, B.D., Kardon, G., Tabin, C.J.: **Autonomous and nonautonomous roles of Hedgehog signaling in regulating limb muscle formation.** *Genes Dev.* 2012, **26**: 2088–2102.

Robson, L.G., Kara, T., Crawley, A., Tickle, C.: **Tissue and cellular patterning of the musculature in chick wings.** *Development* 1994, **120**: 1265–1276.

Schweiger, H., Johnson, R.L., Brand-Sabin, B.: **Characterization of migration behaviour of myogenic precursor cells in the limb bud with respect to Lmx1b expression.** *Anat. Embryol.* 2004, **208**: 7–18.

Shellswell, G.B., Wolpert, L.: **The pattern of muscle and tendon development in the chick wing.** In *Limb and Somite Morphogenesis* (eds Ede, D.A., Hinchliffe, J.R., Balls, M.) Cambridge: Cambridge University Press, 1977.

11.16 The initial development of cartilage, muscles, and tendons is autonomous

Kardon, G.: **Muscle and tendon morphogenesis in the avian hind limb.** *Development* 1998, **125**: 4019–4032.

Ros, M.A., Rivero, F.B., Hinchliffe, J.R., Hurle, J.M.: **Immunohistological and ultrastructural study of the developing tendons of the avian foot.** *Anat. Embryol.* 1995, **192**: 483–496.

11.17 Joint formation involves secreted signals and mechanical stimuli

Hartmann, C., Tabin, C.J.: **Wnt-14 plays a pivotal role in inducing synovial joint formation in the developing appendicular skeleton.** *Cell* 2001, **104**: 341–351.

Kahn, J., Shwartz, Y., Blitz, E., Krief, S., Sharir, A., Breitel, D.A., Rattenbach, R., Relaix, F., Maire, P., Rountree, R.B., Kingsley, D.M., Zelzer, E.: **Muscle contraction is necessary to maintain joint progenitor cell fate.** *Dev. Cell* 2009, **16**: 734–743.

Khan, I.M., Redman, S.N., Williams, R., Dowthwaite, G.P., Oldfield, S.F., Archer, C.W.: **The development of synovial joints.** *Curr. Top. Dev. Biol.* 2007, **79**: 1–36.

Koyama, E., Shibukawa, Y., Nagayama, M., Sugito, H., Young, B., Yuasa, T., Okabe, T., Ochiai, T., Kamiya, N., Rountree, R.B., Kingsley, D.M., Iwamoto, M., Enomoto-Iwamoto, M., Pacifici, M.: **A distinct cohort of progenitor cells participates in synovial joint and articular cartilage formation during mouse limb skeletogenesis.** *Dev Biol.* 2008, **316**: 62–73.

11.18 Separation of the digits is the result of programmed cell death

Garcia-Martinez, V., Macias, D., Gañan, Y., Garcia-Lobo, J.M., Francia, M.V., Fernandez-Teran, M.A., Hurle, J.M.: **Internucleosomal DNA fragmentation and programmed cell death (apoptosis) in the interdigital tissue of the embryonic chick leg bud.** *J. Cell. Sci.* 1993, **106**: 201–208.

Hernandez-Martinez, R., Covarrubias, L.: **Interdigital cell death function and regulation: new insights on an old programmed cell death model.** *Dev. Growth Diff.* 2011, **53**: 245–258.

Zuzarte-Luis, V., Hurle, J.M.: **Programmed cell death in the embryonic vertebrate limb.** *Semin. Cell Dev. Biol.* 2005, **16**: 261–269.

11.19 The adult wing emerges at metamorphosis after folding and evagination of the wing imaginal disc

Morata, G.: **How *Drosophila* appendages develop.** *Nature Rev. Mol. Cell Biol.* 2001, **2**: 89–97.

11.20 A signaling center at the boundary between anterior and posterior compartments patterns the *Drosophila* wing along the antero-posterior axis

Crozatier, M., Glise, B., Vincent, A.: **Patterns in evolution: veins of the *Drosophila* wing.** *Trends Genet.* 2004, **20**: 498–505.

Entchev, E.V., Schwabedissen, A., González-Gaitán, M.: **Gradient formation of the TGF-β homolog Dpp.** *Cell* 2000, **103**: 981–991.

Han, C., Belenkaya, T.Y., Wang, B., Lin, X.: ***Drosophila* glypicans control the cell-to-cell movement of Hedgehog by a dynamin-independent process.** *Development* 2004, **131**: 601–611.

Kruse, K., Pantazis, P., Bollenbach, T., Julicher, F., González-Gáitan, M.: **Dpp gradient formation by dynamin-dependent endocytosis: receptor trafficking and the diffusion model.** *Development* 2004, **131**: 4843–4856.

Moser, M., Campbell, G.: **Generating and interpreting the Brinker gradient in the *Drosophila* wing**. *Dev. Biol.* 2005, **286**: 647–658.

Muller, B., Hartmann, B., Pyrowolakis, G., Affolter, M., Basler, K.: **Conversion of an extracellular Dpp/BMP morphogen gradient into an inverse transcriptional gradient**. *Cell* 2003, **113**: 221–233.

Tabata, T.: **Genetics of morphogen gradients**. *Nature Rev. Genet.* 2001, **2**: 620–630.

11.21 A signaling center at the boundary between dorsal and ventral compartments patterns the *Drosophila* wing along the dorso-ventral axis

Baeg, G.H., Selva, E.M., Goodman, R.M., Dasgupta, R., Perrimon, N.: **The Wingless morphogen gradient is established by the cooperative action of Frizzled and heparan sulfate proteoglycan receptors**. *Dev. Biol.* 2004, **276**: 89–100.

Bollenbach, T., Pantazis, P., Kicheva, A., Bökel, C., González-Gaitán, M., Jülicher, F.: **Precision of the Dpp gradient**. *Development.* 2008, **135**: 1137–1146.

Fujise, M., Takeo, S., Kamimura, K., Matsuo, T., Aigaki, T., Izumi, S., Nakato, H.: **Dally regulates Dpp morphogen gradient formation in the *Drosophila* wing**. *Development* 2003, **130**: 1515–1522.

Hayward, P., Kalmar, T., Martinez Arias, A.: **Wnt/Notch signalling and information processing during development**. *Development* 2008, **135**: 411–424.

Kicheva, A., González-Gaitán, M.: **The Decapentaplegic morphogen gradient: a precise definition**. *Curr. Opin. Cell Biol.* 2008, **20**: 137–143.

Lawrence, P.: **Morphogens: how big is the picture?** *Nat. Cell Biol.* 2001, **3**: E151–E154.

Martinez Arias, A.: **Wnts as morphogens? The view from the wing of *Drosophila***. *Nat. Rev. Mol. Cell. Biol.* 2003, **4**: 321–325.

Milán, M., Cohen, S.M.: **A re-evaluation of the contributions of Apterous and Notch to the dorsoventral lineage restriction boundary in the *Drosophila* wing**. *Development* 2003, **130**: 553–562.

Piddini, E., Vincent, J.P.: **Interpretation of the Wingless gradient requires signaling-induced self-inhibition**. *Cell* 2009, **136**: 296–307.

Rauskolb, C., Correia, T., Irvine, K.D.: **Fringe-dependent separation of dorsal and ventral cells in the *Drosophila* wing**. *Nature* 1999, **401**: 476–480.

Zecca, M., Basler, K., Struhl, G.: **Sequential organizing activities of engrailed, hedgehog and decapentaplegic in the *Drosophila* wing**. *Development* 1995, **121**: 2265–2278.

11.22 Vestigial is a key regulator of wing development that acts to specify wing identity and control wing growth

Kim, J., Sebring, A., Esch, J.J., Kraus, M.E., Vorwerk, K., Magee, J., Carroll, S.B.: **Integration of positional signals and regulation of wing formation and identity by *Drosophila* vestigial gene**. *Nature* 1996, **382**: 133–138.

Klein, T., Martinez Arias, A.: **The vestigial gene product provides a molecular context for the interpretation of signals during the development of the wing in *Drosophila***. *Development* 1999, **126**: 913–925.

11.23 How the proximo-distal axis of the *Drosophila* wing is patterned is not yet clear

Klein, T., Martinez Arias, A.: **Different spatial and temporal interactions between *Notch*, *wingless* and *vestigial* specify proximal and distal pattern elements of the wing in *Drosophila***. *Dev. Biol.* 1998, **194**: 196–212.

11.24 The leg disc is patterned in a similar manner to the wing disc, except for the proximo-distal axis

Emerald, B.S., Cohen, S.M.: **Spatial and temporal regulation of the homeotic selector gene *Antennapedia* is required for the establishment of leg identity in *Drosophila***. *Dev. Biol.* 2004, **267**: 462–472.

Estella, C., Voutev, R., Mann, R.S.: **A dynamic network of morphogens and transcription factors patterns the fly leg**. *Curr. Top. Dev. Biol.* 2012, **98**: 173–198.

Galindo, M.I., Bishop, S., Greig, S., Couso, J.P.: **Leg patterning driven by proximal-distal interactions and EGFR signaling**. *Science* 2002, **297**: 258–259.

Kojima, T.: **The mechanism of *Drosophila* leg development along the proximodistal axis**. *Dev. Growth Differ.* 2004, **46**: 115–129.

11.25 Butterfly wing markings are organized by additional positional fields

Brakefield, P.M., French, V.: **Butterfly wings: the evolution of development of colour patterns**. *BioEssays* 1999, **21**: 391–401.

French, V., Brakefield, P.M.: **Pattern formation: a focus on Notch in butterfly spots**. *Curr. Biol.* 2004, **14**: R663–R665.

11.26 Different imaginal discs can have the same positional values

Carroll, S.B.: **Homeotic genes and the evolution of arthropods and chordates**. *Nature* 1995, **376**: 479–485.

Morata, G.: **How *Drosophila* appendages develop**. *Nature Rev. Mol. Cell Biol.* 2001, **2**: 89–97.

Si Dong, P.D., Chu, J., Panganiban, G.: **Coexpression of the homeobox genes *Distal-less* and *homothorax* determines *Drosophila* antennal identity**. *Development* 2000, **127**: 209–216.

11.27 The vertebrate eye develops mainly from the neural tube and the ectoderm of the head

Adler, R., Canto-Soler V.: **Molecular mechanisms of optic vesicle development: complexities, ambiguities and controversies**. *Dev. Biol.* 2007, **305**: 1–13.

Eiraku, M., Takata, N., Ishibashi, H., Kawada, M., Sakakura, E., Okuda, S., Sekiguchi, K., Adachi, T., Sasai, Y.: **Self-organizing optic-cup morphogenesis in three-dimensional culture**. *Nature* 2011, **427**: 51–56.

Franze, K., Grosche, J., Skatchkov, S.N., Schinkinger, S., Foja, C., Schild, D., Uckermann, O., Travis, K., Reichenbach, A., Guck, J.: **Müller cells are living optical fibers in the vertebrate retina**. *Proc. Natl Acad. Sci. USA* 2007, **104**: 8287–8292.

Gehring, W.J.: **New perspectives on eye development and the evolution of eyes and photoreceptors**. *J. Hered.* 2005, **96**: 171–184.

Heavner, W., Pevney, L.: **Eye development and retinogenesis**. *Cold Spring Harb. Perspect. Biol.* 2012, **4**: a008391.

Jeffery, W.R.: **Evolution and development in the cavefish** *Astyanax*. *Curr. Top. Dev. Biol.* 2009, **86**: 191–221.

Sasai, Y., Eiraku, M., Suga, H.: *In vitro* **organogenesis in three-dimensional self-organizing stem cells.** *Development* 2012, **139**: 4111–4121.

Streit, A.: **The preplacodal region: an ectodermal domain with multipotential progenitors that contribute to sense organs and cranial sensory ganglia.** *Int. J. Dev. Biol.* 2007, **51**: 447–461.

Takahashi, S., Asashima, M., Kurata, S., Gehring, W.J.: **Conservation of** *Pax6* **function and upstream activation by Notch signaling in eye development of frogs and flies.** *Proc. Natl Acad. Sci. USA* 2002, **99**: 2020–2025.

11.28 Patterning of the *Drosophila* **eye involves cell–cell interactions**

Baonza, A., Casci, T., Freeman, M.: **A primary role for the epidermal growth factor receptor in ommatidial spacing in the** *Drosophila* **eye.** *Curr. Biol.* 2001, **11**: 396–404.

Chou, W.H., Huber, A., Bentrop, J., Schulz, S., Schwab, K., Chadwell, L.V., Paulsen, R., Britt, S.G.: **Patterning of the R7 and R8 photoreceptor cells of** *Drosophila***: evidence for induced and default cell-fate specification.** *Development* 1999, **126**: 607–616.

Frankfort, B.J., Mardon, G.: **R8 development in the** *Drosophila* **eye: a paradigm for neural selection and differentiation.** *Development* 2002, **129**: 1295–1306.

Freeman, M.: **Cell determination strategies in the** *Drosophila* **eye.** *Development* 1997, **124**: 261–270.

Sahin, H.B., Çelik A.: *Drosophila* **eye development and photoreceptor specification.** *eLS* 2013.

Strutt, H., Strutt, D.: **Polarity determination in the** *Drosophila* **eye.** *Curr. Opin. Genet. Dev.* 1999, **9**: 442–446.

Tomlinson, A., Struhl, G.: **Delta/Notch and Boss/Sevenless signals act combinatorially to specify the** *Drosophila* **R7 photoreceptor.** *Mol. Cell* 2001, **7**: 487–495.

11.29 The vertebrate lung develops by branching of epithelial tubes

Metzger, R.J., Klein, O.D., Martin, G.R.Krasnow, M.A.: **The branching program of mouse lung development.** *Nature* 2008, **453**: 745–750.

Morrisey, E.E., Hogan, B.L.M.: **Preparing for the first breath: genetic and cellular mechanisms in lung development.** *Dev. Cell* 2010, **18**: 8–23.

Ochoa-Espinosa, A., Affolter, M.: **Branching morphogenesis: from organs to cells and back.** *Cold Spring Harb. Perspect. Biol.* 2012, **4**: a008243.

Pepicelli, C.V., Lewis, P.M., McMahon, A.P.: **Sonic hedgehog regulates branching morphogenesis in the mammalian lung.** *Curr. Biol.* 1998, **8**: 1083–1086.

Sakakura, T., Nishizuka, Y., Dawe, C.J.: **Mesenchyme-dependent morphogenesis and epithelium-specific cytodifferentiation in mouse mammary gland.** *Science* 1976, **194**: 1439–1441.

11.30 The *Drosophila* **tracheal system is a prime example of branching morphogenesis**

Affolter, M., Caussinus, E.: **Branching morphogenesis in** *Drosophila***: new insights into cell behaviour**

and organ architecture. *Development* 2008, **135**: 2055–2064.

Ribeiro, C., Neumann, M., Affolter, M.: **Genetic control of cell intercalation during tracheal morphogenesis in** *Drosophila*. *Cell* 2004, **14**: 2197–2207.

11.31 The vascular system develops by vasculogenesis followed by sprouting angiogenesis

Carmeliet, P., Tessier-Lavigne, M.: **Common mechanisms of nerve and blood vessel wiring.** *Nature* 2005, **436**: 193–200.

Coultas, L., Chawengsaksophak, K., Rossant, J.: **Endothelial cells and VEGF in vascular development.** *Nature* 2005, **438**: 937–945.

Geudens, I., Gerhardt, H.: **Coordinating cell behavior during blood vessel formation.** *Development* 2011, **138**: 4569–4583.

Harvey, N.L., Oliver, G.: **Choose your fate: artery, vein or lymphatic vessel?** *Curr. Opin. Genet. Dev.* 2004, **14**: 499–505.

Jain, R.K., Carmeliet, P.: **Snapshot: tumor angiogenesis.** *Cell* 2012, **149**: 1408.e1.

Larrivée, B., Freitas, C., Suchting, S., Brunet, I., Eichmann, A.: **Guidance of vascular development: lessons from the nervous system.** *Circ. Res.* 2009, **104**: 428–441.

Reese, D.E., Hall, C.E., Mikawa, T.: **Negative regulation of midline vascular development by the notochord.** *Dev. Cell* 2004, **6**: 699–708.

Rossant, J., Hirashima, M.: **Vascular development and patterning: making the right choices.** *Curr. Opin. Genet. Dev.* 2003, **13**: 408–412.

Sato, Y., Poynter, G., Huss, D., Filla, M.B., Czirok, A., Rongish, B.J., Little, C.D., Fraser, S.E., Lansford, R.: **Dynamic analysis of vascular morphogenesis using transgenic quail embryos.** *PLoS One* 2010, **5**: e12674.

Wang, H.U., Chen, Z.-F., Anderson, D.J.: **Molecular distinction and angiogenic interaction between embryonic arteries and veins revealed by ephrin-B2 and its receptor Eph-B4.** *Cell* 1998, **93**: 741–753.

11.32 The development of the vertebrate heart involves morphogenesis and patterning of a mesodermal tube

Bruneau, B.G.: **The developmental genetics of congenital heart disease.** *Nature* 2008, **451**: 943–948.

Linask, K.K., Yu, X., Chen, Y., Han, M.D.: **Directionality of heart looping: effects of Pitx2c misexpression on flectin asymmetry and midline structures.** *Dev. Biol.* 2002, **246**: 407–417.

Meilhac, S.M., Esner, M., Kelly, R.G., Nicolas, J.F., Buckingham, M.E.: **The clonal origin of myocardial cells in different regions of the embryonic mouse heart.** *Dev. Cell* 2004, **6**: 685–698.

Perez-Pomares, J.M., de la Pompa J.L.: **Signalling during epicardium and coronary vessel development.** *Circ. Res.* 2011, **109**: 1429–1442.

Ray, H.J., Niswander, L.: **Mechanisms of tissue fusion during development.** *Development* 2012, **139**: 1701–1711.

Srivastava, D., Olson, E.N.: **A genetic blueprint for cardiac development.** *Nature* 2000, **407**: 221–226.

Staudt, D., Stainier, D.: **Uncovering the molecular and cellular mechanisms of heart development using the zebrafish**. *Annu. Rev. Genet.* 2012, **46**: 397–418.

Vincent, S.D., Buckingham, M.E.: **How to make a heart: the origin and regulation of cardiac progenitor cells**. *Curr. Top. Dev. Biol.* 2010, **90**: 1–41.

Zaffran, S., Kelly, R.G.: **New developments in the second heart field**. *Differentiation* 2012, **84**: 17–24.

11.33 Tooth development involves epithelial-mesenchymal interactions and a homeobox gene code specifies tooth identity

Cobourne, M.T., Sharpe, P.T.: **Making up the numbers: the molecular control of mammalian dental formula**. *Semin. Cell Dev. Biol.* 2010, **21**: 314–324.

Jernvall, J., Thesleff, I.: **Tooth shape formation and tooth renewal: evolving with the same signals**. *Development* 2012, **139**: 3487–3497.

Jussila, M., Thesleff, I.: **Signaling networks regulating tooth organogenesis and regeneration, and the specification of dental mesenchymal and epithelial cell lineages**. *Cold Spring Harb. Perspect. Biol.* 2012, **4**: a008425.

Mitsiadis, T.A., Angeli, I., James, C., Lendahl, U., Sharpe, P.T.: **Role of Islet1 in the patterning of murine dentition**. *Development* 2003, **130**: 4451–4460.

Tucker, A., Sharpe, P.: **The cutting-edge of mammalian development: how the embryo makes teeth**. *Nat. Rev. Genet.* 2004, **5**: 499–508.

12

Development of the nervous system

- Specification of cell identity in the nervous system
- The formation and migration of neurons
- Axon navigation
- Synapse formation and refinement

The nervous system is the most complex of all the organ systems in the animal embryo. In mammals, for example, billions of nerve cells, or neurons, develop a highly organized pattern of connections, creating the neuronal networks that make up the functioning brain and the rest of the nervous system. There are many hundreds of different types of neurons, differing in their identity and the connections they make, even though they may look quite similar. Understanding the developmental processes that shape the nervous system is not simple, as they involve neuronal cell specification, differentiation, morphogenesis, and migration. Neurons send out long processes from the cell body and these must be guided to find their targets. The picture is further complicated by the electrical activity of the nerve cells, which can refine the pattern of connections they make.

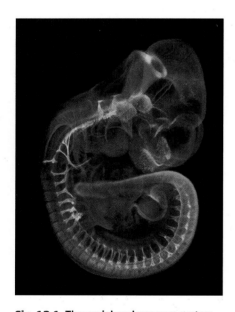

Fig. 12.1 The peripheral nervous system of a mouse embryo at 11 days of gestation. Parts of the nervous system of an intact mid-term mouse embryo have been revealed by immunocytochemical staining with of a pan-neural antigen (red) and the expression of a somatic motor neuron marker gene (Hb9, green). The cranial sensory ganglia and the segmental array of spinal nerves are particularly prominent.

Photograph courtesy of I. Lieberam and T. Jessell.

The nervous system is the most complex of all the organ systems in the animal embryo. In mammals, for example, billions of nerve cells, or neurons, develop a highly organized pattern of connections, creating the neuronal networks that make up the functioning brain and the rest of the nervous system. The nervous system is broadly divided into the **central nervous system**—the brain and spinal cord—and the **peripheral nervous system** of nerves and ganglia outside the brain and spinal cord, which connects the rest of the body with the central nervous system (Fig. 12.1).

All nervous systems, vertebrate and invertebrate, provide a system of communication through a network of electrically excitable cells—the nerve cells or **neurons**—of varying sizes, shapes, and functions (Fig. 12.2). The nervous system also contains non-neuronal cells known collectively as **glia**, which have a variety of supporting roles. The Schwann cells that provide the myelin sheaths of neurons in the peripheral nervous system are **glial cells**, as are the oligodendrocytes that serve the same function in the central nervous system. Neurons connect with each other and with other target cells, such as muscle, at specialized junctions known as **synapses**. A neuron receives input from other neurons through its highly branched **dendrites** and, if the signals are strong enough to activate the neuron, it generates a new electrical

Fig. 12.2 Neurons come in many shapes and sizes. Upper panel: a montage of the various types of neurons found in the optic lobe of the avian brain. In the brain there are many more nervous connections that are not shown here. Lower panel: a single neuron consists of a small, rounded cell body, which contains the nucleus, and from which extend a single axon and a 'tree' of much-branched dendrites. Signals from other neurons are received at the dendrites, processed, and integrated in the cell body, and an outgoing signal—a nerve impulse—is generated and sent down the axon. We shall consider in more detail how neurons signal and communicate with each other in the last part of the chapter.

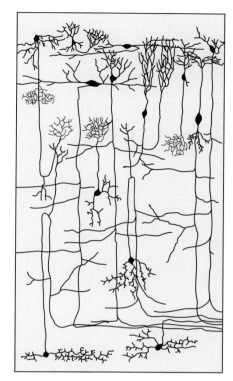

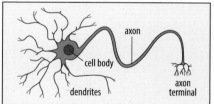

signal—a **nerve impulse**, or **action potential**—at the **cell body**. This electrical signal is then conducted along the **axon** to the **axon terminal**, or nerve ending, which makes a synapse with the dendrites or cell body of another neuron or with the surface of a muscle cell. Some neurons in the central nervous system are connected in this way to as many as 500 other neurons.

The dendrites and axon terminals of individual neurons can be extensively branched, and a single neuron in the central nervous system can receive inputs through as many as 100,000 synapses. At a synapse, the electrical signal is converted into a chemical signal, in the form of a chemical neurotransmitter, which is released from the nerve ending and acts on receptors in the membrane of the opposing target cell to generate or suppress a new electrical signal. The nervous system can only function properly if the neurons are correctly connected to one another, and so a central question in nervous-system development is how the connections between neurons develop with the appropriate specificity. The number of neurons in the human brain seems generally to be estimated at around 100 billion. How many of them have unique or similar identities is not known.

In this chapter, we shall use examples from both invertebrate and vertebrate central nervous systems and this raises the question: do the central nervous systems of vertebrates and invertebrates trace back to a common precursor or are they of independent evolutionary origin? Anatomically, a central nervous system is a spatially delimited nervous tissue composed of neurons with specialized functions that are interconnected by highly organized tracts of axons. The central nervous system receives sensory inputs from the body, processes them, integrates them centrally, and produces outputs that guide behavior, such as movement. Such an organization broadly defines the central nervous systems of animals as far apart evolutionarily as annelid worms and mammals, and is in contrast to the diffuse nervous system, or nerve net, of cnidarians (*Hydra* and sea anemones), which receives sensory input and processes output without central integration.

One important aspect of nervous-system development in both vertebrates and invertebrates is the early segregation of the ectoderm into a non-neural epidermal portion and a neural portion—the neuroectoderm, which gives rise to both the central nervous system (brain and spinal cord) and the peripheral nervous system. In the Bilateria—all animals with bilateral symmetry—the brain and associated sensory organs develop at the anterior end of the embryo, in the head region. Similarities can be traced between the embryonic central nervous systems of simple invertebrates and those of vertebrates. In the annelid worm *Platynereis*, for example, the neuroectoderm is subdivided longitudinally into partially overlapping domains of progenitor neurons that correspond to similar domains in the vertebrate neural tube and give rise to similar cell types as those of vertebrates. In vertebrates, as discussed later in this chapter, the genes that pattern the central nervous system are sensitive to signaling by members of the bone morphogenetic protein (BMP) family, and this is also the case in *Platynereis*. All this evidence suggests that a central-nervous-system architecture was present in the last common ancestor of the bilaterians, and supports a common origin for the nervous system in Bilateria.

For all its complexity, the nervous system is the product of the same kind of cellular and developmental processes as those involved in the development of other organs.

The overall process of nervous-system development can be divided into six major stages: the regionalization of neurectoderm into brain, brain subregions, and spinal cord; the specification of individual neural cell identity; the formation and migration of neurons; the outgrowth of axons to their targets; the formation of synapses with target cells, which can be other neurons, muscle or gland cells; and the refinement of synaptic connections through the elimination of axon branches and neuronal death.

We have already considered some aspects of the development of the vertebrate nervous system, including neural induction (Chapters 4 and 5), the formation of the neural tube (Chapter 9), and the migration of neural crest cells to differentiate into neurons and populate dorsal root ganglia (Chapter 9). We will revisit some of these events when we consider the specification of neural cell identity. A single chapter, such as this one, cannot cover all aspects of nervous-system development, and the general focus here is on the mechanisms that regionalize the nervous system, control the identity of neurons, and pattern their synaptic connections.

Specification of cell identity in the nervous system

Prospective neural tissue—the neuroectoderm—is distinguished from the rest of the ectoderm as part of the early patterning of invertebrate and vertebrate embryos, as discussed in Chapters 2, 4, and 5. We will start this chapter by considering the next stage in the process of generating a nervous system: how the neuroectoderm acquires major regions, each with its characteristic size and shape, and repertoire of neuronal cell types. How, for example, does the brain become subdivided into fore-, mid- and hindbrain regions? And then, how do the hundreds of distinct neuronal types present in any one of these regions acquire their correct individual identities, each in its correct position with respect to the whole? Specification is a process of continuous and progressive refinement of neural precursors, from initial coarse regional pattern to final cell identity, and is complete before a precursor cell undergoes its last cell division and produces a post-mitotic neuron.

12.1 Initial regionalization of the vertebrate brain involves signals from local organizers

We have seen how, during vertebrate neural induction, the neuroectoderm acquires antero-posterior polarity (see Chapters 4 and 5), with an expanded anterior region that becomes the brain. The developing brain next becomes divided into three main regions—**forebrain** (prosencephalon), **midbrain** (mesencephalon), and **hindbrain** (rhombencephalon)—illustrated here for 2-day and 4.5-day chick embryos (Fig. 12.3). Considerable lateral expansions develop from the embryonic forebrain and these are the rudiments of the optic vesicles, from which the eyes will form (see Section 11.27).

The embryonic forebrain gives rise to the **pallium** and basal ganglia (the **telencephalon**), the **thalamus**, and the **hypothalamus**. The pallium of mammalian embryos develops into the outer layers of the bilateral cerebral hemispheres, the **cerebral cortex**, where the highest-level processing centers for sensory information, motor control, learning, and memory are located. The thalamus is a major relay station that distributes incoming sensory information to the appropriate region of the cortex, while the hypothalamus is the brain's link with the endocrine system, and controls the release of hormones from the pituitary.

The embryonic midbrain gives rise to the **tectum**, which is the site of integration and relay centers for signals coming to and from the hindbrain and also for inputs from the sensory organs, such as the eye and ear. The embryonic hindbrain is characterized by a series of seven or eight transient swellings, known as **rhombomeres**, and gives rise to the **cerebellum** (rhombomere 1), the pons (rhombomeres 2 and 3) and the

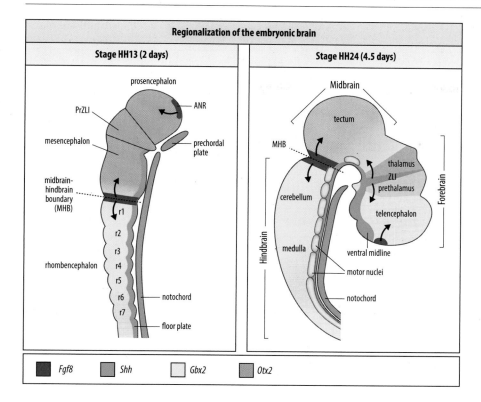

Regionalization of the embryonic brain

Stage HH13 (2 days)

prosencephalon

PrZLI

ANR

mesencephalon

prechordal plate

midbrain–hindbrain boundary (MHB)

r1
r2
r3
r4
r5
r6
r7

rhombencephalon

notochord

floor plate

Stage HH24 (4.5 days)

Midbrain

tectum

MHB

thalamus
ZLI
prethalamus

cerebellum

telencephalon

Hindbrain

Forebrain

medulla

ventral midline

motor nuclei

notochord

Fgf8   Shh   Gbx2   Otx2

Fig. 12.3 Local signaling centers in the developing vertebrate brain. Lateral view of the brain of a 2-day chick embryo (HH stage 13; left panel) and of a 4.5-day chick embryo (HH stage 24: right panel). The three local signaling centers—the isthmus at the midbrain–hindbrain boundary (MHB), the zona limitans intrathalamica (ZLI) in the posterior forebrain, and the anterior neural ridge (ANR) at the anterior pole—are indicated by coloured lines. Bi-directional signaling from MHB and ZLI, and unidirectional signaling from the ANR (indicated by black arrows) is responsible for patterning the adjacent brain regions, and involves secretion of FGF-8 (red) from the MHB and ANR, and Sonic hedgehog (Shh, green) from the ZLI. Wnt-1 is also expressed on the anterior side of the MHB. The boundary between *Otx2* expression in the midbrain (violet) and *Gbx2* expression in the hindbrain (yellow) is linked to the formation of the MHB signaling center. PrZLI, presumptive zona limitans intrathalamica.

Adapted from Kiecker, C., Lumsden, A.: **Compartments and their boundaries in vertebrate brain development.** Nat. Rev. Neurosci. *2005, **6**: 553-564.*

medulla oblongata, which are concerned with the regulation of basic body activities that are independent of conscious control, such as regulation of muscle tone and posture (by the cerebellum), and heartbeat, breathing, and blood pressure.

12.2 Local signaling centers pattern the brain along the antero-posterior axis

Three local signaling centers, or **local organizers**, in the developing brain are responsible for the specification of the brain's major subregions (see Fig. 12.3). One, known as the **midbrain–hindbrain boundary** (MHB), is located at the isthmus between these brain regions and regulates patterning of the posterior midbrain (tectum) and the anterior hindbrain (cerebellum); another, known as the **zona limitans intrathalamica** (ZLI), is located in the posterior forebrain and regulates patterning of the thalamus; a third, known as the **anterior neural ridge** (ANR) lies immediately adjacent to and within the anterior-most forebrain and is responsible for telencephalic patterning. The midbrain–hindbrain boundary and the anterior neural ridge are already major signaling centers in the 2-day chick brain and are still active in the brain of 4.5-day chick embryos, whereas the zona limitans intrathalamica is formed later and acts as a signaling center in the 4.5-day chick brain.

The signaling properties of the midbrain–hindbrain boundary were first shown by grafting experiments in 2-day chick embryos. If the isthmus region is grafted into the anterior midbrain or posterior forebrain, it converts presumptive thalamus into tectum, whereas if grafted to the posterior hindbrain, it converts presumptive medulla into cerebellum. *Fgf8* is expressed at the isthmus and FGF-8 is a good candidate for a bidirectional signal controlling the patterning of posterior midbrain and anterior hindbrain. Induced expression of *Fgf8* in the chick anterior midbrain, for example, results in the expression of genes characteristic of the posterior midbrain. The zebrafish mutant *acerebellar* has a mutation that abolishes expression of FGF-8 almost completely, and as the name suggests, mutant fish lack the cerebellum and its associated organizer.

Fig 12.4 The regulation of cortical pattern by local signals. First panel: schematic front view of the forebrain of a mid-gestation mouse embryo, showing various signal centers and their signals: the anterior neural ridge (ANR) and its derivative, the commissural plate, in the anterior forebrain (FGFs, red); the basal forebrain (Shh, green); and the cortical midline (Wnts, blue). Second panel: top view of the FGF source in the anterior telencephalon. Third panel: the graded expression domains of three prominent transcription factors, Emx2, Coup-tf1, and Pax6. Fourth panel: when any of these factors is knocked out there is a shift in areal identity, implying that, for example, visual cortex (V) requires Emx2 and Coup-tf1, whereas motor cortex (M) requires Pax6. S, sensory cortex.

First panel adapted from Sur, M., Rubenstein, J.L.R.: *Patterning and plasticity of the cerebral cortex.* Science 1973, **310**: 805-810.

Second panel from O'Leary, D., Nakagawa, Y.: *Patterning centers, regulatory genes and extrinsic mechanisms controlling arealization of the neocortex.* Curr. Opin. Neurobiol. 2002, **12**: 14-25.

Third panel from Rakic, P., et al.: *Decision by division: making cortical maps.* Trends Neurosci. 2009, **32**: 291-301.

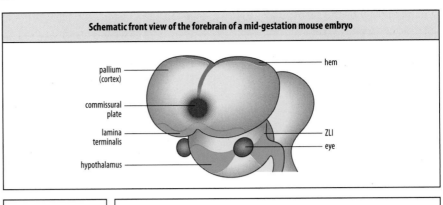

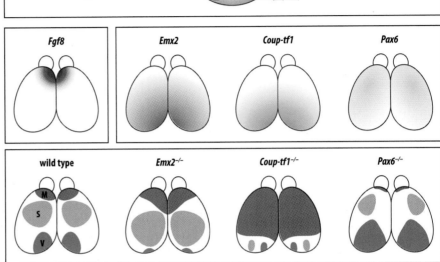

The midbrain–hindbrain organizer region develops at the border between the expression domains of transcription factors Otx2 (on the midbrain side) and Gbx2 (on the hindbrain side). Initial division into an Otx2-expressing anterior territory and a Gbx2-expressing posterior territory occurs in the neural plate, and is one of the earliest patterning events in brain development. Later, in the neural tube, FGF-8 is expressed on the hindbrain side of the border and the signal molecule Wnt-1 on the midbrain side. The zona limitans intrathalamica signals bi-directionally via the signal molecule Sonic hedgehog (Shh), and is essential for formation of the pre-thalamus anteriorly and the thalamus posteriorly (see Fig. 12.3).

A key feature of local organizer function is the release of a diffusible molecule that induces the expression of transcription factors in a spatial array. Because the same molecule, for example FGF-8, can induce different gene expression at different positions, such molecules are known as morphogens (see Box 11B). A morphogen confers positional information: by diffusing from a localized source and forming a concentration gradient across adjacent tissue, the morphogen functions to set the size of a responding field and to induce the expression of different transcription factors at different concentrations (see Box 11B). These factors encode a cell's positional value and are involved in translating this positional value into the cell's final identity. We will see this principle in operation when comparing the different responses of the midbrain–hindbrain region and the cerebral cortex to the same morphogen, FGF-8.

12.3 The cerebral cortex is patterned by signals from the anterior neural ridge

It was long thought that the embryonic cortex was a blank slate (a 'tabula rasa') with its highly complex pattern of **functional areas** being imposed late in development and principally by the spatial organization of incoming axons from the thalamus. It is now clear, however, that the precursor of the cortex, the pallium, acquires considerable areal identity much earlier in development. The genes for many transcription factors, including Emx2 (the vertebrate ortholog of the fly head gene, *empty spiracles*), Pax6 (see Section 11.27), Coup-tf1, and Lhx2, are expressed at different antero-posterior and medio-lateral subdomains of the pallium, under the inductive control of morphogens such as FGF-8 from the anterior neural ridge and Wnt proteins from the midline (Fig. 12.4). Mouse mutants for *Emx2*, *Pax6*, and *Coup-tf1* show how these transcription factors regulate areal identity: loss of Emx2 or Coup-tf1 result in the expansion of anterior areas, representing motor and sensory cortex, at the expense of posterior, visual, cortical areas, whereas loss of Pax6 results in a complementary change.

12.4 The hindbrain is segmented into rhombomeres by boundaries of cell-lineage restriction

Patterning of the hindbrain (and of the face and head through the neural crest cells that emerge from the hindbrain), involves **segmentation** along the antero-posterior axis. This type of patterning of the nervous system does not occur in the midbrain and forebrain, where pattern is controlled by local organizers, or along the spinal cord, where the pattern of dorsal root ganglia and ventral motor nerves at regular intervals—one pair per somite—is imposed by the somites (see Section 9.16). In the chick embryo, two segmented systems can be seen in the posterior head region by 3 days of development (Fig. 12.5): the hindbrain (the rhombencephalon) is divided into eight rhombomeres (r1 to r8); and ventral to the hindbrain lie a series of **branchial arches** (b1 to b4) that are populated by neural crest cells migrating from the hindbrain. The branchial arches give rise to the tissues of the head and throat other than the brain. A similar division of the hindbrain into rhombomeres occurs in all model vertebrates.

Development of the hindbrain region of the head involves several interacting components. The neural tube in this region produces both the segmentally arranged cranial motor nerves that innervate the face and neck (see Fig. 12.5), and also neural crest cells, which give rise to cranial sensory and parasympathetic ganglia, the Schwann cells of cranial nerves, and the **ectomesenchyme** that forms most of the facial skeleton and connective tissues. The sensory ganglia also have contributions from placodes forming in the surface ectoderm. In addition, the otic vesicle, which develops opposite rhombomeres 5 and 6, gives rise to the ear.

The main skeletal elements of the face develop from the first two branchial arches. For example, the first arch gives rise to the jaws (maxilla and mandible) and two middle ear bones (incus and malleus), while the second arch develops into the hyoid bone and the stapes of the middle ear. This region of the head is a particularly valuable model for studying patterning along the antero-posterior axis because of the presence of the numerous different structures regularly ordered along it.

Immediately after the neural tube of the chick embryo closes, the future hindbrain becomes transversely constricted to define eight rhombomeres (see Fig. 12.5). The cellular basis for these constrictions is not entirely understood, but is likely to involve differential cell division and changes in cell shape. The boundaries between rhombomeres subsequently become differentiated from the adjacent neuroepithelium, with a marked increase in the extracellular space between cells. Cell-lineage tracing experiments show that once formed, individual rhombomeres are areas of cell-lineage restriction; that is, cells and their descendants are confined within the rhombomere in which they originate and do not cross from one side of the boundary to the other. Tracking the descendants of individual labeled cells shows that before constrictions become visible, the descendants of a given cell can form a clone that spans a boundary. After the constrictions appear, however, clones of labeled cells are confined to a single rhombomere (Fig. 12.6). This lineage restriction suggests that

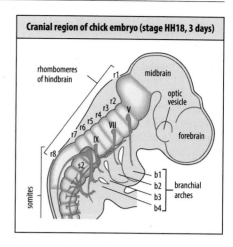

Cranial region of chick embryo (stage HH18, 3 days)

Fig. 12.5 The nervous system in a 3-day chick embryo (HH stage 18). At this stage in development the hindbrain is divided into eight rhombomeres (r1 to r8). The positions of three (V, VII, IX) of the nine cranial nerves that arise from the hindbrain are shown in green; b1 to b4 are the four branchial arches; b1 gives rise to the jaws; s = somite.

*Illustration adapted, with permission, from Lumsden, A.: **Cell lineage restrictions in the chick embryo hindbrain**. Phil. Trans. R. Soc. Lond. B 1991, **331**: 281–286.*

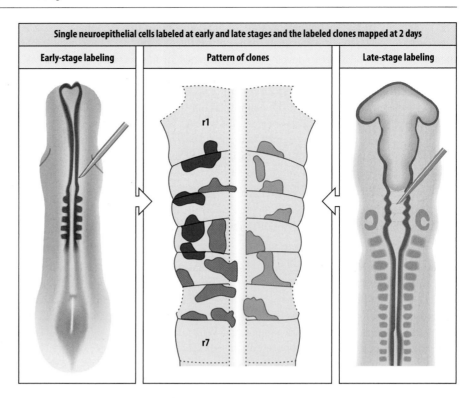

Single neuroepithelial cells labeled at early and late stages and the labeled clones mapped at 2 days

Early-stage labeling | **Pattern of clones** | **Late-stage labeling**

r1

r7

Fig. 12.6 Lineage restriction in rhombomeres of the embryonic chick hindbrain. Single neural precursor cells are injected with a label (rhodamine-labeled dextran) at an early stage (left panel) or a later stage (right panel) of neurulation, and their descendants are mapped 2 days later. Cells injected before rhombomere boundaries form give rise to some clones that span two rhombomeres (dark red) as well as those that do not cross boundaries (red). Clones marked after rhombomere formation do not cross the boundary of the rhombomere that they originate in (blue), though their descendent neurons may do so.

*Illustration adapted, with permission, from Lumsden, A.: **Cell lineage restrictions in the chick embryo hindbrain**. Phil. Trans. R. Soc. Lond. B 1991, **331**: 281–286.*

each rhombomere is acting as an independent developmental unit. In this respect a rhombomere is behaving rather like a compartment, which is an important feature of insect development, as we saw in Chapter 2, but seems to be rare in vertebrates (but note the dorsal and ventral compartments in limb buds, Section 11.10). The rhombomeres are cell-tight boxes within which neural precursors acquire their positional values. Once these precursors become differentiated neurons, they may migrate tangentially, away from their rhombomere of origin. The rhombomeres are no longer visible morphologically after about 4.5 days in the chick embryo (see Fig. 12.3).

Clues to how cells from adjacent rhombomeres might be prevented from mixing come from the discovery that the cells of odd-numbered rhombomeres share some adhesive property that is not shared by even-numbered rhombomeres. This was originally revealed by experiments in which an odd-numbered and an even-numbered rhombomere from non-adjacent positions were placed together—a new boundary formed between them. No new boundary formed when two odd-numbered rhombomeres were placed together, however, suggesting that their cells are compatible with each other. The boundary region itself is unlikely to be acting purely as a mechanical barrier, as cells from adjacent rhombomeres do not mix even when the boundary between them is surgically removed. The finding that particular cell-surface proteins of the **ephrin** family and their **Eph receptors** are expressed separately in alternate rhombomeres provides one possible mechanism for preventing cell mixing.

Ephrins and their receptors are transmembrane signaling proteins that interact with each other on adjacent cells and can both signal to their respective cells. They direct either repulsive or attractive cell–cell interactions (see Box 9D). In particular, the EphA4 receptor, which is strongly expressed in rhombomeres r3 and r5, and its ligand ephrin B2, which is expressed in r2, r4, and r6, could provide repulsive interactions at each rhombomere boundary and so prevent cell mixing (see Fig. 12.6). The expression of EphA4 in r3 and r5 is under the control of the transcription factor Krox20, which is itself expressed in two stripes in the neural plate that become r3 and r5. Separate sets of transcription factors have been shown to switch on *Krox20* in r3 and r5, respectively.

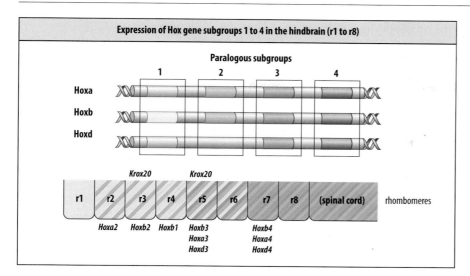

Expression of Hox gene subgroups 1 to 4 in the hindbrain (r1 to r8)

Paralogous subgroups

1   2   3   4

Hoxa

Hoxb

Hoxd

Krox20   Krox20

r1   r2   r3   r4   r5   r6   r7   r8   (spinal cord)   rhombomeres

Hoxa2   Hoxb2   Hoxb1   Hoxb3   Hoxb4
                          Hoxa3   Hoxa4
                          Hoxd3   Hoxd4

Fig. 12.7 Expression of Hox genes in the hindbrain. The expression of genes of three paralogous Hox clusters in the hindbrain (rhombomeres r1 to r8) is shown. *Hoxa1* and *Hoxd1* are not expressed at this stage. Note that there are no Hox genes expressed in r1. The Hox genes whose anterior expression starts in each rhombomere are listed underneath. With the exception of *Hoxb1*, which is only expressed in r4, all the Hox genes listed are also expressed in all rhombomeres posterior to their most anterior expression. The gene encoding the transcription factor Krox20 is expressed in rhombomeres 3 and 5.

Illustration after Krumlauf, R.: **Hox genes and pattern formation in the branchial region of the vertebrate head.** Trends Genet. *1993,* **9***: 106–112.*

The division of the hindbrain into rhombomeres has further functional significance. A similar pattern of neuronal cell types develops along the antero-posterior axis of each rhombomere, and at the same time, each rhombomere acquires a unique identity in respect of the particular cranial nerves and other tissues it gives rise to. The delimitation and future development of individual rhombomeres is under the control of Hox and other transcription factors, and we shall next consider how the Hox gene expression provides positional information and identity both to the developing hindbrain itself and to the neural crest cells that migrate away from it.

12.5 Hox genes provide positional information in the developing hindbrain

Visible constriction of the hindbrain into rhombomeres is preceded by the patterned expression of transcriptional control genes, including *Krox20* and Hox genes, that initially divides the neural tube into the prospective rhombomeric segments. Hox gene expression provides at least part of the molecular basis for the identities of both the rhombomeres and the neural crest at different positions along the hindbrain. No Hox genes are expressed in the most anterior part of the head or in r1, but Hox genes from four paralogous groups (1, 2, 3, and 4) are expressed in the rest of the mouse embryonic hindbrain in a well-defined pattern, which closely correlates with the segmental pattern (Fig. 12.7). The most anterior Hox gene expressed at this stage is *Hoxa2*, which is expressed in r2. In general, the different paralogous groups have different anterior margins of expression. For example, the anterior border of *Hoxb2* expression is at the boundary between r2 and r3, whereas the anterior margin of expression of *Hoxb3* is at the border between r4 and r5.

Studies at the molecular level have provided some indication to how Hox gene expression is controlled in the individual rhombomeres. For example, although the *Hoxb2* gene is expressed in the three contiguous rhombomeres r3, r4, and r5, its expression in r3 and r5 is controlled independently from that in r4. The regulatory region of the *Hoxb2* gene carries two separate *cis*-regulatory modules, one of which regulates its expression in r3 and r5, while expression in r4 is controlled through the other. In r3 and r5, *Hoxb2* is activated in part by the transcription factor Krox20, which is expressed in these rhombomeres but not in r4 (see Fig. 12.7), and there are binding sites for Krox20 in the corresponding regulatory control region of *Hoxb2*.

In the case of *Hoxb4*, whose anterior border of expression is at the boundary between r6 and r7, it seems that the gene is switched on by signals both from within the neural tube and from adjacent somites. One of these signals could be retinoic acid, which is produced by the somites and is involved in patterning the hindbrain. An absence of retinoic acid leads to loss of hindbrain rhombomeres, whereas an excess causes a transformation of neural cell fate from anterior to posterior along the length of the hindbrain.

Evidence that Hox genes determine the fate and future development of cells in the rhombomeres comes from misexpression experiments. Each pair of rhombomeres produces motor axons that project to a single branchial arch: axons from ventral, premotor regions of rhombomeres r2/r3 project to the first branchial arch, whereas those from r4/r5 project to the second arch (see Fig. 12.5). *Hoxb1* is normally expressed in r4 but not in r2. If *Hoxb1* is misexpressed in r2, however, this then sends axons to the second arch. This is yet another example of a homeotic transformation produced by misexpression of Hox genes.

Hox genes are not expressed in the anterior-most neural tissue—the forebrain, midbrain, and r1 of the hindbrain. Instead, homeodomain transcription factors such as Otx and Emx are expressed anterior to the hindbrain and specify antero-posterior pattern in the anterior brain. As we saw earlier, the posterior limit of *Otx2* expression delineates the midbrain–hindbrain boundary in the chick (see Fig. 12.3). The *Drosophila orthodenticle* gene and the vertebrate *Otx* genes are homologous and provide a good example of the conservation of gene function during evolution. *orthodenticle* is expressed in the posterior region of the future *Drosophila* brain, and mutations in the gene result in a greatly reduced brain. In mice, *Otx1* and *Otx2* are expressed in overlapping domains in the developing forebrain and midbrain, and mutation in *Otx1* leads to brain abnormalities and epilepsy. Mice with a defective *Otx* gene can be partly rescued by replacing it with *orthodenticle*, even though the sequence similarity of the two proteins is confined to the homeodomain region. Human *Otx* can even rescue *orthodenticle* mutants in *Drosophila*.

12.6 The pattern of differentiation of cells along the dorso-ventral axis of the spinal cord depends on ventral and dorsal signals

We have seen how a coarse pattern of specification occurs in the developing brain as a result of signals from local organizers activating transcription factors that then act as regional or subregional determinants of fate. The next challenge facing the embryo is to ensure, at a finer level of patterning, that individual neurons of particular types and function differentiate in the correct positions in the neural tube. The developing spinal cord is a particularly good system in which to study both coarse and finer aspects of patterning, as it has a particularly clear dorso-ventral pattern of functionally distinct neurons, which is similar along its length (Fig. 12.8).

Future motor neurons and their associated interneurons are located ventrally, and form the ventral roots of the spinal cord, whereas **commissural neurons**, secondary sensory neurons, and their associated interneurons differentiate primarily in the dorsal region. Each cell type is present symmetrically on either side of the midline. While motor and most sensory neurons differentiating in the right and left halves of the spinal cord extend their axons on the side of the body in which they originated, the axons of commissural neurons will cross the midline of the spinal cord, providing a link between the two sides of the body at the level of the spinal cord (Fig. 12.9). The primary sensory neurons develop from neural crest cells that are located laterally and dorsally in the neural tube, and migrate away from the surface of the tube to form the dorsal root ganglia that lie at regular intervals on each side of the spinal cord (see Fig. 12.9 and Figs. 9.38 and 9.39).

We shall now consider how the dorso-ventral organization of the spinal cord is produced by signals received by the early neural tube, which induce the differentiation of particular types of neurons in the correct positions. In additional to the future nerve-cell precursors, a group of non-neural cells forms a region called the **floor plate** in the ventral midline of the neural tube, while another group of non-neural cells forms the **roof plate** in the dorsal midline. These two regions produce signals that pattern the neural tube along its dorso-ventral axis.

The floor plate is induced by signals secreted by the mesodermal notochord, which lies immediately below the neural tube. The patterning activity of the notochord on the neural tube can be shown by grafting a segment of notochord to a site lateral or dorsal to the neural tube (see Fig. 5.34, where the experimental set-up is described in regard to its effect on somite patterning). A second floor-plate is induced in the neural tube in direct contact with the transplanted notochord. Molecular markers of dorsal-cell differentiation are suppressed in the nearby neural cells, and additional motor neurons, which normally only develop in the ventral region, are generated. *In situ* staining for gene expression and misexpression experiments identified the inductive signal secreted by the notochord as Shh. Shh induces the floor-plate cells themselves to produce Shh, which then forms a gradient of activity from ventral to dorsal in the neural tube, and acts as the ventral patterning signal (Fig. 12.10).

The conversion of neuroepithelial neuronal precursors in the ventral neural tube to committed motor neuron progenitor cells is driven by Shh from the floor plate and by retinoic acid secreted by the adjacent mesoderm. Control of the proliferation of neural progenitor cells by mitogenic Wnt signals is also important in regulating the number of neurons produced. In line with the role of Shh and retinoic acid in neural differentiation *in vivo*, they can direct the development of embryonic stem cells in culture into motor neurons and interneurons.

The roof plate is initially specified by signals from the dorsal epidermal ectoderm, and signals from the roof plate pattern the dorsal half of the neural tube. From genetic knock-outs and cell-fate mapping, it has been shown that selective removal of the roof plate from the neural tube of the mouse embryo results in the loss of the most dorsal interneurons. The dorsalization signal includes several members of the BMP family. BMPs from the ectoderm (see Chapter 5) specify the roof plate, and BMPs subsequently produced in the roof plate then pattern the dorsal region of the neural tube. The BMP signals apparently ensure dorsal differentiation by opposing the action of signals from the ventral region, thus together patterning the dorso-ventral axis of the neural tube (see Fig. 12.10).

The spatial pattern of expression of BMPs in the dorsal neural tube is strongly influenced by the ventral signals, and this system bears a strong similarity to the mechanism for

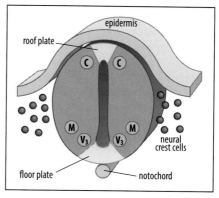

Fig. 12.8 Dorso-ventral organization in embryonic spinal neural tube. In neural tube that develops into spinal cord, a floor plate of non-neuronal cells develops along the ventral midline, and a roof plate of non-neuronal cells along the dorsal midline. Commissural neurons (C) differentiate in the dorsal region, near the roof plate. Motor neurons (M) and V3 interneurons differentiate close to the floor plate (see Fig. 12.11).

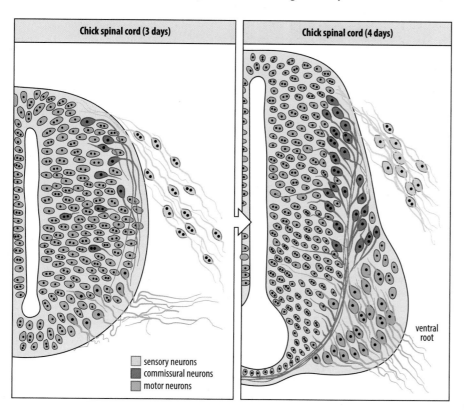

Fig. 12.9 Formation of sensory and motor neurons in the chick spinal cord. Three days after an egg has been laid, motor neurons are beginning to form in the chick embryo neural tube (left panel). A day later, the motor neurons send out axons to form the ventral root (right panel). The primary sensory neurons that migrate from the dorsal part of the spinal cord are derived from neural crest cells and form the segmental dorsal root ganglia (see Section 9.16); their centrally directed axons connect with various types of secondary sensory neurons in the dorsal spinal cord. One type, the commissural neuron, sends axons towards and across the ventral midline.

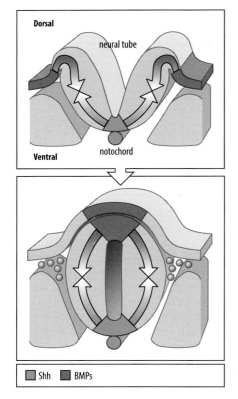

Fig. 12.10 Dorso-ventral patterning of the spinal cord involves both dorsal and ventral signals. Dorsal and ventral signaling operates as the neural tube forms (top panel). The ventral signal from the notochord and floor plate is the Shh protein (green), whereas the dorsal signal (blue) includes BMP-4, which is induced during neural tube formation by the adjacent ectoderm. Ventral and dorsal signals oppose each other's action.

patterning the dorso-ventral body axis in early *Drosophila* development. In that case, the expression of the gene *decapentaplegic* is confined to dorsal regions because its expression is repressed in ventral regions by the transcription factor Dorsal (see Section 2.16). Like its vertebrate homologues, BMP-2 and BMP-4, Decapentaplegic is a member of the TGF-β family of secreted signals. Similarly, in the spinal cord, Shh and BMPs provide positional signals with opposing actions emanating from the two poles of the dorso-ventral axis.

12.7 Neuronal subtypes in the ventral spinal cord are specified by the ventral to dorsal gradient of Shh

The ventral spinal cord produces five different classes of neurons—motor neurons and four classes of interneurons—each distinguished by the expression of specific genes. Experiments *in vitro* have shown that these neuronal subtypes can be specified in a dose-dependent manner by Shh, the signal produced by the notochord and floor plate; the different neuronal subtypes can be generated from chick neural-plate explants in response to a two- to three-fold change in Shh concentration. The incremental changes in Shh concentration are mimicked by smaller changes in the activity of the Gli transcription factors, which are activated by Shh signaling (see Box 11D), and which provide the key intracellular signal in a graded form (Fig. 12.11).

A number of homeobox transcription factor genes are regulated in ventral progenitor cells in response to Shh signaling, and these genes can be divided into two classes—one (class I) containing genes that are generally repressed by Shh, and another (class II) containing genes that are activated (see Fig. 12.11). The threshold concentration levels of Shh that allow expression are different for each gene, so that overall, the responses of cells to the Shh gradient divides the dorso-ventral axis of the spinal cord into at least five regions of different types of progenitor neurons. Boundaries between each region are sharpened by cross-repressive interactions between the class I and class II proteins.

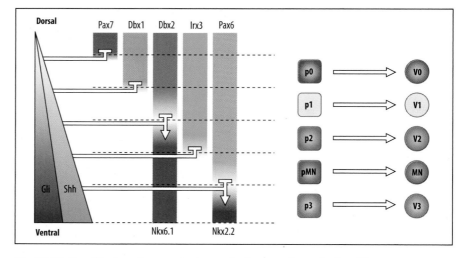

Fig. 12.11 Specification of neuronal subtypes in the ventral neural tube. Different neural subtypes (shown as different colors) are specified along the dorso-ventral axis of the ventral neural tube in response to a gradient of Sonic hedgehog (Shh) emanating from the floor plate. The Shh gradient results in graded activity of Gli transcription factors in cells along the dorso-ventral axis. Shh signaling mediates the repression of the so-called Class I homeodomain protein genes (*Pax7, Pax6, Dbx1, Dbx2,* and *Irx3*), whereas Class II homeodomain protein genes (*Nkx2.2* and *Nkx6.1*), are activated. Interactions between Class I and Class II genes pattern this expression further. Five neuronal subtypes are generated from the five domains. MN, motor neurons; V, ventral interneuron types.

*Adapted from Jessell, T. M.: **Neuronal specification in the spinal cord: inductive signals and transcriptional controls**. Nat. Rev. Genet. 2000, **1**: 20–29*

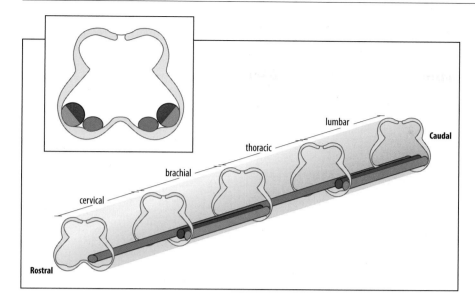

Fig. 12.12 The organization of the motor columns in the 6-day chick embryo spinal cord. The medial motor column is in blue, the medial division of the lateral column in red and the lateral division of the lateral motor column in green. The lateral motor columns are only present in the brachial and lumbar regions, in register with the developing limbs.

The detailed mechanism by which Shh specifies graded Gli activity in the different progenitor neurons *in vivo* is not yet completely worked out. As discussed elsewhere for morphogens in general (see Box 11B), a simple diffusion model assumes that cells at different positions along a concentration gradient of morphogen acquire different positional values, involving the induction of different target genes, according to the ambient concentration of, for example, Shh. However, this model is complicated by the finding that the duration of signaling is also important, as cells are progressively desensitized to Shh signaling. In this 'temporal adaptation' model, different ambient concentrations of Shh would signal for different lengths of time, such that the duration of signaling is proportional to Shh concentration. This idea overcomes the problem that diffusion alone would be an unreliable way of setting up such a gradient because the extracellular concentration of a diffusing molecule at any point changes over time, and its movement is affected by other extracellular molecules.

12.8 Spinal cord motor neurons at different dorso-ventral positions project to different trunk and limb muscles

The motor neurons developing in the ventral region of the spinal cord can be further classified on the basis of the position of their cell bodies along the cord's dorso-ventral axis, and the muscles that their axons innervate. In the chick embryo, motor neurons are subdivided into longitudinal columns on each side of the midline—a **medial column** nearest the midline, which runs the whole length of the spinal cord and sends axons to axial and body-wall muscles, and a **lateral motor column** (LMC), which is only present in the **brachial** and **lumbar** regions, where it innervates the developing fore and hindlimbs (Fig. 12.12). The LMC is divided into two divisions—lateral and medial.

Once designated, each motor neuron subtype expresses a different combination of transcription factors of the LIM homeodomain family, which provides a neuron with its positional identity within the spinal cord. In the LMC, for example, neurons expressing Lim1 settle in the lateral division, whereas those expressing Isl1 settle in the medial division. Experiments tracing the outgrowth of axons from LMC motor neurons show that those in the lateral LMC send axons into the dorsal muscle masses in the limb bud, whereas medial LMC neurons send their axons into the ventral limb bud (Fig. 12.13). We shall look in more detail at the mechanisms that guide the growing axon in the next part of the chapter.

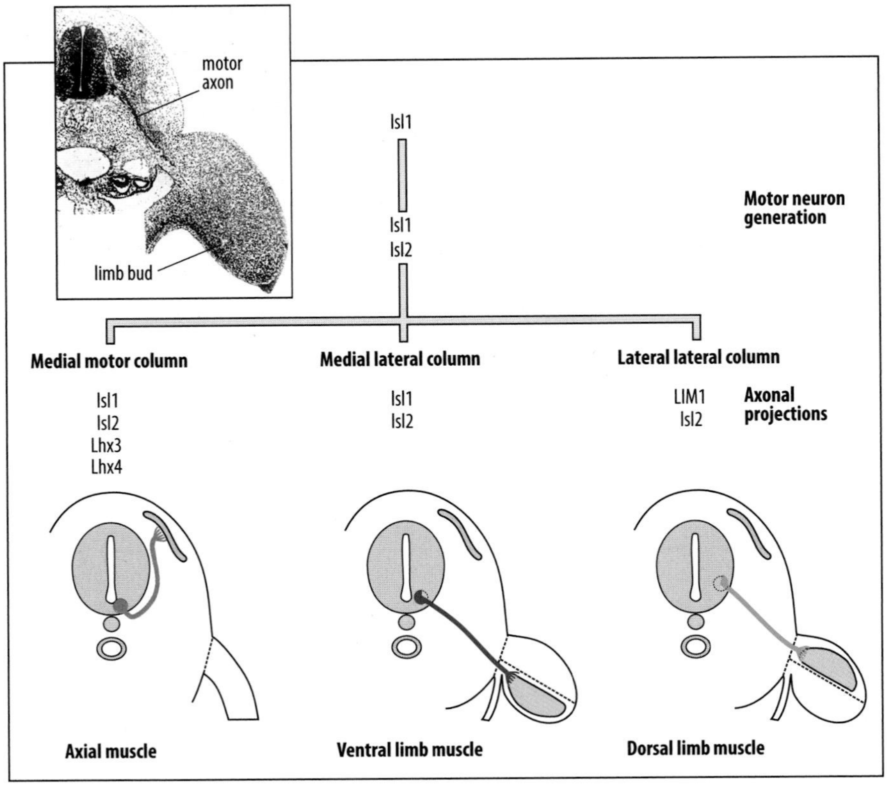

Fig. 12.13 Different motor neurons in the chick spinal cord express different LIM homeodomain proteins. The photograph (inset) shows a transverse section through a chick embryo at the level of the developing wing. Motor neurons can be seen entering the limb bud. Initially, all developing motor neurons express the LIM proteins Isl1 and Isl2. At the time of axon extension, motor neurons that innervate different muscle regions express different combinations of Isl1, Isl2, Lim1, Lhx3, and Lhx4.

Micrograph courtesy of K. Tosney.

12.9 Antero-posterior pattern in the spinal cord is determined in response to secreted signals from the node and adjacent mesoderm

As well as being organized along the dorso-ventral axis, neurons at different positions along the antero-posterior axis of the spinal cord become specified to serve different functions. Antero-posterior specification of neuronal function in the spinal cord was dramatically illustrated some 40 years ago by experiments in which a section of the spinal cord, which would normally innervate wing muscles, was transplanted from one chick embryo into the lumbar region, which normally serves the legs, of another embryo. Chicks developing from the grafted embryos spontaneously activated both legs together as though they were trying to flap their wings, rather than activating each leg alternately as if walking. These studies showed that motor neurons generated at a given antero-posterior level in the spinal cord had intrinsic properties characteristic of that position.

Neuronal fate along the antero-posterior axis of the spinal cord is determined initially by signal molecules secreted by the primary organizer (or node) and later by signals from the adjacent paraxial mesoderm (see Chapter 5). This was first shown by experiments in which quail paraxial mesoderm was grafted to the thoracic level in chick embryos at the time of neural-tube closure. The result was that presumptive chick thoracic neurons became specified as brachial neurons. The signals from the node and the mesoderm form gradients along the antero-posterior axis of the embryo that activate patterns of Hox gene expression in post-mitotic neurons, giving groups of neurons a positional value that specifies their fate. The signals are composed of FGF, growth/differentiation factor (GDF)—a member of the TGF-β family—and retinoic acid. FGF is graded from posterior to anterior, and there is a shallow retinoic acid gradient at the anterior end of the spinal cord and a similar one of GDF at the posterior end.

As in the hindbrain, the spinal cord becomes demarcated into different regions along the antero-posterior axis by combinations of expressed Hox genes. For example, *Hoxc6* is expressed in the motor neurons in the brachial region—adjacent to the forelimb—while

Hoxc9 is expressed in thoracic motor neurons. In some neuronal subtypes, different concentrations of the transcription factor FoxP1 act together with the Hox proteins to specify different neuronal identities. This is in addition to the gene expression specifying dorso-ventral identity described in the previous section. A typical vertebrate limb contains more than 50 muscle groups with which neurons must connect in a precise pattern. Individual neurons express particular combinations of Hox genes, which determines which muscle they will innervate.

So, all together, the expression of signal molecules on orthogonal axes results in the expression of transcriptional control genes that determine regional and cellular fate at antero-posterior and dorso-ventral positions. Cells acquire their unique morphological and functional identities according to their 'grid reference' in this Cartesian system of positional information.

SUMMARY

The developing brain is divided into three regions—forebrain, midbrain, and hindbrain—along the antero-posterior axis. Signaling centers located at the midbrain-hindbrain boundary and within the forebrain produce signals that pattern adjacent regions of the hindbrain, midbrain, and forebrain, respectively. The hindbrain is segmented into rhombomeres, with the cells of each rhombomere respecting their boundaries. A Hox gene code provides positional values for the rhombomeres of the hindbrain, and the neural crest cells that derive from them, while other genes specify more anterior regions. Patterning of neuronal cell types along the dorso-ventral axis of the spinal cord involves signals from both the ventral and dorsal regions. Secretion of Shh by the notochord and floor plate provides a graded signal that leads to specification of different neuronal subtypes in the ventral region of the spinal cord. Neuronal identity along the antero-posterior axis of the cord is specified by the combinatorial expression of Hox genes.

The formation and migration of neurons

We will now consider the next stage in the process of generating a nervous system: how individual precursor cells in the neuroectoderm that have acquired their positional identity now become **neuroblasts**, and subsequently give rise to neurons and glia. We will first look at *Drosophila*, which has revealed some key developmental processes in the formation of neurons, or **neurogenesis**. We then turn to the more complex process of neurogenesis in the vertebrate neural tube. As the epithelial cells of the neural tube proliferate and differentiate into neurons, the wall of the tube develops a multilayered structure, with the neurons migrating outwards to form the different layers. We shall look at the formation of this layered structure in relation to the mammalian cortex, which develops from the forebrain region of the neural tube.

The nervous systems of both invertebrates and vertebrates contain enormous numbers of glial cells as well as neurons. These also develop from the neural stem cells of the neuroectoderm, but we shall not consider the specification and development of glial cells in any detail here.

12.10 Neurons in *Drosophila* arise from proneural clusters

In *Drosophila*, the cells that will give rise to the central nervous system of the larva are specified at an early stage of embryonic development, as part of the patterning process that divides the embryo into different regions along the dorso-ventral and antero-posterior axes (see Chapter 2). In insects, the main nerve cord runs ventrally, rather than dorsally as in vertebrates, and the future central nervous system is specified as two

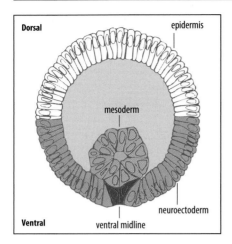

Fig. 12.14 Cross-section of a *Drosophila* early gastrula. At around 3 hours after fertilization, the neuroectoderm (blue) is located on either side of the ventral midline. The mesoderm (red), which originally lies along the ventral midline, has already been internalized.

longitudinal regions of neuroectoderm in the ventral half of the embryo, just above the mesoderm. The neuroectoderm, or neurogenic zone, comprises ectodermal cells that are fated to form either neural cells (neurons or glia) or epidermis. After gastrulation and internalization of the mesoderm, the neuroectoderm remains on the outside of the embryo on either side of the ventral midline (Fig. 12.14). Once individual cells are specified as neuroblasts in the neuroectoderm they move from the surface into the interior of the embryo, where they divide further and differentiate into neurons and glia, forming two tracts of axons running longitudinally on either side of the ventral midline, connected at intervals by neurons whose axons cross the midline.

Despite very great differences between the insect central nervous system and that of vertebrates, there are intriguing parallels between the genes that pattern both systems at a very early stage. *Drosophila* neuroblasts form three longitudinal columns of cells on each side of the ventral midline, and the identity of these columns is specified while the cells are still in the neuroectoderm (Fig. 12.15). The homeodomain transcription factor genes *msh*, *ind*, and *vnd* are expressed in dorsal to ventral order in the *Drosophila* neuroectoderm, in response to the actions of earlier dorso-ventral patterning genes, such as *rhomboid* (see Section 2.16), and the activation of the *Drosophila* EGF signaling pathway. The developing vertebrate neural plate also has three longitudinal domains of prospective neural cells (see Fig. 12.15), which express genes homologous to those in the fly—*Msx*, *Gsh*, and *Nkx2.1*, respectively. This is a striking example of the evolutionary conservation of a mechanism for regional specification.

In response to previous patterning, the *Drosophila* neuroectoderm becomes subdivided along its antero-posterior and dorso-ventral axes into a precise orthogonal pattern of **proneural clusters**, each composed of three to five cells, which can be distinguished by the expression of genes known as **proneural genes** (Fig. 12.16). Initially, all the cells within a proneural cluster are capable of becoming neural precursor cells, or neuroblasts, but one cell, through an apparently random event, produces a signal that promotes its own development as a neuroblast and prevents the adjacent

Fig. 12.15 The neuroepithelium of both *Drosophila* and vertebrates is organized into three columns of neural precursors on either side of the midline. The developing *Drosophila* embryonic central nervous system (left) and the vertebrate embryonic neural plate (right) are both organized into three longitudinal domains (columns) of gene expression on either side of the midline. The cells of each column express a specific homeobox gene and, as is evident before neurulation (upper panels), similar genes are expressed in the same medial-lateral order in *Drosophila* and in vertebrates, with *vnd/Nkx2.1* expressed most medially (nearest the midline), *ind/Gsh* in an intermediate position, and *msh/Msx* expressed most laterally.

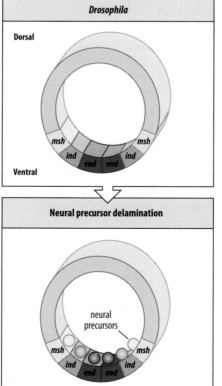
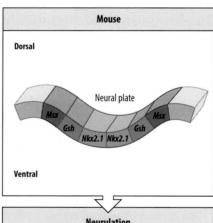
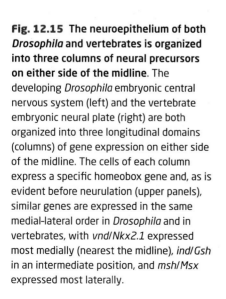
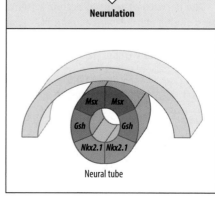

Fig. 12.16 Proneural clusters in the neuroectoderm of the *Drosophila* embryo. At the stage of development shown here, there are eight proneural clusters within each future segment, marked here by expression of the proneural gene *achaete* (black). Half of the clusters are formed in the region of engrailed expression (purple) at the posterior end of the segment, and half in a more anterior region. Some of the clusters have resolved into a single neural cell, the prospective neuroblast. The embryo is viewed here from the ventral surface; the ventral midline runs horizontally along the middle of the photo.

*Photograph from Skeath, J. B., et al. 1996.: **At the nexus between pattern formation and cell-type specification: the generation of individual neuroblast fates in the** Drosophila embryonic central nervous system.* BioEssays, **21**: 11, 922–931.

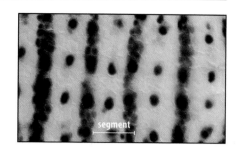

cells from following that fate. This cell will become the single neuroblast in the cluster, and can be distinguished, for example, by a high level of expression of a proneural gene such as *achaete*, whereas the surrounding cells no longer express *achaete* and will develop as epidermis (see Fig. 12.16).

The singling out of the prospective neuroblast is an example of lateral inhibition (Fig. 12.17) and is due to signaling between the prospective neuroblast and its neighbors via the transmembrane protein Notch and its transmembrane ligand Delta (see Box 5D for the Notch–Delta signaling pathway). All the cells in a proneural cluster initially can express both Notch and Delta. One cell will, however, start to express Delta sooner or more strongly than the others. Through the interaction of Delta with the Notch receptors on its less-advanced neighbors, this cell inhibits their further development as neural cells, and also suppresses their expression of Delta (see Fig. 12.17, bottom row). In this way, the initially equivalent cells of the proneural cluster are resolved by lateral inhibition into one neuroblast per cluster. The other cells of the proneural

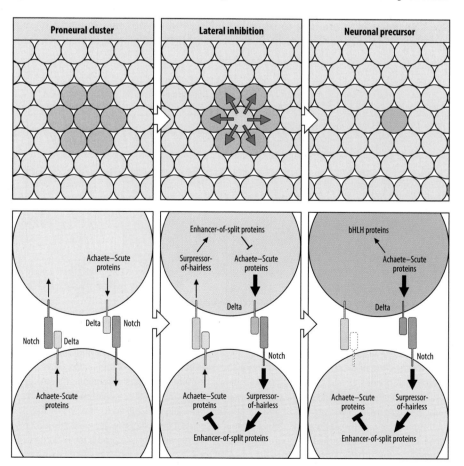

Fig. 12.17 The role of Notch signaling in lateral inhibition in embryonic neural development in *Drosophila*. As shown in the top row, the proneural cluster gives rise to a single neuronal precursor cell, the neuroblast, by means of lateral inhibition. The rest of the cells in the cluster become epidermal cells. As shown in the bottom panels, the presence of Achaete and Scute proteins produces Notch signaling between cells of the proneural cluster. The level of signaling is initially similar and keeps all cells in the proneural state. An imbalance develops when one cell (green) begins to express higher levels of Delta, the ligand for Notch, and thus activates Notch signaling to a higher level in neighboring cells. This imbalance is rapidly amplified by a feedback pathway involving two transcription factors, Suppressor of hairless and Enhancer of split. Their activities repress the production of Achaete-Scute proteins and Delta protein in the affected cells, which prevents the cells from proceeding along the pathway of neuronal development and from delivering inhibitory Delta signals to the prospective neuroblast (dark green). Basic helix–loop–helix (bHLH) transcription factors produced in the prospective neuroblast turn on genes required for neuronal development.

After Kandel, E. R., et al.: Principles of Neural Science (4th edition) New York: McGraw-Hill 2000.

cluster cease expression of Delta and of the proneural genes and go on to become epidermis. In the *Drosophila* ectoderm, if either Notch or Delta function is inactivated, the neuroectoderm makes many more neuroblasts and fewer epidermal cells than normal.

After specification, neuroblasts enlarge, leave the epithelium and move into the interior of the embryo, where they divide repeatedly and give rise to neurons and glia, as described in Section 12.12. In *Drosophila*, neuroblasts are specified and exit the neuroectoderm in five waves over a period of around 90 minutes, to form an invariant pattern of approximately 30 neuroblasts in each hemisegment (a lateral half-segment) arranged in three longitudinal columns. Each neuroblast has a unique identity based on the particular embryonic segment it is part of, its position within the segment, and the time of its formation, and gives rise to a distinct set of neurons and glial cells in the larva.

12.11 The development of neurons in *Drosophila* involves asymmetric cell divisions and timed changes in gene expression

Before it leaves the neuroectoderm, a neuroblast becomes polarized in an apico–basal direction, with the apical and basal ends marked by the accumulation of specific sets of cytoplasmic determinants. The determinants at the apical end help set the plane of cell division by influencing the orientation of the mitotic spindle, and also direct the correct positioning of the basal determinants, which are crucial for determining neural cell fate. Once it has left the surface of the embryo, the neuroblast behaves as a stem cell, dividing asymmetrically to give a larger apical cell, which remains a neural stem cell, and a smaller basal cell, the **ganglion mother cell**, which will divide once more and whose daughter cells will differentiate into neurons or glia (Fig. 12.18). Among the basal cytoplasmic determinants distributed to the ganglion mother cell by this asymmetric division is the transcription factor Prospero, which has been shown by DNA microarray profiling to regulate hundreds of target genes involved in suppressing neural stem-cell behavior and promoting differentiation of the neuroblast.

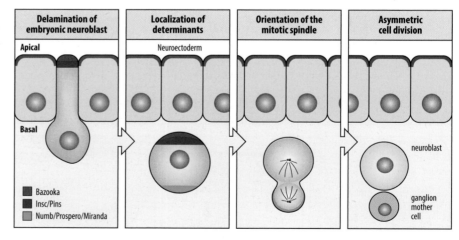

Fig. 12.18 The formation of neuronal cells from neuroblasts in *Drosophila* by asymmetric cell division. Generation of ganglion mother cells of the central nervous system by the asymmetric division of neuroblasts in the *Drosophila* embryo. Once specified, neuroblasts move out of the neuroectodermal epithelium and then behave as neural stem cells. Each neuroblast divides asymmetrically to give an apical cell, which remains a neuroblast stem cell, and a smaller basal cell—the ganglion mother cell. The orientation of cell division and cell fate are specified by localized protein determinants. Numb protein is localized to one end of the neuroblast before division and the daughter that receives it becomes the ganglion mother cell that will give rise to neurons. Prospero and Miranda are required for the proper localization of Numb, and Inscuteable (Insc) and Pins are required for the correct orientation of cell division.

BOX 12A Specification of the sensory organs of adult *Drosophila*

The sensory organs of adult *Drosophila* have provided a classic system for investigating both the patterning of the epithelia in which they arise, and the role of asymmetric localization of cytoplasmic determinants and asymmetric cell division to produce a structure composed of a neuron and non-neuronal cells. Adult flies have several types of sensory organ: sensory bristles (Figure 1) and others that act as mechanoreceptors and chemoreceptors; and internal chordotonal organs that monitor stretching of tissues. Each sensory organ is made up of four cells, one of which is a neuron, and each organ arises from a single **sensory organ precursor (SOP)** cell that is initially specified in the epidermis and in the larval imaginal discs in a constant pattern.

Unlike the neuroblasts in the neuroectoderm, division of an SOP is oriented such that both daughter cells stay in

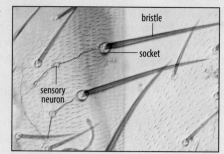
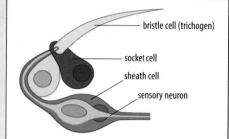

Figure 1

Photograph courtesy of Y. Jan.

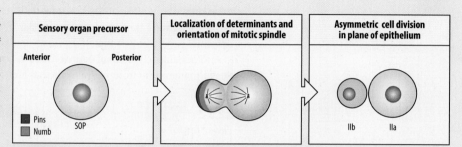

Figure 2

the epidermis (Figure 2), and this is reflected in the location of the Pins protein, which is involved in orienting the plane of cell division (compare Fig. 12.18). The Numb protein is also asymmetrically located in the SOP, such that only one daughter cell receives it. This cell (IIb) will give rise to a sensory neuron and a sheath cell at the next division, with the daughter cell that receives Numb protein becoming the neuron. The SOP daughter cell that did not receive Numb protein (IIa) gives rise at the next division to a bristle cell and a socket cell, which are non-neuronal cells. Notch signaling is also required for neuronal differentiation; if it is absent, a sheath cell develops instead of the neuron.

The pattern of sensory bristles in the wing epidermis is already present in the imaginal disc. As in the embryonic neuroectoderm (see Section 12.11), proneural clusters are first selected by expression of proneural genes—in this case genes of the *achaete-scute* complex—in a precise spatial pattern. This pattern is controlled by the use of different *cis*-regulatory modules in the *achaete-scute* complex DNA to determine gene expression at specific locations in the imaginal disc. The *achaete-scute* complex is an excellent example of how a spatial pattern of gene expression can be precisely controlled by highly complex control regions (Figure 3). The top panel shows the position of various control modules (colored bands) that determine the position-specific expression of the *achaete* and *scute* genes (shown in black). The bottom panel shows the location of the proneural clusters in the imaginal disc (left) and of the corresponding sensory bristles in the adult wing and adjacent thoracic segments (right), with colors corresponding to the enhancer responsible for the site-specific expression of the genes.

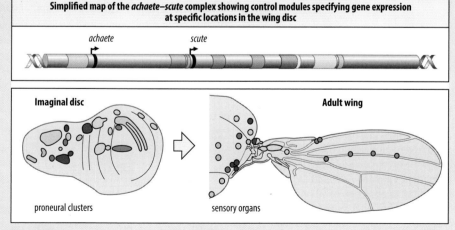

Figure 3

Diagram after Campuzano, S., Molodell, J.: **Patterning of the** Drosophila **nervous system: the achaete-scute gene complex.** Trends Genet. *1992,* **8**: *202-208.*

The neuroblasts at different locations along the body continue to divide according to the fly's developmental program, giving rise to a ganglion mother cell and a neural stem cell at each division. As a neuroblast generates more ganglion mother cells, the older cells are pushed further into the embryo, giving the final nerve cord a layered structure. As neurogenesis proceeds, the neuroblast undergoes successive changes in the expression of a particular set of transcription factors, which give the neurons produced at different times their specific identities.

Because neurons with different functions must develop in the correct positions within the ventral nerve cord, the precise timing of the transition from expression of one transcription factor to the next in the neuroblasts is crucial for the correct formation of the nerve cord. The timing of the first transition seems to be linked to neuroblast division, whereas later transitions are linked to an intrinsic timing mechanism in the neuroblast that is independent of cell division.

The embryonic phase of neuroblast division and neurogenesis provides a functional central nervous system for the first instar *Drosophila* larva, but these neurons comprise only about 10% of the neurons that are eventually required by the adult fly. A further extended period of neurogenesis, involving most of the original embryonic neurons, occurs before metamorphosis and produces the remaining 90%.

Another classic system of neurogenesis in *Drosophila* is the generation of sensory bristles in the adult fly from neural cells that are specified in the imaginal discs (Box 12A).

12.12 The production of vertebrate neurons involves lateral inhibition, as in *Drosophila*

The mechanisms involved in vertebrate neurogenesis are in many ways remarkably similar to those in *Drosophila*. As we have seen, the cells of the vertebrate nervous system derive from the neural plate, a region of columnar epithelium induced from the ectoderm on the dorsal surface of the embryo during gastrulation, and from sensory placodes in the head region that contribute to the cranial sensory ganglia and produce the sensory receptor cells of the eyes (see Section 11.27), ears, and nose.

Neuronal precursors are not generated simultaneously throughout the neural plate and the ventricular zone of its successor neural tube, but, as in *Drosophila*, are initially confined to three longitudinal stripes in the neural plate on each side of the dorsal midline (see Fig. 12.15). In the vertebrate spinal cord, the normal fate of the medial stripe (the one nearest the midline), which becomes the ventral region of the spinal cord, is to give rise to motor neurons. The more lateral stripes, which become the lateral and dorsal regions of the spinal cord, give rise to interneurons and secondary sensory neurons (see Fig. 12.9).

The initial specification of the stripes in the neural plate ectoderm involves lateral inhibition, in which Delta–Notch signaling between adjacent cells restricts the expression

Fig. 12.19 Lateral inhibition specifies single cells as neuronal precursors in the vertebrate nervous system. The *neurogenin* gene is initially expressed in stripes of contiguous cells in the neural plate, and these cells also express the Delta and Notch proteins. The Delta-Notch interaction between adjacent cells mutually inhibits expression of *neurogenin*. When, by chance, one cell expresses more Delta than its neighbors, it inhibits the expression of Delta protein in the neighboring cells and so develops as a neuron, expressing *neurogenin* and *neuroD*.

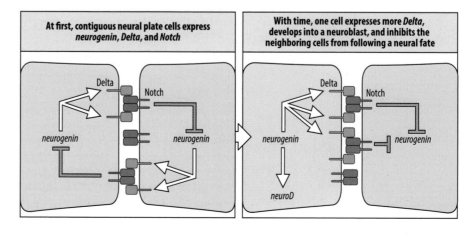

of neuroblast-specifying proneural genes in the same way as it does in *Drosophila* (Fig. 12.19). Prospective neuroblasts in the three stripes come to express Delta more strongly than adjacent cells, and the feedback loop set up by Notch activation in the adjacent cells shuts down proneural gene expression, leaving only the prospective neuroblasts expressing proneural genes. Vertebrate proneural proteins, such as the transcription factor **neurogenin,** are homologs of *Drosophila* proneural transcription factors, while the proteins that inhibit specification as neural cells are vertebrate homologs of the Notch-activated transcription factor Suppressor of hairless and its target gene Enhancer of split (see Fig. 12.17). Cells that express neurogenin go on to express genes characteristic of differentiation into neurons, such as *NeuroD*. Confirmation of the role of Notch and Delta in vertebrates comes from the transgenic overexpression of Delta in the neural plate or expression of a mutant Notch that signals continuously, which both result in a reduction in the number of neurons formed. Inhibition of Delta, on the other hand, leads to an increase in the number of neurons, as does overexpression of *neurogenin*.

The neural tube is formed from the whole of the neural plate in amphibians (see Chapter 4) and will contain both committed neural precursors expressing the *neurogenin* gene and cells in which it is repressed. The latter form a pool of undifferentiated potential neural precursors in the neural tube, which can be brought into play later, as the first wave of neuroblasts to be formed stop dividing and start to differentiate. This ensures that neurogenesis can continue over a relatively long period of time. In mammals and birds, the neural tube that will become the spinal cord is derived from cells of the stem zone as the embryo elongates and the node regresses (see Chapter 5).

12.13 Neurons are formed in the proliferative zone of the vertebrate neural tube and migrate outwards

In vertebrate embryos, all the neurons and glia of the brain and spinal cord arise from a proliferative layer of epithelial cells that lines the lumen of the neural tube and is called the **ventricular zone**. For many years it was thought that no new neurons could be generated in the adult mammalian brain, although adult neurogenesis was known to occur in the brains of songbirds, reptiles, amphibians, and fish. Neural stem cells are now known to exist in limited regions in the adult mammalian brain, and can give rise to new neurons and glial cells, as described in Section 8.13.

As we saw in Chapters 5 and 9, the neural tube is initially made up of a single layer of epithelium. As the cells of the neural tube divide, the ventricular zone develops as a zone of rounded-up dividing cells on the inner surface of the tube. In most regions of the nervous system, proliferation is confined to a single layer but in some regions of the forebrain, proliferation and neurogenesis is supplemented by a **subventricular zone**, which in human brains is exceptionally thick.

Once formed, neurons migrate outward from the ventricular zone and build up the neural tissue in concentric layers. Layers of neuronal cell bodies, the **mantle zone** or **gray matter**, form outside the ventricular zone, with successive waves of new neurons displacing the first formed neurons further outwards. The axons produced by these cells form the outermost layer, the **marginal zone**. Later, as the axons become myelinated, the marginal zone is known as the **white matter**. In this way, the hollow neural tube eventually develops into the hollow spinal cord and brain with its fluid-filled ventricles that represent the remains of the neural tube lumen. This arrangement, with an inner core of gray matter surrounded by a shell of white matter, characterizes all regions of the central nervous system except the cerebral cortex. Here, as we will see, the relationship between gray and white matter is reversed: the cortex has most of its cell bodies located at the outer surface, whereas the axons extend inward to form an intermediate zone of white matter adjacent to the ventricular zone. By keeping the relatively extensible grey matter outside the more rigid white matter, the mammalian cortex has acquired great potential for expansion during evolution.

The neural progenitor cells in the ventricular zone are multipotent neural stem cells, which give rise to many different types of neurons and to glia. In the ventricular zone of the dorsal telencephalon, for example, cells known as **radial glial cells** can act as neural stem cells, generating both radial glia, whose role is described below, and some types of neurons. As with all stem cells, neural stem cells must generate both neural precursors and more stem cells. Two ways of achieving this seem to operate in the developing nervous system. The stem cell can either divide 'vertically' and give rise to two similar cells—both stem cells or both neurons—or it can divide horizontally, parallel to the ventricular surface, with one daughter remaining a stem cell and the other developing into a neuron. Vertical divisions that produce more and more stem cells predominate during the early stages of brain growth, giving way to increasing numbers of horizontal divisions as neurogenesis gathers pace.

Once specified as neuroblasts, the cells migrate from the inner surface of the neural tube towards the outer surface. Here we shall look at the process linking neuron production and neuron migration with reference to the cerebral cortex, which develops from the embryonic dorsal telencephalon (the pallium) and is unique to mammals. The cortex is organized into six layers (I–VI), numbered from the cortical surface inward, and each layer contains neurons with distinctive shapes and connections. For example, large pyramidal cells that project long distances to sub-cortical targets are concentrated in layer V, and smaller neurons that receive input from thalamic afferents predominate in layer IV. All of the cortical excitatory neurons, which use glutamate as their transmitter, have their origin in the pallial ventricular zone and migrate out to their final positions along radial glial cells—greatly elongated cells that extend

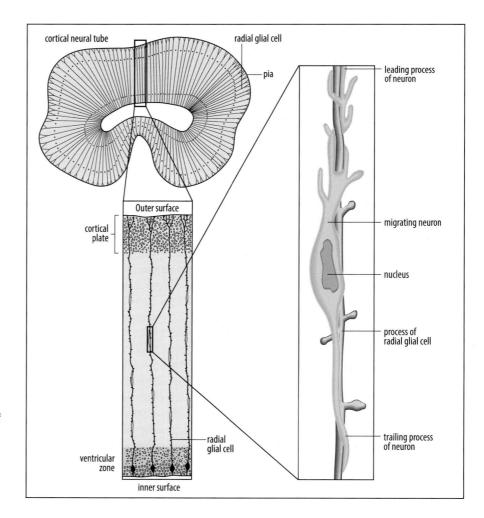

Fig. 12.20 Cortical pyramidal neurons migrate along radial glial cells. Neurons generated in the ventricular zone migrate to their final locations in the cortex along radial glial cells that extend right across the wall of the pallial neural tube.

*Illustration after Rakic, P.: **Mode of cell migration to the superficial layers of fetal monkey neocortex**. J. Comp. Neurol. 1972, **145**: 61-83.*

EXPERIMENTAL BOX 12B Timing the birth of cortical neurons

The time at which a neuron is born determines what cortical layer it becomes part of. A neuron's time of birth is determined as the time that the precursor cell stops dividing and leaves the mitotic cycle. Some of the earliest experiments, done in the 1960s and 1970s, used radiolabeled isotopes to label cells so that their migration could be followed. If neuronal precursors are given a short pulse of a labeled compound that can be incorporated into DNA, those cells that undergo their last round of replication immediately after taking up the compound will have the most heavily labeled DNA. If they continue to divide and replicate their DNA, the label becomes diluted among their progeny.

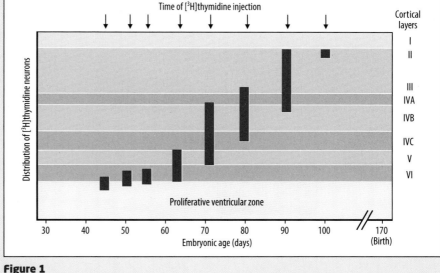

Figure 1

Cells that stop dividing immediately after taking up the label can therefore be distinguished and their final destinations detected.

Figure 1 shows the results of an experiment of this type that originally determined the order in which neurons reached their final layers in the monkey visual cortex. Radioactively labeled thymidine ([³H]thymidine), which becomes incorporated into replicating DNA, was injected into the developing neural tube at different times during development. The red bars represent the distribution of heavily labeled neurons found after each injection.

Neurons born earliest remain closest to their site of birth, the proliferative ventricular zone. Neurons born later become part of successively higher cortical zones. Cortical zones are numbered from the surface of the neural tube inward; that is, layer VI is the layer nearest to the ventricular zone. For this type of experiment nowadays, radioisotopic labeling has been superseded by the labeling of cells with, for example, replication-incompetent retroviruses carrying a marker gene such as *lacZ*, whose product can be detected histochemically.

across the developing pallium (Fig. 12.20). The three-dimensional representation of radial fibers and migrating neurons was painstakingly reconstructed in the 1960s and 70s from electron-microscopic images of successive sections of monkey fetal cortex, and the reconstruction drawn by hand in great detail. The smaller and more numerous interneurons of the cortex, which are inhibitory and use γ-aminobutyric acid (GABA) as their transmitter, are mostly generated in the subpallium, and once formed have to migrate dorsally into their settling positions in the cortex.

The cortical layer to which neurons migrate after their birth in the proliferative layer is related to the time when the neuron is born, and the identity of a cortical neuron is thought to be specified before it begins to migrate. Most of the neurons of the mammalian central nervous system do not divide once they have become specified, which is why newly formed neurons are often called **post-mitotic neurons**. A neuron's time of birth is defined by the last mitotic division that its progenitor cell underwent; the experiments that originally established the relation between the birth time of a neuron and the layer in which it ended up are described in Box 12B. The newly formed neuron is still immature; later it extends an axon and dendrites (see Fig. 12.2), and assumes the morphology of a mature neuron.

Neurons born at early stages of cortical development migrate to layers closest to their site of birth, whereas those born later end up further away, in more superficial layers (Fig. 12.21). The younger neurons must migrate through the older ones on their way to their correct position, giving rise to layers and columns of cell bodies. The exception is the first-formed layer, which remains as the outer layer. Mouse

Scan here

Scan this QR code image with your mobile device to see an online movie showing neuroblast migration or log on to **http://global.oup.com/uk/orc/biosciences/ devbiol/wolpert5e/qr/qr12a/**

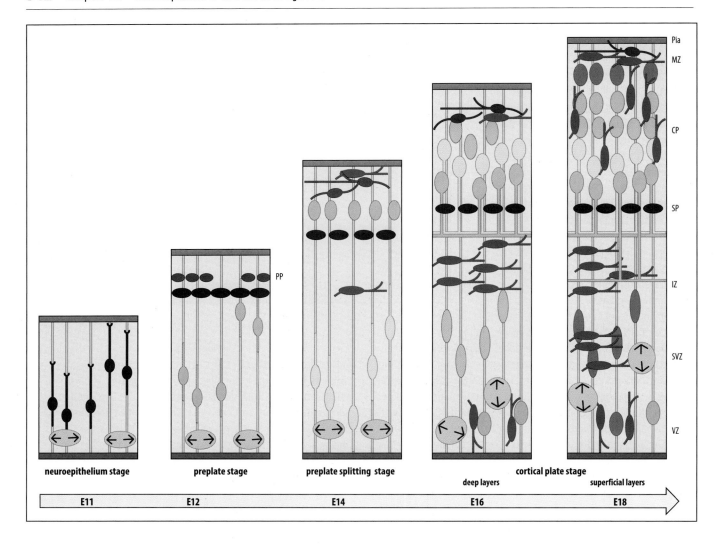

Fig. 12.21 The generation of neuronal layers in the mammalian cortex. Development of the cortex begins (left) with the radial migration of the first wave of post-mitotic neurons from the ventricular zone (VZ, green) towards the outer surface of the neural tube to form the preplate (PP, black). The second wave of newly formed neurons (blue) migrates on radial glial fibers (gray) through the lower layer of the preplate to form the first layer of cortical plate (CP) neurons. The upper layer of the split preplate becomes the superficial marginal zone (MZ) and will become cortical layer I (cortex layers are numbered from the outside inwards), whereas the lower layer of the preplate becomes the subplate (SP), which will later undergo programmed cell death. The ventricular zone releases a third wave of neurons (yellow), which migrate through the first layer of cortical plate neurons to form a second layer of cortical-plate neurons. Increasing numbers of horizontal (asymmetric) precursor divisions in the ventricular zone, together with the late-forming subventricular zone (SVZ), will form successively more superficial layers of the cortical plate (orange, red). Finally, all six layers of the cortical plate are in place and the axons of the major projecting layers (V and VI) grow inwards towards the VZ and turn into the intermediate zone of future white matter. The intermediate zone contains inhibitory interneurons (purple) immigrating mainly from the subpallium but also from the pallial VZ, that now enter the cortical plate. The time line is for mouse development, days *in utero*.

*Adapted from Honda, T., et al.: **Cellular and molecular mechanisms of neuronal migration in neocortical development**. Semin. Cell Dev. Biol. 2003, **14**: 169–174.*

mutations that disrupt cortical neuronal migration have provided considerable insight into the process. The *reeler* mouse, as its name implies, has very poor motor coordination, which is hardly surprising as its cortical layers are arranged in reverse order. The *reeler* gene encodes an extracellular matrix molecule, reelin, that is normally expressed in the first-formed, outermost layer of the cortex. Mutations in *reeler* that cause a loss of reelin protein disrupt radial neuronal migration, with the successive waves of neurons being unable to pass through the layer formed by their predecessors. As one might expect, the migration of neurons in the developing brain is in fact

Fig. 12.22 Origin and tangential migration of cortical interneurons. Schematic transverse section through the left telencephalon of a mouse embryo showing the migration of GABAergic interneurons from the subpallium into the pallium. Major regions of neurogenesis according to neurotransmitter type; glutamatergic neurons are born in the pallium (cortex and hippocampus: Cx, H, purple), whereas GABAergic neurons are born in the subpallium (lateral and medial ganglionic eminences, LGE, MGE, green); cholinergic neurons form in the most ventral region (preoptic area, POA). Green arrows show the migration paths taken by GABAergic neurons from the subpallium into the pallium, reaching as far as the prospective hippocampus (H).
*Adapted from Marin, O., Rubenstein, J.L.R.: **A long, remarkable journey: Tangential migration in the telencephalon.** Nat. Rev. Neurosci. 2001, **2**: 780-790.*

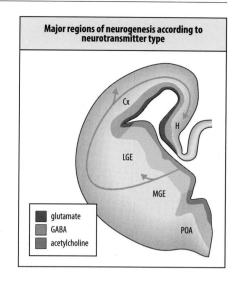

much more complicated than described here; in addition to radial migration, brain neurons also migrate tangentially within the wall of the neural tube in a process that is independent of radial glial fibers.

12.14 Many cortical interneurons migrate tangentially

All immature neurons migrate radially away from their birthplace in the ventricular zone and head towards the surface of the brain. Most neurons then stop moving and extend their axon and dendrites. In a few locations, however, neurons do not settle down at the completion of their radial path but continue moving, now in a tangential direction—parallel to the brain surface. In the case of GABAergic cortical interneurons, this **tangential migration** involves huge numbers of cells moving over a considerable distance, from their place of origin in the subpallium (the medial ganglionic eminence, Fig. 12.22) all the way dorsally into the pallium, where they eventually intersperse among the glutamatergic neurons of the forming cortical plate. Studies on the molecular mechanisms that direct this migration have revealed the existence of both short- and long-range chemoattraction by secreted proteins called **neuregulins**, which diffuse from the lateral ganglionic eminence and cortical plate, respectively. These proteins act on ErbB4 receptors carried by the interneurons. Attraction by molecules released by the pathway and target is complemented by repulsion, which is mediated by another group of proteins, the **semaphorins**, which are secreted by the medial ganglionic eminence. The semaphorins act on the interneurons through receptors called neuropilins.

A simple explanation for this remarkable process of tangential migration is that while both glutamatergic (excitatory) and GABAergic (inhibitory) neurons are needed close to one another for correct network function, the mechanism of cell specification in the telencephalon can only generate these distinct cell types at different overall positions along the dorso-ventral axis. Therefore migration is needed to bring these two types of neurons together.

SUMMARY

In *Drosophila*, prospective neural tissue is specified early in development as a band of neuroectoderm along the dorso-ventral axis. Within the neuroectoderm, expression of proneural genes distinguishes clusters of neuroectoderm cells with the potential to form neural cells. As a result of lateral inhibition involving Notch and Delta, only one cell of the cluster finally gives rise to a neuroblast. Neuroblasts that give rise to the central nervous system behave as stem cells, undergoing repeated asymmetric division, each giving rise to a neuroblast and a daughter cell that can give rise to neurons.

The vertebrate nervous system is derived from the neural plate, which is specified during gastrulation. It contains the cells that form the brain and neural crest cells contribute to the peripheral nervous system. Prospective neural cells are initially specified within the neural plate by a mechanism of lateral inhibition, similar to that in *Drosophila*. Neurons are born in the proliferative zone on the inner surface of the neural tube and then migrate to different locations within the developing brain and spinal cord.

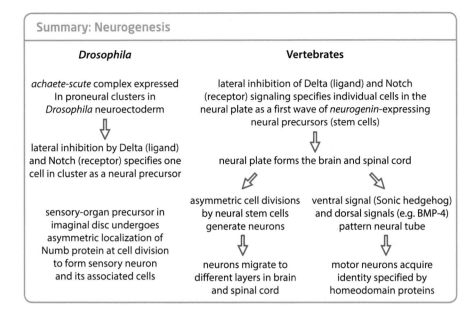

Summary: Neurogenesis

Drosophila	Vertebrates
achaete-scute complex expressed In proneural clusters in *Drosophila* neuroectoderm	lateral inhibition of Delta (ligand) and Notch (receptor) signaling specifies individual cells in the neural plate as a first wave of *neurogenin*-expressing neural precursors (stem cells)
lateral inhibition by Delta (ligand) and Notch (receptor) specifies one cell in cluster as a neural precursor	neural plate forms the brain and spinal cord
sensory-organ precursor in imaginal disc undergoes asymmetric localization of Numb protein at cell division to form sensory neuron and its associated cells	asymmetric cell divisions by neural stem cells generate neurons / ventral signal (Sonic hedgehog) and dorsal signals (e.g. BMP-4) pattern neural tube
	neurons migrate to different layers in brain and spinal cord / motor neurons acquire identity specified by homeodomain proteins

Axon navigation

We shall now look at a feature of development that is unique to the nervous system—the outgrowth and guidance of axons to their final targets. This stage of neuronal differentiation occurs mainly after immature post-mitotic neurons have migrated to their eventual locations. Each neuronal cell body extends an axon and dendrites, through which it will, respectively, send and receive signals. The working of the nervous system depends on the formation of discrete neuronal circuits, in which neurons make numerous and precise connections with each other (Fig. 12.23). In the rest of this chapter we will look at how these connections can be set up in various situations. What controls the growth of axons and the contacts they make with other cells, and how are such precise connections achieved? In this part of the chapter we will first consider axon outgrowth and the evidence that the growing axon is guided specifically to its target.

Fig. 12.23 Neurons make precise connections with their targets. Neurons (green) and their target cells usually develop in different locations. Connections between them are established by axonal outgrowth, guided by the movement of the axon tip (growth cone). Often, the initial set of relatively nonspecific synaptic connections is then refined to produce a more precise pattern of connectivity.

Illustration after Alberts, B., et al.: Molecular Biology of the Cell, 2nd edition. New York: Garland Publishing, 1989.

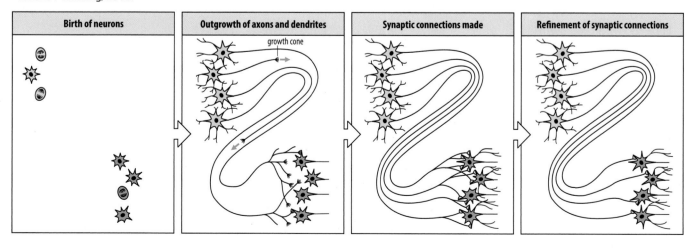

Birth of neurons	Outgrowth of axons and dendrites	Synaptic connections made	Refinement of synaptic connections

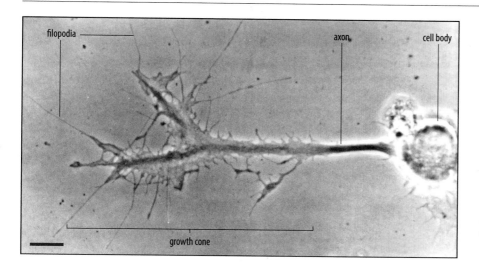

filopodia   axon   cell body

growth cone

Fig. 12.24 A developing axon and growth cone. The axon growing out from a neuron's cell body ends in a motile structure called the growth cone. Many filopodia are continually extended and retracted from the growth cone to explore the surrounding environment. Scale bar = 10 μm.

Photograph courtesy of P. Gordon-Weeks.

12.15 The growth cone controls the path taken by a growing axon

An early event in the differentiation of a neuron is the extension of its axon by the growth and migration of the **growth cone** located at the tip of the axon (Fig. 12.24). The growth cone is specialized both for movement and for sensing its environment for guidance cues. As in other cells capable of migration, such as the primary mesenchyme of the sea-urchin embryo (see Section 9.8), the growth cone continually extends and retracts filopodia at its leading edge, making and breaking connections with the underlying substratum to pull the axon tip forward. Between the filopodia, the edge of the growth cone forms thin ruffles—lamellipodia—similar to those on a moving fibroblast (see Box 9B). Indeed, in its ultrastructure and mechanism of movement, the growth cone closely resembles the leading edge of a fibroblast crawling over a surface. Unlike a moving fibroblast, however, the extending axon also grows in length and diameter, with an accompanying increase in the total surface area of the neuron's plasma membrane. The additional membrane is provided by intracellular vesicles, which fuse with the plasma membrane.

The growth cone guides axon outgrowth, and is influenced by the contacts the filopodia make with other cells and with the extracellular matrix. In general, the growth cone moves in the direction in which its filopodia make the most stable contacts. In addition, its direction of migration can be influenced by diffusible extracellular signal molecules that bind to receptors on the growth-cone surface. Some of these extracellular cues promote axon extension, whereas others inhibit it by causing the retraction of the filopodia and the 'collapse' of the growth cone. The extension and retraction of filopodia and lamellipodia involve the assembly and disassembly of the actin cytoskeleton, and many of the extracellular signals that affect axon guidance are known to have effects on the actin cytoskeleton through the Ras-related GTPase family of intracellular signaling proteins. But precisely how growth cones transduce extracellular signals so as to extend or collapse filopodia is not yet fully understood.

Axon growth cones are guided by two main types of cue—attractive and repulsive. In addition, cues can act either at long or short range, thus giving four ways in which the growth cone can be guided (Fig. 12.25). Long-range attraction involves diffusible **chemoattractant** molecules released from the target cells. In contrast, **chemorepellents** are molecules that repel migrating cells or axons. Both chemoattractant and chemorepellent proteins have been identified in the developing nervous system. Short-range guidance is mediated

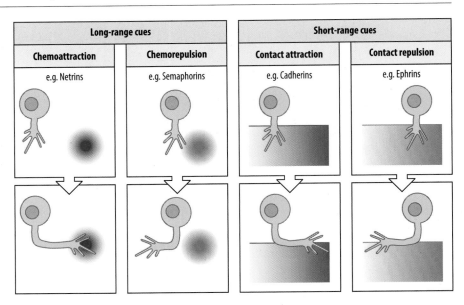

Fig. 12.25 Axon-guidance mechanisms. Four general types of mechanism can contribute to controlling the direction in which a growth cone moves: long-range attraction, long-range repulsion, short-range attraction, and short-range repulsion. Long-range cues are provided by secreted molecules (such as netrins and semaphorins) that form gradients in the extracellular matrix. Short-range cues are often provided by transmembrane proteins, such as the ephrins and their Eph receptors, and the cadherins, which bind to cadherins on other cells.

by contact-dependent mechanisms involving molecules bound to other cells or to the extracellular matrix; again, such interactions can be either attractive or repulsive. Some signals can function as either an attractant or a repellent, depending on the cellular and developmental context.

A number of different extracellular molecules that guide axons have been identified. The semaphorins are among the most prominent of the conserved families of axon-guidance molecules. There are seven different subfamilies of semaphorins, three of which are secreted, whereas the others are associated with the cell membrane. There are two classes of neuronal receptors for semaphorins—the plexins and the neuropilins. Axon repulsion is a major theme of semaphorin action. Some semaphorins can, however, either attract or repel growth cones, depending on the nature of the neuron and on the type of semaphorin receptor present in the neuron membrane. Different regions of the same neuron can respond in opposite ways to the same signal—the apical dendrites of pyramidal neurons in the cortex grow towards a source of semaphorin 3A, whereas their axons are repelled, which enables the receiving end of the neuron and the transmitting end of the neuron to be oriented in opposite directions by the same signal. This difference in response appears to be due to the presence of guanylate cyclase, a component of the pathway that transduces the semaphorin signal, in the dendrites but not in the axon. The role of semaphorins in directing the formation of the simple neural circuit that mediates the knee-jerk reflex is discussed in Box 12C.

Other classes of neuronal-guidance proteins, such as the secreted **netrins**, are also bifunctional, in that they attract some neurons and repel others. The attractant function of netrins is mediated by the DCC (deleted in colorectal cancer) family of receptors on growth cones. The **Slit proteins** are secreted glycoproteins with a repellent function. They repel a variety of axon classes by acting on receptors of the **Robo** family on neurons, but can also stimulate elongation and branching in sensory axons.

Guidance molecules that act by short-range cell–cell contact include the transmembrane **ephrins** and their **Eph receptors** (see Box 9D), which in the nervous system mediate mainly repulsive interactions, and the **cadherins**, which bind to cadherins on adjacent cells (see Box 9A), and which can act as short-range attractors of axons. As we shall see in the next section, ephrins are involved in guiding the axons of spinal cord motor neurons to their correct target muscles in the limb.

BOX 12C The development of the neural circuit for the knee-jerk reflex

Neural circuit that underlies the knee-jerk reflex

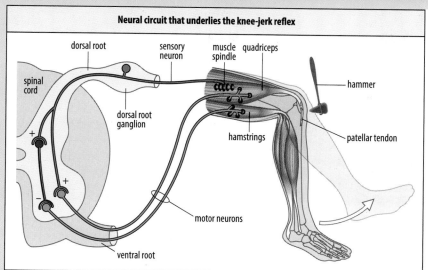

Development of the circuit

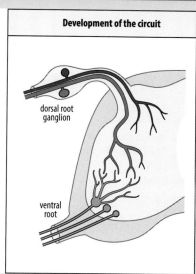

Figure 1

How neural circuits develop is currently a topic of great interest. The formation of a particular circuit will integrate several mechanisms considered in this chapter: neuronal specification and migration, axon navigation, and neuron–neuron recognition being the most prominent. We look here at the formation of the simple circuit constructed of just two neurons—a sensory neuron and a motor neuron—that mediates the familiar knee-jerk reflex.

The knee-jerk reflex is elicited by a tap on the patellar tendon just below the knee that lengthens the quadriceps muscle, causing the leg to jerk upwards. It is an example of a general muscle-stretch reflex that detects and responds to changes in muscle length (proprioception) and is responsible for maintaining limb and body position. In the knee-jerk circuit (see Figure 1, left panel), stretch receptors (spindles, orange) in the quadriceps detect the increase in muscle length. They are innervated by large, rapidly conducting proprioceptive sensory neurons (1a afferents, blue), which have their cell bodies in the dorsal root ganglion (DRG) of spinal cord segment L4. The sensory neuron axons synapse directly with the dendrites of motor neurons (green) in the ventral horn of the same segment, whose axons innervate the quadriceps muscle and cause its contraction. This monosynaptic circuit operates simultaneously with a disynaptic circuit, which includes inhibitory interneurons (red) that reciprocally prevent contraction of the antagonistic muscle, in this case, the hamstrings. This simple circuit enables us to study how the axon of the sensory neuron finds and connects with the correct motor neuron during development.

The axons of all DRG sensory neurons enter the dorsal horn of the spinal cord (see Figure 1, right panel), but whereas those that relay pain and temperature (purple) terminate close to their entry point, where they synapse with interneurons, those that relay proprioception (blue) extend further ventrally to synapse

with the dendrites of the appropriate motor neurons (green), ignoring those motor neurons that innervate antagonistic muscles. Semaphorin III, expressed in the ventral spinal cord, serves to repel all ingrowing sensory axons except the 1a afferents.

Recognition between pre- and post-synaptic partners appears to involve at least two types of mechanism. The sensory axon and the motor dendrites first grow towards a common meeting point known as the termination zone. Like a 'grid reference,' this zone is defined by a coordinate system of positional information across the dorsoventral and mediolateral axes of the spinal cord provided by orthogonal gradients of axon-guidance molecules such as the Semaphorins and Slits.

The ability to detect and respond appropriately to these gradients is part of each neuron's specification. This has been shown by an experiment in which the positions of motor neurons in the mouse lateral motor column (see Fig. 12.12) are scrambled, leaving the sensory neurons unperturbed. In this situation, the 1a sensory axons still terminate in the right place, even connecting with the 'wrong' motor neurons. A useful analogy for considering how both axon and dendrites grow towards the same place in the cord, and how such termination zones are defined, is the "Waterloo Station Clock Model," where two people receive instructions to meet at a specified remote location. Once the axon terminal is in place, the second recognition mechanism comes into play. This is likely to be a specific 'lock-and-key' mechanism furnished by receptor-ligand and adhesive interactions at the pre- and post-synaptic cell surfaces.

Although remarkable progress has been made in understanding how two neurons meet and connect with one another, our understanding of this, and other comparably "simple" circuits, is still far from complete—little is known about the molecular mechanism underlying fine matching.

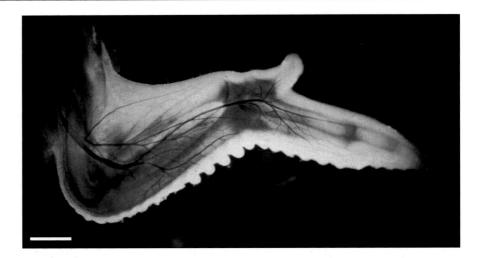

Fig. 12.26 Innervation of the embryonic chick wing. The nerves are stained brown and include both sensory and motor neurons. Scale bar = 1 mm.

Photograph courtesy of J. Lewis.

12.16 Motor neuron axons in the chick limb are guided by ephrin-Eph interactions

The muscular system enables an animal to carry out an enormous range of movements, which all depend on motor neurons from the brainstem and spinal cord having taken the correct path and having made the correct connections with muscles during embryonic development. The pattern of innervation of the chick wing, for example, is precise and always the same (Fig. 12.26). As we saw in Sections 12.8 and 12.9, particular groups of motor neurons that will innervate specific muscles in the limb are already distinguished by the expression of particular combinations of LIM and Hox genes in the spinal cord. To reach their correct targets, however, the growing axons require local guidance cues and these are provided by the limb tissue.

In the chick embryo, the motor neuron axons enter the developing limb bud at around 4.5 days, after cartilage elements, and dorsal and ventral muscle masses, have started to form. When the motor axons growing out from the spinal cord first reach the base of the limb bud, they are all mixed up in a single bundle. At the base of the limb bud, however, the axons separate out and form new nerves containing only those axons that will make connections with particular muscles.

Evidence that local cues in the limb are responsible for this sorting-out comes from experiments in chick embryos in which a section of the spinal cord whose motor neurons will innervate the hindlimb was inverted before the axons grew out. After inversion along the antero-posterior axis, motor axons from lumbosacral segments 1–3, which normally enter anterior parts of the limb, initially grew into posterior limb regions, but then took novel paths to innervate the correct muscles. So, even when the axon bundles entered in reverse antero-posterior order, the correct relationship between motor neurons and muscles was achieved. Motor axons can thus find their appropriate muscles when displaced relative to the limb—provided that the mismatch is small. However, a complete dorso-ventral inversion of the limb bud, equivalent to an antero-posterior reversal of five or more segments, resulted in the axons failing to find their appropriate targets. In such cases, the axons followed paths normally taken by other neurons. The mesoderm of the limb thus provides local cues for the motor axons, and the axons themselves have an identity that allows them to choose the correct pathway.

These local cues include signals such as the transmembrane ephrins. Motor neurons in the medial division of the LMC innervate ventral muscle masses in the limb, while those in the lateral division of the LMC innervate dorsal limb muscles (see Fig. 12.13). EphA4 signaling in the lateral LMC motor axons is a major determinant

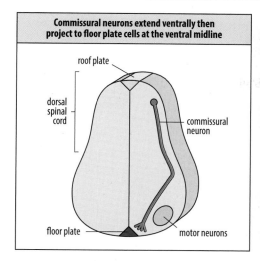

Commissural neurons extend ventrally then project to floor plate cells at the ventral midline

roof plate

dorsal spinal cord

commissural neuron

floor plate

motor neurons

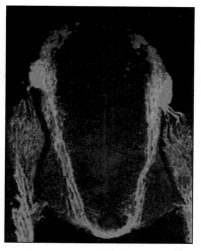

Fig. 12.27 Chemotaxis can guide commissural axons in the spinal cord. In the vertebrate spinal cord, the axons of spinal cord commissural neurons extend ventrally, and then towards the floor-plate cells. They cross the midline at the floor plate and then grow anteriorly along the spinal cord. The photograph shows a section through a rat spinal cord with the axons of the commissural neurons (those inside the spinal cord) stained green and the cell bodies red.

Photograph courtesy of S. Butler.

of the dorsal route they take into the limb. EphA4 is thought to mediate a repulsive interaction with its ligand ephrin A, which is restricted to the ventral limb mesenchyme. The expression of EphA4 is determined by the LIM homeodomain proteins (see Section 12.8), with Lim1 expression in the lateral LMC motor neurons leading to raised levels of EphA4, whereas Isl1 expression in the medial LMC neurons leads to a reduction in EphA4. The concentration of ephrin A in the dorsal limb mesenchyme is apparently kept at a low level by the expression there of the LIM protein Lmx1b (see Fig. 11.16). Migration of medial LMC motor axons into the ventral limb mesenchyme appears to be regulated by another set of receptors that similarly respond to repulsive signals from the dorsal limb mesenchyme.

12.17 Axons crossing the midline are both attracted and repelled

Although the development of the nervous system is largely symmetrical about the midline of the body, and proceeds independently on either side, one half of the body needs to know what the other half is doing. So, although some neurons must remain entirely within the lateral half of the body in which they were formed, many axons must cross the ventral midline of the embryo. Chemoattractant and chemorepellent molecules are involved in these decisions.

In the vertebrate spinal cord, commissural neurons provide the connections between the two sides of the cord. The cell bodies of commissural neurons are located in the dorsal region but they send their axons ventrally along the lateral margin of the cord. Signals from the roof plate prevent the commissural axons extending dorsally. About halfway down the dorso-ventral axis of the spinal cord, the axons make an abrupt turn and project to the floor plate in the ventral midline, bypassing the motor neurons (Fig. 12.27; also see Fig. 12.9). After crossing the floor plate to the other side of the midline, commissural axons make another sharp turn and project anteriorly along the spinal cord towards the head; they do not re-cross the midline. Chemoattractant and chemorepellent molecules present in the ventral midline are involved in both enabling the commissural neurons to cross the midline and in preventing them from crossing back.

Axons approaching the midline of the spinal cord provide good examples of the long-range attraction and repulsion of growth cones. The axons of vertebrate commissural neurons are attracted to the floor plate at the ventral midline by the chemoattractant Netrin-1, which signals attraction via the netrin-1 receptor DCC. Netrin-1 is a key attractant produced in the floor plate and in the midline, and knock-out of the mouse netrin-1 gene, or the gene for its receptor DCC, results in abnormal commissural axon

Fig. 12.28 Effect of *netrin-1* gene knock-out in mice. In mice lacking *netrin-1*, the commissural axons do not migrate towards the floor plate. Scale bar = 0.1 mm.

Photographs courtesy of M. Tessier-Lavigne, reproduced with permission from Serafini, T., et al.: **Netrin-1 is required for commissural axon guidance in the developing vertebrate nervous system.** *Cell 1996, 87: 1001–1014. © 1996 Cell Press.*

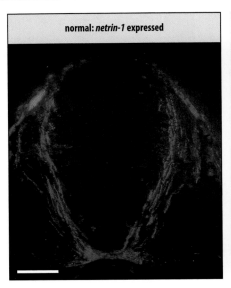
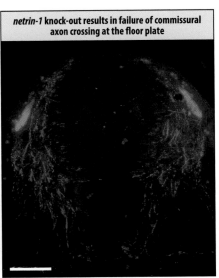

normal: *netrin-1* expressed

netrin-1 knock-out results in failure of commissural axon crossing at the floor plate

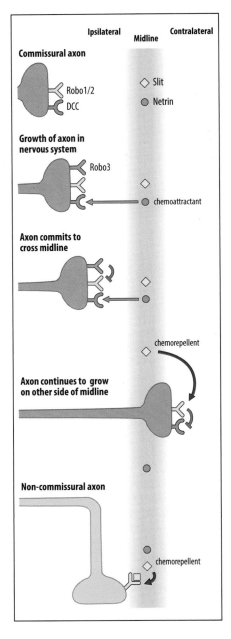

pathways (Fig. 12.28). In both *Drosophila* and vertebrates, the mechanism that repels axons from the midline involves the extracellular protein Slit and its Robo receptors on the axonal growth cones. Slit is present at the midline and, when it binds to receptors Robo1 and Robo2 on the growth cone, it exerts a chemorepellent effect. For vertebrate commissural axons approaching the midline, however, the chemorepellent effect of Slit is overcome by the Robo receptor-related protein Robo3 expressed on the growth cone, which interferes with Slit signaling through Robo1 and Robo2 (Fig. 12.29). Once the commissural axons cross the midline, Robo3 ceases to be expressed and the growth cone can be repelled from the midline again by Slit proteins, which prevents it crossing back. The sharp anterior turn made by mammalian commissural neurons after crossing the midline is thought to be due to a gradient of Wnt proteins in the floor plate with the high point anterior.

12.18 Neurons from the retina make ordered connections with visual centers in the brain

An arguably even more complex task for the developing nervous system is to link up the sensory receptors that receive signals from the outside world with neuronal targets in the brain that enable us to make sense of these signals. A characteristic feature of the vertebrate brain is the existence of **topographic maps**. That is, the neurons from one region of the nervous system project in an ordered manner to another region, so that nearest-neighbor relations are maintained. The vertebrate visual system has

Fig. 12.29 Competing chemoattractant and chemorepellent signals both enable commissural axons to cross the midline and prevent them crossing back. Commissural axons are attracted to the midline by the chemoattractant protein Netrin (green circle) acting through its receptor DCC (green). As they grow toward the midline, the level of Robo1/2 (yellow), the receptors for the chemorepellent protein Slit (yellow diamond), are kept low. Commissural axon growth cones also start to express the cell-surface receptor Robo3 (orange) as they approach the midline, which inhibits the chemorepellent effect of Slit acting at Robo1/2. Once commissural axons reach and cross the midline, alternative splicing of Robo3 produces a cell-surface receptor that is now responsive to Slit and does not inhibit Robo1/2, and so repulsion by Slit can now prevent the axons from re-crossing the midline. Increased expression of Robo1/2 occurs after crossing and helps block attraction to the midline exerted by Netrin. Non-commissural axons do not express Robo3 and so are continuously repelled from the midline by Slit.

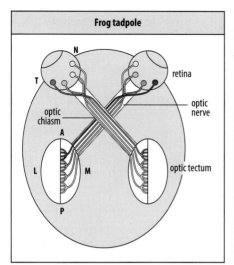
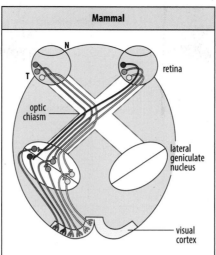

Fig. 12.30 Comparison of amphibian and mammalian visual systems. Left panel: in the frog tadpole, neurons from the retina project directly to the optic tectum, neurons from the left retina connecting to the right optic tectum and those from the right retina connecting to the left tectum, crossing over at the optic chiasm. Right panel: in mammals, the main visual pathway is to the lateral geniculate nucleus (LGN), from which neurons project to the visual cortex. In those mammals with full binocular and stereoscopic vision, neurons from the temporal half of the retina project to the LGN on the same side as the eye, and those from the nasal half cross over to the opposite LGN at the optic chiasm. For the sake of clarity, the neurons from only one half of the retina are shown. N, nasal; T, temporal, L, lateral, M, medial. A, anterior, P, posterior.

Illustration after Goodman, C.S., Shatz, C.J.: **Developmental mechanisms that generate precise patterns of neuronal connectivity.** Cell Suppl. *1993,* **72**: *77-98.*

been the subject of intense investigation for many years, and the highly organized projection of neurons from the eye via the optic nerve and tract to the thalamus and tectum is one of the best models we have to show how topographic neural projections are made. There are around 126 million individual photoreceptor cells in a human retina, and some animals with excellent night vision, such as owls, have many more. Each of those photoreceptor cells is continuously recording a minute part of the eye's visual field and the signals must be sent to the brain in an orderly manner so that they can be assembled into a coherent picture. Small sets of adjacent photoreceptor cells indirectly activate individual neurons—the retinal ganglion cells—whose axons are bundled together and exit the eye as the optic nerve (see Section 11.27 for a brief overview of the development of the eye itself). The optic nerve from each human eye is made up of over a million axons.

In birds and amphibians, the optic nerve connects the retina to a region of the midbrain called the **optic tectum**, where visual signals are processed. In birds and in larval amphibians, the optic nerve from the right eye makes connections with the left optic tectum, while the nerve from the left eye makes connections to the right side of the brain, the axons crossing over each other in a structure called the **optic chiasm** (see Fig. 12.30, left panel). In mammals, the main target for retinal axons are the paired **lateral geniculate nuclei** (**LGN**) in the thalamus, from which other neurons then convey the signals to the **visual cortex**, in which most visual processing occurs. A minor pathway in mammals goes from the retina to the **superior colliculus** in the midbrain (analogous to the avian and amphibian optic tectum), which is involved mainly in aiding the orientation of the eyes and head to a variety of sensory stimuli and in controlling the pupillary reflex. As this pathway provides a simpler map than that to the lateral geniculate nucleus, it is often studied as the mammalian counterpart to the optic tectum.

In most mammals (and in frogs after metamorphosis), retinal axons from one eye do not all go to the same LGN. Instead, axons from the ventro-temporal region of the retina project to the LGN on the same side as the eye (**ipsilateral**), while the remaining retinal axons project to the LGN on the opposite side (**contralateral**) (see Fig. 12.30, right panel). This separation of outputs occurs at the optic chiasm and enables binocular vision in mammals with forward-facing eyes, such as primates, by providing continuity of representation in the visual-processing structures in the brain. Birds, many of which also have excellent binocular vision, integrate the visual signals from each eye in other ways.

At the optic chiasm, retinal ganglion cell axons must make a decision to either avoid or traverse the midline of the brain. How this is achieved is not yet clear, but it

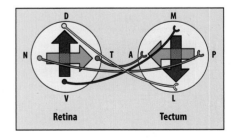

Fig. 12.31 The retina maps onto the tectum. In the amphibian tectum, dorsal (D) retinal neurons connect to the lateral (L) side of the tectum and ventral (V) retinal neurons connect to the medial (M) tectum. Similarly, temporal (T) (or posterior) retinal neurons connect to anterior (A) tectum and nasal (N) (or anterior) retinal neurons connect to the posterior (P) tectum.

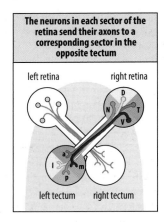

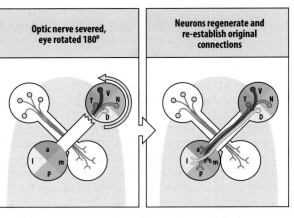

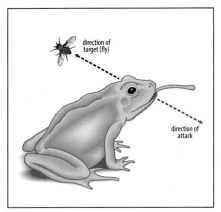

Fig. 12.32 Retino-tectal connections in amphibians are re-established in the original arrangement after severance of the optic nerve and rotation of the eye. First panel: neurons in the optic nerve from the left eye mainly connect to right optic tectum and those from the right eye to the left optic tectum. There is a point-to-point correspondence between neurons from different regions in the retina (nasal (N), temporal (T), dorsal (D), and ventral (V)) and their connections in the tectum (posterior (p), anterior (a), lateral (l), and medial (m), respectively). Center and right panels: if one optic nerve of a frog is cut and the eye rotated dorso-ventrally through 180°, the severed ends of the axons degenerate. When the neurons regenerate, they make connections with their original sites of contact in the tectum. However, because the eye has been rotated, the image falling on the tectum is upside down compared with normal. When the frog sees a fly above its head, it thinks the fly is below it, and moves its head downward to try to catch it (fourth panel).

is likely that short-range guidance molecules, such as ephrins and Eph receptors, are involved. In mice, for example, the ipsilateral retinal projection arises from neurons in the peripheral ventro-temporal crescent of the retina (see Fig. 12.30, right panel). These neurons express the guidance receptor EphB1 (see Box 9D), which interacts with ephrin B2 on radial glia cells at the optic chiasm to repel the axons away from the midline and into the ipsilateral optic tract.

Retinal neurons map in a highly ordered manner onto their target structures, with a point-to-point correspondence between a position on the retina and one on the tectum. Figure 12.31 shows how neurons in different positions along the naso-temporal (antero-posterior) axis of the tadpole retina map in a reverse direction to points along the antero-posterior axis of the optic tectum. A similar reversed polarity is seen in the mapping of neurons along the dorso-ventral axis of the retina: dorsal retinal neurons project to the lateral region of the tectum and neurons from the ventral region of the retina project to the medial region of the tectum.

Remarkably, in some lower vertebrates, such as fish and amphibians, the pattern of connections can be re-established with precision when the optic nerve is cut. The axons distal to the cut die, new growth cones form at the proximal stump, and axon outgrowth reforms connections to the tectum. In frogs, even if the eye is inverted through 180°, the axons still find their way back to their original sites of contact (Fig. 12.32). However, the animals subsequently behave as if their visual world has been turned upside down: if a visual stimulus, such as a fly, is presented above the inverted eye, the frog moves its head downward instead of upward, and can never learn to correct this error.

Such experiments raised the possibility that each retinal neuron carries a chemical label that enables it to connect reliably with an appropriately molecularly labeled cell in the tectum. This is known as the **chemoaffinity hypothesis** of connectivity. In the retinotectal projection, it is thought that graded spatial distributions of a relatively small number of factors on the tectum provide positional information, which can be detected by the retinal axons. The spatially graded expression of another set of factors on the retinal axons would provide them with their own positional information. The development of the retino-tectal projection could thus, in principle, result from the interaction between these two gradients.

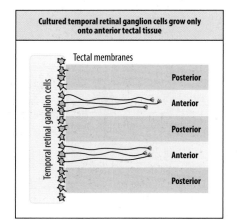

Fig. 12.33 Choice of targets by retinal axons. If pieces of temporal retina are placed next to a 'carpet' of alternating stripes (90 μm wide) of anterior and posterior tectal cell membranes, the temporal retinal cells only extend axons on the anterior tectal membranes.

Fig. 12.34 Complementary expression of the ephrins and their receptors in the mouse retino-collicular projection. The superior colliculus is a structure in the mammalian midbrain to which some retinal neurons project and which is analogous to the tectum of amphibians and birds. The receptor EphA is a receptor tyrosine kinase and ephrin As are its ligands. Temporal retinal neurons, which express high levels of EphA on their surface, are repulsed by the posterior tectum, where high levels of ephrin A are present, but can make connections with the anterior tectum. Nasal retinal neurons make the best contacts in the posterior tectum, as they express low levels of EphA and thus are repulsed only where the levels of ephrins are highest.

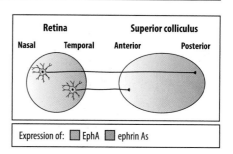

Such gradients were first found in the developing visual system of the chick embryo. An axon-guiding activity based on repulsion has been detected along the antero-posterior axis of the chick tectum. Normally, the temporal (posterior) half of the retina projects to the anterior part of the tectum and the nasal (anterior) retina projects to the posterior tectum. When offered a choice between growing on membranes from posterior or anterior tectal cells, temporal axons from an explanted chick retina show a preference for anterior tectal cell membranes (Fig. 12.33). This choice is mediated by repulsion of the axons—as shown by the collapse of their growth cones—by a factor located on the surface of posterior tectal cells. Ephrins and their receptors are the likely molecules mediating this repulsion as they are expressed in reciprocal gradients in retina and tectum.

In the mouse, the role of ephrins and Ephs in creating a map has been investigated in the retino-collicular pathway, and repulsive interactions between the EphA family of receptors and their ephrin A ligands have been found to pattern this map in the antero-posterior dimension. Mapping of the retina onto the superior colliculus is similar to that from the retina to the optic tectum: retinal neurons from the nasal region map to the posterior colliculus and neurons from the temporal retina map to the anterior colliculus. EphAs are expressed in a gradient in the mouse retina—low in the nasal region, high in the temporal region—whereas ephrin As are graded in the superior colliculus from high posterior to low anterior. Thus retinal neurons with the most EphA receptors connect to collicular targets with the fewest ephrin As. Retinal axons bearing low levels of the receptor move into areas containing high levels of the ligands, whereas retinal axons bearing high levels of receptor stop before they reach areas with high levels of ephrin A (Fig. 12.34). A possible explanation for this behavior is that binding of ligand to the receptor sends a repulsive signal, and the axon ceases to extend when this signal reaches a threshold value. The strength of the signal received by the migrating axon will be proportional to the product of the concentrations of receptor and ligand. Thus, the threshold will be reached when there is a high level of both receptor and ligand, but not when there is a low level of receptor, even when ligand level is high.

Mapping from the dorso-ventral dimension of the retina to the medial-lateral dimension of the superior colliculus or optic tectum also involves graded expression of Ephs and ephrins. In this case, EphB is graded in the retinal neurons from high-ventral to low-dorsal and ephrin B in the tectum from high-medial to low-lateral. EphB and ephrin B mediate adhesive and attractive interactions, rather than repulsion, and this poses a conceptual problem. If their interaction were the only factor at work, all the retinal neurons would be attracted to the area of highest ephrin B, rather than mapping across the whole tectum. In this case, the necessary repulsive interactions in these dimensions seem to be provided by gradients of the signaling protein Wnt3 in the tectum and a receptor, Ryk, on the retinal axons.

SUMMARY

Growth cones at the tip of the extending axon guide it to its destination. Filopodial activity at the growth cone is influenced by environmental factors, such as contact with the substratum and with other cells, and can also be guided by chemotaxis. Guidance involves both attraction and repulsion. In the development of motor neurons that innervate vertebrate limb muscles, the growth cones guide the axons so as to make the correct muscle-specific connections, even when their normal site of entry into the limb is disturbed. Combinations of transcription factors give each motor neuron an identity that determines its pathway. Attraction and repulsion control axons crossing the midline in both *Drosophila* and vertebrates. Gradients in diffusible molecules are probably responsible for the directional growth of commissural axons in the spinal cord. Neuronal guidance of axons from the retina to make the correct connections with the optic tectum involves gradients in cell-surface molecules both on the tectal neurons and on the retinal axons that can promote or repel growth cone approach, and there is competition between neurons for sites.

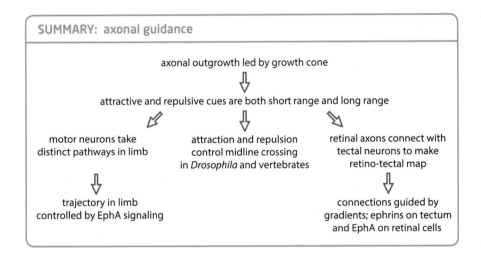

SUMMARY: axonal guidance

axonal outgrowth led by growth cone

⇩

attractive and repulsive cues are both short range and long range

↙   ⇩   ↘

motor neurons take distinct pathways in limb   attraction and repulsion control midline crossing in *Drosophila* and vertebrates   retinal axons connect with tectal neurons to make retino-tectal map

⇩   ⇩

trajectory in limb controlled by EphA signaling   connections guided by gradients; ephrins on tectum and EphA on retinal cells

Synapse formation and refinement

Having seen the outcome of axonal navigation, we now focus on the specialized connections—the **synapses**—that the axons make with their targets and which are essential for signaling between neurons and their target cells. Formation of synapses in the correct pattern is a basic requirement of any developing nervous system. The connections may be made with other nerve cells, with muscles, and also with certain glandular tissues. Here, we focus mainly on the development and stabilization of synapses at the junctions between motor neurons and muscle cells in vertebrates, which are called **neuromuscular junctions** (Fig. 12.35). There are very many neuronal cell types—hundreds if not thousands—and how they match up to make synapses reliably is a central problem. The molecular mechanisms are only beginning to be known. Studies on the vertebrate neuromuscular junction have guided much of the understanding of synapse formation.

Setting up the organization of a complex nervous system in vertebrates involves refining an initially rather imprecise organization by extensive programmed cell death (see Box 6A). The establishment of a connection between a neuron and its target appears to be essential not only for the functioning of the nervous system, but also for the very survival of many neurons. Neuronal death is very common in the developing vertebrate nervous system; too many neurons are produced initially and only those that make appropriate connections survive. Survival depends on the neuron receiving neurotrophic factors or **neurotrophins** such as nerve growth factor (NGF), which are produced by the target tissue and for which neurons compete.

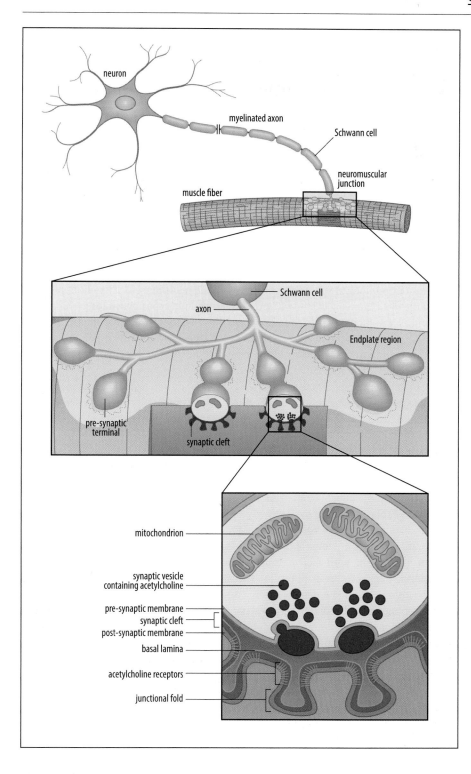

Fig. 12.35 Structure of the vertebrate neuromuscular junction. The motor neuron axon, which is covered in a myelin sheath produced by the glial Schwann cell, innervates a muscle fiber. At the neuromuscular junction, the axon branches in the endplate region, and makes synaptic connections with the muscle membrane. Communication between nerve and muscle is by release of the neurotransmitter acetylcholine from synaptic vesicles into the synaptic cleft. Acetylcholine diffuses across the cleft and binds to acetylcholine receptors on the muscle cell membrane.

Illustration after Kandell, E.R., et al.: Principles of Neural Science, *3rd edition. New York: Elsevier Science Publishing Co., Inc., 1991.*

A special feature of nervous-system development is that fine-tuning of synaptic connections depends on the interaction of the organism with its environment and the consequent neuronal activity. This is particularly true of the vertebrate visual system, where sensory input from the retina in a period immediately after birth modifies synaptic connections so that the animal can perceive fine detail. Again, this refinement seems to involve competition for neurotrophins. We return to this topic after first considering synapse formation.

Fig. 12.36 Development of the neuromuscular junction. Before contact by an axon, acetylcholine receptors (AChR) are gathered in a central region of the muscle fiber. When a motor axon growth cone contacts a muscle fiber it releases the proteoglycan Agrin (blue arrow) into the extracellular material. Agrin binds to and activates the muscle-cell kinase MuSK, which in turn maintains clustering of AChR in the muscle-cell membrane and specialization of the post-synaptic surface. A signal (as yet unknown) from the muscle cell (purple arrow) then helps to induce the differentiation of the pre-synaptic terminal, with the formation of synaptic vesicles. Synaptic signaling by acetylcholine (small red circles) released from the synaptic vesicles can then begin.

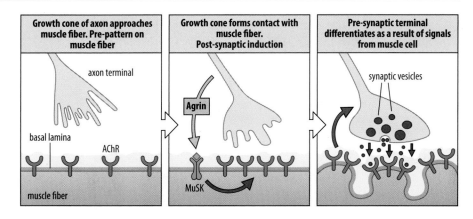

12.19 Synapse formation involves reciprocal interactions

We will start our look at synapse formation with the neuromuscular junction, as this is one of the most intensively studied and best-understood types of synapse. The mature neuromuscular junction is a complex structure involving extensive modification of the nerve ending and the muscle cell membrane (see Fig. 12.35). Just before the junction, the axon terminal branches into a network-like arrangement. Each branch ends in a swelling, which is in contact with a special endplate region on the muscle fiber. The axon's plasma membrane is separated from the muscle-cell's plasma membrane by a narrow cleft (the **synaptic cleft**) filled with extracellular material secreted by both the neuron and the muscle cell. A basal lamina of extracellular material forms around the muscle-cell surface. The synapse comprises the axon plasma membrane, the opposing muscle-cell plasma membrane, and the cleft between.

Electrical signals cannot pass across the synaptic cleft, and for the neuron to signal to the muscle, the electrical impulse propagated down the axon is converted at the terminal into a chemical signal. This is a chemical neurotransmitter that is released into the synaptic cleft from **synaptic vesicles** in the axon terminal. Molecules of the neurotransmitter diffuse across the cleft and interact with receptors on the muscle-cell membrane, causing the muscle fiber to contract. The neurotransmitter used by motor neurons connecting with skeletal muscles is acetylcholine. Because the signal travels from nerve to muscle, the axon terminal is called the **pre-synaptic** part of the junction and the muscle cell is the **post-synaptic** partner (see Fig. 12.35). The growth cone of the developing axon becomes the pre-synaptic terminal.

Development of a neuromuscular junction is progressive. Before the arrival of the axon there is already some pre-patterning on the muscle, with acetylcholine receptors concentrated in the central region of the muscle fiber. When the motor neuron terminal arrives, it releases the proteoglycan Agrin into the basal lamina, which binds to, and activates, the transmembrane muscle-specific kinase (MuSK), whose activity is required to maintain the clustering of acetylcholine receptors and the specialization of the muscle-cell surface at the site (Fig. 12.36). Signals from the muscle in turn induce differentiation of the pre-synaptic zone on the axon terminal and align this zone with the post-synaptic area on the muscle. These signals are not yet clearly identified, but the glycoprotein laminin β_2 in the basal lamina is required for correct alignment and full pre-synaptic differentiation. The growth factor neuregulin-1, which is related to epidermal growth factor and acts through the ErbB receptors, is required for the formation of some synapses, including neuromuscular junctions, where it helps mediate the expression of acetylcholine receptors on the muscle cell.

Neuregulin-1 is also involved in the development of glial cells, including the Schwann cells of the peripheral nervous system. In vertebrates, nerve impulses often have to travel long distances along axons in peripheral nerves—up to several meters

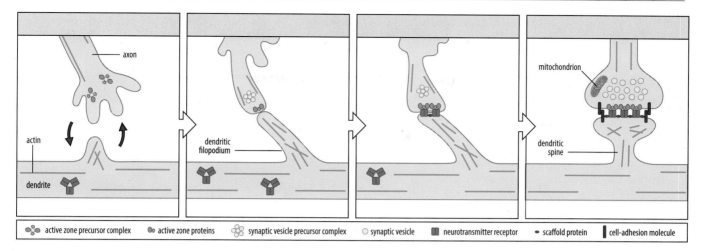

| ⚬:⚬ active zone precursor complex | ●● active zone proteins | ⚬:⚬ synaptic vesicle precursor complex | ○ synaptic vesicle | ▪ neurotransmitter receptor | • scaffold protein | ▮ cell-adhesion molecule |

Fig. 12.37 Interneuronal synapse formation. A dendrite sends out filopodia that contact the axon growth cone. Reciprocal signaling (red arrows) initiates synapse formation aided by cell-adhesion molecules. In the axon, proteins that comprise the pre-synaptic active zone gather at the site of contact along with synaptic vesicle precursors. In the dendrite, neurotransmitter receptors are synthesized and inserted into the post-synaptic membrane along with post-synaptic scaffold proteins. The post-synaptic connection develops into a typical dendritic spine.
Illustration after Li, Z., Sheng, M.: **Some assembly required: the development of neuronal synapses**. Nat. Rev. Mol. Cell Biol. *2003,* **4:** *833–841.*

in a large animal—and rapid propagation of nerve impulses over such distances is achieved by the myelination of axons by glial cells. Schwann cells in the peripheral nervous system and oligodendrocytes in the central nervous system wrap themselves around axons so that their cell membranes form a fatty, insulating layer of myelin. Depolarization of the neuron membrane then only occurs at the small areas of bare neuronal membrane between one insulating cell and the next, and the nerve impulse is able to 'jump' over these areas without loss of signal. The progressive loss of neural function in the human disease multiple sclerosis is due to the destruction of myelin, and eventually leads to severe muscle weakness as a result of the reduction in signal from nerve to muscle.

The neuromuscular junction is a highly specialized type of synapse, but synapse formation within the vast network of neurons that form the central nervous system follows similar principles. Synapses are made between the axon terminal of one neuron and a dendrite, cell body, or more rarely the axon, of another neuron. An individual motor neuron in the spinal cord, for example, will receive input from many other neurons on its dendrites and cell body. As with neuromuscular junctions, interneuronal synapse formation is likely to involve reciprocal signaling between axon and dendrite. Synapses between neurons can form quite rapidly, on a time scale of about an hour, and observations suggest that filopodia on the dendrite initiate synapse formation by reaching out to the axon (Fig. 12.37). Signals from the axon, including release of neurotransmitter, may initially guide the dendrite filopodia. Many other factors have been shown to promote synapse formation, including cell-adhesion molecules and Wnt signaling. Cell-adhesion molecules implicated in synapse formation include the **neuroligins**, which are post-synaptic membrane proteins that bind to proteins called β-**neurexins** on the pre-synaptic axon surface, causing the neurexins to cluster, which induces differentiation in the pre-synaptic terminal. Neuroligins promote the function of excitatory synapses rather than inhibitory synapses. Cadherins, protocadherins, and members of the immunoglobulin superfamily (see Box 9A) have also been found localized to synapses and thereby confer specificity. Defects in inter-neuronal synapse formation and/or plasticity are thought to be a major contributor to autism and autism spectrum disorders (Box 12D).

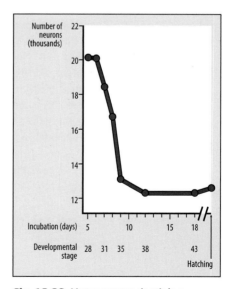

Fig. 12.38 Motor neuron death is a normal part of development in the chick spinal cord. The number of spinal cord motor neurons innervating a chick limb decreases by about half before hatching, as a result of programmed cell death during development. Most of the neurons die over a period of 4 days.

MEDICAL BOX 12D Autism: a developmental disorder that involves synapse dysfunction

The brain works through the release of neurotransmitter molecules at synapses, which interconnect neurons in highly complex networks. Correct synapse formation and refinement during development is therefore of crucial importance to cognitive function. Autism is a common (1 in around 120 live births) and pervasive disorder of developmental origin that is characterized by deficits in communication and social interaction together with repetitive behaviors and restricted interests.

Human genetics studies have identified hundreds of genes that, when mutated or present in an abnormal number of copies, are associated with autism. A large proportion of these genes act in developmental pathways that converge on synapse function, with the mutant genes interfering with synapse formation and/or with synaptic homeostasis—the cell-wide management of synaptic activity to keep a suitable balance between excitation and inhibition. Among this group of genes, those encoding proteins located at the synapse itself predominate; some of these proteins are known to be involved in neurotransmission, others in cell-cell adhesion. For example, neurexins and their partner neuroligins (see Section 12.19) are synaptic cell-adhesion proteins with a crucial role in synaptogenesis, and are also thought to modulate neurotransmission in more specific ways by acting as receptors (Figure 1). They are frequently found mutated in autism. The scaffolding protein SHANK3 is found at the postsynaptic membrane in excitatory neurons, where it is associated with neuroligins and the glutamic acid receptors NMDAR and mGLUR via the adaptor proteins GKAP, PSD-95, and homer. Mutations in the *SHANK3* gene are also frequent in autism.

Whereas the genetics of autism is complex, there are several monogenic human disorders with Mendelian inheritance that include an autistic phenotype. For example, Fragile X mental retardation, a common form of intellectual disability, is caused by mutations in the *FMR1* gene. This encodes a suppressor of translation, FMRP, which targets the mRNAs for multiple genes encoding synaptic proteins that ultimately regulate synaptic plasticity. The loss of these proteins especially affects mGluR-dependent synaptic depression (Figure 1), a weakening of synaptic strength that is thought to be essential to learning and memory, acting alongside complementary processes of synaptic strengthening, or long-term potentiation.

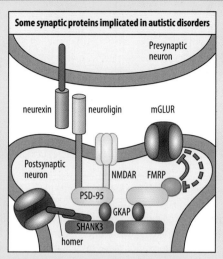

Figure 1

Rett syndrome is a severe X-linked neurodevelopmental disorder that is often associated with autistic behaviors, and almost always results from mutations in the *MeCP2* gene, which encodes a methylcytosine-binding protein. *MeCP2* knock-out mice exhibit a typical Rett's phenotype—including abnormal motor function, tremors, seizures and hind-limb clasping. Remarkably, subsequent reactivation of *MeCP2* in juvenile animals can lead to resolution of the abnormalities—suggesting that the requirement for the MeCP2 protein would be confined to the activity-dependent stages of nervous-system development, during which the brain's wiring can undergo considerable synaptic plasticity. Interestingly, treatment of the *MeCP2* knock-out mice with insulin-like growth factor-1 (IGF-1) also leads to reversal of many aspects of the phenotype, including synaptic abnormalities. Induced pluripotent stem cells (iPS cells; see Box 8D) derived from patients with Rett syndrome, in which *MeCP2* is silenced, have been induced to differentiate into neurons in culture, providing an *in vitro* means of studying the syndrome. The neurons display synaptic abnormalities, particularly disruptions to the balance between excitation and inhibition. As with the mutant mice, application of IGF-1 to these iPS-cell-derived neurons rescues the reduction in the number of excitatory synapses, bringing closer the possibility of an effective molecular therapy.

In *Drosophila*, the cell-adhesion molecule Dscam, a member of the immunoglobulin superfamily, is required both for axon guidance in the embryonic nervous system and in specifying particular neuronal connections. It initiates a signaling pathway that leads to activation of Rho-family GTPases that affect axonal migration by causing changes in the actin cytoskeleton. The Dscam gene is exceptional in that alternative splicing could produce an estimated 38,000 distinct forms of cell-surface receptor, thus raising the possibility that different forms of the protein might be produced in different neurons.

12.20 Many motor neurons die during normal development

During development, some 20,000 motor neurons are formed in the segment of spinal cord that provides innervation to a chick leg, but about half of them die soon after they are formed (Fig. 12.38). Cell death occurs after the axons have grown out from the cell bodies and entered the limb, at about the time that the axon terminals are reaching their potential targets—the skeletal muscles of the limb. The role of the target muscles in preventing cell death is suggested by two experiments. If the leg bud is removed, the number of surviving motor neurons decreases sharply. When an additional limb bud is grafted at the same level as the leg, providing additional targets for the axons, the number of surviving motor neurons increases.

Survival of a motor neuron depends on its establishing contacts with a muscle cell. Once a contact is established, the neuron can activate the muscle, and this is followed by the death of a proportion of the other motor neurons that are approaching the muscle cell. Muscle activation by neurons can be blocked by the drug curare, which prevents neuromuscular signal transmission, and this block results in a large increase in the number of motor neurons that survive.

Even after neuromuscular connections have been made, some are subsequently eliminated. At early stages of development, single muscle fibers are innervated by axon terminals from several different motor neurons. With time, most of these connections are eliminated, until each muscle fiber is innervated by the axon terminals from just one motor neuron (Fig. 12.39). This is due to competition between the synapses, with the most powerful input to the target cell destabilizing the less powerful inputs to the same target. This elimination process and the maturation of a stable neuromuscular junction take up to 3 weeks in the rat.

The well-established matching of the number of motor neurons to the number of appropriate targets by these mechanisms suggests that a general mechanism for the development of nervous-system connectivity in all parts of the vertebrate nervous system is that excess neurons are generated and only those that make the required connections are selected for survival. This mechanism is well suited to regulating cell numbers by matching the size of the neuronal population to its targets. We now look in more detail at the neurotrophic factors that promote neuronal survival, and then consider the role of neural activity in the elimination of neuromuscular synapses.

12.21 Neuronal cell death and survival involve both intrinsic and extrinsic factors

Neuronal death during development takes place as a result of the activation of the apoptotic pathway (see Box 6A). Apoptosis can be triggered or prevented either by an intrinsic developmental program or by external factors. In the development of the nematode, for example, particular neurons and other cells are intrinsically programmed to die (see Chapter 6), and the specificity of cell death can be shown to be under genetic control.

As we saw in Box 6A, members of the Bcl-2 protein family control the apoptotic pathway in vertebrates—some acting to promote apoptosis and some to prevent it. A key family member promoting neuronal apoptosis during development is *Bax*; mice mutant for *Bax* have increased numbers of motor neurons in the face. A protein called Survivin is known to inhibit apoptosis during early neurogenesis. Mutant mice in which Survivin is deleted from embryonic day 12.5 onwards are born with much smaller brains than normal and numerous foci of apoptosis throughout the brain and spinal cord and die soon after birth.

In many developmental situations it seems that cells are programmed to die unless they receive specific survival signals. Extracellular factors play a key role in neuronal survival in vertebrates. The first one identified was NGF, whose

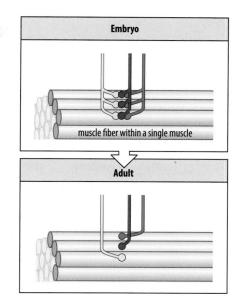

Fig. 12.39 Refinement of muscle innervation by neural activity. Initially, several motor neurons innervate the same muscle fiber. Elimination of synapses means that each fiber is eventually innervated by only one neuron.

Illustration after Goodman, C.S., Shatz, C.J.: ***Developmental mechanisms that generate precise patterns of neuronal connectivity.*** Cell Suppl. *1993, **72**: 77–98.*

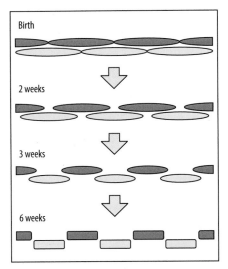

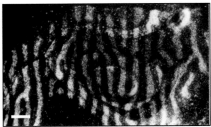

Fig. 12.40 Visualization of ocular dominance columns in the monkey visual cortex. A radioactive tracer is injected into one eye, from where it is transported to the visual cortex through the neurons. Tracer injected at birth is broadly distributed in the cortex. Tracer injected at later times becomes confined to alternating columns of cortical cells (brighter stripes), representing the ocular dominance columns for that eye, as seen in the photograph. Scale = 1 mm.

Illustration after Kandel, E.R., et al.: Essentials of Neural Science and Behavior. *Norwalk, Connecticut: Appleton & Lange, 1991.*

discovery was due to the serendipitous observation that a mouse tumor implanted into a chick embryo evoked extensive growth of nerve fibers towards the tumor. This suggested that the tumor was producing a factor that promoted axonal outgrowth. The factor was eventually identified as a protein, NGF, using axon outgrowth in culture as an assay. NGF is necessary for the survival of a number of types of neurons, particularly those of the sensory and sympathetic nervous systems.

NGF is a member of a family of proteins known as the neurotrophins. Different types of neuron require different neurotrophins for their survival, and the requirement for certain neurotrophins also changes during development. The neurotrophin GDNF, for example, prevents the death of facial motor neurons, and is also expressed in developing limb buds. There is increasing evidence that individual neurotrophins act on a range of neural cell types.

12.22 The map from eye to brain is refined by neural activity

The development and function of the nervous system depends not only on synapse formation, but also on the regulated disassembly of previously functional synaptic connections. We have already considered how axons from the retina make connections with the tectum so that a retino-tectal map is established. This map is initially rather coarse-grained, in that axons from neighboring cells in the retina make contacts over a large area of the tectum. This area is much larger early on than at later stages of development, when the retino-tectal map is more finely tuned. Fine-tuning of the map results, as in muscle, from the withdrawal of axon terminals from most of the initial contacts, and requires neural activity. This requirement is seen particularly clearly in the development of visual connections in mammals. If a mammal, including a human, is deprived of vision during a critical period, then it will suffer from poor visual acuity and will not be able to resolve fine detail.

In mammals with full binocular and stereoscopic vision, such as monkeys and humans, retinal axons first connect to the lateral geniculate nucleus, on to which they map in an ordered manner. Input from half of each eye goes to the opposite side of the brain, whereas input from the other half of each eye goes to the same side of the brain (see Fig. 12.30, right panel).

In the lateral geniculate nucleus, as in the cortex, the neurons are arranged in layers. Each layer receives input from retinal axons from either the right or the left eye, but not both. Thus, inputs from left and right eyes are kept separate. Neurons from the lateral geniculate nucleus then send axons to the visual cortex (see Fig. 12.30, right panel). When there is a visual stimulus, the inputs from the retinal axons activate neurons in the lateral geniculate nucleus, which then activate neurons in the corresponding region of the visual cortex. The adult visual cortex consists of six cell layers, but we need only focus here on layer 4, which is where many axons from the lateral geniculate nucleus make connections. Layers in the lateral geniculate nucleus that receive left or right inputs both make connections with layer 4 of the visual cortex. This means that input from corresponding positions in both eyes, that is from the same part of the visual field, finally arrives at the same location in the cortex (see Fig. 12.30, right panel).

At birth, the nerve endings from the two eyes overlap and are mixed at their common final location, but with time the inputs from left and right eyes become separated into blocks of cortical cells about 0.5 mm wide, which are known as **ocular dominance columns** (Fig. 12.40). Adjacent columns respond to the same stimulus in the visual field; one column responding to signals from the left eye, and the next to signals from the right eye. This arrangement enables good binocular and stereoscopic vision. The columns can be detected and mapped by making electrophysiological recordings. They can also be directly observed by injecting a tracer, such as radioactive proline, into one eye. The tracer is taken up by retinal neurons, transported by the optic nerve to the lateral geniculate, and from there to the visual cortex, where its pattern can be detected by autoradiography. This reveals a striking array of stripes representing the input from one eye (see Fig. 12.40).

Neural activity and visual input are essential for the development and maintenance of the ocular dominance columns. While sensory input is important, spontaneous activity also plays a key role. Initial formation of the stripes in non-human primates occurs before visual experience, and involves spontaneously generated waves of action potentials in the mammalian retina. These waves may cause their effects through the release of neurotrophins that can remodel synaptic connections. If neural activity is blocked during development by the injection of tetrodotoxin (a sodium channel blocker), ocular dominance columns do not develop, and the inputs from the two eyes into the visual cortex remain mixed. If the input from one eye is blocked, the territory in the visual cortex occupied by the other eye's input expands at the expense of the blocked eye. The release of the inhibitory neurotransmitter γ-aminobutyric acid (GABA) is required to initiate the critical period.

The favored explanation for the formation of ocular dominance columns is based on competition between incoming axons (Fig. 12.41). Because of the initial connections that establish the map, individual cortical neurons can initially receive input from both eyes. Within a particular region, there will, therefore, be overlap of stimuli originating from the two eyes, and this overlap has to be resolved. Neighboring cells carrying input from the same eye tend to fire simultaneously in response to a visual stimulus; if they both innervate the same target cell, they can thus cooperate to excite it. As in muscle, stimulation of electrical activity in the target cell tends to strengthen the active synapses and suppress those that are not active at the time—cells that fire together, wire together. As there is competition between neurons for targets, this could generate discrete regions of cortical cells that respond only to one eye or the other, and so form the ocular dominance columns. Such a mechanism explains why experimental exposure of animals to continuous strobe lighting after birth, which causes simultaneous firing of neurons in both eyes, prevents the formation of ocular dominance columns.

In frogs the optic tecta do not normally receive input from both eyes until after morphogenesis and ocular dominance columns do not develop in the tecta under normal circumstances. Tecta that receive inputs from two eyes throughout development can, however, be produced by implanting a third eye primordium into the embryo. Retinal axons from the two eyes then segregate within the tectum and form alternating stripes that resemble the pattern of ocular dominance columns in cats and primates. In these tecta, the dendrites of some types of tectal neurons seem to respond to the stripe boundaries, as evidenced by marked changes in their normal direction of migration or by their termination at a stripe border. The behavior of these neurons in this artificial situation is further evidence that in all vertebrates, not simply mammals, highly localized interactions between incoming axon terminals and the post-synaptic dendrites with which they connect are critical in producing the patterning of a segregated, topographically organized cortex.

One possible mechanism for refining connections in response to neuronal activity involves local release of neurotrophins. A certain level of activation, or activation by two axons simultaneously, could induce the release of neurotrophins from the target cells, and only those axons that have been recently active might be able to respond to them. When neural activity in developing chick or amphibian embryos is blocked with drugs that prevent neuronal firing, no fine-grained retino-tectal map develops.

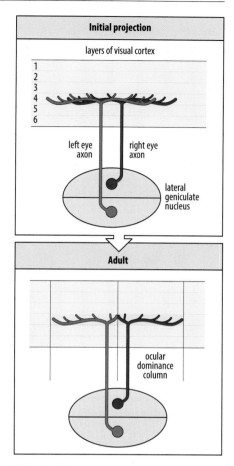

Fig. 12.41 Development of ocular dominance columns. Initially, neurons from the lateral geniculate nucleus, representing projections from both eyes and stimulated by the same visual stimulus, project to the same region of the visual cortex. (Only projection to layer 4 is shown here.) With visual stimulation, the neuronal connections separate out into columns, each representing innervation from only one eye. If stimulation by vision is blocked, ocular dominance columns do not form.

Illustration after Goodman, C.S., Shatz, C.J.: ***Developmental mechanisms that generate precise patterns of neuronal connectivity.*** Cell Suppl. *1993, **72**: 77-98.*

SUMMARY

Neurons communicate with each other and with target cells, such as muscles, by means of specialized junctions called synapses. The formation of a neuromuscular junction involves changes in both pre-synaptic (neuronal) and post-synaptic (muscle cell) membranes once contact is made, and depends on reciprocal signaling between muscle and nerve cell. Reciprocal interactions also occur in the formation of synapses between axons and dendrites, and cell-adhesion molecules are likely to be involved. Initially, most mammalian muscle fibers are innervated by two or more motor axons, but nervous activity results in competition between synapses so that a single fiber is eventually innervated by only one motor neuron. Many neurons produced in the developing nervous system die. About half of the motor neurons that initially innervate the vertebrate limb undergo cell death; those that survive do so because they make functional connections with muscles. Many neurons depend on neurotrophins, such as nerve growth factor, for their survival, with different classes of neurons requiring different neurotrophins. Brain function is based not only on the assembly of synapses, but also on their regulated disassembly. Neural activity has a major role in refining the connections between the eye and the brain. In mammals, input from the left and right eyes is required for the development of ocular dominance columns in the visual cortex. These are adjacent columns of cells responding to the same stimulus from left and right eyes, respectively. Formation of the columns, which are essential for binocular vision, is a result of competition for cortical targets by axons carrying visual input from different eyes.

SUMMARY: synapse formation and refinement

Neuromuscular junction	**Neuron–neuron synapse**
axon of motor neuron releases Agrin at junction with muscle	dendrite extends filopodia which contact axon terminal; cell-adhesion molecules form initial contact between dendrite and axon
acetylcholine receptors cluster at junction and local synthesis of receptors occurs	pre-synaptic active zone assembles; neurotransmitter receptors cluster on pre-synaptic membrane
electrical activity of muscle reduces receptor synthesis elsewhere	
neurotrophins are required for neuronal survival—about 50% of limb motor neurons die	retino-tectal map refined by neuronal activity, leading to ocular dominance columns

---

Summary to Chapter 12

- The processes involved in the development of the nervous system are superficially similar to those found in other developmental systems, but involve the acquisition of individual cell identities such that a vast number of highly specific cell–cell connections can be generated, often over long distances.
- The early neuroectoderm is first patterned into broad regions with distinct and diverse fates by the action of local signaling centers, before region-specific neuronal cell types appear.
- Signal molecules confer positional information on a responding field by forming a diffusion gradient that induces the expression of transcription factors that act as region or cell-type determinants.

- Presumptive neural tissue is specified early in development—during gastrulation in vertebrates, and when the dorso-ventral axis is patterned in *Drosophila*.
- Within the neuroectodermal tissue, the specification of cells that give rise to neural cells involves lateral inhibition.
- The further development of neurons from neuronal precursors involves both asymmetric cell divisions and cell–cell signaling.
- The patterning of different types of neurons within the vertebrate spinal cord is due to both ventral and dorsal signals.
- As they develop, neurons extend axons and dendrites. The axons are guided to their destination by growth cones at their tips.
- Guidance is due to the growth cone's response to attractive and repulsive signals, which may be diffusible or bound to the substratum.
- Gradients in such molecules can guide the axons to their destination, as in the retino-tectal system of vertebrates.
- The functioning of the nervous system depends on the establishment of specific synapses between axons and their targets.
- Specificity appears to be achieved by an initial overproduction of neurons that compete for targets, with many neurons dying during development. Refinement of synaptic connections involves further competition.
- Neural activity plays a major role in refining connections, such as those between the eye and the brain.

■ End of chapter questions

Long answer (concept questions)

1. How is positional information and positional value assigned during regionalization the vertebrate neuroectoderm?

2. The response of posterior midbrain to FGF8 signalling from the isthmus differ from that of the anterior hindbrain. How might this asymmetry be controlled?

3. What are distinguishing features of local signaling centers or organizers? How do they differ from the primary (Spemann) organizer?

4. What are rhombomeres? What are the similarities between rhombomeric segmentation and segmentation in the *Drosophila* larva?

5. Describe an experiment that demonstrates a homeotic transformation of rhombomeres. Outline the strategy used and compare this with other strategies that result in homeotic transformations – drawing your examples from both vertebrates and invertebrates.

6. The formation of motor neurons in the developing spinal cord depends on a gradient of Sonic hedgehog (Shh) protein. What is the source of this Shh gradient? What Class I and Class II homeodomain protein genes are active in response to the precise level of this gradient required for motor neuron progenitors (pMN) to become specified as motor neurons (MN)? What role do the LIM-homeodomain proteins play in motor neuron identity?

7. In *Drosophila*, the central nervous system is ventral, whereas in vertebrates, the central nervous system is dorsal. Despite these differences, what similarities exist that would indicate a common evolutionary origin?

8. Outline the mechanism of lateral inhibition in the proneural cluster of *Drosophila*, in which the expression of Delta in one cell leads to loss of Delta in a neighboring cell. Compare the *Drosophila* pathway to that in vertebrates.

9. How do Numb and Pins proteins collaborate to cause one daughter cell to become a neuron after (a) division of the neuroblast in the *Drosophila* embryo? (b) After division of the sensory organ precursor in the *Drosophila* adult?

10. What are radial glial cells? Include in your answer: the embryonic structure in which they are found, their location in that structure, and their roles in development of the nervous system (name two roles).

11. What is the evolutionary advantage of the distinct pattern of neurogenesis and neuronal migrations that distinguish the mammalian cortex from, e.g. the spinal cord.

12. Although the adult mammalian brain cannot in general produce new neurons, the dentate gyrus of the hippocampus is an exception. Relate this observation to the embryological origin of neurons in the neural tube.

13. Compare and contrast netrins, semaphorins, cadherins, and ephrins: in what way are netrins and cadherins similar in effect, and in what way are they different? In what way are semaphorins and ephrins similar, and in what way do they differ?

14. What is meant by the phrase 'the retina maps onto the tectum' (see Fig. 12.31)? How does the chemoaffinity hypothesis attempt to explain this phenomenon? How do the ephrins and their receptors fit into this hypothesis?

15. Describe the results of the experiment in which the optic nerve to a frog's eye is cut and the eye rotated by 180°. How is the subsequent response of the frog to a visual stimulus explained?

16. What is the role of neurotrophic factors, such as nerve growth factor, in refining the connections between neurons and their targets? Give a specific example of such action.

17. Draw a schematic diagram of a synapse, labeling the pre-synaptic cell and the post-synaptic cell and indicating the direction of signal transmission. Label the synaptic cleft and say why it prevents neurons from passing electrical signals directly to each other? How do neurons communicate across the synaptic cleft?

18. Describe the steps involved in the formation of the neuromuscular junction, starting with the agrin signal from the neuronal growth cone.

19. Cell death (apoptosis) in the developing nervous system is under the control of both intrinsic and extrinsic influences. Give examples of intrinsic (i.e. intracellular) and extrinsic (i.e. cell–cell signaling) proteins that control apoptosis in the nervous system during development.

20. Contrast the retinal inputs to the lateral geniculate nucleus in the mammalian brain with those to the tectum in the amphibian brain. What are ocular dominance columns, and why are they important? What is meant by 'cells that fire together, wire together'?

Multiple choice (factual recall questions)

NB There is only one correct answer to each question.

1. Local organizers of brain pattern

a) are of mesodermal origin
b) cease functioning after the neural plate closes
c) may signal bidirectionally
d) have high rates of cell proliferation

2. Neural crest cells contribute to which of the following?

a) skin melanocytes
b) mesenchymal components of the face and skull
c) peripheral nervous system
d) all of these

3. In the developing cerebral cortex

a) both excitatory and inhibitory neurons arise from the pallial ventricular zone
b) excitatory neurons have local connections whereas inhibitory neurons project over long distances
c) the outermost layer of neurons is born after layer 6
d) areal identity is conferred by incoming axons from the thalamus

4. The genes of the achaete–scute complex encode proteins of which type?

a) cell–cell adhesion molecules
b) cytoskeletal proteins
c) membrane transport proteins
d) transcription factors

5. The vertebrate central nervous system (brain and spinal cord) is derived from

a) cells of the neural tube
b) neural crest cells that migrate along the dorso-lateral route
c) neural crest cells that migrate through the anterior half of the somites
d) somitic mesoderm

6. What would be the consequence of reducing Delta expression in the developing vertebrate neural tube?

a) Excess neurons are produced, because reduced Delta prevents the inhibition of neurogenesis in neighboring cells.
b) Excess neurons are produced, because the reduction in Delta expression increases neurogenin expression.
c) Fewer neurons are produced, because Delta signaling is required for neurogenesis.
d) There is no change in neurogenesis unless Delta expression is eliminated completely.

7. Which of the following is a determined neuronal cell type?

a) premigratory neural crest cells
b) floor plate cells
c) commissural cells
d) roof plate cells

8. Cortical neurons migrate to their final positions in the cortex along

a) Schwann cells
b) commissural neurons
c) radial glia
d) roof-plate cells

9. Axons of spinal cord neurons from the lateral motor column in the chick embryo innervate

a) muscles of the body wall
b) muscles of the limb buds
c) muscles attached to the spine
d) muscles in the gut wall

10. Which of these is a cell-surface adhesion molecule that mainly attracts axons during their outgrowth?

a) neurotrophins
b) cadherins
c) netrins
d) semaphorins

11. Knock-outs of the mouse *netrin-1* gene have a phenotype similar to that of knock-outs of what other gene or gene product?

a) the DCC receptor
b) the Robo1 receptor
c) the semaphorin signaling molecule
d) the Slit signaling molecule

12. Neurons in the anterior (nasal) region of the right retina will project axons to

a) the anterior portion of the left tectum
b) the anterior portion of the right tectum
c) the posterior portion of the left tectum
d) the posterior portion of the right tectum

13. The synapses formed between motor neurons and muscles are called

a) myotomes
b) neuromuscular junctions
c) Nieuwkoop centers
d) retino-tectal projections

14. Schwann cells

a) are glial cells

b) are responsible for myelination of axons in the peripheral nervous system

c) are related to the oligodendrocytes of the central nervous system

d) are all of the above

15. The neurotransmitter used in the communication between motor neurons and muscles is

a) serotonin

b) dopamine

c) epinephrine

d) acetylcholine

Multiple choice answer key

1: c, 2: d, 3: a, 4: d, 5: a, 6: a, 7: c, 8: c, 9: b, 10: b, 11: a, 12: c, 13: b, 14: d, 15: d.

▦ General further reading

Kandel, E.R., Schwartz, J.H., Jessell, T.H., Siegelbaum, S.A., Hudspeth, A.J.: *Principles of Neural Science* (5th edn). New York: McGraw-Hill, 2013.

Kerszberg, M.: **Genes, neurons and codes: remarks on biological communication**. *BioEssays* 2003, **25**: 699–708.

Squire, L.R., Berg, D., Bloom, F.E., Du Lac, S., Ghosh, A., Spitzer, N.C.: *Fundamental Neuroscience* (4th edn). New York: Academic Press, 2013.

▦ Section further reading

12.1 Initial regionalization of the vertebrate brain involves signals from local organizers

Kiecker, C., Lumsden, A.: **The role of organizers in patterning the nervous system**. *Annu. Rev. Neurosci.* 2012, **35**: 347–367.

12.2 Local signaling centers pattern the brain along the antero-posterior axis

Broccoli, V., Boncinelli, E., Wurst, W.: **The caudal limit of *Otx2* expression positions the isthmc organizer**. *Nature*, 1999, **401**: 164–168.

Crossley, P.H., Martinez, S., Martin, G.R.: **Midbrain development induced by FGF8 in the chick embryo**. *Nature*, 1996, **380**: 66–68.

Houart, C., Westerfield, M., Wilson, S.W.: **A small population of anterior cells patterns the forebrain during zebrafish gastrulation**. *Nature*, 1998, **391**: 788–792.

Millet, S.: **A role for *Gbx2* in repression of *Otx2* and positioning the mid/hindbrain organizer**. *Nature*, 1999, **401**: 161–164.

Rhinn, M., Brand, M.: **The midbrain-hindbrain boundary organizer**. *Curr. Opin. Neurobiol.* 2001, **11**: 34–42.

Zeltser, L.M., Larsen, C.W., Lumsden, A.: **A new developmental compartment in the forebrain regulated by *Lunatic fringe***. *Nat. Neurosci.* 2001, **4**: 683–684.

12.3 The cerebral cortex is patterned by signals from the anterior neural ridge

Hamasaki, T., Leingaertner, A., Ringstedt, T., O'Leary, D.D.M.: **Emx2 regulates sizes and positioning of the primary sensory and motor areas in neocortex by direct specification of cortical progenitors**. *Neuron* 2004, **43**: 359–372.

Hanashima, C., Li, S.C., Shen, L., Lai, E., Fishell, G.: **Foxg1 suppresses early cortical cell fate**. *Science* 2004, **303**: 56–59.

Molyneaux, B.J., Arlotta, P., Menezes, J., Macklis, J.D.: **Neuronal subtype specification in the cerebral cortex**. *Nat. Rev. Neurosci.* 2007, **8**: 427–437.

O'Leary, D.D.M., Nakagawa, Y.: **Patterning centers, regulatory genes and extrinsic mechanisms controlling arealization of the neocortex**. *Curr. Opin. Neurobiol.* 2002, **12**: 14–25.

Rakic, P.: **Evolution of the necortex: a perspective from developmental biology**. *Nat. Rev. Neurosci.* 2009, **10**: 724–735.

Shimamura, K., Rubenstein, J.L.R.: **Inductive interactions direct early regionalization of the mouse forebrain**. *Development*, 1997, **124**: 2709–2718.

Toyoda, R., Assimacopoulos, S., Wilcoxon, J., Tayloy, A., Feldman, P., Suzuki-Hirano, A., Shimogori, T., Grove, E.A.: **FGF8 acts as a classic diffusible morphogen to pattern the neocortex**. *Development* 2010, **137**: 3439–3448.

12.4 The hindbrain is segmented into rhombomeres by boundaries of cell-lineage restriction

Fraser, S., Keynes, R.J., Lumsden, A.: **Segmentation in the chick embryo hindbrain is defined by cell lineage restrictions**. *Nature* 1990, **344**: 636–638.

Lumsden, A., Keynes, R.J.: **Segmental patterns of neuronal development in the chick hindbrain**. *Nature* 1989, **337**: 424–428.

Wizenmann, A., Lumsden, A.: **Segregation of rhombomeres by differential cell affinity**. *Mol. Cell Neurosci.* 1997, **9**: 448–459.

Xu, Q., Mellitzer, G., Robinson, V., Wilkinson, D.W.: ***In vivo* cell sorting in complementary segmental domains mediated by Eph receptors and ephrin ligands**. *Nature* 1999, **399**: 267–271.

12.5 Hox genes provide positional information in the developing hindbrain

Bell, E., Wingate, R., Lumsden, A.: **Homeotic transformation of rhombomere identity after localized Hoxb1 misexpression**. *Science* 1999, **284**: 2168–2171.

Gavalas, A., Krumlauf, R.: **Retinoid signalling and hindbrain patterning**. *Curr. Opin. Genet Dev.* 2000, **10**: 380–386.

Grammatopoulos, G.A., Bell, E., Toole, L., Lumsden, A., Tucker, A.S. **Homeotic transformation of branchial arch identity after Hoxa2 overexpression**. *Development* 2000, **127**: 5355–5365.

Krumlauf, R.: **Hox genes and pattern formation in the branchial region of the vertebrate head**. *Trends Genet.* 1993, **9**: 106–112.

Rijli, F.M., Mark, M., Lakkaraju, S., Dierich, A., Dolle, P., Chambon, P.: **A homeotic transformation is generated in the rostral branchial region of the head by disruption of Hoxa2, which acts as a selector gene**. *Cell* 1993, **75**: 1333–1349.

Wassef, M.A., Chomette, D., Pouilhe, M., Stedman, A., Havis, E., Trin-Dihn-Desmarquet, C., Schneider-Maunoury, S., Gilardi-Hebenstreit, P., Charnay, P., Ghislain, J.: **Rostral hindbrain patterning involves the direct activation of a Krox20 transcriptional enhancer by Hox/Pbx and Meis activators.** *Development* 2008, **135**: 3369–3378.

12.6 The pattern of differentiation of cells along the dorso-ventral axis of the spinal cord depends on ventral and dorsal signals

Le Douarin, N., Creuzet, S., Couly, G., Dupin, E.: **Neural crest plasticity and its limits.** *Development* 2004, **131**: 4637–4650.

Keynes, R., Lumsden, A.: **Segmentation and the origin of regional diversity in the vertebrate central nervous system.** *Neuron* 1990, **4**: 1–9.

12.7 Neuronal subtypes in the ventral spinal cord are specified by the ventral to dorsal gradient of Shh

Briscoe, J., Pierani A., Jessell, T.M., Ericson, J.: **A homeodomain protein code specifies progenitor cell identity and neuronal fate in the ventral neural tube.** *Cell* 2000, **101**: 435–445.

Jessell, T.M.: **Neuronal specification in the spinal cord: inductive signals and transcriptional controls.** *Nat. Rev. Genet.* 2000, **1**: 20–29.

Megason, S.G., McMahon, A.P.: **A mitogen gradient of dorsal midline Wnts organizes growth in the CNS.** *Development* 2002, **129**: 2087–2098.

12.8 Spinal cord motor neurons at different dorso-ventral positions project to different trunk and limb muscles

Chamberlain, C.E.Jeong, J., Guo, C., Allen, B.L., McMahon, A.P.: **Notochord-derived Shh concentrates in close association with the apically positioned basal body in neural target cells and forms a dynamic gradient during neural patterning.** *Development* 2008, **135**: 1097–1106.

Dessaud, E., McMahon, A.P., Briscoe, J.: **Pattern formation in the vertebrate neural tube: a sonic hedgehog morphogen-regulated transcriptional network.** *Development* 2008, **135**: 2489–2503.

Dessaud, E., Yang, L.L., Hill, K., Cox, B., Ulloa, F., Ribeiro, A., Mynett, A., Novitch, B.G., Briscoe, J.: **Interpretation of the sonic hedgehog morphogen gradient by a temporal adaptation mechanism.** *Nature* 2007, **450**: 717–720.

Stamataki, D., Ulloa, F., Tsoni, S.V., Mynett, A., Briscoe, J.: **A gradient of Gli activity mediates graded Sonic Hedgehog signaling in the neural tube.** *Genes Dev.* 2005, **19**: 626–641.-

12.9 Antero-posterior pattern in the spinal cord is determined in response to secreted signals from the node and adjacent mesoderm

Dasen, J.S., Jessell, T.M.: **Hox networks and the origins of motor neuron diversity.** *Curr. Top. Dev. Biol.* 2009, **88**: 169–200.

Ensini M., Tsuchida T.N., Belting, H.G., Jessell, T.M.: **The control of rostrocaudal pattern in the developing spinal cord: specification of motor neuron subtype identity is initiated by signals from paraxial mesoderm.** *Development* 1998, **125**: 969–982.

Sockanathan, S., Perlmann, T., Jessell, T.M.: **Retinoid receptor signaling in postmitotic motor neurons regulates rostrocaudal**

positional identity and axonal projection pattern. *Neuron* 2003, **40**: 97–111.

12.10 Neurons in *Drosophila* arise from proneural clusters

Cornell, R.A., Ohlen, T.V.: **Vnd/nkx, ind/gsh, and msh/msx: conserved regulators of dorsoventral neural patterning?** *Curr. Opin. Neurobiol.* 2000, **10**: 63–71.

Skeath, J.B.: **At the nexus between pattern formation and cell-type specification: the generation of individual neuroblast fates in the *Drosophila* embryonic central nervous system.** *BioEssays* 1999, **21**: 922–931.

Weiss, J.B., Von Ohlen, T., Mellerick, D.M., Dressler, G., Doe, C.Q., Scott, M.P.: **Dorsoventral patterning in the *Drosophila* central nervous system: the intermediate neuroblasts defective homeobox gene specifies intermediate column identity.** *Genes Dev.* 1998, **12**: 3591–3602.

12.11 The development of neurons in *Drosophila* involves asymmetric cell divisions and timed changes in gene expression

Brody, T., Odenwald, W.: **Regulation of temporal identities during *Drosophila* neuroblast lineage development.** *Curr. Opin. Cell Biol.* 2005, **17**: 672–675.

Grosskortenhaus, R., Pearson, B.J., Marusich, A., Doe, C.Q.: **Regulation of temporal identity transitions in *Drosophila* neuroblasts.** *Dev. Cell* 2005, **8**: 193–202.

Jan, Y-N., Jan, L.Y.: **Polarity in cell division: what frames thy fearful asymmetry?** *Cell* 2000, **100**: 599–602.

Karcavich, R.E.: **Generating neuronal diversity in the *Drosophila* central nervous system: a view from the ganglion mother cells.** *Dev. Dyn.* 2005, **232**: 609–616.

Box 12A Specification of the sensory organs of adult *Drosophila*

Gómez-Skarmeta, J.L., Campuzano, S., Modolell, J.: **Half a century of neural prepatterning: the story of a few bristles and many genes.** *Nat. Rev. Neurosci.* 2003, **4**: 587–598.

Knoblich, J.A.: **Asymmetric cell division during animal development.** *Nat. Rev. Mol. Cell. Biol.* 2001, **2**: 11–20.

12.12 The production of vertebrate neurons involves lateral inhibition, as in *Drosophila*

Chitins, A., Henrique, D., Lewis, J., Ish-Horowitcz, D., Kintner, C.: **Primary neurogenesis in *Xenopus* embryos regulated by a homologue of the *Drosophila* neurogenic gene *Delta*.** *Nature* 1995, **375**: 761–766.

Ma, Q., Kintner, C., Anderson, D.J.: **Identification of *neurogenin*, a vertebrate neuronal determination gene.** *Cell* 1996, **87**: 43–52.

12.13 Neurons are formed in the proliferative zone of the neural tube and migrate outwards

D'Arcangelo, G., Curran, T.: ***Reeler*: new tales on an old mutant mouse.** *BioEssays* 1998, **20**: 235–244.

Gage, F.H.: **Mammalian neural stem cells.** *Science* 2000, **287**: 1433–1438.

Hansen, D.V., Lui, J.H., Parker, P.R.L., Kriegstein, A.R.: **Neurogenic radial glia in the outer subventricular zone of human neocortex.** *Nature* 2010, **464**: 554–561.

Kriegstein, A., Alvarez-Buylla, A.: **The glial nature of embryonic and adult neural stem cells.** *Annu. Rev. Neurosci.* 2009, **32**: 149–184.

Kriegstein, A.R., Noctor, S., Martinez-Cerdano, V.: **Patterns of neural stem and progenitor cell division may underlie evolutionary cortical expansion.** *Nat. Rev. Neurosci.* 2006, **7**: 883–890.

Lancaster, M.A., Knoblich, J.A.: **Spindle orientation in mammalian cerebral cortical development.** *Curr. Opin. Neurobiol.* 2012, **22**: 737–746.

Lyuksyutova, A.I., Lu, C.C., Milanesio, N., King, L.A., Guo, N., Wang, Y., Nathans, J., Tessier-Lavigne, M., Zou, Y.: **Anterior-posterior guidance of commissural axons by Wnt-frizzled signaling.** *Science* 2003, **302**: 1984–1988.

Box 12B Timing the birth of cortical neurons

Rakic, P.: **Neurons in rhesus monkey visual cortex: systematic relation between time of origin and eventual disposition.** *Science* 1974, **183**: 425–427.

12.14 Many cortical interneurons migrate tangentially

Anderson, S.A., Eisenstat, D.D., Shi, L., Rubenstein, J.L.R.: **Interneuron migration from basal forebrain to neocortex: dependence on *Dlx* genes.** *Science* 1997, **278**: 474–476.

Flames, N., Long, J.E., Garratt, A.N., Fischer, T.M. Gassmann, M., Birchmeier, C., Lai, C., Rubenstein, J.L.R., Marin, O.: **Short- and long-range attraction of cortical GABAergic interneurons by Neuregulin-1.** *Neuron* 2004, **44**: 251–261.

De Marco Garcia, N.V., Karayannis, T., Fishell, G.: **Neuronal activity is required for the development of specific cortical interneuron subtypes.** *Nature* 2011, **472**: 351–355.

Marin, O., Rubenstein, J.L.R.: **A long, remarkable journey: tangential migration in the telencephalon.** *Nat. Rev. Neurosci.* 2001, **2**: 780–790.

Wonders, C.P., Anderson, S.A.: **The origin and specification of cortical interneurons.** *Nat. Rev. Neurosci.* 2006, **7**: 687–696.

12.15 The growth cone controls the path taken by a growing axon

Baudet, M.L., Bellon, A., Holt, C.E.: **Role of microRNAs in Semaphorin function and neural circuit formation.** *Semin. Cell Dev. Biol.* 2013, **24**: 146–155.

Drees, F., Gertler, F.B.: **Ena/VASP: proteins at the tip of the nervous system.** *Curr. Opin. Neurobiol.* 2008, **18**: 53–59.

Dudanova, I., Klein, R.: **Integration of guidance cues: parallel signaling and crosstalk.** *Trends Neurosci.* 2013, **36**: 295–304.

Lowery, L.A., Van Vactor, D.: **The trip of the tip: understanding the growth cone machinery.** *Nat. Rev. Mol. Cell. Biol.* 2009, **10**: 332–343.

Tear, G.: **Neuronal guidance: a genetic perspective.** *Trends Genet.* 1999, **15**: 113–118.

Box 12C The development of the neural circuit for the knee-jerk reflex

Arber, S.: **Motor circuits in action: specification, connectivity, and function.** *Neuron* 2012, **74**: 975–989.

Brierley, D.J., Rathore, K., Vijayraghavan, K., Williams, D.: **Dendritic targeting in the leg neuropil of *Drosophila*: the role**

of midline signalling molecules in generating a myotopic map. *PLoS Biol* 2009, **7**: e100019.

Li, W-C., Cooke, T., Sautois, B., Soffe, S.R., Borisyuk, R., Roberts, A.: **Axon and dendrite geography predict the specificity of synaptic connections in a functioning spinal cord network.** *Neural Dev.* 2007, **2**: 17.

Surmeli, G., Akay, T., Ippolito, G.C., Tucker, P.W., Jessell, T.M.: **Patterns of spinal sensory-motor connectivity prescribed by a dorsoventral positional template.** *Cell* 2011, **147**: 653–665.

12.16 Motor neuron axons in the chick limb are guided by ephrin-Eph interactions

Lance-Jones, C., Landmesser, L.: **Pathway selection by embryonic chick motoneurons in an experimentally altered environment.** *Proc. R. Soc. Lond. B* 1981, **214**: 19–52.

Tosney, K.W, Hotary, K.B., Lance-Jones, C.: **Specifying the target identity of motoneurons.** *BioEssays* 1995, **17**: 379–382.

Xu, N-J., Henkemeyer, M.: **Ephrin reverse signaling in axon guidance and synaptogenesis.** *Semin. Cell Dev. Biol.* 2012, **23**: 58–64.

12.17 Axons crossing the midline are both attracted and repelled

Chédotal, A.: **Further tales of the midline.** *Curr. Opin. Neurobiol.* 2011, **21**: 68–75.

Giger, R.J., Kolodkin, A.L.: **Silencing the siren: guidance cue hierarchies at the CNS midline.** *Cell* 2001, **105**: 1–4.

Simpson, J.H., Bland, K.S., Fetter, R.D., Goodman, C.S.: **Short-range and long-range guidance by Slit and its Robo receptors: a combinatorial code of Robo receptors controls lateral position.** *Cell* 2000, **103**: 1019–1032.

Williams, S.E., Mason, C.A., Herrera, E.: **The optic chiasm as a midline choice point.** *Curr. Opin. Neurobiol.* 2004, **14**: 51–60.

Zou, Y., Lyuksyutova, A.I.: **Morphogens as conserved axon guidance cues.** *Curr. Opin. Neurobiol.* 2007, **17**: 22–28.

12.18 Neurons from the retina make ordered connections with visual centers in the brain

Hansen, M.J., Dallal, G.E., Flanagan, J.G.: **Retinoic axon response to ephrin-As shows a graded concentration-dependent transition from growth promotion to inhibition.** *Neuron* 2004, **42**: 707–730.

Klein, R.: **Eph/ephrin signaling in morphogenesis, neural development and plasticity.** *Curr. Opin. Cell Biol.* 2004, **16**: 580–589.

Löschinger, J., Weth, F., Bonhoeffer, F.: **Reading of concentration gradients by axonal growth cones.** *Phil. Trans. R. Soc. Lond. B* 2000, **355**: 971–982.

McLaughlin, T., Hindges, R., O'Leary, D.D.: **Regulation of axial patterning of the retina and its topographic mapping in the brain.** *Curr. Opin. Neurobiol.* 2003, **13**: 57–69.

Petros, T.J., Shrestha, B.R., Mason, C.: **Specificity and sufficiency of EphB1 in driving the ipsilateral retinal projection.** *J. Neurosci.* 2009, **29**: 3463–3474.

Reber, M., Bursold, P., Lemke, G.: **A relative signalling model for the formation of a topographic neural map.** *Nature* 2004, **431**: 847–853.

Suetterlin, P., Marler, K.M., Drescher, U.: **Axonal ephrinA/EphA interactions, and the emergence of order in topographic projections**. *Semin. Cell Dev. Biol.* 2012, **23**: 1–6.

12.19 Synapse formation involves reciprocal interactions

Buffelli, M., Burgess, R.W., Feng, G., Lobe, C.G., Lichtman, J.W., Sanes, J.R.: **Genetic evidence that relative synaptic efficacy biases the outcome of synaptic competition**. *Nature* 2003, **424**: 430–434.

Clarke, L.E., Barres, B.A.: **Emerging roles of astrocytes in neural circuit development**. *Nat. Rev. Neurosci.* 2013, **14**: 311–321.

Goda, Y., Davis, G.W.: **Mechanisms of synapse assembly and disassembly**. *Neuron* 2003, **40**: 243–264.

Hua, J.Y, Smith, S.J.: **Neural activity and the dynamics of central nervous-system development**. *Nat. Neurosci.* 2004, **7**: 327–332.

Jan, Y-N., Jan, L.Y.: **The control of dendritic development**. *Neuron* 2003, **40**: 229–242.

Jin, Y., Garner, C.C.: **Molecular mechanisms of presynaptic differentiation**. *Annu. Rev. Cell Dev. Biol.* 2008, **24**: 237–262.

Katz, L.C., Constantine-Paton, M.: **Relationships between segregated afferents and postsynaptic neurones in the optic tectum of three-eyed frogs**. *J. Neurosci.* 1988, **8**: 3160–3180.

Kummer, T.T., Misgled, T., Sanes, J.R.: **Assembly of the postsynaptic membrane at the neuromuscular junction**. *Curr. Opin. Neurobiol.* 2006, **16**: 74–82.

Levinson, J.N., El-Husseini, A.: **Building excitatory and inhibitory synapses: balancing neuroligin partnerships**. *Neuron* 2005, **48**: 171–174.

Li, Z., Sheng, M.: **Some assembly required: the development of neuronal synapses**. *Nat. Rev. Mol. Cell Biol.* 2003, **4**: 833–841.

Pasterkamp, R.J.: **Getting neural circuits into shape with semaphorins**. *Nat. Rev. Neurosci.* 2012, **13**: 605–618.

Schmucker, D., Clemens, J.C., Shu, H., Worby, C.A., Xiao, J., Muda, M., Dixon, J.E., Zipursky, S.L.: ***Drosophila* Dscam is an axon guidance receptor exhibiting extraordinary molecular diversity**. *Cell* 2000, **101**: 671–684.

Box 12D Autism: a developmental disorder that involves synapse dysfunction

Ebert, D.H., Greenberg, M.E.: **Activity-dependent neuronal signalling and autism spectrum disorder**. *Nature* 2013, **493**: 327–337.

Geschwind, D.H.: **Autism: many genes, common pathways?** *Cell* 2008, **135**: 391–395.

Grabrucker, A.M.: **Environmental factors in autism**. *Front. Psychiatry* 2012, **3**: 118.

Guy, J.G., Gan, J., Selfridge, J., Cobb, S., Bird, A.: **Reversal of neurological defects in a mouse model of Rett syndrome**. *Science* 2007, **315**: 1143–1147.

Sidorov, M.S., Auerbach, B.D., Bear, M.F.: **Fragile X mental retardation protein and synaptic plasticity**. *Mol. Brain* 2013, **6**: 15.

Sudhof, T.: **Neuroligins and neurexins link synaptic function to cognitive disease**. *Nature* 2008, **455**: 903–911.

Walsh, C.A., Morrow, E.M., Rubenstein, J.L.R.: **Autism and brain development**. *Cell* 2008, **135**: 396–400.

12.20 Many motor neurons die during normal development

Oppenheim, R.W.: **Cell death during development of the nervous system**. *Annu. Rev. Neurosci.* 1991, **14**: 453–501.

Pettmann, B., Henderson, C.E.: **Neuronal cell death**. *Neuron* 1998, **20**: 633–647.

12.21 Neuronal cell death and survival involve both intrinsic and extrinsic factors

Burden, S.J.: **Wnts as retrograde signals for axon and growth cone differentiation**. *Cell* 2000, **100**: 495–497.

Davies, A.M.: **Neurotrophic factors. Switching neurotrophin dependence**. *Curr. Biol.* 1994, **4**: 273–276.

Harrington, A.W., Ginty, D.D.: **Long-distance retrograde neurotrophic factor signalling in neurons**. *Nat. Rev. Neurosci.* 2013, **14**: 177–187.

Jiang, Y., de Bruin, A., Caldas, H., Fangusaro, J., Hayes, J., Conway, E.M., Robinson, M.L., Altura, R.A.: **Essential role for survivin in early brain development**. *J. Neurosci.* 2005, **25**: 6962–6970.

Park, H., Poo, M.M.: **Neurotrophin regulation of neural circuit development and function**. *Nat. Rev. Neurosci.* 2013, **14**: 7–23.

Serafini, T.: **Finding a partner in a crowd: neuronal diversity and synaptogenesis**. *Cell* 1999, **98**: 133–136.

12.22 The map from eye to brain is refined by neural activity

Del Rio, T., Feller, M.B.: **Early retinal activity and visual circuit development**. *Neuron* 2006, **52**: 221–222.

Huberman, A.D.: **Mechanisms of eye-specific visual circuit development**. *Curr. Opin. Neurobiol.* 2007, **7**: 73–80.

Katz, L.C., Crowley, J.C.: **Development of cortical circuits: lessons from ocular dominance columns**. *Nat. Rev. Neurosci.* 2002, **3**: 34–42.

Katz, L.C., Shatz, C.J.: **Synaptic activity and the construction of cortical circuits**. *Science* 1996, **274**: 1133–1138.

13

Growth, post-embryonic development and regeneration

- Growth
- Molting and metamorphosis
- Regeneration
- Aging and senescence

Patterning of the embryo occurs on a small scale and is followed by growth. The control of growth, and therefore of size, is a key problem in development, involving at its core the control of cell proliferation. Such control is crucial in adult life in preventing cancer—the result of uncontrolled cell proliferation. The genetic programs that determine why different species grow to different characteristic sizes are, however, still a complete mystery. In this chapter we look at the balance between intrinsic programs of cell proliferation during embryonic development, and growth, usually at later stages, which is stimulated by extrinsic signals such as circulating hormones. One important aspect of post-embryonic development in many animals is metamorphosis, in which the form of the animal is completely changed between the larval stage and the adult. In this chapter we shall look at some aspects of metamorphosis in insects and in the vertebrate amphibians. Another intriguing feature of some animals is their capacity as adults to regrow various body parts—such as tails, limbs, and even the heart—when injured. Using examples from amphibians, insects, and the zebrafish, we shall examine the adult regeneration process focusing on the source of cells for regeneration and how patterning and growth of the regenerated structures is controlled. The final stage of post-embryonic development in most organisms is aging, which results from the accumulation of cellular damage that outstrips the ability of the body to repair itself. This age-related decline in repair—senescence—is genetically programmed.

Development does not stop once the embryonic phase is complete. Most, but by no means all, of the growth in animals and plants occurs in the post-embryonic period, when the basic form and pattern of the organism has already been established. Growth is a central aspect of all developing systems, determining the final size and shape of the organism and its parts. In many animals, the embryonic stage is immediately succeeded by a free-living larval or immature adult stage, in which growth occurs. In others, such as mammals, considerable growth occurs during a late embryonic or fetal period, while the embryo is still dependent on maternal resources. Growth then continues after birth. In animals with a larval stage, the larva not only grows in size

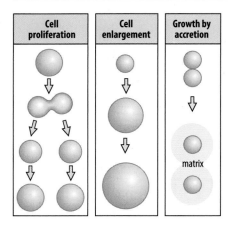

Fig. 13.1 The three main mechanisms for growth in vertebrates. The most common mechanism is cell proliferation—cell growth followed by division. A second mechanism is cell enlargement, in which cells increase their size without dividing. The third mechanism is to increase size by accretionary growth, such as secretion of a matrix.

but eventually undergoes **metamorphosis**, in which it is transformed into the adult form. Metamorphosis often involves a radical change in form and the development of new organs.

A related phenomenon to growth is that of **regeneration**. This is the ability of the fully developed juvenile or adult organism to replace tissues, organs, and even most of the body, by the regrowth or repatterning of somatic tissue. All animals can repair damaged tissues to some extent, but among vertebrates, the ability to regrow lost limbs, for example, or regenerate injured organs such as the heart, is much more restricted. In this chapter we shall look at a few classical examples of regeneration—the regrowth of amphibian limbs, insect appendages, and the zebrafish heart. The regenerative powers of mammals are much more limited. Like all vertebrates, mammals renew some tissues throughout their lives from stem cells (see Chapter 8) and can regenerate liver tissue if not too much is removed (see Section 13.6). But they cannot regrow lost limbs or hearts. Why animals such as amphibians and fishes, with much the same basic anatomy as mammals, are able to do so is an intriguing question. Finally, we also look at the last stage of post-embryonic development—the phenomenon of aging.

Unlike the situation in animals, where the embryo is essentially a miniature version of the free-living larva or adult, plant embryos bear little resemblance to the mature plant. Most of the adult plant structures are generated after germination by the shoot or root meristems, which have a capacity for continual growth. Some aspects of plant growth are discussed in Chapter 7.

We will start this chapter by considering first what we mean by 'growth', and then look at the roles of intrinsic growth programs and of extrinsic factors, such as growth factors and hormones, in controlling both embryonic and post-embryonic growth. In this part of the chapter we shall also touch on the disturbances of growth control that lead to cancer.

Growth

Growth is defined as an increase in the mass or overall size of a tissue or organism; this increase may result from cell proliferation, cell enlargement without division, or by accretion of extracellular material, such as bone matrix or even water (Fig. 13.1). In animals, the basic body pattern is laid down when the embryo is still very small, of the order of millimeters in size. The intrinsic growth program—that is, how large an organism or an individual organ can grow, and how it responds to growth stimulants such as hormones—may also be specified at an early stage in development, as we shall see later in relation to finger length. Overall growth of the organism mainly occurs in the post-embryonic period after basic patterning has been established; there are, however, many examples where earlier organogenesis involves localized growth, as in the vertebrate limb bud, where the number of cell divisions may act as a timing mechanism to provide positional information, and in the vertebrate lung (see Chapter 11), where cell proliferation mediates morphogenesis. Localized growth also underlies some of the size differences in the brain between species. In the embryonic mouse brain, for example, the neuronal progenitor cells that give rise to the neocortex divide only around 11 times before differentiating into neurons, whereas in primates they undergo a greater number of cell divisions before differentiating, expanding this cell population and so enabling it to give rise to the much more extensive primate neocortex. In this way, genetically programmed differences in the rate of growth in different parts of the body, or at different times, during early development profoundly affect the shape of organs and the organism.

Growth will determine the size of the different organs and the overall size of the body of the organism. The size of the organs must also fit the size of the body, and the same principles are involved in determining both organ size and overall body size. One important question is how do individual organs or an organism know when the correct size has been reached and therefore know when to stop growing. The enormous differences in body size in different mammals result mainly from differences in cell number, not in cell size. A 70-kilogram person is composed of some 10^{13} cells, and has a body mass about 3000 times greater than that of a mouse, which is composed of about 3×10^9 cells. In principle, therefore, overall body size could be controlled by counting cell divisions and cell number. But there is also evidence that body size can be determined by monitoring overall dimensions, rather than the absolute number of cells or cell divisions. In species in which the cells are either larger or smaller than normal, as a result of differences in the number of chromosomes, the overall body size is not affected. Thus, for example, some naturally tetraploid salamanders with cells twice as large as diploid cells grow to the same size as their diploid relatives but have only half the number of cells.

Growth depends on both intrinsic and extrinsic factors. Although considerable progress has been made in identifying extrinsic factors, for example, the growth factors and hormones that control growth, the genetic basis for differences in size are completely unknown. A classic illustration of an intrinsic growth program, which is genetically determined, comes from the limb and the results of grafting limb buds between large and small species of salamanders of the genus *Ambystoma*. A limb bud from the larger species grafted to the smaller species initially grows slowly, but eventually ends up at its normal size, which is much larger than any of the limbs of the host animal (Fig. 13.2). This indicates that whatever the circulatory factors, such as hormones, that influence growth, the intrinsic response of a tissue to them is crucial.

13.1 Tissues can grow by cell proliferation, cell enlargement, or accretion

Although growth often occurs through **cell proliferation**, which increases the number of cells by mitotic cell division, this is only one of three main ways by which growth in size of an organ or body structure can occur (see Fig. 13.1). There is also a significant amount of programmed cell death in many growing tissues, and the overall growth rate is therefore determined by the balance between the rates of cell death and cell proliferation.

A second way in which growth occurs is through **cell enlargement**—that is, by individual cells increasing their mass and getting bigger. This is the case in *Drosophila* larval growth—except for the imaginal discs, in which cells proliferate during the larval stages. And differences in size between closely related species of *Drosophila* are partly the result of the larger species having larger cells. In mammals, skeletal and heart muscle cells and neurons divide very rarely, if at all, once differentiated, although they do increase in size. Neurons grow by the extension and growth of axons and dendrites, whereas muscle growth involves an increase in mass, as well as the fusion of satellite cells to pre-existing muscle fibers, providing additional nuclei to support the large cell mass. Much growth involves a combination of cell proliferation and cell enlargement. For example, the cells in the lens of the eye are produced by cell division from a proliferative zone for an extended period, whereas their differentiation involves considerable enlargement.

The third growth mechanism, **accretionary growth**, involves an increase in the volume of the extracellular space, which is achieved by the secretion of large quantities of extracellular matrix by cells. Accretionary growth occurs in both cartilage and bone, where most of the tissue mass is extracellular.

Intracellular signaling pathways controlling growth in cell size and cell proliferation have been discovered (Fig. 13.3). The TOR (Target Of Rapamycin) pathway, whose activity is regulated by nutrient supply, positively controls cell size, whereas the Hippo signaling pathway controls cell number by suppressing proliferation and enhancing cell death. We will come across both pathways later in this chapter and look at the Hippo signaling pathway in more detail.

Fig. 13.2 The size of limbs is genetically programmed in salamanders. An embryonic limb bud from a large species of salamander, *Ambystoma tigrinum*, grafted to the embryo of a smaller species, *Ambystoma punctatum*, grows much larger than the host limbs—to the size it would have grown in *Ambystoma tigrinum*.

Photograph reproduced with permission from Harrison, R.G.: Organization and Development of the Embryo. *New Haven: Yale University Press, 1969.*

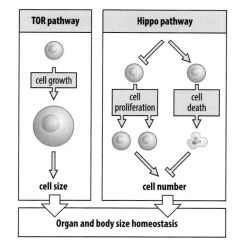

Fig. 13.3 Intracellular pathways controlling cell size and cell number together can determine tissue growth. TOR (target of rapamycin) is a serine-threonine protein kinase that regulates organ size by stimulating cell growth, thereby increasing cell size. Its general function is as a central control hub in cells, integrating information on the availability of energy and nutrients and responding to growth factor signaling. Hippo is a serine-threonine protein kinase that acts in a signaling pathway (outlined in Box 13A) that controls organ size by restricting cell number through the inhibition of proliferation and induction of apoptosis.

Tumaneng, et al.: **Organ Size Control by Hippo and TOR Pathways**. *Curr Biol. 2012, 22, R368-R370.*

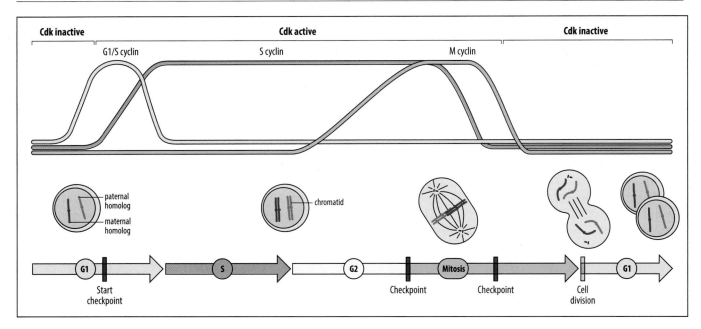

Fig. 13.4 Progression through the cell cycle is regulated by the levels of cyclins. Different cyclins (G1/S, S and M) regulate progression through different phases of the cell cycle. The expression of each cyclin rises and falls throughout the cell cycle. As cyclin protein levels increase, the cyclins bind to their corresponding inactive cyclin-dependent kinases (Cdks) (not shown) forming active enzymes. These active kinases are ultimately responsible for driving cell-cycle events such as the initiation of DNA replication (S phase) and the entry into mitosis (M phase). There are a number of different checkpoints in the cell cycle (red bars) that ensure that the cell does not progress to the next stage until the previous stage has been completed successfully. At the end of mitosis, existing cyclins are targeted to the proteasome and destroyed, so that the new cells formed at cell division are re-set to early G1 phase.

Adapted from D.O. Morgan, The Cell Cycle. Oxford University Press, 2006.

13.2 Cell proliferation is controlled by regulating entry into the cell cycle

When a eukaryotic cell duplicates itself by mitosis it goes through a fixed sequence of phases called the **cell cycle**, which were described in Box 1B. The cell grows in size (phases G_1 and G_2), the DNA is replicated (S phase), and the replicated chromosomes undergo mitosis (M phase) and become segregated into two daughter nuclei. The cell then divides (cytokinesis) to give two daughter cells. The cell cycle shown in Box 1B is the standard cell cycle of a dividing somatic animal cell. At different stages of development, or in specialized cell types, different phases of the cell cycle are absent, or the length of time taken to complete them varies. During cleavage of a fertilized amphibian egg, for example, growth phases G_1 and G_2 are virtually absent.

The timing of events in the cell cycle is controlled by a set of 'central' timing mechanisms (Fig. 13.4). Proteins known as **cyclins** control the passage through key transition points in the cycle. Cyclin concentrations oscillate during the cell cycle, and these oscillations correlate with transitions from one phase of the cycle to the next. Cyclins act by forming complexes with, and helping to activate, protein kinases known as **cyclin-dependent kinases (Cdks)**. These kinases phosphorylate proteins that trigger the events of each phase, such as DNA replication in S phase or mitosis in M phase.

Once a cell has entered the cycle and progressed beyond a point commonly known as 'Start', which occurs in mid- to late G_1, it will continue through and complete the cycle without needing any further external signal. Transitions into successive phases are marked by **cell-cycle checkpoints** at which the cell monitors progress to ensure, for example, that an appropriate size has been reached, that DNA replication is complete, and that any DNA damage has been repaired. If such criteria are not met, progress into the next stage is delayed until all the necessary processes have been completed. If the cell has suffered some damage that cannot be repaired, the cell cycle will be arrested and the cell will usually undergo apoptosis. These mechanisms are intrinsic to all normal eukaryotic cells and, for example, prevent them from continuing to divide with damaged DNA.

Extracellular signals, such as those provided by growth factors, control cell proliferation by stimulating entry into the cell cycle. Studies of animal cells in culture show that growth factors are essential for cells to multiply, with the particular growth factor or factors required depending on the cell type. When somatic cells are not stimulated to proliferate they are usually in a state known as G_0, into which they withdraw after mitosis (see Box 1B). Growth factors enable the cell to proceed out of G_0 and progress through the cell cycle. Most cells in adult organisms are not actively proliferating and so need an external signal before they can re-enter the cell cycle. Many of the extracellular signal molecules we met in earlier chapters in their roles as patterning agents, such as fibroblast growth factors (FGF) and members of the TGF-β family, were initially discovered through their ability to stimulate cell proliferation. They can also act in this way during embryonic and fetal growth, and in the control of cell proliferation in the adult. Other growth factors have specialized roles in directing the proliferation of particular tissues. One example is erythropoietin, which promotes the proliferation of red blood cell precursors (see Section 8.8).

Cells must receive signals, such as growth factors, not only for them to divide, but also simply to survive. In the absence of all such stimulation, cells commit suicide by apoptosis, as a result of activation of an internal death program (see Box 6A).

13.3 Cell division in early development can be controlled by an intrinsic developmental program

Embryonic cells proliferate much more freely compared to the cells of adult organisms, and in many organisms, the pattern of cell division in the earliest embryonic stages is controlled by a cell-autonomous developmental program that does not depend on stimuli such as growth factors. One well-understood example comes from *Drosophila*, in which the early cell cycles are controlled by the proteins that pattern the embryo; these proteins exert their effects by interacting with components of the cell-cycle control system.

The first cell cycles in the *Drosophila* embryo are represented by rapid and synchronous nuclear divisions, without any accompanying cell divisions, which create the syncytial blastoderm (see Fig. 2.2). There are virtually no G phases, just alternations of DNA synthesis (S phase) and mitosis. But at cycle 14, there is a major transition to a different type of cell cycle, a transition similar to the mid-blastula transition in frogs (see Section 4.8). A well-defined G_2 phase is seen in cycle 14, and the blastoderm becomes cellularized. A G_1 phase is seen at the 16th cycle. After the 17th or 18th cycle, cells in the epidermis and mesoderm stop dividing and differentiate. The cessation of cell proliferation is caused by the exhaustion of maternal cyclin E reserves originally laid down in the egg.

At the 14th cell cycle, distinct spatial domains with different cell-cycle times can be seen in the *Drosophila* blastoderm (Fig. 13.5). This patterning of cell cycles is produced by a change in the synthesis and distribution of a protein phosphatase called String, which exerts control on the cell cycle by dephosphorylating and activating a cyclin-dependent kinase and controlling the transition from G_2 to M. In the fertilized egg, the String protein is of maternal origin and is uniformly distributed. It therefore produces a synchronized pattern of nuclear division throughout the embryo. After cycle 13, the maternal String protein disappears and the zygotic expression of String protein becomes the controlling factor for entry into M phase.

Zygotic *string* gene transcription occurs in a complex spatial and temporal pattern. Only cells in which the *string* gene is expressed enter mitosis. This results in variation in the rate of cell division in different parts of the blastoderm, which ensures that the correct number of cells is generated in different tissues. The pattern of zygotic *string* gene expression is controlled by transcription factors encoded by the early patterning genes, such as the gap and pair-rule genes and those patterning the dorso-ventral axis.

One exception to the rule that expression of *string* leads to cell proliferation is the presumptive mesoderm, which is the first domain in which *string* is expressed but the tenth to start cell division. The delay in cell division is due to the expression of the

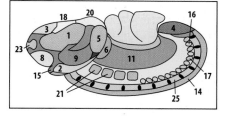

Fig. 13.5 Domains of mitosis in the *Drosophila* blastoderm. Areas composed of cells that divide at the same time are indicated by the various colors, and the numbers indicate the order in which these domains undergo mitosis at the 14th cell cycle, when zygotic String protein is first expressed. The schematic illustrates a lateral view of the embryo corresponding to a stage after mesoderm has been internalized and segmentation has begun (segments are indicated by the black marks along the lower surface). Anterior is to the left and dorsal is up. The gray region on the dorsal surface is the amnioserosa. The internalized mesoderm and some other domains are not visible in this view.

Illustration after Edgar, B.A., et al.: ***Transcriptional regulation of string (cdc25): a link between developmental programming and the cell cycle.*** Development *1994,* ***120****: 3131–3143.*

gene *tribbles* in this region, as Tribbles protein degrades String. The delay is necessary to allow ventral furrow formation and thus mesoderm invagination (see Section 9.9), because cell division inhibits ventral furrow formation. The prospective mesodermal cells start to proliferate after they have been internalized. Thus, the early cell cycles in *Drosophila* development provide a good example of how a pattern of cell division can be under genetic control.

13.4 Extrinsic signals coordinate cell division, cell growth, and cell death in the developing *Drosophila* wing

The *Drosophila* wing has proved an interesting model system for studying many aspects of growth and especially how the size of an organ might be determined. In contrast to the intrinsic embryonic program in the blastoderm, the growth in the wing imaginal disc and other discs is modulated by extracellular signals produced by the previous patterning process, and involves the coordination of cell growth, cell proliferation, and cell death.

At its formation, the wing disc is initially composed of about 40 cells, and normally grows in the larva to about 50,000 cells. Cell division occurs throughout the disc, and then ceases uniformly when the correct size is reached. By the end of the pupal stage, the disc has developed the structure of the adult wing, which is produced at metamorphosis by changes in cell shape with only minimal further cell division (see Fig. 11.27). Early experiments investigating wing growth showed that the final size of the wing does not depend on the imaginal disc undergoing a fixed number of cell divisions, or attaining a particular number of cells. Instead, final size seems to be controlled by some mechanism that monitors the overall size of the developing wing disc and adjusts cell division and cell size accordingly. Experiments show that there is no restriction on how much of the wing the progeny of a given cell can make; the clonal descendants of a single cell can contribute from a tenth to as much as a half of the wing.

Wing growth can also be experimentally decoupled from cell proliferation, as demonstrated by the fact that if cell division is blocked in either the anterior or posterior compartment in the larval wing disc, the final size of the wing is normal but the individual cells are larger. In the same way, the size of a region of the wing need not be determined by the rate of cell division. This can be shown by making mosaics of wild-type cells and slower-growing *Minute* cells (see Box 2H). Within a wing compartment, the faster-dividing wild-type cells contribute more cells than the slow-growing *Minute* cells, but the compartment remains the same size as if all the cells were wild type.

In normal growth, the final size of the wing (and that of any other organ) is likely to be achieved by a balance between cell proliferation and apoptosis. An important coordinator of these cell activities is the Hippo signaling pathway (Box 13A), which was first discovered in *Drosophila* but has since been found to operate in mammals and other vertebrates. The *Drosophila* Hippo pathway involves a cascade of serine-threonine protein kinases, of which Hippo is one, and whose activity results in the inactivation of a transcriptional co-activator called Yorkie (Yki).

When the pathway is inactive, Yki translocates to the nucleus and forms a complex with cell-type-specific transcription factors to induce the expression of genes encoding proteins that promote cell proliferation and suppress apoptosis. In contrast, when Hippo signaling is activated, it phosphorylates Warts, another protein kinase in the cascade, and this, in turn, phosphorylates Yki. Phosphorylated Yki is retained in the cytoplasm and therefore cannot translocate to the nucleus to act as a transcriptional activator. Loss-of-function mutations in either Hippo or Warts or the overexpression of Yki in *Drosophila* wing discs lead to a dramatic increase in cell proliferation and a decrease in apoptosis, resulting in wing discs growing to almost eight times the normal size (Fig. 13.6). The inactivation of Yki by the Hippo signaling network is

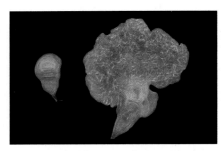

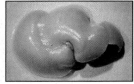

Fig. 13.6 The effect of alterations in the *Hippo* pathway in *Drosophila* and in mouse liver. Upper panel shows the massive growth of a *Drosophila* wing disc (right) when *Yki* is overexpressed. Normal-sized wing disc for comparison on left. Middle panels show overgrowth of clones of *Hippo* mutant cells in the cuticle of *Drosophila* (right) with a normal fly for comparison on the left. Lower panels show the large increase in the size of the liver of a mouse in which both *Mst1* and *Mst2*, the two mammalian *Hippo* homologs, were conditionally inactivated during embryonic development (right). Liver of a normal mouse for comparison on left.

*Top panel from Huang, J., et al.: **The Hippo signaling pathway coordinately regulates cell proliferation and apoptosis by inactivating Yorkie, the** Drosophila **homolog of YAP**. Cell 2005, **122**: 421–434.*

*Middle and bottom panels From Halder, G., Johnson, R.L.: **Hippo signaling; growth control and beyond**. Development 2011, **138**: 9–22.*

CELL BIOLOGY BOX 13A The core Hippo signaling pathways in *Drosophila* and mammals

The outcome of activation of the Hippo pathway is the inactivation of the transcriptional co-activator Yorkie (Yki) in *Drosophila* (see Figure 1, left panels) and the homologous transcriptional co-activators Yap (Yes-associated protein) or Taz in mammals (see Figure 1, right panels). When the pathway is inactive, Yki or Yap/Taz translocate to the nucleus and help to activate expression of genes that promote cell proliferation and suppress apoptosis. When the pathway is active, upstream regulators including Merlin, Expanded and Kibra in *Drosophila* (neurofibromatosis 2 (NF2), FRMD and Kibra, respectively, in mammals), activate the protein kinase Hippo (Mst1 or Mst2 in mammals), which then phosphorylates the protein kinase Warts (Lats 1 or Lats 2 in mammals), which in turn phosphorylates Yki (Yap or Taz in mammals). Phosphorylated Yki (or Yap or Taz) is retained in the cytoplasm and inactivated. Expression of cell-proliferation genes is suppressed and expression of apoptosis genes is promoted. Salvador (Sav) and Mats in *Drosophila* (Mob1 in mammals) are cofactors for the core protein kinases in the Hippo pathway.

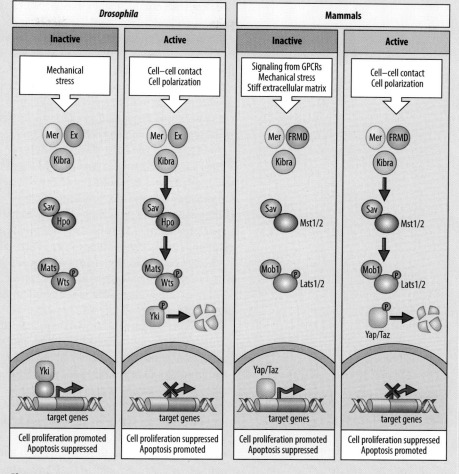

Figure 1

therefore a mechanism for suppressing growth, and the Hippo pathway could be one means of terminating growth when an organ reaches its required size.

It is becoming clear that the Hippo pathway acts as a sensor of the overall state of a tissue and thus could ensure that it does not grow beyond a certain size. How exactly it achieves this is not known, but it is not surprising that there are multiple inputs into the Hippo pathway that are integrated to regulate the activity of the pathway both positively and negatively. Many of these inputs comprise signaling molecules associated with adhesion and the cytoskeleton, which can transmit not only chemical information but also mechanical information resulting from cell–cell contacts.

Inputs that activate the pathway comprise growth-inhibitory signals including signals from molecules in septate junctions (these signals act on the kinase cascade), and signals from molecules localized in apical domains of epithelial cells (for example, Crumbs) that provide information about tissue organization (these signals act on the upstream regulators of the kinase cascade). There is also evidence that the atypical cadherins Fat and Dachsous are activators of the Hippo pathway.

Inputs that inhibit the pathway comprise growth-promoting signals, including signals from the cytoskeleton that provide information about the mechanical forces generated in the tissue (see Chapter 9), and these signals act on Yki. The pathway also interacts with other signaling pathways important in regulating growth and development, such as the Wnt, TGF-β, BMP, Hedgehog, and Notch pathways.

At the output end of the Hippo pathway, Yki acts as a co-activator of genes involved in regulating growth. In the *Drosophila* wing imaginal disc, for example, Yki primarily functions as a co-activator for the transcription factor Scalloped (see Section 11.22), which is essential for normal wing growth. Genes activated by the Yki–Scalloped complex include the cell-cycle-promoting gene *cyclin E*, the gene *diap1*, which encodes an inhibitor of apoptosis, and the gene for a microRNA called *bantam*, which promotes both cell proliferation and cell survival by suppressing production of the pro-apoptotic protein Hid.

The Hippo pathway is highly conserved between invertebrates and vertebrates, with the protein kinases Mst1 and Mst2 in vertebrates being homologs of Hippo, and the proteins Yap and Taz being homologs of Yki (see Box 13A for details). When *Yap* is overexpressed or *Mst1* and *Mst2* are conditionally deleted in mouse liver to mimic inactivation of the pathway, there is a striking three- to four-fold increase in liver size (see Fig. 13.6), consistent with the Hippo pathway functioning in the same way in vertebrate cells as in insect cells and regulating organ size. However, when *Mst1 and Mst2* are conditionally deleted in the mouse limb, there is little effect on limb size, indicating that in mammals the function of the Hippo pathway is not the same in all organs.

The Hippo pathway in vertebrate cells is activated by cell–cell contact, and is therefore responsible for density-dependent inhibition of cell proliferation, a well-known property of normal cells in culture. Cell polarization also activates Hippo signaling. Both these inputs act directly on the transcriptional co-activators, Yap and Taz. The pathway is inhibited by signaling of G-protein-coupled receptors to which extracellular growth-promoting molecules may bind, in addition to mechanical cues transmitted by the cytoskeleton. These cues include those that depend on the stiffness of the extracellular matrix, and thus the Hippo pathway is responsible for anchorage-dependent cell proliferation, another well-known property of normal cells in culture. As in *Drosophila*, the vertebrate Hippo pathway also interacts with other signaling pathways important in regulating growth and development. A picture is emerging of a multitude of signals conveying information about a tissue cell's environment converging on the Hippo pathway to determine whether that cell will divide or not, and whether it will live or die.

13.5 Cancer can result from mutations in genes that control cell proliferation

The majority of cancers—more than 85%—occur in epithelia. This is not surprising when it is recalled that many epithelia (such as the epidermis of the skin and the lining of the gut) are renewed from stem cells that have the potential to divide indefinitely (see Chapter 8). In normal epithelia, stem-cell division is under strict regulation and the cells generated by stem cells soon stop dividing. By contrast, cancerous epithelial cells continue to divide, although not necessarily more rapidly, and usually fail to differentiate. Another feature of cancer cells, unlike normal cells, is that when they divide, they are genetically unstable; the gain or loss of chromosomes is common in solid tumors.

Most cancers derive from a single abnormal cell that has acquired a number of mutations that increase its rate of proliferation. The progression of such a cell to becoming a malignant tumor cell, known as **tumor progression**, is an evolutionary process, involving both further mutation and selection of those cells best able to proliferate. A greater frequency of division increases the cell's chance of acquiring

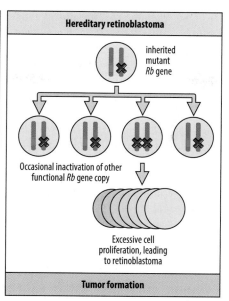

Normal individual	Hereditary retinoblastoma

Occasional deletion of one of the two *Rb* genes

No tumor formation

inherited mutant *Rb* gene

Occasional inactivation of other functional *Rb* gene copy

Excessive cell proliferation, leading to retinoblastoma

Tumor formation

Fig. 13.7 The *retinoblastoma* (*RB*) gene is a human tumor-suppressor gene. If only one copy of the *RB* gene is lost or inactivated by mutation, no tumor develops (left panel). In individuals already carrying an inherited mutant *RB* gene, if the other copy of the gene is also lost or inactivated in a retinal cell, that cell will generate a retinal tumor (right panel). Individuals with an inherited *RB* mutation are thus at a much greater risk of developing retinoblastoma, and usually do so at a young age.

more mutations. In almost all cancers, the cells are found to have a mutation in at least one, and usually many, genes. On average, 63 mutations are found in a pancreatic cancer cell. Genes that when mutated can lead to cancer fall into two classes—**proto-oncogenes** and **tumor-suppressor genes**.

Proto-oncogenes are involved in driving cell proliferation in normal cells. If a proto-oncogene becomes permanently switched on or abnormally expressed as a result of mutation or a chromosomal rearrangement, this can lead to the over-stimulation of cell division and thus contribute to cancer formation. Mutation of one copy of a proto-oncogene in a cell can be sufficient to promote uncontrolled proliferation. At least 70 proto-oncogenes are known in mammals; the first to be identified in humans was Ras, a small GTPase involved in intracellular signaling pathways that promote cell proliferation.

Tumor–suppressor genes, on the other hand, are involved in suppressing cell proliferation in normal cells. Inactivation or deletion of both copies of the gene in a cell removes these restraints to proliferation and is required for a cell to become cancerous. The classic example of a human tumor-suppressor gene is the *retinoblastoma* (*RB*) gene that encodes a protein, RB, which is involved in inhibiting cell-cycle progression.

Retinoblastoma is a childhood tumor of retinal cells, and although it is normally very rare, certain families have an inherited predisposition to it. This led to the discovery of the *RB* gene. In some of these families, the predisposing genetic defect was identified as a deletion of a particular region on one copy of chromosome 13. A deletion on one copy of the chromosome does not, on its own, cause retinal cells to become cancerous. However, if a retinal cell also acquires a deletion of the same region on the other copy of chromosome 13, a retinal tumor develops. The gene in this region that is responsible for susceptibility to retinoblastoma is the *RB* gene. Both copies of *RB* must be lost or inactivated for a cell to become cancerous (Fig. 13.7).

The tumor-suppressor gene *p53* plays a key role in many cancers; about half of all human tumors contain a mutated form of *p53*. This gene is not required for development *per se*, but when cells are exposed to agents that damage DNA, then *p53* is activated and arrests the cell cycle, giving the cell time to repair the DNA. The p53 protein thus prevents the cell from replicating damaged DNA and giving rise to mutant cells. Instead, p53 will cause the cell to die by apoptosis if the damage is too severe to be repaired. The mutant forms of *p53*, found in many cancers, do not promote apoptosis, and so the affected cells are more likely to accumulate mutations.

Genes encoding components of the Wnt and Hedgehog signaling pathways are frequently found mutated in human cancers. Indeed, the first mammalian member of the Wnt family, mouse *Int-1*, which is involved in brain development, was discovered through its ability to act as a proto-oncogene. The Wnt pathway component APC (adenomatous polyposis coli), on the other hand, is a tumor suppressor. Wnt signaling in the stem-cell niche is necessary for the proliferation of intestinal cells, along with those of many other stem-cell pools (see Section 8.12), and APC normally helps to keep the Wnt pathway inhibited in the absence of Wnt signaling (see Box 4B). Mutations that inactivate *APC* result in the constitutive activation of the pathway, leading to unregulated cell proliferation; such mutations are the cause of the inherited syndrome familial adenomatous polyposis coli (hence the gene name), in which precancerous polyps form in the colon. Inactivation of *APC* on its own is not sufficient to produce a malignant tumor, but greatly increases the risk of further mutations leading to polyps developing into colorectal cancer.

Mutations in Hedgehog-pathway genes can also lead to tumors. The link between Hedgehog signaling and cancer was first recognized by the finding that mutations in the gene for the Hedgehog receptor Patched (see Box 2F) are the cause of the rare, inherited Gorlin's syndrome, one feature of which is multiple cancers of epidermal basal cells. Patched acts as a tumor suppressor, keeping the Hedgehog pathway switched off in the absence of Hedgehog. Abnormal activation of the Hedgehog pathway is present in nearly all pancreatic cancers, where it is thought to maintain cancer stem cells and to be involved in cancer progression. The Hippo pathway has been found to be deregulated in many solid tumors, most probably due to mutations in genes involved in the Wnt and Hedgehog signaling pathways.

13.6 Size-control mechanisms differ in different organs

As we have seen, manipulating the Hippo pathway can increase the size of a mouse liver (see Fig. 13.6), but has little effect on the size of the limb. This shows that the molecular mechanisms that determine organ size in vertebrates differ in different organs. Intrinsic developmental programs and extracellular factors that stimulate or inhibit growth can both contribute to organ size, but the relative importance of these two mechanisms in different organs varies a good deal. This is nicely illustrated by comparing what happens to the liver and the pancreas after parts of them are destroyed in the embryo. The liver, which is the main detoxifying organ in mammals, can regrow and restore its mass to its normal size in both the embryo and the adult after damage. The size of the pancreas, on the other hand is intrinsically determined, and if part of it is destroyed in the embryo it ends up smaller than normal.

Experiments that show these different properties of the liver and the pancreas make use of the fact that organ precursor cells can be specifically destroyed in embryonic mice using transgenic techniques (Fig. 13.8). A strain of mice is generated that carries an introduced gene for diphtheria toxin under the control of a tetracycline-repressible promoter. Part of this promoter comes from a gene that is only expressed in the particular organ precursor cells targeted, and so only directs expression of the toxin in those cells. If a pregnant mouse of this strain is continuously fed tetracycline, the toxin is not expressed at all in the embryos she carries. If tetracycline is withdrawn, the toxin is produced in the targeted precursor cells, killing them but leaving other tissues in the embryo untouched.

To investigate the regenerative powers of the embryonic liver, tetracycline was withdrawn for a short period at the appropriate time during gestation, with the result that a proportion of the liver precursor cells in the embryos are destroyed. The longer the period without tetracycline, the greater the number of precursor cells lost. After such treatment,

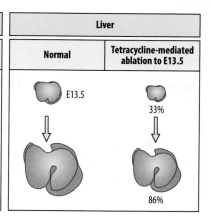

Pancreas		Liver	
Normal	**Tetracycline-mediated ablation to E10.5**	**Normal**	**Tetracycline-mediated ablation to E13.5**

Embryo ↓ **Adult**

E18.5 — weight 36% of normal

weight 40–50% of normal

E13.5 — 33%

86%

Fig. 13.8 Tetracycline-controlled cell killing in embryonic pancreas and liver in mice reveals their different potentials for regeneration. The pancreas does not regenerate when its cells are killed in the embryo and remains small, whereas the liver can regenerate, and rapidly regains almost its normal size when its cells are killed in the embryo.

*Adapted from Liu & Baron.: **Mechanisms Limiting Body Growth in Mammals**. 2011,* Endocrine Reviews *32(3): 422-440*

the embryonic liver grows back to a normal size, indicating that it does not arise from a fixed number of cells. In contrast, if some of the cells of the pancreas in a mouse embryo are destroyed (by the technique described above) after the pancreatic 'bud' has formed, a smaller than normal pancreas develops (see Fig. 13.8).

This finding argues against a mechanism in the pancreas that senses organ size, and instead suggests that the size of the embryonic pancreas is largely under intrinsic control. One intrinsic mechanism could involve counting the number of cell divisions, and each pancreatic precursor cell might only be capable of giving rise to a fixed number of cells. Another organ with intrinsic growth control is the thymus gland. If multiple fetal thymus glands are transplanted into a developing mouse embryo, each one grows to full size. If the same experiment is done with spleens, however, each spleen grows much smaller than normal, so that the final total mass of the spleens is equivalent to one normal spleen. It seems, therefore, that the liver and spleen are regulating their final size by some kind of negative-feedback mechanism, responding to factors in the tissue environment.

Because of the importance of the liver in human health and the relatively high incidence of liver damage, there have been many studies of regeneration of the mammalian liver. Liver cells can be induced to divide by injury to the liver or other stimuli, such as removal of part of the organ. In the regenerating human liver, hepatocytes start to divide a day after surgery (to remove a liver tumor, for example). In adult mammals, the liver can regenerate after as much as two-thirds of it has been surgically removed. In rats and mice, for example, if two of the five liver lobes are removed, the three remaining lobes increase in size to restore the normal mass of liver tissue, but the removed lobes do not regrow.

So what switches on growth of the liver when part of it is removed, and how does it know when it has grown enough? Although various growth-promoting factors are known to be required for liver regeneration, the size-regulating mechanism is still unclear. One negative-feedback mechanism for regulating liver size could be by sensing when the liver is big enough to carry out its function of removing bile acids from the circulation. Bile acids are made in the liver and released into the small intestine to aid digestion. They are potentially toxic and so are subsequently resorbed into the blood vessels draining the small intestine and returned directly to the liver, where they are taken up by hepatocytes and reused. An increase in the concentration of bile acids in the blood produced, for example, by feeding mice a diet containing bile acid, signals that the liver is not large enough to maintain the proper levels of bile in the blood, and thus stimulates regeneration, whereas draining the bile duct before removing liver lobes delays regeneration, because the levels of bile acids in the circulation are low and so signal that the liver is too large.

A similar feedback mechanism may also contribute to regulating the size of the kidney. Removal of a kidney leads to an increase in size of the remaining kidney, perhaps

Belgian blue cow

Whippet and 'bully' whippet

Fig. 13.9 'Double-muscled' breeds of cattle and dogs have inherited inactivating mutations in the *myostatin* gene.

Bull: This file is licensed under the Creative Commons Attribution-Share Alike 3.0 Unported license. Author: Stoolhog.

Bully whippet: reproduced with permission from S. Isett.

in response to a temporary rise in the concentration of creatinine in the serum, which signals the need to increase kidney function. In the kidney, however, the increase in size is mainly the result of cell enlargement rather than cell proliferation.

One hypothesis for the regulation of liver size is that organs such as the liver produce molecules that negatively regulate their growth. Such factors were first postulated some 40 years ago and called 'chalones'. Because the concentration of these factors would depend on the number of cells in an organ, this could provide a mechanism to stop the organ growing once the correct number of cells had been produced. But these factors remained unidentified and the concept of chalones fell out of favor. However, secreted proteins with the properties of the classical chalones are now beginning to be identified, although no chalone-like molecule has so far been identified in the liver.

Skeletal muscle is the best example so far of a tissue in which size is regulated by negative feedback mediated directly by a 'chalone'-like protein. This protein is myostatin, a TGF-β-family protein that is produced and secreted by myoblasts (immature muscle cells, see Section 8.5) and which inhibits muscle growth. The importance of myostatin in regulating muscle size was shown by functionally inactivating the *myostatin* gene in mice. There is a significant increase in muscle mass in the mutant mice; the number of muscle fibers and their size are both increased. Similarly, "double-muscled" domesticated animals, for example Belgian blue cattle (Fig. 13.9) and certain breeds of sheep, have been shown to lack myostatin function. So-called "bully" whippets with greatly enlarged muscles are also homozygous for a loss-of-function *myostatin* mutation. The mutation is maintained in the heterozygous state in the whippet breed, as one copy of the mutant gene makes the dog a faster racer (see Fig. 13.9).

13.7 Overall body size depends on the extent and the duration of growth

Overall body size depends not only on the rate of growth, but on how long growth continues. Human growth during the embryonic, fetal, and post-natal periods is typical of the different phases of growth in mammals. The human embryo increases in length from 150 μm at implantation to about 50 cm over the 9 months of gestation. During the first 8 weeks after conception, the embryonic body does not increase greatly in size, but the basic human form is laid down in miniature. After 8 weeks the human embryo is technically known as a fetus.

The greatest rate of growth occurs at about 4 months, when the fetus grows as much as 10 cm per month. Growth after birth follows a well-defined pattern (Fig. 13.10, left panel). During the first year after birth, growth occurs at a rate of about 2 cm per month. The growth rate then declines steadily until the start of a characteristic adolescent growth spurt at puberty at about 11 years in girls and 13 years in boys (Fig. 13.10, right panel). In pygmies, sexual maturation at puberty is not accompanied by

Fig. 13.10 Normal human growth. Left panel: an average growth curve for a human male after birth. Right panel: comparative growth rates of boys and girls. There is a growth spurt at puberty in both sexes, which occurs earlier in girls.

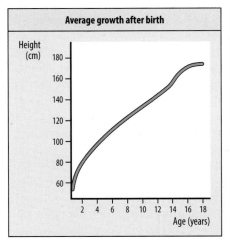

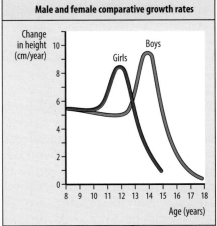

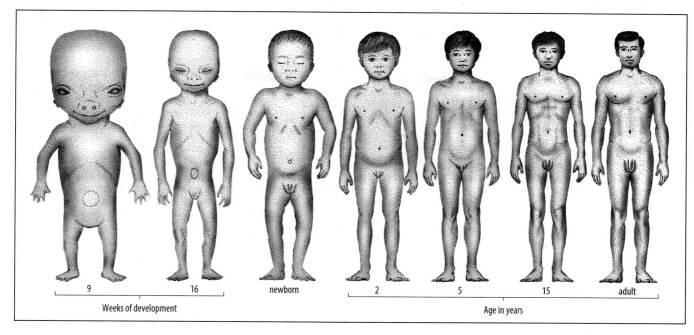

9   16   newborn   2   5   15   adult

Weeks of development   Age in years

Fig. 13.11 Different parts of the human body grow at different rates. At 9 weeks of development the head is relatively large but, with time, other parts of the body grow much more than the head.

Illustration after Gray, H.: Gray's Anatomy. *Edinburgh: Churchill-Livingstone, 1995.*

this adolescent growth spurt, hence their characteristic short stature. There are similar declines in growth rate in other mammals after birth, but in rodents, for example, the decline takes place over a few weeks rather than years.

Different tissues and organs grow at different rates during development, and this affects the way that the proportions of the different parts of the body change in relation to each other during gestation and after birth (Fig. 13.11). At 9 weeks of development, the head of a human embryo, for example, is more than a third of the length of the whole embryo, whereas at birth it is only a quarter. After birth, the rest of the body grows much more than the head, which is only about an eighth of the body length in an adult.

13.8 Hormones and growth factors coordinate the growth of different tissues and organs and contribute to determining overall body size

Circulating hormones, such as growth hormone, thyroid hormone, and the steroid hormones, together with locally acting growth factors, have a major role in coordinating growth and controlling overall body size. Most hormones have multiple effects on cells, but one of these effects is often to stimulate or inhibit cell division or cell differentiation. The response to any given hormone differs from organ to organ, and generally is also different at different stages of development. In this way hormones can act throughout the body as a whole to coordinate the growth of different tissues and organs.

Cell proliferation in the early mammalian embryo, when it is only a few millimeters long, can be controlled by growth factors secreted into the local extracellular environment. The **insulin-like growth factors 1 and 2 (IGF-1 and IGF-2)** are two such growth factors with key roles in promoting cell proliferation and growth during embryonic development. We have encountered other growth-promoting factors, such as FGF, in earlier chapters in their role as patterning molecules. IGF-1 and IGF-2 are single-chain protein growth factors that closely resemble the hormone insulin and each other in their amino-acid sequence, and produce their effects on cells through a shared insulin-signaling pathway. Newborn mice lacking a functional *Igf1* gene develop relatively normally, but weigh only 60% of the normal newborn body weight. Mice in which the *Igf2* gene has been inactivated are also growth retarded. Both the

BOX 13B The major determinant of body size in dogs is the growth hormone–IGF-1 axis

Domestic dogs are said to show the largest range of sizes of any living land animal. The dog (*Canis canis*) became a species distinct from its gray wolf (*Canis lupus*) ancestor at least 15,000 years ago, and archaeological evidence suggests that there were already size differences in the earliest domestic dogs. Some present-day breeds, such as Greyhounds and the Pharaoh Hound, look very similar to dogs depicted in 8000-year-old temple paintings although it is not clear whether our dogs are their direct descendants.

Intensive selective breeding over the past 200–300 years has led to the 350 or so distinct breeds that we know today, with enormous differences in size and shape: compare Toby, a Lurcher (a greyhound cross) with Roly, the Norfolk terrier (Figure 1). A Mastiff, one of the large breeds of dog, is 30 times larger overall than a small breed, the Yorkshire terrier, and is 50 times heavier than a Chihuahua. This size variation is matched by a large amount of genetic variation between breeds, whereas dogs of the same breed have a small amount of genetic variation because they are, effectively, closed breeding populations. This makes dogs ideal subjects for searching for genes that control features such as body size.

So what is the genetic basis for these enormous differences in body size? All domestic dogs possess the same genome of about 2.5 billion base pairs, and the full genome sequence of a female Boxer is now used as the reference dog genome. To identify regions of the genome associated with size, DNA sequences were compared in detail not only between dogs of different sizes within a single breed (Portuguese water dogs), but also between dogs from small and giant breeds. The first success was the identification of a primary region of the genome responsible for difference in size, which was then refined to a single gene, *IGF1*, which encodes insulin-like growth factor 1 (IGF-1), known to be an important growth factor and hormone in controlling growth during embryonic and fetal development and post-natally (discussed in Section 13.8). All small Portuguese water dogs were homozygous for the same *IGF1* allele (a particular version of the gene), whereas larger Portuguese water dogs had a different *IGF1* allele. Because the same characteristic mutation in *IGF1* found in small Portuguese water dogs was found in virtually all the breeds of small dogs tested, even in breeds that are not closely related, and was hardly ever found in giant dog breeds, such as the Saint Bernard or the Irish Wolfhound, this allele is likely to have evolved early in the history of dog domestication. In addition, small Portuguese water dogs were shown to have lower levels of IGF-1 protein in their serum than large dogs of this breed.

Analysis of other locations in the dog genome associated with size has since identified five more genes, which together with *IGF1* account for 50% of the variation of body size in dogs. These genes include the IGF receptor and the growth hormone receptor, consistent with the growth hormone–IGF axis (see Sections 13.8

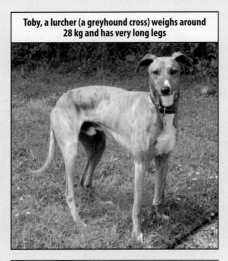

Toby, a lurcher (a greyhound cross) weighs around 28 kg and has very long legs

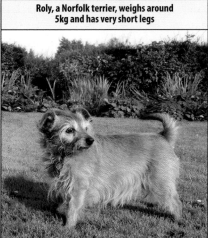

Roly, a Norfolk terrier, weighs around 5kg and has very short legs

Figure 1

Photograph of Toby courtesy M. Welten.
Photograph of Roly courtesy R. Mumford.

and 13.9) being a major determinant of body size. IGF-1 is one of the main effectors of growth hormone action, because growth hormone stimulates its synthesis.

This small number of genes associated with size in dogs should, however, be contrasted with the large number of genes that have been implicated in size determination in humans. In the human genome, 180 locations—potentially 180 different genes—have been found to be significantly associated with height, but even then only account for 10% of the variation in height among people. The reason for this difference in the number of genes involved in size is most likely to be due to the intense artificial selection of dogs for particular characteristics from a very small number of founders. For example, all Golden Retrievers are descended from the puppies that resulted from mating a male Yellow Retriever with a female Water Spaniel in the 1860s.

factors and their receptors can be detected as early as the eight-cell stage of mouse development. *Igf2* is one of the genes that are imprinted in mammals (see Section 10.8); it is inactivated in maternal germ cells and is only expressed in the embryo from the paternal genome. Like many growth factors, the IGFs also have important roles in post-natal growth and in controlling cell proliferation in adult mammals. In humans, IGF-1 is produced after birth mainly by the liver under the stimulation of **growth hormone**, a circulating hormone produced by the pituitary (Fig. 13.12). It is also produced in growth-hormone-responsive tissues, where it can act locally. *IGF1* and genes encoding other components of the growth hormone–IGF-1 axis have been identified as major genetic determinants of body size in dogs (Box 13B).

Growth hormone is crucial to the growth of humans and other mammals after birth. Within the first year of birth, the pituitary gland begins to secrete growth hormone. A child with insufficient growth hormone grows less than normal, but if the synthetic form of the hormone is given regularly, normal growth is restored. In this case, there is a catch-up phenomenon, with a rapid initial response that tends to restore the growth curve to its original trajectory.

Production of growth hormone in the pituitary is under the control of two hormones produced in the hypothalamus: **growth hormone-releasing hormone**, which promotes growth hormone synthesis and secretion, and **somatostatin**, which inhibits its production and release. As we have just noted, growth hormone produces many of its effects by inducing synthesis of IGF-1, but it can also act directly on some tissues such as the growth plates of long bones (see Fig. 13.12). Post-natal growth, in addition to embryonic growth, is therefore largely due to the actions of the insulin-like growth factors, and to the complex hormonal regulatory circuits that control their production.

Puberty is also initiated by activity in the hypothalamus, which governs the intermittent release of **gonadotropin-releasing hormone (GnRH)** by hypothalamic neurons. The mechanism determining this timing is not known. One of the results of a pulse of GnRH is a sharp increase in the secretion of gonadotropins (luteinizing hormone and follicle-stimulating hormone) by the pituitary; these cause increased production of the steroid sex hormones—the estrogens and androgens. These in turn stimulate the production of pulses of growth hormone, which are responsible for the growth spurts at puberty in girls and boys.

13.9 Elongation of the long bones illustrates how growth can be determined by a combination of an intrinsic growth program and extracellular factors

Patterning of the embryo occurs while the organs are still very small. For example, human limbs have their basic pattern established when they are less than 1 cm long. Over the years, the limb grows to be at least one hundred times longer and this growth is driven by elongation of the long bones. The long bones in the arm are the humerus, radius and ulna, and in the leg, the femur, fibula and tibia. So how is this growth controlled?

Each of the skeletal elements in the chick wing has its own intrinsic growth program, which is specified during embryonic development. The cartilaginous elements representing the long bones—the humerus and the ulna, for example—are initially similar in size to the elements in the wrist (Fig. 13.13). Yet, with growth, the humerus and ulna increase many times in length compared with the wrist bones. These growth programs are specified when the elements are initially patterned and involve both cell multiplication and matrix secretion (accretion). Each skeletal element follows its own growth program even when grafted to a developmentally neutral site, provided that a good blood supply is established. It is likely that this differential growth is due at least in part to the different Hox genes that come to be expressed in the different limb 'segments' (see Section 11.13). In addition, the radius/ulna region of the chick wing (and

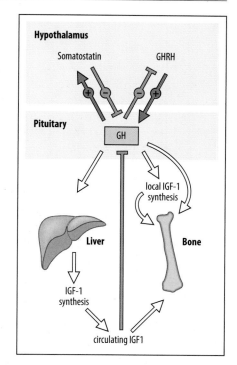

Fig. 13.12 Growth hormone production is under the control of the hypothalamic hormones. Growth hormone is made in the pituitary gland and is secreted. Growth-hormone-releasing hormone (GHRH) from the hypothalamus promotes the synthesis of growth hormone (GH), whereas somatostatin inhibits it. Growth hormone controls its own release by feedback signals to the hypothalamus. Growth hormone stimulates the synthesis of the insulin-like growth factor IGF-1, and this, in turn, has a negative effect on the production of growth hormone in the pituitary. IGF-1 is mainly made in the liver but can also be made locally. Long bone growth is controlled directly by the action of growth hormone and by both circulating and locally produced IGF-1.

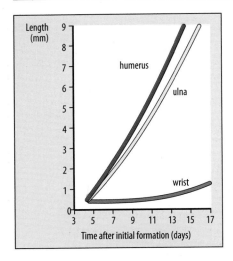

Fig. 13.13 Comparative growth of the cartilaginous elements in the embryonic chick wing. When first laid down, the cartilaginous elements of the humerus, ulna, and wrist are the same size, but the humerus and ulna then grow much more than the wrist element.

the tibia/fibula region of the chick leg) expresses the *Shox* (*Short stature homeobox*) gene. *SHOX* deficiencies in humans are associated with Turner syndrome and various short stature conditions, which are characterized by the forearms and the lower legs being disproportionately short.

One intriguing effect of the sex hormones testosterone and estrogen is on differences in relative finger length between the two sexes. This growth program is apparently established in a narrow developmental window in the embryo as a result of the differing levels of these hormones in male and female embryos (Box 13C).

The growth of the long bones occurs in the **growth plates**. Long bones are initially laid down as cartilaginous elements (see Section 11.1). In both fetal and post-natal growth, the cartilage is replaced by bone in a process known as **endochondral ossification**, in which ossification starts in the centers of the long bones and spreads outward (Fig. 13.14). Secondary ossification centers then develop at each end of the bone, known as the epiphyses. The adult long bones thus have a bony shaft (diaphysis) with cartilage confined to the articulating surfaces at each end, and to two regions near each end—the **growth plates**—in which growth occurs. In the growth plates, the cartilage cells, or **chondrocytes**, are usually arranged in columns, and various zones can be identified. Just next to the bony epiphysis is a narrow germinal zone, which contains stem cells. Next is a proliferative zone of cell division, followed by a zone of maturation, and a hypertrophic zone, in which the cartilage cells increase in size. Finally, there is a zone in which the cartilage cells die and are replaced by bone laid down by cells called **osteoblasts**, which differentiate from cells that form the perichondrium surrounding the cartilage. This stage involves Wnt signaling. There is a marked similarity to the development of skin, where basal stem cells give rise to dividing cells, which differentiate into keratinocytes and finally die (see Section 8.10).

The proliferation of chondrocytes at the ends of the long bones, and later in the growth plate, is controlled so that at a given distance from the end of the bone they stop dividing and enlarge. In mice, the proliferation of chondrocytes is controlled by the secreted signaling proteins parathyroid-hormone-related protein (PHRP) and Indian hedgehog (Ihh). Ihh belongs to the family of Hedgehog signaling proteins. PHRP is secreted by the chondrocytes and cells of the perichondrium at the ends of the prospective bones and stimulates chondrocytes to proliferate, which prevents them from expressing *Ihh*. Once chondrocytes move out of the zone of influence of PHRP they stop proliferating, start to express *Ihh*, and become hypertrophic (Fig. 13.15). Ihh protein diffuses back into the pool of proliferating chondrocytes, where it increases their rate of proliferation. Ihh also, by some mechanism as yet unknown, stimulates production of PHRP by the cells at the ends of the bone and this ensures that the proliferating zone of cells is maintained. Ihh also acts on the adjacent perichondrial cells to form bone-producing osteoblasts. In the absence of the *Ihh* gene in mice there is an accelerated rate of hypertrophy resulting in short, stubby limbs and cell proliferation is also suppressed.

Circulating growth hormone affects bone growth by acting on the growth plates. The cells in the germinal zone have receptors for growth hormone, and growth hormone is probably directly responsible for stimulating these stem cells to proliferate. Further growth, however, is probably mediated by IGF-1, whose production in the growth plate is stimulated by growth hormone. Thyroid hormones are also necessary for optimal bone growth; they act both by increasing the secretion of growth hormone and IGF-1, and by stimulating hypertrophy of the cartilage cells. FGF is also important for bone growth; the genetic defect that gives rise to **achondroplasia** (short-limbed dwarfism) is a dominant mutation in the FGF receptor-3, whose normal function is to limit, rather than promote, bone formation. When it is mutated, the FGF pathway is activated in the absence of the ligand and therefore the inhibition of growth is greatly exaggerated.

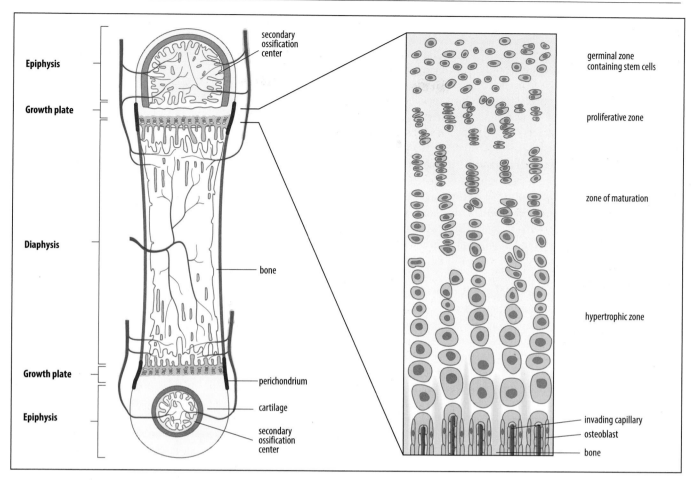

Fig. 13.14 Growth plates and endochondral ossification in the long bone of a vertebrate. The long bones of vertebrate limbs increase in length by growth from cartilaginous growth plates. The growth plates are cartilaginous regions that lie between the epiphysis, the end of the bone, and the shaft of the bone, the diaphysis. In the figure, bone has already replaced cartilage in the diaphysis, and more bone is being added at the growth plates. Within the growth plates, cartilage cells multiply in the proliferative zone, then mature and undergo hypertrophy (cell enlargement). They are then replaced by bone, which is laid down by specialized cells called osteoblasts. These derive from perichondrial cells. Osteoclasts invade the bone along with the blood vessels and are involved in resorbing the cartilage and remodeling bone. Secondary sites of ossification are located within the epiphyses.

*Illustration after Walls, G.A.: **Here today, bone tomorrow**. Curr. Biol. 1993, **3**: 687-689.*

The rate of increase in the length of a long bone is equal to the rate of new cell production per column multiplied by the mean height of an enlarged cell. The rate of new cell production depends both on the time cells take to complete a cycle in the proliferative zone, and the size of this zone. Different bones grow at different rates, and this can reflect the size of the proliferative zone, the rate of proliferation, and the

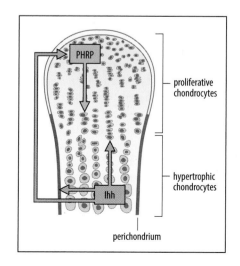

Fig. 13.15 Indian hedgehog (Ihh) and parathyroid-hormone-related protein (PHRP) form feedback loops that maintain chondrocyte proliferation and growth at the ends of developing bones. PHRP secreted by the chondrocytes at the ends of developing long bones acts on proliferating chondrocytes (blue) to maintain proliferation and prevent the expression of the gene for Indian hedgehog. Chondrocytes farther away from the end of the bone (orange) escape the influence of PHRP and express *Ihh*. Ihh protein acts on adjacent chondrocytes to increase the rate of proliferation and also acts on perichondrial cells to form osteoblasts. In some way that is not yet understood, production of Ihh also stimulates PHRP synthesis by the chondrocytes at the end of the bone, thus forming a positive feedback loop that maintains PHRP production.

*Illustration after Kronenberg, H.M.: **Developmental regulation of the growth plate**. Nature (Insight) 2003, **423**: 332-336.*

BOX 13C Digit length ratio is determined in the embryo

More than 100 years ago, it was noticed that the relative lengths of different fingers varies between the sexes. In men, the fourth finger (ring finger) is generally longer than the second finger (index finger), whereas in women, it is shorter or the same length. The effect is stronger in the right hand. These proportions are usually expressed as the **2D:4D digit ratio** (the ratio is less than 1 in men; the same or more than 1 in women) (Figure 1).

In the past decade there has been a resurgence of interest in this ratio and a low 2D:4D ratio has been correlated with various attributes, including greater sporting prowess, and has been linked to developmental disorders such as autism. It has also been suggested that the 2D:4D digit ratio could be a biomarker for some diseases such as osteoarthritis, where the risk is greater in one sex than the other.

Sexual dimorphism in finger length is determined before birth. It has been detected in human fetuses and was proposed to reflect the degree of prenatal exposure to the 'male' hormone testosterone. This proposal has been tested experimentally in mice. Mice also show the same sex differences in the 2D:4D digit ratio as humans and, like humans, the effect is stronger on right limbs. These differences can first be detected in the embryo when the chondrocytes (cartilage cells) in the developing digits have just begun to hypertrophy, the mean 2D:4D digit ratio in the right paws of male embryos being significantly smaller than that in female embryos.

To test whether this ratio is controlled by testosterone, and whether the 'female' sex hormone estrogen was also involved, mice were created in which either the androgen receptor (the receptor for testosterone) or the estrogen receptor were deleted genetically in the developing mouse limbs. When the androgen-receptor gene is deleted, the digit ratio in mutant males is significantly higher, consistent with testosterone being required for a masculine (low) digit ratio. In contrast, when the estrogen-receptor gene is deleted, the digit ratio in mutant males is significantly lower, consistent with estrogen being required for a feminine (high) digit ratio. Thus, estrogen as well as testosterone influences digit growth and their effects are opposite.

There is a narrow window in digit development when the balance of hormones determines whether there is a masculine or feminine digit ratio. Surprisingly, the differences in digit ratios in the hormone-treated offspring are due entirely to effects on the length of digit 4.

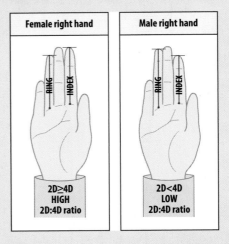

Figure 1

The length of this digit increases in female offspring of mothers treated with testosterone, whereas it decreases in male offspring of mothers treated with estrogen. The reason that digit 4 is more responsive to hormone levels than digit 2 is that there are many more receptors for both testosterone and estrogen in digit 4 (Figure 2).

These experiments explain how the 2D:4D ratio arises during normal embryonic development. In female embryos, which have higher levels of estrogen than of testosterone, there will be higher levels of activated estrogen receptor in digit 4 and its growth will decrease. In male embryos, which have higher levels of testosterone than estrogen, there will be higher levels of activated androgen receptor in digit 4 and its growth will increase.

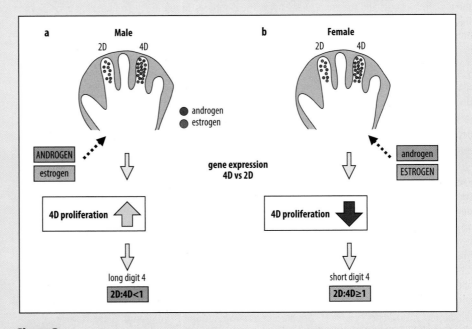

Figure 2

degree of cell enlargement in the growth plate. If bone growth is prevented, for example by heavy loading, then there is some catch-up growth when the load is removed.

In view of the complexity of the growth plate, it is remarkable that human bones in limbs on opposite sides of the body can grow for some 15 years independently of each other, and yet eventually match to an accuracy of about 0.2%. This may be achieved by having many columns of cells in each plate, so that growth variation between cells is averaged out. When growth of a bone ceases the growth plate ossifies, and this occurs at different times for different bones. Ossification of growth plates occurs in a strict order in different bones and can therefore be used to provide a measure of physiological age. The timing of growth cessation in the growth plate appears to be instrinsic to the plate itself rather than to hormonal influence. The cessation of growth is due to the cessation of cell proliferation, and this timing may be programmed in the cells. Cell senescence, and thus growth cessation, may be due to the chondrocyte stem cells having only a finite potential for division.

The length of the long bones dictates the length of the limbs and the growth of other tissues including the muscles. The number of striated (skeletal) muscle fibers in a muscle is determined during embryonic development. Once differentiated, striated muscle cells lose the ability to divide. Post-embryonic growth of muscle tissue results from an increase in individual fiber size, both in length and girth, during which the number of myofibrils within the enlarged muscle fiber can increase more than 10-fold. Additional nuclei to support the functioning of the much-enlarged cell are provided by the fusion of satellite cells with the fiber. Satellite cells, which are undifferentiated cells lying adjacent to the differentiated muscle, also act as a reserve population of stem cells that can replace damaged muscle (see Section 8.13).

The increase in the length of a muscle fiber is associated with an increase in the number of sarcomeres—the functional contractile units—it contains. For example, in the soleus muscle of the mouse leg, as the muscle increases in length, the number of sarcomeres increases from 700 to 2300 at 3 weeks after birth. This increase in number seems to depend on the growth of the long bones putting tension on the muscle through its tendons. If the soleus muscle is immobilized by placing the leg in a plaster cast at birth, sarcomere number increases slowly over the next 8 weeks, but then increases rapidly when the cast is removed. One can, therefore, see how bone and muscle growth are mechanically coordinated.

13.10 The amount of nourishment an embryo receives can have profound effects in later life

Whatever the type of growth program, no animal will reach its full potential size if inadequately nourished as an embryo and during the post-natal growth period. In mammals, inadequate nutrition or poor nutrition of the embryo not only has direct effects on embryonic and fetal growth, but can have serious, and in some ways unexpected, effects in adult life. Population studies in developed countries, such as the United Kingdom, have associated smaller size or relative thinness at birth (due to either maternal undernutrition or premature birth) and during early infancy with an increased risk of developing coronary heart disease, stroke, or type 2 diabetes in adult life.

When undernutrition during early development is followed by improved nutrition later, whether later in gestation or in the early post-natal period, the period of 'catch-up' growth may incur a cost, as the body's resources are diverted from 'repair and maintenance' to growth. Premature birth itself, independent of size for gestational age, has been associated with insulin intolerance and glucose intolerance in pre-pubertal children that may continue into young adulthood and may be accompanied by high blood pressure. Catch-up growth seems to predispose to overweight and even obesity, and this may partly explain the effects on health in later life. One proposal is that in response to undernutrition *in utero* the fetus lays down more fat cells as a precaution. That is a good strategy if conditions after birth are indeed hard and food is short, but in conditions of plentiful food, this leads to undesirable consequences.

Experiments in animals support the effects of undernutrition and an unbalanced maternal diet observed in humans. The embryos of pregnant rats fed a low-protein, but otherwise calorie-sufficient, diet during the preimplantation period (0 to 4.5 days) showed altered development in multiple organ systems. If the pregnancy was allowed to go to full term, the offspring had low birth weight, increased post-natal growth, or adult-onset high blood pressure.

Obesity is associated with numerous diseases in later life, including type 2 diabetes and heart disease. Although much obesity in children and adults is due to overeating and lack of exercise, early developmental nutritional experience and genetic background can contribute. Human fatty tissue comprises some 40 billion adipose cells, with most of the bulk stashed under the skin, and obesity represents both greater numbers of adipose cells compared with lean people and excessive deposition of fat in these cells, which increases their size. Humans are born with a certain number of adipose cells, with females generally having more than males. The number of adipose cells increases throughout late childhood and early puberty, and after this remains fairly constant. However, the number of adipose cells increases more rapidly in genuinely obese children than in lean children, meaning that they end up with more adipose cells. Once fat cells develop in the body, they remain there for life and they seldom die, although there is some turnover. Each month about 1% of adipose cells in the human body die and are replaced. Thus adult obesity is often linked to childhood obesity.

SUMMARY

Growth in animals mainly occurs after the basic body plan has been laid down and the organs are still very small. Final organ size can be controlled both by external signals and by intrinsic growth programs. In some cases, organ size can be determined by monitoring the dimensions of the growing organ rather than being set by cell number or cell size. In animals, growth can occur by cell multiplication, cell enlargement, and secretion of large amounts of extracellular matrix. The Hippo signaling pathway integrates multiple inputs from the tissue environment and suppresses proliferation and promotes apoptosis of excess unwanted cells. Cancer is the result of the loss of growth control and differentiation. In mammals, insulin-like growth factors are required for normal embryonic growth, and they also mediate the effects of growth hormone after birth. Human post-natal growth is largely controlled by growth hormone, which is made in the pituitary gland. Growth in the long bones is both intrinsically programmed and controlled by local factors, and, in addition, occurs in response to growth hormone stimulation of the cartilaginous growth plates at either end of the bone.

Molting and metamorphosis

Many animals do not develop directly from an embryo into an 'adult' form, but into a larva from which the adult eventually develops by metamorphosis. The changes that occur at metamorphosis can be rapid and dramatic, the classic examples being the metamorphosis of a caterpillar into a butterfly, a maggot into a fly, and a tadpole into a frog. Another striking example of metamorphosis is the transformation of the pluteus larva of the sea urchin into the adult (see Fig. 6.19). In some cases it is hard to see any resemblance between the animal before and after metamorphosis. The adult fly does not resemble the larva at all, because adult structures develop from the imaginal discs and so are completely absent from the larval stages (see Chapters 2 and 11). In frogs, the most obvious external changes at metamorphosis are the regression of the tadpole's tail and the development of limbs, although many other structural changes occur. In some insects, the entire body plan is transformed, with most larval tissues undergoing cell death as the adult tissues develop from the imaginal discs and histoblasts. In arthropods and nematodes, increase in size in larval and pre-adult stages requires shedding of the external cuticle, which is rigid, a process known as **molting**.

A number of features distinguish early embryogenesis from molting, metamorphosis, and other aspects of post-embryonic development. Whereas the signal molecules in early development act over a short range and are typically protein growth factors, many signals in post-embryonic development are produced by specialized endocrine cells, and include both protein and non-protein hormones. The synthesis of these hormones is orchestrated by the central nervous system in response to environmental cues, and there is complex feedback between the endocrine glands and their secretions.

13.11 Arthropods have to molt in order to grow

Arthropods have a rigid outer skeleton, the cuticle, which is secreted by the epidermis. This makes it impossible for the animals to increase in size gradually. Instead, increase in body size takes place in steps, associated with the loss of the old outer skeleton and the deposition of a new larger one. This process is known as **ecdysis** or molting. The stages between molts are known as instars. *Drosophila* larvae have three instars and molts. The increase in overall size between molts can be striking, as illustrated in Fig. 13.16 for the tobacco hornworm (*Manduca*).

At the start of a molt, the epidermis separates from the cuticle in a process known as apolysis, and a fluid (molting fluid) is secreted into the space between the two (Fig. 13.17). The epidermis then increases in area by cell multiplication or cell enlargement, and becomes folded. It begins to secrete a new cuticle, and the old cuticle is partly digested away, eventually splits, and is shed.

Molting is under hormonal control. Stretch receptors that monitor body size are activated as the animal grows, and this results in the brain secreting **prothoracicotropic hormone** (**PTTH**). This activates the prothoracic gland to release the steroid hormone **ecdysone**, which is the hormone that causes molting. A similar hormonal circuit controls metamorphosis.

13.12 Insect body size is determined by the rate and duration of larval growth

When an insect larva has reached a particular stage, it does not grow and molt any further but undergoes a more radical metamorphosis into the adult form. Larval tissues, such as gut, salivary glands, and certain muscles, undergo programmed cell death. The imaginal discs now develop into rudimentary adult appendages, such as wings, legs, and antennae. The nervous system is also remodeled.

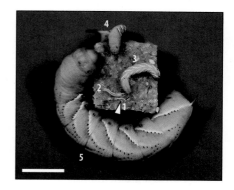

Fig. 13.16 Growth and molting of the caterpillar of the tobacco hawkmoth (*Manduca sexta*). The caterpillar, known as the tobacco hornworm, goes through a series of molts. The tiny hatchling (1)—indicated by the arrow—molts to become a caterpillar (2), and then undergoes three further molts (3, 4, and 5). The increase in size between molts is about twofold. The caterpillars are sitting on a lump of caterpillar food. Scale bar = 1 cm. *Photograph courtesy of S.E. Reynolds.*

Fig. 13.17 Molting and growth of the epidermis in arthropods. The cuticle is secreted by the epidermis. At the start of molting, the cuticle separates from the epidermis—apolysis—and a fluid is secreted between them. The epidermis grows, becomes folded, and begins to secrete a new cuticle. Enzymes weaken the old cuticle, which is shed.

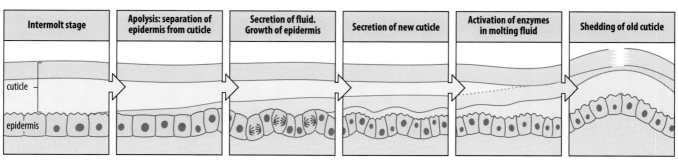

Intermolt stage	Apolysis: separation of epidermis from cuticle	Secretion of fluid. Growth of epidermis	Secretion of new cuticle	Activation of enzymes in molting fluid	Shedding of old cuticle
cuticle					
epidermis					

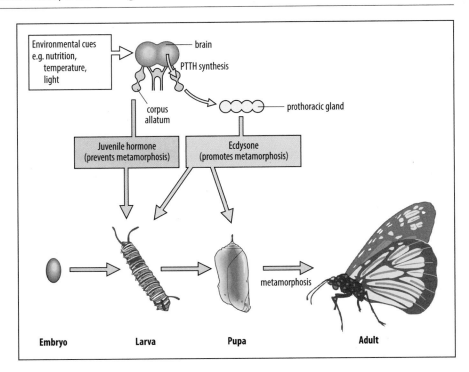

Fig. 13.18 Insect metamorphosis.
The corpora allata of a butterfly larva secrete juvenile hormone, which inhibits metamorphosis. In response to environmental changes, such as an increase in light and temperature, the brain of the final instar larva begins to produce prothoracicotropic hormone (PTTH), which is released from the corpora allata. This acts on the prothoracic gland to stimulate the secretion of ecdysone, the hormone that overcomes the inhibition by juvenile hormone and causes metamorphosis.
*Illustration after Tata, J.R.: **Gene expression during metamorphosis: an ideal model for post-embryonic development**.* BioEssays 1993, **15**: 239–248.

Growth occurs only in the larval stages and therefore the size of the adult is determined by the size the larva reaches before it stops feeding and becomes a **pupa**. In normal circumstances the larva pupates when it reaches a size typical for the species—this is genetically controlled, but if the larval stage is prolonged or shortened experimentally, adults that are, respectively, larger or smaller than usual can be produced. The same hormonal circuit, involving prothoracicotrophic hormone, that controls molting also controls pupation and metamorphosis, with a pulse of ecdysone being the universal signal that terminates larval development (Fig. 13.18).

How body size is monitored by insect larvae and how the timing of the ecdysone signal is determined are still not entirely understood, and the mechanisms are different in different groups of insects. In *Drosophila*, for example, the larva must attain a minimum critical size, as otherwise it will not survive metamorphosis. In normal circumstances this minimum size is attained about half-way through the last instar, coinciding with a pulse of ecdysone, and the larva continues to feed and grow throughout the terminal growth period until pupation (Fig. 13.19). If the larva is starved after it reaches the critical size, it cannot grow because the terminal growth period is dependent on nutrition. Under these circumstances, the larvae still undergo metamorphosis but produce smaller adults than normal. However, if the larvae are kept on a low-nutrient diet throughout the third instar, pupation is delayed, enabling the larvae to grow to nearly normal size (see Fig. 13.19).

The larva of the *Manduca* moth also attains the minimum critical size about half-way through the final instar. In this case however, as in other butterflies and moths, the pulse of ecdysone at the critical size is triggered by a drop in the level of another hormone—**juvenile hormone**—produced by the corpora allata, a pair of endocrine glands located just behind the brain. As its name implies, juvenile hormone maintains the larval state and inhibits metamorphosis. There is also another difference between *Drosophila* and *Manduca*, in that the terminal growth phase in *Manduca* is constant and not dependent on nutrition. Therefore if nutrients are scarce during the final instar in a *Manduca* larva, a smaller adult moth is produced (see Fig. 13.19).

The delay in pupation in *Drosophila* when nutrients are low is accompanied by downregulation of the TOR pathway in the prothoracic gland, and this has the effect

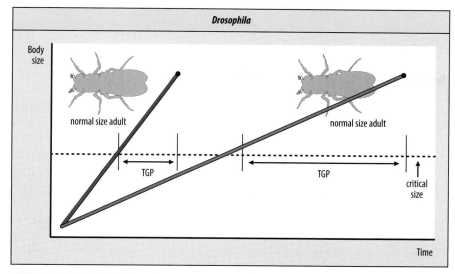

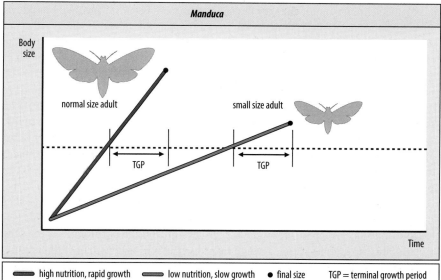

| high nutrition, rapid growth | low nutrition, slow growth | ● final size | TGP = terminal growth period |

Fig. 13.19 Nutrition-dependent regulation of insect body size. The level of nutrition available regulates adult body size in different ways in different insects. Upper panel: in *Drosophila*, once the larva has reached a critical size, if nutrients are scarce the subsequent time the larva spends feeding (the terminal growth period, TGP) is simply prolonged until the larva reaches a normal size. It then undergoes metamorphosis, producing a normal-sized adult. Lower panel: in the tobacco hornworm (in which the adult is a moth), the larval terminal growth period (after reaching a critical size) remains the same whether nutrients are plentiful or scarce. This means that when nutrients are restricted a smaller adult is produced.

*From Nijhout, H.F.: **Size matters (but so does time), and it's OK to be different**. Dev. Cell, 2008, **15**: 491–492.*

of keeping ecdysone production at a low level. As the larvae does eventually pupate and metamorphose, it seems that in *Drosophila*, exposure to low levels of ecdysone over a period of time can have the same effect as the distinct pulse of ecdysone which is the usual signal for pupation.

The TOR pathway is a nutrient-sensing signal-transduction cascade that controls cell size (see Fig 13.3). Its activity in the fat body in *Drosophila* larvae is the way in which nutritional conditions are sensed. The fat body communicates to the central nervous system by releasing a signal, whose identity is as yet unknown. This unidentified signal regulates the release of *Drosophila* insulin-like peptides from neurosecretory cells in the brain. These peptides, which circulate in the hemolymph, are potent activators of growth and their signaling pathway strongly resembles the vertebrate insulin/IGF signaling pathway. Thus the equivalent major hormonal pathways control growth in both *Drosophila* and vertebrates. As we have seen, the *IGF1* gene and a few other functionally related genes are associated with the large differences in body size in dogs of different breeds (see Box 13B). The circulating *Drosophila* insulin-like peptides activate the insulin/TOR signaling pathways in the prothoracic gland and this regulates the synthesis of ecdysone. Other environmental factors such as temperature and light can also regulate synthesis of ecdysone by

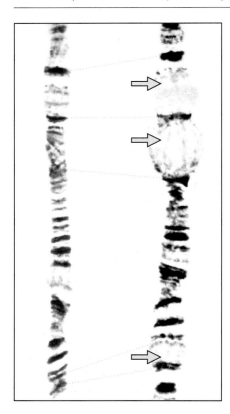

Fig. 13.20 Gene activity seen as puffs on the polytene chromosomes of *Drosophila*. A region of a chromosome is shown from a young third instar (left), and from an older larva (right), after ecdysone has induced puffs at three loci (arrowed).

Photograph courtesy of M. Ashburner.

acting on the neurosecretory cells in the brain to control the release of prothoracicotropic hormone.

Ecdysone crosses the plasma membrane, where it interacts with intracellular ecdysone receptors that belong to the steroid hormone receptor superfamily. These receptors are gene-regulatory proteins, which are activated by binding their hormone ligand, similarly to the receptors for retinoic acid (see Box 5C). The hormone–receptor complex binds to the regulatory regions of a number of different genes, inducing a new pattern of gene activity characteristic of metamorphosis.

There are alterations in the expression of many genes—several hundred at least—during *Drosophila* metamorphosis. In *Drosophila*, because of the special characteristics of polytene chromosomes, changes in gene activity that occur during metamorphosis are actually visible. Cells in some larval tissues (such as the salivary glands) grow and repeatedly pass through S phase without undergoing mitosis and cell division. The cells become very large and can have several thousand times the normal complement of DNA. In salivary gland cells, many copies of each chromosome are packed side by side to form giant polytene chromosomes. When a gene is active, the chromosome at that site expands into a large localized 'puff', which is easily visible (Fig. 13.20). The puff represents the unfolding of chromatin and the associated transcriptional activity. When the gene is no longer active, the puff disappears. During the last days of larval life, a large number of puffs are formed in a precise sequence, a pattern that is under the direct influence of ecdysone.

13.13 Metamorphosis in amphibians is under hormonal control

Metamorphosis occurs in many animal groups other than arthropods, including amphibians. In amphibians, environmental cues, such as nutrition, temperature, and light, as well as the animal's internal developmental program, control metamorphosis through their effects on neurosecretory cells in the brain. The neurosecretory cells are located in the hypothalamus and release **corticotropin-releasing hormone**, which acts on the pituitary gland, causing it to release **thyroid-stimulating hormone** (thyrotropin). (This action of corticotropin-releasing hormone is peculiar to non-mammalian vertebrates, and in *Xenopus* only occurs at the tadpole stage; in adult frogs, as in mammals, the release of thyroid-stimulating hormone is due to thyrotropin-releasing hormone.) Thyroid-stimulating hormone in turn acts on the thyroid gland to stimulate the secretion of the **thyroid hormones** that bring about metamorphosis (Fig. 13.21). The thyroid hormones are the iodo-amino acids thyroxine (T_4), and tri-iodothyronine (T_3). They are signaling molecules of ancient origin, occurring even in plants. Although very different in chemical structure to ecdysone, they too pass through the plasma membrane and interact with intracellular receptors that belong to the same superfamily as the receptors for retinoic acid and steroid hormones (see Box 5C). The pituitary also produces the protein hormone prolactin, which was originally thought to be an inhibitor of metamorphosis; however, overexpression of prolactin does not prolong tadpole life but reduces tail resorption.

A striking feature of the hormones that stimulate metamorphosis is that, in addition to affecting a wide range of tissues, they affect different tissues in different ways, their effects varying from the subtle to the gross. In the tadpole limb, for example,

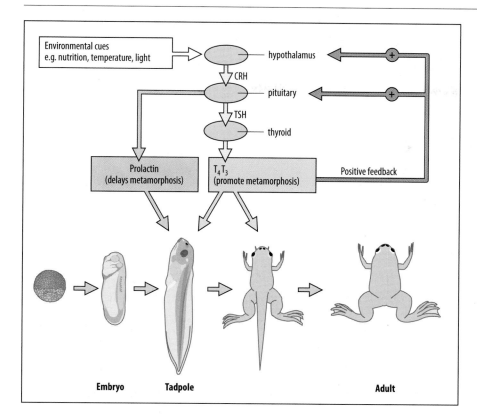

Embryo   **Tadpole**   **Adult**

Fig. 13.21 Amphibian metamorphosis. Changes in the environment, such as an increase in nutritional levels, cause the secretion of corticotropin-releasing hormone (CRH) from the larval hypothalamus, which acts on the pituitary to release thyroid-stimulating hormone (TSH). This in turn acts on the thyroid glands to stimulate secretion of the thyroid hormones thyroxine (T_4) and tri-iodothyronine (T_3), which cause metamorphosis. The thyroid hormones also act on the hypothalamus and pituitary to maintain synthesis of CRH and TSH. Prolactin is also produced by the pituitary and was traditionally thought to delay metamorphosis. *Illustration after Tata, J.R.:* **Gene expression during metamorphosis: an ideal model for post-embryonic development**. BioEssays *1993,* **15***: 239–248.*

thyroid hormones promote development and growth, whereas they cause cell death and degeneration in the tail. Fast muscles are the first to go, and later the notochord collapses. Yet in all these cases, the hormone produces these very different effects by binding to the same intracellular receptor. The difference in outcomes is due to the hormone–receptor complex switching on or off different sets of genes in different tissues, as a result of the different developmental histories of the regions. Each tissue has its own response to the hormones that cause metamorphosis, and some of these effects can be reproduced in culture. When excised *Xenopus* tadpole tails are exposed to thyroid hormones in culture, for example, they cause cell death and complete tissue regression. Metamorphosis also leads to changes in the responsiveness of cells to other signals; for example, in *Xenopus*, estrogen can only induce the synthesis of vitellogenin, a protein required for the yolk of the egg, after metamorphosis and the attainment of sexual maturity.

SUMMARY

Arthropod larvae grow by undergoing a series of molts in which the rigid cuticle is shed. Metamorphosis during the post-embryonic period can result in a dramatic change in the form of an organism. In insects, it is hard to see any resemblance between the animal before and after metamorphosis, whereas in amphibians, the change is somewhat less dramatic. Environmental and hormonal factors control metamorphosis. Thyroid hormones cause metamorphosis in amphibians, and ecdysone does the same in insects. In insects, the size of the adult body is determined by the size of the larva when it begins metamorphosis. In *Drosophila*, gene activity during metamorphosis can be monitored by localized puffing on the giant polytene chromosomes.

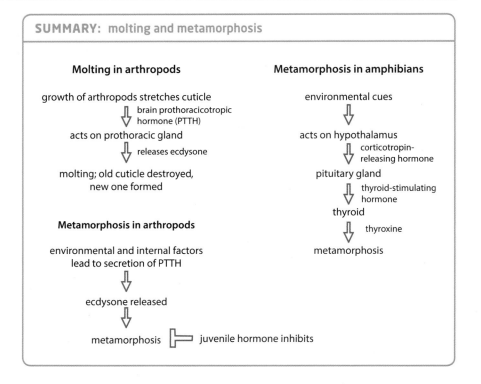

Regeneration

Regeneration is the repair and replacement of missing parts. As we have seen, many animal embryos can regulate (see Chapters 4 and 5), but the ability to regenerate and make good missing tissues and organs as an adult is much rarer. In contrast, plants can regenerate very well—this is the reason that cuttings, for example, can produce an entire new plant (see Chapter 7).

In land-living animals, growth both in the embryo and during the post-natal period determines the final size of the adult body and the shape and size of its constituent parts. Once the adult size has been attained, growth stops. But, as we have seen, some organs, such as the liver in mammals can regenerate even in the adult. But in liver regeneration, the missing parts are not replaced, the remaining parts just get bigger to restore the mass of tissue. In contrast, some insects and other arthropods and even some vertebrates can regenerate new, fully functional organs. Insects can regenerate lost appendages, such as legs, as can urodele, 'tailed,' amphibians (salamanders) such as newts and the axolotl (the Mexican salamander, *Ambystoma mexicanum*). These amphibians have a remarkable capacity for regeneration even as adults, being able to regenerate not only limbs but also new tails and some internal tissues (Fig. 13.22). Regeneration of the lens in the newt eye from the pigmented epithelium of the iris is an example of transdifferentiation (see Fig. 8.32). In contrast, in frogs (anuran amphibians), regeneration occurs in larval stages but is generally lost at metamorphosis. Another adult vertebrate that can regenerate new organs is the zebrafish, which can regenerate fins, and even the heart after removal of part of the ventricle. In contrast, mammals cannot regenerate lost limbs, although they have the capacity to replace the very ends of the digits, and are unable to regenerate the heart.

As a general rule, the efficiency of wound healing decreases with age, and the same was thought to be true of regeneration. But, unexpectedly, it has recently been shown that regeneration of the newt lens can occur repeatedly, and that its efficiency does not diminish with age. In this long-term experiment, the lens was removed 18 times in the same animal, with successful regeneration still occurring when the animal was 30 years old.

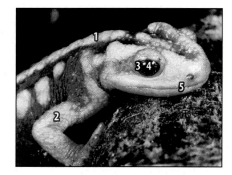

Fig. 13.22 The capacity for regeneration in urodele amphibians. The emperor newt can regenerate its dorsal crest (1), limbs (2), retina and lens (3 and 4), jaw (5), and tail (not shown).

Planarian	*Hydra*	Starfish
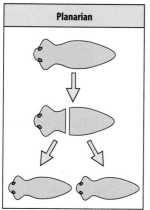	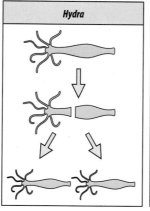	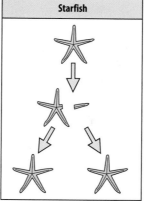

Fig. 13.23 Regeneration in some invertebrate animals. A planarian, *Hydra*, and a starfish all show remarkable powers of regeneration. When parts are removed or a small fragment isolated, a whole animal can be regenerated.

Some invertebrate animals also show great ability to regenerate: small fragments of animals such as starfish, planarians (flatworms), and *Hydra* can give rise to a whole animal (Fig. 13.23). This ability to regenerate a whole animal may be related to the ability of these animals to reproduce asexually; that is, to produce a complete new individual by budding or fission.

The issue of regeneration raises several major questions. What is the origin of the cells that give rise to the regenerated structures? What mechanisms pattern the regenerated tissue and how are these related to the patterning processes that occur in embryonic development?

We will mainly focus here on regeneration of limbs in urodele amphibians (newts and axolotls) and insects. We will also look briefly at regeneration of heart muscle in zebrafish. Understanding regeneration in these systems could lead to progress in stimulating it in mammalian tissues and so help the development of medical ways of repairing tissues such as the heart. There is also the question of whether stem cells are involved in regeneration, and so whether stem cells might be used to help repair damaged tissues (see Chapter 8).

13.14 There are two types of regeneration—morphallaxis and epimorphosis

Regeneration requires, at the minimum, the production of a population of cells whose growth, differentiation, and patterning is regulated. A distinction has been drawn between two types of regeneration. In one—**morphallaxis**—there is little new cell division and growth, and regeneration of structure occurs mainly by the repatterning of existing tissue and the re-establishment of boundaries. Regeneration of the whole body in the freshwater cnidarian *Hydra* is a good example of morphallaxis (Box 13D). By contrast, regeneration of a newt limb depends on the growth of completely new, correctly patterned structures, and this is known as **epimorphosis**. Both types of regeneration can be illustrated with reference to the French flag pattern (Fig. 13.24). In morphallaxis, new boundary regions are first established and new positional values are specified in relation to them; in epimorphosis, new positional values are linked to growth from the cut surface. In planarians, a population of adult stem cells that exists throughout the body produces the regenerated tissue (Box 13E), and regeneration involves both morphallaxis and epimorphosis.

13.15 Regeneration of amphibian and insect limbs involves epimorphosis

Urodele amphibians show a remarkable capacity for regenerating body structures such as tails, limbs, jaws, and the lens of the eye (see Fig. 13.22). Regeneration of these structures involves cell proliferation and new growth, and is therefore of the epimorphic type. Regeneration of a structure such as an adult vertebrate limb, which contains a range of fully differentiated cell types in a highly organized arrangement, raises the question of the origin of the cells that give rise to the regenerated structure.

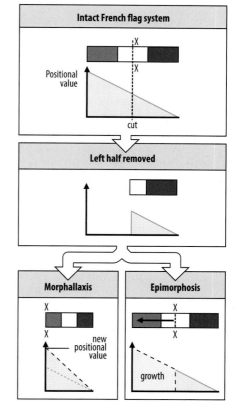

Fig. 13.24 Morphallaxis and epimorphosis. A pattern such as the French flag may be specified by a gradient in positional value (see Fig. 1.26). If the system is cut in half, it can regenerate in one of two ways. In regeneration by morphallaxis, a new boundary is established at the cut and the positional values are changed throughout. In regeneration by epimorphosis, new positional values are linked to growth from the cut surface.

BOX 13D Regeneration in *Hydra*

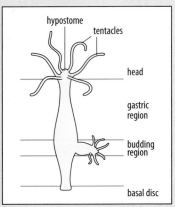

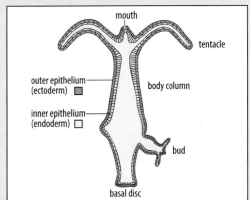

Figure 1

*Photograph reproduced with permission from Müller, W.A.: **Diacylglycerol-induced multihead formation in Hydra**. Development. 1989, **105**: 309–316. Published by permission of The Company of Biologists Ltd.*

The remarkable ability of the small freshwater cnidarian *Hydra* to regenerate the missing region of the body when cut transversely was discovered more than 200 years ago. Regeneration in *Hydra* does not require growth or cell proliferation and is a good example of morphallaxis (see Section 13.14). In morphallaxis, the remaining tissue is repatterned so that the missing region is replaced. Regeneration in *Hydra* has also been an important model for testing the principle of positional information.

Hydra consists of a hollow tubular body about 0.5 cm long. It has distinct antero-posterior polarity, with a head at one end and a basal region at the other, by which it sticks to a surface (Figure 1). The head consists of a conical hypostome with an opening from the gut surrounded by a number of tentacles. Unlike most of the animals discussed in this book, which have three germ layers, *Hydra* has only two. The body wall is composed of an outer epithelium, which corresponds to the ectoderm, and an inner epithelium lining the gut cavity, which corresponds to the endoderm. These two layers are separated by a gel-like basement membrane or mesoglea. There is also a population of multipotent stem cells, called interstitial cells, which give rise to about 20 different specialized cell types, including neurons, secretory cells, and the stinging cells, or nematocytes, that are used to capture prey. *Hydra* has no central nervous system but has a network of neurons distributed throughout the body.

Hydra grows continuously in that there is dynamic turnover of cells. Both endoderm and ectoderm cells in the gastric region of the body divide every 3–4 days, and daughter cells get displaced towards the head or the tail where they are either lost

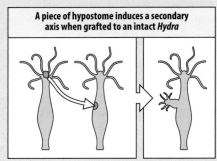

A piece of hypostome induces a secondary axis when grafted to an intact *Hydra*

Figure 2

or participate in forming buds. The interstitial cells in the gastric region also self-renew but at a faster rate. This means that cells in the adult animal are continually changing their relative positions and are forming new structures as they move up or down the body column. Cells are repatterned during this dynamic process, an ability that gives *Hydra* its remarkable capacity for regeneration.

If the body column of a *Hydra* is cut transversely, the lower piece regenerates a head and the upper piece a foot. Two small *Hydra* are produced, each half the size of the original. Even more dramatically, a mixture of dissociated *Hydra* cells will reform an animal. Heavily irradiated *Hydra*, in which no cell divisions occur, can still regenerate more or less normally, showing that regeneration does not require growth. The ectoderm and endoderm cells are the main contributors to regeneration, as *Hydra* lacking interstitial cells can still regenerate.

Do the differentiated cells dedifferentiate and start dividing? Are there special reserve cells? Do existing cells change their character after dedifferentiation? Damaged insect legs and other appendages can also regenerate. Insect regeneration has been studied mainly in the larval cockroach, which has relatively large legs that are easy to manipulate, and we shall also briefly discuss this system.

Hydra has a well-defined overall polarity and this determines that the appropriate structure is regenerated. A head regenerates at an anterior-facing cut surface whereas a foot regenerates at a posterior-facing cut surface. The basis for this polarity was revealed by grafting experiments. When a small fragment of the hypostome region is grafted into the gastric region of another *Hydra*, a new body axis is induced complete with head and tentacles (Figure 2). Similarly, transplantation of a fragment of the basal region induces a new body column with a basal disc at its end. *Hydra*, therefore, has two organizing regions, one at each end, which give the animal its overall polarity.

Figure 3

Additional grafting experiments revealed that the hypostome is also the source of a long-range diffusible inhibitor of head formation, which prevents the inappropriate formation of extra heads by lateral inhibition (see Section 1.16). The basal disc is similarly thought to produce an inhibitor of foot formation. The organizing regions at the ends of the body therefore each produce a pair of signals, which form opposing gradients along the body column: one signal from each organizer specifies a gradient in positional values and the other signal gradient inhibits formation of either a head or a foot.

A simple model for head regeneration in *Hydra* is based on the pair of signal gradients produced by the head organizer. Both gradients are linear, and their values decrease at a constant rate with distance from the head (see Figure 3, graph 1). The model proposes that, in the intact animal, provided the level of inhibitor (indicated by the gradient labeled I) at every point along the body is greater than the threshold set by the positional value at those points (indicated by the gradient labeled P), head regeneration is inhibited. When the head is removed, however, the concentration of inhibitor falls at the cut surface (see Figure 3, graph 2), and when the inhibitor falls below the threshold set by the positional value at that point, the positional value of cells at the cut surface increases to that of a normal head region (see Figure 3, graph 3). Thus, the first key step is the specification of a new head region at the cut surface. The new head then starts to make inhibitor and the inhibitory gradient is re-established. The gradient in positional value also returns to normal, but this can take more than 24 hours.

Genes homologous with those associated with vertebrate embryonic organizers are expressed in the adult *Hydra* head organizer, suggesting an ancient evolutionary origin for such organizers. For example, the *Wnt3* homolog is expressed at the tip of the hypostome in an adult *Hydra*. Within an hour after amputation of the head region, *Wnt3* is expressed at high levels throughout the cut surface (Figure 4). This local Wnt signaling seems to be involved in setting up the head organizer and is required for regeneration, because in its absence the head does not regenerate. Furthermore, when Wnt signaling is activated throughout the body of a transgenic *Hydra*, multiple heads are generated all along the body column.

Hydra has also pointed to the possible importance of apoptosis in regeneration. Bisection of a *Hydra* in the mid-gastric region also leads to upregulation of *Wnt3* expression within a few hours in endodermal cells at the head-regenerating surface. This upregulation is in response to Wnt3 signals released by interstitial cells, which undergo apoptosis at the amputation site. When apoptosis is inhibited, endodermal *Wnt3* expression is not induced and regeneration is blocked. Apoptosis is also required for regeneration of tadpole tails, raising the possibility that signals from dying cells are a general trigger for regeneration.

Figure 4

Photographs courtesy of B. Hobmayer and T.W. Holstein, from Hobmayer, et al.: **WNT signalling molecules act in axis formation in the dipoblastic metazoan Hydra.** *Nature 2000,* **407**: *186–189.*

13.16 Amphibian limb regeneration involves cell dedifferentiation and new growth

Amputation of a newt limb is followed by a rapid migration of epidermal cells from the edges of the wound over the wound surface. The wound epidermis (also known as the **apical epidermal cap**) is essential for subsequent regeneration; if its formation

BOX 13E Planarian regeneration

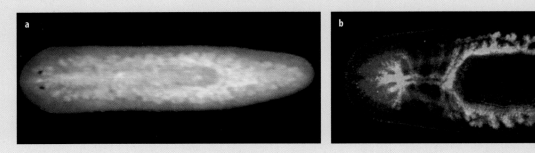

Figure 1

a from Adell, et al.: **Gradients in planarian regeneration and homeostasis.** *CSH Perspect., 2012,* **2***: a000505. b from Newmark, P.A., Sánchez-Alvarado, A.:* **Regeneration in planaria.** *Encyclopedia of Life Sciences. Wiley, 2001.*

Planarians are free-living flatworms with three body layers—ectoderm, mesoderm and endoderm—but no coelom. They have a complex anatomy with various organ systems, including a nervous system, kidneys, reproductive organs, and gut. Some species of planarians are unique among bilaterian animals in having a large pool of adult stem cells that enables them to regenerate any part of the body, including the brain. Planarian regeneration is unparalleled among animals. When one of these animals is cut into many small pieces, each piece will regenerate and give rise to a complete new animal. The minimum size of fragment that can regenerate is 1/279th (0.3%) of an adult animal—about 1000 cells.

The species most used for studying regeneration is the freshwater planarian *Schmidtea mediterranea,* which is about 0.1–2 cm long (Figure 1). The body is bilaterally symmetrical and has distinct antero-posterior polarity, with a well-developed head at one end and a tail at the other, and also distinct dorso-ventral polarity, with, for example, two eye spots on the dorsal side of the head (see Figure 1a). The mouth is in the middle of the ventral surface, through which a muscular pharynx protrudes during feeding.

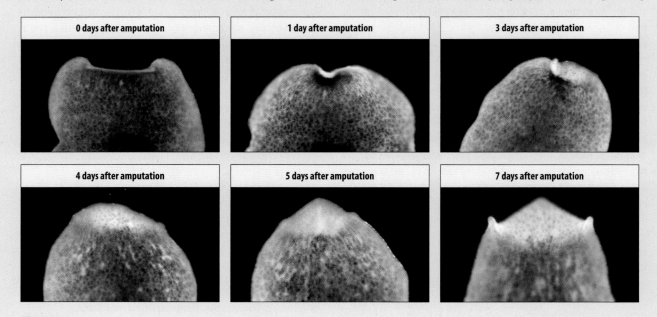

Figure 2

From Newmark, P.A., Sánchez-Alvarado, A.: **Regeneration in planaria.** *Encyclopedia of Life Sciences. Wiley, 2001.*

is prevented by suturing the edges of the stump together, there is no regeneration. The apical cap may have a role similar to that of the apical ectodermal ridge in embryonic limb development (see Section 11.3). A mass of undifferentiated cells called the **blastema** then forms under the apical cap and this mass of cells gives rise to the regenerated limb (Fig. 13.25). The blastema cells start to divide, eventually forming

Figure 1b shows a ventral view with the gut (green), the pharynx in the center, and the nerve cords (red) coming from the head end.

Adult planarians constantly replace their tissues. In planarians, the source of new cells is a population of small, undifferentiated cells known as **neoblasts**, which are the only cells in the adult animal that are mitotically active. Neoblasts constitute about 25–30% of the total number of cells and are scattered throughout the body. They not only self-renew but also can differentiate into all the different cell types—epidermis, muscle, neurons, and germ cells.

Both morphallaxis and epimorphosis (see Section 13.14) are involved in planarian regeneration. Morphallaxis is clearly seen by the remodeling of fragments of the body that regenerate to form a complete small animal. But heavily irradiated planarians are unable to regenerate, and regeneration of the head, for example, following amputation involves epimorphosis, with the early proliferation and migration of neoblasts to form a blastema. The blastema resembles that of an amphibian limb blastema in that it consists of a mass of undifferentiated cells covered by a wound epithelium. The head blastema provides at least some of the replacement parts needed for the new head, including the brain, which forms over the next 1 to 2 weeks (Figure 2). Neoblasts are particularly sensitive to radiation and this explains why irradiated planarians are unable to regenerate.

A key issue in planarian regeneration is whether the neoblasts are pluripotent and able to give rise to all the cell types in the regenerated tissue, or whether the population of neoblasts is heterogeneous, with individual cells being lineage restricted. We will come across a similar issue about multipotency in relation to blastema cells in amphibian limb regeneration (see Fig. 13.36). In planarians, this question can be addressed by transplanting a single neoblast cell into an adult planarian that has been irradiated to destroy its own population of neoblasts, and following the fate of the transplanted cell's progeny. Some transplanted neoblasts were found to be pluripotent, and produced colonies

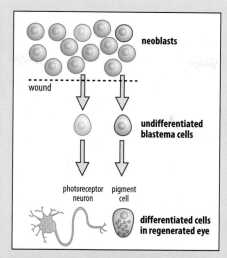

Figure 3

From Reddien.: **Specialized progenitors and regeneration**. *Development 2013,* **140**: *951-957.*

of neoblasts that differentiated to form neurons, intestinal cells, and other cell types. Others did not produce colonies, however, and so it is not clear whether all neoblasts are pluripotent. Indeed, during eye regeneration, specialized neoblasts give rise to pigment cells and other specialized neoblasts give rise to neurons (Figure 3). This lineage restriction of planarian blastema cells for the eye resembles the lineage restriction of blastema cells in the amphibian limb.

The planarian body has intrinsic polarity. Antero-posterior polarity is maintained even in small fragments during regeneration, so that a head regenerates at an anterior-facing cut surface whereas a tail regenerates at the posterior cut surface. Around the beginning of the 20th century it was suggested that this polarity could be based on opposing gradients of "head stuffs" and "tail stuffs". More than 100 years later, a gradient of Wnt signaling that is critical for maintaining antero-posterior polarity has been discovered. A planarian Wnt signaling protein (Wnt-P1) is expressed in a small number of cells in the tail in the adult animal, and diffuses anteriorly to form a gradient. The Wnt signaling suppresses head formation along this gradient except at the anterior end where it is weakest. When Wnt signaling is reduced in an uninjured planarian by knocking down β-catenin expression for an extended period of time, ectopic heads emerge all over the body, and in a regenerating planarian, a head instead of a tail develops from a posterior-facing cut surface, giving a two-headed animal. When Wnt signaling was reduced in non-regenerative species of planarians, they were able to regenerate a fully functional head.

Planarians can also regenerate the missing lateral half of the body if cut in half longitudinally. This regeneration requires BMP-4, which is normally expressed in a stripe along the dorsal midline. BMP signaling is also necessary for re-establishing a dorso-ventral axis during regeneration. All in all, the remarkable ability of adult planarians to regenerate seems to depend on using the signals that normally maintain tissue homeostasis.

an elongated cone. Over a period of weeks, as the limb regenerates, the blastema cells differentiate into cartilage, muscle, and connective tissue. The undifferentiated cells of the early blastema are derived locally from the differentiated tissues of the stump, close to the site of amputation. They come from the dermis in particular, but also from cartilage and muscle.

The ability of newt skeletal muscle cells to dedifferentiate and return to a proliferating state in the blastema is particularly intriguing, as vertebrate skeletal muscle cells do

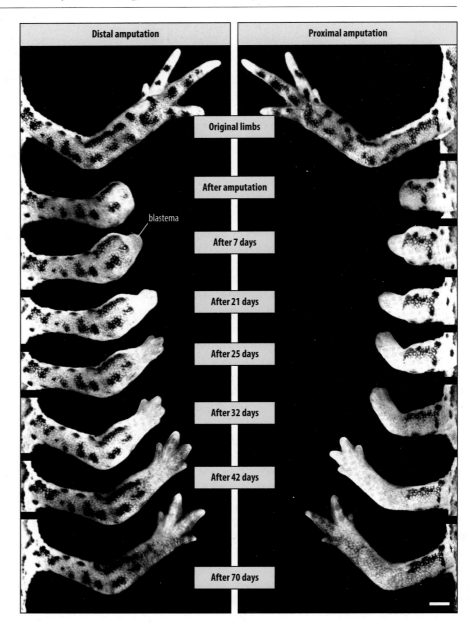

Distal amputation

Proximal amputation

Original limbs

After amputation

blastema

After 7 days

After 21 days

After 25 days

After 32 days

After 42 days

After 70 days

Fig. 13.25 Regeneration of the forelimb in the red-spotted newt *Notophthalmus viridescens*. The left panel shows the regeneration of a forelimb after amputation at a distal (mid-radius/ulna) site. The right panel shows regeneration after amputation at a proximal (mid-humerus) site. At the top, the limbs are shown before amputation. Successive photographs were taken at the times shown after amputation. Note that the blastema gives rise to structures distal to the cut. Scale bar = 1 mm.

not normally divide after becoming fully differentiated. A general feature of vertebrate muscle differentiation is the withdrawal from the cell cycle after myoblast fusion to produce myotubes (see Section 8.5). This withdrawal involves the dephosphorylation of the cell-cycle control protein RB (retinoblastoma protein) (see Section 13.5). Cultured mouse muscle cells lacking RB can re-enter the cell cycle and in myotubes in the regenerating newt limb, RB protein is inactivated by phosphorylation. The myotube then fragments to form mononucleate cells that divide. Re-entry into the cell cycle is also associated with the activation of thrombin in the blastema. Indeed, thrombin could be a key factor in regeneration in several systems, as the regeneration of the newt lens from the dorsal margin of the iris also correlates with thrombin activity in that region.

Thrombin is a proteolytic enzyme that is more familiar as part of the blood-clotting cascade, but seems also to be involved in providing an environment for dedifferentiation. Multinucleate post-mitotic newt muscle cells can be returned to the cell cycle in culture in the presence of thrombin in the culture medium. Dedifferentiation of the muscle cells also involves renewed expression of the homeodomain transcription factor Msx1. This multifunctional transcriptional factor is known to prevent myogenic

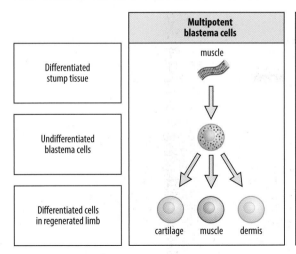

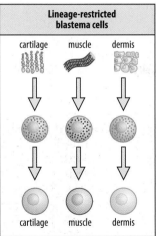

Fig. 13.26 Are blastema cells multipotent or are they lineage restricted? Multipotent blastema cells derived from differentiated tissues would be able to give rise to all the tissues in the limb—for example, cartilage, muscle and dermis. Lineage-restricted blastema cells would only be able to give rise to their own tissue of origin.

differentiation in mammalian cells (see Section 8.6), and its expression is characteristic of undifferentiated mesenchymal cells that can undergo regeneration. Cells that express Pax7, functionally similar to the satellite stem cells that regenerate skeletal muscle fibers in adult mammals (see Section 8.13), have also been identified in adult newts and axolotols.

One of the key questions in limb regeneration is whether the limb blastema cells are multipotent or whether they are lineage restricted (Fig. 13.26). Can blastema cells derived from muscle, for example, give rise to all the differentiated cell types in the regenerated limb? In other words, can they transdifferentiate (see Section 8.17)? Or do they just give rise to muscle?

Although early work suggested that cells in the amphibian blastema are multipotent, more recent work on the regenerating limb of the axolotl *Ambystoma mexicanum* revealed that cells in the blastema are indeed lineage restricted. In these experiments, the blastema cells did not revert to a multipotent state but retained a restricted developmental potential related to their origin. The fate of individual tissue types in a regenerating limb was traced by transplanting tissue from a transgenic axolotl line expressing green fluorescent protein (GFP). A patch of tissue of a particular type, such as muscle or epidermis, was transplanted into the forelimbs of non-transgenic juveniles of the same species. The forelimb of the juvenile axolotl was amputated across the site of the transplant, and the fate of the glowing green transplanted cells could then be traced over time as the limb regenerated (Fig. 13.27). The tissue type made by the fluorescent cells after redifferentiation was then identified by testing for tissue-specific markers.

These experiments showed that transplanted muscle tissue produced blastema cells that only gave rise to muscle (Fig. 13.28). Additional cell-lineage tracing experiments have shown that the regenerated muscle in axolotl limbs is derived from resident *Pax7*-expressing satellite cells, whereas in newt limbs it is derived from as a result of dedifferentiation of muscle cells. Cartilage and epidermal cells were also restricted in their further development, as were the glial Schwann cells, which provide the myelin coating of peripheral nerve fibers.

One exception was the cells of the dermis, which could contribute to both new dermis and new cartilage skeleton. This investigation also found that cartilage-derived blastema cells retained a "memory" of their original positional identity along the proximo-distal axis of the limb. Thus cartilage cells transplanted from a proximal position in the original limb give rise to more distal regions in the regenerating limb, whereas cartilage cells transplanted from distal regions do not give rise to more proximal structures. In contrast, muscle cells do not seem to have any positional identity, suggesting that, as in limb development (see Section 11.15), the connective tissues guide muscle patterning.

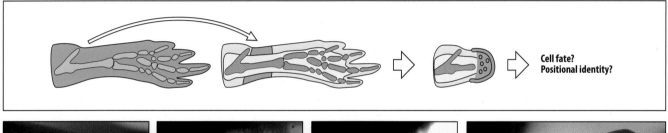

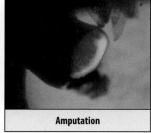
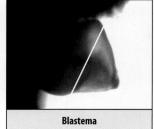
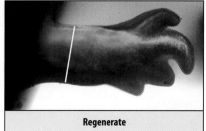

| Graft | Amputation | Blastema | Regenerate |

Fig. 13.27 Cells that regenerate the axolotl limb have restricted developmental potential. The experimental procedure is shown in the top row and photographs of the limb at various stages are shown in the bottom row. Cells of a particular tissue type (for example, dermis as shown here) and position (for example, proximal as shown here) in the limb of a juvenile axolotl transgenic for green fluorescent protein, were transplanted into the limb of a non-transgenic animal and the limb amputated across the graft. In the regenerating limb that grew from the blastema, the transplanted cells and their progeny can be detected by their green fluorescence. These cells could then be tested for their tissue type. In the case illustrated, the transplanted dermis cells could give rise to new dermis and to the cartilage skeleton, but not to muscle. Figure 13.28 shows the lineage restrictions discovered. Cells taken from a proximal position in the limb could give rise to distal structures (as seen here by the labeled cells in the digits), but transplanted cartilage cells would not give rise to more-proximal structures.

From Kragl, M., et al.: ***Cells keep a memory of their tissue origin during axolotl limb regeneration***. Nature, *2009,* ***460****: 60-65.*

It has recently been discovered that macrophages are required for limb regeneration in adult axolotls. This may be part of the injury response that triggers initiation of regeneration. It is well known that macrophages are recruited to wounds in mammals, and that this leads to scarring. It was surprising, therefore, that within the first 24 hours after limb amputation, macrophages were seen in and around the wound epithelium. Furthermore, depletion of macrophages in the axolotl during the few days before limb amputation blocks regeneration completely. However, these stumps can regenerate when re-amputated, once macrophages have been replenished. Expression of several genes normally expressed in the early blastema, including *Msx1*, is reduced in the absence of macrophages, suggesting that macrophages may be involved in promoting dedifferentiation.

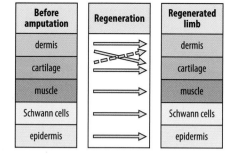

Before amputation	Regeneration	Regenerated limb
dermis		dermis
cartilage		cartilage
muscle		muscle
Schwann cells		Schwann cells
epidermis		epidermis

Fig. 13.28 Lineage restriction of cells in the axolotl blastema. Blastema cells derived from a particular tissue are in general lineage restricted and only contribute to the same tissue in the regenerated limb. Blastema cells derived from the dermis contribute to both dermis and cartilage and there is a small contribution by blastema cells derived from cartilage to the dermis.

13.17 Limb regeneration in amphibians is dependent on the presence of nerves

Growth and development of the amphibian limb blastema not only depends on the overlying wound epidermis (the apical cap), but also on its nerve supply (Fig. 13.29). In amphibian limbs in which the nerves have been cut before amputation, a blastema forms but fails to grow. The nerves have no influence on the character or pattern of the regenerated structure; it is the amount of innervation—not the type of nerve—that matters. In regenerating adult newt limbs, the nerves provide an essential growth factor called anterior gradient protein (AG). This is secreted initially by the glial Schwann cells of the incoming nerves and later by the glandular cells of the wound epidermis. Treatment of the blastema of a denervated limb with AG is sufficient to enable it to regenerate completely.

An interesting phenomenon is that if embryonic newt limbs are denervated very early in their development, and so do not become exposed to the influence of nerves, they can regenerate in the complete absence of any nerve supply (see Fig. 13.29, fourth panel). If an aneurogenic limb is subsequently innervated, however, it rapidly

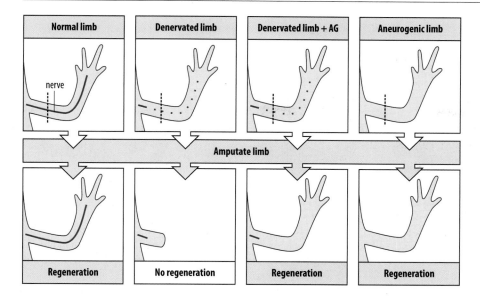

Fig. 13.29 Innervation and limb regeneration. Normal limbs require a nerve supply to regenerate (first panels). Limbs denervated before amputation will not regenerate (second panels). However, if the limb is amputated and the cut surface is electroporated with plasmid DNA encoding newt AG, regeneration is rescued (third panels). Limbs that have never been innervated, because the nerve was removed during development, can regenerate normally in the absence of innervation (fourth panels).

becomes dependent on nerves for regeneration. This suggests that dependence on innervation is imposed on the limb only after the nerves grow into it. It is now known that these phenomena can be attributed to the levels of AG. During normal limb development, AG is highly expressed in the epidermis, but its expression is reduced when the nerves grow in, and remains low in the adult. In denervated limbs, this reduction in expression does not occur, and high levels of AG persist in the adult, which explains why these limbs can regenerate even though they are not innervated. When an aneurogenic limb is subsequently innervated, AG is downregulated in the epidermis and thus the limb once more becomes dependent on nerves for regeneration.

A striking instance of the influence of nerves on regeneration of amphibian limbs is that if a major peripheral nerve, such as the sciatic nerve, is cut and the branch inserted into a wound on a limb or on the surface of the adjacent flank, a supernumerary limb develops at that site. This experimental system provides an opportunity to study limb regeneration in the absence of the considerable cell damage caused by amputation.

13.18 The limb blastema gives rise to structures with positional values distal to the site of amputation

Regeneration always proceeds in a direction distal to the cut surface, enabling replacement of the lost part. If the hand is amputated at the wrist, only the carpals and digits are regenerated, whereas if the limb is amputated through the middle of the humerus, everything distal to the cut (including the distal humerus) is regenerated. Positional value along the proximo-distal axis is therefore of great importance, and is at least partly retained in the blastema. The blastema has considerable morphogenetic autonomy. If it is transplanted to a neutral location that permits growth, such as the dorsal crest of a newt larva or even the anterior chamber of the eye, it gives rise to a regenerated structure appropriate to the position from which it was taken.

The growth of the blastema and the nature of the structures it gives rise to depend on the site of the amputation and not on the nature of the more proximal tissues. The limb is not, however, simply 'trying' to replace missing parts. This was shown in a classic experiment in which the distal end of a newt limb that had been amputated at the wrist was inserted into the belly of the same animal, so as to establish a blood supply to it. The limb was then cut mid-humerus. Both surfaces regenerated distally, even though the part attached to the belly already had a radius and ulna (Fig. 13.30).

The blastema is much larger than the embryonic limb bud—it can be ten times larger in terms of cell numbers. This size makes it most unlikely that signals can

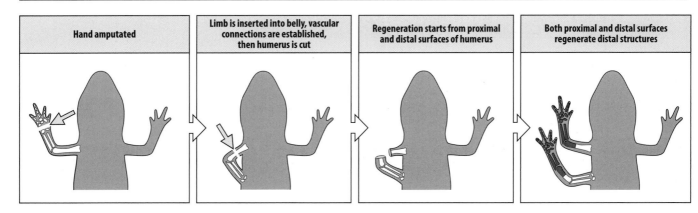

| Hand amputated | Limb is inserted into belly, vascular connections are established, then humerus is cut | Regeneration starts from proximal and distal surfaces of humerus | Both proximal and distal surfaces regenerate distal structures |

Fig. 13.30 Limb regeneration is always in the distal direction. The distal end of a limb is amputated and the limb inserted into the belly. Once vascular connections are established, a cut is made through the humerus. Both cut surfaces regenerate the same distal structures even though, in the case of one of the regenerating limbs, distal structures are already present.

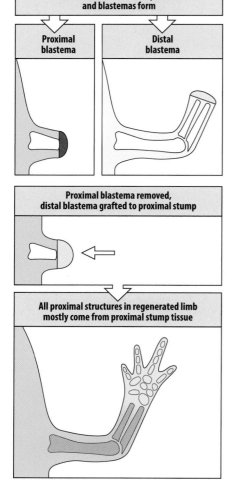

diffuse right across the blastema. Instead, regeneration can best be understood in terms of an adult limb having a set of positional values along its proximo-distal axis, which have been set up during embryonic development (see Section 11.6). The regenerating limb in some way reads the positional value at the site of the amputation and then regenerates all positional values distal to it. The ability of cells to recognize a discontinuity in positional values is illustrated by grafting a distal blastema to a proximal stump. In this experiment, the forelimb's stump and blastema have different positional values, corresponding to shoulder and wrist, respectively. The result is a normal limb in which structures between the shoulder and wrist have been generated by **intercalary growth**, predominantly from the proximal stump, whereas the cells from the wrist blastema mostly give rise to the hand (Fig. 13.31).

A fundamental question in any discussion of pattern formation is how the proposed positional information is encoded molecularly. A major advance on this front has been the identification of Prod1, a cell-surface protein that is expressed in a graded manner along the proximo-distal axis of the newt limb, with higher levels of expression proximally. Proximo-distal values in the blastema can be altered by treatment with retinoic acid, as discussed in more detail in the next section, and retinoic-acid treatment was used to compare gene expression in blastemas whose proximo-distal values had been altered. Treatment with retinoic acid increases the level of expression of the *Prod1* gene and expression is higher in "proximal" blastemas compared with "distal" blastemas. Prod1 might therefore encode positional values. The growth factor AG, which is essential for limb regeneration (see Section 13.17), is a ligand for Prod1. It is thought that AG has no input into generating proximo-distal pattern, but acts through Prod1 to promote cell division.

The gradient in Prod1 fits very well with the observation that cell-surface properties are involved in regeneration. When mesenchyme from two blastemas from different proximo-distal sites are confronted in culture, the more proximal mesenchyme engulfs the distal (Fig. 13.32, left panels), whereas mesenchyme from two blastemas from similar sites maintains a stable boundary. This behavior suggests that there is a graded difference in cell adhesiveness along the axis, with adhesiveness being highest distally. We have seen such behavior in mixtures of explants of ectoderm, mesoderm and endoderm from amphibian blastulas, in which tissues normally adjacent and adherent to each other, but with different degrees of cohesion, will envelop each other (see Section 9.1). In the blastema explants, the cells in the distal explant remain more tightly bound to each other than do the cells from the proximal blastema, which therefore spreads to a greater extent. Prod1 is involved in this process because when a blocking antibody to Prod1 is added to the culture, the proximal explant fails to spread over the distal explant.

Fig. 13.31 Proximo-distal intercalation in limb regeneration. A distal blastema grafted to a proximal stump results in intercalation of all the structures proximal to the distal blastema. Almost all of the intercalated tissue comes from the proximal stump.

Donor and host limbs amputated and blastemas form

Proximal blastema

Distal blastema

Proximal blastema removed, distal blastema grafted to proximal stump

All proximal structures in regenerated limb mostly come from proximal stump tissue

Fig. 13.32 Cell-surface properties vary along the proximo-distal axis. Left panels: when mesenchyme from distal and proximal blastemas is placed in contact in culture, the proximal mesenchyme engulfs the distal mesenchyme, which has greater adhesion between its cells. Center panels: if an anti-Prod1 antibody is added to the cultured blastemas, the proximal tissue does not envelop the distal tissue. Right panels: if a distal blastema (in this case from a cut wrist) is grafted to the dorsal surface of a more proximal blastema, the regenerating wrist blastema will move distally to a position on the host limb that corresponds to its original level and regenerates a hand.

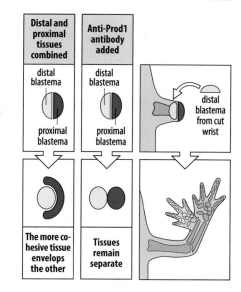

A difference in the adhesiveness of tissues along the proximo-distal axis is also suggested by the behavior of a distal blastema when grafted to the dorsal surface of a proximal blastema so that their mesenchymal cells are in contact. Under these conditions, the distal blastema moves during limb regeneration to end up at the site from which it originated (see Fig. 13.32, right panels). This suggests that its cells adhere more strongly to the regenerated wrist region than they do to the proximal region to which it was grafted. When a shoulder-level blastema is grafted to a shoulder stump, it does not move but leads to a normal distal outgrowth from the shoulder blastema.

All these experiments suggest that proximo-distal positional values in limb regeneration in urodele amphibians are encoded as a graded property, probably in part at the cell surface, and that cell behavior—growth, movement, and adhesion—relevant to axial specification is a function of the expression of this property, relative to neighboring cells.

Maintaining the continuity of positional values by intercalation is a fundamental property of regenerating epimorphic systems, and later in the chapter we will consider it in relation to the cockroach leg. It is not clear to what extent all blastemal cells inherit a particular positional value from their differentiated precursors, and to what extent they are subject to signals that induce the appropriate expression of positional value. In the axolotl experiment outlined in Fig. 13.27, cartilage cells retained their original positional identity whereas muscle cells did not. Normal regeneration by outgrowth from a blastema could be considered to be the result of intercalation between the cells at the level of amputation and those with the most distal positional values, as specified by the wound epidermis. But although intercalation was originally thought to be involved, it has now been shown that cells in the regenerating blastema are specified in a proximo-distal direction, as has been suggested to occur in developing limbs (see Chapter 11).

13.19 Retinoic acid can change proximo-distal positional values in regenerating limbs

In Chapter 11 we saw how experimental treatment with retinoic acid can alter positional values in the developing chick limb (see Section 11.8). It also has striking effects on regenerating amphibian limbs.

When a regenerating limb is exposed to retinoic acid, the blastema becomes proximalized; that is, the limb regenerates as if it had originally been amputated at a more proximal site. For example, if a limb is amputated through the wrist, treatment with retinoic acid can result not only in the regeneration of the elements distal to the cut, but also in the production of an extra complete radius and ulna and humerus (Fig. 13.33). The effect of retinoic acid is dose-dependent, and with a high dose it is possible to regenerate a whole extra limb, including part of the shoulder girdle, on a limb from which only the hand has been amputated. Retinoic acid therefore alters the proximo-distal positional value of the blastema, making it more proximal. It probably does this by affecting various pathways for specifying positional information, in particular increasing the expression of *Prod1* (see Section 13.18), as overexpression of *Prod1* in distal blastema cells also causes them to translocate to more proximal positions. Retinoic acid also appears to shift positional values in the blastema in a proximal direction through the activation of *Meis* homeobox genes, which are involved in specifying proximal positional values in the developing limb and are normally repressed in distal regions. Retinoic acid can also,

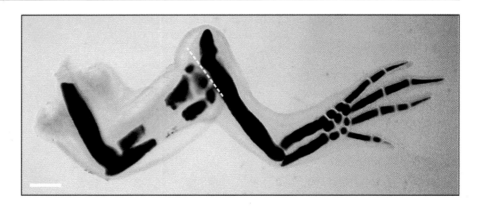

Fig. 13.33 Retinoic acid can proximalize positional values. A forelimb amputated at the level of the hand, as indicated by the dotted line, and then treated with retinoic acid, regenerates structures that would normally arise from amputating at the level of the proximal end of the humerus. Scale bar = 1 mm.

Photograph courtesy of M. Maden.

under some experimental conditions, shift positional values along the antero-posterior axis in a posterior direction and duplicated digits can be produced.

Retinoic acid can act through several different nuclear receptors, but only one (RAR δ2) is involved in changes in proximo-distal positional value in the regenerating limb. By constructing a chimeric receptor from a RAR δ2 DNA-binding portion and a thyroxine receptor hormone-binding portion, the retinoic acid receptor can be selectively activated by thyroxine, so that the effects of its activation can be studied experimentally (Fig. 13.34). Cells in the distal blastema are first transfected with the chimeric receptor gene and then grafted to a proximal stump and treated with thyroxine. The transfected cells behave as if they have been treated with retinoic acid. The result is a movement of the transfected blastemal cells to more proximal regions in the intercalating regenerating blastema, which shows that activation of the retinoic-acid pathway can proximalize the positional values of the cells, which then respond by movement to more proximal sites. Prod1 is a very strong candidate for mediating this change, as it too can cause cells to move proximally when its concentration is increased.

Another remarkable effect of retinoic acid is its ability to bring about a homeotic transformation of tails into limbs in tadpoles of the frog *Rana temporaria*. If the tail of a tadpole is removed, it will regenerate. But treatment of regenerating tails with retinoic acid at the same time as the hindlimbs are developing results in the appearance of additional hindlimbs in place of a regenerated tail (Fig. 13.35). There is, as yet, no satisfactory explanation for this result, but it has been speculated that the retinoic acid alters the antero-posterior positional value of the regenerating tail blastema to that of the site along the antero-posterior axis where hindlimbs would normally develop.

Fig. 13.34 Retinoic acid proximalizes the positional value of individual cells. Some of the cells of a newt distal blastema are transfected by a chimeric receptor, through which retinoic acid receptor function can be activated by thyroxine. This blastema is grafted to a proximal stump and treated with thyroxine. During intercalary growth, the transfected cells, which have been labeled, move proximally because their positional values have been proximalized by the activation of retinoic acid receptor function. The photographs illustrate proximalization of the transfected cells. Scale bar = 0.5 mm.

*Photographs reproduced with permission from Pecorino, L.T., et al.: **Activation of a single retinoic acid receptor isoform mediates proximodistal respecification**. Curr. Biol. 1996, **6**: 563-569. Copyright Elsevier.*

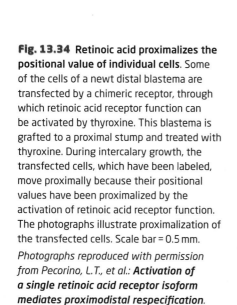

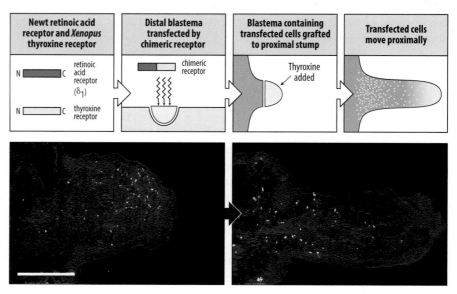

Fig. 13.35 Retinoic acid can induce additional limbs in the regenerating tail of a frog tadpole. After amputation, the regenerating tail stump of a tadpole of *Rana temporaria* is treated with retinoic acid at the time when hindlimbs are developing. This results in the appearance of additional hindlimbs in place of a new tail. Scale bar = 5 mm.

Photograph courtesy of M. Maden.

(In the photograph: label "normal limb")

13.20 Mammals can regenerate the tips of the digits

Although mammals cannot regenerate whole limbs, many mammals, including young children, can regenerate the tips of their digits, provided the nail-generating tissue—the nail organ—is still present. In mice and children, the level from which digits are able to regenerate is limited to the base of the claw or nail, respectively. This probably reflects the presence of connective-tissue cells under the nail organ that express the transcription factor *Msx1*, which regulates BMP-4 expression and is associated with undifferentiated mesenchymal cells, as noted earlier. In mammalian limb development, Msx1 is expressed at the tip of the embryonic limb bud, and in mice it continues to be expressed in the tips of the digits even after birth. The region in which digit regeneration in mice can occur corresponds with the domain of *Msx1* expression.

The cells that proliferate to regenerate the digit tip in the mouse seem to be lineage-restricted, as in the regenerating amphibian limb. For example, in transgenic mice expressing GFP-labeled *Sox9*, which would label skeletal-cell precursors, only the bone cells in the skeleton of the regenerated digit tip are labeled with GFP, whereas cells in other tissues such as tendons and dermis are unlabeled. Wnt signaling is activated during digit-tip regeneration and promotes the growth of nerves, which could be essential for regeneration, as in the amphibian limb. The significance of the regenerative capacity of the tips of mammalian digits is not clear. It is important to consider how the ability to regenerate appendages may have evolved because this will help us to answer the question of why we can't regenerate our limbs (Box 13F).

13.21 Insect limbs intercalate positional values by both proximo-distal and circumferential growth

The legs of some insects, such as the cockroach and cricket, can regenerate. Unlike *Drosophila*, these insects do not undergo a complete metamorphosis, and the juvenile stages (nymphs) are more similar to miniature adults. When the tibia of a third-instar nymph of the cricket *Gryllus bimaculatus* is amputated, it takes about 40 days to restore the adult leg. Wingless and Decapentaplegic signaling activate epidermal growth factor receptor (EGFR) signaling in a distal-to-proximal gradient in the cricket blastema, which patterns the regenerating distal leg.

In *Drosophila*, as we saw in earlier in the chapter, signaling through the cadherins Fat and Dachsous and the associated Hippo pathway coordinates cell proliferation and apoptosis in imaginal discs, determining their final size and thus the size of adult wings and legs (see Section 13.4). Experiments using RNA interference to investigate the role of Fat, Dachsous, and the Hippo pathway in cricket-limb regeneration suggest that this is also the case for an insect whose legs develop directly. Knockdown of either Fat or Dachsous results in a shorter than normal regenerated limb, whereas knockdown of another set of genes in the Hippo pathway produces a longer leg than normal. It is proposed that a Dachsous/Fat gradient exists along the limb and that the steepness of the gradient could control cell proliferation.

Regenerating insect legs follow an epimorphic process of blastema formation and outgrowth that intercalates missing positional values. This contrasts with the current view of the way in which the proximo-distal axis of the regenerating amphibian limb is specified (see Section 13.18). However, in amphibian limbs when cells with disparate positional values are placed next to one another, intercalary growth occurs in order to regenerate the missing positional values. Such intercalation of positional values seems to be a general property of epimorphically regenerating systems and is particularly clearly illustrated by limb regeneration in the cockroach. It also occurs in regenerating *Drosophila* leg and wing imaginal discs after amputation.

A cockroach leg is made up of a number of distinct segments, arranged along the proximo-distal axis in the order: coxa, femur, tibia, tarsus. Each segment seems to contain a similar set of proximo-distal and circumferential positional values, and will intercalate missing positional values during regrowth. When a distally amputated tibia is grafted onto a host tibia that has been cut at a more proximal site, localized growth occurs at the junction between graft and host, and the missing central regions of the tibia are intercalated (Fig. 13.36, left panels). In contrast to amphibian regeneration, there is a predominant contribution from the distal piece. As in the amphibian, however, regeneration is a local phenomenon and the cells are indifferent to the overall pattern of the tibia. Thus, when a proximally cut tibia is grafted onto a more distal site, making an abnormally long tibia, regenerative intercalation again restores the missing positional values, making the tibia even longer (Fig. 13.36, right panels). The regenerated portion

Fig. 13.36 Intercalation of positional values by growth in the regenerating cockroach leg. Left panels: when a distally amputated tibia (5) is grafted to a proximally amputated host (1), intercalation of the positional values 2–4 occurs, irrespective of the proximo-distal orientation of the grafts, and a normal tibia is regenerated. Right panels: when a proximally amputated tibia (1) is grafted to a distally amputated host (4), however, the regenerated tibia is longer than normal and the regenerated portion is in the reverse orientation to normal, as judged by the orientation of surface bristles. The reversed orientation of regeneration is due to the reversal in positional value gradient. The proposed gradient in positional value is shown under each figure.

*Illustration after French, V., et al.: **Pattern regulation in epimorphic fields**. Science 1976, **193**: 969-981.*

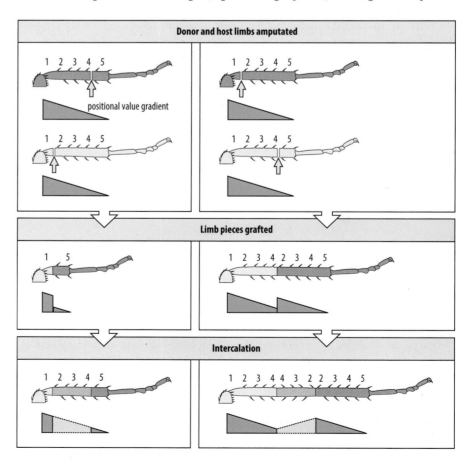

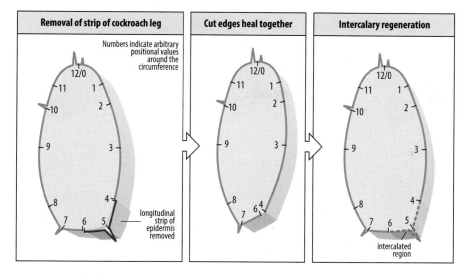

| Removal of strip of cockroach leg | Cut edges heal together | Intercalary regeneration |

Numbers indicate arbitrary positional values around the circumference

longitudinal strip of epidermis removed

intercalated region

Fig. 13.37 Circumferential intercalation in the cockroach leg. The leg is seen in transverse section. When a piece of cockroach ventral epidermis is removed (left panel), the cut edges heal together (center panel). When the insect molts and the cuticle regrows, circumferential positional values are intercalated (right panel). The positional values are arranged around the circumference of the leg, rather like the hours on a clock face.

*Illustration after French, V., et al.: **Pattern regulation in epimorphic fields**. Science 1976, **193**: 969–981.*

is in the reverse orientation to the rest of the limb, as indicated by the direction in which the bristles point, suggesting that the gradient in positional values also specifies cell polarity, as in insect body segments (see Chapter 2). These results show that when cells with non-adjacent positional values are placed next to each other, the missing values are intercalated by growth to establish a set of continuous positional values.

A similar set of positional values is present in each segment of the limb. Thus, a mid-tibia amputation, when grafted to the mid-femur of a host, will heal without intercalation. In contrast, a distally amputated femur when grafted onto a proximally amputated host tibia results in intercalation, largely femur in type. There must be other factors making each segment different, rather like the segments of the insect larva.

Intercalary regeneration also occurs in a circumferential direction. When a longitudinal strip of epidermis is removed from the leg of a cockroach, normally non-adjacent cells come into contact with one another, and intercalation in a circumferential direction occurs after molting (Fig. 13.37). Cell division occurs preferentially at sites of mismatch around the circumference. One can treat positional values in the circumferential direction as a clock face, with values going continuously 12, 1, 2, 3 … 6 … 9 … 11. As in the proximo-distal axis, there is intercalation of the missing positional values.

13.22 Heart regeneration in zebrafish involves the resumption of cell division by cardiomyocytes

Some vertebrates have the ability to regenerate the heart. For example, if 20% of the ventricle of an adult zebrafish heart is removed, it will regrow, and the regenerated tissue will become functionally integrated with the existing heart tissue (Fig. 13.38). This phenomenon is also seen in newts, but is of particular interest in the zebrafish, as in this model organism regeneration can be studied genetically.

One of the key questions in zebrafish heart regeneration, as in the regenerating limb, is the origin of the cells that make up the regenerated tissue. Genetic fate-mapping experiments show that existing mature muscle cells of the adult myocardium, the **cardiomyocytes**, are the main source of proliferating cells for regeneration, rather than undifferentiated progenitor cells. Cardiomyocytes proliferate at a very low rate in the adult zebrafish heart but after removal of the apex of the heart ventricle, cardiomyocytes in and around the wounded area in the ventricle dedifferentiate and start to divide. The new cardiomyocytes then migrate into the wound area and at the same time the regenerated tissue becomes vascularized. The new myocardial tissue becomes electrically coupled to the cardiomyocytes in the undamaged region of the heart between 2 and 4 weeks after injury.

Fig. 13.38 Time course of regeneration after the amputation of the apex of the ventricle of the adult zebrafish heart. Endocardial activation occurs throughout the heart within a few hours of amputation, epicardial activation a few days later marked by expression of the retinoic acid synthesizing enzyme gene, *raldh2* and genes involved in other signaling pathways. By 7 days after amputation, activation of the epicardium is localized to the wound area. Cardiomyocytes are stimulated to activate regulatory sequences controlling expression of *gata4*, a transcription factor involved in heart development and to proliferate. The regenerating muscle is vascularized over the next 7 days. By about 30 days, the myocardium has regenerated and is electrically coupled with the rest of the heart.

*Adapted from Figure 1 Gemberling, M., et al.: **The zebrafish as a model for complex tissue regeneration**. Trends Genet. 2013, 29: 611-620.*

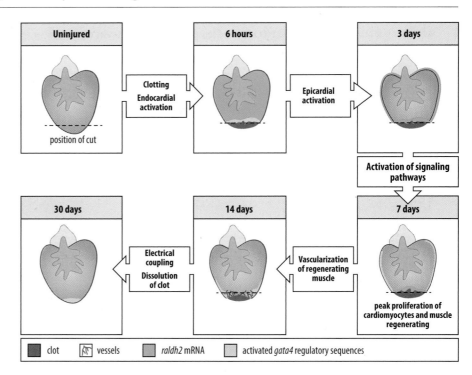

In an embryonic model of zebrafish heart regeneration, however, damage to ventricular muscle can also be repaired by transdifferentiation of cells from atrial muscle. Following damage to the ventricle, atrial cells at the border between the atrium and the ventricle dedifferentiate and start to divide. They then migrate to the damaged region where they differentiate into ventricular muscle.

Regeneration of the myocardium in the adult zebrafish heart involves interactions with cells in the other layers of the heart, the endocardium and epicardium (see Section 11.32). As early as 1 hour after injury, the entire endocardium, the endothelial lining of the heart, begins to express *raldh2*, which encodes a retinoic-acid-synthesizing enzyme (see Figure 13.38). Within the next day or so, *raldh2* is also expressed throughout the entire epicardium, the thin outer layer covering the myocardium, and then later becomes confined to the wound area. The epicardial cells proliferate and surround the regenerating muscle, releasing retinoic acid and other signals, including Shh and IGF-2, that aid cardiomyocyte proliferation. However, the signal that initiates adult cardiomyocyte proliferation has not yet been identified. The epicardial cells give rise to vascular supporting cells in the regenerated tissue just as they do during normal heart development. The peak of cell proliferation occurs 7 days after injury and at this time, regulatory sequences controlling expression of *gata4*, which encodes a transcription factor involved in heart development, are activated throughout the outer layer of the myocardium. The new cardiac muscle is made up largely of the progeny of cells in which these *gata4*-associated sequences were activated. Other genes are expressed that are also involved in the normal development of cardiac muscle, such as *nkx2.5*, an early marker of prospective heart cells (see Section 11.34). In addition, expression of the zebrafish homeodomain-containing transcription factors MsxB and MsxC (related to Msx1) increases during regeneration, whereas these factors are not expressed during embryonic heart development.

In adult mammals, growth of the heart after birth is due to cell enlargement, and the rates of cell proliferation are extremely low. Unlike the adult zebrafish heart, the mammalian heart cannot regenerate and produce fully functional heart tissue after injury. If the adult mammalian heart muscle is damaged, it forms a fibrous scar. However, the cardiomyocytes in an adult mouse heart can resume cell division when specific signaling pathways are activated. Even then, because very few cells are

BOX 13F Why can't we regenerate our limbs?

Examples of adult animals that are able to regenerate complex organs are found throughout the animal kingdom. Regenerative capacity is not confined to animals with simple body plans, and not all simple animals can regenerate. The nematodes, for example, have no regenerative capacity, probably because their bodies have high internal hydrostatic pressure, which makes healing impossible.

Among vertebrates, adult salamanders such as newts and the axolotl can regrow multiple body parts, including limbs, the lens of the eye, and the heart whereas mammals can only regrow the very tips of the digits and repair the heart only in the first week of postnatal life. So why do newts and other salamanders show these remarkable regenerative capacities whereas mammals, including humans, have such a limited capacity? The view that prevailed until recently was that the ability of some adult vertebrates to regenerate is a fundamental vertebrate property that we mammals and other vertebrates that cannot regenerate have lost during our evolutionary history. Indeed, studies in invertebrates have shown that the variability in regenerative ability in a group of closely related annelids is most likely to be due to the evolutionary loss of regenerative ability in some species.

In adult salamanders, however, there is growing evidence that regenerative ability evolved locally in this group of animals. The first indication of this was the finding that Prod1, the key protein that encodes proximo-distal positional values in regenerating amphibian limbs (see Section 13.18), is a salamander-specific protein and has no mammalian homolog. Recent analysis of the transcriptome of newts has identified several more genes encoding novel proteins that have no apparent homologs in mammals. The levels of transcripts of some of these genes have already been shown to change during regeneration (Figure 1).

This emerging focus on novel newt proteins involved in regeneration contrasts with much of the previous research on regeneration, which concentrated on molecules and pathways already known to be conserved during embryonic development. So how do novel proteins interact with these pathways during limb regeneration? Some insights into this question have recently come from studies on Prod1, as it has been shown that expression of the newt *Prod1* gene is regulated by the homeoprotein Meis. Meis was already implicated in specifying proximal positional values in chick and mouse developing limbs (see Section 11.6). Like *Prod1*, it is expressed at higher levels in proximal blastemas compared with distal blastemas in regenerating amphibian limbs, although unlike *Prod1* it is not expressed in the intact limb. The interaction between *Prod1* and Meis that occurs

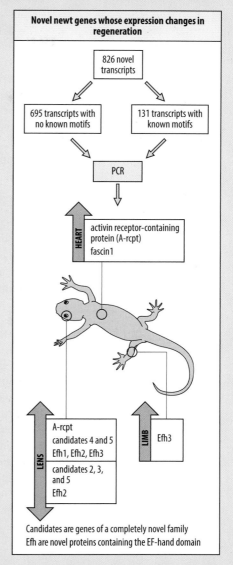

Figure 1

Adapted from Mihaylova, Y., Aboobaker, A.A.: **What is it about eye of newt?** Genome Biol. *2013,* **14***: 106.*

following limb amputation could be involved in specifying proximal positional values in regenerating limb blastemas.

This new perspective on the question of why newts have such great regenerative abilities has implications for thinking about ways of tackling our inability to regenerate.

involved, this is not sufficient to restore adult function. There is, however, in the mouse a brief period just after birth when a fully functioning heart can regenerate after surgical removal of the apex of the left ventricle (15% of the ventricular myocardium is removed). Genetic-lineage analysis has shown that just as in the adult zebrafish, the new cardiomyocytes are derived from pre-existing ones. Another similarity to adult heart regeneration in zebrafish is that epicardial activation occurs. Thus, although the zebrafish heart differs from the mammalian heart in having only one ventricle, understanding the mechanisms involved in regeneration of the zebrafish heart will be relevant to mammalian heart regeneration.

The ability of the mouse heart to regenerate is lost when the mouse is 7 days old. One of the molecules that blocks cardiomyocytes from entering the cell cycle is Meis1, a homeodomain transcription factor required for embryonic heart development. When *Meis1* function is deleted specifically in adult mouse cardiomyocytes, they enter the cell cycle and proliferate without undergoing hypertrophy. Conversely, postnatal overexpression of *Meis1* reduces cardiomyocyte proliferation and prevents the regeneration normally seen in the first week after birth. There is evidence that Meis1 acts by directly controlling expression of genes encoding cell-cycle inhibitors.

SUMMARY

Urodele amphibians can regenerate amputated limbs and tails. Stump tissues at the site of amputation first dedifferentiate to form a blastema, which then grows and gives rise to a regenerated structure. The dedifferentiated cells of the blastema give rise to the various cell types within the regenerated limbs and are generally lineage restricted with cells giving rise to their own tissue of origin. Regeneration is usually dependent on the presence of nerves which provide an essential an essential growth factor, but limbs that have never been innervated can regenerate. Regeneration always gives rise to structures with positional values more distal than those at the site of amputation. When a blastema is grafted to a stump with different positional values, proximo-distal intercalation of the missing positional values occurs. Positional values may be related to a proximo-distal gradient in a cell-surface protein. Retinoic acid changes the positional values of the cells of the blastema, giving them more proximal values. Insect limbs can also regenerate, with intercalation of missing positional values occurring in both proximo-distal and circumferential directions. The adult heart can regenerate in amphibians and fish. In the first week of birth, the mouse heart can also regenerate. In both adult zebrafish and neonatal mice, the new myocardial heart tissue is produced by proliferation of pre-existing cardiomyocytes and this is facilitated by interactions with epicardial cells in the outer covering of the heart.

SUMMARY: regeneration of an amphibian limb

amputation of a newt limb

local dedifferentiation of stump tissue to form a blastema

the blastema grows to form distal structures
provided nerves are present

graft of distal blastema to proximal stump

intercalation of missing proximo-distal positional values from stump tissue growth

Aging and senescence

Organisms are not immortal, even if they escape disease or accidents. With aging—which occurs over the passage of time—comes an increasing impairment of physiological functions, which reduces the body's ability to deal with various stresses, and an increased susceptibility to disease. This age-related decline in function is known as **senescence**. Aging is observed in most multicellular animals, but there are notable exceptions, such as cnidarians (sea anemones, *Hydra*, and their relations) and planarians. All sexually reproducing animals age, whereas most asexual animals do not. *Hydra*, whose capacity for regeneration is discussed in Box 13D, does not age unless it undergoes sexual differentiation. The phenomenon of senescence raises many questions as to the underlying mechanisms, and these are still largely unanswered, but we can at least consider some general questions, such as whether senescence is part of an organism's post-embryonic developmental program or whether it is simply the result of wear and tear. Germ cells do not age; if they did, the species would die out.

Although individuals may vary in the time at which particular aspects of aging appear, the overall effect is summed up as an increased probability of dying in most animals, including humans, with increased age. This life pattern, which is illustrated in relation to *Drosophila* (Fig. 13.39), is typical of many animals, but there is little evidence that aging contributes to mortality in the wild; more than 90% of wild mice die during their first year. There are, however, exceptions, such as the Pacific salmon, in which death does not come after a process of gradual aging, but is linked to a certain stage in the life cycle, in this case to spawning.

One view of aging is that it is due to an accumulation of damage that eventually outstrips the ability of the body to repair itself, and so leads to the loss of essential functions. For example, some old elephants die of starvation because their teeth have worn out. The nematode *Caenorhabditis elegans* lives for an average or around 20 days and the major cellular change that occurs with age is the progressive deterioration of muscle. This deterioration has a random effect on the life span, which varies from 10 to 30 days. Nevertheless, there is clear evidence that senescence is under genetic control, as different species age at vastly different rates, as shown by their different life spans (Fig. 13.40). An elephant, for example, is born after 21 months' embryonic development, and at that point shows few, if any, signs of aging, whereas a 21-month-old mouse is already well into middle age and beginning to show signs of senescence.

The genetic control of aging can be understood in terms of the 'disposable soma' theory, which puts it into the context of evolution. The disposable soma theory proposes that natural selection tunes the **life history** of the organism so that sufficient resources are invested in maintaining the repair mechanisms that prevent aging, at least until the organism has reproduced and cared for its young. Thus mice, which start reproducing when just a few months old, need to maintain their repair mechanisms for much less time than do elephants, which only start to reproduce when around 13 years old. In most species of animals in the wild, few individuals live long enough to show obvious signs of senescence and senescence need only be delayed until reproduction is complete. Cells have numerous mechanisms to delay aging, which are quite similar to the mechanisms used to prevent malignant transformation. These cellular mechanisms protect the cell from internal damage by reactive chemicals and routinely repair damage to DNA, which is occurring continually in living cells even when they are not actively dividing. They are particularly active in germ cells.

13.23 Genes can alter the timing of senescence

The maximum recorded life spans of animals show dramatic differences (see Fig. 13.40). Humans can live as long as 120 years, some owls 68 years, cats 28 years,

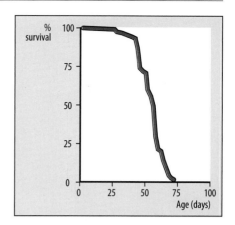

Fig. 13.39 Aging in *Drosophila*. The probability of dying increases rapidly at older ages.

Longevity and time to attain reproductive maturity at puberty for various mammals			
	Maximum lifespan (months)	Length of gestation (months)	Age at puberty (months)
Human	1440	9	144
Fin whale	960	12	~72
Indian elephant	840	21	156
Horse	744	11	15
Chimpanzee	720	8	120
Brown bear	468	7	72
Dog	348	2	7
Cattle	588	9	15
Rhesus monkey	480	5.5	36
Cat	486	2	8
Pig	276	4	5
Squirrel monkey	180	5	30
Sheep	276	5	7
Gray squirrel	180	1.5	12
European rabbit	156	1	4
Guinea-pig	90	2	2
House rat	56	0.7	2
Golden hamster	48	0.5	2
Mouse	48	0.7	1.5

Fig. 13.40 Table showing life span, length of gestation, and age at puberty for various mammals. The numbers for the lifespans are the maximum lifespan recorded for the species; average lifespans are usually much shorter.

Xenopus 15 years, mice 3.5 years, and the nematode about 25 days. Mutations in genes that affect life span have been identified in *C. elegans*, *Drosophila*, mice, and humans, and may give clues to the mechanisms involved; the ability to resist damage to DNA and the effects of oxygen radicals are important.

A *C. elegans* worm that hatches as a first instar larva in an uncrowded environment with ample food grows to adulthood and can survive for 25 days. In crowded conditions and when food is short, however, the animal enters a quiescent third-instar larval state known as the **dauer** state, where it neither eats nor grows until food becomes available again. When conditions become favorable, the dauer larva molts and becomes a fourth instar larva. The dauer state can last for 60 days and has no effect on the post-dauer life span: it is therefore considered to be a state in which the larva does not age. An insulin/IGF-1 signaling system plays an important role both in controlling entry into the dauer state in response to stress and in the overall control of fertility, life span, and metabolism in *C. elegans* and also in *Drosophila*. Reduction in insulin/IGF-I signaling increases longevity (Fig. 13.41). This is the same hormonal system that is responsible for controlling metabolism and promoting growth in *Drosophila* and vertebrates, as we discussed earlier.

In *C. elegans*, mutations that cause a strong reduction in the expression of *daf-2*, which encodes a receptor in the insulin-signaling pathway, arrest development in the dauer state. The normal role of DAF-2 is to antagonize the activity of DAF-16, a transcription factor whose activity lengthens life span and increases resistance to some types of stress. A partial loss of DAF-2 function results in longer adult life-span after the dauer state but with reduced fertility and viability of the progeny, while a further reduction in DAF-2 at the larval stage by RNA interference increases life span even more, without shortening the dauer state. The presence of a germline seems to have a negative effect on life-span extension. Removal of germline precursor cells in *daf-2* mutant embryos resulted in their having a mean life span of 125 days and remaining quite healthy—in humans, this would equate to a life span of 500 years. But when *daf-2* mutants were cultured alongside wild-type worms, the mutants became extinct in just a few generations, partly due to their reduced fertility. Microarray analysis of the effects of DAF-16 shows that it activates stress-response and antimicrobial genes. In laboratory conditions it seems that aging animals in which DAF-16 is inhibited are killed by the bacteria on which they feed.

A similar system regulates aging in *Drosophila*. Mutations disabling the insulin/IGF-1 pathway almost double the life span in this animal, and calorie restriction also extends life span. The mutant flies, rather like the nematode dauer larvae, enter a state of reproductive diapause or quiescence. These effects on life span may be due to resistance to oxidative stress. Ablation of germline cells in *Drosophila* also extends lifespan, probably through the effects of the germline on insulin signaling. Aging in mice can be retarded by reduction in pituitary activity, and there is also evidence that female mice with a mutation in the gene for the IGF-1 receptor (which is the receptor activated by IGF-1 and IGF-2) can live 33% longer.

There is evidence that oxidative damage accelerates aging and that reactive oxygen radicals are key players in causing cell damage. Reduction in food intake increases life span in other animals; rats on a minimal diet live about 40% longer than rats allowed to eat as much as they like. This is thought to be partly due to a reduced exposure to free radicals, which are formed during the oxidative breakdown of food. Free radicals are highly reactive, and can damage both DNA and proteins. A long-lived rodent species generates less reactive oxygen than the laboratory mouse. Oxidative stress also exerts its effects by damaging mitochondria.

Humans who are homozygous for a recessive gene defect leading to Werner syndrome age prematurely. There is growth retardation at puberty, and by their early twenties those affected by this syndrome have gray hair and suffer from various illnesses, such as heart disease, that are typical of old age. Most people affected die before the age of 50. The gene affected in Werner syndrome has been identified, and encodes a protein involved in unwinding DNA. Such unwinding is required for DNA replication, DNA repair, and gene expression. The inability to carry out DNA repair properly in Werner-syndrome patients could subject the genetic material to a much higher level of damage than normal. The link between Werner syndrome and DNA thus fits with the possibility that aging is linked to the accumulation of damage in DNA. Aging may also be related to cell senescence. Hutchinson–Gilford progeria syndrome (progeria means premature aging) is even more severe in its effects. The gene affected in this syndrome encodes the intranuclear protein lamin A, and the primary cellular defect is an instability in the structure of the nuclear envelope, which means cells are more likely to die prematurely.

13.24 Cell senescence blocks cell multiplication

One might think that when cells are isolated from an animal, placed in culture, and provided with adequate medium and growth factors, they would continue to proliferate almost indefinitely. But this is not the case. For example, mammalian fibroblasts—connective tissue cells—will only go through a limited number of cell doublings in culture; the cells then stop dividing, however long they are cultured (Fig. 13.42). For normal fibroblasts, the number of cell doublings depends both on the species and the age of the animal from which they are taken. Fibroblasts taken from a human fetus go through about 60 doublings, those from an 80-year-old about 30, and those from an adult mouse about 12–15 doublings. When the cells stop dividing, they appear to be healthy, but are stuck at some point in the cell cycle, often G_0; this phenomenon is known as **cell senescence**. Cells taken from patients with Werner's syndrome, who show an acceleration of many features of normal aging, make significantly fewer divisions in culture than normal cells. However, it is far from clear how this behavior of cells in culture is significant for the aging of the organism and to what extent it reflects the way the cells are cultured.

A feature shared by senescent cells in culture and *in vivo* is shortening of the **telomeres**. Telomeres are the repetitive DNA sequences at the ends of the chromosomes that preserve chromosome integrity and ensure that chromosomes replicate themselves completely without loss of information-encoding DNA at the ends. The length of the

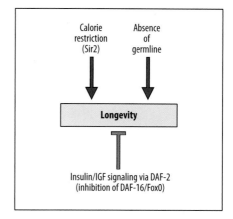

Fig. 13.41 Potential inputs to aging in *Caenorhabditis elegans* and *Drosophila*. Calorie restriction and the absence of a germline promote an extended life-span in both animals. In *Drosophila*, the calorie-restriction effect has been shown to require the histone deacetylase Sir2, which is likely to act on gene expression through its ability to modifiy chromatin (see Box 8A). In *C. elegans*, insulin/IGF signaling has been shown to regulate life span via its inhibitory effect on expression of the gene for the transcription factor DAF-16 (a member of the FoxO family of transcription factors).

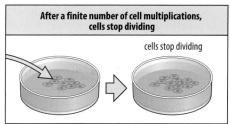

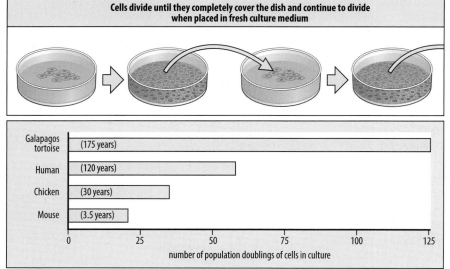

Fig. 13.42 Vertebrate fibroblasts can only go through a limited number of divisions in culture. Fibroblasts placed in culture are subcultured until they stop growing (top panels). The number of cell doublings in culture before they stop dividing is related to their maximum age, as indicated by the figures in brackets on the graph (bottom panel).

telomeres is reduced in older human cells. Telomere length is found to decrease at each DNA replication, suggesting that they are not completely replicated at each cell division and this may be related to senescence. If the enzyme telomerase, which maintains telomere length and is normally absent from cells in culture, is expressed in those cells, senescence in culture does not occur. ES cells also express telomerase and proliferate indefinitely. However, it is not yet clear whether telomere shortening is a major cause of aging in somatic cells, as certain rodent cells such as Schwann cells can, under appropriate conditions, proliferate indefinitely and their telomeres do not control replication.

Senescent cells that have stopped dividing have changes in the expression of proteins that are involved in cell-cycle control. The proteins p21 and p16 are often expressed; they are part of the tumor-suppressor pathways governed by p53 and RB in response to DNA damage (see Section 13.5). They act as inhibitors of cyclin-dependent kinases and prevent a cell with damaged DNA from entering the cell cycle. This helps to prevent senescent cells, with their DNA damage, from becoming cancerous.

SUMMARY

Aging is largely caused by damage to cells, particularly by reactive oxygen. Senescence is the age-related decline in the ability to carry out repairs and is under genetic control. Many normal cells in culture can only undergo a limited number of cell divisions, which correlate with their age at isolation and the normal life-span of the animal from which they came. Genes that can increase life span have been identified in *C. elegans* and *Drosophila* and may act by increasing the animals' resistance to oxidative stress.

Summary to Chapter 13

- The form of many animals, such as mammals and birds, is laid down in miniature during embryonic development, and they then grow in size, keeping the basic body form, although different regions grow at different rates.
- Growth may involve cell multiplication, cell enlargement, and the laying down of extracellular material.
- Cancer can be viewed as an aberration of growth, as it usually results from mutations that lead to excessive cell proliferation and failure of cells to differentiate.

- In vertebrates, structures have an intrinsic growth program, which is under hormonal control.

- The larvae of arthropods and some other groups, such as frogs among the vertebrates do not resemble the adult and undergo metamorphosis to reach the adult form. Arthropod larvae are covered by a relatively inextensible chitin cuticle and so must undergo successive molts as they grow in size.

- Regeneration is the ability of an adult organism to replace a lost part of its body and can involve growth. The capacity to regenerate varies greatly among different groups of organisms and even between different species within a group. Mammals have very limited powers of regeneration, whereas newts can regenerate limbs, jaws, and the lens of the eye.

- Regeneration of amphibian limbs involves dedifferentiation of cells and the formation of a blastema, which grows and forms the regenerate. Regeneration requires the presence of nerves.

- Evidence exists for a proximo-distal gradient in positional values in the regenerating limb, because intercalation of positional values occurs when normally non-adjacent values are placed next to each other in amphibian and insect limbs.

- The adult zebrafish heart can regenerate as a result of dedifferentiation of cardiomyocytes and their proliferation to form new tissue. The neonatal mouse heart in the first week after birth can also regenerate by the same mechanism.

- The symptoms of aging (senescence) appear mainly to be caused by damage to cells that accumulates over time. Aging is also under genetic control, as shown by aging syndromes in humans and by the effect of mutations in the nematode.

■ End of chapter questions

Long answer (concept questions)

1. Define 'growth.' What are the main mechanisms of growth? How does cell death fit into a discussion of growth?

2. Why is it inaccurate to say that the cell cycle starts when cytokinesis produces two cells after M phase? According to molecular analysis of the cell cycle, each round of the cycle 'starts' at what point in the cycle? What are the molecular events that define this start point? (The cell-cycle diagram in Box 1B might also be helpful.)

3. In what way is the standard cell cycle modified during the first 13 pre-cellular divisions in *Drosophila* embryogenesis, to allow very rapid cell divisions? What is the maternally supplied protein that drives these early divisions; and what is this protein's function?

4. The final size of some organs, such as liver and spleen, is regulated by signaling between cells, whereas the size of other organs, such as pancreas and thymus, is under the control of a cell-intrinsic developmental program. Describe the experiments that reveal these two different mechanisms of growth control.

5. In *Drosophila*, the Hippo protein and the *bantam* microRNA have opposing functions, as their names imply. What are the functions of these two regulators of growth? Outline the pathway that initiates signaling to Hippo, and signaling to the *bantam* gene.

6. Address the following questions about human growth hormone: What is growth hormone? Where is it produced? How is its synthesis and secretion controlled? What other growth factor is a major mediator of growth hormone's effect on cellular proliferation?

7. Review the various signaling molecules and signaling pathways discussed in the chapter, and indicate which could be classified as oncogenes, and which as tumor suppressors.

8. In one experiment, a frog tadpole is exposed to thyroxine in the water in which it is being reared. In another experiment, the thyroid gland is removed from a tadpole. What results would you predict in each case. Justify your predictions.

9. Discuss the evidence for and against the statement: 'The changes that occur with aging are an inevitable part of an animal's genetically determined postembryonic developmental program.'

10. Contrast morphallaxis and epimorphosis. Give an example of each.

11. What is the effect of denervation on regeneration of a salamander limb? What is the role of the nerve supply? What protein can substitute for the nerves in enabling regeneration? Why can a limb that has never been innervated regenerate?

12. Refer to Fig. 13.30. Describe in your own words what the experiment illustrated was intended to investigate, what was observed, and what interpretation can be drawn from results of the experiment.

13. The dedifferentiated cells of the salamander blastema will regenerate only structures distal to the site of the amputation, which requires some kind of memory of proximo-distal position in the intact limb. Summarize the information that Prod1 may be involved in registering proximo-distal position; be sure to include the results of retinoic acid treatment on Prod1 expression.

14. Describe the effect of retinoic acid on regeneration; for example, if a limb is cut at the wrist level and treated with retinoic acid, how will its regeneration be altered, compared to regeneration without such treatment?

15. In the regeneration of the cockroach leg, proximo-distal values will be restored through intercalation if two pieces of leg are apposed by grafting (see Fig. 13.36). How does this process of intercalation explain the results of the experiments shown in the figure?

16. Discuss the importance to biomedical science of the ability of portions of the adult zebrafish heart to regenerate (you may wish to review some of the advantages of the zebrafish as a model organism from Chapter 3).

17. How is muscle size regulated? Discuss whether the same principles apply to regulation of the size of other organs such as the liver.

Multiple choice (factual recall questions)

NB There is only one correct answer to each question.

1. What will be the result of grafting a limb bud from a large species of the salamander *Ambystoma* onto a smaller species?

a) The grafted bud will be unable to grow in a smaller animal, and will be lost.

b) The grafted bud will grow to a size appropriate to its host, and will thus be of the same size as the other limbs.

c) The grafted bud will grow to become a larger limb than the others, reflecting its origin in the larger species.

d) The grafted bud will initially grow larger than the others, but with time will regulate and become a limb of the same size as the other limbs of its host.

2. Growth is stimulated by which signaling molecules?

a) insulin
b) IGF-1 and -2
c) FGF
d) all of these may stimulate growth

3. Growth in the length of long bones in vertebrates occurs in the

a) diaphysis
b) epiphysis
c) growth plate
d) ossification centers

4. Which of the following statements is true?

a) The genes involved in cancer are all classed as oncogenes.
b) All genes involved in cancer are of a class called tumor suppressor genes.
c) Both oncogenes and tumor suppressor genes are involved in cancer.
d) No unifying classifications can be applied to the genes involved in cancer.

5. Molting in insects and other arthropods is triggered by

a) auxin
b) ecdysone
c) Hedgehog
d) juvenile hormone

6. Which signaling pathway seems to play a role in aging in several different animals, from worms to mice?

a) ecdysone
b) FGF
c) Hippo
d) insulin/IGF-1

7. Which of the following is consistent with a model for aging in which stresses leading to DNA damage cause senescence?

a) Werner's syndrome is a premature aging illness, possibly caused by a defect in DNA repair.
b) Dietary restriction in mammals reduces the production of DNA-damaging free radicals in the mitochondria.
c) Senescent cells in culture that have stopped dividing often express p21 and p16, proteins that are normally activated in response to DNA damage.
d) All of these facts are consistent with a model for aging based on DNA damage.

8. Epimorphosis is regeneration through

a) repatterning of existing cells without a requirement for new growth, as occurs in *Hydra*
b) repatterning of existing cells without a requirement for new growth, as occurs in newts
c) the reinitiation of division in existing cells and new growth, followed by patterning, as occurs in newts

d) the reinitiation of division in existing cells and new growth, followed by patterning, as occurs in *Hydra*.

9. If the nerve supply to a newt limb is severed before amputation, how will this affect regeneration?

a) A blastema will form but will not grow, and regeneration will fail.
b) No regeneration occurs, and the stump heals over as it would in a mammal.
c) Outgrowth will occur, but the identity of the limb will be lost and normal proximo-distal patterning will not occur.
d) Regeneration of most tissues will occur normally, but regeneration of the nerves will not occur.

10. Which is the result of grafting a blastema derived from a distal amputation onto a stump that has been amputated at a proximal position?

a) The distal blastema is unable to properly interpret its positional values in this circumstance, and regeneration fails.
b) The distal blastema regenerates the entire limb, supplying all the cells required for this regeneration.
c) The distal blastema regenerates the limb regions removed in the distal amputation, resulting in a shortened limb missing the regions between the proximal and distal amputation sites.
d) The limb regions between the distal and proximal amputation sites are filled in by intercalary growth from the proximal stump, and the regions distal to the distal amputation are regenerated from the distal blastema.

Multiple choice answer key

1: c, 2: d, 3: c, 4: c, 5: b, 6: d, 7: a, 8: c, 9: a, 10: d.

■ General further reading

Birnbaum, K.D., Sánchez Alvarado, A.: **Slicing across kingdoms: regeneration in plants and animals**. *Cell* 2008, **132**: 697–710.
Brockes, J.P., Kumar, A.: **Principles of appendage regeneration in adult vertebrates and their implications for regenerative medicine**. *Science* 2005, **310**: 1919–1923.
Brockes, J.P., Kumar, A.: **Comparative aspects of animal regeneration**. *Annu. Rev. Cell Dev. Biol.* 2008, **24**: 525–549.
Galliot, B., Tanaka, E., Simon, A.: **Regeneration and tissue repair**. *Cell. Mol. Life Sci.* 2008, **65**: 3–7.
Goss, R.J.: *The Physiology of Growth*. New York: Academic Press, 1978.
Kirkwood, T.B.L.: **Understanding the odd science of aging**. *Cell* 2005, **120**: 437–447.
Nature Insight: **Cell division and cancer**. *Nature* 2004, **432**: 293–341.
Partridge, L.: **The new biology of ageing**. *Phil. Trans R. Soc. B* 2010, **365**: 147–154.

■ Section further reading

13.1 Tissues can grow by cell proliferation, cell enlargement, or accretion

Tumaneng, K., Russell, R.C.Guan, K.-L.: **Organ size control by TOR and Hippo pathways**. *Curr. Biol.* 2012, **22**: R368–R379.

13.2 Cell proliferation is controlled by regulating entry into the cell cycle

Morgan, D.O.: *The Cell Cycle: Principles of Control.* Oxford University Press: 2007.

13.3 Cell division in early development can be controlled by an intrinsic developmental program

Edgar, B., Lehner, C.F.: **Developmental control of cell cycle regulators: a fly's perspective.** *Science* 1996, **274**: 1646–1652.

Follette, P.J., O'Farrell, P.H.: **Connecting cell behavior to patterning: lessons from the cell cycle.** *Cell* 1997, **88**: 309–314.

13.4 Extrinsic signals coordinate cell division, cell growth, and cell death in the developing *Drosophila* wing

de la Cova, C., Abril, M., Bellosta, P., Gallant, P., Johnston, L.A.: ***Drosophila* myc regulates organ size by inducing cell competition.** *Cell* 2004, 117: 107–116.

Halder, G., Johnson, R.L.: **Hippo signaling: growth control and beyond.** *Development* 2011, **138**: 9–22.

Huang, J., Wu, S., Barrera, J., Matthews, K., Pan, D.: **The Hippo signaling pathway coordinately regulates cell proliferation and apoptosis by inactivating Yorkie, the Drosophila Homolog of YAP.** *Cell* 2005, **122**: 421–434.

Martin, F.A., Herrera, S.C., Morata, G.: **Cell competition, growth and size control in the *Drosophila* wing imaginal disc.** *Development* 2009, **136**: 3747–3756.

Willecke, M., Hamaratoglu, F., Sansores-Garcia, L., Tao, C., Halder, G.: **Boundaries of Dachsous cadherin activity modulate the Hippo signaling pathway to induce cell proliferation.** *Proc. Natl Acad. Sci USA* 2008, **105**: 14897–14902.

Box 13A The core Hippo signaling pathways in *Drosophila* and mammals

Tumaneng, K., Russell, R.C.Guan, K.-L.: **Organ size control by TOR and Hippo pathways.** *Curr. Biol.* 2012, **22**: R368–R379.

13.5 Cancer can result from mutations in genes that control cell proliferation

Beachy, P.A., Karhadkar, S.S., Berman, D.M.: **Tissue repair and stem cell renewal in carcinogenesis.** *Nature* 2004, **432**: 324–331.

Harvey, K.F., Zhang, X., Thomas, D.M.: **The hippo pathway and human cancer.** *Nature Rev Cancer* 2013, **13**: 246–257.

Hunter, T.: **Oncoprotein networks.** *Cell* 1997, **88**: 333–346.

Jones, S., Zhang, X., Parsons, D.W., Lin, J.C., Leary, R.J., Angenendt, P., Mankoo, P., Carter, H., Kamiyama, H., Jimeno, A., *et al.*: **Core signaling pathways in human pancreatic cancers revealed by global genomic analyses.** *Science* 2008, **321**: 1801–1806.

Shilo, B.Z.: **Tumor suppressors. Dispatches from patched.** *Nature* 1996, **382**: 115–116.

Taipale, J., Beachy, P.A.: **The hedgehog and Wnt signalling pathways in cancer.** *Nature* 2001, **411**: 349–353.

Van Dyke, T.: **p53 and tumor suppression.** *N. Engl. J. Med.* 2007, **356**: 79–92.

13.6 Size-control mechanisms differ in different organs

Amthor, H., Huang, R., McKinnell, I., Christ, B., Kambadur, R., Sharma, M., Patel, K.: **The regulation and action of myostatin as a negative regulator of muscle development during avian embryogenesis.** *Dev. Biol.* 2002, **251**: 241–257.

Lui, J.C., Baron, J.: **Mechanisms limiting body growth in mammals.** *Endocr. Rev.* 2011, **32**: 422–440.

Shingleton, A.W.: **Body-size regulation: combining genetics and physiology.** *Curr. Biol.* 2005, **15**: R825–R827.

Stanger, B.Z.: **The biology of organ size determination.** *Diabetes Obes. Metab.* 2008, **10**: 16–26.

Stanger, B.Z., Tanaka, A.J., Melton, D.A.: **Organ size is limited by the number of embryonic progenitor cells in the pancreas but not in the liver.** *Nature* 2007, **445**: 886–891.

Taub, R.: **Liver regeneration: from myth to mechanism.** *Nat. Rev. Mol. Cell Biol.* 2004, **5**: 836–847.

13.8 Hormones and growth factors coordinate the growth of different tissues and organs and contribute to determining overall body size

Efstratiadis, A.: **Genetics of mouse growth.** *Int. J. Dev Biol.* 1998, **42**: 955–976.

Lui, J.C., Baron, J.: **Mechanisms limiting body growth in mammals.** *Endocr. Rev.* 2011, **32**: 422–440.

Lupu, F., Terwilliger, J.D., Lee, K., Segre, G.V., Efstratiadis, A.: **Roles of growth hormone and insulin-like growth factor 1 in mouse postnatal growth.** *Dev Biol.* 2001, **229**: 141–162.

Sanders, E.J., Harvey S.: **Growth hormone as an early embryonic growth and differentiation factor.** *Anat. Embryol.* 2004, **209**: 1–9.

Box 13B The major determinant of body size in dogs is the growth hormone–IGF-1 axis

Ostrander, E.A. **Genetics and the shape of dogs.** *Am. Sci.* 2007, **95**: 406–413.

Sutter, N.B., Sutter, N.B., Bustamante, C.D., Chase, K., Gray, M.M., Zhao, K., Zhu, L., Padhukasahasram, B., Karlins, E., Davis, S., *et al.*: **A single *IGF-1* allele is a major determinant of small size in dogs.** *Science* 2007, **316**: 112–115.

Shearin, A.L., Ostrander, E.A.: **Canine morphology: hunting for genes and tracking mutations.** *PloS Biol.* 2010, **8**: e1000310.

Rimbault, M., Ostrander, E.A.: **So many doggone traits: mapping genetics of multiple phenotypes in the domestic dog.** *Hum. Mol. Genet.* 2012, **21**: R52–R57

Rimbault, M., Beale, H.C., Schoenebeck, J.J., Hoopes, B.C., Allen, J.J., Kilroy-Glynn, P., Wayne, R.K., Sutter, N.B., Ostrander, E.A.: **Derived variants at six genes explain nearly half of size reduction in dog breeds.** *Genome Res.* 2013, **23**: 1985–1995.

13.9 Elongation of the long bones illustrates how growth can be determined by a combination of an intrinsic growth program and extracellular factors

Kember, N.F.: **Cell kinetics and the control of bone growth.** *Acta Paediatr. Suppl.* 1993, **391**: 61–65.

Kronenberg, H.M.: **Developmental regulation of the growth plate.** *Nature* 2003, **423**: 332–336.

Nilsson, O., Baron, J.: **Fundamental limits on longitudinal bone growth: growth plate senescence and epiphyseal function.** *Trends Endocrinol. Metab.* 2004, **8**: 370–374.

Roush, W.: **Putting the brakes on bone growth.** *Science* 1996, **273**: 579.

Schultz, E.: **Satellite cell proliferative compartments in growing skeletal muscles**. *Dev. Biol.* 1996, **175**: 84–94.

Williams, P.E., Goldspink, G.: **Changes in sarcomere length and physiological properties in immobilized muscle**. *J. Anat.* 1978, **127**: 450–468.

Box 13C Digit length ratio is determined in the embryo

Zheng, Z., Cohn, M.J.: **Developmental basis of sexually dimorphic digit ratios** *Proc. Natl Acad. Sci. USA* 2011, **108**: 16289–16294.

Manning, J.T.: *The Finger Book*. London: Faber and Faber, 2008.

Manning, J.T.: **Resolving the role of prenatal sex steroids in the development of digit ratio**. *Proc. Natl Acad. Sci. USA* 2011, **108**: 16143–16144.

13.10 The amount of nourishment an embryo receives can have profound effects in later life

Barker, D.J.: The Wellcome Foundation Lecture. 1994: **The fetal origins of adult disease**. *Proc. R. Soc. Lond.* 1995, **262**: 37–43.

Gluckman, P.D., Hanson, M.A., Cooper, C., Thornburg, K.L.: **Effect of *in utero* and early-life conditions on adult health and disease**. *N. Engl. J. Med.* 2008, **359**: 61–73.

13.12 Insect body size is determined by the rate and duration of larval growth

Andersen, D.S., Colombani, J., Leopold, P.: **Coordination of organ growth: principles and outstanding questions from the world of insects**. *Trends Cell Biol.* 2013, **23**: 336–344.

De Loof, A.: **Ecdysteroids, juvenile hormone and insect neuropeptides: Recent successes and remaining major challenges**. *Gen. Comp. Endocrinol.* 2008, **155**: 3–13.

Mirth, C.K., Riddiford, L.M.: **Size assessment and growth control: how adult size is determined in insects**. *BioEssays* 2007, **29**: 344–355.

Mirth C.K., Shingelton, A.W. **Integrating body and organ size in *Drosophila*: recent advances and outstanding problems**. *Front. Endocr.* 2012, **3**: 1–13.

Nijhout, H.F.: **Size matters (but so does time), and it's OK to be different**. *Dev. Cell* 2008, **15**: 491–492.

Stern, D.: **Body-size control: how an insect knows it has grown enough**. *Curr. Biol.* 2003, **13**: R267–R269.

Thummel, C.S.: **Flies on steroids—*Drosophila* metamorphosis and the mechanisms of steroid hormone action**. *Trends Genet.* 1996, **12**: 306–310.

13.13 Metamorphosis in amphibians is under hormonal control

Brown, D.D., Cai, L.: **Amphibian metamorphosis**. *Dev. Biol.* 2007, **306**: 20–33.

Huang, H., Brown, D.D.: **Prolactin is not juvenile hormone in *Xenopus laevis* metamorphosis**. *Proc. Natl Acad. Sci. USA* 2000, **97**: 195–199.

Tata, J.R.: *Hormonal Signaling and Postembryonic Development*. Heidelberg: Springer, 1998.

13.14 There are two types of regeneration–morphallaxis and epimorphosis

Box 13D Regeneration in *Hydra*

Bode, H.R.: **The head organizer in *Hydra***. *Int. J. Dev. Biol.* 2012, **56**: 473–478.

Broun, M., Gee, L., Reinhardt, B., Bode, H.R.: **Formation of the head organizer in Hydra involves the canonical Wnt pathway**. *Development* 2005, **132**: 2907–2916.

Chera, S., Ghila, L., Dobretz, K., Wenger, Y., Bauer, C., Buzgariu, W., Martinou, J.C., Galliot, B.: **Apoptotic cells provide an unexpected source of Wnt3 signaling to drive hydra head regeneration**. *Dev. Cell* 2009, **17**: 279–289.

Galliot, B.: **Regeneration in Hydra**. eLS (Wiley, 2013). 10.1002/9780470015902.a0001096.pub3.

Hicklin, J., Wolpert, L.: **Positional information and pattern regulation in *Hydra*: the effect of gamma-radiation**. *J. Embryol. Exp. Morph.* 1973, **30**: 741–752.

Hobmayer, B., Rentzsch, F., Kuhn, K., Happel, C.M., von Laue, C.C., Snyder, P., Rothbächer, U., Holstein, T.W.: **WNT signalling molecules act in axis formation in the diploblastic metazoan *Hydra***. *Nature* 2000, **407**: 186–189.

Müller, W.A.: **Pattern formation in the immortal Hydra**. *Trends Genet.* 1996, **12**: 91–96.

Takahashi, T., Fujisawa, T.: **Important roles for epithelial cell peptides in hydra development**. *BioEssays* 2009, **31**: 610–619.

Wolpert, L., Hornbruch, A., Clarke, M.R.B.: **Positional information and positional signaling in *Hydra***. *Am. Zool.* 1974, **14**: 647–663.

Box 13E Planarian regeneration

Adell, T., Cebrià, F., Saló, E.: **Gradients in planarian regeneration and homeostasis**. *Cold Spring Harb Perspect Biol.* 2010, **2**: a000505.

Adell, T., Saló, E., Boutros, M., Bartscherer, K.: **Smed-Evi/Wntless is required for beta-catenin-dependent and -independent processes during planarian regeneration**. *Development* 2009, **136**: 905–910.

De Robertis, E.M.: **Wnt signaling in axial patterning and regeneration: lessons from planaria**. *Sci. Signal.* 2010, **3**: pe21.

Newmark, P.A., Sánchez-Alvarado, A.: **Regeneration in planaria**. *eLS* (Wiley, 2001). doi: 10.1038/npg.els.0001097

Reddien, P.W.: **Specialized progenitors and regeneration**. *Development* 2013, **140**: 951–957.

Reddien, P.W., Sánchez Alvarado, A.: **Fundamentals of planarian regeneration**. *Annu. Rev. Cell Dev. Biol.* 2004, **20**: 725–757.

Simon, A.: **On with their heads**. *Nature*, 2013, **500**: 32–33.

Wagner, D.E., Wang, I.E., Reddien, P.W.: **Clonogenic neoblasts are pluripotent adult stem cells that underlie planarian regeneration**. *Science*, 2011, **332**: 811–816.

13.15 Regeneration of amphibian and insect limbs involves epimorphosis

Ferretti, P., **Regeneration of vertebrate appendages**. eLS. (Wiley, 2013) 10.1002/9780470015902.a0001099.pub3

Hopkins, P.M., **Regeneration in insects and crustaceans**. eLS. (Wiley, 2001) doi: 10.1038/npg.els.0001098

13.16 Amphibian limb regeneration involves cell dedifferentiation and new growth

Alvarado, A.S.: **A cellular view of regeneration**. *Nature News and Views*, 2009, **460**: 39–40.

Brockes, J.P., Kumar, A.: **Plasticity and reprogramming of differentiated cells in amphibian regeneration**. *Nat. Rev. Mol. Cell Biol.* 2002, **3**: 566–574.

Godwin, J.W., Pinto, A.R., Rosenthal, N.A.: **Macrophages are required for adult salamander limb regeneration**. *Proc. Natl Acad. Sci. USA* 2013, **110**: 9415–9420.

Imokawa, Y., Simon, A., Brockes, J.P.: **A critical role for thrombin in vertebrate lens regeneration**. *Phil. Trans. R. Soc. Lond. B Biol. Sci.* 2004, **359**: 765–776.

Kragl, M., Knapp, D., Nacu, E., Khattak, S., Maden, M., Epperlein, H.H., Tanaka, E.M.: **Cells keep a memory of their tissue origin during axolotl limb regeneration**. *Nature* 2009, **460**: 60–65.

Kumar, A., Velloso, C.P., Imokawa, Y., Brockes, J.P.: **The regenerative plasticity of isolated urodele myofibers and its dependence on MSX1**. *PLoS Biol.* 2004, **2**: E218.

Nacu, E., Glausch, M., Le, H. Q., Damanik, R.F.F., Schuez, M., Knapp, D., Khattak, S., Richter, T., Tanaka, E.M.: **Connective tissue cells, but not muscle cells, are involved in establishing the proximo-distal outcome of limb regeneration in the axolotl**. *Development* 2013, **140**: 513–518.

Sandoval-Guzmán, T., Wang, H., Khattak, S., Schuez, M., Roensch, K., Nacu, E., Tazaki, A., Joven, A., Tanaka, E.M., Simon, A.: **Fundamental differences in dedifferentiation and stem cell recruitment during skeletal muscle regeneration in two salamander species**. *Cell Stem Cell* 2013, **14**: 174–187.

Satoh, A., Graham, G.M.C., Bryant, S.V., Gardiner, D.M.: **Neurotrophic regulation of epidermal dedifferentiation during wound healing and limb regeneration in the axolotl (*Ambystoma mexicanum*)**. *Dev. Biol.* 2008, **319**: 321–355.

Tanaka, E.M., Drechel, D.N., Brockes, J.P.: **Thrombin regulates S-phase re-entry by cultured newt myotubes**. *Curr. Biol.* 1999, **9**: 792–799.

13.17 Limb regeneration in amphibians is dependent on the presence of nerves

Kumar, A., Godwin, J.W., Gates, P.B., Garza-Garcia, A.A., Brockes, J.P.: **Molecular basis for the nerve dependence of limb regeneration in an adult vertebrate**. *Science* 2007, **318**: 772–777.

Kumar, A., Delgado, J.P., Gates, P.B., Neville, G., Forge, A., Brockes, J.P.: **The aneurogenic limb identifies developmental cell interactions underlying vertebrate limb regeneration**. *Proc. Natl Acad. Sci. USA* 2011, **108**: 13588–13593.

13.18 The limb blastema gives rise to structures with positional values distal to the site of amputation

Da Silva, S., Gates, P.B., Brockes, J.P.: **New ortholog of CD59 is implicated in proximodistal identity during amphibian limb regeneration**. *Dev. Cell* 2002, **3**: 547–551.

Echeverri, K., Tanaka, E.M.: **Proximodistal patterning during limb regeneration**. *Dev. Biol.* 2005, **279**: 391–401.

Roensch, K., Tazaki, A., Chara, O., Tanaka, E.M.: **Progressive specification rather than intercalation of segments during limb regeneration**. *Science* 2013, **342**: 1375–1379.

13.19 Retinoic acid can change proximo-distal positional values in regenerating limbs

Maden, M.: **The homeotic transformation of tails into limbs in *Rana temporaria* by retinoids**. *Dev. Biol.* 1993, **159**: 379–391.

Mercader, N., Tanaka, E.M., Torres, M.: **Proximodistal identity during limb regeneration is regulated by Meis homeodomain proteins**. *Development* 2005, **132**: 4131–4142.

Pecorino, L.T., Entwistle, A., Brockes, J.P.: **Activation of a single retinoic acid receptor isoform mediates proximo-distal respecification**. *Curr. Biol.* 1996, **6**: 563–569.

Scadding, S.R., Maden, M.: **Retinoic acid gradients during limb regeneration**. *Dev. Biol.* 1994, **162**: 608–617.

13.20 Mammals can regenerate the tips of the digits

Han, M., Yang, X., Farrington, J.E., Muneoka, K.: **Digit regeneration is regulated by *Msx1* and BMP4 in fetal mice**. *Development* 2003, **130**: 5123–5132.

Han, M., Yang, X., Lee, J., Allan, C.H., Muneoka, K.: **Development and regeneration of the neonatal digit tip in mice**. *Dev. Biol.* 2008, **315**: 125–135.

Rinkevich, Y., Lindau, P., Ueno, H., Longaker, M.T., Weissman, I.L.: **Germ-layer and lineage-restricted stem/progenitors regenerate the mouse digit tip**. *Nature* 2011, **476**: 409–413.

Takeo, M., Chou, W.C., Sun, Q., Lee, W., Rabbani, P., Loomis, C., Taketo, M.M., Ito, M.: **Wnt activation in nail epithelium couples nail growth to digit regeneration**. *Nature* 2013, **499**: 228–232.

Box 13F Why can't we regenerate our limbs?

Bely, A.E.: **Evolutionary loss of animal regeneration: pattern and process**. *Integr.Comp. Biol.* 2010, **50**: 515–527.

Garza-Garcia, A.A., Driscoll P.C., Brockes, J.P.: **Evidence for the local evolution of mechanisms underlying limb regeneration in salamanders**. *Integr.Comp. Biol.* 2010, **50**: 528–535.

Mihaylova, Y., Aboobaker, A.A.: **What is it about eye of newt?** *Genome Biol.* 2013, **14**: 106.

Shaikh, N., Gates, P.B., Brockes, J.P.: **The Meis homeoprotein regulates the axolotl *Prod 1* promoter during limb regeneration**. *Gene* 2011, **484**: 69–74.

13.21 Insect limbs intercalate positional values by both proximo-distal and circumferential growth

Bando, T., Mito, T., Maeda, Y., Nakamura, T., Ito, F., Watanabe, T., Ohuchi, H., Noji, S.: **Regulation of leg size and shape by the Dachsous/Fat signalling pathway during regeneration**. *Development* 2009, **136**: 2235–2245.

French, V.: **Pattern regulation and regeneration**. *Phil. Trans. R. Soc. Lond. B Biol. Sci.* 1981, **295**: 601–617.

Nakamura, T., Mito, T., Miyawaki, K., Ohuchi, H., Noji, S.: **EGFR signaling is required for re-establishing the proximodistal axis during distal leg regeneration in the cricket *Gryllus bimaculatus* nymph**. *Dev. Biol.* 2008, **319**: 46–55.

13.22 Heart regeneration in zebrafish involves the resumption of cell division by cardiomyocytes

Ahmed I. Mahmoud, A.I., Fatih Kocabas, F., Shalini A., Muralidhar, S.A., Wataru Kimura, W., Ahmed S. Koura, A.S., Suwannee Thet, S., Enzo R. Porrello, E.R., Hesham A. Sadek, H.A.: **Meis1 regulates postnatal cardiomyocyte cell-cycle arrest**. *Nature* 2013, **497**: 249–253.

Gemberling, M., Bailey, T.J., Hyde, D.R., Poss, K.D.: **The zebrafish as a model for complex tissue regeneration**. *Trends Genet.* 2013, **29**: 611–620.

Itou, J., Oishi, I., Kawakami, H., Glass, T.J., Richter, J., Johnson, A., Lund, T.C., Kawakami, Y.: **Migration of cardiomyocytes is essential for heart regeneration in zebrafish**. *Development* 2012, **139**: 4133–4142.

Oyama, K., El-Nachef, D., MacLellan, WR.: **Regeneration potential of adult cardiac myocytes**. *Cell Res.* 2013, **23**: 978–979.

Porrello, E.R., Mahmoud, A.I., Simpson, E., Hill, J.A., Richardson, J.A., Olson, E.N., Sadek, H.A.: **Transient regenerative potential of the neonatal mouse heart**. *Science* 2011, **331**: 1078–1080.

Poss, K.D., Wilson, L.G., Keating, M.T.: **Heart regeneration in zebrafish**. *Science* 2002, **298**: 2188–2190.

Zhang, R., Han, P., Yang, H., Ouyang, K., Lee, D., Lin, Y.-F., Ocorr, K., Kang, G., Chen, J., Stainier, D.Y.R., Yelon, D., Chi, N.C.: **In vivo cardiac reprogramming contributes to zebrafish heart regeneration**. *Nature* 2013, **498**: 497–501.

13.23 Genes can alter the timing of senescence

Arantes-Oliviera, N., Berman, J.R., Kenyon, C.: **Healthy animals with extreme longevity**. *Science* 2003, **302**: 611.

Campisi, J., d'Adda di Fagagna, F.: **Cellular senescence: when bad things happen to good cells**. *Nat. Rev. Mol. Cell Biol.* 2007, **8**: 729–740.

Finkel, T., Serrano, M., Blasco, M.A.: **The common biology of cancer and ageing**. *Nature* 2007, **448**: 767–774.

Harper, M.E., Bevilacqua, L., Hagopian, K., Weindruch, R., Ramsey, J.J.: **Ageing, oxidative stress, and mitochondrial uncoupling**. *Acta Physiol. Scand.* 2004, **182**: 321–331.

Kenyon, C.: **The plasticity of aging: insights from long-lived mutants**. *Cell* 2005, **120**: 449–460.

Kipling, D., Davis, T., Ostler, E.L., Faragher, R.G.: **What can progeroid syndromes tell us about human aging?** *Science* 2004, **305**: 1426–1431.

Kudlow, B.A., Kennedy, B.K., Monnat, R.J.: **Werner and Hutchinson-Gilford progeria syndromes: mechanistic basis of progerial diseases**. *Nat. Rev. Mol. Cell Biol.* 2007, **8**: 394–404.

Murphy, C.T., Partridge, L., Gems, D.: **Mechanisms of ageing: public or private**. *Nat Rev. Genet.* 2002, **3**: 165–175.

Partridge, L.: **Some highlights of research on aging with invertebrates**. *Aging Cell* 2008, **7**: 605–608.

Tatar, M., Bartke, A., Antebi, A.: **The endocrine regulation of aging by insulin-like signals**. *Science* 2003, **299**: 1346–1351.

Weindruch, R.: **Caloric restriction and aging**. *Science* 1996, **274**: 46–52.

13.24 Cell senescence blocks cell multiplication

Shay, J.W., Wright, W.E.: **When do telomeres matter?** *Science* 2001, **291**: 839–840.

Sherr, C.J., DePinho, R.A.: **Cellular senescence: mitotic clock or culture shock**. *Cell* 2000, **102**: 407–410.

14

Evolution and development

- **The evolution of development**

- **The evolutionary modification of embryonic development**

- **Changes in the timing of developmental processes**

The evolution of multicellular organisms is fundamentally linked to embryonic development, for it is through development that genetic changes cause changes in body form that can be passed on to future generations. This raises important questions as to how evolutionary and developmental processes are linked. How did development itself evolve? How has embryonic development been modified during animal evolution? How important are changes in the timing of developmental processes? In this chapter, we shall discuss these general questions using particular examples from animal evolution.

From paleontological, molecular, and cellular evidence, the multicellular animals—the Metazoa—are presumed to descend from a common ancestor that was itself multicellular and that had, in turn, evolved from a unicellular organism. A simplified view of the evolution of animals is shown in Fig. 14.1. The history of evolution is a long one, taking place over thousands of millions of years, and it can only be accessed indirectly—through the study of fossils and by comparisons between living organisms. As Charles Darwin was the first to realize, evolution is the result of heritable changes in life forms and the selection of those that reproduce best (Box 14A).

Development is a fundamental process in evolution. The evolution of multicellular forms of life is the result of changes in embryonic development, and these in turn are entirely due to heritable genetic changes that control cell fates and cell behavior in the embryo. It is also true, however, as the evolutionary biologist Theodosius Dobzhansky once said, that nothing in biology makes sense unless viewed in the light of evolution. Certainly, it would be very difficult to make sense of many aspects of development without an evolutionary perspective. For example, in our consideration of vertebrate development we have seen how, despite different modes of very early development, all vertebrate embryos develop through a rather similar stage, after which their development diverges again (see Fig. 3.2). This shared stage, the phylotypic or tailbud stage, which is the embryonic stage after neurulation and the formation of the somites, is a stage through which some distant ancestor of the vertebrates passed. It has persisted ever since, to become a fundamental characteristic of the development of all vertebrates, whereas the stages before and after this stage have evolved differently in different organisms.

Some major events in metazoan evolution			
Period	**Million years ago (Mya)**	**Paleontology (fossil evidence)**	**Genetic, cellular and molecular evolution**
Quaternary	0.2 2.6	Earliest fossils of modern humans (*Homo sapiens*)	
Neogene (Tertiary)	4–5 ~8 23	Earliest hominin fossils (e.g. *Ardipithecus* (debated); *Australopithecus*)	Estimated date of chimpanzee–human last common ancestor from DNA analysis
Paleogene (Tertiary)	66	Radiations of mammals, birds and insects	
Cretaceous	145	Cretaceous–Tertiary mass extinction; extinction of the dinosaurs Earliest birds	
Jurassic	200	Dinosaurs dominant	
Triassic	252	End-Triassic mass extinction Earliest mammals Earliest dinosaurs	
Permian	298	Permian–Triassic mass extinction Radiation of reptiles	
Carboniferous	359	Earliest reptile (first amniote) Earliest tetrapods on land	Evolution of keratin (the material of scales, feathers, and hair)
Devonian	419	Late Devonian mass extinction Earliest amphibians Diversification of bony fish Earliest tetrapods Earliest insects Earliest bony fish Invasion of land by arthropods	Evolution of limbs from fins Evolution of bone
Silurian	445	Jawed fish	
Ordovician	485	Ordovician–Silurian mass extinction	
Cambrian	541	Agnathan (jawless fish) fossils Earliest vertebrate fossil known—a jawless craniate (an animal with a backbone and distinct skull) The Cambrian explosion: fossils from China (524 Mya), the Burgess Shale (505 Mya) and elsewhere include early arthropods, echinoderms, chordates, and possible jawless vertebrates Earliest definitive arthropod fossils, echinoderms, shelled molluscs Widespread biomineralization	Hemoglobin duplication into α and β chains Genome duplication in the lineage leading to vertebrates Evolution of neural crest Body plans of all extant phyla established during the Cambrian Phylotypic organization of metazoans Segmentation Hox gene clusters
Precambrian	~600 1200 1500 2000 3500 ~4600	Fossil reef-building animals with mineralized skeletons Earliest fossil of a bilaterian organism Earliest cnidarian fossils Abundant 'Ediacaran' fossils, the earliest evidence for complex multicellular life. Experts still disagree about what many of these enigmatic remains represent Earliest sponge fossils First evidence for sexual reproduction (in a red alga) Major radiation of single-celled, colonial and simple multicellular eukaryotes Earliest eukaryotes Earliest reasonable evidence for life (microbial mats) Earth formed	Hox genes Evolution of collagen (characteristic of multicellular animals)

Fig. 14.1 Some selected events in metazoan evolution. The lengths of the geological periods are not to scale. Paleontological events within each period are listed in the order of their occurrence according to the fossil record.

Modified from Gerhardt, J. and Kirschner, M.: Cells, Embryos, and Evolution: Towards a Cellular and Developmental Understanding of Phenotypic Variation and Evolutionary Adaptability. Blackwell, 1997

Genetically based changes in development that generated more successful modes of reproduction or adult forms better adapted to their environment, have been selected for during evolution. Changes in vertebrate development occurring before neurulation are more often associated with changes in reproduction; those occurring after neurulation are more associated with the evolution of animal form. Genetic variability resulting from mutations, sexual reproduction, and genetic recombination is present in the populations of all the organisms we have looked at in this book, and provides new phenotypes upon which selection can act.

Both changes in the regulation of gene expression and changes in protein structure that generate novel protein functions have played a part in evolution. There is ample evidence for changes in protein structure having a direct impact on animal biochemistry, and playing an important role in producing key physiological differences between

BOX 14A Darwin's finches

'Darwin's finches' are a good example of the evolutionary role of development and of changes in gene expression. Charles Darwin visited the Galapagos Islands in 1835 and collected a group of finches, distinguishing at the time 13 closely related species. What he found particularly striking was the variation in their beaks. He wrote in *The Voyage of the Beagle*: 'It is very remarkable that a nearly perfect gradation of structure in this one group can be traced in the form of the beak, from one exceeding in dimensions that of the largest gros-beak, to another differing but little from that of a warbler.' The shapes of the beaks reflected differences in the birds' diets and how they got their food. And Darwin later commented: 'Seeing this gradation and diversity of structure in one small, intimately related group of birds, one might really fancy that from an original paucity of birds in this archipelago, one species had been taken and modified for different ends.' Figure 1 shows the great variation in the shape and size of their beaks of closely related Galapagos finches as a result of adaptation to different sources of food. The drawings are by the English ornithologist John Gould from specimens collected by Darwin on the Galapagos Islands.

Modern investigations have uncovered clues to the developmental basis of these differences in beak form. The species with broader, deeper beaks relative to length express higher levels of the bone morphogenetic protein BMP-4 in the beak growth zone compared with species with long, pointed beaks. Experiments in

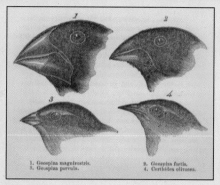

1. *Geospiza magnirostris.*   2. *Geospiza fortis.*
3. *Geospiza parvula.*   4. *Certhidea olivacea.*

Figure 1

which BMP-4 was injected into the developing beak of a chick embryo resulted in the beak growing broader and deeper. A subsequent DNA microarray analysis of differences in the levels of all the mRNAs expressed in the different finch beaks revealed another key player in shaping beak form. The calcium-binding protein calmodulin, a key component of a calcium-dependent signal transduction pathway, is expressed at higher levels in the long, pointed beaks than in the short, broad ones, and was also able to cause the upper beak of a chick embryo to elongate. So beaks are indeed an excellent example of how basic developmental processes and pathways can be varied, producing differences in form and function on which natural selection can act.

species. Changes in the control regions of genes that alter the level, the timing, or the tissue-specificity of their expression, are particularly important in evolution. Many developmental genes have acquired a large number of separate modules in their control regions during their evolutionary history, each of which drives expression of the gene in a different spatial pattern or in a different tissue at various times in development—for example, the *eve* gene in *Drosophila* (see Sections 2.22 and 8.1). Therefore, it is easy to see how mutations in regulatory regions, by changing where and when a gene is expressed, could be a major force for evolutionary change.

The animal kingdom is conventionally divided into three main groups in terms of basic body structure—the bilaterally symmetrical Bilateria, the radially symmetrical Cnidaria and the Ctenophora, and the Parazoa (the sponges (Porifera) and the Placozoa), which are the simplest in terms of their morphology and cellular differentiation and lack body symmetry. The Cnidaria and the Bilateria are 'sister groups'—that is, both groups are considered to arise from a common ancestor. The order of origin of the other groups—the Placozoa, Ctenophora and Porifera—is still a matter of debate. Together, all these groups constitute the Metazoa—the multicellular animals—and are considered to ultimately derive from a common ancestor (Fig. 14.2).

The largest group of animals is the Bilateria, which have bilateral symmetry across the main body axis in at least some stage of development, and have a characteristic pattern of Hox gene expression along the antero-posterior axis. The Bilateria includes vertebrates and other chordates, echinoderms (despite the apparent radial symmetry of the adults), arthropods, annelids, molluscs, and nematodes. These animals are **triploblasts**, as they have the three germ layers—endoderm, mesoderm, and ectoderm. The Cnidaria (corals, jellyfish, hydras and their relatives) are **diploblasts**,

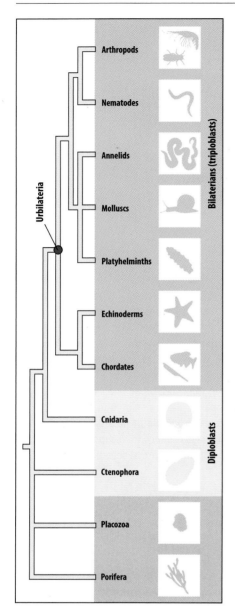

Fig. 14.2 A metazoan family tree. The tree shown here is based on ribosomal DNA and Hox gene sequences and on comparisons of whole-genome sequences. The position of the Urbilateria is represented here by a red circle.

as are the Ctenophora (the comb-jellies). They have only two germ layers (ectoderm and endoderm), and generally have radial symmetry. Some cnidarians, such as *Nematostella*, do however show evidence of bilateral symmetry, and studies of Hox gene expression in these animals have suggested to some researchers that the common ancestor of the Bilateria and the Cnidaria might have had bilateral symmetry.

Throughout this book we have emphasized the conservation of some developmental mechanisms at the cellular and molecular level among distantly related organisms. The widespread use of the Hox gene complexes and of the same few families of protein-signaling molecules provides excellent examples of this, and has become a hallmark of the animal kingdom. It is these basic similarities in molecular mechanisms that have made developmental biology so exciting in recent years; it has meant that discoveries of genes in one animal have had important implications for understanding development in other animals. It seems that when a useful developmental mechanism evolved, it was retained and redeployed in very different organisms, and at different times and places in the same organism.

In the rest of this book we have looked at the early development of quite a wide range of different organisms and found some similarities, as well as a number of differences. In this chapter we mainly confine our attention to two phyla—the chordates (which include the vertebrates) and the arthropods (which include the insects and the crustaceans (crabs, lobsters, and woodlice, for example)). We focus on those differences that distinguish the members of a large group of related animals, such as the vertebrates or the insects, from each other. We will look at the relationship between the development of the individual organism (its **ontogeny**) and the evolutionary history of the species or group (the **phylogeny**): why, for example, do mammalian embryos pass through an apparently fish-like stage that has structures resembling gill slits? We discuss the many variations that occur on the theme of a basic, segmented, body plan: what determines the different numbers and positions of paired appendages, such as legs and wings, in different groups of segmented organisms? We consider the timing of developmental events, and how variations in growth can have major effects on the shape and form of an organism. In all cases, we ultimately want to understand the changes in the developmental processes and their genetic basis that have resulted in the extraordinary range of different multicellular animals. This is an exciting area of study in which many problems remain to be solved. But first we will briefly consider how development itself might have evolved. What is the origin of the egg, and how might processes such as pattern formation and gastrulation have evolved?

The evolution of development

To look for the origins of development we must consider what we know about the ancestors of multicellular animals. The origin of the ancestor of animals is a tough problem. From genomic evidence and studies of the simplest animals, the last common ancestors of the bilaterians—the Urbilateria—must already have been complex creatures, possessing most of the developmental gene pathways used by existing bilaterians. What Urbilateria might have been like and how they might have been constructed are now key questions in the evolution–development (**evo-devo**) field. Another central challenge is to explain how conserved gene networks already present in the archetypal ancestor were modified to generate the wonderful diversity of animal life on Earth today.

14.1 Genomic evidence is throwing light on the origin of metazoans

About 35 different animal phyla with distinct body plans currently exist, and almost 30 of these phyla are bilaterians. Bilaterians are traditionally subdivided into the protostomes and the deuterostomes according to the origin of the mouth with regard to gastrulation. The protostomes develop the mouth during gastrulation, close to the invaginating endoderm and mesoderm. They have a ventral nerve cord which emanates from two dorsal ganglia in the head that serve as a 'brain'. The two branches from the ganglia extend ventrally on either side of the foregut and then fuse to form the single ventral cord. Of the animals covered in this book, the arthropods, such as *Drosophila* (see Chapter 2), and the nematodes, such as *Caenorhabditis* (see Chapter 6), are protostomes.

In deuterostomes the mouth develops independently of the gastrulation process, which defines the anus, and the nervous system develops dorsally. Deuterostomes include the chordates, such as the vertebrates (see Chapters 3–5), the ascidians, and the echinoderms, such as the sea urchin (sea-urchin development is described in Chapter 6 and ascidian development online).

Fossils of adult bilaterians make their appearance suddenly, early in the Cambrian period, around 540 million years ago, or even earlier, a time called the 'Cambrian explosion' because of the sudden appearance of an enormous range of animal life. Fossils from this time reveal the existence of segmented organisms with bilateral symmetry and a well differentiated head.

The remaining, non-bilaterian, metazoans mostly belong to the phyla Cnidaria, Ctenophora and Porifera; they are the simplest and oldest known animals possessing a gut. The genome sequences of species from all these groups have been determined: the cnidarians *Hydra magnipapillata* and the starlet sea anemone *Nematostella vectensis*, the sponge *Amphimedon queenslandica*, *Trichoplax adhaerens* from the Placozoa, and the comb-jelly *Pleurobrachia bachei*, representing the ctenophores. These genome sequences are filling gaps in our knowledge of the evolution of key developmental genes such as the Hox genes, and the genes for signaling proteins such as those of the Wnt and Hedgehog pathways, and reveal that the full complement of proteins characteristic of animal cell types emerged very early in the evolution of multicellular animals. Some typical developmental signaling molecules present in these phyla and the Bilateria are shown in Fig. 14.3.

The genome of *Nematostella* reveals an extensive gene repertoire more similar to that of vertebrates than to flies or nematodes, implying that the genome of the ancestor of all animals was similarly complex. Nearly one-fifth of the inferred genes in the *Nematostella* genome are novelties—not known from other animal groups—and the genome is rich in genes for animal functions such as cell–cell signaling, cell adhesion, and synaptic transmission.

Taking all the evidence from bilaterians and cnidarians together, we can speculate that the ancestor of all extant animals lived perhaps 700 million years ago and would have had flagellate sperm, development through a process of gastrulation, multiple germ layers, true epithelia lying on a basement membrane, a lined gut, neuromuscular and sensory systems, and fixed body axes. *Nematostella*'s genome contains about 18,000 protein-coding genes, a similar number to that of *Caenorhabditis*. Genes representing 56 families of homeodomain-containing gene-regulatory proteins are present in *Nematostella*, including three putative Hox genes, and also including representatives of four different classes of Pax genes, genes that in bilaterians are involved in many different aspects of development such as eye development, segmentation, and neural patterning. More remarkably for an animal with only two germ layers, at least seven genes associated with the formation of mesoderm in bilaterians are expressed in the developing endoderm of *Nematostella*. Thus the machinery for germ-layer specification was already present in the ancestors of both bilaterians and cnidarians.

The sponges are very different from the Cnidaria and the Bilateria in their morphology and level of structural differentiation. They have a few different cell types, some of them sensory and others digestive, but they share many features with other animal groups at the molecular level, such as the presence of signaling molecules (Notch,

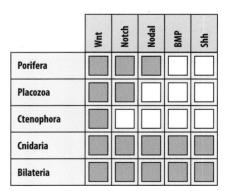

Fig. 14.3 Signaling molecules present in different animal groups. The signaling pathways that are present in each phylum, as judged by their genome sequences.

Wnt and Nodal, see Figure 14.3), some homeobox genes, but not Hox genes, and neural-like cell types. All of this suggests that the ancestor of all metazoans already had, at the very least, some of the cell types and molecules familiar from modern animals.

There is one very simple free-living marine animal called *Trichoplax*, which is so unusual it has been placed in a phylum of its own, the Placozoa. Its relationship to other animals is not yet clear, but genomic comparisons place it as a very simple member of the Metazoa. *Trichoplax* is composed of two layers of epithelium, which form a flat disc with no gut, and it has just four different cell types. It mainly reproduces by fission. Nevertheless, in line with the genomes of other animals, the *Trichoplax* genome contains an estimated 11,500 protein-coding genes, which encode a rich array of transcription factors and signaling proteins, including homeodomain-containing proteins, T-box proteins similar to Brachyury, and components of the Wnt/β-catenin signaling pathway.

14.2 Multicellular organisms evolved from single-celled ancestors

If recognizable animals had evolved by some 700 million years ago, the question still remains of how they evolved from a unicellular ancestor, which from genomic and other evidence is thought to have been an organism rather similar to the modern choanoflagellates. These unicellular and colonial protozoa have a single flagellum surrounded by a 'collar' of villi, and are very similar in morphology to the collared cells (choanocytes) of sponges. What had to be invented for the transition to multicellularity? And how did embryonic development from an egg evolve? The key requirements for embryonic development, as we have seen, are a program of gene activity, cell differentiation, signal transduction (so cells can communicate with each other), and cell motility and cohesion (so that overall form can change).

Judging by modern unicellular eukaryotes, such as the choanoflagellates, the single-celled organism ancestral to animals would have possessed all these features in primitive form, and little new would have had to be invented, although much had to evolve further and acquire different functions for multicellularity to evolve. The genome sequence of the modern choanoflagellate *Monosiga brevicollis* shows evidence of some cell adhesion and signaling proteins that are otherwise only present in metazoans. Unicellular eukaryotes have the ability to move, to adhere, and to respond to signals. At a minimum, the ancestor of multicellular organisms would have already possessed the characteristic eukaryotic cell cycle, which entails a complex program of gene activity, the ability to undergo cellular differentiation, signaling by receptor tyrosine kinases, and cadherin-mediated cell adhesion.

What then was the origin of multicellularity and the embryo? One possibility, and this is highly speculative, is that mutations resulted in the progeny of a single-celled organism not separating after cell division, leading to a loose colony of identical cells that occasionally fragmented to give new 'individuals.' One advantage of a colony might originally have been that when food was in short supply, the cells could feed off each other, and so the colony survived. This could have been the origin of both multicellularity and the requirement for cell death in multicellular organisms. The egg might subsequently have evolved as the cell fed by other cells; in sponges, for example, the egg phagocytoses neighboring cells. The evolutionary advantage of an organism developing from a single cell—the egg—is that all the cells of the organism will have the same genes. This is a prerequisite for the development of complex patterns and body plans, as patterning requires communication between cells, and that can only be reliable if the cells obey the same rules as a result of having the same genetic instructions. There are 'multicellular' organisms that develop from more than one cell, such as slime molds, but they have never evolved numerous complex forms.

Once multicellularity evolved, it opened up all sorts of new possibilities, such as cell specialization for different functions. Some cells could specialize in providing

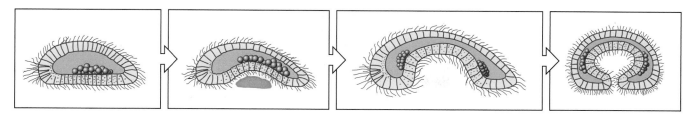

motility, for example, and others in feeding, as we see in the nematocytes, the sting-ing cells of *Hydra* that are used to capture prey. The origin of pattern formation—that is, cell differentiation in a spatially organized arrangement—is not known, but may have depended on gradients set up by external influences such as inside–outside differences.

How gastrulation evolved is also unknown, but it is not implausible to consider a scenario in which a hollow sphere of cells, the common ancestor of all multicel-lular animals, changed its form to assist feeding (Fig. 14.4). This ancestor may, for example, have sat on the ocean floor, ingesting food particles by phagocytosis. A small invagination developing in the body wall could have promoted feeding by forming a primitive gut. Movement of cilia could have swept food particles more efficiently into this region, where they would be phagocytosed. Once the invagination formed, it is not too difficult to imagine how it could eventually extend right across the sphere to fuse with the other side and form a continuous gut, which would be the endoderm. At a later stage in evolution, cells migrating inside, between the gut and the outer epithelium, would give rise to the mesoderm.

Gastrulation provides a good example of developmental change during evolution. While there is considerable similarity in the process of gastrulation in many different animals, there are also significant differences, as discussed throughout this book in relation to our model animals. But how these evolved and what could have been the adaptive nature of the intermediate forms is a hard problem.

Fig. 14.4 A possible scenario for the development of the gastrula. A hollow multicellular sphere of cells, perhaps derived from a colonial protozoan, could have settled on the sea bottom and developed a gut-like invagination to aid feeding.

Based on Jaegerstern G.: **The early phylogeny of the metazoa. The bilatero-gastrea theory.** Zool. Bidrag. (Uppsala) *1956, 30: 321-354.*

SUMMARY

The embryo arose during the evolution of multicellular organisms from single-celled organisms, which have most of the cellular properties required by embryonic develop-ment. Multicellularity, having arisen, might have persisted originally because a colony of cells could provide a source of food for some of the members in times of scarcity. The egg might have evolved as the cell that is fed by other cells and provides the basis for the development of complex structures. The development of an embryo provides insights into the evolutionary origin of the animal.

The evolutionary modification of embryonic development

Comparisons of embryos of related species have suggested an important generaliza-tion about development: the more-general characteristics of a group of animals (that is, those shared by all members of the group) usually appear earlier in their embryos than the more-specialized ones, and arose earlier in evolution. In the vertebrates, a good example of a general characteristic would be the notochord, which is common to all vertebrates and is also found in other chordate embryos. The phylum Chordata comprises three subphyla—the vertebrates; the cephalochordates, such as the amphi-oxus (*Branchiostoma*) (Fig. 14.5); and the urochordates, such as the ascidians (see online material). They all have, at some embryonic stage, a notochord, flanked by muscle, and a dorsal neural tube. The amphioxus in particular has many features of

Fig. 14.5 The cephalochordate amphioxus compared with a hypothetical primitive vertebrate similar to a present-day lamprey. The overall construction of the two organisms is very similar, with a dorsal nerve cord, the axial rod-like notochord below it, and a ventral digestive tract. Both animals have gills in the pharyngeal region, structures that are designed to capture food and also take in oxygen from the water. The notochord extends right to the anterior end of the amphioxus, but the vertebrate has a prominent head at the anterior end, extending beyond the notochord. In more advanced vertebrates, the notochord is present only in the embryo, being replaced by the vertebral column.

*Modified from Finnerty J. R.: **Evolutionary developmental biology: Head start.** Nature, 2000, **408**: 6814.*

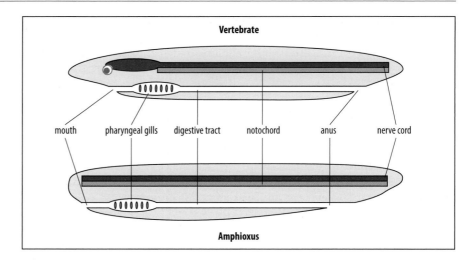

a primitive vertebrate, even when adult: a dorsal hollow nerve cord, in this case supported by the persistent notochord, rather than a bony spine, and segmental muscles that derive from somites.

Among the chordates, paired appendages, such as limbs, are special characters that have evolved only in the vertebrates, and they differ in form among different vertebrates. All vertebrate embryos pass through a phylotypic stage at which they all more or less resemble each other and show the specific embryonic features of the phylum to which they belong (see Fig. 3.2). After the phylotypic stage, vertebrate embryos diversify and give rise to the diverse forms of the different vertebrate classes. The phylotypic stage itself is somewhat variable, however, as can be seen in Fig. 14.6, which shows that the number of somites can be very different, and that some organs, such as limb buds, are at different stages of development at this time. The development of the different vertebrate classes before the phylotypic stage is also highly divergent, because of their very different modes of reproduction; some developmental features that precede the phylotypic stage are unique to their class, such as the formation of a trophoblast and inner cell mass in mammals. This is an example of a special character that developed late in vertebrate evolution, and is related to the nutrition of the embryo through a placenta, rather than a yolky egg.

The reason that vertebrate embryos pass through a phylotypic stage is still a matter of debate. It may be related to that being the stage at which the Hox genes, which control general pattern along the main body axis, are expressed. Also, it is the stage at which structures common to all vertebrates are developed, such as somites, notochord, and neural tube. We will now consider the modifications that have occurred to various embryonic structures during evolution, including the basic body plan and the limbs.

14.3 Hox gene complexes have evolved through gene duplication

Hox genes play a key role in the development of both vertebrates and insects—and in most other animals. By comparing the organization and structure of the Hox genes in insects and vertebrates, we can determine how one set of important developmental genes has changed during evolution. A major general mechanism of evolutionary change has been gene duplication and divergence. Tandem duplication of a gene, which can occur by various mechanisms during DNA replication, provides the embryo with an additional copy of the gene. This copy can diverge in the nucleotide sequences of both its protein-coding region and its control region, including the promoter and *cis*-regulatory regions such as enhancers, thus changing the gene's function and its pattern

Fig. 14.6 Vertebrate embryos at the phylotypic (tailbud) stage show considerable variation in somite number. Not to scale.

Modified from Richardson M. K., et al.: **There is no highly conserved embryonic stage in the vertebrates: implications for current theories of evolution and development**. Anat. Embryol. 1997, **196**: 91–106.

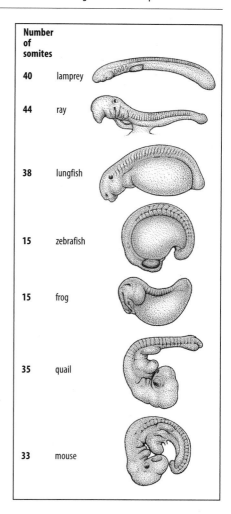

Number of somites	
40	lamprey
44	ray
38	lungfish
15	zebrafish
15	frog
35	quail
33	mouse

of expression without depriving the organism of the function of the original gene (Fig. 14.7). The process of gene duplication has been fundamental in the evolution of new proteins and new patterns of gene expression; it is clear, for example, that the different hemoglobins in humans (see Fig. 8.17) have arisen as a result of gene duplication. In the case of genes encoding transcription factors, alterations in their coding regions, and thus in the structure and binding properties of the protein they encode, can lead to the factor being able to regulate a different set of downstream target genes.

The Hox gene complexes provide one of the clearest examples of the importance of gene duplication in developmental evolution. The Hox genes specify pattern along the antero-posterior axis, as we have seen in previous chapters, and are among the trademark genes that characterize the Metazoa. They are members of the homeobox gene superfamily, which is characterized by a short 180-base-pair motif, the homeobox, which encodes a DNA-binding helix–turn–helix domain called the homeodomain that is involved in transcriptional regulation (see Box 5E). Two features characterize all vertebrate Hox genes: the individual genes are organized into one or more gene clusters, or complexes, and the order of expression of individual genes along the antero-posterior axis is usually the same as their sequential order in the gene complex (see Section 5.10).

The simplest Hox gene complexes are found in invertebrates, and comprise a small number of sequence-related genes carried on one chromosome. Higher vertebrates typically have four sets of Hox genes, carried on four different chromosomes, suggesting two rounds of wholesale duplication of an ancestral Hox gene complex, in line with the generally accepted idea that large-scale duplications of the genome have occurred during vertebrate evolution. There have also been further duplications of Hox genes within each complex. In the ancestor of fish there was one further round of genome duplication; thus, for example, the zebrafish and the Japanese puffer fish *Takifugu rubripes* have seven Hox clusters each, the result of three rounds of duplication and the subsequent loss of one cluster. It seems that in the ancestral chordates that gave rise to the vertebrates, the spatial co-linearity between the order of the Hox genes along the chromosome and their expression along the main body axis was confined to the ectoderm-derived neural tube, and possibly the epidermis, and that it was only later extended to other tissues, thus allowing increased complexity of body plan.

Comparisons of Hox genes of various arthropod and chordate species indicate that the common ancestor of arthropods and vertebrates most probably had a simple Hox

Fig. 14.7 Gene duplication and diversification. Once a gene has become duplicated, the second copy can evolve new functions and/or new expression patterns. Illustrated is a hypothetical example of a gene (purple) with two control regions (blue and green) that each confer expression in a different tissue. Left pathway: after duplication of the complete gene and its control regions, mutation in the coding region of one of the copies can generate a gene with a new and useful function that is selected for. Both copies will be retained in the genome as they both fulfill useful functions. Subsequent mutation in a control region could lead to the new gene acquiring a new expression pattern. Right pathway: alternatively, a mutation in one control region in one copy and the other control region in the other copy can generate two copies of the gene that retain the original function but are now only expressed in a single tissue each. Both will be retained in the genome as they are both now needed to fulfill the function of the original gene. Examples of all these types of duplication and divergence are found in animal genomes.

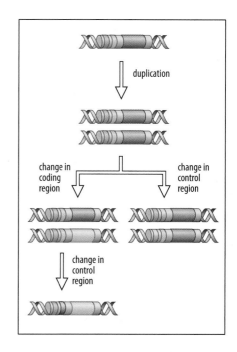

Fig. 14.8 Gene duplication and Hox gene evolution. A suggested evolutionary relationship between the Hox genes of a hypothetical common ancestor and *Drosophila* (an arthropod), amphioxus (a cephalochordate), and the mouse (a vertebrate). Duplications of genes of the ancestral set (red) could have given rise to the additional genes in *Drosophila* and amphioxus. Two duplications of the whole cluster in a chordate ancestor of the vertebrates could have given rise to the four separate Hox gene complexes in vertebrates. There has also been a loss of some of the duplicated genes in vertebrates. In *Drosophila*, the two Hox gene clusters comprising the HOM-C complex are called the Antennapedia complex and the bithorax complex and their co-linearity has been broken as a result of a chromosomal rearrangement. Hox14 has been found in the coelacanth and the horn shark as well as in amphioxus, but Hox15 has only been found so far in amphioxus species, and has no relationship to the other paralagous subgroups.

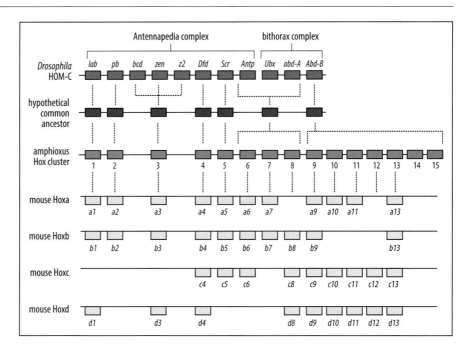

cluster of seven genes (Fig. 14.8). As we get nearer to the vertebrates, the cephalo-chordate amphioxus has just one Hox cluster, containing 15 genes, and one can think of this as most closely resembling the ancestor of the four vertebrate Hox gene clusters: Hoxa, Hoxb, Hoxc, and Hoxd (see Fig. 14.8). Possible ways in which both the vertebrate and *Drosophila* Hox clusters could have evolved from a simpler ancestral gene cluster by gene duplication have been reconstructed. In *Drosophila*, for example, successive tandem duplications of an ancestral *Antp*-like gene could have given rise to the genes *abdominal-A* (*abd-A*), *Ultrabithorax* (*Ubx*), and *Antennapedia* (*Antp*) (see Fig. 14.8). *Antennapedia* is associated with the specification of the second thoracic segment, T2, and if the later-evolved posteriorly expressed Hox genes of the bithorax complex—namely *Ubx*, *abd-A* and *Abd-B*—are deleted in *Drosophila*, all the thoracic and abdominal segments posterior to T2 revert to the T2 type, and express *Antp* (see Fig. 2.51, second panel).

In vertebrates, the Hox genes are arranged in four separate clusters, each of which is on a different chromosome. The separate gene complexes probably arose from duplications of whole chromosomal regions, in which new Hox genes had already been generated by tandem duplication. For example, sequence comparisons suggest that the multiple mouse Hox genes most similar in sequence to the *Drosophila Abd-B* gene do not have direct homologs within the *Drosophila* HOM-C complex (see Fig. 14.8); they probably arose by tandem duplication from an ancestral gene after the split of the insect and vertebrate lineages, but before duplication of the whole cluster in vertebrates. The advantage of duplication was that the embryo had more Hox genes to control downstream targets and so could make a more complicated body. We now consider the role of these genes in the evolution of the axial body plan.

14.4 Changes in both Hox genes and their target genes generated the elaboration and diversification of bilaterian body plans

The broad similarities in the pattern of Hox gene expression in vertebrates and arthropods, whose evolution diverged hundreds of millions of years ago, are taken as good supporting evidence for the idea that all multicellular animals descend from a common ancestor. Hox genes are key genes in the control of development and are expressed regionally along the antero-posterior axis of the bilaterian embryo (see

Chapters 2 and 5). The apparently conserved nature of the Hox genes (and of certain other developmentally important genes) in animal development has led to the concept of the **zootype,** a general pattern of expression of these key genes that is present in all animal embryos.

The highly conserved organization of Hox gene expression along the body axis suggests that changes in the pattern have important consequences for the morphology of the animal. The Hox genes exert their influence by controlling the activity of target genes that determine, for example, how the cells in a region develop the morphological features characteristic of a particular segment—producing segment-specific structures such as legs and wings in insects, or the different types of vertebrae in vertebrates, for example. These target genes are the 'effector' genes that interpret the positional information encoded by the Hox genes, and any individual Hox gene will control many downstream target genes.

Mutations in the Hox target genes, especially changes that affect how positional information is interpreted, are equally likely to be a major source of morphological change in evolution. A number of genes targeted by the Hox transcription factors are now known, but many still remain to be identified.

One easily visible modification of the body plan involving Hox genes that has taken place during vertebrate evolution is in the pattern of the axial skeleton—the vertebral column and ribs. This pattern is characterized by the number and type of vertebrae in the main anatomical regions—cervical (neck), thoracic (rib-bearing), lumbar, sacral, and caudal (Fig. 14.9). The number of vertebrae of a particular type varies considerably among the different vertebrate classes—mammals, with rare exceptions, have seven cervical vertebrae, whereas birds can have between 13 and 15, and certain reptiles, such as snakes, can have no cervical vertebrae but more than 200 thoracic vertebrae.

How do these differences arise? A comparison between the mouse and the chick shows that the patterns of Hox gene expression along the body differ. The anterior limit of expression of, for example, *Hoxc6*, one of the Hox genes that has been shown experimentally to specify thoracic regional identity, is at somite 12 in the mouse but at somite 19 in the chick (see Fig. 5.30; the first five somites do not give rise to vertebrae). Moreover, the anterior limit of *Hoxc6* expression is at somite 22/23 in geese, which have three more cervical vertebrae than chickens. In snakes, most of the body consists of an elongated trunk, and the trunk axial skeleton consists of a very large number of morphologically similar thoracic vertebrae, which all have ribs, as can be seen in the skeleton of the python embryo in Fig. 14.10. In pythons, *Hoxc6* and *Hoxc8* are expressed in the somites throughout the trunk, suggesting that alterations in the pattern of Hox gene expression could have been involved in the evolution of the python body plan.

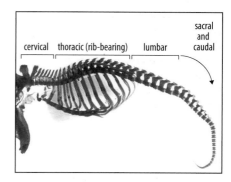

Fig. 14.9 Vertebrae come in many different types. Lateral view of an E18.5 mouse skeleton stained with Alcian blue and alizarin red (for cartilage and bone respectively), showing the pattern of vertebrae in the main anatomical regions: cervical (neck) 7; thoracic (rib-bearing) 13; lumbar (lower back) 6; sacral (hip region) 4; caudal (tail), approximately 30 (although not all of these are easy to make out in this specimen).

*Photograph courtesy D. M. Wellik, from Wellik, D.M.: **Hox genes and vertebrate axial pattern**. Curr. Topics Dev. Biol. 2009, **88**: 257-278.*

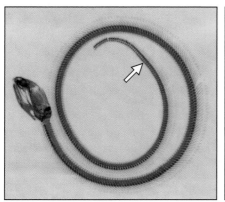

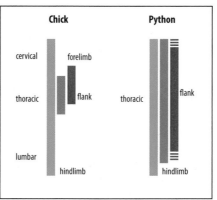

Fig. 14.10 Comparison of Hox gene expression in python and chick embryos. The photograph shows a skeleton of a python embryo at 24 days' incubation stained with Alcian blue and Alizarin red. The arrow marks the position of hindlimb rudiments, which have been removed in this preparation. Note how similar the vertebrae are in the trunk, which is anterior to the arrow. The right panel shows a schematic comparison of domains of expression of *Hoxb5* (green), *Hoxc8* (blue) and *Hoxc6* (red) in chick and python embryos. The expansion of the *Hoxc8* and *Hoxc6* domains in the python correlates with the expansion of thoracic identity in the axial skeleton and flank identity in the lateral plate mesoderm.

*Adapted from Cohn, M.J. and Tickle C.: **Developmental basis of limblessness and axial patterning in snakes**. Nature 1999, **399**: 474-479.*

An example of a change in a Hox target gene, rather than in the Hox gene itself, that affects morphology has been found in the North American corn snakes. In corn snakes, a more comprehensive analysis of Hox gene expression patterns revealed that these are not as homogeneous as in the python. Some Hox10 paralogs, *Hoxa10* for example, which specify lumbar vertebral identity in mice (see Section 5.11), are unexpectedly expressed in the somites in the posterior region of the corn snake trunk. Experiments have shown that the reason that rib-bearing thoracic vertebrae still develop in this region in these snakes is not because the functioning of Hoxa10 protein is changed. Just as when mouse *Hoxa10* is expressed conditionally in pre-somitic mesoderm in a transgenic mouse, corn snake *Hoxa10* can also repress rib formation on thoracic vertebrae. Instead, what has happened in the corn snake is a change in the interpretation of the Hox code. A single nucleotide change in the enhancer of one of the target genes of Hoxa10 has been identified. The gene is required for rib formation, and in mice its expression is inhibited in the lumbar region by Hoxa10 protein. As a result of the mutation in corn snakes, Hoxa10 is unable to bind to the enhancer and repress expression of the rib-forming gene, and so ribs develop even when Hoxa10 is present. Remarkably, the identical genetic change is also found in some afrotherian vertebrates, such as elephants, which have elongated rib cages. These findings point up how important changes in Hox target genes can be in animal evolution.

In vertebrates, Hox gene expression also specifies positional differences along the antero-posterior axis in the lateral plate mesoderm from which limbs develop (see Section 11.2). Limb loss during evolution is quite common, and the python provides a good example of the developmental changes involved. Pythons, like other snakes, lack forelimbs but have a pair of hindlimb rudiments at the junction between the rib-bearing thoracic vertebrae and the vertebrae with shorter forked ribs. In the lateral plate mesoderm of four-limbed vertebrates, *Hoxb5*, *Hoxc6*, and *Hoxc8* are expressed in a short trunk region between the limb-forming regions, whereas in pythons these genes are expressed along the whole body as far as the pelvic rudiment (see Fig. 14.10). In the chick embryo, *Hoxc8* can repress the action of a control region in the T-box gene *Tbx5*, which normally drives its expression in the prospective forelimb region of the lateral plate mesoderm. The transcription factor Tbx5 is essential for forelimb development (see Section 11.2), and so repression of its gene by Hoxc8, which is expressed more widely in snakes, could provide the mechanism that leads to the loss of forelimbs. Python hindlimb buds begin to develop, but the signals associated with the apical ridge and the polarizing region are not activated, and therefore the limbs are very rudimentary.

14.5 Differences in Hox gene expression determine the variation in position and type of paired appendages in arthropods

Changes in patterns of Hox gene expression can also help explain the evolution of the great diversity of arthropod body plans, examples of which are shown in Fig. 14.11. A segmented body is distinguishable in all arthropod embryos, even though in some classes, such as the spiders, the segmented structure is not easily discernible in the adult. Efforts to determine the homology of different types of segments between the different arthropod groups, and thus to identify the nature of the changes that have occurred during their evolution, have proved difficult on anatomical grounds alone.

We shall focus here on insects and crustaceans, in which most is known about patterns of Hox gene expression. Insects and crustaceans evolved from a common ancestor, which probably had an adult body composed of more or less uniform segments, all bearing similar paired leg-like appendages (Fig. 14.12). Evolution has restricted the legs to particular segments. Differences in the type and distribution of the paired appendages along the body in crustaceans have provided a good test of the role of

Drosophila

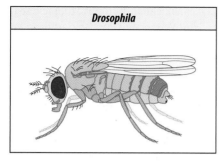

Insecta

Head						Thorax			Abdomen									
Oc	An	In	Mn	Mx	La	T1	T2	T3	A1	A2	A3	A4	A5	A6	A7	A8	A9	A10
						leg	leg	leg										
							wing	wing										

Shrimp

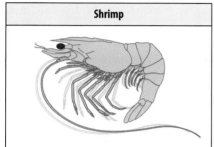

Crustacea (Malacostraca: Decapoda)

Head						Thorax (Pereon)								Abdomen (Pleon)						
Oc	An1	An2	Mn	Mx1	Mx2	T1	T2	T3	T4	T5	T6	T7	T8	A1	A2	A3	A4	A5	A6	A7
						mxp	mxp	mxp	leg	leg	leg	leg	leg	sw	sw	sw	sw	sw		

Centipede

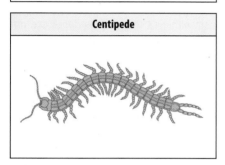

Myriapoda

Head							Trunk											
Oc	An	In	Mn	Mx1	Mx2	Mxp	T1	T2	T3	T4	T5	T6	T7	T8	T9			
						mxp	leg	leg	leg	leg	leg	leg	leg	leg	leg			

more posterior segments added with growth (all with legs)

Spider

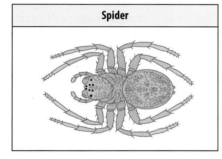

Chelicerata

Prosoma							Opisthosoma								
Oc	Ch	Pp	L1	L2	L3	L4	O1	O2	O3	O4	O5	O6	O7	O8	O9
			leg	leg	leg	leg									

Hox genes in controlling this process and thus their probable role in the evolution of arthropod body plans. If the product of the Hox gene is actively 'instructing' the identity of a particular segment or contiguous group of segments by regulating the expression of the target genes that specify the structural features of that segment, then the same change in segmental morphology in different groups should be accompanied by a consistent change in the expression pattern of that Hox gene.

Crustaceans are particularly well suited for such a study. Their thoracic segments vary in number between different groups, but each segment bears one of two kinds of appendages: legs (which can be of several types) or specialized 'feeding' legs (maxillipeds), which resemble mouthparts in structure and are used to manipulate food. Different orders of crustaceans have a different complement of legs and maxillipeds on the thorax, and a comparison of the patterns of Hox gene expression in different species reveals a clear relationship between Hox expression and segment identity and function, as judged by the appendages they carry. Figure 4.13 shows the four different

Fig. 14.11 The variety of arthropod body plans. Shown here are examples from the four main arthropod classes and their segment pattern. Segments that bear maxillipeds (mxp), legs, swimmerets (sw), and wings are noted. The head segments bear a variety of specialized appendages such as antennae and mouthparts. An, antenna; Ch, chelicera; In, intercalary segment; Lb, labium; Mn, mandible; Mx, maxilla; Oc, ocular; Pp, pedipalp.

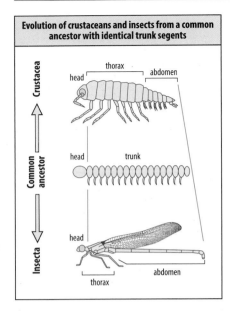

Evolution of crustaceans and insects from a common ancestor with identical trunk segents

Fig. 14.12 Insects and crustaceans evolved from a common ancestor.

From Abzhanov, A., Kaufman, T.C.:
Crustacean (malacostracan) Hox genes and the evolution of the arthropod trunk. Development *2000,* **127***: 2239-2249.*

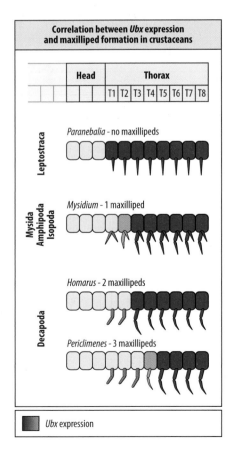

patterns found among the malacostracan crustaceans, which have eight thoracic segments, and the correlation with expression of the Hox gene *Ubx*. The other main groups of crustaceans either carry legs on all the thoraic segments (the branchiopds, such as the brine shrimp *Artemia*) or have one maxilliped (the maxillipods, such as the American tadpole shrimp *Triops*), and show the same correlation between the loss of *Ubx* expression in a thoracic segment and the presence of a maxilliped instead of a leg.

The presence of maxillipeds on a segment is always associated with an absence of *Ubx* expression. However, correlations of this sort indicate causality, but do not definitively prove it. As we saw in *Drosophila*, causality can be tested by deliberately altering the spatial pattern of Hox gene expression in an individual species and analyzing the effects (see Fig. 2.51). The results of manipulating *Ubx* expression in the amphipod *Parhyale* (a malacostracan crustacean) support the morphological comparison. Lowering *Ubx* expression allowed the emergence of maxilliped-like structures instead of legs, whereas expressing *Ubx* in a thoracic segment in which it is normally not expressed led to the appearance of legs instead of maxillipeds (Fig. 14.14). Notably, the transformations only affected the first three thoracic segments—in both cases, the rest of the appendage distribution was normal. Even when *Ubx* expression was silenced throughout the thorax, maxillipeds only appeared on the first three segments; a leg developed as usual on the fourth and following thoracic segments. This reflects the natural distribution of maxillipeds in crustaceans, as these appendages are not found further back than thoracic segment 3 (see Fig. 14.13). In this case, therefore, we can definitely say that changes in the regulation of *Ubx* that resulted in a change in its spatial pattern of expression have driven evolutionary change in crustaceans.

Despite the enormous variety in appearance among extant insects, their basic body plans are less diverse than those of crustaceans. All adult insects have just three thoracic segments, bearing a pair of legs on each. Wings, when present, appear on the second and third segments, or just the second in the case of flies. Insect fossils, however, display a much wider range of patterns in the position and number of their legs and wings. Some insect fossils have legs on every segment, whereas others only have legs in a distinct thoracic region. The number of abdominal segments bearing legs varies, as does the size and shape of the legs. Wings arose later than legs in insect evolution. Wing-like appendages are present on all the thoracic and abdominal segments of some insect fossils, but are restricted to the thorax in others. To see in principle how these different patterns of appendages could have arisen during insect evolution we will compare two orders of modern insects, the Lepidoptera (butterflies and moths) and the Diptera (flies, including *Drosophila*).

The basic pattern of Hox gene expression along the antero-posterior axis is the same in all present-day insect species that have been studied. Yet caterpillars (larvae) of Lepidoptera have small walking appendages (pro-legs) on the abdomen and the thorax, and the adults have two pairs of wings, whereas the more recently evolved Diptera have no legs on the abdomen in the larva, and only one pair of wings in the

Fig. 14.13 The presence and position of maxillipeds (feeding legs) in different species of crustacean depends on evolutionary changes in the pattern of *Ubx* and *abdA* expression. Four examples from the class Malacostraca are shown here, as this class shows the greatest diversity. The number and position of maxillipeds (blue) on thoracic segments varies between different orders (for example, the Decapoda and the Amphipoda), and their presence clearly correlates with the anterior boundary of expression of the Hox genes *Ubx* and *abdA*. A monoclonal antibody that recognizes both the Ubx and abdA proteins was used to stain the larval stages of different species. Red indicates higher levels of the two proteins, which correlates with the presence of walking legs, and orange indicates lower-level expression, which correlates with grasping legs. Maxillipeds develop where *Ubx* and *abdA* are not expressed.

Adapted from Averof, M., Patel, N.H.: ***Crustacean appendage evolution associated with changes in Hox gene expression***. Nature *1997,* **388***: 682-685.*

Fig. 14.14 Manipulation of *Ubx* expression in the amphipod *Parhyale hawaiensis*.
Wild-type *P. hawaiensis* bears a single pair of maxillipeds on the first thoracic segment (top panel), a type of leg called a grasping leg on the next two segments, and walking legs on the remaining five thoracic segments (only thoracic segments 1–4 are shown here). When *Ubx* expression was lowered in the embryo by injection of short-interfering RNAs (siRNAs; see Box 6B), maxillipeds appeared on the first three thoracic segments (second panel). When transgenic *Ubx* was expressed ectopically in all thoracic segments under the control of a heat-shock gene promoter, legs of the type normally found on T2 appeared on T1 instead of a maxilliped (third panel), reflecting the transformation of the segment to a more posterior fate. Because uniform high-level *Ubx* transgene expression throughout the body in addition to the underlying wild-type expression was lethal, the transgene was expressed for only short periods each day. The outcome illustrated here was observed in around 10% of hatchlings. In a smaller proportion of hatchlings (3%) legs of the T4 type developed on all thoracic segments. As in Fig. 14.13, red indicates higher levels of *Ubx* expression, and shades of orange lower levels.

*Data from Liubicich, D.M., et al.: **Knockdown of Parhyale Ultrabithorax recapitulates evolutionary changes in crustacean appendage morphology.** Proc. Natl Acad. Sci USA 2009, **106**: 13892-13896; Pavlopoulos, A., et al.: **Probing the evolution of appendage specialization by Hox gene misexpression in an emerging model crustacean.** Proc. Natl Acad. Sci USA 2009, **106**: 13897-13902.*

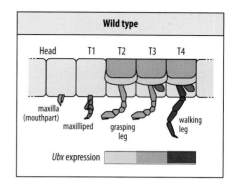

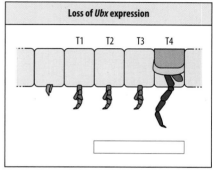

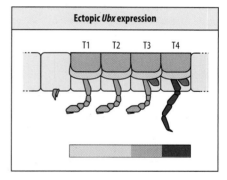

adult, with the second pair of wings having been modified into small balacing organs, the halteres. How are these differences related to differences in Hox gene activity in the two groups of insects?

In *Drosophila*, the *Distal-less* gene, which is required for leg formation (see Section 11.24) is a target of the Hox genes of the bithorax complex, which are expressed in the abdomen. The Hox proteins repress *Distal-less* expression and thus suppress leg formation in the abdomen.

If one removes the bithorax complex completely, every abdominal segment in the embryo bears a leg disc and a wing disc (this mutation is lethal so larvae do not develop). This suggests that the potential for appendage development is still present in every segment in insects, even in the recently evolved Diptera, and is under active repression in the fly abdomen. This evidence supports the idea that the ancestral arthropod from which insects evolved had appendages on all its segments (see Fig. 14.12).

During the embryonic development of lepidopterans, by contrast, the genes *Ubx* and *abd-A* (a member of the bithorax complex) are turned off in the ventral parts of the embryonic abdominal segments; this results in *Distal-less* being expressed and pro-legs developing on the abdomen in the caterpillar. The presence or absence of legs on the abdomen is thus determined by whether or not particular Hox genes are expressed there, which is yet more evidence that changes in the pattern of Hox gene expression have played a key role in evolution.

Nevertheless, differences in the downstream targets of the Hox genes are likely to be even more important in the evolution of novel body plans. Members of the bithorax complex repress appendage development in the insect abdomen, but in crustaceans these genes are expressed in the thorax (see Fig. 14.13) and in the abdomen, yet appendages (legs on the thorax and swimmerets on the abdomen) still develop. The difference must lie in the genes targeted by the bithorax group genes in crustaceans and insects.

Although the sequence encoding the DNA-binding homeodomain is highly conserved between Hox genes in different species, the coding sequences outside the homedomain have undergone considerable change during evolution. This will, in turn, affect which regulatory regions and gene-regulatory proteins the Hox proteins can interact with in different species, and thus which genes they can target.

Evolutionary changes in the targets of Hox genes also help to explain why flies have two wings whereas butterflies have four. In *Drosophila*, expression of *Ubx* in the third

Fig. 14.15 The vertebrate and *Drosophila* dorso-ventral axes are related but inverted. Left panels: in arthropods, the nerve cord is ventral, whereas in vertebrates it is dorsal–dorsal and ventral being defined by the position of the mouth. Right panels: in arthropods, illustrated here by *Drosophila* (transverse section of blastoderm), and vertebrates, *Xenopus (surface view of blastula),* the signals specifying the dorso-ventral axis are similar, but are expressed in inverted positions. The protein Chordin, a dorsal specifier in vertebrates, is related to Sog, which is a ventral specifier in *Drosophila,* and the vertebrate ventral specifier BMP-4 is related to *Drosophila* Decapentaplegic (Dpp), which specifies dorsal.

Illustration after Ferguson, E.L.: Conservation of dorsal-ventral patterning in arthropods and chordates. Curr. Opin. Genet. Dev. *1996, 6: 424–431.*

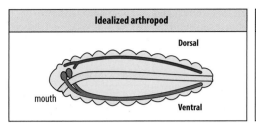

Idealized arthropod

Dorsal

mouth

Ventral

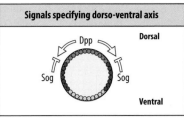

Signals specifying dorso-ventral axis

Dpp

Dorsal

Sog   Sog

Ventral

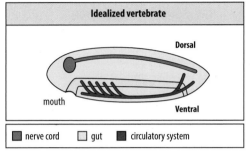

Idealized vertebrate

Dorsal

mouth

Ventral

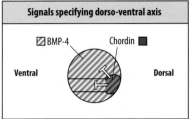

Signals specifying dorso-ventral axis

☑ BMP-4   Chordin ■

Ventral   Dorsal

■ nerve cord   □ gut   ■ circulatory system

thoracic segment (T3) specifies the appendage on that segment as a haltere, rather than the wing that develops on the neighbouring T2, in which *Ubx* is not expressed at the time that the wing is specified (see Sections 2.29 and 11.26). Genes known to be involved in patterning the wing are repressed by *Ubx* and are therefore not expressed in segment T3, explaining the differences between the appendages on neighboring segments. But in butterflies, because of changes in the regulatory regions of these genes during evolution, only some of them are repressed by Ubx, and this results in the development of a hindwing on segment T3 rather than a haltere (see Fig. 11.40).

Genes other than Hox genes are involved in the basic specification and patterning of the arthropod head. The head regions of different types of arthropods are very different, and indeed spiders do not seem to have a head at all (see Fig. 14.11). Their body is divided into a front and a back region (the prosoma and the opisthoma), the front end having segments that bear the spider's fangs (the chelicerae after which the class is named), pedipalps, and four pairs of walking legs; the front region thus acts as a head and thorax combined. Despite these differences, typical 'head genes' are expressed in the anterior-most region in both spiders and insects, including the homeobox gene *orthodenticle*, which is involved in specifying the larval head and anterior brain in *Drosophila* (see Section 2.33), and the Hox gene *labial*, which in *Drosophila* is expressed in the imaginal discs specifying the labial palps in the adult fly. This comparison shows how much morphology can change as a result of changes in the downstream targets of genes. It is interesting to note that in vertebrates, the *Otx* genes of the orthodenticle gene family are required for specifying anterior structures, including the brain (see Sections 4.14 and 12.2).

14.6 The basic body plan of arthropods and vertebrates is similar, but the dorso-ventral axis is inverted

A comparison between the body plans of arthropods and chordates reveals an intriguing difference. In spite of many similarities in their basic body plan—both have an anterior head, a nerve cord running anterior to posterior, a gut, and appendages—the dorso-ventral axis of vertebrates is inverted when compared to that of arthropods. The most obvious manifestation of this is that the main nerve cord runs ventrally in arthropods and dorsally in vertebrates (Fig. 14.15, left panels).

One explanation for this, first proposed in the nineteenth century, is that during the evolution of the vertebrates from their common ancestor with the arthropods, the dorso-ventral axis was turned upside-down, so that the ventral nerve cord of the ancestor became dorsal. This startling idea has recently found some support from molecular evidence showing that the same genes are expressed along the dorso-ventral axis in both insects and vertebrates, but in inverse directions. This inversion may have been dictated by the position of the mouth. The mouth defines the ventral side, and a change in the position of the mouth away from the side of the nerve cord would have resulted in the reversal of the dorso-ventral axis in relation to the mouth. The position of the mouth is specified during gastrulation in protostomes but not in deuterostomes, where as we have said above, it develops as a secondary independent opening, and thus it is not difficult to imagine how its position could have moved, but other changes in body structure and the nervous system were also involved.

We have seen in Chapters 2 and 4 that the patterning of the dorso-ventral axis in insects and vertebrates involves cell–cell signaling. In *Xenopus*, the protein Chordin is one of the signals that specifies the dorsal region, whereas the growth factor BMP-4 specifies a ventral fate. In *Drosophila*, the pattern of gene expression is reversed: the protein Decapentaplegic, which is closely related to BMP-4, is the dorsal signal, and the protein short gastrulation (Sog), which is related to Chordin, is the ventral signal (Fig. 14.15, right panels, and see Chapter 2). These signaling molecules are experimentally interchangeable between insects and frogs. Chordin can promote ventral development in *Drosophila*, and Decapentaplegic protein promotes ventral development in *Xenopus*. The molecules and mechanisms that set up the dorso-ventral axes in the two groups of animals are thus homologous, strongly suggesting that the divergence in the body plans of the arthropods and vertebrates involved an inversion of this axis, by movement of the mouth during the evolution of the vertebrates. Thus our image of the common ancestor of chordates and arthropods is an animal in which the establishment of the two major body axes, antero-posterior and dorso-ventral, was similar to that in present-day animals.

14.7 Limbs evolved from fins

The limbs of tetrapod vertebrates are special characters that develop after the phylotypic stage. Amphibians, reptiles, birds, and mammals have limbs, whereas fish have fins. The limbs of the first land vertebrates evolved from the pelvic and pectoral fins of their fish-like ancestors. The basic limb pattern is highly conserved in both the forelimbs and hindlimbs of all tetrapods, although there are differences both between forelimbs and hindlimbs, and between different vertebrates. Limbs evolved from fins, but how fins evolved is far less clear, and is a complex problem. Several scenarios have been proposed, including the evolution of fins by the splitting of a lateral fin fold. The evolution of fins made use of signaling molecules, such as Sonic hedgehog (Shh) and fibroblast growth factor (FGF), and of transcription factors, such as the Hox proteins, which were already being used to pattern the body.

Other key transcription factors that have been co-opted for use in the development of paired appendages are those encoded by the *Tbx* genes *Tbx4* and *Tbx5*, whose ancestral role is likely to have been in heart development (see Section 11.32). *Tbx5* is expressed in the anterior pair of appendages and *Tbx4* in the posterior pair of appendages in both fish and tetrapods (see Section 11.2). The amphioxus, which lacks paired appendages, has only a single *Tbx4/5* gene but it can function in the same way as the vertebrate genes in limb formation. A dramatic demonstration of this is that expression of the amphioxus *Tbx4/5* gene in transgenic mice in which *Tbx5* was knocked out rescues forelimb development.

The fossil record suggests that the transition from fins to limbs occurred in the Devonian period, between 400 and 360 million years ago. The transition probably occurred when the fish ancestors of the tetrapod vertebrates living in shallow waters moved onto the land. The fins of Devonian lobe-finned fishes, such as *Panderichthys*, are probably ancestral to tetrapod limbs, an early example of which is the limb of the Devonian tetrapod *Tulerpeton* (Fig. 14.16). The proximal skeletal elements

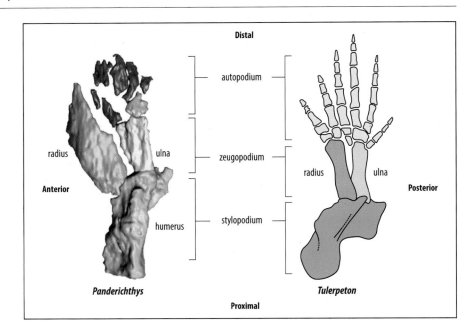

Fig. 14.16 The fin-to-limb transition.
In the lobe-like fin of the Devonian fish *Panderichthys*, there were proximal elements corresponding to the humerus (the stylopodium), radius and ulna (the zeugopodium), and distal elements (the autopodium). The Devonian tetrapod *Tulerpeton* has similar proximal elements, and the distal elements are recognizable digits.

*From Boisvert, C.A., et al.: **The pectoral fin of Panderichthys and the origin of digits**. Nature 2008, **456**: 636-638.*

corresponding to the humerus, radius, and ulna of the tetrapod limb are present in the ancestral fish, and a recent analysis of a fossil *Panderichthys* by computed tomography scanning has shown that the distal region of the pectoral fin contains separate skeletal elements. So fingers may have evolved from skeletal elements present in fins.

To gain insights into the genetic basis of the transition from fin to limb, researchers have turned to the zebrafish, the model organism in which fin development can be followed in detail and the genes involved can be identified. The fin buds of the zebrafish embryo are initially similar to tetrapod limb buds, but important differences soon arise during development. The proximal part of the fin bud gives rise to skeletal elements, which may be homologous to the proximal skeletal elements of the tetrapod limb. There are four main proximal skeletal elements in a zebrafish fin, which arise from the subdivision of a cartilaginous sheet, with six to eight smaller nodular skeletal elements developing more distally (Fig. 14.17). These skeletal elements support a flattened distal region containing a large number of fine bony fin rays. The essential difference between fin and limb development is that in the zebrafish fin bud, an ectodermal fin fold develops at the distal end of the bud and it is within this fold that the fin rays form. So the question is whether any of the supporting skeletal elements that develop from the zebrafish fin bud can be considered to be equivalent to tetrapod digits.

As in the tetrapod limb bud (see Chapter 11), the key gene *Shh* is expressed at the posterior margin of the zebrafish fin buds. The fin buds also express Hoxd and Hoxa

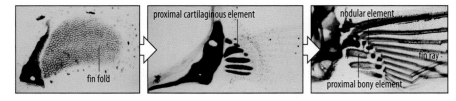

Fig. 14.17 The development of the pectoral fin of the zebrafish *Danio*. Left panel: the pectoral girdle and fin fold. Middle panel: four proximal cartilaginous elements and several more distal nodular cartilaginous elements have developed and fin rays have formed in the fin fold at the tip (the rays are very faint in this image). Right panel: four proximal bony elements and 6–8 smaller distal nodular elements support the fin ray region in the adult fish.

*Photographs courtesy of D. Dubole, from Sordino P., et al.: **Hox gene expression in teleost fins and the origin of vertebrate digits**. Nature 1995, **375**: 678-681.*

Fig. 14.18 Acquisition of a new control region for Hox gene expression could have driven the fin to limb transition. Regulatory regions on both sides of Hoxa and Hoxd clusters in fish and tetrapods control Hox gene expression in the developing fin and limb, respectively. Experiments in mice show that in tetrapod limbs, illustrated by a human arm, enhancers on the 3' (telomeric) side (orange) drive the proximal expression and patterning of this part of the limb (the humerus and the radius and the ulna), whereas enhancers on the 5' (centromeric) side (purple) drive Hox gene expression involved in patterning the distal part of the limb (the hand). In teleost fins, 3' and 5' enhancers (orange and yellow, respectively) interact in the same way with genes in the fish Hox clusters as the enhancers do in tetrapods, and drive expression in different proximo-distal regions of the fin bud. When tested in transgenic mice, however, both sets of fish enhancers drive expression only in the proximal limb. In the coelacanth, the 5' regulatory region contains a new enhancer that is also present in tetrapods but not in teleosts (thin yellow or purple band). It is speculated that acquisition of this enhancer and subsequent further changes in this regulatory region led to the transition from fin to limb and the evolution of fingers.

*Reproduced from Woltering, J.M., et al.: **Conservation and divergence of regulatory strategies at Hox loci and the origin of tetrapod digits**. PLoS Biol. 2014, **12**: e1001773.*

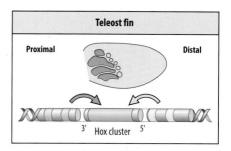

Teleost fin

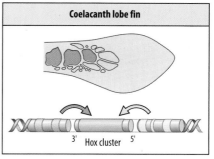

Coelacanth lobe fin

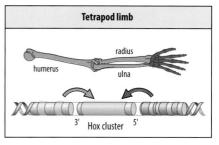

Tetrapod limb

genes. Initial studies suggested that the late phase of Hoxd gene expression which characterizes the digit-forming region in distal tetrapod limb buds (see Section 11.13) does not occur in zebrafish fin buds. This led to the proposal that digits are a tetrapod novelty rather than evolving from structures already present in an ancestral fish fin. Subsequent more comprehensive studies of Hox gene expression, however, pointed to greater similarities between the expression patterns of the Hoxa and Hoxd genes in zebrafish pectoral fins and tetrapod limb buds. More recent comparative analysis of the regulatory organization of Hoxa and Hoxd complexes showed that in zebrafish, as in mice, genes towards the 3' end of the gene complex interact with enhancers on the adjacent telomeric side (3') of the complex and genes towards the 5' end of the complex interact with enhancers on the adjacent centromeric side (5'); genes in the centers of the complexes can interact with enhancers on both sides.

In mouse limbs, enhancers in the centromeric side of the Hox clusters drive expression of the Hox genes in the digit-forming region (see Fig. 11.20). But when the functioning of the fish regulatory regions—both those on the 3' and the 5' sides of the Hox gene clusters—was tested in transgenic mice, only proximal Hox gene expression was obtained. This suggests that the digits of the tetrapod limb evolved by the acquisition of new control sites in the region 5' to the Hoxa and Hoxd clusters, thereby establishing a late phase of Hox gene expression in the digit-forming region at the tip of the limb bud. Interestingly, the genome of the coelacanth, an ancient lobe-finned fish and the closest living ancestor to tetrapods, contains one regulatory sequence 5' to the Hoxd cluster that is not found in ray-finned fishes but is present in tetrapods. This regulatory sequence can drive Hox gene expression in limb buds in transgenic mice, and it has been speculated that the acquisition of this new sequence may have contributed to the evolution of the 5' regulatory region in tetrapods and the formation of digits (Fig. 14.18).

14.8 Limbs have evolved to fulfill different specialized functions

The great range of anatomical specializations in the limbs of mammals (Fig. 14.19) is due to changes both in limb patterning and in the differential growth of parts of the limbs during embryonic development, but the basic underlying pattern of skeletal elements is the same. This is an excellent example of the modularity of the skeletal elements. If one compares the forelimb of a bat and a horse, one can see that, although both retain the basic pattern of limb bones, the limb has been modified to provide a specialized function in each animal. In the bat, the limb is adapted for flying: the digits are greatly lengthened to support a membranous wing. In the horse, the limb is adapted for running: in the forelimb, lateral digits are reduced, the central metacarpal (a hand bone in humans) is lengthened, and the radius and ulna are fused for greater strength. The

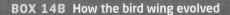

BOX 14B How the bird wing evolved

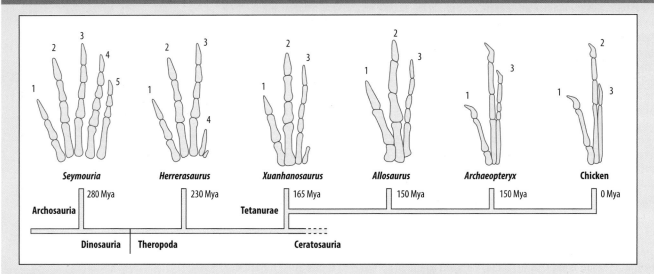

Figure 1

From Wieschampel, D.B. Dodson, P., Osmolska, H. (eds) The Dinosauria 2nd edition, University of California Press, 2004.

The wings of birds have three digits—the reduction in digit number being an adaptation that enabled the evolution of flight—and there is a long-running controversy about whether these are digits 1 2 3 or digits 2 3 4 of the ancestral pentadactyl limb plan. This issue is important for tracing the evolution of the wing from the birds' dinosaur ancestors. A group of dinosaurs known as theropods bridge the transition between a five-digit hand found in the earlier dinosaurs and the three-digit bird wing (Figure 1).

Paleontologists identify the three digits in theropods such as *Allosaurus* as 1 2 3, with digits 5 and then 4 being first reduced, and subsequently lost, during theropod evolution. According to fossil evidence, the three chick wing digits are 1 2 3.

The pattern of digit loss seen in theropods is unusual. In other cases of digit loss, it is the lateral digits that are lost (see Section 14.12 for the horse and other hoofed mammals). Indeed, classical morphologists comparing the order and the positions in which the

whale flipper is modified for balance and steering, with the elongated flippers having digits with many phalanges.

A feature of vertebrate limb evolution is that, while reduction in digit number is common (there are only three in the chick wing and reduction is common in lizards), there are no modern species with more than five digits. The giant panda's 'thumb',

Fig. 14.19 Diversification of vertebrate limbs. The basic pattern of bones in the forelimb is conserved throughout the vertebrates, but there are changes in the proportions of the different bones, as well as fusion and loss of bones. In the horse limb, the radius and ulna have become fused into a single bone, and the central metacarpal (a hand bone in humans) is greatly elongated. In addition, there has been loss and reduction of the digits in the horse. In the bat wing, by contrast, the digits have become greatly elongated to support the membranous wing. In the whale flipper, there has been digit reduction, and elongation of two of the digits is accompanied by an increase in the number of phalanges.

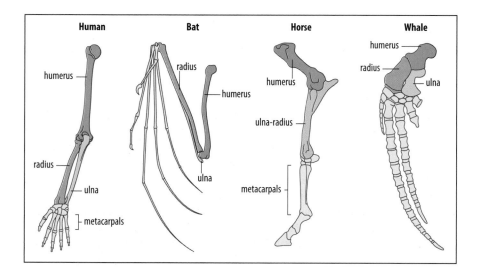

digits develop in the embryonic chick wing with the embryonic forelimbs of other vertebrates—alligator and turtle—inferred that the digits are 2 3 4, and digit 1 has been lost; a cartilage condensation representing digit 5 appears transiently (Figure 2). An anterior expression domain of *Sox9* (the transcription factor required for cartilage differentiation) recently detected in the chick wing bud could represent the remnants of digit 1. So how can the different views be reconciled? One suggestion is that the three bird wing digits are 1 2 3, like the three digits of dinosaur hands, but that during evolution they shifted to the positions in which digits 2 3 4 normally form in tetrapod limbs—the frame-shift theory. Comparative analysis of Hox gene-expression patterns in the chick wing, chick leg, and mouse limbs supports the 1 2 3 theory and microarray analysis shows that the transcriptomes of the first chick wing digit and digit I in the chick leg are highly conserved. Fate maps made by grafting GFP-expressing tissue from transgenic chickens into chick wing and leg buds or by genetically labeling Shh-producing cells in mouse limb buds show that the origin of the chick wing digits is the same as the origin of digits 1 2 3 in the chick leg and the mouse limb.

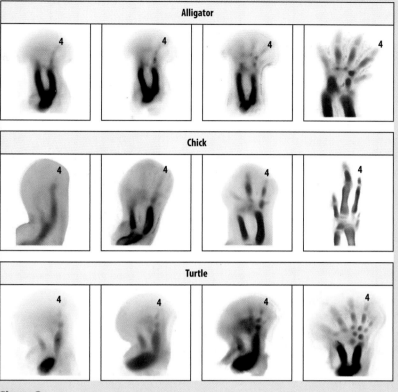

Figure 2

*Photos from Burke, A.C., Feduccia, A.: **Developmental patterns and the identification of homologies in the avian hand**. Science 1977, 278: 666-668.*

The debate seems likely to continue: domains of *Sox-9* expression might not necessarily represent digits, and fate maps only give a retrospective view of evolution. This underscores the problems of integrating evidence from the fossil record with evidence from the embryology of present-day animals. By contrast, the study of recent evolutionary changes, such as the loss of pelvic fins in different forms of sticklebacks co-existing today, can uncover the basis of the evolutionary change (see Section 14.9).

which appears to be a sixth digit, is in fact a modified wrist bone. Species with more than five digits appear in the fossil record of the earliest tetrapods but it is not clear whether there were more than five morphologically distinct types of digits. In addition, in modern limbs with polydactyly due to mutations, at least two of the digits are the same (see Box 11C). It seems that there is a **developmental constraint** on evolving more than five different kinds of digits. It was originally suggested that the five-digit pattern might be due to there being five Hoxd genes expressed in the distal part of the early limb bud, providing only five discrete genetic programs for giving a digit an identity. While this is an attractive idea, it is likely to be an over-simplification.

Reduction in the number of digits has occurred repeatedly in tetrapod evolution. In limbs of different species of the Australian lizard *Hemigeris*, digit loss is correlated with reduction in the extent and duration of Shh signaling by the polarizing region. In a recent study of limb development in bovine embryos, changes in Shh signaling in early limb buds were also detected and a change in a *cis*-regulatory region of the gene involved identified. In contrast, in camels and horses, digit loss appears to be due to extensive cell death at later stages in limb development. Reduction in the number of phalanges in the digits is also common. Whales and dolphins are exceptional in having elongated digits with additional phalanges (see Fig. 14.19) and it has been suggested that this is due to persistence of the apical ectodermal ridge which prolongs outgrowth of the bud.

EXPERIMENTAL BOX 14C Pelvic reduction in sticklebacks is based on mutations in a gene control region

Marine three-spine sticklebacks have a substantial pair of serrated pelvic spines on the ventral side—their scientific name, *Gasterosteus aculeatus*, means 'bony stomach with spines'. But populations of three-spine stickleback which spread into freshwater environments in different parts of the world, including Alaska, Canada, and Scotland, after the last Ice Age, have evolved into a form in which the pelvic structures—the pelvic

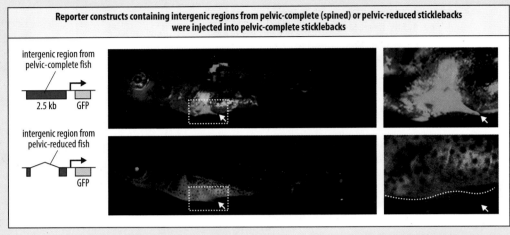

Figure 1 *From Chan, Y.F., et al.: **Adaptive evolution of pelvic reduction in sticklebacks by recurrent deletion of a** Pitx1 **enhancer**. Science 2010, **327**: 302-305.*

girdle and associated spines—are reduced or completely absent. Genetic crosses between fish from these different natural populations identified a region in the genome containing the gene *Pitx1* that is linked with the variance in pelvic spine size, but no mutations in the *Pitx1* protein-coding region were found. However, no *Pitx1* expression could be detected in the developing pelvic region of pelvic-reduced fish, and this suggested that mutations in a control region of the *Pitx1* gene that activates its expression in the pelvic region could explain the missing spines.

To locate this control region, reporter transgenes composed of the coding sequence for green fluorescent protein (GFP) linked to a 2.5-kilobase DNA sequence representing the intergenic region upstream of the *Pitx1* coding sequence were made (this region would be expected to contain the *Pitx1* regulatory modules). The reporter DNA was injected into fertilized eggs of marine sticklebacks, which were then allowed to develop. When intergenic DNA

from marine fish was used in the reporter transgene, GFP was strongly expressed in the developing pelvic region, whereas with the intergenic DNA from spine-reduced fish, which has a large deletion, no GFP was expressed in the region (outlined here with a white dotted line) (Figure 1). The square outlined with a white dotted line in the left panels indicates the region shown at higher magnification in the panels on the right. These experiments suggested that the DNA upstream of this *Pitx1* gene contains an enhancer that normally activates expression of the gene in the pelvic region, but that this enhancer was not working, or was deleted, in pelvic-reduced fish. To test this hypothesis, the *Pitx1* control region from marine sticklebacks was linked to the *Pitx1* coding region and this construct was injected into fish with pelvic reduction. This resulted in the dramatic rescue of pelvic development and provided direct evidence that pelvic reduction is due to mutation in the control region of the *Pitx1* gene (Figure 2).

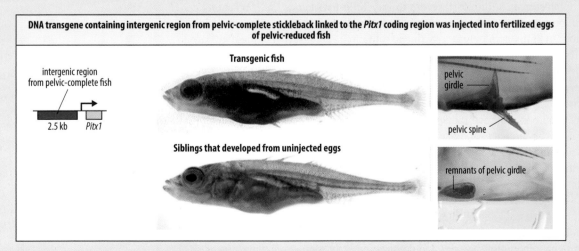

Figure 2 *From Chan, Y.F., et al.: **Adaptive evolution of pelvic reduction in sticklebacks by recurrent deletion of a** Pitx1 **enhancer**. Science 2010, **327**: 302-305.*

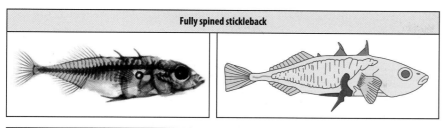

Fig. 14.20 Loss of the pelvic spines in three-spine sticklebacks (*Gasterosteus aculeatus*). Upper row: marine populations and many freshwater populations of three-spine sticklebacks have a pair of pelvic spines (red in the drawing). Fish in marine populations also have bony plates along their sides (yellow in the drawing) and three dorsal spines (blue in the drawing). Lower row: in some freshwater populations, the pelvic spines have been lost and the pelvic girdle very reduced as a result of a mutation in the *Pitx1* gene.

*From Tickle, C., Cole, N.J.: **Morphological diversity: Taking the spine out of three-spine stickleback**. Curr. Biol. 2004, **14**: R422–R424.*

14.9 Adaptive evolution within the same species provides a way of studying the developmental basis for evolutionary change

Because of the existence of very different morphological forms within the same species due to adaptive evolution, the three-spine stickleback fish, *Gasterosteus aculeatus*, is a good model in which to study the developmental mechanisms involved in evolutionary changes. Fish in populations that became isolated when they colonized new lakes and streams formed as the glaciers melted after the ice ages about 20,000 years ago have undergone reduction or loss of the pelvic spine (a modified pelvic fin) together with other morphological changes. Paired pelvic spines are present in all marine populations of three-spine sticklebacks, but the pelvic spine has been lost repeatedly in several freshwater populations, as a result of failure of pelvic fin bud development (Fig. 14.20). From genetic crosses between sticklebacks from marine and freshwater populations, the cause of pelvic reduction was identified as mutations in the control region of the gene for the homeodomain transcription factor Pitx1 (Box 14C).

Pelvic-spine loss in sticklebacks is an interesting example of **convergent evolution**, because different populations have different mutations in the *Pitx1* control region, which have been selected independently in each population. Pitx1 is known to have a key role in vertebrate hindlimb development (see Section 11.2). When the *Pitx1* gene is knocked-out in mice, hindlimb development is reduced, with the right hindlimb being more affected than the left. This asymmetry in limb reduction is likely to be due to the expression of a closely related gene, *Pitx2*, on the left-hand side of the body (see Section 5.16), partially compensating for lack of Pitx1. Interestingly, pelvic spine reduction in sticklebacks is also asymmetrical, being more marked on the right than on the left (see online practical on stickleback evolution).

But why have pelvic spines been lost in freshwater sticklebacks? Two environmental factors that may allow spine-reduced sticklebacks to be more successful in freshwater environments are the limited availability of calcium required for bone formation and the reduced need for protection against predation.

We have already seen another intriguing example of developmental evolution within another species of fish, the teleost fish *Astyanax mexicanus* (see Section 11.27). *Astyanax mexicanus* has two radically different forms: a sighted, pigmented, surface-dwelling form and a blind, unpigmented, cave-dwelling form known as a cavefish (see Fig. 11.49). At least 30 different populations of *Astyanax* cavefish are present in limestone caverns in Mexico, having been isolated from their surface relations for the past few million years. Although functional eyes are lacking in cavefish adults, the embryos begin to develop eye primordia, which subsequently degenerate. The major cause of eye degeneration is apoptosis of the developing lens cells,

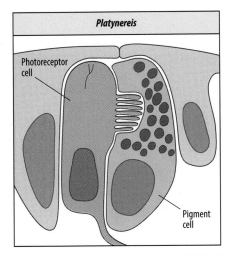

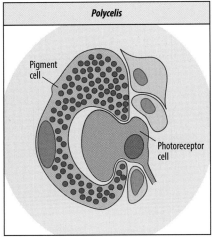

Fig. 14.21 Simple eyes like those of some invertebrates are thought to resemble the prototypical bilaterian eye. Illustrated here are the larval eye of the polychaete worm *Platynereis* (upper panel), and one of the numerous simple eyes of the planarian *Polycelis auricularia* (lower panel). They both consist of a photoreceptor cell shielded on one side by a pigment cell.

*From Gehring, W.J.: **Chance and necessity in eye evolution**. Genome Biol. Evol. 2011, 3: 1053-1056, and Arendt, D.: **Evolution of eyes and photoreceptor cell types**. Int. J. Dev. Biol. 2003, 47: 563-571.*

which then prevents the growth of other optic tissues, including the retina. The non-development of the eye is, in turn, the cause of other craniofacial differences between cavefish and surface fish. Lens apoptosis is induced by increased activity of the Shh signaling system along the cavefish embryonic midline compared to the embryos of surface-dwelling *Astyanax*. However, the genetic basis for this increased activity is unknown. Recent work suggests that, in cavefish, natural selection has acted on pre-existing variation in the population. In surface-dwelling forms of cavefish, variations in eye size appear to be buffered by the heat-shock protein HSP60. The proposal is that the environment in caves disturbs this system, thereby unmasking variations in eye size on which natural selection can act.

Eyes are initially formed and then degraded during larval or adult development in all sightless cave-dwelling vertebrates. It is surprising that eyes begin to develop at all. This may be because early steps in eye development are required for other essential steps in development, and the elimination of these steps would be fatal. Because of this strong developmental constraint, it is thus unlikely that cave vertebrates lacking embryonic eyes will be discovered. But why have eyes been lost in cavefish? The answer is not known. A recent suggestion is that loss of eyes is an indirect consequence of selection for genes required for the development of the sensory receptors involved in adaptive behavior that makes foraging in the dark easier.

14.10 Evolution of different types of eyes in different animal groups is an example of parallel evolution using an ancient genetic circuitry

Charles Darwin considered the evolution of the vertebrate eye as one of the most difficult problems in evolution. Modern molecular biology has thrown some light on the problem and in the process has uncovered an example of conservation of gene function that is likely even to pre-date the evolution of animals.

As we have seen, the remarkable conservation of *Pax6/eyeless* function is demonstrated by the fact that *Pax6* from the mouse can substitute for *eyeless* and induce ectopic eye structures in *Drosophila* if expressed in the imaginal disc for the antenna (see Fig. 11.48). Whatever the source of the *Pax6* gene, the eye structures that form are always *Drosophila* ommatidia, not eyes like those of the animal from which the *Pax6* gene came. *Pax6* is acting as a master gene that turns on *Drosophila*'s own eye-development program.

The discovery of this underlying conservation of function now suggests to many biologists that the very different eyes found in different groups of animals are examples of parallel evolution that has adapted the same genetic circuitry and certain cell specializations that were present in the ancestors of all bilaterian animals, or even earlier. Animals are thought to have evolved from unicellular organisms, probably most resembling flagellate protozoa (see Section 14.2), which are proposed to have had a light-sensitive organelle similar to the 'eyespot' in some modern flagellates. The eyespot transmits signals to the flagella and enables the cell to move towards light. The evolution of multicellularity and of differentiated cell types eventually led to simple types of photoreceptor cells, such as those in present-day jellyfish larva, which have individual photosensitive cells in the ectoderm. Jellyfish photoreceptor cells not only use opsins as the photoreceptor proteins, as do all animals and some unicellular organisms, but also contain the pigment melanin, which is the pigment in the retinal pigmented epithelium of the vertebrate eye. The function of the pigment is to absorb any light that passes through the photoreceptor layer and so prevent it reflecting back onto the photoreceptors and producing an imprecise signal.

The next step in the evolution of the eye would have been the differentiation of the single light-sensing cell into two cell types: a sensory photoreceptor cell and a pigment cell (Fig. 14.21). The idea that the origin of the eye involved two such cells goes back to Darwin. Simple two-celled 'pigment cup' light-sensitive organs are present in the larvae of many bilaterian animals, (animals with bilateral symmetry) while

Fig. 14.22 Modification of the branchial arches during the evolution of jaws in vertebrates. The ancestral jawless fish had a series of at least seven gill slits-branchial clefts-supported by cartilaginous or bony arches —branchial arches. Jaws developed from a modification of the first arch to give the mandibular arch, with the mandibular cartilage of the lower jaw and the hyoid arch behind it. The spiracle is a respiratory opening behind the eye in present day cartilaginous fish.

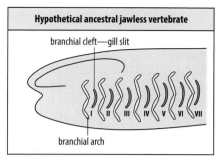

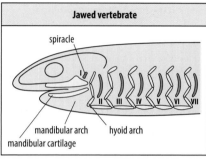

the more complex frontal eye of the larval amphioxus (a cephalochordate) consists of a pigment cup and two rows of photoreceptor cells connected by neurons to the central nervous system, foreshadowing the eyes of vertebrates. When and how *Pax6* became such a crucial gene in eye development is unknown.

The crystallin proteins that give the eye lenses of cephalopods (octopuses and squids) and vertebrates their transparency also have an intriguing evolutionary history. The crystallins were originally thought to be unique to the lens and to have evolved for this special function, but more recent research indicates that they are co-opted proteins that are not structurally specialized for lens function, and in other contexts act as enzymes.

14.11 Embryonic structures have acquired new functions during evolution

If two groups of animals that differ greatly in their adult structure and habits (such as fishes and mammals) pass through a very similar embryonic stage, this could indicate that they are descended from a common ancestor and, in evolutionary terms, are closely related. Thus, an embryo's development reflects the evolutionary history of its ancestors. Division of the body into segments, which then diverge from each other in structure and function, is a common feature in the evolution of both vertebrates and arthropods. In the vertebrates, one example of segmented structure is the branchial arches and clefts that are present in all vertebrate embryos, including humans, located just behind the head on either side.

These structures are not the relics of the gill arches and gill slits of an adult fish-like ancestor, but represent structures that would have been present in the embryo of the fish-like ancestor of vertebrates as developmental precursors to gill slits and gill arches. During evolution, the branchial arches gave rise both to the gill arches of the primitive jawless fishes and, in a later modification, to gills and jaw elements in later-evolved fishes (Fig. 14.22). With time the arches became further modified, and in mammals they now give rise to various structures in the face and neck (Fig. 14.23), many of which are derived from the neural crest cells that migrate into the branchial

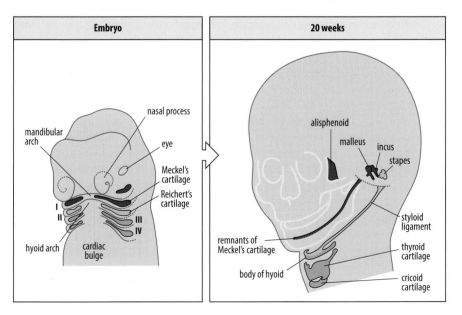

Fig. 14.23 Fate of branchial arch cartilage in humans. In the embryo, cartilage develops in the branchial arches, which gives rise to elements of the three auditory ossicles, the hyoid, and the pharyngeal skeleton. The fate of the various elements is shown by the color coding.

Illustration after Larsen, W.J.: Human Embryology. *New York: Churchill Livingstone, 1993.*

Fig. 14.24 Evolution of the bones of the mammalian middle ear. The articular and quadrate bones of ancestral reptiles (left panel) were part of the lower jaw articulation. Sound was transmitted to the inner ear via these bones and their connection to the stapes. When the lower jaw of mammals became a single bone (the dentary), the articular bone became the malleus, and the quadrate bone the incus of the middle ear, acquiring a new function in transmitting sound to the inner ear from the tympanic membrane (right panel). The eustachian tube forms between branchial arches I and II.

Illustration after Romer, A.S.: The Vertebrate Body. Philadelphia: W.B. Saunders, 1949.

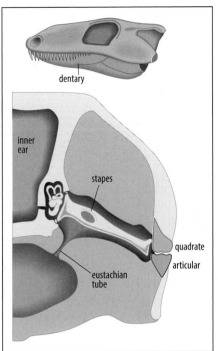

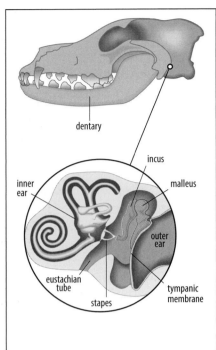

arches early in development (see Section 5.15). The cleft between the first and second branchial arches provides the opening for the Eustachian tube, and endodermal cells in the clefts give rise to various glands, such as the thyroid and thymus.

Evolution rarely, if ever, generates a completely novel structure out of the blue. New anatomical features usually arise from modification of an existing structure. One can therefore think of much of evolution as a 'tinkering' with existing structures, which gradually fashions something different. It is possible because many structures are modular: that is, animals have anatomically distinct parts that can evolve independently. Vertebrae are modules, for example, and can evolve independently of each other; so too are limbs; and so are digits.

A nice example of a modification of an existing structure is provided by the evolution of the mammalian middle ear. This is made up of three bones—malleus, incus, and stapes—that transmit sound from the eardrum (the tympanic membrane) to the inner ear. In the reptilian ancestors of mammals, the joint between the skull and the lower jaw was between the quadrate bone of the skull and the articular bone of the lower jaw, which were also involved in transmitting sound (Fig. 14.24). The vertebrate lower jaw was originally composed of several bones, but during mammalian evolution one of these bones, the dentary, increased in size and came to comprise the whole lower jaw, and the other bones were lost. The articular bone was no longer attached to the lower jaw, and by changes in their development, the articular and the quadrate in mammals were modified into two bones, the malleus and incus, respectively, whose function was now to help transmit sound from the tympanic membrane. The quadrate is evolutionarily and developmentally homologous with the dorsal cartilage of the first branchial arch in the ancestral vertebrate, and the stapes with the dorsal cartilage of the second.

These examples provide evidence of a key relationship between evolution and development, the gradual change of a structure into a different form. In many cases, however, we do not understand how intermediate forms were adaptive and gave a selective advantage to the animal. Consider, for example, the intermediate forms in the transition of the first branchial arch to jaws; what was the adaptive advantage? We do not know, and because of the passage of time and our current ignorance of the ecology of ancient organisms, we may never know.

SUMMARY

Groups of animals that pass through a similar embryonic stage are descended from a common ancestor and have evolved modifications of gene-regulatory circuits. The basic body plan of all animals is defined by patterns of Hox gene expression that encode positional information, the interpretation of which has changed in evolution. The Hox genes themselves have undergone considerable evolution by gene duplication and divergence and are a good example of Darwin's description of evolution as 'descent with modification.' The limbs of tetrapod vertebrates evolved from fins, with the digits evolving in tetrapods due to the acquisition of novel regulatory circuits controlling the expression of Hox genes. Comparison of patterns of dorso-ventral gene expression in specification of the body plan suggests that during the evolution of the vertebrates, the dorso-ventral axis of an invertebrate ancestor was inverted. Sticklebacks are a good model for studying evolutionary mechanisms, and mutations in a regulatory region of a key developmental gene underlie changes in the body plan. During evolution, the development of structures can be altered so that they acquire new functions, as has happened in the evolution of the mammalian middle ear from a reptilian jaw bone, which itself evolved from a skeletal element in an ancestral branchial arch.

SUMMARY: evolution of structures by the modification of development

articular and quadrate bones in reptile jaw fins of ancestral fish

malleus and incus bones of mammalian middle ear

limbs of tetrapods, with digits as novel structures

Hox gene duplication, diversification, and interpretation

different body plans

signals for *Xenopus* and *Drosophila* dorso-ventral axes are homologous but axes are inverted: BMP-4–Decapentaplegic; Chordin–Short gastrulation

Changes in the timing of developmental processes

In the previous part of the chapter we focused on changes in spatial patterning that have occurred during evolution. But changes in the timing of developmental processes can also have major effects. In this part of the chapter, we look at some examples of how changes in the timing of growth and sexual maturation can affect animal form and behavior.

14.12 Changes in growth can modify the basic body plan

Differential growth rates during embryonic development are a major way in which parts of the body can be modified during evolution for specialized functions. Many of the changes that occur during evolution reflect changes in the relative dimensions of parts of the body. We have seen how growth can alter the proportions of the human baby after birth, as the head grows much less than the rest of the body (see Fig. 1.17). The various face shapes in the different breeds of dogs, which are all members of the same species, descended from the gray wolf, also provides a good example of the

Fig. 14.25 Evolution of the forelimb in horses. *Hyracotherium*, the first true equid, was about the size of a large dog. Its forefeet had four digits, of which one (the third digit in anatomical terms) was slightly longer, as a result of a faster growth rate. All digits were in contact with the ground. As equids increased in size, the lateral digits lost contact with the ground, as the result of the relatively greater increase in length of metacarpal 3 (the proximal bone of the third digit). At a later stage, the lateral digits became even shorter, because of a separate genetic change.

Illustration after Gregory, W.K.: Evolution Emerging. New York: Macmillan, 1957.

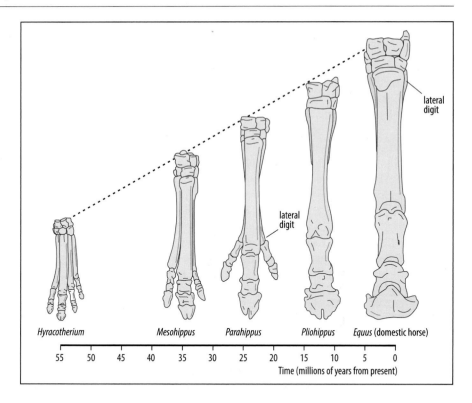

lateral digit

lateral digit

Hyracotherium   Mesohippus   Parahippus   Pliohippus   Equus (domestic horse)

55   50   45   40   35   30   25   20   15   10   5   0

Time (millions of years from present)

effects of differential growth after birth. All dogs are born with rounded faces; some, such as bulldogs and pugs, keep this shape, but in others such as greyhounds and pointers the nasal regions and jaws elongate during growth. The elongated face of the baboon is also the result of growth of this region after birth.

Because individual structures, such as bones, can grow at different rates, the overall shape of an organism can be changed substantially during evolution by heritable changes in the duration of growth, which also leads to an increase in overall size of the organism. In the horse, for example, the central digit of the ancestral horse grew faster than the digits on either side, so that it ended up longer than the lateral digits (Fig. 14.25). As horses continued to increase in overall size during evolution, this discrepancy in growth rates resulted in the relatively smaller lateral digits no longer touching the ground because of the much greater length of the central digit. At a later stage in evolution, the now-redundant lateral digits became reduced even further in size because of a separate genetic change that led to extensive cell death on both sides of the digital-plate.

We have already seen that the bones in the digits in the bat wing are greatly lengthened (see Fig. 14.19). In contrast, the relative lengths of the different bones of the bat hindlimb are similar to those in the mouse hind limb. In bats, there are differences in size of fore- and hindlimbs from a very early embryonic stage, although the large increase in the length of the digits in the bat wing occurs relatively late once the main pattern of the skeleton is laid down. In the flightless kiwi, the wing buds are relatively small compared with the leg buds even at early embryonic stages (Fig. 14.26).

The homeobox transcription factor Prx1 is expressed at high levels in the distal region of the developing bat wing and this may contribute to digit elongation. When the enhancer from the mouse *Prx1* gene is replaced by the bat *Prx1* enhancer in a transgenic mouse, there is a small but significant increase in the length of the mouse forelimb. Elongation of the bat wing digits may also promoted by a second wave of Shh signaling as *Shh* expression has been detected at later stages in developing bat wings after the primordia of the digits have been laid down. FGF-8 is also produced and could act to prolong outgrowth of the digits. This postulated mechanism is similar

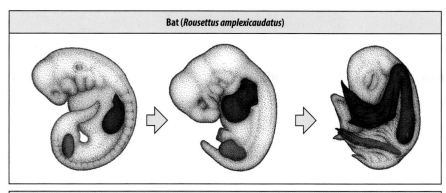

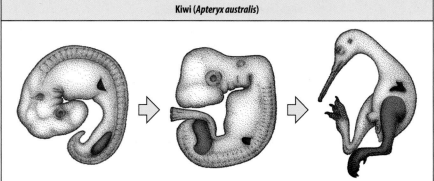

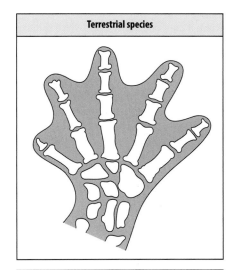

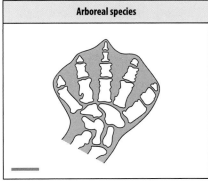

Fig. 14.26 Developmental expression of adult limb size. Top row: the bat *Rousettus amplexicaudatus* has large forelimbs and smaller hindlimbs when adult. The forelimb bud is large at all stages, relative to the hindlimb. Bottom row: the flightless kiwi (*Apteryx australis*) has relatively small forelimbs and large hind limbs at all stages of development, even in the early embryo.

*Adapted from Richardson MK.: **Vertebrate evolution: the developmental origins of adult variation**. Bioessays 1999, 21: 604-613.*

to that invoked to explain elongation of the digits in the whale flipper (see Section 14.8) although in the case of the flipper, prolonging outgrowth of the bud has been proposed to lead to the formation of additional phalanges rather than their elongation.

14.13 Evolution can be due to changes in the timing of developmental events

Differences among species in respect of the time at which developmental processes occur relative to one another, and relative to their timing in an ancestor, can have dramatic effects on both morphology and behavior. Differences in the feet of members of a genus of tropical salamanders illustrate the effect of a change in developmental timing on both the morphology and ecology of different species. Many species of the salamander genus *Bolitoglossa* are arboreal (tree living), rather than typically terrestrial, and their feet are modified for climbing on smooth surfaces. The feet of the arboreal species are smaller and more webbed than those of terrestrial species, and their digits are shorter (Fig. 14.27). These differences seem to be mainly the result of the development and growth of the foot ceasing at an earlier stage in the arboreal species than in the terrestrial species. The term used to describe such differences in timing is **heterochrony**.

Some of the clearest examples of heterochrony come from alterations in the timing of onset of sexual maturity in organisms with larval stages. The acquisition of sexual maturity by an animal while still in the larval stage is a process that goes under the name **neoteny**. The development of the animal, although not its growth, is retarded in relation to the maturation of the reproductive organs. This occurs in the axolotl, *Ambystoma mexicanum*, a type of salamander; the larva grows in size and matures sexually, but does not undergo metamorphosis. The sexually mature form remains aquatic and looks like an overgrown larva. However, the axolotl can be induced to undergo metamorphosis by treatment with the hormone thyroxine (Fig. 14.28).

Many animals have evolved free-swimming larval forms that have an advantage when it comes to dispersal and feeding, and then undergo a dramatic change in morphology to reach the adult state at metamorphosis. The essence of development is gradual change;

Fig. 14.27 Heterochrony in salamanders. In terrestrial species of the salamander *Bolitoglossa* (top panel), the foot is larger, has longer digits, and is less markedly webbed than in those that live in trees (bottom panel). This difference can be accounted for by foot growth ceasing at an earlier stage in the arboreal species. Scale bar = 1mm.

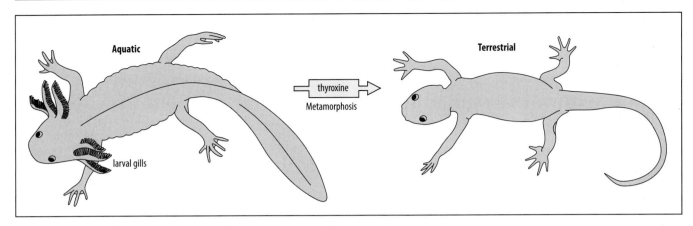

Fig. 14.28 Neoteny in salamanders.
The sexually mature axolotl (*Ambystoma mexicanum*) retains larval features, such as gills, and remains aquatic. This neotenic form metamorphoses into a typical terrestrial adult salamander if treated with thyroxine, which it does not produce, and which is the hormone that causes metamorphosis in other amphibians (see Section 13.13).

yet at metamorphosis there is no gradual continuity between larva and adult. Metamorphosis makes more evolutionary sense, however, if it is assumed that all larval forms evolved by the insertion of the larval stage into the pre-existing development program of a directly developing animal, possibly as a result of heterochrony. In many invertebrates, the larva initially resembles the late gastrula stage, which could have given rise to the free-swimming larval form. Larval stages could not have been the original state because the metamorphosis that brings a larva back to the developmental program into a sexually mature adult is a highly complex process, and it is difficult to see how it could have evolved other than to re-enter the developmental pathway.

In vertebrates that undergo metamorphosis, such as frogs, if we assume that frog ancestors developed directly into adults, a change in the timing of events in the post-neurula stages, such as a delay in hindlimb development, could have led to acquisition of mobility and evolution of the feeding tadpole stage. Becoming mobile would have aided dispersal, but to become a reproducing adult, the larva had to return to the normal developmental program. That, in essence is what metamorphosis is for.

Larval stages can be lost as well as acquired during evolution. Some modern frogs have re-evolved direct development to the adult by a loss of the tadpole stage and the acceleration of the development of adult features. Frogs of the genus *Eleutherodactylus*, unlike the more typical amphibians *Rana* and *Xenopus*, develop directly into an adult frog and there is no aquatic tadpole stage, the eggs being laid on land. Typical tadpole features, such as gills and cement glands, do not develop, and prominent limb buds appear shortly after the formation of the neural tube (Fig. 14.29). In the embryo, the tail is modified into a respiratory organ. Such direct development requires a large supply of yolk to the egg in order to support development through to an adult without a tadpole feeding stage. This increase in yolk may itself be an example of heterochrony, involving a longer or more rapid period of yolk synthesis in the development of the egg.

Most sea urchins have a larval stage that takes a month or more to become adult. But there are some species in which direct development has evolved, so that they no longer go through a functional larval stage. Such species have large eggs and, as a result of their very rapid development, become juvenile urchins within four days. This has involved changes in early development such that the directly developing embryo

Fig. 14.29 Development of the frog
Eleutherodactylus. Typical frogs, such as *Xenopus* or *Rana*, lay their eggs in water and develop through an aquatic tadpole stage, which undergoes metamorphosis into the adult. Frogs of the genus *Eleutherodactylus* lay their eggs on land and the frog hatches from the egg as a miniature adult, without going through an aquatic free-living larval stage. The embryonic tail is modified as a respiratory organ.

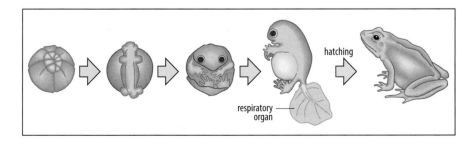

gives rise to a 'larval' stage that lacks a gut and cannot feed, and which metamorphoses rapidly into the adult form.

14.14 The evolution of life histories has implications for development

Animals and plants have very diverse life histories: small birds breed in the spring following their birth, and continue to do so each year until their death; Pacific salmon breed in a suicidal burst at 3 years of age; and oak trees require 30 years of growth before producing acorns, which they eventually produce by the thousand. In order to understand the evolution of such life histories, evolutionary ecologists consider them in terms of probabilities of survival, rates of reproduction, and optimization of reproductive effort. These factors have important implications for the evolution of developmental strategies, particularly in relation to the rate of development. For example, as we have just seen, a characteristic feature of many animal life histories is the presence of a free-living larval stage that is distinct from, and usually simpler in form than, the sexually mature adult, and which feeds in a different way.

Life histories help us to understand the evolution of long germ-band insects, such as *Drosophila*, which are of more recent origin than short germ-band insects, such as grasshoppers (see Section 2.23). Unlike *Drosophila*, short germ-band insects do not have a larval stage and develop directly into small, immature adult-like forms. In contrast, *Drosophila* develops very rapidly into a feeding larva, taking only 24 hours to start feeding, compared with the grasshopper's 5–6 days. It is easy to imagine conditions in which there would have been a selective advantage to insects whose larvae begin to feed as quickly as possible. It is also likely that embryos are more vulnerable than adults and so there would be selection for making the embryonic stage shorter. Thus it is likely that the complex developmental mechanisms of long-band insects—the system for setting up the complete antero-posterior axis in the egg—evolved as a result of selection pressure for rapid development. In the case of fruit flies like *Drosophila*, this was perhaps related to the ability to feed on fast-disappearing fruit.

Egg size can also be best understood within the context of life histories. If we assume that the parent has limited energy resources to put into reproduction, the question is how should these resources best be invested in making gametes, particularly eggs; is it more advantageous to make lots of small eggs or a few large ones? In general, it seems that the larger the egg, and thus the larger the offspring at birth, the better are the chances of the offspring surviving. This would seem to suggest that in most circumstances an embryo needs to give rise to a hatchling as large as possible. Why then, do some species lay many small eggs capable of rapid development? One possible answer is that a parental investment in large eggs may reduce the parents' own chance of surviving and laying more eggs. This strategy of laying many small eggs may be especially successful in variable environmental conditions, where populations can suddenly crash. The rapid development of a larval stage enables early feeding and dispersal to new sites in such circumstances.

SUMMARY

Changes in the timing of developmental processes that have occurred during evolution can alter the form of the body, for if different regions grow faster than others, their size is proportionately increased if the animal gets larger. The increase in size of the central digit of horses is partly due to this type of developmental change. Speed of development and egg size also have important evolutionary implications. Changing the time at which an animal becomes sexually mature can result in adults with larval characteristics. Some animals, such as frogs and sea urchins, which usually have a larval form, have evolved species that develop directly into the adult without a larval stage. The life histories of animals also have implications for evolution and development.

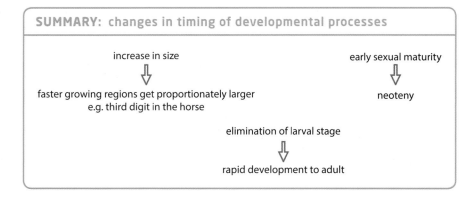

Summary to Chapter 14

- Many developmental processes have been conserved during evolution. While many questions relating to evolution and development remain unanswered, it is clear that development reflects the evolutionary history of ancestral embryos.

- The origins of multicellular animals (the Metazoa) and their embryos from a single-celled ancestor are still highly speculative. Analysis of complete genomes of the earliest evolved metazoans (such as sponges, cnidarians and ctenophores) show that the earliest ancestor of multicellular animals must have already had many of the molecular signaling pathways and other cell biological features present in modern animals.

- All vertebrate embryos pass through a conserved phylotypic developmental stage, where they share many morphological features, although there can be considerable divergence both earlier and later in development.

- The signals involved in patterning the dorso-ventral axis of the body in vertebrates and arthropods show remarkable similarity and conservation, and show that the dorso-ventral axis became inverted in the evolution of vertebrates from their common ancestor with arthropods.

- Comparison of the pattern of Hox gene expression between snakes and other vertebrates correlates with their very different morphologies, notably with the number of vertebrae with thoracic identity. Changes in a control region of a Hox target gene have also been identified in snakes. Changes in both Hox gene regulation and the interpretation of a Hox code could have led to major morphological changes during evolution.

- The limbs of tetrapods evolved from the fins of their ancient fish ancestors, and comparative analysis of the regulation of Hox gene expression in the paired appendages of teleosts, the coelacanth, and mammals have identified changes in regulatory regions that may be significant in this evolution.

- The pattern of Hox gene expression along the body axis of the different types of arthropods is also conserved, and variations in expression of Hox genes between different arthropod classes correlate closely with differences in body plan and the numbers and positions of appendages.

- Genetic analysis of recent evolutionary changes in natural populations of organisms such as sticklebacks and cavefish throw light on the mechanisms of evolution. In sticklebacks, the evolution of a major morphological change is based on changes in the control region of a key developmental gene.

- Alterations in the timing of developmental events have played an important role in evolution. Such changes can alter the overall form of an organism as a result of differences in growth rates of different structures, and can also result in sexual maturation at larval stages.

■ End of chapter questions

Long answer (concept questions)

1. Trace the possible steps in the evolution of the Metazoa from a single-celled ancestor. What were the key innovations that led to the evolution of the Metazoa, and especially of bilaterally symmetric animals?

2. Why has the phylotypic stage been so strongly conserved during evolution? What events and processes are occurring at this point in development that make this stage so central to animal development?

3. Give an example of a change in protein structure and function that occurred after gene duplication. Give an example of a change in gene expression that occurred after gene duplication.

4. Tetrapod limbs evolved from fins. What conclusions about how this might have occurred emerge from (1) studying the fossil record and (2) studying fin development in zebrafish. If you could obtain embryos of an early tetrapod such as *Panderichthys*, what experiments would you carry out to investigate the molecular basis of the fin-to-limb transition?

5. *Drosophila* larvae have no legs, yet lepidopteran larvae have legs on each thoracic and abdominal segment. Explain how Hox genes and the *Distal-less* gene mediate this difference.

6. Snakes have evolved an elongated body with many morphologically similar vertebrae and no limbs. How are these changes related to changes in Hox gene function? Consider both changes in Hox gene expression and changes in how the information encoded by Hox genes is interpreted.

7. 'Darwin's finches' are a classical example of the evolutionary role of development. What is currently known about the developmental basis for the adaptive evolution of beak shape. What are the remaining outstanding questions and what approaches could you would use to find answers?

8. Synthesize the information in Figs 14.22, 14.23, and 14.24 and provide a narrative for the evolution of the incus and malleus bones in the inner ear of mammals.

9. Describe what is meant by the term heterochrony and give an example of an evolutionary change that can be described as being due to heterochrony.

10. Using examples from the text, discuss how probabilities of survival, rates of reproduction, and optimization of reproductive effort can drive evolutionary changes in development.

Multiple choice (factual recall questions)

NB There is only one correct answer to each question.

1. Ontogeny is

a) the development of an individual organism
b) the development of the ear
c) the evolutionary history of a species or group
d) the study of cancer

2. An example of a triploblast is

a) *Hydra*
b) sponges
c) *Xenopus*
d) triploid plants such as seedless watermelons

3. Considering the ancient evolutionary split between protostomes and deuterostomes, which of the following pairs of animals is most closely related evolutionarily?

a) *Caenorhabditis elegans* and amphioxus
b) humans and *Drosophila*
c) sea urchins and humans
d) snails and starfish

4. The evolution of the vertebrate Hox clusters probably arose through

a) duplications of an ancestral cluster to give four ancestral clusters, followed by tandem duplications which have apparently occurred differently in each of the four clusters
b) random changes in DNA sequences followed by natural selection, leading to the independent evolution of each of the Hox genes from unrelated sequences
c) tandem duplications of an ancestral gene, followed by duplications of the whole cluster
d) the generation of four ancestral Hox genes on separate chromosomes, followed by tandem duplications to give rise to the four Hox clusters

5. An animal's ventral surface is defined by

a) the expression of BMP family members
b) the position of the mouth
c) the side of the animal that faces the ground
d) the side opposite the central nervous system

6. The horse's hoof is homologous to human

a) fingers or toes
b) humerus
c) radius and ulna
d) wrist bones

7. In humans, the branchial arches are

a) embryonic gill slits, reflecting our evolution from fish
b) embryonic precursors to gills, which do not continue to develop in humans
c) embryonic precursors to structures predominantly in the jaw
d) embryonic structures that develop into the ribs

8. The evolutionary acquisition of sexual maturity by an animal while still in a larval stage is called

a) apoptosis
b) metamorphosis
c) neoteny
d) ontogeny

9. Which is true?

a) Bat wings evolved from bird wings
b) Ears evolved from jaws
c) Insects evolved from crustaceans
d) Limbs evolved from fins

10. The loss of pelvic spines during adaptive evolution in stickleback is based on

a) a mutation in the protein-coding region of the *Pitx1* gene
b) a mutation in the protein-coding region of the *Pitx2* gene
c) a mutation in the control region of the *Pitx1* gene
d) the same mutation in *Pitx1* in geographically distinct populations of spine-deficient fish

Multiple choice answer key

1: a, 2: c, 3: c, 4: c, 5: b, 6: a, 7: c, 8: c, 9: d, 10: c.

■ General further reading

Carroll, S.B.: **Evo-devo and an expanding evolutionary synthesis: a genetic theory of morphological evolution.** *Cell* 2008, **134**: 25–36.

Carroll, S.B., Grenier, J.K., Weatherbee, S.D.: *From DNA to Diversity*, 2nd edn. Malden, MA: Blackwell Science, 2005.

Finnerty, J.R., Pang, K., Burton, P., Paulson, D., Martindale, M.Q.: **Origins of bilateral symmetry: Hox and *dpp* expression in a sea anemone.** *Science* 2004, **304**: 1335–1337.

Hoekstra, H.E., Coyne, J.A.: **The locus of evolution: evo devo and the genetics of adaptation.** *Evolution* 2007, **61**: 995–1016.

Kirschner, M., Gerhart, J.: **Evolvability.** *Proc. Natl Acad. Sci. USA* 1998, **95**: 8420–8427.

Kirschner, M., Gerhart, J.: *The Plausibility of Life*. Yale University Press, 2005.

Raff, R.A.: *The Shape of Life*. University of Chicago Press, 1996.

Richardson, M.K.: **Vertebrate evolution: the developmental origins of adult variation.** *BioEssays* 1999, **21**: 604–613.

Wray, G.A.: **The evolutionary significance of *cis*-regulatory mutations.** *Nat. Rev. Genet.* 2007, **8**: 206–216.

Box 14A 'Darwin's finches'

Abzhanov, A., Kuo, W.P., Hartmann, C., Grant, B.R., Grant, P.R., Tabin, C.J.: **The calmodulin pathway and evolution of elongated beak morphology in Darwin's finches.** *Nature* 2006, **442**: 563–567.

■ Section further reading

14.1 Genomic evidence is throwing light on the origin of metazoans

Petersen, C.P., Reddien, P.W.: **Wnt signaling and the polarity of the primary body axis.** *Cell* 2009, **139**: 1056–1068.

Putnam, N.H., Srivastava, M., Hellsten, U., Dirks, B., Chapman, J., Salamov, A., Terry, A., Shapiro, H., Lindquist, E., Kapitonov, V.V., *et al.*: **Sea anemone genome reveals ancestral eumetazoan gene repertoire and genomic organization.** *Science* 2007, **317**: 86–94.

Richards, G.S., Degnan, B.M.: **The dawn of developmental signaling in the metazoa.** *Cold Spring Harb. Symp. Quant. Biol.* 2009, **74**: 81–90.

Ryan, J.F., Pang, K., Schnitzler, C.E., Nguyen, A.D., Moreland, R.T., Simmons, D.K., Koch, B.J., Francis, W.R., Havlak, P., NISC Comparative Sequencing Program, Smith, *et al.*: **The genome of the ctenophore *Mnemiopsis leidyi* and its implications for cell type evolution.** *Science* 2013, **342**: 1242592.

Srivastava, M., Begovic, E., Chapman, J., Putnam, N.H., Hellsten, U., Kawashima, T., Kuo, A., Mitros, T., Salamov, A., Carpenter, M.L., *et al.*: **The *Trichoplax* genome and the nature of placozoans.** *Nature* 2008, **454**: 955–960.

14.2 Multicellular organisms evolved from single-celled ancestors

Brooke, N.M., Holland P.W.H.: **The evolution of multicellularity and early animal genomes.** *Curr. Opin. Genet. Dev.* 2003, **6**: 599–603.

Jaegerstern, G.: **The early phylogeny of the metazoa. The bilaterogastrea theory.** *Zool. Bidrag. (Uppsala)* 1956, **30**: 321–354.

King, N.: **The unicellular ancestry of animal development.** *Dev. Biol.* 2004, **7**: 313–325.

King, N., Westbrook, M.J., Young, S.L., Kuo, A., Abedin, M., Chapman, J., Fairclough, S., Hellsten, U., Isogai, Y., Letunic, I., *et al.*: **The genome of the choanoflagellate *Monosiga brevicollis* and the origin of metazoans.** *Nature* 2008, **451**: 783–788.

Martindale, M.Q.: **The evolution of metazoan axial properties.** *Nat. Rev. Genet.* 2005, **6**: 917–927.

Miller, D.J., Ball, E.E.: **Animal evolution: the enigmatic phylum Placozoa revisited.** *Curr. Biol.* 2005, **15**: R26–R28.

Rudel, D., Sommer, R.J.: **The evolution of developmental mechanisms.** *Dev. Biol.* 2003, **264**: 15–37.

Szathmary, E., Wolpert, L.: **The transition from single cells to multicellularity.** In *Genetic and Cultural Evolution of Cooperation* (ed.Hammersteen, P.) 271–289. Cambridge, MA: MIT Press, 2004.

Wolpert, L.: **Gastrulation and the evolution of development.** *Development (Suppl.)* 1992, 7–13.

Wolpert, L., Szathmary, E.: **Evolution and the egg.** *Nature* 2002, **420**: 745.

14.3 Hox gene complexes have evolved through gene duplication

Brooke, N.M., Gacia-Fernandez, J., Holland, P.W.H.: **The ParaHox gene cluster is an evolutionary sister of the Hox gene cluster.** *Nature* 1998, **392**: 920–922.

Duboule, D.: **The rise and fall of Hox gene clusters.** *Development* 2007, **134**: 2549–2560.

Ferrier, D.E.K., Holland, P.W.H.: **Ancient origin of the Hox gene cluster.** *Nat. Rev. Genet.* 2001, **2**: 33–34.

Holland, L.Z. *et al.*: **The amphioxuss genome illuminates vertebrate origins and cephalochordate biology.** *Genome Res.* 2008, **18**: 1100–1111.

Pascual-Anaya, J., D'Aniello, S., Kuratani, S., Garcia-Fernandez, J.: **Evolution of Hox gene clusters in deuterostomes.** *BMC Dev. Biol.* 2013, **13**: 26.

Prince, V.E., Pickett, F.B.: **Splitting pairs: the diverging fates of duplicated genes.** *Nat. Rev. Genet.* 2002, **3**: 827–837.

Valentine, J.W., Erwin, D.H., Jablonski, D.: **Developmental evolution of metazoan bodyplans: the fossil evidence.** *Dev. Biol.* 1996, **173**: 373–381.

14.4 Changes in both Hox genes and their target genes generated the elaboration and diversification of bilaterian body plans

Akam, M.: **Hox genes and the evolution of diverse body plans.** *Phil. Trans. R. Soc. Lond. B* 1995, **349**: 313–319.

Cohn, M.J., Tickle, C.: **Developmental basis of limblessness and axial patterning in snakes.** *Nature* 1999, **399**: 474–479.

Duboule, D.: **A Hox by any other name.** *Nature* 2000, **403**: 607–610.

Galant, R., Carroll, S.B.: **Evolution of a transcriptional repression domain in an insect Hox protein.** *Nature* 2003, **415**: 910–913.

Guerreiro, I., Nunes, A., Woltering, J.M., Casaca, A., Nóvoa, A., Vinagre, T., Hunter M. E., Duboule, D., Mallo, M.: **Role of a polymorphism in a Hox/Pax-responsive enhancer in the evolution of the vertebrate spine.** *Proc. Natl Acad. Sci. USA* 2013, **110**: 10682–10686.

Mansfield, J.H.: ***cis*-regulatory change associated with snake body plan evolution.** *Proc. Natl Acad. Sci. USA* 2013, **110**: 10473–10474.

Nishimoto, S., Minguillon, C., Wood, S., Logan, M.P.: **A combination of activation and repression by a colinear Hox code controls forelimb-restricted expression of Tbx5 and reveals Hox protein specificity.** *PLoS Genet.* 2014, **10**: e1004245.

Pavlopoulos, A., Averof, M.: **Developmental evolution: Hox proteins ring the changes.** *Curr. Biol.* 2002, **12**: R291–R293.

Woltering, J.M., Vonk, F.J., Müller, H., Bardine, N., Tuduce, I.L., de Bakker, M.A., Knöchel, W., Sirbu, I.O., Durston, A.J., Richardson, M.K.: **Axial patterning in snakes and caecilians: evidence for an alternative interpretation of the Hox code.** *Dev. Biol.* 2009, **332**: 82–89.

14.5 Differences in Hox gene expression determine the variation in position and type of paired appendages in arthropods

Angelini, D.R., Kaufman, T.C.: **Comparative developmental genetics and the evolution of arthropod body plans.** *Annu. Rev. Genet.* 2005, **39**: 95–119.

Averof, M.: **Origin of the spider's head.** *Nature* 1998, **395**: 436–437.

Carroll, S.B., Weatherbee, S.D., Langeland, J.A.: **Homeotic genes and the regulation and evolution of insect wing number.** *Nature* 1995, **375**: 58–61.

Levine, M.: **How insects lose their wings.** *Nature* 2002, **415**: 848–849.

Liubicich, D.M., Serano, J.M., Pavlopoulos, A., Kontarakis, Z., Protas, M.E., Kwan, E., Chatterjee, S., Tran, K.D., Averof, M., Patel, N.H.: **Knockdown of *Parhyale Ultrabithorax* recapitulates evolutionary changes in crustacean appendage morphology.** *Proc. Natl Acad. Sci. USA* 2009, **106**: 13892–13896.

Pavlopoulos, A., Akam, M.: **Hox gene Ultrabithorax regulates distinct sets of target genes at successive stages of *Drosophila* haltere morphogenesis.** *Proc. Natl Acad. Sci. USA* 2011, **108**: 2855–2860.

Pavlopoulos, A., Kontarakis, Z., Liubicich, D.M., Serano, J.M., Akam, M., Patel, N.P., Averof, M.: **Probing the evolution of appendage specialization by Hox gene misexpression in an emerging model crustacean.** *Proc. Natl Acad. Sci. USA* 2009, **106**: 13897–13902.

Weatherbee, S.D., Carroll, S.: **Selector genes and limb identity in arthropods and vertebrates.** *Cell* 1999, **97**: 283–286.

Weatherbee, S.D., Nijhout, H.F, Grunnert, L.W., Halder, G., Galant, R., Selegue, J., Carroll, S.: **Ultrabithorax function in butterfly wings and the evolution of insect wing patterns.** *Curr. Biol.* 1999, **9**: 109–115.

14.6 The basic body plan of arthropods and vertebrates is similar, but the dorso-ventral axis is inverted

Arendt, D., Nübler-Jung, K.: **Dorsal or ventral: similarities in fate maps and gastrulation patterns in annelids, arthropods and chordates.** *Mech. Dev.* 1997, **61**: 7–21.

Davis, G.K., Patel, N.H.: **The origin and evolution of segmentation.** *Trends Biochem. Sci.* 1999, **24**: M68–M72.

Gerhart, J., Lowe, C., Kirschner, M.: **Hemichordates and the origins of chordates.** *Curr. Opin. Genet. Dev.* 2005, **15**: 461–467.

Holley, S.A., Jackson, P.D., Sasai, Y., Lu, B., De Robertis, E., Hoffman, F.M., Ferguson, E.L.: **A conserved system for dorso-ventral patterning in insects and vertebrates involving *sog* and *chordin*.** *Nature* 1995, **376**: 249–253.

Mizutani, C.M., Bier, E.: **EvoD/Vo: the origins of BMP signalling in the neuroectoderm.** *Nat. Rev. Genet.* 2008, **9**: 663–677.

14.7 Limbs evolved from fins

Ahn, D., Ho, R.K.: **Tri-phasic expression of posterior Hox genes during development of pectoral fins in zebrafish: implications for the evolution of vertebrate paired appendages.** *Dev. Biol.* 2008, **322**: 220–233.

Amemiya, C.T. *et al.*: **The African coelacanth genome provides insights into tetrapod evolution.** *Nature*, 2013, **496**: 311–316.

Boisvert, C.A., Mark-Kurik, E., Ahlberg, P.E.: **The pectoral fin of *Panderichthys* and the origin of digits.** *Nature* 2008, **456**: 636–638.

Cooper, K.L., Sears, K.E., Uygur, A., Maier, J., Baczkowski, K.S., Brosnahan, M., Antczak, D., Skidmore, J.A., Tabin, C.J.: **Patterning and post-patterning modes of evolutionary digit loss in mammals.** *Nature* 2014, **511**: 41–45.

Cooper, L.N., Berta, A., Dawson, S.D., Reidenberg, J.S.: **Evolution of hyperphalangy and digit reduction in the cetacean manus.** *Anat. Rec.* 2007, **290**: 654–672.

Hoff, M.: **A footnote to the evolution of digits.** *PLoS Biol.* 2014, **12**: e1001774.

Lopez-Rios, J. *et al.*: **Attenuated sensing of SHH by Ptch1 underlies evolution of bovine limbs.** *Nature* 2014, **511**: 46–51.

Minguillon, C., Gibson-Brown, J.J., Logan, M. P.: **Tbx4/5 gene duplication and the origin of vertebrate paired appendages.** *Proc. Natl Acad. Sci. USA* 2009, **106**: 21726–21730.

Richardson, M. K., Oelschläger, H.H.: **Time, pattern, and heterochrony: a study of hyperphalangy in the dolphin embryo flipper.** *Evol Dev.* 2002 **4**: 435–444.

Sordino, P., van der Hoeven, F., Duboule, D.: **Hox gene expression in teleost fins and the origin of vertebrate digits.** *Nature* 1995, **375**: 678–681.

Tabin, C.J.: **Why we have (only) five fingers per hand: hox genes and the evolution of paired limbs.** *Development* 1992, **116**: 289–296.

Woltering, J.M., Noordermeer, D., Leleu, M., Duboule, D.: **Conservation and divergence of regulatory strategies at *Hox*** loci and the origin of tetrapod digits. *PLoS Biol.* 2014, **12**: e1001773.

Yano, T., Tamura, T.: **The making of differences between fins and limbs.** *J. Anat.* 2013, **222**: 100–113.

14.9 Adaptive evolution within the same species provides a way of studying the developmental basis for evolutionary change

Chan, Y.F., Marks, M.E., Jones, F.C., Villarreal, G., Jr., Shapiro, M.D., Brady, S.D., Southwick, A.M., Absher, D.M., Grimwood, J., Schmutz, J., *et al.*: **Adaptive evolution of pelvic reduction in sticklebacks by recurrent deletion of a *Pitx1* enhancer.** *Science* 2010, **327**: 302–305.

Gompel, N., Carrol, S.B.: **Genetic mechanisms and constraints governing the evolution of correlated traits in drosophilid flies.** *Nature* 2003, **424**: 931–935.

Gompel, N., Prud'homme, B., Wittkopp, P.J., Kassner, V.A., Carroll, S.B.: **Chance caught on the wing: cis-regulatory evolution and the origin of pigment patterns in *Drosophila*.** *Nature* 2005, **433**: 481–487.

Jeffery, W.R.: **Evolution and development in the cavefish *Astyanax*.** *Curr. Top. Dev. Biol.* 2009, **86**: 191–221.

Rohner, N., Jarosz, D.F., Kowalko, J.E., Yoshizawa, M., Jeffery, W.R., Borowsky, R.L., Lindquist, S., Tabin, C.J.: **Cryptic variation in morphological evolution: HSP90 as a capacitor for loss of eyes in cavefish.** *Science* 2013, **342**: 1372–1375.

Shapiro, M.D., Marks, M.E., Peichel, C.L., Blackman, B.K., Nereng, K.S., Jonsson, B., Schluter, D., Kingsley, D.M.: **Genetic and developmental basis of evolutionary pelvic reduction in threespine sticklebacks.** *Nature* 2004, **428**: 717–723.

Yamamoto, Y., Stock, D.W., Jeffery, W.R.: **Hedgehog signaling controls eye degeneration in blind cavefish.** *Nature* 2004, **431**: 844–847.

Yoshizawa, M., Yamamoto, Y., O'Quin, K. E, Jeffery, W.R.: **Evolution of an adaptive behavior and its sensory receptors promotes eye regression in blind cavefish.** *BMC Biol.* 2012, **10**: 108.

14.10 Evolution of different types of eyes in different animal groups is an example of parallel evolution using an ancient genetic circuitry

Arendt, D.: **Evolution of eyes and photoreceptor cell types.** *Int. J. Dev. Biol.* 2003, 563–571.

Gehring, W.J.: **Chance and necessity in eye development.** *Genome Biol. Evol.* 2011, **3**: 1053–1056.

Halder, G., Callaerts, P., Gehring, W.J.: **Induction of ectopic eyes by targeted expression of the eyeless gene in *Drosophila*.** *Science* 1995, **267**: 1788–1792.

Shubin, N., Tabin, C., Carroll, S.: **Deep homology and the origins of evolutionary novelty.** *Nature* 2009, **457**: 818–823.

Wawersik S., Maas, R.L.: **Vertebrate eye development as modeled in *Drosophila*.** *Hum. Mol. Genet.* 2000, **9**: 917–925.

14.11 Embryonic structures have acquired new functions during evolution

Cerny, R., Lwigale, P., Ericsson, R., Meulemans, D., Epperlein, H.H.Bronner-Fraser, M.: **Developmental origins and evolution of jaws: new interpretation of 'maxillary' and 'mandibular'.** *Dev Biol.* 2004, **276**: 225–236.

Cohn, M.J.: **Lamprey Hox genes and the origin of jaws.** *Nature* 2002, **416**: 386–387.

De Robertis, E.M.: **Evo-devo:variations on ancestral themes.** *Cell* 2008, **132**: 185–195.

Erwin, D.H.: **Early origin of the bilaterian developmental toolkit.** *Philos. Trans. R. Soc. Lond. B Biol. Sci.* 2009, **364**: 2253–2261.

Romer, A.S.: *The Vertebrate Body*. Philadelphia: W.B. Saunders, 1949.

14.12 Changes in growth can modify the basic body plan

Behringer, R.R., Rasweiler, J.J., Chen, C.H., Cretekos, C.J.: **Genetic regulation of mammalian diversity.** *Cold Spring Harb Symp Quant Biol.* 2009, **74**: 297–302.

Hockman, D., Cretekos, C.J., Mason, M.K, Behringer, R.R, Jacobs, D.S., Illing, N.: **A second wave of Sonic hedgehog expression during the development of the bat limb.** *Proc. Natl Acad. Sci. USA.* 2008, **105** : 16982–16987.

Richardson, M.K.: **Vertebrate evolution: the developmental origins of adult variation.** *BioEssays* 1999, **21**: 604–613.

14.13 Evolution can be due to changes in the timing of developmental events

Alberch, P., Alberch, J.: **Heterochronic mechanisms of morphological diversification and evolutionary change in the neotropical salamander *Bolitoglossa occidentales* (Amphibia: Plethodontidae).** *J. Morphol.* 1981, **167**: 249–264.

Lande, R.: **Evolutionary mechanisms of limb loss in tetrapods.** *Evolution* 1978, **32**: 73–92.

Raynaud, A.: **Developmental mechanism involved in the embryonic reduction of limbs in reptiles.** *Int. J. Dev. Biol.* 1990, **34**: 233–243.

Wray, G.A., Raff, R.A.: **The evolution of developmental strategy in marine invertebrates.** *Trends Evol. Ecol.* 1991, **6**: 45–56.

14.14 The evolution of life histories has implications for development

Partridge, L., Harvey, P.: **The ecological context of life history evolution.** *Science* 1988, **241**: 1449–1455.

Box 14B How the bird wing evolved

Burke, A.C., Feduccia, A.: **Developmental patterns and the identification of homologies in the avian hand.** *Science* 1997, **278**: 666–669.

Feduccia, A., Lingham-Soliar, T., Hinchliffe, J. R.: **Do feathered dinosaurs exist? Testing the hypothesis on neontological and paleontological evidence.** *J. Morphol.* 2005, **226**: 125–166.

Tamura, K., Nomura, N., Seki, R., Yonei-Tamura, S., Yokoyama, H.: **Embryological evidence identifies wing digits in birds as digits 1, 2, and 3.** *Science* 2011, **331**: 753–757.

Towers, M., Signolet, J., Sherman, A., Sang, H., Tickle, C.: **Insights into bird wing evolution and digit specification from polarizing region fate maps.** *Nat. Commun.* **2**: 426.

Vargas, A.O., Fallon, J.F.: **The digits of the wing of birds are 1, 2, and 3. A review.** *J. Exp. Zool. B Mol. Dev. Evol.* 2005, **304**: 206-219.

Xu, X., Mackem, S.: **Tracing the evolution of avian wing digits.** *Curr. Biol.* 2013, **23**: R538–R544.

Glossary

2D:4D digit ratio the ratio of the lengths of the second and fourth fingers in humans, which differs between the sexes, being less than 1 in men; the same or more than 1 in women.

abdomen in insects, the most posterior of the three distinct parts of the body (head, thorax, and abdomen).

abembryonic pole *see embryonic–abembryonic axis.*

aboral refers to the side of a sea urchin embryo opposite to the mouth.

accretionary growth increase in mass due to secretion of large quantities of extracellular matrix by cells, which enlarges the volume of extracellular space. It occurs, for example, in cartilage and bone, where most of the tissue mass is extracellular.

acheiropodia rare inherited human condition in which structures below the elbow or knee do not develop, but there is no other defect.

achondroplasia form of dwarfism in which the limbs are short in relation to the rest of the body.

acrosomal reaction the release of enzymes and other proteins from the **acrosome** of the sperm head that occurs once a sperm has bound to the zona pellucida surrounding the mammalian egg. It helps the sperm to penetrate the zona pellucida enabling it to approach the egg plasma membrane.

acrosome vesicular structure on the sperm head that contains enzymes and other proteins. *See also acrosomal reaction.*

actin filament one of the three principal types of protein filaments in the cytoskeleton. Actin filaments are involved in cell movement and changes in cell shape, and are also part of the contractile apparatus of muscle cells. Also called microfilaments.

action potential the electrical signal generated in a neuron when it receives sufficient stimulatory inputs. It propagates the length of the axon and stimulates the release of neurotransmitter from the axon terminal. Also called a nerve impulse.

activator gene-regulatory protein that helps to turn a gene on when it binds to a particular site in a gene's control region.

actomyosin an assembly of actin filaments and the motor protein myosin that can undergo contraction.

adaxial–abaxial axis in plants, the axis that runs from the center of the plant stem to the circumference. In plant leaves it runs from the upper surface to the lower surface (dorsal to ventral).

adherens junction type of adhesive cell junction in which the adhesion molecules linking the two cells together are transmembrane cadherins that are linked intracellularly to the actin cytoskeleton.

adhesion molecule *see cell-adhesion molecule.*

affinity the strength of binding of a molecule at a single site on another molecule.

allantois a set of extra-embryonic membranes that develops in many vertebrate embryos. In bird and reptile embryos the allantois collects liquid waste and acts as a respiratory surface, whereas in mammals its blood vessels carry blood to and from the placenta.

allele a particular version of a gene. In diploid organisms, two alleles of each gene are present (one on each chromosome of a homologous pair), which may or may not be identical.

alternative splicing RNA splicing that generates an alternative form of an mRNA by using different splicing.

amnion extra-embryonic membrane in birds, reptiles, and mammals that forms a fluid-filled sac which encloses and protects the embryo. It is derived from extra-embryonic ectoderm and mesoderm.

amnioserosa extra-embryonic membrane on the dorsal side of the *Drosophila* embryo.

amniotes vertebrates whose embryos have an amnion. They comprise the mammals, birds, and reptiles.

anamniotes vertebrates whose embryos do not have an amnion. They comprise the fish and amphibians.

androgenetic describes an embryo in which the two sets of homologous chromosomes are both paternal in origin.

angioblast mesodermal precursor cells that will give rise to the endothelial cells of blood vessels.

angiogenesis the process by which small blood vessels sprout from the larger vessels.

animal cap, animal pole *see animal region.*

animal pole *see animal region.*

animal region in eggs of amphibians, the hemispherical end of the egg where the nucleus resides, away from the yolk. The most terminal part of this region is the **animal pole**, which is directly opposite the vegetal pole at the other end of the egg. In *Xenopus*, the pigmented animal half is called the **animal cap**.

animal–vegetal axis axis that runs from the animal pole to the vegetal pole in an egg or early embryo.

aniridia the absence of an iris in the eye.

Antennapedia complex a cluster of genes that comprises one part of the Hox gene clusters (the HOM-C complex) in *Drosophila*.

anterior at, or in the direction of, the head end of an embryo or animal.

anterior neural ridge a signaling center in the developing brain of the early chick embryo that lies immediately adjacent to and within the anterior-most forebrain, and is responsible for patterning of the telencephalon (the part of the forebrain that gives rise to the pallium and the basal ganglia).

anterior visceral endoderm (AVE) an extra-embryonic tissue in the early mouse embryo that is involved in inducing anterior regions of the embryo.

antero-posterior axis the axis running from the head end to the tail end of an animal. The head is anterior and the tail posterior. In the vertebrate limb, this axis runs from the thumb to little finger.

anticlinal describes cell divisions in planes at right angles to the outer surface of a tissue.

antisense RNA an RNA complementary in nucleotide sequence to an mRNA, which blocks expression of a protein by binding to its mRNA and blocking translation.

apical–basal axis the axis running from shoot tip to root tip in a plant. The shoot end is apical and the root end is basal.

apical dominance the phenomenon in plants that buds in the axils of nodes behind the tip of a plant shoot will not form side stems when the tip is intact. It is due to the production of the hormone auxin by the shoot tip which suppresses outgrowth of axillary buds.

apical ectodermal ridge, apical ridge a thickening of the ectoderm at the distal end of the developing chick and mammalian limb bud, which is essential for limb bud outgrowth and for correct patterning along the proximo-distal axis of the limb.

apical epidermal cap the epidermis that forms over the cut surface of an amputated amphibian limb. Its formation is essential for subsequent regeneration. Also called wound epidermis.

apical meristem the region of dividing cells at the tip of a growing plant shoot or root, from which all new tissue is produced.

apico-basal polarity cell polarity seen in an epithelium, in which the apical and basal faces of the cells have different properties.

apoptosis type of cell death that occurs widely during development. Also called programmed cell death. The cell is induced by an external or internal signal to commit 'cell suicide,' which involves fragmentation of the DNA and shrinkage of the cell, without release of the cell contents. Apoptotic cells are removed by the body's scavenger cells and, unlike necrosis, their death does not cause damage to surrounding cells.

archenteron the cavity formed inside the embryo when the endoderm and mesoderm invaginate during gastrulation. It forms the gut.

area opaca the outer dark area of the chick blastoderm.

area pellucida the central clear area of the chick blastoderm.

arterial pole the anterior end of the developing heart tube, which will become the outflow (arterial) end.

aster structure made of microtubules radiating away from the centrosome and anchored in the cortex, found at each pole of the mitotic or meiotic spindle.

asymmetric cell division, asymmetric division a cell division in which the daughter cells are different from each other because some cytoplasmic determinant(s) has been distributed unequally between them.

atrial chamber one of the chambers into which the vertebrate heart tube is first divided by a transverse septum, the other being the ventricular chamber. In fish, the heart has only two chambers; in the four-chambered heart of other vertebrates each chamber is divided longitudinally to give right and left atria and right and left ventricles.

autonomous describes any developmental process that can continue without a requirement for extracellular signals to be continuously present. *See also* **cell-autonomous**.

Aux/IAA protein any of a family of proteins that, by binding to auxin-response factors, block the expression of auxin-responsive genes in the absence of the plant hormone auxin.

auxin small organic molecule that is an important plant hormone in almost all aspects of plant development. Auxin acts by regulating the expression of auxin-responsive genes by stimulating the degradation of Aux/IAA proteins and thus enabling gene expression.

auxin-responsive gene any plant gene whose expression is controlled by the plant hormone auxin.

auxin-response factor (ARF) DNA-binding transcriptional activator (or in a few cases a repressor) that binds to auxin-responsive elements in the control region of auxin-responsive genes.

AVE *see* **anterior visceral endoderm**.

axial structures those structures that form along the main axis of the body, such as the notochord, vertebral column, somites and neural tube in vertebrates.

axon long cell process of a neuron that conducts nerve impulses away from the cell body. The end of an axon—the axon terminal—forms contacts (synapses) with other neurons, muscle cells, or glandular cells.

axon terminal *see* **axon**.

β-catenin protein that functions both as a transcriptional co-activator and as a protein that links adhesive cell junctions to the cytoskeleton. As a transcriptional co-activator, β-catenin is activated in early development in many embryos as the end result of a Wnt signaling pathway.

β-neurexin transmembrane protein in the pre-synaptic membrane of some synapses, which interacts with the protein neuroligin in post-synaptic membranes to promote the functional differentiation of the synapse.

basal lamina *see* **basement membrane**.

basal layer the deepest layer of the epidermis of the skin, which contains epidermal stem cells from which the epidermis is renewed.

basement membrane sheet of extracellular matrix that separates an epithelial layer from the underlying tissues. For example, the epidermis of the skin is separated from the dermis by a basement membrane. Also called the basal lamina.

bilateral symmetry type of symmetry possessed by animals with a main axis of symmetry running from head to tail and the two sides of the body being mirror images of each other.

bithorax complex a cluster of genes that comprises one part of the Hox gene complex (HOM-C complex) in *Drosophila*.

blastema the mass of cells that develops at the site of amputation of the amphibian limb. It is formed from the dedifferentiation and proliferation of cells beneath the wound epidermis, and gives rise to the regenerated limb.

blastocoel fluid-filled cavity that develops in the interior of a blastula.

blastocyst the stage of a mammalian embryo that corresponds in form to the blastula stage of other animal embryos, and is the stage at which the embryo implants in the uterine wall.

blastoderm a post-cleavage embryo composed of a solid layer of cells rather than a spherical blastula, as found in early chick, zebrafish, and *Drosophila* embryos.

blastodisc alternative name for the blastoderm in the chick embryo.

blastomeres any of the cells formed by the cleavage of the fertilized egg.

blastopore slit-like or circular invagination on the surface of amphibian and sea-urchin embryos, at which the mesoderm and endoderm move inside the embryo at gastrulation.

blastula early stage in the development of some embryos (e.g. amphibians, sea urchin), which is the outcome of cleavage. It is a hollow ball of cells, composed of an epithelial layer of cells enclosing a fluid-filled cavity—the blastocoel.

blastula organizer *see* **Nieuwkoop center**.

block to polyspermy the mechanism in some animals that prevents fertilization of an egg by more than one sperm. In sea urchins and *Xenopus* it is composed of two stages—an immediate rapid electrical block to polyspermy that involves depolarization of the egg plasma membrane, and a subsequent slow block to polyspermy which involves the formation of an impenetrable fertilization membrane around the egg.

body plan the overall organization of an organism; for example, the relative positions of the head and tail, and the plane of bilateral symmetry, where it exists. The body plan of most animals is organized around two main axes: the antero-posterior axis and the dorso-ventral axis.

bone marrow stromal cells cells in bone marrow that form the stem-cell niche that supports and maintains the hematopoietic stem cells.

brachial describes the region of a vertebrate embryo trunk at the level of the forelimb.

branchial arches structures that develop on each side of the embryonic head and give rise to the gill arches in fishes, and to the jaws and other facial structures in other vertebrates.

branching morphogenesis describes the development and growth of stuctures such as blood vessels or the bronchi and bronchioles of the lung, which develop by the successive branching of a tube of epithelium.

cadherin any of a family of cell-adhesion molecules with important roles in development. Cadherins are the adhesion molecules in adherens junctions.

cambium a ring of meristem in the stems of plants that gives rise to new stem tissue, which increases the diameter of the stem.

canonical Wnt/β-catenin pathway an intracellular signaling pathway stimulated by members of the Wnt family of signaling proteins that leads to the stabilization of β-catenin and its entry into nuclei, where it acts as a transcriptional co-activator.

capacitation the functional maturation of sperm after they have been deposited in the female reproductive tract.

cardia bifida condition in which two laterally positioned hearts are formed. It results from a failure of cells of the cardiac crescent to migrate to form a centrally positioned heart tube. Found in zebrafish, as a result of a mutation.

cardiac crescent regions of heart precursor cells located in the splanchnic lateral plate mesoderm on either side of the body underlying the head folds.

cardiogenic crescent *see* **cardiac crescent**.

cardiomyocyte a mature fully differentiated heart muscle cell.

caspase any of a family of intracellular proteases, some of which are involved in apoptosis.

catenin any of a family of proteins that are part of adherens junctions, involved in linking cadherins to the cytoskeleton. One member of the family, β-catenin, also acts as a transcriptional activator in the canonical Wnt/β-catenin signaling pathway.

cell-adhesion molecule any of several different families of proteins that bind cells to each other and to the extracellular matrix. The main classes of adhesion molecules important in development are the cadherins, the immunoglobulin superfamily, and the integrins.

cell adhesiveness the property of cells that causes them to stick together. It is mediated by adhesive cell junctions composed of cell-adhesion molecules that bind to similar or different types of cell-adhesion molecules on the neighboring cell.

cell-autonomous describes the effects of a gene that only affects the cell the gene is expressed in.

cell body the part of a neuron that contains the nucleus, and from which the axon and dendrites extend.

cell–cell interaction, cell–cell signaling general terms that describe various types of intercellular communication by which one cell influences the behavior of another cell. Cells can communicate with each other via cell contact and interaction of membrane-bound signaling molecules with receptors on the contacting cell, or by the secretion of signaling molecules that act on specific receptors on (or in) other cells nearby or at a distance to influence their behavior.

cell cycle the sequence of events by which a cell duplicates its DNA and divides in two.

cell-cycle checkpoint any one of several points that occur during the cell cycle at which the cell monitors progress through the cycle and ensures that a previous stage has been completed before embarking on the next.

cell differentiation the process by which cells become functionally and structurally different from one another and become distinct cell types, such as muscle or blood cells.

cell enlargement an increase in cell size without division, which is one way in which a structure or embryo can grow in size. Growth by cell enlargement is especially common in plants.

cell-lineage restriction the situation where all the descendants of a particular group of cells remain within a 'boundary' and never mix with an adjacent group of cells of a different lineage. The area within such a boundary is known as a compartment.

cell migration the active movement of cells from one location to another in the embryo by 'crawling' along the extracellular matrix, often under the guidance of molecules previously laid down in the matrix.

cell motility the ability of cells to move, to contract, or to change shape, mediated by changes in the cytoskeleton.

cell proliferation the growth and division of cells to form new cells, which is one way in which a structure or embryo can grow in size.

cell-replacement therapy the correction of disease by replacing damaged cells with new healthy cells, including stem cells. So far, mainstream cell-replacement therapy using stem cells is limited to hematopoietic stem cell transplants (such as bone marrow transplants), which use naturally occurring sources of donor stem cells, but therapies using other types of stem cells are being researched.

cell senescence *see* **senescence**.

central nervous system the brain and spinal cord.

centrosome the organizing center for microtubule growth within a cell. It duplicates before mitosis or meiosis, and each centrosome forms one end of the microtubule spindle. It consists of a pair of centrioles, one of which gives rise to the basal body of cilia, including primary cilia which are found on most vertebrate cells.

cerebellum part of the hindbrain concerned with the routine regulation of muscle tone and posture independent of conscious control.

cerebral cortex the outer layers of the cerebral hemispheres, which develop from the pallium of the embryonic forebrain and in which the highest-level processing centers for sensory information, motor control, learning, and memory are located.

chemoaffinity hypothesis the idea that each retinal neuron carries a chemical label that enables it to connect reliably with an appropriately labeled cell in the optic tectum.

chemoattractant a molecule that attracts cells to move towards it.

chemorepellent a molecule that repels cells, causing them to move away from it.

chimera an organism or tissue that is composed of cells from two or more sources with different genetic constitutions.

ChIP-chip and ChIP-seq techniques for determining which sites in chromosomal DNA are bound by particular proteins *in vivo*. They involve immunoprecipitation of chromatin fragments by antibodies against the protein of interest followed by analysis of the bound DNA by microarrays (ChIP-chip) or DNA sequencing (ChIP-seq).

chondrocyte a differentiated cartilage cell.

chorion the outermost of the extra-embryonic membranes in birds, reptiles, and mammals. It is involved in respiratory gas exchange. In birds and reptiles it lies just beneath the shell. In mammals it is part of the placenta and is also involved in nutrition and waste removal. The chorion of insect eggs has a different structure.

chromatin the material of which chromosomes are made. It is composed of DNA packaged into nucleosomes by histone proteins, and is also associated with many other proteins that influence its structure. Chemical modifications to DNA and histone proteins affect the structure of chromatin and the proteins that are associated with it and determine whether a given gene is available for transcription or has been packed away so that it is inaccessible. Such chemical modifications are known as epigenetic modifications as, unlike mutation, they affect gene expression without changing the nucleotide sequence of the DNA.

chromatin-remodeling complex enzyme complex that acts on chromatin to modify it and alter the ability of the DNA to be transcribed.

ciliopathy any of a number of human syndromes, which may affect many different organs, in which the function of cilia is defective. Some ciliopathies, such as Kartagener's syndrome, show abnormal right–left asymmetry of organs such as heart and liver (situs inversus) in a proportion of cases.

circadian clock an internal 24-hour timer present in living organisms that causes many metabolic and physiological processes, including the expression of some genes, to vary in a regular manner throughout the day.

cis-**regulatory module** *see* **cis-regulatory region**.

cis-**regulatory region** the non-coding sequences flanking a gene and containing sites at which the expression of that gene can be controlled by the binding of transcription factors (activators and repressors). Also called control regions. Many regulatory regions in developmental genes contain a series of different *cis*-**regulatory modules**, which are short regions containing multiple binding sites for different transcription factors; the combination of factors bound determines whether the gene is switched on or off. Different regulatory modules control gene expression in different locations or times during development.

cleavage a series of rapid cell divisions without cell growth that occurs after fertilization and divides the embryo up into a number of small cells.

clock and wavefront model a model that has been proposed to explain the periodicity of somite formation. It proposes that a 'clock' represented by regularly cycling gene expression in the pre-somitic mesoderm interacts with a wavefront of somite determination moving anterior to posterior to specify the formation of each somite.

clone (1) a collection of genetically identical cells derived from a single cell by repeated cell division. (2) The genetically identical offspring of a single individual produced by asexual reproduction or artifical cloning techniques.

cloning procedure by which an individual genetically identical to a 'parent' is produced by transplantation of a parental somatic cell nucleus into an unfertilized oocyte.

co-activator gene-regulatory protein that promotes gene expression but does not bind to DNA itself. It binds specifically to other DNA-binding transcription factors, enabling them to act as activators.

coding region that part of a gene that encodes a polypeptide or functional RNA.

co-linearity the correspondence between the order of Hox genes on a chromosome and their temporal and spatial order of expression in the embryo.

colony-stimulating factor any protein that drives the differentiation of blood cells.

commissural neuron neuron in the spinal cord of vertebrates, or the ventral nerve cord of insects, whose axon crosses the midline, forming connections with neurons in the contralateral side of the cord.

community effect the phenomenon that induction of cell differentiation in some tissues depends on there being a sufficient number of responding cells present.

compaction in the mouse embryo, the stage in early cleavage at which the blastomeres flatten against each other and microvilli become confined to the outer surface of the ball of cells.

compartment a discrete area of an embryo that contains all the descendants of a small group of founder cells and shows cell-lineage restriction. Cells in compartments respect the compartment boundary and do not cross over into an adjacent compartment. Compartments tend to act as discrete developmental units.

competence the ability of a tissue to respond to an inducing signal. Embryonic tissues only remain competent to respond to a particular signal for a limited period of time.

condensation the increased packing together of cells to form structures such as cartilage elements (precursors of bones).

conditional mutation a mutation whose effects only become manifest in a particular condition, e.g. at a higher temperature than normal.

contractile ring contractile structure composed of actomyosin that develops around the circumference of an animal cell at the plane of cleavage. Contraction of the ring forms the cleavage furrow that divides the cell in two.

contralateral referring to the opposite side of the body.

control region a non-coding region of a gene to which regulatory proteins bind and determine whether or not the gene is transcribed. also called *cis*-regulatory region.

convergent evolution evolution in two different lineages or populations that independently produces a similar outcome in both.

convergent extension the process by which a sheet of cells changes shape by extending in one direction and narrowing—converging—in a direction at right angles to the extension, caused by the cells intercalating between each other. It contributes, for example, to elongation of the notochord, elongation of the body in amphibian and zebrafish embryos, and extension of the germ band in *Drosophila* embryos.

cooperativity the phenomenon in which the binding of a ligand to one site on another molecule (such as the binding of a transcription factor to a site in DNA) makes the binding of subsequent ligands to other sites on the same molecule more likely to occur.

co-repressor gene-regulatory protein that acts to repress gene expression but does not bind to DNA itself. It acts in a complex with DNA-binding transcription factors.

cortex (1) the outer layer of the cytoplasm of an animal cell, just underneath the plasma membrane, which is more gel-like in consistency than the rest of the cytoplasm and is rich in actin and myosin. (2) The tissue between the epidermis and central vascular tissue in plant stems and roots.

cortical granules granules present in the cortex of some eggs, which release their contents on fertilization to form the fertilization membrane.

cortical rotation a movement of the cortex of the fertilized amphibian egg that occurs immediately after fertilization. The cortex rotates with respect to underlying cytoplasm away from the point of sperm entry.

corticotropin-releasing hormone peptide hormone released by neurosecretory cells in the hypothalamus. In non-mammalian vertebrates it acts on the pituitary gland to cause the release of thyroid-stimulating hormone and consequent metamorphosis.

cotyledon the part of the plant embryo that acts as a food storage organ.

Cre/*loxP* system a transgenic modification of mice that enables a gene to be expressed in a particular tissue or at a particular time in development.

CRISPR/Cas9 system a gene-editing system derived from bacteria that can be introduced into animal cells to make precise targeted mutations.

critical period the period during which a teratogen can interfere with development of a specific organ.

cumulus cells somatic cells that surround the developing mammalian oocyte and are shed with it at ovulation.

cyclin any of a family of proteins that periodically rise and fall in concentration during the cell cycle and are involved in controlling progression through the cycle. Cyclins act by binding to and activating **cyclin-dependent kinases (Cdks)**.

cyclin-dependent kinases (Cdks) protein kinases that have specific roles in particular phases of the cell cycle and are activated by the binding of cyclins. Because cyclin concentration varies periodically in the cell cycle, so do the activities of the cyclin-dependent kinases.

cytoplasmic determinant a cytoplasmic protein or mRNA that has a specific influence on development. The unequal distribution of cytoplasmic determinants to daughter cells by an asymmetric cell division is one way of producing cells with different developmental fates.

cytoplasmic localization the non-uniform distribution of some protein factor or determinant in a cell's cytoplasm, so that when the cell divides, the protein is unequally distributed to the daughter cells.

cytoskeleton the network of protein filaments—microfilaments, microtubules, and intermediate filaments—that gives cells their shape, enables them to move, and provides tracks for transport of materials in the cell.

dauer a larval state in *Caenorhabditis elegans* that arises in response to starvation conditions and in which the larvae neither eat nor grow until food is again available.

dedifferentiation loss of the structural characteristics of a differentiated cell, which may result in the cell then differentiating into a new cell type.

deep layer the layer several cells deep under the outer enveloping layer of the zebrafish blastoderm. It gives rise to the embryo.

delamination the process by which epithelial cells leave an epithelium as individual cells. It occurs, for example, in the primitive streak and in the movement of neural crest cells out of the neural tube.

dendrite an extension from the body of a nerve cell that receives stimuli from other nerve cells.

dental lamina band of thickened oral epithelium in the jaw within which individual teeth will develop.

dental papilla condensation of ectomesenchyme cells that gives rise to dentin and the soft connective tissue in the centre of the tooth (the dental pulp).

dental placode specific region of the dental lamina that develops into a tooth.

denticle small tooth-like outgrowth of the cuticle on insect larvae.

dermis the connective tissue layer of the skin lying beneath the epidermis, from which it is separated by a basement membrane.

dermomyotome the region of the somite that will give rise to both muscle and dermis.

desmosome type of cell junction in which the cadherin cell-adhesion molecules are linked intracellularly to keratin filaments of the cytoskeleton.

determinant cytoplasmic factor (e.g. a protein or RNA) in the egg and in embryonic cells that can be asymmetrically distributed at cell division and so influence how the daughter cells develop.

determination a stable change in the internal state of a cell such that its fate is now fixed, or **determined**. A determined cell will follow that fate if grafted into other regions of the embryo.

deuterostomes animals, such as chordates and echinoderms, that have radial cleavage of the egg, and in which the primary invagination of the gut at gastrulation forms the anus, with the mouth developing independently.

developmental biology the study of the development of a multicellular organism from its origin in a single cell—the fertilized egg—until it is fully adult.

developmental constraint a pre-existing developmental process that constrains the evolution of new forms.

developmental gene any gene that specifically controls a developmental process.

differential adhesion hypothesis a hypothesis to explain the movement of cells in developing embryos, e.g. at gastrulation, in terms of the differences in the strength of adhesion of different types of cells for each other.

differential gene expression the turning on and off of different genes in different cells in a multicellular organism, thus generating cells with different developmental and functional properties.

differentiation factor any extracellular signaling protein that promotes cellular differentiation.

digital-plate *see* **hand-plate**.

diploblasts animals with two germ layers (endoderm and ectoderm) only; they include cnidarians such as *Hydra* and jellyfish.

diploid describes cells that contain two sets of homologous chromosomes, one from each parent, and thus two copies of each gene.

directed dilation the extension of a tube-like structure at each end due to hydrostatic pressure, the direction of extension reflecting greater circumferential resistance to expansion.

distal describes the end of a structure, such as a limb, that is furthest away from the point of attachment to the body.

distal visceral endoderm (DVE) the endoderm located at the distal end of the cup-shaped mouse epiblast, which moves and extends away from the posterior side of the embryo, upon which it becomes the anterior visceral endoderm (AVE).

DNA methylase enzyme that adds methyl groups to DNA.

DNA methylation covalent attachment of methyl groups to DNA, which alters the ability of the DNA to be transcribed.

DNA microarray analysis technique for detecting and measuring the expression of large numbers of genes simultaneously, by hybridization of cellular RNA or cDNA to arrays of oligonucleotides representing gene sequences.

dominant describes an allele that determines the phenotype even when present in only a single copy.

dominant-negative describes a mutation that inactivates a particular cellular function by the production of a defective RNA or protein molecule that blocks the normal function of the gene product.

dorsal describes the back of an embryo or animal, the side opposite to the belly or underside.

dorsal closure in *Drosophila*, the bringing together of the two edges of the dorsal epidermis over the amnioserosa, to fuse along the dorsal midline of the embryo.

dorsal convergence in zebrafish, directed migration of individual mesodermal cells from lateral regions towards the midline to form the notochord.

dorsal ectoderm one of the four main regions into which the dorso-ventral axis of the early *Drosophila* embryo is divided before gastrulation. This region will give rise to epidermis.

dorsalized describes embryos that have developed greatly increased dorsal regions at the expense of ventral regions, as a result of experimental manipulation.

dorsalizing factor in the early vertebrate embryo, a protein or RNA that promotes the formation of dorsal structures, such as maternal components of the Wnt signaling pathway in *Xenopus*.

dorso-lateral hinge point site at each side of the developing neural tube at which it bends to bring the tips of the neural folds towards each other, to eventually fuse and close the neural tube.

dorso-ventral axis the body axis running from the upper surface or back (dorsal) to the under surface (ventral) of an organism or structure. The mouth is always on the ventral side.

dosage compensation the mechanism that ensures that, although the number of X chromosomes in males and females is different, the level of expression of X-chromosome genes is the same in both sexes. Mammals, insects, and nematodes all have different dosage-compensation mechanisms.

ecdysis molting in arthropods, in which the external cuticle is shed to allow for growth.

ecdysone a steroid hormone in insects that is responsible for initiating molting and also the transition to pupation and metamorphosis.

ectoderm the embryonic germ layer that gives rise to the epidermis and the nervous system.

ectomesenchyme cells of ectodermal origin (neural crest) that give rise to bone, cartilage, and connective tissue of the face.

egg chamber the structure in a female *Drosophila* within which an oocyte develops surrounded by its nurse cells and follicle cells.

egg cylinder the cylindrical structure comprising the epiblast covered by visceral endoderm in early post-implantation mouse embryogenesis.

electroporation technique for introducing DNA into cells within a small region of an embryo by subjecting the area to pulses of electric current delivered through fine electrodes, which makes neighboring cell membranes more permeable to the injected DNA.

embryogenesis the process of development of the embryo from the fertilized egg.

embryology the study of the development of an embryo.

embryonic–abembryonic axis in the mammalian blastocyst the axis running from the site of attachment of the inner cell mass (the embryonic pole), to the opposite pole (the abembryonic pole).

embryonic endoderm the layer of cells in the mammalian embryo that gives rise to the endoderm of the embryo.

embryonic-lethal mutation any mutation that causes the death of the embryo.

embryonic organizer alternative name for the Spemann organizer in amphibians and similar organizing regions in other vertebrates (such as the node in chick) that can direct the development of a complete embryo.

embryonic stem cells (ES cells) pluripotent cells derived from the inner cell mass of a mammalian embryo, most commonly mouse embryos, that can be indefinitely maintained in culture. When injected into another blastocyst, they combine with the inner cell mass and can potentially contribute to all the tissues of the embryo.

endoblast in the chick embryo, the layer of cells that grows out from the posterior marginal zone prior to primitive streak formation, and replaces the hypoblast underlying the epiblast.

endocardial cushions localized bulges on opposite sides of the inner wall of the heart tube that develop into the septa that divide the developing heart tube transversely into atrial and ventricular chambers.

endocardium the inner endothelial layer of the developing heart.

endochondral ossification the replacement of cartilage with bone in the growth plates of vertebrate embryonic skeletal elements, such as the long bones of the limbs.

endoderm the embryonic germ layer that gives rise to the gut and associated organs, such as the lungs and liver in vertebrates.

endodermis a tissue layer in plant roots interior to the cortex and outside the vascular tissue.

endomesoderm *see* **mesendoderm**.

endosperm nutritive tissue in higher plant seeds that serves as a source of food for the embryo.

endothelial cell cell type that makes up the endothelium lining blood vessels.

endothelium the epithelium lining blood vessels.

enhancer site in a gene control region to which activating proteins bind to switch on the gene, especially in respect of highly regulated tissue-specific genes.

enhancer-trap technique used in *Drosophila* to turn on the expression of a specific gene in a particular tissue or stage in development.

Eph receptor cell-surface protein that is the receptor for an **ephrin** protein. Interactions of ephrins and their receptors can cause repulsion of cells or attraction and adhesion, and are involved, for example, in delimiting compartments in rhombomeres and in axonal guidance.

ephrin *see* **Eph receptor**.

epiblast in mouse and chick embryos, a group of cells within the blastocyst or blastoderm, respectively, that gives rise to the embryo proper. In the mouse, it develops from cells of the inner cell mass.

epiblast stem cells (epiSCs) stem cells derived from a later-stage mouse epiblast than the inner cell mass when injected into a blastocyst. They can be maintained indefinitely in culture and can give rise to a wide range of cell types, but unlike embryonic stem cells they cannot give rise to chimeras when injected into another blastocyst.

epiboly the process during gastrulation in which the ectoderm extends to cover the whole of the embryo.

epicardium the outer lining of the heart, overlying the heart muscle.

epidermis in vertebrates, insects, and plants the outer layer of cells that forms the interface between the organism and its environment. Its structure is quite different in the different organisms.

epigenetic describes mechanisms of gene regulation that involve the modification of chromatin by, e.g. DNA methylation, histone methylation, and histone acetylation.

epimorphosis a type of regeneration in which the regenerated structures are formed by new growth.

epiSCs *see* *epiblast stem cells*.

epithelial-to-mesenchymal transition (EMT) the process by which epithelial cells lose adhesiveness and detach from the epithelium as single cells to form mesenchyme.

epithelium (plural **epithelia**) a sheet of cells tightly bound to each other by adhesive cell junctions.

erythroid lineage the lineage of blood cells that gives rise to red blood cells (erythrocytes) and megakaryocytes.

ES cells *see* *embryonic stem cells*.

euchromatin chromatin in which genes are available for transcription.

evagination the turning of a pouch or tubular structure inside out and its protrusion from the surface of a structure or the body.

evo-devo an informal term for the study of the evolution of development.

extra-embryonic ectoderm in mammals, embryonic tissue that contributes to the formation of the placenta.

extra-embryonic membranes membranes external to the embryo proper that are involved in its protection and nutrition. In mammals they include the amnion, chorion, and placental tissues.

Fallopian tubes the name given to the oviducts in mammals.

fate describes what a cell will normally develop into. Having a particular fate does not, however, imply that a cell could not develop differently if placed in a different environment.

fertilization the fusion of sperm and egg to form the zygote.

fertilization membrane a membrane formed around the eggs of some species after fertilization to prevent entry of further sperm.

fetus the name given to a human embryo after 8 weeks gestation.

file in a plant root, a vertical column of cells that originates from a single initial in the root meristem.

first heart field the cardiac crescent, a region of prospective heart cells located in the splanchnic mesoderm on either side of the body underlying the head folds. These regions come together to form the heart tube. Cells from the cardiac crescent are the first to differentiate into heart muscle cells.

floor plate a small region of the developing neural tube at the ventral midline that is composed of non-neural cells. It is involved in patterning the ventral part of the neural tube.

floral meristem a region of dividing cells at the tip of a shoot that gives rise to a flower.

floral organ identity gene any of a number of genes in plants that give the different parts of a flower (petal, stamen, etc) their individual identities.

floral organ primordium parts of a floral meristem that have become specified to give rise to the individual parts of a flower (petal, stamen, etc).

follicle the structure in the vertebrate ovary that contains an egg cell and its supporting somatic cells (follicle cells). In *Drosophila*, follicle cells are the somatic cells that surround the oocyte and nurse cells in the egg chamber during egg development.

follicle cells *see* *follicle*

forebrain the anterior part of the vertebrate embryonic brain that will give rise to the cerebral hemispheres, the thalamus, and the hypothalamus. Also called the prosencephalon.

forward genetics type of genetic analysis in which a mutant organism is first identified by its unusual phenotype and then genetic experiments are done to discover which gene is responsible for the mutant phenotype.

founder cells the small number of cells deriving from the apical meristem that give rise to a plant organ such as a flower or leaf.

functional area any of the areas of the cerebral cortex that are dedicated to a particular function, such as the visual cortex, motor cortex, or sensory cortex.

gamete cell that carries the genes into the next generation—in animals it is the egg or sperm.

ganglion mother cell a cell formed by division of a neuroblast in *Drosophila* and which gives rise to neurons.

gap gene any of a number of zygotic genes coding for transcription factors expressed in early *Drosophila* development that subdivide the embryo into regions along the antero-posterior axis.

gap junction type of cell junction in which there is direct communication between the cytoplasms of adjoining cells, and through which ions and small molecules can pass from cell to cell.

gastrula the stage in animal development at which prospective endodermal and mesodermal cells of the blastula or blastoderm move inside the embryo.

gastrulation the process in animal embryos in which prospective endodermal and mesodermal cells move from the outer surface of the embryo to the inside, where they give rise to internal organs.

gene cluster, gene complex a group of related genes closely linked on a chromosome that usually constitute a functional unit. Examples are the Hox gene clusters in animals, in which sequential expression of genes of each cluster specify positional values along the antero-posterior axis of the embryo.

gene expression the process of gene activation, transcription, and translation that produces a functional protein or RNA from a gene. Transcription of the gene can be detected by *in situ* nucleic acid hybridization while production of the protein is detected by *in situ* immunostaining.

gene knockdown *see* *gene silencing*.

gene knock-in the introduction of a new functional gene into the genome using techniques involving homologous recombination and transgenesis.

gene knock-out the complete and permanent inactivation or deletion of a particular gene in an organism by means of genetic manipulation.

gene-regulatory network the interactions of genes, especially transcription factor genes, expressed during a particular developmental process. Such networks can be represented as flow diagrams.

gene-regulatory protein general term for any protein involved in switching genes on and off.

gene silencing the switching off of a gene (or block of genes) by epigenetic modifications to chromatin, by microRNAs, or by RNA interference. Unlike gene knock-out, it does not affect the structure of the gene itself but prevents its transcription or the translation of the mRNA.

general transcription factor gene-regulatory protein involved in the expression of many different genes. These transcription factors form part of the transcription initiation complex; they bind to RNA polymerase and position it in the correct place on the DNA to start transcription.

generative program the instructions for development contained in the genome that determine when and where the proteins that control cell behavior, and thus development, are produced.

genetic equivalence the fact that all the somatic cells in the body of a multicellular organism contain the same set of genes.

genital ridge the region of mesoderm lining the abdominal cavity from which the gonads develop in vertebrates.

genome the hereditary information of a particular organism including its complete set of genes and non-coding DNA sequences.

genomic imprinting the process in which certain genes are switched off in either the egg or the sperm during their development, and remain silenced in the genome of the early embryo. This imprinting occurs during gamete formation, and is probably due to differential DNA methylation of the gene during egg or sperm production.

genotype the exact genetic constitution of a cell or organism in terms of the alleles it possesses for any given gene.

germ band the ventral blastoderm of the early *Drosophila* embryo, from which most of the trunk region of the embryo will develop.

germ-band extension the extension of the germ band of the *Drosophila* embryo towards the posterior and onto the dorsal side, which occurs during gastrulation. The germ band later retracts as development is completed.

germ cells cells in an animal that gives rise to eggs or sperm.

germ layers the regions of the early animal embryo that will give rise to distinct types of tissue. Most animals have three germ layers—ectoderm, mesoderm, and endoderm.

germ ring in zebrafish embryos, the thickened edge of the blastoderm when it has spread out to cover the upper half of the yolk.

germarium reproductive structure in the adult female *Drosophila* containing stem cells that give rise to a succession of egg chambers, each containing an oocyte.

germline cells the cells that give rise to the gametes—eggs and sperm.

germline cyst the 16-cell structure in adult female *Drosophila* that contains an oocyte precursor and nurse cells, before meiosis occurs.

germplasm the special cytoplasm in some animal eggs, such as those of *Drosophila*, that is involved in the specification of germ cells.

glia, glial cells the non-neuronal supporting cells of the nervous system, such as the Schwann cells of the peripheral nervous system and the astrocytes of the brain.

globular stage the stage at which a plant embryo is a ball of around 32 cells.

glycosylation the addition of carbohydrate side-chains to a protein, a common modification in membrane and extracellular proteins that occurs after their translation.

gonad reproductive organ in animals. The gonads comprise the ovary (female) and the testis (male).

gonadotropin-releasing hormone (GnRH) protein hormone secreted by the hypothalamus and acts on the pituitary to stimulate the release of gonadotropins, which in turn increase the production of the steroid sex hormones at puberty.

granulocyte colony-stimulating factor (G-CSF) hematopoietic growth factor that helps promote the differentiation of granulocytes from the granulocyte–macrophage progenitor cell.

granulocyte–macrophage colony-stimulating factor (GM-CSF) hematopoietic growth factor that is required for the development of most myeloid cells from the earliest progenitors that can be identified.

gray matter regions of the brain and spinal cord that are mainly composed of the cell bodies of neurons.

growth an increase in size, which can occur by cell multiplication, increase in cell size and deposition of extracellular material.

growth cone the region at the end of the extending axon of a developing neuron. The growth cone crawls forward on the substratum and senses its environment by means of filopodia.

growth hormone protein hormone produced by the pituitary gland that is essential for the post-embryonic growth of humans and other mammals.

growth hormone-releasing hormone protein hormone produced by the hypothalamus that stimulates the production of growth hormone by the pituitary.

growth plate cartilaginous region in the long bones of vertebrates at which growth of the bone occurs. The cartilage grows in size and is eventually replaced by bone by the process of endochondral ossification.

gynogenetic an embryo in which the two sets of homologous chromosomes are both maternal in origin.

hand-plate the region at the end of the developing vertebrate limb from which the digits will develop. Also called the digital-plate.

haploid in diploid organisms, describes cells that contain a single set of chromosomes and that are derived from diploid cells by meiosis. In most animals the only haploid cells are the gametes—the sperm or egg.

head the structure located at the anterior end of a bilaterally symmetrical animal, such as an arthropod or a vertebrate, that typically houses the brain (or equivalent), various sense organs, and the mouth.

head fold an infolding of the three germ layers in the head region of the gastrula in chick and mammalian embryos that indicates the start of the development of the pharynx and the foregut.

head process the anterior end of the notochord that projects into the head of mammalian and avian embryos.

heart stage a stage in embryogenesis in dicotyledonous plants in which the cotyledons and embryonic root are starting to form, giving a heart-shaped embryo.

Hedgehog a secreted signaling protein first identified in *Drosophila*, that is a member of an important family of developmental signaling proteins that includes Sonic hedgehog in vertebrates.

hematopoiesis the formation of blood cells from a multipotent stem cell. In adult vertebrates this occurs mainly in the bone marrow.

hematopoietic growth factor any of various secreted signaling proteins that induce differentiation of blood-cell progenitors into the various blood cell types.

hematopoietic stem cell a multipotent stem cell in the bone marrow of vetebrates that gives rise to all the blood cells.

hemidesmosome type of adhesive cell junction through which cells adhere to the extracellular matrix. Integrins in the cell membrane bind to proteins such as laminin in the extracellular matrix.

Hensen's node a condensation of cells at the anterior end of the primitive streak in chick embryos. It corresponds to the Spemann organizer in amphibians. The equivalent region in mammals is just called the node. Cells of the node give rise to the prechordal plate and the notochord in chick embryos and to the notochord in mammals.

hermaphrodite an organism that possesses both male and female gonads and produces both male and female gametes.

heterochromatin the state of chromatin in which transcription of the DNA is not possible.

heterochronic *see* **heterochrony**

heterochrony an evolutionary change in the timing of developmental events. A mutation that changes the timing of a developmental event is called a **heterochronic** mutation.

heterozygous describes the state in which a diploid organism carries two different alleles of a given gene, one inherited from the father and one from the mother.

hindbrain the most posterior part of the embryonic brain, which gives rise to the cerebellum, the pons, and the medulla oblongata. It is also called the rhomboencephalon.

histone protein any of a family of proteins that are packaged with DNA to form chromatin. Chemical modifications of histones affect the ability of the DNA in the modified chromatin to be transribed.

HOM-C complex the name of the Hox gene clusters in *Drosophila*.

homeobox a short region of DNA in certain genes that encodes a DNA-binding domain called the homeodomain. Genes containing this motif are known generally as homeobox genes. The homeodomain is present in a large number of transcription factors that are important in development, such as the products of the Hox genes and the Pax genes.

homeobox gene *see* **homeobox**.

homeodomain *see* **homeobox**.

homeosis the phenomenon in which one structure is transformed into another, homologous, structure as a result of a mutation. An example of such a homeotic transformation is the development of legs in place of antennae in *Drosophila* as a result of mutation.

homeotic gene a gene which when mutated can result in a homeotic transformation. Examples are the Hox genes.

homeotic selector genes in *Drosophila*, genes that specify the identity and developmental pathway of a group of cells. They encode homeodomain transcription factors and act by controlling the expression of other genes. Their expression is required throughout development. The *Drosophila* gene *engrailed* is an example of a homeotic selector gene.

homeotic transformation *see* **homeosis**.

homologous genes genes in different species that share significant similarity in their nucleotide sequence and are derived from a common ancestral gene.

homologous recombination the recombination of two DNA molecules at a specific site of sequence similarity.

homology morphological or structural similarity due to common ancestry.

homozygous describes the state in which a diploid organism carries two identical alleles of a given gene.

Hox clusters, Hox complex *see* **Hox genes**

Hox code the combinations of Hox genes that specify different regions along the antero-posterior axis.

Hox genes a family of homeobox-containing genes that are present in all animals except sponges (as far as is known) and are involved in patterning the antero-posterior axis. In many animals they are clustered on the chromosomes in one or more gene clusters (gene complexes). Combinatorial expression of different Hox genes characterizes different regions or structures along the axis.

hypoblast a sheet of cells in the early chick embryo that covers the yolk under the blastoderm and gives rise to extra-embryonic structures such as the stalk of the yolk sac.

hypocotyl the seedling stem that develops from the region between the embryonic root and the future shoot.

hypodermis the name given to the epidermis in nematodes.

hypophysis a cell in some plant embryos that is recruited from the suspensor and contributes to the embryonic root meristem and root cap.

hypothalamus region of the brain containing neurosecretory cells that produce and secrete peptide hormones such as growth hormone-releasing hormone, which act on cells in the pituitary to stimulate the release of the relevant hormone.

IGF *see* **insulin-like growth factors**

imaginal discs small sacs of epithelium present in the larva of *Drosophila* and other insects, which at metamorphosis give rise to adult structures such as wings, legs, antennae, eyes, and genitalia.

immunoglobulin superfamily large family of molecules that contain immunoglobulin-like domains, and which includes some cell-adhesion molecules such as N-CAM.

imprinting *see genomic imprinting*.

indeterminate describes growth in plants that do not make a fixed number of leaves or flowers.

induced pluripotent stem cells (iPS cells) pluripotent stem cells that are produced by the conversion of differentiated somatic cells by the introduction and expression of a few specific transcription factor genes that induce the cell to revert to a state of pluripotency.

induced transdifferentiation potential technique of regenerative medicine in which differentiated cells would be induced to change into another differentiated cell type. This has been achieved experimentally in the conversion of mouse liver cells to insulin-producing cells.

induction is the process whereby one group of cells signals to another group of cells in the embryo and so affects how they will develop.

inflorescence a flowerhead or flowering shoot in plants.

inflorescence meristem meristems that develop from vegetative shoot meristems and produce flowering shoots.

ingression the movement of individual cells from the outside of the embryo into the interior during gastrulation. In mammalian and avian embryos cells detach from the epiblast surface and ingress through the primitive streak.

initial any cell in the meristem of a plant that is able to divide continuously, giving rise both to dividing cells that stay within the meristem and to cells that leave the meristem and go on to differentiate.

initiation of transcription a crucial step in the expression of a gene. In most developmental genes it is under tight control so that genes are not expressed at the wrong time and place.

inner cell mass a discrete mass of cells in the blastocyst of the early mammalian embryo which is derived from the inner cells of the morula, and which will give rise to the embryo proper and some extra-embryonic membranes.

instar the larval phase between each molt in animals in which the larva goes through successive phases of growth and molting before developing into an adult.

instructive describes induction in which cells respond differently to different concentrations of the inducing signal.

insulin-like growth factors 1 and 2 (IGF-1 and 2) polypeptide growth factors that stimulate cell division in the early embryo and which after birth mediate many of the effects of growth hormone and are essential for post-natal growth in mammals.

integrin any of a family of cell-adhesion molecules by which cells attach to the extracellular matrix.

intercalary growth growth that occurs in animals capable of epimorphic regeneration when two pieces of tissue with different positional values are placed next to each other. The intercalary growth replaces the intermediate positional values.

interdigital regions the regions between the forming digits in the developing hand or foot.

interfollicular epidermis the epidermis of the skin between hair follicles.

intermediate filament one of the three principal types of protein filaments of the cytoskeleton. Intermediate filaments are involved in strengthening tissues such as epithelia.

intermediate mesoderm the region of mesoderm between the dorsal mesoderm that develops into notochord and somites and the lateral plate mesoderm. The kidneys develop from the intermediate mesoderm.

interneuron a type of neuron in the central nervous system that relays signals from one neuron to another.

internode that portion of a plant stem between two nodes (sites at which a leaf or leaves form).

intracellular signaling the process by which an extracellular signal received by a receptor at the cell's surface is relayed onward to its final destination inside the cell by a series of intracellular signaling proteins that interact with each other.

invagination the local inward deformation of a sheet of embryonic epithelial cells to form a bulge-like structure, as in early gastrulation in the sea-urchin embryo.

involution a type of cell movement that occurs at the beginning of amphibian gastrulation, when a sheet of cells enters the interior of the embryo at the blastopore by rolling in under itself.

iPS cells *see induced pluripotent stem cells*.

ipsilateral the same side of the body.

juvenile hormone hormone in insects that maintains the larval state and prevents premature metamorphosis. Metamorphosis follows a drop in the level of juvenile hormone.

keratinocyte differentiated epidermal skin cell that produces keratin, eventually dies, and is shed from the skin surface.

Koller's sickle a crescent-shaped region of small cells lying at the front of the posterior marginal zone in the chick blastoderm.

lateral geniculate nuclei in mammals, paired regions (one on each side) in the brain at which most of the axons from the retina terminate.

lateral inhibition the mechanism by which cells inhibit neighboring cells from developing in a similar way to themselves.

lateral line cells sensory organ precursor cells that migrate from the cranial ectoderm to form the lateral-line organs in fishes.

lateral-line primordium group of cells arising from the cranial ectoderm in fishes that migrates collectively along the route of the future lateral line, depositing sensory-organ precursor cells, that will form the lateral-line organs, as it moves.

lateral motor column longitudinal column of motor neurons, present in both sides of the spinal cord in the brachial and lumbar regions. Its neurons innervate the fore- and hindlimbs.

lateral plate mesoderm mesoderm in vertebrate embryos that lies lateral and ventral to the somites and gives rise to the tissues of the heart, kidney, gonads, blood, and the limb connective tissues.

lateral shoot meristem meristem that arises from the apical shoot meristem and gives rise to lateral shoots.

leaf primordium small set of cells at the edge of the shoot apical meristem from which a leaf develops.

left–right asymmetry the bilateral asymmetry of arrangement and structure of most internal organs in vertebrates. In mice and humans, for example, the heart is on the left side, the right lung has more lobes than the left, and the stomach and spleen lie to the left.

life history the life cycle of an organism or species viewed in terms of its reproductive strategy and its unique ecology or interaction with the environment.

ligand general term for any molecule that binds to another molecule.

ligand-mediated apoptosis programmed cell death induced by an external signal molecule interacting with a cell-surface receptor.

limb bud small structures that grow out from the flank of the vertebrate embryo and develop into limbs.

lineage (1) the ancestry of a given cell, i.e. the progenitor cells and the sequence of cell divisions that gave rise to the cell in question. (2) The genealogy of an organism or species.

local organizer general term for any signaling center with localized effects on neighboring tissue (compared with the embryonic organizer), e.g. the signaling centers at the midbrain–hindbrain boundary, the anterior neural ridge, and the zona limitans intrathalamica in the developing vertebrate brain.

locus control region *cis*-regulatory region that controls the sequential expression of the genes in a multigene locus, such as a globin gene cluster or a Hox cluster, and which is located quite far from the genes that it controls.

long-germ development type of insect development in which the embryonic blastoderm gives rise to the whole of the future embryo, as in *Drosophila*.

lymphoid lineage the lymphocytes of the immune system, which derive from the multipotent hematopoietic stem cell via the lymphoid progenitor cell.

macromere the larger of the cells that result from an unequal cleavage division in certain embryos, such as those of sea urchins.

macrophage colony-stimulating factor (M-CSF) hematopoietic differentiation factor that promotes the differentiation of macrophages from the granulocyte–macrophage progenitor cell.

mantle zone the layers of neuronal cell bodies outside the ventricular zone, that are formed by neuronal migration.

marginal zone the belt-like region of presumptive mesoderm at the equator of the late blastula of an amphibian embryo.

master regulatory gene gene, usually encoding a transcription factor, whose expression is necessary and sufficient to trigger activation of many other genes in a specific program leading to the development of a particular cell type, tissue or organ.

maternal-effect gene see *maternal-effect mutation*.

maternal-effect mutation a mutation in the mother that has no effect on the phenotype of the mother but affects the development of the egg and later the embryo. Genes affected by such mutations are called maternal-effect genes.

maternal factor protein or RNA that is deposited in the egg by the mother during oogenesis. The production of these maternal proteins and RNAs is under the control of so-called **maternal genes**.

maternal gene in general terms, any gene inherited from the mother. *See also* **maternal-effect gene**.

medial at or towards the midline.

medial column longitudinal column of motor neurons, running along the ventral midline of the spinal cord.

median hinge point the midline of the neural plate at which the plate bends to start to form the neural tube.

medio-lateral axis the axis in vertebrates that runs from the midline to the periphery.

medio-lateral intercalation cell movements that occur during convergent extension in amphibian gastrulation. Cells push in sideways between their neighbors, resulting in the extension and narrowing of the cell sheet.

medulla oblongata brain region that develops from part of the hindbrain and which is involved in regulating unconscious activities such as heartbeat, breathing, and blood pressure.

meiosis a special type of cell division that occurs during formation of sperm and eggs, and in which the number of chromosomes is halved from diploid to haploid.

mericlinal chimera plant in which a genetically marked cell gives rise to a sector of an organ or of the whole plant.

meristem a group of undifferentiated, dividing cells that persist at the growing tips of plants. They give rise to all the adult structures— shoots, leaves, flowers, and roots.

meristem identity genes in plants, genes that specify whether a meristem is a vegetative or an inflorescence meristem.

mesectoderm tissue that can give rise to both ectoderm and mesoderm.

mesenchymal stem cell cell derived from bone marrow that can be differentiated in culture into a variety of cells such as chondrocytes, osteoblasts, and adipocytes.

mesenchymal-to-epithelial transition (MTE) the aggregation of mesenchyme cells to form an epithelium. Occurs, for example, in the formation of tubular structures such as kidney tubules and blood capillaries.

mesenchyme loose connective tissue, usually of mesodermal origin, whose cells are capable of migration; some epithelia of ectodermal origin, such as the neural crest, undergo an epithelial-to-mesenchymal transition.

mesendoderm tissue that gives rise to both endoderm and mesoderm.

mesoderm germ layer that gives rise to the skeleto-muscular system, connective tissues, the blood, and internal organs such as the kidney and heart.

mesonephros an embryonic kidney in mammals that contributes to the male and female reproductive organs.

messenger RNA (mRNA) the RNA molecule that specifies the sequence of amino acids in a protein. It is produced by transcription from DNA.

metamorphosis the process by which a larva is transformed into an adult. It often involves a radical change in form and the development of new organs, such as wings in butterflies and limbs in frogs.

metastasis the movement of cancer cells from their site of origin to invade underlying tissues and to spread to other parts of the body.

microfilament see *actin filament*.

micromere the smaller cell that results from an unequal cleavage division during early animal development.

microRNAs (miRNAs) small RNAs that suppress the expression of specific genes.

microtubule one of the three principal protein filaments of the cytoskeleton. They are involved in the transport of proteins and RNAs within cells.

mid-blastula transition in amphibian embryos, the stage in development when the embryo's own genes begin to be transcribed, cleavages become asynchronous, and the cells of the blastula become motile.

midbrain the middle section of the embryonic vertebrate brain that gives rise to the tectum (in amphibians and birds) and similar structures in mammals, which are the sites of integration and relay centers for signals coming to and from the hindbrain, and also for inputs from the sensory organs. Also alled the mesencephalon.

midbrain–hindbrain boundary the site of a signaling center in the embryonic brain that helps to pattern the brain along the antero-posterior axis.

miRNAs see *microRNAs*.

mitosis the nuclear division that occurs during the proliferation of somatic diploid cells, and results in both daughter cells having the same diploid complement of chromosomes as the parent cell.

mitotic spindle the microtubule-based apparatus in a dividing cell that partitions the chromosomes equally to daughter nuclei at mitosis.

model organisms the small number of species that are commonly studied in developmental biology and about whose development most is known.

molting the shedding of an external cuticle when arthropods grow, and its replacement with a new one.

morphallaxis a type of regeneration that involves repatterning of existing tissues without growth.

morphogen any substance active in pattern formation whose spatial concentration varies and to which cells respond differently at different threshold concentrations.

morphogenesis the processes involved in bringing about changes in form in the developing embryo.

morphogenetic furrow in *Drosophila* eye development, a furrow that moves across the eye disc and initiates the development of the ommatidia.

morpholino antisense RNA a type of antisense RNA composed of stable morpholino nucleotide analogs.

morula the very early stage in a mammalian embryo when cleavage has resulted in a solid ball of cells.

mosaic a term used historically to describe the development of organisms that appeared to develop mainly by distribution of localized cytoplasmic determinants.

motor neuron neuron that innervates muscle and controls muscle ontraction and thus movement.

Müllerian ducts tubules that runs adjacent to the Wolffian ducts in the mammalian embryo and become the oviducts in females.

Müllerian-inhibiting substance protein secreted by the developing testis that induces regression of the Müllerian ducts in males.

multipotent describes a cell that can give rise to many different types, but not all types, of differentiated cell.

mural trophectoderm the trophectoderm in the blastocyst that is not in contact with the cells of the inner cell mass.

muscle fiber fully differentiated muscle cell, a multinucleate cell that produces muscle specific proteins such as musle actin and myosin that enable the muscle cell to undergo contraction.

myeloid lineage blood cells such as granulocytes, monocytes/macrophages, and mast cells which derive from the multipotent hematopoietic stem cell via the erythroid–myeloid progenitor cell.

myoblast a committed but still undifferentiated muscle cell.

myocardium the contractile layer of the developing heart, composed of muscle cells called cardiomyocytes.

myoplasm a special cytoplasm in ascidian eggs that is involved in the specification of muscle cells.

myotome that part of the somite that gives rise to muscle.

myotube multinucleate cell that is an intermediate stage in muscle cell differentiation. It develops from the fusion of myoblasts and will develop into a muscle fiber.

necrosis a type of cell death due to pathological damage in which cells break up, releasing their contents.

negative feedback loop a type of regulation in which the end-product of a pathway or process inhibits an earlier stage.

neoblasts a population of small, undifferentiated cells in planarians, which are the only cells in the adult that are mitotically active. They not only self-renew but also can differentiate into all the different cell types—epidermis, muscle, neurons, and germ cells.

neoteny the phenomenon in which an animal acquires sexual maturity while still in larval form.

nerve impulse see *action potential*.

netrin a secreted protein that act as a guidance molecule for neurons in the nervous system; netrins can be either attractant or repellent.

neural crest cells in vertebrates, a population of cells derived from the edge of the neural plate. They migrate from the dorsal neural tube to different regions of the body and give rise to a wide variety of tissues, including the autonomic nervous system, the sensory nervous system, the pigment cells of the skin, and some cartilage in the head.

neural folds the two folds that rise up at each edge of the neural plate at the beinning of neurulation and will eventually fuse to form the neural tube, which gives rise to the nervous system.

neural furrow a longitudinal infolding along the midline of the neural plate epithelium, caused by localized changes in cell shape, which is one of the first signs of neurulation.

neural groove in neural tube formation, the longitudinal depression in the neural plate between the two neural folds.

neural plate an area of thickened dorsal ectodermal epithelium at the anterior of a vertebrate embryo that gives rise to the nervous system through the process of neurulation.

neural tube the ectodermally derived tubular structure that forms along the dorsal midline of a vertebrate embryo and gives rise to the nervous system.

neural tube closure the coming together and fusion of the dorsal tips of the neural folds to form the neural tube that occurs during neurulation.

neural tube defect any of a number of congenital defects, such as spinal bifida, that are caused by localized failures of the neural tube to close properly.

neuregulin any of a family of chemoattractant proteins that are involved in guiding the tangential migration of neurons in the developing brain.

neuroblast an embryonic cell that will give rise to neural tissue (neurons and glia).

neuroectoderm embryonic ectoderm with the potential to form neural cells.

neurogenesis the formation of neurons from their precursor cells.

neurogenin a transcription factor specifically expressed in prospective neurons.

neuroligin protein present in the post-synapptic membrane of some synapses, which interacts with the transmembrane protein β-neurexin in pre-synaptic membranes to promote the functional differentiation of the synapse.

neuromuscular junction the specialized synapse made between a motor neuron and a muscle fiber, at which the neuron stimulates muscle activity.

neuron the electrically excitable cell type in the nervous system, also called a nerve cell, which conveys information in the form of electrical signals to other neurons, to muscles and to some glandular cells. Neurons communicate with other cells through specialized structures called synapses, at which the electrical signal stimulates production of a chemical signal (a neurotransmitter) by the pre-synaptic cell. The neurotransmitter diffuses across a narrow space between the two cell membranes and binds to receptors on the post-synaptic cell.

neurotrophic factor, neurotrophin general name for a protein that is necessary for neuronal survival, such as nerve growth factor.

neurula the stage of vertebrate embryonic development at the end of gastrulation when the neural tube is forming.

neurulation the process in vertebrates in which the future brain and spinal cord are formed from the ectodermal neural plate. As a result largely of localized changes in cell shape, the neural plate develops a central groove (the neural groove) with folds rising up on either side (neural folds). The folds eventually meet and fuse along the midline to form a tubular structure (the neural tube) that develops into the brain and spinal cord. In birds and mammals, the neural plate gives rise to the brain and the spinal cord is formed from the stem zone.

Nieuwkoop center a signaling center on the dorsal side of the early *Xenopus* embryo. It forms in the dorsal vegetal region of the blastula as a result of cortical rotation, and helps to specify the position of the Spemann organizer, which is formed just above it.

Nodal, Nodal-related proteins a subfamily of the TGF-β family of signaling proteins of vetebrates. They are involved in all stages of development, but particularly in early mesoderm induction and patterning.

node (1) in avian and mammalian embryos, the embryonic organizing center analogous to the Spemann organizer of amphibians. It is also known as Hensen's node in birds. (2) In plants, that part of the stem at which leaves and lateral buds form.

non-autonomous, non-cell-autonomous describes gene expression in a cell that affects cells other than the cell in which the gene is expressed. Genes for secreted signaling proteins, for example, have non-autonomous effects.

non-coding RNA any of a large number of RNAs encoded in the genome that have no protein-coding function. Many act as gene-regulatory RNAs.

notochord a transient stiff, rod-like cellular structure in vertebrate embryos that runs from head to tail and lies centrally beneath the neural tube. It is derived from mesoderm and its cells eventually become incorporated into the vertebral column.

in situ **nucleic acid hybridization** technique used to detect where in the embryo particular genes are being transcribed. The mRNA is detected by its hybridization to a labeled single-stranded complementary DNA probe.

nucleosome a typical structural unit in chromatin, which consists of a core of histone proteins with DNA wrapped round it.

nurse cells germline cells surrounding the developing oocyte in *Drosophila* which synthesize proteins and RNAs that are deposited in the oocyte.

octant stage the eight-cell stage of a plant embryo.

ocular dominance columns columns of neurons in the visual cortex that respond to the same visual stimulus from either the left or the right eye.

ommatidia (singular **ommatidium**) the individual units in an insect compound eye, which each consist of a lens and a small number of photoreceptor cells and supporting cells. There are hundreds of ommatidia in a compound eye.

ontogeny the development of an individual organism.

oocyte an immature egg.

oogenesis the process of egg formation in the female.

oogonia diploid germ cells that divide by mitosis within the ovary before entering meiosis to produce the oocytes.

optic chiasm the site at which the retinal nerve fibers from the right eye cross over fibers from the left eye on their way to opposite sides of the optic tectum (chicks and amphibians). In mammals, the optic chiasm is the site at which the nerve fibers from an eye divide, some taking a contralateral course to the opposite side of the lateral geniculate nucleus (LGN), whereas some connect with the ipsilateral side of the LGN.

optic tectum the region of the brain in amphibians and birds where the axons from the retina terminate.

optic vesicle the precursor of the retina of the vertebrate eye and is derived from the wall of the forebrain.

oral referring to the mouth, or to the side of the body on which the mouth is located.

oral–aboral axis in sea urchins and other radially symmetrical organisms, an axis that runs from the centrally situated mouth to the opposite side of the body.

organizer, organizing region, organizing center a signaling center that directs the development of the whole embryo or of part of the embryo, such as a limb. In amphibians, the organizer usually refers to the Spemann organizer. The organizing center in plants refers to the cells underlying the central zone of the meristem, which maintains the stem cells of the central zone.

organogenesis the development of specific organs such as limbs, eyes, and heart.

osteoblast precursor cell from which differentiated bone cells are formed.

outer enveloping layer in zebrafish, the outermost layer, one cell thick, in the blastoderm stage of the early embryo. It disappears as the embryo develops.

ovariole the string of egg chambers produced in the *Drosophila* ovary.

ovary the internal reproductive structure in female animals; it produces the female germ cells, the oocytes.

oviduct the tube in female birds and mammals that transports the eggs from the ovaries to the uterus.

ovule the structure in plants that contains an egg cell.

P element transposable DNA element found in *Drosophila*. It is a short sequence of DNA that can become inserted in different positions within a chromosome and can also move to other chromosomes. This property has been exploited in the technique of P-element-mediated transformation for making transgenic flies.

P granules granules that become localized to the posterior end of the fertilized egg of *Caenorhabditis elegans*.

pair-rule gene any of a number of genes in *Drosophila* that are involved in delimiting parasegments. They are expressed in transverse stripes in the blastoderm, each pair-rule gene being expressed in alternate parasegments.

pallium part of the embryonic forebrain that develops into the cerebral cortex, the outer layers of the cerebral hemispheres, where the highest-level processing centers for sensory information, motor control, learning, and memory are located.

paralogous genes genes within a species that have arisen by duplication and divergence. Examples are the Hox genes in vertebrates, which comprise several paralogous gene subgroups made up of paralogous genes.

parasegments developmental units arranged along the body of the developing *Drosophila* embryo and which give rise to the segments of the larva and adult.

paraxial mesoderm mesoderm lying on either side of the dorsal midline and which gives rise to the somites.

parthenogenetic describes eggs that are able to develop into an embryo without fertilization.

PAR (partitioning) proteins a group of proteins, initially discovered in *Caenorhabditis elegans*, that are required for the correct positioning and orientation of mitotic spindles in the early cleavages to ensure division at the required place in the cell and in the required plane.

pattern formation the process by which cells in a developing embryo acquire identities that lead to a well-ordered spatial pattern of cell activities.

Pax genes family of genes encoding transcriptional regulatory proteins that contain both a homeodomain and another protein motif, the paired motif.

periclinal describes cell divisions in a plane parallel to the surface of the tissue.

periclinal chimera in plants, chimera in which one of the three meristem layers has a genetic marker that distinguishes it from the other two.

peripheral nervous system all of the nervous system apart from the brain and spinal cord.

perivitelline space the space between the vitelline membrane lining the egg case and the egg plasma membrane in the fertilized eggs and early embryos of insects and other animals.

permissive describes induction in which a cell makes only one kind of response to an inducing signal, and makes it when a given level of signal is reached.

phenotype the observable or measurable characters and features of a cell or an organism.

phocomelia a congenital abnormality in which proximal structures in a limb are absent, whereas distal ones, such as the digits, are present and can be almost normal.

photoperiodism the response of an organism to relative day length; in plants it is responsible for promoting flowering as days become longer.

phyllotaxis the way the leaves are arranged along a shoot.

phylogeny the evolutionary history of a species or group.

phylotypic stage the stage at which the embryos of the different vertebrate groups closely resemble each other. This is the stage at which the embryo possesses a distinct head, a neural tube, and somites.

placenta a structure that forms in the uterine wall at which the blood systems of mother and embryo form an interface with each other. Mammalian embryos (with the exception of the monotremes such as the egg-laying duck-billed platypus and echidna) are nourished by the mother by the passage of nutrients through the placenta.

placode a region of thickened epithelium, usually on the surface of the embryo, that gives rise to a particular structure. An example is the lens placode, which gives rise to the lens of the eye.

planar cell polarity the polarization of cells in the plane of the tissue.

plasmodesmata (singular **plasmodesma**) threads of cytoplasm that run through the cell wall and interconnect adjacent plant cells.

pluripotent describes a stem cell, such as an embryonic stem cell, that can give rise to all the types of cells in the body.

pluteus the larval stage of the sea urchin.

polar body a small cell that is a product of meiosis during egg development. Polar bodies take no part in embryonic development.

polar follicle cells cells that are specified at the anterior and posterior ends of a *Drosophila* egg follicle. They induce stalk formation between egg chambers and help position the oocyte at the posterior end of the follicle.

polar trophectoderm the trophectoderm in contact with the inner cell mass in the mammalian blastocyst.

polarity the property of a cell, structure or organism in which one end is different from the other.

polarizing region in developing chick and mouse limb buds, an area at the posterior margin of the bud which produces a signal specifying position along the antero-posterior axis. Also called the zone of polarizing activity (ZPA).

pole cells cells formed at the posterior end of the *Drosophila* blastoderm that are precursors of the germ cells.

pole plasm the cytoplasm at the posterior end of the *Drosophila* egg that is involved in specifying germ cells.

polydactyly the occurrence of extra digits on hands or feet.

polyspermy the entry of more than one sperm into the egg.

polytene chromosome giant chromosome that is formed by repeated DNA replication in the absence of cell division.

populational asymmetry a pattern of stem-cell division that has been proposed to explain the dynamics of the renewal of the epidermis.

positional information molecular information, in the form, for example, of a gradient of an extracellular signaling molecule, that can be interpreted by cells to provide the basis for pattern formation in the embryo.

positional value the property acquired by a cell by virtue of its position with respect to the boundaries of a field of positional information.

The cell then interprets this positional value according to its genetic constitution and developmental history, and develops accordingly.

positive-feedback loop a type of regulation in which the end-product of a pathway or process can activate an earlier stage.

posterior at, or in the direction of, the tail end of an embryo or animal.

posterior dominance, posterior prevalence the fact that the more posteriorly expressed Hox genes can inhibit the action of more anteriorly expressed Hox genes when they are expressed in the same region.

posterior marginal zone a dense region of cells at the edge of the blastoderm of the chick embryo that will give rise to the primitive streak.

post-mitotic neuron a neuron after its formation from a precursor cell, so-called because most neurons do not divide further once they are formed.

post-synaptic describes the side of a synapse that receives the signal.

post-translational modification any modification that occurs to a protein after it has been synthesized. The protein can, for example, be enzymatically cleaved, glycosylated, or acetylated.

preaxial polydactyly the occurrence of extra digits on the anterior side of the hands or feet, that is on the thumb or big toe side.

prechordal plate mesoderm the anterior-most mesoderm in the vertebrate embryo, located anterior to the notochord. It gives rise to various ventral tissues of the head.

preimplantation genetic diagnosis a means of determining the genotype of embryos produced by IVF before implantation without harming the embryo, which can be used if the embryo is known to be at risk of carrying a mutation for a genetic disease. One blastomere is removed from the embryo during its early cleavage *in vitro*, which does not affect its subsequent development. The DNA from this blastomere can be amplified and tested for the presence or absence of mutation.

prepattern (1) a basic pattern generated automatically in a structure. It may subsequently be modified during development. (2) a pattern laid down in a structure before its final differentiation.

pre-somitic mesoderm the unsegmented mesoderm between the node (in chick and mouse) and the already formed somites. It will form somites from its anterior end.

pre-synaptic describes that side of a synapse that generates the signal.

primary body formation in avian and mammalian embryos, the formation of the head and trunk.

primary cilium an immotile microtubule-based structure found on most vertebrate cells; it is the site of the Sonic hedgehog signaling.

primary mesenchyme in sea urchin embryos, the first mesodermal cells to enter the blastula at gastrulation, and which migrate along the interior wall and eventually lay down the skeletal rods of the sea urchin larval endoskeleton by secretion of matrix proteins

primary neurulation the folding of the neural plate to form the neural tube.

primary oocyte female germ cell that has entered meiosis.

primary trophoblast the layer of giant cells that invade the uterus wall at the implantation of the mammalian blastocyst.

primitive endoderm in mammalian embryos that part of the inner cell mass that contributes to extra-embryonic membranes.

primitive streak the site of gastrulation in avian and mammalian embryos and the forerunner of the antero-posterior axis. It is a strip of ingressing cells that extends into the epiblast from the posterior margin. Epiblast cells move through the streak into the interior of the embryo to form mesoderm and endoderm.

primordial germ cells precursor cells in the early embryo that represent the germline and will produce the germ cells.

primordium (plural primordia) minute undifferentiated growth that will give rise to a structure such as a tooth, leaf, flower or floral organ.

proembryo the two-celled stage in a plant embryo.

programmed cell death see **apoptosis**.

progress zone in the timing model of vertebrate limb development, an area at the tip of the limb bud where cells acquire positional values.

promeristem the central region of the meristem that contains cells capable of continued division—the initials.

promoter a region of DNA immediately preceding the coding sequence to which RNA polymerase binds to begin transcription of a gene.

pronephros the first primitive kidney that arises from the intermediate mesoderm in vertebrate embryos. It degenerates in the fetus in mammals, in which the adult kidney develops from the metanephros, but, together with the mesonephros, it functions as the adult kidney in some amphibians and fish.

proneural cluster small cluster of cells within the neuroectoderm in which one cell will eventually become a neuroblast.

proneural gene a gene that promoted a neural fate in neuroectoderm cells.

pronucleus the haploid nucleus of sperm or egg after fertilization but before nuclear fusion and the first mitotic division.

prothoracicotropic hormone (PTTH) protein hormone secreted by the insect brain that causes the secretion of the steroid hormone ecdysone, and the initiation of molting or pupation and metamorphosis.

proto-oncogene a gene that is involved in the regulation of cell proliferation and that can cause cancer when mutated or expressed under abnormal control.

protostomes those animals, such as insects, in which cleavage of the zygote is not radial and in which gastrulation primarily forms the mouth.

proximo-distal axis the axis of a limb or other appendage (such as a leaf in a plant) that runs from the point of attachment to the body or stem (proximal) to the tip of the appendage (distal).

pupa in *Drosophila* and other insects that undergo metamorphosis, a stage following the larval stages in which the organism can remain dormant for long periods and in which metamorphosis occurs.

quiescent center a central group of cells in a plant root-tip meristem that divide rarely but are essential for meristem function.

radial axis the axis running from the center of a structure to the circumference.

radial cleavage a type of cleavage that occurs at right angles to the egg surface and produces blastomeres sitting directly over each other.

radial glial cell elongated glial cell that spans the whole width of the neural tube wall and provides tracks for migrating neurons.

radial intercalation process that occurs in the multilayered ectoderm of an amphibian gastrula in which cells intercalate in a direction perpendicular to the surface, thus thinning and extending the cell sheet.

radial symmetry the symmetry around the central axis in cylindrical structures such as plant stems and roots.

rapid block to polyspermy see **block to polyspermy**.

reaction–diffusion mechanism that can produce self-organizing patterns of chemical concentrations, which has been proposed to underlie some types of patterning in development.

recessive describes a mutation that only changes the phenotype when both copies of a gene carry the mutation.

recombination exchange of DNA between homologous chromosomes during meiosis, which shuffles parental genes into new combinations in the haploid gametes.

redundancy the existence of different genes or pathways that can substitute for each other during development.

regeneration the ability of a fully developed organism to replace lost parts.

regenerative medicine an approach that aims to use stem cells and their derivatives to replace diseased tissue with healthy tissue.

regulation the ability of the embryo to develop normally even when parts are removed or rearranged. Embryos that can undergo regulation are described as **regulative**.

repressor gene-regulatory protein that helps to suppress gene activity when it binds to a particular site in a gene's control region.

retinoic acid small non-protein secreted signaling molecule with many roles in development.

reverse genetics an approach to genetic analysis that starts with the nucleotide sequence of a gene or amino acid sequence of a protein, and then uses that information to determine the gene's function.

rhombomere any of a sequence of compartments of cell-lineage restriction along the hindbrain of chick and mice embryos.

RNA interference (RNAi) a means of suppressing gene expression by promoting the destruction of a given mRNA by targeting a nuclease to it by means of a short complementary RNA called a short interfering RNA (siRNA).

RNA processing the process in eukaryotic cells in which newly transcribed RNAs are modified in various ways to make a functional messenger RNA or structural RNA. It includes RNA splicing, which removes introns from the transcript to leave a continuous coding messenger RNA.

RNA seq a technique used to determine which genes are expressed in a particular tissue or at a particular time in development, in which the total RNA in the cells is converted into cDNA and then sequenced.

Robo family of receptors on neurons that bind the chemorepellent Slit proteins.

robust the characteristic of development that it can withstand large variations in conditions and still continue normally.

roof plate a small strip of non-neural cells in the dorsal-most part of the neural tube around the midline, which is involved in patterning the dorsal part of the tube.

satellite cell undifferentiated stem cell present in adult muscle that can be reactivated to produce more skeletal muscle cells if the muscle is damaged.

sclerotome that part of a somite that will give rise to the cartilage of the vertebrae.

second heart field a second lineage of cells that gives rise to the heart, and exclusively to the outflow tract. Thee cells lie behind the cardiac crescent (the first heart field) in the pharyngeal mesoderm.

secondary body formation formation of the most posterior region of the body from the tailbud in mouse and chick embryos.

secondary neurulation in vertebrates, the formation of the posterior neural tube, beyond the lumbo-sacral region, as a solid rod of cells, derived from stem-like cells in the tailbud, which then develops an interior cavity or lumen that connects with that of the anterior neural tube.

segmentation the division of the body of an organism, or a particular structure, into a succession of morphological units—the segments—along the antero-posterior axis.

segmentation gene any of a number of genes in *Drosophila* involved in patterning the parasegments and segments.

selector gene a gene whose expression determines the behavior or properties of a group of cells, and whose continued expression is required to maintain that behavior.

semaphorin any of a family of secreted proteins that act as guidance cues for neurons.

semi-dominant describes a mutation that affects the phenotype when just one allele carries the mutation, but where the effect on the phenotype is much greater when both alleles carry the mutation.

senescence the impairment of function associated with aging.

sensory neuron a type of neuron that is triggered by signals from within the body (e.g. muscle stretch) or from the environment (e.g. touch, heat), and carries those signals to the central nervous system.

sensory organ precursor (SOP) an ectodermal cell that will give rise to a sensory bristle in the adult *Drosophila* epidermis.

septate junction a type of cell junction in invertebrate cells with a similar function to the vertebrate tight junction.

sex chromosome any chromosome that specifically determines sexual phenotype (e.g. the X and Y chromosomes in mammals).

sex determination the genetic and developmental process by which an organism's sex is specified.

sex-determining region of the Y chromosome (SRY) a genetic locus on the Y chromosome that determines maleness by specifying the gonad as a testis.

shield the name given to the shield-shaped embryonic organizer region in zebrafish embryos.

shield stage the stage in in zebrafish embryos in which the embryonic organizer (the shield) has been formed on one side of the blastoderm.

short interfering RNA (siRNA) in the phenomenon of RNA interference, a short single-stranded RNA that is exactly complementary to the mRNA of a given gene, and which becomes incorporated into a nuclease that can then target the mRNA for destruction.

short-germ development type of insect embryonic development in which the blastoderm only gives rise to the anterior segments of the embryo and the remaining segments are added sequentially during growth.

signal transduction the process by which a cell converts an extracellular signal received in one form, such as the binding of a molecule to a cell-surface receptor, into an intracellular signal of a different form, such as the phosphorylation of a cytoplasmic protein.

signaling center a localized region of the embryo that exerts a special influence on surrounding cells, usually by means of secreted signaling proteins, and thus determines how those cells develop.

silenced *see gene silencing.*

situs inversus in humans a rare condition in which there is complete mirror-image reversal of the positions of the internal organs.

Slit proteins family of secreted proteins that act as guidance cues to repel growing axons in the developing nervous system.

slow block to polyspermy reaction that follows fertilization in some animals, in which an impenetrable membrane (the fertilization membrane) is formed around the egg, preventing the entry of sperm.

somatic cells any cell in the body other than germ cells. In most animals, the somatic cells are diploid.

somatic cell nuclear transfer cloning of an animal by the transfer of a somatic cell nucleus into an enucleated egg, and the subsequent development of the egg into a new individual. The embryo that develops will have an identical nuclear genetic constitution to the somatic cell donor.

somatic transgenesis the introduction of an mRNA or a DNA construct directly into somatic cells to test the effect of overexpression of that gene. Also called transient transgenesis.

somatostatin a protein hormone produced by the hypothalamus that inhibits the production and release of growth hormone by the pituitary.

somites blocks of mesoderm that segment from the mesoderm on either side of the notochord. They give rise to trunk and limb muscles, the vertebral column and ribs, and the dermis.

specified describes the developmental status of a group of cells that when isolated and cultured in a minimal medium will develop according to their normal fate.

Spemann organizer, Spemann–Mangold organizer a signaling center on the dorsal side of the amphibian early embryo that acts as the main embryonic organizer. Signals from this center can organize new antero-posterior and dorso-ventral axes.

sperm the haploid male gamete in animals.

spermatogenesis the production of sperm.

sphere stage in zebrafish embryos the developmental stage at which the embryo consists of a hemispherical blastoderm of around 1000 cells lying over a spherical yolk.

spindle midzone the structure formed by the overlap of the microtubules of the two halves of a mitotic spindle; its position along the long axis of the cell marks the location of cell division.

spiral cleavage type of cleavage typical of molluscs and annelid worms, in which the plane of cell division is at a slight angle to the egg surface and blastomeres end up in a spiral arrangement.

stem cells undifferentiated cells that are both self renewing and also give rise to differentiated cell types. Stem cells are found in most adult tissues and may contribute to their repair. *See also **embryonic stem cells***.

stem-cell niche the cells adjacent to stem cells in tissues, and which provide a specialized environment that supports stem-cell maintenance and self-renewal.

stem zone in avian and mammalian embryos, an arc of self-renewing epiblast cells on either side of the primitive streak immediately posterior to the regressing node that give rise to the trunk neural tube and the medial parts of the somites.

subgerminal space the cavity that develops under the area pellucida in the early chick blastoderm.

subventricular zone a layer of proliferating cells under the ventricular layer in the forebrain neural tube which also gives rise to neurons and glial cells.

superior colliculus region of the brain in mammals to which some retinal neurons project. It corresponds to the optic tectum of amphibians and birds.

suspensor structure that attaches the plant embryo to maternal tissue and is a source of nutrients.

symmetric division a cell division in which the daughter cells are identical to each other and have the same fates.

synapse specialized cell–cell junction at which the axon terminal of a neuron transmits a signal to another neuron, or to a muscle cell, by means of a chemical neurotransmitter that diffuses across a small space (synaptic cleft) between the apposed cell membranes where it binds to receptors on the receiving cell. The transmitting cell is called the pre-synaptic neuron and the receiving cells are called the post-synaptic cells.

synaptic cleft the small space between the cell membrane of the transmitting cell and that of the receiving cell at a synapse.

synaptic vesicle small neurotransmitter-filled membrane-bounded vesicle in the axon terminals of neurons, which releases its contents into the synaptic cleft when an action potential (nerve impulse) generated at the cell body arrives at the terminal.

syncytial blastoderm the very early embryo in *Drosophila*, in which nuclear division after fertilization is not accompanied by cytoplasmic division, resulting in a blastoderm containing many nuclei in a common cytoplasm. The nuclei are arranged around the periphery of the embryo.

syncytium a cell with many nuclei in a common cytoplasm.

syndrome a collection of symptoms, often affecting different functions of the body, that characterize a particular pathological condition.

tailbud the structure at the posterior end of vertebrate embryos containing stem-like cells that give rise to the post-anal tail.

TALENS transcription activator-like effector nucleases. Nucleases that can be engineered to make targeted mutations in DNA.

tangential migration the migration of brain neurons parallel to the brain surface.

tectum brain structure, deriving from the embryonic midbrain, that is the site of integration and relay center for signals coming to and from the hindbrain, and also for inputs from the sensory organs, such as the eye.

telencephalon that part of the embryonic forebrain that gives rise to the pallium and the basal ganglia.

telomere structure composed of repetitive non-coding DNA at each end of a chromosome that prevents chromosomes sticking to each other and prevents gene loss at DNA replication. Telomeres typically become shorter at each cell division.

temperature-sensitive mutation a mutation that only causes a change in phenotype at a different temperature than normal, most commonly a higher temperature.

teratocarcinoma a solid malignant tumor that arises from germ cells, and which typically contains a mixture of differentiated cell types characteristic of more than one germ layer, as well as embryonal cancer cells. Undifferentiated mouse ES cells injected under the skin of an adult mouse have also been shown to give rise to teratocarcinomas.

teratogen any environmental agent (e.g. a chemical or an infectious agent) that causes abnormal embryonic development.

teratoma a capsulated solid tumor containing a mixture of tissues, often with differentiated cells, that are characteristic of more than one germ layer. It may arise from germ cells or pluripotent embryonic cells. Undifferentiated human ES cells injected under the skin of an immunocompromised mouse form teratomas, and this property has been used as a test of pluripotency.

terminally differentiated describes a cell that has become fully structurally and functionally differentiated and will not undergo any further differentiation. Terminally differentiated cells often do not undergo further cell division.

testis the internal male reproductive organ in animals; it produces the male germ cells, the sperm.

tetraploid describes a cell that contains four sets of chromosomes, or a species whose somatic cells contain four sets of chromosomes.

thalamus brain region, derived from the embryonic forebrain, which is a major relay station that distributes incoming sensory information to the appropriate region of the cortex.

therapeutic cloning the potential use of somatic-cell nuclear transfer to generate embryos from which ES cells can be derived. The goal is to create cells that are an exact genetic match to those of a patient and that could be used to alleviate disease.

thorax in insects, the middle part of the distinctly three-part body (head, thorax and abdomen). It follows the head, and in adults carries the legs and the wings, where the latter are present. In vertebrates, the thorax is the chest region.

threshold concentration that concentration of a chemical signal that can elicit a particular response from a cell.

thyroid hormones thyroxine (T4), and tri-iodothyronine (T3), produced by the thyroid gland, which are required generally for growth, and for metamorphosis in *Xenopus*.

thyroid-stimulating hormone hormone released by the pituitary gland that acts on the thyroid gland to stimulate production of thyroid hormones.

tight junction a type of adhesive cell junction that binds epithelial cells very tightly together to form an epithelium and that seals off the environment on one side of the epithelium from the other.

TILLING targeting-induced local lesions in genomes: a technique for detecting mutations by hybridizing the mutated DNA against unmutated DNA and detecting mismatched bases, which indicate the site of a mutation.

tissue engineering the building of replacement tissues *in vitro* for repair of damaged tissues and body parts.

tissue-specific describes the expression of a gene only in a particular tissue(s) or cell type(s).

tooth primordium the mesenchymal structure from which a tooth develops. Also called a tooth germ.

totipotent the capacity of a cell to develop into a new organism. The only truly totipotent cell in most animals is the fertilized egg, which gives rise to the embryo and any extra-embryonic membranes, including the placenta in mammals. In plants, somatic cells remain potentially totipotent, being able to form a complete new plant if cultured under the appropriate conditions.

tracheal system in insects, a system of fine tubules that deliver air (and thus oxygen) to the tissues.

transcription the copying of the DNA sequence of a gene into a complementary RNA sequence. In some genes, the RNA is the end-product, but in protein-coding genes, the RNA is then translated to produce a protein.

transcription factor a protein required to initiate or regulate the transcription of a gene into RNA. Transcription factors act within the nucleus of a cell by binding to specific sites in the regulatory regions of the gene.

transcription initiation complex a protein complex consisting of a set of 'general' transcription factors and the RNA polymerase that assembles on the promotor and ensures that the polymerase is positioned to start transcription at the appropriate place.

transdetermination the process by which a committed, but not yet differentiated, cell can become redetermined as a different cell type. The best known examples occur in regenerating *Drosophila* imaginal discs, where rare transdetermination events occur, resulting in the homeotic transformation of one type of adult structure into another.

transdifferentiation the process in which a differentiated cell dedifferentiates and redifferentiates into a different cell type.

transfection technique by which mammalian and other animal cells are induced to take up foreign DNA molecules. The introduced DNA sometimes becomes inserted permanently into the host cell's DNA.

transgene a gene, sometimes from another organism, that has been introduced into a cell or organism by genetic engineering techniques.

transgenic describes an organism whose genetic make-up is the result of the deliberate introduction of new DNA; e.g. new genes; genes with specific mutations; or DNAs that cause the inactivation of specific genes. The term is also used to describe the various genetic engineering techniques that can be used to accomplish this.

transient transgenesis see *somatic transgenesis*.

transit-amplifying cells in continually renewing tissues such as epidermis and the gut lining, rapidly dividing cells that are produced from stem cells and after several divisions will differentiate into the specialized cell types of the tissue.

translation the process by which messenger RNA directs the order of amino acids in a protein during protein synthesis on ribosomes.

transposon a DNA sequence that can become inserted into a different site on the chromosome, either by the insertion of a copy of the original sequence or by excision and reinsertion of the original sequence.

triploblast an animal with three germ layers—endoderm, mesoderm, and ectoderm.

trisomy the presence of three copies of a chromosome, rather than the normal two, which is characteristic of some conditions, such as Down's syndrome.

trophectoderm the outer layer of cells of the early mammalian embryo. It gives rise to extra-embryonic structures such as the placenta.

tumor progression the transition of a tumor through a series of stages that transform cells from normal to cancerous.

tumor-suppressor gene gene that can cause a cell to become cancerous when both copies of the gene have been inactivated. In non-cancerous cells, tumor-suppressor genes are involved in suppressing cell proliferation.

Turing mechanisms *see reaction–diffusion*.

vas deferens the duct that connects the sperm-producing testis to the penis.

vasculogenesis the initial stages in the formation of blood vessels, comprising the condensation of angioblasts to form a tubular vessel.

vegetal pole is the central point on the surface of the vegetal region, directly opposite the animal pole. *See vegetal region*.

vegetal region the yolky 'lower' hemisphere of amphibian eggs and blastulas, and the region from which the endoderm will develop. The

venous pole the posterior end of the developing heart tube, which will become the inflow (venous) end.

ventral closure the coming together and closing of the sides of the chick or mammalian embryo on the ventral side of the body to form the gut.

ventral ectoderm one of the four main regions into which the dorso-ventral axis of the early *Drosophila* embryo is divided before gastrulation. This region will give rise to the nervous system and some epidermis.

ventralized describes embryos that are deficient in dorsal regions and have much increased ventral regions.

ventricular chamber one of the chambers into which the vertebrate heart tube is first divided by a transverse septum, the other being the atrial chamber. In fish, the heart has only two chambers; in the four-chambered heart of other vertebrates each chamber is divided longitudinally to give right and left atria and right and left ventricles.

ventricular zone a layer of proliferating cells lining the lumen of the vertebrate neural tube, from which neurons and glia are formed.

vernalization the phenomenon by which flowering is accelerated after the plant has been exposed to a long period of cold temperature.

vertebral column the backbone or spine of vertebrates, composed of a succession of vertebrae.

visceral endoderm tissue derived from the primitive endoderm that develops on the surface of the egg cylinder in the mammalian blastocyst.

visual cortex that part of the mammalian cerebral cortex to which visual signals are sent after the LGN and are processed to produce a visual perception.

vitelline envelope, vitelline membrane extracellular layer surrounding the eggs of animals. In the sea urchin it gives rise to the fertilization membrane.

white matter regions of the brain and spinal cord that are mainly composed of the axons of neurons.

Wingless secreted signaling protein in *Drosophila* with a role in segment patterning and wing development. It is a member of the Wnt family of signaling proteins.

Wnt family family of secreted signaling proteins present in all Metazoa, including the sponges, members of which act in many aspects of development. It includes the Wingless protein in *Drosophila* and the Wnt proteins of vertebrates. Wnt proteins can signal through various pathways including the canonical Wnt/β-catenin pathway and the Wnt planar polarity pathway.

Wnt/β-catenin pathway *see canonical Wnt/β-catenin pathway*.

Wolffian ducts paired ducts associated with the mesonephros in mammalian embryos. They become the vas deferens in males.

X-chromosome inactivation the random inactivation of one copy of the X chromosome that occurs in the somatic cells of female mammals.

yolk sac an extra-embryonic structure with an internal cavity in birds and mammals. In the chick embryo the cavity contains yolk.

yolk syncytial layer a continuous layer of multinucleate non-yolky cytoplasm underlying the blastoderm in zebrafish embryos.

zinc-finger nucleases (ZFNs) nucleases that are used in a technique to specifically cleave and disrupt genes *in vivo* in a targeted fashion.

zona limitans intrathalamica a signaling center located in the forebrain of the vertebrate embryo, which helps pattern the brain along the antero-posterior axis.

zona pellucida a layer of glycoprotein surrounding the mammalian egg that serves to prevent polyspermy.

zone of polarizing activity (ZPA) *see polarizing region*.

zootype the expression of Hox genes and certain other genes along the antero-posterior axis of the embryo that is characteristic of all animal embryos.

zygote the fertilized egg. It is diploid and contains chromosomes from both the male and female parents.

zygotic gene any gene present in the fertilized egg and which is expressed in the embryo itself.

Index